# The Properties of Gases and Liquids

## Other McGraw-Hill Books in Chemical Engineering

*Chopey and Hicks* • HANDBOOK OF CHEMICAL ENGINEERING CALCULATIONS (1984)
*Considine* • PROCESS INSTRUMENTS AND CONTROLS HANDBOOK, 3D ED. (1985)
*Dean* • LANGE'S HANDBOOK OF CHEMISTRY, 13TH ED. (1985)
*Grant* • GRANT & HACKH'S CHEMICAL DICTIONARY, 5TH ED. (1987)
*Hicks* • STANDARD HANDBOOK OF ENGINEERING CALCULATIONS, 2D ED. (1985)
*Hopp and Hennig* • HANDBOOK OF APPLIED CHEMISTRY (1983)
*Lyman* • HANDBOOK OF CHEMICAL PROPERTY ESTIMATION METHODS (1982)
*Meyers* • HANDBOOK OF CHEMICALS PRODUCTION PROCESSES (1986)
*Meyers* • HANDBOOK OF PETROLEUM REFINING PROCESSES (1986)
*Meyers* • HANDBOOK OF SYNFUELS TECHNOLOGY (1984)
*Perry* • ENGINEERING MANUAL (1976)
*Perry and Green* • PERRY'S CHEMICAL ENGINEERS' HANDBOOK, 6TH ED. (1984)
*Rohsenow et al.,* • HANDBOOK OF HEAT TRANSFER APPLICATIONS (1985)
*Rohsenow et al.,* • HANDBOOK OF HEAT TRANSFER FUNDAMENTALS (1985)
*Rosaler and Rice* • STANDARD HANDBOOK OF PLANT ENGINEERING (1983)
*Sandler and Luckiewicz* • PRACTICAL PROCESS ENGINEERING (1987)
*Schweitzer* • HANDBOOK OF SEPARATION TECHNIQUES FOR CHEMICAL ENGINEERS (1979)
*Shinskey* • DISTILLATION CONTROL, 2D ED. (1983)
*Shinskey* • PROCESS CONTROL SYSTEMS, 2D ED. (1979)

# The Properties of Gases and Liquids

**Robert C. Reid**
Professor of Chemical Engineering
Massachusetts Institute of Technology

**John M. Prausnitz**
Professor of Chemical Engineering
University of California at Berkeley

**Bruce E. Poling**
Professor of Chemical Engineering
University of Missouri at Rolla

**Fourth Edition**

**McGraw-Hill Book Company**
New York St. Louis San Francisco Auckland Bogotá
Hamburg London Madrid Mexico
Milan Montreal New Delhi Panama
Paris São Paulo Singapore
Sydney Tokyo Toronto

Library of Congress Cataloging-in-Publication Data

Reid, Robert C.
The properties of gases and liquids.

Includes bibliographies and index.
1. Gases 2. Liquids. I. Prausnitz, J. M.
II. Poling, Bruce E. III. Title.
TP242.R4 1987 660′.042 86-21358
ISBN 0-07-051799-1



2 3 4 5 6 7 8 9 0 DOC/DOC 8 9 3 2 1 0 9 8

ISBN 0-07-051799-1

*The editors for this book were Betty Sun and Galen H. Fleck, the designer was Naomi Auerbach, and the production supervisor was Thomas G. Kowalczyk. It was set in Century Schoolbook by University Graphics, Inc.*

*Printed and bound by R. R. Donnelley & Sons Company.*

# Contents

# Preface

Reliable values of the properties of materials are necessary for the design of industrial processes. An enormous amount of data has been collected and correlated over the years, but the rapid advance of technology into new fields seems always to maintain a significant gap between demand and availability. The engineer is still required to rely primarily on common sense, experience, and a variety of methods for estimating physical properties.

This book presents a critical review of various estimation procedures for a limited number of properties of gases and liquids: critical and other pure component properties, *PVT* and thermodynamic properties of pure components and mixtures, vapor pressures, and phase-change enthalpies, standard enthalpies of formation, standard Gibbs energies of formation, heat capacities, surface tensions, viscosities, thermal conductivities, diffusion coefficients, and phase equilibria. Comparisons of experimental and estimated values are normally shown in tables to indicate reliability. Most methods are illustrated by examples.

The procedures described are necessarily limited to those which appear to the authors to have the greatest validity and practical use. Wherever possible, we have included recommendations delineating the best methods for estimating each property and the most reliable techniques for extrapolating or interpolating available data. Recommended methods are often illustrated by detailed examples.

Although the book is intended to serve primarily the practicing engineer, especially the process or chemical engineer, other engineers and scientists concerned with gases and liquids may find it useful.

The first edition of this book was published in 1958, the second in 1966, and the third in 1977. Each revision is essentially a new book, because many estimation methods are proposed each year and, over an 8- to 10-year span, most earlier methods are modified or displaced by more accurate or more general techniques. Most new methods are still empirical in nature, although there are often theoretical bases for the correlation;

whenever possible, the theory is outlined to provide the user with the foundation of the proposed estimation method.

The data bank, Appendix A, is now about 15 percent larger than that in the third edition. More important, many estimated quantities in the earlier version have now been replaced by experimental results.

Many colleagues and students have contributed data, advice, examples, or illustrative calculations; we are grateful to them all. For their helpful contributions, we want to record our thanks to C. Baroncini, M. R. Brulé, R. Lopes Cardozo, T. -H. Chung, J. H. Dymond, W. Hayduk, J. B. Irving, J. D. Isdale, D. T. Jamieson, K. G. Joback, J. Kestin, D. Klingenberg, D. Kyser, G. Latini, L. L. Lee, K. Lucas, D. Reichenberg, R. L. Rowley, A. S. Teja, T. P. Thinh, R. Topliss, A. Vetere, and J. Wong.

Special thanks are due to D. Ambrose for contributions to the data bank and to estimation methods for critical constants and to J. McGarry for extensive help in regression of vapor-pressure data.

ROBERT C. REID
JOHN M. PRAUSNITZ
BRUCE E. POLING

Chapter

# 1

# The Estimation of Physical Properties

## 1-1 Introduction

The structural engineer cannot design a bridge without knowing the properties of steel and concrete. Similarly, scientists and engineers often require the properties of gases and liquids. The chemical or process engineer, in particular, finds knowledge of physical properties of fluids essential to the design of many kinds of industrial equipment. Even the theoretical physicist must occasionally compare theory with measured properties.

The physical properties of every substance depend directly on the nature of the molecules of the substance. Therefore, the ultimate generalization of physical properties of fluids will require a complete understanding of molecular behavior, which we do not yet have. Though its origins are ancient, the molecular theory was not generally accepted until about the beginning of the nineteenth century, and even then there were setbacks until experimental evidence vindicated the theory early in the twentieth century. Many pieces of the puzzle of molecular behavior have now fallen into place, but as yet it has not been possible to develop a complete generalization.

In the nineteenth century, the laws of Charles and Gay-Lussac were combined with Avogadro's hypothesis to form the gas law, $PV = NRT$, which was perhaps the first important correlation of properties. Deviations from the ideal-gas law, though often small, were tied to the fundamental nature of the molecules. The equation of van der Waals, the virial equation, and other equations of state express these quantitatively. These extensions of the ideal-gas law have not only facilitated progress in the development of a molecular theory but, more important for our purposes here, have provided a framework for correlating physical properties of fluids.

The original "hard-sphere" kinetic theory of gases was a significant contribution to progress in understanding the statistical behavior of a system containing a large number of molecules. Thermodynamic and transport properties were related quantitatively to molecular size and speed. Deviations from the hard-sphere kinetic theory led to studies of the interaction of molecules based on the realization that molecules attract at intermediate separations and repel when they come very close. The semiempirical potential functions of Lennard-Jones and others describe attraction and repulsion in an approximate quantitative fashion. More recent potential functions allow for the shapes of molecules and for asymmetric charge distribution in polar molecules.

Although allowance for the forces of attraction and repulsion between molecules is primarily a development of the twentieth century, the concept is not new. In about 1750, Boscovich suggested that molecules (which he referred to as atoms) are "endowed with potential force, that any two atoms attract or repel each other with a force depending on their distance apart. At large distances the attraction varies as the inverse square of the distance. The ultimate force is a repulsion which increases without limit as the distance decreases without limit, so that the two atoms can never coincide" [3].

From the viewpoint of mathematical physics, the development of a comprehensive molecular theory would appear to be complete. J. C. Slater [4] observes that, while we are still seeking the laws of nuclear physics, "in the physics of atoms, molecules and solids, we have found the laws and are exploring the deductions from them." However, the suggestion that, in principle (the Schrödinger equation of quantum mechanics), everything is known about molecules is of little comfort to the engineer who needs to know the properties of some new chemical to design a commercial plant.

Paralleling the continuing refinement of the molecular theory has been the development of thermodynamics and its application to properties. The two are intimately related and interdependent. Carnot was an engineer interested in steam engines, but the second law of thermodynamics was shown by Clausius, Kelvin, Maxwell, and Gibbs to have broad applications in all branches of science.

Thermodynamics by itself cannot provide physical properties; only molecular theory or experiment can do that. But thermodynamics reduces experimental or theoretical efforts by relating one physical property to another. For example, the Clausius-Clapeyron equation provides a useful method for obtaining enthalpies of vaporization from more easily measured vapor pressures.

The second law led to the concept of chemical potential which is basic to an understanding of chemical and phase equilibria, and Maxwell's equations provide ways to obtain important thermodynamic properties of a substance from $PVTx$ relations. Since derivatives are often required, the $PVTx$ function must be known accurately.

In spite of impressive developments in molecular theory, the engineer frequently finds a need for physical properties which have not been measured and which cannot be calculated from existing theory. The *International Critical Tables,* Beilstein, Landolt-Börnstein, and many other handbooks provide convenient data sources, and there exists an increasing number of journals devoted to compilation and critical review of physical property data. Further, computerized data banks are now becoming routine components of computer-aided process design. But it is inconceivable that all desired experimental data will ever be available for the hundreds or thousands of compounds of interest in science and industry: while the number of possibly interesting compounds is already very large, the number of mixtures formed by these compounds is much larger.

While the need for accurate design data is increasing, the rate of accumulation of new data is not increasing fast enough. Data on multicomponent mixtures are particularly scarce. The process engineer who is frequently called upon to design a plant to produce a new chemical (or a well-known chemical in a new way) often finds that the required physical property data are not available. It may be possible to obtain the desired properties from new experimental measurements, but that is often not practical because such measurements tend to be expensive and time-consuming. To meet budgetary and deadline requirements, the process engineer almost always must estimate at least some of the properties required for design.

## 1-2 Estimation of Properties

In the all-too-frequent situation in which no experimental value of the needed property is at hand, the value must be estimated or predicted. "Estimation" and "prediction" are often used as if they were synonymous, although the former properly carries the frank implication that the result may be only approximate. Estimates may be based on theory, on correlations of experimental values, or on a combination of both. A theoretical relation, although not generally valid, may nevertheless serve adequately in specific cases.

To relate mass and volumetric flow rates of air through an air-conditioning unit, the engineer is justified in using $PV = NRT$. Similarly, he or she may properly use Dalton's law and the vapor pressure of water to calculate the mass fraction of water in saturated air. However, the engineer must be able to judge the operating pressure when such simple calculations lead to unacceptable error.

Completely empirical correlations are often useful, but one must avoid the temptation to use them outside the narrow range of properties on which they are based. In general, the stronger the theoretical basis, the more reliable the correlation.

Most of the better estimation methods use equations based on the form of an incomplete theory with empirical correlations of the constants that are not provided by that theory. Introduction of empiricism into parts of a theoretical relation provides a powerful method for developing a reliable correlation. For example, the van der Waals equation of state is a modification of the simple $PV = NRT$; setting $N = 1$,

$$\left(P + \frac{a}{V^2}\right)(V - b) = RT \tag{1-2.1}$$

Equation (1-2.1) is based on the idea that the pressure on a container wall, exerted by the impinging molecules, is decreased because of the attraction by the mass of molecules in the bulk gas; that attraction rises with density. Further, the available space in which the molecules move is less than the total volume by the excluded volume $b$ due to the size of the molecules themselves. Therefore, the "constants" $a$ and $b$ have some theoretical basis. The correlation of $a$ and $b$ in terms of other properties of a substance is an example of the use of an empirically modified theoretical form.

Empirical extension of theory can often lead to a correlation useful for estimation purposes. For example, several methods for estimating diffusion coefficients in low-pressure binary gas systems are empirical modifications of the equation given by the simple kinetic theory. Almost all the better estimation procedures are based on correlations developed in this way.

## 1-3 Types of Estimation

An ideal system for the estimation of a physical property would (1) provide reliable physical and thermodynamic properties for pure substances and for mixtures at any temperature, pressure, and composition, (2) indicate the state (solid, liquid, or gas), (3) require a minimum of input data,

(4) choose the least-error route (i.e., the best estimation method), (5) indicate the probable error, and (6) minimize computation time. Few of the available methods approach this ideal, but some serve remarkably well.

In numerous practical cases, the most accurate method may not be the best for the purpose. Many engineering applications properly require only approximate estimates, and a simple estimation method requiring little or no input data is often preferred over a complex but more accurate correlation. The simple gas law is useful at low to modest pressures, although more accurate correlations are available. Unfortunately, it is often not easy to provide guidance on when to reject the simpler in favor of the more complex (but more accurate) method.

Although a variety of molecular theories may be useful for data correlation, there is one theory which is particularly helpful. This theory, called the theory of corresponding states, was originally based on macroscopic arguments, but in its modern form it has a molecular basis.

### The law of corresponding states

Proposed by van der Waals in 1873, the law of corresponding states expresses the generalization that equilibrium properties which depend on intermolecular forces are related to the critical properties in a universal way. Corresponding states provides the single most important basis for the development of correlations and estimation methods. In 1873, van der Waals showed it to be theoretically valid for all pure substances whose *PVT* properties could be expressed by a two-constant equation of state such as Eq. (1-2.1). As shown by Pitzer in 1939, it is similarly valid if the intermolecular potential function requires only two characteristic parameters. Corresponding states holds well for fluids containing simple molecules and, upon semiempirical extension, it also holds for many other substances where molecular orientation is not important, i.e., for molecules that are not strongly polar or hydrogen-bonded.

The relation of pressure to volume at constant temperature is different for different substances; however, corresponding states theory asserts that if pressure, volume, and temperature are related to the corresponding critical properties, the function relating reduced pressure to reduced volume becomes the same for all substances. The reduced property is commonly expressed as a fraction of the critical property: $P_r = P/P_c$; $V_r = V/V_c$; and $T_r = T/T_c$.

To illustrate corresponding states, Fig. 1-1 shows the law of corresponding states for *PVT* data for methane and nitrogen. In effect, the critical point is taken as the origin. The data for saturated liquid and saturated vapor coincide well for the two substances. The isotherms (constant $T_r$), of which only one is shown, agree equally well.

Successful application of the law of corresponding states for correlation

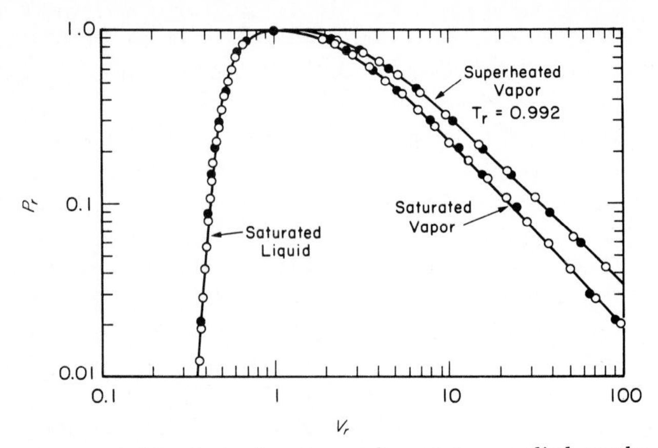


**Figure 1-1** The law of corresponding states applied to the *PVT* properties of methane and nitrogen. Experimental values [3]: ○ methane, ● nitrogen.

of *PVT* data has encouraged similar correlations of other properties which depend primarily on intermolecular forces. Many of these have proved valuable to the practicing engineer. Modifications of the law are common to improve accuracy or ease of use. Good correlations of high-pressure gas viscosity have been obtained by expressing $\eta/\eta_c$ as a function of $P_r$ and $T_r$. But since $\eta_c$ is seldom known and not easily estimated, this quantity has been replaced in other correlations by $\eta_c^\circ$, $\eta_T^\circ$, or the group $M^{1/2}P_c^{2/3}T_c^{1/6}$, where $\eta_c^\circ$ is the viscosity at $T_c$ and low pressure, $\eta_T^\circ$ is the viscosity at the temperature of interest, again at low pressure, and the group containing $M$, $P_c$, and $T_c$ is suggested by dimensional analysis. Other alternatives to the use of $\eta_c$ might be proposed, each modeled on the law of corresponding states but essentially empirical as applied to transport properties.

The law of corresponding states can be derived from statistical mechanics when severe simplifications are introduced into the partition function. Sometimes other useful results can be obtained by introducing less severe simplifications into statistical mechanics toward providing a framework for the development of estimation methods. Fundamental equations describing various properties (including transport properties) can sometimes be derived, provided that an expression is available for the potential-energy function for molecular interactions. This function may be, at least in part, empirical; but the fundamental equations for properties are often insensitive to details in the potential function from which they stem, and two-constant potential functions frequently serve remarkably well for some systems. Statistical mechanics may at present be far removed from engineering practice, but there is good reason to believe that it will become increasingly useful, especially when combined with computer simulations.

### Nonpolar and polar molecules

Spherically symmetric molecules (for example, $CH_4$) are well fitted by a two-constant law of corresponding states. Nonspherical and weakly polar molecules do not fit poorly, but deviations are often great enough to encourage the development of correlations using a third parameter, e.g., the acentric factor. The acentric factor is obtained from the deviation of the experimental vapor pressure–temperature function from that which might be expected for a similar substance consisting of spherically symmetric molecules. Typical corresponding states correlations express the dimensionless property as a function of $P_r$, $T_r$, and the chosen third parameter.

Unfortunately, the properties of strongly polar molecules are often not satisfactorily represented by the two- or three-constant correlations which do so well for nonpolar molecules. An additional parameter based on the dipole moment has often been suggested but with limited success, since polar molecules are not easily characterized by using only the dipole moment and critical constants. As a result, although good correlations exist for properties of nonpolar fluids, similar correlations for polar fluids are often not available or else are of restricted reliability.

### Structure

All macroscopic properties are related to molecular structure, which determines the magnitude and predominant type of the intermolecular forces. For example, structure determines the energy storage capacity of a molecule and thus the molecule's heat capacity.

The concept of structure suggests that a macroscopic property can be calculated from group contributions. The relevant characteristics of structure are related to the atoms, atomic groups, bond type, etc.; to them we assign weighting factors and then determine the property, usually by an algebraic operation which sums the contributions from the molecule's parts. Sometimes the calculated sum of the contributions is not for the property itself but instead is for a correction to the property as calculated by some simplified theory or empirical rule. For example, Lydersen's method for estimating $T_c$ starts with the loose rule that the ratio of the normal boiling temperature to the critical temperature is about 2:3. Additive structural increments based on bond types are then used to obtain empirical corrections to that ratio.

Some of the better correlations of ideal-gas heat capacities employ theoretical values of $C_p^\circ$ (which are intimately related to structure) to obtain a polynomial expressing $C_p^\circ$ as a function of temperature; the constants in the polynomial are determined by contributions from the constituent atoms, atomic groups, and types of bonds.

## 1-4 Organization of the Book

Reliable experimental data are always to be preferred over values obtained by estimation methods. But all too often reliable data are not available.

In this book, the various estimation methods are correlations of experimental data. The best are based on theory, with empirical corrections for the theory's defects. Others, including those stemming from the law of corresponding states, are based on generalizations which are partly empirical but which nevertheless have application to a remarkably wide range of properties. Totally empirical correlations are useful only when applied to situations very similar to those used to establish the correlations.

The text includes a large number of numerical examples to illustrate the estimation methods, especially those which are recommended. Almost all of them are designed to explain the calculation procedure for a single property. However, most engineering design problems require estimation of several properties; the error in each contributes to the overall result, but some individual errors are more important than others. Fortunately, the result is often found adequate for engineering purposes, in spite of the large measure of empiricism incorporated in so many of the estimation procedures.

As an example, consider the case of a chemist who has synthesized a new compound which has the chemical formula $CCl_2F_2$ and boils at $-20.5°C$ at atmospheric pressure. Using only this information, is it possible to obtain a useful prediction of whether or not the substance has the thermodynamic properties which might make it a practical refrigerant?

Figure 1-2 shows portions of a Mollier diagram developed by the prediction methods described in later chapters. The dashed curves and points are obtained from estimates of liquid and vapor heat capacities, critical properties, vapor pressure, enthalpy of vaporization, and pressure corrections to ideal-gas enthalpies and entropies. The substance is, of course, a well-known refrigerant, and its known properties are shown by the solid curves.

For a standard refrigeration cycle operating between 48.9 and $-6.7°C$, the evaporator and condenser pressures are estimated to be 2.4 and 12.4 bar, vs. the known values 2.4 and 11.9 bar. The estimate of the heat absorption in the evaporator checks closely, and the estimated volumetric vapor rate to the compressor also shows good agreement: 2.39 versus 2.45 $m^3/hr$ per kW of refrigeration. (This number indicates the size of the compressor.) Constant-entropy lines are not shown in Fig. 1-2, but it is found that the constant-entropy line through the point for the low-pressure vapor essentially coincides with the saturated vapor curve. The estimated coefficient of performance (ratio of refrigeration rate to isentropic compression power) is estimated to be 3.8; the value obtained from the

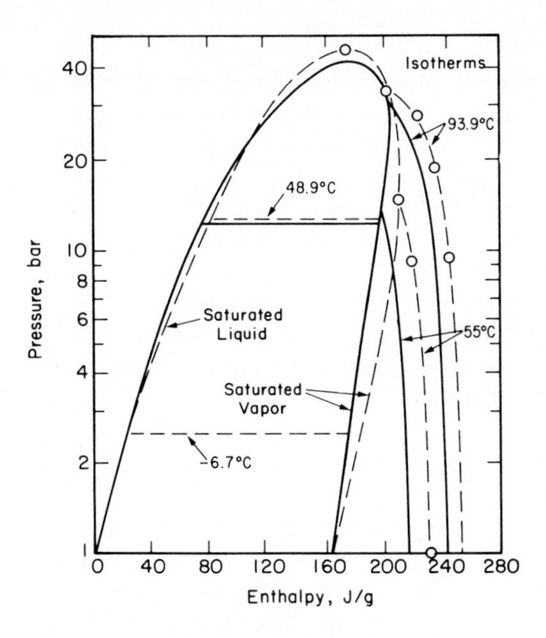


**Figure 1-2** Mollier diagram for dichlorodifluoromethane. The solid lines represent measured data. Dashed lines and points represent results obtained by estimation methods when only the chemical formula and the normal boiling temperature are known.

data is 3.5. This is not a very good check, but it is nevertheless remarkable because the only data used for the estimate were the normal boiling point and the chemical formula.

Most estimation methods require parameters which are characteristic of single pure components or of constituents of a mixture of interest. The more important of these are considered in Chap. 2, and tables of values for common substances are provided in Appendix A.

Thermodynamic properties (such as enthalpy and heat capacity) are discussed in Chaps. 3 to 6. Although the more accurate equations of state are employed, the basic thermodynamic relations are developed in a general way so that other equations of state can be introduced whenever they are more applicable for a particular purpose.

Chapters 5 and 6 discuss heat capacities; Chap. 6 discusses Gibbs energy and enthalpy of formation; and Chap. 7 discusses vapor pressures and enthalpies of vaporization of pure substances. Chapter 8 presents techniques for estimation and correlation of phase equilibria in mixtures. Chapters 9 to 11 describe estimation methods for viscosity, thermal conductivity, and diffusion coefficients. Surface tension is considered briefly in Chap. 12.

The literature searched was voluminous, and the lists of references following each chapter represent but a fraction of the material examined. Of the many estimation methods available, only a few were selected for detailed discussion. These were selected on the basis of their generality, accuracy, and availability of required input data. Tests of all methods were more extensive than those suggested by the abbreviated tables comparing experimental with estimated values. However, no comparison is adequate to indicate expected errors for new compounds. The arithmetic average errors given in the comparison tables represent but a crude overall evaluation; the inapplicability of a method for a few compounds may so increase the average error as to distort judgment of the method's merit.

Many estimation methods are of such complexity that a computer is desirable. This is less of a handicap than it once was, since computers and efficient computer programs have become widely available. Electronic desk computers, which have become so popular in recent years, have made the more complex correlations practical. However, accuracy is not necessarily enhanced by greater complexity.

The scope of the book is inevitably limited. The properties discussed were selected arbitrarily because they are believed to be of wide interest, especially to chemical engineers. Electrical properties are not included, nor are the properties of salts, metals, or alloys or chemical properties other than some thermodynamically derived properties such as enthalpy and the Gibbs energy of formation. The difficult areas of organic polymers and crystals are not included here; introductions are given by Bondi [1] and van Krevelen [5].

This book is intended to provide estimation methods for a limited number of physical properties of fluids. Hopefully, the need for such estimates, and for a book of this kind, may diminish as more experimental values become available and as the continually developing molecular theory advances beyond its present incomplete state. In the meantime, estimation methods are essential for most process design calculations and for many other purposes in engineering and applied science.

## REFERENCES

1. Bondi, A.: *Physical Properties of Molecular Crystals, Liquids and Glasses,* Wiley, New York, 1968.
2. Din, F., (ed.): *Thermodynamic Functions of Gases,* vol. 3, Butterworth, London, 1961.
3. Maxwell, James Clerk: "Atoms," *Encyclopaedia Britannica,* 9th ed., A. & C. Black, Edinburgh, 1875–1888.
4. Slater, J. C.: *Modern Physics,* McGraw-Hill, New York, 1955.
5. van Krevelen, D. W.: *Properties of Polymers: Their Estimation and Correlations with Chemical Structure,* 2d ed., Elsevier, Amsterdam, 1976.

Chapter

# 2

# Pure Component Constants

## 2-1 Scope

Chemical engineers normally deal with mixtures rather than pure materials. However, the chemical compositions of many mixtures are known. Even for poorly characterized mixtures in the petroleum industry, one assigns pseudo pure component constants to fractions of the mixture where the fractions are normally specified by a volatility range. Also, few mixture-property correlations to date have incorporated true mixture parameters. Instead, the techniques employ parameters which are functions only of the properties of the pure components. These *pure component constants* are then used with the state variables such as temperature, pressure, and composition to generate property estimation methods.

In this chapter, we introduce the more common and useful pure component constants and show how they can be estimated if no experimental data are available. A few useful pure component constants, however, are not covered, because it is more convenient and appropriate to discuss them in subsequent chapters. For example, the liquid specific volume (or density) at some reference state (e.g., as a saturated liquid at 1 bar) is a useful pure component constant, yet it is more easily introduced in Chap. 3, where volumetric properties are covered.

The correlations presented in this chapter are primarily of the *group contribution* type. When the molecular structure is not well known, or when one is interested in, say, characterizing pseudocritical properties of petroleum or coal liquid fractions, other types of estimation methods would then be necessary. For example, Lin and Chao [25] present a technique to estimate critical properties and the acentric factor for quite complex organic compounds (and mixtures) from the average molecular weight, boiling point, and the specific gravity (at 20°C). Similar techniques have also been suggested by others [8, 43, 50].

## 2-2 Critical properties

Critical temperature, pressure, and volume represent three widely used pure component constants, yet recent experimental measurements are almost nonexistent. In Appendix A we have tabulated the critical properties of many materials. In all cases the values given were measured. An excellent compilation of critical properties is available in a National Physical Laboratory Report [4] by Ambrose. Earlier reviews were given by Kudchadker et al. [22] for organic compounds and Mathews [27] for inorganic compounds. The Design Institute for Physical Property Data [10] also presents a detailed discussion of critical properties and their estimation.

### Estimation techniques

**Ambrose method [2, 3].** In this method, the three critical properties $T_c$, $P_c$, and $V_c$ are estimated by a group contribution technique using the following relations:

$$T_c = T_b[1 + (1.242 + \Sigma\Delta_T)^{-1}] \qquad (2\text{-}2.1)$$

$$P_c = M(0.339 + \Sigma\Delta_P)^{-2} \qquad (2\text{-}2.2)$$

$$V_c = 40 + \Sigma\Delta_V \qquad (2\text{-}2.3)$$

The units employed are kelvins, bars, and cubic centimeters per mole.† For perfluorinated compounds or for compounds containing only halogens (including fluorine), the constant 1.242 in Eq. (2-2.1) would be replaced by 1.570 and the constant 0.339 in Eq. (2-2.2) by 1.000. The $\Delta$ quantities are evaluated by summing contributions for various atoms or groups of atoms as shown in Table 2-1. To employ these relations, the normal boiling point $T_b$ (at 1 atm) and the molecular weight $M$ are needed. The technique is illustrated in Example 2-1, and the expected errors are noted later after alternate methods are introduced.

**Joback modification of Lydersen's method.** One of the first very successful group contribution methods to estimate critical properties was developed

†Only the gram mole is used in this book.

**TABLE 2-1 Ambrose Group Contributions for Critical Constants**

| | Δ values for | | |
|---|---|---|---|
| | $T_c$ | $P_c$ | $V_c$ |
| Carbon atoms in alkyl groups | 0.138 | 0.226 | 55.1 |
| Corrections: | | | |
| >CH− (each) | −0.043 | −0.006 | −8 |
| >C< (each) | −0.120 | −0.030 | −17 |
| Double bonds (nonaromatic) | −0.050 | −0.065 | −20 |
| Triple bonds | −0.200 | −0.170 | −40 |
| Delta Platt number,[1] multiply by | −0.023 | −0.026 | — |
| Aliphatic functional groups: | | | |
| −O− | 0.138 | 0.160 | 20 |
| >CO | 0.220 | 0.282 | 60 |
| −CHO | 0.220 | 0.220 | 55 |
| −COOH | 0.578 | 0.450 | 80 |
| −CO−O−OC− | 1.156 | 0.900 | 160 |
| −CO−O− | 0.330 | 0.470 | 80 |
| $-NO_2$ | 0.370 | 0.420 | 78 |
| $-NH_2$ | 0.208 | 0.095 | 30 |
| −NH− | 0.208 | 0.135 | 30 |
| >N− | 0.088 | 0.170 | 30 |
| −CN | 0.423 | 0.360 | 80 |
| −S− | 0.105 | 0.270 | 55 |
| −SH | 0.090 | 0.270 | 55 |
| $-SiH_3$ | 0.200 | 0.460 | 119 |
| $-O-Si(CH_3)_2$ | 0.496 | — | — |
| −F | 0.055 | 0.223 | 14 |
| −Cl | 0.055 | 0.318 | 45 |
| −Br | 0.055 | 0.500 | 67 |
| −I | 0.055 | — | 90 |
| Halogen correction in aliphatic compounds: | | | |
| F is present | 0.125 | — | — |
| F is absent, but Cl, Br, I are present | 0.055 | | |
| Aliphatic alcohols[2] | [3] | [4] | 15 |
| Ring compound increments (listed only when different from aliphatic values): | | | |
| $-CH_2-$ | 0.090 | 0.182 | 44.5 |
| >CH− in fused ring | 0.030 | 0.182 | 44.5 |
| Double bond | −0.030 | — | −15 |
| −O− | 0.090 | — | 10 |
| −NH− | 0.090 | — | — |
| −S− | 0.090 | — | 30 |
| Aromatic compounds: | | | |
| Benzene | 0.448 | 0.924 | [5] |
| Pyridine | 0.448 | 0.850 | |
| $C_4H_4$ (fused as in naphthalene) | 0.220 | 0.515 | |
| −F | 0.080 | 0.183 | |
| −Cl | 0.080 | 0.318 | |
| −Br | 0.080 | 0.600 | |
| −I | 0.080 | 0.850 | |
| −OH | 0.198 | −0.025 | |
| Corrections for nonhalogenated substitutions: | | | |
| First | 0.010 | 0 | |
| Each subsequent | 0.030 | 0.020 | |
| Ortho pairs containing −OH | −0.080 | −0.050 | |
| Ortho pairs with no −OH | −0.040 | −0.050 | |

**TABLE 2-1 Ambrose Group Contributions for Critical Constants (*Continued*)**

| | Δ values for | | |
|---|---|---|---|
| | $T_c$ | $P_c$ | $V_c$ |
| Highly fluorinated aliphatic compounds: | | | |
| $-CF_3$, $-CF_2-$, $>CF-$ | 0.200 | 0.550 | |
| $-CF_2$, $>CF-$ (ring) | 0.140 | 0.420 | |
| $>CF-$ (in fused ring) | 0.030 | — | |
| $-H$ (monosubstitution) | −0.050 | −0.350 | |
| Double bond (nonring) | −0.150 | −0.500 | |
| Double bond (ring) | −0.030 | — | |
| (Other increments as in nonfluorinated compounds) | | | |

[1]The delta Platt number is defined as the Platt number of the isomer minus the Platt number of the corresponding alkane. (For $n$-alkanes, the Platt number is $n - 1$.) The Platt number is the total number of carbon atoms three bonds apart [40, 41]. This correction is used only for branched alkanes.

[2]Includes naphthenic alcohols and glycols but not aromatic alcohols such as xylenol.

[3]First determine the hydrocarbon homomorph, i.e., substitute $-CH_3$ for each $-OH$ and calculate $\Sigma\Delta_T$ for this compound. Subtract 0.138 from $\Sigma\Delta_T$ for each $-OH$ substituted. Next, add $0.87 - 0.11n + 0.003n^2$ where $n = [T_b \text{ (alcohol, K)} - 314]/19.2$. Exceptions include methanol ($\Sigma\Delta_T = 0$), ethanol ($\Sigma\Delta_T = 0.939$), and any alcohol whose value of $n$ exceeds 10.

[4]Determine the hydrocarbon homomorph as in footnote 3. Calculate $\Sigma\Delta_P$ and subtract 0.226 for each $-OH$ substituted. Add $0.100 - 0.013n$, where $n$ is computed as in footnote 3.

[5]When estimating the critical volumes of aromatic substances, use alkyl group values.

by Lydersen [26] in 1955. Since that time, more experimental values have been reported and efficient statistical techniques have been developed to determine the optimum group contributions. Joback [19] reevaluated Lydersen's scheme, added several functional groups, and determined the values of the group contributions. His proposed relations are

$$T_c = T_b[0.584 + 0.965\Sigma\Delta_T - (\Sigma\Delta_T)^2]^{-1} \tag{2-2.4}$$

$$P_c = (0.113 + 0.0032n_A - \Sigma\Delta_P)^{-2} \tag{2-2.5}$$

$$V_c = 17.5 + \Sigma\Delta_V \tag{2-2.6}$$

As with the Ambrose method, the units are kelvins, bars, and cubic centimeters per mole. $n_A$ is the number of atoms in the molecule. The Δ values are given in Table 2-2, and the procedure is illustrated in Example 2-2.

**Fedors method [13].** The Fedors group contribution method is valid only for critical temperatures. As will be seen later, it is less accurate than the methods of Ambrose and Joback, but it has the advantage of not requiring the normal boiling point to calculate $T_c$. The Fedors equation may be written as

$$T_c = 535 \log \Sigma\Delta_T \tag{2-2.7}$$

**TABLE 2-2 Joback Group Contributions for Critical Properties, the Normal Boiling Point, and the Freezing Point**

| | Δ | | | | |
|---|---|---|---|---|---|
| | $T_c$ | $P_c$ | $V_c$ | $T_b$ | $T_f$ |
| Nonring increments: | | | | | |
| $-CH_3$ | 0.0141 | −0.0012 | 65 | 23.58 | −5.10 |
| $>CH_2$ | 0.0189 | 0 | 56 | 22.88 | 11.27 |
| $>CH-$ | 0.0164 | 0.0020 | 41 | 21.74 | 12.64 |
| $>C<$ | 0.0067 | 0.0043 | 27 | 18.25 | 46.43 |
| $=CH_2$ | 0.0113 | −0.0028 | 56 | 18.18 | −4.32 |
| $=CH-$ | 0.0129 | −0.0006 | 46 | 24.96 | 8.73 |
| $=C<$ | 0.0117 | 0.0011 | 38 | 24.14 | 11.14 |
| $=C=$ | 0.0026 | 0.0028 | 36 | 26.15 | 17.78 |
| $\equiv CH$ | 0.0027 | −0.0008 | 46 | 9.20 | −11.18 |
| $\equiv C-$ | 0.0020 | 0.0016 | 37 | 27.38 | 64.32 |
| Ring increments: | | | | | |
| $-CH_2-$ | 0.0100 | 0.0025 | 48 | 27.15 | 7.75 |
| $>CH-$ | 0.0122 | 0.0004 | 38 | 21.78 | 19.88 |
| $>C<$ | 0.0042 | 0.0061 | 27 | 21.32 | 60.15 |
| $=CH-$ | 0.0082 | 0.0011 | 41 | 26.73 | 8.13 |
| $=C<$ | 0.0143 | 0.0008 | 32 | 31.01 | 37.02 |
| Halogen increments: | | | | | |
| $-F$ | 0.0111 | −0.0057 | 27 | −0.03 | −15.78 |
| $-Cl$ | 0.0105 | −0.0049 | 58 | 38.13 | 13.55 |
| $-Br$ | 0.0133 | 0.0057 | 71 | 66.86 | 43.43 |
| $-I$ | 0.0068 | −0.0034 | 97 | 93.84 | 41.69 |
| Oxygen increments: | | | | | |
| $-OH$ (alcohol) | 0.0741 | 0.0112 | 28 | 92.88 | 44.45 |
| $-OH$ (phenol) | 0.0240 | 0.0184 | −25 | 76.34 | 82.83 |
| $-O-$ (nonring) | 0.0168 | 0.0015 | 18 | 22.42 | 22.23 |
| $-O-$ (ring) | 0.0098 | 0.0048 | 13 | 31.22 | 23.05 |
| $>C=O$ (nonring) | 0.0380 | 0.0031 | 62 | 76.75 | 61.20 |
| $>C=O$ (ring) | 0.0284 | 0.0028 | 55 | 94.97 | 75.97 |
| $O=CH-$ (aldehyde) | 0.0379 | 0.0030 | 82 | 72.24 | 36.90 |
| $-COOH$ (acid) | 0.0791 | 0.0077 | 89 | 169.09 | 155.50 |
| $-COO-$ (ester) | 0.0481 | 0.0005 | 82 | 81.10 | 53.60 |
| | | | | — | |
| $=O$ (except as above) | 0.0143 | 0.0101 | 36 | −10.50 | 2.08 |
| Nitrogen increments: | | | | | |
| $-NH_2$ | 0.0243 | 0.0109 | 38 | 73.23 | 66.89 |
| $>NH$ (nonring) | 0.0295 | 0.0077 | 35 | 50.17 | 52.66 |
| $>NH$ (ring) | 0.0130 | 0.0114 | 29 | 52.82 | 101.51 |
| $>N-$ (nonring) | 0.0169 | 0.0074 | 9 | 11.74 | 48.84 |
| $-N=$ (nonring) | 0.0255 | −0.0099 | — | 74.60 | — |
| $-N=$ (ring) | 0.0085 | 0.0076 | 34 | 57.55 | 68.40 |
| $-CN$ | 0.0496 | −0.0101 | 91 | 125.66 | 59.89 |
| $-NO_2$ | 0.0437 | 0.0064 | 91 | 152.54 | 127.24 |
| Sulfur increments: | | | | | |
| $-SH$ | 0.0031 | 0.0084 | 63 | 63.56 | 20.09 |
| $-S-$ (nonring) | 0.0119 | 0.0049 | 54 | 68.78 | 34.40 |
| $-S-$ (ring) | 0.0019 | 0.0051 | 38 | 52.10 | 79.93 |

**TABLE 2-3 Fedors Group Contributions for Critical Temperature**

| Group | $\Delta_T$ | Group | $\Delta_T$ |
|---|---|---|---|
| $-CH_3$ | 1.79 | $-NH-$ (aromatic) | 7.64 |
| $-CH_2-$ | 1.34 | $>N-$ | 0.89 |
| $>CH-$† | 0.45 | $>N-$ (aromatic) | 4.74 |
| $>C<$ | −0.22 | $-N=$ | 4.51 |
| $=CH_2$ | 1.59 | $-S-S-$ | 9.83 |
| $=CH-$ | 1.40 | $-S-$ | 4.91 |
| $>C=$ | 0.89 | $-SH$ | 5.36 |
| $\equiv CH$ | 1.79 | $-F$ | 2.10 |
| $\equiv C-$ | 2.46 | $-F$ (aromatic) | 0.45 |
| $=C=$ | 1.03 | $-F$ (perfluoro) | 0.54 |
| $-COOH$ | 10.72 | $-Cl$ | 4.20 |
| $-CO-O-OC-$ (anhydride) | 7.95 | $-Cl$ (disubstituted) | 3.71 |
| $-CO-O-$ | 5.32 | $-Cl$ (trisubstituted) | 3.17 |
| $-O-OC-CO-O-$ (oxalate) | 6.25 | $-Br$ | 5.58 |
| $-CO-$ | 5.36 | $-I$ | 8.04 |
| $-O-$ | 1.56 | $-I$ (aromatic) | 10.77 |
| $-O-$ (aromatic) | 2.68 | Three-membered ring | 0.45 |
| $-OH$ | 5.63 | Five-membered ring | 2.23 |
| $-OH$ (aromatic) | 9.65 | Six-membered ring | 2.68 |
| $-CHO$ | 5.49 | Heteroatom in ring | 0.45 |
| $-C\equiv N$ | 8.49 | Substitution on carbon in a double bond (nonaromatic) | 0.58 |
| $-C\equiv N$ (aromatic) | 9.38 | | |
| $-NH_2$ | 4.56 | Orthosubstitution in a benzene ring | 1.16 |
| $-NH_2$ (aromatic) | 9.20 | | |
| $-NH-$ | 3.04 | Conjugation, per double bond | 0.13 |

†Except for adjacent pairs of $>CH-$; then add 0.76 for each.

with $T_c$ in kelvins. The $\Delta$ values are shown in Table 2-3, and the method is illustrated in Example 2-3.

**Example 2-1** Since the Ambrose method is somewhat more complicated than other methods, this example treats several types of compounds to illustrate some of the less obvious features.

Estimate the critical properties of the following compounds; for most, the experimental critical values shown are from Appendix A (when available) as well as the values of $T_b$ and $M$ necessary to use Ambrose's method.

| Compound | $T_b$, K | $M$ | $T_c$, K | $P_c$, bar | $V_c$, cm$^3$/mol |
|---|---|---|---|---|---|
| (*a*) 2,2,3-Trimethylpentane | 383.0 | 114.23 | 563.5 | 27.3 | 436 |
| (*b*) 1,*trans*-3,5-Trimethylcyclohexane | 413.7 | 126.24 | 602.2 | — | — |
| (*c*) 1,2,3-Trimethylbenzene | 449.3 | 120.20 | 664.5 | 34.5 | 430 |
| (*d*) 3-Methylbutan-2-one | 367.5 | 86.13 | 553.4 | 38.5 | 310 |
| (*e*) *N*-Methylaniline | 469.4 | 107.16 | 701 | 52.0 | — |
| (*f*) 2-Methylpentan-2-ol | 394.2 | 102.18 | 559.5 | — | — |
| (*g*) 1,2,3,4-Tetrafluorobenzene | 367.5 | 150.08 | 550.8 | 37.9 | 313 |
| (*h*) 1-Chloro-2,2-difluoroethylene | 254.6 | 98.48 | 400.6 | 44.6 | 197 |
| (*i*) Perfluorocyclohexene | 325.2 | 262.06 | 461.8 | — | — |

**solution**

(*a*) 2,2,3-TRIMETHYLPENTANE. There are eight aliphatic carbons, one $>$CH$-$ and one $>$C$<$. To obtain the delta Platt number, the Platt number for the isomer and alkane (*n*-octane) are:

$$\begin{array}{ccccccccc} & & C_6 & & C_8 & & & & \\ & & | & & | & & & & \\ C_1 & - & C_2 & - & C_3 & - & C_4 & - & C_5 \\ & & | & & & & & & \\ & & C_7 & & & & & & \end{array} \qquad C_1-C_2-C_3-C_4-C_5-C_6-C_7-C_8$$

The carbon pairs three bonds apart are:

For 2,2,3-trimethylpentane: 1-4, 2-5, 4-6, 4-7, 1-8, 6-8, 7-8, and 5-8, and the Platt number is 8.

For *n*-octane: 1-4, 2-5, 3-6, 4-7, and 5-8, and the Platt number is 5.

The delta Platt number = 8 − 5 = 3.

$$\Sigma\Delta_T = (8)\ (0.138) - 0.043 - 0.120 + (3)\ (-0.023) = 0.872$$

$$\Sigma\Delta_P = (8)\ (0.226) - 0.006 - 0.030 + (3)\ (-0.026) = 1.694$$

$$\Sigma\Delta_V = (8)\ (55.1) - 8 - 17 = 416$$

With Eqs. (2-2.1) to (2-2.3)

$$T_c = (383.0)\ [1 + (1.242 + 0.872)^{-1}] = 564.2 \text{ K}$$

$$P_c = (114.23)\ (0.339 + 1.694)^{-2} = 27.6 \text{ bar}$$

$$V_c = 40 + 416 = 456 \text{ cm}^3/\text{mol}$$

The percent errors in $T_c$, $P_c$, and $V_c$ are 0.1, 1.1, and 4.6%.

(*b*) 1,*trans*-3,5-TRIMETHYLCYLOHEXANE. There are nine carbon atoms. Six have the normal contribution, and three have ring $-CH_2-$. In addition, there are three corrections for $>$CH$-$. No delta Platt number is involved.

$$\Sigma\Delta_T = (6)\ (0.138) + (3)\ (0.090) + (3)\ (-0.043) = 0.969$$

$$\Sigma\Delta_P = (6)\ (0.226) + (3)\ (0.182) + (3)\ (-0.006) = 1.884$$

$$\Sigma\Delta_V = (6)\ (55.1) + (3)\ (44.5) + (3)\ (-8) = 440$$

With Eqs. (2-2.1) to (2-2.3)

$$T_c = (413.7)\ [1 + (1.242 + 0.969)^{-1}] = 600.8 \text{ K}$$

$$P_c = (126.24)\ (0.339 + 1.884)^{-2} = 25.5 \text{ bar}$$

$$V_c = 40 + 440 = 480 \text{ cm}^3/\text{mol}$$

The percent error for $T_c$ is −0.2%. No experimental values exist for $P_c$ and $V_c$. The fact that the configuration is trans does not enter into the calculations.

(*c*) 1,2,3-TRIMETHYLBENZENE. This is an aromatic compound. There is one contribution for the benzene ring; there are three alkyl contributions for $CH_3-$; and there is a correction for substitution on an aromatic ring. Also, since there are two ortho groups, there are two further corrections.

$$\Sigma\Delta_T = 0.448 + (3)\ (0.138) + 0.010 + (2)\ (0.030) + (2)\ (-0.040) = 0.852$$

$$\Sigma\Delta_P = (0.924) + (3)\ (0.226) + 0 + (2)\ (0.020) + (2)\ (-0.050) = 1.542$$

$$\Sigma\Delta_V = (9)\ (55.1) + (3)\ (-20) = 436 \text{ cm}^3/\text{mol}$$

Note that, for $\Sigma\Delta_V$, Table 2-1 suggests that one should use alkyl group values when determining $V_c$ for aromatic compounds. Thus we assume nine alkyl carbons but also include three double bonds.

$$T_c = (449.3)\,[1 + (1.242 + 0.852)^{-1}] = 663.9 \text{ K}$$

$$P_c = (120.20)\,(0.339 + 1.542)^{-2} = 34.0 \text{ bar}$$

$$V_c = 40 + 436 = 476 \text{ cm}^3\text{/mol}$$

The percent errors are −0.1, −1.5, and 11%.

(*d*) 3-METHYLBUTAN-2-ONE. There are four aliphatic carbons; there is one >CH− correction; and there is one >CO.

$$\Sigma\Delta_T = (4)\,(0.138) - 0.043 + 0.220 = 0.729$$

$$\Sigma\Delta_P = (4)\,(0.226) - 0.006 + 0.282 = 1.180$$

$$\Sigma\Delta_V = (4)\,(55.1) - 8 + 60 = 272$$

Then,

$$T_c = (367.5)\,[1 + (1.242 + 0.729)^{-1}] = 554.0 \text{ K}$$

$$P_c = (86.13)\,(0.339 + 1.180)^{-2} = 37.3 \text{ bar}$$

$$V_c = 40 + 272 = 312 \text{ cm}^3\text{/mol}$$

The percent errors are 0.1, −3.1, and 0.6%.

(*e*) *N*-METHYLANILINE. There is one benzene ring; there is one aliphatic carbon; there is one >NH; and there is an aromatic substitution.

$$\Sigma\Delta_T = 0.448 + 0.138 + 0.01 + 0.208 = 0.804$$

$$\Sigma\Delta_P = 0.924 + 0.226 + 0.135 = 1.285$$

$$\Sigma\Delta_V = (7)\,(55.1) + (3)\,(-20) + 30 = 356$$

Note that for $\Sigma\Delta_V$, we treated the seven carbon atoms as aliphatic and corrected with three double bonds.

$$T_c = (469.4)\,[1 + (1.242 + 0.804)^{-1}] = 699 \text{ K}$$

$$P_c = (107.16)\,(0.339 + 1.285)^{-2} = 40.6 \text{ bar}$$

$$V_c = 40 + 356 = 396 \text{ cm}^3\text{/mol}$$

The percent errors in $T_c$ and $P_c$ are −0.3 and −22%. The estimate for critical pressure in this case is quite poor, and it seems probable that the experimental value is in error.

(*f*) 2-METHYLPENTAN-2-OL. The hydrocarbon homomorph of 2-methylpentan-2-ol is 2,2-dimethylpentane, and we need to determine $\Sigma\Delta_T$ and $\Sigma\Delta_P$ for the latter.

$$\Sigma\Delta_T = (7)\,(0.138) - 0.120 = 0.846$$

$$\Sigma\Delta_P = (7)\,(0.226) - 0.030 = 1.552$$

For the critical temperature we need to correct $\Sigma\Delta_T$ as indicated in the footnotes of Table 2-1, i.e., subtract 0.138 for the −OH. Also, we calculate

$$n = \frac{T_b\,(\text{alcohol}) - 314.1}{19.2} = \frac{394.2 - 314.1}{19.2} = 4.17$$

and determine the correction $= (0.87 - 0.11n + 0.003n^2) = 0.463$. Then,

$$\Sigma\Delta_T\,(\text{alcohol}) = 0.846 - 0.138 + 0.463 = 1.171$$

In a similar manner, for $\Sigma\Delta_P$, with $n = 4.17$

$$\Sigma\Delta_P\,(\text{alcohol}) = 1.552 - 0.226 + [0.100 - (0.13)\,(4.17)] = 0.884$$

For $\Sigma\Delta_V$, we do not require these corrections, so

$$\Sigma\Delta_V = (6)(55.1) - 17 + 15 = 329 \text{ cm}^3/\text{mol}$$

Then we can use Eqs. (2-2.1) to (2-2.3)

$$T_c = (394.2)[1 + (1.242 + 1.171)^{-1}] = 557.6 \text{ K}$$

$$P_c = (102.18)(0.339 + 0.884)^{-2} = 68.3 \text{ bar}$$

$$V_c = 40 + 329 = 369 \text{ cm}^3/\text{mol}$$

Only the critical temperature is known; for this property, the percent error is $-0.3\%$

(*g*) 1,2,3,4-TETRAFLUOROBENZENE. Since tetrafluorobenzene is not a completely fluorinated aliphatic compound, the estimation method uses the benzene ring plus four aromatic $-F$ contributions.

$$\Sigma\Delta_T = 0.448 + (4)(0.080) = 0.768$$

$$\Sigma\Delta_P = 0.924 + (4)(0.183) = 1.656$$

$$\Sigma\Delta_V = (6)(55.1) + (3)(-20) + (4)(14) = 327$$

$$T_c = (367.5)[1 + (1.242 + 0.768)^{-1}] = 550.3 \text{ K}$$

$$P_c = (150.08)(0.339 + 1.656)^{-2} = 37.7 \text{ bar}$$

$$V_c = 40 + 327 = 367 \text{ cm}^3/\text{mol}$$

The percent errors are $-0.1$, $-0.5$, and 17%.

(*h*) 1-CHLORO-2,2-DIFLUOROETHYLENE. Here we have two aliphatic carbons, a double bond, a fluorine correction, two $-F$, and one $-Cl$.

$$\Sigma\Delta_T = (2)(0.138) - 0.050 + 0.125 + (2)(0.055) + 0.055 = 0.516$$

$$\Sigma\Delta_P = (2)(0.226) - 0.065 + (2)(0.223) + 0.318 = 1.151$$

$$\Sigma\Delta_V = (2)(55.1) - 20 + (2)(14) + 45 = 163$$

$$T_c = (254.6)[1 + (1.242 + 0.516)^{-1}] = 399.4 \text{ K}$$

$$P_c = (98.48)(0.339 + 1.151)^{-2} = 44.4 \text{ bar}$$

$$V_c = 40 + 163 = 203 \text{ cm}^3/\text{mol}$$

The percent errors are $-0.3$, $-0.4$, and 3.0%.

(*i*) PERFLUOROCYCLOHEXENE. In this final case, we have a perfluorinated compound, so that constants 1.242 and 0.339 in Eqs. (2-2.1) and (2-2.2) are changed to 1.570 and 1.00. We have four $(-CF_2)_{\text{ring}}$, two $(>CF-)_{\text{ring}}$, and one ring double bond.

$$\Sigma\Delta_T = (4)(0.140) + (2)(0.140) - 0.030 = 0.810$$

$$\Sigma\Delta_P = (4)(0.420) + (2)(0.420) - 0 = 2.520$$

$\Sigma\Delta_V$ cannot be calculated

$$T_c = (325.2)[1 + (1.570 + 0.810)^{-1}] = 461.9 \text{ K}$$

$$P_c = (262.06)(1.00 + 2.52)^{-2} = 21.1 \text{ bar}$$

Only $T_c$ is known; the error is 0.1%.

**Example 2-2** Using the Joback modification of Lydersen's method, estimate the critical properties of 3-methylbutan-2-one.

**solution** With Table 2-2, for this compound

| | $\Delta_T$ | $\Delta_P$ | $\Delta_V$ |
|---|---|---|---|
| (3) $-CH_3$ | (3) (0.0141) | (3) (−0.0012) | (3) (65) |
| $>$CH$-$ | 0.0164 | 0.0020 | 41 |
| C=0 | 0.0380 | 0.0031 | 62 |
| | 0.0967 | 0.0015 | 298 |

Thus $\Sigma\Delta_T = 0.0976$, $\Sigma\Delta_P = 0.0015$, $\Sigma\Delta_V = 298$, and the number of atoms $n_A$ is equal to 16 ($C_5H_{10}O$). With Eqs. (2-2.4) to (2-2.6) and with $T_b = 367.5$ K from the table in Example 2-1,

$$T_c = (367.5)\,[0.584 + (0.965)\,(0.0967) - (0.0967)^2]^{-1} = 550.2 \text{ K}$$

$$P_c = [0.113 + (0.0032)\,(16) - 0.0015]^{-2} = 37.8 \text{ bar}$$

$$V_c = 17.5 + 298 = 315.5 \text{ cm}^3\text{/mol}$$

The percent errors are −0.6, 1.9, and 1.7% for $T_c$, $P_c$, and $V_c$. Shown below are the percent errors found for Joback's technique with the other compounds tested in Example 2-1.

| Compound | Percent error in $T_c$ | $P_c$ | $V_c$ | Compound | Percent error in $T_c$ | $P_c$ | $V_c$ |
|---|---|---|---|---|---|---|---|
| (*a*) | 0.4 | −4.4 | 7.0 | (*f*) | −1.2 | — | — |
| (*b*) | 1.4 | — | — | (*g*) | 3.6 | −7.4 | 7.0 |
| (*c*) | −0.3 | −8.6 | 0.3 | (*h*) | −0.1 | 2.4 | 8.3 |
| (*d*) | −0.6 | 1.9 | 1.7 | (*i*) | −2.3 | — | — |
| (*e*) | 0 | −20 | — | | | | |

**Example 2-3** Using Fedors method, estimate the critical temperature of 2,2,3-trimethylpentane.

**solution** Using Table 2-3,

$$\Sigma\Delta_T = (5)\,(-CH_3) + -CH_2 + >CH- + >C< = (5)\,(1.79) + 1.34 + 0.45 - 0.22$$
$$= 10.52$$

With Eq. (2-2.7),

$$T_c = (535)\,(\log_{10} 10.52) = 546.8 \text{ K}$$

The percent error is −3.0%. For the various compounds tested in Example 2-1, the Fedors method was found to give percent errors of (*a*) −3.0%, (*b*) 0.2%, (*c*) −0.4%, (*d*) 1.3%, (*e*) −0.7%, (*f*) 8.0%, (*g*) 12%, (*h*) 46%, and (*i*) 17%.

## Discussion

For *critical temperatures,* a comparison was made between experimental values and the values estimated by the methods of Ambrose and Joback.

Over 400 compounds from Appendix A were included. The results are shown below.

| | Ambrose | Joback |
|---|---|---|
| Mean of the absolute error, K | 4.3 | 4.8 |
| Average of the absolute percent error | 0.7 | 0.8 |

The Ambrose method yields the smaller error, but it is somewhat more complicated in its use than Joback's. Other recent group contribution estimation methods were also evaluated [14, 18, 21, 47]. The Fishtine procedure [14] shows errors comparable to those found above, but it involves a reference substance of "similar" structure whose properties must be known. The other references cited showed higher errors in the testing than either the method of Ambrose or that of Joback, or they were limited to hydrocarbons.

The Design Institute for Physical Property Data [10] recommends either Fishtine's or Ambrose's method for nonhydrocarbons and the API technique for hydrocarbons [6].

We suggest either the Ambrose or the Joback form to estimate $T_c$ when a reliable value of $T_b$ is known. The Fedors relation, Eq. (2-2.7), is of value when $T_b$ is not known. For many compounds, it yields a relatively small error, but for some materials (see Example 2-3), the accuracy is poor.

Finally, we note a deceptively simple relation derived by Klincewicz [20, 21] which involves no group contributions.

$$T_c = 50.2 - 0.16M + T_b \tag{2-2.8}$$

where $T_b$ is the normal boiling temperature (at 1 atm) in kelvins and $M$ is the molecular weight. When Eq. (2-2.8) was used to estimate $T_c$ for the compounds in Example 2-1, the percent errors were (*a*) 1.5%, (*b*) 1.8%, (*c*) 0%, (*d*) 0.2%, (*e*) −0.9%, (*f*) 5.4%, (*g*) −1.2%, (*h*) −1.8%, and (*i*) 1.1%. Except for case (*f*), the errors found range from about 1 to 2 percent. This is typical of the range found in more extensive testing.

For *critical pressure,* all the estimation methods tested yielded higher percent errors than for critical temperature, and the database employed was somewhat smaller. About 390 compounds were tested. The results for the Ambrose and Joback methods are shown below.

| | Ambrose | Joback |
|---|---|---|
| Mean of the absolute error, bar | 1.8 | 2.1 |
| Average of the absolute percent error | 4.6 | 5.2 |

These errors are typical of those shown in Examples 2-1 and 2-2. The very large error for $N$-methylaniline is suspect. The experimental value for this material was listed by Kudchadker et al. [22].

A number of other estimation methods also were examined [14, 20, 21, 26, 34, 44, 47]. The Fishtine procedure [14] is similar in accuracy to the methods of Ambrose and Joback but, as noted earlier in the discussion on critical temperature, it requires the choice of a reference material of similar structure. It was tested, along with Ambrose's, by Danner and Daubert [10] and found to be about equal in accuracy for nonhydrocarbons. For hydrocarbons, these authors recommend the method in [6].

We recommend the method of either Ambrose or Joback to estimate critical pressures. In the former the molecular weight is employed as a secondary parameter, whereas in the latter, the number of atoms is used instead.

For *critical volume,* the Ambrose and Joback methods were tested with data from 310 compounds. The associated errors were found to be

| | Ambrose | Joback |
|---|---|---|
| Mean of the absolute error, $cm^3/mol$ | 8.5 | 7.5 |
| Average of the absolute percent error | 2.8 | 2.3 |

Several other recent estimation methods for critical volume also were evaluated. The methods of Fishtine [14], Vetere [48], Fedors [11, 12], and Klincewicz [20, 21] are also group contribution methods. They are not recommended relative to the Ambrose or the Joback technique either because the overall accuracy was less or because they could not treat as wide a range of compounds. Nath [33] has also proposed that $V_c$ be estimated from $T_c$, $P_c$, and the acentric factor, but the correlation is limited to nonpolar substances.

We recommend either the Joback or the Ambrose method to estimate critical volumes. Expected errors are in the range of 5 to 10 percent. Danner and Daubert [10], however, recommend the Fedors method [12].

As a final comment relative to the estimation of critical properties, we note that occasionally one may "back out" critical properties from experimental data if a corresponding states correlation is believed to be applicable. That is, if the property were some function of reduced temperature ($T/T_c$), reduced pressure ($P/P_c$), and/or reduced volume ($V/V_c$), then by knowing the values of the property over a range of temperatures, pressures, and/or volumes, it is possible to estimate "characteristic" critical properties. See, for example, Refs. 5, 15, and 24.

## 2-3 Acentric Factor

One of the more common pure component constants is the acentric factor [38, 39], which is defined as

$$\omega = -\log P_{vp_r} \text{ (at } T_r = 0.7) - 1.000 \tag{2-3.1}$$

To obtain values of $\omega$, the reduced vapor pressure ($P_r = P/P_c$) at $T_r = T/T_c = 0.7$ is required.

As originally proposed, $\omega$ represented the acentricity or nonsphericity of a molecule. For monatomic gases, $\omega$ is, therefore, essentially zero. For methane, it is still very small. However, for higher-molecular-weight hydrocarbons, $\omega$ increases. It also rises with polarity. At present, $\omega$ is very widely used as a parameter which in some manner is supposed to measure the complexity of a molecule with respect to both the geometry and polarity, but the large values of $\omega$ for some polar compounds ($\omega > 0.4$) are not meaningful in the context of the original meaning of this property.

We show in Appendix A the acentric factor for many materials. The values were obtained, in most cases, from experimental data on $T_c$, $P_c$, and vapor pressures.

If acentric factors are needed for a material not shown in Appendix A, the usual technique is to locate (or estimate) the critical constants $T_c$ and $P_c$ and then determine the vapor pressure at $T_r = 0.7$. This latter estimation would normally be made by using one of the reduced vapor pressure correlations given later in Chap. 7. As an example, if the vapor pressure correlation chosen were

$$\log P_{vp} = A + \frac{B}{T} \tag{2-3.2}$$

with $A$ and $B$ found, say, from the sets ($T_c$, $P_c$; $T_b$, $P = 1$ atm), then

$$\omega = \frac{3}{7}\frac{\theta}{1-\theta}\log P_c - 1 \tag{2-3.3}$$

where $P_c$, in this case, must be expressed in atmospheres and $\theta = T_b/T_c$.

Similarly, if the Lee-Kesler vapor pressure relations (7-2.6) to (7-2.8) were used,

$$\omega = \frac{\alpha}{\beta} \tag{2-3.4}$$

where $\alpha = -\ln P_c - 5.97214 + 6.09648\theta^{-1} + 1.28862 \ln \theta - 0.169347\theta^6$

$\beta = 15.2518 - 15.6875\theta^{-1} - 13.4721 \ln \theta + 0.43577\theta^6$

and $P_c$ is in atmospheres. Lee and Kesler [23] report that Eq. (2-3.4) yields values of $\omega$ very close to those selected by Passut and Danner [36] and Henry and Danner [16] in their critical reviews.

**Example 2-4** Estimate the acentric factor of isopropylbenzene by using Eqs. (2-3.3) and (2-3.4). The accepted value is 0.326.

**solution** From Appendix A, $T_c$ = 631.1 K, $T_b$ = 425.6 K, $P_c$ = 32.1 bar = 31.7 atm. Thus, $\Theta$ = (425.6/631.1) − 0.674. With Eq. (2-3.3),

$$\omega = \frac{3}{7}\frac{0.674}{1-0.674}\log 31.7 - 1$$
$$= 0.330$$

Using Eq. (2-3.4),

$$\alpha = -\ln(31.7) - 5.92714 + (6.09648)(0.674)^{-1} + (1.28862)\ln(0.674)$$
$$-(0.169347)(0.674)^6$$
$$= -0.8669$$
$$\beta = 15.2518 - (15.6875)(0.674)^{-1} - (13.4721)\ln(0.674) + (0.43577)(0.674)^6$$
$$= -2.662$$
$$\omega = \frac{\alpha}{\beta} = \frac{-0.8669}{-2.662} = 0.326$$

In many instances in the literature, one finds $\omega$ related to $Z_c$ by

$$Z_c = \frac{P_c V_c}{RT_c} = 0.291 - 0.080\omega \tag{2-3.5}$$

This equation results from applying a $PVT$ correlation that employs $\omega$ at the critical point, where $Z = Z_c$. Equation (2-3.5) is only very approximate, as the reader can readily show from the values in Appendix A.

In other recent papers dealing with the acentric factor, Nath [32] relates $\omega$ to the enthalpy of vaporization and to the reduced temperature; Hoshino et al. [17] propose a group contribution method to estimate $\omega$ that is applicable to saturated hydrocarbons.

Chappelear [9] makes an observation that, over the years, the "accepted" values of the acentric factor may change due to new vapor pressure data or critical constants. However, if one is using a correlation that was developed from earlier values of $\omega$, then these acentric factors should be employed and not the newer, updated values. She notes the problem of carbon dioxide in particular. In Appendix A, we show $\omega$ = 0.225; others have quoted a value of 0.267 [35]. The differences result from the extrapolation technique used to extend the liquid region past the freezing point to a reduced temperature of 0.7. Neither $\omega$ = 0.225 nor $\omega$ = 0.267 can be considered a firm value, but if one were to use a correlation such as the Peng-Robinson equation of state [37], the acentric factor value of $CO_2$ should be 0.225, since that was the value used by Peng and Robinson to develop their correlation.

## 2-4 Boiling and Freezing Points

Ordinarily, when one refers to a freezing or boiling point, there is an implied condition that the pressure is 1 atm. A more exact terminology for these temperatures might be the *normal freezing* and *boiling points.* In Appendix A, values for $T_f$ and $T_b$ are given for many substances.

A number of methods to estimate the normal boiling point have been proposed. Many are reviewed in a previous edition of this book [42]. More recent techniques are usually specific for a given homologous series as, for example, the work of Ambrose [1] on alkanols. Others, e.g., [30, 31], attempt to use London's theory to relate $T_b$ to basic molecular parameters such as ionization potential, molar refraction, and shape. None in this latter class yield accurate estimations, but they can provide useful guidelines.

To obtain a very approximate guess of $T_b$, one may use the group contributions for $T_b$ in Table 2-2 with the relation:

$$T_b = 198 + \Sigma\Delta_b \qquad (2\text{-}4.1)$$

where $T_b$ is in kelvins. The group increments were developed by Joback [19], and these, with Eq. (2-4.1) were tested on 438 diverse organic compounds. The average absolute error found was 12.9 K, and the standard deviation of the error was 17.9 K. The average of the absolute percent errors was 3.6%. Whereas these errors are not small, this simple technique may be useful as a guide in obtaining approximate values of $T_b$ should no experimental value be available.

**Example 2-5** Estimate the normal boiling temperature of isopentylmercaptan. The reported value of $T_b$ is 393 K [4].

**solution** From Table 2-2, for this compound we have two $-CH_3$, two $-CH_2$, one $>CH-$, and one $-SH$. $\Sigma\Delta_b = (2)(23.6) + (2)(22.9) + 21.7 + 63.6 = 178.3$

$$T_b = 198 + 178 = 376 \text{ K}$$

The error is $376 - 393 = -17$ K.

The estimation of the normal freezing point is complicated by the fact that $T_f = \Delta H_{\text{fus}}/\Delta S_{\text{fus}}$ and, whereas $\Delta H_{\text{fus}}$ depends primarily upon intermolecular forces, $\Delta S_{\text{fus}}$ is a function of the molecular symmetry. As noted by Bondi [7], $\Delta S_{\text{fus}}$ is larger when the molecule can assume many orientations in the liquid phase relative to the solid. Thus, $\Delta S_{\text{fus}}$ is smaller for spherical, rigid molecules and $T_f$ is higher than for molecules of the same size which are not spherical and are flexible.

No reliable methods are now available to estimate $T_f$. For a *very* approximate guess, one may use Eq. (2-4.2) with the group contributions developed by Joback [19] and shown in Table 2-2 under $T_f$.

$$T_f = 122 + \Sigma\Delta_f \qquad (2\text{-}4.2)$$

When tested with 388 simple and complex organic compounds, the average absolute error was 23 K with a standard deviation of 25 K. The average of the absolute percent errors was 11%.

**Example 2-6** Estimate the normal freezing temperature of 2-bromobutane by using Eq. (2-4.2) and the group contributions in Table 2-2. The experimental value of $T_f$ is 161 K [49].

**solution** Here we have two $-CH_3$, one $-CH_2-$, one $>CH-$, and one $-Br$. Thus,

$$\Sigma\Delta_f = (2)(-5.1) + 11.3 + 12.6 + 43.4 = 57$$

$$T_f = 122 + 57 = 179 \text{ K}$$

The error is 179 − 161 = 18 K.

## 2-5 Dipole Moments

Dipole moments of molecules are often required in property correlations for polar materials. The best source of this constant is the compilation by McClellan [28], which has, to a large degree, superseded prior summaries such as those given by Smith [45] and Smyth [46]. For those rare occasions when one may be forced to estimate a value, there are vector group contribution methods, although they ordinarily require considerable effort. Most such methods are summarized in the text by Minkin et al. [29].

Dipole moments for many materials are listed in Appendix A. No temperature effect is shown, because dipole moments are insensitive to that variable. Also, we have not noted whether the values were measured in the gas phase or in a solvent, because differences between such measurements are ordinarily small.

Dipole moments are expressed in debye units, 1 debye being equivalent to $10^{-18}$ $(\text{dyn}\cdot\text{cm}^4)^{1/2} = 3.162 \times 10^{-25}$ $(\text{J}\cdot\text{m}^3)^{1/2}$. Thus, the physical unit for this property is $[(\text{energy})(\text{volume})]^{1/2}$.

## Notation

| | |
|---|---|
| $M$ | molecular weight |
| $n_A$ | number of atoms in a molecule |
| $P$ | pressure, bar; $P_c$, critical pressure; $P_r$, reduced pressure, $P/P_c$; $P_{vp}$, vapor pressure; $P_{vpr}$, $P_{vp}/P_c$ |
| $R$ | gas constant |
| $T$ | temperature, kelvins; $T_c$, critical temperature; $T_r$, reduced temperature, $T/T_c$; $T_b$, normal boiling point (at 1 atm); $T_f$, freezing point |
| $V$ | molar volume, $cm^3$/mol; $V_c$, critical volume |
| $Z$ | compressibility factor; $Z_c$, at the critical point |

Greek

Δ group contribution. When subscripted by $T$, for $T_c$; by $P$ for $P_c$; by $V$ for $V_c$; by $b$ for $T_b$; and by $f$ for $T_f$

θ $T_b/T_c$

## REFERENCES

1. Ambrose, D.: "Correlation of the Boiling Points of Alkanols," National Physical Laboratory, Teddington, *NPL Rep. Chem. 57,* December 1976; *Appl. Chem. Biotechnol.,* **26**: 711 (1976).
2. Ambrose, D.: "Correlation and Estimation of Vapour-Liquid Critical Properties. I. Critical Temperatures of Organic Compounds," National Physical Laboratory, Teddington, *NPL Rep. Chem. 92,* September 1978, corrected March 1980.
3. Ambrose, D.: "Correlation and Estimation of Vapour-Liquid Critical Properties. II. Critical Pressures and Volumes of Organic Compounds," National Physical Laboratory, Teddington, *NPL Rep. Chem. 98,* 1979.
4. Ambrose, D.: "Vapour-Liquid Critical Properties," National Physical Laboratory, Teddington, *NPL Rep. Chem. 107,* February 1980.
5. Ambrose, D., and N. C. Patel: *J. Chem. Thermodyn.,* **16**: 459 (1984).
6. Am. Petrol. Inst.: *Tech. Data Book—Petroleum Refining,* 3rd ed., API, Washington, D.C., 1977 (extant 1981), chap. 4.
7. Bondi, A.: *Physical Properties of Molecular Crystals, Liquids and Glasses,* Wiley, New York, 1968, chap. 6.
8. Brulé, M. R., C. T. Lin, L. L. Lee, and K. E. Starling: *AIChE J.,* **28**: 616 (1982).
9. Chappelear, P. S.: *Fluid Phase Equil.* **9**: 319 (1982).
10. Danner, R. P., and T. E. Daubert: *Manual for Predicting Chemical Process Design Data,* Design Institute for Physical Property Data, AIChE, 1983, chap. 2.
11. Fedors, R. F.: *Polymer Lett. Ed.,* **11**: 767 (1973).
12. Fedors, R. F.: *AIChE J.,* **25**: 202 (1979).
13. Fedors, R. F.: *Chem. Eng. Commun.,* **16**: 149 (1982).
14. Fishtine, S. H.: *Z. Physik. Chem. Neue Folge,* **123**: 39 (1980).
15. Gunn, R. D., and V. J. Mahajan: "Corresponding States Theory for High Boiling Compounds," paper presented at *Natl. Mtg. AIChE, New Orleans, La., March 1974.*
16. Henry, W. P., and R. P. Danner: *Ind. Eng. Chem. Process Design Develop.,* **17**: 373 (1978).
17. Hoshino, D., K. Nagahama, and M. Hirata: *J. Chem. Eng. Japan,* **15**: 153 (1982).
18. Jalowka, J. W.: M.S. thesis, Dept. Chem. Eng., Pennsylvania State University, University Park, Pa., August 1984.
19. Joback, K. G.: S.M. thesis in chemical engineering, Massachusetts Institute of Technology, Cambridge, Mass., June 1984.
20. Klincewicz, K. M.: S.M. thesis in chemical engineering, Massachusetts Institute of Technology, Cambridge, Mass., June 1982.
21. Klincewicz, K. M., and R. C. Reid: *AIChE J.,* **30**: 137 (1984).
22. Kudchadker, A. P., G. H. Alani, and B. J. Zwolinski: *Chem. Rev.,* **68**: 659 (1968).
23. Lee, B. I., and M. G. Kesler: *AIChE J.,* **21**: 510 (1975).
24. Lewis, G. N., and M. Randall: *Thermodynamics,* 2d ed., rev. by K. S. Pitzer and L. Brewer, app. 1, McGraw-Hill, New York, 1961.
25. Lin, H.-M., and K.-C. Chao: *AIChE J.,* **30**: 981 (1984).
26. Lydersen, A. L.: "Estimation of Critical Properties of Organic Compounds," *Univ. Wisconsin Coll. Eng., Eng. Exp. Stn. rept.3,* Madison, Wis., April 1955.
27. Mathews, J. F.: *Chem. Rev.,* **72**: 71 (1972).
28. McClellan, A. L.: *Tables of Experimental Dipole Moments,* Freeman, San Francisco, 1963.
29. Minkin, V. I., O. A. Osipov, and Y. A. Zhdanov: *Dipole Moments in Organic Chemistry,* trans. from the Russian by B. J. Hazard, Plenum, New York, 1970.

30. More, R., and A. L. Capparelli: *J. Phys. Chem.*, **84**: 1870 (1980).
31. Myers, R. T.: *J. Phys. Chem.*, **83**: 294 (1979).
32. Nath, J.: *Ind. Eng. Chem. Fundam.*, **18**: 297 (1979).
33. Nath, J.: *Ind. Eng. Chem. Fundam.*, **21**: 325 (1982).
34. Nath, J.: *Ind. Eng. Chem. Fundam.*, **22**: 358 (1983).
35. Nat. Gas. Processors Assoc.: *Engineering Data Book,* 1981.
36. Passut, C. A., and R. P. Danner: *Ind. Eng. Chem. Process Design Develop.*, **12**: 365 (1973).
37. Peng, D.-Y., and D. B. Robinson: *Ind. Eng. Chem. Fundam.*, **15**: 59 (1976).
38. Pitzer, K. S.: *J. Am. Chem. Soc.*, **77**: 3427 (1955).
39. Pitzer, K. S., D. Z. Lippmann, R. F. Curl, C. M. Huggins, and D. E. Petersen: *J. Am. Chem. Soc.*, **77**: 3433 (1955).
40. Platt, J. R.: *J. Chem. Phys.*, **15**: 419 (1947).
41. Platt, J. R.: *J. Phys. Chem.*, **56**: 328 (1952).
42. Reid, R. C., and T. K. Sherwood: *The Properties of Gases and Liquids,* 2d ed., McGraw-Hill, New York, 1966, chap. 2.
43. Riazi, M. R., and T. E. Daubert: *Hydrocarbon Process. Petrol. Refiner,* **59**(3): 115 (1980).
44. Riedel, L.: *Chem. Ing. Tech.*, **26**: 83 (1954).
45. Smith, J. W.: *Electric Dipole Moments,* Butterworth, London, 1955, chap. 3.
46. Smyth, C. P.: *Dielectric Behavior and Structure,* McGraw-Hill, New York, 1955, pp. 16–50.
47. Soulie, M. A., and J. Rey: *Bull. Soc. Chim. France,* **1980**(3-4): I-117.
48. Vetere, A.: *AIChE J.*, **22**: 950 (1976); ibid, **23**: 406 (1977).
49. Weast, R. C., (ed.): *The CRC Handbook of Chemistry and Physics,* CRC Press, Cleveland, Ohio, 1977.
50. Wilson, G. M., R. H. Johnson, S. C. Hwang, and C. Tsonopoulos: *Ind. Eng. Chem. Process Design Develop.*, **20**: 94 (1981).

# Chapter 3

# Pressure-Volume-Temperature Relations of Pure Gases and Liquids

## 3-1 Scope

Methods are presented in this chapter for estimating the volumetric behavior of pure gases and liquids as functions of temperature and pressure. Extension to mixtures is given in Chap. 4. Emphasis is placed on equations of state which are applicable to computer-based property-estimation systems.

The equations of state described in this chapter are employed in Chap. 5 to determine thermodynamic departure functions and partial molar properties.

## 3-2 Two-Parameter Correlations

The nonideality of a gas is conveniently expressed by the compressibility factor $Z$:

$$Z = \frac{PV}{RT} \tag{3-2.1}$$

where $V$ = molar volume
$P$ = absolute pressure
$T$ = absolute temperature
$R$ = universal gas constant

The gas constant $R$ assumes different values for different sets of units. Common values are shown in Table 3-1. For the remainder of this book, unless otherwise noted, pressure will be in bars, volume in $cm^3$/mol, and the term mol will refer to gram moles. Note that 1 bar = $10^5$ Pa = $10^5$ N/$m^2$ and 1 atm = 1.01325 bar. For an ideal gas $Z$ = 1.0. For real gases, $Z$ is normally less than 1 except at high reduced temperatures and pressures. Equation (3-2.1) can also be used to define $Z$ for a liquid; in this case $Z$ is normally much less than unity.

The compressibility factor is often correlated with the reduced temperature $T_r$ and pressure $P_r$ as

$$Z = f(T_r, P_r) \tag{3-2.2}$$

where $T_r = T/T_c$ and $P_r = P/P_c$. The function $f(\ )$ has been obtained from experimental $PVT$ data by Nelson and Obert [73], and the final curves are shown in Figs. 3-1 to 3-3. Except as noted below, the use of these figures to obtain $Z$ at a given $T_r$ and $P_r$ should lead to errors of less than 4 to 6 percent except near the saturation curve or near the critical point, where $Z$ is very sensitive to both $T_r$ and $P_r$.

Figures 3-1 and 3-3 should not be used for strongly polar fluids, nor are they recommended for helium, hydrogen, or neon unless special, modified critical constants are used [28, 62, 69, 74]. For very high pressures or very high temperatures, the reduced pressure-temperature-density charts of Breedveld and Prausnitz [9] are useful.

Many graphs similar to those in Figs. 3-1 to 3-3 have been published. All differ somewhat, because each reflects the author's choice of experimental data and how the data are smoothed. Those shown are as accurate as any two-parameter plots published, and they have the added advantage that volumes can be found directly. Note, however, that in these figures

**TABLE 3-1 Values of the Gas Constant *R***

| Value of $R$ | Unit† for $R$ |
|---|---|
| 83.144 | bar·cm³/(mol·K) |
| 8.3144 | J/(mol·K) |
| 10.732 | psia·ft³/(lb-mol·°R) |
| 82.057 | atm·cm³/(mol·K) |

†The unit mol refers to gram moles.

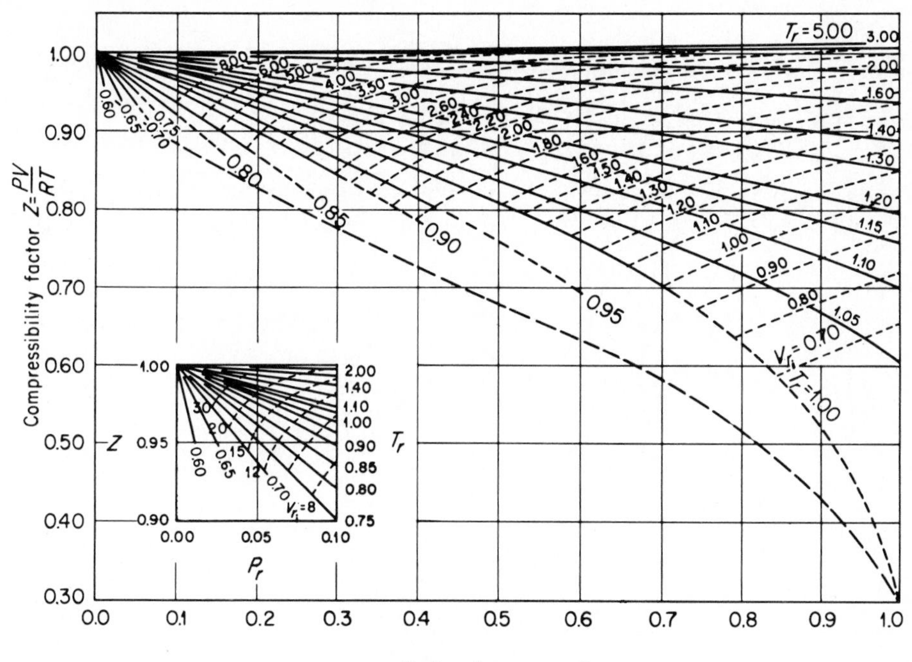


**Figure 3-1** Generalized compressibility chart. $V_{r_i}$ is $V/(RT_c/P_c)$. *(From Ref. 73.)*

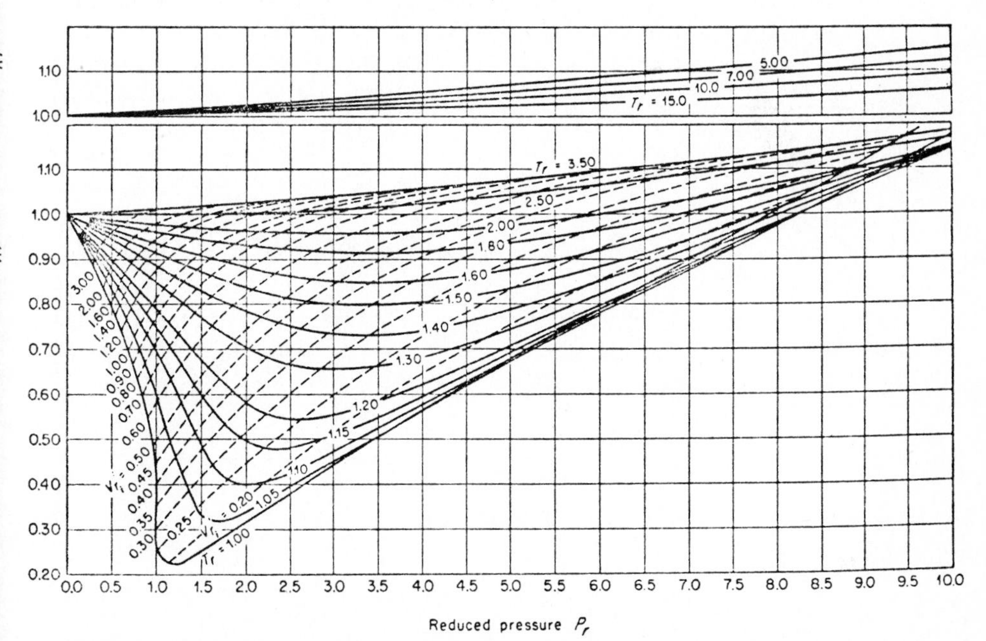


**Figure 3-2** Generalized compressibility chart. $V_{r_i}$ is $V/(RT_c/P_c)$. *(From Ref. 73.)*

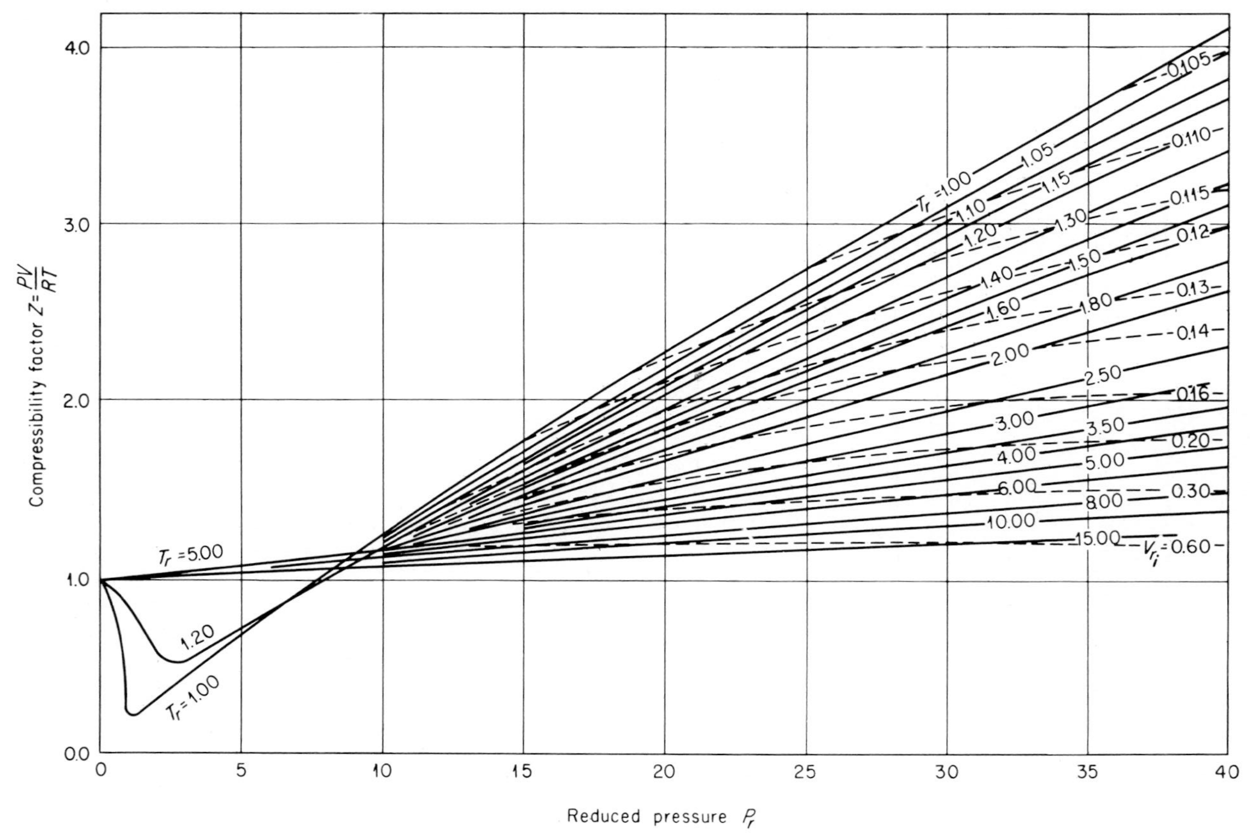


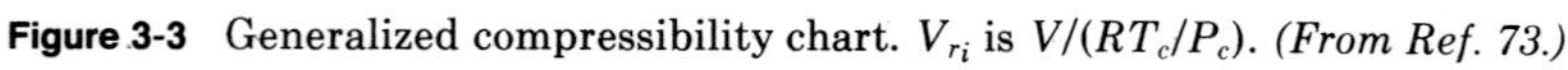

**Figure 3-3** Generalized compressibility chart. $V_{ri}$ is $V/(RT_c/P_c)$. *(From Ref. 73.)*

$V_{r_i}$ is not defined in the usual manner, that is, $V/V_c$, but instead is an "ideal reduced volume" given by

$$V_{r_i} = \frac{V}{RT_c/P_c} \tag{3-2.3}$$

Equation (3-2.2) is an example of the law of corresponding states. This law, though not exact, suggests that reduced properties of all fluids are essentially the same if compared at equal reduced temperatures and pressures. For $PVT$ properties, this law gives

$$V_r = \frac{V}{V_c} = \frac{(Z/Z_c)(T/T_c)}{P/P_c} = f_1(T_r, P_r)$$

$$\text{or} \qquad Z = Z_c f_2(T_r, P_r) \tag{3-2.4}$$

Except for monatomic gases, highly polar fluids, and fluids composed of large molecules, values of $Z_c$ for most organic compounds range from 0.27 to 0.29. If it is assumed to be a constant, Eq. (3-2.4) reduces to Eq. (3-2.2). In Sec. 3-3, $Z_c$ is introduced as a third correlating parameter (in addition to $T_c$ and $P_c$) to estimate $Z$, but not in the form of Eq. (3-2.4).

In Eq. (3-2.2), $T_c$ and $P_c$ are scaling factors to reduce $T$ and $P$; i.e., to make them nondimensional. Other scaling factors have been proposed, but none have been widely accepted. A tabulation of $T_c$ and $P_c$ for a number of elements and compounds is given in Appendix A, and methods for estimating them are described in Sec. 2-2.

## 3-3 Three-Parameter Correlations

Equation (3-2.2) is a two-parameter equation of state, the two parameters being $T_c$ and $P_c$. That is, by knowing $T_c$ and $P_c$ for a given fluid, it is possible to estimate the volumetric properties at various temperatures and pressures. The calculation may involve the use of Figs. 3-1 to 3-3, or one may employ an analytical function for $f(\;)$ in Eq. (3-2.2). Both methods are only approximate. Many suggestions which retain the general concept yet allow an increase in accuracy and applicability have been offered. In general, the more successful modifications have involved the inclusion of an additional third parameter into the function expressed by Eq. (3-2.2). Most often, this third parameter is related to the reduced vapor pressure at some specified reduced temperature or to some volumetric property at or near the critical point, although one correlation employs the molar polarizability as the third parameter [93]. Two common and well-tested three-parameter correlations are described below.

Assume that there are different, but unique, functions $Z = f(T_r, P_r)$ for each group of pure substances with the same $Z_c$. Then, for each $Z_c$ we have a different set of Figs. 3-1 to 3-3. All fluids with the same $Z_c$ values then follow the $Z$-$T_r$-$P_r$ behavior shown on charts drawn for that particular $Z_c$. Such a structuring indeed leads to a significant increase in accuracy. This is exactly what was done in the development of the Lydersen-Greenkorn-Hougen tables, which first appeared in 1955 [53] and were later modified [37]. There $Z$ is tabulated as a function of $T_r$ and $P_r$ with separate tables for various values of $Z_c$. Edwards and Thodos [20] have also utilized $Z_c$ in a correlation to estimate saturated vapor densities of nonpolar compounds.

An alternate third parameter is the Pitzer acentric factor [86–89], defined in Sec. 2-3. This factor is an indicator of the nonsphericity of a molecule's force field; e.g., a value of $\omega = 0$ denotes rare-gas spherical symmetry. Deviations from simple-fluid behavior are evident when $\omega > 0$. Within the context of the present discussion, it is assumed that all molecules with equal acentric factors have identical $Z = f(T_r, P_r)$ functions, as in Eq. (3-2.2). However, rather than prepare separate $Z$, $T_r$, $P_r$ tables for different values of $\omega$, it was suggested that a linear expansion could be employed:

$$Z = Z^{(0)}(T_r, P_r) + \omega Z^{(1)}(T_r, P_r) \tag{3-3.1}$$

Thus, the $Z^{(0)}$ function would apply to spherical molecules, and the $Z^{(1)}$ term is a deviation function.

Pitzer et al. tabulated $Z^{(0)}$ and $Z^{(1)}$ as functions of $T_r$ and $P_r$ [89], and Edmister has shown the same values graphically [18]. Several modifications as well as extensions to wider ranges of $T_r$ and $P_r$ have been published [38, 51, 99]. Tables 3-2 and 3-3 list those prepared by Lee and Kesler [47]. The method of calculation is described in Sec. 3-7. With Tables 3-2 and 3-3, $Z$ can be determined for both gases and liquids.* The $Z^{(0)}$ table agrees well with that presented originally by Pitzer et al. over the range of $T_r$ and $P_r$ common to both. The deviation function table of Lee and Kesler (Table 3-3) differs somewhat from that of Pitzer and Curl, but extensive testing [47, 114] indicates the new table is the more accurate.

Tables 3-2 and 3-3 were not intended to be applicable to strongly polar fluids, though they are often so used with surprising accuracy except at low temperatures near the saturated vapor region. Though none have been widely adopted, special techniques have been suggested to modify Eq. (3-3.1) for polar materials [21, 31, 50, 75, 111, 120].

Considerable emphasis has been placed on the Pitzer-Curl generalized relation. It has proved to be accurate and general when applied to pure gases. Only the acentric factor and critical temperature and pressure need

*For mixtures, see Table 4.3.

be known. It is probably the most successful and useful result of corresponding states theory [48, 109, 110].

**Example 3-1** Estimate the specific volume of dichlorodifluoromethane vapor at 20.67 bar and 366.5 K.

**solution** From Appendix A, $T_c = 385.0$ K, $P_c = 41.4$ bar, and $\omega = 0.204$.

$$T_r = \frac{366.5}{385.0} = 0.952 \qquad P_r = \frac{20.67}{41.4} = 0.499$$

From Fig. 3-1, $Z = 0.77$ and

$$V = \frac{ZRT}{P} = \frac{(0.77)(83.14)(366.5)}{20.67} = 1134 \text{ cm}^3/\text{mol}$$

The value reported in the literature is 1109 cm$^3$/mol [4].

If the Pitzer-Curl method were to be used, from Tables 3-2 and 3-3, $Z^{(0)} = 0.761$ and $Z^{(1)} = -0.082$. From Eq. (3-3.1),

$$Z = 0.761 + (0.204)(-0.082) = 0.744$$

$$V = \frac{ZRT}{P} = 1097 \text{ cm}^3/\text{mol}$$

## 3-4 Analytical Equations of State

An analytical equation of state is an algebraic relation between pressure, temperature, and molar volume. Three classes of equations of state are presented in the next three sections. The virial equation is discussed in Sec. 3-5. In its truncated form, it is a simple equation, and it can represent only modest deviations in the vapor phase from ideal-gas behavior. In Sec. 3-6, equations which are cubic in volume are discussed. These equations can represent both liquid and vapor behavior of nonpolar molecules over limited ranges of temperature and pressure, and they remain relatively simple from a computational point of view. Section 3-7 describes the Lee-Kesler generalized version of the Benedict-Webb-Rubin equation, which is applicable over broader ranges of temperatures and pressure than are the cubic equations. But it is also computationally more complex.

## 3-5 Virial Equation

The virial equation of state is a polynomial series in inverse volume which is explicit in pressure and can be derived from statistical mechanics:

$$P = \frac{RT}{V} + \frac{RTB}{V^2} + \frac{RTC}{V^3} + \cdots \tag{3-5.1}$$

The parameters $B, C, \ldots$ are called the second, third, ... virial coefficients and are functions only of temperature for a pure fluid. Much has been written about this particular equation, and several reviews have been pub-

**TABLE 3-2 Values of $Z^{(0)}$**

| $T_r$ | $P_r$ | | | | | | |
|---|---|---|---|---|---|---|---|
| | 0.010 | 0.050 | 0.100 | 0.200 | 0.400 | 0.600 | 0.800 |
| 0.30 | 0.0029 | 0.0145 | 0.0290 | 0.0579 | 0.1158 | 0.1737 | 0.2315 |
| 0.35 | 0.0026 | 0.0130 | 0.0261 | 0.0522 | 0.1043 | 0.1564 | 0.2084 |
| 0.40 | 0.0024 | 0.0119 | 0.0239 | 0.0477 | 0.0953 | 0.1429 | 0.1904 |
| 0.45 | 0.0022 | 0.0110 | 0.0221 | 0.0442 | 0.0882 | 0.1322 | 0.1762 |
| 0.50 | 0.0021 | 0.0103 | 0.0207 | 0.0413 | 0.0825 | 0.1236 | 0.1647 |
| 0.55 | 0.9804 | 0.0098 | 0.0195 | 0.0390 | 0.0778 | 0.1166 | 0.1553 |
| 0.60 | 0.9849 | 0.0093 | 0.0186 | 0.0371 | 0.0741 | 0.1109 | 0.1476 |
| 0.65 | 0.9881 | 0.9377 | 0.0178 | 0.0356 | 0.0710 | 0.1063 | 0.1415 |
| 0.70 | 0.9904 | 0.9504 | 0.8958 | 0.0344 | 0.0687 | 0.1027 | 0.1366 |
| 0.75 | 0.9922 | 0.9598 | 0.9165 | 0.0336 | 0.0670 | 0.1001 | 0.1330 |
| 0.80 | 0.9935 | 0.9669 | 0.9319 | 0.8539 | 0.0661 | 0.0985 | 0.1307 |
| 0.85 | 0.9946 | 0.9725 | 0.9436 | 0.8810 | 0.0661 | 0.0983 | 0.1301 |
| 0.90 | 0.9954 | 0.9768 | 0.9528 | 0.9015 | 0.7800 | 0.1006 | 0.1321 |
| 0.93 | 0.9959 | 0.9790 | 0.9573 | 0.9115 | 0.8059 | 0.6635 | 0.1359 |
| 0.95 | 0.9961 | 0.9803 | 0.9600 | 0.9174 | 0.8206 | 0.6967 | 0.1410 |
| 0.97 | 0.9963 | 0.9815 | 0.9625 | 0.9227 | 0.8338 | 0.7240 | 0.5580 |
| 0.98 | 0.9965 | 0.9821 | 0.9637 | 0.9253 | 0.8398 | 0.7360 | 0.5887 |
| 0.99 | 0.9966 | 0.9826 | 0.9648 | 0.9277 | 0.8455 | 0.7471 | 0.6138 |
| 1.00 | 0.9967 | 0.9832 | 0.9659 | 0.9300 | 0.8509 | 0.7574 | 0.6353 |
| 1.01 | 0.9968 | 0.9837 | 0.9669 | 0.9322 | 0.8561 | 0.7671 | 0.6542 |
| 1.02 | 0.9969 | 0.9842 | 0.9679 | 0.9343 | 0.8610 | 0.7761 | 0.6710 |
| 1.05 | 0.9971 | 0.9855 | 0.9707 | 0.9401 | 0.8743 | 0.8002 | 0.7130 |
| 1.10 | 0.9975 | 0.9874 | 0.9747 | 0.9485 | 0.8930 | 0.8323 | 0.7649 |
| 1.15 | 0.9978 | 0.9891 | 0.9780 | 0.9554 | 0.9081 | 0.8576 | 0.8032 |
| 1.20 | 0.9981 | 0.9904 | 0.9808 | 0.9611 | 0.9205 | 0.8779 | 0.8330 |
| 1.30 | 0.9985 | 0.9926 | 0.9852 | 0.9702 | 0.9396 | 0.9083 | 0.8764 |
| 1.40 | 0.9988 | 0.9942 | 0.9884 | 0.9768 | 0.9534 | 0.9298 | 0.9062 |
| 1.50 | 0.9991 | 0.9954 | 0.9909 | 0.9818 | 0.9636 | 0.9456 | 0.9278 |
| 1.60 | 0.9993 | 0.9964 | 0.9928 | 0.9856 | 0.9714 | 0.9575 | 0.9439 |
| 1.70 | 0.9994 | 0.9971 | 0.9943 | 0.9886 | 0.9775 | 0.9667 | 0.9563 |
| 1.80 | 0.9995 | 0.9977 | 0.9955 | 0.9910 | 0.9823 | 0.9739 | 0.9659 |
| 1.90 | 0.9996 | 0.9982 | 0.9964 | 0.9929 | 0.9861 | 0.9796 | 0.9735 |
| 2.00 | 0.9997 | 0.9986 | 0.9972 | 0.9944 | 0.9892 | 0.9842 | 0.9796 |
| 2.20 | 0.9998 | 0.9992 | 0.9983 | 0.9967 | 0.9937 | 0.9910 | 0.9886 |
| 2.40 | 0.9999 | 0.9996 | 0.9991 | 0.9983 | 0.9969 | 0.9957 | 0.9948 |
| 2.60 | 1.0000 | 0.9998 | 0.9997 | 0.9994 | 0.9991 | 0.9990 | 0.9990 |
| 2.80 | 1.0000 | 1.0000 | 1.0001 | 1.0002 | 1.0007 | 1.0013 | 1.0021 |
| 3.00 | 1.0000 | 1.0002 | 1.0004 | 1.0008 | 1.0018 | 1.0030 | 1.0043 |
| 3.50 | 1.0001 | 1.0004 | 1.0008 | 1.0017 | 1.0035 | 1.0055 | 1.0075 |
| 4.00 | 1.0001 | 1.0005 | 1.0010 | 1.0021 | 1.0043 | 1.0066 | 1.0090 |

lished [e.g., 63, 123]. One reason for the equation's popularity is that the coefficients $B$, $C$, . . . can be related to parameters characterizing the intermolecular potential function. Little information is available for the third and higher virial coefficients. Two methods have, however, appeared recently for predicting $C$, the third virial coefficient [15, 77]. More often the virial equation is truncated to contain only the second virial coeffi-

| $P_r$ | | | | | | | |
|---|---|---|---|---|---|---|---|
| 1.000 | 1.200 | 1.500 | 2.000 | 3.000 | 5.000 | 7.000 | 10.000 |
| 0.2892 | 0.3470 | 0.4335 | 0.5775 | 0.8648 | 1.4366 | 2.0048 | 2.8507 |
| 0.2604 | 0.3123 | 0.3901 | 0.5195 | 0.7775 | 1.2902 | 1.7987 | 2.5539 |
| 0.2379 | 0.2853 | 0.3563 | 0.4744 | 0.7095 | 1.1758 | 1.6373 | 2.3211 |
| 0.2200 | 0.2638 | 0.3294 | 0.4384 | 0.6551 | 1.0841 | 1.5077 | 2.1338 |
| 0.2056 | 0.2465 | 0.3077 | 0.4092 | 0.6110 | 1.0094 | 1.4017 | 1.9801 |
| 0.1939 | 0.2323 | 0.2899 | 0.3853 | 0.5747 | 0.9475 | 1.3137 | 1.8520 |
| 0.1842 | 0.2207 | 0.2753 | 0.3657 | 0.5446 | 0.8959 | 1.2398 | 1.7440 |
| 0.1765 | 0.2113 | 0.2634 | 0.3495 | 0.5197 | 0.8526 | 1.1773 | 1.6519 |
| 0.1703 | 0.2038 | 0.2538 | 0.3364 | 0.4991 | 0.8161 | 1.1241 | 1.5729 |
| 0.1656 | 0.1981 | 0.2464 | 0.3260 | 0.4823 | 0.7854 | 1.0787 | 1.5047 |
| 0.1626 | 0.1942 | 0.2411 | 0.3182 | 0.4690 | 0.7598 | 1.0400 | 1.4456 |
| 0.1614 | 0.1924 | 0.2382 | 0.3132 | 0.4591 | 0.7388 | 1.0071 | 1.3943 |
| 0.1630 | 0.1935 | 0.2383 | 0.3114 | 0.4527 | 0.7220 | 0.9793 | 1.3496 |
| 0.1664 | 0.1963 | 0.2405 | 0.3122 | 0.4507 | 0.7138 | 0.9648 | 1.3257 |
| 0.1705 | 0.1998 | 0.2432 | 0.3138 | 0.4501 | 0.7092 | 0.9561 | 1.3108 |
| 0.1779 | 0.2055 | 0.2474 | 0.3164 | 0.4504 | 0.7052 | 0.9480 | 1.2968 |
| 0.1844 | 0.2097 | 0.2503 | 0.3182 | 0.4508 | 0.7035 | 0.9442 | 1.2901 |
| 0.1959 | 0.2154 | 0.2538 | 0.3204 | 0.4514 | 0.7018 | 0.9406 | 1.2835 |
| 0.2901 | 0.2237 | 0.2583 | 0.3229 | 0.4522 | 0.7004 | 0.9372 | 1.2772 |
| 0.4648 | 0.2370 | 0.2640 | 0.3260 | 0.4533 | 0.6991 | 0.9339 | 1.2710 |
| 0.5146 | 0.2629 | 0.2715 | 0.3297 | 0.4547 | 0.6980 | 0.9307 | 1.2650 |
| 0.6026 | 0.4437 | 0.3131 | 0.3452 | 0.4604 | 0.6956 | 0.9222 | 1.2481 |
| 0.6880 | 0.5984 | 0.4580 | 0.3953 | 0.4770 | 0.6950 | 0.9110 | 1.2232 |
| 0.7443 | 0.6803 | 0.5798 | 0.4760 | 0.5042 | 0.6987 | 0.9033 | 1.2021 |
| 0.7858 | 0.7363 | 0.6605 | 0.5605 | 0.5425 | 0.7069 | 0.8990 | 1.1844 |
| 0.8438 | 0.8111 | 0.7624 | 0.6908 | 0.6344 | 0.7358 | 0.8998 | 1.1580 |
| 0.8827 | 0.8595 | 0.8256 | 0.7753 | 0.7202 | 0.7761 | 0.9112 | 1.1419 |
| 0.9103 | 0.8933 | 0.8689 | 0.8328 | 0.7887 | 0.8200 | 0.9297 | 1.1339 |
| 0.9308 | 0.9180 | 0.9000 | 0.8738 | 0.8410 | 0.8617 | 0.9518 | 1.1320 |
| 0.9463 | 0.9367 | 0.9234 | 0.9043 | 0.8809 | 0.8984 | 0.9745 | 1.1343 |
| 0.9583 | 0.9511 | 0.9413 | 0.9275 | 0.9118 | 0.9297 | 0.9961 | 1.1391 |
| 0.9678 | 0.9624 | 0.9552 | 0.9456 | 0.9359 | 0.9557 | 1.0157 | 1.1452 |
| 0.9754 | 0.9715 | 0.9664 | 0.9599 | 0.9550 | 0.9772 | 1.0328 | 1.1516 |
| 0.9865 | 0.9847 | 0.9826 | 0.9806 | 0.9827 | 1.0094 | 1.0600 | 1.1635 |
| 0.9941 | 0.9936 | 0.9935 | 0.9945 | 1.0011 | 1.0313 | 1.0793 | 1.1728 |
| 0.9993 | 0.9998 | 1.0010 | 1.0040 | 1.0137 | 1.0463 | 1.0926 | 1.1792 |
| 1.0031 | 1.0042 | 1.0063 | 1.0106 | 1.0223 | 1.0565 | 1.1016 | 1.1830 |
| 1.0057 | 1.0074 | 1.0101 | 1.0153 | 1.0284 | 1.0635 | 1.1075 | 1.1848 |
| 1.0097 | 1.0120 | 1.0156 | 1.0221 | 1.0368 | 1.0723 | 1.1138 | 1.1834 |
| 1.0115 | 1.0140 | 1.0179 | 1.0249 | 1.0401 | 1.0747 | 1.1136 | 1.1773 |

cient. The virial equation may also be written as a power series in either $V$ or $P$, so that truncation leads to two forms. These are

$$Z = 1 + \frac{BP}{RT} \tag{3-5.2a}$$

and

$$Z = 1 + \frac{B}{V} \tag{3-5.2b}$$

**TABLE 3-3 Values of $Z^{(1)}$**

| | $P_r$ | | | | | | |
|---|---|---|---|---|---|---|---|
| $T_r$ | 0.010 | 0.050 | 0.100 | 0.200 | 0.400 | 0.600 | 0.800 |
| 0.30 | -0.0008 | -0.0040 | -0.0081 | -0.0161 | -0.0323 | -0.0484 | -0.0645 |
| 0.35 | -0.0009 | -0.0046 | -0.0093 | -0.0185 | -0.0370 | -0.0554 | -0.0738 |
| 0.40 | -0.0010 | -0.0048 | -0.0095 | -0.0190 | -0.0380 | -0.0570 | -0.0758 |
| 0.45 | -0.0009 | -0.0047 | -0.0094 | -0.0187 | -0.0374 | -0.0560 | -0.0745 |
| 0.50 | -0.0009 | -0.0045 | -0.0090 | -0.0181 | -0.0360 | -0.0539 | -0.0716 |
| 0.55 | -0.0314 | -0.0043 | -0.0086 | -0.0172 | -0.0343 | -0.0513 | -0.0682 |
| 0.60 | -0.0205 | -0.0041 | -0.0082 | -0.0164 | -0.0326 | -0.0487 | -0.0646 |
| 0.65 | -0.0137 | -0.0772 | -0.0078 | -0.0156 | -0.0309 | -0.0461 | -0.0611 |
| 0.70 | -0.0093 | -0.0507 | -0.1161 | -0.0148 | -0.0294 | -0.0438 | -0.0579 |
| 0.75 | -0.0064 | -0.0339 | -0.0744 | -0.0143 | -0.0282 | -0.0417 | -0.0550 |
| 0.80 | -0.0044 | -0.0228 | -0.0487 | -0.1160 | -0.0272 | -0.0401 | -0.0526 |
| 0.85 | -0.0029 | -0.0152 | -0.0319 | -0.0715 | -0.0268 | -0.0391 | -0.0509 |
| 0.90 | -0.0019 | -0.0099 | -0.0205 | -0.0442 | -0.1118 | -0.0396 | -0.0503 |
| 0.93 | -0.0015 | -0.0075 | -0.0154 | -0.0326 | -0.0763 | -0.1662 | -0.0514 |
| 0.95 | -0.0012 | -0.0062 | -0.0126 | -0.0262 | -0.0589 | -0.1110 | -0.0540 |
| 0.97 | -0.0010 | -0.0050 | -0.0101 | -0.0208 | -0.0450 | -0.0770 | -0.1647 |
| 0.98 | -0.0009 | -0.0044 | -0.0090 | -0.0184 | -0.0390 | -0.0641 | -0.1100 |
| 0.99 | -0.0008 | -0.0039 | -0.0079 | -0.0161 | -0.0335 | -0.0531 | -0.0796 |
| 1.00 | -0.0007 | -0.0034 | -0.0069 | -0.0140 | -0.0285 | -0.0435 | -0.0588 |
| 1.01 | -0.0006 | -0.0030 | -0.0060 | -0.0120 | -0.0240 | -0.0351 | -0.0429 |
| 1.02 | -0.0005 | -0.0026 | -0.0051 | -0.0102 | -0.0198 | -0.0277 | -0.0303 |
| 1.05 | -0.0003 | -0.0015 | -0.0029 | -0.0054 | -0.0092 | -0.0097 | -0.0032 |
| 1.10 | -0.0000 | 0.0000 | 0.0001 | 0.0007 | 0.0038 | 0.0106 | 0.0236 |
| 1.15 | 0.0002 | 0.0011 | 0.0023 | 0.0052 | 0.0127 | 0.0237 | 0.0396 |
| 1.20 | 0.0004 | 0.0019 | 0.0039 | 0.0084 | 0.0190 | 0.0326 | 0.0499 |
| 1.30 | 0.0006 | 0.0030 | 0.0061 | 0.0125 | 0.0267 | 0.0429 | 0.0612 |
| 1.40 | 0.0007 | 0.0036 | 0.0072 | 0.0147 | 0.0306 | 0.0477 | 0.0661 |
| 1.50 | 0.0008 | 0.0039 | 0.0078 | 0.0158 | 0.0323 | 0.0497 | 0.0677 |
| 1.60 | 0.0008 | 0.0040 | 0.0080 | 0.0162 | 0.0330 | 0.0501 | 0.0677 |
| 1.70 | 0.0008 | 0.0040 | 0.0081 | 0.0163 | 0.0329 | 0.0497 | 0.0667 |
| 1.80 | 0.0008 | 0.0040 | 0.0081 | 0.0162 | 0.0325 | 0.0488 | 0.0652 |
| 1.90 | 0.0008 | 0.0040 | 0.0079 | 0.0159 | 0.0318 | 0.0477 | 0.0635 |
| 2.00 | 0.0008 | 0.0039 | 0.0078 | 0.0155 | 0.0310 | 0.0464 | 0.0617 |
| 2.20 | 0.0007 | 0.0037 | 0.0074 | 0.0147 | 0.0293 | 0.0437 | 0.0579 |
| 2.40 | 0.0007 | 0.0035 | 0.0070 | 0.0139 | 0.0276 | 0.0411 | 0.0544 |
| 2.60 | 0.0007 | 0.0033 | 0.0066 | 0.0131 | 0.0260 | 0.0387 | 0.0512 |
| 2.80 | 0.0006 | 0.0031 | 0.0062 | 0.0124 | 0.0245 | 0.0365 | 0.0483 |
| 3.00 | 0.0006 | 0.0029 | 0.0059 | 0.0117 | 0.0232 | 0.0345 | 0.0456 |
| 3.50 | 0.0005 | 0.0026 | 0.0052 | 0.0103 | 0.0204 | 0.0303 | 0.0401 |
| 4.00 | 0.0005 | 0.0023 | 0.0046 | 0.0091 | 0.0182 | 0.0270 | 0.0357 |

For the evaluation of $B$ from experimental data, however, Eq. (3-5.1) should be used, i.e.,

$$B = \lim_{1/V \to 0} \left( \frac{\partial Z}{\partial 1/V} \right)_T \tag{3-5.3}$$

| | | | $P_r$ | | | | |
|---|---|---|---|---|---|---|---|
| 1.000 | 1.200 | 1.500 | 2.000 | 3.000 | 5.000 | 7.000 | 10.000 |
| -0.0806 | -0.0966 | -0.1207 | -0.1608 | -0.2407 | -0.3996 | -0.5572 | -0.7915 |
| -0.0921 | -0.1105 | -0.1379 | -0.1834 | -0.2738 | -0.4523 | -0.6279 | -0.8863 |
| -0.0946 | -0.1134 | -0.1414 | -0.1879 | -0.2799 | -0.4603 | -0.6365 | -0.8936 |
| -0.0929 | -0.1113 | -0.1387 | -0.1840 | -0.2734 | -0.4475 | -0.6162 | -0.8606 |
| -0.0893 | -0.1069 | -0.1330 | -0.1762 | -0.2611 | -0.4253 | -0.5831 | -0.8099 |
| -0.0849 | -0.1015 | -0.1263 | -0.1669 | -0.2465 | -0.3991 | -0.5446 | -0.7521 |
| -0.0803 | -0.0960 | -0.1192 | -0.1572 | -0.2312 | -0.3718 | -0.5047 | -0.6928 |
| -0.0759 | -0.0906 | -0.1122 | -0.1476 | -0.2160 | -0.3447 | -0.4653 | -0.6346 |
| -0.0718 | -0.0855 | -0.1057 | -0.1385 | -0.2013 | -0.3184 | -0.4270 | -0.5785 |
| -0.0681 | -0.0808 | -0.0996 | -0.1298 | -0.1872 | -0.2929 | -0.3901 | -0.5250 |
| -0.0648 | -0.0767 | -0.0940 | -0.1217 | -0.1736 | -0.2682 | -0.3545 | -0.4740 |
| -0.0622 | -0.0731 | -0.0888 | -0.1138 | -0.1602 | -0.2439 | -0.3201 | -0.4254 |
| -0.0604 | -0.0701 | -0.0840 | -0.1059 | -0.1463 | -0.2195 | -0.2862 | -0.3788 |
| -0.0602 | -0.0687 | -0.0810 | -0.1007 | -0.1374 | -0.2045 | -0.2661 | -0.3516 |
| -0.0607 | -0.0678 | -0.0788 | -0.0967 | -0.1310 | -0.1943 | -0.2526 | -0.3339 |
| -0.0623 | -0.0669 | -0.0759 | -0.0921 | -0.1240 | -0.1837 | -0.2391 | -0.3163 |
| -0.0641 | -0.0661 | -0.0740 | -0.0893 | -0.1202 | -0.1783 | -0.2322 | -0.3075 |
| -0.0680 | -0.0646 | -0.0715 | -0.0861 | -0.1162 | -0.1728 | -0.2254 | -0.2989 |
| -0.0879 | -0.0609 | -0.0678 | -0.0824 | -0.1118 | -0.1672 | -0.2185 | -0.2902 |
| -0.0223 | -0.0473 | -0.0621 | -0.0778 | -0.1072 | -0.1615 | -0.2116 | -0.2816 |
| -0.0062 | 0.0227 | -0.0524 | -0.0722 | -0.1021 | -0.1556 | -0.2047 | -0.2731 |
| 0.0220 | 0.1059 | 0.0451 | -0.0432 | -0.0838 | -0.1370 | -0.1835 | -0.2476 |
| 0.0476 | 0.0897 | 0.1630 | 0.0698 | -0.0373 | -0.1021 | -0.1469 | -0.2056 |
| 0.0625 | 0.0943 | 0.1548 | 0.1667 | 0.0332 | -0.0611 | -0.1084 | -0.1642 |
| 0.0719 | 0.0991 | 0.1477 | 0.1990 | 0.1095 | -0.0141 | -0.0678 | -0.1231 |
| 0.0819 | 0.1048 | 0.1420 | 0.1991 | 0.2079 | 0.0875 | 0.0176 | -0.0423 |
| 0.0857 | 0.1063 | 0.1383 | 0.1894 | 0.2397 | 0.1737 | 0.1008 | 0.0350 |
| 0.0864 | 0.1055 | 0.1345 | 0.1806 | 0.2433 | 0.2309 | 0.1717 | 0.1058 |
| 0.0855 | 0.1035 | 0.1303 | 0.1729 | 0.2381 | 0.2631 | 0.2255 | 0.1673 |
| 0.0838 | 0.1008 | 0.1259 | 0.1658 | 0.2305 | 0.2788 | 0.2628 | 0.2179 |
| 0.0816 | 0.0978 | 0.1216 | 0.1593 | 0.2224 | 0.2846 | 0.2871 | 0.2576 |
| 0.0792 | 0.0947 | 0.1173 | 0.1532 | 0.2144 | 0.2848 | 0.3017 | 0.2876 |
| 0.0767 | 0.0916 | 0.1133 | 0.1476 | 0.2069 | 0.2819 | 0.3097 | 0.3096 |
| 0.0719 | 0.0857 | 0.1057 | 0.1374 | 0.1932 | 0.2720 | 0.3135 | 0.3355 |
| 0.0675 | 0.0803 | 0.0989 | 0.1285 | 0.1812 | 0.2602 | 0.3089 | 0.3459 |
| 0.0634 | 0.0754 | 0.0929 | 0.1207 | 0.1706 | 0.2484 | 0.3009 | 0.3475 |
| 0.0598 | 0.0711 | 0.0876 | 0.1138 | 0.1613 | 0.2372 | 0.2915 | 0.3443 |
| 0.0565 | 0.0672 | 0.0828 | 0.1076 | 0.1529 | 0.2268 | 0.2817 | 0.3385 |
| 0.0497 | 0.0591 | 0.0728 | 0.0949 | 0.1356 | 0.2042 | 0.2584 | 0.3194 |
| 0.0443 | 0.0527 | 0.0651 | 0.0849 | 0.1219 | 0.1857 | 0.2378 | 0.2994 |

As the density approaches zero, Eqs. (3-5.2$a$) and (3-5.2$b$) become identical, and at low densities, both forms of Eq. (3-5.2) approximate true behavior. But neither Eq. (3-5.2$a$) nor (3-5.2$b$) should be used if $\rho > \rho_c/2$ or if $V_{r_i}$ as defined in Figs. 3-1 to 3-3 is less than about 0.5. When $\rho = \rho_c/2$, Eq. (3-5.2$a$) predicts a $Z$ which is too high and Eq. (3-5.2$b$) predicts a $Z$ which is too low [128]; Eq. (3-5.2*a*) is easier to use and is thus preferred over Eq. (3-5.2*b*). If two virial coefficients are retained, the preferred form is that given in Eq. (3-5.1).

Since $B$ is a function only of temperature, Eq. (3-5.2*a*) predicts that $Z$ is a linear function of pressure along an isotherm. Examination of Fig. 3-1 shows that this is not a bad assumption at low values of $P_r$.

A compilation of second virial coefficients is given by Dymond and Smith [17]. To estimate values, a number of techniques are available. Most are based on the integration of a theoretical expression relating intermolecular energy to the distance of separation between molecules. With our present limited ability to determine such energies, however, it is more common to employ corresponding states relations to estimate $B$.

For nonpolar molecules [128],

$$\frac{BP_c}{RT_c} = B^{(0)} + \omega B^{(0)} \tag{3-5.4}$$

$$B^{(0)} = 0.083 - \frac{0.422}{T_r^{1.6}} \tag{3-5.5}$$

$$B^{(1)} = 0.139 - \frac{0.172}{T_r^{4.2}} \tag{3-5.6}$$

In Fig. 3-4, Eq. (3-5.3) is shown to correlate well experimental second virial coefficient data for 14 nonpolar fluids. Except at high temperatures, $B$ is negative; by Eq. (3-5.2), the compressibility factor is less than unity.

Equation (3-5.2) should be considered applicable only for nonpolar or slightly polar materials. For polar molecules, Tsonopoulos [121] recom-

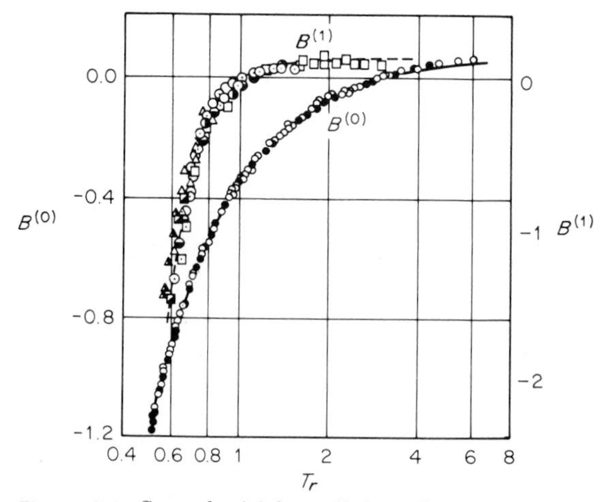


**Figure 3-4** Second virial coefficient for 14 nonpolar fluids; Eqs. (3-5.5) and (3-5.6) *(From Ref. 128.)*

**TABLE 3-4 Values of *a* and *b* for Eq. (3-5.7) [121, 122, 124]**

| Compound class | $a$ | $b$ |
|---|---|---|
| Ketones, aldehydes, nitriles, ethers, $NH_3$, $H_2S$, HCN, esters | $-2.112 \times 10^{-4}\,\mu_r$† $- 3.877 \times 10^{-21}\,\mu_r^{8}$ | 0 |
| Mercaptans | 0 | 0 |
| Monoalkyhalides | $2.078 \times 10^{-4}\,\mu_r - 7.048 \times 10^{-21}\,\mu_r^{8}$ | |
| Alcohols | 0.0878 | 0.04−0.06‡ |
| Phenol | −0.0136 | 0 |

†$\mu_r$ is defined in Eq. (3-5.8).
‡See Ref. 113 for specific values.

mends that Eq. (3-5.4) be modified by the addition of another term $B^{(2)}$, where

$$B^{(2)} = \frac{a}{T_r^6} - \frac{b}{T_r^8} \tag{3-5.7}$$

Neither $a$ nor $b$ can be estimated with much accuracy; Tsonopoulos [121, 122, 124], however, has correlated $a$ and $b$ for several compound classes. $b$ is zero for nonhydrogen-bonded materials, and $a$ is given by the values or expressions in Table 3-4. These expressions are functions of the reduced dipole moment,

$$\mu_r = \frac{10^5 \mu^2 P_c}{T_c^2} \tag{3-5.8}$$

where $\mu$ = dipole moment, debyes
$P_c$ = critical pressure, bars
$T_c$ = critical temperature, K

**Example 3-2** Estimate the second virial coefficient of methyl isobutyl ketone at 120°C.

**solution** From Appendix A, $T_c$ = 571 K, $P_c$ = 32.7 bar, $\omega$ = 0.385, and $\mu$ = 2.8 debyes. With $T_r = (120 + 273)/571 = 0.689$ and with Eqs. (3-5.5) and (3-5.6), $B^{(0)} = -0.684$ and $B^{(1)} = -0.684$, and with Eq. (3-5.8) $\mu_r = (10^5)(2.8)^2(32.7/571^2) = 78.6$. From Table 3-4 and Eq. (3-5.7) $a = 0.0166$ and $B^{(2)} = -0.155$. Since methyl isobutyl ketone is nonhydrogen-bonded, $b = 0$. Then, using Eqs. (3-5.4) and (3-5.7),

$$\frac{(B)(32.7)}{(83.14)(571)} = -0.684 + (0.385)(-0.686) + (-0.155)$$

$$B = -1601 \text{ cm}^3/\text{mol}$$

The experimental value is $-1580$ cm$^3$/mol [35].

Several other methods have been proposed for estimating $B$ for polar compounds [36, 91, 108, 113]. Values of the parameters used in the Hayden-O'Connell method for a number of compounds and a summary of the equations are given in Ref. 94. A technique for treating compounds that associate, such as acetic acid, is also given in Ref. 94. More recently, McCann and Danner [66] have developed a rather involved group contribution method for second virial coefficients. Their method requires only the critical temperature, and except for organic acids it reproduces all existing data to an accuracy equivalent to that of the Tsonopoulos method. The McCann-Danner method can be used for new compounds for which very little information is available, but it has not been tested for mixtures.

## 3-6 Cubic Equations of State

The term "cubic equation of state" implies an equation which, if expanded, would contain volume terms raised to either the first, second, or third power. Many of the common two-parameter cubic equations can be expressed by the equation

$$P = \frac{RT}{V - b} - \frac{a}{V^2 + ubV + wb^2} \tag{3-6.1}$$

An equivalent form of Eq. (3-6.1) is

$$Z^3 - (1 + B^* - uB^*)Z^2 + (A^* + wB^{*2} - uB^* - uB^{*2})Z - A^*B^* - wB^{*2} - wB^{*3} = 0 \tag{3-6.2}$$

where $$A^* = \frac{aP}{R^2T^2} \tag{3-6.3}$$

and $$B^* = \frac{bP}{RT} \tag{3-6.4}$$

Four well-known cubic equations are the van der Waals, Redlich-Kwong (RK) [96], Soave (SRK) [101], and Peng-Robinson (PR) [84] equations. For these four equations, $u$ and $w$ take on the integer values listed in Table 3-5. There are several approaches which have been used to set the values of the two parameters, $a$ and $b$, that appear in Eq. (3-6.1). One approach is to choose $a$ and $b$ so that the two critical point conditions

$$\left(\frac{\partial P}{\partial V}\right)_{T_c} = 0 \tag{3-6.5}$$

$$\left(\frac{\partial^2 P}{\partial V^2}\right)_{T_c} = 0 \tag{3-6.6}$$

**TABLE 3-5 Constants for Four Common Cubic Equations of State**

| Equation | $u$ | $w$ | $b$ | $a$ |
|---|---|---|---|---|
| van der Waals | 0 | 0 | $\frac{RT_c}{8P_c}$ | $\frac{27}{64}\frac{R^2T_c^2}{P_c}$ |
| Redlich-Kwong | 1 | 0 | $\frac{0.08664RT_c}{P_c}$ | $\frac{0.42748R^2T_c^{2.5}}{P_cT^{1/2}}$ |
| Soave | 1 | 0 | $\frac{0.08664RT_c}{P_c}$ | $\frac{0.42748R^2T_c^2}{P_c}[1+f\omega(1-T_r^{1/2})]^2$ where $f\omega = 0.48 + 1.574\omega - 0.176\omega^2$ |
| Peng-Robinson | 2 | −1 | $\frac{0.07780RT_c}{P_c}$ | $\frac{0.45724R^2T_c^2}{P_c}[1+f\omega(1-T_r^{1/2}]^2$ where $f\omega = 0.37464 + 1.54226\omega - 0.26992\omega^2$ |

are satisfied. Eqs. (3-6.5) and (3-6.6) are applicable only to pure components. Both Soave and Peng and Robinson used Eqs. (3-6.5) and (3-6.6) to find $a$ and $b$ at the critical point. They then made the parameter $a$ a function of temperature and acentric factor so as to reproduce hydrocarbon vapor pressures. The expresssions for $a$ and $b$ which result from this procedure are listed in Table 3-5. Graboski and Daubert [26] have given a slightly different expression for $f\omega$ in the SRK equation.

An alternative approach to the determination of $a$ and $b$ in Eq. (3-6.1) is that first suggested by Joffe et al. [39]. In this approach, both $a$ and $b$ are functions of temperature and can be set so as to reproduce some selected pure component data. This technique has been extended by various workers, and temperature-dependent values of $a$ and $b$ have been published [12, 32, 134]. If the two conditions, saturated liquid volume and vapor pressure, are used to set $a$ and $b$, then Eq. (3-6.1) will reproduce exactly these two pure component properties. For this case, Panagiotopoulos and Kumar [80] have shown that $a$ and $b$ can be calculated approximately by

$$b = \frac{RTZ^L}{P}\frac{\sum_i A_i(\ln Z^L)^i}{1+\sum_i A_i(\ln Z^L)^i} \tag{3-6.7}$$

$$a = bRT\sum_i B_i(\ln Z^L)^i \tag{3-6.8}$$

$Z^L$ is the compressibility factor for the pure saturated liquid. The coefficients $A_i$ and $B_i$ are given in Table 3-6. Morris and Turek [70] have used the vapor pressure and volumetric data over a range of pressures (at a fixed temperature) to determine optimal values of $a$ and $b$ for eight sub-

**TABLE 3-6 Coefficients for Eqs. (3-6.7) and (3-6.8)**

| | $6.91 \times 10^{-13} \leq Z^L \leq 0.011$ | | $0.011 \leq Z^L \leq Z_c$ | |
|---|---|---|---|---|
| $i$ | $A_i$ | $B_i$ | $A_i$ | $B_i$ |
| | Redlich-Kwong (or Soave) Form ($u = 1, w = 0$) | | | |
| 0 | $-0.874084$ | 1.50479 | $-3.66182 \times 10^{-2}$ | 5.88848 |
| 1 | $-0.827262$ | $-1.66630$ | $-0.203841$ | 2.07104 |
| 2 | $-1.74216 \times 10^{-3}$ | $-6.30280 \times 10^{-3}$ | 0.147671 | 1.31927 |
| 3 | — | $-7.68315 \times 10^{-5}$ | $1.19456 \times 10^{-2}$ | 0.226604 |
| 4 | — | — | — | $1.52897 \times 10^{-2}$ |
| | Peng-Robinson Form ($u = 2, w = -1$) | | | |
| 0 | $-1.21190$ | 1.82378 | $2.14469 \times 10^{-2}$ | 7.09646 |
| 1 | $-0.918850$ | $-1.84430$ | $-5.87391 \times 10^{-2}$ | 2.41021 |
| 2 | $-1.91042 \times 10^{-3}$ | $-5.75993 \times 10^{-3}$ | 0.195961 | 1.41294 |
| 3 | — | $-2.90784 \times 10^{-5}$ | $1.53575 \times 10^{-2}$ | 0.229028 |
| 4 | — | $8.43108 \times 10^{-7}$ | — | $1.47396 \times 10^{-2}$ |

stances. Their results are shown in Fig. 3-5 for $CO_2$. The volumetric behavior predicted with the $a$ and $b$ from Table 3-5 for the RK and SRK equations is shown in Figs. 3-6 and 3-7. Figure 3-7 indicates that the original RK equation gives more accurate volume behavior for $T_r$ and $P_r$ greater than 1. This is expected because the Soave $a$ parameter was fit to vapor pressure data which don't exist above $T_r = 1$. Thus, the Soave equation should not be used at large reduced temperatures. Results with the Morris-Turek parameters shown in Fig. 3-5 demonstrate better volume behavior than either the RK or SRK equation because the area with errors greater than 10 percent (located at $P_r \sim 1.4$ and $T_r \sim 1.01$) is very small.

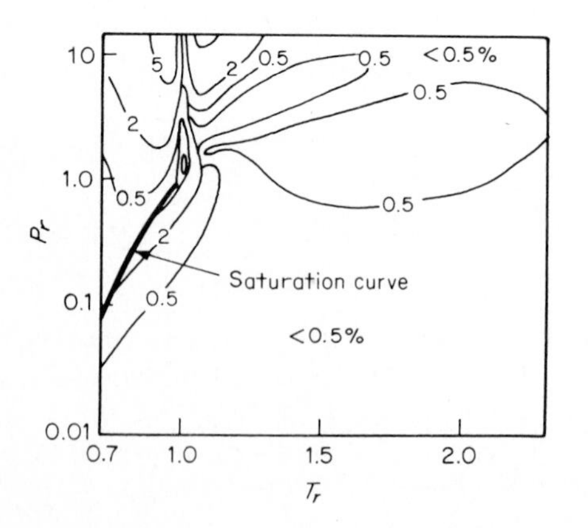


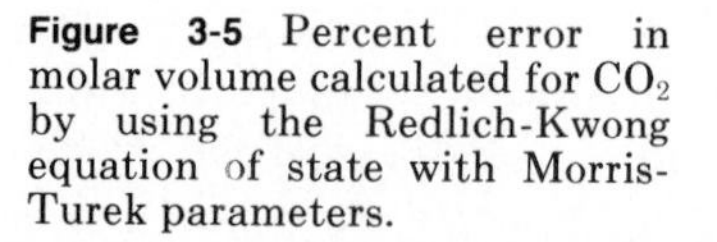

**Figure 3-5** Percent error in molar volume calculated for $CO_2$ by using the Redlich-Kwong equation of state with Morris-Turek parameters.

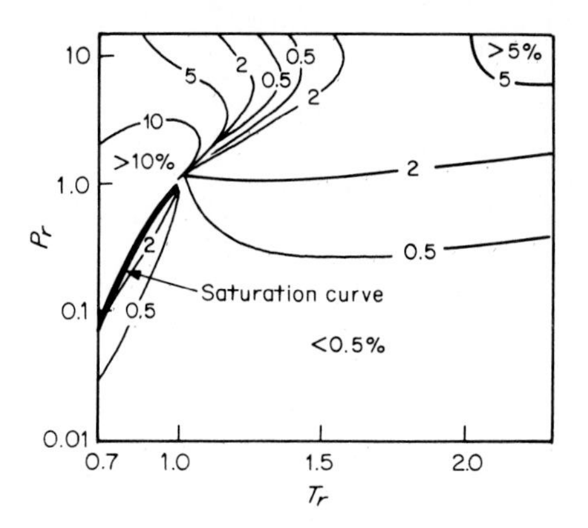


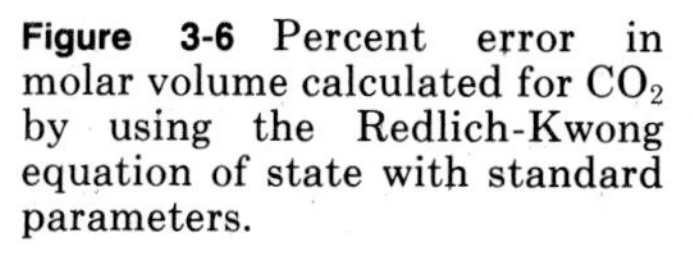

**Figure 3-6** Percent error in molar volume calculated for $CO_2$ by using the Redlich-Kwong equation of state with standard parameters.

There have been numerous modifications of cubic equations to improve their predictions of liquid volumes, most of which involve a third parameter [24, 34, 100, 127] in addition to the $a$ and $b$ already in Eq. (3-6.1). Peneloux and Rauzy [83] have correlated the errors in the liquid volumes as predicted by the Soave equation at $T_r = 0.7$ and recommend that the following correction be subtracted from the volume obtained in the SRK equation

$$c = 0.40768(0.29441 - Z_{RA})\frac{RT_c}{P_c} \tag{3-6.9}$$

$Z_{RA}$ is the Rackett compressibility factor appearing in Spencer and Danner's [106] modification of the Rackett equation (Sec. 3-11). In terms of a new equation of state, Eq. (3-6.9) represents a volume translation which has no effect on the VLE behavior predicted by the SRK equation. This concept would allow one to fit $a$ and $b$ in the original cubic equation with

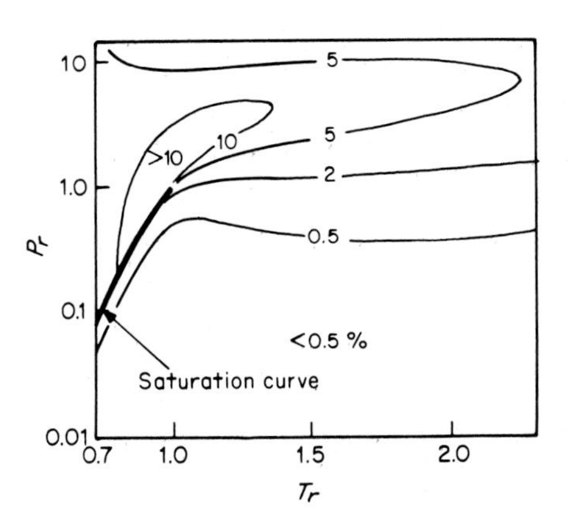


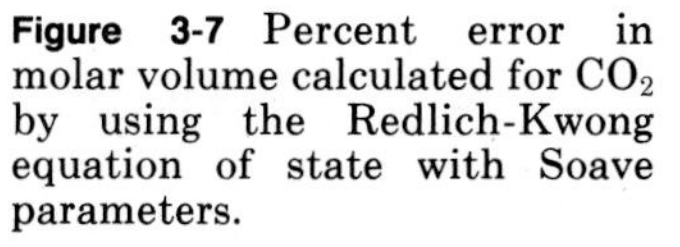

**Figure 3-7** Percent error in molar volume calculated for $CO_2$ by using the Redlich-Kwong equation of state with Soave parameters.

vapor pressures and vapor volumes. The parameter $c$ could then be fit so as to reproduce liquid volumes [54].

**Example 3-3** Calculate the saturated liquid and vapor volumes and vapor pressure of isobutane at 300 K by the following methods:
(*a*) The SRK equation with $a$ and $b$ from Table 3-5
(*b*) Peneloux's correction to the results from part (a)
(*c*) The PR equation with $a$ and $b$ from Table 3-5
(*d*) The SRK equation with $a$ and $b$ from Eqs. (3-6.7) and (3-6.8)

Literature values and data for isobutane at 300 K are [14, 107]

$$\text{Vapor pressure} = 3.704 \text{ bar}$$
$$V^L = 105.9 \text{ cm}^3/\text{mol}$$
$$V^V = 6031 \text{ cm}^3/\text{mol}$$

From Appendix A, $T_c = 408.2$ K, $P_c = 36.5$ bar, and $\omega = 0.183$. Also, $Z_{RA} = 0.27569$ [105].

**solution**

(*a*) The vapor pressure is the pressure at which the fugacity coefficient for the liquid, $\phi^L$, is equal to the fugacity coefficient for the vapor, $\phi^V$. An iterative process whereby one guesses $P$, solves Eq. (3-6.2) for $Z^L$ and $Z^V$, calculates $\phi^L$ and $\phi^V$ with the equations in Chap. 5, and adjusts $P$ by $P_{\text{new}} = P_{\text{old}}\phi^L/\phi^V$ leads to the result that the predicted vapor pressure is 3.706 bar. At that pressure, Eq. (3-6.2) is $Z^3 - Z^2 + 0.08668Z - 0.0011825 = 0$. This may be solved analytically for the largest and smallest values of $Z$ to give $Z^L = 0.01687$ and $Z^V = 0.9057$. Then

$$V^L = (0.01687)\frac{(83.14)(300)}{3.706}$$

$$V^L = 113.5 \text{ cm}^3/\text{mol } (7.2\% \text{ error})$$

Similarly

$$V^V = 6096 \text{ cm}^3/\text{mol } (1.1\% \text{ error})$$

(*b*) Equation (3-6.9) gives

$$c = (0.40768)(83.14)(408.2)(0.29441 - 0.27569)/(36.5)$$
$$c = 7.1 \text{ cm}^3/\text{mol}$$
$$V^L = 113.5 - 7.1$$
$$V^L = 106.4 \text{ cm}^3/\text{mol } (0.5\% \text{ error})$$
$$V^V = 6089 \text{ cm}^3/\text{mol } (0.6\% \text{ error})$$

(*c*) The PR equation, by the procedure described in part (*a*), predicts a vapor pressure for isobutane at 300 K of 3.683 bar. At that pressure, Eq. (3-6.2) is $Z^3 - 0.9893Z^2 + 0.08025Z - 0.0009738 = 0$. Solution of this equation gives $Z^L = 0.01479$ and $Z^V = 0.9015$. Thus

$$V^L = 100.2 \text{ cm}^3/\text{mol } (5.4\% \text{ error})$$
$$V^V = 6105 \text{ cm}^3/\text{mol } (1.2\% \text{ error})$$

(*d*) Equations (3-6.7) and (3-6.8) require the experimental values of $Z^L$ and the vapor pressure

$$Z^L = \frac{(3.704)(105.9)}{(83.14)(300)} = 0.01573$$

Equations (3-6.7) and (3-6.8) give $b = 75.66\ \text{cm}^3/\text{mol}$ and $a = 1.577 \times 10^7\ \text{bar} \cdot \text{cm}^6/\text{mol}^2$. These constants, along with the procedure described in part (*a*), predict a vapor pressure for isobutane at 300 K of 3.711 bar. The liquid and vapor compressibility factors are 0.01577 and 0.9104 respectively. Thus,

$V^L = 106.0\ \text{cm}^3/\text{mol}\ (0.1\%\ \text{error})$

$V^V = 6119\ \text{cm}^3/\text{mol}\ (1.5\%\ \text{error})$

## 3-7 Generalized Benedict-Webb-Rubin Equations

The Benedict-Webb-Rubin (BWR) equation of state is more complicated than cubic equations, and it has been used with success over wide ranges of temperature and pressure. Constants for the BWR equation for a number of pure compounds have been tabulated [13, 78, 85]. The success of the original Benedict-Webb-Rubin equation has led to a number of studies wherein the equation or a modification of it has been generalized to apply to many types of compounds [19, 76, 112, 132].

Lee and Kesler [47] developed a modified BWR equation within the context of Pitzer's three-parameter correlation. To employ the analytical form, care must be taken in the method of solution. The compressibility factor of a real fluid is related to properties of a simple fluid ($\omega = 0$) and those of *n*-octane as a reference fluid. Assume that $Z$ is to be calculated for a fluid at some temperature and pressure. First, using the critical properties of this fluid, determine $T_r$ and $P_r$. Then determine an ideal reduced volume of a simple fluid with Eq. (3-7.1).

$$\frac{P_r V_r^{(0)}}{T_r} = 1 + \frac{B}{V_r^{(0)}} + \frac{C}{(V_r^{(0)})^2} + \frac{D}{(V_2^{(0)})^5} + \frac{c_4}{T_r^3 (V_r^{(0)})^2}\left[\beta + \frac{\gamma}{(V_r^{(0)})^2}\right] \exp\left[-\frac{\gamma}{(V_r^{(0)})^2}\right] \tag{3-7.1}$$

where $\quad B = b_1 - \dfrac{b_2}{T_r} - \dfrac{b_3}{T_r^2} - \dfrac{b_4}{T_r^3} \qquad C = c_1 - \dfrac{c_2}{T_r} + \dfrac{c_3}{T_r^3} \qquad D = d_1 + \dfrac{d_2}{T_r}$

$V_r^{(0)} = P_c V^{(0)}/RT_c$, and the constants are given in Table 3-7 for a simple fluid. With $V_r^{(0)}$, the simple fluid compressibility factor is calculated.

$$Z^{(0)} = \frac{P_r V_r^{(0)}}{T_r} \tag{3-7.2}$$

Next, using the same reduced temperature and pressure as determined above, Eq. (3-7.1) is again solved for $V_r^{(0)}$ but with the reference fluid constants in Table 3-7; call this value $V_r^{(R)}$. Then

$$Z^{(R)} = \frac{P_r V_r^{(R)}}{T_r} \tag{3-7.3}$$

**TABLE 3-7 Lee-Kesler Constants for Eq. (3-7.1)**

| Constant | Simple fluid | Reference fluid | Constant | Simple fluid | Reference fluid |
|---|---|---|---|---|---|
| $b_1$ | 0.1181193 | 0.2026579 | $c_3$ | 0.0 | 0.016901 |
| $b_2$ | 0.265728 | 0.331511 | $c_4$ | 0.042724 | 0.041577 |
| $b_3$ | 0.154790 | 0.027655 | $d_1 \times 10^4$ | 0.155488 | 0.48736 |
| $b_4$ | 0.030323 | 0.203488 | $d_2 \times 10^4$ | 0.623689 | 0.0740336 |
| $c_1$ | 0.0236744 | 0.0313385 | $\beta$ | 0.65392 | 1.226 |
| $c_2$ | 0.0186984 | 0.0503618 | $\gamma$ | 0.060167 | 0.03754 |

The compressibility factor $Z$ for the fluid of interest is then calculated

$$Z = Z^{(0)} + \left(\frac{\omega}{\omega^{(R)}}\right)(Z^{(R)} - Z^{(0)}) \tag{3-7.4}$$

where $\omega^{(R)} = 0.3978$.

Equation (3-7.4) was used to generate the $Z^{(0)}$ and $Z^{(1)}$ values in Tables 3-2 and 3-3. Although tested primarily on hydrocarbons, average errors normally were less than 2 percent for both the vapor and liquid phases. The reduced-temperature range is 0.3 to 4, and the reduced-pressure range is 0 to 10. The application of Lee-Kesler equations to mixtures is covered in Sec. 4-6 and to thermodynamic properties in Sec. 5-4.

Starling and coworkers [7, 129, 130] have also developed a generalized BWR equation. As in the Lee-Kesler equation, Starling's version retains the density dependence given in the original BWR equation. Starling uses a third parameter, an "orientation parameter," which is similar to but not identical with the Pitzer acentric factor. Although Starling's equation and its performance are generally similar to the Lee-Kesler equation, it has been more extensively tested for coal-derived compounds. McFee et al. [67] have compared the Starling and Lee-Kesler equations and found they gave similar results. For one-time calculations the Lee-Kesler method is easier to use because Tables 3-2 and 3-3 are available. Twu [125] has recast the Starling equation into a form such that only the normal boiling point is required as a parameter and has successfully reproduced properties for the paraffins methane through $n$-hectane ($C_{100}H_{202}$).

The Pitzer approach to three-parameter corresponding states theory assumes that the compressibility factor $Z$ depends linearly on the acentric factor $\omega$. In Eq. (3-7.4), the two points used to establish this straight line are the $Z$ values of a simple fluid and the reference fluid, $n$-octane. There is nothing special about these two reference fluids and, in fact, any two fluids could be used as reference fluids. In this case, Eq. (3-7.4) takes the form [117]

$$Z = Z^{(R1)} + \frac{\omega - \omega^{(R1)}}{\omega^{(R2)} - \omega^{(R1)}}(Z^{(R2)} - Z^{(R1)}) \tag{3-7.5}$$

In the application of Eq. (3-7.5) the $Z$ values for the two reference fluids and the fluid of interest are all evaluated at the same reduced temperature and pressure. In the Lee-Kesler method, the reference fluid properties were expressed by Eqs. (3-7.1) to (3-7.3), but, in general, any equation which satisfactorily represented the reference fluid behavior could be used. Thus, Teja et al. [118] used a cubic equation of state while others [65, 81] have used a 35-term extended BWR equation.

## 3-8 Discussion of Equations of State

In this section, the equations of state previously presented are compared. In the low-density limit, all reduce to the ideal-gas law. In the critical region, none of the equations satisfactory. The primary differences occur with respect to computational simplicity and to the quality of the results at high pressures, for the liquid phase, and for polar molecules. Historically, equations of state have been used to represent the behavior of only vapor phases. More recently, they have been used for the liquid phase as well. Thus, it is desirable to have an equation of state that represents the *PVT* behavior of both vapor and liquid phases. It is necessary to strive for an equation that can be extended to mixtures and generate accurate mixture enthalpies and phase equilibria results. Finally, it is desirable that the equation remain relatively simple from a computational point of view. Of course, to have all of these is not possible.

The truncated virial equation, Eq. (3-5.2) is simple but can be used only for the vapor phase, and then only for modest deviations from ideal-gas behavior, i.e., if the reduced density is less than $\frac{1}{2}$. Temperatures and pressures for which that condition applies can be identified as the regions in Figs. 3-1 to 3-3 for which $V_{ri}$ is greater than but about 0.5.

Both the cubic equations in Sec. 3-6 and the generalized BWR equations in Sec. 3-7 are capable of representing liquid phase behavior. The BWR-based equations are applicable over broader temperature and pressure ranges; cubic equations are less complex. It is possible, and generally recommended, to solve cubic equations for volume (or equivalently, the compressibility factor) noniteratively. The BWR equations must be solved iteratively. Specific computational problems encountered with both types of equations have been discussed [27, 40, 90, 92]; see also Sec. 8-12. All the cubic equations of state in Table 3-5 are of the van der Waals form

$$P = P_{\text{repulsive}} + P_{\text{attractive}} \tag{3-8.1}$$

This idea, that the pressure of a fluid results from the sum of repulsive and attractive forces, was first expressed by van der Waals in his equation. In cubic equations, the repulsive part is represented by $RT/(V-b)$. This gives only a qualitative description of the repulsive behavior of molecules. But cubic equations can give good results, which means that the repulsive and attractive contributions are not truly separated. Thus, even though cubic equations are effective for curve fitting *PVT* behavior of fluids, they should not be extrapolated outside the regions of fitted data, especially to high pressures. Efforts to extend the range of applicability of cubic equations have met with some success [34, 41, 43]. However, there are certain limitations that result because the density dependence can be no more complicated than cubic; these limitations have been discussed by Abbott [1, 2] and Martin [55–61].

The modified BWR equations are similar to the cubic equations in the sense that they should not be extrapolated outside the temperature and pressure ranges for which they have been tested. Within those ranges, they do very well, however; in fact, one of their primary advantages is that they have been tested extensively for many hydrocarbons and coal-derived compounds and with respect to almost all thermodynamic properties.

Neither cubic equations nor BWR equations can be used to *predict*, with confidence, the *PVT* behavior of polar molecules. That is consistent with the fact that polar molecules often do not obey three-parameter corresponding states theory. An equation of state presented by Gmehling et al. [25, 26*a*] has been used with some success, but the equation requires four pure component parameters which must be determined from experimental data. The equation allows pure components to dimerize and predicts changes in pure component molar volumes because the apparent number of moles changes (with reaction). This approach is particularly effective for compounds like acetic acid which are known to dimerize, even in the vapor phase.

Wu and Stiel [131*a*] have extended the Lee-Kesler method to polar compounds by the addition of a term to Eq. (3-3.1)

$$Z = Z^{(0)} + \omega Z^{(1)} + YZ^{(2)} \qquad (3\text{-}8.2)$$

$Y$ is a polar parameter unique to each compound, and water is used as a reference fluid to determine values of $Z^{(2)}$. Wu and Stiel tabulate values of $Y$ for several compounds, and they give values of $Z^{(2)}$ as functions of $T_r$ and $P_r$. Errors in compressibility factor predictions for several polar compounds were reduced from 10 to 1 percent by the addition of the term $YZ^{(2)}$. To date, Eq. (3-8.2) has not been tested for highly associated compounds (acetic acid, for example), nor has it been tested for mixtures. A different form is required for alcohols [131*a*].

### Recommendations

To characterize small deviations from ideal-gas behavior, use the truncated virial equation, Eq. (3-5.2). Do not use the virial equation for the liquid phase.

For nonpolar molecules near saturated conditions, use the Soave or Peng-Robinson equation of state; use Peneloux's correction to the Soave equation to obtain more accurate liquid volumes. The Peng-Robinson equation is essentially equivalent to the Soave equation for saturated vapors.

For expanded ranges of temperature and pressure, use the Lee-Kesler method or Starling's equation. The Gmehling equation is recommended for polar molecules if the parameters are available. All three of these equations are capable of representing the liquid phase accurately. However, if one wishes to calculate only liquid volumes, one of the correlations in the following sections is recommended.

## 3-9 *PVT* Properties of Liquids—General Considerations

Liquid specific volumes are relatively easy to measure. For most common organic liquids, at least one experimental value is available. There are a number of references in which these experimental volumes (or densities) are tabulated or in which constants are given to allow one to calculate them rapidly with an empirical equation [3, 16, 23, 49, 64, 68, 72, 115, 131]. References 30 and 116 are extensive bibliographies for liquid *PVT* data published to 1978 and 1983 respectively. Ritter, Lenoir, and Schweppe [97] have published convenient nomographs to estimate saturated liquid densities as functions of temperature for some 90 liquids covering, primarily, hydrocarbons and hydrocarbon derivatives. In Appendix A, single-liquid densities are tabulated for many compounds at a given temperature.

## 3-10 Estimation of the Liquid Molar Volume at the Normal Boiling Point

A number of additive methods are discussed by Partington [82]. Each element and certain bond linkages are assigned numerical values, so that the molar volume at the normal boiling point can be calculated by the addition of these values in a manner similar to that described in Chap. 2 for estimating the critical volume.

### Additive methods

Schroeder [82] has suggested a novel and simple additive method for estimating molar volumes at the normal boiling point. His rule is to count the number of atoms of carbon, hydrogen, oxygen, and nitrogen, add 1 for each double bond, and multiply the sum by 7. This gives the volume in cubic centimeters per mole. His rule is surprisingly good: it gives results within 3 to 4 percent except for highly associated liquids. Table 3-8 gives the values to be used with these and other atoms and functional groups. The accuracy of Schroeder's method is shown in Table 3-9, where molar volumes at the normal boiling point are compared with experimental values for a wide range of materials. The average error for the compounds

**TABLE 3-8 Volume Increments for the Calculation of Molar Volumes $V_b$**

| | Increment, $cm^3/mol$ | |
|---|---|---|
| | Schroeder | Le Bas |
| Carbon | 7 | 14.8 |
| Hydrogen | 7 | 3.7 |
| Oxygen (except as noted below) | 7 | 7.4 |
| In methyl esters and ethers | — | 9.1 |
| In ethyl esters and ethers | — | 9.9 |
| In higher esters and ethers | — | 11.0 |
| In acids | — | 12.0 |
| Joined to S, P, or N | — | 8.3 |
| Nitrogen | 7 | |
| Doubly bonded | — | 15.6 |
| In primary amines | — | 10.5 |
| In secondary amines | — | 12.0 |
| Bromine | 31.5 | 27 |
| Chlorine | 24.5 | 24.6 |
| Fluorine | 10.5 | 8.7 |
| Iodine | 38.5 | 37 |
| Sulfur | 21 | 25.6 |
| Ring, three-membered | −7 | −6.0 |
| Four-membered | −7 | −8.5 |
| Five-membered | −7 | −11.5 |
| Six-membered | −7 | −15.0 |
| Naphthalene | −7 | −30.0 |
| Anthracene | −7 | −47.5 |
| Double bond between carbon atoms | 7 | — |
| Triple bond between carbon atoms | 14 | — |

tested is 3.0 percent. (Schroeder's original rule has been expanded to include halogens, sulfur, and triple bonds.)

Additive volumes published by Le Bas [44] represent a refinement of Schroeder's rule. Volume increments from Le Bas are shown in Table 3-8, and calculated values of $V_b$ are also compared with experimental values in Table 3-9. The average error for the compounds tested is 4.0 percent. Although the average error in this case is greater than that found by Schroeder's increments, the method appears to be more general and as accurate as Schroeder's for most of the compounds tested; i.e., the average error is not particularly representative.

Other additive methods are discussed by Fedors [22].

### Tyn and Calus method [126]

$V_b$ is related to the critical volume by

$$V_b = 0.285\, V_c^{1.048} \tag{3-10.1}$$

**TABLE 3-9 Comparison of Calculated and Experimental Liquid Molar Volumes at the Normal Boiling Point**

| Compound | Molar volume, $cm^3/mol$ | | Percent error† when calculated by method of | | |
|---|---|---|---|---|---|
| | Exp. $V_b$ | Ref. | Tyn and Calus | Schroeder | Le Bas |
| Methane | 37.7 | 46 | −6.7 | −7.2 | −21.5 |
| Propane | 74.5 | 104 | 0.2 | 3.3 | −0.7 |
| Heptane | 162 | 46 | 1.8 | −0.6 | 0.5 |
| Cyclohexane | 117 | 46 | −1.2 | 1.7 | 1.0 |
| Ethylene | 49.4 | 46 | −6.0 | −0.8 | −10 |
| Benzene | 96.5 | 46 | −0.1 | 1.6 | −0.5 |
| Fluorobenzene | 102 | 46 | −0.9 | −0.5 | −1.0 |
| Bromobenzene | 120 | 46 | 1.6 | 2.1 | −1.6 |
| Chlorobenzene | 115 | 46 | 0.0 | 0.0 | 1.7 |
| Iodobenzene | 130 | 46 | 1.9 | −0.4 | −0.5 |
| Methanol | 42.5 | 46 | −0.5 | −1.2 | −13 |
| *n*-Propyl alcohol | 81.8 | 46 | −1.4 | 2.7 | −0.5 |
| Dimethyl ether | 63.8 | 46 | 2.0 | −1.3 | −4.5 |
| Ethyl propyl ether | 129 | 46 | | −2.3 | −0.5 |
| Acetone | 77.5 | 46 | −0.6 | −0.6 | −4.5 |
| Acetic acid | 64.1 | 46 | −2.7 | −1.7 | 6.7 |
| Isobutyric acid | 109 | 46 | 0.3 | −3.7 | 3.5 |
| Methyl formate | 62.8 | 46 | 0.0 | 0.3 | −0.3 |
| Ethyl acetate | 106 | 46 | 0.9 | −0.9 | 2.5 |
| Diethylamine | 109 | 46 | 3.5 | 2.8 | 2.7 |
| Acetonitrile | 57.4 | 46 | 10 | −2.4 | |
| Methyl chloride | 50.6 | 46 | −0.8 | 3.7 | −0.2 |
| Carbon tetrachloride | 102 | 46 | 1.0 | 2.8 | 11 |
| Diochlorodifluoromethane | 80.7 | 42 | −0.8 | −4.6 | 0.9 |
| Ethyl mercaptan | 75.5 | 46 | 0.9 | 2.0 | 2.5 |
| Diethyl sulfide | 118 | 46 | 1.3 | 0.9 | 3.2 |
| Phosgene | 69.5 | 46 | 0.2 | 0.7 | 2.7 |
| Ammonia | 25.0 | 46 | 1.5 | 12 | |
| Chlorine | 45.5 | 46 | −2.1 | 7.7 | 8.1 |
| Water | 18.7 | 45 | 3.5 | 12 | |
| Hydrochloric acid | 30.6 | 46 | −6.8 | 2.9 | −7.5 |
| Sulfur dioxide | 43.8 | 102 | 0.0 | −12 | −3.7 |
| Average error | | | 1.9 | 3.1 | 3.9 |

†Percent error = [calc. − exp.)/exp.] × 100

where both $V_b$ and $V_c$ are expressed in cubic centimeters per mole. This simple relation is generally accurate within 3 percent except for the low-boiling permanent gases (He, $H_2$, Ne, Ar, Kr) and some polar nitrogen and phosphorus compounds (HCN, $PH_3$). A similar relation was suggested earlier by Benson [8], but in that case the critical pressure also was employed in the correlation.

### Recommendation

The Tyn and Calus method is recommended for estimating liquid molar volumes at the boiling point. The average error for 32 compounds is only 2 percent, as shown in Table 3-9. A reliable value of the critical volume must be available, however.

**Example 3-4** Estimate the molar volume of liquid chlorobenzene at its normal boiling point. The critical volume is 308 $cm^3$/mol (Appendix A). The experimental value is 115 $cm^3$/mol.

**solution** SCHROEDER METHOD. From Table 3-8, C = 7, H = 7, Cl = 24.5, the ring = −7, and each double bond = 7. Therefore, for $C_6H_5Cl$

$$V_b = (6)(7)+(5)(7)+24.5-7+(3)(7) = 115 \text{ cm}^3/\text{mol}$$

$$\text{Error} = \frac{115-115}{115} \times 100 = 0\%$$

LE BAS METHOD. From Table 3-8, C = 14.8, H = 3.7, Cl = 24.6, and the ring = −15.0. Therefore,

$$V_b = (6)(14.8)+(5)(3.7)+24.6-15.0 = 117 \text{ cm}^3/\text{mol}$$

$$\text{Error} = \frac{117-115}{115} \times 100 = +1.7\%$$

TYN AND CALUS METHOD. With Eq. (3-10.1),

$$V_b = 0.285 V_c^{1.048} = (0.285)(308^{1.048}) = 115 \text{ cm}^3/\text{mol}$$

$$\text{Error} = \frac{115-115}{115} \times 100 = 0\%$$

## 3-11 Estimation of Liquid Densities

Even if no data are available, there are a number of techniques for estimating pure liquid specific volumes or densities. Three techniques are presented to estimate *saturated* liquid densities; one is presented for compressed liquids.

### Hankinson-Brobst-Thomson (HBT) technique

Hankinson and Thomson [33] present the following correlation for saturated densities of liquids:

$$\frac{V_s}{V^*} = V_R^{(0)}\,[1 - \omega_{\text{SRK}} V_R^{(\delta)}] \qquad (3\text{-}11.1)$$

$$V_R^{(0)} = 1 + a(1-T_r)^{1/3} + b(1-T_r)^{2/3} + c(1-T_r) + d(1-T_r)^{4/3} \qquad 0.25 < T_r < 0.95 \qquad (3\text{-}11.2)$$

$$V_R^{(\delta)} = [e + fT_r + gT_r^2 + hT_r^3]/(T_r - 1.00001) \qquad 0.25 < T_r < 1.0 \qquad (3\text{-}11.3)$$

When computing $T_r$, $T_c$ should be obtained from Table 3-10. Values of the constants are:

| | | | |
|---|---|---|---|
| $a$ | −1.52816 | $b$ | 1.43907 |
| $c$ | −0.81446 | $d$ | 0.190454 |
| $e$ | −0.296123 | $f$ | 0.386914 |
| $g$ | −0.0427258 | $h$ | −0.0480645 |

$V^*$ is a pure component characteristic volume generally within 1 to 4 percent of the critical volume; $\omega_{SRK}$ is the acentric factor which forces the Soave equation to give a best fit of existing vapor pressure data. Values of $V^*$ and $\omega_{SRK}$ for over 400 compounds given in Refs. 33, 71, and 102 are listed in Table 3-10.

If a value of $V^*$ is not available, it may be estimated by:

$$V^* = \frac{RT_c}{P_c}(a + b\omega_{SRK} + c\omega_{SRK}^2) \tag{3-11.4}$$

Values for the constants in Eq. (3-11.4) are given in the accompanying tables.

| Constant | Paraffins | Olefins and diolefins | Cycloparaffins | Aromatics | All hydrocarbons |
|---|---|---|---|---|---|
| $a$ | 0.2905331 | 0.3070619 | 0.6564296 | 0.2717636 | 0.2851686 |
| $b$ | −0.08057958 | −0.2368581 | −3.391715 | −0.05759377 | −0.06379110 |
| $c$ | 0.02276965 | 0.2834693 | 7.442388 | 0.05527757 | 0.01379173 |
| Avg. abs. % error in $V_s$ | 1.23% | 1.43% | 1.00% | 0.58% | 1.89% |

| Constant | Sulfur compounds | Fluorocarbons | Cryogenic liquids | Condensable gases |
|---|---|---|---|---|
| $a$ | 0.3053426 | 0.5218098 | 0.2960998 | 0.2828447 |
| $b$ | −0.1703247 | −2.346916 | −0.05468500 | −0.1183987 |
| $c$ | 0.1753972 | 5.407302 | −0.1901563 | 0.1050570 |
| Avg. abs. % error in $V_s$ | 1.98% | 0.82% | 0.85% | 3.65% |

If no data are available for a compound, $\omega_{SRK}$ should be replaced by the true acentric factor. The change in $V_s$ will often be less than 1 percent but can be as high as 4 percent. If a compound does not fit into the categories described for Eq. (3-11.4), $V^*$ may be replaced by the true critical volume. Again the resulting error will often be less than 1 percent but can be as high as 4 percent. The temperature-dependent criticals suggested by Gunn et al. [28] were used for hydrogen and helium.

**TABLE 3-10 Pure Component Parameters for the Hankinson-Brobst-Thomson and the Rackett Liquid Volume Correlations† [33, 71, 102]**

| Paraffins | $T_c$, K‡ | $\omega_{SRK}$ | $V^*$, L/mol | $Z_{RA}$ |
|---|---|---|---|---|
| Methane | 190.58 | 0.0074 | 0.0994 | 0.2892 |
| Ethane | 305.42 | 0.0983 | 0.1458 | 0.2808 |
| Propane | 369.82 | 0.1532 | 0.2001 | 0.2766 |
| $n$-Butane | 425.18 | 0.2008 | 0.2544 | 0.2730 |
| Isobutane | 408.14 | 0.1825 | 0.2568 | 0.2754 |
| $n$-Pentane | 469.65 | 0.2522 | 0.3113 | 0.2684 |
| Isopentane | 460.43 | 0.2400 | 0.3096 | 0.2717 |
| Neopentane | 433.78 | 0.1975 | 0.3126 | 0.2756 |
| $n$-Hexane | 507.43 | 0.3007 | 0.3682 | 0.2635 |
| 2-Methylpentane | 497.50 | 0.2791 | 0.3677 | 0.2672 |
| 3-Methylpentane | 504.43 | 0.2741 | 0.3633 | 0.2690 |
| 2,2-Dimethylbutane | 488.78 | 0.2330 | 0.3634 | 0.2733 |
| 2,3-Dimethylbutane | 499.98 | 0.2477 | 0.3610 | 0.2705 |
| $n$-Heptane | 540.26 | 0.3507 | 0.4304 | 0.2604 |
| 2,2-Dimethylpentane | 520.50 | 0.2882 | 0.4225 | 0.2684 |
| 2,4-Dimethylpentane | 519.79 | 0.3040 | 0.4251 | 0.2671 |
| 3,3-Dimethylpentane | 536.40 | 0.2681 | 0.4137 | 0.2707 |
| 2,3-Dimethylpentane | 537.35 | 0.2973 | 0.4127 | 0.2703 |
| 2-Methylhexane | 530.37 | 0.3310 | 0.4274 | 0.2638 |
| 3-Methylhexane | 535.25 | 0.3243 | 0.4231 | 0.2654 |
| 3-Ethylpentane | 540.64 | 0.3118 | 0.4163 | 0.2658 |
| 2,2,3-Trimethylbutane | 531.17 | 0.2511 | 0.4125 | 0.2727 |
| $n$-Octane | 568.83 | 0.3998 | 0.4904 | 0.2571 |
| Isooctane | 543.96 | 0.3045 | 0.4790 | 0.2684 |
| 2,2,3,3-Tetramethylbutane | 567.93 | 0.2513 | 0.4569 | 0.2738 |
| 2-Methylheptane | 559.57 | 0.3780 | 0.4889 | |
| 3-Methylheptane | 563.60 | 0.3699 | 0.4837 | |
| 4-Methylheptane | 561.67 | 0.3708 | 0.4841 | |
| 2,2-Dimethylhexane | 549.80 | 0.3374 | 0.4829 | |
| 2,3-Dimethylhexane | 563.42 | 0.3458 | 0.4765 | |
| 2,4-Dimethylhexane | 553.45 | 0.3425 | 0.4811 | |
| 2,5-Dimethylhexane | 549.99 | 0.3556 | 0.4858 | |
| 3,4-Dimethylhexane | 568.78 | 0.3376 | 0.4722 | |
| 2,2,3-Trimethylpentane | 563.43 | 0.2965 | 0.4679 | |
| 2,3,3-Trimethylpentane | 573.49 | 0.2889 | 0.4632 | |
| 2,3,4-Trimethylpentane | 566.34 | 0.3144 | 0.4689 | |
| $n$-Nonane | 594.64 | 0.4478 | 0.5529 | 0.2543 |
| 2-Methyloctane | 586.60 | 0.4225 | 0.5524 | |
| 2,3-Dimethylheptane | 589.60 | 0.3848 | 0.5383 | |
| 2,6-Dimethylheptane | 577.90 | 0.4006 | 0.5500 | |
| 3-Ethylheptane | 590.40 | 0.4083 | 0.5415 | |
| 2,2,3-Trimethylhexane | 588.00 | 0.3349 | 0.5283 | |
| 2,2,4-Trimethylhexane | 573.70 | 0.3481 | 0.5361 | |
| 2,2,5-Trimethylhexane | 568.00 | 0.3569 | 0.5408 | |
| 2,2,3,4-Tetramethylpentane | 592.70 | 0.3122 | 0.5198 | |
| 2,3,3,4-Tetramethylpentane | 607.60 | 0.3117 | 0.5127 | |
| 3,3-Diethylpentane | 610.00 | 0.3365 | 0.5175 | |
| $n$-Decane | 617.65 | 0.4916 | 0.6192 | 0.2507 |
| 4-Methylnonane | 610.50 | 0.4572 | 0.6104 | |
| 2,7-Dimethyloctane | 602.90 | 0.4432 | 0.6135 | |
| 3,3,5-Trimethylheptane | 609.60 | 0.3827 | 0.5895 | |

**TABLE 3-10 Pure Component Parameters for the Hankinson-Brobst-Thomson and the Rackett Liquid Volume Correlations† [33, 71, 102] (*Continued*)**

| Paraffins | $T_c$, K‡ | $\omega_{SRK}$ | $V^*$, L/mol | $Z_{RA}$ |
|---|---|---|---|---|
| 2,2,3,3,-Tetramethylhexane | 623.10 | 0.3646 | 0.5737 | |
| 2,2,5,5-Tetramethylhexane | 581.50 | 0.3760 | 0.5992 | |
| *n*-Undecane | 638.73 | 0.5422 | 0.6865 | 0.2499 |
| *n*-Dodecane | 658.26 | 0.5807 | 0.7558 | 0.2466 |
| *n*-Tridecane | 675.76 | 0.6340 | 0.8317 | 0.2473 |
| *n*-Tetradecane | 691.87 | 0.6821 | 0.9022 | |
| *n*-Pentadecane | 706.76 | 0.7254 | 0.9772 | |
| *n*-Hexadecane | 720.54 | 0.7667 | 1.0539 | 0.2388 |
| *n*-Heptadecane | 733.37 | 0.7946 | 1.1208 | 0.2343 |
| *n*-Octadecane | 745.26 | 0.8124 | 1.1989 | 0.2275 |
| *n*-Nonadecane | 755.93 | 0.8328 | 1.2715 | 0.2236 |
| *n*-Eicosane | 767.04 | 0.9239 | 1.3754 | 0.2281 |
| Heneicosane | 776.77 | 1.0505 | 1.5081 | 0.2363 |
| Docosane | 788.99 | 1.0561 | 1.5839 | 0.2350 |
| Tricosane | 801.29 | 1.0477 | 1.6507 | 0.2341 |
| Tetracosane | 813.65 | 1.0316 | 1.7104 | |
| Pentacosane | 826.24 | 1.0014 | 1.7887 | |
| Hexacosane | 839.30 | 0.9498 | 1.8133 | 0.2302 |
| Heptacosane | 852.66 | 0.9001 | 1.9420 | |
| Octacosane | 866.44 | 0.8455 | 1.8972 | 0.2277 |
| Triacontane | 840.35 | 0.9919 | 2.0455 | |
| Dotriacontane | 850.57 | 1.0045 | 2.2453 | |
| Tetracontane | 882.18 | 0.9781 | 2.7094 | |
| Tritetracontane | 891.12 | 0.9214 | 2.9342 | |
| **Halogenated Paraffins** | | | | |
| Fluoromethane (R-41) | 317.70 | 0.1851 | 0.1054 | 0.2491 |
| Chloromethane (R-40) | 416.25 | 0.1472 | 0.1363 | 0.2679 |
| Bromomethane | 464.00 | 0.2032 | 0.1506 | |
| Nitromethane | 588.00 | 0.3295 | 0.1626 | |
| Dichloromethane (R-30) | 510.00 | 0.1959 | 0.1767 | 0.2618 |
| Chloroform (R-20) | 536.40 | 0.2181 | 0.2245 | 0.2750 |
| Trifluorobromomethane (R-13B1) | 340.20 | 0.1700 | 0.1970 | |
| Carbon tetrachloride (R-10) | 556.40 | 0.1875 | 0.2754 | 0.2722 |
| Trichlorofluoromethane (R-11) | 471.15 | 0.1871 | 0.2460 | 0.2745 |
| Dichlorodifluoromethane (R-12) | 385.15 | 0.1699 | 0.2147 | 0.2757 |
| Chlorotrifluoromethane (R-13) | 302.00 | 0.1747 | 0.1807 | 0.2771 |
| Dichloromonofluoromethane (R-21) | 451.65 | 0.2102 | 0.1958 | 0.2705 |
| Monochlorodifluoromethane (R-22) | 369.15 | 0.2215 | 0.1637 | 0.2663 |
| Trichlorotrifluoroethane (R-113) | 487.26 | 0.2560 | 0.3263 | 0.2721 |
| Dichlorotetrafluoroethane (R-114) | 418.87 | 0.2582 | 0.2954 | 0.2737 |
| Fluoroethane | 375.31 | 0.2150 | 0.1577 | |
| Chloroethane (R-160) | 460.40 | 0.1880 | 0.1858 | 0.2654 |
| Bromoethane | 503.80 | 0.2266 | 0.2064 | 0.2896 |
| 1,1-Dichloroethane (R-150A) | 523.00 | 0.2365 | 0.2369 | |
| 1,2-Dichloroethane | 561.00 | 0.2754 | 0.2302 | |
| 2-Fluoropropane | 417.26 | 0.1822 | 0.2130 | |
| 1-Chloropropane | 503.00 | 0.2263 | 0.2434 | |
| 2-Chloropropane | 485.00 | 0.2474 | 0.2494 | |
| 1,2,3-Trichloropropane | 651.00 | 0.3282 | 0.3279 | |
| 1-Chlorobutane | 542.00 | 0.2265 | 0.2969 | |

| Paraffins | $T_c$, K‡ | $\omega_{SRK}$ | $V^*$, L/mol | $Z_{RA}$ |
|---|---|---|---|---|
| Halogenated Paraffins (*Continued*) | | | | |
| 2-Chlorobutane | 520.60 | 0.2920 | 0.3019 | |
| tert-Butyl chloride | 507.00 | 0.1999 | 0.3031 | |
| Cycloparaffins | | | | |
| Cyclopropane | 397.81 | 0.1305 | 0.1610 | 0.2716 |
| Cyclopentane | 511.76 | 0.1969 | 0.2600 | 0.2745 |
| Methylcyclopentane | 532.79 | 0.2322 | 0.3181 | 0.2711 |
| Cyclohexane | 553.54 | 0.2128 | 0.3090 | 0.2729 |
| Methylcyclohexane | 572.19 | 0.2371 | 0.3709 | 0.2704 |
| 1,1-Dimethylcyclopentane | 547.04 | 0.2691 | 0.3754 | 0.2768 |
| 1-*trans*-3-Dimethylcyclopentane | 553.15 | 0.2676 | 0.3796 | 0.2768 |
| 1-*trans*-2-Dimethylcyclopentane | 553.15 | 0.2689 | 0.3784 | 0.2763 |
| 1-*cis*-3-Dimethylcyclopentane | 550.93 | 0.2825 | 0.3825 | 0.2823 |
| 1-*cis*-2-Dimethylcyclopentane | 564.82 | 0.2685 | 0.3706 | 0.2699 |
| Ethylcyclopentane | 569.46 | 0.2689 | 0.3740 | |
| 1,1,2-Trimethylcyclopentane | 579.50 | 0.2527 | 0.4255 | |
| 1,1,3-Trimethylcyclopentane | 569.50 | 0.2173 | 0.4330 | |
| 1,1-Dimethylcyclohexane | 591.00 | 0.2351 | 0.4216 | |
| *trans*-1,3-Dimethylcyclohexane | 598.00 | 0.2374 | 0.4213 | |
| *trans*-1,4-Dimethylcyclohexane | 590.00 | 0.2395 | 0.4319 | |
| Ethylcyclohexane | 609.00 | 0.2497 | 0.4227 | |
| Propylcyclohexane | 639.00 | 0.2654 | 0.4812 | |
| Cyclodecane | 709.00 | 0.2803 | 0.5100 | |
| 1,2-Dicyclohexylethane | 756.92 | 0.4986 | 0.7482 | |
| Olefins | | | | |
| Ethylene | 282.36 | 0.0882 | 0.1310 | 0.2815 |
| Propylene | 364.76 | 0.1455 | 0.1829 | 0.2779 |
| 1-Butene | 419.59 | 0.1921 | 0.2377 | 0.2736 |
| *cis*-2-Butene | 435.58 | 0.2039 | 0.2311 | 0.2701 |
| *trans*-2-Butene | 428.63 | 0.2153 | 0.2367 | 0.2720 |
| Isobutene | 417.90 | 0.1959 | 0.2369 | 0.2728 |
| 1-Pentene | 464.78 | 0.2824 | 0.2951 | 0.2899 |
| *cis*-2-Pentene | 475.93 | 0.2426 | 0.2875 | 0.2671 |
| *trans*-2-Pentene | 475.37 | 0.2399 | 0.2929 | 0.2704 |
| 2-Methyl-1-Butene | 465.37 | 0.2355 | 0.2887 | 0.2627 |
| 3-Methyl-1-Butene | 450.37 | 0.2266 | 0.2940 | 0.2739 |
| 2-Methyl-2-Butene | 470.37 | 0.2852 | 0.2883 | 0.2592 |
| 1-Hexene | 504.03 | 0.2850 | 0.3509 | 0.2658 |
| *cis*-2-Hexene | 518.00 | 0.2509 | 0.3447 | |
| *trans*-3-Hexene | 516.00 | 0.2532 | 0.3491 | |
| *cis*-3-Hexene | 517.00 | 0.2318 | 0.3465 | |
| *trans*-3-Hexene | 519.00 | 0.2153 | 0.3473 | |
| 2,3-Dimethyl-1-Butene | 501.00 | 0.2242 | 0.3425 | |
| 3,3,-Dimethyl-1-Butene | 490.00 | 0.1499 | 0.3466 | |
| 2-Ethyl-1-Butene | 515.46 | 0.2100 | 0.3395 | |
| 2-Methyl-1-Pentene | 511.54 | 0.2099 | 0.3433 | |
| 1-Heptene | 533.29 | 0.3639 | 0.4113 | 0.2611 |
| 2,3,3-Trimethyl-1-Butene | 533.00 | 0.1981 | 0.3914 | |
| 2,3-Dimethyl-1-Pentene | 537.57 | 0.2318 | 0.3955 | |
| 1-Octene | 566.65 | 0.3876 | 0.4710 | 0.2600 |

**TABLE 3-10 Pure Component Parameters for the Hankinson-Brobst-Thomson and the Rackett Liquid Volume Correlations† [33, 71, 102] (*Continued*)**

| Paraffins | $T_c$, K‡ | $\omega_{SRK}$ | $V^*$, L/mol | $Z_{RA}$ |
|---|---|---|---|---|
| | Olefins (*Continued*) | | | |
| *trans*-2-Octene | 580.00 | 0.3431 | 0.4674 | |
| 2,3-Dimethyl-2-hexene | 577.15 | 0.3256 | 0.4510 | |
| 2,4,4-Trimethyl-1-pentene | 560.15 | 0.2207 | 0.4510 | |
| 2,4,4-Trimethyl-2-pentene | 563.15 | 0.2363 | 0.4493 | |
| 1-Nonene | 592.04 | 0.4327 | 0.5333 | 0.2539 |
| 1-Decene | 614.82 | 0.4975 | 0.6013 | 0.2546 |
| 1-Dodecene | 657.00 | 0.5638 | 0.7340 | |
| | Halogenated Olefin | | | |
| Vinyl chloride (R-1140) | 429.70 | 0.1293 | 0.1722 | |
| | Diolefins | | | |
| Propadiene | 393.15 | 0.1430 | 0.1470 | 0.2584 |
| 1,2-Butadiene | 443.71 | 0.2492 | 0.2183 | 0.2675 |
| 1,3-Butadiene | 425.37 | 0.1934 | 0.2202 | 0.2712 |
| 1,2-Pentadiene | 503.15 | 0.1760 | 0.2692 | 0.2677 |
| *cis*-1,3-Pentadiene | 499.15 | 0.1849 | 0.2691 | |
| *trans*-1,3-Pentadiene | 496.00 | 0.1830 | 0.2742 | |
| Isoprene | 484.26 | 0.1700 | 0.2691 | 0.2652 |
| 1,5-Hexadiene | 507.00 | 0.2444 | 0.3306 | |
| | Cyclic Olefins | | | |
| Cyclopentene | 506.00 | 0.1007 | 0.2375 | |
| Cyclohexene | 560.41 | 0.2091 | 0.2903 | |
| 4-Vinylcyclohexene | 604.03 | 0.2883 | 0.3902 | |
| 1,5-Cyclooctadiene | 647.23 | 0.2672 | 0.3717 | |
| | Acetylenes | | | |
| Acetylene | 308.32 | 0.2049 | 0.1128 | 0.2709 |
| Methylacetylene | 402.39 | 0.2184 | 0.1609 | 0.2706 |
| Vinyl acetylene | 455.00 | 0.1335 | 0.1961 | |
| Dimethylacetylene | 488.15 | 0.1581 | 0.2106 | 0.2693 |
| 1-Butyne | 463.71 | 0.0986 | 0.2154 | 0.2711 |
| | Aromatics | | | |
| Benzene | 562.16 | 0.2137 | 0.2564 | 0.2698 |
| Toluene | 591.79 | 0.2651 | 0.3137 | 0.2644 |
| *o*-Xylene | 630.37 | 0.3118 | 0.3673 | 0.2620 |
| *m*-Xylene | 617.05 | 0.3270 | 0.3731 | 0.2625 |
| *p*-Xylene | 616.26 | 0.3216 | 0.3740 | 0.2592 |
| 1,2,3-Trimethylbenzene | 664.45 | 0.3642 | 0.4183 | |
| 1,2,4-Trimethylbenzene | 649.17 | 0.3745 | 0.4279 | |
| 1,3,5-Trimethylbenzene | 637.28 | 0.3974 | 0.4337 | |
| Ethylbenzene | 617.17 | 0.3048 | 0.3702 | |
| *n*-Propylbenzene | 638.30 | 0.3432 | 0.4298 | 0.2599 |

| Paraffins | $T_c$, K‡ | $\omega_{SRK}$ | $V^*$, L/mol | $Z_{RA}$ |
|---|---|---|---|---|
| | Aromatics (*Continued*) | | | |
| Styrene | 647.59 | 0.2420 | 0.3482 | 0.2634 |
| Cumene | 631.15 | 0.3277 | 0.4271 | 0.2617 |
| *n*-Butylbenzene | 660.40 | 0.3921 | 0.4921 | |
| Isobutylbenzene | 650.00 | 0.3921 | 0.4944 | |
| *sec*-Butylbenzene | 664.00 | 0.2817 | 0.4778 | |
| *tert*-Butylbenzene | 660.00 | 0.2710 | 0.4733 | |
| Pentylbenzene | 679.93 | 0.4406 | 0.5561 | 0.2547 |
| Heptylbenzene | 713.54 | 0.5441 | 0.6906 | 0.2508 |
| Nonylbenzene | 740.93 | 0.6583 | 0.8369 | 0.2471 |
| Undecylbenzene | 764.26 | 0.7659 | 0.9919 | 0.2444 |
| Tridecylbenzene | 783.15 | 0.9001 | 1.1693 | 0.2434 |
| Heptadecylbenzene | 818.76 | 0.9404 | 1.4564 | |
| Tricosylbenzene | 847.43 | 1.1399 | 1.9952 | |
| Heavy aromatic | 895.75 | 0.7619 | 2.2323 | |
| Naphthalene | 748.35 | 0.3000 | 0.3834 | |
| 1-Methylnaphthalene | 772.00 | 0.3422 | 0.4504 | |
| 2-Methylnaphthalene | 761.00 | 0.3669 | 0.4591 | |
| Tetralin | 719.00 | 0.3209 | 0.4304 | |
| Biphenyl | 789.00 | 0.3633 | 0.4890 | 0.2743 |
| Phenanthrene | 878.00 | 0.4316 | 0.5711 | |
| | Aromatic Derivatives | | | |
| Fluorobenzene | 560.09 | 0.2434 | 0.2702 | 0.2662 |
| Chlorobenzene | 632.40 | 0.2461 | 0.3056 | 0.2651 |
| Bromobenzene | 670.00 | 0.2481 | 0.3204 | 0.2637 |
| Nitrobenzene | 718.86 | 0.4348 | 0.3339 | |
| Iodobenzene | | | | 0.2645 |
| | Heterocycles | | | |
| Ethylene oxide | 468.15 | 0.2114 | 0.1345 | 0.2569 |
| 1,2-Propylene oxide | 482.20 | 0.2593 | 0.1910 | |
| Furan | 490.25 | 0.2061 | 0.1968 | |
| Tetrahydrofuran | 540.15 | 0.2227 | 0.2308 | |
| Pyrrole | 640.00 | 0.3305 | 0.2130 | |
| Pyrrolidine | 568.55 | 0.2718 | 0.2423 | |
| *n*-Methylpyrrolidone | 723.59 | 0.3654 | 0.3070 | |
| Thiophene | 579.40 | 0.1934 | 0.2279 | |
| Thiophane | 622.15 | 0.2299 | 0.2626 | |
| Sulfolane | 854.88 | 0.3591 | 0.3136 | 0.2384 |
| 3-Methyl sulfolane | 817.38 | 0.4132 | 0.4127 | |
| 1,4-Dioxane | 587.00 | 0.2779 | 0.2523 | |
| Pyridine | 620.00 | 0.2398 | 0.2400 | |
| | Nitriles | | | |
| Acetonitrile | 548.00 | 0.3076 | 0.1606 | 0.1987 |
| Acrylonitrile | 536.00 | 0.3369 | 0.1918 | 0.2275 |
| Benzonitrile | 699.40 | 0.3566 | 0.3257 | |

**TABLE 3-10 Pure Component Parameters for the Hankinson-Brobst-Thomson and the Rackett Liquid Volume Correlations† [33, 71, 102] (*Continued*)**

| Paraffins | $T_c$, K‡ | $\omega_{SRK}$ | $V^*$, L/mol | $Z_{RA}$ |
|---|---|---|---|---|
| | | Acids and Anhydrides | | |
| Hydrogen cyanide | 456.80 | 0.3838 | 0.1076 | |
| Formic acid | 580.00 | 0.4700 | 0.1170 | 0.1880 |
| Acetic acid | 594.45 | 0.4310 | 0.1741 | 0.2225 |
| Trifluoroacetic acid | 491.30 | 0.5335 | 0.2285 | |
| Adipic acid | 809.11 | 1.1045 | 0.4844 | |
| Stearic acid | 798.83 | 1.2312 | 1.3430 | |
| Acetic anhydride | 569.00 | 0.9057 | 0.3287 | |
| | | Esters | | |
| Methyl acetate | 506.80 | 0.3205 | 0.2262 | 0.2552 |
| Ethyl acetate | 523.20 | 0.3595 | 0.2853 | 0.2539 |
| Vinyl acetate | 525.00 | 0.3362 | 0.2669 | 0.2573 |
| Methyl acrylate | 536.00 | 0.3373 | 0.2640 | |
| Ethyl acrylate | 552.00 | 0.3908 | 0.3245 | |
| Methyl methacrylate | 563.15 | 0.2890 | 0.3112 | |
| | | Amides | | |
| Formamide | 765.33 | 0.4061 | 0.1305 | |
| *n*-Methylformamide | 727.30 | 0.3965 | 0.1893 | |
| *n,n*-Dimethylformamide | 643.15 | 0.3672 | 0.2399 | |
| Acetamide | 760.79 | 0.4292 | 0.1830 | |
| Propionamide | 730.31 | 0.4559 | 0.2406 | |
| | | Aldehydes | | |
| Formaldehyde | 408.00 | 0.2656 | 0.1001 | |
| Acetaldehyde | 460.93 | 0.2647 | 0.1519 | 0.2269 |
| Furfural | 670.15 | 0.3847 | 0.2622 | |
| | | Ketones | | |
| Ketene | 380.00 | 0.0967 | 0.1450 | |
| Acetone | 508.15 | 0.3149 | 0.2080 | 0.2477 |
| Methyl ethyl ketone | 536.78 | 0.3188 | 0.2523 | |
| Diethyl ketone | 561.00 | 0.3465 | 0.3034 | |
| Methyl isopropyl ketone | 553.00 | 0.3323 | 0.3156 | |
| Methyl isobutyl ketone | 571.00 | 0.3743 | 0.3758 | |
| Cyclopentanone | 626.00 | 0.2949 | 0.2686 | |
| Cyclohexanone | 629.00 | 0.4409 | 0.3271 | 0.2465 |
| | | Alcohols | | |
| Methanol | 513.15 | 0.5536 | 0.1198 | 0.2334 |
| Ethanol | 516.16 | 0.6378 | 0.1752 | 0.2502 |
| 1-Propanol | 537.04 | 0.6249 | 0.2305 | 0.2541 |
| Isopropanol | 508.76 | 0.6637 | 0.2313 | 0.2493 |
| 1-Butanol | 562.93 | 0.5928 | 0.2841 | 0.2538 |
| Isobutanol | 547.73 | 0.5883 | 0.2730 | |
| *sec*-Butanol | 535.95 | 0.5792 | 0.2803 | |

| Paraffins | $T_c$, K‡ | $\omega_{SRK}$ | $V^*$, L/mol | $Z_{RA}$ |
|---|---|---|---|---|
| | Alcohols (*Continued*) | | | |
| *tert*-Butanol | 506.20 | 0.6134 | 0.2876 | |
| 1-Pentanol | 586.00 | 0.5975 | 0.3437 | 0.2596 |
| 2-Methyl-1-butanol | 571.00 | 0.6108 | 0.3407 | |
| 3-methyl-1-butanol | 579.40 | 0.5629 | 0.3413 | |
| 2-methyl-2-butanol | 545.00 | 0.5007 | 0.3313 | |
| 3-Pentanol | 542.62 | 0.7094 | 0.3434 | |
| 1-Dodecanol | 679.00 | 1.1256 | 0.8283 | |
| Allyl alcohol | 508.31 | 0.6663 | 0.2273 | |
| Furfuryl alcohol | 597.30 | 0.5887 | 0.4900 | |
| Cyclohexanol | 625.00 | 0.5296 | 0.3377 | |
| Benzyl alcohol | 677.00 | 0.7231 | 0.3591 | |
| | Glycols | | | |
| Ethylene glycol | 647.15 | 1.2280 | 0.2120 | 0.2488 |
| Diethylene glycol | 680.15 | 1.0713 | 0.3522 | 0.2489 |
| Triethylene glycol | 710.15 | 1.2540 | 0.5347 | 0.2462 |
| Tetraethylene glycol | 747.59 | 1.7224 | 0.8966 | |
| Glycerol | 726.00 | 1.9845 | 0.4119 | |
| | Phenol | | | |
| Phenol | 694.20 | 0.4297 | 0.2809 | 0.2780 |
| | Amines | | | |
| Methyl amine | 430.70 | 0.2872 | 0.1223 | |
| Dimethyl amine | 437.22 | 0.3044 | 0.1812 | |
| Ethyl amine | 456.00 | 0.2871 | 0.1772 | 0.2642 |
| Diethyl amine | 496.60 | 0.3045 | 0.2906 | 0.2568 |
| Triethyl amine | 535.00 | 0.3196 | 0.4026 | 0.2693 |
| Hexamethylene diamine | 666.53 | 0.6418 | 0.4260 | |
| Aniline | 699.00 | 0.3809 | 0.2901 | 0.2616 |
| Monoethanolamine | 636.76 | 0.8271 | 0.2135 | |
| Diethanolamine | 706.54 | 1.5299 | 0.4543 | 0.2527 |
| Diisopropanol amine | 672.35 | 1.4690 | 0.6151 | |
| | Ethers | | | |
| Dimethyl ether | 400.00 | 0.1972 | 0.1692 | 0.2742 |
| Ethyl ether | 466.76 | 0.2800 | 0.2812 | 0.2632 |
| Methyl ethyl ether | 437.80 | 0.2401 | 0.2216 | 0.2673 |
| Methyl *n*-butyl ether | 512.78 | 0.3137 | 0.3372 | |
| Methyl isobutyl ether | 496.31 | 0.3049 | 0.3379 | |
| Methyl *tert*-butyl ether | 497.10 | 0.2670 | 0.3249 | |
| Diisopropyl ether | 500.32 | 0.3300 | 0.3995 | |
| Methyl vinyl ether | 436.00 | 0.2489 | 0.2011 | |
| Ethyl vinyl ether | 475.00 | 0.2673 | 0.2477 | |
| | Mercaptans | | | |
| Methyl mercaptan | 469.93 | 0.1567 | 0.1508 | 0.2781 |
| Ethyl mercaptan | 499.26 | 0.1915 | 0.2023 | 0.2704 |

**TABLE 3-10 Pure Component Parameters for the Hankinson-Brobst-Thomson and the Rackett Liquid Volume Correlations† [33, 71, 102] (*Continued*)**

| Paraffins | $T_c$, K‡ | $\omega_{SRK}$ | $V^*$, L/mol | $Z_{RA}$ |
|---|---|---|---|---|
| | | Mercaptans (*Continued*) | | |
| *n*-Propyl mercaptan | 535.64 | 0.2380 | 0.2572 | 0.2685 |
| Isopropyl mercaptan | 517.41 | 0.2105 | 0.2606 | 0.2810 |
| *n*-Butyl mercaptan | 569.11 | 0.2781 | 0.3135 | 0.2644 |
| *sec*-Butyl mercaptan | 554.01 | 0.2494 | 0.3139 | 0.2731 |
| Isobutyl mercaptan | 559.44 | 0.2496 | 0.3159 | 0.2725 |
| *tert*-Butyl mercaptan | 530.14 | 0.1966 | 0.3162 | 0.2831 |
| *n*-amyl mercaptan | 597.79 | 0.3235 | 0.3728 | 0.2613 |
| 2-Pentanethiol | 581.90 | 0.2932 | 0.3709 | 0.2685 |
| 3-Pentanethiol | 584.17 | 0.2931 | 0.3676 | 0.2666 |
| Isoamyl mercaptan | 591.88 | 0.2935 | 0.3663 | 0.2641 |
| 3-Methyl-1-butanethiol | 590.89 | 0.2934 | 0.3697 | 0.2658 |
| *tert*-amyl mercaptan | 570.13 | 0.2390 | 0.3674 | 0.2722 |
| *sec*-Isoamyl mercaptan | 583.53 | 0.2641 | 0.3650 | 0.2698 |
| *tert*-Butyl methyl mercaptan | 577.10 | 0.2393 | 0.3661 | |
| *n*-Hexyl mercaptan | 622.73 | 0.3726 | 0.4335 | 0.2583 |
| *sec*-Hexyl mercaptan | 607.84 | 0.3413 | 0.4310 | 0.2646 |
| *tert*-Hexyl mercaptan | 595.39 | 0.2861 | 0.4233 | |
| Diisopropyl mercaptan | 602.64 | 0.2562 | 0.4160 | 0.2664 |
| *n*-Heptyl mercaptan | 644.82 | 0.4253 | 0.4971 | 0.2557 |
| *sec*-Heptyl mercaptan | 630.80 | 0.3932 | 0.4939 | 0.2614 |
| *n*-Octyl mercaptan | 663.77 | 0.4808 | 0.5616 | 0.2532 |
| *sec*-Octyl mercaptan | 650.80 | 0.4477 | 0.5579 | 0.2583 |
| *n*-Nonyl mercaptan | 680.86 | 0.5388 | 0.6301 | 0.2514 |
| *sec*-Nonyl mercaptan | 669.47 | 0.5048 | 0.6258 | 0.2559 |
| *n*-Decyl mercaptan | 696.37 | 0.5986 | 0.7007 | 0.2493 |
| *sec*-Decyl mercaptan | 685.81 | 0.5640 | 0.6957 | 0.2534 |
| Benzenethiol | 689.52 | 0.2677 | 0.3157 | |
| Benzyl mercaptan | 710.17 | 0.3101 | 0.3704 | |
| | | Sulfides | | |
| Carbonyl sulfide | 375.37 | 0.1021 | 0.1410 | 0.2709 |
| Dimethyl sulfide | 503.04 | 0.1936 | 0.2010 | 0.2692 |
| Methyl ethyl sulfide | 533.15 | 0.2435 | 0.2569 | 0.2689 |
| Methyl *n*-propyl sulfide | 564.82 | 0.2770 | 0.3129 | 0.2653 |
| Diethyl sulfide | 557.04 | 0.2938 | 0.3137 | 0.2671 |
| Methyl isopropyl sulfide | 553.71 | 0.2494 | 0.3133 | 0.2728 |
| Methyl *n*-butyl sulfide | 593.15 | 0.3220 | 0.3716 | 0.2620 |
| Ethyl *n*-propyl sulfide | 585.37 | 0.3250 | 0.3728 | 0.2643 |
| Methyl *sec*-butyl sulfide | 581.48 | 0.2946 | 0.3715 | 0.2688 |
| Methyl isobutyl sulfide | 582.04 | 0.2933 | 0.3705 | 0.2683 |
| Ethyl isopropyl sulfide | 574.26 | 0.2940 | 0.3730 | 0.2713 |
| Methyl *tert*-butyl sulfide | 569.82 | 0.2387 | 0.3666 | 0.2720 |
| Ethyl *n*-butyl sulfide | 610.37 | 0.3730 | 0.4355 | 0.2611 |
| Di-*n*-propyl sulfide | 608.15 | 0.3741 | 0.4332 | 0.2615 |
| *n*-Propyl isopropyl sulfide | 597.59 | 0.3428 | 0.4328 | 0.2677 |
| Ethyl *sec*-butyl sulfide | 600.37 | 0.3398 | 0.4288 | 0.2658 |
| Ethyl isobutyl sulfide | 600.93 | 0.3421 | 0.4316 | 0.2665 |
| Ethyl *tert*-butyl sulfide | 588.71 | 0.2848 | 0.4276 | 0.2704 |
| Diisopropyl sulfide | 585.37 | 0.3098 | 0.4327 | 0.2747 |

| Paraffins | $T_c$, K‡ | $\omega_{SRK}$ | $V^*$, L/mol | $Z_{RA}$ |
|---|---|---|---|---|
| | | Sulfides (*Continued*) | | |
| Di-*n*-butyl sulfide | 649.26 | 0.4824 | 0.5616 | 0.2561 |
| Diisoamyl sulfide | 664.26 | 0.6181 | 0.7013 | 0.2589 |
| Diallyl sulfide | 653.15 | 0.1031 | 0.3732 | 0.2525 |
| | | Disulfides | | |
| Dimethyl disulfide | 605.74 | 0.2610 | 0.2638 | 0.2751 |
| Diethyl disulfide | 642.04 | 0.3424 | 0.3803 | 0.2694 |
| Ethyl *n*-propyl disulfide | 657.07 | 0.3876 | 0.4403 | 0.2662 |
| Ethyl isopropyl disulfide | 650.71 | 0.3556 | 0.4392 | 0.2711 |
| Di-*n*-propyl disulfide | 675.31 | 0.4391 | 0.5041 | 0.2634 |
| Propyl isopropyl disulfide | 666.40 | 0.4059 | 0.5023 | 0.2680 |
| Ethyl *tert*-butyl disulfide | 659.92 | 0.3482 | 0.4950 | 0.2696 |
| Diisopropyl disulfide | 659.36 | 0.3734 | 0.4999 | 0.2727 |
| Di-*n*-butyl disulfide | 699.64 | 0.5507 | 0.6359 | 0.2589 |
| Di-*tert*-butyl disulfide | 691.68 | 0.2939 | 0.4933 | |
| Di-*n*-amyl disulfide | 721.73 | 0.6707 | 0.7803 | 0.2552 |
| Di-*n*-hexyl disulfide | 741.59 | 0.7928 | 0.9399 | 0.2525 |
| Di-*tert*-dodecyl disulfide | 728.14 | 1.0299 | 1.7258 | |
| | | Inorganic Gases | | |
| Hydrogen | 33.15 | − 0.2324 | 0.0642 | 0.3060 |
| Oxygen | 154.09 | 0.0298 | 0.0738 | 0.2905 |
| Nitrogen | 126.25 | 0.0358 | 0.0901 | 0.2900 |
| Air | 132.41 | − 0.0031 | 0.0875 | 0.2692 |
| Carbon monoxide | 133.15 | 0.0295 | 0.0921 | 0.2896 |
| Carbon dioxide | 304.15 | 0.2373 | 0.0938 | 0.2722 |
| Fluorine | 144.30 | 0.0493 | 0.0669 | 0.2887 |
| Chlorine | 417.11 | 0.0822 | 0.1223 | 0.2767 |
| Hydrogen fluoride | 461.00 | 0.3281 | 0.0586 | 0.1451 |
| Hydrogen chloride | 324.54 | 0.1254 | 0.0838 | 0.2653 |
| Hydrogen bromide | 363.20 | 0.0779 | 0.0992 | |
| Sulfur dioxide | 430.59 | 0.2645 | 0.1204 | 0.2661 |
| Sulfur trioxide | 491.37 | 0.5025 | 0.1222 | 0.2515 |
| Hydrogen sulfide | 373.65 | 0.0930 | 0.0994 | 0.2855 |
| Carbon disulfide | 552.04 | 0.1035 | 0.1690 | 0.2808 |
| Ammonia | 405.43 | 0.2620 | 0.0701 | 0.2465 |
| Hydrazine | 653.15 | 0.3410 | 0.0984 | 0.2640 |
| Nitrous oxide | 309.58 | 0.1691 | 0.0980 | 0.2758 |
| Nitric oxide | 180.37 | 0.5896 | 0.0665 | 0.2668 |
| Nitrogen dioxide | 431.37 | 0.8634 | 0.0912 | 0.2413 |
| Nitrogen tetroxide | 431.15 | 0.8573 | 0.1894 | 0.3665 |
| Helium | 5.37 | − 0.4766 | 0.0546 | 0.2981 |
| Neon | 44.40 | − 0.0362 | 0.0425 | 0.3085 |
| Argon | 150.80 | − 0.0092 | 0.0754 | 0.2922 |
| Krypton | 209.40 | − 0.0050 | 0.0917 | 0.2901 |
| Xenon | 289.73 | − 0.0023 | 0.1135 | 0.2829 |
| | | Oils and Petroleum Products | | |
| Jet fuel naphtha (44.4 grav.) | 622.18 | 0.3982 | 0.5083 | |
| Aromatic naphtha (34.5 grav.) | 593.46 | 0.3120 | 0.3522 | |

**TABLE 3-10 Pure Component Parameters for the Hankinson-Brobst-Thomson and the Rackett Liquid Volume Correlations† [33, 71, 102] (*Continued*)**

| Paraffins | $T_c$, K‡ | $\omega_{SRK}$ | $V^*$, L/mol | $Z_{RA}$ |
|---|---|---|---|---|
| Oils and Petroleum Products (*Continued*) | | | | |
| Low-boiling naphtha (59.9 grav.) | 571.39 | 0.3708 | 0.4686 | |
| High-boiling naphtha (54.2 grav.) | 608.31 | 0.4212 | 0.5562 | |
| Kerosene (43.5 grav.) | 664.30 | 0.4783 | 0.6674 | |
| Fuel oil (33.0 grav.) | 745.21 | 0.6009 | 0.9150 | |
| Gas oil (35.3 grav.) | 738.70 | 0.6007 | 0.9256 | |
| Miscellaneous | | | | |
| Water | 647.37 | 0.3852 | 0.0436 | 0.2338 |
| Sulfuric acid | 899.79 | 0.9302 | 0.1230 | |
| Dimethyl sulfoxide | 727.56 | 0.2985 | 0.2288 | |
| Di-*n*-propyl sulfone | 763.52 | 0.5771 | 0.4937 | |
| Di-*n*-butyl sulfone | 766.74 | 0.6928 | 0.6302 | |
| Dowtherm A | 770.37 | 0.4084 | 0.5185 | 0.2542 |

†The Phillips Petroleum Company is gratefully acknowledged for providing many of the values in this table.

‡$T_c$ values in many cases are not experimental values and should be used only in Eqs. (3-11.2) and (3-11.3). More reliable values of the true $T_c$ are given in Appendix A.

More recently, Thomson et al. [119] have extended the HBT method to allow prediction of compressed liquid volumes by generalizing the constants in the Tait equation. Thus

$$V = V_s\left(1 - c \ln \frac{\beta + P}{\beta + P_{vp}}\right) \tag{3-11.5}$$

$V_s$, the saturated liquid volume at the vapor pressure $P_{vp}$, should be obtained from Eq. (3-11.1). $\beta$ and $c$ are obtained from

$$\beta/P_c = -1 + a(1 - T_r)^{1/3} + b(1 - T_r)^{2/3} + d(1 - T_r) + e(1 - T_r)^{4/3} \tag{3-11.6}$$

where $$e = \exp(f + g\omega_{SRK} + h\omega_{SRK}^2) \tag{3-11.7}$$

and $$c = j + k\omega_{SRK} \tag{3-11.8}$$

Values of $P_c$ should be obtained from Appendix A. The constants $a$ through $k$ for Eqs. (3-11.6) to (3-11.8) are:

| | | | |
|---|---|---|---|
| $a$ | −9.070217 | $b$ | 62.45326 |
| $d$ | −135.1102 | $f$ | 4.79594 |
| $g$ | 0.250047 | $h$ | 1.14188 |
| $j$ | 0.0861488 | $k$ | 0.0344483 |

### Modified Rackett technique

An equation to estimate saturated volumes which was developed by Rackett [95] and later modified by Spencer and Danner [102] is

$$V_s = \frac{RT_c}{P_c} Z_{\mathrm{RA}}^{[1 + (1-T_r)^{2/7}]} \tag{3-11.9}$$

$Z_{\mathrm{RA}}$ is a unique constant for each compound, and sample values are listed in Table 3-10. If a value of $Z_{\mathrm{RA}}$ is not available, it may be estimated by [133]:

$$Z_{\mathrm{RA}} = 0.29056 - 0.08775\omega \tag{3-11.10}$$

If one experimental density is available at a reference temperature $T^R$, the recommended form of the Rackett equation is

$$V_s = V_s^R Z_{\mathrm{RA}}^{\phi} \tag{3-11.11}$$

where $$\phi = (1-T_r)^{2/7} - (1-T_r^R)^{2/7} \tag{3-11.12}$$

Note that Eq. (3-11.9) does not predict the correct volume at the critical point unless $Z_{\mathrm{RA}} = Z_c$.

### Bhirud's method

Bhirud [5] has presented the following corresponding states equation for the saturated liquid volume of normal (nonpolar) fluids

$$\ln \frac{P_c V_s}{RT} = \ln V^{(0)} + \omega \ln V^{(1)} \tag{3-11.13}$$

$$\ln V^{(0)} = 1.39644 - 24.076T_r + 102.615T_r^2 - 255.719T_r^3 + 355.805T_r^4 - 256.671T_r^5 + 75.1088T_r^6 \tag{3-11.14}$$

$$\ln V^{(1)} = 13.4412 - 135.7437T_r + 533.380T_r^2 - 1091.453T_r^3 + 1231.43T_r^4 - 728.227T_r^5 + 176.737T_r^6 \tag{3-11.15}$$

Above $T_r = 0.98$, values of $\ln V^{(0)}$ and $\ln V^{(1)}$ from tables in [5] should be used. Equation (3-11.13) gave an average percent deviation of 0.76 percent for 752 data points for hydrocarbons for reduced temperatures between 0.3 and 1.0 [5]. Bhirud [6] has extended his method to polar compounds, but the extension requires an experimental density.

Other methods that can be used to estimate saturated liquid densities include those of Gunn and Yamada [29] and Yen and Woods [135]. They were reviewed in the third edition of this book. Hankinson and Thomson [33] have reviewed shortcomings of the two methods. The methods of Yen and Woods [135] as well as those of Chueh and Prausnitz [11] and Lyckman, Eckert, and Prausnitz [52] can be used for compressed liquids. These methods are, however, more complicated than the HBT method [Eqs. (3-11.5) to (3-11.8)] and appear to be less accurate [119]. Campbell and Thodos [10] have presented a method for predicting saturated liquid densities for both polar and nonpolar substances. The method requires, as input parameters, critical properties, the normal boiling point, and, for polar compounds, the dipole moment. The method has not been tested for mixtures, but it appears to give pure component results with an accuracy equivalent to the Hankinson and modified Rackett methods. Ouyang [79] has presented a group contribution type method based on "contact volume." The method requires the melting temperature, the critical temperature, and one experimental density.

## Recommendation

To estimate the molar volume of saturated liquids, use either the Hankinson or Rackett equation with parameters from Table 3-10. If parameters are not listed in Table 3-10, $Z_{RA}$ may be estimated with Eq. (3-11.10). For the Hankinson parameters, $\omega_{SRK}$ may be replaced with $\omega$ and $V^*$ may be estimated with Eq. (3-11.4) or replaced with $V_c$. If a single experimental density is available, $V^*$ may be calculated directly or for the Rackett equation, Eq. (3-11.11) may be used. For compressed liquids, use the HBT method, Eq. (3-11.5).

**Example 3-5** For isobutane at 310.93 K, calculate the volume of the saturated liquid and compressed liquid at 137.9 bar. Experimental values [98] are 108.2 and 102.7 cm$^3$/mol, respectively.

**solution** From Table 3-10, $\omega_{SRK} = 0.1825$, $V^* = 256.8$, and $Z_{RA} = 0.2754$. $T_r = 0.762$. Using the Hankinson method, Eqs. (3-11.1) to (3-11.3) give $V_R^{(0)} = 0.4399$, $V_R^{(\delta)} = 0.1990$, $V_s = (0.4399)(256.8)[(1-(0.1825)(0.1990)] = 108.9$ cm$^3$/mol (error = 0.6%).

For the compressed liquid case, Eqs. (3-11.6) to (3-11.8) give

$$c = 0.09244$$

$$e = 131.58$$

$$\beta/P_c = 4.615$$

$$\beta = (4.615)(36.5) = 168.3 \text{ bar}$$

From Ref. 107, $P_{vp} = 4.958$ bar; Eq. (3-11.5) gives

$$V = (108.9)(1 - 0.09244\left(\ln \frac{168.3+137.9}{168.3+4.958}\right) = 103.2 \text{ cm}^3/\text{mol} \qquad (\text{error} = 0.5\%)$$

For saturated liquid, the Rackett equation (3-11.9) gives

$$V_s = \frac{(83.14)(408.2)}{36.5}(0.2754)^{[1+(1-0.762)^{2/7}]} = 108.9 \text{ cm}^3/\text{mol} \qquad (\text{error} = 0.6\%)$$

## Notation

In many equations in this chapter, special constants are defined and usually denoted $a, b, \ldots, A, B, \ldots$ They are not defined in this notation; for they apply to the specific equation and do not occur elsewhere in the chapter.

| | |
|---|---|
| $A^*, B^*$ | dimensionless parameters for cubic equations of state, see Eqs. (3-6.3), (3-6.4) |
| $B$ | second virial coefficient, $cm^3/mol$ |
| $C$ | third virial coefficient, Eq. (3-5.1), $(cm^3/mol)^2$ |
| $M$ | molecular weight |
| $P$ | pressure; $P_{vp}$, vapor pressure; $P_c$, critical pressure; $P_r$, reduced pressure, $P/P_c$, bar |
| $R$ | gas constant, see Table 3-1 |
| $T$ | absolute temperature; $T_c$, critical temperature; $T_r$, reduced temperature, $T/T_c$; $T_b$, normal boiling point at 1 atm, kelvins |
| $V$ | molar volume; $V_c$, critical volume; $V_r$, reduced volume, $V/V_c$; $V_b$, at normal boiling point; $V_{ri}$, ideal reduced volume, $V/(RT_c/P_c)$; $V_s$, saturated liquid volume, $cm^3/mol$ |
| $V_R^{(0)}, V_R^{(\delta)}$ | functions of $T_r$ for HBT correlations, Eqs. (3-11.2) and (3-11.3) |
| $V^*, \omega_{SRK}$ | parameters for HBT correlations, Eq. (3-11.1) and Table 3-10 |
| $Z$ | compressibility factor, $PV/RT$; $Z_c$, at the critical point; $Z^{(0)}$, simple fluid compressibility factor, Eq. (3-3.1) and Table 3-2; $Z^{(1)}$, deviation compressibility factor, Eq. (3-3.1) and Table 3-3; $Z_{RA}$, Rackett compressibility factor, Eq. (3-11.9) |

GREEK

| | |
|---|---|
| $\mu$ | dipole moment, debyes; $\mu_r$, reduced dipole moment, see Eq. (3-5.8) |
| $\rho$ | molar density; $\rho_c$, critical density; $\rho_r$, reduced density, $\rho/\rho_c$; $\rho_b$, at normal boiling point; $\rho_s$, saturated liquid density |
| $\phi$ | fugacity coefficient |
| $\omega$ | Pitzer acentric factor, see Sec. 2-3 |

SUPERSCRIPTS

| | |
|---|---|
| $R$ | reference fluid or reference state |
| $L$ | liquid phase property |
| $V$ | vapor phase property |

## REFERENCES

1. Abbott, M. M.: *AIChE J.*, **19:** 596 (1973).
2. Abbott, M. M.: "Equations of State in Engineering and Research," *Advan. Chem. Ser.*, **182:** (1979).

3. American Chemical Society: "Physical Properties of Chemical Compounds," *Advan. Chem. Ser.*, vols. 15, 22, and 29, R. R. Dreisbach.
4. *ASHRAE Thermodynamic Properties of Refrigerants,* 1969, p. 45.
5. Bhirud, V. L.: *AIChE J.,* **24:** 1127 (1978).
6. Bhirud, V. L.: *AIChE J.,* **24:** 880 (1978).
7. Brulé, M. R., C. T. Lin, L. L. Lee, and K. E. Starling: *AIChE J.,* **28:** 616 (1982).
8. Benson, S. W.: *J. Phys. Colloid Chem.,* **52:** 1060 (1948).
9. Breedveld, G. J. F., and J. M. Prausnitz: *AIChE J.,* **19:** 783 (1973).
10. Campbell, S. W., and G. Thodos: *J. Chem. Eng. Data,* **30:** 102 (1985), *Ind. Eng. Chem. Fundam.,* **23:** 500 (1984).
11. Chueh, P. L., and J. M. Prausnitz: *AIChE J.,* **13:** 1099 (1967); **115:** 471 (1969).
12. Chung, W. K., S. E. M. Hamam, and B. C.-Y. Lu: *Ind. Eng. Chem. Fund.,* **16:** 494 (1977).
13. Cooper, H. W., and J. C. Goldfrank: *Hydrocarbon Process. Petrol. Refiner,* **46**(12): 141 (1967).
14. Das, T. R., C. O. Reed, Jr., and P. T. Eubank: *J. Chem. Eng. Data,* **18:** 253 (1973).
15. DeSantis, R., and B. Grande: *AIChE J.,* **25:** 937 (1979).
16. Doolittle, A. K.: *AIChE J.,* **6:** 150, 153, 157 (1960).
17. Dymond, J. H., and E. B. Smith: *The Second Virial Coefficient of Pure Gases and Mixtures: A Critical Compilation,* Clarendon Press, Oxford, 1969.
18. Edmister, W. C.: *Petrol. Refiner,* **37**(4): 173 (1958).
19. Edmister, W. C., J. Vairogs, and A. J. Klekers: *AIChE J.,* **14:** 479 (1968).
20. Edwards, M. N. B., and G. Thodos: *J. Chem. Eng. Data,* **19:** 14 (1974).
21. Eubank, P. T., and J. M. Smith: *AIChE J.,* **8:** 117 (1962).
22. Fedors, R. F.: *Polymer Eng. Sci.,* **14:** 147, 153 (1974).
23. Francis, A. W.: *Chem. Eng. Sci.,* **10:** 37 (1959): *Ind. Eng. Chem.,* **49:** 1779 (1957).
24. Fuller, G. G.: *Ind. Eng. Chem. Fundam.,* **15:** 254 (1976).
25. Gmehling, J., D. D. Liu, and J. M. Prausnitz: *Chem. Eng. Sci.* **34:** 951 (1979).
26. Graboski, M. S., and T. E. Daubert: *Ind. Eng. Chem. Process Design Develop.,* **17:** 443 (1978).

26*a*. Grenzheuser, P., and J. Gmehling: *Fluid Phase Equil.,* **25:** 1 (1986).

27. Gunderson, T.: *Computers Chem. Eng.,* 6: 245 (1982).
28. Gunn, R. D., P. L. Chueh, and J. M. Prausnitz: *AIChE J.,* **12:** 937 (1966).
29. Gunn, R. D., and T. Yamada: *AIChE J.,* **17:** 1341 (1971).
30. Hales, J. L., "Bibliography of Fluid Density," *Natl. Phys. Lab. Gt. Brit. Rept. Chem.,* **106:** February 1980.
31. Halm, R. L., and L. I. Stiel: *AIChE J.,* **13:** 351 (1967).
32. Hamam, S. E. M., W. K. Chung, I. M. Elshayal, and B. C.-Y Lu: *Ind. Eng. Chem. Process Design Develop.,* **16:** 50 (1977).
33. Hankinson, R. W., and G. H. Thomson: *AIChE J.,* **25:** 653 (1979).
34. Harmens, A., and H. Knapp: *Ind. Eng. Chem. Fundam.,* **19:** 291 (1980).
35. Hauthal, W. H., and H. Sackman: *Proc. 1st Intern. Conf. Calorimetry Thermodyn., Warsaw, 1969,* p. 625.
36. Hayden, J. G., and J. P. O'Connell: *Ind. Eng. Chem. Fundam.,* **14:** 209 (1975).
37. Hougen, O., K. M. Watson, and R. A. Ragatz: *Chemical Process Principles,* pt. II, Wiley, New York, 1959.
38. Hsiao, Y-J., and B. C.-Y. Lu: *Can. J. Chem. Eng.,* **57:** 102 (1979).
39. Joffe, J., G. M. Schroeder, and D. Zudkevitch: *AIChE J.,* **16:** 496 (1970).
40. Johnson, D. W., and C. P. Colver: *Hydrocarbon Process. Petrol. Refiner,* **47:** 79 (1968).
41. Kim, H., T. M. Guo, and K. C. Chao: paper presented at *3rd Intern. Conf. on Fluid Properties and Phase Equil. for Chem. Process. Design, Calloway Gardens, Ga., April 10–15, 1983.*
42. Knoebel, D. H., and W. C. Edmister: *J. Chem. Eng. Data,* **13:** 312 (1968).
43. Kumar, K. H., and K. E. Starling: *Ind. Eng. Chem. Fundam.,* **21:** 255 (1982).
44. Le Bas, G.: *The Molecular Volumes of Liquid Chemical Compounds,* Longmans, Green, New York, 1915.

45. Lee, B. I., and W. C. Edmister: *AIChE J.*, **17:** 1412 (1971).
46. Lee, B. I., J. H. Erbar, and W. C. Edmister: *AIChE J.*, **19:** 349 (1973).
47. Lee, B. I., and M. G. Kesler: *AIChE J.*, **21:** 510 (1975).
48. Leland, T. W., Jr., and P. S. Chappelear: *Ind. Eng. Chem.*, **60**(7): 15 (1968). See also G. D. Fisher and T. W. Leland, Jr., *Ind. Eng. Chem. Fundam.*, **9:** 537 (1970).
49. Li, K., R. L. Arnett, M. B. Epstein, R. B. Ries, L. P. Butler, J. M. Lynch, and F. D. Rossini: *J. Phys. Chem.*, **60:** 1400 (1956).
50. Lo, H. Y., and L. I. Stiel: *Ind. Eng. Chem. Fundam.*, **8:** 713 (1969).
51. Lu, B.C.-Y., J. A. Ruether, C. Hsi, and C.-H. Chiu: *J. Chem. Eng. Data*, **18:** 241 (1973).
52. Lyckman, E. W., C. A. Eckert, and J. M. Prausnitz: *Chem. Eng. Sci.*, **20:** 703 (1965).
53. Lydersen, A. L., R. A. Greenkorn, and O. A. Hougen: "Generalized Thermodynamic Properties of Pure Fluids," *Univ. Wisconsin Coll. Eng., Eng. Exp. Stn. Rept. 4*, Madison, Wis., October 1955.
54. Manley, D. B.: personal communication (1984).
55. Martin, J. J.: *Ind. Eng. Chem. Fundam.*, **18:** 81 (1979).
56. Martin, J. J.: *Ind. Eng. Chem. Fundam.*, **23:** 454 (1984).
57. Martin, J. J.: *Chem. Eng. Progr. Symp. Ser.*, **59:**(44): 120 (1963).
58. Martin, J. J.: *Ind. Eng. Chem.*, **59**(12): 34 (1967).
59. Martin, J. J., and Y.-C. Hou: *AIChE J.*, **1:** 142 (1955).
60. Martin, J. J., R. M. Kapoor, and N. DeNevers: *AIChE J.*, **5:** 159 (1959).
61. Martin, J. J., and T. G. Stanford: *Chem. Eng., Progr. Symp. Ser.*, **70**(140): 1 (1974).
62. Maslan, F. D., and T. M. Littman: *Ind. Eng. Chem.*, **45:** 1566 (1953).
63. Mason, E. A., and T. H. Spurling: *The Virial Equation of State*, Pergamon, New York, 1968.
64. *Matheson Gas Data Book*, 5th ed., 1971.
65. McCarty, R. D.: "Four Mathematical Models for the Prediction of LNG Densities," *NBS Tech. Note 1030*, Washington, D.C., 1980.
66. McCann, D. W., and R. P. Danner: *Ind. Eng. Chem. Process Design Develop.*, **23:** 529 (1984).
67. McFee, D. G., K. H. Mueller, and J. Lielmezs: *Thermochimica Acta*, **54:** 9 (1982).
68. Morecroft, D. W.: *J. Inst. Petrol.*, **44:** 433 (1958).
69. Morgan, R. A., and J. A. Childs: *Ind. Eng. Chem.*, **37:** 667 (1945).
70. Morris, R. W., and E. A. Turek: paper presented at *189th ACS Mg., Miami, Fla., April 29–May 2, 1985.* See also R. W. Morris and E. A. Turek in K. C. Chao and R. L. Robinson (eds.), *Equations of State*, ACS Symp. Ser. 300, ACS, 1986.
71. Murrieta-Guevara, F., and A. T. Rodriguez: *J. Chem. Eng. Data*, **29:** 204 (1984).
72. Nakanishi, K., M. Kurata, and M. Tamura: *J. Chem. Eng. Data*, **5:** 210 (1960).
73. Nelson, L. C., and E. F. Obert: *Trans. ASME*, **76:** 1057 (1954).
74. Newton, R. H.: *Ind. Eng. Chem.*, **27:** 302 (1935).
75. O'Connell, J. P., and J. M. Prausnitz: *Ind. Chem. Process Design Develop.*, **6:** 245 (1967).
76. Opfell, J. B., B. H. Sage, and K. S. Pitzer: *Ind. Eng. Chem.*, **48:** 2069 (1956).
77. Orbey, H., and J. H. Vera: *AIChE J.*, **29:** 107 (1983).
78. Orye, R. V.: *Ind. Eng. Chem. Process Design Develop.*, **8:** 579 (1969).
79. Ouyang, R: *J. Chem. Eng. Japan*, **9:** 171 (1976).
80. Panagiotopoulos, A. Z., and S. K. Kumar: *Fluid Phase Equil.*, **22:** 77 (1985).
81. Parrish, W. R.: "Liquid Densities of Ethane-Propane Mixtures," *Tech. Pub. TP-12, GPA*, Tulsa, Okla., 1985.
82. Partington, J.: *An Advanced Treatise on Physical Chemistry*, Vol. I, *Fundamental Principles: The Properties of Gases*, Longmans, Green, New York, 1949.
83. Peneloux, A., and E. Rauzy: *Fluid Phase Equil.*, **8:** 7 (1982).
84. Peng, D. Y., and D. B. Robinson: *Ind. Eng. Chem. Fundam.*, **15:** 59 (1976).
85. Perry, R. H., D. W. Green, and J. O. Maloney, *Perry's Chemical Engineers' Handbook*, 6th ed., McGraw-Hill, New York, 1984, p. 3-271.
86. Pitzer, K. S., and R. F. Curl: *J. Am. Chem. Soc.*, **77:** 3427 (1955).

87. Pitzer, K. S., and R. F. Curl: *J. Am. Chem. Soc.*, **79:** 2369 (1957).
88. Pitzer, K. S., and R. F. Curl: *The Thermodynamic Properties of Fluids,* Inst. Mech. Eng., London, 1957.
89. Pitzer, K. S., D. Z. Lippmann, R. F. Curl, C. M. Huggins, and D. E. Petersen: *J. Am. Chem. Soc.*, **77:** 3433 (1955).
90. Plocker, U. J., and H. Knapp: *Hydrocarbon Process. Petrol. Refiner.*, **55:** 199 (1976).
91. Polak, J., and B. C.-Y. Lu: *Can. J. Chem. Eng.*, **50:** 553 (1972).
92. Poling, B. E., E. A. Grens II, and J. M. Prausnitz: *Ind. Eng. Chem. Process Design Develop.*, **20:** 127 (1981).
93. Prasad, D. H. L., and D. S. Viswanath, Dept. of Chemical Engineering, Indian Institute of Science, Bangalore, India: personal communication, 1974.
94. Prausnitz, J. M., T. F. Anderson, E. A. Grens II, C. A. Eckert, R. Hsieh, and J. P. O'Connell: *Computer Calculations for Multicomponent Vapor-Liquid and Liquid-Liquid Equilibria,* Prentice-Hall, Englewood Cliffs, N.J., 1980, pp. 130–137, 160–178.
95. Rackett, H. G: *J. Chem. Eng. Data,* **15:** 514 (1970).
96. Redlich, O., and J. N. S. Kwong: *Chem. Rev.*, **44:** 233 (1949).
97. Ritter, R. B., J. M. Lenoir, and J. L. Schweppe: *Petrol. Refiner,* **37**(11): 225 (1958).
98. Sage, B. H., and W. N. Lacey: *Ind. Eng. Chem.*, **30:** 673 (1938).
99. Satter, A., and J. M. Campbell: *Soc. Petrol. Engrs. J.*, December 1963: 333.
100. Schmidt, G., and H. Wenzel: *Chem. Eng. Sci.*, **35:** 1503 (1980).
101. Soave, G.: *Chem. Eng. Sci.*, **27:** 1197 (1972).
102. Spencer, C. F., and R. P. Danner: *J. Chem. Eng. Data,* **17:** 236 (1972).
103. Starling, K. E.: paper presented at *NGPA Annual Conv., Denver, Colo., March 17–19, 1970.*
104. Starling, K. E., and M. S. Han: *Hydrocarbon Process. Petrol. Refiner,* **51**(5): 129, **51**(6): 107 (1972).
105. Spencer, C. F., and S. B. Adler: *J. Chem. Eng. Data,* **23:** 82 (1978).
106. Spencer, C. F., and R. P. Danner: *J. Chem. Eng. Data,* **18:** 230 (1973).
107. Steele, K., B. E. Poling, and D. B. Manley: *J. Chem. Eng. Data,* **21:** 399 (1976).
108. Stein, F. P., and E. J. Miller: *Ind. Eng. Chem. Process Design Develop.*, **19:** 123 (1980).
109. Stiel, L. I.: *Chem. Eng. Sci.*, **27:** 2109 (1972).
110. Stiel, L. I.: *Ind. Eng. Chem.*, **60**(5): 50 (1968).
111. Storvick, T. S., and T. H. Spurling: *J. Phys. Chem.*, **72:** 1821; Suh, K. W., and T. S. Storvick, *J. Phys. Chem.*, **71:** 1450 (1967).
112. Su, G. S., and D. S. Viswanath: *AIChE J.*, **11:** 205 (1965).
113. Tarakad, R. R., and R. P. Danner: *AIChE J.*, **23:** 685 (1977).
114. Tarakad, R. R., and T. E. Daubert, Pennsylvania State University, University Park: private communication, Sept. 23, 1974.
115. *Technical Data Book, Petroleum Refining,* 2d ed., American Petroleum Institute, Washington, D.C., 1971.
116. Tekac, V., I. Cibulka, and R. Holub: *Fluid Phase Equil.*, **19:** 33 (1985).
117. Teja, A. S.: *AIChE J.*, **26:** 337 (1980).
118. Teja, A. S., S. I. Sander, and N. C. Patel: *Chem. Eng. J.*, **21:** 21 (1981).
119. Thomson, G. H., K. R. Brobst, and R. W. Hankinson: *AIChE J.*, **28:** 671 (1982).
120. Tseng, J. K., and L. I. Stiel: *AIChE J.*, **17:** 1283 (1971).
121. Tsonopoulos, C.: *AIChE J.*, **20:** 263 (1974).
122. Tsonopoulos, C.: *AIChE J.*, **21:** 827 (1975).
123. Tsonopoulos, C., and J. M. Prausnitz: *Cryogenics,* October 1969: 315.
124. Tsonopoulos, C.: *AIChE J.*, **24:** 1112 (1978).
125. Twu, C. H.: *Fluid Phase Equil.*, **11:** 65 (1983).
126. Tyn, M. T., and W. F. Calus: *Processing,* **21**(4): 16 (1975).
127. Usdin E., and J. C. McAuliffe: *Chem. Eng. Sci.*, **31:** 1077 (1976).
128. Van Ness, H. C., and M. M. Abbott: *Classical Thermodynamics of Nonelectrolyte Solutions,* McGraw-Hill, New York, 1982 pp. 126–33.
129. Watanasiri, S., M. R. Brulé, and K. E. Starling: *AIChE J.*, **28:** 626 (1982).

130. Watanasiri, S., H. Kumar, and K. E. Starling: paper presented at *AIChE Annual Mtg., Los Angeles, Calif., Nov. 14–18, 1982.*
131. Weast, R. C.: *Handbook of Chemistry and Physics,* 49th ed., Chemical Rubber Co., Cleveland, 1968.
131*a*. Wu, G. Z. A., L. I. Stiel: *AIChE J.,* **31:** 1632 (1985).
132. Yamada, T.: AIChE J., **19:** 286 (1973).
133. Yamada, T., and R. D. Gunn: *J. Chem. Eng. Data,* **18:** 234 (1973).
134. Yarborough, L.: "Equations of State in Engineering Research," *Advan. Chem. Ser.,* **182:** 385 (1979).
135. Yen, L. C., and S. S. Woods: *AIChE J.,* **12:** 95 (1966).

Chapter

# 4

# Volumetric Properties of Mixtures

## 4-1 Scope

In Chap. 3 we reviewed methods for calculating the *PVT* properties of gases and liquids. To extend the methods to mixtures, they must be modified to include the additional variable of composition. In essentially all cases, the inclusion is accomplished by averaging pure component constants to obtain constants which hopefully characterize the mixtures. Equations which do this are called mixing rules. Many algebraic relations have been suggested, although it can be shown (Sec. 4-2) that essentially all can be derived from a single general expression. One must keep in mind that the rules covered in this chapter are, with a single exception [Eq. (4-4.1)], essentially empirical and have resulted from many trials and comparisons of calculated mixture properties with experimental data. Our treatment is to present first a general discussion and then follow with the recommended mixing rules for all gas PVT estimation methods covered in Chap. 3. Rules applicable to liquid mixtures are covered in Sec. 4-9.

This chapter is concerned primarily with the volumetric properties of mixtures. But mixing rules are used in the estimation of many other properties as well. Thus mixing rules appear in Chaps. 5, 8, 9, 10, 11, and 12. In Chap. 5 we examine the thermodynamic properties of both pure components and mixtures; for the former, one may simply use the constants given with each method outlined in Chap. 3. For the latter, the equation-of-state parameters must be determined separately for each mixture with a different composition. In Chap. 8, and in particular, in Sec. 8-12, equations of state are used to calculate vapor-liquid equilibria. In this application derivatives of the mixing rules with respect to composition must accurately represent the true behavior. In Chaps. 9 to 12, mixing rules are used in the estimation of transport properties and the surface tension of mixtures.

## 4-2 Mixing Rules—General Discussion

Typically a mixing rule expresses a mixture parameter $Q_m$ in terms of composition and pure component parameters according to

$$Q_m = \sum_i \sum_j y_i y_j Q_{ij} \tag{4-2.1}$$

In Eq. (4-2.1), the mole fractions $y_i$ and $y_j$ may apply to a liquid or vapor phase. $Q_{ii}$ and $Q_{jj}$ involve properties of the pure compounds $i$ and $j$. For example, $Q_{ii}$ might be the critical temperature of pure $i$. The difficulty arises when interaction terms $Q_{ij}$ are considered. Rules used to find $Q_{ij}$ are called combining rules. If $Q_{ij}$ is set equal to either the arithmetic average or geometric mean, the double summation in Eq. (4-2.1) reduces to a single summation.

$$\text{For} \quad Q_{ij} = \frac{Q_{ii} + Q_{jj}}{2} \qquad Q_m = \sum_i y_i Q_i \tag{4-2.2}$$

$$\text{For} \quad Q_{ij} = (Q_{ii}Q_{jj})^{1/2} \qquad Q_m = \left(\sum_i y_i Q_i^{1/2}\right)^2 \tag{4-2.3}$$

where $Q_{ii}$ and $Q_i$ are the same quantity. Equations (4-2.2) and (4-2.3) correspond to the two combining and mixing rules van der Waals originally proposed for his equation. To improve performance, it is customary to add binary interaction parameters $k_{ij}$ to the expressions for $Q_{ij}$. One possibility would be

$$Q_{ij} = \frac{k_{ij}(Q_{ii} + Q_{jj})}{2} \tag{4-2.4}$$

Values of $k_{ii}$ would be taken as unity and values of $k_{ij}$ for all possible binary pairs would be determined by regressing experimental mixture data. Values of $k_{ij}$ are more sensitive to derivative, or partial properties (such as fugacity coefficients), than to total properties (such as mixture molar volumes). For that reason, values of $k_{ij}$ have most often been determined from VLE data, although Brulé and Starling [4] suggest multiproperty regression. $k_{ij}$ is usually assumed to be independent of temperature, pressure, and composition. As defined in Eq. (4-2.4), one would hope for values of $k_{ij}$ close to unity, which would suggest the original model was reasonable and only a small correction was necessary to give the best fit to data. Other ways in which $k_{ij}$ and $Q_{ij}$ could be defined include

$$Q_{ij} = k_{ij}(Q_{ii}Q_{jj})^{1/2} \qquad (k_{ii} = 1) \tag{4-2.5}$$

$$Q_{ij} = \frac{(1 - k_{ij})(Q_{ii} + Q_{jj})}{2} \qquad (k_{ii} = 0) \tag{4-2.6}$$

$$Q_{ij} = (1 - k_{ij})(Q_{ii}Q_{jj})^{1/2} \qquad (k_{ii} = 0) \tag{4-2.7}$$

In Secs. 4-3 to 4-6, four different binary interaction parameters, $k_{ij}$, $k^*_{ij}$, $\bar{k}_{ij}$, and $k'_{ij}$ are introduced. Four symbols are used to emphasize that these four quantities are not numerically the same. They are, however, related to each other so that it would, in theory, be possible to calculate any three of the interaction parameters from the fourth. In practice this is usually not done.

The theory of mixing rules is discussed elsewhere [21, 31, 48] and will not be dealt with to any extent here. However, it is worth noting that, more often than not, an arithmetic average is used for size parameters and a geometric mean is used for energy parameters.

## 4-3 Corresponding States: The Pseudocritical Method

To apply the corresponding states correlations in Secs. 3-2 and 3-3 to mixtures, one must determine appropriate scaling factors. For mixtures, these scaling factors are called *pseudocritical properties*. Values of pseudocritical properties are not the same as those of true critical properties, but Eqs. (3-6.5) and (3-6.6) are satisfied at the pseudocritical point. The assumption in applying corresponding states is that the *PVT* behavior of the *mixture* will be the same as that of a *pure component* whose critical temperature and pressure are equal to the pseudocritical temperature and pressure of the mixture.

For the pseudocritical temperature $T_{c_m}$ a simple mole fraction average method is usually satisfactory. This rule, often called Kay's rule [27], is

$$T_{c_m} = \sum_j y_j T_{c_j} \tag{4-3.1}$$

Comparison of $T_{c_m}$ from Eq. (4-3.1) with values determined from other, more complicated rules shows that the differences are usually less than 2 percent if, for all components [48],

$$0.5 < \frac{T_{c_i}}{T_{c_j}} < 2 \qquad \text{and} \qquad 0.5 < \frac{P_{c_i}}{P_{c_j}} < 2$$

For the pseudocritical pressure, a simple mole fraction average of the pure component critical pressures is normally not satisfactory unless all components have similar critical pressures or critical volumes. The simplest rule which gives acceptable results is the modified Prausnitz and Gunn combination [44]

$$P_{c_m} = \frac{R\left(\sum_j y_j Z_{c_j}\right) T_{c_m}}{\sum_j y_j V_{c_j}} \tag{4-3.2}$$

The mixture acentric factor is usually given by [24]:

$$\omega_m = \sum_j y_j \omega_j \tag{4-3.3}$$

although Brulé et al. [5] use a different rule.

No binary (or higher) interaction parameters are included in Eqs. (4-3.1) to (4-3.3); thus these mixing rules cannot truly reflect mixture properties. Yet surprisingly good results are often obtained when these simple pseudomixture parameters are used in corresponding states calculations to determine mixture properties.

Less satisfactory results are found for mixtures of dissimilar components, especially if one or more of the components is polar or shows any tendency to associate into dimers, etc.

However, if one has available some experimental data for any of the possible binaries in the mixture, it is frequently worthwhile to use those data to modify the pseudocritical rules. Though many options are open, one which has often proved successful is to change Eq. (4-3.1) from a linear to a quadratic form:

$$T_{c_m} = \sum_i \sum_j y_i y_j T_{c_{ij}} \tag{4-3.4}$$

with

$$T_{c_{ii}} = T_{c_i} \qquad \text{and} \qquad T_{c_{ij}} = k_{ij}^* \frac{T_{c_i} + T_{c_j}}{2} \tag{4-3.5}$$

In some instances, it is assumed that $T_{c_{ij}} = (T_{c_i} T_{c_j})^{1/2}(1 - k_{ij})$. It is clear that $k_{ij}^*$ and $k_{ij}$ are not the same, although they are easily related to one another. Also, the arithmetic and geometric means are essentially the same unless critical temperatures $i$ and $j$ are greatly different.

From the available data, the best values of the binary constants $k_{ij}^*$ are back-calculated by trial and error. If $k_{ij}^*$ is assumed equal to unity, Eq. (4-3.4) reduces to Eq. (4-3.1). With $k_{ij}^*$ values for all possible binary sets,

**TABLE 4-1 Barner and Quinlan $k_{ij}^*$ Values [2]**

| Component *i* | Component *j* | $k_{ij}^*$ | Component *i* | Component *j* | $k_{ij}^*$ |
|---|---|---|---|---|---|
| Methane | Ethylene | 1.01 | *n*- or Isobutane | Isobutane | 1.00 |
| | Ethane | 1.03 | | *n*-Pentane | 1.00 |
| | Propylene | 1.06 | | Isopentane | 1.00 |
| | Propane | 1.07 | | *n*-Hexane | 1.02 |
| | *n*-Butane | 1.11 | | *n*-Heptane | 1.03 |
| | Isobutane | 1.11 | | Cyclohexane | 1.01 |
| | *n*-Pentane | 1.15 | *n*- or Isopentane | Isopentane | 1.00 |
| | Isopentane | 1.15 | | *n*-Hexane | 1.00 |
| | *n*-Hexane | 1.19 | | *n*-Heptane | 1.01 |
| | *n*-Heptane | 1.22 | | *n*-Octane | 1.02 |
| | Cyclohexane | 1.16 | | Cyclohexane | 1.00 |
| | Naphthalene | 1.23 | *n*-Hexane | *n*-Heptane | 1.00 |
| Ethylene | Ethane | 1.00 | | *n*-Octane | 1.01 |
| | Propylene | 1.02 | | Toluene | 0.98 |
| | Propane | 1.02 | Cyclohexane | *n*-Heptane | 1.00 |
| | *n*-Butane | 1.05 | | *n*-Octane | 1.00 |
| | Isobutane | 1.05 | | Toluene | 0.99 |
| | *n*-Pentane | 1.08 | *n*-Heptane | *n*-Octane | 1.01 |
| | Isopentane | 1.08 | Nitrogen | Methane | 0.97 |
| | *n*-Hexane | 1.11 | | Ethylene | 1.01 |
| | Cyclohexane | 1.09 | | Ethane | 1.02 |
| | *n*-Heptane | 1.13 | | *n*-Butane | 1.13 |
| | Benzene | 1.07 | | 1-Pentene | 1.13 |
| | Naphthalene | 1.15 | | 1-Hexene | 1.25 |
| Ethane | Propylene | 1.01 | | *n*-Hexane | 1.26 |
| | Propane | 1.01 | | *n*-Heptane | 1.31 |
| | *n*-Butane | 1.03 | | *n*-Octane | 1.34 |
| | Isobutane | 1.03 | Argon | Oxygen | 0.99 |
| | *n*-Pentane | 1.05 | | Nitrogen | 0.99 |
| | Isopentane | 1.05 | Carbon dioxide | Ethylene | 0.94 |
| | *n*-Hexane | 1.08 | | Ethane | 0.92 |
| | *n*-Heptane | 1.10 | | Propylene | 0.93 |
| | Cyclohexane | 1.06 | | Propane | 0.93 |
| | Benzene | 1.04 | | *n*-Butane | 0.93 |
| | Naphthalene | 1.11 | | Naphthalene | 1.07 |
| Propylene | Propane | 1.00 | Hydrogen sulfide | Methane | 0.93 |
| | *n*-Butane | 1.01 | | Ethane | 0.92 |
| | Isobutane | 1.01 | | Propane | 0.92 |
| | *n*-Pentane | 1.02 | | *n*-Pentane | 0.96 |
| | Isopentane | 1.03 | | Carbon dioxide | 0.92 |
| | Benzene | 1.03 | Acetylene | Ethylene | 0.94 |
| Propane | *n*-Butane | 1.01 | | Ethane | 0.92 |
| | Isobutane | 1.01 | | Propylene | 0.95 |
| | *n*-Pentane | 1.01 | | Propane | 0.94 |
| | Isopentane | 1.02 | Hydrogen chloride | Propane | 0.88 |
| | Benzene | 1.00 | | | |

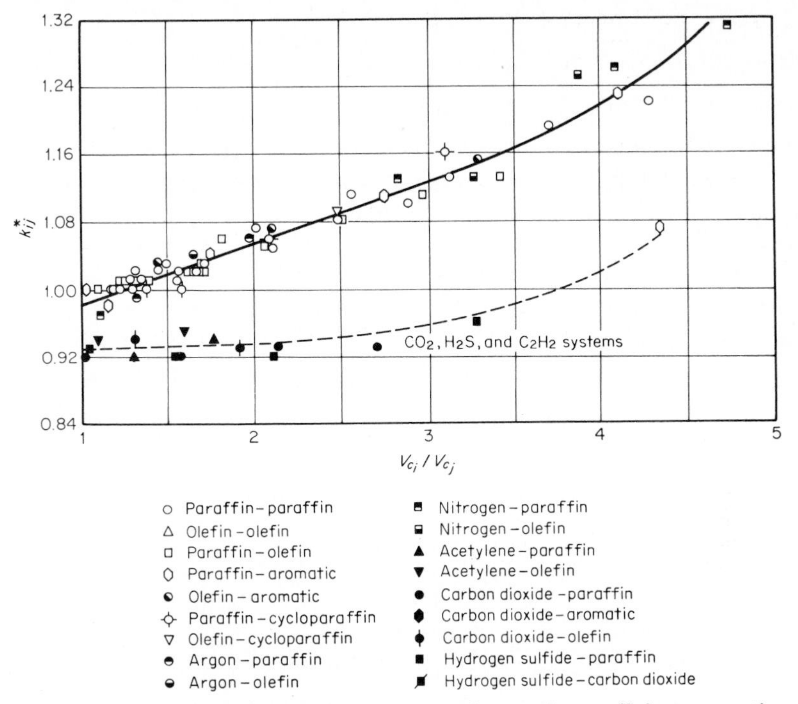


**Figure 4-1** Plot of pseudocritical temperature-interaction coefficients vs. ratios of molar critical volumes. *(From Ref. 2.)*

multicomponent mixture properties can then be estimated. In Table 4-1, $k_{ij}^*$ values are tabulated for many binary pairs [2]. For simple binaries, $k_{ij}^*$ can be approximately correlated with the ratio of the critical volumes of pure $i$ and $j$, as shown in Fig. 4-1.

There is nothing rigorous in this approach. First, Eq. (4-3.5) may not be the best form for expressing $T_{c_{ij}}$. From molecular theory, the geometric mean of $T_{c_i}$ and $T_{c_j}$ could more easily be defended. Second, no ternary or higher interaction parameters are considered. Third, it is probably too much to expect that, for a given binary, there is but a single value of $k_{ij}^*$ that is not a function of composition, temperature, or pressure. Only by experience have we obtained some confidence in approaches of this sort.

We consider next how the analytical equations of state mentioned in Chap. 3 can be modified for mixtures.

## 4-4 Second Virial Coefficients for Mixtures

The truncated virial equation shown in Sec. 3-5 is the only equation for real gases for which an exact relation is known for mixture coefficients:

$$B_m = \sum_i \sum_j y_i y_j B_{ij} \tag{4-4.1}$$

For example, for a ternary mixture of 1, 2, and 3,

$$B_m = y_1^2B_1 + y_2^2B_2 + y_3^2B_3 + 2y_1y_2B_{12} + 2y_1y_3B_{13} + 2y_2y_3B_{23} \tag{4-4.2}$$

The pure component second virial coefficients $B_1$, $B_2$, and $B_3$ can often be found as described in Sec. 3-5 [e.g., Eqs. (3-5.4) to (3-5.6)] from the critical temperatures, critical pressures, and acentric factors of the pure materials. If one wishes to employ Eq. (3-5.4) to calculate the mixture interaction virials $B_{ij}$, combination rules must be devised to obtain $T_{c_{ij}}$, $P_{c_{ij}}$, and $\omega_{ij}$. This problem has been discussed from a theoretical point of view by Leland and Chappelear [31] and Ramaiah and Stiel [45]. For typical engineering calculations involving normal fluids, the following simple rules are useful [43]:

$$T_{c_{ij}} = (T_{c_i}T_{c_j})^{1/2}(1 - k_{ij}) \tag{4-4.3}$$

$$V_{c_{ij}} = \left[\frac{V_{ci}^{1/3} + V_{cj}^{1/3}}{2}\right]^3 \tag{4-4.4}$$

$$Z_{c_{ij}} = \frac{Z_{c_i} + Z_{c_j}}{2} \tag{4-4.5}$$

$$\omega_{ij} = \frac{\omega_i + \omega_j}{2} \tag{4-4.6}$$

$$P_{c_{ij}} = \frac{Z_{c_{ij}}RT_{c_{ij}}}{V_{c_{ij}}} \tag{4-4.7}$$

For molecules which do not differ greatly in size or chemical structure the binary constant $k_{ij}$ can be set equal to zero. $k_{ij}$ values for a variety of binary systems have been published [10]. Tsonopoulos [54] has provided additional $k_{ij}$ values and has discussed their prediction and correlation. Tarakad and Danner [51] have also given guidelines for the estimation of $k_{ij}$. For binaries where both components fall into one of these categories (hydrocarbons, rare gases, permanent gases, carbon monoxide, perhalocarbons), $k_{ij}$ may be estimated by

$$k_{ij} = 1 - \frac{8(V_{c_i}V_{c_j})^{1/2}}{(V_{ci}^{1/3} + V_{cj}^{1/3})^3} \tag{4-4.8}$$

Also, if the molecule pairs contain a quantum gas ($H_2$, He, or Ne), modified critical constants are recommended [43]. Note that the $T_{c_{ij}}$ in Eq. (4-4.3) is not the same quantity as the $T_{c_{ij}}$ in Eq. (4-3.5).

A technique for estimating second virial coefficients of gas mixtures of simple fluids and heavy hydrocarbons has been given by Kaul and Prausnitz [26]. Maris and Stiel [35] have presented a method for estimating second virial coefficients of mixtures of polar compounds.

**Example 4-1** Estimate the second virial coefficient for an equimolar methane (1)–$n$-butane (2) mixture at 444.3 K. $B_1 = -8.1$ cm$^3$/mol; $B_2 = -293.4$ cm$^3$/mol. Reference 62 gives $B_{12} = -63.4$ cm$^3$/mol, so for the mixture, $B_m = -107$ cm$^3$/mol.

**solution**

| Component | $T_c$ | $V_c$ | $Z_c$ | $\omega$ |
|---|---|---|---|---|
| 1 | 190.6 | 99.2 | 0.288 | 0.012 |
| 2 | 425.2 | 255 | 0.274 | 0.199 |

Equation (4-4.8) gives $k_{12} = 0.036$ (Ref. 10 gives $k_{12} = 0.04$); Eqs. (4-4.3) through (4-4.7) give

$$T_{c12} = [(190.6)(425.2)]^{1/2}(1 - 0.036) = 274.4 \text{ K}$$

$$\omega_{12} = (0.5)(.012) + (0.5)(0.199) = 0.1055$$

$$Z_{c12} = (0.5)(0.288) + (0.5)(0.274) = 0.281$$

$$V_{c12} = \frac{(99.2^{1/3} + 255^{1/3})^3}{8} = 165.0 \text{ cm}^3/\text{mol}$$

$$P_{c12} = \frac{(0.281)(83.14)(273.3)}{165} = 38.7 \text{ bar}$$

$$T_{r12} = 444.3/274.4 = 1.619$$

Equations (3-5.5) and (3-5.6) give

$$B_{12}^{(0)} = -0.112 \qquad B_{12}^{(1)} = 0.116$$

Equation (3-5.4) gives

$$B_{12} = \frac{(83.14)(273.3)}{38.7}[-0.112 + (0.1055)(0.116)]$$

$$= -58.6 \text{ cm}^3/\text{mol}$$

Equation (4-4.1) gives

$$B_m = (0.5)^2(-8.1) + (2)(0.5)^2(-58.6) + (0.5)^2(-293.4)$$

$$= -105 \text{ cm}^3/\text{mol} \qquad (\text{error} = 2\%)$$

With $k_{12} = 0.036$, the predicted $B_{12}$ is 4.8 cm$^3$/mol too high; with $k_{12} = 0$, the predicted $B_{12}$ would be 6.3 cm$^3$/mol too low. Note that, in this example, the binary interaction parameter $k_{12}$ was used in the estimation of the reduced temperature at which $B_{12}$ was evaluated. This temperature was neither the pseudocritical nor the true critical temperature of the mixture, and it was not a function of the mixture composition. The pseudocritical approach offers an alternate solution method to Example 4-1. In theory, this approach is equivalent, but in practice it is less accurate.

**Example 4-2** Repeat Example 4-1 by using the pseudocritical method.

**solution** From Table 4-1, $k_{12}^* = 1.11$. Equations (4-3.4) and (4-3.5) give

$$T_{c12} = \frac{(1.11)(190.6 + 425.2)}{2} = 341.8 \text{ K}$$

$$T_{cm} = (0.5)^2(190.6) + (2)(0.5)^2(341.8) + (0.5)^2(425.2)$$
$$= 324.8 \text{ K}$$
$$T_r = \frac{444.3}{324.8} = 1.3678$$

Equations (3-5.5) and (3-5.6) give $B^{(0)} = -0.1727$ and $B^{(1)} = 0.0928$; Eq. (4-3.2) gives

$$P_{cm} = \frac{(83.14)[(0.5)(0.288) + (0.5)(0.274)](324.8)}{(0.5)(99.2) + (0.5)(255)}$$
$$= 42.85 \text{ bar}$$

Equation (3-5.4) gives

$$B_m = \frac{(83.14)(324.8)}{42.85}[-0.1727 + (0.1055)(0.0928)]$$
$$= -103 \text{ cm}^3/\text{mol} \qquad (\text{error} = 4\%)$$

If $k^*_{12}$ were taken as zero, the answer would be $B_m = -89$ cm$^3$/mol, for an error of 17%. This example illustrates how $k^*_{12}$ is related to $k_{12}$.

## 4-5 Mixing Rules for Redlich-Kwong-Type Equations of State

The mixing rules recommended for all two-constant cubic equations of state (i.e., van der Waals, Redlich-Kwong, Soave, and Peng-Robinson) are

$$a_m = \sum_i \sum_j y_i y_j (a_i a_j)^{1/2}(1 - \bar{k}_{ij}) \tag{4-5.1}$$

$$b_m = \sum_i y_i b_i \tag{4-5.2}$$

$a_i$ and $b_i$ are given in Table 3-5. Some values of the binary interaction coefficient $\bar{k}_{ij}$ for the Soave and Peng-Robinson equations are given in Table 4-2. A more extensive tabulation is given in Ref. 28. Values for $\bar{k}_{ij}$ for specific systems and as a function of temperature are given in [9, 16, 23, 25, 33]. For hydrocarbon pairs, $\bar{k}_{ij}$ is usually taken as zero. If all $\bar{k}_{ij}$ are zero, Eq. (4-5.1) reduces to

$$a_m = \left(\sum_i y_i a_i^{1/2}\right)^2 \tag{4-5.3}$$

It can be shown that the second virial coefficient $B$ is given by

$$B = \frac{b - a}{RT} \tag{4-5.4}$$

This provides a relation between $\bar{k}_{ij}$ and the $k_{ij}$ in Eq. (4-4.3). In fact,

TABLE 4-2 $\bar{k}_{ij}$ Values for Soave (SRK) and Peng-Robinson (PR) Equations [28]†

| | Carbon dioxide | | Hydrogen sulfide | | Nitrogen | | Carbon monoxide | |
|---|---|---|---|---|---|---|---|---|
| | SRK | PR | SRK | PR | SRK | PR | SRK | PR |
| Methane | 0.093 | 0.092 | | | 0.028 | 0.031 | 0.032 | 0.030 |
| Ethylene | 0.053 | 0.055 | 0.085 | 0.083 | 0.080 | 0.086 | | |
| Ethane | 0.136 | 0.132 | | | 0.041 | 0.052 | −0.028 | −0.023 |
| Propylene | 0.094 | 0.093 | | | 0.090 | 0.090 | | |
| Propane | 0.129 | 0.124 | 0.088 | 0.088 | 0.076 | 0.085 | 0.016 | 0.026 |
| Isobutane | 0.128 | 0.120 | 0.051 | 0.047 | 0.094 | 0.103 | | |
| *n*-Butane | 0.143 | 0.133 | | | 0.070 | 0.080 | | |
| Isopentane | 0.131 | 0.122 | | | 0.087 | 0.092 | | |
| *n*-Pentane | 0.131 | 0.122 | 0.069 | 0.063 | 0.088 | 0.100 | | |
| *n*-Hexane | 0.118 | 0.110 | | | 0.150 | 0.150 | | |
| *n*-Heptane | 0.110 | 0.100 | | | 0.142 | 0.144 | | |
| *n*-Decane | 0.130 | 0.114 | | | | | | |
| Carbon dioxide | | | 0.099 | 0.097 | −0.032 | −0.017 | | |
| Cyclohexane | 0.129 | 0.105 | | | | | | |
| Benzene | 0.077 | 0.077 | | | 0.153 | 0.164 | | |
| Toluene | 0.113 | 0.106 | | | | | | |

†For a more complete list, see Ref. 28, pages 771–793.

setting $\bar{k}_{ij}$ equal to zero is equivalent to using Eq. (4-4.8) to calculate $k_{ij}$ in a cross-virial-coefficient calculation [52]. Note that, once $a$ and $b$ for the mixture ($a_m$ and $b_m$) are determined, computations proceed as though $a$ and $b$ were for a pure component (unless derivatives with respect to composition are required as with fugacity coefficients). In this approach, a pseudocritical temperature and pressure are not determined, although they could be. The pseudocritical temperature and pressure for the Soave equation are, for example, given by

$$T_{cm} = \frac{\sum_i \sum_j y_i y_j (a_i a_j)^{1/2}(1 - \bar{k}_{ij})}{\Omega_a R^2 \sum_i y_i T_{ci}/P_{ci}} \tag{4-5.5}$$

and

$$\frac{T_{cm}}{P_{cm}} = \frac{\sum_i y_i T_{ci}}{P_{ci}} \tag{4-5.6}$$

where

$$a_i = \frac{\Omega_a\{1 + fw_i[1 - (T_{cm}/T_{ci})^{1/2}]\}^2 R^2 T_{ci}^2}{P_{ci}} \tag{4-5.7}$$

$fw_i$ is given in Table 3-5 and $\Omega_a = 0.42748$. Equation (4-5.5) must be solved iteratively, since $a_i$ depends on $T_{cm}$.

Gray et al. [19] have reviewed some of the methods [14, 18, 20, 32, 38, 39, 63] for using the Soave equation for hydrogen-containing mixtures, and Abbott [1] has presented mixing rules for cubic equations with more than the two parameters $a$ and $b$.

Peneloux's correction to the mixture volumes is given by

$$c_m = \sum_i y_i c_i \tag{4-5.8}$$

The procedure for using this correction is the same as for pure components (which was dicussed in Sec. 3-6):

$$V = V_{\mathrm{SRK}} - c_m \tag{4-5.9}$$

$V_{\mathrm{SRK}}$ is the volume predicted by the Soave equation; $c_m$ is the correction for the mixture from Eq. (4-5.8); and $V$ is the corrected volume.

## 4-6 Mixing Rules for the Lee-Kesler Equation

Numerous sets of mixing rules have been proposed for the Lee-Kesler method [2, 5, 29, 44, 47, 57, 60]. Those shown in Table 4-3 are recommended by Knapp et al. [28, 42]. These rules yield values of the mixture pseudocritical properties ($T_{cm}$, $P_{cm}$, $V_{cm}$, and $\omega_m$) from which reduced mixture properties can be determined. Computations then proceed as for a pure component.

Sample values of the binary interaction coefficient $k'_{ij}$ are given in Table 4-4. Values of $k'_{ij}$ have been correlated with pure component properties as shown in Figs. 4-2 to 4-5 [42]. Values of $k'_{ij}$ determined from experimental data, such as those in Table 4-4, are preferred to those estimated from Figs. 4-2 to 4-5, but estimates from the figures are better than taking $k_{ij}$ to be unity. Figure 4-4 often provides reasonable estimates for hydrogen sulfide–hydrocarbon systems [42]. More detailed figures and additional values are given is Refs. 42 and 28. For hydrogen and helium, modified critical properties [43] should be used.

**TABLE 4-3 Mixing Rules for the Lee-Kesler Equation [42]**

$$T_{cm} = \frac{1}{V_{cm}^{1/4}} \sum_i \sum_j y_i y_j V_{cij}^{1/4} T_{cij} \tag{4-6.1}$$

$$V_{cm} = \sum_i \sum_j y_i y_j V_{cij} \tag{4-6.2}$$

$$\omega_m = \sum_i y_i \omega_i \tag{4-6.3}$$

$$T_{cij} = (T_{ci} T_{cj})^{1/2} k'_{ij} \tag{4-6.4}$$

$$V_{cij} = \tfrac{1}{8} (V_{ci}^{1/3} + V_{cj}^{1/3})^3 \tag{4-6.5}$$

$$P_{cm} = (0.2905 - 0.085\omega_m) R T_{cm} / V_{cm} \tag{4-6.6}$$

**TABLE 4-4 Binary Parameters $k'_{ij}$ for the Lee-Kesler Method [42]**

| System | $k'_{ij}$ | System | $k'_{ij}$ |
|---|---|---|---|
| Methane-ethane | 1.052 | Ethylene-$n$-butane | 0.998 |
| -ethylene | 1.014 | -benzene | 1.094 |
| -propane | 1.113 | -$n$-heptane | 1.163 |
| -propylene | 1.089 | Acetylene-ethylene | 0.948 |
| -$n$-butane | 1.171 | Propane-propylene | 0.992 |
| -isobutane | 1.155 | -$n$-butane | 1.003 |
| -$n$-pentane | 1.240 | -isobutane | 1.003 |
| -isopentane | 1.228 | -$n$-pentane | 1.006 |
| -$n$-hexane | 1.304 | -isopentane | 1.009 |
| -cyclohexane | 1.269 | -$n$-hexane | 1.047 |
| -benzene | 1.234 | -cyclohexane | 1.037 |
| -$n$-heptane | 1.367 | -benzene | 1.011 |
| -$n$-octane | 1.423 | -$n$-heptane | 1.067 |
| -$n$-nonane | 1.484 | -$n$-octane | 1.090 |
| -$n$-decane | 1.533 | -$n$-nonane | 1.115 |
| Ethane-ethylene | 0.991 | -$n$-decane | 1.139 |
| -propane | 1.010 | Propylene-$n$-butane | 1.010 |
| -propylene | 1.002 | -isobutane | 1.009 |
| -$n$-butane | 1.029 | -isobutene | 1.006 |
| -isobutane | 1.036 | $n$-Butane-isobutane | 1.001 |
| -$n$-pentane | 1.064 | -$n$-pentane | 0.994 |
| -isopentane | 1.070 | -isopentane | 0.998 |
| -$n$-hexane | 1.106 | -$n$-hexane | 1.018 |
| -cyclohexane | 1.081 | -cyclohexane | 1.008 |
| -benzene | 1.066 | -benzene | 0.999 |
| -$n$-heptane | 1.143 | -$n$-heptane | 1.027 |
| -$n$-octane | 1.165 | -$n$-octane | 1.046 |
| -$n$-nonane | 1.214 | -$n$-nonane | 1.064 |
| -$n$-decane | 1.237 | -$n$-decane | 1.078 |

## 4.7 Interaction Parameters—General Discussion

A frequently asked question is "Can values of $k_{ij}$ from one equation of state be used in another equation of state?" To answer the question, one must first remember that all interaction parameters are empirical. If a $k_{ij}$

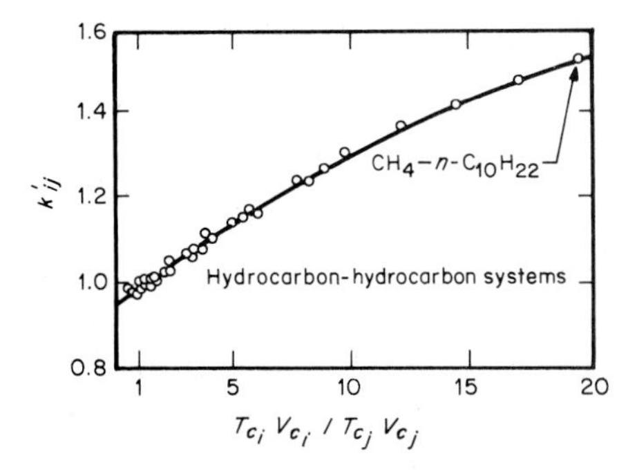


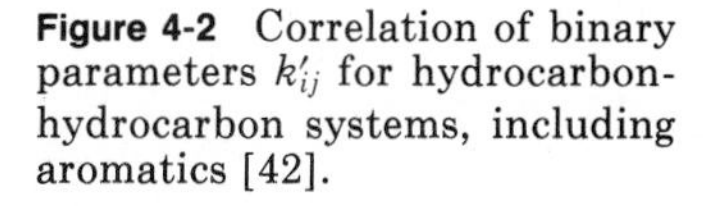
**Figure 4-2** Correlation of binary parameters $k'_{ij}$ for hydrocarbon-hydrocarbon systems, including aromatics [42].

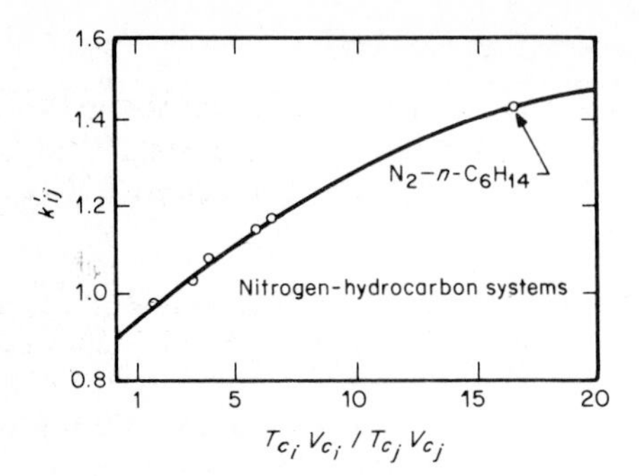


**Figure 4-3** Correlation of binary parameters $k'_{ij}$ for nitrogen-hydrocarbon systems [42].

for one equation is available, then mixture data are probably available. To use a different equation, these data should be regressed to find the best interaction parameter for the new equation. If data at different temperatures and pressures or data of a different type (*PVT* data instead of *VLE* data, for example) are used, a slightly different $k_{ij}$ will likely be obtained. The important thing to remember is that $k_{ij}$ values have no theoretical basis; they are empirical, and their role is to help overcome deficiencies in corresponding states theory, or deficiencies in a particular model's ability to describe what we perceive to be corresponding states behavior.

But even when one does not have access to a regression program, it is often possible to estimate an interaction parameter for one model from a value for another model. If all models were exact, they would predict identical pseudocritical temperatures and second virial coefficients. Thus, the $T_{c_m}$'s in Eqs. (4-3.4), (4-5.5), and (4-6.1) are all about the same number. The three equations contain, respectively, $k^*_{ij}$, $\bar{k}_{ij}$, and $k'_{ij}$. By equating $T_{c_m}$ from Eqs. (4-3.4) and (4-5.5), one could, for example, estimate $k^*_{ij}$ from $\bar{k}_{ij}$. The second virial coefficient expressions, Eqs. (4-4.1) and (4-5.4), provide a relation between $k_{ij}$ and $\bar{k}_{ij}$. These interaction parameters are numerically different, but the mixture virial coefficients they predict should be about the same number [36]. $k^*_{ij}$ and $k_{ij}$ can be related by the technique illustrated by Examples 4-1 and 4-2.

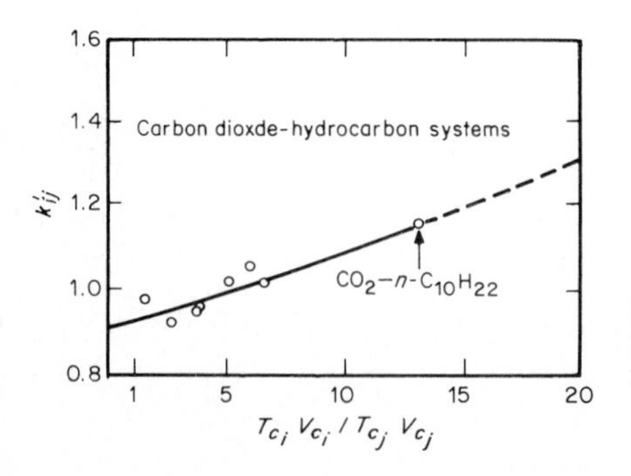


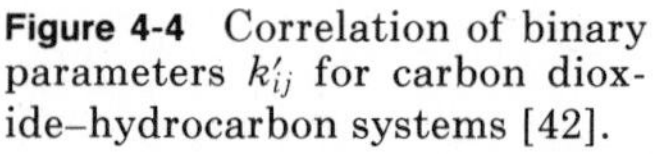
**Figure 4-4** Correlation of binary parameters $k'_{ij}$ for carbon dioxide–hydrocarbon systems [42].

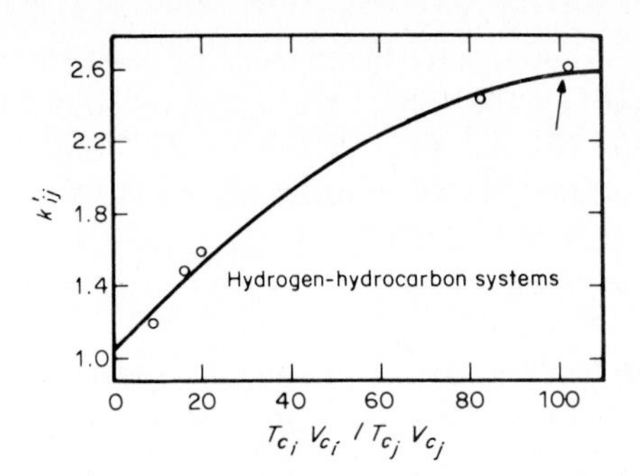


**Figure 4-5** Correlation of binary parameters $k'_{ij}$ for hydrogen-hydrocarbon systems. The arrow indicates data for the $H_2$–$C_7H_{16}$ system [42].

## 4-8 Recent Developments in Mixing Rules

Numerous modifications to traditional mixing rules have been suggested. Some are the result of empirical studies [42]; others have resulted from a theoretical analysis [34]. Still others have been suggested to overcome shortcomings of the traditional mixing rules which arise in specific applications. Two such applications are for mixtures which contain polar molecules or which contain a large number of components.

For systems with even modest deviations from ideal solution behavior, a single binary parameter [for example, $\bar{k}_{ij}$ in Eq. (4-5.1)] often is a weak function of composition and temperature. Often this dependence is small and can be ignored, but for polar molecules, the variation in the interaction parameter with temperature and composition can become significant. Techniques for modifying equations of state to account for this behavior fall roughly into three categories. The categories include reaction models, local composition models, and empirical extensions. All the methods require more than one adjustable parameter per binary.

A method developed by Gmehling et al. [17] assumes that reactions occur among all the polar molecules in a mixture. The model also assumes pure polar compounds exist as an equilibrium mixture of monomers and dimers. Disadvantages of the model are (1) there are a large number of monomer-dimer species, (2) computations are complex because a combined reaction and phase equilibria problem must be solved iteratively, and (3) three or four parameters are required for each pure component. Advantages of the model are that only two adjustable parameters per binary are required and the model does work for mixtures of polar molecules. Only binary mixtures were examined in the original paper, but Buck [6] has extended the model to a seven-component system.

Local composition models, also called two-fluid theories, have as their underlying premise that the composition is not uniform throughout a binary mixture. For example, in a mixture of components A and B, if A molecules are more attracted to other A molecules than to B molecules, the "local mole fraction" of A around another A molecule will be greater than the bulk mole fraction of A. This idea has been used by various authors [30, 49, 56, 58, 59, 61] to modify equation-of-state mixing rules. The mixing rule Mathias and Copeman [37] have proposed for the Peng-Robinson equation is typical. The constant $b$ is given as before by Eq. (4-5.2), but

$$a_m = \sum_i \sum_j y_i y_j (a_i a_j)^{1/2}(1 - \bar{k}_{ij})$$
$$+ \frac{1}{2bRT\sqrt{2}} \ln \left[ \frac{V + b + b\sqrt{2}}{V + b - b\sqrt{2}} \right] \sum_i y_i a_{ci}^2 \left[ \sum_j y_j d_{ji}^2 - \left( \sum_j y_j d_{ji} \right)^2 \right] \tag{4-8.1}$$

where $a_{ci} = 0.45724 R^2 T_c^2 P_c$ and $a_i$ is given in Table 3-5. $\bar{k}_{ij}$ is symmetric, but $d_{ij}$ is not. That is, $d_{ij} \neq d_{ji}$, but $\bar{k}_{ij} = \bar{k}_{ji}$. Also, $\bar{k}_{ii} = d_{ii} = 0$. Thus, in the above formulation, there are three parameters per binary, $\bar{k}_{12}$, $d_{12}$, and $d_{21}$. Local compositon mixing rules as applied by other authors vary in the number of parameters used and in the equation of state to which the mixing rules are applied. Some forms do not reduce to a quadratic mixing rule at low densities as is required by the virial equation. Equation (4-8.1) does not have this shortcoming because, as $V$ approaches infinity, the second term in Eq. (4-8.1) goes to zero. With this additional volume dependence, the equation is no longer cubic in volume.

All the local composition models require at least two parameters per binary. Several empirical forms which are simpler than Eq. (4-8.1) have been presented, but they still allow the inclusion of an additional binary parameter. Panagiotopoulos [41] suggests that Eq. (4-5.1) be replaced by

$$a_m = \sum_i \sum_j y_i y_j \left[ (a_i a_j)^{1/2}(1 - \bar{k}_{ij}) + \frac{b_m}{VRT}(y_i \lambda_{ij} + y_i \lambda_{ji}) \right] \tag{4-8.2}$$

where $$\lambda_{ij} = -\lambda_{ji} \tag{4-8.3}$$

In Eq. (4-8.1), there are three parameters per binary; in Eq. (4-8.2) there are two. Both of these equations make $\bar{k}_{ij}$, as defined in Eq. (4-5.1), a linear function of composition. The composition dependence expressed by Eq. (4-8.2) is obtained from Eq. (4-8.1) if $d_{ij}$ is set equal to minus $d_{ji}$. Burcham et al. [7] have examined the case when one of the two $d$ parameters in Eq. (4-8.1) is set equal to zero.

Several authors have added a second binary interaction parameter $\epsilon_{ij}$ in the $b$ constant [8, 15, 49, 55], i.e.

$$b_m = \sum_i \sum_j y_i y_j \frac{(b_i + b_j)}{2}(1 - \epsilon_{ij}) \tag{4-8.4}$$

(See also Sec. 8-12.) Efforts to apply equations of state to polar molecules are still in the development stage. It is not yet clear which form is best.

All the mixing rules presented thus far in this chapter are for a discrete number of components. For mixtures with a very large number of components, such as crude oil, the computations required to evaluate the summations many times can be prohibitively expensive. Traditionally these complex mixtures have been split into fractions of pseudocomponents. Several authors [3, 12, 13, 46] have developed the idea whereby these complex mixtures can be described by a single, continuous distribution function. Summations are replaced by integrals with a corresponding increase in computational efficiency. Cotterman et al. [12, 13] have used this idea along with a cubic equation of state and have presented the associated equations.

## 4-9 Densities of Liquid Mixtures

Mixing rules for both the modified Rackett and Hankinson equations have been published and are given below. The modified Rackett equation for mixtures at their bubble points is [50]

$$V_m = R\left(\sum_i \frac{x_i T_{ci}}{P_{ci}}\right) Z_{\mathrm{RA}_m}^{[1 + (1 - T_r)^{2/7}]} \tag{4-9.1}$$

$$Z_{\mathrm{RA}_m} = \sum_i x_i Z_{\mathrm{RA}_i} \tag{4-9.2}$$

where $T_r = T/T_{cm}$ and the Chueh-Prausnitz rules [11, 54] are recommended [50] for $T_{cm}$:

$$T_{cm} = \sum_i \sum_j \phi_i \phi_j T_{cij} \tag{4-9.3}$$

$$\phi_i = \frac{x_i V_{ci}}{\sum_i x_i V_{ci}} \tag{4-9.4}$$

$$T_{cij} = (1 - k_{ij})(T_{c_i} T_{c_j})^{1/2} \tag{4-9.5}$$

$$1 - k_{ij} = \frac{8(V_{c_i} V_{c_j})^{1/2}}{(V_{c_i}^{1/3} + V_{c_j}^{1/3})^3} \tag{4-9.6}$$

Mixing rules recommended [22] for the Handinson-Brobst-Thomson equation are

$$T_{cm} = \frac{\sum_i \sum_j x_i x_j V_{ij}^* T_{cij}}{V_m^*} \tag{4-9.7}$$

$$V_m^* = \frac{1}{4}\left[\sum_i x_i V_i^* + 3\left(\sum_i x_i V_i^{*2/3}\right)\left(\sum_i x_i V_i^{*1/3}\right)\right] \tag{4-9.8}$$

$$V_{ij}^* T_{cij} = (V_i^* T_{ci} V_j^* T_{cj})^{1/2} \tag{4-9.9}$$

$$\omega_{\text{SRK}m} = \sum_i x_i \omega_{\text{SRK}i} \tag{4-9.10}$$

$$P_{cm} = \frac{(0.291 - 0.080\omega_{\text{SRK}m})RT_{cm}}{V_m^*} \tag{4-9.11}$$

In the HBT compressed liquid correlation Eq. (3-11.5), $P_{\text{vp}m}$ for a mixture is calculated by

$$P_{\text{vp}m} = P_{cm}P_{rm} \tag{4-9.12}$$

$P_{cm}$ is from Eq. (4-9.11), and $P_{rm}$ is calculated from the generalized Riedel vapor pressure equation

$$\log_{10} P_{rm} = P_{rm}^{(0)} + \omega_{\text{SRK}m} P_{rm}^{(1)} \tag{4-9.13}$$

$$P_{rm}^{(0)} = 5.8031817 \log_{10} T_{rm} + 0.07608141\alpha \tag{4-9.14}$$

$$P_{rm}^{(1)} = 4.86601 \log_{10} T_{rm} + 0.03721754\alpha \tag{4-9.15}$$

$$\alpha = 35.0 - \frac{36.0}{T_{rm}} - 96.736 \log_{10} T_{rm} + T_{rm}^6 \tag{4-9.16}$$

and

$$T_{rm} = \frac{T}{T_{cm}} \tag{4-9.17}$$

Teja [52] and Teja and Sandler [53] have calculated densities of LNG mixtures and $CO_2$–crude oil systems with Eq. (3-7.5) in which the two reference fluids were methane and either butane or decane. While they obtained very accurate results, their particular approach did require that binary interaction parameters be determined from binary density data.

**Example 4-3** Calculate the volume of a saturated liquid mixture of 50.2 mole percent $CO_2$ (1) and 49.8 mole percent $n$-butane (2) at 344.26 K (160°F). Repeat for compressed liquid at 344.26 K and 345 bar (5000 psia). Experimental values [40] are $V^L$ = 99.13 cm$^3$/mol at the bubble point pressure of 64.8 bar and $V^L$ = 78.22 cm$^3$/mol at 345 bar.

**solution** With $T_c$, $\omega_{\text{SRK}}$, $V^*$, and $Z_{\text{RA}}$ from Table 3-10 and $V_c$ and $P_c$ from Appendix A,

| | $T_c$, K | $P_c$, bar | $\omega_{\text{SRK}}$ | $V^*$, cm$^3$/mol | $Z_{\text{RA}}$ | $V_c$, cm$^3$/mol |
|---|---|---|---|---|---|---|
| $CO_2$ | 304.15 | 73.8 | 0.2373 | 93.8 | 0.2722 | 93.9 |
| $n-C_4H_{10}$ | 425.18 | 38.0 | 0.2008 | 254.4 | 0.2730 | 255.0 |

HBT METHOD FOR SATURATED VOLUME Equations (4-9.7) to (4-9.10) give

$\omega_{\text{SRK}m} = 0.2191 \qquad V_m^* = 167.3 \text{ cm}^3/\text{mol} \qquad \text{and} \qquad T_{cm} = 369.56 \text{ K}$

Thus $\quad T_r = \dfrac{344.26}{369.56} = 0.9315$

From Eqs. (3-11.2) and (3-11.3),

$V_r^{(0)} = 0.5653 \qquad \text{and} \qquad V_r^{(\delta)} = 0.1698.$

Equation (3-11.1) then gives

$$\begin{aligned} V_s &= (167.3)(0.5653)[1 - (0.219)(0.1698)] \\ &= 91.0 \text{ cm}^3/\text{mol} \qquad (\text{error} = 8\%) \end{aligned}$$

MODIFIED RACKETT. Equations (4-9.2) to (4-9.6) give

$1 - k_{12} = 0.959 \qquad T_{c12} = 345.0 \text{ K} \qquad \phi_1 = 0.2704$

$\phi_2 = 0.7296 \qquad Z_{\text{RA}m} = 0.2726 \qquad \text{and} \qquad T_{cm} = 384.7 \text{ K}$

Thus, $T_r = 0.895$, and Eq. (4-9.1) then gives

$V = 87.5 \text{ cm}^3/\text{mol} \qquad (\text{error} = 12\%)$

HBT METHOD FOR COMPRESSED LIQUID. Equation (4-9.11) gives

$$\begin{aligned} P_{cm} &= \frac{[0.291 - (0.080)(0.2191)](83.14)(369.56)}{167.3} \\ &= 50.23 \text{ bar} \end{aligned}$$

Equations (3-11.6) to (3-11.8) give

$$\begin{aligned} c &= 0.0861488 + (0.0344483)(0.2191) = 0.09370 \\ e &= 135.0 \\ \frac{\beta}{P_c} &= 0.2727 \\ \beta &= 13.72 \end{aligned}$$

Equations (4-9.12) to (4-9.17) give

$$\begin{aligned} \alpha &= 35.0 - \frac{36.0}{0.9314} - 96.736 \log_{10} 0.9314 + (0.9314)^6 \\ &= -0.01298 \\ P_{rm}^{(1)} &= -0.1522 \\ P_{rm}^{(0)} &= -0.1797 \\ \log_{10} P_{rm} &= -0.1796 - (0.2191)\,(0.1522) \\ P_{rm} &= 0.612 \\ P_{vpm} &= (0.612)(50.22) \\ &= 30.75 \end{aligned}$$

Equation (3-11.5) gives

$$\begin{aligned} V &= (91.1)(1 - 0.0937 \ln \left(\frac{13.69 + 345}{13.69 + 30.75}\right) \\ &= 73.2 \text{ cm/mol} \qquad (\text{error } 6\%) \end{aligned}$$

The errors in this example are atypical. Tests with hydrocarbon mixtures and mixtures of hydrocarbons with other light gases gave much smaller errors for both the modified Rackett and HBT methods.

## Recommendation

To estimate the volume of both saturated and compressed liquid mixtures, use the HBT correlations, Eqs. (3-11.1) and (3-11.5). Mixing rules are given in Eqs. (4-9.7) to (4-9.17).

## Notation

| | |
|---|---|
| $a$ | attractive parameter in cubic equations of state, bar·cm$^6$/mol$^2$ |
| $b$ | repulsive parameter in cubic equations of state, cm$^3$/mol |
| $B$ | second virial coefficient, cm$^3$/mol |
| $c$ | Peneloux's volume correction, Eq. (4-5.8), cm$^3$/mol |
| $d$ | local composition binary parameter, Eq. (4-8.1) |
| $k_{ij}$, $k^*_{ij}$, $\bar{k}_{ij}$, $k'_{ij}$ | binary interaction parameter |
| $P$ | pressure; $P_c$, critical pressure; $P_r$, reduced pressure, $P/P_c$; $P_{cm}$, pseudocritical mixture pressure, bar |
| $Q$ | generalized property |
| $R$ | gas constant, see Table 3-1 |
| $T$ | temperature; $T_c$, critical temperature; $T_r$, reduced temperature, $T/T_c$; $T_{cT}$, true critical temperature of a mixture; $T_{cm}$, pseudocritical mixture temperature, K |
| $V$ | volume; $V_c$, critical volume; $V_r$, reduced volume, $V/V_c$; $V_{cT}$, true critical volume for a mixture; $V_{cm}$, pseudocritical mixture volume, cm$^3$/mol |
| $V^*$ | parameter in HBT correlation |
| $x$ | mole fraction in the liquid phase |
| $y$ | mole fraction in the vapor phase, or in some cases in both the liquid and vapor phases |
| $Z_c$ | critical compressibility factor |
| $Z_{RA}$ | Rackett compressibility factor |

### Greek

| | |
|---|---|
| $\alpha$ | function of $T_r$; see Eq. (4-9.16) |
| $\phi$ | volume fraction, Eqs. (4-9.3) and (4-9.4) |
| $\omega_{SRK}$ | parameter in HBT correlation, Eq. (4-9.10) |

SUPERSCRIPTS

| | |
|---|---|
| $L$ | liquid |
| $V$ | vapor |

SUBSCRIPTS

| | |
|---|---|
| $c$ | critical |
| $j$ | component $j$ |
| $ij$ | interaction between $i$ and $j$ |
| $m$ | mixture |

## REFERENCES


1. Abbott, M. M.: "Equations of State in Engineering and Research," *Advan. Chem. Ser.*, **182:** (1979).
2. Barner, H. E., and C. W. Quinlan: *Ind. Eng. Chem. Process Design Develop.*, **9:** 407 (1969).
3. Briano, J. G., and E. D. Glandt: *Fluid Phase Equil.*, **14:** 91 (1983).
4. Brulé, M. R., and K. E. Starling: *Ind. Eng. Chem. Process Design Develop.*, **23:** 833 (1984).
5. Brulé, M. R., C. T. Lin, L. L. Lee, and K. E. Starling: *AIChE J.*, **28:** 616 (1982).
6. Buck, E: *CHEM. TECH.*, **14:** 570 (1984).
7. Burcham, A. F., Trampe, M. D., Poling, B. E., and D. B. Manley: *ACS Symp. Ser.*, **300:** 86 (1986).
8. Chaback, J. J., and E. A. Turek: *ACS Symp. Ser.*, **300:** 406 (1986).
9. Chang, T., R. W. Rousseau, and J. K. Ferrell: *Ind. Eng. Chem. Process Design Develop.* **22:** 462 (1983).
10. Chueh, P. L., and J. M. Prausnitz: *Ind. Eng. Chem. Fundam.*, **6:** 492 (1967).
11. Chueh, P. L., and J. M. Prausnitz: *AIChE J.*, **13:** 1099 (1967).
12. Cotterman, R. L., Bender, R., and J. M. Prausnitz: *Ind. Eng. Chem. Process Design Develop.*, **24:** 194 (1985).
13. Cotterman, R. L., and J. M. Prausnitz: *Ind. Eng. Chem. Process Design Develop.*, **24:** 434 (1985), **24:** 891 (1985).
14. El-Twaty, A. I., and J. M. Prausnitz: *Chem. Eng. Sci.*, **35:** 1765 (1980).
15. Evelein, K. A., R. G. Moore, and R. A. Heidemann: *Ind. Eng. Chem. Process Design Develop.*, **15:** 423 (1976).
16. Fu, C.-T., S. Yoshimura, and B. C.-Y. Lu: *Advan. Cryog. Eng.*, **27:** 875 (1982).
17. Gmehling, J., D. D. Liu, and J. M. Prausnitz: *Chem. Eng. Sci.*, **34:** 951 (1979).
18. Graboski, M. S., and T. E. Daubert: *Ind. Eng. Chem. Process Design Develop.*, **17:** 448 (1979).
19. Gray, R. D., Jr., J. L. Heideman, S. C. Hwang, and C. Tsonopoulos: *Fluid Phase Equil.*, **14:** 59 (1983).
20. Gray, R. D., Jr.: "Correlation of $H_2$/Hydrocarbon VLE Using Redlich-Kwong Variants," paper presented at *Annual AIChE Mtg., New York, Nov. 13–17, 1977.*
21. Gunn, R. D.: *AIChE J.*, **18:** 183 (1972).
22. Hankinson, R. W., and G. H. Thomson: *AIChE J.*, **25:** 653 (1979).
23. Ishikawa, T., W. K. Chung, and B. C.-Y. Lu: *Advan. Cryog. Eng.*, **25:** 671 (1980).
24. Joffe, J.: *Ind. Eng Chem. Fundam.*, **10:** 532 (1971).
25. Kato, K: *Fluid Phase Equil.*, **7:** 219 (1981).
26. Kaul, B. K., and J. M. Prausnitz: *Ind. Eng. Chem. Fundam.*, **16:** 336 (1977).
27. Kay, W. B.: *Ind. Eng. Chem.*, **28:** 1014 (1936).

28. Knapp, H., R. Doring, L. Oellrich, U. Plocker, and J. M. Prausnitz: "Vapor-Liquid Equilibria for Mixtures of Low Boiling Substances," *Chem. Data. Ser.*, vol. VI (1982), DECHEMA.
29. Lee, B. I., and M. G. Kesler: *AIChE J.*, **21:** 510 (1975).
30. Lee, L. L., T. H. Chung, and K. E. Starling: *Fluid Phase Equil.*, **12:** 105 (1983).
31. Leland, T. W., Jr., and P. S. Chappelear: *Ind. Eng. Chem.*, **60**(7): 15 (1968).
32. Liu, N. M.: *Ind. Eng. Chem. Process Design Develop.*, **19:** 501 (1980).
33. Lu, B. C.-Y., W. K. Chung, M. Kato, and Y.-J. Hsiao: *Advan. Cryog. Eng.*, **23:** 580 (1978).
34. Mansoori, G. A.: *ACS Symp. Ser.*, **300:** 314 (1986).
35. Maris, N. J. P., and L. I. Stiel: *Ind. Eng. Chem. Process Design Develop.*, **24:** 183 (1985).
36. Martin, J. J.: *Ind. Eng. Chem. Fundam.*, **23:** 454 (1984).
37. Mathias, P. M., and T. W. Copeman: *Fluid Phase Equil.*, **13:** 91 (1983).
38. Moysan, J. M., M. J. Huron, H. Paradowski, and J. Vidal: *Chem. Eng. Sci*, **38:** 1085 (1983).
39. Moysan, J. M., H. Paradowski, and J. Vidal: *Hydrocarbon Process. Petrol. Refiner*, **64**(7): 73 (1985).
40. Olds, R. H., H. H. Reamer, B. H. Sage, and W. N. Lacey: *Ind Eng. Chem.*, **41:** 475 (1949).
41. Panagiotopoulos, A. Z., and R. C. Reid: *Fluid Phase Equil.*, **29:** 525 (1986); see also *ACS Symp. Ser.*, **300:** 571 (1986).
42. Plocker, Ulf, H. Knapp, and J. M. Prausnitz: *Ind. Eng. Chem. Process Design Develop.* **17:** 324 (1978).
43. Prausnitz, J. M.: *Molecular Thermodynamics of Fluid-Phase Equilibria*, Prentice-Hall, Englewood Cliffs, N.J., 1969, pp. 128, 129.
44. Prausnitz, J. M., and R. D. Gunn: *AIChE J.*, **4:** 430, 494 (1958).
45. Ramaiah, V., and L. I. Stiel; *Ind Eng. Chem. Process Design Develop.*, **11:** 501 (1972), **12:** 305 (1973).
46. Ratzsch, M. T., and Kehlen, H.: *Fluid Phase Equil.*, **14:** 225 (1983).
47. Reed, T. M., and K. E, Gubbins: *Applied Statistical Mechanics*, McGraw-Hill, New York (1973).
48. Reid, R. C., and T. W. Leland, Jr.: *AIChE J.*, **11:** 228 (1965), **12:** 1227 (1966).
49. Soave, G: *Chem. Eng. Sci.*, **39:** 357 (1984).
50. Spencer, C. F., and R. P. Danner: *J. Chem. Eng. Data*, **18:** 230 (1973).
51. Tarakad, R. R., and R. P. Danner: *AIChE J.*, **23:** 685 (1977).
52. Teja, A. S.: *Chem. Eng. Sci.*, **33:** 609 (1978).
53. Teja, A. S., and S. I. Sandler: *AIChE J.*, **26:** 341 (1980).
54. Tsonopoulos, C.: "Equations of State in Engineering and Research," *Advan. Chem. Ser.*, **182:** 143, (1979).
55. Turek, E. A., R. S. Metcalfe, L. Yarborough, and R. L. Robinson: *Soc. Petrol. Engrs. J.*, **24:** 308 (1984).
56. Vachhani, H. N., and T. F. Anderson: "Mixing Rules Based on the Two-Fluid Theory for Representing Highly Nonideal Systems with an Equation of State," paper presented at *Annual AIChE Mtg., Los Angeles, November 1982.*
57. Watanasiri, S., M. R. Brulé, and K. E. Starling: *AIChE J.*, **28:** 626 (1982).
58. Whiting, W. B., and J. M. Prausnitz: *Fluid Phase Equil.*, **9:** 119 (1982).
59. Won, K. W.: *Fluid Phase Equil.*, **10:** 191 (1983).
60. Wong, D. S., and S. I. Sandler: *Ind. Eng. Chem. Fundam.*, **23:** 348 (1984).
61. Wong, J. M., and K. P. Johnston: *Ind. Eng. Chem. Fundam.*, **23:** 320 (1984).
62. Yarborough, L.: "Equations of State in Engineering and Research," *Advan. Chem. Ser.*, **182:** (1979).
63. Zudkevitch, D., and J. Joffe: *AIChE J.*, **16:** 112 (1970).

Chapter

# 5

# Thermodynamic Properties

## 5-1 Scope

In this chapter we first develop relations to calculate the Helmholtz and Gibbs energies, enthalpies, entropies, and fugacity coefficients. These relations are then used with equation-of-state correlations from Chap. 3 to develop estimation techniques for enthalpy and entropy departure functions and fugacity-pressure ratios for pure components and mixtures. In Sec. 5-5 methods are presented for determining the heat capacities of real gases. The true critical properties of mixtures are discussed in Sec. 5-6, and heat capacities of liquids are treated in Sec. 5-7. Fugacity coefficients of components in gas mixtures are considered in Sec. 5-8.

## 5-2 Fundamental Thermodynamic Principles

Enthalpy, internal energy, entropy, fugacity, etc., are useful thermodynamic properties. In analyzing or designing process equipment, a variation in these properties can often be related to operating variables, e.g., the temperature rise of a fluid in a heat exchanger. It is therefore important to estimate such property variations as the temperature, pressure, and other independent variables of a system change.

The variation of any thermodynamic property between two states is independent of the path chosen to pass from one state to the other. For example, with a pure fluid or a mixture of a fixed composition, if the difference in enthalpy between states $P_1$, $T_1$ and $P_2$, $T_2$ is desired, there are an infinite number of possible calculational paths, all of which give the same numerical result. Two of the most obvious are illustrated in Eqs. (5-2.2) and (5-2.3):

$$H = f(P, T)$$

$$dH = \left(\frac{\partial H}{\partial P}\right)_T dP + \left(\frac{\partial H}{\partial T}\right)_P dT \tag{5-2.1}$$

$$H_2 - H_1 = \int_{P_1}^{P_2} \left(\frac{\partial H}{\partial P}\right)_{T_1} dP + \int_{T_1}^{T_2} \left(\frac{\partial H}{\partial T}\right)_{P_2} dT \tag{5-2.2}$$

$$H_2 - H_1 = \int_{P_1}^{P_2} \left(\frac{\partial H}{\partial P}\right)_{T_2} dP + \int_{T_1}^{T_2} \left(\frac{\partial H}{\partial T}\right)_{P_1} dT \tag{5-2.3}$$

In the first method, a stepwise process is visualized whereby the temperature is held constant at $T_1$ and the isothermal variation in $H$ is determined from $P_1$ to $P_2$; this change is then added to the isobaric variation in $H$ with $T$ from $T_1$ to $T_2$ at pressure $P_2$. The second method is similar; but now the variation in $H$ is first determined at $P_1$ from $T_1$ to $T_2$, and then the variation of $H$ from $P_1$ to $P_2$ is determined at $T_2$. These paths are shown schematically in Fig. 5-1, where Eq. (5-2.2) is illustrated by path $ADC$, whereas Eq. (5-2.3) refers to path $ABC$. The net $\Delta H = H_2 - H_1$ represents the change $AC$. Obviously, any other convenient path is possible, for example, $AEFGHC$, but to calculate $\Delta H$ by this path, values of $(\partial H/\partial P)_T$ and $(\partial H/\partial T)_P$ must be available for the various isotherms and isobars.

The partial derivatives of enthalpy (or other thermodynamic properties) can be visualized as slopes of the isotherms or isobars in Fig. 5-1. To obtain numerical values of such derivatives, both a pressure and a tem-

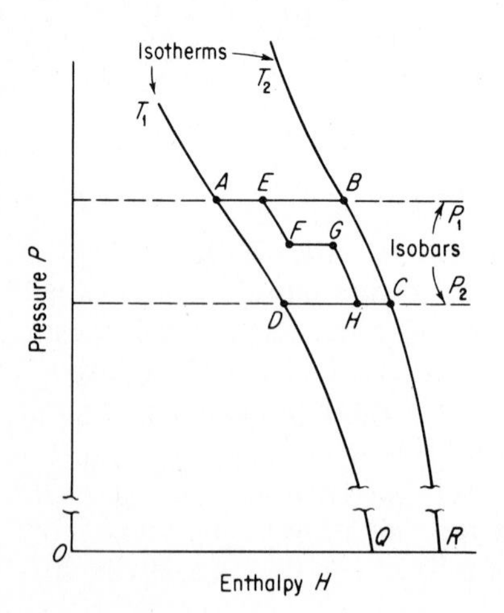


**Figure 5-1** Schematic diagram showing possible isotherms and isobars for changes in enthalpy.

perature must be specified; i.e., values of these derivatives are also functions of two independent intensive variables in the same way that $H$, $U$, ... are functions of two such variables.

In Chap. 6, the quantity $(\partial H/\partial T)_{P^\circ}$ is considered. This derivative is called the *constant-pressure heat capacity* $C_p^\circ$, and the superscript indicates that the pressure level is zero absolute pressure; i.e., the fluid is an ideal gas.

Rarely are heat capacities available at high pressures. Therefore, the usual path for determining $H_2 - H_1$ in Fig. 5-1 with values of $C_p^\circ$ is $AQRC$, that is,

$$\Delta H = \int_{P_1}^{P^\circ} \left(\frac{\partial H}{\partial P}\right)_{T_1} dP + \int_{T_1}^{T_2} C_p^\circ \, dT + \int_{P^\circ}^{P_2} \left(\frac{\partial H}{\partial P}\right)_{T_2} dP \tag{5-2.4}$$

or

$$\Delta H = (H^\circ - H_{P_1})_{T_1} + \int_{T_1}^{T_2} C_p^\circ \, dT - (H^\circ - H_{P_2})_{T_2} \tag{5-2.5}$$

The terms $(H^\circ - H_{P_1})_{T_1}$ and $(H^\circ - H_{P_2})_{T_2}$ are called *departure functions*. They relate a thermodynamic property (enthalpy in this case) at some $P$, $T$ to a reference state ($P = 0$, for enthalpy) *at the same temperature*. It is shown in Sec. 5-3 that departure functions can be calculated solely from $PVT$ data or, equivalently, from an equation of state. On the other hand, the term $\int C_p^\circ \, dT$ is evaluated in the ideal-gas state and values of $C_p^\circ$ are estimated as described in Chap. 6.

## 5-3 Departure Functions

Let $\mathcal{L}$ be the value of some thermodynamic property of a pure component (or a mixture with a fixed composition) at some $P$, $T$. If $\mathcal{L}^\circ$ is defined to be the value of $\mathcal{L}$ at the same temperature (and at the same composition if a mixture) but in an *ideal-gas state* and *at a reference pressure* $P^\circ$, then a *departure function* is defined as $\mathcal{L} - \mathcal{L}^\circ$ or $\mathcal{L}^\circ - \mathcal{L}$. In the reference state at $T$, $P^\circ$, the molal volume $V^\circ$ would be given by

$$V^\circ = \frac{RT}{P^\circ} \tag{5-3.1}$$

As shown below, departure functions can be expressed in terms of the $PVT$ properties of a fluid. Two general approaches are used. The first is more convenient if the $PVT$ properties of a fluid are characterized by an equation of state explicit in pressure. All the analytical equations of state described in Chap. 3 are of this form. The second is more useful when temperature and pressure are the independent variables. The corresponding states correlations in Chap. 3, for example, are expressed as $Z = f(T_r, P_r)$.

For the first form, we develop the departure function for the Helmholtz energy $A$, and from this result all other departure functions can readily be obtained. At constant temperature and composition, the variation in the Helmholtz energy with molar volume $V$ is

$$dA = -P\,dV \tag{5-3.2}$$

Integrating at constant temperature and composition from the reference volume $V°$ to the system volume $V$ gives

$$A - A° = -\int_{V°}^{V} P\,dV \tag{5-3.3}$$

The evaluation of Eq. (5-3.3) is inconvenient, since one limit of the integration refers to the real state but the other to the reference, ideal-gas state. Thus, we break the integral into two parts:

$$A - A° = -\int_{\infty}^{V} P\,dV - \int_{V°}^{\infty} P\,dV \tag{5-3.4}$$

The first integral requires real-gas properties, that is, $P = f(V)$ at constant temperature $T$, whereas the second is written for an ideal gas and can be integrated immediately. Before doing so, however, to avoid the difficulty introduced by the infinity limit, we add and subtract $\int_{\infty}^{V}(RT/V)\,dV$ from the right-hand side. Then

$$A - A° = -\int_{\infty}^{V}\left(P - \frac{RT}{V}\right)dV - RT\ln\frac{V}{V°} \tag{5-3.5}$$

The departure function for $A$ [Eq. (5-3.5)] depends upon the choice of $V°$. Note that $A - A°$ does *not* vanish even for an ideal gas unless $V°$ is chosen to equal $V$. Other departure functions are readily obtained from Eq. (5-3.5):

$$S - S° = \frac{-\partial}{\partial T}(A - A°)_V \tag{5-3.6}$$

$$= \int_{\infty}^{V}\left[\left(\frac{\partial P}{\partial T}\right)_V - \frac{R}{V}\right]dV + R\ln\frac{V}{V°} \tag{5-3.7}$$

$$H - H° = (A - A°) + T(S - S°) + RT(Z - 1) \tag{5-3.8}$$

$$U - U° = (A - A°) + T(S - S°) \tag{5-3.9}$$

$$G - G° = (A - A°) + RT(Z - 1) \tag{5-3.10}$$

Also, although not strictly a departure function, the fugacity-pressure ratio can be expressed in a similar manner:

$$\ln \frac{f}{P} = \frac{A - A^\circ}{RT} + \ln \frac{V}{V^\circ} + (Z - 1) - \ln Z$$
$$= -\frac{1}{RT}\int_\infty^V \left(P - \frac{RT}{V}\right) dV + (Z - 1) - \ln Z \qquad (5\text{-}3.11)$$

where $Z = PV/RT$.

Therefore, from any pressure-explicit equation of state and a definition of the reference state ($P^\circ$ or $V^\circ$), all departure functions can readily be found.

**Example 5-1** Derive the departure functions for a pure material or for a mixture of constant composition by using the Redlich-Kwong equation of state (3-6.1).

**solution** Using Eq. (3-6.1) and Table 3-5,

$$P = \frac{RT}{V - b} - \frac{a}{V(V + b)}$$

Then, with Eq. (5-3.5),

$$A - A^\circ = -\int_\infty^V \left[\frac{RT}{V - b} - \frac{RT}{V} - \frac{a}{V(V + b)}\right] dV - RT \ln \frac{V}{V^\circ}$$
$$= -RT \ln \frac{V - b}{V} - \frac{a}{b} \ln \frac{V + b}{V} - RT \ln \frac{V}{V^\circ}$$

For entropy, enthalpy, and internal energy with Eqs. (5-3.7) to (5-3.9),

$$S - S^\circ = -\left[\frac{\partial(A - A^\circ)}{\partial T}\right]_V$$
$$= R \ln \frac{V - b}{V} - \frac{a}{2bT} \ln \frac{V + b}{V} + R \ln \frac{V}{V^\circ}$$
$$H - H^\circ = (A - A^\circ) + T(S - S^\circ) + RT(Z - 1)$$
$$= PV - RT - \frac{3a}{2b} \ln \frac{V + b}{V}$$
$$= \frac{bRT}{V - b} - \frac{a}{(V + b)} - \frac{3a}{2b} \ln \frac{V + b}{V}$$
$$U - U^\circ = (A - A^\circ) + T(S - S^\circ) = -\frac{3a}{2b} \ln \frac{V + b}{V}$$

For $\ln(f/P)$, with Eq. (5-3.11), where

$$Z = \frac{PV}{RT} = \frac{V}{V - b} - \frac{a}{RT(V + b)}$$

we have

$$\ln \frac{f}{P} = \frac{b}{V - b} - \frac{a}{RT(V + b)} - \ln \frac{V - b}{V} - \frac{a}{bRT} \ln \frac{V + b}{V} - \ln \left[\frac{V}{V - b} - \frac{a}{RT(V + b)}\right]$$

Finally, from Eq. (5-3.10)

$$G - G^\circ = \frac{bRT}{V - b} - \frac{a}{(V + b)} - RT \ln \frac{V - b}{V} - \frac{a}{b} \ln \frac{V + b}{V} - RT \ln \frac{V}{V^\circ}$$

From Eqs. (5-3.5) to (5-3.11), or with Example 5-1, one can see that the departure functions $H - H^\circ$, $U - U^\circ$, and $\ln(f/P)$ do not depend upon the value of the reference state pressure $P^\circ$ (or $V^\circ$). In contrast, $A - A^\circ$, $S - S^\circ$, and $G - G^\circ$ do depend upon $P^\circ$ (or $V^\circ$). Either of two common reference states is normally chosen. In the first, $P^\circ$ is set equal to a unit pressure, for example, 1 bar if that is the pressure unit chosen. Then, from Eq. (5-3.1), $V^\circ = RT$, but it is necessary to express $R$ in the same units of pressure. In the second reference state, $P^\circ = P$, the system pressure. Then $V/V^\circ = Z$, the compressibility factor. Other reference states can be defined, for example, $V^\circ = V$, but the two noted above are the more common.

An alternate calculational path to obtain departure functions is more convenient if the equation of state is explicit in volume or if pressure and temperature are the independent variables. In such cases, we again choose as a reference state an ideal gas at the same temperature and composition as that of the system under study. The reference pressure is $P^\circ$, and Eq. (5-3.1) applies. We begin, however, with the Gibbs energy rather than the Helmholtz energy. The analog to Eq. (5-3.4) is

$$G - G^\circ = \int_{P^\circ}^{P} V\,dP = \int_0^P V\,dP + \int_{P^\circ}^{0} V\,dP$$

$$= \int_0^P \left(V - \frac{RT}{P}\right) dP + RT \ln \frac{P}{P^\circ} \tag{5-3.12}$$

$$= RT \int_0^P (Z - 1)\,d \ln P + RT \ln \frac{P}{P^\circ} \tag{5-3.13}$$

For entropy,

$$\begin{aligned} S - S^\circ &= \frac{-\partial}{\partial T}(G - G^\circ)_P \\ &= R \int_0^P \left[1 - Z - T\left(\frac{\partial Z}{\partial T}\right)_P\right] d \ln P - R \ln \frac{P}{P^\circ} \end{aligned} \tag{5-3.14}$$

and for the enthalpy and internal-energy departure functions,

$$H - H^\circ = (G - G^\circ) + T(S - S^\circ) \tag{5-3.15}$$

$$U - U^\circ = (G - G^\circ) + T(S - S^\circ) - RT(Z - 1) \tag{5-3.16}$$

The Helmholtz-energy departure function is

$$A - A^\circ = (G - G^\circ) - RT(Z - 1) \tag{5-3.17}$$

and

$$\ln \frac{f}{P} = \left(\frac{G - G^\circ}{RT}\right) - \ln \frac{P}{P^\circ} \tag{5-3.18}$$

Again, simple algebraic substitution shows that the departure functions $H - H°$, $U - U°$, and $\ln(f/P)$ do not depend upon the choice of $P°$ (or $V°$).

## 5-4 Evaluation of Departure Functions

The departure functions shown in Eqs. (5-3.5) to (5-3.11) or in Eqs. (5-3.12) to (5-3.18) can be evaluated with $PVT$ data and, when necessary, a definition of the reference state. Generally, either an analytical equation of state or some form of the law of corresponding states is used to characterize $PVT$ behavior, although, if available, experimental $PVT$ data for a pure substance or a given mixture can be employed.

### Departure functions from equations of state

Several analytical equations of state were introduced in Chap. 3. All are pressure-explicit. In the same manner as that shown in Example 5-1, departure functions can be determined. $A - A°$ and $S - S°$ for three equations of state are shown in Table 5-1. With $A - A°$ and $S - S°$, $H - H°$, $U - U°$, etc., are readily found from Eqs. (5-3.8) to (5-3.11). In each case reference is made to the appropriate equations or tables in Chaps. 3 and 4, where the characteristic parameters of the equation are defined.

The Lee-Kesler correlation [Eqs. (3-7.1) to (3-7.4)] is not shown in Table 5-1. To employ it to calculate thermodynamic properties, care must be exercised in following the procedure recommended by the authors. The method is illustrated with the enthalpy departure function, and analogous rules apply to the entropy departure function and fugacity-pressure ratio.

Given a pressure and temperature, the first step involves the calculation of the reduced temperature and pressure. If the fluid is pure, $T_c$ and $P_c$ values can be found in Appendix A or estimated by relations given in Chap. 2. For mixtures, the appropriate pseudocritical properties are determined from Eqs. (4-6.1) to (4-6.6). With $T_r$ and $P_r$ following the procedure given in Sec. 3-7, $V_r^{(0)}$, $V_r^{(R)}$, $Z^{(0)}$, and $Z^{(R)}$ are determined. From $T_r$, $V_r^{(0)}$, and $Z^{(0)}$ the simple fluid enthalpy departure function can then be found:

$$\left(\frac{H° - H}{RT_c}\right)^{(0)} = -T_r\left[Z^{(0)} - 1 - \frac{b_2 + 2b_3/T_r + 3b_4/T_r^2}{T_r(V_r^{(0)})} - \frac{c_2 - 3c_3/T_r^2}{2T_r(V_r^{(0)})^2} + \frac{d_2}{5T_r(V_r^{(0)})^5} + 3E\right] \qquad (5\text{-}4.1)$$

where

$$E = \frac{c_4}{2T_r^3\gamma}\left\{\beta + 1 - \left[\beta + 1 + \frac{\gamma}{(V_r^{(0)})^2}\right]\exp\left[-\frac{\gamma}{(V_r^{(0)})^2}\right]\right\} \qquad (5\text{-}4.2)$$

**TABLE 5-1 Departure Functions for the Virial and Cubic Equations of State**
Only $A - A°$ and $S - S°$ departure functions are given; $H - H°$, $U - U°$, $G - G°$, and $\ln (f/P)$ are readily obtained from Eqs. (5-3.8) to (5-3.11).

Virial, Secs. 3-5 and 4-4

$$Z = 1 + \frac{BP}{RT}$$

$$A - A° = RT \ln \frac{V}{V - B} - RT \ln \frac{V}{V°}$$

$$S - S° = -\frac{RT}{V - B}\frac{dB}{dT} - R \ln \frac{V}{V - B} + R \ln \frac{V}{V°}$$

$B$ is given in Eqs. (3-5.3) and (4-4.1)

Cubic, Secs. 3-6 and 4-5

$$P = \frac{RT}{V - b} - \frac{a}{V^2 + uVb + wb^2}$$

$$A - A° = \frac{a}{b\sqrt{u^2 - 4w}} \ln \frac{2Z + B^*(u - \sqrt{u^2 - 4w})}{2Z + B^*(u + \sqrt{u^2 - 4w})} - RT \ln \frac{Z - B^*}{Z} - RT \ln \frac{V}{V°}$$

$$S - S° = R \ln \frac{Z - B^*}{Z} + R \ln \frac{V}{V°} - \frac{1}{b\sqrt{u^2 - 4w}} \frac{\partial a}{\partial T} \ln \frac{2Z + B^*(u - \sqrt{u^2 - 4w})}{2Z + B^*(u + \sqrt{u^2 - 4w})}$$

where

$$\frac{\partial a}{\partial T} = -\frac{R}{2}\left(\frac{\Omega_a}{T}\right)^{1/2} \sum_i \sum_j y_i y_j (1 - \bar{k}_{ij}) \left[ f\omega_j \left(\frac{a_i T_{cj}}{P_{cj}}\right)^{1/2} + f\omega_i \left(\frac{a_j T_{ci}}{P_{ci}}\right)^{1/2} \right]$$

For the Soave Equation

$$f\omega_i = 0.480 + 1.574\omega_i - 0.176\omega_i^2$$
$$\Omega_a = 0.42748$$

For the Peng-Robinson Equation

$$f\omega_i = 0.37464 + 1.54226\omega_i - 0.26992\omega_i^2$$
$$\Omega_a = 0.45724$$

Note: $u$, $w$, $a_i$, $a$, and $b$ are given in Table 3-5 and Eqs. (4-5.1) and (4-5.2).

Next, using the same $T_r$ but $V_r^{(R)}$ and $Z^{(R)}$, recompute Eq. (5-4.1) using the reference fluid constants in Table 3-7; call this departure function $[(H° - H)/RT_c]^{(R)}$. The departure function for the real fluid is then

$$\frac{H° - H}{RT_c} = \left(\frac{H° - H}{RT_c}\right)^{(0)} + \frac{\omega}{\omega^R}\left[\left(\frac{H° - H}{RT_c}\right)^{(R)} - \left(\frac{H° - H}{RT_c}\right)^{(0)}\right] \quad (5\text{-}4.3)$$

Equation (5-4.3) may be rewritten as

$$\frac{H° - H}{RT_c} = \left(\frac{H° - H}{RT_c}\right)^{(0)} + \omega\left(\frac{H° - H}{RT_c}\right)^{(1)} \quad (5\text{-}4.4)$$

if the deviation function for enthalpy is defined as

$$\left(\frac{H^\circ - H}{RT_c}\right)^{(1)} = \frac{1}{\omega^R}\left[\left(\frac{H^\circ - H}{RT_c}\right)^{(R)} - \left(\frac{H^\circ - H}{RT_c}\right)^{(0)}\right] \tag{5-4.5}$$

With Eqs. (5-4.1) and (5-4.5), Tables 5-2 and 5-3 were developed [47]. In these calculations $\omega^R$ was set equal to 0.3978. Tables 5-2 and 5-3 are presented graphically in [84].

### Entropy

Analogous expressions for the entropy departure functions are

$$\left(\frac{S^\circ - S}{R}\right)^{(0)} = -\ln\frac{P^\circ}{P} - \ln Z^{(0)} + \frac{b_1 + b_3/T_r^2 + 2b_4/T_r^3}{V_r^{(0)}} + \frac{c_1 - 2c_3/T_r^3}{2(V_r^{(0)})^2} + \frac{d_1}{5(V_r^{(0)})^5} - 2E \tag{5-4.6}$$

$$\frac{S^\circ - S}{R} = \left(\frac{S^\circ - S}{R}\right)^{(0)} + \frac{\omega}{\omega^R}\left[\left(\frac{S^\circ - S}{R}\right)^{(R)} - \left(\frac{S^\circ - S}{R}\right)^{(0)}\right] \tag{5-4.7}$$

If Eq. (5-4.6) is written as

$$\frac{S^\circ - S}{R} = -\ln\frac{P^\circ}{P} + \left(\frac{S^\circ - S}{R}\right)^{(0)} + \omega\left(\frac{S^\circ - S}{R}\right)^{(1)} \tag{5-4.8}$$

the simple fluid entropy departure $(\ )^{(0)}$ and the deviation function $(\ )^{(1)}$ can be found in Tables 5-4 and 5-5.

### Fugacity-pressure ratio

As with the enthalpy departure calculation,

$$\left(\ln\frac{f}{P}\right)^{(0)} = Z^{(0)} - 1 - \ln Z^{(0)} + \frac{B}{V_r^{(0)}} + \frac{C}{2(V_r^{(0)})^2} + \frac{D}{5(V_r^{(0)})^5} + E \tag{5-4.9}$$

$$\ln\frac{f}{P} = \left(\ln\frac{f}{P}\right)^{(0)} + \frac{\omega}{\omega^R}\left[\left(\ln\frac{f}{P}\right)^{(R)} - \left(\ln\frac{f}{P}\right)^{(0)}\right] \tag{5-4.10}$$

and when Eq. (5-4.10) is written as

$$\ln\frac{f}{P} = \left(\ln\frac{f}{P}\right)^{(0)} + \omega\left(\ln\frac{f}{P}\right)^{(1)} \tag{5-4.11}$$

the simple fluid and deviation functions can be found from Tables 5-6 and 5-7. (Note that $\log\frac{f}{P}$, not $\ln\frac{f}{P}$, is obtained from these tables.)

**TABLE 5-2 Lee-Kesler Residual Enthalpy [47]**

Simple fluid $\left(\frac{H^\circ - H}{RT_c}\right)^{(0)}$

| $T_r$ | $P_r$ = 0.010 | 0.050 | 0.100 | 0.200 | 0.400 | 0.600 | 0.800 |
|---|---|---|---|---|---|---|---|
| 0.30 | 6.045 | 6.043 | 6.040 | 6.034 | 6.022 | 6.011 | 5.999 |
| 0.35 | 5.906 | 5.904 | 5.901 | 5.895 | 5.882 | 5.870 | 5.858 |
| 0.40 | 5.763 | 5.761 | 5.757 | 5.751 | 5.738 | 5.726 | 5.713 |
| 0.45 | 5.615 | 5.612 | 5.609 | 5.603 | 5.590 | 5.577 | 5.564 |
| 0.50 | 5.465 | 5.463 | 5.459 | 5.453 | 5.440 | 5.427 | 5.414 |
| 0.55 | 0.032 | 5.312 | 5.309 | 5.303 | 5.290 | 5.278 | 5.265 |
| 0.60 | 0.027 | 5.162 | 5.159 | 5.153 | 5.141 | 5.129 | 5.116 |
| 0.65 | 0.023 | 0.118 | 5.008 | 5.002 | 4.991 | 4.980 | 4.968 |
| 0.70 | 0.020 | 0.101 | 0.213 | 4.848 | 4.838 | 4.828 | 4.818 |
| 0.75 | 0.017 | 0.088 | 0.183 | 4.687 | 4.679 | 4.672 | 4.664 |
| 0.80 | 0.015 | 0.078 | 0.160 | 0.345 | 4.507 | 4.504 | 4.499 |
| 0.85 | 0.014 | 0.069 | 0.141 | 0.300 | 4.309 | 4.313 | 4.316 |
| 0.90 | 0.012 | 0.062 | 0.126 | 0.264 | 0.596 | 4.074 | 4.094 |
| 0.93 | 0.011 | 0.058 | 0.118 | 0.246 | 0.545 | 0.960 | 3.920 |
| 0.95 | 0.011 | 0.056 | 0.113 | 0.235 | 0.516 | 0.885 | 3.763 |
| 0.97 | 0.011 | 0.054 | 0.109 | 0.225 | 0.490 | 0.824 | 1.356 |
| 0.98 | 0.010 | 0.053 | 0.107 | 0.221 | 0.478 | 0.797 | 1.273 |
| 0.99 | 0.010 | 0.052 | 0.105 | 0.216 | 0.466 | 0.773 | 1.206 |
| 1.00 | 0.010 | 0.051 | 0.103 | 0.212 | 0.455 | 0.750 | 1.151 |
| 1.01 | 0.010 | 0.050 | 0.101 | 0.208 | 0.445 | 0.728 | 1.102 |
| 1.02 | 0.010 | 0.049 | 0.099 | 0.203 | 0.434 | 0.708 | 1.060 |
| 1.05 | 0.009 | 0.046 | 0.094 | 0.192 | 0.407 | 0.654 | 0.955 |
| 1.10 | 0.008 | 0.042 | 0.086 | 0.175 | 0.367 | 0.581 | 0.827 |
| 1.15 | 0.008 | 0.039 | 0.079 | 0.160 | 0.334 | 0.523 | 0.732 |
| 1.20 | 0.007 | 0.036 | 0.073 | 0.148 | 0.305 | 0.474 | 0.657 |
| 1.30 | 0.006 | 0.031 | 0.063 | 0.127 | 0.259 | 0.399 | 0.545 |
| 1.40 | 0.005 | 0.027 | 0.055 | 0.110 | 0.224 | 0.341 | 0.463 |
| 1.50 | 0.005 | 0.024 | 0.048 | 0.097 | 0.196 | 0.297 | 0.400 |
| 1.60 | 0.004 | 0.021 | 0.043 | 0.086 | 0.173 | 0.261 | 0.350 |
| 1.70 | 0.004 | 0.019 | 0.038 | 0.076 | 0.153 | 0.231 | 0.309 |
| 1.80 | 0.003 | 0.017 | 0.034 | 0.068 | 0.137 | 0.206 | 0.275 |
| 1.90 | 0.003 | 0.015 | 0.031 | 0.062 | 0.123 | 0.185 | 0.246 |
| 2.00 | 0.003 | 0.014 | 0.028 | 0.056 | 0.111 | 0.167 | 0.222 |
| 2.20 | 0.002 | 0.012 | 0.023 | 0.046 | 0.092 | 0.137 | 0.182 |
| 2.40 | 0.002 | 0.010 | 0.019 | 0.038 | 0.076 | 0.114 | 0.150 |
| 2.60 | 0.002 | 0.008 | 0.016 | 0.032 | 0.064 | 0.095 | 0.125 |
| 2.80 | 0.001 | 0.007 | 0.014 | 0.027 | 0.054 | 0.080 | 0.105 |
| 3.00 | 0.001 | 0.006 | 0.011 | 0.023 | 0.045 | 0.067 | 0.088 |
| 3.50 | 0.001 | 0.004 | 0.007 | 0.015 | 0.029 | 0.043 | 0.056 |
| 4.00 | 0.000 | 0.002 | 0.005 | 0.009 | 0.017 | 0.026 | 0.033 |

| | | | $P_r$ | | | | |
|---|---|---|---|---|---|---|---|
| 1.000 | 1.200 | 1.500 | 2.000 | 3.000 | 5.000 | 7.000 | 10.000 |
| 5.987 | 5.975 | 5.957 | 5.927 | 5.868 | 5.748 | 5.628 | 5.446 |
| 5.845 | 5.833 | 5.814 | 5.783 | 5.721 | 5.595 | 5.469 | 5.278 |
| 5.700 | 5.687 | 5.668 | 5.636 | 5.572 | 5.442 | 5.311 | 5.113 |
| 5.551 | 5.538 | 5.519 | 5.486 | 5.421 | 5.288 | 5.154 | 4.950 |
| 5.401 | 5.388 | 5.369 | 5.336 | 5.270 | 5.135 | 4.999 | 4.791 |
| 5.252 | 5.239 | 5.220 | 5.187 | 5.121 | 4.986 | 4.849 | 4.638 |
| 5.104 | 5.091 | 5.073 | 5.041 | 4.976 | 4.842 | 4.704 | 4.492 |
| 4.956 | 4.945 | 4.927 | 4.896 | 4.833 | 4.702 | 4.565 | 4.353 |
| 4.808 | 4.797 | 4.781 | 4.752 | 4.693 | 4.566 | 4.432 | 4.221 |
| 4.655 | 4.646 | 4.632 | 4.607 | 4.554 | 4.434 | 4.303 | 4.095 |
| 4.494 | 4.488 | 4.478 | 4.459 | 4.413 | 4.303 | 4.178 | 3.974 |
| 4.316 | 4.316 | 4.312 | 4.302 | 4.269 | 4.173 | 4.056 | 3.857 |
| 4.108 | 4.118 | 4.127 | 4.132 | 4.119 | 4.043 | 3.935 | 3.744 |
| 3.953 | 3.976 | 4.000 | 4.020 | 4.024 | 3.963 | 3.863 | 3.678 |
| 3.825 | 3.865 | 3.904 | 3.940 | 3.958 | 3.910 | 3.815 | 3.634 |
| 3.658 | 3.732 | 3.796 | 3.853 | 3.890 | 3.856 | 3.767 | 3.591 |
| 3.544 | 3.652 | 3.736 | 3.806 | 3.854 | 3.829 | 3.743 | 3.569 |
| 3.376 | 3.558 | 3.670 | 3.758 | 3.818 | 3.801 | 3.719 | 3.548 |
| 2.584 | 3.441 | 3.598 | 3.706 | 3.782 | 3.774 | 3.695 | 3.526 |
| 1.796 | 3.283 | 3.516 | 3.652 | 3.744 | 3.746 | 3.671 | 3.505 |
| 1.627 | 3.039 | 3.422 | 3.595 | 3.705 | 3.718 | 3.647 | 3.484 |
| 1.359 | 2.034 | 3.030 | 3.398 | 3.583 | 3.632 | 3.575 | 3.420 |
| 1.120 | 1.487 | 2.203 | 2.965 | 3.353 | 3.484 | 3.453 | 3.315 |
| 0.968 | 1.239 | 1.719 | 2.479 | 3.091 | 3.329 | 3.329 | 3.211 |
| 0.857 | 1.076 | 1.443 | 2.079 | 2.807 | 3.166 | 3.202 | 3.107 |
| 0.698 | 0.860 | 1.116 | 1.560 | 2.274 | 2.825 | 2.942 | 2.899 |
| 0.588 | 0.716 | 0.915 | 1.253 | 1.857 | 2.486 | 2.679 | 2.692 |
| 0.505 | 0.611 | 0.774 | 1.046 | 1.549 | 2.175 | 2.421 | 2.486 |
| 0.440 | 0.531 | 0.667 | 0.894 | 1.318 | 1.904 | 2.177 | 2.285 |
| 0.387 | 0.466 | 0.583 | 0.777 | 1.139 | 1.672 | 1.953 | 2.091 |
| 0.344 | 0.413 | 0.515 | 0.683 | 0.996 | 1.476 | 1.751 | 1.908 |
| 0.307 | 0.368 | 0.458 | 0.606 | 0.880 | 1.309 | 1.571 | 1.736 |
| 0.276 | 0.330 | 0.411 | 0.541 | 0.782 | 1.167 | 1.411 | 1.577 |
| 0.226 | 0.269 | 0.334 | 0.437 | 0.629 | 0.937 | 1.143 | 1.295 |
| 0.187 | 0.222 | 0.275 | 0.359 | 0.513 | 0.761 | 0.929 | 1.058 |
| 0.155 | 0.185 | 0.228 | 0.297 | 0.422 | 0.621 | 0.756 | 0.858 |
| 0.130 | 0.154 | 0.190 | 0.246 | 0.348 | 0.508 | 0.614 | 0.689 |
| 0.109 | 0.129 | 0.159 | 0.205 | 0.288 | 0.415 | 0.495 | 0.545 |
| 0.069 | 0.081 | 0.099 | 0.127 | 0.174 | 0.239 | 0.270 | 0.264 |
| 0.041 | 0.048 | 0.058 | 0.072 | 0.095 | 0.116 | 0.110 | 0.061 |

**TABLE 5-3 Lee-Kesler Residual Enthalpy [47]**

Deviation function $\left(\frac{H^{\circ} - H}{RT_c}\right)^{(1)}$

| $T_r$ | $P_r$ = 0.010 | 0.050 | 0.100 | 0.200 | 0.400 | 0.600 | 0.800 |
|---|---|---|---|---|---|---|---|
| 0.30 | 11.098 | 11.096 | 11.095 | 11.091 | 11.083 | 11.076 | 11.069 |
| 0.35 | 10.656 | 10.655 | 10.654 | 10.653 | 10.650 | 10.646 | 10.643 |
| 0.40 | 10.121 | 10.121 | 10.121 | 10.120 | 10.121 | 10.121 | 10.121 |
| 0.45 | 9.515 | 9.515 | 9.516 | 9.517 | 9.519 | 9.521 | 9.523 |
| 0.50 | 8.868 | 8.869 | 8.870 | 8.872 | 8.876 | 8.880 | 8.884 |
| 0.55 | 0.080 | 8.211 | 8.212 | 8.215 | 8.221 | 8.226 | 8.232 |
| 0.60 | 0.059 | 7.568 | 7.570 | 7.573 | 7.579 | 7.585 | 7.591 |
| 0.65 | 0.045 | 0.247 | 6.949 | 6.952 | 6.959 | 6.966 | 6.973 |
| 0.70 | 0.034 | 0.185 | 0.415 | 6.360 | 6.367 | 6.373 | 6.381 |
| 0.75 | 0.027 | 0.142 | 0.306 | 5.796 | 5.802 | 5.809 | 5.816 |
| 0.80 | 0.021 | 0.110 | 0.234 | 0.542 | 5.266 | 5.271 | 5.278 |
| 0.85 | 0.017 | 0.087 | 0.182 | 0.401 | 4.753 | 4.754 | 4.758 |
| 0.90 | 0.014 | 0.070 | 0.144 | 0.308 | 0.751 | 4.254 | 4.248 |
| 0.93 | 0.012 | 0.061 | 0.126 | 0.265 | 0.612 | 1.236 | 3.942 |
| 0.95 | 0.011 | 0.056 | 0.115 | 0.241 | 0.542 | 0.994 | 3.737 |
| 0.97 | 0.010 | 0.052 | 0.105 | 0.219 | 0.483 | 0.837 | 1.616 |
| 0.98 | 0.010 | 0.050 | 0.101 | 0.209 | 0.457 | 0.776 | 1.324 |
| 0.99 | 0.009 | 0.048 | 0.097 | 0.200 | 0.433 | 0.722 | 1.154 |
| 1.00 | 0.009 | 0.046 | 0.093 | 0.191 | 0.410 | 0.675 | 1.034 |
| 1.01 | 0.009 | 0.044 | 0.089 | 0.183 | 0.389 | 0.632 | 0.940 |
| 1.02 | 0.008 | 0.042 | 0.085 | 0.175 | 0.370 | 0.594 | 0.863 |
| 1.05 | 0.007 | 0.037 | 0.075 | 0.153 | 0.318 | 0.498 | 0.691 |
| 1.10 | 0.006 | 0.030 | 0.061 | 0.123 | 0.251 | 0.381 | 0.507 |
| 1.15 | 0.005 | 0.025 | 0.050 | 0.099 | 0.199 | 0.296 | 0.385 |
| 1.20 | 0.004 | 0.020 | 0.040 | 0.080 | 0.158 | 0.232 | 0.297 |
| 1.30 | 0.003 | 0.013 | 0.026 | 0.052 | 0.100 | 0.142 | 0.177 |
| 1.40 | 0.002 | 0.008 | 0.016 | 0.032 | 0.060 | 0.083 | 0.100 |
| 1.50 | 0.001 | 0.005 | 0.009 | 0.018 | 0.032 | 0.042 | 0.048 |
| 1.60 | 0.000 | 0.002 | 0.004 | 0.007 | 0.012 | 0.013 | 0.011 |
| 1.70 | 0.000 | 0.000 | 0.000 | -0.000 | -0.003 | -0.009 | -0.017 |
| 1.80 | -0.000 | -0.001 | -0.003 | -0.006 | -0.015 | -0.025 | -0.037 |
| 1.90 | -0.001 | -0.003 | -0.005 | -0.011 | -0.023 | -0.037 | -0.053 |
| 2.00 | -0.001 | -0.003 | -0.007 | -0.015 | -0.030 | -0.047 | -0.065 |
| 2.20 | -0.001 | -0.005 | -0.010 | -0.020 | -0.040 | -0.062 | -0.083 |
| 2.40 | -0.001 | -0.006 | -0.012 | -0.023 | -0.047 | -0.071 | -0.095 |
| 2.60 | -0.001 | -0.006 | -0.013 | -0.026 | -0.052 | -0.078 | -0.104 |
| 2.80 | -0.001 | -0.007 | -0.014 | -0.028 | -0.055 | -0.082 | -0.110 |
| 3.00 | -0.001 | -0.007 | -0.014 | -0.029 | -0.058 | -0.086 | -0.114 |
| 3.50 | -0.002 | -0.008 | -0.016 | -0.031 | -0.062 | -0.092 | -0.122 |
| 4.00 | -0.002 | -0.008 | -0.016 | -0.032 | -0.064 | -0.096 | -0.127 |

| | | | $P_r$ | | | | |
|---|---|---|---|---|---|---|---|
| 1.000 | 1.200 | 1.500 | 2.000 | 3.000 | 5.000 | 7.000 | 10.000 |
| 11.062 | 11.055 | 11.044 | 11.027 | 10.992 | 10.935 | 10.872 | 10.781 |
| 10.640 | 10.637 | 10.632 | 10.624 | 10.609 | 10.581 | 10.554 | 10.529 |
| 10.121 | 10.121 | 10.121 | 10.122 | 10.123 | 10.128 | 10.135 | 10.150 |
| 9.525 | 9.527 | 9.531 | 9.537 | 9.549 | 9.576 | 9.611 | 9.663 |
| 8.888 | 8.892 | 8.899 | 8.909 | 8.932 | 8.978 | 9.030 | 9.111 |
| 8.238 | 8.243 | 8.252 | 8.267 | 8.298 | 8.360 | 8.425 | 8.531 |
| 7.596 | 7.603 | 7.614 | 7.632 | 7.669 | 7.745 | 7.824 | 7.950 |
| 6.980 | 6.987 | 6.997 | 7.017 | 7.059 | 7.147 | 7.239 | 7.381 |
| 6.388 | 6.395 | 6.407 | 6.429 | 6.475 | 6.574 | 6.677 | 6.837 |
| 5.824 | 5.832 | 5.845 | 5.868 | 5.918 | 6.027 | 6.142 | 6.318 |
| 5.285 | 5.293 | 5.306 | 5.330 | 5.385 | 5.506 | 5.632 | 5.824 |
| 4.763 | 4.771 | 4.784 | 4.810 | 4.872 | 5.008 | 5.149 | 5.358 |
| 4.249 | 4.255 | 4.268 | 4.298 | 4.371 | 4.530 | 4.688 | 4.916 |
| 3.934 | 3.937 | 3.951 | 3.987 | 4.073 | 4.251 | 4.422 | 4.662 |
| 3.712 | 3.713 | 3.730 | 3.773 | 3.873 | 4.068 | 4.248 | 4.497 |
| 3.470 | 3.467 | 3.492 | 3.551 | 3.670 | 3.885 | 4.077 | 4.336 |
| 3.332 | 3.327 | 3.363 | 3.434 | 3.568 | 3.795 | 3.992 | 4.257 |
| 3.164 | 3.164 | 3.223 | 3.313 | 3.464 | 3.705 | 3.909 | 4.178 |
| 2.471 | 2.952 | 3.065 | 3.186 | 3.358 | 3.615 | 3.825 | 4.100 |
| 1.375 | 2.595 | 2.880 | 3.051 | 3.251 | 3.525 | 3.742 | 4.023 |
| 1.180 | 1.723 | 2.650 | 2.906 | 3.142 | 3.435 | 3.661 | 3.947 |
| 0.877 | 0.878 | 1.496 | 2.381 | 2.800 | 3.167 | 3.418 | 3.722 |
| 0.617 | 0.673 | 0.617 | 1.261 | 2.167 | 2.720 | 3.023 | 3.362 |
| 0.459 | 0.503 | 0.487 | 0.604 | 1.497 | 2.275 | 2.641 | 3.019 |
| 0.349 | 0.381 | 0.381 | 0.361 | 0.934 | 1.840 | 2.273 | 2.692 |
| 0.203 | 0.218 | 0.218 | 0.178 | 0.300 | 1.066 | 1.592 | 2.086 |
| 0.111 | 0.115 | 0.108 | 0.070 | 0.044 | 0.504 | 1.012 | 1.547 |
| 0.049 | 0.046 | 0.032 | -0.008 | -0.078 | 0.142 | 0.556 | 1.080 |
| 0.005 | -0.004 | -0.023 | -0.065 | -0.151 | -0.082 | 0.217 | 0.689 |
| -0.027 | -0.040 | -0.063 | -0.109 | -0.202 | -0.223 | -0.028 | 0.369 |
| -0.051 | -0.067 | -0.094 | -0.143 | -0.241 | -0.317 | -0.203 | 0.112 |
| -0.070 | -0.088 | -0.117 | -0.169 | -0.271 | -0.381 | -0.330 | -0.092 |
| -0.085 | -0.105 | -0.136 | -0.190 | -0.295 | -0.428 | -0.424 | -0.255 |
| -0.106 | -0.128 | -0.163 | -0.221 | -0.331 | -0.493 | -0.551 | -0.489 |
| -0.120 | -0.144 | -0.181 | -0.242 | -0.356 | -0.535 | -0.631 | -0.645 |
| -0.130 | -0.156 | -0.194 | -0.257 | -0.376 | -0.567 | -0.687 | -0.754 |
| -0.137 | -0.164 | -0.204 | -0.269 | -0.391 | -0.591 | -0.729 | -0.836 |
| -0.142 | -0.170 | -0.211 | -0.278 | -0.403 | -0.611 | -0.763 | -0.899 |
| -0.152 | -0.181 | -0.224 | -0.294 | -0.425 | -0.650 | -0.827 | -1.015 |
| -0.158 | -0.188 | -0.233 | -0.306 | -0.442 | -0.680 | -0.874 | -1.097 |

## Discussion

In testing the departure functions presented in this chapter, only estimated values of $H^\circ - H$ can be compared with experimental data; and even here, most reliable data are limited to simple hydrocarbons and the permanent light gases. Also, it is convenient to summarize at this point the methods for estimating molar volumes of gas mixtures.

Recommended methods for estimating the molar volumes of pure gases are given in Sec. 3-8 and those for pure liquids in Sec. 3-11. Liquid mixtures are treated in Sec. 4-9. The Lee-Kesler equation also has been used to calculate molar volumes of pure liquids and liquid mixtures of light hydrocarbons, but the methods described in Sec. 4-9 are generally more accurate.

### Molar volumes for gas mixtures

No comprehensive test has been made of all estimation techniques. Authors of individual methods have made limited comparisons, as have the authors of this book. Generally, errors were found to be less than 2 to 3 percent, except near the critical point or for mixtures containing highly polar components in significant concentrations.

Recommendations to estimate the molar volume of hydrocarbon gas mixtures (including those with components associated with natural gas, for example, $CO_2$ and $H_2S$), are essentially the same as for pure gases. For small deviations from ideal-gas behavior, the truncated virial equation is satisfactory. Otherwise, use any of the other equations of state in Chaps. 3 and 4. Values of binary interaction coefficients ($k_{ij}$'s) determined from VLE data can be used in volume calculations; mixture volumes are much less sensitive to $k_{ij}$ values than are VLE calculations.

To estimate the molar volumes of gas mixtures containing nonhydrocarbons, the Lee-Kesler equation may be used, even though it has not been critically evaluated for nonhydrocarbons; in most cases, binary interaction parameters are required for high accuracy. Equations such as those proposed by Gmehling et al. [26] and Vimalchand et al. [85, 86] are more complicated but have demonstrated some ability to describe the behavior of polar molecules.

The equations of state noted above for hydrocarbon and nonhydrocarbon mixtures are complex and normally require a computer for efficient utilization. Cubic equations of state may be slightly less accurate, but they are simpler in form. The truncated virial equation is the easiest of all equations of state to use, although for mixtures its accuracy may be poor unless care is taken to estimate the cross-coefficient $B_{12}$. Also, it should not be used if $\rho/\rho_c > 0.5$.

### Enthalpy departures for pure gases and mixtures—recommendations

For hydrocarbons and hydrocarbon gas mixtures (including light gases such as $N_2$, $CO_2$, and $H_2S$) calculate $H° - H$ from the Soave, Peng-Robinson, or Lee-Kesler equation. Errors should be less than 4 J/g [14, 62, 80, 89]. The truncated virial equation may be used at low to moderate densities.

For gas mixtures containing nonhydrocarbons, the Lee-Kesler correlation [Eqs. (5-4.1) to (5-4.5) or Tables 5-2 and 5-3] is recommended to estimate $H° - H$. Though errors vary, for nonpolar gas mixtures, differences between calculated and experimental values of $H° - H$ should be only a few joules per gram.

The cubic or truncated virial equations of state are simple to employ and may be useful when extensive iterative calculations must be performed. However, the truncated virial equation can be used only at low or moderate densities.

All the correlations noted above may be used up to the saturated vapor envelope except the truncated virial, which is limited in range to about one-half the critical density. As this envelope is approached, and especially in the critical region, errors are expected to increase. At very high pressures, the correlation of Breedveld and Prausnitz [9] can be used.

### Enthalpy departures for pure liquids and liquid mixtures

To estimate the enthalpy departure of pure liquids, it is generally preferable to break the computation into several steps, i.e.,

$$H^L - H° = (H^L - H^{SL}) + (H^{SL} - H^{SV}) + (H^{SV} - H°) \qquad (5\text{-}4.12)$$

where $H^L$ = liquid enthalpy at $T$ and $P$

$H°$ = ideal-gas enthalpy at $T$ and $P°$

$H^{SL}$ = saturated liquid enthalpy at $T$ and $P_{vp}$

$H^{SV}$ = saturated vapor enthalpy at $T$ and $P_{vp}$

The vapor contribution $H^{SV} - H°$ can be estimated by methods described earlier in this chapter. The term $H^{SL} - H^{SV}$ is simply $-\Delta H_v$, and it can be obtained from enthalpy of vaporization correlations given in Chap. 7. Finally, $H^L - H^{SL}$ represents the effect of pressure on liquid enthalpy. It is normally small relative to the other two terms. The Lee-Kesler correlation [Eqs. (5-4.1) to (5-4.5) or Tables 5-2 and 5-3] may be applied to the liquid phase and differences taken between $(H° - H)^{SL}$ and $(H° - H)^{SCL}$, where $SL$ and $SCL$ refer to saturated and subcooled liquid.

In the Lee-Kesler method, the enthalpy departure term is $H° - H$, that is, the difference between the enthalpy of the fluid in an ideal-gas state at $P°$ and $T$ and that of the fluid at $P$, $T$ (liquid or gas). It is not generally recommended that enthalpies of liquids be calculated directly from this difference. The corresponding states methods are usually not sufficiently accurate to estimate $\Delta H$ for a phase change. It is preferable to determine phase change $\Delta H$ values separately, as indicated in Eq. (5-4.12), and to use other methods for $H^L - H^{SL}$ and $H^{SV} - H°$. (See, however, p. 220.)

When enthalpy departures are desired for liquid mixtures, no completely satisfactory recommendations can be formulated. One method employs a modified form of Eq. (5-4.12). $H^{SV} - H°$ is calculated as described earlier for gas mixtures, and $H^{SL} - H^{SV}$ is set equal to the mole fraction average of the pure component values of $-\Delta H_v$,

$$H^{SL} - H^{SV} = -\sum_j x_j \, \Delta H_{vj} \qquad (5\text{-}4.13)$$

Finally, $H^L - H^{SL}$ can be neglected at low pressure, or it can be estimated from Tables 5-2 and 5-3 with pseudocritical constants determined from Eqs. (4-3.1) to (4-3.3), (4-3.4), or (4-6.1) to (4-6.6). This approach is useful only if all components are subcritical. Further, it neglects any heat of mixing in the liquid phase, an assumption which is often warranted unless the liquid phase contains polar components [75].

Good results have been reported when the Soave modification of the Redlich-Kwong equation of state has been used to calculate liquid mixture enthalpy departures for hydrocarbon mixtures not containing hydrogen [89]. This relation is preferred to earlier modified Redlich-Kwong equations [38, 91]. A theoretical equation-of-state method applicable for cryogenic mixtures also is available [58].

If Tables 5-2 and 5-3 are used directly to calculate liquid mixture enthalpy departures, good results are often reported [21, 36, 80, 91].

Mixture enthalpies of vaporization can also be obtained from phase equilibrium measurements [46, 79], but very accurate data are needed to yield reasonable enthalpy values.

To estimate enthalpies (and densities) of hydrocarbon mixtures containing hydrogen, the generalized method of Chueh and Deal [11] is recommended. Huang and Daubert [34] illustrate how one can predict enthalpies of liquid (and vapor) mixtures of petroleum fractions, and Ghormley and Lenoir treat saturated liquid and saturated vapor enthalpies for aliphatic hydrocarbon mixtures [22].

### Departure entropies and fugacity coefficients

To estimate departure functions for entropy and fugacity coefficients of pure gases or mixtures, when possible, follow the recommendations made earlier for enthalpies. Table 5-1 may be used for $S - S°$ if an analytical equation of state is selected and $\ln(f/P)$ is determined from Eq. (5-3.11). If the Lee-Kesler method is chosen, use Eqs. (5-4.6) to (5-4.11) or Tables 5-4 to 5-7.

Finally, it should be pointed out that enthalpy departures, entropy departures, and fugacity coefficients are not independent and are related by

$$\frac{H° - H}{RT} = \frac{S° - S}{R} - \ln \frac{f}{P°} \qquad (5\text{-}4.14)$$

where $P°$ is the pressure in the ideal-gas reference state (see Sec. 5-3).

**Example 5-2** Estimate the enthalpy and entropy departures for propylene at 398.15 K and 100 bar. Bier et al. [6] report experimental values as

$$H° - H = 244.58 \text{ J/g} \qquad S° - S = 1.4172 \text{ J/(g·K)}$$

(The ideal-gas reference pressure for entropy $P°$ is 1 bar.)

**solution** From Appendix A, $M = 42.081$, $T_c = 364.9$ K, $P_c = 46.0$ bar, $Z_c = 0.274$, and $\omega = 0.144$. Thus, $T_r = (398.15)/364.9 = 1.09$ and $P_r = 100/46.0 = 2.17$.

LEE-KESLER METHOD Equations (5-4.4) and (5-4.8), along with Tables 5-2 to 5-5, give

$$\left(\frac{H° - H}{RT_c}\right)^{(0)} = 3.11 \qquad \left(\frac{H° - H}{RT_c}\right)^{(1)} = 1.62$$

$$\left(\frac{S° - S}{R}\right)^{(0)} = 2.18 \qquad \left(\frac{S° - S}{R}\right)^{(1)} = 1.57$$

$$H° - H = \frac{RT_c}{M}\left[\left(\frac{H° - H}{RT_c}\right)^{(0)} + \left(\frac{H° - H}{RT_c}\right)^{(1)}\right]$$

$$= \frac{(8.314)(364.9)}{42.081}[3.11 + (0.144)(1.62)]$$

$$= 241 \text{ J/g}$$

$$S° - S = \frac{R}{M}\left[\left(\frac{S° - S}{R}\right)^{(0)} + \left(\frac{S° - S}{R}\right)^{(1)} - \ln \frac{P°}{P}\right]$$

$$= \frac{8.314}{42.081}\left[2.18 + (0.144)(1.57) - \ln \frac{1}{100}\right]$$

$$= 1.39 \text{ J/(g·K)}$$

The method underpredicts $H° - H$ by 3 J/g and $S° - S$ by 0.03 J/(g·K).

SOAVE EQUATION (3-6.1). From Eq. (5-3.8) and Table 5-1

$$S° - S = -R \ln \frac{Z - B^*}{Z} - R \ln \frac{V}{V°} + \frac{1}{b}\frac{\partial a}{\partial T} \ln \frac{Z}{Z + B^*}$$

**TABLE 5-4 Lee-Kesler Residual Entropy [47]**

Simple fluid $\left(\frac{S^\circ - S}{R}\right)^{(0)}$

| $T_r$ | $P_r$ 0.010 | 0.050 | 0.100 | 0.200 | 0.400 | 0.600 | 0.800 |
|---|---|---|---|---|---|---|---|
| 0.30 | 11.614 | 10.008 | 9.319 | 8.635 | 7.961 | 7.574 | 7.304 |
| 0.35 | 11.185 | 9.579 | 8.890 | 8.205 | 7.529 | 7.140 | 6.869 |
| 0.40 | 10.802 | 9.196 | 8.506 | 7.821 | 7.144 | 6.755 | 6.483 |
| 0.45 | 10.453 | 8.847 | 8.157 | 7.472 | 6.794 | 6.404 | 6.132 |
| 0.50 | 10.137 | 8.531 | 7.841 | 7.156 | 6.479 | 6.089 | 5.816 |
| 0.55 | 0.038 | 8.245 | 7.555 | 6.870 | 6.193 | 5.803 | 5.531 |
| 0.60 | 0.029 | 7.983 | 7.294 | 6.610 | 5.933 | 5.544 | 5.273 |
| 0.65 | 0.023 | 0.122 | 7.052 | 6.368 | 5.694 | 5.306 | 5.036 |
| 0.70 | 0.018 | 0.096 | 0.206 | 6.140 | 5.467 | 5.082 | 4.814 |
| 0.75 | 0.015 | 0.078 | 0.164 | 5.917 | 5.248 | 4.866 | 4.600 |
| 0.80 | 0.013 | 0.064 | 0.134 | 0.294 | 5.026 | 4.649 | 4.388 |
| 0.85 | 0.011 | 0.054 | 0.111 | 0.239 | 4.785 | 4.418 | 4.166 |
| 0.90 | 0.009 | 0.046 | 0.094 | 0.199 | 0.463 | 4.145 | 3.912 |
| 0.93 | 0.008 | 0.042 | 0.085 | 0.179 | 0.408 | 0.750 | 3.723 |
| 0.95 | 0.008 | 0.039 | 0.080 | 0.168 | 0.377 | 0.671 | 3.556 |
| 0.97 | 0.007 | 0.037 | 0.075 | 0.157 | 0.350 | 0.607 | 1.056 |
| 0.98 | 0.007 | 0.036 | 0.073 | 0.153 | 0.337 | 0.580 | 0.971 |
| 0.99 | 0.007 | 0.035 | 0.071 | 0.148 | 0.326 | 0.555 | 0.903 |
| 1.00 | 0.007 | 0.034 | 0.069 | 0.144 | 0.315 | 0.532 | 0.847 |
| 1.01 | 0.007 | 0.033 | 0.067 | 0.139 | 0.304 | 0.510 | 0.799 |
| 1.02 | 0.006 | 0.032 | 0.065 | 0.135 | 0.294 | 0.491 | 0.757 |
| 1.05 | 0.006 | 0.030 | 0.060 | 0.124 | 0.267 | 0.439 | 0.656 |
| 1.10 | 0.005 | 0.026 | 0.053 | 0.108 | 0.230 | 0.371 | 0.537 |
| 1.15 | 0.005 | 0.023 | 0.047 | 0.096 | 0.201 | 0.319 | 0.452 |
| 1.20 | 0.004 | 0.021 | 0.042 | 0.085 | 0.177 | 0.277 | 0.389 |
| 1.30 | 0.003 | 0.017 | 0.033 | 0.068 | 0.140 | 0.217 | 0.298 |
| 1.40 | 0.003 | 0.014 | 0.027 | 0.056 | 0.114 | 0.174 | 0.237 |
| 1.50 | 0.002 | 0.011 | 0.023 | 0.046 | 0.094 | 0.143 | 0.194 |
| 1.60 | 0.002 | 0.010 | 0.019 | 0.039 | 0.079 | 0.120 | 0.162 |
| 1.70 | 0.002 | 0.008 | 0.017 | 0.033 | 0.067 | 0.102 | 0.137 |
| 1.80 | 0.001 | 0.007 | 0.014 | 0.029 | 0.058 | 0.088 | 0.117 |
| 1.90 | 0.001 | 0.006 | 0.013 | 0.025 | 0.051 | 0.076 | 0.102 |
| 2.00 | 0.001 | 0.006 | 0.011 | 0.022 | 0.044 | 0.067 | 0.089 |
| 2.20 | 0.001 | 0.004 | 0.009 | 0.018 | 0.035 | 0.053 | 0.070 |
| 2.40 | 0.001 | 0.004 | 0.007 | 0.014 | 0.028 | 0.042 | 0.056 |
| 2.60 | 0.001 | 0.003 | 0.006 | 0.012 | 0.023 | 0.035 | 0.046 |
| 2.80 | 0.000 | 0.002 | 0.005 | 0.010 | 0.020 | 0.029 | 0.039 |
| 3.00 | 0.000 | 0.002 | 0.004 | 0.008 | 0.017 | 0.025 | 0.033 |
| 3.50 | 0.000 | 0.001 | 0.003 | 0.006 | 0.012 | 0.017 | 0.023 |
| 4.00 | 0.000 | 0.001 | 0.002 | 0.004 | 0.009 | 0.013 | 0.017 |

| $P_r$ | | | | | | | |
|---|---|---|---|---|---|---|---|
| 1.000 | 1.200 | 1.500 | 2.000 | 3.000 | 5.000 | 7.000 | 10.000 |
| 7.099 | 6.935 | 6.740 | 6.497 | 6.182 | 5.847 | 5.683 | 5.578 |
| 6.663 | 6.497 | 6.299 | 6.052 | 5.728 | 5.376 | 5.194 | 5.060 |
| 6.275 | 6.109 | 5.909 | 5.660 | 5.330 | 4.967 | 4.772 | 4.619 |
| 5.924 | 5.757 | 5.557 | 5.306 | 4.974 | 4.603 | 4.401 | 4.234 |
| 5.608 | 5.441 | 5.240 | 4.989 | 4.656 | 4.282 | 4.074 | 3.899 |
| 5.324 | 5.157 | 4.956 | 4.706 | 4.373 | 3.998 | 3.788 | 3.607 |
| 5.066 | 4.900 | 4.700 | 4.451 | 4.120 | 3.747 | 3.537 | 3.353 |
| 4.830 | 4.665 | 4.467 | 4.220 | 3.892 | 3.523 | 3.315 | 3.131 |
| 4.610 | 4.446 | 4.250 | 4.007 | 3.684 | 3.322 | 3.117 | 2.935 |
| 4.399 | 4.238 | 4.045 | 3.807 | 3.491 | 3.138 | 2.939 | 2.761 |
| 4.191 | 4.034 | 3.846 | 3.615 | 3.310 | 2.970 | 2.777 | 2.605 |
| 3.976 | 3.825 | 3.646 | 3.425 | 3.135 | 2.812 | 2.629 | 2.463 |
| 3.738 | 3.599 | 3.434 | 3.231 | 2.964 | 2.663 | 2.491 | 2.334 |
| 3.569 | 3.444 | 3.295 | 3.108 | 2.860 | 2.577 | 2.412 | 2.262 |
| 3.433 | 3.326 | 3.193 | 3.023 | 2.790 | 2.520 | 2.362 | 2.215 |
| 3.259 | 3.188 | 3.081 | 2.932 | 2.719 | 2.463 | 2.312 | 2.170 |
| 3.142 | 3.106 | 3.019 | 2.884 | 2.682 | 2.436 | 2.287 | 2.148 |
| 2.972 | 3.010 | 2.953 | 2.835 | 2.646 | 2.408 | 2.263 | 2.126 |
| 2.178 | 2.893 | 2.879 | 2.784 | 2.609 | 2.380 | 2.239 | 2.105 |
| 1.391 | 2.736 | 2.798 | 2.730 | 2.571 | 2.352 | 2.215 | 2.083 |
| 1.225 | 2.495 | 2.706 | 2.673 | 2.533 | 2.325 | 2.191 | 2.062 |
| 0.965 | 1.523 | 2.328 | 2.483 | 2.415 | 2.242 | 2.121 | 2.001 |
| 0.742 | 1.012 | 1.557 | 2.081 | 2.202 | 2.104 | 2.007 | 1.903 |
| 0.607 | 0.790 | 1.126 | 1.649 | 1.968 | 1.966 | 1.897 | 1.810 |
| 0.512 | 0.651 | 0.890 | 1.308 | 1.727 | 1.827 | 1.789 | 1.722 |
| 0.385 | 0.478 | 0.628 | 0.891 | 1.299 | 1.554 | 1.581 | 1.556 |
| 0.303 | 0.372 | 0.478 | 0.663 | 0.990 | 1.303 | 1.386 | 1.402 |
| 0.246 | 0.299 | 0.381 | 0.520 | 0.777 | 1.088 | 1.208 | 1.260 |
| 0.204 | 0.247 | 0.312 | 0.421 | 0.628 | 0.913 | 1.050 | 1.130 |
| 0.172 | 0.208 | 0.261 | 0.350 | 0.519 | 0.773 | 0.915 | 1.013 |
| 0.147 | 0.177 | 0.222 | 0.296 | 0.438 | 0.661 | 0.79[illegible] | 0.908 |
| 0.127 | 0.153 | 0.191 | 0.255 | 0.375 | 0.570 | 0.70[illegible] | 0.815 |
| 0.111 | 0.134 | 0.167 | 0.221 | 0.325 | 0.497 | 0.62[illegible] | 0.733 |
| 0.087 | 0.105 | 0.130 | 0.172 | 0.251 | 0.388 | 0.492 | 0.599 |
| 0.070 | 0.084 | 0.104 | 0.138 | 0.201 | 0.311 | 0.399 | 0.496 |
| 0.058 | 0.069 | 0.086 | 0.113 | 0.164 | 0.255 | 0.329 | 0.416 |
| 0.048 | 0.058 | 0.072 | 0.094 | 0.137 | 0.213 | 0.277 | 0.353 |
| 0.041 | 0.049 | 0.061 | 0.080 | 0.116 | 0.181 | 0.236 | 0.303 |
| 0.029 | 0.034 | 0.042 | 0.056 | 0.081 | 0.126 | 0.166 | 0.216 |
| 0.021 | 0.025 | 0.031 | 0.041 | 0.059 | 0.093 | 0.123 | 0.162 |

**TABLE 5-5 Lee-Kesler Residual Entropy [47]**

Deviation function $\left(\frac{S^{\circ} - S}{R}\right)^{(1)}$

| $T_r$ | $P_r$ 0.010 | 0.050 | 0.100 | 0.200 | 0.400 | 0.600 | 0.800 |
|---|---|---|---|---|---|---|---|
| 0.30 | 16.782 | 16.774 | 16.764 | 16.744 | 16.705 | 16.665 | 16.626 |
| 0.35 | 15.413 | 15.408 | 15.401 | 15.387 | 15.359 | 15.333 | 15.305 |
| 0.40 | 13.990 | 13.986 | 13.981 | 13.972 | 13.953 | 13.934 | 13.915 |
| 0.45 | 12.564 | 12.561 | 12.558 | 12.551 | 12.537 | 12.523 | 12.509 |
| 0.50 | 11.202 | 11.200 | 11.197 | 11.192 | 11.182 | 11.172 | 11.162 |
| 0.55 | 0.115 | 9.948 | 9.946 | 9.942 | 9.935 | 9.928 | 9.921 |
| 0.60 | 0.078 | 8.828 | 8.826 | 8.823 | 8.817 | 8.811 | 8.806 |
| 0.65 | 0.055 | 0.309 | 7.832 | 7.829 | 7.824 | 7.819 | 7.815 |
| 0.70 | 0.040 | 0.216 | 0.491 | 6.951 | 6.945 | 6.941 | 6.937 |
| 0.75 | 0.029 | 0.156 | 0.340 | 6.173 | 6.167 | 6.162 | 6.158 |
| 0.80 | 0.022 | 0.116 | 0.246 | 0.578 | 5.475 | 5.468 | 5.462 |
| 0.85 | 0.017 | 0.088 | 0.183 | 0.408 | 4.853 | 4.841 | 4.832 |
| 0.90 | 0.013 | 0.068 | 0.140 | 0.301 | 0.744 | 4.269 | 4.249 |
| 0.93 | 0.011 | 0.058 | 0.120 | 0.254 | 0.593 | 1.219 | 3.914 |
| 0.95 | 0.010 | 0.053 | 0.109 | 0.228 | 0.517 | 0.961 | 3.697 |
| 0.97 | 0.010 | 0.048 | 0.099 | 0.206 | 0.456 | 0.797 | 1.570 |
| 0.98 | 0.009 | 0.046 | 0.094 | 0.196 | 0.429 | 0.734 | 1.270 |
| 0.99 | 0.009 | 0.044 | 0.090 | 0.186 | 0.405 | 0.680 | 1.098 |
| 1.00 | 0.008 | 0.042 | 0.086 | 0.177 | 0.382 | 0.632 | 0.977 |
| 1.01 | 0.008 | 0.040 | 0.082 | 0.169 | 0.361 | 0.590 | 0.883 |
| 1.02 | 0.008 | 0.039 | 0.078 | 0.161 | 0.342 | 0.552 | 0.807 |
| 1.05 | 0.007 | 0.034 | 0.069 | 0.140 | 0.292 | 0.460 | 0.642 |
| 1.10 | 0.005 | 0.028 | 0.055 | 0.112 | 0.229 | 0.350 | 0.470 |
| 1.15 | 0.005 | 0.023 | 0.045 | 0.091 | 0.183 | 0.275 | 0.361 |
| 1.20 | 0.004 | 0.019 | 0.037 | 0.075 | 0.149 | 0.220 | 0.286 |
| 1.30 | 0.003 | 0.013 | 0.026 | 0.052 | 0.102 | 0.148 | 0.190 |
| 1.40 | 0.002 | 0.010 | 0.019 | 0.037 | 0.072 | 0.104 | 0.133 |
| 1.50 | 0.001 | 0.007 | 0.014 | 0.027 | 0.053 | 0.076 | 0.097 |
| 1.60 | 0.001 | 0.005 | 0.011 | 0.021 | 0.040 | 0.057 | 0.073 |
| 1.70 | 0.001 | 0.004 | 0.008 | 0.016 | 0.031 | 0.044 | 0.056 |
| 1.80 | 0.001 | 0.003 | 0.006 | 0.013 | 0.024 | 0.035 | 0.044 |
| 1.90 | 0.001 | 0.003 | 0.005 | 0.010 | 0.019 | 0.028 | 0.036 |
| 2.00 | 0.000 | 0.002 | 0.004 | 0.008 | 0.016 | 0.023 | 0.029 |
| 2.20 | 0.000 | 0.001 | 0.003 | 0.006 | 0.011 | 0.016 | 0.021 |
| 2.40 | 0.000 | 0.001 | 0.002 | 0.004 | 0.008 | 0.012 | 0.015 |
| 2.60 | 0.000 | 0.001 | 0.002 | 0.003 | 0.006 | 0.009 | 0.012 |
| 2.80 | 0.000 | 0.001 | 0.001 | 0.003 | 0.005 | 0.008 | 0.010 |
| 3.00 | 0.000 | 0.001 | 0.001 | 0.002 | 0.004 | 0.006 | 0.008 |
| 3.50 | 0.000 | 0.000 | 0.001 | 0.001 | 0.003 | 0.004 | 0.006 |
| 4.00 | 0.000 | 0.000 | 0.001 | 0.001 | 0.002 | 0.003 | 0.005 |

| $P_r$ | | | | | | | |
|---|---|---|---|---|---|---|---|
| 1.000 | 1.200 | 1.500 | 2.000 | 3.000 | 5.000 | 7.000 | 10.000 |
| 16.586 | 16.547 | 16.488 | 16.390 | 16.195 | 15.837 | 15.468 | 14.925 |
| 15.278 | 15.251 | 15.211 | 15.144 | 15.011 | 14.751 | 14.496 | 14.153 |
| 13.896 | 13.877 | 13.849 | 13.803 | 13.714 | 13.541 | 13.376 | 13.144 |
| 12.496 | 12.482 | 12.462 | 12.430 | 12.367 | 12.248 | 12.145 | 11.999 |
| 11.153 | 11.143 | 11.129 | 11.107 | 11.063 | 10.985 | 10.920 | 10.836 |
| 9.914 | 9.907 | 9.897 | 9.882 | 9.853 | 9.806 | 9.769 | 9.732 |
| 8.799 | 8.794 | 8.787 | 8.777 | 8.760 | 8.736 | 8.723 | 8.720 |
| 7.810 | 7.807 | 7.801 | 7.794 | 7.784 | 7.779 | 7.785 | 7.811 |
| 6.933 | 6.930 | 6.926 | 6.922 | 6.919 | 6.929 | 6.952 | 7.002 |
| 6.155 | 6.152 | 6.149 | 6.147 | 6.149 | 6.174 | 6.213 | 6.285 |
| 5.458 | 5.455 | 5.453 | 5.452 | 5.461 | 5.501 | 5.555 | 5.648 |
| 4.826 | 4.822 | 4.820 | 4.822 | 4.839 | 4.898 | 4.969 | 5.082 |
| 4.238 | 4.232 | 4.230 | 4.236 | 4.267 | 4.351 | 4.442 | 4.578 |
| 3.894 | 3.885 | 3.884 | 3.896 | 3.941 | 4.046 | 4.151 | 4.300 |
| 3.658 | 3.647 | 3.648 | 3.669 | 3.728 | 3.851 | 3.966 | 4.125 |
| 3.406 | 3.391 | 3.401 | 3.437 | 3.517 | 3.661 | 3.788 | 3.957 |
| 3.264 | 3.247 | 3.268 | 3.318 | 3.412 | 3.569 | 3.701 | 3.875 |
| 3.093 | 3.082 | 3.126 | 3.195 | 3.306 | 3.477 | 3.616 | 3.796 |
| 2.399 | 2.868 | 2.967 | 3.067 | 3.200 | 3.387 | 3.532 | 3.717 |
| 1.306 | 2.513 | 2.784 | 2.933 | 3.094 | 3.297 | 3.450 | 3.640 |
| 1.113 | 1.655 | 2.557 | 2.790 | 2.986 | 3.209 | 3.369 | 3.565 |
| 0.820 | 0.831 | 1.443 | 2.283 | 2.655 | 2.949 | 3.134 | 3.348 |
| 0.577 | 0.640 | 0.618 | 1.241 | 2.067 | 2.534 | 2.767 | 3.013 |
| 0.437 | 0.489 | 0.502 | 0.654 | 1.471 | 2.138 | 2.428 | 2.708 |
| 0.343 | 0.385 | 0.412 | 0.447 | 0.991 | 1.767 | 2.115 | 2.430 |
| 0.226 | 0.254 | 0.282 | 0.300 | 0.481 | 1.147 | 1.569 | 1.944 |
| 0.158 | 0.178 | 0.200 | 0.220 | 0.290 | 0.730 | 1.138 | 1.544 |
| 0.115 | 0.130 | 0.147 | 0.166 | 0.206 | 0.479 | 0.823 | 1.222 |
| 0.086 | 0.098 | 0.112 | 0.129 | 0.159 | 0.334 | 0.604 | 0.969 |
| 0.067 | 0.076 | 0.087 | 0.102 | 0.127 | 0.248 | 0.456 | 0.775 |
| 0.053 | 0.060 | 0.070 | 0.083 | 0.105 | 0.195 | 0.355 | 0.628 |
| 0.043 | 0.049 | 0.057 | 0.069 | 0.089 | 0.160 | 0.286 | 0.518 |
| 0.035 | 0.040 | 0.048 | 0.058 | 0.077 | 0.136 | 0.238 | 0.434 |
| 0.025 | 0.029 | 0.035 | 0.043 | 0.060 | 0.105 | 0.178 | 0.322 |
| 0.019 | 0.022 | 0.027 | 0.034 | 0.048 | 0.086 | 0.143 | 0.254 |
| 0.015 | 0.018 | 0.021 | 0.028 | 0.041 | 0.074 | 0.120 | 0.210 |
| 0.012 | 0.014 | 0.018 | 0.023 | 0.035 | 0.065 | 0.104 | 0.180 |
| 0.010 | 0.012 | 0.015 | 0.020 | 0.031 | 0.058 | 0.093 | 0.158 |
| 0.007 | 0.009 | 0.011 | 0.015 | 0.024 | 0.046 | 0.073 | 0.122 |
| 0.006 | 0.007 | 0.009 | 0.012 | 0.020 | 0.038 | 0.060 | 0.100 |

**TABLE 5-6 Lee-Kesler Fugacity-Pressure Ratio [47]**

Simple fluid $\left(\log \frac{f}{P}\right)^{(0)}$

| $T_r$ | $P_r$ 0.010 | 0.050 | 0.100 | 0.200 | 0.400 | 0.600 | 0.800 |
|---|---|---|---|---|---|---|---|
| 0.30 | -3.708 | -4.402 | -4.696 | -4.985 | -5.261 | -5.412 | -5.512 |
| 0.35 | -2.471 | -3.166 | -3.461 | -3.751 | -4.029 | -4.183 | -4.285 |
| 0.40 | -1.566 | -2.261 | -2.557 | -2.848 | -3.128 | -3.283 | -3.387 |
| 0.45 | -0.879 | -1.575 | -1.871 | -2.162 | -2.444 | -2.601 | -2.707 |
| 0.50 | -0.344 | -1.040 | -1.336 | -1.628 | -1.912 | -2.070 | -2.177 |
| 0.55 | -0.008 | -0.614 | -0.911 | -1.204 | -1.488 | -1.647 | -1.755 |
| 0.60 | -0.007 | -0.269 | -0.566 | -0.859 | -1.144 | -1.304 | -1.413 |
| 0.65 | -0.005 | -0.026 | -0.283 | -0.576 | -0.862 | -1.023 | -1.132 |
| 0.70 | -0.004 | -0.021 | -0.043 | -0.341 | -0.627 | -0.789 | -0.899 |
| 0.75 | -0.003 | -0.017 | -0.035 | -0.144 | -0.430 | -0.592 | -0.703 |
| 0.80 | -0.003 | -0.014 | -0.029 | -0.059 | -0.264 | -0.426 | -0.537 |
| 0.85 | -0.002 | -0.012 | -0.024 | -0.049 | -0.123 | -0.285 | -0.396 |
| 0.90 | -0.002 | -0.010 | -0.020 | -0.041 | -0.086 | -0.166 | -0.276 |
| 0.93 | -0.002 | -0.009 | -0.018 | -0.037 | -0.077 | -0.122 | -0.214 |
| 0.95 | -0.002 | -0.008 | -0.017 | -0.035 | -0.072 | -0.113 | -0.176 |
| 0.97 | -0.002 | -0.008 | -0.016 | -0.033 | -0.067 | -0.105 | -0.148 |
| 0.98 | -0.002 | -0.008 | -0.016 | -0.032 | -0.065 | -0.101 | -0.142 |
| 0.99 | -0.001 | -0.007 | -0.015 | -0.031 | -0.063 | -0.098 | -0.137 |
| 1.00 | -0.001 | -0.007 | -0.015 | -0.030 | -0.061 | -0.095 | -0.132 |
| 1.01 | -0.001 | -0.007 | -0.014 | -0.029 | -0.059 | -0.091 | -0.127 |
| 1.02 | -0.001 | -0.007 | -0.014 | -0.028 | -0.057 | -0.088 | -0.122 |
| 1.05 | -0.001 | -0.006 | -0.013 | -0.025 | -0.052 | -0.080 | -0.110 |
| 1.10 | -0.001 | -0.005 | -0.011 | -0.022 | -0.045 | -0.069 | -0.093 |
| 1.15 | -0.001 | -0.005 | -0.009 | -0.019 | -0.039 | -0.059 | -0.080 |
| 1.20 | -0.001 | -0.004 | -0.008 | -0.017 | -0.034 | -0.051 | -0.069 |
| 1.30 | -0.001 | -0.003 | -0.006 | -0.013 | -0.026 | -0.039 | -0.052 |
| 1.40 | -0.001 | -0.003 | -0.005 | -0.010 | -0.020 | -0.030 | -0.040 |
| 1.50 | -0.000 | -0.002 | -0.004 | -0.008 | -0.016 | -0.024 | -0.032 |
| 1.60 | -0.000 | -0.002 | -0.003 | -0.006 | -0.012 | -0.019 | -0.025 |
| 1.70 | -0.000 | -0.001 | -0.002 | -0.005 | -0.010 | -0.015 | -0.020 |
| 1.80 | -0.000 | -0.001 | -0.002 | -0.004 | -0.008 | -0.012 | -0.015 |
| 1.90 | -0.000 | -0.001 | -0.002 | -0.003 | -0.006 | -0.009 | -0.012 |
| 2.00 | -0.000 | -0.001 | -0.001 | -0.002 | -0.005 | -0.007 | -0.009 |
| 2.20 | -0.000 | -0.000 | -0.001 | -0.001 | -0.003 | -0.004 | -0.005 |
| 2.40 | -0.000 | -0.000 | -0.000 | -0.001 | -0.001 | -0.002 | -0.003 |
| 2.60 | -0.000 | -0.000 | -0.000 | -0.000 | -0.000 | -0.001 | -0.001 |
| 2.80 | 0.000 | 0.000 | 0.000 | 0.000 | 0.000 | 0.000 | 0.001 |
| 3.00 | 0.000 | 0.000 | 0.000 | 0.000 | 0.001 | 0.001 | 0.002 |
| 3.50 | 0.000 | 0.000 | 0.000 | 0.001 | 0.001 | 0.002 | 0.003 |
| 4.00 | 0.000 | 0.000 | 0.000 | 0.001 | 0.002 | 0.003 | 0.004 |

| $P_r$ | | | | | | | |
|---|---|---|---|---|---|---|---|
| 1.000 | 1.200 | 1.500 | 2.000 | 3.000 | 5.000 | 7.000 | 10.000 |
| -5.584 | -5.638 | -5.697 | -5.759 | -5.810 | -5.782 | -5.679 | -5.461 |
| -4.359 | -4.416 | -4.479 | -4.547 | -4.611 | -4.608 | -4.530 | -4.352 |
| -3.463 | -3.522 | -3.588 | -3.661 | -3.735 | -3.752 | -3.694 | -3.545 |
| -2.785 | -2.845 | -2.913 | -2.990 | -3.071 | -3.104 | -3.063 | -2.938 |
| -2.256 | -2.317 | -2.387 | -2.468 | -2.555 | -2.601 | -2.572 | -2.468 |
| -1.835 | -1.897 | -1.969 | -2.052 | -2.145 | -2.201 | -2.183 | -2.096 |
| -1.494 | -1.557 | -1.630 | -1.715 | -1.812 | -1.878 | -1.869 | -1.795 |
| -1.214 | -1.278 | -1.352 | -1.439 | -1.539 | -1.612 | -1.611 | -1.549 |
| -0.981 | -1.045 | -1.120 | -1.208 | -1.312 | -1.391 | -1.396 | -1.344 |
| -0.785 | -0.850 | -0.925 | -1.015 | -1.121 | -1.204 | -1.215 | -1.172 |
| -0.619 | -0.685 | -0.760 | -0.851 | -0.958 | -1.046 | -1.062 | -1.026 |
| -0.479 | -0.544 | -0.620 | -0.711 | -0.819 | -0.911 | -0.930 | -0.901 |
| -0.359 | -0.424 | -0.500 | -0.591 | -0.700 | -0.794 | -0.817 | -0.793 |
| -0.296 | -0.361 | -0.437 | -0.527 | -0.637 | -0.732 | -0.756 | -0.735 |
| -0.258 | -0.322 | -0.398 | -0.488 | -0.598 | -0.693 | -0.719 | -0.699 |
| -0.223 | -0.287 | -0.362 | -0.452 | -0.561 | -0.657 | -0.683 | -0.665 |
| -0.206 | -0.270 | -0.344 | -0.434 | -0.543 | -0.639 | -0.666 | -0.649 |
| -0.191 | -0.254 | -0.328 | -0.417 | -0.526 | -0.622 | -0.649 | -0.633 |
| -0.176 | -0.238 | -0.312 | -0.401 | -0.509 | -0.605 | -0.633 | -0.617 |
| -0.168 | -0.224 | -0.297 | -0.385 | -0.493 | -0.589 | -0.617 | -0.602 |
| -0.161 | -0.210 | -0.282 | -0.370 | -0.477 | -0.573 | -0.601 | -0.588 |
| -0.143 | -0.180 | -0.242 | -0.327 | -0.433 | -0.529 | -0.557 | -0.546 |
| -0.120 | -0.148 | -0.193 | -0.267 | -0.368 | -0.462 | -0.491 | -0.482 |
| -0.102 | -0.125 | -0.160 | -0.220 | -0.312 | -0.403 | -0.433 | -0.426 |
| -0.088 | -0.106 | -0.135 | -0.184 | -0.266 | -0.352 | -0.382 | -0.377 |
| -0.066 | -0.080 | -0.100 | -0.134 | -0.195 | -0.269 | -0.296 | -0.293 |
| -0.051 | -0.061 | -0.076 | -0.101 | -0.146 | -0.205 | -0.229 | -0.226 |
| -0.039 | -0.047 | -0.059 | -0.077 | -0.111 | -0.157 | -0.176 | -0.173 |
| -0.031 | -0.037 | -0.046 | -0.060 | -0.085 | -0.120 | -0.135 | -0.129 |
| -0.024 | -0.029 | -0.036 | -0.046 | -0.065 | -0.092 | -0.102 | -0.094 |
| -0.019 | -0.023 | -0.028 | -0.036 | -0.050 | -0.069 | -0.075 | -0.066 |
| -0.015 | -0.018 | -0.022 | -0.028 | -0.038 | -0.052 | -0.054 | -0.043 |
| -0.012 | -0.014 | -0.017 | -0.021 | -0.029 | -0.037 | -0.037 | -0.024 |
| -0.007 | -0.008 | -0.009 | -0.012 | -0.015 | -0.017 | -0.012 | 0.004 |
| -0.003 | -0.004 | -0.004 | -0.005 | -0.006 | -0.003 | 0.005 | 0.024 |
| -0.001 | -0.001 | -0.001 | -0.001 | 0.001 | 0.007 | 0.017 | 0.037 |
| 0.001 | 0.001 | 0.002 | 0.003 | 0.005 | 0.014 | 0.025 | 0.046 |
| 0.002 | 0.003 | 0.003 | 0.005 | 0.009 | 0.018 | 0.031 | 0.053 |
| 0.004 | 0.005 | 0.006 | 0.008 | 0.013 | 0.025 | 0.038 | 0.061 |
| 0.005 | 0.006 | 0.007 | 0.010 | 0.016 | 0.028 | 0.041 | 0.064 |

**TABLE 5-7 Lee-Kesler Fugacity-Pressure Ratio [47]**

Deviation function $\left(\log \dfrac{f}{P}\right)^{(1)}$

| | $P_r$ | | | | | | |
|---|---|---|---|---|---|---|---|
| $T_r$ | 0.010 | 0.050 | 0.100 | 0.200 | 0.400 | 0.600 | 0.800 |
| 0.30 | -8.778 | -8.779 | -8.781 | -8.785 | -8.790 | -8.797 | -8.804 |
| 0.35 | -6.528 | -6.530 | -6.532 | -6.536 | -6.544 | -6.551 | -6.559 |
| 0.40 | -4.912 | -4.914 | -4.916 | -4.919 | -4.929 | -4.937 | -4.945 |
| 0.45 | -3.726 | -3.728 | -3.730 | -3.734 | -3.742 | -3.750 | -3.758 |
| 0.50 | -2.838 | -2.839 | -2.841 | -2.845 | -2.853 | -2.861 | -2.869 |
| 0.55 | -0.013 | -2.163 | -2.165 | -2.169 | -2.177 | -2.184 | -2.192 |
| 0.60 | -0.009 | -1.644 | -1.646 | -1.650 | -1.657 | -1.664 | -1.671 |
| 0.65 | -0.006 | -0.031 | -1.242 | -1.245 | -1.252 | -1.258 | -1.265 |
| 0.70 | -0.004 | -0.021 | -0.044 | -0.927 | -0.934 | -0.940 | -0.946 |
| 0.75 | -0.003 | -0.014 | -0.030 | -0.675 | -0.682 | -0.688 | -0.694 |
| 0.80 | -0.002 | -0.010 | -0.020 | -0.043 | -0.481 | -0.487 | -0.493 |
| 0.85 | -0.001 | -0.006 | -0.013 | -0.028 | -0.321 | -0.327 | -0.332 |
| 0.90 | -0.001 | -0.004 | -0.009 | -0.018 | -0.039 | -0.199 | -0.204 |
| 0.93 | -0.001 | -0.003 | -0.007 | -0.013 | -0.029 | -0.048 | -0.141 |
| 0.95 | -0.001 | -0.003 | -0.005 | -0.011 | -0.023 | -0.037 | -0.103 |
| 0.97 | -0.000 | -0.002 | -0.004 | -0.009 | -0.018 | -0.029 | -0.042 |
| 0.98 | -0.000 | -0.002 | -0.004 | -0.008 | -0.016 | -0.025 | -0.035 |
| 0.99 | -0.000 | -0.002 | -0.003 | -0.007 | -0.014 | -0.021 | -0.030 |
| 1.00 | -0.000 | -0.001 | -0.003 | -0.006 | -0.012 | -0.018 | -0.025 |
| 1.01 | -0.000 | -0.001 | -0.003 | -0.005 | -0.010 | -0.016 | -0.021 |
| 1.02 | -0.000 | -0.001 | -0.002 | -0.004 | -0.009 | -0.013 | -0.017 |
| 1.05 | -0.000 | -0.001 | -0.001 | -0.002 | -0.005 | -0.006 | -0.007 |
| 1.10 | -0.000 | -0.000 | 0.000 | 0.000 | 0.001 | 0.002 | 0.004 |
| 1.15 | 0.000 | 0.000 | 0.001 | 0.002 | 0.005 | 0.008 | 0.011 |
| 1.20 | 0.000 | 0.001 | 0.002 | 0.003 | 0.007 | 0.012 | 0.017 |
| 1.30 | 0.000 | 0.001 | 0.003 | 0.005 | 0.011 | 0.017 | 0.023 |
| 1.40 | 0.000 | 0.002 | 0.003 | 0.006 | 0.013 | 0.020 | 0.027 |
| 1.50 | 0.000 | 0.002 | 0.003 | 0.007 | 0.014 | 0.021 | 0.028 |
| 1.60 | 0.000 | 0.002 | 0.003 | 0.007 | 0.014 | 0.021 | 0.029 |
| 1.70 | 0.000 | 0.002 | 0.004 | 0.007 | 0.014 | 0.021 | 0.029 |
| 1.80 | 0.000 | 0.002 | 0.003 | 0.007 | 0.014 | 0.021 | 0.028 |
| 1.90 | 0.000 | 0.002 | 0.003 | 0.007 | 0.014 | 0.021 | 0.028 |
| 2.00 | 0.000 | 0.002 | 0.003 | 0.007 | 0.013 | 0.020 | 0.027 |
| 2.20 | 0.000 | 0.002 | 0.003 | 0.006 | 0.013 | 0.019 | 0.025 |
| 2.40 | 0.000 | 0.002 | 0.003 | 0.006 | 0.012 | 0.018 | 0.024 |
| 2.60 | 0.000 | 0.001 | 0.003 | 0.006 | 0.011 | 0.017 | 0.023 |
| 2.80 | 0.000 | 0.001 | 0.003 | 0.005 | 0.011 | 0.016 | 0.021 |
| 3.00 | 0.000 | 0.001 | 0.003 | 0.005 | 0.010 | 0.015 | 0.020 |
| 3.50 | 0.000 | 0.001 | 0.002 | 0.004 | 0.009 | 0.013 | 0.018 |
| 4.00 | 0.000 | 0.001 | 0.002 | 0.004 | 0.008 | 0.012 | 0.016 |

| $P_r$ | | | | | | | |
|---|---|---|---|---|---|---|---|
| 1.000 | 1.200 | 1.500 | 2.000 | 3.000 | 5.000 | 7.000 | 10.000 |
| -8.811 | -8.818 | -8.828 | -8.845 | -8.880 | -8.953 | -9.022 | -9.126 |
| -6.567 | -6.575 | -6.587 | -6.606 | -6.645 | -6.723 | -6.800 | -6.919 |
| -4.954 | -4.962 | -4.974 | -4.995 | -5.035 | -5.115 | -5.195 | -5.312 |
| -3.766 | -3.774 | -3.786 | -3.806 | -3.845 | -3.923 | -4.001 | -4.114 |
| -2.877 | -2.884 | -2.896 | -2.915 | -2.953 | -3.027 | -3.101 | -3.208 |
| -2.199 | -2.207 | -2.218 | -2.236 | -2.273 | -2.342 | -2.410 | -2.510 |
| -1.677 | -1.684 | -1.695 | -1.712 | -1.747 | -1.812 | -1.875 | -1.967 |
| -1.271 | -1.278 | -1.287 | -1.304 | -1.336 | -1.397 | -1.456 | -1.539 |
| -0.952 | -0.958 | -0.967 | -0.983 | -1.013 | -1.070 | -1.124 | -1.201 |
| -0.700 | -0.705 | -0.714 | -0.728 | -0.756 | -0.809 | -0.858 | -0.929 |
| -0.499 | -0.504 | -0.512 | -0.526 | -0.551 | -0.600 | -0.645 | -0.709 |
| -0.338 | -0.343 | -0.351 | -0.364 | -0.388 | -0.432 | -0.473 | -0.530 |
| -0.210 | -0.215 | -0.222 | -0.234 | -0.256 | -0.296 | -0.333 | -0.384 |
| -0.146 | -0.151 | -0.158 | -0.170 | -0.190 | -0.228 | -0.262 | -0.310 |
| -0.108 | -0.114 | -0.121 | -0.132 | -0.151 | -0.187 | -0.220 | -0.265 |
| -0.075 | -0.080 | -0.087 | -0.097 | -0.116 | -0.149 | -0.180 | -0.223 |
| -0.059 | -0.064 | -0.071 | -0.081 | -0.099 | -0.132 | -0.162 | -0.203 |
| -0.044 | -0.050 | -0.056 | -0.066 | -0.084 | -0.115 | -0.144 | -0.184 |
| -0.031 | -0.036 | -0.042 | -0.052 | -0.069 | -0.099 | -0.127 | -0.166 |
| -0.024 | -0.024 | -0.030 | -0.038 | -0.054 | -0.084 | -0.111 | -0.149 |
| -0.019 | -0.015 | -0.018 | -0.026 | -0.041 | -0.069 | -0.095 | -0.132 |
| -0.007 | -0.002 | 0.008 | 0.007 | -0.005 | -0.029 | -0.052 | -0.085 |
| 0.007 | 0.012 | 0.025 | 0.041 | 0.042 | 0.026 | 0.008 | -0.019 |
| 0.016 | 0.022 | 0.034 | 0.056 | 0.074 | 0.069 | 0.057 | 0.036 |
| 0.023 | 0.029 | 0.041 | 0.064 | 0.093 | 0.102 | 0.096 | 0.081 |
| 0.030 | 0.038 | 0.049 | 0.071 | 0.109 | 0.142 | 0.150 | 0.148 |
| 0.034 | 0.041 | 0.053 | 0.074 | 0.112 | 0.161 | 0.181 | 0.191 |
| 0.036 | 0.043 | 0.055 | 0.074 | 0.112 | 0.167 | 0.197 | 0.218 |
| 0.036 | 0.043 | 0.055 | 0.074 | 0.110 | 0.167 | 0.204 | 0.234 |
| 0.036 | 0.043 | 0.054 | 0.072 | 0.107 | 0.165 | 0.205 | 0.242 |
| 0.035 | 0.042 | 0.053 | 0.070 | 0.104 | 0.161 | 0.203 | 0.246 |
| 0.034 | 0.041 | 0.052 | 0.068 | 0.101 | 0.157 | 0.200 | 0.246 |
| 0.034 | 0.040 | 0.050 | 0.066 | 0.097 | 0.152 | 0.196 | 0.244 |
| 0.032 | 0.038 | 0.047 | 0.062 | 0.091 | 0.143 | 0.186 | 0.236 |
| 0.030 | 0.036 | 0.044 | 0.058 | 0.086 | 0.134 | 0.176 | 0.227 |
| 0.028 | 0.034 | 0.042 | 0.055 | 0.080 | 0.127 | 0.167 | 0.217 |
| 0.027 | 0.032 | 0.039 | 0.052 | 0.076 | 0.120 | 0.158 | 0.208 |
| 0.025 | 0.030 | 0.037 | 0.049 | 0.072 | 0.114 | 0.151 | 0.199 |
| 0.022 | 0.026 | 0.033 | 0.043 | 0.063 | 0.101 | 0.134 | 0.179 |
| 0.020 | 0.023 | 0.029 | 0.038 | 0.057 | 0.090 | 0.121 | 0.163 |

$$H^\circ - H = -[A - A^\circ + T(S - S^\circ) + RT(Z - 1)]$$

$$= -\frac{a}{b}\ln\frac{Z}{Z + B^*} + \frac{T}{b}\frac{\partial a}{\partial T}\ln\frac{Z}{Z + B^*} - RT(Z - 1)$$

$$R = 83.14\ \text{cm}^3\cdot\text{bar/(mol}\cdot\text{K)}$$

$$b = 0.08664\frac{RT_c}{P_c}$$

$$= \frac{(0.08664)(83.14)(364.9)}{46.0} = 57.14\ \text{cm}^3/\text{mol}$$

$$B^* = \frac{bP}{RT}$$

$$= \frac{(57.14)(100)}{(83.14)(398.15)} = 0.1726$$

$$f\omega = 0.48 + 1.574\omega - 0.176\omega^2$$

$$= 0.703$$

$$a = 0.42748\frac{[1 + f\omega(1 - T_r^{0.5})]^2R^2T_c^2}{P_c}$$

$$= 0.42748\frac{[1 + 0.703(1 - 1.09^{1/2})]^2(83.14)^2(364.9)^2}{46.0}$$

$$= 8.03 \times 10^6\ \text{bar}\cdot(\text{cm}^3/\text{mol})^2$$

For a pure component,

$$\frac{\partial a}{\partial T} = -Rf\omega\left(\frac{0.42748aT_c}{TP_c}\right)^{1/2}$$

$$= -(83.14)(0.703)\left[\frac{(0.42748)(8.03 \times 10^6)}{(1.09)(46.0)}\right]^{1/2}$$

$$= -15{,}281\ \text{bar}\cdot(\text{cm}^3/\text{mol})^2/\text{K}$$

If Eq. (3-6.2) is solved for $Z$, the result is $Z = 0.4470$, $V = 148.0\ \text{cm}^3/\text{mol}$.

$$S^\circ - S = -83.14\ln\frac{0.4470 - 0.1726}{0.4470} - 83.14\ln\frac{148.0(1)}{(83.14)(398.15)} - \frac{15{,}281}{57.14}\ln\frac{0.4470}{0.4470 + 0.1726}$$

$$= 578\ \frac{\text{bar}\cdot\text{cm}^3}{\text{mol}\cdot\text{K}}$$

$$S^\circ - S = \frac{578}{42.081}\frac{8.314}{83.14} = 1.37\ \text{J/(g}\cdot\text{K)}$$

$$H^\circ - H = \left[-\frac{8.03 \times 10^6}{56.91} + \frac{398.15}{56.91}(-15{,}281)\right]\ln\frac{0.447}{0.447 + 0.1726} - (83.14)(398.15)(0.447 - 1)$$

$$= 9.89 \times 10^4\ \text{bar}\cdot\text{cm}^3/\text{mol}$$

$$H^\circ - H = \frac{9.89 \times 10^4}{42.081}\frac{8.314}{83.14} = 235\ \text{J/g}$$

PENG-ROBINSON EQUATION. For the Peng-Robinson equation, the corresponding equation for $H^\circ - H$ is

$$H^\circ - H = \left(-\frac{a}{2b\sqrt{2}} + \frac{T}{2b\sqrt{2}}\frac{\partial a}{\partial T}\right)\ln\left(\frac{Z - 0.414B^*}{Z + 2.414B^*}\right) - RT(Z - 1)$$

The Peng-Robinson results are

$$S° - S = 1.36 \text{ J/(g·K)}$$
$$H° - H = 235 \text{ J/g}$$

## 5-5 Heat Capacities of Real Gases

In Chap. 6 we present methods for estimating the heat capacity of pure gases in the ideal-gas state as a function of temperature. Also in this ideal-gas state, for a gas mixture,

$$C_{pm}^{\circ} = \sum_i y_i C_{pi}^{\circ} \tag{5-5.1}$$

The heat capacity of a real gas is related to the value in the ideal-gas state, at the same temperature and composition:

$$C_p = C_p^{\circ} + \Delta C_p \tag{5-5.2}$$

where this relation applies to either a pure gas or gas mixture at constant composition. $\Delta C_p$ is a residual heat capacity. For a pressure-explicit equation of state, $\Delta C_p$ is most conveniently determined by (see p. 159, Ref. 56)

$$\Delta C_p = T \int_{\infty}^{V} \left(\frac{\partial^2 P}{\partial T^2}\right) dV - \frac{T(\partial P/\partial T)_V^2}{(\partial P/\partial V)_T} - R \tag{5-5.3}$$

For the Lee-Kesler method

$$C_p - C_p^{\circ} = \Delta C_p = (\Delta C_p)^{(0)} + \omega(\Delta C_p)^{(1)} \tag{5-5.4}$$

The simple fluid contribution $(\Delta C_p)^{(0)}$ is given in Table 5-8, and the deviation function $(\Delta C_p)^{(1)}$ is given in Table 5-9 as a function of $T_r$ and $P_r$. If Eq. (5-5.4) and Tables 5-8 and 5-9 are employed for mixtures, the pseudocritical rules given in Eqs. (4-6.3) to (4-6.7) should be used. These rules have been developed primarily from hydrocarbon mixture data, but they should be satisfactory unless highly polar components are present. Tables 5-8 and 5-9 differ somewhat from an earlier correlation by Edmister [17] especially in the critical region, where high accuracy is difficult to achieve in any case.

## 5-6 True Critical Points of Mixtures

In Chap. 4, emphasis was placed upon the estimation of pseudocritical constants for mixtures. Such constants are necessary if one is to use most corresponding states correlations to estimate mixture $PVT$ or derived properties. However, these pseudocritical constants often differ considerably from the true critical points for mixtures. Estimation techniques

**TABLE 5-8 Residual Heat Capacities [47]**

Simple fluid $\left(\frac{C_p - C_p^\circ}{R}\right)^{(0)}$

| $T_r$ | $P_r$ 0.010 | 0.050 | 0.100 | 0.200 | 0.400 | 0.600 | 0.800 |
|---|---|---|---|---|---|---|---|
| 0.30 | 2.805 | 2.807 | 2.809 | 2.814 | 2.830 | 2.842 | 2.854 |
| 0.35 | 2.808 | 2.810 | 2.812 | 2.815 | 2.823 | 2.835 | 2.844 |
| 0.40 | 2.925 | 2.926 | 2.928 | 2.933 | 2.935 | 2.940 | 2.945 |
| 0.45 | 2.989 | 2.990 | 2.990 | 2.991 | 2.993 | 2.995 | 2.997 |
| 0.50 | 3.006 | 3.005 | 3.004 | 3.003 | 3.001 | 3.000 | 2.998 |
| 0.55 | 0.118 | 3.002 | 3.000 | 2.997 | 2.990 | 2.984 | 2.978 |
| 0.60 | 0.089 | 3.009 | 3.006 | 2.999 | 2.986 | 2.974 | 2.963 |
| 0.65 | 0.069 | 0.387 | 3.047 | 3.036 | 3.014 | 2.993 | 2.973 |
| 0.70 | 0.054 | 0.298 | 0.687 | 3.138 | 3.099 | 3.065 | 3.033 |
| 0.75 | 0.044 | 0.236 | 0.526 | 3.351 | 3.284 | 3.225 | 3.171 |
| 0.80 | 0.036 | 0.191 | 0.415 | 1.032 | 3.647 | 3.537 | 3.440 |
| 0.85 | 0.030 | 0.157 | 0.336 | 0.794 | 4.404 | 4.158 | 3.957 |
| 0.90 | 0.025 | 0.131 | 0.277 | 0.633 | 1.858 | 5.679 | 5.095 |
| 0.93 | 0.023 | 0.118 | 0.249 | 0.560 | 1.538 | 4.208 | 6.720 |
| 0.95 | 0.021 | 0.111 | 0.232 | 0.518 | 1.375 | 3.341 | 9.316 |
| 0.97 | 0.020 | 0.104 | 0.217 | 0.480 | 1.240 | 2.778 | 9.585 |
| 0.98 | 0.019 | 0.101 | 0.210 | 0.463 | 1.181 | 2.563 | 7.350 |
| 0.99 | 0.019 | 0.098 | 0.204 | 0.447 | 1.126 | 2.378 | 6.038 |
| 1.00 | 0.018 | 0.095 | 0.197 | 0.431 | 1.076 | 2.218 | 5.156 |
| 1.01 | 0.018 | 0.092 | 0.191 | 0.417 | 1.029 | 2.076 | 4.516 |
| 1.02 | 0.017 | 0.089 | 0.185 | 0.403 | 0.986 | 1.951 | 4.025 |
| 1.05 | 0.016 | 0.082 | 0.169 | 0.365 | 0.872 | 1.648 | 3.047 |
| 1.10 | 0.014 | 0.071 | 0.147 | 0.313 | 0.724 | 1.297 | 2.168 |
| 1.15 | 0.012 | 0.063 | 0.128 | 0.271 | 0.612 | 1.058 | 1.670 |
| 1.20 | 0.011 | 0.055 | 0.113 | 0.237 | 0.525 | 0.885 | 1.345 |
| 1.30 | 0.009 | 0.044 | 0.089 | 0.185 | 0.400 | 0.651 | 0.946 |
| 1.40 | 0.007 | 0.036 | 0.072 | 0.149 | 0.315 | 0.502 | 0.711 |
| 1.50 | 0.006 | 0.029 | 0.060 | 0.122 | 0.255 | 0.399 | 0.557 |
| 1.60 | 0.005 | 0.025 | 0.050 | 0.101 | 0.210 | 0.326 | 0.449 |
| 1.70 | 0.004 | 0.021 | 0.042 | 0.086 | 0.176 | 0.271 | 0.371 |
| 1.80 | 0.004 | 0.018 | 0.036 | 0.073 | 0.150 | 0.229 | 0.311 |
| 1.90 | 0.003 | 0.016 | 0.031 | 0.063 | 0.129 | 0.196 | 0.265 |
| 2.00 | 0.003 | 0.014 | 0.027 | 0.055 | 0.112 | 0.170 | 0.229 |
| 2.20 | 0.002 | 0.011 | 0.021 | 0.043 | 0.086 | 0.131 | 0.175 |
| 2.40 | 0.002 | 0.009 | 0.017 | 0.034 | 0.069 | 0.104 | 0.138 |
| 2.60 | 0.001 | 0.007 | 0.014 | 0.028 | 0.056 | 0.084 | 0.112 |
| 2.80 | 0.001 | 0.006 | 0.012 | 0.023 | 0.046 | 0.070 | 0.093 |
| 3.00 | 0.001 | 0.005 | 0.010 | 0.020 | 0.039 | 0.058 | 0.078 |
| 3.50 | 0.001 | 0.003 | 0.007 | 0.013 | 0.027 | 0.040 | 0.053 |
| 4.00 | 0.000 | 0.002 | 0.005 | 0.010 | 0.019 | 0.029 | 0.038 |

| $P_r$ | | | | | | | |
|---|---|---|---|---|---|---|---|
| 1.000 | 1.200 | 1.500 | 2.000 | 3.000 | 5.000 | 7.000 | 10.000 |
| 2.866 | 2.878 | 2.896 | 2.927 | 2.989 | 3.122 | 3.257 | 3.466 |
| 2.853 | 2.861 | 2.875 | 2.897 | 2.944 | 3.042 | 3.145 | 3.313 |
| 2.951 | 2.956 | 2.965 | 2.979 | 3.014 | 3.085 | 3.164 | 3.293 |
| 2.999 | 3.002 | 3.006 | 3.014 | 3.032 | 3.079 | 3.135 | 3.232 |
| 2.997 | 2.996 | 2.995 | 2.995 | 2.999 | 3.019 | 3.054 | 3.122 |
| 2.973 | 2.968 | 2.961 | 2.951 | 2.938 | 2.934 | 2.947 | 2.988 |
| 2.952 | 2.942 | 2.927 | 2.907 | 2.874 | 2.840 | 2.831 | 2.847 |
| 2.955 | 2.938 | 2.914 | 2.878 | 2.822 | 2.753 | 2.720 | 2.709 |
| 3.003 | 2.975 | 2.937 | 2.881 | 2.792 | 2.681 | 2.621 | 2.582 |
| 3.122 | 3.076 | 3.015 | 2.928 | 2.795 | 2.629 | 2.537 | 2.469 |
| 3.354 | 3.277 | 3.176 | 3.038 | 2.838 | 2.601 | 2.473 | 2.373 |
| 3.790 | 3.647 | 3.470 | 3.240 | 2.931 | 2.599 | 2.427 | 2.292 |
| 4.677 | 4.359 | 4.000 | 3.585 | 3.096 | 2.626 | 2.399 | 2.227 |
| 5.766 | 5.149 | 4.533 | 3.902 | 3.236 | 2.657 | 2.392 | 2.195 |
| 7.127 | 6.010 | 5.050 | 4.180 | 3.351 | 2.684 | 2.391 | 2.175 |
| 10.011 | 7.451 | 5.785 | 4.531 | 3.486 | 2.716 | 2.393 | 2.159 |
| 13.270 | 8.611 | 6.279 | 4.743 | 3.560 | 2.733 | 2.395 | 2.151 |
| 21.948 | 10.362 | 6.897 | 4.983 | 3.641 | 2.752 | 2.398 | 2.144 |
| ****** | 13.281 | 7.686 | 5.255 | 3.729 | 2.773 | 2.401 | 2.138 |
| 22.295 | 18.967 | 8.708 | 5.569 | 3.821 | 2.794 | 2.405 | 2.131 |
| 13.184 | 31.353 | 10.062 | 5.923 | 3.920 | 2.816 | 2.408 | 2.125 |
| 6.458 | 20.234 | 16.457 | 7.296 | 4.259 | 2.891 | 2.425 | 2.110 |
| 3.649 | 6.510 | 13.256 | 9.787 | 4.927 | 3.033 | 2.462 | 2.093 |
| 2.553 | 3.885 | 6.985 | 9.094 | 5.535 | 3.186 | 2.508 | 2.083 |
| 1.951 | 2.758 | 4.430 | 6.911 | 5.710 | 3.326 | 2.555 | 2.079 |
| 1.297 | 1.711 | 2.458 | 3.850 | 4.793 | 3.452 | 2.628 | 2.077 |
| 0.946 | 1.208 | 1.650 | 2.462 | 3.573 | 3.282 | 2.626 | 2.068 |
| 0.728 | 0.912 | 1.211 | 1.747 | 2.647 | 2.917 | 2.525 | 2.038 |
| 0.580 | 0.719 | 0.938 | 1.321 | 2.016 | 2.508 | 2.347 | 1.978 |
| 0.475 | 0.583 | 0.752 | 1.043 | 1.586 | 2.128 | 2.130 | 1.889 |
| 0.397 | 0.484 | 0.619 | 0.848 | 1.282 | 1.805 | 1.907 | 1.778 |
| 0.336 | 0.409 | 0.519 | 0.706 | 1.060 | 1.538 | 1.696 | 1.656 |
| 0.289 | 0.350 | 0.443 | 0.598 | 0.893 | 1.320 | 1.505 | 1.531 |
| 0.220 | 0.265 | 0.334 | 0.446 | 0.661 | 0.998 | 1.191 | 1.292 |
| 0.173 | 0.208 | 0.261 | 0.347 | 0.510 | 0.779 | 0.956 | 1.086 |
| 0.140 | 0.168 | 0.210 | 0.278 | 0.407 | 0.624 | 0.780 | 0.917 |
| 0.116 | 0.138 | 0.172 | 0.227 | 0.332 | 0.512 | 0.647 | 0.779 |
| 0.097 | 0.116 | 0.144 | 0.190 | 0.277 | 0.427 | 0.545 | 0.668 |
| 0.066 | 0.079 | 0.098 | 0.128 | 0.187 | 0.289 | 0.374 | 0.472 |
| 0.048 | 0.057 | 0.071 | 0.093 | 0.135 | 0.209 | 0.272 | 0.350 |

**TABLE 5-9 Residual Heat Capacities [47]**

Deviation function $\left(\frac{C_p - C_p^\circ}{R}\right)^{(1)}$

| $T_r$ | $P_r$ 0.010 | 0.050 | 0.100 | 0.200 | 0.400 | 0.600 | 0.800 |
|---|---|---|---|---|---|---|---|
| 0.30 | 8.462 | 8.445 | 8.424 | 8.381 | 8.281 | 8.192 | 8.102 |
| 0.35 | 9.775 | 9.762 | 9.746 | 9.713 | 9.646 | 9.568 | 9.499 |
| 0.40 | 11.494 | 11.484 | 11.471 | 11.438 | 11.394 | 11.343 | 11.291 |
| 0.45 | 12.651 | 12.643 | 12.633 | 12.613 | 12.573 | 12.532 | 12.492 |
| 0.50 | 13.111 | 13.106 | 13.099 | 13.084 | 13.055 | 13.025 | 12.995 |
| 0.55 | 0.511 | 13.035 | 13.030 | 13.021 | 13.002 | 12.981 | 12.961 |
| 0.60 | 0.345 | 12.679 | 12.675 | 12.668 | 12.653 | 12.637 | 12.620 |
| 0.65 | 0.242 | 1.518 | 12.148 | 12.145 | 12.137 | 12.128 | 12.117 |
| 0.70 | 0.174 | 1.026 | 2.698 | 11.557 | 11.564 | 11.563 | 11.559 |
| 0.75 | 0.129 | 0.726 | 1.747 | 10.967 | 10.995 | 11.011 | 11.019 |
| 0.80 | 0.097 | 0.532 | 1.212 | 3.511 | 10.490 | 10.536 | 10.566 |
| 0.85 | 0.075 | 0.399 | 0.879 | 2.247 | 9.999 | 10.153 | 10.245 |
| 0.90 | 0.058 | 0.306 | 0.658 | 1.563 | 5.486 | 9.793 | 10.180 |
| 0.93 | 0.050 | 0.263 | 0.560 | 1.289 | 3.890 | ****** | 10.285 |
| 0.95 | 0.046 | 0.239 | 0.505 | 1.142 | 3.215 | 9.389 | 9.993 |
| 0.97 | 0.042 | 0.217 | 0.456 | 1.018 | 2.712 | 6.588 | ****** |
| 0.98 | 0.040 | 0.207 | 0.434 | 0.962 | 2.506 | 5.711 | ****** |
| 0.99 | 0.038 | 0.198 | 0.414 | 0.911 | 2.324 | 5.027 | ****** |
| 1.00 | 0.037 | 0.189 | 0.394 | 0.863 | 2.162 | 4.477 | 10.511 |
| 1.01 | 0.035 | 0.181 | 0.376 | 0.819 | 2.016 | 4.026 | 8.437 |
| 1.02 | 0.034 | 0.173 | 0.359 | 0.778 | 1.884 | 3.648 | 7.044 |
| 1.05 | 0.030 | 0.152 | 0.313 | 0.669 | 1.559 | 2.812 | 4.679 |
| 1.10 | 0.024 | 0.123 | 0.252 | 0.528 | 1.174 | 1.968 | 2.919 |
| 1.15 | 0.020 | 0.101 | 0.205 | 0.424 | 0.910 | 1.460 | 2.048 |
| 1.20 | 0.016 | 0.083 | 0.168 | 0.345 | 0.722 | 1.123 | 1.527 |
| 1.30 | 0.012 | 0.058 | 0.116 | 0.235 | 0.476 | 0.715 | 0.938 |
| 1.40 | 0.008 | 0.042 | 0.083 | 0.166 | 0.329 | 0.484 | 0.624 |
| 1.50 | 0.006 | 0.030 | 0.061 | 0.120 | 0.235 | 0.342 | 0.437 |
| 1.60 | 0.005 | 0.023 | 0.045 | 0.089 | 0.173 | 0.249 | 0.317 |
| 1.70 | 0.003 | 0.017 | 0.034 | 0.068 | 0.130 | 0.187 | 0.236 |
| 1.80 | 0.003 | 0.013 | 0.027 | 0.052 | 0.100 | 0.143 | 0.180 |
| 1.90 | 0.002 | 0.011 | 0.021 | 0.041 | 0.078 | 0.111 | 0.140 |
| 2.00 | 0.002 | 0.008 | 0.017 | 0.032 | 0.062 | 0.088 | 0.110 |
| 2.20 | 0.001 | 0.005 | 0.011 | 0.021 | 0.040 | 0.057 | 0.072 |
| 2.40 | 0.001 | 0.004 | 0.007 | 0.014 | 0.028 | 0.039 | 0.049 |
| 2.60 | 0.001 | 0.003 | 0.005 | 0.010 | 0.020 | 0.028 | 0.035 |
| 2.80 | 0.000 | 0.002 | 0.004 | 0.008 | 0.014 | 0.021 | 0.026 |
| 3.00 | 0.000 | 0.001 | 0.003 | 0.006 | 0.011 | 0.016 | 0.020 |
| 3.50 | 0.000 | 0.001 | 0.002 | 0.003 | 0.006 | 0.009 | 0.012 |
| 4.00 | 0.000 | 0.001 | 0.001 | 0.002 | 0.004 | 0.006 | 0.008 |

| $P_r$ | | | | | | | |
|---|---|---|---|---|---|---|---|
| 1.000 | 1.200 | 1.500 | 2.000 | 3.000 | 5.000 | 7.000 | 10.000 |
| 8.011 | 7.920 | 7.785 | 7.558 | 7.103 | 6.270 | 5.372 | 4.020 |
| 9.430 | 9.360 | 9.256 | 9.080 | 8.728 | 8.013 | 7.290 | 6.285 |
| 11.240 | 11.188 | 11.110 | 10.980 | 10.709 | 10.170 | 9.625 | 8.803 |
| 12.451 | 12.409 | 12.347 | 12.243 | 12.029 | 11.592 | 11.183 | 10.533 |
| 12.964 | 12.933 | 12.886 | 12.805 | 12.639 | 12.288 | 11.946 | 11.419 |
| 12.939 | 12.917 | 12.882 | 12.823 | 12.695 | 12.407 | 12.103 | 11.673 |
| 12.589 | 12.574 | 12.550 | 12.506 | 12.407 | 12.165 | 11.905 | 11.526 |
| 12.105 | 12.092 | 12.060 | 12.026 | 11.943 | 11.728 | 11.494 | 11.141 |
| 11.553 | 11.536 | 11.524 | 11.495 | 11.416 | 11.208 | 10.985 | 10.661 |
| 11.024 | 11.022 | 11.013 | 10.986 | 10.898 | 10.677 | 10.448 | 10.132 |
| 10.583 | 10.590 | 10.587 | 10.556 | 10.446 | 10.176 | 9.917 | 9.591 |
| 10.297 | 10.321 | 10.324 | 10.278 | 10.111 | 9.740 | 9.433 | 9.075 |
| 10.349 | 10.409 | 10.401 | 10.279 | 9.940 | 9.389 | 8.999 | 8.592 |
| 10.769 | 10.875 | 10.801 | 10.523 | 9.965 | 9.225 | 8.766 | 8.322 |
| 11.420 | 11.607 | 11.387 | 10.865 | 10.055 | 9.136 | 8.621 | 8.152 |
| 13.001 | ****** | 12.498 | 11.445 | 10.215 | 9.061 | 8.485 | 7.986 |
| ****** | ****** | ****** | 11.856 | 10.323 | 9.037 | 8.420 | 7.905 |
| ****** | ****** | ****** | 12.388 | 10.457 | 9.011 | 8.359 | 7.826 |
| ****** | ****** | ****** | 13.081 | 10.617 | 8.990 | 8.293 | 7.747 |
| ****** | ****** | ****** | ****** | 10.805 | 8.973 | 8.236 | 7.670 |
| ****** | ****** | ****** | ****** | 11.024 | 8.960 | 8.182 | 7.595 |
| 7.173 | 2.277 | ****** | ****** | 11.852 | 8.939 | 8.018 | 7.377 |
| 3.877 | 4.002 | 3.927 | ****** | ****** | 8.933 | 7.759 | 7.031 |
| 2.587 | 2.844 | 2.236 | 7.716 | 12.812 | 8.849 | 7.504 | 6.702 |
| 1.881 | 2.095 | 1.962 | 2.965 | 9.494 | 8.508 | 7.206 | 6.384 |
| 1.129 | 1.264 | 1.327 | 1.288 | 3.855 | 6.758 | 6.365 | 5.735 |
| 0.743 | 0.833 | 0.904 | 0.905 | 1.652 | 4.524 | 5.193 | 5.035 |
| 0.517 | 0.580 | 0.639 | 0.666 | 0.907 | 2.823 | 3.944 | 4.289 |
| 0.374 | 0.419 | 0.466 | 0.499 | 0.601 | 1.755 | 2.871 | 3.545 |
| 0.278 | 0.312 | 0.349 | 0.380 | 0.439 | 1.129 | 2.060 | 2.867 |
| 0.212 | 0.238 | 0.267 | 0.296 | 0.337 | 0.764 | 1.483 | 2.287 |
| 0.164 | 0.185 | 0.209 | 0.234 | 0.267 | 0.545 | 1.085 | 1.817 |
| 0.130 | 0.146 | 0.166 | 0.187 | 0.217 | 0.407 | 0.812 | 1.446 |
| 0.085 | 0.096 | 0.110 | 0.126 | 0.150 | 0.256 | 0.492 | 0.941 |
| 0.058 | 0.066 | 0.076 | 0.089 | 0.109 | 0.180 | 0.329 | 0.644 |
| 0.042 | 0.048 | 0.056 | 0.066 | 0.084 | 0.137 | 0.239 | 0.466 |
| 0.031 | 0.036 | 0.042 | 0.051 | 0.067 | 0.110 | 0.187 | 0.356 |
| 0.024 | 0.028 | 0.033 | 0.041 | 0.055 | 0.092 | 0.153 | 0.285 |
| 0.015 | 0.017 | 0.021 | 0.026 | 0.038 | 0.067 | 0.108 | 0.190 |
| 0.010 | 0.012 | 0.015 | 0.019 | 0.029 | 0.054 | 0.085 | 0.146 |

for the latter can be evaluated by comparison with experimental data. For the former, evaluation is indirect, since the pseudocritical state does not exist in a physical sense.

In this section, we briefly discuss methods of estimating the true critical properties of mixtures. Most techniques are limited to hydrocarbon mixtures or to mixtures of hydrocarbons with $CO_2$, $H_2S$, CO, and the permanent gases. A summary of experimental values is given in Ref. 33.

Estimation methods fall into two categories: empirical methods and rigorous methods in which an equation of state is used to solve the Gibbs criteria for a critical point in a mixture. The empirical methods are simpler and are reviewed first.

### Mixture critical temperature

The true mixture critical temperature is usually not a linear mole fraction average of the pure component critical temperatures. Li [48] has suggested that if the composition is expressed as

$$\phi_j = \frac{y_j V_{cj}}{\sum_i y_i V_{ci}} \tag{5-6.1}$$

the true mixture critical temperature can be estimated by

$$T_{cT} = \sum_j \phi_j T_{cj} \tag{5-6.2}$$

where $y_j$ = mole fraction of component $j$
$V_{cj}$ = critical volume of $j$
$T_{cj}$ = critical temperature of $j$
$T_{cT}$ = true mixture critical temperature

Chueh and Prausnitz [12] have proposed a similar technique. By defining a *surface fraction* $\theta_j$,

$$\theta_j = \frac{y_j V_{cj}^{2/3}}{\sum_i y_i V_{ci}^{2/3}} \tag{5-6.3}$$

they then relate $\theta_j$ and $T_{cT}$ by

$$T_{cT} = \sum_j \theta_j T_{cj} + \sum_i \sum_j \theta_i \theta_j \tau_{ij} \tag{5-6.4}$$

where $\tau_{ij}$ is an interaction parameter. $\tau_{ii}$ is considered to be zero, and $\tau_{ij}$ $(i \neq j)$ can be estimated for several different binary types by

$$\psi_T = A + B\,\delta_T + C\,\delta_T^2 + D\,\delta_T^3 + E\,\delta_T^4 \tag{5-6.5}$$

where $$\psi_T = \frac{2\tau_{ij}}{T_{c_i} + T_{c_j}} \tag{5-6.6}$$

and $$\delta_T = \left| \frac{T_{c_i} - T_{c_j}}{T_{c_i} + T_{c_j}} \right| \tag{5-6.7}$$

The coefficients for Eq. (5-6.5) are shown below for a few binary types [73], where $0 \leq \delta_T \leq 0.5$:

| Binary | $A$ | $B$ | $C$ | $D$ | $E$ |
|---|---|---|---|---|---|
| Containing aromatics | −0.0219 | 1.227 | −24.277 | 147.673 | −259.433 |
| Containing $H_2S$ | −0.0479 | −5.725 | 70.974 | −161.319 | |
| Containing $CO_2$ | −0.0953 | 2.185 | −33.985 | 179.068 | −264.522 |
| Containing $C_2H_2$ | −0.0785 | −2.152 | 93.084 | −722.676 | |
| Containing CO | −0.0077 | −0.095 | −0.225 | 3.528 | |
| All other systems | −0.0076 | 0.287 | −1.343 | 5.443 | −3.038 |

In Fig. 5-2 the critical temperature of the methane–$n$-pentane binary is shown plotted as a function of mole fraction, surface fraction, and volume fraction. It is clear that the use of a volume fraction provides essentially a linear relation between $\phi_j$ and $T_{cT}$, as predicted by Eq. (5-6.2). When the surface fraction is employed, $T_{cT}$ is slightly nonlinear with $\theta_j$ and the interaction term in Eq. (5-6.4) compensates for the nonlinearity.

Spencer et al. [74] reviewed and evaluated a number of proposed methods for estimating true critical temperatures for mixtures. They recommended either the Li or Chueh-Prausnitz correlations described above; 135 binary hydrocarbon mixtures were tested, and the average deviation noted for both methods was less than 4 K. For multicomponent hydro-

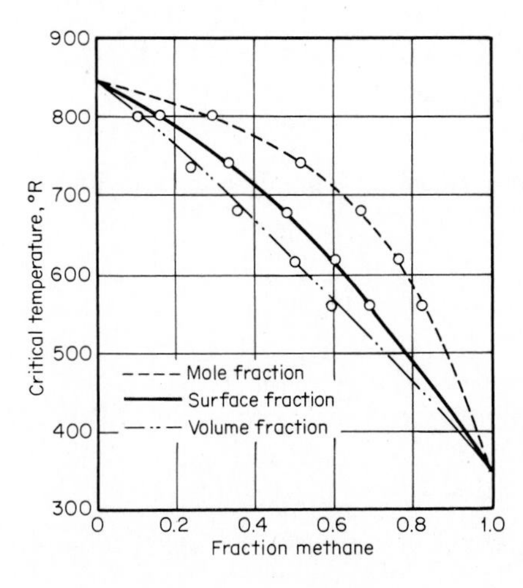


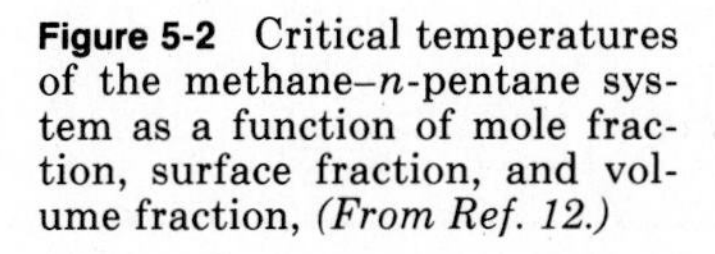

**Figure 5-2** Critical temperatures of the methane–$n$-pentane system as a function of mole fraction, surface fraction, and volume fraction, *(From Ref. 12.)*

carbon systems, larger errors were found (average deviation about 11 K). When hydrocarbon-nonhydrocarbon mixtures were evaluated, the Chueh-Prausnitz method yielded a smaller average deviation [70].

Of the two methods recommended by Spencer et al. [74], the Li relation (5-6.2) is the easier to use, and unless one of the components is a nonhydrocarbon, it is slightly more accurate.

**Example 5-3** Estimate the true critical temperature for a mixture of methane, ethane, and $n$-butane with mole fractions:

$$y_{C1} = 0.193$$
$$y_{C2} = 0.470$$
$$y_{C4} = \underline{0.337}$$
$$1.000$$

Li [48] indicates that the experimental value is 354 K.

**solution** From Appendix A:

| | $T_c$, K | $V_c$, cm$^3$/mol |
|---|---|---|
| Methane | 190.4 | 99.2 |
| Ethane | 305.4 | 148.3 |
| $n$-Butane | 425.2 | 255 |

LI METHOD. Using Eq. (5-6.1),

| | Volume fraction $\phi_j$ |
|---|---|
| Methane | 0.110 |
| Ethane | 0.398 |
| $n$-Butane | 0.492 |
| | 1.000 |

With Eq. (5-6.2)

$$T_{cT} = (0.110)(190.4) + (0.398)(305.4) + (0.492)(425.2) = 352 \text{ K}$$

Deviation = 2 K

CHUEH-PRAUSNITZ METHOD. First, surface fractions $\theta_j$ are determined with Eq. (5-6.3):

| | Surface fraction $\theta_j$ |
|---|---|
| Methane | 0.134 |
| Ethane | 0.427 |
| $n$-Butane | 0.439 |
| | 1.000 |

Next, $\tau_{ij}$ is found from Eq. (5-6.5) by using $A = -0.0076$, $B = 0.287$, $C = -1.343$, $D = 5.443$, and $E = -3.038$:

| $i$ | $j$ | $\delta_T$ | $\psi_T$ | $\tau_{ij}$ |
|---|---|---|---|---|
| Methane | Ethane | 0.232 | 0.046 | 11.4 |
| Methane | $n$-Butane | 0.381 | 0.144 | 44.3 |
| Ethane | $n$-Butane | 0.164 | 0.025 | 9.1 |

Then, with Eq. (5-6.4),

$$T_{cT} = (0.134)(190.4) + (0.427)(305.4) + (0.439)(425.2)$$
$$+ (2)(0.134)(0.427)(11.4) + (2)(0.134)(0.439)(44.3)$$
$$+ (2)(0.427)(0.439)(9.1)$$
$$= 353 \text{ K}$$
$$\text{Deviation} = 1 \text{ K}$$

### Mixture critical volumes

Only a few experimental values are available for mixture critical volumes. Thus the range and accuracy of estimation methods are not clearly established. Grieves and Thodos [29] have suggested an approximate graphical method for hydrocarbon mixtures, but an analytical technique by Chueh and Prausnitz [12], modified by Schick and Prausnitz [70], appears to be more accurate. When the surface fraction $\theta_j$ is defined as in Eq. (5-6.3), the mixture critical volume is given by a relation analogous to Eq. (5-6.4)

$$V_{cT} = \sum_j \theta_j V_{cj} + \sum_i \sum_j \theta_i \theta_j \nu_{ij} \tag{5-6.8}$$

$V_{cj}$ is the critical volume of $j$, and $\nu_{ij}$ is an interaction parameter such that $\nu_{ii} = 0$ and $\nu_{ij}$ $(i \neq j)$ can be estimated as follows:

$$\psi_v = A + B\delta_v + C\delta_v^2 + D\delta_v^3 + E\delta_v^4 \tag{5-6-9}$$

$$\psi_v = \frac{2\nu_{ij}}{V_{ci} + V_{cj}} \tag{5-6.10}$$

$$\delta_v = \left| \frac{V_{ci}^{2/3} - V_{cj}^{2/3}}{V_{ci}^{2/3} + V_{cj}^{2/3}} \right| \tag{5-6.11}$$

The coefficients for Eq. (5-6.9) are given below for a few binary types [73] when $0 \leq \delta_v \leq 0.5$.

| Binary | $A$ | $B$ | $C$ | $D$ | $E$ |
|---|---|---|---|---|---|
| Aromatic-aromatic | 0 | 0 | 0 | 0 | 0 |
| Containing at least one cycloparaffin | 0 | 0 | 0 | 0 | 0 |
| Paraffin-aromatic | 0.0753 | −3.332 | 2.220 | 0 | 0 |
| System with $CO_2$ or $H_2S$ | −0.4957 | 17.1185 | −168.56 | 587.05 | −698.89 |
| All other systems | 0.1397 | −2.9672 | 1.8337 | −1.536 | 0 |

Spencer et al. [74] evaluated Eq. (5-6.8) for 23 binary hydrocarbon mixtures and 8 binaries consisting of a hydrocarbon and a nonhydrocarbon. They report an average error of 10.5 percent, with particularly poor results for ethane-cyclohexane, ethylene-propylene, and several of the systems containing nonhydrocarbons. However, this large error may in part be explained by experimental inaccuracies. It is much more difficult to measure $V_{cT}$ than to measure $T_{cT}$ or $P_{cT}$. Spencer et al. [74] note that, when $V_{cT}$ is correlated by

$$V_{cT} = y_1 V_{c1} + y_2 V_{c2} + V_c^{EX} \tag{5-6.12}$$

for a binary of 1 and 2, $V_c^{EX}$ values for some systems are positive and for others are negative. No generalized methods are available to estimate either the sign or the magnitude.

**Example 5-4** Estimate the true critical volume of a mixture of toluene and $n$-hexane containing 50.5 mole percent $n$-hexane. The experimental value is 325 cm$^3$/mol [88].

**solution** Equation (5-6.8) will be used. From Appendix A, $V_c$ ($n$-hexane) = 370 cm$^3$/mol and $V_c$ (toluene) = 316 cm$^3$/mol. Thus with Eq. (5-6.3),

$$\theta_{n-C6} = \frac{(0.505)(370^{2/3})}{(0.505)(370^{2/3}) + (0.495)(316^{2/3})} = 0.531$$

$$\theta_{tol} = 0.469$$

With Eq. (5-6.11)

$$\delta_v = \left| \frac{370^{2/3} - 316^{2/3}}{370^{2/3} + 316^{2/3}} \right| = 0.0525$$

Then with Eq. (5-6.9),

$$\begin{aligned}
\psi_v &= A + B\delta_v + C\delta_v^2 + D\delta_v^3 + E\delta_v^4 \\
&= 0.0753 + (-3.332)(0.0525) + (2.220)(0.0525)^2 \\
&= -0.094 \\
&= \frac{2\nu_{12}}{370 + 316}
\end{aligned}$$

$$\nu_{12} = -32$$

Then, with Eq. (5-6.8),

$$\begin{aligned}
V_{cT} &= (0.531)(370) + (0.469)(316) + (2)(-32)(0.531)(0.469) \\
&= 328 \text{ cm}^3/\text{mol}
\end{aligned}$$

$$\text{Error} = \frac{328 - 325}{325} \times 100 = 1\%$$

### Mixture critical pressure

The dependence of mixture critical pressures on mole fraction is often nonlinear, and estimation of $P_{cT}$ is often unreliable. Two approaches are illustrated below.

Kreglewski and Kay [45] derived an approximate expression for $P_{cT}$ by using conformal solution theory. $T_{cT}$ is required in the calculation, and Kreglewski and Kay also suggest how this property may be determined. However, Spencer et al. [74] found better results in determining $P_{cT}$ if $T_{cT}$ is found from Eq. (5-6.2). To find $P_{cT}$, the liquid molal volumes for each component at $T_r = 0.6$, $V_i^*$, must first be obtained. Kreglewski [44] tabulates this volume for many pure liquids; alternatively, it can be estimated by methods presented in Chap. 3. The sequential set of equations to employ then follow for a binary of 1 and 2:

$$V_{12}^* = \frac{[(V_1^*)^{1/3} + (V_2^*)^{1/3}]^3}{8} \tag{5-6.13}$$

$$V^* = V_1^* y_1 + V_2^* y_2 + (2V_{12}^* - V_1^* - V_2^*) y_1 y_2 \tag{5-6.14}$$

where $y_1$ and $y_2$ are mole fractions. Next, *surface* fractions are defined as

$$\theta_1 = \frac{y_1 V_1^{*2/3}}{y_1 V_1^{*2/3} + y_2 V_2^{*2/3}} \tag{5-6.15}$$

$$\theta_2 = 1 - \theta_1 \tag{5-6.16}$$

$$T_{12}^* = \frac{2 V_{12}^{*1/3}}{V_1^{*1/3}/T_{c1} + V_2^{*1/3}/T_{c2}} \tag{5-6.17}$$

$$T^* = V^{*1/3} \left[ \frac{T_{c1}\theta_1}{V_1^{*1/3}} + \frac{T_{c2}\theta_2}{V_2^{*1/3}} + \left( \frac{2T_{12}^*}{V_{12}^{*1/3}} - \frac{T_{c1}}{V_1^{*1/3}} - \frac{T_{c2}}{V_2^{*1/3}} \right) \theta_1 \theta_2 \right] \tag{5-6.18}$$

$$\omega_{12} = \frac{2}{1/\omega_1 + 1/\omega_2} \tag{5-6.19}$$

$$\omega = \omega_1\theta_1 + \omega_2\theta_2 + (2\omega_{12} - \omega_1 - \omega_2)\theta_1\theta_2 \tag{5-6.20}$$

where $\omega$ is the acentric factor. Finally,

$$P^* = \frac{T^*}{V^{*1/3}} \frac{P_{c1}\theta_1 + P_{c2}\theta_2}{T_{c1}\theta_1/V_1^{*1/3} + T_{c2}\theta_2/V_2^{*1/3}} \tag{5-6.21}$$

and

$$P_{cT} = P^* \left[ 1 + (5.808 + 4.93\omega) \left( \frac{T_{cT}}{T^*} - 1 \right) \right] \tag{5-6.22}$$

with $T_{cT}$ from Eq. (5-6.2) as noted before.

Spencer et al. [73, 74] evaluated this approach and found an average deviation of about 1 bar when calculated values of $P_{cT}$ were compared against experimental data; 967 mixture data points were tested. Methane systems were not included, because all available correlations give significant error for such mixtures. The method is illustrated in Example 5-5. In most cases, Eq. (5-6.22) can be simplified by approximating $P^*$, $T^*$, and $\omega$ by mole fraction averages of the pure component properties $P_{ci}$, $T_{ci}$, and $\omega_i$ [40, 73, 74]. Average errors increase to only about 1.3 bar.

**Example 5-5** Estimate the critical pressure for a mixture of ethane and benzene which contains 39.2 mole percent ethane. The true critical pressure and temperature are reported to be 84.96 bar and 499.0 K [39].

**solution** Properties of the pure components from Appendix A are:

| Property | Ethane | Benzene |
|---|---|---|
| $T_c$, K | 305.4 | 562.2 |
| $P_c$, bar | 48.8 | 48.9 |
| $\omega$ | 0.099 | 0.212 |
| $V_c$, cm$^3$/mol | 148.3 | 259 |

From Kreglewski [44] $V^*$ (ethane) = 54.87 cm$^3$/mol and $V^*$ (benzene) = 93.97 cm$^3$/mol. Following Eqs. (5-6.13) to (5-6.21), with ethane = 1 and benzene = 2, we have

$$V_{12}^* = \frac{[(54.87)^{1/3} + (93.97)^{1/3}]^3}{8} = 72.7 \text{ cm}^3/\text{mol}$$

$$V^* = (54.87)(0.392) + (93.97)(0.608) + [(2)(72.7) - 54.87 - 93.97](0.392)(0.608) = 77.8 \text{ cm}^3/\text{mol}$$

$$\theta_1 = \frac{(0.392)(54.87)^{2/3}}{(0.392)(54.87)^{2/3} + (0.608)(93.97)^{2/3}} = 0.310$$

$$\theta_2 = 1 - 0.310 = 0.690$$

$$T_{12}^* = \frac{(2)(72.7)^{1/3}}{(54.87)^{1/3}/305.4 + (93.97)^{1/3}/562.2} = 406 \text{ K}$$

$$T^* = (77.8)^{1/3}\left\{\frac{(305.4)(0.310)}{(54.87)^{1/3}} + \frac{(562.2)(0.690)}{(93.97)^{1/3}} + \left[\frac{(2)(406)}{(72.7)^{1/3}} - \frac{305.4}{(54.87)^{1/3}} - \frac{562.2}{(93.97)^{1/3}}\right](0.310)(0.690)\right\}$$
$$= 462 \text{ K}$$

$$\omega_{12} = \frac{2}{1/0.099 + 1/0.212} = 0.135$$

$$\omega = (0.099)(0.310) + (0.212)(0.690) + [(2)(0.135) - 0.099 - 0.212](0.310)(0.690)$$
$$= 0.168$$

$$P^* = \frac{462}{(77.8)^{1/3}} \frac{(488)(0.310) + (489)(0.690)}{\dfrac{(305.4)(0.310)}{(54.87)^{1/3}} + \dfrac{(562.2)(0.690)}{(93.97)^{1/3}}}$$
$$= 48.0 \text{ bar}$$

Before using Eq. (5-6.22), $T_{cT}$ must be estimated. With Eqs. (5-6.1) and (5-6.2),

$$\theta_1 = \frac{(0.392)(148.3)}{(0.392)(148.3) + (0.608)(259)} = 0.270$$

$$T_{cT} = (0.270)(305.4) + (0.730)(562.2) = 493 \text{ K}$$

Then, with Eq. (5-6.22),

$$P_{cT} = 48.0 \left\{ 1 + [5.808 + (4.93)(0.168)] \left( \frac{493}{462} - 1 \right) \right\} = 69.4 \text{ bar}$$

$$\text{Error} = \frac{69.4 - 83.8}{83.8} \times 100 = -18\%$$

Spencer et al. [73] report an average deviation of −13 percent when the Kreglewski and Kay method is applied to this very nonlinear system. Tests in most other systems led to much less error.

An alternate method for predicting the critical pressures for mixtures was developed by Chueh and Prausnitz [12]. $P_{c_m}$ was related to $T_{cm}$ and $V_{c_m}$ by a modified Redlich-Kwong equation of state (see Secs. 3-6 and 4-5).

$$P_{cT} = \frac{RT_{cT}}{V_{cT} - b_m} - \frac{a_m}{T_{cT}^{1/2} V_{cT}(V_{cT} + b_m)} \tag{5-6.23}$$

where $T_{cT}$ and $V_{cT}$ are calculated from methods described earlier in this section. The mixture coefficients for determining $P_{cT}$ are defined as

$$b_m = \sum_j y_j b_j = \sum_j \frac{y_j \Omega_{b_j}^* RT_{c_j}}{P_{c_j}} \tag{5-6.24}$$

$$a_m = \sum_i \sum_j y_i y_j a_{ij} \tag{5-6.25}$$

with

$$\Omega_{b_j}^* = 0.0867 - 0.0125\omega_j + 0.011\omega_j^2 \tag{5-6.26}$$

$$a_{ii} = \frac{\Omega_{a_i}^* R^2 T_{c_i}^{2.5}}{P_{c_i}} \tag{5-6.27}$$

$$a_{ij} = \frac{(\Omega_{a_i}^* + \Omega_{a_j}^*) RT_{c_{ij}}^{1.5} (V_{c_i} + V_{c_j})}{4[0.291 - 0.04(\omega_i + \omega_j)]} \tag{5-6.28}$$

$$T_{c_{ij}} = (1 - k_{ij}) \sqrt{T_{c_i} T_{c_j}} \tag{5-6.29}$$

$$\Omega_{a_j}^* = \left( \frac{RT_{c_j}}{V_{c_j} - b_j} - P_{c_j} \right) \frac{P_{c_j} V_{c_j} (V_{c_j} + b_j)}{(RT_{c_j})^2} \tag{5-6.30}$$

The interaction parameter $k_{ij}$ usually ranges from 0.1 to 0.01. Values for a large number of binary systems have been tabulated [12].

Spencer et al. [74] have reported that the average deviation between $P_{cT}$ estimated from the Chueh-Prausnitz correlation and experimental values was 2 bar unless methane was one of the components. In such cases, the deviation was often much larger. Other techniques for estimating $P_{c_m}$ for systems containing methane have been suggested [2, 29, 77], but they either are limited to aliphatic hydrocarbons or are graphical with a trial-and-error solution. A method developed by Teja et al. [83] gave very large errors in some cases.

**Example 5-6** Repeat Example 5-5 by using the Chueh-Prausnitz method to estimate $P_{cT}$.

**solution** The properties of pure ethane and benzene are given in Example 5-5. Using the Chueh-Prausnitz methods to calculate $T_{cT}$ and $V_{cT}$ for this mixture gives $T_{cT}$ = 493 K and $V_{cT}$ = 185 cm³/mol; $k_{12}$ = 0.03 for this binary [12]. Then

$$\Omega_{b1}^* = 0.0867 - (0.0125)(0.099) + (0.011)(0.099)^2 = 0.0856$$

$$\Omega_{b2}^* = 0.0846$$

$$b_1 = \frac{(0.0856)(83.14)(305.4)}{48.8} = 44.52 \text{ cm}^3/\text{mol}$$

$$b_2 = 80.87 \text{ cm}^3/\text{mol}$$

$$b_m = (0.392)(44.52) + (0.608)(80.87) = 66.62 \text{ cm}^3/\text{mol}$$

$$\Omega_{a1}^* = \left(\frac{RT_{c1}}{V_{c1} - b_1} - P_{c1}\right)\frac{P_{c1}V_{c1}(V_{c1} + b_1)}{(RT_{c1})^2}$$

$$= \left[\frac{(83.14)(305.4)}{148.3 - 44.52} - 48.8\right]\left\{\frac{(48.8)(148.3)(148.3 + 44.52)}{[(83.14)(305.4)]^2}\right\}$$

$$= 0.424$$

$$\Omega_{a2}^* = 0.421$$

$$a_1 = \frac{\Omega_{a1}R^2T_{c1}^{2.5}}{P_{c1}} = \frac{(0.424)(83.14)^2(305.4)^{2.5}}{48.8}$$

$$= 97.88 \times 10^6 \text{ cm}^6\cdot\text{bar}\cdot\text{K}^{0.5}/\text{mol}^2$$

$$a_2 = 446.0 \times 10^6 \text{ cm}^6\cdot\text{bar}\cdot\text{K}^{0.5}/\text{mol}^2$$

$$T_{c12} = [(305.4)(562.2)]^{1/2}(1 - 0.03) = 401.9 \text{ K}$$

$$a_{12} = \frac{(\tfrac{1}{4})(0.424 + 0.421)(83.14)(401.9)^{1.5}(148.3 + 259)}{0.291 - (0.04)(0.099 + 0.212)}$$

$$= 206.9 \times 10^6 \text{ cm}^6\cdot\text{bar}\cdot\text{K}^{0.5}/\text{mol}^2$$

$$a_m = 10^6\,[(0.392)^2(97.88) + (0.608)^2(446.0) + (2)(0.392)(0.608)(206.9)]$$

$$= 278.5 \times 10^6 \text{ cm}^6\cdot\text{bar}\cdot\text{K}^{0.5}/\text{mol}^2$$

Then

$$P_{cT} = \frac{RT_{cT}}{V_{cT} - b_m} - \frac{a_m}{T_{cT}^{1/2}V_{cT}(V_{cT} + b_m)}$$

$$= \frac{(83.14)(493)}{185 - 66.62} - \frac{278.5 \times 10^6}{(493)^{1/2}(185)(185 + 66.62)}$$

$$= 76.8 \text{ bar}$$

$$\text{Error} = \frac{76.8 - 84.9}{84.9} \times 100 = -10\%$$

## Rigorous methods

The Gibbs criteria for the true critical point of a mixture with $n$ components may be expressed in various forms, but the most convenient (see pp. 250 and 251, Ref. 56) when using a pressure explicit equation of state is

$$L = \begin{vmatrix} A_{11} & A_{12} \ldots A_{1n} \\ A_{12} & A_{22} \\ \vdots & \\ A_{n1} & \ldots\ldots A_{nn} \end{vmatrix} = 0 \tag{5-6.31}$$

$$M = \begin{vmatrix} A_{11} & A_{12} \ldots A_{1n} \\ A_{21} & A_{22} \\ \vdots & \\ A_{n-1,1} & \ldots A_{n-1,n} \\ \dfrac{\partial L}{\partial N_1} & \cdots \dfrac{\partial L}{\partial N_n} \end{vmatrix} = 0 \tag{5-6.32}$$

where $$A_{12} = \left(\frac{\partial^2 \underline{A}}{\partial N_1 \partial N_2}\right)_{T,\underline{V}} \tag{5-6.33}$$

Thus, all the $A$ terms in Eqs. (5-6.31) and (5-6.32) are second derivatives of the total Helmholtz energy $\underline{A}$ with respect to moles at constant temperature and total volume $\underline{V}$. The determinants expressed by Eqs. (5-6.31) and (5-6.32) are solved simultaneously for the critical volume and critical temperature. The critical pressure is then found from the original equation of state. Peng and Robinson [61] used their equation of state to calculate mixture critical points; later Heidemann and Khalil [32], Michelsen and Heidemann [54], and Michelsen [53] used the Soave equation. Both of these equations of state gave more accurate estimates for critical temperatures and pressures than did the other empirical methods in this section. For example, the Soave equation predicts a critical pressure of 83.9 bar for the mixture in Example 5-5 for an error of only 0.1 percent. These equations of state should not be used for the prediction of critical volumes.

### Recommendations

To estimate the true critical temperature and pressure of a mixture, use Eqs. (5-6.31) and (5-6.32) along with either the Soave or the Peng-Robinson equation. Heidemann's [32, 54] method for solving these equations is particularly efficient, but it must still be carried out on a computer. For more rapid estimates of the true critical temperature of a hydrocarbon mixture, use the method of Li [Eq. (5-6.2)]. If the mixture contains nonhydrocarbons, the Chueh-Prausnitz correlation is preferred [Eq. (5-6.4)], although the interaction parameter $\tau_{ij}$ can be evaluated only for relatively simple binary pairs. For rapid estimates of the true critical pressure of a mixture, either the Kreglewski-Kay or the Chueh-Prausnitz method may be used. Neither is particularly applicable for systems containing methane. Errors found are usually considerably larger for $P_{cT}$ estimations than for $T_{cT}$.

To estimate the true critical volume of a mixture, Eq. (5-6.8) is recommended. However, as above, the interaction parameter $\nu_{ij}$ can be determined for only a limited number of binary types. The Soave and Peng-Robinson equations do not give accurate estimates of mixture critical volumes.

The empirical methods in this section have been developed and tested primarily with hydrocarbon mixtures. When these estimation methods are applied to nonhydrocarbon mixtures, no reliable estimate of the error can be given.

## 5-7 Heat Capacities of Liquids

There are three liquid heat capacities in common use: $C_{pL}$, $C_{\sigma L}$, and $C_{\text{sat}L}$. The first represents the change in enthalpy with temperature at constant pressure; the second shows the variation in enthalpy of a saturated liquid with temperature; the third indicates the energy required to effect a temperature change while maintaining the liquid in a saturated state. The three heat capacities are related as follows:

$$C_{\sigma L} = \frac{dH_{\sigma L}}{dT} = C_{pL} + \left[ V_{\sigma L} - T\left(\frac{\partial V}{\partial T}\right)_p \right] \left(\frac{dP}{dT}\right)_{\sigma L} = C_{\text{sat}L} + V_{\sigma L}\left(\frac{dP}{dT}\right)_{\sigma L} \tag{5-7.1}$$

The term $(dP/dT)_{\sigma L}$ represents the change in vapor pressure with temperature. Except at high reduced temperatures, all three forms of the liquid heat capacity are in close numerical agreement. Most estimation techniques yield either $C_{pL}$ or $C_{\sigma L}$, although $C_{\text{sat}L}$ is often the quantity measured experimentally.

Liquid heat capacities are not strong functions of temperature except above $T_r = 0.7$ to 0.8. In fact, a shallow minimum is often reported at temperatures slightly below the normal boiling point. At high reduced temperatures, liquid heat capacities are large and are strong functions of temperature. The general trend is illustrated in Fig. 5-3 for propylene.

Near the normal boiling point, most liquid organic compounds have heat capacities between 1.2 and 2 J/(g·K). In this temperature range, there is essentially no effect of pressure [20].

Experimentally reported liquid heat capacities of hydrocarbons have been correlated in a nomograph [20] and expressed in analytical form [30, 31, 78]. A tabulation of available data is given by San José [69].

Estimation methods applicable for liquid heat capacities fall into four general categories: theoretical, group contribution, corresponding states, and Watson's thermodynamic cycle [65, 87]. Group contribution and cor-

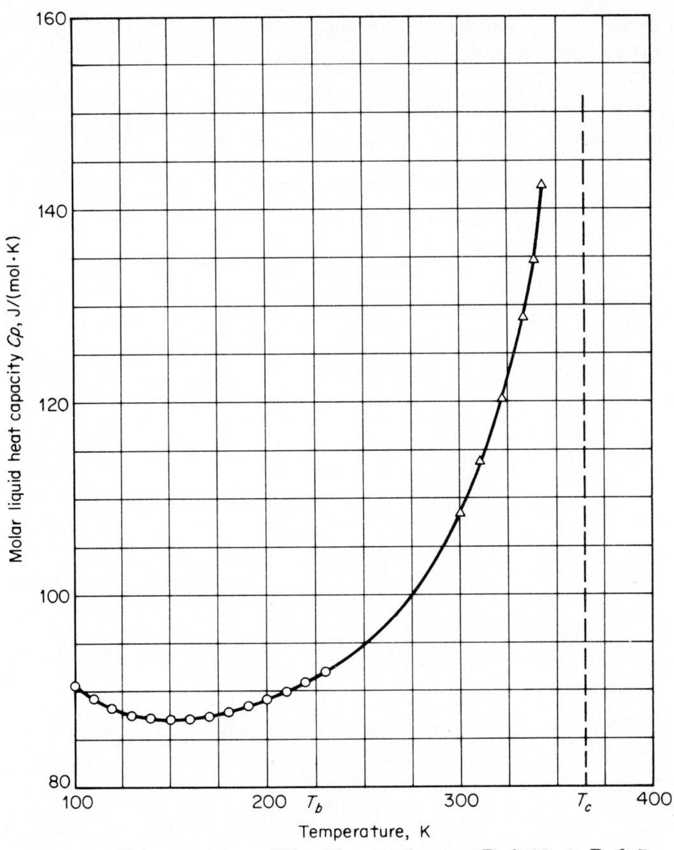


**Figure 5-3** Heat capacity of liquid propylene. ○, Ref. 63; Δ, Ref. 5.

responding states methods are described below, and recommendations are presented at the end of the section. Theoretical methods are based on the estimation of liquid heat capacities at constant volume by considering each mode of energy storage separately. Reliable estimation procedures have not yet been developed for engineering use, although Bondi [7, 8], for high-molecular-weight liquids and polymers, has suggested some useful approximations that are particularly valuable. A similar, earlier treatment was published by Sakiadis and Coates [68]. Ogiwara et al. [57] have given a group contribution method for constant-volume liquid heat capacities.

### Group contribution methods

The assumption is made that various groups in a molecule contribute a definite value to the total molar heat capacity that is independent of other

**TABLE 5-10 Group Contributions for Molar Liquid Heat Capacity at 293 K for the Chueh-Swanson Method, J/(mol·K) [10]**

| Group | Value† |
|---|---|
| **Alkane** | |
| $-CH_3$ | 36.8 |
| $-CH_2-$ | 30.4 |
| $-\overset{\vert}{C}H-$ | 21.0 |
| $-\overset{\vert}{\underset{\vert}{C}}-$ | 7.36 |
| **Olefin** | |
| $=CH_2$ | 21.8 |
| $=\overset{\vert}{C}-H$ | 21.3 |
| $=\overset{\vert}{C}-$ | 15.9 |
| **Alkyne** | |
| $-C{\equiv}H$ | 24.7 |
| $-C{\equiv}$ | 24.7 |
| **In a Ring** | |
| $-\overset{\vert}{C}H-$ | 18 |
| $-\overset{\vert}{C}=$ or $-\overset{\vert}{\underset{\vert}{C}}-$ | 12 |
| $-CH=$ | 22 |
| $-CH_2-$ | 26 |
| **Nitrogen** | |
| $H-\overset{\overset{H}{\vert}}{N}-$ | 58.6 |
| $-\overset{\overset{H}{\vert}}{N}-$ | 43.9 |
| $-\overset{\vert}{N}-$ | 31 |
| $-N=$ (in a ring) | 19 |
| $-C{\equiv}N$ | 58.2 |

| Group | Value† |
|---|---|
| **Oxygen** | |
| $-O-$ | 35 |
| $>C=O$ | 53.0 |
| $-\overset{\overset{H}{\vert}}{C}=O$ | 53.0 |
| $-\overset{\overset{O}{\Vert}}{C}-OH$ | 79.9 |
| $-\overset{\overset{O}{\Vert}}{C}-O-$ | 60.7 |
| $-CH_2OH$ | 73.2 |
| $-\overset{\vert}{C}HOH$ | 76.1 |
| $-\overset{\vert}{\underset{\vert}{C}}COH$ | 111.3 |
| $-OH$ | 44.8 |
| $-ONO_2$ | 119.2 |
| **Halogen** | |
| $-Cl$ (first or second on a carbon) | 36 |
| $-Cl$ (third or fourth on a carbon) | 25 |
| $-Br$ | 38 |
| $-F$ | 17 |
| $-I$ | 36 |
| **Sulfur** | |
| $-SH$ | 44.8 |
| $-S-$ | 33 |
| **Hydrogen** | |
| $H-$ (for formic acid, formates, hydrogen cyanide, etc.) | 15 |

†Add 18.8 for any carbon group which fulfills the following criterion: a carbon group which is joined by a single bond to a carbon group connected by a double or triple bond with a third carbon group. In some cases a carbon group fulfills the above criterion in more ways than one. In these cases 18.8 should be added each time the group fulfills the criterion. The following are exceptions to the 18.8 addition rule:

1. No 18.8 additions for $-CH_3$ groups.
2. For a $-CH_2-$ group fulfilling the 18.8 addition criterion, add 10.5 instead of 18.8. However, when the $-CH_2-$ group fulfills the addition criterion in more ways than one, the addition should be 10.5 the first time and 18.8 for each subsequent addition. (See Example 5-7.)
3. No 18.8 addition for any carbon group in a ring.

groups present. Johnson and Huang [38], Shaw [72], and Chueh and Swanson [10] have all proposed values for different molecular groups to estimate $C_{pL}$ at room temperature. Shaw's method is applicable at 298 K, and the others are to be used at 293 K. Both the Shaw and Chueh-Swanson methods are accurate, but the latter is more general; group contributions for this technique are shown in Table 5-10. Another additive method in which structural increments are given from −25 to 100°C has been suggested by Missenard [55]; group contributions for this scheme are given in Table 5-11. The Chueh-Swanson and Missenard methods are illustrated in Examples 5-7 and 5-8. Missenard's method cannot be used for compounds with double bonds, nor should it be used if the temperature corresponds to a reduced temperature in excess of 0.75. When $T_r <$ 0.75, the estimated value may be considered to be $C_{pL}$, $C_{\sigma L}$, or $C_{satL}$, because the three are essentially identical at low reduced temperatures.

**TABLE 5-11 Group Contributions for Missenard Method, J/(mol·K) [55]**

| Group | Temperature, K | | | | | |
|---|---|---|---|---|---|---|
| | 248 | 273 | 298 | 323 | 348 | 373 |
| −H | 12.5 | 13.4 | 14.6 | 15.5 | 16.7 | 18.8 |
| $-CH_3$ | 38.5 | 40.0 | 41.6 | 43.5 | 45.8 | 48.3 |
| $-CH_2$ | 27.2 | 27.6 | 28.2 | 29.1 | 29.9 | 31.0 |
| −CH− | 20.9 | 23.8 | 24.9 | 25.7 | 26.6 | 28.0 |
| −C− (with two additional vertical bonds) | 8.4 | 8.4 | 8.4 | 8.4 | 8.4 | |
| −C≡C− | 46.0 | 46.0 | 46.0 | 46.0 | | |
| −O− | 28.9 | 29.3 | 29.7 | 30.1 | 30.5 | 31.0 |
| −CO− (ketone) | 41.8 | 42.7 | 43.5 | 44.4 | 45.2 | 46.0 |
| −OH | 27.2 | 33.5 | 43.9 | 52.3 | 61.7 | 71.1 |
| −COO− (ester) | 56.5 | 57.7 | 59.0 | 61.1 | 63.2 | 64.9 |
| −COOH | 71.1 | 74.1 | 78.7 | 83.7 | 90.0 | 94.1 |
| $-NH_2$ | 58.6 | 58.6 | 62.8 | 66.9 | | |
| −NH− | 51.0 | 51.0 | 51.0 | | | |
| −N− (with one additional vertical bond) | 8.4 | 8.4 | 8.4 | | | |
| −CN | 56.1 | 56.5 | 56.9 | | | |
| $-NO_2$ | 64.4 | 64.9 | 65.7 | 66.9 | 68.2 | |
| −NH−NH− | 79.5 | 79.5 | 79.5 | | | |
| $C_6H_5-$ (phenyl) | 108.8 | 113.0 | 117.2 | 123.4 | 129.7 | 136.0 |
| $C_{10}H_7-$ (naphthyl) | 179.9 | 184.1 | 188.3 | 196.6 | 205. | 213. |
| −F | 24.3 | 24.3 | 25.1 | 25.9 | 27.0 | 28.2 |
| −Cl | 28.9 | 29.3 | 29.7 | 30.1 | 30.8 | 31.4 |
| −Br | 35.1 | 35.6 | 36.0 | 36.4 | 37.2 | 38.1 |
| −I | 39.3 | 39.7 | 40.4 | 41.0 | | |
| −S− | 37.2 | 37.7 | 38.5 | 39.3 | | |

Errors for the Chueh-Swanson method rarely exceed 2 to 3 percent and those for Missenard's method ±5 percent.

**Example 5-7** Estimate the liquid heat capacity of 1,4-pentadiene at 20°C by using the Chueh-Swanson group contribution method.

**solution** From Table 5-10

$$C_{pL}\,(20°\mathrm{C}) = 2(\mathrm{CH_2{=}}) + 2(-\mathrm{CH{=}}) + -\mathrm{CH_2^-} + \text{corrections noted in Table 5-10}$$
$$= (2)(21.8) + (2)(21.3) + 30.4 + 10.5 + 18.8$$
$$= 146\ \mathrm{J/(mol\cdot K)}$$

Tamplin and Zuzic [78] indicate that $C_{pL} = 147$ J/(mol·K) at 293 K.

**Example 5-8** By using Missenard's group contribution method, estimate the liquid heat capacity of isopropyl alcohol at 273 K.

**solution** With Table 5-11

$$C_{pL}\,(0°\mathrm{C}) = 2(-\mathrm{CH_3}) + -\overset{|}{\underset{|}{\mathrm{C}}}\mathrm{H} + -\mathrm{OH}$$
$$= (2)(40.0) + 23.8 + 33.5$$
$$= 137.2\ \mathrm{J/(mol\cdot K)}$$

The experimental value is 135.8 J/(mol·K) [25].

### Corresponding states methods

Several corresponding states methods for liquid heat capacity estimation have been cast in the form of Eq. (5-5.4). For example, with Tables 5-8 and 5-9, one can estimate the heat capacity departure function $C_{pL} - C_p^\circ$ for liquids as well as for gases. Good results have also been reported by using an analytical form of the Lee-Kesler heat capacity departure function [47] for calculating liquid heat capacities for hydrocarbons [13]. Bondi [7] has reviewed many forms. A modification† of one of his equations that was suggested originally by Rowlinson [66] is

$$\frac{C_{pL} - C_p^\circ}{R} = 1.45 + 0.45(1 - T_r)^{-1} + 0.25\omega[17.11 + 25.2(1 - T_r)^{1/3}T_r^{-1} + 1.742(1 - T_r)^{-1}] \tag{5-7.2}$$

$C_{pL}$, $C_{\sigma L}$, and $C_{\mathrm{sat}L}$ can also be related to each other by the approximate corresponding states relations

$$\frac{C_{pL} - C_{\sigma L}}{R} = \exp(20.1T_r - 17.9) \tag{5-7.3}$$

and

$$\frac{C_{\sigma L} - C_{\mathrm{sat}L}}{R} = \exp(8.655T_r - 8.385) \tag{5-7.4}$$

†This equation has been given incorrectly in Refs. 7, 8, 64, and 81.

Equations (5-7.3) and (5-7.4) are valid for $T_r < 0.99$. Below $T_r \sim 0.8$, $C_{\sigma L}$, $C_{pL}$, and $C_{\text{sat}L}$ may be considered to be the same number.

**Example 5-9** Estimate the liquid heat capacity of *cis*-2-butene at 349.8 K by using the Rowlinson-Bondi corresponding states correlation.

**solution** From Appendix A, $T_c = 435.6$ K and $\omega = 0.202$. From [76], $C_p^\circ = 91.00$ J/(mol·K). The reduced temperature is 349.8/435.6 = 0.803. Equation (5-7.2) gives

$$\frac{C_{pL} - C_p^\circ}{R} = 1.45 + \frac{0.45}{1 - 0.803}$$

$$+ (0.25)(0.202)\left[17.11 + \frac{(25.2)(1 - 0.803)^{1/3}}{0.803} + \frac{1.742}{1 - 0.803}\right]$$

$$= 5.97$$

Equation (5-7.3) gives

$$\frac{C_{pL} - C_{\sigma L}}{R} = \exp(20.1T_r - 17.9)$$

$$= 0.172$$

$$C_{\sigma L} = 8.314(5.97 - 0.172) + 91$$

$$= 139.2 \text{ J/(mol·K)}$$

The experimental value is 152.7 J/(mol·K) [71], so error = (139.2 − 152.7)/152.7 × 100 = −8.8%.

## Discussion

Three methods for estimating liquid heat capacities have been described. The Chueh-Swanson method is a group contribution method and is applicable only at 20°C. The Missenard method also is a group contribution method; values for the groups are given in Table 5-11 for temperatures between 248 and 373 K. The Missenard method should not be used when $T_r > 0.75$. The Rowlinson-Bondi method, Eq. (5-7.2), is a corresponding states method and is applicable at low values of $T_r$ as well as at values approaching unity. The Rowlinson-Bondi method requires $C_p^\circ$, $T_c$, and $\omega$. A comparison with experimental values is given in Table 5-12. Errors are generally less than 5 percent except for alcohols at low temperatures, for which Eq. (5-7.2) is not applicable. Other methods for estimating liquid heat capacities have been published [1, 30, 31, 49, 50, 81, 92]. These methods usually require additional parameters which result in only marginal improvement or are applicable to limited classes of compounds.

For liquid mixtures, Teja [82] has examined the extension of the multiple reference fluid method [see Eq. (3-7.5)] to liquid heat capacity prediction. This method avoids the assumption that the mixture molar heat capacity is a mole fraction average of the pure component values [15], an assumption which neglects any contribution due to the effect of temperature on heats of mixing.

**TABLE 5-12 Percent Error in Liquid Heat Capacity Calculated by Eq. (5-7.2)†**

| Compound | $T$, K | $C_{\sigma L}$(exp.), J/(mol·K) | Ref. | Percent error‡ |
|---|---|---|---|---|
| Methane | 102.3 | 54.8 | 90 | 0 |
| | 140.5 | 59.8 | | 0.8 |
| | 180.9 | 92.9 | | 0.7 |
| Propane | 100 | 84.9 | 43 | 7.4 |
| | 150 | 87.9 | | 2.3 |
| | 200 | 93.3 | | 0.9 |
| | 305.5 | 115.1 | 67 | 4.6 |
| | 344.7 | 150.2 | | −1.1 |
| *n*-Pentane | 200 | 144.3 | 52 | −3.1 |
| | 250 | 153.6 | | 1.5 |
| | 300 | 167.8 | | 0.8 |
| | 363 | 193.3 | 4 | −0.2 |
| | 443 | 273.2 | | −8.3 |
| *n*-Heptane | 300 | 225.5 | 16,52 | −0.9 |
| | 400 | 270.3 | 16 | −0.1 |
| | 503 | 355.6 | 4 | −4.8 |
| *n*-Decane | 250 | 297.5 | 52 | −0.9 |
| | 320 | 325.5 | | 1.2 |
| Cyclohexane | 305.3 | 156.6 | 5 | −1.3 |
| | 360.9 | 179.9 | | −0.9 |
| *cis*-2-Butene | 305.3 | 131.4 | 71 | −5.1 |
| | 349.8 | 152.7 | | −8.9 |
| Isopropylbenzene | 305.3 | 212.1 | 71 | 2.3 |
| | 333.1 | 225.5 | | 1.6 |
| | 366.4 | 243.1 | | 0.1 |
| Ethyl alcohol | 208 | 91.2 | 41 | 84 |
| | 294 | 110.0 | | 44 |
| | 383 | 159.0 | 19 | 4 |
| Acetone | 297 | 124.7 | 42 | 3.9 |
| Ethylene oxide | 170 | 82.4 | 23 | 5.0 |
| | 230 | 82.4 | | 2.7 |
| | 280 | 86.6 | | 2.5 |
| Ethyl ether | 290 | 170.7 | 59 | −5.3 |
| Ethyl chloride | 150 | 96.2 | 27 | −1.6 |
| | 200 | 95.8 | | −2.1 |
| | 250 | 98.7 | | −1.2 |
| | 290 | 103.3 | | −0.2 |
| Ethyl mercaptan | 208 | 111.7 | 51 | −7.5 |
| | 275 | 115.5 | | −3.2 |
| | 315 | 120.1 | | −2.2 |
| Chlorine | 200 | 66.5 | 24 | −5.7 |
| | 240 | 65.7 | | −3.0 |

†When $C_{pL}$ or $C_{\mathrm{sat}L}$ was reported, it was converted to $C_{\sigma L}$.
‡Percent error = [(calc. − exp.)/exp.] × 100. Ideal-gas heat capacities were taken from Refs. 3, 18, 28, 35, 76, Appendix A, and the ASPEN data bank.

### Recommendations

Use the Rowlinson-Bondi method Eq. (5-7.2) to estimate the liquid heat capacity of nonpolar or slightly polar compounds. For compounds for which the critical temperature is unknown, or for alcohols at low temperatures ($T_r < 0.75$), use the Missenard group contribution method.

## 5-8 Vapor Phase Fugacity of a Component in a Mixture

From thermodynamics, the chemical potential or fugacity can be related to the Gibbs or Helmoltz energies. Using the latter gives

$$\mu_i \equiv \left(\frac{\partial \underline{A}}{\partial N_i}\right)_{T,\ \underline{V},\ N_j[i]} \tag{5-8.1}$$

The subscripts on the partial derivative indicate that the temperature, total system volume, and all mole numbers (except $i$) are to be held constant. Thus, one may take the Helmholtz energy function for a particular equation of state, as shown in Table 5-1, and find $\mu_i$ by differentiation. Since the functions so given are expressed as the difference in specific Helmholtz energy between the real state and the chosen reference state, one must multiply the entire expression by $N$, the total moles, before differentiating. Then

$$\mu_i - \mu_i^\circ = \frac{\partial}{\partial N_i}(\underline{A} - \underline{A}^\circ)_{T,\underline{V},N_j[i]} \tag{5-8.2}$$

But this difference in chemical potential is related to fugacity by

$$\mu_i - \mu_i^\circ = RT \ln \frac{\hat{f}_i}{\hat{f}_i^\circ} \tag{5-8.3}$$

The fugacity $\hat{f}_i$ refers to the value for component $i$ in the mixture and $\hat{f}_i^\circ$ to the reference state at $T$, $P^\circ$, $V^\circ$, $N$. However, the reference state was previously chosen as an ideal-gas state. For such cases, as a part of the definition of fugacity,

$$\hat{f}_i^\circ = P^\circ y_i \tag{5-8.4}$$

Thus
$$RT \ln \frac{\hat{f}_i}{P^\circ y_i} = \frac{\partial}{\partial N_i}(\underline{A} - \underline{A}^\circ)_{T,\underline{V},N_j[i]} \tag{5-8.5}$$

and from Eq. (5-3.5), multiplying by $N$, we have

$$\underline{A} - \underline{A}^\circ = -\int_\infty^{\underline{V}} \left(P - \frac{NRT}{\underline{V}}\right) d\underline{V} - NRT \ln \frac{\underline{V}}{\underline{V}^\circ} \tag{5-8.6}$$

With Eqs. (5-8.5) and (5-8.6), noting that $\underline{V} = ZNRT/P$, $\underline{V}^\circ = NRT/P^\circ$, we have

$$RT \ln \phi_i = -\int_\infty^{\underline{V}} \left[ \left( \frac{\partial P}{\partial N_i} \right)_{T,\underline{V},N_j[i]} - \frac{RT}{\underline{V}} \right] d\underline{V} - RT \ln Z \tag{5-8.7}$$

or

$$RT \ln \phi_i = \left( \frac{\partial(\underline{A} - \underline{A}^\circ)}{\partial N_i} \right)_{T,\underline{V},N_j[i]} - RT \ln \frac{P}{P^\circ} \tag{5-8.8}$$

$\phi_i$ is the fugacity coefficient of $i$ in the gas mixture.

To obtain a usable relation for $\phi_i$, Eq. (5-8.7) must be integrated; but before it can be, the derivative of $P$ with respect to $N_i$ must be found. Thus any pressure-explicit equation of state is convenient provided that the composition dependence of all the parameters can be expressed in analytical form.

For the analytical equations of state covered in Chap. 3, mixture combining rules are given in Chap. 4; thus evaluation of the integral in Eq. (5-8.7) is possible. For example, the original Redlich-Kwong equation is given as Eq. (3-6.1), and this same relation expressed in terms of total volume would be

$$P = \frac{NRT}{\underline{V} - Nb} - \frac{aN^2}{\underline{V}(\underline{V} + Nb)} \tag{5-8.9}$$

In the differentiation indicated in Eq. (5-8.7), the variables are $N$, $a$, and $b$, where the parameters $a$ and $b$ are shown as functions of composition in Eqs. (4-5.1), (4-5.2), and (3-6.1). The final result is

$$\ln \phi_i = \ln \frac{V}{V - b} + \frac{b_i}{V - b} - \ln Z + \frac{ab_i}{RTb^2} \left( \ln \frac{V + b}{V} - \frac{b}{V + b} \right) - \frac{2 \sum_j y_i a_{ij}}{RTb} \ln \frac{V + b}{V} \tag{5-8.10}$$

For all the analytical equations of state, the working equations for $\ln \phi_i$ are given in Table 5-13.

It is difficult to evaluate these expressions, since fugacity coefficients themselves are difficult to determine. Presumably, if the mixture equation of state is a valid representation, the derived property $\phi_i$ should also be accurate. This assumes, of course, that the equation of state not only yields accurate predictions of volumetric properties but also that it yields accurate derivatives of pressure with respect to mole numbers. These two attributes are not necessarily compatible.

Vapor phase fugacity expressions are of value in vapor-liquid equilibrium calculations, as described in Chap. 8.

$$\phi_i \equiv \frac{\hat{f}_i}{y_i P}$$

Virial, $Z = 1 + \frac{BP}{RT}$ [Eqs. (3-5.2$a$) and (4-4.1)]

$$\ln \phi_i = \left(2 \sum_j y_j B_{ij} - B\right) \frac{P}{RT} \tag{5-8.11}$$

Cubic, $P = \frac{RT}{V - b} - \frac{a}{V^2 + ubV + wb^2}$ [Eqs. (3-6.1), (4-5.1), and (4-5.2) and Table 3-5]

$$\ln \phi_i = \frac{b_i}{b}(Z - 1) - \ln(Z - B^*) + \frac{A^*}{B^* \sqrt{u^2 - 4w}} \left(\frac{b_i}{b} - \delta_i\right) \ln \frac{2Z + B^*(u + \sqrt{u^2 - 4w})}{2Z + B^*(u - \sqrt{u^2 - 4w})} \tag{5-8.12}$$

where $$\frac{b_i}{b} = \frac{T_{ci}/P_{ci}}{\sum_j y_j T_{cj}/P_{cj}} \tag{5-8.13}$$

$$\delta_i = \frac{2a_j^{1/2}}{a} \sum_j x_j a_j^{1/2}(1 - \overline{k}_{ij}) \tag{5-8.14}$$

If all $\overline{k}_{ij} = 0$, this reduces to

$$\delta_i = 2\left(\frac{a_i}{a}\right)^{1/2} \tag{5-8.15}$$

Lee-Kesler [Eqs. (3-7.1) to (3-7.4) and Table 4-3]

$$\ln \phi_i = \ln\left(\frac{f}{P}\right)_m + \frac{H^\circ - H}{TRT_{cm}} \sum_{j \neq i} y_j \left(\frac{dT_{cm}}{dy_j}\right)_{y_k} + \frac{Z_m - 1}{P_{cm}} \sum_{j \neq i} y_j \left(\frac{dP_{cm}}{dy_j}\right)_{y_k} - \ln\left(\frac{f}{P}\right)_m^{(1)} \sum_{j \neq i} y_j \left(\frac{d\omega_m}{dy_j}\right)_{y_k} \tag{5-8.16}$$

$$\left(\frac{dT_{cm}}{dy_j}\right)_{y_k} = \left[2 \sum_l y_l (V_{clj}^{1/4} T_{clj} - V_{cli}^{1/4} T_{cli}) - \frac{0.25}{V_{cm}^{3/4}} \left(\frac{dV_{cm}}{dy_j}\right)_{y_k} T_{cm}\right] \Big/ V_{cm}^{1/4} \tag{5-8.17}$$

$$\left(\frac{dV_{cm}}{dy_j}\right)_{y_k} = 2 \sum_l y_l (V_{clj} - V_{cli}) \tag{5-8.18}$$

$$\left(\frac{dP_{cm}}{dy_j}\right)_{y_k} = P_{cm}\left[-\frac{0.085(\omega_j - \omega_i)}{Z_{cm}} + \frac{1}{T_{cm}}\left(\frac{dT_{cm}}{dy_j}\right)_{y_k} - \frac{1}{V_{cm}}\left(\frac{dV_{cm}}{dy_j}\right)_{y_k}\right] \tag{5-8.19}$$

Note: In Eqs. (5-8.16) to (5-8.19), $k \neq i, j$. $\ln (f/P)_m$ is given in Eq. (5-4.11); $(H^\circ - H)/RT_{cm}$ is given by Eq. (5-4.3); and $T_{cm}$, $V_{cm}$, $\omega_m$, and $P_{cm}$ are given in Table 4-3. $\log (f/P)_m^{(1)}$ [not $\ln (f/P)_m^{(1)}$] is given in Table 5-7.

## Notation

| | |
|---|---|
| $a$ | cubic-equation-of-state constant, Table 3-5 |
| $A$ | Helmholtz energy, J/mol |
| $\underline{A}$ | Helmholtz energy, J |
| $b$ | cubic-equation-of-state constant, Table 3-5 |
| $C$ | heat capacity, J/(mol·K); $C_p$, at constant pressure; $C_v$, at constant volume; $C_{\sigma L}$, variation of saturated liquid enthalpy with temperature; $C_{\text{sat}L}$, $(dQ/dT)_{SL}$ |
| $f$ | fugacity, bar; $\hat{f}_i$ fugacity of $i$ in a mixture |
| $G$ | Gibbs energy, J/mol |
| $H$ | enthalpy, J/mol |
| $\Delta H_v$ | enthalpy of vaporization, J/mol |
| $M$ | molecular weight |
| $N$ | total moles |
| $P$ | pressure, usually bar; $P_c$, at the critical point; $P_r$, $P/P_c$ |
| $Q$ | heat, J |
| $R$ | gas constant, 8.314 J/(mol·K) |
| $S$ | entropy, J/(mol·K) |
| $T$ | temperature, K; $T_c$, at the critical point; $T_r$, $T/T_c$; $T_b$, at the normal boiling point |
| $U$ | internal energy, J/mol |
| $V$ | volume, $cm^3$/mol; $V_c$, at the critical point; $V_r$, $V/V_c$ |
| $\underline{V}$ | volume, $cm^3$ |
| $x_i$ | mole fraction $i$ |
| $y_i$ | mole fraction $i$ |
| $Z$ | compressibility factor; $Z_c$, at the critical point |

GREEK

| | |
|---|---|
| $\delta_T$, $\delta_v$ | parameters in Eqs. (5-6.7) and 5-6.11) |
| $\theta$ | surface fraction, Eqs. (5-6.3) and (5-6.15) |
| $\mu$ | chemical potential |
| $\nu_{ij}$ | interaction parameter in Eq. (5-6.8) |
| $\tau_{ij}$ | interaction parameter in Eq. (5-6.4) |
| $\phi_i$ | fugacity coefficient, $\hat{f}_i/Py_i$ |
| $\phi$ | volume fraction defined in Eq. (5-6.1) |
| $\psi_T$, $\psi_v$ | parameter in Eqs. (5-6.6) and (5-6.10) |
| $\omega$ | Pitzer acentric factor |

SUPERSCRIPTS

| | |
|---|---|
| ° | reference state or an ideal-gas state |
| (0) | simple fluid function |
| $(R)$ | simple fluid function for the reference fluid |
| $R$ | reference fluid |
| $SL$ | saturated liquid |
| $SV$ | saturated vapor |
| $L$ | liquid |
| $EX$ | excess function |
| (1) | deviation function |

SUBSCRIPTS

| | |
|---|---|
| $b$ | normal boiling point |
| $c$ | critical state |
| $SCL$ | subcooled liquid |
| $SL$ | saturated liquid |
| $SV$ | saturated vapor |
| $m$ | mixture |
| $T$ | true critical property of mixture |
| $\sigma_L$ | saturated liquid state |

## References


1. Akhmedov, A. G.: *Russ. J. Phys. Chem.*, **54:** 1341 (1980).
2. Akiyama, T., and G. Thodos: *Can. J. Chem. Eng.*, **48:** 311 (1970).
3. American Petroleum Institute: *Selected Values of Physical and Thermodynamic Properties of Hydrocarbons and Selected Compounds*, project 44, Carnegie Press, Pittsburgh, Pa., 1953, and supplements.
4. Amirkhanov, Kh. I., B. G. Alibekov, D. I. Vikhrov, V. A. Mirskaya, and L. N. Levina: *Teplofiz. Vys. Temp.*, **9:** 1211 (1971).
5. Auerbach, C. E., B. H. Sage, and W. N. Lacey: *Ind. Eng. Chem.*, **42:** 110 (1950).
6. Bier, K., G. Ernst, J. Kunze, and G. Maurer: *J. Chem. Thermodyn.*, **6:** 1039 (1974).
7. Bondi, A.: *Ind. Eng. Chem. Fundam.*, **5:** 443 (1966).
8. Bondi, A.: *Physical Properties of Molecular Crystals, Liquids and Glasses*, Wiley, New York, 1968.
9. Breedveld, G. J. F., and J. M. Prausnitz: *AIChE J.*, **19:** 783 (1973).
10. Chueh, C. F., and A. C. Swanson: (*a*) *Chem. Eng. Progr.*, **69**(7): 83 (1973); (*b*) *Can. J. Chem. Eng.*, **51:** 596 (1973).
11. Chueh, P. L., and C. H. Deal: *AIChE J.*, **19:** 138 (1973).
12. Chueh, P. L., and J. M. Prausnitz: *AIChE J.*, **13:** 1099 (1967).
13. Daubert, T. E.: private communication, 1975.
14. Dillard, D. D., W. C. Edmister, J. H. Erbar, and R. L. Robinson, Jr.: *AIChE J.*, **14:** 923 (1968).
15. Dimoplon, W.: *Chem. Eng.*, **74**(22): 64 (1972).

16. Douglas, T. B., G. T. Furukawa, R. E. McCoskey, and A. F. Ball: *J. Res. Natl. Bur. Stand.,* **53:** 139 (1954).
17. Edmister, W. C.: *Hydrocarbon Process. Petrol. Refiner,* **46**(5): 187 (1967).
18. Eucken, Von A., and E. U. Franck: *Z. Electrochem.,* **52:** 195 (1948).
19. Fick, E. F., D. C. Ginnings, and W. B. Holten: *J. Res. Natl. Bur. Stand.,* **6:** 881 (1931).
20. Gambill, W. R.: *Chem. Eng.,* **64**(5): 263, **64**(6): 243, **64**(7): 263, **64**(8): 257 (1957).
21. Garcia-Rangel, S., and L. C. Yen: "Evaluation of Generalized Correlations for Mixture Enthalpy Predictions," paper presented at *159th Natl. Mtg. Am. Chem. Soc., Houston, Tex., February 1970.*
22. Ghormley, E. L., and J. M. Lenoir: *Can. J. Chem. Eng.,* **50:** 89 (1972).
23. Giauque, W. F., and J. Gordon: *J. Am. Chem. Soc.,* **71:** 2176 (1949).
24. Giauque, W. F., and T. M. Powell: *J. Am. Chem. Soc.,* **61:** 1970 (1939).
25. Ginnings, D. C., and R. J. Corruccini: *Ind. Eng. Chem.,* **40:** 1990 (1948).
26. Gmehling, J., D. D. Liu, and J. M. Prausnitz: *Chem. Eng. Sci.,* **34:** 951 (1979).
27. Gordon, J., and W. F. Giauque: *J. Am. Chem. Soc.,* **70:** 1506 (1948).
28. Green, J. H. S.: *Trans. Faraday Soc.,* **57:** 2132 (1961).
29. Grieves, R. B., and G. Thodos: *AIChE J.,* **9:** 25 (1963).
30. Hadden, S. T.: *Hydrocarbon Process. Petrol. Refiner,* **45**(7): 137 (1966).
31. Hadden, S. T.: *J. Chem. Eng. Data,* **15:** 92 (1970).
32. Heidemann, R. A., and A. M. Khalil: *AIChE J.,* **26:** 769 (1980).
33. Hicks, C. P., and C. L. Young: *Chem. Rev.,* **75:** 119 (1975).
34. Huang, P. K., and T. E. Daubert: *Ind. Eng. Chem. Process Design Develop.,* **13:** 359 (1974).
35. "JANAF Thermochemical Tables," 2d ed., *NSRDS-NBS 37,* June 1971.
36. Joffe, J.: *Ind. Eng. Chem. Fundam.,* **12:** 259 (1973).
37. Joffe, J., and D. Zudkevitch: paper presented at *159th Natl. Mtg., Am. Chem. Soc., Houston, Tex., February 1970.*
38. Johnson, A. I., and C. J. Huang: *Can. J. Technol.,* **33:** 421 (1955).
39. Kay, W. B., and T. D. Nevens: *Chem. Eng. Progr. Symp. Ser.,* **48**(3): 108 (1952).
40. Kay, W. B., D. W. Hissong, and A. Kreglewski: *Ohio State Univ. Res. Found. Rept. 3,* API Proj. PPC 15.8, 1968.
41. Kelley, K. K.: *J. Am. Chem. Soc.,* **51:** 770 (1929).
42. Kelley, K. K.: *J. Am. Chem. Soc.,* **51:** 1145 (1929).
43. Kemp, J. D., and C. J. Egan: *J. Am. Chem. Soc.,* **60:** 1521 (1938).
44. Kreglewski, A.: *J. Phys. Chem.,* **73:** 608 (1969).
45. Kreglewski, A., and W. B. Kay: *J. Phys. Chem.,* **73:** 3359 (1969).
46. Lee, B. I., and W. C. Edmister: *AIChE J.,* **15:** 615 (1969).
47. Lee, B. I., and M. G. Kesler: *AIChE J.,* **21:** 510 (1975).
48. Li, C. C.: *Can. J. Chem. Eng.,* **19:** 709 (1971).
49. Luria, M., and S. W. Benson: *J. Chem. Eng. Data,* **20:** 90 (1977).
50. Lyman, T. J., and R. P. Danner: *AIChE J.,* **22:** 759 (1976).
51. McCullough, J. P., D. W. Scott, H. L. Finke, M. E. Gross, K. D. Williamson, R. E. Pennington, G. Waddington, and H. M. Huffman: *J. Am. Chem. Soc.,* **74:** 2801 (1952).
52. Messerly, J. F., G. B. Gutherie, S. S. Todd, and H. L. Finke: *J. Chem. Eng. Data,* **12:** 338 (1967).
53. Michelsen, M. L.: *Fluid Phase Equil.,* **16:** 57 (1984).
54. Michelsen, M. L., and R. A. Heidemann: *AIChE J.,* **27:** 521 (1981).
55. Missenard, F.-A.: *C. R.,* **260:** 5521 (1965).
56. Modell, M., and R. C. Reid: *Thermodynamics and Its Applications in Chemical Engineering,* 2d ed. Prentice-Hall, Englewood Cliffs, N.J., 1983, chap. 5.
57. Ogiwara, K., A. Yasuhiko, and S. Shozaburo: *J. Chem. Eng. Japan,* **14:** 156 (1981).
58. Orentlicher, M., and J. M. Prausnitz: *Can. J. Chem. Eng.,* **45:** 78 (1967).
59. Parks, G. S., and H. M. Huffman: *J. Am. Chem. Soc.,* **48:** 2788 (1926).
60. Passut, C. A., and R. P. Danner: *Ind. Eng. Chem. Process Design Develop.,* **11:** 543 (1972).
61. Peng, D. Y., and D. B. Robinson: *AIChE J.,* **23:** 137 (1977).
62. Peng, D. Y., and D. B. Robinson: *Ind. Eng. Chem. Fundam.,* **15:** 59 (1976).

63. Powell, T. M., and W. F. Giauque: *J. Am. Chem. Soc.,* **61:** 2366 (1939).
64. Reid, R. C., J. M. Prausnitz, and T. K. Sherwood: *Properties of Gases and Liquids,* 3d ed., McGraw-Hill, New York, 1977.
65. Reid, R. C., and J. E. Sobel: *Ind. Eng. Chem. Fundam.,* **4:** 328 (1965).
66. Rowlinson, J. S.: *Liquids and Liquid Mixtures,* 2d ed., Butterworth, London, 1969.
67. Sage, B. H., and W. N. Lacey: *Ind. Eng. Chem.,* **27:** 1484 (1935).
68. Sakiadis, B. C., and J. Coates: *AIChE J.,* **2:** 88 (1956).
69. San José, J.: Ph.D. thesis, Department of Chemical Engineering, Massachusetts Institute of Technology, Cambridge, Mass., 1975.
70. Schick, L. M., and J. M. Prausnitz: *AIChE J.,* **14:** 673 (1968).
71. Schlinger, W. G., and B. H. Sage: *Ind. Eng. Chem.,* **44:** 2454 (1952).
72. Shaw, R.: *J. Chem. Eng. Data,* **14:** 461 (1969).
73. Spencer, C. F., T. E. Daubert, and R. P. Danner: "Critical Properties" in *Technical Data Book,* American Petroleum Institute, OP72, 539, Xerox University Microfilm, chap. 4.
74. Spencer, C. F., T. E. Daubert, and R. P. Danner: *AIChE J.,* **19:** 522 (1973).
75. Stein, F. P., and J. J. Martin: *Chem. Eng. Progr. Symp. Ser.,* **59**(44): 112 (1963).
76. Stull, D. R., E. F. Westrum, Jr., and G. C. Sinke: *The Chemical Thermodynamics of Organic Compounds,* Wiley, New York (1969).
77. Sutton, J. R.: *Advances in Thermophysical Properties of Extreme Temperatures and Pressures,* ASME, New York, 1965, p. 76.
78. Tamplin, W. S., and D. A. Zuzic: *Hydrocarbon Process. Petrol. Refiner,* **46**(8): 145 (1967).
79. Tao, L. C.: *AIChE J.,* **15:** 362, 469 (1969).
80. Tarakad, R., and T. E. Daubert: *API-5-74,* Pennsylvania State University, University Park, Sept. 23, 1974.
81. Tarakad, R. R., and R. P. Danner: *AIChE J.,* **23:** 944 (1977), **23:** 685 (1977).
82. Teja, A. S.: *J. Chem. Eng. Data,* **28:** 83 (1983).
83. Teja, A. S., K. B. Garg, and R. L. Smith: *Ind. Eng. Chem. Process Design Develop.,* **22:** 672 (1983).
84. Valdes-Krieg, E., J. A. Renuncio, and J. M. Prausnitz: *Ind. Eng. Chem. Process Design Develop.,* **15:** 429 (1976).
85. Vimalchand, P., M. D. Donohue, and I. C. Celmins: *ACS Symp. Ser.,* **300:** 297 (1986).
86. Vimalchand, P., and M. D. Donohue: *Ind. Eng. Chem. Fundam.,* **24:** 246 (1985).
87. Watson, K. M.: *Ind. Eng. Chem.,* **35:** 398 (1943).
88. Watson, L. M., and B. F. Dodge: *Chem. Eng. Progr. Symp. Ser.,* **48**(3): 73 (1952).
89. West, E. W., and J. H. Erbar: "An Evaluation of Four Methods of Predicting Properties of Light Hydrocarbon Systems," paper presented at *NGPA 52nd Annual Meet., Dallas, Tex.,* March 1973.
90. Wiebe, R., and M. J. Brevoort: *J. Am. Chem. Soc.,* **52:** 622 (1930).
91. Wilson, G. M.: *Advan. Cryog. Eng.,* **11:** 392 (1966).
92. Yuan, T.-F., and L. I. Stiel: *Ind. Eng. Chem. Fundam.,* **9:** 393 (1970).

Chapter

# 6

# Thermodynamic Properties of Ideal Gases

## 6-1 Scope and Definitions

Methods are described to estimate the enthalpy and Gibbs energy of formation as well as the entropy for organic compounds in the ideal-gas state. The reference temperature is 298.16 K, and the reference pressure is *one atmosphere*. In addition, ideal-gas heat capacity estimation techniques are presented to allow one to determine $C_p^\circ$ as a function of temperature.

The enthalpy of formation is defined as the isothermal enthalpy change in a synthesis reaction from the elements in their standard states. In such a reaction scheme, the elements are assumed initially to be at reaction temperature, at 1 atm, and in their most stable phase, e.g., diatomic oxygen as an ideal gas, carbon as a solid in the form of $\beta$-graphite, etc. Ordinarily one need not be concerned with the numerical values of the enthalpy of formation of the constituent elements, since, to obtain a standard enthalpy of *reaction,* the enthalpies of formation of all elements cancel. For a general reaction

$$aA + bB = cC + dD$$

the standard enthalpy change to form products C and D from A and B in stoichiometric amounts when reactants and products are pure, at $T$ and 1 atm, is given by

$$\Delta H^\circ = c\,\Delta H_f^\circ\,(C) + d\,\Delta H_f^\circ\,(D) - a\,\Delta H_f^\circ\,(A) - b\,\Delta H_f^\circ\,(B) \qquad (6\text{-}1.1)$$

At 298 K, if A, B, C, and D are elements in their most stable phase configuration, the value of $\Delta H_f^\circ$ is set equal to zero.

Reported enthalpies of formation are normally available only at 298 K. At other temperatures,

$$\Delta H_f^\circ\ (T) = \Delta H_f^\circ\ (298\text{ K}) + \int_{298}^{T} \Delta C_p^\circ\ dT \tag{6-1.2}$$

where $$\Delta C_p^\circ = \sum_j \nu_j C_{pj}^\circ \tag{6-1.3}$$

that is, $\Delta C_p^\circ$ is the sum of the heat capacities of the compound and the constituent elements, each element in its standard state, and each multiplied by the appropriate stoichiometric multiplier $\nu_j$. For most elements, there is no difficulty in obtaining $C_p^\circ$. As examples, oxygen at 298 K has a standard state as an ideal gas and $C_p^\circ$ ($O_2$) is then the heat capacity of diatomic oxygen as an ideal gas. Or, for carbon, one would need the heat capacity $C_p^\circ$ ($\beta$-graphite) as a function of temperature. However, in some instances, care is necessary to include phase change enthalpies in addition to heat capacities when calculating $\Delta H_f^\circ$ as a function of temperature. A case in point would be for compounds containing bromine. The standard state for diatomic bromine at 298 K is the pure liquid at its vapor pressure. For this state, $\Delta H_f^\circ$ ($Br_2$) $=$ 0. Then, in Eq. (6-1.2), to determine $\Delta H_f^\circ\ (T)$ for a compound containing bromine, one must integrate the heat capacity of the liquid bromine up to its normal boiling point at 1 atm, add the enthalpy of vaporization at $T_b$, and, if $T > T_b$, add the integral of the ideal-gas heat capacity between $T_b$ and $T$.

Another way to write Eq. (6-1.1) is to combine Eq. (6-1.2) and Eq. (6-1.3):

$$\Delta H^\circ\ (T) = \sum_j \nu_j\ \Delta H_{fj}^\circ\ (298\text{ K}) + \sum_j \int_{298}^{T} \nu_j C_{pj}^\circ\ dT \tag{6-1.4}$$

where $\nu_j$ is the stoichiometric multiplier in the synthesis reaction (negative for reactants and positive for products). Thus, to determine enthalpies of reaction, one needs to know the enthalpy of formation at 298 K as well as the ideal-gas (or element) heat capacities of all reactants and products.

The Gibbs energy of formation $\Delta G_f^\circ$ is defined in a manner analogous to that for $\Delta H_f^\circ$, and the standard Gibbs energy of reaction $\Delta G^\circ\ (T)$ is written in a form similar to that of Eq. (6-1.1). Unless otherwise stated, all reactants and products are pure, ideal gases at temperature $T$ and 1 atm. (As noted earlier, elements which are not normally gases at $T$ are defined somewhat differently [10, 24].)

Finally, the standard entropies of most elements and all compounds, $S^\circ\ (T)$, apply to materials in an ideal-gas state at 1 atm and at a temperature $T$. This entropy is often termed an *absolute* entropy because it is relative to a value of zero at absolute zero temperature in a perfectly ordered solid state. The entropy of formation could also have been introduced, but no estimation scheme yields this property directly. For use in chemical reactions, since the standard-state entropies of the elements cancel, one may employ $S^\circ\ (T)$ directly; i.e., for the general reaction introduced above,

$$\Delta S^\circ\ (T) = cS_c^\circ\ (T) + dS_d^\circ\ (T) - aS_a^\circ\ (T) - bS_b^\circ\ (T) \tag{6-1.5}$$

$\Delta S^\circ\ (T)$ may be expressed in terms of $S^\circ$ (298 K) for reactants and products in a manner similar to Eq. (6-1.4),

$$\Delta S^\circ\ (T) = \sum_j \nu_j S_j^\circ\ (298\ \text{K}) + \sum_j \int_{298}^{T} \nu_j C_p^\circ\ d\ \ln T \tag{6-1.6}$$

The Gibbs energy change for a chemical reaction $\Delta G^\circ\ (T)$ may be written as

$$\Delta G^\circ\ (T) = \Delta H^\circ\ (T) - T\,\Delta S^\circ\ (T) \tag{6-1.7}$$

where $\Delta H^\circ\ (T)$ and $\Delta S^\circ\ (T)$ are found from Eqs. (6-1.4) and (6-1.6). The Gibbs energy change is related to the chemical equilibrium constant by

$$\ln K = \frac{-\Delta G^\circ}{RT} \tag{6-1.8}$$

Because of the exponential character of this equation, small errors in estimating $\Delta G^\circ\ (T)$ are amplified when calculating $K$. We illustrate this in Fig. 6-1. Here we define $Y$ = true value of $\Delta G^\circ\ (T)/RT$ and show how the equilibrium constant is affected when we overestimate (or underestimate) $\Delta G^\circ\ (T)/RT$. For example, if the true value of $\Delta G^\circ\ (T)/RT$ were 6 and we overestimated by 15 percent, then $\ln K_{est} = -(6)(1.15)$, and $K_{est} = 10^{-3}$, whereas $K_{true}$ is $2.5 \times 10^{-3}$. Note that estimation errors are more serious when $|Y| \gg 0$ because the slope on the plot is proportional to $Y$.

## 6-2 Estimation Methods

Since the properties of compounds treated in this chapter refer to those of an *ideal gas,* intermolecular forces play no role in their estimation. Also, by the same reasoning, the law of corresponding states, used so widely in other chapters, is inapplicable. All estimation methods for $C_p^\circ$, $\Delta H_f^\circ$, $\Delta G_f^\circ$, and $S^\circ$ involve some group estimation scheme related to the molecular structure. Benson [3] and Benson and Buss [4] have pointed out a hierarchy of such methods. The most simple methods would be those which assign contributions based on the *atoms* present in the

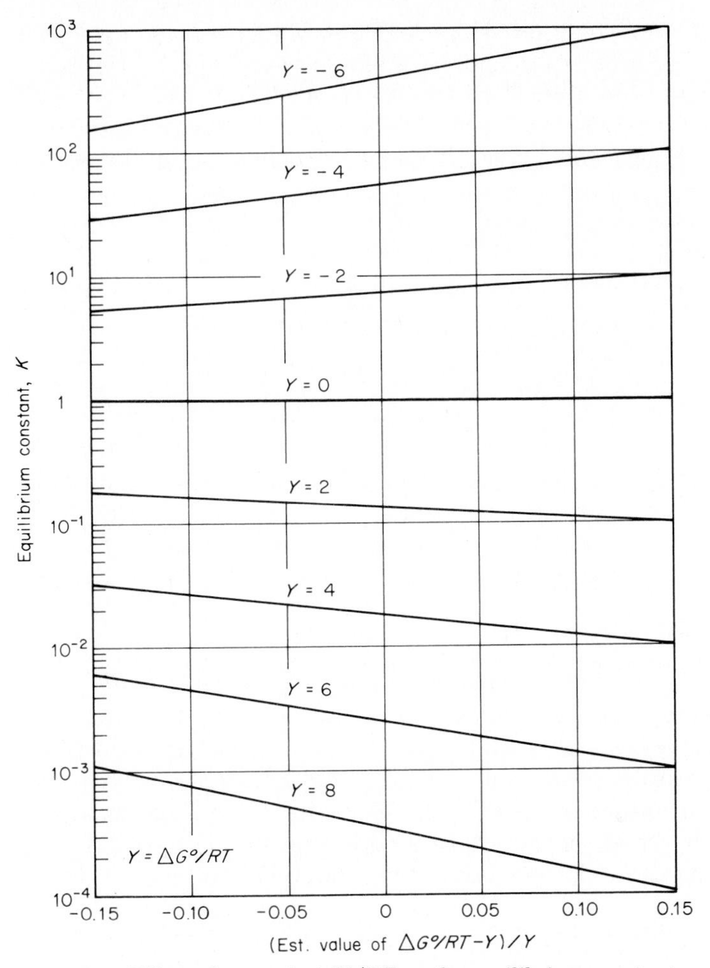


**Figure 6-1** Effect of errors in $\Delta G°/RT$ on the equilibrium constant.

molecule. Although exact for molecular weights and even reasonable for a few other properties, e.g., the liquid molar volume at the normal boiling point, such methods are completely inadequate for the properties discussed in this chapter.

Only slightly more complicated are methods which assign contributions to various chemical bonds. Such techniques are easy to use but again do not normally yield very accurate results. A more successful method assigns contributions to common molecular groupings, for example, $-CH_3$, $-NH_2-$, $-COOH$, and, by simple additivity, one can then estimate ideal-gas properties from a table of group values. The Joback method discussed in Sec. 6-3 employs this approach.

Proceeding to more complicated, and usually more accurate, methods, atoms or molecular groups are chosen, and allowance is made for next-nearest neighbors to this atom or group. The methods of both Yoneda and Benson and his coworkers, discussed in Secs. 6-4 and 6-6, are illustrative of this type of development. Allowance could be made for atoms or groups two or more atoms removed from the atom or group being considered. But such effects are normally so small as to be insignificant, and estimation techniques incorporating them would be quite cumbersome as well as suspect because there are insufficient data to develop a reliable table of group contributions including second next-nearest neighbor effects.

In this chapter, we present five group contribution methods. Three allow an estimation of $\Delta H_f^\circ$ at 298 K and $C_p^\circ\ (T)$, i.e., those of Joback (Sec. 6-3), Yoneda (Sec. 6-4), and Benson et al. (Sec. 6-6). Joback's procedure also provides a scheme for estimating $\Delta G_f^\circ$, whereas the others yield methods for $S^\circ$. Thinh and coworkers proposed a method applicable only to hydrocarbons which estimates $\Delta H_f^\circ$, $\Delta G_f^\circ$, $S^\circ$, and $C_p^\circ$ (Sec. 6-5). Cardozo has suggested a technique for the enthalpy of combustion (as well as the enthalpy of formation) applicable to quite complicated molecules (Sec. 6-7). All methods are evaluated and discussed in Sec. 6-8.

## 6-3 Method of Joback

Choosing the same atomic and molecular groups as in Table 2-2 to estimate critical properties, Joback has used the values given in Stull et al. [25] to obtain group contributions for $\Delta H_f^\circ$ (298 K), $\Delta G_f^\circ$ (298 K), and polynomial coefficients to relate $C_p^\circ$ to temperature. His group values are shown in Table 6-1, and they are to be used in Eqs. (6-3.1) to (6-3.3).

With $\Delta H_f^\circ$ and $\Delta G_f^\circ$ in kJ/mol and $C_p^\circ$ in J/(mol·K),

$$\Delta H_f^\circ\ (298\ \text{K}) = 68.29 + \sum_j n_j\,\Delta_H \tag{6-3.1}$$

$$\Delta G_f^\circ\ (298\ \text{K}) = 53.88 + \sum_j n_j\,\Delta_G \tag{6-3.2}$$

$$C_p^\circ = \left(\sum_j n_j\,\Delta_a - 37.93\right) + \left(\sum_j n_j\,\Delta_b + 0.210\right)T + \left(\sum_j n_j\,\Delta_c - 3.91\times 10^{-4}\right)T^2 + \left(\sum_j n_j\,\Delta_d + 2.06\times 10^{-7}\right)T^3 \tag{6-3.3}$$

where $n_j$ is the number of groups of the $j$th type and the $\Delta$ contributions are for the $j$th atomic or molecular group. The temperature $T$ is in kelvins.

**Example 6-1** Estimate $\Delta H_f^\circ$ (298 K), $\Delta G_f^\circ$ (298 K), and the isobaric ideal-gas heat capacity of butyronitrile at 500 K by using Joback's additive-group method.

**TABLE 6-1 Joback Group Contributions for Ideal-Gas Properties**

| | | | Δ Values | | | |
|---|---|---|---|---|---|---|
| | $\Delta_H$ | $\Delta_G$ | $\Delta_a$ | $\Delta_b$ | $\Delta_c$ | $\Delta_d$ |
| | kJ/mol | kJ/mol | .............. | J/mol K | .................. | |
| Non-ring increments | | | | | | |
| $-CH_3$ | -76.45 | -43.96 | 1.95E+1 | -8.08E-3 | 1.53E-4 | -9.67E-8 |
| $>CH_2$ | -20.64 | 8.42 | -9.09E-1 | 9.50E-2 | -5.44E-5 | 1.19E-8 |
| >CH- | 29.89 | 58.36 | -2.30E+1 | 2.04E-1 | -2.65E-4 | 1.20E-7 |
| >C< | 82.23 | 116.02 | -6.62E+1 | 4.27E-1 | -6.41E-4 | 3.01E-7 |
| $=CH_2$ | -9.63 | 3.77 | 2.36E+1 | -3.81E-2 | 1.72E-4 | -1.03E-7 |
| =CH- | 37.97 | 48.53 | -8.00 | 1.05E-1 | -9.63E-5 | 3.56E-8 |
| =C< | 83.99 | 92.36 | -2.81E+1 | 2.08E-1 | -3.06E-4 | 1.46E-7 |
| =C= | 142.14 | 136.70 | 2.74E+1 | -5.57E-2 | 1.01E-4 | -5.02E-8 |
| ≡CH | 79.30 | 77.71 | 2.45E+1 | -2.71E-2 | 1.11E-4 | -6.78E-8 |
| ≡C- | 115.51 | 109.82 | 7.87 | 2.01E-2 | -8.33E-6 | 1.39E-9 |
| Ring increments | | | | | | |
| $-CH_2-$ | -26.80 | -3.68 | -6.03 | 8.54E-2 | -8.00E-6 | -1.80E-8 |
| >CH- | 8.67 | 40.99 | -2.05E+1 | 1.62E-1 | -1.60E-4 | 6.24E-8 |
| >C< | 79.72 | 87.88 | -9.09E+1 | 5.57E-1 | -9.00E-4 | 4.69E-7 |
| =CH- | 2.09 | 11.30 | -2.14 | 5.74E-2 | -1.64E-6 | -1.59E-8 |
| =C< | 46.43 | 54.05 | -8.25 | 1.01E-1 | -1.42E-4 | 6.78E-8 |
| Halogen increments | | | | | | |
| -F | -251.92 | -247.19 | 2.65E+1 | -9.13E-2 | 1.91E-4 | -1.03E-7 |
| -Cl | -71.55 | -64.31 | 3.33E+1 | -9.63E-2 | 1.87E-4 | -9.96E-8 |
| -Br | -29.48 | -38.06 | 2.86E+1 | -6.49E-2 | 1.36E-4 | -7.45E-8 |
| -I | 21.06 | 5.74 | 3.21E+1 | -6.41E-2 | 1.26E-4 | -6.87E-8 |

**TABLE 6-1 Joback Group Contributions for Ideal-Gas Properties**
***(Continued)***

| | | | Δ Values | | | |
|---|---|---|---|---|---|---|
| | $\Delta_H$ | $\Delta_G$ | $\Delta_a$ | $\Delta_b$ | $\Delta_c$ | $\Delta_d$ |
| | kJ/mol | kJ/mol | ............. | J/mol K | ............. | ............. |
| Oxygen increments | | | | | | |
| -OH (alcohol) | -208.04 | -189.20 | 2.57E+1 | -6.91E-2 | 1.77E-4 | -9.88E-8 |
| -OH (phenol) | -221.65 | -197.37 | -2.81 | 1.11E-1 | -1.16E-4 | 4.94E-8 |
| -O- (nonring) | -132.22 | -105.00 | 2.55E+1 | -6.32E-2 | 1.11E-4 | -5.48E-8 |
| -O- (ring) | -138.16 | -98.22 | 1.22E+1 | -1.26E-2 | 6.03E-5 | -3.86E-8 |
| >C=O (nonring) | -133.22 | -120.50 | 6.45 | 6.70E-2 | -3.57E-5 | 2.86E-9 |
| >C=O (ring) | -164.50 | -126.27 | 3.04E+1 | -8.29E-2 | 2.36E-4 | -1.31E-7 |
| O=CH- (aldehyde) | -162.03 | -143.48 | 3.09E+1 | -3.36E-2 | 1.60E-4 | -9.88E-8 |
| -COOH (acid) | -426.72 | -387.87 | 2.41E+1 | 4.27E-2 | 8.04E-5 | -6.87E-8 |
| -COO- (ester) | -337.92 | -301.95 | 2.45E+1 | 4.02E-2 | 4.02E-5 | -4.52E-8 |
| =O (except as above) | -247.61 | -250.83 | 6.82 | 1.96E-2 | 1.27E-5 | -1.78E-8 |
| Nitrogen Increments | | | | | | |
| $-NH_2$ | -22.02 | 14.07 | 2.69E+1 | -4.12E-2 | 1.64E-4 | -9.76E-8 |
| >NH (nonring) | 53.47 | 89.39 | -1.21 | 7.62E-2 | -4.86E-5 | 1.05E-8 |
| >NH (ring) | 31.65 | 75.61 | 1.18E+1 | -2.30E-2 | 1.07E-4 | -6.28E-8 |
| >N- (nonring) | 123.34 | 163.16 | -3.11E+1 | 2.27E-1 | -3.20E-4 | 1.46E-7 |
| -N= (nonring) | 23.61 | - | - | - | - | - |
| -N= (ring) | 55.52 | 79.93 | 8.83 | -3.84E-3 | 4.35E-5 | -2.60E-8 |
| =NH | 93.70 | 119.66 | 5.69 | -4.12E-3 | 1.28E-4 | -8.88E-8 |
| -CN | 88.43 | 89.22 | 3.65E+1 | -7.33E-2 | 1.84E-4 | -1.03E-7 |
| $-NO_2$ | -66.57 | -16.83 | 2.59E+1 | -3.74E-3 | 1.29E-4 | -8.88E-8 |
| Sulfur increments | | | | | | |
| -SH | -17.33 | -22.99 | 3.53E+1 | -7.58E-2 | 1.85E-4 | -1.03E-7 |
| -S- (nonring) | 41.87 | 33.12 | 1.96E+1 | -5.61E-3 | 4.02E-5 | -2.76E-8 |
| -S- (ring) | 39.10 | 27.76 | 1.67E+1 | 4.81E-3 | 2.77E-5 | -2.11E-8 |

**solution** Butyronitrile contains one $-CH_3$, two $-CH_2-$, and one $-CN$. From Table 6-1

| Group | $n_j$ | $n_j\ \Delta_H$ | $n_j\ \Delta_G$ |
|---|---|---|---|
| $-CH_3$ | 1 | $-76.45$ | $-43.96$ |
| $-CH_2-$ | 2 | $-41.28$ | 16.84 |
| $-CN$ | 1 | 88.43 | 89.22 |
| | | $\Sigma n_j\ \Delta_H = -29.30$ | $\Sigma n_j\ \Delta_G = 62.10$ |

With Eqs. (6-3.1) and (6-3.2)

$$\Delta H_f^\circ\ (298\text{ K}) = 68.29 - 29.30 = 38.99\text{ kJ/mol}$$

$$\Delta G_f^\circ\ (298\text{ K}) = 53.88 + 62.10 = 115.98\text{ kJ/mol}$$

The literature values for these two properties are 34.08 and 108.73 kJ/mol, respectively [25], thus the differences, estimated − literature, are

$$\Delta H_f^\circ\ (298\text{ K}) \qquad \text{Difference} = 38.99 - 34.08 = 4.91\text{ kJ/mol}$$

$$\Delta G_f^\circ\ (298\text{ K}) \qquad \text{Difference} = 115.98 - 108.73 = 7.25\text{ kJ/mol}$$

For the heat capacity,

| Group | $n_j$ | $n_j\ \Delta_a$ | $n_j\ \Delta_b$ | $n_j\ \Delta_c$ | $n_j\ \Delta_d$ |
|---|---|---|---|---|---|
| $-CH_3$ | 1 | $1.95\ E{+}1$ | $-8.08\ E{-}3$ | $1.53\ E{-}4$ | $-9.67\ E{-}8$ |
| $-CH_2-$ | 2 | $-1.82$ | $1.90\ E{-}1$ | $-1.09\ E{-}4$ | $2.38\ E{-}8$ |
| $-CN$ | 1 | $3.65\ E{+}1$ | $7.33\ E{-}2$ | $1.84\ E{-}4$ | $-1.03\ E{-}7$ |
| | | 54.18 | 0.109 | $2.28\ E{-}4$ | $-1.76\ E{-}7$ |

Then, with Eq. (6-3.3),

$$C_p^\circ\ (500\text{ K}) = (54.18 - 37.93) + (0.109 + 0.210)\ (500) + (2.28\ E{-}4 - 3.91\ E{-}4)\ (500)^2 + (-1.76\ E{-}7 + 2.06\ E{-}7)\ (500)^3 = 138.75\text{ J/(mol}\cdot\text{K)}$$

The literature value is 138.37 J/(mol·K), so the error is

$$\text{Error} = \frac{138.75 - 138.37}{138.37} \times 100 = 0.3\%$$

Comparisons between estimated and literature values of $\Delta H_f^\circ$ (298 K) are shown in Table 6-7, for $\Delta G_f^\circ$ (298 K) in Table 6-9, and for $C_p^\circ\ (T)$ in Table 6-6. A comparison of this technique with others is presented in Sec. 6-8.

## 6-4 Method of Yoneda

Yoneda modified a group contribution technique for ideal-gas properties that was originally proposed by Anderson et al. [2, 14]. In this method, one begins with a *base* molecule and sequentially modifies the structure by substituting other groups to arrive at the final structure. Each substitution has a group contribution value, and the values are summed to arrive at the final value of the property. The mechanics of the method are described in detail below. With it one can estimate $\Delta H_f^\circ$ (298 K), $S^\circ$ (298 K), and the polynomial constants for $C_p^\circ$. Equations (6-4.1) to (6-4.3) are used.

$$\Delta H_f^\circ \ (298 \text{ K}) = \sum_j n_j \Delta_H \tag{6-4.1}$$

$$S^\circ \ (298 \text{ K}) = \sum_j n_j \Delta_S \tag{6-4.2}$$

$$C_p^\circ \ (T) = \sum_j n_j \Delta_a + \left(\sum_j n_j \Delta_b\right)\left(\frac{T}{1000}\right) + \left(\sum_j n_j \Delta_c\right)\left(\frac{T}{1000}\right)^2 \tag{6-4.3}$$

$\Delta$ is the contribution (Table 6-2), and $n_j$ is the number of times the contribution is required. $\Delta H_f^\circ$ (298 K) is estimated in kJ/mol, and both $C_p^\circ$ $(T)$ and $S^\circ$ (298 K) are estimated in J/(mol·K). The temperature is in kelvins.

To use this method, one must first select a *base* group from which to synthesize the desired molecule. The base groups allowed are:

Methane
Cyclopentane
Cyclohexane
Benzene
Naphthalene

Compounds which cannot be synthesized from these groups, e.g., thiophene, cannot be treated. There are contributions for each base group (Table 6-2*a*), and for subsequent substitutions (Tables 6-2*b* to 6-2*e*). Corrections are also necessary in some cases (Table 6-2*f*).

To introduce the method, let us initially limit the discussion to hydrocarbons. Later we will demonstrate how functional groups containing atoms other than carbon and hydrogen are inserted.

We begin with the appropriate base group and proceed in a series of steps:

1. The desired structure is built up by substituting $-CH_3$ groups for hydrogen atoms. The *first* substitution on the base group is termed a *primary methyl substitution.* For methane, only a single type of primary substitution is possible and, of course, ethane is formed. For ring compounds as base groups, one may have several types of primary methyl substitutions for hydrogens on the ring. For example, if the final compound were to consist of a 1,3-*trans* form of cyclohexane, we would require a *first* primary methyl substitution as well as a *second* primary methyl substitution of the 1,3-*trans* type. (At this stage, our compound would be 1,3-

**TABLE 6-2 Yoneda Group Contributions for Ideal-Gas Properties[a]**

Table 6-2a Contributions for Base Groups

| Base Group | ΔH | ΔS | Δa | Δb | Δc |
|---|---|---|---|---|---|
| Methane | -74.90 | 186.31 | 16.71 | 65.65 | -9.96 |
| Cyclopentane | -77.29 | 293.08 | -41.95 | 474.03 | -182.71 |
| Cyclohexane | -123.22 | 298.44 | -52.25 | 600.18 | -231.07 |
| Benzene | 82.98 | 269.38 | -22.52 | 402.81 | -171.53 |
| Naphthalene | 151.06 | 335.87 | -28.43 | 623.67 | -269.09 |

Table 6-2b Contributions for Primary Methyl Substitutions

| Base Group | ΔH | ΔS | Δa | Δb | Δc |
|---|---|---|---|---|---|
| Methane | -9.84 | 43.33 | -9.92 | 103.87 | -43.54 |
| Cyclopentane | | | | | |
| First primary | -34.46 | 49.28 | 8.75 | 68.29 | -23.19 |
| Second primary to form: | | | | | |
| 1,1 | -26.63 | 17.17 | -6.03 | 116.43 | -55.60 |
| 1,2 (cis) | -17.88 | 24.03 | -3.64 | 110.53 | -53.26 |
| 1,2 (trans) | -25.04 | 24.70 | -2.47 | 107.64 | -52.17 |
| 1,3 (cis) | -24.20 | 24.70 | -2.47 | 107.64 | -52.17 |
| 1,3 (trans) | -21.94 | 24.70 | -2.47 | 107.64 | -52.17 |
| Cyclohexane | | | | | |
| First primary | -33.66 | 46.35 | 11.60 | 81.27 | -39.61 |
| Second primary to form: | | | | | |
| 1,1 | -24.24 | 20.47 | 13.52 | 111.49 | -41.03 |
| 1,2(cis) | -15.41 | 29.98 | -8.00 | 100.06 | -38.73 |
| 1,2(trans) | -23.24 | 26.38 | -5.82 | 103.37 | -43.25 |
| 1,3(cis) | -28.01 | 25.92 | -6.32 | 95.21 | -33.03 |
| 1,3(trans) | -19.80 | 31.69 | -4.31 | 88.49 | -32.20 |
| 1,4(cis) | -19.89 | 25.92 | -4.31 | 88.47 | -32.30 |
| 1,4(trans) | -27.84 | 20.26 | -8.42 | 107.68 | -44.05 |
| Benzene | | | | | |
| First primary | -35.50 | 47.94 | 5.78 | 64.68 | -19.51 |
| Second primary to form: | | | | | |
| 1,2 | -27.80 | 36.43 | 12.48 | 50.03 | -11.97 |
| 1,3 | -29.14 | 41.66 | 5.02 | 64.81 | -19.64 |
| 1,4 | -28.72 | 36.22 | 5.48 | 60.33 | -16.16 |
| 1,2,3 | -30.40 | 42.87 | 14.15 | 29.27 | 9.67 |
| 1,2,4 | -33.49 | 43.63 | 16.41 | 18.63 | 16.24 |
| 1,3,5 | -34.22 | 26.84 | 6.20 | 58.41 | -14.74 |

*trans*-dimethylcyclohexane.) The various primary methyl substitution group values are shown in Table 6-2*b*. These substitutions must be made before any secondary methyl substitutions.

2. After making the primary methyl substitutions, we continue to replace hydrogen atoms in the synthesis procedure with additional $-CH_3$ groups. Each of these steps is termed a *secondary methyl substitution.* Contributions for such substitutions depend on the type of carbon upon which the substitution is made as well as upon the type of adjacent carbon

**TABLE 6-2 Yoneda Group Contributions for Ideal-Gas Properties[a] (*Continued*)**

Table 6-2b Contributions for Primary Methyl Substitutions (continued)

| Base Group | $\Delta_H$ | $\Delta_S$ | $\Delta_a$ | $\Delta_b$ | $\Delta_c$ |
|---|---|---|---|---|---|
| Naphthalene | | | | | |
| First primary | | | | | |
| Position 1 | -34.12 | 41.83 | 6.36 | 37.39 | -32.11 |
| Position 2 | -34.88 | 44.42 | 10.68 | 61.80 | -20.18 |
| Second primary to form: | | | | | |
| 1,2 | -26.42 | 30.31 | 13.06 | 64.81 | -24.58 |
| 1,3 | -27.76 | 35.59 | 5.61 | 79.59 | -32.24 |
| 1,4 | -27.34 | 30.10 | 6.07 | 75.11 | -28.76 |
| 2,3 | -26.42 | 30.31 | 13.06 | 64.81 | -24.58 |

Table 6-2c Contributions for Secondary Methyl Substitutions

| Type Number A | Type Number B | $\Delta_H$ | $\Delta_S$ | $\Delta_a$ | $\Delta_b$ | $\Delta_c$ |
|---|---|---|---|---|---|---|
| 1 | 1 | -21.10 | 43.71 | -3.68 | 98.22 | -42.29 |
| 1 | 2 | -20.60 | 38.90 | 1.47 | 81.48 | -31.48 |
| 1 | 3 | -15.37 | 36.63 | -0.96 | 91.69 | -38.98 |
| 1 | 4 | -15.37 | 36.63 | -0.96 | 91.69 | -38.98 |
| 1 | 9 | -19.68 | 45.34 | 1.55 | 88.59 | -37.68 |
| 2 | 1 | -28.76 | 21.48 | -2.09 | 95.75 | -41.70 |
| 2 | 2 | -26.59 | 27.34 | -0.63 | 90.73 | -37.56 |
| 2 | 3 | -22.23 | 27.38 | -4.90 | 97.68 | -41.66 |
| 2 | 4 | -20.68 | 27.51 | -1.21 | 92.11 | -38.02 |
| 2 | 9 | -24.37 | 28.09 | -3.18 | 90.43 | -36.34 |
| 3 | 1 | -31.48 | 11.76 | -2.76 | 107.77 | -49.28 |
| 3 | 2 | -28.64 | 18.00 | -6.91 | 111.79 | -51.71 |
| 3 | 3 | -20.77 | 25.96 | -6.91 | 111.79 | -51.75 |
| 3 | 4 | -23.70 | 4.56 | -4.19 | 129.62 | -66.36 |
| 3 | 9 | -26.13 | 28.09 | -3.18 | 90.43 | -36.34 |

atoms. We designate by the letter A the carbon atom where the substitution is made and by the letter B the *highest type number* (see below) of carbon adjacent to A. The type numbers are as follows:

| Type | | | | |
|---|---|---|---|---|
| 1 | 2 | 3 | 4 | 9 |
| $-CH_3$ | $-CH_2-$ | $>CH-$ | $>C<$ | C in aromatic ring |

Contributions for secondary methyl substitutions are given in Table 6-2*c*. As an example, assume we were interested in 2-methylbutane. The sequential steps would be:

Table 6-2d Multiple Bond Contributions Replacing Single Bonds

| Bond type | $\Delta_H$ | $\Delta_S$ | $\Delta_a$ | $\Delta_b$ | $\Delta_c$ |
|---|---|---|---|---|---|
| 1=1 | 137.08 | -10.05 | 0.50 | -32.78 | 3.73 |
| 1=2 | 126.23 | -5.99 | 3.81 | -50.95 | 16.33 |
| 1=3 | 116.98 | 0.71 | 12.81 | -71.43 | 27.93 |
| 2=2 (cis) | 118.49 | -6.32 | -6.41 | -37.60 | 11.30 |
| 2=2 (trans) | 114.51 | -11.39 | 9.17 | -67.57 | 26.80 |
| 2=3 | 114.72 | 0.59 | -1.05 | -54.09 | 21.23 |
| 3=3 | 115.97 | -2.09 | 5.90 | -95.92 | 57.57 |
| 1≡1 | 311.62 | -28.68 | 19.18 | -98.81 | 22.99 |
| 1≡2 | 290.98 | -20.81 | 16.54 | -117.15 | 40.74 |
| 2≡2 | 274.40 | -23.95 | 12.85 | -127.11 | 51.71 |
| Adjacent double bonds | 41.41 | -13.27 | 9.76 | -7.79 | 2.14 |
| Conjugated double bonds | -15.32 | -17.00 | -6.70 | 37.30 | -27.51 |
| Double bond conjugated with aromatic ring | -7.20 | -9.50 | 5.36 | -9.09 | 5.19 |
| Triple bond conjugated with aromatic ring | 8.79 | -20.10 | -3.77 | 4.61 | 0.42 |
| Conjugated triple bonds | 17.58 | -20.52 | 3.35 | 14.65 | -14.65 |
| Conjugated double and triple bonds | 13.82 | -5.86 | 12.56 | 22.19 | 9.63 |

Table 6-2e Fundamental Contributions (Replacing $CH_n$ Groups)

| Functional group | $\Delta_H$ | $\Delta_S$ | $\Delta_a$ | $\Delta_b$ | $\Delta_c$ |
|---|---|---|---|---|---|
| =O (aldehyde) | -10.13 | -54.43 | 17.12 | -214.20 | 84.32 |
| =O (ketone) | -29.68 | -84.53 | 6.32 | -148.59 | 36.68 |
| -OH | -119.07 | 8.62 | 7.29 | -65.73 | 24.45 |
| *-OH | -146.58 | -1.26 | 12.02 | -49.82 | 24.28 |
| -O- | -85.54 | -5.28 | 13.27 | -85.37 | 38.60 |
| *-O- | -97.85 | -15.07 | 18.00 | -69.50 | 38.10 |
| -OOH | -103.41 | | | | |
| -OO- | -21.86 | | | | |
| -COOH | -350.39 | 53.05 | 7.91 | 29.22 | -26.67 |
| *-COOH | -337.87 | 51.92 | -8.04 | 25.20 | -4.56 |
| -COO- | -306.14 | 54.85 | -17.58 | 1.26 | 7.95 |
| *-COO- | -317.90 | 54.85 | -17.58 | 1.26 | 7.95 |

Base group → methane

Primary methyl substitution → ethane

Secondary methyl substitutions

A=1, B=1 → propane

A=1, B=2 → *n*-butane

A=2, B=2 → 2-methylbutane

3. Next, we insert any necessary double or triple bonds in the molecule. Contributions for such substitutions are shown in Table 6-2*d*, and they

**TABLE 6-2 Yoneda Group Contributions for Ideal-Gas Properties**[a] **(*Continued*)**

Table 6-2e Fundamental Contributions (Replacing $CH_n$ Groups)

| Functional group | $\Delta_H$ | $\Delta_S$ | $\Delta_a$ | $\Delta_b$ | $\Delta_c$ |
|---|---|---|---|---|---|
| *-OOC- | -310.33 | | | | |
| -COOCO- | -470.26 | 116.94 | -5.28 | 124.72 | -69.29 |
| $-COO_2CO-$ | -392.30 | | | | |
| -OOCH | -276.04 | 71.80 | 7.91 | 29.22 | -26.67 |
| $-CO_3-$ | -490.57 | | | | |
| -F | -154.28 | -16.62 | 4.23 | -76.62 | 24.58 |
| *-F | -165.34 | -18.05 | 6.49 | -59.54 | 18.38 |
| *-F (ortho) | -143.36 | -12.85 | 5.90 | -78.92 | 32.45 |
| -COF | -355.71 | 57.36 | 14.24 | -18.00 | 4.61 |
| *-COF | -351.69 | | | | |
| -Cl | 2.05 | -5.90 | 7.45 | -64.90 | 14.95 |
| *-Cl | 9.88 | -3.98 | 10.72 | -83.40 | 31.07 |
| -COCl | -159.35 | 53.97 | 22.65 | -23.57 | -2.43 |
| *-COCl | -155.41 | | | | |
| -Br | 49.57 | 13.10 | 11.14 | -49.95 | 13.06 |
| *-Br | 57.61 | 7.29 | 12.31 | -70.38 | 28.93 |
| -COBr | -105.80 | 68.66 | 20.93 | -43.54 | 9.21 |
| *-COBr | -98.56 | | | | |
| -I | 101.19 | 14.57 | 11.39 | -72.56 | 18.30 |
| *-I | 115.18 | 8.79 | 12.56 | -92.95 | 34.33 |
| -COI | -38.06 | 88.34 | 23.45 | -33.08 | 9.63 |
| *-COI | -30.98 | | | | |
| -SH | 60.37 | 24.07 | 14.40 | -65.98 | 28.43 |
| *-SH | 64.14 | 19.76 | 12.14 | -42.45 | 19.43 |
| -S- | 69.67 | 21.65 | 17.12 | -83.65 | 46.05 |
| *-S- | 71.01 | 17.17 | 15.07 | -60.29 | 38.64 |
| -SS- | 79.88 | 63.51 | 35.63 | -58.45 | 20.43 |
| -SO- | -43.17 | | | | |
| *-SO- | -39.77 | | | | |
| $-SO_2-$ | -280.10 | | | | |
| $-*SO_2-$ | -276.66 | | | | |
| $-SO_3H$ | 1183.61 | | | | |
| $-OSO_2-$ | -379.78 | | | | |
| $-OSO_3-$ | -583.56 | | | | |

depend upon the type numbers of the carbon atoms involved. As an example, for 2-methyl-2-butene, we would first synthesize 2-methylbutane as above and insert a double bond of the type 2=3. Corrections are also given in Table 6-2*d* for adjacent and conjugated double and triple bonds.

With steps 1 to 3 we can synthesize hydrocarbons and estimate their ideal-gas properties.

For nonhydrocarbons, one must first prepare a suitable hydrocarbon and then insert the desired functional groups by substituting for $-CH_n-$

Table 6-2e Fundamental Contributions (Replacing $CH_n$ Groups) (continued)

| Functional group | $\Delta_H$ | $\Delta_S$ | $\Delta_a$ | $\Delta_b$ | $\Delta_c$ |
|---|---|---|---|---|---|
| $-NH_2$ | 61.50 | 13.10 | 7.49 | -37.68 | 13.19 |
| $*-NH_2$ | 39.44 | 2.05 | 8.83 | -14.40 | 4.40 |
| -NH- | 87.00 | -0.21 | 1.38 | -24.62 | 7.75 |
| *-NH- | 57.61 | -11.30 | 2.51 | -1.26 | -0.84 |
| -N< | 110.74 | -5.86 | 0.04 | -18.59 | 4.40 |
| *-N< | 80.72 | -16.75 | 1.26 | 4.61 | -4.19 |
| =N- (keto) | 187.15 | | | | |
| -N=N- | 266.28 | | | | |
| $-NHNH_2$ | 170.15 | 49.24 | | | |
| $*-NHNH_2$ | 153.61 | | | | |
| $-N(NH_2)-$ | 187.86 | 31.53 | | | |
| $*-N(NH_2)-$ | 171.24 | | | | |
| -NHNH- | 195.86 | 39.10 | | | |
| *-NHNH- | 179.20 | | | | |
| -CN | 172.66 | 6.70 | 14.32 | -53.42 | 14.70 |
| *-CN | 171.49 | 3.94 | 17.79 | -47.60 | 20.18 |
| -NC | 235.05 | 17.29 | 17.58 | -47.73 | 20.10 |
| =NOH | 92.11 | | | | |
| $-CONH_2$ | -153.74 | 77.46 | 15.07 | 23.86 | -12.56 |
| $*-CONH_2$ | -141.22 | | | | |
| -CONH- | -128.12 | | | | |
| *-NHCO- | -158.39 | | | | |
| -CON< | 87.92 | | | | |
| $-NO_2$ | 11.51 | 45.55 | 4.77 | 4.65 | -14.57 |
| $*-NO_2$ | 18.00 | 45.64 | 4.61 | 4.61 | -14.65 |
| -ONO- | 20.68 | 54.85 | 10.34 | 6.32 | -16.08 |
| $-ONO_2$ | -36.72 | 72.43 | 17.25 | 31.86 | -29.14 |
| -NCS | 234.46 | 61.55 | | | |

Table 6-2f Corrections for Type-Number and Multiple Substitutions

| Functional Group | $\Delta_H$ | $\Delta_S$ | $\Delta_a$ | $\Delta_b$ | $\Delta_c$ |
|---|---|---|---|---|---|
| =O (aldehyde) | -22.69 | 18.84 | -3.60 | 6.74 | -4.81 |
| =O (ketone) | -13.82 | 30.94 | 6.66 | -47.31 | 34.37 |
| -OH | -11.10 | 0.84 | 0.42 | 0.00 | -0.42 |
| -O- | -9.55 | -2.30 | 2.14 | -5.02 | 3.31 |
| *-O- | -11.76 | -2.51 | 2.09 | -5.02 | 3.35 |

groups (not hydrogen atoms). Functional groups are of three general types characterized by one, two, or three bonds. Examples are $-Br$, $-O-$, $>N-$. In the first case, the $-Br$ substitutes for a $-CH_3$; in the second, the $-O-$ replaces a $-CH_2-$; and in the third, the $>N-$ replaces a $>CH-$. An aldehyde or ketone group (=O) replaces two $-CH_3$ groups.

Values for such *fundamental contributions* are shown in Table 6-2*e*.

**TABLE 6-2 Yoneda Group Contributions for Ideal-Gas Properties[a] (*Continued*)**

Table 6-2f Corrections for Type-Number and Multiple Substitutions (continued)

| Functional Group | $\Delta_H$ | $\Delta_S$ | $\Delta_a$ | $\Delta_b$ | $\Delta_c$ |
|---|---|---|---|---|---|
| -OOH | 8.37 | | | | |
| -OO- | -10.47 | | | | |
| -COOH | 6.45 | 35.92 | 0.00 | 0.00 | 0.00 |
| +-COO- | -5.07 | 35.92 | 0.00 | 0.00 | 0.00 |
| +-OOC- | -11.72 | -2.50 | 2.09 | -5.02 | 3.35 |
| *-COO- | 7.49 | -2.51 | 2.09 | -5.02 | 3.35 |
| -COOCO- | -5.07 | 36.01 | 0.00 | 0.00 | 0.00 |
| $-COO_2CO-$ | -21.35 | | | | |
| -OOCH | 33.45 | -2.51 | 2.09 | -5.02 | 3.35 |
| $-CO_3-$ | -1.21 | | | | |
| -F | -6.15 | 4.14 | 1.59 | -0.54 | 1.59 |
| -F,-F | -15.37 | -3.81 | -2.01 | -0.75 | -1.76 |
| -F,-Cl | 11.01 | -0.67 | 7.20 | -13.98 | 18.34 |
| -F,-Br | 17.54 | 6.82 | 4.14 | -16.79 | 4.40 |
| -F,-I | 17.25 | -0.38 | 7.03 | -6.49 | 4.23 |
| -COF | 1.67 | | | | |
| -Cl | -2.60 | 5.19 | 3.77 | -12.56 | 8.04 |
| -Cl,-Cl | 17.79 | -6.24 | -2.60 | 6.49 | -3.77 |
| -Cl,-Br | 21.52 | 6.20 | 7.24 | -29.10 | 12.64 |
| -Cl,-I | 20.52 | 5.19 | 7.03 | -27.59 | 18.92 |
| -COCl | 1.88 | | | | |
| -Br | -7.24 | -5.23 | 1.63 | -26.59 | 9.67 |
| -Br,-Br | 17.63 | 9.92 | 4.69 | -35.96 | 19.68 |
| -Br,-I | 20.52 | 7.95 | -1.59 | -32.41 | 16.08 |
| -COBr | 1.67 | | | | |
| -I | -4.31 | 3.94 | 2.76 | -10.13 | 7.29 |
| -I,-I | 23.40 | -3.06 | 0.50 | 0.75 | -1.51 |
| -COI | 1.67 | | | | |
| -SH | -1.13 | 1.59 | 1.47 | -1.21 | -1.59 |
| -S- | -3.56 | -0.17 | -0.17 | 4.52 | -3.77 |
| *-S- | -1.17 | -0.42 | -0.42 | 4.61 | -3.77 |
| -SS- | -3.43 | 0.08 | -1.76 | 11.14 | -9.59 |
| -SO- | -8.25 | | | | |
| *-SO- | -8.37 | | | | |

There are, in addition, two further contributions which may be necessary when nonhydrocarbons are synthesized. These are termed *corrections*, and they are shown in Table 6-2*f*.

1. *Type number corrections.* These corrections are never necessary if a functional group is attached to an aromatic ring, e.g., when $-Cl$ replaces $-CH_3$ in toluene to form chlorobenzene. In other cases, the general rules to follow are:

- Make *all* substitutions of functional groups first.

Table 6-2f Corrections for Type-Number and Multiple Substitutions (continued)

| Functional Group | $\Delta_H$ | $\Delta_S$ | $\Delta_a$ | $\Delta_b$ | $\Delta_c$ |
|---|---|---|---|---|---|
| $-SO_2-$ | -1.13 | | | | |
| $*-SO_2-$ | 25.87 | | | | |
| $-SO_3H$ | -11.72 | | | | |
| $-OSO_2-$ | -11.76 | | | | |
| $-OSO_3-$ | -10.76 | | | | |
| $-NH_2$ | -5.44 | -1.42 | 0.67 | 1.97 | -2.55 |
| -NH- | -9.76 | -1.26 | 0.84 | 2.09 | -2.51 |
| *-NH- | -8.71 | -1.26 | 0.84 | 2.09 | -2.51 |
| -N< | -7.12 | -1.26 | 0.84 | 2.09 | -2.51 |
| *-N< | -4.19 | -1.26 | 0.84 | 2.09 | -2.51 |
| ←=N- | 0.84 | | | | |
| ←-N= | -3.77 | | | | |
| -N=N- | -3.77 | | | | |
| $-NHNH_2$ | -5.44 | -1.26 | | | |
| $-N(NH_2)-$ | -5.44 | -1.26 | | | |
| $*-N(NH_2)-$ | -5.44 | | | | |
| -NHNH- | -5.44 | -1.26 | | | |
| *-NHNH- | -5.44 | | | | |
| -CN | -12.90 | 2.34 | 4.27 | -20.43 | 18.76 |
| -NC | -12.98 | 2.51 | 4.19 | -20.52 | 18.84 |
| =NOH | 0.84 | | | | |
| $-CONH_2$ | 0.13 | 36.01 | 0.00 | 0.00 | 0.00 |
| ←-CONH- | -5.02 | | | | |
| ←-NHCO- | -9.63 | | | | |
| *-NHCO- | -5.02 | | | | |
| $-NO_2$ | -9.46 | 0.00 | 0.00 | 0.00 | 0.00 |
| -ONO- | -26.54 | 0.00 | 0.00 | 0.00 | 0.00 |
| $-ONO_2$ | -10.34 | 2.76 | -1.55 | 3.43 | -2.30 |
| -NCS | -3.77 | -1.26 | | | |

a. Values of $\Delta_H$ are used to estimate the enthalpy of formation at 298 K in kJ/mol; values of $\Delta_S$ are used to estimate the entropy in the ideal-gas state at one atmosphere and 298 K in J/mol K; values of $\Delta_a$, $\Delta_b$, and $\Delta_c$ are used to estimate the ideal-gs heat capacity in Eq. (6-4.3) in J/mol K.

The symbol * indicates an aromatic nucleus; the symbol ← indicates the directionality of the bond when making type-number corrections.

- For *each* functional group, focus on the carbon atoms bonded to the functional group. For each bonded carbon, count the number of adjacent, bonded atoms other than hydrogen. The sum is the multiplier for the type number correction in Table 6-2*f*. If the functional group has multiple bonds, this procedure is repeated for each *single* bond. For example, consider 1-chloropropane. Here the —Cl is attached to a carbon which itself has only a single nonhydrogen atom attached. Thus, only one type number correction is applied for —Cl in Table 6-2*f*. For 2-chloropropane, there would be two type number corrections for the —Cl. For methyl ethyl ketone, the =O is bonded to a carbon which itself is connected to two other carbons. Thus the number of type number corrections for the =O is two. Finally, for perfluoroethane, each fluorine is bonded to a carbon which is attached to three other nonhydrogen atoms. With six —F, each with three type number corrections, the total number of type number corrections for —F in Table 6-2*f* is $3 \times 6 = 18$.

2. *Multiple corrections.* If certain functional groups should be connected to the *same* carbon atom, corrections given in Table 6-2*f* are necessary. For example, with 1,1,1-trichloroethane, there would be three —Cl,—Cl multiple-bond corrections to account for the three binary interactions between the chlorine atoms.

We illustrate Yoneda's method in Example 6-2. Calculated values of $\Delta H_f^\circ$ (298 K), $S^\circ$ (298 K), and $C_p^\circ(T)$ are compared with literature values in Tables 6-7, 6-8, and 6-6, respectively. A comparison of the reliability of the method in comparison with others is discussed in Sec. 6-8.

**Example 6-2** By using Yoneda's method, estimate $\Delta H_f^\circ$ (298 K), $S^\circ$ (298 K), and $C_p^\circ$ (800 K) for isopropyl ether. The literature values are −319.0 kJ/mol, 390.5 J/(mol·K), and 311.5 J/(mol·K) [25].

**solution** The synthesis strategy is to form 2,4-dimethylpentane and then insert the functional group —O— for the number 3 carbon atom (—$CH_2$—). The base group is methane.

| | $\Delta_H$ | $\Delta_S$ | $\Delta_a$ | $\Delta_b$ | $\Delta_c$ |
|---|---|---|---|---|---|
| Base group → methane | −74.90 | 186.31 | 16.71 | 65.65 | −9.96 |
| Primary methyl: | | | | | |
| Substitution → ethane | −9.84 | 43.33 | −9.92 | 103.87 | −43.54 |
| Secondary methyl: | | | | | |
| Substitutions | | | | | |
| A=1, B=1 → propane | −21.10 | 43.71 | −3.68 | 98.22 | −42.29 |
| A=1, B=2 → *n*-butane | −20.60 | 38.90 | 1.47 | 81.48 | −31.48 |
| A=1, B=2 → *n*-pentane | −20.60 | 38.90 | 1.47 | 81.48 | −31.48 |
| A=2, B=2 → 2-methylpentane | −26.59 | 27.34 | −0.63 | 90.73 | −37.56 |
| A=2, B=2 → 2,4-dimethylpentane | −26.59 | 27.34 | −0.63 | 90.73 | −37.56 |
| Fundamental contribution: | | | | | |
| —O— → isopropyl ether | −85.54 | −5.28 | 13.27 | −85.37 | 38.60 |
| Type number correction: | | | | | |
| —O— four times | −38.20 | −9.20 | 8.56 | −20.08 | 13.24 |
| | −323.96 | 391.35 | 26.62 | 506.71 | −182.03 |

With Eqs. (6-4.1) to (6-4.3)

$$\Delta H_f^\circ\ (298\ \text{K}) = -323.96\ \text{kJ/mol}$$
$$\text{Difference} = -323.96 - (-319.0) = -4.9\ \text{kJ/mol}$$
$$S^\circ\ (298\ \text{K}) = 391.35\ \text{J/(mol}\cdot\text{K)}$$
$$\text{difference} = 391.35 - 390.5 = 0.9\ \text{J/(mol}\cdot\text{K)}$$
$$C_p^\circ\ (800\ \text{K}) = 26.62 + (506.71)\left(\frac{800}{1000}\right) - (182.03)\left(\frac{800}{1000}\right)^2$$
$$= 315.5\ \text{J/(mol}\cdot\text{K)}$$
$$\text{Error} = \frac{315.5 - 311.5}{311.5} \times 100 = 1.3\%$$

## 6-5 Method of Thinh et al. [26, 27]

An additive group method for estimating $\Delta H_f^\circ$ (298 K), $\Delta G_f^\circ$ (298 K), $S^\circ$ (298 K), and $C_p^\circ$ $(T)$ for *hydrocarbons* employs the following equations:

$$\Delta H_f^\circ\ (298\ \text{K}) = \sum_j n_j\, \Delta_H \tag{6-5.1}$$

$$\Delta G_f^\circ\ (298\ \text{K}) = \sum_j n_j\, \Delta_G \tag{6-5.2}$$

$$S^\circ\ (298\ \text{K}) = \sum_j n_j\, \Delta_S \tag{6-5.3}$$

$$C_p^\circ\ (T) = \sum_j n_j\left[A + B_1 \exp\left(\frac{-C_1}{T^{n_1}}\right) - B_2 \exp\left(\frac{-C_2}{T^{n_2}}\right)\right] \qquad \text{6-5.4)}$$

Values of $\Delta_H$, $\Delta_G$, $\Delta_S$, $A$, $B_1$, $B_2$, $C_1$, $C_2$, $n_1$, and $n_2$ are shown in Table 6-3 for many hydrocarbon groups. Equation (6-5.4) is a modified form suggested earlier by Yuan and Mok [30]. It is sometimes inconvenient to use because it cannot be integrated analytically and numerical techniques must be employed.

This method is illustrated in Example 6-3, and estimated values of the ideal-gas properties are compared to literature values in Tables 6-6 through 6-9. It is compared with other estimation methods in Sec. 6-8.

**Example 6-3** Estimate the heat capacity of 2-methyl-1,3-butadiene at 800 K and the enthalpy of formation at 298 K by using the group contribution method suggested by Thinh and his coworkers. The literature values for these two properties are 201.0 J/(mol·K) and 75.78 kJ/mol, respectively [25].

**solution** 2-Methyl-1,3-butadiene may be broken down to $-CH_3$, $-HC{=}CH_2$, and $>C{=}CH_2$. With Table 6-3, for this compound,

$$\Delta H_f^\circ = -42.36 + 55.12 + 67.81 = 80.54\ \text{kJ/mol}$$
$$\text{Difference} = \text{estimated} - \text{literature} = 80.54 - 75.78 = 4.8\ \text{kJ/mol}$$

**TABLE 6-3 Thinh et al., Group Contributions for Ideal-Gas Properties**

| GROUP | $\Delta_H$ | $\Delta_S$ | $\Delta_G$ | A | $B_1$ | $C_1$ | $n_1$ | $B_2$ | $C_2$ | $n_2$ |
|---|---|---|---|---|---|---|---|---|---|---|
| $-CH_3$ | -42.362 | 114.823 | -16.454 | 19.8312 | 85.5824 | 1013.8229 | 1.0489 | 0 | 0 | 0 |
| $-CH_2-$ | -20.641 | 38.979 | 8.374 | 12.9037 | 573.0482 | 788.7739 | 1.0380 | 503.6800 | 832.8313 | 1.0452 |
| >CH- | -7.519 | -49.634 | 28.428 | -2.6017 | 318.1767 | 601.8911 | 0.9953 | 256.7467 | 1013.8229 | 1.0489 |
| >C< | 3.358 | -152.693 | 50.576 | -10.3372 | 375.4710 | 733.9538 | 1.0396 | 342.3287 | 1013.8229 | 1.0489 |
| $=CH_2$ | 26.159 | 109.799 | 34.085 | 14.3951 | 64.4428 | 527.1308 | 0.9644 | 0 | 0 | 0 |
| ≡C- | 114.467 | 32.9773 | 105.687 | 10.4586 | 100.8906 | 1133.5426 | 1.0774 | 85.5824 | 1013.8229 | 1.0489 |
| ≡CH | 113.450 | 100.4769 | 104.670 | 6.9241 | 52.1081 | 21.3585 | 0.5 | 0 | 0 | 0 |
| =C= | 139.940 | 24.49 | 134.346 | 6.4087 | 152.4962 | 321.4962 | 0.9048 | 128.8860 | 527.1308 | 0.9644 |
| $-HC=CH_2$ | 55.119 | 139.46 | 75.383 | 6.9526 | 131.7946 | 134.8699 | 0.8030 | 0 | 0 | 0 |
| $>C=CH_2$ | 67.809 | 64.14 | 91.021 | 20.6066 | 262.0187 | 652.9594 | 0.9990 | 171.1643 | 1013.8229 | 1.0489 |
| >C=C< | 102.836 | -96.42 | 134.982 | -0.0063 | 430.5353 | 511.5828 | 0.9573 | 342.3287 | 1013.8229 | 1.0489 |
| >C=CH- | 84.506 | -5.67 | 109.066 | 11.0372 | 329.0795 | 874.4157 | 1.0464 | 256.7467 | 1013.8229 | 1.0489 |
| -HC=CH-(cis) | 77.732 | 71.38 | 98.808 | 13.3291 | 260.7753 | 1110.9532 | 1.0843 | 171.1643 | 1013.8229 | 1.0489 |
| -HC=CH-(trans) | 73.545 | 67.03 | 95.920 | 24.7813 | 252.2304 | 990.4639 | 1.0590 | 171.1643 | 1013.8229 | 1.0489 |
| $>C=C=CH_2$ | 214.515 | 90.23 | 231.656 | 31.4039 | 279.1155 | 578.8631 | 0.9870 | 171.1643 | 1013.8229 | 1.0489 |
| $-HC=C=CH_2$ | 204.684 | 178.38 | 215.034 | 35.8315 | 208.9812 | 700.3515 | 1.0163 | 85.5824 | 1013.8229 | 1.0489 |
| -HC=C=CH- | 223.307 | 95.25 | 238.982 | 33.7481 | 270.4279 | 949.4988 | 1.0596 | 171.1643 | 1013.8229 | 1.0489 |
| HC↔ | 13.8303 | 44.8963 | 21.6240 | 6.0060 | 45.5189 | 1174.9378 | 1.1387 | 0 | 0 | 0 |
| -C↔ | 23.2422 | -19.3501 | 30.7052 | 5.7087 | 333.3065 | 1269.0479 | 1.1387 | 313.6947 | 986.1985 | 1.0928 |
| ↔C↔ | 18.7849 | -10.6483 | 23.6500 | 9.7540 | 199.4973 | 1587.2948 | 1.1915 | 182.0760 | 1174.9378 | 1.1387 |

Note: ↔ indicates that the carbon is bonded to other carbons in an aromatic ring

| GROUP | $\Delta_H$ | $\Delta_S$ | $\Delta_G$ | A | $B_1$ | $C_1$ | $n_1$ | $B_2$ | $C_2$ | $n_2$ |
|---|---|---|---|---|---|---|---|---|---|---|
| RING FORMATION CORRECTIONS | | | | | | | | | | |
| 3-membered ring | 103.117 | 129.9386 | 64.393 | -7.2109 | 126.0424 | 4304.6037 | 1.3222 | 137.8135 | 1237.7501 | 1.1173 |
| 5-mem.ring-cyclopentane | 25.916 | 98.18 | -3.224 | -23.2619 | 364.2466 | 1281.9257 | 1.1221 | 346.3003 | 545.5184 | 0.9912 |
| 5-mem.ring-cyclopentene | 17.141 | 101.53 | -13.063 | -9.8327 | 295.5738 | 1877.9329 | 1.1852 | 295.5236 | 813.0059 | 1.0604 |
| 5-mem.ring-cyclohexane | 0.628 | 64.56 | -18.464 | -7.8410 | 400.4109 | 4371.2289 | 1.3210 | 415.5604 | 545.5184 | 0.9912 |
| 6-mem.ring-cyclohexene | -2.286 | 83.65 | -27.130 | -19.9091 | 381.0306 | 972.7163 | 1.1001 | 364.4877 | 756.9254 | 1.0481 |
| BRANCHING IN FIVE-MEMBERED RINGS | | | | | | | | | | |
| Single branching | 0.234 | 20.85 | -6.448 | 11.2315 | 430.9482 | 1040.8363 | 1.0906 | 438.4517 | 1062.1344 | 1.1013 |
| Double branching | | | | | | | | | | |
| 1,1 | -2.604 | 14.03 | -6.783 | 23.0044 | 482.8142 | 1362.2334 | 1.1394 | 511.9707 | 951.9636 | 1.0909 |
| 1,2 (cis) | 6.146 | 20.89 | -0.084 | 21.7090 | 488.0603 | 1170.5749 | 1.1148 | 511.9707 | 951.9636 | 1.0909 |
| 1,2 (trans) | -1.013 | 21.56 | -7.452 | 21.0747 | 490.8382 | 1084.4398 | 1.1022 | 511.9707 | 951.9636 | 1.0909 |
| 1,3 (cis) | -0.176 | 21.56 | -6.615 | 21.0747 | 490.8382 | 1084.4398 | 1,1022 | 511.9707 | 951.9636 | 1.0909 |
| 1,3 (trans) | 2.085 | 21.56 | -4.354 | 21.0747 | 490.8382 | 1084.4398 | 1.1022 | 511.9707 | 951.9636 | 1.0909 |

**TABLE 6-3 Thinh et al., Group Contributions for Ideal-Gas Properties** *(Continued)*

| GROUP | ΔH | ΔS | ΔG | A | $B_1$ | $C_1$ | $n_1$ | $B_2$ | $C_2$ | $n_2$ |
|---|---|---|---|---|---|---|---|---|---|---|
| BRANCHING IN SIX-MEMBERED RINGS | | | | | | | | | | |
| Single branching | -2.412 | 18.92 | -8.080 | 6.5147 | 472.5457 | 2470.4850 | 1.2344 | 473.6389 | 3061.5552 | 1.2706 |
| Double branching | | | | | | | | | | |
| 1,1 | 0.578 | 14.40 | -3.726 | 18.9943 | 521.8143 | 3819.4942 | 1.3044 | 546.4984 | 2405.6012 | 1.2375 |
| 1,2 (cis) | 9.412 | 23.91 | -2.261 | 15.2973 | 530.8527 | 2640.9182 | 1.2462 | 546.4984 | 2405.6012 | 1.2375 |
| 1,2 (trans) | 1.583 | 20.31 | -4.480 | 12.5169 | 536.5874 | 2189.0807 | 1.2193 | 546.4984 | 2405.6012 | 1.2375 |
| 1,3 (cis) | -3.190 | 19.85 | -9.127 | 19.7734 | 527.3798 | 3032.2880 | 1.2661 | 546.4984 | 2405.6012 | 1.2375 |
| 1,3 (trans) | 5.016 | 25.62 | -2.638 | 16.1489 | 533.1098 | 2403.6929 | 1.2294 | 546.4984 | 2405.6012 | 1.2375 |
| 1,4 (cis) | 4.932 | 19.85 | -1.005 | 16.1489 | 533.1098 | 2403.6929 | 1.2294 | 546.4984 | 2405.6012 | 1.2375 |
| 1,4 (trans) | -3.023 | 14.19 | -7.243 | 14.5722 | 530.9130 | 2616.9819 | 1.2473 | 546.4984 | 2405.6012 | 1.2375 |
| CORRECTIONS FOR BRANCHING IN AROMATICS | | | | | | | | | | |
| Double branching | | | | | | | | | | |
| 1,2 | 1.926 | 17.54 | 7.159 | 9.2562 | 390.1846 | 1137.6065 | 1.1076 | 393.0425 | 1381.8801 | 1.1442 |
| 1,3 | 0.167 | -12.60 | 3.927 | 1.7204 | 394.2571 | 1266.8220 | 1.1287 | 393.0425 | 1381.8801 | 1.1442 |
| 1,4 | 0.879 | -17.88 | 6.217 | 4.4970 | 390.6682 | 1414.2979 | 1.1428 | 393.0425 | 1381.8801 | 1.1442 |
| Triple branching | | | | | | | | | | |
| 1,2,3 | 6.280 | -29.64 | 15.127 | 13.5058 | 446.9279 | 1416.3668 | 1.1311 | 452.4386 | 1496.1894 | 1.1513 |
| 1,2,4 | 1.926 | -24.49 | 9.236 | 15.8357 | 442.3262 | 1513.4331 | 1.1414 | 452.4386 | 1496.1894 | 1.1513 |
| 1,3,5 | -0.209 | -35.29 | 10.325 | 5.9959 | 450.0936 | 1482.6086 | 1.1443 | 452.4386 | 1496.1894 | 1.1513 |

| GROUP | $\Delta H$ | $\Delta S$ | $\Delta G$ | A | $B_1$ | $C_1$ | $n_1$ | $B_2$ | $C_2$ | $n_2$ |
|---|---|---|---|---|---|---|---|---|---|---|
| SPECIAL CORRECTIONS FOR FIRST FEW ($-CH_2-$) IN NORMAL SERIES | | | | | | | | | | |
| For the first three $-CH_2-$ in normal paraffins | | | | | | | | | | |
| first $-CH_2-$ | 1.45 | 1.47 | 1.03 | -6.7303 | 248.7780 | 653.8562 | 0.9992 | 242.1603 | 779.2506 | 1.0174 |
| first and second $-CH_2-$ | -0.23 | 2.72 | -1.01 | -2.4878 | 310.2695 | 745.5631 | 1.0220 | 312.2675 | 694.6245 | 1.0053 |
| first and second and third $-CH_2-$ | 0.11 | 2.60 | -0.59 | -1.3214 | 376.7655 | 750.0491 | 1.0256 | 382.0195 | 652.7659 | 0.9996 |
| For the first three ($-CH_2-$) in normal alkyl benzenes | | | | | | | | | | |
| first $-CH_2-$ | 0.42 | 1.76 | -0.08 | -2.6088 | 405.3002 | 972.3475 | 1.0980 | 401.3810 | 1115.4125 | 1.1162 |
| first and second $-CH_2-$ | -0.92 | 3.02 | -1.77 | 3.8192 | 463.3222 | 1065.8703 | 1.1084 | 469.7326 | 1014.5118 | 1.0997 |
| first and second and third $-CH_2-$ | -1.93 | 2.89 | -2.71 | 4.9593 | 529.7412 | 1024.6444 | 1.1005 | 538.2738 | 943.4540 | 1.0870 |
| For the first two ($-CH_2-$) in normal monoolefins | | | | | | | | | | |
| first $-CH_2-$ | 7.758 | 12.54 | 4.250 | 7.8264 | 268.1097 | 616.9205 | 0.9973 | 287.2903 | 313.8732 | 0.8983 |
| first and second $-CH_2-$ | 7.591 | 13.80 | 3.705 | 12.6199 | 328.8149 | 721.9341 | 1.0225 | 355.8680 | 352.8265 | 0.9180 |
| For the first two ($-CH_2-$) in normal acetylenes | | | | | | | | | | |
| first $-CH_2-$ | 0.38 | 3.77 | -0.054 | -8.1701 | 226.4510 | 333.3507 | 0.9005 | 209.3387 | 504.9147 | 0.9628 |
| first and second $-CH_2-$ | 0.17 | 3.77 | -0.641 | -1.0002 | 280.4281 | 533.4160 | 0.9750 | 278.6772 | 510.4336 | 0.9686 |

**TABLE 6-3 Thinh et al., Group Contributions for Ideal-Gas Properties** *(Continued)*

| GROUP | $\Delta_H$ | $\Delta_S$ | $\Delta_G$ | A | $B_1$ | $C_1$ | $n_1$ | $B_2$ | $C_2$ | $n_2$ |
|---|---|---|---|---|---|---|---|---|---|---|
| For the first $-CH_2-$ outside the ring, in normal alkyl cyclopentanes | | | | | | | | | | |
| first $-CH_2-$ | 0.251 | -0.54 | 0.460 | -30.4929 | 587.9033 | 312.1951 | 0.9010 | 499.7695 | 956.9970 | 1.0777 |
| For the first two $-CH_2-$ outside the ring, in normal alkyl cyclohexanes | | | | | | | | | | |
| first $-CH_2-$ | 3.642 | 0.29 | 3.601 | 0.8016 | 536.6820 | 2165.9117 | 1.2143 | 539.4721 | 2097.6983 | 1.2081 |
| first and second $-CH_2-$ | 2.721 | -1.72 | 3.308 | -2.1981 | 615.9164 | 1465.2225 | 1.1521 | 606.7502 | 1839.8190 | 1.1870 |
| CORRECTIONS FOR BRANCHING IN PARAFFIN CHAINS | | | | | | | | | | |
| side chain with two or more C atoms | 6.699 | -0.21 | 6.573 | 0 | 0 | 0 | 0 | 0 | 0 | 0 |
| three adjacent $-CH_2-$ groups | -12.460 | 29.35 | 19.510 | 0 | 0 | 0 | 0 | 0 | 0 | 0 |
| adjacent >HC-C<- groups | 10.986 | 11.72 | 7.536 | 0 | 0 | 0 | 0 | 0 | 0 | 0 |

For $C_p^\circ$, one cannot add the contributions for $A$, $B_1$, etc., but rather, each group must be evaluated separately. At 800 K,

$$-\mathrm{CH_3} = 19.8312 + 85.5824 \exp \frac{-1013.8229}{(800)^{1.0489}} = 54.15$$

$$-\mathrm{HC{=}CH_2} = 6.9526 + 131.7946 \exp \frac{-134.8699}{(800)^{0.8030}} = 77.21$$

$$>\mathrm{C{=}CH_2} = 20.6066 + 262.0187 \exp \frac{-652.9594}{(800)^{0.9990}}$$

$$- 171.1643 \exp \frac{-1013.8229}{(800)^{1.0489}} = 67.19$$

$$C_p^\circ\ (800\ \mathrm{K}) = 54.15 + 77.21 + 67.19 = 198.55\ \mathrm{J/(mol \cdot K)}$$

$$\mathrm{Error} = \frac{198.55 - 201.0}{201.0} \times 100 = -1.2\%$$

## 6-6 Method of Benson

For the estimation of $\Delta H_f^\circ$ (298 K), $S^\circ$ (298 K), and $C_p^\circ$ ($T$), an accurate group contribution method was developed by Benson and his colleagues. It is thoroughly described in a book [3] and in a comprehensive review [5]. Contributions are given only for atoms with valences greater than unity. For each group, the key atom is given but followed by a notation specifying other atoms bonded to the key atom. For example, $\mathrm{C{-}(C)(H)_3}$ refers to a carbon atom bonded to another carbon and three hydrogens, that is, $-\mathrm{CH_3}$. The contributions for this method are shown in Table 6-4.

To employ this technique, one must become acquainted with the shorthand notation introduced; for example, $\mathrm{C}_d$ refers to a carbon atom which also participates in a double bond with another carbon atom; it is assumed to have a valence of 2. Notes at the bottoms of the tables define terms which are not immediately obvious.

Table 6-4 is in a continual state of change as new groups are added and older values are updated [9, 18, 22, 23]. It has been adapted for computer use [16, 19], and many new values were added by Shell Development Co. [8] and Olson [17].

Bures et al. [6] have attempted to fit the tabular values of $C_p^\circ$ presented by Benson with various equation forms. In many tests of Bures et al.'s equations, it was found that the calculated group values of $C_p^\circ$ agreed well with Benson's tabulated values. In other cases, however, serious errors were noted; thus, Bures et al.'s equations are not given.

### Corrections for symmetry

When determining $S^\circ$ (298 K) by the Benson group contribution method, one must make corrections for molecular symmetry.

**TABLE 6-4 Benson Group Contributions to Ideal-Gas Properties**

| GROUP | $\Delta H_f^o$(298 K) kJ/mol | $S^o$(298 K) J/mol K | $C_p^o$, J/mol K | | | | | |
|---|---|---|---|---|---|---|---|---|
| | | | 300 K | 400 K | 500 K | 600 K | 800 K | 1000 K |
| **Hydrocarbon Groups** | | | | | | | | |
| C-(C)(H)$_3$ | -42.20 | 127.32 | 25.92 | 32.82 | 39.36 | 45.18 | 54.51 | 61.84 |
| C-(C)$_2$(H)$_2$ | -20.72 | 39.44 | 23.03 | 29.10 | 34.54 | 39.15 | 46.35 | 51.67 |
| C-(C)$_3$(H) | -7.95 | -50.53 | 19.01 | 25.12 | 30.02 | 33.70 | 38.98 | 42.08 |
| C-(C)$_4$ | 2.09 | -146.96 | 18.30 | 25.67 | 30.81 | 34.00 | 36.72 | 36.68 |
| $C_d$-(H)$_2$ | 26.21 | 115.60 | 21.35 | 26.63 | 31.44 | 35.59 | 42.16 | 47.19 |
| $C_d$-(C)(H) | 35.96 | 33.37 | 17.42 | 21.06 | 24.33 | 27.21 | 32.03 | 35.38 |
| $C_d$-(C)$_2$ | 43.29 | -53.17 | 17.17 | 19.30 | 20.89 | 22.02 | 24.28 | 25.46 |
| $C_d$-($C_d$)(H) | 28.39 | 26.71 | 18.67 | 24.24 | 28.26 | 31.07 | 34.96 | 37.64 |
| $C_d$-($C_d$)(C) | 37.18 | -61.13 | (18.42) | (22.48) | (24.83) | (25.87) | (27.21) | (27.72) |
| $C_d$-($C_d$)$_2$ | 19.26 | | | | | | | |
| $C_d$-($C_B$)(H) | 28.39 | 26.80 | 18.67 | 24.24 | 28.26 | 31.07 | 34.96 | 37.64 |
| $C_d$-($C_B$)(C) | 36.17 | (-61.13) | (18.42) | (22.48) | (24.83) | (25.87) | (27.21) | (27.72) |
| $C_d$-($C_B$)$_2$ | 33.49 | | | | | | | |
| $C_d$-($C_t$)(H) | 28.39 | 26.80 | 18.67 | 24.24 | 28.26 | 31.07 | 34.96 | 37.64 |
| $C_d$-($C_t$)(C) | 35.71 | | 18.42 | 22.48 | 24.83 | 25.87 | 27.21 | 27.72 |
| C-($C_d$)(H)$_3$ | -42.20 | 127.32 | 25.92 | 32.82 | 39.36 | 45.18 | 54.51 | 61.84 |
| C-($C_d$)$_2$(H)$_2$ | -17.96 | (42.71) | (19.68) | (28.47) | (35.17) | (40.14) | (47.31) | (52.75) |
| C-($C_d$)$_2$(C)$_2$ | 4.86 | | 14.95 | 25.04 | 31.44 | 35.04 | 37.68 | 37.76 |
| C-($C_d$)(C)$_3$ | 7.03 | (-145.37) | 16.71 | (25.29) | (31.11) | (34.58) | (37.35) | (37.51) |
| C-($C_d$)(C)(H)$_2$ | -19.93 | 41.03 | 22.69 | 28.72 | 34.83 | 39.73 | 46.98 | 52.25 |
| C-($C_d$)(C)$_2$(H) | -6.20 | (-48.99) | (17.42) | (24.74) | (30.73) | 34.29 | (39.61) | (42.66) |
| C-($C_d$)$_2$(C)(H) | -5.19 | | 15.66 | 24.49 | 30.65 | 34.75 | 39.94 | 43.17 |

| GROUP | $\Delta H_f^o$(298 K) kJ/mol | $S^o$(298 K) J/mol K | $C_p^o$, J/mol K 300 K | 400 K | 500 K | 600 K | 800 K | 1000 K |
|---|---|---|---|---|---|---|---|---|
| Hydrocarbon Groups (continued) | | | | | | | | |
| $C-(C_t)(H)_3$ | -42.20 | 127.32 | 25.92 | 32.82 | 39.36 | 45.18 | 54.51 | 61.84 |
| $C-(C_t)(C)(H)_2$ | -19.80 | 43.12 | 20.72 | 27.47 | 33.20 | 38.02 | 45.47 | 51.04 |
| $C-(C_t)(C)_2(H)$ | -7.20 | (-46.89) | (16.71) | (23.49) | (28.68) | (32.57) | (38.10) | (41.45) |
| $C-(C_B)(H)_3$ | -42.20 | 127.32 | 25.92 | 32.82 | 39.36 | 45.18 | 54.51 | 61.84 |
| $C-(C_B)(C)(H)_2$ | -20.35 | 38.94 | 24.45 | 31.86 | 37.60 | 41.91 | 48.11 | 52.50 |
| $C-(C_B(C)_2(H)$ | -4.10 | (-51.08) | (20.43) | (27.88) | (33.08) | (36.63) | (40.74) | (42.91) |
| $C-(C_B)(C)_3$ | 11.76 | (-147.29) | (18.30) | (28.43) | (33.87) | (36.76) | (38.48) | (37.51) |
| $C-(C_B)_2(C)(H)$ | -5.19 | | 15.66 | 24.49 | 30.65 | 34.75 | 39.94 | 43.17 |
| $C-(C_B)_2(C)_2$ | -4.86 | | 14.95 | 25.04 | 31.44 | 35.04 | 37.68 | 37.76 |
| $C-(C_B)(C_d)(H)_2$ | -17.96 | (42.71) | (19.68) | (28.47) | (35.17) | (40.19) | (47.31) | (52.75) |
| $C_t-(H)$ | 112.75 | 103.41 | 22.06 | 25.08 | 27.17 | 28.76 | 31.28 | 33.33 |
| $C_t-(C)$ | 115.35 | 26.59 | 13.10 | 14.57 | 15.95 | 17.12 | 19.26 | 20.60 |
| $C_t-(C_d)$ | 122.25 | (26.92) | (10.76) | (14.82) | (14.65) | (20.60) | (22.36) | (23.03) |
| $C_t-(C_B)$ | (122.25) | 26.92 | 10.76 | 14.82 | 14.65 | 20.60 | 22.36 | 23.03 |
| $C_B-(H)$ | 13.82 | 48.27 | 13.57 | 18.59 | 22.86 | 26.38 | 31.57 | 35.21 |
| $C_B-(C)$ | 23.07 | -32.20 | 11.18 | 13.15 | 15.41 | 17.38 | 20.77 | 22.78 |
| $C_B-(C_d)$ | 23.78 | -32.66 | 15.03 | 16.62 | 18.34 | 19.76 | 22.11 | 23.49 |
| $C_B-(C_t)$ | 23.86 | -32.66 | 15.03 | 16.62 | 18.34 | 19.76 | 22.11 | 23.49 |
| $C_B-(C_B)$ | 20.77 | -36.17 | 13.94 | 17.67 | 20.47 | 22.06 | 24.12 | 24.91 |
| $C_a$ | 143.19 | 25.12 | 16.33 | 18.42 | 19.68 | 20.93 | 22.19 | 23.03 |
| $C_{BF}-(C_b)_2(C_{BF})$ | 20.10 | -20.93 | 12.52 | 15.32 | 17.67 | 19.43 | 21.90 | 23.24 |
| $C_{BF}-(C_B)(C_{BF})_2$ | 15.49 | -20.93 | 12.52 | 15.32 | 17.67 | 19.43 | 21.90 | 23.24 |
| $C_{BF}-(C_{BF})_3$ | 6.07 | 7.62 | 8.71 | 11.93 | 14.65 | 16.87 | 19.89 | 21.52 |

**TABLE 6-4 Benson Group Contributions to Ideal-Gas Properties** *(Continued)*

| GROUP | $\Delta H_f^o$(298 K) kJ/mol | $S^o$(298 K) J/mol K | $C_p^o$, J/mol K 300 K | 400 K | 500 K | 600 K | 800 K | 1000 K |
|---|---|---|---|---|---|---|---|---|
| | | Next-Nearest-Neighbor Correction | | | | | | |
| Alkane gauche | 3.35 | | | | | | | |
| Alkene gauche | 2.09 | | | | | | | |
| Cis | 4.19 | d | -5.61 | -4.56 | -3.39 | -2.55 | -1.63 | -1.09 |
| Ortho | 2.39 | -6.74 | 4.69 | 5.65 | 5.44 | 4.90 | 3.68 | 2.76 |
| Ring ($\sigma$)[e,f] | | Corrections to be Applied for Ring Compounds | | | | | | |
| Cyclopropane (6) | 115.56 | 134.40 | -12.77 | -10.59 | -8.79 | -7.95 | -7.41 | -6.78 |
| Cyclopropene (2) | 224.83 | 140.68 | | | | | | |
| Cyclobutane (8) | 109.69 | 124.77 | -19.30 | -16.29 | -13.15 | -11.05 | -7.87 | -5.78 |
| Cyclobutene (2) | 124.77 | 121.42 | -10.59 | -9.17 | -7.91 | -7.03 | -6.20 | -5.57 |
| Cyclopentane (10) | 26.38 | 114.30 | -27.21 | -23.03 | -18.84 | -15.91 | -11.72 | -7.95 |
| Cyclopentene (2) | 24.70 | 108.02 | -25.04 | -22.40 | -20.47 | -17.33 | -12.27 | -9.46 |
| Cyclopentadiene | 25.12 | 117.23 | -18.00 | | | | | |
| Cyclohexane (6) | 0 | 78.71 | -24.28 | -17.17 | -12.14 | -5.44 | 4.61 | 9.21 |
| Cyclohexene (2) | 5.86 | 90.02 | -17.92 | -12.73 | -8.29 | -5.99 | -1.21 | 0.33 |
| Cycloheptane (1) | 26.80 | 66.57 | | | | | | |
| Cyclooctane (8) | 41.45 | 69.08 | | | | | | |
| Naphthalene | -- | 33.91 | | | | | | |

| GROUP | $\Delta H_f^o$(298 K) kJ/mol | $S^o$(298 K) J/mol K | $C_p^o$, J/mol K | | | | | |
|---|---|---|---|---|---|---|---|---|
| | | | 300 K | 400 K | 500 K | 600 K | 800 K | 1000 K |
| Oxygen-Containing Compounds | | | | | | | | |
| CO-(CO)(H) | -108.86 | | 28.14 | 32.78 | 37.26 | 41.41 | 47.86 | 50.74 |
| CO-(CO)(C) | -122.25 | | 22.86 | 26.46 | 29.98 | 32.95 | 37.68 | 40.86 |
| CO-(O)($C_d$) | -136.07 | | 25.00 | 28.05 | 31.02 | 33.58 | 37.14 | 39.19 |
| CO-(O)($C_B$) | -136.07 | | 9.13 | 11.51 | 16.66 | 21.06 | 26.33 | 29.56 |
| CO-(O)(C) | -146.96 | 20.01 | 25.00 | 28.05 | 30.98 | 33.58 | 37.14 | 39.19 |
| CO-(O)(H) | -134.40 | 146.24 | 29.43 | 32.95 | 36.93 | 40.53 | 46.72 | 51.08 |
| CO-($C_d$)(H) | -132.72 | | 29.43 | 32.95 | 36.93 | 40.53 | 46.72 | 51.08 |
| CO-($C_B$)$_2$ | -159.52 | | 22.02 | 28.34 | 32.11 | 35.50 | 40.28 | 41.24 |
| CO-($C_B$)(C) | -129.37 | | 23.78 | 28.97 | 32.24 | 35.00 | 39.31 | 40.86 |
| CO-($C_B$)(H) | -144.86 | | 26.80 | 32.32 | 37.30 | 41.24 | 48.11 | 50.62 |
| CO-(C)$_2$ | -131.47 | 62.84 | 23.40 | 26.46 | 29.68 | 32.49 | 37.22 | 40.24 |
| CO-(C)(H) | -121.84 | 146.24 | 29.43 | 32.95 | 36.93 | 40.53 | 46.72 | 51.08 |
| CO-(H)$_2$ | -108.86 | 224.71 | 35.46 | 39.27 | 43.79 | 48.23 | 55.98 | 62.01 |
| O-($C_B$)(CO) | -136.07 | | 8.62 | 11.30 | 13.02 | 14.32 | 16.24 | 17.50 |
| O-(CO)$_2$ | -213.11 | | -1.72 | 7.45 | 13.40 | 16.75 | 21.48 | 24.49 |
| O-(CO)(O) | -79.55 | | 15.49 | 15.49 | 15.49 | 15.49 | 17.58 | 17.58 |
| O-(CO)($C_d$) | -196.36 | | 6.03 | 12.48 | 16.66 | 18.80 | 20.81 | 21.77 |
| O-(CO)(C) | -185.48 | 35.13 | 16.33 | 15.11 | 17.54 | 19.34 | 20.89 | 20.18 |
| O-(CO)(H) | -243.25 | 102.66 | 15.95 | 20.85 | 24.28 | 26.54 | 30.10 | 32.45 |
| O-(O)(C) | (-18.84) | (39.36) | (15.49) | (15.49) | (15.49) | (15.49) | (17.58) | (17.58) |
| O-(O)$_2$ | (-79.55) | (39.36) | (15.49) | (15.49) | (15.49) | (15.49) | (17.58) | (17.58) |
| O-(O)(H) | -68.12 | 116.60 | 21.65 | 24.24 | 26.29 | 27.88 | 29.94 | 31.44 |
| O-($C_d$)$_2$ | -137.33 | 42.29 | 14.24 | 15.49 | 15.49 | 15.91 | 18.42 | 19.26 |
| O-($C_d$)(C) | -133.56 | 40.61 | 14.24 | 15.49 | 15.49 | 15.91 | 18.42 | 19.26 |
| O-($C_B$)$_2$ | -88.34 | | 4.56 | 5.11 | 6.28 | 8.33 | 11.93 | 14.70 |
| O-($C_B$)(C) | -94.62 | | 14.24 | 15.49 | 15.49 | 15.91 | 18.42 | 19.26 |
| O-($C_B$)(H) | -158.68 | 121.84 | | 18.84 | 20.10 | 21.77 | 25.12 | 27.63 |

**TABLE 6-4 Benson Group Contributions to Ideal-Gas Properties** *(Continued)*

| GROUP | $\Delta H_f^o$(298 K) kJ/mol | $S^o$(298 K) J/mol K | $C_p^o$, J/mol K 300 K | 400 K | 500 K | 600 K | 800 K | 1000 K |
|---|---|---|---|---|---|---|---|---|
| Oxygen-containing Compounds, continued | | | | | | | | |
| O-(C)$_2$ | -99.23 | 36.34 | 14.24 | 15.49 | 15.49 | 15.91 | 18.42 | 19.26 |
| O-(C)(H) | -158.68 | 121.71 | 18.13 | 18.63 | 20.18 | 21.90 | 25.20 | 27.67 |
| $C_d$-(CO)(O) | 37.68 | | 23.40 | 29.31 | 31.32 | 32.45 | 33.58 | 34.04 |
| $C_d$-(CO)(C) | 39.36 | | 15.62 | 18.76 | 21.02 | 22.61 | 24.91 | 26.67 |
| $C_d$-(CO)(H) | 35.59 | | 15.87 | 20.52 | 24.45 | 27.80 | 32.66 | 36.59 |
| $C_d$-(O)($C_d$) | 37.26 | | (18.42) | (22.48) | (24.83) | (25.87) | (27.21) | (27.72) |
| $C_d$-(O)(C) | 43.12 | | 17.17 | 19.30 | 20.89 | 22.02 | 24.28 | 25.46 |
| $C_d$-(O)(H) | 36.01 | | 17.42 | 21.06 | 24.33 | 27.21 | 32.03 | 35.38 |
| $C_B$-(CO) | 40.61 | | 11.18 | 13.15 | 15.41 | 17.38 | 20.77 | 22.78 |
| $C_B$-(O) | -3.77 | -42.71 | 16.33 | 22.19 | 25.96 | 27.63 | 28.89 | 28.89 |
| C-(CO)$_2$(H)$_2$ | -31.82 | | 23.45 | 29.52 | 35.13 | 40.53 | 48.48 | 53.88 |
| C-(CO)(C)$_2$(H) | -7.54 | -50.24 | 26.00 | 31.65 | 33.49 | 34.37 | 38.43 | 40.32 |
| C-(CO)(C)(H)$_2$ | -21.77 | 40.19 | 25.96 | 32.24 | 36.43 | 39.77 | 46.47 | 51.08 |
| C-(CO)(C)$_3$ | 6.70 | | 21.23 | 28.81 | 32.70 | 34.62 | 36.84 | 36.09 |
| C-(CO)(H)$_3$ | -42.29 | 127.32 | 25.92 | 32.82 | 39.36 | 45.18 | 54.51 | 61.84 |
| C-(O)$_2$(C)$_2$ | -77.87 | | 6.66 | 16.54 | 25.96 | 30.94 | 31.90 | 35.50 |
| C-(O)$_2$(C)(H) | -68.24 | | 21.19 | 30.48 | 37.81 | 39.40 | 43.17 | 45.01 |
| C-(O)$_2$(H)$_2$ | -63.22 | | 11.85 | 21.19 | 31.48 | 38.18 | 43.21 | 47.27 |
| C-(O)($C_B$)(H)$_2$ | -33.91 | 40.61 | 15.53 | 26.25 | 34.67 | 40.99 | 49.36 | 55.27 |
| C-(O)($C_B$)(C)(H) | -25.46 | | 21.52 | 30.56 | 36.97 | 39.48 | 42.83 | 44.38 |
| C-(O)($C_d$)(H)$_2$ | -28.89 | | 19.51 | 29.18 | 36.22 | 41.37 | 48.32 | 53.30 |
| C-(O)(C)$_3$ | -27.63 | -140.51 | 18.13 | 25.92 | 30.35 | 32.24 | 34.33 | 34.50 |
| C-(O)(C)$_2$(H) | -30.14 | -46.05 | 20.10 | 27.80 | 33.91 | 36.55 | 41.07 | 43.54 |
| C-(O)(C)(H)$_2$ | -33.91 | 41.03 | 20.89 | 28.68 | 34.75 | 39.48 | 46.52 | 51.62 |
| C-(O)(H)$_3$ | -42.29 | 127.32 | 25.92 | 32.82 | 39.36 | 45.18 | 54.55 | 61.84 |

| GROUP | $\Delta H_f^o$(298 K) kJ/mol | $S^o$(298 K) J/mol K | $C_p^o$, J/mol K 300 K | 400 K | 500 K | 600 K | 800 K | 1000 K |
|---|---|---|---|---|---|---|---|---|
| Strain and Ring Corrections for Oxygen-containing Compounds | | | | | | | | |
| Ether oxygen, gauche | 1.3 | | -0.42 | -3.73 | -4.61 | -3.06 | -2.51 | -0.96 |
| Ditertiary ethers | 32.7 | | -16.50 | -23.61 | -29.94 | -36.97 | -50.41 | -62.38 |
| Ethylene oxide | 115.6 | 131.5 | -8.4 | -11.7 | -12.6 | -10.9 | -9.6 | -9.6 |
| Trimethylene oxide | 110.5 | 116.0 | -19.3 | -20.9 | -17.6 | -14.7 | -10.9 | 0.8 |
| Tetrahydrofuran | 28.1 | | -17.8 | -19.01 | -17.04 | -14.86 | -12.94 | -10.93 |
| Tetrahydropyran | 9.2 | | -17.92 | -12.73 | -8.29 | -5.99 | -1.21 | 0.33 |
| 1,3-Dioxane | 3.8 | | -10.51 | -12.06 | -9.55 | -6.24 | -1.09 | 2.34 |
| 1,4-Dioxane | 22.6 | | -17.42 | -19.13 | -13.02 | -7.87 | -4.56 | -1.97 |
| 1,3,5-Trioxane | 21.4 | | 7.49 | 2.34 | -2.55 | -2.72 | -5.02 | -10.17 |
| Furan | -24.3 | | -17.54 | -15.20 | -12.23 | -10.01 | -8.33 | -7.20 |
| Dihydropyran | 5.0 | | -18.59 | -13.40 | -6.53 | -1.88 | 1.76 | 2.76 |
| Cyclopentanone | 21.8 | | -35.71 | -30.10 | -22.23 | -15.57 | -9.46 | -5.11 |
| Cyclohexanone | 9.2 | | -33.91 | -27.51 | -17.75 | -8.00 | 2.93 | 8.25 |
| Succinic anhydride | 18.8 | | -33.08 | -25.20 | -18.80 | -14.99 | -14.03 | -12.81 |
| Glutaric anhydride | 3.3 | | -33.20 | -25.29 | -18.84 | -15.03 | -14.03 | -12.85 |
| Maleic anhydride | 15.1 | | -21.44 | -14.15 | -8.46 | -9.17 | -1.55 | -0.04 |

**TABLE 6-4 Benson Group Contributions to Ideal-Gas Properties** *(Continued)*

| GROUP | $\Delta H_f^o$(298 K) kJ/mol | $S^o$(298 K) J/mol K | $C_p^o$, J/mol K 300 K | 400 K | 500 K | 600 K | 800 K | 1000 K |
|---|---|---|---|---|---|---|---|---|
| Nitrogen-containing Compounds | | | | | | | | |
| $C-(N)(H)_3$ | -42.20 | 127.32 | 25.92 | 32.82 | 39.36 | 45.18 | 54.51 | 61.84 |
| $C-(N)(C)(H)_2$ | -27.6 | 41.0 | 21.98 | 28.89 | 34.57 | 39.31 | 46.43 | 51.67 |
| $C-(N)(C)_2(H)$ | -21.8 | -49.0 | 19.55 | 26.46 | 31.99 | 35.13 | 40.03 | 42.83 |
| $C-(N)(C)_3$ | -13.4 | -142.8 | 18.21 | 25.79 | 30.61 | 33.12 | 35.55 | 35.59 |
| $N-(C)(H)_2$ | 20.1 | 124.4 | 23.95 | 27.26 | 30.65 | 33.79 | 39.40 | 43.84 |
| $N-(C)_2(H)$ | 64.5 | 37.4 | 17.58 | 21.81 | 25.67 | 28.60 | 33.08 | 36.22 |
| $N-(C)_3$ | 102.2 | -56.4 | 14.57 | 19.09 | 22.73 | 25.00 | 27.47 | 27.93 |
| $N-(N)(H)_2$ | 47.7 | 122.0 | 25.54 | 30.90 | 35.29 | 38.81 | 44.13 | 48.23 |
| $N-(N)(C)(H)$ | 87.5 | 40.2 | 20.18 | 24.28 | 27.21 | 29.31 | 32.66 | 34.75 |
| $N-(N)(C)_2$ | 122.3 | -57.8 | 6.53 | 10.47 | 13.86 | 16.20 | 19.34 | 20.89 |
| $N-(N)(C_B)(H)$ | 92.5 | | 13.73 | 16.96 | 19.89 | 22.23 | 26.29 | 28.93 |
| $N_I-(H)$ | (68.2) | (51.5) | 12.35 | 19.18 | 27.00 | 32.28 | 38.23 | 41.53 |
| $N_I-(C)$ | 89.2 | | 10.38 | 13.98 | 16.54 | 17.96 | 19.22 | 19.26 |
| $N_I-(C_B)$ | 69.9 | | 10.89 | 13.48 | 15.95 | 17.67 | 20.05 | 21.44 |
| $N_A-(H)$ | 105.1 | 112.2 | 18.34 | 20.47 | 22.78 | 24.87 | 28.34 | 31.07 |
| $N_A-(C)$ | 136.1 | 33.5 | 11.30 | 17.17 | 20.60 | 22.36 | 23.82 | 23.91 |
| $N-(C_B)(H)_2$ | 20.1 | 124.4 | 23.95 | 27.26 | 30.65 | 33.79 | 39.40 | 43.84 |
| $N-(C_B)(C)(H)$ | 62.4 | | 15.99 | 20.47 | 23.91 | 26.29 | 30.10 | 32.36 |
| $N-(C_B)(C)_2$ | 109.7 | | 2.60 | 8.46 | 13.69 | 17.29 | 21.90 | 23.40 |
| $N-(C_B)_2(H)$ | 68.2 | | 9.04 | 13.06 | 17.29 | 21.35 | 28.30 | 32.99 |
| $C_B-(N)$ | -2.1 | -40.6 | 16.54 | 21.81 | 24.87 | 26.46 | 27.34 | 27.47 |
| $N_A-(N)$ | 96.3 | | 8.88 | 17.50 | 23.07 | 28.34 | 28.72 | 29.52 |
| $CO-(N)(H)$ | -123.9 | 146.2 | 29.43 | 32.95 | 36.93 | 40.53 | 46.72 | 51.08 |
| $CO-(N)(C)$ | -137.3 | 67.8 | 22.48 | 25.83 | 29.60 | 32.07 | 40.28 | 46.85 |
| $N-(CO)(H)_2$ | -62.4 | 103.37 | 17.04 | 24.03 | 29.85 | 34.71 | 41.70 | 46.98 |
| $N-(CO)(C)(H)$ | -18.4 | 16.3 | 16.20 | 21.27 | 24.91 | 28.30 | 28.76 | 27.38 |
| $N-(CO)(C)_2$ | 19.7 | | 7.66 | 15.87 | 21.94 | 25.92 | 29.77 | 31.07 |
| $N-(CO)(C_B)(H)$ | 1.7 | | 12.69 | 16.37 | 19.26 | 23.36 | 26.08 | 26.46 |

| GROUP | $\Delta H_f^o$(298 K) kJ/mol | $S^o$(298 K) J/mol K | $C_p^o$, J/mol K 300 K | 400 K | 500 K | 600 K | 800 K | 1000 K |
|---|---|---|---|---|---|---|---|---|
| Nitrogen-containing Compounds (continued) | | | | | | | | |
| N-$(CO)_2$(H) | -77.5 | | 15.03 | 23.19 | 28.05 | 30.94 | 33.29 | 34.29 |
| N-$(CO)_2$(C) | -24.7 | | 4.48 | 12.98 | 18.05 | 20.93 | 22.94 | 27.09 |
| N-$(CO)_2$(C) | -2.1 | | 4.10 | 12.81 | 17.71 | 20.31 | 22.11 | 22.15 |
| C-(CN)(C)$(H)_2$ | 94.2 | 168.31 | 46.47 | 56.10 | 64.90 | 72.01 | 82.5 | 89.18 |
| C-(CN)$(C)_2$(H) | 108.0 | 82.90 | 46.05 | 53.17 | 59.03 | 64.48 | 72.43 | 77.87 |
| C-(CN)$(C)_3$ | 121.4 | -11.72 | 36.22 | 46.72 | 53.97 | 58.82 | 64.94 | 67.78 |
| C-$(CN)_2(C)_2$ | | -118.91 | 61.63 | 74.78 | 83.74 | 90.48 | 99.56 | 104.50 |
| $C_d$-(CN)(H) | 156.6 | 153.15 | 41.03 | 48.89 | 55.68 | 60.71 | 68.24 | 72.43 |
| $C_d$-(CN)(C) | 163.91 | 66.61 | 40.78 | 47.23 | 52.25 | 55.52 | 60.50 | 62.51 |
| $C_d$-$(CN)_2$ | 352.1 | | 56.94 | 69.29 | 78.21 | 84.78 | 93.53 | 98.77 |
| $C_d$-$(NO)_2$(H) | | 185.9 | 51.5 | 63.2 | 72.9 | 80.4 | 90.4 | 97.1 |
| $C_B$-(CN) | 149.9 | 85.83 | 41.0 | 46.9 | 51.5 | 54.9 | 59.5 | 62.4 |
| $C_t$-(CN) | 267.1 | 148.21 | 43.12 | 47.31 | 50.66 | 53.17 | 56.94 | 59.87 |
| C-$(NO_2)$(C)$(H)_2$ | -63.2 | 202.6 | 52.71 | 66.24 | 77.54 | 86.50 | 99.60 | 108.44 |
| C-$(NO_2)(C)_2$(H) | -66.2 | 112.6 | 50.20 | 63.68 | 74.19 | 82.10 | 92.86 | 99.23 |
| C-$(NO_2)(C)_3$ | | 16.3 | 41.41 | 55.85 | 66.40 | 73.77 | 81.27 | 87.34 |
| C-$(NO_2)_2$(C)(H) | -62.4 | | 72.52 | 95.54 | 113.34 | 126.48 | 143.82 | 154.20 |
| O-(NO)(C) | -24.7 | 175.4 | 38.10 | 43.12 | 46.9 | 50.2 | 55.7 | 58.2 |
| O-$(NO_2)$(C) | -81.2 | 203.06 | 39.94 | 48.32 | 55.52 | 65.31 | 68.62 | 72.77 |
| Ring Corrections for Nitrogen-containing Compounds | | | | | | | | |
| Ethyleneimine | 116.0 | 132.3 | -8.67 | -9.13 | -9.09 | -8.58 | -8.12 | -7.87 |
| Azetidine | 109.7 | 122.7 | -19.80 | -18.92 | -17.08 | -15.11 | -11.14 | 0.04 |
| Pyrrolidine | 28.5 | 118.8 | -25.83 | -23.36 | -20.10 | -16.75 | -12.02 | -9.09 |
| Piperidine | 4.2 | | -2.34 | 1.55 | 4.52 | 6.53 | 7.16 | -1.93 |
| Succinimide | 35.6 | | 9.04 | 17.08 | 25.71 | 33.54 | 38.14 | 40.91 |

**TABLE 6-4 Benson Group Contributions to Ideal-Gas Properties** *(Continued)*

| GROUP | $\Delta H_f^o$(298 K) kJ/mol | $S^o$(298 K) J/mol K | $C_p^o$, J/mol K 300 K | 400 K | 500 K | 600 K | 800 K | 1000 K |
|---|---|---|---|---|---|---|---|---|
| Halogen Groups | | | | | | | | |
| C-(F)$_3$(C) | -663.2 | 177.9 | 53.2 | 62.8 | 68.7 | 74.9 | 80.8 | 83.7 |
| C-(F)$_2$(H)(C) | -457.6 | 163.7 | 41.4 | 50.2 | 57.4 | 63.2 | 69.9 | 74.5 |
| C-(F)(H)$_2$(C) | -215.6 | 148.2 | 33.9 | 41.87 | 50.2 | 54.43 | 63.6 | 69.5 |
| C-(F)$_2$(C)$_2$ | -406.1 | 74.5 | 41.4 | 49.4 | 56.5 | 60.3 | 67.4 | 69.5 |
| C-(F)(H)(C)$_2$ | -205.2 | 58.6 | 30.56 | 37.85 | 43.84 | 48.40 | 54.85 | 58.66 |
| C-(F)(C)$_3$ | -203.1 | - | 28.47 | 37.10 | 42.71 | 46.72 | 52.04 | 53.26 |
| C-(F)$_2$(Cl)(C) | -445.1 | 169.6 | 57.4 | 67.4 | 73.3 | 77.9 | 82.9 | 85.4 |
| C-(Cl)$_3$(C) | -86.7 | 211.0 | 68.2 | 75.4 | 80.0 | 82.9 | 86.2 | 87.9 |
| C-(Cl)$_2$(H)(C) | (-79.1) | 183.0 | 50.7 | 58.6 | 64.5 | 69.1 | 74.9 | 78.3 |
| C-(Cl)(H)$_2$(C) | -69.1 | 158.3 | 37.3 | 44.8 | 51.5 | 56.1 | 64.1 | 69.9 |
| C-(Cl)$_2$(C)$_2$ | -92.1 | 93.8 | 51.1 | 62.30 | 66.78 | 69.00 | 71.01 | 71.26 |
| C-(Cl)(H)(C)$_2$ | -62.0 | 73.7 | 37.7 | 41.4 | 44.0 | 46.9 | 58.2 | 61.1 |
| C-(Cl)(C)$_3$ | -53.6 | -22.6 | 38.9 | 44.0 | 46.1 | 47.3 | 51.9 | 53.2 |
| C-(Br)$_3$(C) | - | 233.2 | 69.9 | 75.4 | 78.7 | 81.2 | 83.3 | 85.0 |
| C-(Br)(H)$_2$(C) | -22.6 | 170.8 | 38.1 | 46.1 | 52.8 | 57.4 | 64.9 | 70.3 |
| C-(Br)(H)(C)$_2$ | -14.2 | - | 37.39 | 44.63 | 50.07 | 53.76 | 58.82 | 61.63 |
| C-(Br)(C)$_3$ | -1.7 | -8.4 | 38.9 | 46.1 | 48.1 | 51.5 | 55.7 | 55.7 |
| C-(I)(H)$_2$(C) | 33.5 | 180.0 | 38.5 | 46.1 | 54.0 | 58.2 | 66.2 | 72.0 |
| C-(I)(H)(C)$_2$ | 44.0 | 89.2 | 38.5 | 45.6 | 51.1 | 54.4 | 59.5 | 62.0 |
| C-(I)(C)(C$_d$)(H) | 55.77 | - | 34.04 | 41.95 | 44.49 | 52.8 | 58.6 | 62.4 |
| C-(I)(C$_d$)(H)$_2$ | 34.29 | - | 36.93 | 45.68 | 54.30 | 58.78 | 66.78 | 72.60 |
| C-(I)(C)$_3$ | 54.4 | 0 | 41.16 | 49.19 | 54.09 | 56.31 | 57.74 | 56.94 |

| GROUP | $\Delta H_f^o$(298 K) kJ/mol | $S^o$(298 K) J/mol K | $C_p^o$, J/mol K 300 K | 400 K | 500 K | 600 K | 800 K | 1000 K |
|---|---|---|---|---|---|---|---|---|
| Halogen Groups (continued) | | | | | | | | |
| C-(Cl)(Br)(H)(C) | - | 191.3 | 51.9 | 58.6 | 65.3 | 68.2 | 74.9 | 79.5 |
| N-$(F)_2$(C) | -32.7 | - | 34.54 | 42.41 | 48.23 | 53.59 | 60.16 | 62.72 |
| C-(Cl)(C)(O)(H) | -90.4 | 63. | 41.24 | 43.50 | 46.26 | 48.44 | 52.13 | 55.01 |
| C-$(I)_2$(C)(H) | (108.9) | (228.6) | 53.13 | 61.88 | 67.87 | 71.68 | 76.66 | 79.67 |
| C-(I)(O)$(H)_2$ | 15.9 | 170.4 | 34.42 | 43.92 | 51.20 | 56.73 | 64.27 | 69.38 |
| $C_d$-$(F)_2$ | -324.5 | 156.2 | 40.6 | 46.1 | 50.2 | 53.2 | 57.8 | 60.7 |
| $C_d$-$(Cl)_2$ | -7.5 | 176.3 | 47.7 | 52.3 | 55.7 | 58.2 | 61.1 | 62.8 |
| $C_d$-$(Br)_2$ | - | 199.3 | 51.5 | 55.3 | 58.2 | 59.9 | 62.4 | 63.6 |
| $C_d$-(F)(Cl) | - | 166.6 | 43.1 | 49.0 | 52.8 | 55.7 | 59.5 | 61.5 |
| $C_d$-(F)(Br) | - | 177.9 | 45.2 | 50.2 | 53.6 | 56.5 | 59.9 | 61.5 |
| $C_d$-(Cl)(Br) | - | 188.8 | 50.7 | 53.2 | 56.5 | 59.0 | 61.5 | 61.5 |
| $C_d$-(F)(H) | -157.4 | 137.3 | 28.5 | 35.2 | 39.8 | 44.0 | 49.4 | 53.2 |
| $C_d$-(Cl)(H) | -5.0 | 148.2 | 33.1 | 38.5 | 43.1 | 46.9 | 51.5 | 54.8 |
| $C_d$-(Br)(H) | 46.1 | 160.4 | 33.9 | 39.8 | 44.4 | 47.7 | 51.9 | 55.3 |
| $C_d$-(I)(H) | 102.6 | 169.6 | 36.8 | 41.9 | 45.6 | 48.6 | 52.8 | 55.7 |
| $C_d$-(C)(Cl) | -8.8 | 62.8 | 33.5 | 35.2 | 35.6 | 37.7 | 38.5 | 39.4 |
| $C_d$-(C)(I) | 98.8 | - | 37.3 | 38.5 | 38.1 | 39.4 | 39.8 | 40.2 |
| $C_d$-$(C_d)$(Cl) | -14.91 | - | 34.8 | 38.5 | 39.4 | 41.4 | 41.4 | 41.4 |
| $C_d$-$(C_d)$(I) | 92.70 | - | 38.5 | 41.4 | 41.9 | 43.1 | 43.1 | 42.3 |
| $C_t$-(Cl) | - | 139.8 | 33.1 | 35.2 | 36.4 | 37.7 | 39.4 | 40.2 |
| $C_t$-(Br) | - | 151.1 | 34.8 | 36.4 | 37.7 | 38.5 | 39.8 | 40.6 |
| $C_t$-(I) | - | 158.7 | 35.2 | 36.8 | 38.1 | 38.9 | 40.2 | 41.0 |
| $C_B$-(F) | -179.20 | 67.4 | 26.4 | 31.8 | 35.6 | 38.1 | 41.0 | 42.7 |
| $C_B$-(Cl) | -15.9 | 79.1 | 31.0 | 35.2 | 38.5 | 40.6 | 42.7 | 43.5 |
| $C_B$-(Br) | 44.8 | 90.4 | 32.7 | 36.4 | 39.4 | 41.4 | 43.1 | 44.0 |
| $C_B$-(I) | 100.5 | 99.2 | 33.5 | 37.3 | 40.2 | 41.4 | 43.1 | 44.0 |

**TABLE 6-4 Benson Group Contributions to Ideal-Gas Properties** *(Continued)*

| GROUP | $\Delta H_f^o$(298 K) kJ/mol | $S^o$(298 K) J/mol K | $C_p^o$, J/mol K | | | | | |
|---|---|---|---|---|---|---|---|---|
| | | | 300 K | 400 K | 500 K | 600 K | 800 K | 1000 K |
| **Halogen Groups (continued)** | | | | | | | | |
| $C-(C_B)(F)_3$ | -681.2 | 179.2 | 52.3 | 64.1 | 72.0 | 77.5 | 84.2 | 87.9 |
| $C-(C_B)(Br)(H_2)$ | -28.9 | | 38.90 | 46.47 | 52.51 | 57.32 | 65.27 | 69.96 |
| $C-(C_B)(I)(H)_2$ | 35.2 | | 40.95 | 48.40 | 54.01 | 58.95 | 66.49 | 70.80 |
| $C-(Cl)_2(CO)(H)$ | -74.5 | | 53.6 | 61.76 | 66.36 | 69.71 | 75.07 | 77.71 |
| $C-(Cl)_3(CO)$ | -82.1 | | 71.2 | 78.50 | 81.85 | 83.53 | 86.37 | 87.34 |
| $CO-(Cl)(C)$ | -126.4 | | 37.14 | 39.52 | 42.87 | 46.39 | 52.46 | 56.90 |
| **Corrections for Next-Nearest-Neighbor Halogen Compounds** | | | | | | | | |
| Ortho (F)(F) | 20.9 | | 0 | 0 | 0 | 0 | 0 | 0 |
| Ortho (Cl)(Cl) | 9.2 | | -2.09 | -1.84 | -2.30 | -2.22 | -1.17 | -0.08 |
| Ortho (alkane)(halogen) | 2.5 | | 1.76 | 1.84 | 1.17 | 0.80 | 0.50 | 0.59 |
| Cis (halogen)(halogen) | 1.3 | | -0.75 | -0.04 | -0.13 | -0.71 | 0 | -0.13 |
| Cis (halogen)(alkane) | -3.3 | | -4.06 | -2.93 | -2.22 | -1.97 | -1.00 | -0.54 |
| **Organosulfur Groups** | | | | | | | | |
| $C-(H)_3(S)$ | -42.20 | 127.32 | 25.92 | 32.82 | 39.36 | 45.18 | 54.51 | 61.84 |
| $C-(C)(H)_2(S)$ | -23.66 | 41.37 | 22.52 | 29.64 | 36.01 | 41.74 | 51.33 | 59.24 |
| $C-(C)_2(H)(S)$ | -11.05 | -47.39 | 20.31 | 27.26 | 32.57 | 36.38 | 41.45 | 44.25 |
| $C-(C)_3(S)$ | -2.30 | -144.07 | 19.13 | 26.25 | 31.19 | 34.12 | 36.51 | 33.91 |
| $C-(C_B)(H)_2(S)$ | -19.80 | | 17.21 | 28.26 | 36.43 | 42.50 | 49.95 | 54.85 |
| $C-(C_d)(H)_2(S)$ | -27.00 | | 20.93 | 29.27 | 36.30 | 42.16 | 51.9[illegible] | 59.83 |
| $C_B-(S)$ | -7.5 | 42.71 | 16.33 | 22.19 | 25.96 | 27.63 | 28.89 | 28.89 |

a. Data were obtained largely from References 3 and 5. Some $\Delta H_f^o$ values were from References 9 and 23. Many $C_p^o$ values are from Reference 17. Grateful acknowledgment is extended to Shell Development Co., and particularly to the late Dr. P. Chueh for supplying missing values as well as additional contributions. Finally, Dr. S.W. Benson provided an up-to-date errata list that showed recent modifications of the original tables.

b. $C_d$ represents a carbon atom that is joined to another carbon atom by a double bond. It is considered divalent. For example, 2-pentene would have the groups C-($C_d$)(H)$_3$, $C_d$-(C)(H) twice, C-($C_d$)(H)$_2$, and C-(C)(H)$_3$. $C_t$ represents a carbon atom that is joined to another carbon atom by a triple bond. It is considered monovalent. For example, propyne would have the groups $C_t$-(H), $C_t$-(C), and C-($C_t$)(H)$_3$. $C_B$ represents a carbon atom in an aromatic ring. It is considered monovalent. For example, p-ethyltoluene would have the groups C-(C)(H)$_3$, C-($C_B$)(C)(H)$_2$, C-($C_B$)(H)$_3$, $C_B$-(C) twice, and $C_B$-(H) four times. $C_a$ represents the allene group, >C=C=C<; the end carbons are treated as normal $C_d$ atoms. For example, 1,2-butadiene would have the groups $C_a$, $C_d$-(H)$_2$, $C_d$-(C)(H), and C-($C_d$)(H)$_3$. $C_{BF}$) represents a carbon atom at the border of two or three fused aromatic rings. For example, benzo[a]pyrene would consist of 12 [$C_B$-(H)], 4[$C_{BF}$-($C_B$)$_2$($C_{BF}$)], 2[$C_{BF}$-($C_B$)($C_{BF}$)$_2$] and 2[$C_{BF}$-($C_{BF}$)$_3$].

c. When one of the groups is t-butyl, the cis correction = 4.0; when both are t-butyl, the cis correction = 10.0; and when there are two cis corrections around one double bond, the total correction = 3.0.

d. Value is 1.2 for but-2-ene but zero for other dienes and 0.6 for trienes.

e. The number in parentheses beside each ring is the symmetry number.

f. For $\Delta H_f^o$(298 K) contributions for other ring structures, see Chem. Rev. 69, 279 (1969).

g. $N_I$ represents a double-bonded nitrogen in imines; $N_I$-($C_B$) represents a pyridine nitrogen. $N_A$ represents a double-bonded nitrogen in azo compounds. For ortho or para substitution in pyridine add -6.3 kJ/mol per group to $\Delta H_f^o$(298 K).

| GROUP | $\Delta H_f^o$(298 K) kJ/mol | $S^o$(298 K) J/mol K | $C_p^o$, J/mol K 300 K | 400 K | 500 K | 600 K | 800 K | 1000 K |
|---|---|---|---|---|---|---|---|---|
| Organosulfur Groups (continued) | | | | | | | | |
| $C_d$-(H)(S) | 35.84 | 33.49 | 17.42 | 21.06 | 24.33 | 27.21 | 32.03 | 35.38 |
| $C_d$-(C)(S) | 45.76 | -51.96 | 14.65 | 14.95 | 16.04 | 17.12 | 18.46 | 20.93 |
| S-(C)(H) | 19.34 | 132.03 | 24.53 | 25.96 | 27.26 | 28.39 | 30.56 | 32.28 |
| S-($C_B$)(H) | 50.07 | 53.00 | 21.44 | 22.02 | 23.32 | 25.25 | 29.27 | 32.82 |
| S-(C)$_2$ | 48.19 | 55.06 | 20.89 | 20.77 | 21.02 | 21.23 | 22.65 | 23.99 |
| S-(C)($C_d$) | 41.74 | | 17.67 | 21.27 | 23.28 | 24.16 | 24.58 | 24.58 |
| S-($C_d$)$_2$ | -19.01 | 69.00 | 20.05 | 23.36 | 23.15 | 26.33 | 33.24 | 40.74 |
| S-($C_B$)(C) | 80.22 | | 12.64 | 14.19 | 15.53 | 16.91 | 19.34 | 20.93 |
| S-($C_B$)$_2$ | 108.44 | | 8.37 | 8.42 | 9.38 | 11.47 | 15.91 | 19.72 |
| S-(S)(C) | 29.52 | 51.79 | 21.90 | 22.69 | 23.07 | 23.07 | 22.52 | 21.44 |
| S-(S)($C_B$) | 60.7 | | 12.10 | 14.19 | 15.57 | 17.38 | 20.01 | 21.35 |
| S-(S)$_2$ | 12.73 | 55.94 | 19.7 | 20.9 | 21.4 | 21.8 | 22.2 | 22.6 |
| C-(SO)(H)$_3$ | -42.20 | 127.32 | 25.92 | 32.82 | 39.36 | 45.18 | 54.51 | 61.84 |
| C-(C)(SO)(H)$_2$ | -32.32 | | 19.05 | 26.88 | 33.29 | 38.35 | 48.85 | 51.16 |
| C-(C)$_3$(SO) | -12.77 | | 12.81 | 19.18 | 20.26 | 27.63 | 31.57 | 33.33 |
| C-($C_d$)(SO)(H)$_2$ | -30.77 | | 18.42 | 26.63 | 29.06 | 38.73 | 45.93 | 51.29 |
| $C_B$-(SO) | 9.6 | | 11.18 | 13.15 | 15.41 | 17.38 | 20.77 | 22.78 |
| SO-(C)$_2$ | -60.33 | 75.78 | 37.18 | 41.99 | 43.96 | 45.18 | 45.97 | 46.77 |
| SO-($C_B$)$_2$ | -50.2 | | 23.95 | 38.06 | 40.61 | 47.94 | 47.98 | 47.10 |
| C-($SO_2$)(H)$_3$ | -42.20 | 127.32 | 25.92 | 32.82 | 39.36 | 45.18 | 54.51 | 61.84 |
| C-(C)($SO_2$)(H)$_2$ | -32.15 | | 22.52 | 29.64 | 36.01 | 41.74 | 51.33 | 35.65 |
| C-(C)$_2$($SO_2$)(H) | -10.97 | | 18.51 | 26.17 | 31.65 | 35.50 | 40.36 | 43.12 |
| C-(C)$_3$($SO_2$) | -2.55 | | 9.71 | 18.34 | 23.86 | 27.17 | 30.44 | 31.23 |
| C-($C_d$)($SO_2$)(H)$_2$ | -29.89 | | 20.93 | 29.27 | 36.30 | 42.16 | 51.96 | 59.83 |
| C-($C_B$)($SO_2$)(H)$_2$ | -23.19 | | 15.53 | 27.51 | 34.57 | 40.99 | 49.78 | 55.27 |
| $C_B$-($SO_2$) | 9.6 | | 11.18 | 13.15 | 15.41 | 17.38 | 20.77 | 22.78 |

TABLE 6-4 **Benson Group Contributions to Ideal-Gas Properties** *(Continued)*

| GROUP | $\Delta H_f^o$(298 K) kJ/mol | $S^o$(298 K) J/mol K | $C_p^o$, J/mol K 300 K | 400 K | 500 K | 600 K | 800 K | 1000 K |
|---|---|---|---|---|---|---|---|---|
| Organosulfur Groups (continued) | | | | | | | | |
| $C_d$-(H)($SO_2$) | 52.46 | | 12.73 | 19.55 | 24.83 | 28.64 | 32.95 | 36.30 |
| $C_d$-(C)($SO_2$) | 60.58 | | 7.75 | 13.02 | 16.66 | 19.26 | 22.32 | 23.74 |
| $SO_2$-($C_d$)($C_B$) | -287.13 | | 41.41 | 48.15 | 55.89 | 61.17 | 65.82 | 66.65 |
| $SO_2$-($C_d$)$_2$ | -308.06 | | 48.23 | 50.12 | 55.89 | 59.79 | 64.39 | 66.49 |
| $SO_2$-(C)$_2$ | -291.99 | 87.50 | 42.62 | 49.15 | 54.09 | 57.65 | 63.35 | 66.99 |
| $SO_2$-(C)($C_B$) | -302.66 | | 41.62 | 48.15 | 56.31 | 60.75 | 65.40 | 66.65 |
| $SO_2$-($C_B$)$_2$ | -287.13 | | 35.00 | 46.18 | 56.73 | 62.55 | 66.40 | 66.82 |
| $SO_2$-($SO_2$)($C_B$) | -319.24 | | 41.07 | 48.15 | 56.61 | 61.67 | 65.77 | 67.11 |
| CO-(S)(C) | -132.14 | 64.60 | 23.40 | 26.46 | 29.68 | 32.49 | 37.22 | 40.24 |
| S-(H)(CO) | -5.90 | 130.63 | 31.95 | 33.87 | 34.00 | 34.21 | 35.59 | 34.50 |
| C-(S)(F)$_3$ | | 162.9 | 41.37 | 54.47 | 62.09 | 68.54 | 76.07 | 80.01 |
| CS-(N)$_2$ | -132.14 | 64.60 | 23.40 | 26.46 | 29.68 | 32.49 | 37.22 | 40.24 |
| N-(CS)(H)$_2$ | 53.51 | 122.21 | 25.41 | 30.48 | 34.25 | 37.30 | 42.24 | 45.97 |
| S-(S)(N) | -20.52 | | 15.5 | 15.5 | 15.5 | 15.5 | 17.6 | 17.6 |
| N-(S)(C)$_2$ | 125.19 | | 16.62 | 21.65 | 26.00 | 29.06 | 30.94 | 38.69 |
| SO-(N)$_2$ | -132.14 | | 23.40 | 26.46 | 29.68 | 32.49 | 37.22 | 40.24 |
| N-(SO)(C)$_2$ | 66.99 | | 17.58 | 24.62 | 25.62 | 27.34 | 28.60 | 34.92 |





The correction originates in the fact that, from statistical mechanics, the entropy is given by $R \ln W$, where $W$ is the number of distinguishable configurations of a compound. The rotational entropy contributions must be corrected, since by rotating a molecule one often finds indistinguishable configurations and $W$ must be reduced by this factor. If $\sigma$ is the symmetry number (see below for a more exact definition), the rotational entropy is to be corrected by subtracting $R \ln \sigma$ from the calculated value.

Benson [3] defines $\sigma$ as "the total number of *independent* permutations of identical atoms (or groups) in a molecule that can be arrived at by simple rigid rotations of the entire molecule." Inversion is not allowed.

It is often convenient to separate $\sigma$ into two parts, $\sigma_{ext}$ and $\sigma_{int}$, and then

$$\sigma = \sigma_{ext}\sigma_{int} \tag{6-6.1}$$

For example, propane has two terminal $-CH_3$ groups, and each group has a threefold axis of symmetry. Rotation of these *internal* groups yields $\sigma_{int} = (3)(3)$ as the number of permutations. Also, the entire molecule has a single twofold axis of symmetry, so $\sigma_{ext} = 2$. Then $\sigma = (2)(3^2) = 18$. Some additional examples are shown in the accompanying table. Benson et al. [5] show many other examples in their comprehensive review.

| | $\sigma_{ext}$ | $\sigma_{int}$ | $\sigma$ |
|---|---|---|---|
| Benzene | (6)(2) | 1 | 12 |
| Methane | 4 | 3 | 12 |
| *p*-Cresol | 2 | 3 | 6 |
| 1,3,5-Trimethylbenzene | 2 | $3^4$ | 162 |
| 1,2,4-Trimethylbenzene | 1 | $3^3$ | 27 |
| Cyclohexane | 6 | 1 | 6 |
| Methanol | 1 | 3 | 3 |
| *t*-Butyl alcohol | 1 | $3^4$ | 81 |
| Acetone | 2 | $3^2$ | 18 |
| Acetic acid | 1 | 3 | 3 |
| Aniline | 2 | 1 | 2 |
| Trimethylamine | 3 | $3^3$ | 81 |

### Corrections for isomers

In addition to the symmetry corrections noted above, if a molecule has optical isomers, i.e., contains one or more completely asymmetric carbon atoms (as in 3-methylhexane), the number of spatial orientations is increased and a correction of $+R \ln \eta$ must be added to the calculated absolute entropy, $\eta$ being the number of such isomers. The number of possible optical isomers is $2^m$, where $m$ is the number of asymmetric carbons. However, in some molecules with more than one asymmetric carbon atom, there exist planes of symmetry which negate the optical activity of some forms. An example is the meso form of tartaric acid.

In a similar manner, Benson [3] indicates that, in molecules of the type ROOH and R—OO—R, the O—H and O—R bonds are at approximately right angles and exist in right- and left-hand forms with a higher entropy by $R \ln 2$.

In summary, when using Benson's method to estimate $S°$ (298 K), one first sums the contributions for the various groups ($\Delta_S$ in Table 6-4) and then corrects this sum by

$$S° \text{ (298 K)} = \sum_j n_j \Delta_S - R \ln \sigma + R \ln \eta \tag{6-6.2}$$

where $\sigma$ is the symmetry number from Eq. (6-6.1) and $\eta$ is the number of possible optical isomers.

For $\Delta H_f°$ (298 K) and $C_p°$ $(T)$, no symmetry or optical isomer corrections are necessary. For $\Delta H_f°$ (298 K),

$$\Delta H_f° \text{ (298 K)} = \sum_j n_j \Delta_H \tag{6-6.3}$$

and $C_p°$ $(T)$ is obtained by summing the necessary group contributions at the system temperature from Table 6-4.

The method is illustrated in Example 6-4, and estimated values of $C_p°$ $(T)$, $\Delta H_f°$ (298 K), and $S°$ (298 K) are compared with literature values in Tables 6-6 to 6-8. The method is discussed in Sec. 6-8.

**Example 6-4** Use Benson's method to estimate $\Delta H_f°$ (298 K), $S°$ (298 K), and $C_p°$ (800 K) for 2-methyl-2-butanethiol. Literature values for these three properties are −127.11 kJ/mol, 387.20 J/(mol·K), and 277.50 J/(mol·K), respectively [25].

**solution** 2-Methyl-2-butanethiol is composed of the following groups with the contributions from Table 6-4.

| Group | Number | $\Delta_H$ | $\Delta_S$ | $C_p°$ (800 K) |
|---|---|---|---|---|
| $C-(C)(H)_3$ | 3 | (3)(−42.20) | (3)(127.32) | (3)(54.51) |
| $C-(C)_2(H)_2$ | 1 | −20.72 | 39.44 | 46.35 |
| $C-(C)_3(S)$ | 1 | −2.30 | −144.07 | 36.51 |
| $S-(C)(H)$ | 1 | 19.34 | 137.03 | 30.56 |
| | | −130.88 | 414.36 | 276.95 |

With Eq. (6-6.3),

$$\Delta H_f° \text{ (298 K)} = -130.88 \text{ kJ/mol}$$

$$\text{Difference} = -130.88 - (-127.11) = -3.8 \text{ kJ/mol}$$

For $S°$ (298 K), the symmetry number of 2-methyl-2-butanethiol is computed with $\sigma_{ext} = 1$, $\sigma_{int} = 3^3$, and, by Eq. (6-6.1), $\sigma = (1)(3^3) = 27$. Then, with Eq. (6-6.2) and $\eta = 1$,

$$S° \text{ (298 K)} = 414.36 - (8.314)(\ln 27) + (8.314)(\ln 1)$$
$$= 386.96 \text{ J/(mol·K)}$$
$$\text{Difference} = 386.96 - 387.20 = -0.3 \text{ J/(mol·K)}$$
$$C_p° \text{ (800 K)} = 276.95 \text{ J/(mol·K)}$$
$$\text{Error} = \frac{276.95 - 277.50}{277.50} \times 100 = -0.2\%$$

## 6-7 Method of Cardozo [7]

Cardozo has proposed a rather simple method to estimate the enthalpy of combustion and, as shown later, the enthalpy of formation for a wide variety of organic compounds in the gaseous, liquid, and solid states.

The enthalpy of combustion is defined as the difference in enthalpy of a compound and that of its products of combustion in the gaseous state, all at 298 K and 1 atm. The products of combustion are assumed to be $H_2O(g)$, $CO_2(g)$, $SO_2(g)$, $N_2(g)$, and $HX(g)$, where X is a halogen atom. Since product water is in the gaseous state, this enthalpy of combustion would be termed the *lower* enthalpy of combustion.

In the calculations an *equivalent chain length* $N$ is defined as

$$N = N_C + \sum_i \Delta N_i \tag{6-7.1}$$

where $N_C$ is the *total* number of carbon atoms in the compound and $\Delta N_i$ are corrections for various structures and phases as shown in Table 6-5. Once $N$ has been determined, the enthalpy of combustion (in kJ/mol) is determined from Eqs. (6-7.2) through (6-7.4).

$$\Delta H_c°(g) = -198.42 - 615.14N \tag{6-7.2}$$

$$\Delta H_c°(l) = -196.98 - 610.13N \tag{6-7.3}$$

$$\Delta H_c°(s) = -206.21 - 606.56N \tag{6-7.4}$$

For the normal alkanes, $N = N_C$. Since the technique applies to quite complex compounds, some examples are shown in Table 6-5 to illustrate its wide applicability.

If one has the enthalpy of combustion, it is possible to obtain an enthalpy of formation by using Eq. (6-7.5),

$$\Delta H_f° \text{ (298 K)} = -393.78N_C - 121.00(N_H - N_X) - 271.81N_F - 92.37N_{Cl} - 36.26N_{Br} + 24.81N_I - 297.26N_S - \Delta H_c° \tag{6-7.5}$$

where $N_C$, $N_H$, $N_F$, $N_{Cl}$, $N_{Br}$, $N_I$, and $N_S$ are the numbers of atoms of carbon, hydrogen, fluorine, chlorine, bromine, iodine, and sulfur in the compound and $N_X$ is the total number of halogen atoms. Enthalpies of formation estimated with Eqs. (6-7.1), (6-7.2), and (6-7.5) are compared with literature values in Table 6-7.

## 6-8 Discussion and Recommendations

### Enthalpy of formation

Six methods were evaluated. Values estimated from these methods are compared with literature values in Table 6-7. The methods of Benson (Sec. 6-6) and Yoneda (Sec. 6-4) yield the smallest errors with only a few large deviations. Both techniques allow for effects of next-nearest neighbors and require some patience to master the method of calculation. Thinh et al.'s procedure (Sec. 6-5) is also quite accurate but is limited to hydrocarbons. These authors [26, 27], in a much wider test, found errors normally less than 1 kJ/mol. Joback's technique is broadly applicable and, on the average, is only slightly less accurate than Benson's or Yoneda's. The large errors encountered for some compounds should, however, be noted.

The procedure suggested by Cardozo (Sec. 6-7) determines the enthalpy of combustion at 298 K, and from this value the enthalpy of formation can be found. The errors in Table 6-7 are comparable to those of Joback's. Cardozo's procedure is, however, most useful for quite complex organic compounds especially when the material is in a condensed phase at the temperature of interest.

It is recommended that, for highest accuracy, the Benson or Yoneda method be selected, but the procedures of Joback and Cardozo also normally provide reliable estimates—and are much simpler to use.

Literature values of $\Delta H_f^\circ$ (298 K) are given for many compounds in Appendix A.

### Ideal-gas heat capacities

Four estimation methods were presented, and calculated results are compared with literature values in Table 6-6. All techniques are similar in accuracy, and except for quite unusual structures, errors are less than 1 to 2 percent. Joback's method is the easiest to use and has wide applicability. In this case, the coefficients of a third-order polynomial are estimated and used to determine $C_p^\circ$ in an equation form. Yoneda's approach is similar, but only a second-order polynomial is used. Thinh et al.'s equations for heat capacity are more complex and cannot be integrated analytically. Benson tabulates $C_p^\circ$ group values at increments of 100 K.

None of the estimation procedures except the procedure of Thinh et al. should be used outside the approximate range of 280 to 1100 K. The

**TABLE 6-5 Cardozo Correction Factors for the Enthalpy of Combustion**

A   $n$-Alkanes   Gas $\Delta H_c^\circ = -198.42 - 615.14N$

Liquid $\Delta H_c^\circ = -196.98 - 610.13N$

Solid $\Delta H_c^\circ = -206.21 - 606.56N$

| | | Per | $\Delta N_i$ gas | $\Delta N_i$ liquid | $\Delta N_i$ solid | Remark |
|---|---|---|---|---|---|---|
| $B_1$ | Carbon-to-carbon branch, alkanes | Branch | For g, l, and s: $-0.031 + 0.012 \ln N_C$ | | | 1 |
| $B_2$ | Carbon-to-carbon branch, all other compounds | Branch | −0.02 | −0.02 | −0.02 | |
| $C_1$ | Cyclopropanes | | −0.06 | −0.102 | — | 2 |
| $C_2$ | Cyclobutanes | | −0.16 | −0.17 | — | 2 |
| $C_3$ | Cyclopentanes | | −0.277 | −0.283 | −0.25 | 2 |
| $C_4$ | Cyclohexanes | | −0.311 | −0.311 | −0.278 | 2 |
| $C_5$ | Cycloheptanes | | −0.29 | −0.297 | — | 2 |
| $C_6$ | Cyclooctanes and higher | | −0.256 | −0.269 | −0.271 | 2 |
| $D_1$ | 1-Alkenes | Double bond | −0.189 | −0.189 | −0.189 | 3 |
| $D_2$ | 1-Alkenes ($i \neq 1$) | Double bond | −0.205 | −0.208 | −0.218 | 3 |
| $D_3$ | Cis | | +0.004 | +0.003 | +0.003 | 3 |
| $D_4$ | Trans | | −0.003 | −0.002 | −0.002 | 3 |
| $E_1$ | 1-Alkynes | Triple bond | −0.314 | −0.342 | — | 3 |
| $E_2$ | 1-Alkynes ($i \neq 1$) | Triple bond | −0.34 | −0.347 | — | 3 |
| $F_1$ | Alcohols, primary | −OH | −0.246 | −0.297 | −0.30 | |
| $F_2$ | Alcohols, secondary | −OH | −0.27 | −0.32 | −0.33 | |
| $F_3$ | Alcohols, tertiary | −OH | −0.30 | −0.36 | −0.33 | |
| G | Aldehydes | =O | −0.525 | −0.551 | −0.52 | |
| H | Ketones | =O | −0.576 | −0.609 | −0.57 | |
| I | Carboxylic acids | −OOH | −0.94 | −1.033 | −1.038 | |
| J | Esters | −OO− | −0.857 | −0.93 | −0.90 | |
| K | Lactones | | −1.08 | −1.13 | −1.19 | 2 |
| L | Ethers | −O− | −0.197 | −0.212 | −0.25 | |
| $M_1$ | Amines, primary | $-NH_2$ | +0.24 | +0.21 | +0.18 | |
| $M_2$ | Amines, secondary | =NH | +0.30 | +0.27 | +0.16 | |
| $M_3$ | Amines, tertiary | =N− | +0.32 | +0.33 | +0.14 | |
| N | Amides | $-ONH_2$ | — | −0.542 | −0.542 | |

| | | | | | | |
|---|---|---|---|---|---|---|
| O | Lactams | | — | — | −0.80 | 2 |
| P | Amino acids | | — | — | Additional −0.043 | 4 |
| Q | Dipeptides | | — | — | Σ amino acids +0.44 | 4 |
| R | Diketopiperazines | | — | — | Σ amino acids +0.59 | 4 |
| $S_1$ | 1-Nitro- | $-NO_2$ | −0.22 | −0.27 | — | |
| $S_2$ | 2-Nitro- | $-NO_2$ | −0.26 | −0.27 | −0.28 | |
| $S_3$ | Dinitro- | $=(NO_2)_2$ | — | −0.50 | −0.50 | |
| $S_4$ | Trinitro- | $\equiv(NO_2)_3$ | — | — | −0.64 | |
| T | Nitriles | $\equiv N$ | −0.322 | −0.36 | — | |
| U | Sulfides | −S− | +0.553 | +0.535 | — | |
| V | Disulfides | −S−S− | +1.049 | — | — | |
| W | Thiols, primary | −SH | +0.546 | +0.524 | — | |
| $X_1$ | Fluoro- | −F | −0.26 | −0.26 | — | |
| $X_2$ | Chloro- | −Cl | −0.28 | −0.30 | −0.30 | |
| $X_3$ | Bromo- | −Br | −0.30 | −0.33 | — | |
| $X_4$ | Iodo- | −I | −0.31 | −0.34 | −0.34 | |
| $Y_1$ | Benzenes | | −1.167 | −1.173 | −1.173 | 2 |
| $Y_2$ | Ortho | | −0.006 | −0.006 | −0.006 | |
| $Y_3$ | Meta | | −0.002 | −0.002 | −0.002 | |
| $Y_4$ | Para | | −0.001 | −0.001 | −0.001 | |
| Z | Linear polynuclear aromatic hydrocarbons | | — | — | $0.248\text{-}0.236N_c$ | 5 |
| AA | Quinones | | — | — | −0.86 | |
| BB | Pyridines | | −0.914 | −0.95 | — | 2 |
| CC | Anilides | | — | — | −0.50 | 6 |
| DD | Tetrazoles | | — | — | +0.12 | 2 |
| EE | Pyrroles | | −0.60 | −0.65 | −0.69 | 2 |
| FF | Thiophenes | | −0.303 | −0.327 | — | 2 |
| $GG_1$ | Monosaccharides | Furanose ring | — | — | −0.52 | 2 |
| $GG_2$ | Monosaccharides | Pyranose ring | — | — | −0.50 | 2 |
| $HH_1$ | Di- and oligosaccharides | Furanose ring | — | — | −0.50 | 2, 7 |
| $HH_2$ | Di- and oligosaccharides | Pyranose ring | — | — | −0.47 | 2, 7 |

**TABLE 6-5 Cardozo Correction Factors for the Enthalpy of Combustion** ***(Continued)***

Examples of the Cardozo Method to Estimate the Enthalpy of Combustion

| Compound | Structure | Physical state | Calculation of $N$ | | $\Delta H_c^\circ$ (kJ/mol) Lit. | Calc. |
|---|---|---|---|---|---|---|
| Vinyl chloride | $CH_2{=}CHCl$ | Gas | $(2 + D_1) + X_2$ | = 1.531 | −1157 | −1140 |
| Ethane | $CH_3CH_3$ | Gas | (2) | = 2 | −1429 | −1429 |
| Acrolein | $CH_2{=}CHCHO$ | Liquid | $(3 + D_1) + G$ | = 2.260 | −1560 | −1576 |
| Asparagine | $H_2NCOCH_2CH(NH_2)COOH$ | Solid | $(4) + I + M_1 + N + P$ | = 2.557 | −1754 | −1757 |
| Succinamide | $H_2NCOCH_2{-}CH_2CONH_2$ | Solid | $(4) + 2 \times N$ | = 2.916 | −1962 | −1975 |
| Crotononitrile | $CH_3CH{=}CHCN$ | Liquid | $(4 + D_2 + D_4) + T$ | = 3.430 | −2286 | −2290 |
| Vinylacetylene | $CH_2{=}CH{-}C{\equiv}CH$ | Gas | $(4 + D_1 + E_1)$ | = 3.497 | −2364 | −2350 |
| L-Gulonic acid- $\gamma$-lactone | $CH_2OH{-}CHOH{-}CH{-}CHOH{-}CHOH{-}CO$ (ring closed through –O– between CH and CO) | Solid | $(6 + K) + F_1 + 3 \times F_2$ | = 3.52 | −2352 | −2341 |
| 2,4-Dinitrophenol | $C_6H_3(OH)(NO_2)_2$ | Solid | $(6 + Y_1 + Y_2 + Y_4) + F_2 + S_3$ | = 3.990 | −2614 | −2626 |
| Methyl propyl sulfide | $CH_3{-}S{-}C_3H_7$ | Gas | $(4) + U$ | = 4.553 | −3000 | −2999 |
| Phthalic acid | $C_6H_4(COOH)_2$ | Solid | $(8 + 2 \times B_2 + Y_1 + Y_2) + 2 \times I$ | = 4.705 | −3093 | −3060 |
| Isopentylamine | $CH_3CH(CH_3)CH_2CH_2NH_2$ | Liquid | $(5 + B_2) + M_1$ | = 5.19 | −3346 | −3364 |
| Anisole | $C_6H_5{-}O{-}CH_3$ | Liquid | $(7 + Y_1) + L$ | = 5.615 | −3613 | −3623 |
| 5-Phenylaminotetrazole | $(CN_4H){-}NH{-}C_6H_5$ | Solid | $(7 + Y_1 + DD) + M_2$ | = 6.107 | −3909 | −3910 |
| 2,4-Dimethyl-3-pentanone | $CH_3CH(CH_3)COCH(CH_3)CH_3$ | Liquid | $(7 + 2 \times B_2) + H$ | = 6.351 | −4070 | −4072 |
| 2,2,3-Trimethylbutane | $CH_3C(CH_3)_2CH(CH_3)CH_3$ | Liquid | $(7 + 3 \times B_1)$ | = 6.977 | −4455 | −4454 |
| $\alpha$-Naphthol | $C_{10}H_7OH$ | Solid | $(10 + Z) + F_2$ | = 7.558 | −4787 | −4791 |
| 1,2,4-Trimethylbenzene | $C_6H_3(CH_3)_3$ | Gas | $(9 + 3 \times B_2 + Y_1 + Y_2 + Y_4)$ | = 7.766 | −4982 | −4976 |
| Eugenol | $C_6H_3(OH)(OCH_3)(CH_2{-}CH{=}CH_2)$ | Liquid | $(10 + B_2 + D_1 + Y_1 + Y_2 + Y_4) + F_2 + L$ | = 8.079 | −5123 | −5126 |
| Sucrose | $[CH(CH_2OH){-}CHOH{-}CHOH{-}CHOH{-}CH]{-}O{-}[(CH_2OH)CH{-}CHOH{-}CHOH{-}C(CH_2OH)]$ (each ring closed through –O–) | Solid | $(12 + 3 \times B_2 + HH_1 + HH_2) + 3 \times F_1 + 5 \times F_2 + L$ | = 8.17 | −5160 | −5162 |
| 2,2′-Difluorobiphenyl | $FC_6H_4{-}C_6H_4F$ | Gas | $(12 + 2 \times B_2 + 2 \times Y_1) + 2 \times X_1$ | = 9.106 | −5794 | −5800 |
| Ethyl 4-ethyl-3,5-dimethyl-pyrrole-2-carboxylate | $(C_4HN)(C_2H_5)(CH_3)_2(CO_2C_2H_5)$ | Solid | $(11 + 4 \times B_2 + EE) + J$ | = 9.33 | −5865 | −5865 |
| Benzanilide | $C_6H_5{-}NH{-}CO{-}C_6H_5$ | Solid | $(13 + B_2 + 2 \times Y_1) + CC$ | = 10.134 | −6357 | −6353 |
| Diphenylacetylene | $C_6H_5C{\equiv}CC_6H_5$ | Solid | $(14 + 2 \times B_2 + E_2 + 2 \times Y_1)$ | = 11.267 | −7043 | −7040 |

| Compound | Formula | State | Contributions | Sum | Calc. | Exp. |
|---|---|---|---|---|---|---|
| Valylphenylalanine | $(C_6H_5)CH_2CH(NH)COOH$ / $CH_3CH(CH_3)CH(NH_2)CO$ | Solid | Valine : $(5 + B_2) + I + M_1 + P$ | = 4.079 | | |
| | | | Phen.al. : $(9 + B_2 + Y_1) + I + M_1 + P$ | = 6.906 | | |
| | | | Dipept. : +Q | = 0.44 + | | |
| | | | | = 11.425 | −7164 | −7136 |
| Dicyclohexylmethane | $(C_6H_{11})_2CH_2$ | Liquid | $(13 + 2 \times B_2 + 2 \times C)$ | = 12.38 | −7724 | −7750 |
| Pentacene-6,13-quinone | $C_{22}H_{12}O_2$ | Solid | (22+Z) + AA | = 16.202 | −10041 | −10034 |
| 5-Butyldocosane | $CH_3(CH_2)_3-CH(C_4H_9)-(CH_2)_{16}CH_3$ | Liquid | $(26 + B_1)$ | = 25.997 | −16056 | −16059 |
| Dotriacontane | $CH_3(CH_2)_{30}CH_3$ | Solid | (32) | = 32.0 | −19616 | −19616 |
| Glyceroltribrassidate | $CH_2(OCOR)CH(OCOR)CH_2(OCOR)$ with R = $CH_3(CH_2)_7CH{=}CH(CH_2)_{10}CH_2$-(trans) | Solid | $(69 + 3 \times D_2 + 3 \times D_4) + 3 \times J$ | = 65.640 | −40041 | −40021 |

Remarks

1. The correction factor for branched alkanes has an upper limit of −0.003.

2. The carbon atoms forming the ring in cyclic compounds count in the determination of $N_C$. Functional groups, connected to a carbon atom that is part of a ring and that is connected to two other carbon atoms, are considered as secondary groups. Furthermore, for a carbon atom branched to a carbon atom that is part of a ring, a branch correction should be applied.

3. The correction factors are of general nature and can be used for all compounds with multiple bonds. The cis/trans corrections are valid for all compounds having such structural isomers.

4. The correction for amino acids is calculated by the normal procedure with the addition of the group corrections for amine and carboxylic acid plus an additional constant equal to −0.043. For the amino acid derivatives, the correction consists of the correction factors of the basic amino acids plus an additional structural correction.

5. The correction for linear polynuclear hydrocarbons is calculated for the number of carbon atoms that form the basic structure. It is observed that nonlinear polynuclear hydrocarbons like phenanthrene and chrysene fit well, but for condensed polynuclear hydrocarbons like pyrene and fluoranthrene, the calculated enthalpy of combustion may be up to 2 percent too high.

6. For anilides, the correction factor given is valid for the structural components (NH− and =O). For the rest of the molecule, the normal rules apply.

7. The oxygen bridge is considered as an ether bond. For the rest, the normal procedures apply.

**TABLE 6-6 Comparison of Estimated and Literature Values for the Ideal-Gas Heat Capacity**

| Compound | T,K | $C_p^o$ J/mol K [25] | Per Cent Error[a] Calculated by the Method of: Joback Table 6-1 | Benson Table 6-4 | Yoneda Table 6-2 | Thinh et al. Table 6-3 |
|---|---|---|---|---|---|---|
| Propane | 298 | 73.94 | 1.1 | 0.9 | 0.6 | -0.5 |
| | 800 | 155.25 | -0.4 | 0.6 | 0.5 | 0.2 |
| n-Heptane | 298 | 166.09 | -0.2 | 0.4 | 0.1 | -0.1 |
| | 800 | 340.93 | -0.2 | 0.6 | 0.3 | 0.8 |
| 2,2,3-Trimethylbutane | 298 | 164.67 | 0.9 | 0.8 | 0.3 | 0.7 |
| | 800 | 346.37 | 0.1 | 0.7 | 0.5 | 1.0 |
| trans-2-Butene | 298 | 87.88 | -4.6 | -1.8 | 1.0 | 0 |
| | 800 | 173.75 | -0.3 | -0.1 | 0.6 | 0.1 |
| 3,3-Dimethyl-1-butene | 298 | 126.57 | 4.7 | 4.6 | 4.5 | 6.6 |
| | 800 | 266.28 | 2.5 | 3.5 | 2.6 | 4.3 |
| 2-Methyl-1,3-butadiene | 298 | 104.7 | -4.4 | 0.7 | 1.7 | -2.0 |
| | 800 | 201.0 | -2.2 | 0.5 | 1.0 | -1.2 |
| 2-Pentyne | 298 | 98.77 | 2.0 | -0.2 | 1.1 | -0.3 |
| | 800 | 192.17 | 0.7 | 0.8 | 0.6 | 0.4 |
| p-Ethyltoluene | 298 | 151.65 | 0.9 | 0.8 | 0.4 | -1.3 |
| | 800 | 324.90 | 0.3 | 0.5 | 0.5 | -0.5 |
| 2-Methylnaphthalene | 298 | 159.89 | -2.3 | 0 | 0.6 | -2.2 |
| | 800 | 343.44 | -1.6 | 0.9 | 0.6 | -0.4 |
| cis-1,3-Dimethylcyclopentane | 298 | 134.56 | 1.3 | -5.9 | 0.4 | 0 |
| | 800 | 317.53 | 3.5 | -1.1 | 0.5 | 0.1 |
| 2-Butanol | 298 | 113.38 | -1.9 | -0.8 | -2.1 | |
| | 800 | 220.56 | -0.1 | 0.5 | 0.2 | |
| p-Cresol | 298 | 124.56 | 0.9 | 0.1 | 1.9 | |
| | 800 | 255.86 | 0.1 | -0.1 | 3.9 | |
| Isopropyl ether | 298 | 158.39 | 0 | -0.8 | 1.9 | |
| | 800 | 311.46 | 1.3 | 2.3 | 1.3 | |
| p-Dioxane | 298 | 94.12 | -0.4 | -1.0 | - | |
| | 800 | 218.34 | 0.1 | 0 | - | |
| Methyl ethyl ketone | 298 | 102.95 | -5.1 | -2.5 | -4.1 | |
| | 800 | 192.93 | 0.6 | -0.1 | -0.3 | |

| Compound | T,K | $C_p^o$ J/mol K [25] | Per Cent Error[a] Calculated by the Method of: Joback Table 6-1 | Benson Table 6-4 | Yoneda Table 6-2 | Thinh et al. Table 6-3 |
|---|---|---|---|---|---|---|
| Ethyl acetate | 298 | 113.71 | -0.4 | -0.5 | -28. | |
| | 800 | 213.57 | 0.2 | 0.1 | -11. | |
| Trimethyl amine | 298 | 91.82 | -0.2 | 0.3 | 0.2 | |
| | 800 | 191.00 | 0.1 | 0.5 | 0.4 | |
| Propionitrile | 298 | 73.10 | 1.9 | -1.4 | 1.1 | |
| | 800 | 134.56 | 0.8 | 2.1 | 1.7 | |
| 2-Nitrobutane | 298 | 123.55 | 2.0 | 1.0 | 1.4 | |
| | 800 | 248.86 | -0.3 | 0.1 | 0.3 | |
| 3-Picoline | 298 | 99.65 | 2.2 | - | - | |
| | 800 | 222.40 | -0.4 | - | - | |
| 1,1-Difluoroethane | 298 | 67.99 | -0.5 | -1.3 | 4.1 | |
| | 800 | 124.31 | 0.3 | 0.6 | 0.9 | |
| Octafluorocyclobutane | 298 | 156.25 | -12. | -6.8 | - | |
| | 800 | 245.56 | -1.5 | 5.5 | - | |
| Bromobenzene | 298 | 97.76 | -0.1 | 2.8 | 1.7 | |
| | 800 | 200.05 | 0 | 1.1 | 4.6 | |
| Trichloroethylene | 298 | 80.26 | 1.4 | 0.3 | -5.1 | |
| | 800 | 112.79 | -2.8 | 0.1 | -2.9 | |
| Butyl methyl sulfide | 298 | 140.84 | 0.2 | -0.1 | 0.3 | |
| | 800 | 278.55 | -2.1 | -0.1 | -0.5 | |
| 2-Methyl-2-butanethiol | 298 | 146.31 | -1.5 | 0.4 | -0.9 | |
| | 800 | 277.50 | -0.1 | -0.2 | 0.2 | |
| Propyl disulfide | 298 | 185.48 | -0.2 | 0.4 | 1.2 | |
| | 800 | 350.44 | -1.8 | 0.4 | -1.3 | |
| 3-Methylthiophene | 298 | 94.91 | 1.8 | 0.3 | - | |
| | 800 | 192.38 | 1.0 | 2.5 | - | |
| Number of Compounds | | | 28 | 27 | 24 | 10 |
| Average Absolute Error, Per Cent | | | 1.4 | 1.1 | 1.4[b] | 1.1 |

a. [(calc.-lit.)/lit.]x100;
b. Values do not include the large anomalous errors for ethyl acetate

**TABLE 6-7 Comparison between Estimated and Literature Values for the Enthalpy of Formation at 298 K**

| Compound | $\Delta H_f^o$ kJ/mol [25] | Difference[a], kJ/mole, as Calculated by the Method of: Joback Table 6-1 | Benson Table 6-4 | Yoneda Table 6-2 | Thinh et al. Table 6-3 | Cardozo Table 6-5 |
|---|---|---|---|---|---|---|
| Propane | -103.92 | -1.4 | -1.2 | -1.9 | 0 | -1.5 |
| n-Heptane | -187.90 | 0.1 | -0.1 | 0.3 | 0.1 | -0.1 |
| 2,2,3-Trimethylbutane | -204.94 | 3.0 | 1.5 | 2.6 | -0.1 | 2.8 |
| trans-2-Butene | -11.18 | 2.5 | -1.3 | -0.8 | 0 | -0.8 |
| 3,3-Dimethyl-1-butene | -43.17 | -7.4 | -14. | -12. | -25. | -23. |
| 2-Methyl-1,3-butadiene | 75.78 | 19. | 0 | -0.9 | 4.8 | 6.8 |
| 2-Pentyne | 128.95 | -1.5 | -2.5 | -1.6 | -5.0 | -0.8 |
| p-Ethyltoluene | -3.27 | -1.9 | -0.1 | 2.3 | 0.6 | -1.1 |
| 2-Methylnaphthalene | 116.18 | 30. | 2.9 | 0 | -0.9 | -4.2 |
| cis-1,3-Dimethylcyclopentane | -135.95 | 15. | -0.2 | 0 | 0 | -17. |
| 2-Butanol | -292.49 | 9.4 | -1.5 | -1.8 | | 0.3 |
| p-Cresol | -125.48 | -3.1 | -0.8 | -2.3 | | 8.6 |
| Isopropyl ether | -319.03 | 9.0 | -2.1 | -4.9 | | 5.9 |
| p-Dioxane | 315.27 | 0 | 8.0 | - | | - |
| Methyl ethyl ketone | -238.52 | 0 | 0.8 | -0.5 | | 0.1 |
| Ethyl acetate | -443.21 | 0 | 9.5 | -1.1 | | 32. |
| Methyl methacrylate | -332.0 | 16. | -9.3 | -20. | | 14. |
| Trimethyl amine | -23.86 | -14. | -0.6 | 0 | | -5.8 |
| Propionitrile | 50.66 | 9.0 | 1.3 | 3.3 | | 8.8 |
| 2-Nitrobutane | -163.7 | 22. | -7.5 | 3.3 | | -1.3 |

| Compound | $\Delta H_f^o$ kJ/mol [25] | Difference[a], kJ/mole, as Calculated by the Method of: Joback Table 6-1 | Benson Table 6-4 | Yoneda Table 6-2 | Thinh et al. Table 6-3 | Cardozo (Table 6-5) |
|---|---|---|---|---|---|---|
| 3-Picoline | 106.22 | -4.0 | - | - | | -1.1 |
| 1,1-Difluoroethane | -494. | 13. | -4. | 11. | | 30. |
| Octafluorocyclobutane | -1529. | - | 14. | - | | - |
| Bromobenzene | 105.1 | -9.3 | 8.8 | 0 | | -1.2 |
| Trichloroethylene | -5.86 | -19. | 7.1 | 22. | | - |
| Butyl methyl sulfide | -102.24 | -2.4 | 0.9 | 0.7 | | -1.6 |
| 2-Methyl-2-butanethiol | -127.11 | 10. | -3.2 | 2.4 | | 6.7 |
| Propyl disulfide | -117.27 | 34. | 3.1 | 2.1 | | 0.7 |
| 3-Methylthiophene | 82.86 | 0.8 | 0.4 | - | | 0.4 |

a. Difference = (Calculated - Literature), kJ/mol

**TABLE 6-8 Comparison between Estimated and Literature Values of the Entropy at 298 K**

| Compound | $S^o$(298 K) J/mol K [25] | Difference[a], J/mol K, as Calculated by the Method of: Benson Table 6-4 | Yoneda Table 6-2 | Thinh et al. Table 6-3 |
|---|---|---|---|---|
| Propane | 270.09 | 0 | 3.3 | 0 |
| n-Heptane | 428.18 | -0.4 | 1.0 | -1.0 |
| 2,2,3-Trimethylbutane | 383.55 | 0.8 | 1.5 | -12. |
| trans-2-Butene | 296.68 | 0.7 | 4.2 | 0 |
| 3,3-Dimethyl-1-butene | 343.99 | 5.0 | 7.6 | -13. |
| 2-Methyl-1,3-butadiene | 315.85 | -0.9 | 1.5 | 2.6 |
| 2-Pentyne | 332.01 | 0.7 | -4.8 | 6.3 |
| p-Ethyltoluene | 399.17 | -0.9 | -0.3 | -7.5 |
| 2-Methylnaphthalene | 380.29 | 5.1 | 0 | 8.2 |
| cis-1,3-Dimethylcyclopentane | 367.06 | 0.9 | 0 | 0 |
| 2-Butanol | 359.27 | 2.1 | -7.3 | |
| p-Cresol | 347.88 | 4.6 | 4.4 | |
| Isopropyl ether | 390.50 | 21. | 0.9 | |
| Methyl ethyl ketone | 338.34 | 1.1 | -3.4 | |
| Ethyl acetate | 363.00 | 13. | 1.2 | |
| Trimethyl amine | 288.97 | 0.1 | 0 | |
| Propionitrile | 286.80 | -0.4 | -4.4 | |
| 2-Nitrobutane | 383.59 | 11. | 2.0 | |
| 1,1-Difluoroethane | 282.69 | -0.9 | -8.3 | |
| Octafluorocyclobutane | 400.63 | 4.9 | - | |
| Bromobenzene | 324.60 | 1.4 | 0 | |
| Trichloroethylene | 325.02 | -0.5 | 17. | |
| Butyl methyl sulfide | 412.11 | -0.4 | -0.8 | |
| 2-Methyl-2-butanethiol | 387.20 | -0.3 | -0.8 | |
| Propyl disulfide | 495.30 | 0.5 | -2.7 | |
| 3-Methylthiophene | 321.50 | -3.1 | - | |

a. Difference = Calculated - Literature, J/mol K

Thinh et al. form has been demonstrated (for hydrocarbons) to extend from about 200 to 1500 K. Appendix A presents specific polynomial constants for many compounds. Values given in this data bank should be used in preference to those estimated by methods in this chapter.

Other authors have tabulated polynomial constants for $C_p^\circ$ [20, 21], and new equations to express $C_p^\circ$ as a function of temperature have been suggested [1, 11, 13, 28].

### Entropies

Three procedures were evaluated to estimate the absolute entropy of 298 K. Calculated results are compared with literature values in Table 6-8. All three have about the same degree of accuracy, but Thinh et al.'s is limited to hydrocarbons. A disadvantage of Benson's method is the necessity of determining the symmetry number of the compound, a task often requiring the construction of three-dimensional models to determine the symmetry planes. Neither Yoneda's nor Thinh et al.'s procedure requires such a correction.

### Gibbs energy of formation

Two techniques were tested, Joback's and Thinh et al.'s. The latter is significantly more accurate, although it is limited to hydrocarbons. For nonhydrocarbons we recommend the method of Joback, although it may lead to relatively large errors for complex materials. No symmetry or optical isomer corrections are necessary. Estimated values are compared with those reported in the literature in Table 6-9. Generally, Joback's method is within 5 to 10 kJ/mol of the literature value. An interesting suggestion was made by Fredenslund and Rasmussen [12] to estimate $\Delta G_f^\circ$ from UNIFAC group contributions.

Some literature values of $\Delta G_f^\circ$ (298 K) are tabulated in Appendix A.

## Notation

| | |
|---|---|
| $C_p^\circ$ | ideal-gas heat capacity at constant pressure, J/(mol·K) |
| $\Delta G_f^\circ$ | standard Gibbs energy of formation at $T$ and 1 atm, kJ/mol; $\Delta G^\circ\ (T)$, standard Gibbs energy change in a reaction at $T$ |
| $\Delta H_c^\circ$ | standard (lower) enthalpy of combustion at 298 K, kJ/mol |
| $\Delta H_f^\circ$ | standard enthalpy of formation at $T$, kJ/mol; $\Delta H^\circ\ (T)$, standard enthalpy of reaction at $T$ |
| $n_j$ | number of groups of type $j$ |
| $N$ | equivalent chain length; $\Delta N_i$, corrections for the equivalent chain length, Table 6-5; $N_C$, number of carbon atoms in a compound; simi- |

**TABLE 6-9 Comparison between Estimated and Literature Gibbs Energy of Formation at 298 K (Reference state is 1 atmosphere.)**

| Compound | $\Delta G_f^o$(298 K) kJ/mol [25] | Difference[a], kJ/mol, as Calculated by the Method of: Joback Table 6-1 | Thinh et al. Table 6-3 |
|---|---|---|---|
| Propane | -23.49 | -2.2 | 0 |
| *n*-Heptane | 8.00 | 0 | 0.4 |
| 3-Methylhexane | 4.61 | 0.9 | -0.4 |
| 2,4-Dimethylpentane | 3.10 | 0 | -5.2 |
| 2,2,3-Trimethylbutane | 4.27 | 4.1 | -7.5 |
| Cyclopentane | 38.64 | -3.2 | 0 |
| Cyclohexane | 31.78 | 0 | 0 |
| Methylcyclopentane | 35.80 | 0.4 | 0 |
| Ethylene | 68.16 | -6.7 | 0 |
| 1-Butene | 71.34 | -0.7 | -4.0 |
| *cis*-2-Butene | 65.90 | -2.9 | 0 |
| *trans*-2-Butene | 63.01 | 0 | 0 |
| 1,3-Butadiene | 150.77 | 7.7 | 0 |
| Acetylene | 209.34 | 0 | 0 |
| Methylacetylene | 194.56 | 2.9 | -0.6 |
| Benzene | 129.75 | -8.0 | 0 |
| Ethylbenzene | 130.67 | -1.8 | 0 |
| *o*-Xylene | 122.17 | -2.9 | -7.2 |
| *m*-Xylene | 118.95 | 0.3 | -4.0 |
| Ethyl mercaptan | -4.69 | 0 | |
| Dimethyl sulfide | 17.79 | -1.9 | |
| Thiophene | 126.86 | 0 | |
| Aniline | 166.80 | 12. | |
| Ethyl amine | 37.30 | -4.9 | |
| Pyridine | 190.33 | 0 | |
| Dimethyl ether | -113.00 | -26. | |
| Acetaldehyde | -133.39 | -0.2 | |
| Acetone | -153.15 | -1.4 | |
| Methyl formate | -297.39 | 0 | |
| Acetic acid | -376.94 | -1.0 | |
| Ethyl acetate | -327.62 | 0 | |
| *n*-Propyl alcohol | -163.08 | -0.6 | |
| Isopropyl alcohol | -173.71 | 8.5 | |
| Phenol | -32.91 | 0 | |
| Phosgene | -206.91 | 12. | |
| Methyl chloride | -62.93 | 8.5 | |
| Methylene chloride | -68.91 | 2.6 | |
| Chloroform | -68.58 | -12. | |
| Carbon tetrachloride | -58.28 | -29. | |
| Ethyl bromide | -26.33 | 6.6 | |
| Dichlorodifluoromethane | -394.8 | -11. | |
| Fluorobenzene | -69.08 | -14. | |
| Chlorobenzene | 99.23 | 1.0 | |

a. Difference = Calculated - Literature, kJ/mol

larly for $N_F$, $N_{Cl}$, $N_{Br}$, $N_I$, $N_S$, $N_X$ for fluorine, chlorine, bromine, iodine, sulfur, and halogen atoms

$R$ gas constant, 8.314 J/(mol·K)

$S°$ absolute entropy at $T$, 1 atm, J/(mol·K); $\Delta S°\ (T)$, standard entropy change of reaction at $T$; $\Delta S_f°$, standard entropy of formation at $T$, 1 atm

$T$ temperature, K; $T_b$, normal boiling temperature at 1 atm

$W$ number of distinguishable configurations of a molecule

GREEK

$\eta$ number of optical isomers

$\nu_j$ stoichiometric multiplier, positive for products, negative for reactants

$\sigma$ symmetry number; $\sigma_{ext}$, for rigid body rotation; $\sigma_{int}$, for rotation of subgroups constituting the molecule

$\Delta_H$ group contribution for $\Delta H_f°$; $\Delta_S$, for $S°$; $\Delta_G$, $\Delta G_f°$; $\Delta_a$, $\Delta_b$, $\Delta_c$, $\Delta_d$ for polynomial coefficients in $C_p°$ equations

## REFERENCES

1. Aly, F. A., and L. L. Lee: *Fluid Phase Equil.*, **6**:169 (1981).
2. Anderson, J. W., G. H. Beyer, and K. M. Watson: *Natl. Petrol. News Tech. Sec.*, **36**: R476 (July 5, 1944).
3. Benson, S. W.: *Thermochemical Kinetics,* Wiley, New York, 1968, chap. 2.
4. Benson, S. W., and J. H. Buss: *J. Chem. Phys.*, **29**: 279 (1969).
5. Benson, S. W., F. R. Cruickshank, D. M. Golden, G. R. Haugen, H. E. O'Neal, A. S. Rodgers, R. Shaw, and R. Walsh: *Chem. Rev.*, **69**: 279 (1969).
6. Bures, M., V. Majer, and M. Zabransky: *Chem. Eng. Sci.*, **36**: 529 (1981).
7. Cardozo, R. L.: private communication, *Akzo Zout Chemie Nederland bv,* January 1983; *AIChE J.*, **32:** 844 (1986).
8. Chueh, P. L.: private communication, Shell Development Co., 1974.
9. Eigenmann, H. K., D. M. Golden, and S. W. Benson: *J. Phys. Chem.*, **77**: 1687 (1973).
10. Evans, W. H., and D. D. Wagman: *J. Res. Natl. Bur. Stand.*, **55**: 147 (1955).
11. Fakeeha, A., A. Kache, Z. U. Rehman, Y. Shoup, and L. L. Lee: *Fluid Phase Equil.*, **11**: 225 (1983).
12. Fredenslund, A., and P. Rasmussen: *AIChE J.*, **25**: 203 (1979).
13. Harmens, A.: "Correl. Thermodyn. Data Fluids Fluid Mixtures: Their Estimation, Correlation, Use," *Proc. NPL Cong.*, 1978 (pub. 1979) p. 112–20.
14. Hougen, O. A., K. M. Watson, and R. A. Ragatz: *Chemical Process Principles,* 2d. ed. Pt. II, *Thermodynamics,* Wiley, New York, 1959.
15. Joback, K. G.: thesis, Massachusetts Institute of Technology, Cambridge, Mass., June, 1984.
16. Majer, V., M. Bures, and M. Zabransky: *Chem. Prumysl.*, **29**(9): 462 (1979).
17. Olson, B. A.: thesis, Rutgers University, New Brunswick, N.J., August 1973.
18. O'Neal, H. E., and S. W. Benson: *J. Chem. Eng. Data,* **15**: 266 (1970).
19. Seaton, W. H., and E. Freedman: "Computer Implementation of a Second Order Additivity Method for the Estimation of Chemical Thermodynamic Data," paper presented at 65th Annual Mtg., AIChE, New York, November 1972.
20. Seres, L., L. Zalotal, and F. Marta, *Acta. Phys. Chem.*, **23**(4): 433 (1977).
21. Seres, L.: *Acta. Phys. Chem.*, **27**(1–4): 31 (1981).
22. Shaw, R.: *J. Phys. Chem.*, **75**: 4047 (1971).
23. Stein, S. E., D. M. Golden, and S. W. Benson: *J. Phys. Chem.*, **81**: 314 (1977).

24. Stull, D. R., and G. C. Sinke: "Thermodynamic Properties of the Elements," *Advan. Chem. Ser.*, **8**: (1956).
25. Stull, D. R., E. F. Westrum, and G. C. Sinke: *The Chemical Thermodynamics of Organic Compounds,* Wiley, New York, 1969.
26. Thinh, T.-P., J.-L. Duran, and R. S. Ramalho: *Ind. Eng. Chem. Process Design Develop.,* **10**: 576 (1971).
27. Thinh, T.-P., and T. K. Trong: *Can. J. Chem. Eng.,* **54**: 344 (1976).
28. Thompson, P. A.: *J. Chem. Eng. Data,* **22**: 431 (1977).
29. Yoneda, Y.: *Bull. Chem. Soc. Japan,* **52**: 1297 (1979).
30. Yuan, S. C., and Y. I. Mok: *Hydrocarbon Process. Petrol. Refiner,* **47**(3): 133 (1968); **47**(7): 153 (1968).

Chapter

# 7

# Vapor Pressures and Enthalpies of Vaporization of Pure Fluids

## 7-1 Scope

This chapter covers methods for estimating and correlating vapor pressures of pure liquids. Since enthalpies of vaporization are often derived from vapor pressure–temperature data, the estimation of this property also is included.

## 7-2 Theory and Corresponding States Correlations

When the vapor phase of a pure fluid is in equilibrium with the liquid phase, the equality of chemical potential, temperature, and pressure in both phases leads to the Clausius-Clapeyron equation

$$\frac{dP_{\mathrm{vp}}}{dT} = \frac{\Delta H_v}{T\,\Delta V_v} = \frac{\Delta H_v}{(RT^2/P_{\mathrm{vp}})\,\Delta Z_v} \tag{7-2.1}$$

$$\frac{d\ln P_{\mathrm{vp}}}{d(1/T)} = -\frac{\Delta H_v}{R\,\Delta Z_v} \tag{7-2.2}$$

In this equation, $\Delta H_v$ and $\Delta Z_v$ refer to differences in the enthalpies and compressibility factors of saturated vapor and saturated liquid.

Most vapor pressure estimation and correlation equations stem from an integration of Eq. (7-2.2). When this is done, an assumption must be made regarding the dependence of the group $\Delta H_v/\Delta Z_v$ on temperature. Also, a constant of integration which must be evaluated by using one vapor pressure–temperature point is obtained.

The simplest approach is to assume that the group $\Delta H_v/R\,\Delta Z_v$ is constant and independent of temperature. Then, with the constant of integration denoted as $A$, Eq. (7-2.2) becomes

$$\ln P_{vp} = A - \frac{B}{T} \tag{7-2.3}$$

where $B = \Delta H_v/R\,\Delta Z_v$. Equation (7-2.3) is sometimes called the Clapeyron equation. Surprisingly, it is a fairly good relation for approximating vapor pressure over small temperature intervals. Except near the critical point, both $\Delta H_v$ and $\Delta Z_v$ are weak functions of temperature, and since both decrease with rising temperature, the result is a compensatory effect. However, over large temperature ranges, Eq. (7-2.3) normally represents vapor pressure data poorly. This is shown in Fig. 7-1. The ordinate is the ratio $(P_{exp} - P_{calc})/P_{exp}$, and the abscissa $T_r = T/T_c$. $P_{calc}$ is obtained from Eq. (7-2.3), where the constants $A$ and $B$ are found from experimental data at $T_r = 0.7$ and 1.0. At high reduced temperatures, the fit is reasonably good for oxygen and a typical hydrocarbon, 2,2,4-trimethylpentane, but poor for an associating liquid, $n$-butanol. Ambrose [2] points out the complexity of the curves in this figure and notes that, to represent the changes in curvature that are evident, at least a four-constant vapor pressure equation would be necessary. Also, it is important to note that there is usually a change in curvature at a $T_r$ between 0.8 and 0.85; this fact is utilized in several later developments.

Extending our consideration of Eq. (7-2.3) one step further, a common practice is to use both the normal boiling point and the critical point to obtain generalized constants. Expressing pressures in bars and temperatures on the absolute scale (kelvins or degrees Rankine), with $P = P_c$, $T = T_c$, and $P = 1.01325$, $T = T_b$ = boiling temperature at 1 atm = 1.01325 bar, Eq. (7-2.3) becomes

$$\ln P_{vp_r} = h\left(1 - \frac{1}{T_r}\right) \tag{7-2.4}$$

$$h = T_{b_r}\frac{\ln (P_c/1.01325)}{1 - T_{b_r}} \tag{7-2.5}$$

Figure 7-1 shows that the linear form of $\ln P_{vp}$ versus $1/T$ is not satisfactory for associating materials. Equation (7-2.3) generally overpredicts vapor pressures below $T_b$ (see Fig. 7-1 or Table 7-2).

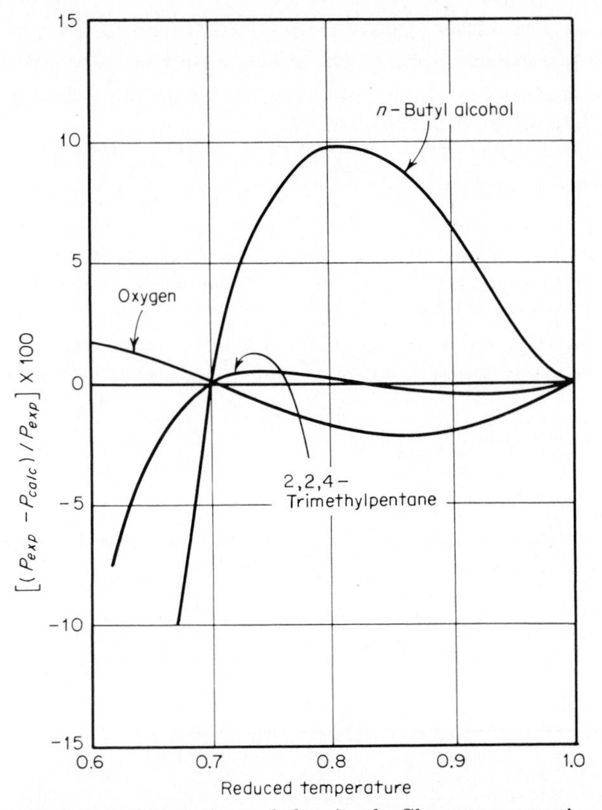


**Figure 7-1** Comparison of the simple Clapeyron equation with experimental vapor pressure data. (Adapted from Ref. 2.)

Equation (7-2.4) is a two-parameter corresponding states equation for vapor pressure. To achieve more accuracy, several investigators have proposed three-parameter forms. The Pitzer expansion is one of the more successful:

$$\ln P_{vp_r} = f^{(0)}(T_r) + \omega f^{(1)}(T_r) \tag{7-2.6}$$

The functions $f^{(0)}$ and $f^{(1)}$ have been expressed in tabular form by several authors [18, 70]; they have been expressed in analytical form by Lee and Kesler [56]:

$$f^{(0)} = 5.92714 - \frac{6.09648}{T_r} - 1.28862 \ln T_r + 0.169347 T_r^6 \tag{7-2.7}$$

$$f^{(1)} = 15.2518 - \frac{15.6875}{T_r} - 13.4721 \ln T_r + 0.43577 T_r^6 \tag{7-2.8}$$

Values of the acentric factor $\omega$ are tabulated in Appendix A for many fluids; but when Eq. (7-2.6) is employed, it is recommended that Eq. (2-

3.4) be used to compute $\omega$. The latter equation was obtained from Eq. (7-2.6) with $T_r = T_{b_r}$. The Lee-Kesler form of the Pitzer equation generally predicts vapor pressures within 1 to 2 percent between $T_b$ and $T_c$. Below $T_b$, it may underpredict $P_{vp}$ by several percent.

As can readily be verified, Eqs. (7-2.6) to (7-2.8) satisfy the definition of the acentric factor [Eq. (2-3.1)].

**Example 7-1** Estimate the vapor pressure of ethylbenzene at 347.2 and 460 K by using both Eq. (7-2.4) and the Lee-Kesler relations. Experimental values are 0.133 bar [101] and 3.325 bar [3], respectively.

**solution** From Appendix A, $T_b = 409.3$ K, $T_c = 617.1$ K, and $P_c = 36.0$ bar.
EQUATION (7-2.4). First $h$ is determined from Eq. (7-2.5), with $T_{b_r} = 409.3/617.1 = 0.663$,

$$h = 0.663 \frac{\ln 36/1.01325}{1 - 0.663} = 7.024$$

Then $\ln P_{vp_r} = 7.024(1 - T_r^{-1})$

LEE-KESLER With $T_{b_r} = 0.663$, from Eq. (2-3.4), $\omega = 0.299$. Then, with Eq. (7-2.6),

$$\ln P_{vp_r} = f^{(0)}(T_r) + 0.299 f^{(1)}(T_r)$$

For the cases considered,

**Vapor Pressure, bar**

| $T$, K | $T_r$ | Exp. | Eq. (7-2.4) | Percent error | Lee-Kesler | Percent error |
|---|---|---|---|---|---|---|
| 347.2 | 0.563 | 0.133 | 0.155 | 16 | 0.132 | −0.8 |
| 460 | 0.745 | 3.325 | 3.252 | −2.2 | 3.353 | 0.8 |

The error was calculated as [(calc. − exp.)/exp.] × 100.

## 7-3 Antoine Vapor Pressure Correlation

Antoine [10] proposed a simple modification of Eq. (7-2.3) which has been widely used over limited temperature ranges.

$$\ln P_{vp} = A - \frac{B}{T + C} \tag{7-3.1}$$

When $C = 0$, Eq. (7-3.1) reverts to the Clapeyron equation (7-2.3).

Simple rules have been proposed [30, 89] to relate $C$ to the normal boiling point for certain classes of materials; but these rules are not reliable, and the only way to obtain values of the constants is to regress experimental data [15, 46, 58, 63, 74, 88].

Values of $A$, $B$, and $C$ are tabulated for a number of materials in Appendix A with $P_{vp}$ in bars and $T$ in kelvins. The applicable temperature range is not large and in most instances corresponds to a pressure interval of about 0.01 to 2 bar. The Antoine equation should never be used outside the stated temperature limits. Extrapolation beyond these limits may lead to absurd results. The constants $A$, $B$, and $C$ form a set. Never use one constant from one tabulation and the other constants from a different tabulation.

Cox [23] suggested a graphical correlation in which the ordinate, representing $P_{vp}$, is a log scale, and a straight line (with a positive slope) is drawn. The sloping line is taken to represent the vapor pressure of water (or some other reference substance). Since the vapor pressure of water is accurately known as a function of temperature, the abscissa scale can be marked in temperature units. When the vapor pressure and temperature scales are prepared in this way, vapor pressures for other compounds are often found to be nearly straight lines, especially for homologous series. Calingaert and Davis [17] have shown that the temperature scale on this Cox chart is nearly equivalent to the function $(T + C)^{-1}$, where $C$ is approximately $-43$ K for many materials boiling between 273 and 373 K. Thus the Cox chart closely resembles a plot of the Antoine vapor pressure equation. Also, for homologous series, a useful phenomenon is often noted on Cox charts. The straight lines for each member of the homologous series often converge to a single point when extrapolated. This point, called the infinite point, is useful for providing one value of vapor pressure for a new member of the series. Dreisbach [25] presents a tabulation of these infinite points for several homologous series.

**Example 7-2** Calculate the vapor pressure of acrylonitrile at 293.15 K by using the Antoine equation.

**solution** From Appendix A, $A = 9.3051$, $B = 2782.21$, and $C = -51.15$. With Eq. (7-3.1),

$$\ln P_{vp} = 9.3051 - \frac{2782.21}{293.15 - 51.15}$$

$$P_{vp} = 0.112 \text{ bar}$$

The literature value is 0.117 bar [15] and

$$\text{Error} = \frac{0.112 - 0.117}{0.117} \times 100 = -4.3\%$$

Usually, in the range 0.01 to 2 bar, the Antoine equation provides an excellent correlating equation for vapor pressures. When Antoine parameters are determined from data in this pressure range (as they usually are), the equation underpredicts vapor pressures at higher pressures.

## 7-4 Gomez-Thodos Vapor Pressure Equation

Gomez-Nieto and Thodos [34 to 36] have presented the following equation for estimating vapor pressures:

$$\ln P_{vp_r} = \beta\left[\frac{1}{T_r^m} - 1\right] + \gamma[T_r^7 - 1] \tag{7-4.1}$$

Equation (7-4.1) is necessarily satisfied at the critical point. The normal boiling point provides an additional equation which relates the constants $\beta$, $\gamma$, and $m$ to each other. This leads to

$$\gamma = ah + b\beta \tag{7-4.2}$$

where
$$a = \frac{1 - 1/T_{b_r}}{T_{b_r}^7 - 1} \tag{7-4.3}$$

and
$$b = \frac{1 - 1/T_{b_r}^m}{T_{b_r}^7 - 1} \tag{7-4.4}$$

$h$ is given by Eq. (7-2.5). Compounds are divided into three classes: nonpolar, polar, and hydrogen-bonded compounds. The procedure for determining $m$, $\beta$, and $\gamma$ is different for each class. For nonpolar compounds (both organic and inorganic) [34]

$$\beta = -4.26700 - \frac{221.79}{h^{2.5} \exp 0.0384h^{2.5}} + \frac{3.8126}{\exp(2272.44/h^3)} + \Delta^* \tag{7-4.5}$$

$$m = 0.78425 \exp(0.089315h) - \frac{8.5217}{\exp(0.74826h)} \tag{7-4.6}$$

where $\Delta^* = 0$ except for He ($\Delta^* = 0.41815$), $H_2$ ($\Delta^* = 0.19904$) and Ne ($\Delta^* = 0.02319$). $\gamma$ is obtained from Eq. (7-4.2).

For polar compounds that do not hydrogen-bond [35] (this class includes ammonia and acetic acid),

$$m = 0.466T_c^{0.166} \tag{7-4.7}$$

$$\gamma = 0.08594 \exp(7.462 \times 10^{-4}\, T_c) \tag{7-4.8}$$

For hydrogen-bonding compounds [36] (water and alcohols)

$$m = 0.0052M^{0.29}T_c^{0.72} \tag{7-4.9}$$

$$\gamma = \frac{2.464}{M} \exp(9.8 \times 10^{-6}\, MT_c) \tag{7-4.10}$$

For these last two categories of compounds, $\beta$ is obtained from Eq. (7-4.2) i.e.,

$$\beta = \frac{\gamma}{b} - \frac{ah}{b} \tag{7-4.11}$$

**Example 7-3** Repeat Example 7-1 by using the Gomez-Thodos vapor-pressure equation.

**solution** From Example 7-1, $T_b$ = 409.3 K, $T_c$ = 617.1 K, $T_{b_r}$ = 0.663, $h$ = 7.024. Eqs. (7-4.5) and (7-4.6) should be used for ethylbenzene:

$$\beta = 4.26700 - \frac{221.79}{(7.024)^{2.5}\exp[(0.03848)(7.024)^{2.5}]} + \frac{3.8126}{\exp(2272.44/7.024^3)}$$
$$= -4.2727$$

$$m = 0.78425\exp[(0.089315)(7.024)] - \frac{8.5217}{\exp[(0.74826)(7.024)]}$$
$$= 1.4242$$

Equations (7-4.2) to (7-4.4) give

$$a = \frac{1 - 1/0.663}{(0.663)^7 - 1} = 0.53863$$

$$b = \frac{1 - 1/(0.663)^{1.4242}}{0.663^7 - 1} = 0.8430$$

$$\gamma = (0.53863)(7.024) + (0.8430)(-4.2727)$$
$$= 0.18143$$

Equation (7-4.1) becomes

$$\ln P_{vp_r} = -4.2727\left(\frac{1}{T_r^{1.4242}} - 1\right) + (0.18143)(T_r^7 - 1)$$

| | | Vapor pressure, bar | | |
|---|---|---|---|---|
| $T$, K | $T_r$ | Calc. | Exp. | $\frac{P_{calc} - P_{exp}}{P_{exp}} \times 100$ |
| 347.2 | 0.563 | 0.135 | 0.133 | 1.0 |
| 460 | 0.745 | 3.320 | 3.325 | −0.2 |

**Example 7-4** Use the Gomez-Thodos method to estimate the vapor pressure of isopropanol at 450 K. The literature value is 16.16 bar [6].

**solution** For isopropanol, from Appendix A, $M$ = 60.096, $T_b$ = 355.4 K, $T_c$ = 508.3 K, and $P_c$ = 47.6 bar. From Eqs. (7-2.5), (7-4.9), and (7-4.10), with $T_{b_r}$ = 355.4/508.3 = 0.699,

$$h = 0.699\,\frac{\ln(47.6/1.01325)}{1 - 0.699} = 8.948$$

$$m = 0.0052(60.096)^{0.29}(508.3)^{0.72} = 1.515$$

$$\gamma = \frac{2.464}{60.096}\exp[(9.8 \times 10^{-6})(60.096)(508.3)] = 0.0553$$

From Eqs. (7-4.2), (7-4.3), and (7-4.11),

$$a = \frac{1 - 1/0.699}{(0.699)^7 - 1} = 0.469$$

$$b = \frac{1 - 1/(0.699)^{1.515}}{(0.699)^7 - 1} = 0.785$$

$$\beta = 0.0553/0.785 - \frac{(0.469)(8.948)}{0.785} = -5.276$$

From Eq. (7-4.1), with $T_r = 450/508.3 = 0.885$,

$$\ln P_{vpr} = -5.276\left(\frac{1}{0.885^{1.515}} - 1\right) + (0.0553)(0.885^7 - 1) = -1.101$$

$$P_{vp} = 47.6 \exp(-1.101) = 15.83 \text{ bar}$$

$$\text{Error} = \frac{15.83 - 16.16}{16.16} \times 100 = -2.1\%$$

## 7-5 Vapor Pressure Estimation with Two Reference Fluids

In the Lee-Kesler method, a fluid's properties are obtained by interpolating between the properties of a simple fluid ($\omega = 0$) and a reference fluid (octane with $\omega = 0.3978$). Several authors [11, 84] have suggested that the simple fluid be replaced by a real fluid. This leads to Eq. (3-7.5) and, when written for vapor pressures, gives

$$\ln P_{vpr} = \ln P_{vpr}^{(R1)} + (\ln P_{vpr}^{(R2)} - \ln P_{vpr}^{(R1)}) \frac{\omega - \omega^{(R1)}}{\omega^{(R2)} - \omega^{(R1)}} \tag{7-5.1}$$

The superscripts R1 and R2 refer to the two reference substances. Ambrose and Patel [8] used either propane and octane or benzene and pentafluorotoluene as the reference fluids. In Eq. (7-5.1), all vapor pressures are calculated at the reduced temperature of the substance whose vapor pressure is being predicted. Reduced vapor pressures for the reference substances were calculated with the Wagner equation in the form

$$\ln P_{vpr} = \frac{a\tau + b\tau^{1.5} + c\tau^3 + d\tau^6}{T_r} \tag{7-5.2}$$

where $\tau = 1 - T_r$. Values of the constants $a$, $b$, $c$, and $d$ for these four reference fluids are given in Table 7-1 as well as Appendix A. Equation (7-5.1) is written to estimate vapor pressures. However, if two or more vapor pressures are known in addition to $T_c$ and $\omega$, Eq. (7-5.1) can be used to estimate the critical pressure. Ambrose and Patel [8] have examined this application and report average errors in $P_c$ of about 2 percent for 65 compounds. This is as good as the methods in Chapter 2. When at least three vapor pressures are known, it is mathematically possible to estimate

**TABLE 7-1 Constants for Eq. (7-5.2) for Four Reference Fluids [8]**

| Substance | $T_c$, K | $P_c$, bar | $\omega$ | $a$ | $b$ | $c$ | $d$ |
|---|---|---|---|---|---|---|---|
| Propane | 369.85 | 42.4535 | 0.153 | −6.72219 | 1.33236 | −2.13868 | −1.38551 |
| Octane | 568.81 | 24.8617 | 0.398 | −7.91211 | 1.38007 | −3.80435 | −4.50132 |
| Benzene | 562.16 | 48.9794 | 0.212 | −6.98273 | 1.33213 | −2.62863 | −3.33399 |
| Pentafluoro-toluene | 566.52 | 31.2481 | 0.415 | −8.05688 | 1.46673 | −3.82439 | −2.78727 |

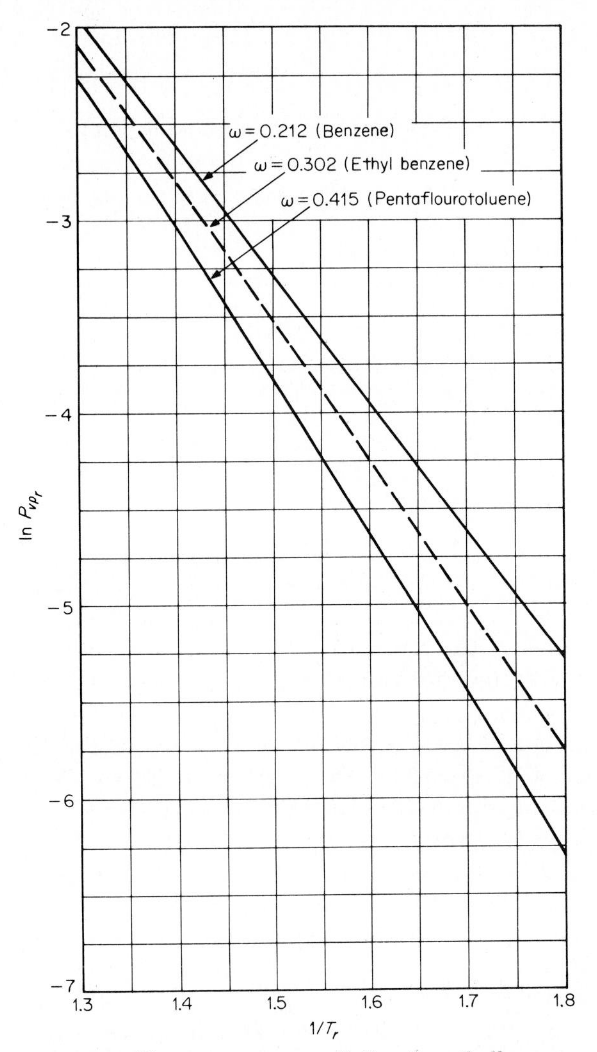


**Figure 7-2** Vapor pressure prediction for ethylbenzene by the two-reference-fluid method; —— from Eq. (7-5.2); — — from Eq. (7-5.1).

$T_c$ also, but Ambrose and Patel [8] recommended this not be done. It should be obvious that, when using Eq. (7-5.1), more reliable estimates will be obtained when

$$\omega^{(R1)} < \omega < \omega^{(R2)} \tag{7-5.3}$$

This case represents an interpolation in the acentric factor rather than an extrapolation. Use of Eq. (7-5.1) to estimate vapor pressures is illustrated by Example 7-5 and Fig. 7-2. Within the accuracy of the graph, the dashed line in Fig. 7-2 coincides with the literature data for ethylbenzene in [3] and [102].

**Example 7-5** Repeat Example 7-1 by using Eq. (7-5.1) and benzene and pentafluorotoluene as reference fluids.

**solution** For ethylbenzene, $T_c$ = 617.1 K, $\omega$ = 0.302, and $P_c$ = 36.0 bar. Using benzene as R1 and pentafluorotoluene as R2, Eqs. (7-5.1) and (7-5.2) along with the values in Table 7-1 lead to the following results.

| T, K | $1/T_r$ | $\ln P_{vpr}^{(R1)}$ | $\ln P_{vpr}^{(R2)}$ | $\ln P_{vpr}$ | $P_{vpcalc}$, bar | $P_{vpexp}$, bar | $\frac{P_{calc} - P_{exp}}{P_{exp}} \times 100$ |
|---|---|---|---|---|---|---|---|
| 347.2 | 1.777 | −5.174 | −6.110 | −5.589 | 0.1346 | 0.1333 | 0.96 |
| 460 | 1.342 | −2.216 | −2.586 | −2.380 | 3.333 | 3.325 | 0.24 |

## 7-6 Correlation and Extrapolation of Vapor Pressure Data

Most vapor pressure equations presented in the preceding sections have been introduced primarily as estimating equations. That is, given some data such as a boiling point and the critical properties, it is possible to develop the constants so that vapor pressures can be estimated as functions of temperature.

Upon occasion, however, one is favored with experimental vapor pressure data over a wide range of temperatures (see [15] for a compilation of vapor pressure data), and it may be useful to store this information in an analytical form. Using standard regression techniques, one can determine the best values of the constants by employing any one of the equation forms introduced earlier in this chapter. This is routinely done to obtain the Antoine constants in Eq. (7-3.1). A number of studies have evaluated different equations in terms of their ability to correlate and/or predict vapor pressures over wide ranges of temperature [1, 5, 37, 53, 57, 62, 76, 92*a*, 95, 96, 103, 106]. One equation which has been particularly successful is the Wagner equation. In the procedure originally proposed by Wagner [96], the form of the equation was variable. Equation (7-5.2) is the simplest form which arises from Wagner's analysis; it has come to be referred to as the Wagner equation, even though a more complex form is superior for water [97, 98], ammonia [12], and oxygen [95]. In all situations in which a form of it was compared to other vapor pressure equations [5, 53, 62, 95, 96], the Wagner equation was selected as the best equation. McGarry [62] has determined values for $a$, $b$, $c$, and $d$ for a number of compounds, and many of those values are listed in Appendix A. For compounds for which ample data are available, $P_c$ was treated as an adjustable parameter; and in some of these cases, this is the value listed for $P_c$ in Appendix A. When Wagner constants were not available for a particular compound, Antoine constants, or constants for the Frost-Kalkwarf-Thodos equation

$$\ln P_{vp} = A - \frac{B}{T} + C \ln T + \frac{DP_{vp}}{T^2} \tag{7-6.1}$$

have been tabulated in Appendix A.

An equation's ability to correlate data depends directly on the number of adjustable parameters in the equation. When one has only a small number of data, or requires computational simplicity, one might use only three adjustable parameters. The Antoine equation (7-3.1) is the most common, but not the best, three-parameter equation. McGarry [62] found that the modified Miller equation

$$\ln P_{vp_r} = -\frac{A}{T_r}[1 - T_r^2 + B(3 + T_r)(1 - T_r)^3] \tag{7-6.2}$$

gave a better fit to experimental data than did the Antoine equation. The three fitted constants in Eq. (7-6.2) are $A$, $B$, and $P_c$ ($P_{vp_r} = P_{vp}/P_c$).

Sometimes vapor pressure data are available over a limited temperature range and one wishes to extrapolate the data to higher or lower temperatures. The Antoine equation does not reproduce the correct shape of a vapor pressure curve over the entire temperature range and should not be used for extrapolation [49]. If it is used to calculate vapor pressures outside the range of temperatures listed in Appendix A, serious errors may result. The Wagner equation may be used to extrapolate data because of the manner in which the constants have been determined. In addition to requiring a best fit of existing vapor pressure data, the equation is constrained so as to generate a "reasonable shape" for the vapor pressure curve from a reduced temperature of 0.5 up to the critical point. The correct shape of a vapor pressure curve is well documented [4, 20, 62, 99], and several authors have demonstrated the use of the constrained fit method [4, 62]. Ambrose et al. [4] have shown that the constrained fit technique may be used to estimate critical pressures with an accuracy comparable to that obtained with Eq. (7-5.1). The Wagner constants in Appendix A are also from a constrained fit [62]. The Wagner equation may not extrapolate well to reduced temperatures below 0.5.

Vapor pressure data below reduced temperatures of about 0.5 are sparse, although Carruth and Kobayashi [19] give vapor pressure data for the normal paraffins ethane through $n$-decane down to their triple points. Low-temperature vapor pressures are usually obtained by integrating the Clausius-Clapeyron equation. King and Al-Najjar [50] have done this integration along the saturation curve for eight normal alkanes, and Ambrose and Davies [7] have used the method for nine other organic compounds. These studies establish reliable values for 17 compounds that may be used as reference fluid values in the two-reference-fluid method, Eq. (7-5.1), to estimate vapor pressures of additional compounds. Mosselman et al. [65] have presented the Clausius-Clapeyron equation in the form of an exact differential so that integration is path-independent. One integrated form they present is

$$\frac{1}{T_1}\int_{P_0}^{P_1} \Delta V_v(P, T_1)\, dP + \int_{1/T_0}^{1/T_1}\left[\int_{T_0}^{T} \Delta C_{p_v}(P_0, T)\, dT\right] dT^{-1} = \Delta H_v(P_0, T_0)\,(T_0^{-1} - T_1^{-1}) \qquad (7\text{-}6.3)$$

In Eq. (7-6.3), $\Delta H_v$ and $P_0$ are the enthalpy of vaporization and vapor pressure at $T_0$. $P_1$ is the vapor pressure to be calculated at $T_1$. $\Delta V_v$ is the vapor volume minus the liquid volume at $P$ and $T_1$; $\Delta C_{p_v}$ is the vapor heat capacity minus the liquid heat capacity at $P_0$ and $T$. If $\Delta V_v$ and $\Delta C_{p_v}$ data are available or can be estimated, Eq. (7-6.3) provides a thermodynamically sound extrapolation method. Equation (7-6.3) is particularly useful for low temperatures and pressures when deviations from ideal-gas behavior are small for the vapor phase.

## 7-7 Discussion and Recommendations for Vapor Pressure Estimation and Correlation

Starting from the Clausius-Clapeyron equation (7-2.2), we have shown only a few of the many vapor pressure equations which have been published. We have emphasized those which appear to be most accurate and general. Properties required for the different estimation equations are $T_b$, $M$, and $T_c$ for Gomez-Thodos and $\omega$ and $T_c$ for the two-reference-fluid and Lee-Kesler methods. It is amazing how well these techniques predict vapor pressures over wide ranges of temperature with this little input. We show in Table 7-2 a detailed comparison of calculated and experimental vapor pressures for acetone for the estimation techniques described in this chapter. The range shown is from 0.04 bar to the critical point, 47 bar. The least accurate is, as expected, the Clapeyron equation, especially at lower temperatures.

The Antoine equation should not be used above 2.0 to 2.7 bar when the constants are obtained from experimental data below that pressure. In the range for which the constants are applicable, it is very accurate. Of the methods shown in Table 7-2 the Wagner equation is the most accurate, although all the predictive methods, i.e., Lee-Kesler, Gomez-Thodos, and the two-reference fluid methods, perform satisfactorily.

The methods presented in this chapter cannot be used to estimate vapor pressures of high-molecular-weight compounds if the normal boiling points cannot be determined. When no vapor pressure data exist but the molecular structure of a compound is known, the UNIFAC group con-

**TABLE 7-2 Comparison between Calculated and Experimental Vapor Pressures for Acetone**

| | | | Percent error[a] | | | | | |
|---|---|---|---|---|---|---|---|---|
| $T$, K | $T_r$ | $P_{vp}$ (exp.), bar[b] | Clapeyron, Eq. (7-2.4) | Antoine, Eq. (7-3.1)[c] | Wagner Eq. (7-5.2) | Gomez and Thodos, Eq. (7-4.1) | Two reference fluids[d] Eq. (7-5.1) | Lee and Kesler, Eq. (7-2.6) |
| 259.2 | 0.510 | 0.04267 | 24 | 1.6 | 0.2 | 1.1 | −5.5 | −7.9[e] |
| 273.4 | 0.538 | 0.09497 | 15 | 0.3 | −0.1 | 0.2 | −3.9 | −6.2 |
| 290.1 | 0.571 | 0.21525 | 8.3 | −0.1 | 0.2 | 0.1 | −2.0 | −3.7 |
| 320.5 | 0.631 | 0.74449 | 1.5 | −0.7 | 0.1 | 0.1 | 0 | −0.2 |
| 350.9 | 0.691 | 2.01571 | −1.0 | −0.6 | 0 | 0.3 | 0.8 | 1.9 |
| 390.3 | 0.768 | 5.655 | −1.5 | 0.1 | 0.2 | 0.8 | 1.3 | 2.2 |
| 446.4 | 0.878 | 17.682 | −0.3 | −0.5 | 0 | 0.6 | 0.6 | 0.7 |
| 470.6 | 0.926 | 26.628 | 0.5 | 1.5[f] | −0.1 | 0.3 | −0.2 | 0.3 |
| 508.1 | 1.00 | 47.000 | | −4.5[f] | | | | |

[a]Percent error = [(calc. − exp.)/exp.] × 100.
[b]Experimental data from Ref. 9.
[c]$\ln P_{vp} = 10.0311 - 2940.45/(T - 35.93)$, $P_{vp}$ is in bars.
[d]Reference fluids are propane and octane.
[e]$\omega$ calculated from Eq. (2-3.4) ($\omega = 0.301$) rather than obtained from Appendix A ($\omega = 0.304$).
[f]Constants not applicable at high reduced temperature.

tribution method [19a, 45, 104] or the group contribution methods [16, 43, 78] similar to the method developed by Prausnitz and coworkers [26, 59] can be used. White [100] has used the molecular connectivity [48] to correlate the normal boiling point and $\Delta H_{vb}$ for polycyclic aromatic hydrocarbons, and Willman and Teja [102*a*] have correlated constants in the Wagner equation with effective carbon numbers. When no vapor pressure data are available and when one is uncertain of the molecular structure of a compound, or a petroleum fraction, the SWAP equation [27, 60, 81] may be used.

### Recommendations

If constants for a particular compound are available in Appendix A, use them along with the appropriate equation. The Antoine equation should not be used for temperatures outside the range listed in Appendix A. The Wagner equation may be extrapolated to higher temperatures with confidence. It may be used down to a reduced temperature of 0.5 or to the value of $T_{\min}$ listed in Appendix A; extrapolations to lower temperatures may lead to unacceptable errors. For reduced temperatures below 0.5, the Lee-Kesler equation (7-2.6) or the two-reference-fluid method, Eq. (7-5.1), is recommended. Except for alcohols, errors should be less than 30 percent. If higher accuracy is required, at low temperatures, vapor pressures may be calculated with Eq. (7-6.2) or as in Refs. 7 and/or 50. For predictions for polar compounds at reduced temperatures between 0.5 and 1.0, the two-reference-fluid or Gomez-Thodos methods are recommended. If no data are available for a compound, and the compound's normal boiling point is unknown, one of the group contribution methods mentioned above may be used. In that event, however, calculated results are not likely to be highly accurate.

## 7-8 Enthalpy of Vaporization of Pure Compounds

The enthalpy of vaporization $\Delta H_v$ is sometimes referred to as the latent heat of vaporization. It is the difference between the enthalpy of the saturated vapor and that of the saturated liquid at the same temperature.

Because of the forces of attraction between the molecules of the liquid, the molecules escaping are those of higher than average energy. The average energy of the remaining molecules in the liquid is reduced, and energy must be supplied to maintain the temperature constant. This is the internal energy of vaporization $\Delta U_v$. Work is done on the vapor phase as vaporization proceeds, since the vapor volume increases if the pressure is maintained constant at $P_{vp}$. This work is $P_{vp}(V_g - V_L)$. Thus

$$\begin{aligned}\Delta H_v &= \Delta U_v + P_{vp}(V_g - V_L) = \Delta U_v + RT(Z_g - Z_L) \\ &= \Delta U_v + RT\,\Delta Z_v\end{aligned} \tag{7-8.1}$$

Many estimation methods for $\Delta H_v$ can be traced to Eq. (7-2.2), where it is shown that $\Delta H_v$ is related to the slope of the vapor pressure–temperature curve. Other methods are based on the law of corresponding states. We review the more accurate techniques in Secs. 7-9 to 7-12; recommendations are presented in Sec. 7-13. Literature references to experimental and calculated values of $\Delta H_v$ have been compiled in [83].

It is often difficult to trace the origin of many "experimental" enthalpies of vaporization. A few have been determined from calorimetric measurements, but in a large number of cases the so-called experimental value was obtained directly from Eq. (7-2.2). Some technique was employed to determine $\Delta Z_v$ separately, and also $(d \ln P_{vp})/dT$ was found by numerical differentiation of experimental vapor pressure data or by differentiating some $P_{vp}$-$T$ correlation analytically. An example of this latter approach can be found in the recent reissue of the API Tables [101]. Enthalpies of vaporization were determined by using Eq. (7-2.2), where $dP_{vp}/dT$ was found from the Antoine vapor pressure equation (Sec. 7-3); the saturated vapor compressibility factor was estimated from a virial equation of state (Sec. 3-5); and experimental data were employed for saturated liquid compressibility factors.

An element of uncertainty is introduced in using any analytical vapor pressure–temperature equation to obtain accurate values of slopes $dP_{vp}/dT$. The constants in the equation may be optimum for correlating vapor pressures, but it does not necessarily follow that these same constants give the best fit for computing slopes. With the Antoine equation, for example, different sets of the constants $A$, $B$, and $C$ can represent a set of experimental data satisfactorily. But differentiation eliminates $A$ and increases the importance of $C$. Uncertainties in $C$ lead directly to uncertainties in $\Delta H_v$.

Since so few calorimetric measurements of $\Delta H_v$ are available, there is little that can be done to rectify the problem. A critical survey of reported $\Delta H_v$ values would, nevertheless, be of value, since one would like to avoid the logical pitfalls of comparing estimated values of $\Delta H_v$ with values estimated by other approximate methods and then making recommendations based on such a comparison.

## 7-9 Estimation of $\Delta H_v$ from the Law of Corresponding States

Equation (7-2.1), in reduced form, becomes

$$d \ln P_{vp_r} = \frac{-\Delta H_v}{RT_c \, \Delta Z_v} \, d \, \frac{1}{T_r} \qquad (7\text{-}9.1)$$

The reduced enthalpy of vaporization $-\Delta H_v/RT_c$ is a function of $(d \ln P_{vp_r})/d(1/T_r)$ and $\Delta Z_v$; both these parameters are commonly assumed to be functions of $T_r$ or $P_{vp_r}$ and some third parameter such as $\omega$, $Z_c$, or $h$. A number of correlations based on this approach have been suggested.

### Pitzer acentric factor correlation [70]

Pitzer et al. have shown that $\Delta H_v$ can be related to $T$, $T_r$, and $\omega$ by an expansion similar to that used to estimate compressibility factors, Eq. (3-3.1), i.e.,

$$\frac{\Delta H_v}{T} = \Delta S_v^{(0)} + \omega\,\Delta S_v^{(1)} \tag{7-9.2}$$

where $\Delta S_v^{(0)}$ and $\Delta S_v^{(1)}$ are expressed in entropy units, for example, J/(mol·K), and are functions only of $T_r$. Multiplying Eq. (7-9.2) by $T_r/R$ gives

$$\frac{\Delta H_v}{RT_c} = \frac{T_r}{R}\,(\Delta S_v^{(0)} + \omega\,\Delta S_v^{(1)}) \tag{7-9.3}$$

Thus $\Delta H_v/RT_c$ is a function only of $\omega$ and $T_r$. From the tabulated $\Delta S_v^{(0)}$ and $\Delta S_v^{(1)}$ functions given by Pitzer et al. and extended to low reduced temperatures by Carruth and Kobayashi [18], Fig. 7-3 was constructed. For a close approximation, an analytical representation of this correlation for $0.6 < T_r \leq 1.0$ is

$$\frac{\Delta H_v}{RT_c} = 7.08(1 - T_r)^{0.354} + 10.95\omega(1 - T_r)^{0.456} \tag{7-9.4}$$

The effect of temperature on $\Delta H_v$ is similar to that suggested by Watson (see Sec. 7-12). Nath [68] has presented an equation similar to Eq. (7-9.4) for $0.5 < T_r < 0.7$. To use the Lee-Kesler method to obtain values for $\Delta S_v^{(0)}$ and $\Delta S_v^{(1)}$, the vapor pressure is determined with Eq. (7-2.6) and $H - H°$ is evaluated for both liquid and vapor. The difference in these two enthalpy departure functions is $\Delta H_v$. Gupte and Daubert [38] report good results with this approach; the approach is given graphically in [47].

**Example 7-6** By using the Pitzer et al. corresponding states correlation, estimate the enthalpy of vaporization of propionaldehyde at 321 K. The literature value is 28280 J/mol [22].

**solution** From Appendix A, $T_c$ = 496 K and $\omega$ = 0.313. $T_r$ = 321/496 = 0.647, and from Eq. (7-9.4)

$$\frac{\Delta H_v}{RT_c} = (7.08)(1-0.647)^{0.354} + (10.95)(0.313)(1 - 0.647)^{0.456}$$

$$= 7.03$$

$$\Delta H_v = (7.03)(8.314)(496) = 29{,}000 \text{ J/mol}$$

$$\text{Error} = \frac{29{,}000 - 28{,}280}{28{,}280} \times 100 = 2.5\%$$

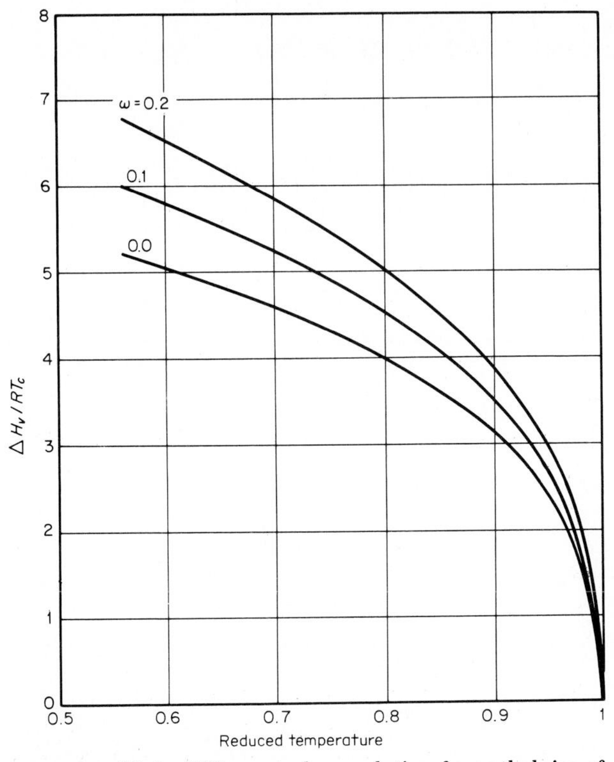


**Figure 7-3** Plot of Pitzer et al. correlation for enthalpies of vaporization.

### Two-reference-fluid method

Sivaraman et al. [80] have used the two-reference-fluid method to calculate enthalpies of vaporization of coal compounds between their freezing points and critical points. An equation developed by Torquato and Stell [90] was used for the reference fluids, benzene and carbazole. These equations are

$$\frac{\Delta H_v}{RT_c} = \left(\frac{\Delta H_v}{RT_c}\right)^{(R1)} + \left(\frac{\omega - \omega^{(R1)}}{\omega^{(R2)} - \omega^{(R1)}}\right)\left[\left(\frac{\Delta H_v}{RT_c}\right)^{(R2)} - \left(\frac{\Delta H_v}{RT_c}\right)^{(R1)}\right] \tag{7-9.5}$$

For benzene,

$$\left(\frac{\Delta H_v}{RT_c}\right)^{(R1)} = 6.537\tau^{1/3} - 2.467\tau^{5/6} - 77.521\tau^{1.208} + 59.634\tau + 36.009\tau^2 - 14.606\tau^3 \tag{7-9.6}$$

The quantity $(\Delta H_v/RT_c)^{(R2)} - (\Delta H_v/RT_c)^{(R1)}$ in Eq. (7-9.5) is the difference in the reduced enthalpies of vaporization of carbazole and benzene and is given by

$$\left(\frac{\Delta H_v}{RT_c}\right)^{(R2)} - \left(\frac{\Delta H_v}{RT_c}\right)^{(R1)} = -0.133\tau^{1/3} - 28.215\tau^{5/6}$$

$$-82.958\tau^{1.208} + 99.000\tau + 19.105\tau^2 - 2.796\tau^3 \quad (7\text{-}9.7)$$

In Eqs. (7-9.6) and (7-9.7), $\tau = 1 - T_r$, $\omega^{(R1)}$ is 0.21, and $\omega^{(R2)}$ is 0.46. The same value of $\tau$ is used in Eqs. (7-9.6) and (7-9.7), where $\tau$ is determined at the reduced temperature of the substance whose enthalpy of vaporization is being estimated.

**Example 7-7** Calculate $\Delta H_v$ for naphthalene at 553.15 K with Eqs. (7-9.5) to (7-9.7). The experimental value [80] is 39.82 kJ/mol.

**solution** For naphthalene, $T_c = 748.4$ and $\omega = 0.302$.

$$\tau = 1 - \frac{553.15}{748.4} = 0.2609$$

From Eqs. (7-9.6 and (7-9.7)

$$\left(\frac{\Delta H_v}{RT_c}\right)^{(R1)} = 5.840 \quad \text{and} \quad \left(\frac{\Delta H_v}{RT_c}\right)^{(R2)} - \left(\frac{\Delta H_v}{RT_c}\right)^{(R1)} = 1.426$$

Equation (7-9.5) gives

$$\frac{\Delta H_v}{RT_c} = 5.840 + \left(\frac{0.302 - 0.21}{0.46 - 0.21}\right)(1.426) = 6.365$$

$$\Delta H_v = (6.365)(8.314)(748.4) = 39{,}600 \text{ J/mol} = 39.6 \text{ kJ/mol}$$

$$\text{Error} = \frac{39.6 - 39.82}{39.82} \times 100 = -0.6\%$$

## 7-10 Estimation of $\Delta H_v$ from Vapor Pressure Equations

The vapor pressure correlations covered in Secs. 7-2 to 7-6 can be used to estimate enthalpies of vaporization. From Eq. (7-9.1), we can define a dimensionless group $\psi$ as

$$\psi \equiv \frac{\Delta H_v}{RT_c\,\Delta Z_v} = \frac{-d\ln P_{vp_r}}{d(1/T_r)} \quad (7\text{-}10.1)$$

By differentiating the vapor pressure equations discussed earlier, we can obtain various expressions for $\psi$. These are shown in Table 7-3. To use the expressions, one must refer back to the reference vapor pressure equation earlier in the chapter for the definition of the various parameters.

In Fig. 7-4, we show experimental values of $\psi$ for propane. These were calculated from smoothed values tabulated in Refs. 24 and 105. Note the pronounced minimum in the curve around $T_r = 0.8$. Since

$$\psi = \frac{-d\ln P_{vp_r}}{d(1/T_r)} \quad (7\text{-}10.7)$$

**TABLE 7-3** **$\psi$ Values from Vapor Pressure Equations**

| Vapor pressure equation | $\psi = \dfrac{\Delta H_v}{RT_c\,\Delta Z_v}$ | |
|---|---|---|
| Clapeyron, Eq. (7-2.4) | $h$, defined in Eq. (7-2.5) | (7-10.2) |
| Lee-Kesler, Eq. (7-2.6) | $6.09648 - 1.28862T_r + 1.016T_r^7 + \omega(15.6875 - 13.4721T_r + 2.615T_r^7)$ | (7-10.3) |
| Antoine, Eq. (7-3.1) | $\dfrac{B}{T_c}\left(\dfrac{T_r}{T_r + C/T_c}\right)^2$ | (7-10.4) |
| Gomez-Thodos, Eq. (7-4.1) | $7\gamma T_r^6 - \dfrac{\beta m}{T_r^{m-1}}$ | (7.10.5) |
| Wagner, Eq. (7-5.2) | $-a + b\tau^{0.5}(0.5\tau - 1.5) + c\tau^2(2\tau - 3) + d\tau^5(5\tau - 6)$ | (7-10.6) |

we have

$$\frac{d\psi}{dT_r} = \frac{1}{T_r^2}\frac{d^2 \ln P_{\text{vp}_r}}{d(1/T_r)^2} \tag{7-10.8}$$

At low values of $T_r$, $d\psi/dT_r < 0$, so that $(d^2 \ln P_{\text{vp}_r})/d(1/T_r)^2$ is also $< 0$. At high values of $T_r$, the signs reverse. When $d\psi/dT_r = 0$, an inflection point results. Thus the general (though exaggerated) shape of a log vapor pressure–inverse temperature curve is that shown in Fig. 7-5.

Figure 7-6 illustrates how the various vapor pressure equations can predict the shape of Fig. 7-4. Except for the Clapeyron equation (where $\psi$ is a constant equal to $h$), the other equations show a remarkable fit to the value of $\psi$ calculated from literature data for propane. The Antoine equation does not predict the $\psi$–$T_r$ minimum, and the calculated curve was drawn only to $T_r = 0.83$. The curve generated with the Wagner constants

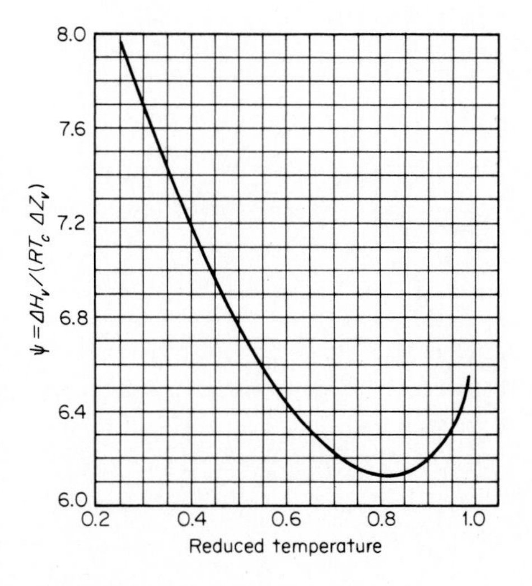


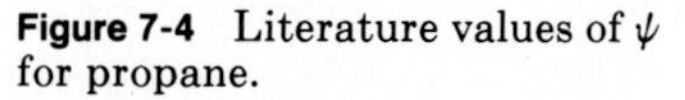

**Figure 7-4** Literature values of $\psi$ for propane.

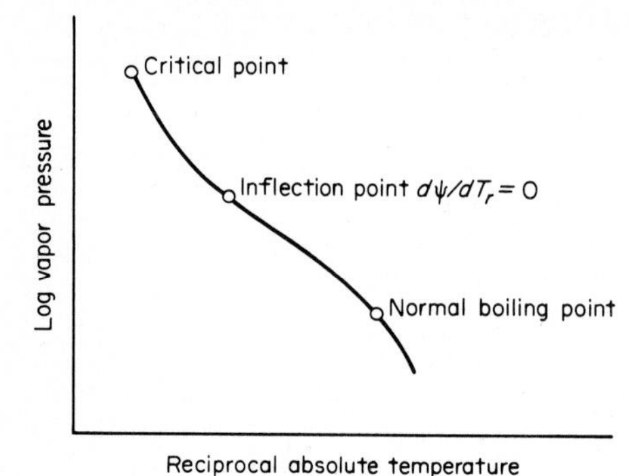


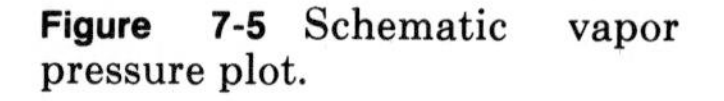
**Figure 7-5** Schematic vapor pressure plot.

in Appendix A and Eq. (7-10.6) is not shown in Fig. 7-6 because it coincides with the literature values over the $T_r$ range shown in the figure.

Other comparisons also produced good results. We may, therefore, recommend these vapor pressure correlations to predict $\psi$ and thus $\Delta H_v$. However, accurate values of $\Delta Z_v$ must be available. To date, available correlations for $\Delta Z_v$ are not accurate. $\Delta Z_v$ is determined best as a difference in the $Z$ values of saturated vapor and saturated liquid. For saturated liquid, the methods in Chap. 3 are appropriate. For the specific saturated

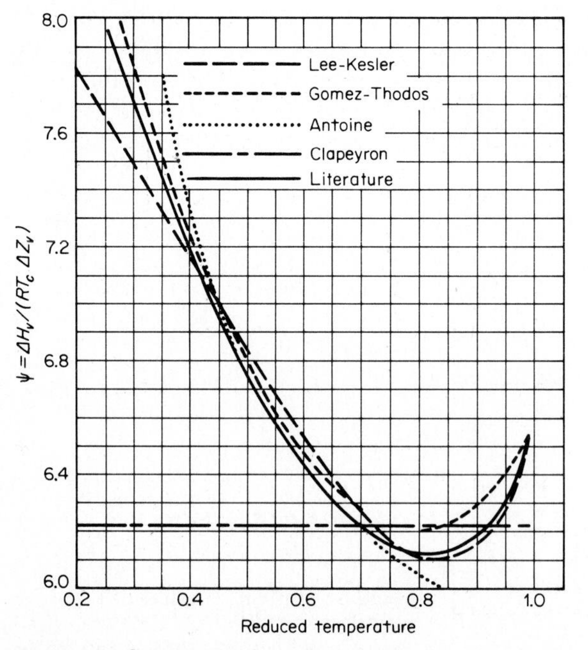


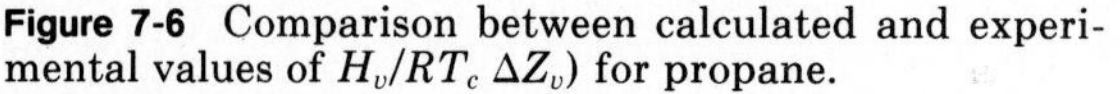
**Figure 7-6** Comparison between calculated and experimental values of $H_v/RT_c\ \Delta Z_v)$ for propane.

volumes of nonpolar materials, the following equation presented by Thompson and Sullivan [86] may be used from the triple point to the critical point

$$(1 - P_{vp_r})^{\beta} = |1 - V_r^{-\lambda}| \tag{7-10.9}$$

$V_r$ is the reduced saturated vapor volume; values of $\beta$ and $\lambda$ are listed in [86] or may be estimated by

$$\beta = 0.346 - 0.067\omega \tag{7-10.10}$$

$$\lambda = 1 + 0.085(1 + 0.19\omega)^{-8} \tag{7-10.11}$$

Equations (7-10.9) to (7-10.11) gave errors which were usually less than 3 percent but could be as high as 10 percent.

## 7-11 $\Delta H_v$ at the Normal Boiling Point

A pure component constant that is occasionally used in property correlations is the enthalpy of vaporization at the normal boiling point $\Delta H_{vb}$. Any one of the correlations discussed in Secs. 7-8 to 7-10 can be used for this state where $T = T_b$, $P = 1.01$ bar. We discuss some of the techniques below. In addition, several special estimation methods are suggested.

### $\Delta H_{vb}$ from vapor pressure relations

In Table 7-3, we show equations for $\psi \equiv \Delta H_v/(RT_c\, \Delta Z_v)$ as determined from a number of the more accurate vapor pressure equations. Each can be used to determine $\psi\ (T_b)$. With $\psi\ (T_b)$ and $\Delta Z_v\ (T_b)$, $\Delta H_{vb}$ can be determined.

Special mention should be made when the Clapeyron equation is used to calculate $\psi$ [see Eq. (7-10.2) in Table 7-3]. Here, $\psi$ is equal to $h$ regardless of $T_r$, that is,

$$\psi\ (T_r) = \psi\ (T_b) = T_{b_r} \frac{\ln (P_c/1.01325)}{1 - T_{b_r}} \tag{7-11.1}$$

and

$$\Delta H_{vb} = RT_c\, \Delta Z_{vb} \left( T_{b_r} \frac{\ln (P_c/1.01325)}{1 - T_{b_r}} \right) \tag{7-11.2}$$

Equation (7-11.2) has been widely employed to make rapid estimates of $\Delta H_{vb}$; usually, in such cases, $\Delta Z_{vb}$ is set equal to unity. In this form, it has been called the Giacalone equation [32]; extensive testing of this simplified form indicates that it normally overpredicts $\Delta H_{vb}$ by a few percent. Correction terms have been suggested [30, 52] to improve the accuracy of the Giacalone equation, but better results are obtained with other relations, noted below.

**TABLE 7-4 Comparison between Calculated and Experimental Values of $\Delta H_{vb}$**

| | | Average absolute percent error | | | |
|---|---|---|---|---|---|
| Class of compounds | Number of compounds | Giacalone, Eq. (7-11.2) | Riedel, Eq. (7-11.3) | Chen, Eq. (7-11.4) | Vetere, Eq. (7-11.5) |
| Saturated hydrocarbons | 22 | 2.9 | 0.9 | 0.4 | 0.4 |
| Unsaturated hydrocarbons | 8 | 2.4 | 1.4 | 1.2 | 1.2 |
| Cycloparaffins and aromatics | 12 | 1.1 | 1.3 | 1.2 | 1.1 |
| Alcohols | 7 | 3.6 | 4.0 | 4.0 | 3.8 |
| Nitrogen and sulfur compounds (organic) | 10 | 1.6 | 1.7 | 1.7 | 1.9 |
| Halogenated compounds | 10 | 1.3 | 1.6 | 1.5 | 1.5 |
| Rare gases | 5 | 8.4 | 2.1 | 2.2 | 2.5 |
| Nitrogen and sulfur compounds (inorganic) | 4 | 3.0 | 2.7 | 2.7 | 2.1 |
| Inorganic halides | 4 | 0.6 | 1.4 | 1.4 | 0.9 |
| Oxides | 6 | 6.9 | 4.4 | 4.9 | 4.6 |
| Other polar compounds | 6 | 2.2 | 1.5 | 1.8 | 1.6 |
| Total | 94 | 2.8 | 1.8 | 1.7 | 1.6 |

### Riedel method

Riedel [77] modified Eq. (7-11.2) slightly and proposed that

$$\Delta H_{vb} = 1.093RT_c \left[ T_{b_r} \frac{\ln P_c - 1.013}{0.930 - T_{b_r}} \right] \tag{7-11.3}$$

When tested as shown in Table 7-4, errors are almost always less than 2 percent.

### Chen method

Chen [21] used Eq. (7-9.3) and a similar expression proposed by Pitzer et al. to correlate vapor pressures so that the acentric factor is eliminated. He obtained a relation between $\Delta H_v$, $P_{vp_r}$, and $T_r$. When applied to the normal boiling point,

$$\Delta H_{vb} = RT_cT_{b_r} \frac{3.978T_{b_r} - 3.958 + 1.555 \ln P_c}{1.07 - T_{b_r}} \tag{7-11.4}$$

This correlation was tested with the results shown in Table 7-4. The accuracy is similar to that of the Riedel equation (7-11.3). In a more complete test, Chen compared estimated $\Delta H_{vb}$ with literature values for 169 materials and found an average error of 2.1 percent.

### Vetere method

Vetere [91] proposed a relation similar to the one suggested by Chen:

$$\Delta H_{vb} = RT_cT_{b_r}\frac{0.4343 \ln P_c - 0.69431 + 0.89584T_{b_r}}{0.37691 - 0.37306T_{b_r} + 0.15075P_c^{-1}\,T_{b_r}^{-2}} \tag{7-11.5}$$

As shown in Table 7-4, this empirical equation is capable of providing a good estimate of $\Delta H_{vb}$; errors are normally less than 2 percent.

### Other methods

A number of other methods for estimating $\Delta H_{vb}$ have been proposed. None appears to offer any significant advantages over those given above. Miller [64] used an earlier version of the Riedel-Plank-Miller vapor pressure equation to determine $\psi$ and then, with $\Delta Z_{vb} \approx 1 - 0.97/P_cT_{b_r}$, he obtained a relation for $\Delta H_{vb}$ in terms of $T_{b_r}$ and $P_c$. The final result is more complex than Eqs. (7-11.3) to (7-11.5), but the results are similar.

Ibrahim and Kuloor [44] presented two correlations for $\Delta H_{vb}$ in which the molecular weight or liquid volume is employed as the independent variable. Specific constants are required for various homologous series. Ogden and Lielmezs [69] proposed a similar relation, but in this case the Altenburg quadratic mean radius was used. The latter term reflects the mass distribution in a molecule. Procopio and Su [73] and Viswanath and Kuloor [93] have suggested modifications to the Giacalone equation, Eq. (7-11.2). Narsimhan [67] relates $\Delta H_{vb}$ to a density function. Various group contribution methods have been proposed to estimate $\Delta H_{vb}$ or $\Delta H_v$ at 298 K [28, 33, 42, 55, 61].

### Comparison of estimated with literature values of $\Delta H_{vb}$

Table 7-4 compares calculated and experimental values of $\Delta H_{vb}$ by using the estimation methods described in this section. The Riedel, Pitzer-Chen, and Vetere relations [Eqs. (7-11.3) to (7-11.5)] are convenient and are generally accurate. In every case, $T_b$, $T_c$, and $P_c$ must be known or estimated.

**Example 7-8** Estimate the enthalpy of vaporization of propionaldehyde at the normal boiling point. The experimental value is 28280 J/mol [22].

**solution** From Appendix A, $T_b = 321$ K, $T_c = 496$ K, and $P_c = 47.6$ bar. Thus $T_{b_r} = 0.647$.

RIEDEL METHOD. With Eq. (7-11.3)

$$\Delta H_{vb} = (1.093)(8.314)(496)\frac{(0.647)(\ln 47.6 - 1.013)}{0.930 - 0.647}$$

$$= 29{,}370 \text{ J/mol}$$

$$\text{Error} = \frac{29{,}370 - 28{,}280}{28{,}280} \times 100 = 3.8\%$$

GIACALONE METHOD. With Eq. (7-11.2) and assuming $\Delta Z_{vb} = 1.0$,

$$\Delta H_{vb} = (8.314)(496)\left(\frac{0.647 \ln 47.6/1.01325}{1 - 0.647}\right)$$

$$= 29{,}100 \text{ J/mol}$$

$$\text{Error} = \frac{29{,}100 - 28{,}280}{28{,}280} \times 100 = 2.9\%$$

CHEN METHOD. From Eq. (7-11.4)

$$\Delta H_{vb} = \frac{(8.314)(496)(0.647)[(3.978)(0.647) - 3.958 + 1.555 \ln 47.6]}{1.07 - 0.647}$$

$$= 29{,}160 \text{ J/mol}$$

$$\text{Error} = \frac{29{,}160 - 28{,}280}{28{,}280} \times 100 = 3.1\%$$

VETERE METHOD. From Eq. (7-11.5)

$$\Delta H_{vb} = (8.314)(496)(0.647)\frac{0.4343 \ln 47.6 - 0.69431 + (0.89584)(0.647)}{0.37691 - (0.37306)(0.647) + (0.15075)(47.6)^{-1}(0.647)^{-2}}$$

$$= 29{,}140 \text{ J/mol}$$

$$\text{Error} = \frac{29{,}140 - 28{,}280}{28{,}280} \times 100 = 3.0\%$$

## 7-12 Variation of $\Delta H_v$ with Temperature

The latent heat of vaporization decreases steadily with temperature and is zero at the critical point. Typical data are shown in Fig. 7-7. The shapes of these curves agree with most other enthalpy-of-vaporization data.

The variation of $\Delta H_v$ with temperature could be determined from any of the $\psi$ relations shown in Table 7-3, although the variation of $\Delta Z_v$ with temperature would also have to be specified.

A widely used correlation between $\Delta H_v$ and $T$ is the Watson relation [85]

$$\Delta H_{v2} = \Delta H_{v1}\left(\frac{1 - T_{r2}}{1 - T_{r1}}\right)^n \tag{7-12.1}$$

where subscripts 1 and 2 refer to temperatures 1 and 2. A common choice for $n$ is 0.375 or 0.38 [87].

Viswanath and Kuloor [94] recommend that $n$ be obtained by

$$n = \left(0.00264\frac{\Delta H_{vb}}{RT_b} + 0.8794\right)^{10} \tag{7-12.2}$$

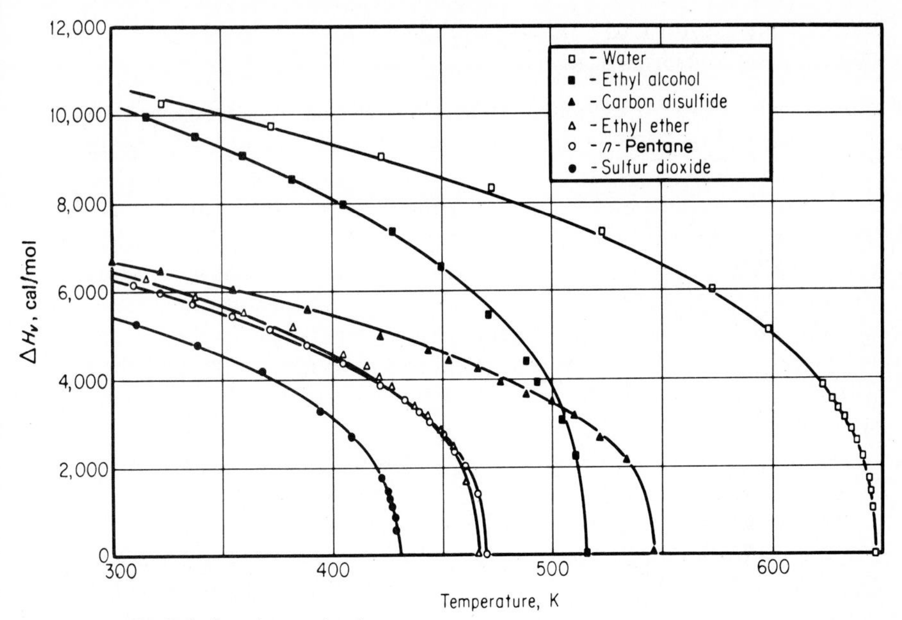


**Figure 7-7** Enthalpies of vaporization.

Fish and Lielmezs [29, 79] suggest a different way to correlate $\Delta H_v$ with $T$:

$$\Delta H_v = \Delta H_{vb} \frac{T_r}{T_{b_r}} \frac{X + X^q}{1 + X^p} \tag{7-12.3}$$

where
$$X = \frac{T_{b_r}}{T_r} \frac{1 - T_r}{1 - T_{b_r}} \tag{7-12.4}$$

and the parameters $q$ and $p$ are:

| | $q$ | $p$ |
|---|---|---|
| Liquid metals | 0.20957 | −0.17467 |
| Quantum liquids* | 0.14543 | 0.52740 |
| Inorganic and organic liquids | 0.35298 | 0.13856 |

*Helium, hydrogen, deuterium, neon.

Compared with Eq. (7-12.1), with $n = 0.38$, Eq. (7-12.3) predicts slightly smaller values of $\Delta H_v$ when $T < T_b$ and somewhat higher when $T > T_b$. For liquid metals and quantum liquids, Eq. (7-12.3) is more accurate than Eq. (7-12.1), but for inorganic and organic liquids, the average errors reported by Fish and Lielmezs for Eqs. (7-12.1) and (7-12.3) were not significantly different.

## 7-13 Discussion and Recommendations for Enthalpy of Vaporization

Three techniques for estimating enthalpies of vaporization of pure liquids have been proposed. The first is based on Eq. (7-2.1) and requires finding $dP_{vp}/dT$ from a vapor pressure–temperature correlation (Sec. 7-10). A separate estimation of $\Delta Z_v$ must be made before $\Delta H_v$ can be obtained. This procedure is inherently accurate, especially if $\Delta Z_v$ is obtained from reliable *PVT* correlations discussed in Chap. 3. Any number of modifications could be programmed for use in a machine computation system. We recommend the Wagner equation (when constants are available), the Lee-Kesler, or Gomez-Thodos vapor pressure equations for use over the entire liquid range, although the accuracy is lower near the freezing point and the critical point.

In the second category are the techniques from the law of corresponding states. The Pitzer et al. form is one of the most accurate and convenient. In an analytical form, this equation for $\Delta H_v$ is approximated by Eq. (7-9.4). Thompson and Braun [87] also recommend the Pitzer et al. form for hydrocarbons. The critical temperature and acentric factor are required. A second corresponding states method is the two-reference fluid method of Sivaraman et al., Eqs. (7-9.5) to (7-9.7).

The third method is to estimate $\Delta H_{vb}$ as recommended in Sec. 7-11 and then scale with temperature with the Watson or Fish-Lielmezs functions discussed in Sec. 7-12.

All three of these techniques are satisfactory, and all yield approximately the same error when averaged over many types of fluids and over large temperature ranges. Fishtine [31], in a separate review of predictive methods of $\Delta H_v$, did not consider the first of the methods discussed above but claimed comparable accuracy for the latter two.

All the methods noted above can be improved in accuracy for special cases. For example, Halm and Stiel [39, 40] have modified the Pitzer et al. correlation for polar fluids at lower reduced temperatures.

Finally, for all correlations discussed, $T_c$ and $P_c$ are required either directly or indirectly. Although these constants are available for many fluids—and can be estimated for most others—there are occasions when one would prefer not to use critical properties. (For example, for some high-molecular-weight materials or for polyhydroxylated compounds, it is difficult to assign reliable values to $T_c$ and $P_c$.) In such cases, one may have to use an approximate rule for the entropy of vaporization at the normal boiling point and then scale with temperature with the Watson relation (described in Sec. 7-12) with an estimated $T_c$. Several entropy-of-vaporization rules are described in an earlier edition of this book [75]. One of the more useful of such rules was suggested by Kistiakowsky [51].

$$\frac{\Delta H_{vb}}{T_b} = \Delta S_{vb} = 36.6 + R \ln T_b \tag{7-13.1}$$

**TABLE 7-5 Vetere's Modification of the Kistiakowsky Equation**
$\Delta H_{vb}/T_b = \Delta S_{vb}$, J/(mol·K); $T_b$ in kelvins; $M$ = molecular weight

| Type of compound | Correlation |
|---|---|
| Alcohols, acids, methylamine | $\Delta S_{vb} = 81.119 + 13.083 \log T_b - 25.769 \frac{T_b}{M} + 0.146528 \frac{T_b^2}{M} - 2.1362 \times 10^{-4} \frac{T_b^3}{M}$ |
| Other polar compounds† | $\Delta S_{vb} = 44.367 + 15.33 \log T_b + 0.39137 \frac{T_b}{M} + 4.330 \times 10^{-3} \frac{T_b^2}{M} - 5.627 \times 10^{-6} \frac{T_b^3}{M}$ |
| Hydrocarbons | $\Delta S_{vb} = 58.20 + 13.7 \log M + 6.49 \frac{[T_b - (263M)^{0.581}]^{1.037}}{M}$ |

†For esters, multiply the calculated value of $\Delta S_{vb}$ by 1.03.

where $T_b$ is in kelvins and $\Delta S_{vb}$ is in J/(mol·K). Fishtine [30] improved the accuracy of Eq. (7-13.1) by employing a multiplicative correction factor that is a function of the compound class. More recently, Vetere [92] proposed a form wherein $\Delta S_{vb}$ is correlated as a function of $T_b$ and $M$; his relations are given in Table 7-5. $\Delta H_{vb}$ predicted from these equations is almost always within ±5 percent of the experimental value, and for most cases the error is below 3 percent.

**Example 7-9** Repeat Example 7-8 by using the Vetere equation for $\Delta S_{vb}$.

**solution** From Appendix A, $T_b$ = 321 K and $M$ = 58.08. Using the second equation in Table 7-5 gives

$$\Delta S_{vb} = 44.376 + 15.33 \log 321 + 0.39137 \frac{321}{58.08} + 4.33 \times 10^{-3} \frac{321^2}{58.08} - 5.627 \times 10^{-6} \frac{321^3}{58.08}$$

$$= 89.43 \text{ J/(mol·K)}$$

$$\Delta H_{vb} = (89.43)(321) = 28{,}710 \text{ J/mol}$$

$$\text{Error} = \frac{28{,}710 - 28{,}280}{28{,}280} \times 100 = 1.5\%$$

## 7-14 Enthalpy of Fusion

The enthalpy change on fusion or melting is commonly referred to as the latent heat of fusion. It depends in part on the crystal form of the solid phase, and attempts to obtain general correlations have been unsuccessful. The Clausius-Clapeyron equation is applicable, but its use to calculate $\Delta H_m$ requires data on the variation of melting point with pressure, which is seldom available. References for literature values are listed in [83].

Sutra [82] and Mukherjee [66] have indicated how the prediction of $\Delta H_m$ might be attained by using the hole theory of liquids, but their calculated values compare poorly with experimental values. Kuczinski [54] presents a theory relating $\Delta H_m$ and the modulus of rigidity of the solid.

Good agreement between theoretical and experimental values is shown for eight metallic elements. For monatomic substances, $\Delta S_m$ is about equal to the gas constant $R$ [41]. The best theoretical treatment has been presented by Bondi [14], who related the entropy of fusion of molecular crystals to molecular structure.

The difficulty in obtaining a general correlation of $\Delta H_m$ in terms of other physical properties is suggested by the selected values tabulated for some hydrocarbons in Table 7-6. It is evident that the simple introduction of a methyl group often has a marked effect; it either increases or decreases $\Delta H_m$. Results for optical and stereoisomers differ markedly. The variation of $\Delta H_m$ and the entropy of fusion $\Delta H_m/T_m$ is as great as that of the melting points, which have never been correlated with other properties. It appears that there is no simple correlation between $\Delta H_m$ and melting point. Table 7-6 shows, for comparison, the relative constancy of the latent heat of vaporization at the normal boiling point. The ratio of $\Delta H_m/\Delta H_{vb}$, however, varies greatly; in the table, values are found between 0.02 and 1.12.

## 7-15 Enthalpy of Sublimation; Vapor Pressures of Solids

Solids vaporize without melting (sublime) at temperatures below the triple-point temperature. Sublimation is accompanied by an enthalpy increase, or latent heat of sublimation. This may be considered to be the sum of a latent heat of fusion and a hypothetical latent heat of vaporization, even though liquid cannot exist at the pressure and temperature in question.

The latent heat of sublimation $\Delta H_s$ is best obtained from solid vapor pressure data. For this purpose, the Clausius-Clapeyron equation (7-2.1) is applicable.

In only a very few cases is the sublimation pressure at the melting point known with accuracy. It can be calculated from a liquid vapor pressure correlation, with input such as $T_b$ and $T_c$, by extrapolating to the melting point. Such a method is not recommended generally, because none of the vapor pressure correlations are accurate in the very low pressure range. Even if $P_{vp}$ at $T_m$ is known, at least one other value of the vapor pressure of the solid is necessary to calculate $\Delta H_s$ from the integrated form of the Clausius-Clapeyron equation. Thus, no useful generalized correlations appear to be available for enthalpies of sublimation for solids.

Another difficulty arises from transitions in the solid phase. Whereas the freezing point is a first-order transition between vapor and solid, the solid may have some liquid-like characteristics such as free rotation. (In a first-order phase transition, there is a discontinuity in volume, enthalpy, and entropy between the phases in equilibrium. Vaporization and melting are first-order transitions. Second-order transitions are relatively rare and are characterized by the fact that the volume, enthalpy, and entropy are continuous between the phases but there is a discontinuity in the heat capacities and compressibilities between the phases.) In many cases, at temperatures somewhat below the melting point, there is another first-order solid-solid transition between the crystalline-like solid and the more mobile liquid-like solid. The enthalpy of fusion and enthalpy of sublimation are different for these different solid phases. Ordinarily the literature values for $\Delta H_m$ and $\Delta H_s$ refer to the solid phase very near its melting point. Bondi [13] suggests that $\Delta H_m$ or $\Delta H_s$ correlations would be much better if the "most crystalline like" solid phase were considered; i.e., the correlations would be based on the lowest first-order transition temperature, where there is a pronounced semblance of order in the solid structure. Solid-phase transitions are also discussed by Preston et al. [72]. References to literature values of $\Delta H_s$ are listed in [83].

In some cases it is possible to obtain $\Delta H_s$ from thermochemical data by subtracting known values of the enthalpies of formation of solid and vapor. This is hardly a basis for estimation of an unkown $\Delta H_s$, however, since the enthalpies of formation tabulated in the standard references are often based in part on measured values of $\Delta H_s$. If the enthalpies of dissociation of both solid and gas phases are known, it is possible to formulate a cycle including the sublimation of the solid, the dissociation of the vapor, and the recombination of the elements to form the solid compound.

Bondi [13] has suggested an additive-group technique to estimate $\Delta H_s$ at the lowest first-order transition temperature for molecular crystals of organic substances and inorganic hydrides, perhalides, and percarbonyls. Usually the lowest first-order transition temperature is not much less than the melting point; e.g., for paraffins the two are almost identical. The exception appears to be $n$-butane; the usually quoted freezing point is 135 K, whereas the lowest first-order transition temperature is near 107 K. For some other molecules, however, there is a considerable difference. Cyclohexane melts at 279.6 K, but the lowest first-order transition temperature is 186 K. There appear to be no general rules for predicting any of the first-order transition temperatures. Bondi's method, however, is only approximate, and great care must be taken in deciding what contributions are necessary in many cases.

Finally, as a rough engineering rule, one might estimate $\Delta H_v$ and $\Delta H_m$ separately and obtain $\Delta H_s$ as the sum. The latent heat of fusion is usually less than one-quarter of the sum; therefore, the estimate may be fair even though that for $\Delta H_m$ is crude.

If enthalpy-of-fusion information is available, Eq. (7-2.1) may be used to estimate vapor pressures of solids in addition to liquid vapor pressures. The technique, along with its limitations, is illustrated with Example 7-10 [71].

**TABLE 7-6 Sample Tabulation of Some Enthalpies and Entropies of Fusion for Simple Hydrocarbons†**

| | $T_m$ | | | $\Delta H_m$ | | $\Delta S_m = \Delta H_m/T$, | $\Delta H_{vb}$, | $\frac{\Delta H_m}{\Delta H_{vb}}$ |
|---|---|---|---|---|---|---|---|---|
| | °C | K | $M$ | J/g | J/mol | J/(mol·K) | J/g | |
| Methane | −182.5 | 90.7 | 16.04 | 58.70 | 942 | 10.4 | 510 | 0.12 |
| Ethane | −183.3 | 89.9 | 30.07 | 95.10 | 2,860 | 31.8 | 490 | 0.19 |
| Propane | −187.7 | 85.5 | 44.09 | 79.91 | 3,523 | 41.2 | 427 | 0.19 |
| $n$-Butane | −138.4 | 134.8 | 58.12 | 80.21 | 4,662 | 34.6 | 385 | 0.21 |
| Isobutane | −159.6 | 113.6 | 58.12 | 78.12 | 4,541 | 40.0 | 367 | 0.21 |
| $n$-Pentane | −129.7 | 143.4 | 72.15 | 116.4 | 8,395 | 58.5 | 357 | 0.33 |
| Isopentane | −159.9 | 113.3 | 72.15 | 71.38 | 5,150 | 45.5 | 339 | 0.21 |
| Neopentane | −16.6 | 256.6 | 72.15 | 45.15 | 3,257 | 12.7 | 315 | 0.14 |
| $n$-Hexane | −95.4 | 177.8 | 86.18 | 151.2 | 13,030 | 73.3 | 335 | 0.45 |
| 2-Methylpentane | −153.7 | 119.5 | 86.18 | 72.84 | 6,278 | 52.5 | 323 | 0.23 |
| 2,2-Dimethylbutane | −99.9 | 173.3 | 86.18 | 6.74 | 581 | 3.3 | 305 | 0.02 |
| 2,3-Dimethylbutane | −128.5 | 144.6 | 86.18 | 9.41 | 811 | 5.6 | 317 | 0.03 |
| $n$-Heptane | −90.6 | 182.6 | 100.2 | 140.0 | 14,030 | 76.8 | 316 | 0.44 |
| 2-Methylhexane | −118.3 | 154.9 | 100.2 | 91.67 | 9,185 | 59.3 | 306 | 0.30 |
| 3-Ethylpentane | −118.6 | 154.6 | 100.2 | 95.31 | 9,550 | 61.8 | 309 | 0.31 |
| 2,2-Dimethylpentane | −123.8 | 149.4 | 100.2 | 58.12 | 5,823 | 39.0 | 291 | 0.20 |
| 2,4-Dimethylpentane | −119.2 | 154.0 | 100.2 | 68.28 | 6,842 | 44.4 | 295 | 0.23 |
| 3,3-Dimethylpentane | −134.5 | 138.7 | 100.2 | 70.54 | 7,068 | 51.0 | 296 | 0.24 |
| 2,2,3-Trimethylbutane | −24.9 | 248.3 | 100.2 | 22.55 | 2,260 | 9.1 | 289 | 0.08 |
| $n$-Octane | −56.8 | 216.4 | 114.2 | 181.6 | 20,740 | 95.8 | 301 | 0.60 |
| 3-Methylheptane | −120.5 | 152.7 | 114.2 | 99.62 | 11,380 | 74.5 | 286 | 0.33 |
| 4-Methylheptane | −120.9 | 152.2 | 114.2 | 94.89 | 10,840 | [illegible] | [illegible] | [illegible] |

| | | | | | | | | |
|---|---|---|---|---|---|---|---|---|
| *n*-Nonane | −53.5 | 219.7 | 128.3 | 120.6 | 15,460 | 70.4 | 286 | 0.42 |
| *n*-Decane | −29.6 | 243.5 | 142.3 | 201.8 | 28,720 | 118.0 | 276 | 0.73 |
| *n*-Dodecane | −9.6 | 263.6 | 170.3 | 216.3 | 36,830 | 139.7 | 256 | 0.84 |
| *n*-Octadecane | 28.2 | 301.3 | 254.5 | 241.2 | 61,390 | 203.7 | 215 | 1.12 |
| *n*-Nonadecane | 32 | 305 | 268.6 | 170.6 | 45,830 | 150.3 | 210 | 0.81 |
| *n*-Eicosane | 36.8 | 310 | 282.6 | 247.3 | 69,890 | 225.5 | 204 | 1.21 |
| Benzene | 5.5 | 278.7 | 78.1 | 125.9 | 9,833 | 35.3 | 394 | 0.32 |
| Toluene | −95 | 178 | 92.1 | 71.55 | 6,589 | 37.0 | 363 | 0.20 |
| Ethylbenzene | −94.9 | 178.2 | 106.2 | 86.32 | 9,167 | 51.4 | 339 | 0.25 |
| *o*-Xylene | −25.1 | 248.0 | 106.2 | 128.1 | 13,600 | 54.8 | 347 | 0.37 |
| *m*-Xylene | −47.9 | 225.3 | 106.2 | 109.0 | 11,570 | 51.4 | 343 | 0.32 |
| *p*-Xylene | 13.3 | 286.4 | 106.2 | 161 | 17,100 | 59.7 | 340 | 0.47 |
| *n*-Propylbenzene | −99.5 | 173.7 | 120.2 | 71.00 | 8,534 | 49.1 | 318 | 0.22 |
| Isopropylbenzene | −96.0 | 177.1 | 120.2 | 59.20 | 7,116 | 40.2 | 312 | 0.19 |
| 1,2,3-Trimethylbenzene | −25.4 | 247.7 | 120.2 | 69.5 | 8,350 | 33.7 | 333 | 0.21 |
| 1,2,4-Trimethylbenzene | −46 | 227 | 120.2 | 106.9 | 12,840 | 56.6 | 326 | 0.33 |
| 1,3,5-Trimethylbenzene | −44.7 | 228.4 | 120.2 | 80.08 | 9,626 | 42.1 | 325 | 0.25 |
| Cyclohexane | 6.5 | 279.6 | 84.2 | 31.7 | 2,670 | 9.5 | 357 | 0.09 |
| Methylcyclohexane | −126.5 | 146.6 | 98.2 | 68.6 | 6,740 | 46.0 | 323 | 0.21 |
| Ethylcyclohexane | −111.3 | 161.8 | 112.2 | 74.18 | 8,323 | 51.4 | 309 | 0.24 |
| 1,1-Dimethylcyclohexane | −33.5 | 239.7 | 112.2 | 5.52 | 620 | 2.6 | 294 | 0.02 |
| 1,*cis*-2-Dimethylcyclohexane | −50.0 | 223.1 | 112.2 | 14.6 | 1,640 | 7.4 | 305 | 0.05 |
| 1,*trans*-2-Dimethylcyclohexane | −88.1 | 185.0 | 112.2 | 93.47 | 10,490 | 56.7 | 297 | 0.31 |

†Data taken primarily from R. R. Dreisbach, "Physical Properties of Chemical Compounds," *Advan. Chem. Ser., ACS Monogr. 15* and *22*, 1955, 1959.

**Example 7-10** Use information at the triple point and Eq. (7-2.1) to estimate the vapor pressure of ice at 263 K.

**solution** Eq. (7-2.1) may be written for the solid in equilibrium with vapor

$$\frac{dP_{vp}^S}{dT} = \frac{\Delta H_{sub}}{T\,\Delta V^S} \tag{i}$$

and for hypothetical subcooled liquid in equilibrium with vapor

$$\frac{dP_{vp}^L}{dT} = \frac{\Delta H_v}{T\,\Delta V^L} \tag{ii}$$

In this example $\Delta V$ in both (i) and (ii) may be taken as

$$\Delta V = V^G - V^L \text{ (or } V^S) \simeq V^G = \frac{RT}{P}$$

Subtracting Eq. (i) from (ii), and using $\Delta H_v - \Delta H_{sub} = \Delta H_{fus}$, we get

$$\ln\left(\frac{P_{263}}{P_{273}}\right)^L - \ln\left(\frac{P_{263}}{P_{273}}\right)^S = \int_{273}^{263} \frac{\Delta H_{fus}}{RT^2}\,dT \tag{iii}$$

$\Delta H_{fus}$ is given as a function of temperature by

$$\Delta H_{fus} = (\Delta H_{fus} \text{ at } T_1) + \int_{T1}^{T} \Delta C_p\,dT \tag{iv}$$

For $H_2O$, $\Delta H_{fus}$ at 273 K is 6008 J/mol; for liquid, $C_p = 75.3$ J/(mol·K); and for ice, $C_p \simeq 37.7$ J/(mol·K).

$$\begin{aligned}\Delta H_{fus} &= 6008 + (37.7)(T - 273)\\ &= 37.7T - 4284\end{aligned} \tag{v}$$

Substitution into (iv) and integration gives

$$\ln\frac{P_{263}^S}{P_{273}^S} - \ln\frac{P_{263}^L}{P_{273}^L} = \frac{37.7}{8.314}\ln\frac{263}{273} + \frac{4284}{8.314}\left(\frac{1}{263} - \frac{1}{273}\right) \tag{vi}$$

$P_{273}^S = P_{273}^L$; $P_{263}^L$ is the vapor pressure of subcooled liquid at 263 K. An extrapolation based on the assumption that $\ln P_{vp}^L$ versus $1/T$ is linear gives $P_{263}^L = 0.00288$ bar. Solving Eq. (vi) gives $P_{263}^S = 0.00261$ bar, which is the same as the literature value.

One should be cautious of the technique used in Example 7-10. Discontinuities can occur in $\Delta H_{fus}$ (see Sec. 7-14). Also, the extrapolation of vapor pressures for hypothetical subcooled liquid over large temperature ranges is uncertain (see Sec. 7-6).

## Notation

| | |
|---|---|
| $a$ | parameter in Eqs. (7-4.2) and (7-4.3), constant in Eq. (7-5.2) |
| $A$ | constant in Eqs. (7-2.3), (7-3.1), (7-6.1), and (7-6.2) |
| $b$ | parameter in Eq. (7-4.2) and (7-4.4), constant in Eq. (7-5.2) |
| $B$ | constant in Eqs. (7-2.3) and (7-3.1) |
| $c$ | constant in Eq. (7-5.2) |

| | |
|---|---|
| $C$ | constant in Eq. (7-3.1) |
| $\Delta C_{p_v}$ | heat capacity of vapor minus heat capacity of liquid, J/(mol·K); see Eq. (7-6.3) |
| $d$ | constant in Eq. (7-5.2) |
| $f^{(0)}, f^{(1)}$ | functions of reduced temperature in Eq. (7-2.6) and defined in Eqs. (7-2.7) and (7-2.8) |
| $h$ | parameter defined in Eq. (7-2.5) |
| $\Delta H_m$ | enthalpy change on melting, J/mol |
| $\Delta H_v$ | enthalpy of vaporization, J/mol; $\Delta H_{vb}$ at $T_b$ |
| $\Delta H_s$ | enthalpy of sublimation, J/mol |
| $m$ | parameter in Eq. (7-4.1) |
| $M$ | molecular weight |
| $n$ | exponent in Eq. (7-12.1); usually chosen as 0.38 |
| $P$ | pressure in bars unless stated otherwise; $P_r = P/P_c$ |
| $P_c$ | critical pressure, bars |
| $P_{vp}$ | vapor pressure; $P_{vp_r} = P_{vp}/P_c$ |
| $R$ | gas constant, 8.314 J/(mol·K) |
| $\Delta S_v$ | entropy of vaporization, J/(mol·K); $\Delta S^{(0)}$ and $\Delta S^{(1)}$, Pitzer parameters in Eq. (7-9.2) |
| $T$ | temperature, K; $T_r = T/T_c$; $T_b$, normal boiling point; $T_m$, melting point; $T_c$, critical temperature, K |
| $\Delta U_v$ | internal energy of vaporization, J/mol |
| $V$ | volume, cm$^3$/mol; $V_g$, saturated vapor; $V_L$, saturated liqud |
| $V_c$ | critical volume, cm$^3$/mol |
| $\Delta V_v$ | volume change on vaporization, cm$^3$/mol |
| $X$ | temperature parameter defined in Eq. (7-12.4) |
| $Z$ | compressibility factor, $PV/RT$; $Z_g$ or $Z_{SV}$, saturated vapor; $Z_L$ or $Z_{SL}$, saturated liquid |
| $\Delta Z_v$ | $Z_g - Z_L$; $\Delta Z_{vb}$, at normal boiling point |
| $Z_c$ | $Z$ at the critical point |

GREEK

| | |
|---|---|
| $\beta$ | parameter in Eqs. (7-4.1) and (7-10.9) |
| $\gamma$ | parameter in Eq. (7-4.1) |
| $\Delta^*$ | quantum gas correction in Eq. (7-4.5) |
| $\tau$ | $1 - T_r$; see Eqs. (7-5.2), (7-9.6), (7-9.7) |
| $\psi$ | $\Delta H_v/(RT_c\,\Delta Z_v)$ |
| $\omega$ | acentric factor |

SUPERSCRIPTS

(0) simple fluid property

(1) deviation function

(R1) property for reference fluid 1

(R2) property for reference fluid 2

## REFERENCES


1. Abrams, D. S., H. A. Massaldi, and J. M. Prausnitz: *Ind. Eng. Chem. Fundam.*, **13:** 259 (1974).
2. Ambrose, D.: "Vapor-Pressure Equations," *Natl. Phys. Lab. Rep. Chem.*, **19:** November 1972.
3. Ambrose, D., B. E. Broderick, and R. Townsend: *J. Chem. Soc.*, **1967A:** 633.
4. Ambrose, D., J. F. Counsell, and C. P. Hicks: *J. Chem. Thermodyn.*, **10:** 771 (1978).
5. Ambrose, D.: *J. Chem. Thermodyn.*, **10:** 765 (1978).
6. Ambrose, D., J. F. Counsell, I. J. Lawrenson, and G. B. Lewis: *J. Chem. Thermodyn.*, **10:** 1033 (1978).
7. Ambrose, D., and R. H. Davies: *J. Chem. Thermodyn.*, **12:** 871 (1980).
8. Ambrose, D., and N. C. Patel: *J. Chem. Thermodyn.*, **16:** 459 (1984).
9. Ambrose, D., C. H. S. Sprake, and R. Townsend: *J. Chem. Thermodyn.*, **6:** 693 (1974).
10. Antoine, C.: *C.R.*, **107:** 681, 836 (1888).
11. Armstrong, B.: *J. Chem. Eng. Data*, **26:** 168 (1981).
12. Baehr, H. D., H. Garnjost, and R. Pollak: *J. Chem. Thermodyn.*, **8:** 113 (1976).
13. Bondi, A.: *J. Chem. Eng. Data*, **8:** 371 (1963).
14. Bondi, A.: *Chem. Rev.*, **67:** 565 (1967).
15. Boublik, T., V. Fried, and E. Hala: *The Vapor Pressures of Pure Substances*, 2d ed., Elsevier, New York, 1973.
16. Burkhard, L. P.: *Ind. Eng. Chem. Fundam.*, **24:** 119 (1985).
17. Calingaert, G., and D. S. Davis: *Ind. Eng. Chem.*, **17:** 1287 (1925).
18. Carruth, G. F., and R. Kobayashi: *Ind. Eng. Chem. Fundam.*, **11:** 509 (1972).
19. Carruth, G. F., and R. Kobayashi: *J. Chem. Eng. Data*, **18:** 115 (1973).

19*a*. Chandar, S. C. R., and R. P. Singh: *Fluid Phase Equil.*, **24:** 177 (1985).

20. Chase, J. D.: *Chem. Eng. Progr.*, **80:** 63 (April 1984).
21. Chen, N. H.: *J. Chem. Eng. Data*, **10:** 207 (1965).
22. Counsell, J. F., and D. A. Lee: *J. Chem. Thermodyn.*, **4:** 915 (1972).
23. Cox, E. R.: *Ind. Eng. Chem.*, **15:** 592 (1923).
24. Das, T. R., and P. T. Eubank: *Advan. Cryog. Eng.*, **18:** 208 (1973).
25. Dreisbach, R. R.: *Pressure-Volume-Temperature Relationships of Organic Compounds*, 3d ed., McGraw-Hill, New York, 1952.
26. Edwards, D. R., and J. M. Prausnitz: *Ind. Eng. Chem. Fundam.*, **20:** 280 (1981).
27. Edwards, D., C. G. Van de Rostyne, J. Winnick, and J. M. Prausnitz: *Ind. Eng. Chem. Process Design Develop.*, **20:** 138 (1981).
28. Fedors, R. F.: *Polymer Eng. Sci.*, **14:** 147 (1974).
29. Fish, L. W., and J. Lielmezs: *Ind. Eng. Chem. Fundam.*, **14:** 248 (1975).
30. Fishtine, S. H.: *Ind. Eng. Chem.*, **55**(4): 20, **55**(5): 49, **55**(6): 47 (1963); *Hydrocarbon Process. Petrol. Refiner*, **42**(10): 143 (1963).
31. Fishtine, S. H.: *Hycrocarbon Process. Petrol. Refiner*, **45**(4): 173 (1966).
32. Giacalone, A.: *Gazz. Chim. Ital.*, **81:** 180 (1951).
33. Guthrie, J. P., and K. F. Taylor: *Can. J. Chem.*, **61:** 602 (1983).
34. Gomez-Nieto, M., and Thodos, G.: *Ind. Eng. Chem. Fundam.*, **17:** 45 (1978).
35. Gomez-Nieto, M., and Thodos, G.: *Can. J. Chem. Eng.*, **55:** 445 (1977).
36. Gomez-Nieto, M., and Thodos, G.: *Ind. Eng. Chem. Fundam.*, **16:** 254 (1977).

37. Gupte, P. A., and T. E. Daubert: *Ind. Eng. Chem. Process Design Develop.*, **24:** 674 (1985).
38. Gupte, P. A., and T. E. Daubert: *Ing. Eng. Chem. Process Design Develop.*, **24:** 761 (1985).
39. Halm, R. L.: Ph.D. thesis, Syracuse University, Syracuse, N.Y., 1968.
40. Halm, R. L., and L. I. Stiel: *AIChE J.*, **13:** 351 (1967).
41. Hirschfelder, J. O., C. F. Curtiss, and R. B. Bird: *Molecular Theory of Gases and Liquids*, Wiley, New York, 1954.
42. Hoshino, D., K. Nagahama, and M. Hirata: *Ind. Eng. Chem. Fundam.*, **22:** 430 (1983).
43. Hoshino, D., X.-R. Zhu, K. Nagahama, and M. Hirata: *Ind. Eng. Chem. Fundam.*, **24:** 112 (1985).
44. Ibrahim, S. H., and N. R. Kuloor: *Chem. Eng.*, **73**(25): 147 (1966).
45. Jensen, T., A. Fredenslund, and P. Rasmussen: *Ind. Eng. Chem. Fundam.*, **20:** 239 (1981).
46. Karapet'yants, M. Kh., and K. Ch'ung: *Ssu Ch'uan Ta Hseuh Hseuh Pao—Tzu Jan K'o Hseuh*, **1958:** 91; *Chem. Abstr.*, **53:** 17613 (1959).
47. Katinas, T. G., and R. P. Danner: *Hydrocarbon Process. Petrol. Refiner*, **56**(3): 157 (1977).
48. Kier, L. B., and L. H. Hall: *Molecular Connectivity in Chemistry and Drug Research*, Academic, New York, 1976.
49. King, M. B.: *Trans. Instr. Chem. Engrs.*, **54:** 54 (1976).
50. King, M. G., and H. Al-Najjar: *Chem. Eng. Sci.*, **29:** 1003 (1974).
51. Kistiakowsky, W.: *Z. Phys. Chem.*, **107:** 65 (1923).
52. Klein, V. A.: *Chem. Eng. Progr.*, **45:** 675 (1949).
53. Kratzke, H.: *J. Chem. Thermodyn.*, **12:** 305 (1980).
54. Kuczinski, G. C.: *J. Appl. Phys.*, **24:** 1250 (1953).
55. Lawson, D. D.: *Appl. Energy*, **6:** 241 (1980).
56. Lee, B. I., and M. G. Kesler: *AIChE J.*, **21:** 510 (1975).
57. Lielmezs, J., K. G. Astley, and J. A. McEvoy: *Thermochimica Acta*, **52:** 9 (1982).
58. Lu, B. C.-Y.: *Can. J. Chem. Eng.*, **38:** 33 (1960).
59. Macknick, A. B., and J. M. Prausnitz: *Ind. Eng. Chem. Fundam.*, **18:** 348 (1979).
60. Macknick, A. B., J. Winnick, and J. M. Prausnitz: *AIChE J.*, **24:** 731 (1978).
61. McCurdy, K. G., and K. J. Laidler: *Can. J. Chem.*, **41:** 1867 (1963).
62. McGarry, J: *Ind. Eng. Chem. Process Design Develop.*, **22:** 313 (1983).
63. Miller, D. G.: *Ind. Eng. Chem. Fundam.*, **2:** 68 (1963).
64. Miller, D. G.: personal communication, April 1964.
65. Mosselman, C., W. H. van Vugt, and H. Vos.: *J. Chem. Eng. Data*, **27:** 246 (1982).
66. Mukerjee, N. R.: *J. Chem. Phys.*, **19:** 502, 1431 (1951).
67. Narsimhan, G.: *Brit. Chem. Eng.*, **10**(4): 253 (1965), **12**(6): 897 (1967).
68. Nath, J.: *Ind. Eng. Chem. Fundam.*, **18:** 297 (1979).
69. Ogden, J. M., and J. Lielmezs: *AIChE J.*, **15:** 469 (1969).
70. Pitzer, K. S., D. Z. Lippmann, R. F. Curl, C. M. Huggins, and D. E. Petersen: *J. Am. Chem. Soc.*, **77:** 3433 (1955).
71. Prausnitz, J. M.: *Molecular Thermodynamics of Fluid-Phase Equilibria*, Prentice Hall, Englewood Cliffs, N.J., 1969, pp. 387–390.
72. Preston, G. T., E. W. Funk, and J. M. Prausnitz: *J. Phys. Chem.*, **75:** 2345 (1971).
73. Procopio, J. M., and G. J. Su: *Chem. Eng.*, **75**(12): 101 (1968).
74. Rehberg, C. E.: *Ind. Eng. Chem.*, **42:** 829 (1950).
75. Reid, R. C., and T. K. Sherwood: *The Properties of Gases and Liquids*, 2d ed., McGraw-Hill, New York, 1966, pp. 149–153.
76. Reynes, E. G., and G. Thodos: *AIChE J.*, **8:** 357 (1962).
77. Riedel, L.: *Chem. Ing. Tech.*, **26:** 679 (1954).
78. Ruzicka, V.: *Ind. Eng. Chem. Fundam.*, **22:** 266 (1983).
79. Santrach, D., and J. Lielmezs: *Ind. Eng. Chem. Fundam.*, **17:** 93 (1978).
80. Sivaraman, A., J. W. Magee, and R. Kobayashi: *Ind. Eng. Chem. Fundam.*, **23:** 97 (1984).

81. Smith, G., J. Winnick, D. S. Abrams, and J. M. Prausnitz: *Can. J. Chem. Eng.*, **54:** 337 (1976).
82. Sutra, G.: *C. R.*, **233:** 1027, 1186 (1951).
83. Tamir, A., E. Tamir, and K. Stephan: *Heats of Phase Change of Pure Components and Mixtures*, Elsevier, Amsterdam, 1983.
84. Teja, A. S., S. I. Sandler, and N. C. Patel: *Chem. Eng. J. (Lausanne)*, **21:** 21 (1981).
85. Thek, R. E., and L. I. Stiel: *AIChE J.*, **12:** 599 (1966), **13:** 626 (1967).
86. Thompson, P. A., and D. A. Sullivan: *Ind. Eng. Chem. Fundam.*, **18:** 1 (1979).
87. Thompson, W. H., and W. G. Braun; *29th Midyear Mtg., Am. Petrol. Inst., Div. Refining, St. Louis, Mo., May 11, 1964*, prepr. 06-64.
88. Thomson, G. W.: *Chem. Rev.*, **38:** 1 (1946).
89. Thomson, G. W.: in A. Weissberger (ed.), *Techniques of Organic Chemistry*, 3d. ed., vol. I, pt. I, Interscience, New York, 1959, p. 473.
90. Torquato, S., and G. R. Stell: *Ind. Eng. Chem. Fundam.*, **21:** 202 (1982).
91. Vetere, A.: *New Generalized Correlations for Enthalpy of Vaporization of Pure Compounds*, Laboratori Ricerche Chimica Industriale, SNAM PROGETTI, San Donato Milanese, 1973.
92. Vetere, A.: *Modification of the Kistiakowsky Equation for the Calculation of the Enthalpies of Vaporization of Pure Compounds*, Laboratori Ricerche Chimica Industriale, SNAM PROGETTI, San Donato Milanese, 1973.
92*a*. Vetere, A.: *Chem. Eng. J.*, **32:** 77 (1986).
93. Viswanath, D. S., and N. R. Kuloor: *J. Chem. Eng. Data*, **11:** 69, 544 (1966).
94. Viswanath, D. S., and N. R. Kuloor: *Can. J. Chem. Eng.*, **45:** 29 (1967).
95. Wagner, W., J. Evers, and W. Pentermann: *J. Chem. Thermodyn.*, **8:** 1049 (1976).
96. Wagner, W.: *Cryogenics*, **13:** 470 (1973).
97. Wagner, W.: *Bull. Inst. Froid. Annexe*, no. 4, 1973, p. 65.
98. Wagner, W.: *A New Correlation Method for Thermodynamic Data Applied to the Vapor-Pressure Curve of Argon, Nitrogen, and Water*, J. T. R. Watson (trans. and ed.), IUPAC Thermodynamic Tables Project Centre, London, 1977.
99. Waring, W.: *Ind. Eng. Chem.*, **46:** 762 (1954).
100. White, C. M.: personal communication, 1985.
101. Wilhoit, R. C., and B. J. Zwolinski: *Handbook of Vapor Pressures and Heats of Vaporization of Hydrocarbons and Related Compounds*, Texas A&M University, Thermodynamics Research Center, College Station, 1971.
102. Willingham, C. B., W. J. Taylor, J. M. Pignorco, and F. D. Rossini: *J. Res. Natl. Bur. Stand.*, **35:** 219 (1945).
102*a*. Willman, B., and A. S. Teja: *Ind. Eng. Chem. Process Design Develop.*, **24:** 1033 (1985).
103. Wilsak, R. A., and G. Thodos: *Ind. Eng. Chem. Fundam.*, **23:** 75 (1984).
104. Yair, O. B., and A. Fredenslund: *Ind. Eng. Chem. Process Design Develop.*, **22:** 433 (1983).
105. Yarbrough, D. W., and C.-H. Tsai: *Advan. Cryog. Eng.:* **23:** 602 (1978).
106. Zhong, Xu: *Ind. Eng. Chem. Process Design Develop.*, **23:** 7 (1984).
107. Zhong, Xu: *Chem. Eng. Commun.*, **29:** 257 (1984).

Chapter

# 8

# Fluid Phase Equilibria in Multicomponent Systems

## 8-1 Scope

In the chemical process industries, fluid mixtures are often separated into their components by diffusional operations such as distillation, absorption, and extraction. Design of such separation operations requires quantitative estimates of the partial equilibrium properties of fluid mixtures. Whenever possible, such estimates should be based on reliable experimental data for the particular mixture at conditions of temperature, pressure, and composition corresponding to those of interest. Unfortunately, such data are often not available. In typical cases, only fragmentary data are at hand and it is necessary to reduce and correlate the limited data to make the best possible interpolations and extrapolations. This chapter discusses some techniques that are useful toward that end. Although primary attention is given to nonelectrolytes, a few paragraphs are devoted to aqueous solutions of electrolytes. Emphasis is given to the calculation of fugacities in liquid solutions; fugacities in gaseous mixtures are discussed in Sec. 5-8.

The scientific literature on fluid phase equilibria goes back well over 100 years and has reached monumental proportions, including thousands of articles and hundreds of books and monographs. Table 8-1*a* and 8-1*b* give

**TABLE 8-1*a* Some Useful Books on Fluid-Phase Equilibria**

| Book | Remarks |
|---|---|
| Chao, K. C., and R. A. Greenkorn: *Thermodynamics of Fluids,* Dekker, New York, 1975. | Introductory survey including an introduction to statistical thermodynamics of fluids; also gives a summary of surface thermodynamics. |
| Francis, A. W.: *Liquid-Liquid Equilibrium,* Wiley-Interscience, New York, 1963 | Phenomenological discussion of liquid-liquid equilibria with extensive data bibliography. |
| Hala, E., et al.: *Vapour-Liquid Equilibrium,* 2d English ed., trans. by George Standart, Pergamon, Oxford, 1967. | Comprehensive survey, including a discussion of experimental methods. |
| Hildebrand, J. H., and R. L. Scott: *Solubility of Nonelectrolytes,* 3d ed., Reinhold, New York, 1950. (Reprinted by Dover, New York, 1964.) | A classic in its field, it gives a survey of solution chemistry from a chemist's point of view. Although out of date, it nevertheless provides physical insight into how molecules "behave" in mixtures. |
| Hildebrand, J. H., and R. L. Scott: *Regular Solutions,* Prentice Hall, Englewood Cliffs, N.J., 1962. | Updates some of the material in Hildebrand's 1950 book. |
| Hildebrand, J. H., J. M. Prausnitz, and R. L. Scott: *Regular and Related Solutions,* Van Nostrand Reinhold, New York, 1970. | Further updates some of the material in Hildebrand's earlier books. |
| King, M. B.: *Phase Equilibrium in Mixtures,* Pergamon, Oxford, 1969. | A general text covering a variety of subjects in mixture thermodynamics. |
| Modell, M., and R. C. Reid, *Thermodynamics and Its Applications:* 2d ed., Prentice Hall, Englewood Cliffs, N.J., 1983. | This semiadvanced text emphasizes the solution of practical problems through application of fundamental concepts of chemical engineering thermodynamics and discusses surface thermodynamics and systems in potential fields. |
| Murrell, J. M., and E. A. Boucher: *Properties of Liquids and Solutions,* Wiley, New York, 1982. | A short introduction to the physics and chemistry of the liquid state. |
| Null, H. R.: *Phase Equilibrium in Process Design,* Wiley-Interscience, New York, 1970. | An engineering-oriented monograph with a variety of numerical examples. |
| Prausnitz, J. M., R. N. Lichtenthaler, and E. G. Azevedo: *Molecular Thermodynamics of Fluid-Phase Equilibria,* 2d ed., Prentice Hall, Englewood Cliffs, N.J., 1986. | A text which attempts to use molecular-thermodynamic concepts useful for engineering. Written from a chemical engineering point of view. |
| Prausnitz, J. M., T. F. Anderson, E. A. Grens, C. A. Eckert, R. Hsieh, and J. P. O'Connell: *Computer Calculations for Vapor-Liquid and Liquid Equilibria,* Prentice Hall, Englewood Cliffs, N.J., 1980. | A monograph with detailed computer programs and a (limited) data bank. |
| Prigogine, I., and R. Defay, *Chemical Thermodynamics,* trans. and rev. by D. H. Everett, Longmans, Green, London, 1954. | A semiadvanced text from a European chemist's point of view. It offers many examples and discusses molecular |

| Book | Remarks |
|---|---|
| | principles. Although out of date, it contains much useful information not easily available in standard American texts. |
| Renon, H., L. Asselineu, G. Cohen, and C. Raimbault, *Calcul sur Ordinateur des Equilibres Liquide-Vapeur et Liquid-Liquid,* Editions Technip, Paris, 1971. | Discusses (in French) the thermodynamic basis for computer calculations for vapor-liquid and liquid-liquid equilibria. Computer programs are given. |
| Rowlinson, J. S., and F. L. Swinton: *Liquids and Liquid Mixtures,* 3d ed., Butterworth, London, 1982. | Presents a thorough treatment of the physics of fluids and gives some statistical mechanical theories of the equilibrium properties of simple pure liquids and liquid mixtures; contains data bibliography. Primarily for research-oriented readers. |
| Van Ness, H. C., and M. M. Abbott: *Classical Thermodynamics of Nonelectrolyte Solutions,* McGraw-Hill, New York, 1982. | Systematic, comprehensive, and clear exposition of the principles of classical thermodynamics applied to solutions of nonelectrolytes. Discusses phase equilibria in fluid systems with numerous examples. |

**TABLE 8-1*b* Some Useful Books on Fluid Phase Equilibria Data Sources**

| Book | Remarks |
|---|---|
| API Research Project 42: *Properties of Hydrocarbons of High Molecular Weight,* American Petroleum Institute, New York, 1966. | A compilation of physical properties (vapor pressure, liquid density, transport properties) for 321 hydrocarbons with carbon number 11 or more. |
| API Research Project 44: *Handbook of Vapor Pressures and Heats of Vaporization of Hydrocarbons and Related Compounds,* Thermodynamics Research Center, College Station, Tex., 1971. | A thorough compilation of the vapor pressures and enthalpies of vaporization of alkanes (up to $C_{100}$), aromatics, and naphthenes (including some with heteroatoms). Other API-44 publications include data on a variety of thermodynamic properties of hydrocarbons and related compounds. |
| Barton, A. F. M.: *Handbook of Solubility Parameters and other Cohesion Parameters,* CRC Press, Boca Raton, Fla., 1983. | Contains an extensive compilation of data for solubility parameters, cohesive energies, and molar volumes for a variety of substances including polymers. Some correlations (group contributions) are also presented. |
| Behrens, D., and R. Eckermann: *Chemistry Data Series,* DECHEMA, Frankfurt a.M., Vol. I, *VLE Data Collection,* by J. Gmehling, U. Onken, W. Arlt, P. Grenzheuser, U. Weidlich, and B. Kolbe, | An extensive compilation (in six volumes, some of them consisting of several parts) of thermodynamic property data for pure compounds and mixtures, *PVT* data, heat capacity, enthalpy, and entropy data, |

**TABLE 8-1*b* Some Useful Books on Fluid Phase Equilibria Data Sources (*Continued*)**

| Book | Remarks |
| --- | --- |
| 1977–1984; Vol. II, *Critical Data,* by K. H. Simmrock, 1984; Vol. III, *Heats of Mixing Data Collection,* by C. Christensen, J. Gmehling, P. Rasmussen, and U. Weidlich, 1984; Vol. IV, *Vapor Pressures,* by M. Schönberg and S. Müller, in prep.; Vol. V, *LLE-Data Collection,* by J. M. Sorensen and W. Arlt, 1979–1980; Vol. VI, *VLE for Mixtures of Low Boiling Substances,* by H. Knapp, R. Döring, L. Oellrich, U. Plöcker, and J. M. Prausnitz, 1982. | phase equilibrium data, and transport and interfacial tension data for a variety of inorganic and organic compounds including aqueous mixtures. |
| Boublik, T., V. Fried, and E. Hala: *The Vapour Pressures of Pure Substances,* 2d ed., Elsevier, Amsterdam, 1984. | Experimental and smoothed data are given for the vapor pressures of pure substances in the normal and low-pressure region; Antoine constants are reported. |
| Brandrup, J., and E. H. Immergut (eds.): *Polymer Handbook,* 2d ed., Wiley-Interscience, New York, 1975. | A thorough compilation of polymerization reactions and of solution and physical properties of polymers and their associated oligomers and monomers. |
| Broul, M., J. Nyult, and O. Söhnel: *Solubility in Inorganic Two-Component Systems,* Elsevier, Amsterdam, 1981. | An extensive compilation of data on solubility of inorganic compounds in water. |
| Christensen, J. J., L. D. Hansen, and R. M. Izatt: *Handbook of Heats of Mixing,* Wiley, New York, 1982. | Experimental heat-of-mixing data for a variety of binary mixtures. |
| Danner, R. P., and T. E. Daubert (eds.): *Technical Data Book—Petroleum Refining,* American Petroleum Institute, Washington, D.C., 1983. | A two-volume data compilation of the physical, transport, and thermodynamic properties of petroleum fractions and related model compound mixtures of interest to the petroleum-refining industry. |
| Dymond, J. H., and E. B. Smith: *The Virial Coefficients of Pure Gases and Mixtures,* Oxford, Oxford, 1980. | A critical compilation of data for virial coefficients of pure gases and binary mixtures published to 1979. |
| Hicks, C. P., K. N. Marsh, A. G. Williamson, I. A. McLure, and C. L. Young: *Bibliography of Thermodynamic Studies,* Chemical Society, London, 1975. | Literature references for vapor-liquid equilibria, enthalpies of mixing, and volume changes of mixing of selected binary systems. |
| Hirata, M., S. Ohe, and K. Nagahama: *Computer-Aided Data Book of Vapor-Liquid Equilibria,* Elsevier, Amsterdam, 1975. | A compilation of binary experimental data reduced with the Wilson equation and, for high pressures, with a modified Redlich-Kwong equation. |
| Hiza, M. J., A. J. Kidnay, and R. C. Miller: *Equilibrium Properties of Fluid Mixtures,* 2 vols., IFL/Plenum, New York, 1975, 1982. | Volume 1 contains references for experimental phase equilibria and thermodynamic properties of fluid mixtures of cryogenic interest. Volume 2 updates to January 1980 the references given in Vol. 1. Includes mixtures |

| Book | Remarks |
|---|---|
| | containing pentane and some aqueous mixtures. |
| IUPAC: *Solubility Data Series,* Pergamon, Oxford, 1974. | A multivolume compilation of the solubilities of inorganic gases in pure liquids, liquid mixtures, aqueous solutions, and miscellaneous fluids and fluid mixtures. |
| Kehiaian, H. V., (ed.-in-chief), and B. J. Kwolinski (exec. officer), *International Data Series: Selected Data on Mixtures,* Thermodynamics Research Center, Chemistry Department, Texas A&M University, College Station, TX 77843, continuing since 1973. | Presents a variety of measured thermodynamic properties of binary mixtures. These properties are often represented by empirical equations. |
| Van Krevelen, D. W.: *Properties of Polymers,* 2d ed., Elsevier, Amsterdam, 1976. | Presents methods to correlate and predict thermodynamic, transport, and chemical properties of polymers as a function of chemical structure. |
| Maczynski, A.: *Verified Vapor-Liquid Equilibrium Data,* Polish Scientific Publishers, Warszawa, 1976. | A four-volume compilation of binary vapor-liquid equilibrium data for mixtures of hydrocarbons with a variety of organic compounds; includes many data from the East European literature. |
| Ohe, S.: *Computer-Aided Data Book of Vapor Pressure,* Data Book Publishing Company, Tokyo, 1976. | Literature references for vapor pressure data for about 2000 substances are given. The data are presented in graphical form, and Antoine constants also are given. |
| Seidell, A.: *Solubilities of Inorganic and Organic Compounds,* 3d ed., Van Nostrand, New York, 1940 (1941, 1952). | A two-volume (plus supplement) data compilation. The first volume concerns the solubilities of inorganic and metal organic compounds in single compounds and in mixtures; the second concerns organics; and the supplement updates the solubility references to 1949. |
| Silcock, H., (ed.): *Solubilities of Inorganic and Organic Compounds,* trans. from Russian, Pergamon, Oxford, 1979. | A systematic compilation of data to 1965 on solubilities of ternary and multicomponent systems of inorganic compounds. |
| Stephen, H., and T. Stephen (eds): *Solubilities of Inorganic and Organic Compounds.* trans. from Russian, Pergamon, Oxford, 1963. | A five-volume data compilation of solubilities for inorganic, metal-organic and organic compounds in binary, ternary and multicomponent systems. |
| Tamir, A., E. Tamir, and K. Stephan: *Heats of Phase Change of Pure Components and Mixtures,* Elsevier, Amsterdam, 1983. | An extensive compilation of data published to 1981 on the enthalpy of phase change for pure compounds and mixtures. |

**TABLE 8-1*b* Some Useful Books on Fluid Phase Equilibria Data Sources (*Continued*)**

| Book | Remarks |
|---|---|
| Timmermans, J.: *Physico-chemical Constants of Pure Organic Compounds,* Elsevier, Amsterdam, 1950. | A compilation of data on vapor pressure, density, melting and boiling point, heat-capacity constants, and transport properties of organic compounds (2 vols.). |
| *Constants of Binary Systems,* Interscience, New York, 1959. | A four-volume compilation of vapor-liquid and liquid-liquid equilibria, densities of the coexisting phases, transport properties, and enthalpy data for binary concentrated solutions. |
| Vergaftik, N. B.: *Handbook of Physical Properties of Liquids and Gases: Pure Substances and Mixtures,* 2d ed., Hemisphere, Washington, 1981. | A compilation of thermal, caloric, and transport properties of pure compounds (including organic compounds, $SO_2$, and halogens) and mainly transport properties of binary gas mixtures, liquid fuels, and oils. |
| Wichterle, I., J. Linek, and E. Hala: *Vapor-Liquid Equilibrium Data Bibliography,* Elsevier, Amsterdam, 1973 (plus 3 supplements: 1976, 1979, 1982). | A thorough compilation of literature sources for binary and multicomponent data; includes many references to the East European literature. |
| Wisniak, J.: *Phase Diagrams,* Elsevier, Amsterdam, 1981. | A literature-source book for published data to 1980 on phase diagrams for a variety of inorganic and organic compounds (2 vols.). |
| Wisniak, J., and A. Tamir: *Liquid-Liquid Equilibrium and Extraction,* Elsevier, Amsterdam, 1980. | A two-volume literature source book for the equilibrium distribution between two immiscible liquids for data published to 1980. |
| Wisniak, J., and A. Tamir: *Mixing and Excess Thermodynamic Properties,* Elsevier, Amsterdam, 1978 (supplement 1982). | An extensive bibliographic compilation of data references on mixing and excess properties published between 1900 and 1982. |

the authors and titles of some books which are useful for obtaining data and for more detailed discussions. The lists are not exhaustive; they are restricted to publications which are likely to be useful to the practicing engineer in the chemical process industries.

There is an important difference between calculating phase equilibrium compositions and calculating typical volumetric, energetic, or transport properties of fluids of known composition. In the latter case we are interested in the property of the mixture as a whole, whereas in the former we are interested in the *partial* properties of the individual components which constitute the mixture. For example, to find the pressure drop of a liquid mixture flowing through a pipe, we need the viscosity and the density of that liquid mixture at the particular composition of interest. But if we ask for the composition of the vapor which is in equilibrium with the liquid mixture, it is no longer sufficient to know the properties of the liquid mixture at that particular composition; we must now know, in addition, how certain of its properties (in particular the Gibbs energy) *depend on composition.* In phase equilibrium calculations, we must know *partial* properties, and to find them, we typically differentiate data with respect to composition. Whenever experimental data are differentiated, there is a loss of accuracy, often a serious loss. Since partial, rather than total, properties are needed in phase equilibria, it is not surprising that phase equilibrium calculations are often more difficult and less accurate than those for other properties encountered in chemical process design.

In one chapter it is not possible to present a complete review of a large subject. Also, since this subject is so wide in its range, it is not possible to recommend to the reader simple, unambiguous rules for obtaining quantitative answers to a particular phase equilibrium problem. Since the variety of mixtures is extensive, and since mixture conditions (temperature, pressure, and composition) cover many possibilities, and, finally, since there are large variations in the availability, quantity, and quality of experimental data, the reader cannot escape responsibility for using judgment, which, ultimately, is obtained only by experience.

This chapter, therefore, is qualitatively different from the others in this book. It does not give specific advice on how to calculate specific quantities. It provides only an introduction to some (by no means all) of the tools and techniques which may be useful for an efficient strategy toward calculating particular phase equilibria for a particular process design.

## 8-2 Thermodynamics of Vapor-Liquid Equilibria

We are concerned with a liquid mixture which, at temperature $T$ and pressure $P$, is in equilibrium with a vapor mixture at the same temperature and pressure. The quantities of interest are the temperature, the pressure, and the compositions of both phases. Given some of these quantities, our task is to calculate the others.

For every component $i$ in the mixture, the condition of thermodynamic equilibrium is given by

$$f_i^V = f_i^L \qquad (8\text{-}2.1)$$

where $f$ = fugacity
$V$ = vapor
$L$ = liquid

The fundamental problem is to relate these fugacities to mixture composition. In the subsequent discussion, we neglect effects due to surface forces, gravitation, electric or magnetic fields, semipermeable membranes, or any other special conditions.

The fugacity of a component in a mixture depends on the temperature, pressure, and composition of that mixture. In principle, any measure of composition can be used. For the vapor phase, the composition is nearly always expressed by the mole fraction $y$. To relate $f_i^V$ to temperature, pressure, and mole fraction, it is useful to introduce the fugacity coefficient $\phi_i$

$$\phi_i = \frac{f_i^V}{y_i P} \tag{8-2.2}$$

which can be calculated from vapor phase $PVTy$ data, usually given by an equation of state as discussed in Sec. 5-8. For a mixture of ideal gases $\phi_i = 1$.

The fugacity coefficient $\phi_i$ depends on temperature and pressure and, in a multicomponent mixture, on *all* mole fractions in the vapor phase, not just $y_i$. The fugacity coefficient is, by definition, normalized such that as $P \to 0$, $\phi_i \to 1$ for all $i$. At low pressures, therefore, it is usually a good assumption to set $\phi_i = 1$. But just what "low" means depends on the composition and temperature of the mixture. For typical mixtures of nonpolar (or slightly polar) fluids at a temperature near or above the normal boiling point of the least volatile component, "low" pressure means a pressure less than a few bars. However, for mixtures containing a strongly associating carboxylic acid, e.g., acetic acid–water at 25°C, fugacity coefficients may differ appreciably from unity at pressures much less than 1 bar.† For mixtures containing one component of very low volatility and another of high volatility, e.g., decane-methane at 25°C, the fugacity coefficient of the light component may be close to unity for pressures up to 10 or 20 bar while at the same pressure the fugacity coefficient of the heavy component is typically much less than unity. A detailed discussion is given in chap. 5 of Ref. 97.

The fugacity of component $i$ in the liquid phase is related to the composition of that phase through the activity coefficient $\gamma_i$. In principle, any composition scale may be used; the choice is strictly a matter of convenience. For some aqueous solutions, frequently used scales are molality (moles of solute per 1000 g of water) and molarity (moles of solute per liter of solution); for polymer solutions, a useful scale is the volume fraction, discussed briefly in Sec. 8-14. However, for typical solutions containing nonelectrolytes of normal molecular weight (including water), the

---

†For moderate pressures, fugacity coefficients can often be estimated with good accuracy as discussed in Ref. 48.

most useful measure of concentration is the mole fraction $x$. Activity coefficient $\gamma_i$ is related to $x_i$ and to standard-state fugacity $f_i^\circ$ by

$$\gamma_i \equiv \frac{a_i}{x_i} = \frac{f_i^L}{x_i f_i^\circ} \tag{8-2.3}$$

where $a_i$ is the activity of component $i$. The standard-state fugacity $f_i^\circ$ is the fugacity of component $i$ at the temperature of the system, i.e., the mixture, and at some arbitrarily chosen pressure and composition. The choice of standard-state pressure and composition is dictated only by convenience, but it is important to bear in mind that the numerical values of $\gamma_i$ and $a_i$ have no meaning unless $f_i^\circ$ is clearly specified.

While there are some important exceptions, activity coefficients for most typical solutions of nonelectrolytes are based on a standard state where, for every component $i$, $f_i^\circ$ is the fugacity of *pure* liquid $i$ at system temperature and pressure; i.e., the arbitrarily chosen pressure is the total pressure $P$, and the arbitrarily chosen composition is $x_i = 1$. Frequently, this standard-state fugacity refers to a hypothetical state, since it may happen that component $i$ cannot physically exist as a pure liquid at system temperature and pressure. Fortunately, for many common mixtures it is possible to calculate this standard-state fugacity by modest extrapolations with respect to pressure; and since liquid phase properties remote from the critical region are not sensitive to pressure (except at high pressures), such extrapolation introduces little uncertainty. In some mixtures, however, namely, those which contain supercritical components, extrapolations with respect to temperature are required, and these, when carried out over an appreciable temperature region, may lead to large uncertainties. We briefly return to this problem in Sec. 8-12.

Whenever the standard-state fugacity is that of the pure liquid at system temperature and pressure, we obtain the limiting relation that $\gamma_i \to 1$ as $x_i \to 1$.

## 8-3 Fugacity of a Pure Liquid

To calculate the fugacity of a pure liquid at a specified temperature and pressure, we require two primary thermodynamic properties: the saturation (vapor) pressure, which depends only on temperature, and the liquid density, which depends primarily on temperature and to a lesser extent on pressure. Unless the pressure is very large, it is the vapor pressure which is by far the more important of these two quantities. In addition, we require volumetric data (equation of state) for pure vapor $i$ at system temperature, but unless the vapor pressure is high or unless there is strong dimerization in the vapor phase, this requirement is of minor, often negligible, importance.

The fugacity of pure liquid $i$ at temperature $T$ and pressure $P$ is given by

$$f_i^L(T, P, x_i = 1) = P_{\mathrm{vp}i}(T)\phi_i^s(T) \exp \int_{P_{\mathrm{vp}i}}^{P} \frac{V_i^L(T, P)}{RT}\, dP \tag{8-3.1}$$

where $P_{\mathrm{vp}}$ is the vapor pressure (see Chap. 7) and superscript $s$ stands for saturation. The fugacity coefficient $\phi_i^s$ is calculated from vapor phase volumetric data, as discussed in Sec. 5-8; for typical nonassociated fluids at temperatures well below the critical, $\phi_i^s$ is close to unity.

The molar liquid volume $V_i^L$ is the ratio of the molecular weight to the density, where the latter is expressed in units of mass per unit volume.† At a temperature well below the critical, a liquid is nearly incompressible. In that case the effect of pressure on liquid phase fugacity is not large unless the pressure is very high or the temperature is very low. The exponential term in Eq. (8-3.1) is called the *Poynting factor.*

To illustrate Eq. (8-3.1), the fugacity of liquid water is shown in Table 8-2. Since $\phi^s$ for a pure liquid is always less than unity, the fugacity at saturation is always lower than the vapor pressure. However, at pressures well above the saturation pressure, the product of $\phi^s$ and the Poynting factor may easily exceed unity, and then the fugacity is larger than the vapor pressure.

At 260 and 316°C, the vapor pressure exceeds 40 bar, and therefore pure liquid water cannot exist at these temperatures and 40 bar. Nevertheless, the fugacity can be calculated by a mild extrapolation: in the Poynting factor we neglect the effect of pressure on molar liquid volume.

Table 8-2 indicates that the vapor pressure is the primary quantity in Eq. (8-3.1). When data are not available, the vapor pressure can be estimated, as discussed in Chap. 7. Further, for nonpolar (or weakly polar) liquids, the ratio of fugacity to pressure can be estimated from a generalized (corresponding states) table for liquids, as discussed in Sec. 5-4.

## 8-4 Simplifications in the Vapor-Liquid Equilibrium Relation

Equation (8-2.1) gives the rigorous, fundamental relation for vapor-liquid equilibrium. Equations (8-2.2), (8-2.3), and (8-3.1) are also rigorous, without any simplifications beyond those indicated in the paragraph following Eq. (8-2.1). Substitution of Eqs. (8-2.2), (8-2.3), and (8-3.1) into Eq. (8-2.1) gives

$$y_i P = \gamma_i x_i P_{\mathrm{vp}i} \mathcal{F}_i \tag{8-4.1}$$

where $$\mathcal{F}_i = \frac{\phi_i^s}{\phi_i} \exp \int_{P_{\mathrm{vp}i}}^{P} \frac{V_i^L\, dP}{RT} \tag{8-4.2}$$

†For volumetric properties of liquids, see Sec. 3-11.

**TABLE 8-2 Fugacity of Liquid Water, bar**

| Temp., °C | $P_{vp}$ | Fugacity | | |
|---|---|---|---|---|
| | | Saturation | 41.4 bar | 345 bar |
| 37.7 | 0.06544 | 0.0654 | 0.0674 | 0.0834 |
| 149 | 4.620 | 4.41 | 4.50 | 5.32 |
| 260 | 46.94 | 39.2 | 39.2† | 45.7 |
| 316 | 106.4 | 79.9 | 77.2† | 90.6 |

†Hypothetical because $P < P_{vp}$.

For subcritical components, the correction factor $\mathcal{F}_i$ is often near unity when the total pressure $P$ is sufficiently low. However, even at moderate pressures, we are nevertheless justified in setting $\mathcal{F}_i = 1$ if only approximate results are required and, as happens so often, if experimental information is sketchy, giving large uncertainties in $\gamma$.

If, in addition to setting $\mathcal{F}_i = 1$, we assume that $\gamma_i = 1$, Eq. (8-4.1) reduces to the familiar relation known as *Raoult's law*.

In Eq. (8-4.1), $\gamma_i$ depends on temperature, composition, and pressure. However, remote from critical conditions, and unless the pressure is large, the effect of pressure on $\gamma_i$ is usually small. [See Eq. (8-12.1).]

## 8-5 Activity Coefficients; Gibbs-Duhem Equation and Excess Gibbs Energy

In typical mixtures, Raoult's law provides no more than a rough approximation; only when the components in the liquid mixture are similar, e.g., a mixture of $n$-butane and isobutane, can we assume that $\gamma_i$ is essentially unity for all components at all compositions. The activity coefficient, therefore, plays a key role in the calculation of vapor-liquid equilibria.

Classical thermodynamics has little to tell us about the activity coefficient; as always, thermodynamics does not give us the experimental quantity we desire but only relates it to other experimental quantities. Thus thermodynamics relates the effect of pressure on the activity coefficient to the partial molar volume, and it relates the effect of temperature on the activity coefficient to the partial molar enthalpy, as discussed in any thermodynamics text (see, for example, chap. 6 of Ref. 97). These relations are of limited use because good data for the partial molar volume and for the partial molar enthalpy are rare.

However, there is one thermodynamic relation which provides a useful tool for correlating and extending limited experimental data: the Gibbs-Duhem equation. This equation is not a panacea, but, given some experimental results, it enables us to use these results efficiently. In essence, the Gibbs-Duhem equation says that, in a mixture, the activity coefficients of the individual components are not independent of one another but are related by a differential equation. In a binary mixture the Gibbs-Duhem relation is

$$x_1 \left( \frac{\partial \ln \gamma_1}{\partial x_1} \right)_{T,P} = x_2 \left( \frac{\partial \ln \gamma_2}{\partial x_2} \right)_{T,P} \tag{8-5.1}†$$

Equation (8-5.1) has several important applications.

1. If we have experimental data for $\gamma_1$ as a function of $x_1$, we can integrate Eq. (8-5.1) and calculate $\gamma_2$ as a function of $x_2$. That is, in a binary mixture, activity coefficient data for one component can be used to predict the activity coefficient of the other component.
2. If we have extensive experimental data for *both* $\gamma_1$ and $\gamma_2$ as a function of composition, we can test the data for thermodynamic consistency by determining whether or not the data obey Eq. (8-5.1). If the data show serious inconsistencies with Eq. (8-5.1), we may conclude that they are unreliable.
3. If we have limited data for $\gamma_1$ and $\gamma_2$, we can use an integral form of the Gibbs-Duhem equation; the integrated form provides us with thermodynamically consistent equations which relate $\gamma_1$ and $\gamma_2$ to $x$. These equations contain a few adjustable parameters which can be determined from the limited data. It is this application of the Gibbs-Duhem equation which is of particular use to chemical engineers. However, there is no *unique* integrated form of the Gibbs-Duhem equation; many forms are possible. To obtain a particular relation between $\gamma$ and $x$, we must assume some model which is consistent with the Gibbs-Duhem equation.

For practical work, the utility of the Gibbs-Duhem equation is best realized through the concept of excess Gibbs energy, i.e., the observed Gibbs energy of a mixture above and beyond what it would be for an ideal solution at the same temperature, pressure, and composition. By definition, an ideal solution is one where all $\gamma_i = 1$. The *total* excess Gibbs energy $G^E$ for a binary solution, containing $n_1$ moles of component 1 and $n_2$ moles of component 2, is defined by

$$G^E = RT(n_1 \ln \gamma_1 + n_2 \ln \gamma_2) \tag{8-5.2}$$

Equation (8-5.2) gives $G^E$ as a function of *both* $\gamma_1$ and $\gamma_2$. Upon applying the Gibbs-Duhem equation, we can relate the *individual* activity coefficients $\gamma_1$ or $\gamma_2$ to $G^E$ by differentiation

†Note that the derivatives are taken at constant temperature $T$ and constant pressure $P$. In a binary, two-phase system, however, it is not possible to vary $x$ while holding *both* $T$ and $P$ constant. At ordinary pressures $\gamma$ is a very weak function of $P$, and therefore it is often possible to apply Eq. (8-5.1) to isothermal data while neglecting the effect of changing pressure. This subject has been amply discussed in the literature; see, for example, chap. 6 and app. IV in Ref. 97.

$$RT \ln \gamma_1 = \left(\frac{\partial G^E}{\partial n_1}\right)_{T,P,n_2} \tag{8.5.3}$$

$$RT \ln \gamma_2 = \left(\frac{\partial G^E}{\partial n_2}\right)_{T,P,n_1} \tag{8-5.4}$$

Equations (8-5.2) to (8-5.4) are useful because they enable us to interpolate and extrapolate limited data with respect to composition. To do so, we must first adopt some mathematical expression for $G^E$ as a function of composition. Second, we fix the numerical values of the constants in that expression from the limited data; these constants are independent of $x$, but they usually depend on temperature. Third, we calculate activity coefficients at any desired composition by differentiation, as indicated by Eqs. (8-5.3) and (8-5.4).

To illustrate, consider a simple binary mixture. Suppose that we need activity coefficients for a binary mixture over the entire composition range at a fixed temperature $T$. However, we have experimental data for only one composition, say $x_1 = x_2 = \frac{1}{2}$. From that one datum we calculate $\gamma_1(x_1 = \frac{1}{2})$ and $\gamma_2(x_2 = \frac{1}{2})$; for simplicity, let us assume symmetrical behavior, that is, $\gamma_1(x_1 = \frac{1}{2}) = \gamma_2(x_2 = \frac{1}{2})$.

We must adopt an expression relating $G^E$ to the composition subject to the conditions that at fixed composition $G^E$ is proportional to $n_1 + n_2$ and that $G^E = 0$ when $x_1 = 0$ or $x_2 = 0$. The simplest expression we can construct is

$$G^E = (n_1 + n_2)g^E = (n_1 + n_2)Ax_1x_2 \tag{8-5.5}$$

where $g^E$ is the excess Gibbs energy per mole of mixture and $A$ is a constant depending on temperature. The mole fraction $x$ is simply related to mole number $n$ by

$$x_1 = \frac{n_1}{n_1 + n_2} \tag{8-5.6}$$

$$x_2 = \frac{n_2}{n_1 + n_2} \tag{8-5.7}$$

The constant $A$ is found from substituting Eq. (8-5.5) into Eq. (8-5.2) and using the experimentally determined $\gamma_1$ and $\gamma_2$ at the composition midpoint:

$$A = \frac{RT}{(\frac{1}{2})(\frac{1}{2})}\left[\tfrac{1}{2} \ln \gamma_1(x_1 = \tfrac{1}{2}) + \tfrac{1}{2} \ln \gamma_2(x_2 = \tfrac{1}{2})\right] \tag{8-5.8}$$

Upon differentiating Eq. (8-5.5) as indicated by Eqs. (8-5.3) and (8-5.4), we find

$$RT \ln \gamma_1 = Ax_2^2 \tag{8-5.9}$$

$$RT \ln \gamma_2 = Ax_1^2 \tag{8-5.10}$$

With these relations we can now calculate activity coefficients $\gamma_1$ and $\gamma_2$ at any desired $x$ even though experimental data were obtained only at one point, namely, $x_1 = x_2 = \frac{1}{2}$.

This simplified example illustrates how the concept of excess function, coupled with the Gibbs-Duhem equation, can be used to interpolate or extrapolate experimental data with respect to composition. Unfortunately, the Gibbs-Duhem equation tells nothing about interpolating or extrapolating such data with respect to temperature or pressure.

Equations (8-5.2) to (8-5.4) indicate the intimate relation between activity coefficients and excess Gibbs energy $G^E$. Many expressions relating $g^E$ (per mole of mixture) to composition have been proposed, and a few are given in Table 8-3. All these expressions contain adjustable constants which, at least in principle, depend on temperature. That dependence may in some cases be neglected, especially if the temperature interval is not large. In practice, the number of adjustable constants per binary is typically two or three; the larger the number of constants, the better the representation of the data but, at the same time, the larger the number of reliable experimental data points required to determine the constants. Extensive and highly accurate experimental data are required to justify more than three empirical constants for a binary mixture at a fixed temperature.†

For many moderately nonideal binary mixtures, all equations for $g^E$ containing two (or more) binary parameters give good results; there is little reason to choose one over another except that the older ones (Margules, van Laar) are mathematically easier to handle than the newer ones (Wilson, NRTL, UNIQUAC). The two-suffix (one-parameter) Margules equation is applicable only to simple mixtures where the components are similar in chemical nature and in molecular size.

For strongly nonideal binary mixtures, e.g., solutions of alcohols with hydrocarbons, the equation of Wilson is probably the most useful because, unlike the NRTL equation, it contains only two adjustable parameters and it is mathematically simpler than the UNIQUAC equation. For such mixtures, the three-suffix Margules equation and the van Laar equation are likely to represent the data with significantly less success, especially in the region dilute with respect to alcohol, where the Wilson equation is particularly suitable.

The four-suffix (three-parameter) Margules equation has no significant advantages over the three-parameter NRTL equation.

†The models shown in Table 8-3 are not applicable to solutions of electrolytes; such solutions are not considered here. However, brief attention is given to aqueous solutions of volatile weak electrolytes in a later section of this chapter.

Numerous articles in the literature use the Redlich-Kister expansion [see Eq. (8-9.20)] for $g^E$. This expansion is mathematically identical to the Margules equation.

The Wilson equation is not applicable to a mixture which exhibits a miscibility gap; it is inherently unable, even qualitatively, to account for phase splitting. Nevertheless, Wilson's equation may be useful even for those mixtures where miscibility is incomplete provided attention is confined to the one-phase region.

Unlike Wilson's equation, the NRTL and UNIQUAC equations are applicable to *both* vapor-liquid and liquid-liquid equilibria.† Therefore, mutual solubility data [see Sec. 8-10] can be used to determine NRTL or UNIQUAC parameters but not Wilson parameters. While UNIQUAC is mathematically more complex than NRTL, it has three advantages: (1) it has only two (rather than three) adjustable parameters, (2) UNIQUAC's parameters often have a smaller dependence on temperature, and (3) because the primary concentration variable is a surface fraction (rather than mole fraction), UNIQUAC is applicable to solutions containing small or large molecules, including polymers.

### Simplifications: one-parameter equations

It frequently happens that experimental data for a given binary mixture are so fragmentary that it is not possible to determine two (or three) *meaningful* binary parameters; limited data can often yield only one significant binary parameter. In that event, it is tempting to use the two-suffix (one-parameter) Margules equation, but this is usually an unsatisfactory procedure because activity coefficients in a real binary mixture are rarely symmetric with respect to mole fraction. In most cases better results are obtained by choosing the van Laar, Wilson, NRTL, or UNIQUAC equation and reducing the number of adjustable parameters through reasonable physical approximations.

To reduce the van Laar equation to a one-parameter form, for mixtures of nonpolar fluids, the ratio $A/B$ can often be replaced by the ratio of molar liquid volumes: $A/B = V_1^L/V_2^L$. This simplification, however, is not reliable for binary mixtures containing one (or two) polar components.

To simplify the Wilson equation, we first note that

$$\Lambda_{ij} = \frac{V_j^L}{V_i^L} \exp\left(-\frac{\lambda_{ij} - \lambda_{ii}}{RT}\right) \tag{8-5.11}$$

†Wilson [130] has given a three-parameter form of his equation which is applicable also to liquid-liquid equilibria; the molecular significance of the third parameter has been discussed by Renon and Prausnitz [106]. The three-parameter Wilson equation has not received much attention, primarily because it is not readily extended to multicomponent systems.

**TABLE 8-3 Some Models for the Excess Gibbs Energy and Subsequent Activity Coefficients for Binary Systems[a]**

| Name | $g^E$ | Binary parameters | $\ln \gamma_1$ and $\ln \gamma_2$ |
|---|---|---|---|
| Two-suffix[b] Margules | $g^E = Ax_1x_2$ | $A$ | $RT \ln \gamma_1 = Ax_2^2$ |
| | | | $RT \ln \gamma_2 = Ax_1^2$ |
| Three-suffix[b] Margules | $g^E = x_1x_2[A + B(x_1 - x_2)]$ | $A, B$ | $RT \ln \gamma_1 = (A + 3B)x_2^2 - 4Bx_2^3$ |
| | | | $RT \ln \gamma_2 = (A - 3B)x_1^2 + 4Bx_1^3$ |
| van Laar | $g^E = \dfrac{Ax_1x_2}{x_1(A/B) + x_2}$ | $A, B$ | $RT \ln \gamma_1 = A\left(1 + \dfrac{A}{B}\dfrac{x_1}{x_2}\right)^{-2}$ |
| | | | $RT \ln \gamma_2 = B\left(1 + \dfrac{B}{A}\dfrac{x_2}{x_1}\right)^{-2}$ |
| Wilson | $\dfrac{g^E}{RT} = -x_1 \ln(x_1 + \Lambda_{12}x_2) - x_2 \ln(x_2 + \Lambda_{21}x_1)$ | $\Lambda_{12}, \Lambda_{21}$ | $\ln \gamma_1 = -\ln(x_1 + \Lambda_{12}x_2) + x_2\left(\dfrac{\Lambda_{12}}{x_1 + \Lambda_{12}x_2} - \dfrac{\Lambda_{21}}{\Lambda_{21}x_1 + x_2}\right)$ |
| | | | $\ln \gamma_2 = -\ln(x_2 + \Lambda_{21}x_1) - x_1\left(\dfrac{\Lambda_{12}}{x_1 + \Lambda_{12}x_2} - \dfrac{\Lambda_{21}}{\Lambda_{21}x_1 + x_2}\right)$ |
| Four-suffix[b] Margules | $g^E = x_1x_2[A + B(x_1 - x_2) + C(x_1 - x_2)^2]$ | $A, B, C$ | $RT \ln \gamma_1 = (A + 3B + 5C)x_2^2 - 4(B + 4C)x_2^3 + 12Cx_2^4$ |
| | | | $RT \ln \gamma_2 = (A - 3B + 5C)x_1^2 + 4(B - 4C)x_1^3 + 12Cx_1^4$ |

| | | | |
|---|---|---|---|
| NRTL[c] | $\frac{g^E}{RT} = x_1 x_2 \left( \frac{\tau_{21} G_{21}}{x_1 + x_2 G_{21}} + \frac{\tau_{12} G_{12}}{x_2 + x_1 G_{12}} \right)$<br>where $\tau_{12} = \frac{\Delta g_{12}}{RT} \qquad \tau_{21} = \frac{\Delta g_{21}}{RT}$<br>$\ln G_{12} = -\alpha_{12}\tau_{12} \qquad \ln G_{21} = -\alpha_{12}\tau_{21}$ | $\Delta g_{12}, \Delta g_{21}, \alpha_{12}$[d] | $\ln \gamma_1 = x_2^2 \left[ \tau_{21} \left( \frac{G_{21}}{x_1 + x_2 G_{21}} \right)^2 + \frac{\tau_{12} G_{12}}{(x_2 + x_1 G_{12})^2} \right]$<br>$\ln \gamma_2 = x_1^2 \left[ \tau_{12} \left( \frac{G_{12}}{x_2 + x_1 G_{12}} \right)^2 + \frac{\tau_{21} G_{21}}{(x_1 + x_2 G_{21})^2} \right]$ |
| UNIQUAC[e] | $g^E = g^E \text{ (combinatorial)} + g^E \text{ (residual)}$<br>$\frac{g^E \text{ (combinatorial)}}{RT} = x_1 \ln \frac{\Phi_1}{x_1} + x_2 \ln \frac{\Phi_2}{x_2} + \frac{z}{2} \left( q_1 x_1 \ln \frac{\theta_1}{\Phi_1} + q_2 x_2 \ln \frac{\theta_2}{\Phi_2} \right)$<br>$\frac{g^E \text{ (residual)}}{RT} = -q_1 x_1 \ln [\theta_1 + \theta_2 \tau_{21}] - q_2 x_2 \ln [\theta_2 + \theta_1 \tau_{12}]$<br>$\Phi_1 = \frac{x_1 r_1}{x_1 r_1 + x_2 r_2} \qquad \theta_1 = \frac{x_1 q_1}{x_1 q_1 + x_2 q_2}$<br>$\ln \tau_{21} = -\frac{\Delta u_{21}}{RT} \qquad \ln \tau_{12} = -\frac{\Delta u_{12}}{RT}$<br>$r$ and $q$ are pure-component parameters and coordination number $z = 10$ | $\Delta u_{12}$ and $\Delta u_{21}$[f] | $\ln \gamma_i = \ln \frac{\Phi_i}{x_1} + \frac{z}{2} q_i \ln \frac{\theta_i}{\Phi_i} + \Phi_j \left( \ell_i - \frac{r_i}{r_j} \ell_j \right) - q_i \ln (\theta_i + \theta_j \tau_{ji}) + \theta_j q_i \left( \frac{\tau_{ji}}{\theta_i + \theta_j \tau_{ji}} - \frac{\tau_{ij}}{\theta_j + \theta_i \tau_{ij}} \right)$<br>where $i = 1 \quad j = 2$ or $i = 2 \quad j = 1$<br>$\ell_i = \frac{z}{2}(r_i - q_i) - (r_i - 1)$<br>$\ell_j = \frac{z}{2}(r_j - q_j) - (r_j - 1)$ |

[a] Reference 97 discusses the Margules, van Laar, Wilson, and NRTL equations. The UNIQUAC equation is discussed in Ref. 4.

[b] Two-suffix signifies that the expansion for $g^E$ is quadratic in mole fraction. Three-suffix signifies a third-order, and four-suffix signifies a fourth-order equation.

[c] NRTL = Non Random Two Liquid.

[d] $\Delta g_{12} = g_{12} - g_{22}$; $\Delta g_{21} = g_{21} - g_{11}$.

[e] UNIQUAC = *Uni*versal *Qua*si *C*hemical. Parameters $q$ and $r$ can be calculated from Eq. (8-10.52').

[f] $\Delta u_{12} = u_{12} - u_{22}$; $\Delta u_{21} = u_{21} - u_{11}$.

where $V_i^L$ is the molar volume of pure liquid $i$ and $\lambda_{ij}$ is an energy parameter characterizing the interaction of molecule $i$ with molecule $j$.

The Wilson equation can be reduced to a one-parameter form by assuming that $\lambda_{ij} = \lambda_{ji}$† and that

$$\lambda_{ii} = -\beta(\Delta H_{v_i} - RT) \tag{8-5.12}$$

where $\beta$ is a proportionality factor and $\Delta H_{vi}$ is the enthalpy of vaporization of pure component $i$ at $T$. A similar equation is written for $\lambda_{jj}$. When $\beta$ is fixed, the only adjustable binary parameter is $\lambda_{ij}$.

Tassios set $\beta = 1$, but from theoretical considerations it makes more sense to assume that $\beta = 2/z$, where $z$ is the coordination number (typically, $z = 10$). This assumption, used by Wong and Eckert (135) and Schreiber and Eckert (109), gives good estimates for a variety of binary mixtures. Hiranuma and Honma have had some success in correlating $\lambda_{ij}$ with energy contributions from intermolecular dispersion and dipole-dipole forces.

Ladurelli et al. [65] have suggested the $\beta = 2/z$ for component 2, having the smaller molar volume, while for component 1, having the larger molar volume, $\beta = (2/z)(V_2^L/V_1^L)$. This suggestion follows from the notion that a larger molecule has a larger area of interaction; parameters $\lambda_{ii}$, $\lambda_{jj}$, and $\lambda_{ij}$ are considered as interaction energies per segment rather than per molecule. In this particular case the unit segment is that corresponding to one molecule of component 2.

Using similar arguments, Bruin and Prausnitz [23] have shown that it is possible to reduce the number of adjustable binary parameters in the NRTL equation by making a reasonable assumption for $\alpha_{12}$ and by substituting NRTL parameter $g_{ii}$ for Wilson parameter $\lambda_{ii}$ in Eq. (8-5.12). Bruin gives some correlations for $g_{ij}$, especially for aqueous systems.

Finally, Abrams and Prausnitz [4] have shown that the UNIQUAC equation can be simplified by assuming that

$$u_{11} = \frac{-\Delta U_1}{q_1} \qquad \text{and} \qquad u_{22} = \frac{-\Delta U_2}{q_2} \tag{8-5.13}$$

and that

$$u_{12} = u_{21} = (u_{11}u_{22})^{1/2}(1 - c_{12}) \tag{8-5.14}$$‡

†The simplifying assumption that cross-parameter $\lambda_{ij} = \lambda_{ji}$ (or, similarly, $g_{ij} = g_{ji}$ or $u_{ij} = u_{ji}$) is often useful but is not required by theory unless severe simplifying assumptions concerning liquid structure are made.

‡See preceding footnote.

where, remote from the critical temperature, energy $\Delta U_i$ is given very nearly by $\Delta U_i \approx \Delta H_{v_i} - RT$. The only adjustable binary parameter is $c_{12}$, which, for mixtures of nonpolar liquids, is positive and small compared with unity. For some mixtures containing polar components, however, $c_{12}$ is of the order of 0.5; and when the unlike molecules in a mixture are attracted more strongly than like molecules, $c_{12}$ may be negative, e.g., in acetone-chloroform.

For mixtures of nonpolar liquids, a one-parameter form (van Laar, Wilson, NRTL, UNIQUAC) often gives results nearly as good as those obtained by using two, or even three, parameters. However, if one or both components are polar, significantly better results are usually obtained by using two parameters, provided that the experimental data used to determine the parameters are of sufficient quantity and quality.

## 8-6 Calculation of Binary Vapor-Liquid Equilibria

First consider the isothermal case. At some constant temperature $T$, we wish to construct two diagrams; $y$ vs. $x$ and $P$ vs. $x$. We assume that, since the pressure is low, we can use Eq. (8.4.1) with $\mathcal{F}_i = 1$. The steps toward that end are:

1. Find the pure liquid vapor pressures $P_{vp1}$ and $P_{vp2}$ at $T$.
2. Suppose a few experimental points for the mixture are available at temperature $T$. Arbitrarily, to fix ideas, suppose there are five points; i.e., for five values of $x$ there are five corresponding experimental equilibrium values of $y$ and $P$. For each of these points calculate $\gamma_1$ and $\gamma_2$ according to

$$\gamma_1 = \frac{y_1 P}{x_1 P_{vp1}} \tag{8-6.1}$$

$$\gamma_2 = \frac{y_2 P}{x_2 P_{vp2}} \tag{8-6.2}$$

3. For each of the five points, calculate the molar excess Gibbs energy $g^E$:

$$g^E = RT(x_1 \ln \gamma_1 + x_2 \ln \gamma_2) \tag{8-6.3}$$

4. Choose one of the equations for $g^E$ given in Table 8.3. Adjust the constants in that equation to minimize the deviation between $g^E$ calculated from the equation and $g^E$ found from experiment in step 3.
5. Using Eqs. (8-5.3) and (8-5.4), find $\gamma_1$ and $\gamma_2$ at arbitrarily selected values of $x_1$ from $x_1 = 0$ to $x_1 = 1$.
6. For each selected $x_1$ find the corresponding $y_1$ and $P$ by solving Eqs.

(8-6.1) and (8-6.2) coupled with the mass balance relations $x_2 = 1 - x_1$ and $y_2 = 1 - y_1$. The results obtained give the desired $y$-vs.-$x$ and $P$-vs.-$x$ diagrams.

The simple steps outlined above provide a rational, thermodynamically consistent procedure for interpolation and extrapolation with respect to composition. The crucial step is 4. Judgment is required to obtain the best, i.e., the most representative, constants in the expression chosen for $g^E$. To do so, it is necessary to decide on how to weight the five individual experimental data, some of which may be more reliable than others. For determining the constants, the experimental points which give the most information are those at the ends of the composition scale, that is, $y_1$ when $x_1$ is small and $y_2$ when $x_2$ is small. Unfortunately, however, these experimental data are often the most difficult to measure. Thus it frequently happens that the data which are potentially most valuable are also the ones which are likely to be least accurate.

Now let us consider the more complicated isobaric case. At some constant pressure $P$, we wish to construct two diagrams: $y$ vs. $x$ and $T$ vs. $x$. Assuming that the pressure is low, we again use Eq. (8-4.1) with $\mathcal{F}_i = 1$. The steps toward construction of these diagrams are:

1. Find pure component vapor pressures $P_{vp_1}$ and $P_{vp_2}$. Prepare plots (or obtain analytical representation) of $P_{vp_1}$ and $P_{vp_2}$ vs. temperature in the region where $P_{vp_1} \approx P$ and $P_{vp_2} \approx P$. (See Chap 7.)

2. Suppose there are available a few experimental data points for the mixture at pressure $P$ or at some other pressure not far removed from $P$ or, perhaps, at some constant temperature such that the total pressure is in the general vicinity of $P$. As in the previous case, to fix ideas, we arbitrarily set the number of such experimental points at five. By experimental point we mean, as before, that for some value of $x_1$ we have the corresponding experimental equilibrium values of $y_1$, $T$, and total pressure.

For each of the five points, calculate activity coefficients $\gamma_1$ and $\gamma_2$ according to Eqs. (8-6.1) and (8-6.2). For each point the vapor pressures $P_{vp_1}$ and $P_{vp_2}$ are evaluated at the experimentally determined temperature for that point. In these equations, the experimentally determined total pressure is used for $P$; the total pressure measured is not necessarily the same as the pressure for which we wish to construct the equilibrium diagrams.

3. For each of the five points, calculate the molar excess Gibbs energy according to Eq. (8-6.3).

4. Choose one of the equations for $g^E$ given in Table 8-3. As in step 4 of the previous (isothermal) case, find the constants in that equation which give the smallest deviation between calculated values of $g^E$ and those found in step 3. When the experimental data used in Eq. (8-6.3) are isobaric rather than isothermal, it may be advantageous to choose an

expression for $g^E$ which contains the temperature as one of the explicit variables. Such a choice, however, complicates the calculations in step 6.

5. Find $\gamma_1$ and $\gamma_2$ as functions of $x$ by differentiation according to Eqs. (8-5.3) and (8-5.4).†

6. Select a set of arbitrary values for $x_1$ for the range $x_1 = 0$ to $x_1 = 1$. For each $x_1$, by iteration, solve simultaneously the two equations of phase equilibrium [Eqs. (8-6.1) and (8-6.2)] for the two unknowns, $y_1$ and $T$. In these equations the total pressure $P$ is now the one for which the equilibrium diagrams are desired.

Simultaneous solution of Eqs. (8-6.1) and (8-6.2) requires trial and error because, at a given $x$, both $y$ and $T$ are unknown and both $P_{vp1}$ and $P_{vp2}$ are strong, nonlinear functions of $T$. In addition, $\gamma_1$ and $\gamma_2$ may also vary with $T$ (as well as $x$), depending on which expression for $g^E$ has been chosen in step 4. For simultaneous solution of the two equilibrium equations, the best procedure is to assume a reasonable temperature for each selected value of $x_1$. Using this assumed temperature, calculate $y_1$ and $y_2$ from Eqs. (8-6.1) and (8-6.2). Then check if $y_1 + y_2 = 1$. If not, assume a different temperature and repeat the calculation. In this way, for fixed $P$ and for each selected value of $x$, find corresponding equilibrium values $y$ and $T$.

Calculation of isothermal or isobaric vapor-liquid equilibria can be efficiently performed with a computer as discussed, for example, in Ref. 99. Further, it is possible in such calculations to include the correction factor $\mathcal{F}_i$ [Eq. (8-4.1)] when necessary. In that event, the calculations are more complex in detail but not in principle.

When the procedures outlined above are followed, the accuracy of any vapor-liquid equilibrium calculation depends primarily on the extent to which the expression for $g^E$ accurately represents the behavior of the mixture at the particular conditions (temperature, pressure, composition) for which the calculation is made. This accuracy of representation often depends not so much on the algebraic form of $g^E$ as on the reliability of the constants appearing in that expression. This reliability, in turn, depends on the quality and quantity of the experimental data used to determine the constants.

Some of the expressions for $g^E$ shown in Table 8-3 have a better theoretical foundation than others, but all have a strong empirical flavor. Experience has indicated that the more recent equations for $g^E$ (Wilson, NRTL, and UNIQUAC) are more consistently reliable than the older equations in the sense that they can usually reproduce accurately even

†Some error is introduced here because Eqs. (8-5.3) and (8-5.4) are based on the isobaric *and* isothermal Gibbs-Duhem equation. For most practical calculations this error is not serious. See chap. 6 and app. IV of Ref. 97.

highly nonideal behavior by using only two or three adjustable parameters.

The oldest equation of $g^E$, that of Margules, is a power series in mole fraction. With a power series it is always possible to increase accuracy of representation by including higher terms, where each term is multiplied by an empirically determined coefficient. (The van Laar equation, as shown by Wohl [133], is also a power series in effective volume fraction, but in practice this series is almost always truncated after the quadratic term.) However, inclusion of higher-order terms in $g^E$ is dangerous because subsequent differentiation to find $\gamma_1$ and $\gamma_2$ can then lead to spurious maxima or minima. Also, inclusion of higher-order terms in binary data reduction often leads to serious difficulties when binary data are used to estimate multicomponent phase equilibria.

It is desirable to use an equation for $g^E$ which is based on a relatively simple model and which contains only two (or at most three) adjustable binary parameters. Experimental data are then used to find the "best" binary parameters. Since experimental data are always of limited accuracy, it often happens that several sets of binary parameters may equally well represent the data within experimental uncertainty [4]. Only in rare cases, when experimental data are both plentiful and highly accurate, is there any justification for using more than three adjustable binary parameters.

## 8-7 Effect of Temperature on Vapor-Liquid Equilibria

A particularly troublesome question is the effect of temperature on the molar excess Gibbs energy $g^E$. This question is directly related to $s^E$, the molar excess entropy of mixing about which little is known.† In practice, either one of two approximations is frequently used.

**(*a*) Athermal solution.** This approximation sets $g^E = -Ts^E$, which assumes that the components mix at constant temperature without change of enthalpy ($h^E = 0$). This assumption leads to the conclusion that, at constant composition, $\ln \gamma_i$ is independent of $T$ or, its equivalent, that $g^E/RT$ is independent of temperature.

**(*b*) Regular solution.** This approximation sets $g^E = h^E$, which is the same as assuming that $s^E = 0$. This assumption leads to the conclusion that, at constant composition, $\ln \gamma_i$ varies as $1/T$ or, its equivalent, that $g^E$ is independent of temperature.

†From thermodynamics, $s^E = -(\partial g^E/\partial T)_{P,x}$ and $g^E = h^E - Ts^E$.

Neither one of these extreme approximations is valid, although the second one is often better than the first. Good experimental data for the effect of temperature on activity coefficients are rare, but when such data are available, they suggest that, for a moderate temperature range, they can be expressed by an empirical equation of the form

$$(\ln \gamma_i)_{\substack{\text{constant}\\ \text{composition}}} = c + \frac{d}{T} \tag{8-7.1}$$

where $c$ and $d$ are empirical constants that depend on composition. In most cases constant $d$ is positive. It is evident that, when $d = 0$, Eq. (8-7.1) reduces to assumption ($a$) and, when $c = 0$, it reduces to assumption ($b$). Unfortunately, in typical cases $c$ and $d/T$ are of comparable magnitude.

Thermodynamics relates the effect of temperature on $\gamma_i$ to the partial molar enthalpy $\bar{h}_i$

$$\left[\frac{\partial \ln \gamma_i}{\partial (1/T)}\right]_{x,P} = \frac{\bar{h}_i - h_i^\circ}{R} \tag{8-7.2}$$

where $h_i^\circ$ is the enthalpy of liquid $i$ in the standard state, usually taken as pure liquid $i$ at system temperature and pressure. Sometimes (but rarely) experimental data for $\bar{h}_i - h_i^\circ$ may be available; if so, they can be used to provide information on how the activity coefficient changes with temperature. However, even if such data are at hand, Eq. (8-7.2) must be used with caution because $\bar{h}_i - h_i^\circ$ depends on temperature and often strongly so.

Some of the expressions for $g^E$ shown in Table 8-3 contain $T$ as an explicit variable. However, one should not therefore conclude that the constants appearing in those expressions are independent of temperature. The explicit temperature dependence indicated provides only an approximation. This approximation is usually, but not always, better than approximation ($a$) or ($b$), but, in any case, it is not exact.

Fortunately, the primary effect of temperature on vapor-liquid equilibria is contained in the pure component vapor pressures or, more precisely, in the pure component liquid fugacities [Eq. (8-3.1)]. While the activity coefficients depend on temperature as well as composition, the temperature dependence of the activity coefficient is usually small when compared with the temperature dependence of the pure liquid vapor pressures. In a typical mixture, a rise of 10°C increases the vapor pressures of the pure liquids by a factor of 1.5 or 2, but the change in activity coefficient is likely to be only a few percent, often less than the experimental uncertainty. Therefore, unless there is a large change in temperature, it is frequently satisfactory to neglect the effect of temperature on $g^E$ when calculating vapor-liquid equilibria. However, in calculating liquid-liquid equilibria, vapor pressures play no role at all, and therefore the effect of temperature on $g^E$, although small, may seriously affect liquid-liquid equilibria. Even small changes in activity coefficients can have a large effect on multicomponent liquid-liquid equilibria, as briefly discussed in Sec. 8.13.

## 8-8 Binary Vapor-Liquid Equilibria: Examples

To introduce the general ideas, we present first two particularly simple methods for reduction of vapor-liquid equilibria. These are followed by a brief introduction to more accurate, but also mathematically more complex, procedures.

**Example 8-1** Given five experimental vapor-liquid equilibrium data for the binary system methanol (1)–1,2-dichloroethane (2) at 50°C, calculate the $P$-$y$-$x$ diagram at 50°C and predict the $P$-$y$-$x$ diagram at 60°C.

**Experimental Data at 50°C [126]**

| $100x_1$ | $100y_1$ | $P$, bar |
|---|---|---|
| 30 | 59.1 | 0.6450 |
| 40 | 60.2 | 0.6575 |
| 50 | 61.2 | 0.6665 |
| 70 | 65.7 | 0.6685 |
| 90 | 81.4 | 0.6262 |

**solution** To interpolate in a thermodynamically consistent manner, we must choose an algebraic expression for the molar excess Gibbs energy. For simplicity, we choose the van Laar equation (see Table 8-3). To evaluate the van Laar constants $A'$ and $B'$, we rearrange the van Laar equation in a linear form†

$$\frac{x_1 x_2}{g^E/RT} = D + C(2x_1 - 1) \qquad \text{where} \qquad \begin{aligned} A' &= (D - C)^{-1} \\ B' &= (D + C)^{-1} \end{aligned} \tag{8-8.1}$$

Constants $D$ and $C$ are found from a plot of $x_1x_2(g^E/RT)^{-1}$ vs. $x_1$. The intercept at $x_1 = 0$ gives $D - C$, and the intercept at $x_1 = 1$ gives $D + C$.
The molar excess Gibbs energy is calculated from the definition

$$\frac{g^E}{RT} = x_1 \ln \gamma_1 + x_2 \ln \gamma_2 \tag{8-8.2}$$

For the five available experimental points, activity coefficients $\gamma_1$ and $\gamma_2$ are calculated from Eq. (8-4.1) with $\mathcal{F}_i = 1$ and from pure component vapor pressure data.
Table 8-4 gives $x_1x_1(g^E/RT)^{-1}$ as needed to obtain van Laar constants. Figure 8-1 shows the linearized van Laar equation. The results shown are obtained with Antoine constants given in Appendix A.

†From Table 8-3, $A' = A/RT$ and $B' = B/RT$.

**TABLE 8-4 Experimental Activity Coefficients for Linearized van Laar Plot, Methanol (1)–1,2-Dichloroethane (2) at 50°C**

| $x_1$ | $\gamma_1$ | $\gamma_2$ | $\dfrac{x_1x_2}{g^E/RT}$ |
|---|---|---|---|
| 0.3 | 2.29 | 1.21 | 0.550 |
| 0.4 | 1.78 | 1.40 | 0.555 |
| 0.5 | 1.47 | 1.66 | 0.560 |
| 0.7 | 1.12 | 2.46 | 0.601 |
| 0.9 | 1.02 | 3.75 | 0.604 |

From Fig. 8-1 we obtain the van Laar constants

$$A' = 1.94 \qquad B' = 1.61 \qquad \frac{A'}{B'} = 1.20$$

We can now calculate $\gamma_1$ and $\gamma_2$ at any mole fraction:

$$\ln \gamma_1 = 1.94 \left(1 + 1.20\frac{x_1}{x_2}\right)^{-2} \tag{8-8.3}$$

$$\ln \gamma_2 = 1.61 \left(1 + \frac{x_2}{1.20x_1}\right)^{-2} \tag{8-8.4}$$

By using Eqs. (8-8.3) and (8-8.4) and the pure component vapor pressures, we can now find $y_1$, $y_2$, and total pressure $P$. There are two unknowns: $y_1$ (or $y_2$) and $P$. To find them, we must solve simultaneously the two equations of vapor-liquid equilibrium

$$y_1 = \frac{x_1\gamma_1 P_{vp1}}{P} \tag{8-8.5}$$

$$1 - y_1 = \frac{x_2\gamma_2 P_{vp2}}{P} \tag{8-8.6}$$

Calculated results at 50°C are shown in Table 8-5.

To predict vapor-liquid equilibria at 60°C, we assume that the effect of temper-

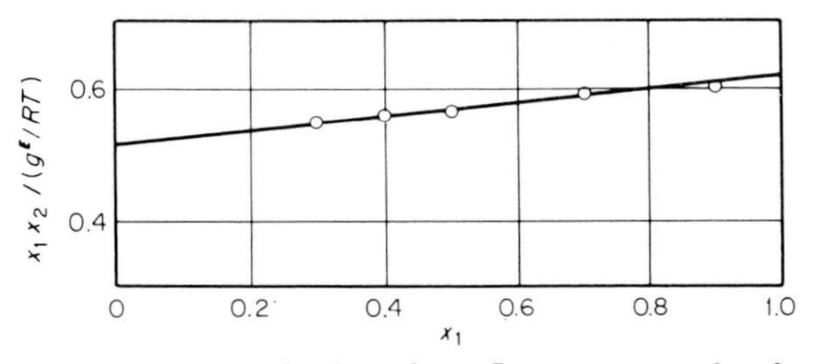


**Figure 8-1** Determination of van Laar constants for the system methanol (1)–1,2-dichloroethane (2) at 50°C.

**TABLE 8-5 Calculated Vapor-Liquid Equilibria in the System Methanol (1)–1,2-Dichloroethane (2) at 50 and 60°C**

| | $\gamma_1$ | | $\gamma_2$ | | $100y_1$ | | $P$, bar | | $100\ \Delta y$ | | $10^3\ \Delta P$, bar | |
|---|---|---|---|---|---|---|---|---|---|---|---|---|
| $100x_1$ | 50°C | 60°C | 50°C | 60°C | 50°C | 60°C | 50°C | 60°C | 50°C | 60°C | 50°C | 60°C |
| 5 | 5.56 | 5.28 | 1.01 | 1.01 | 34.1 | 33.9 | 0.4528 | 0.6589 | | | | |
| 10 | 4.53 | 4.33 | 1.02 | 1.02 | 46.9 | 46.8 | 0.5370 | 0.7830 | −0.9 | 0.4 | −1.9 | 1.5 |
| 20 | 3.15 | 3.04 | 1.09 | 1.09 | 56.4 | 56.5 | 0.6211 | 0.9102 | 0.2 | 0.9 | 11.1 | 20.5 |
| 40 | 1.82 | 1.79 | 1.37 | 1.36 | 61.3 | 62.1 | 0.6601 | 0.9762 | 1.1 | 2.2 | 2.6 | 26.4 |
| 60 | 1.28 | 1.27 | 1.95 | 1.91 | 63.8 | 65.0 | 0.6693 | 0.9914 | 1.3 | 1.8 | −1.7 | 22.9 |
| 80 | 1.06 | 1.06 | 3.01 | 2.91 | 71.6 | 73.1 | 0.6585 | 0.9815 | 0.5 | 1.2 | 1.5 | 34.8 |
| 90 | 1.01 | 1.01 | 3.85 | 3.70 | 80.9 | 92.1 | 0.6249 | 0.9369 | −0.5 | −0.1 | −1.3 | 30.3 |

ature on activity coefficients is given by the regular solution approximation (see Sec. 8-7):

$$\frac{\ln \gamma_i\ (60°\text{C})}{\ln \gamma_i\ (50°\text{C})} = \frac{273 + 50}{273 + 60} \tag{8-8.7}$$

Pure component vapor pressures at 60°C are found from the Antoine relations. The two equations of equilibrium [Eqs. (8-8.5) and (8-8.6)] are again solved simultaneously to obtain $y$ and $P$ as a function of $x$. Calculated results at 60°C are shown in Table 8-5 and in Fig. 8-2.

Predicted $y$'s are in good agreement with experiment [126], but predicted pressures are too high. This suggests that Eq. (8-8.7) is not a good approximation for this system.

Equation (8-8.7) corresponds to approximation (*b*) in Sec. 8-7. If approximation (*a*) had been used, the predicted pressures would have been even higher.

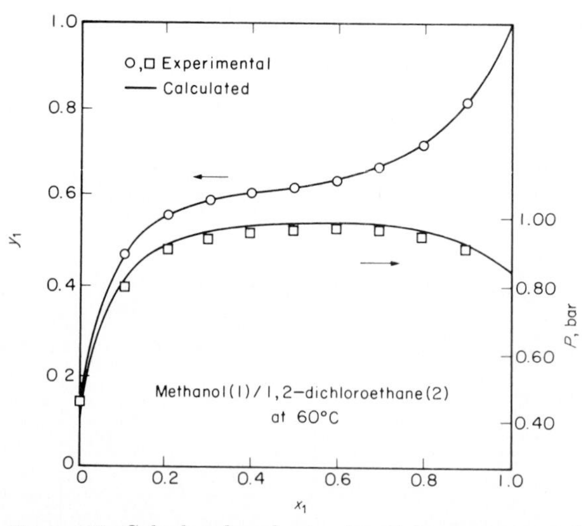


**Figure 8-2** Calculated and experimental vapor compositions and total pressures of the system methanol (1)–1,2-dichloroethane (2) at 60°C.

**Example 8-2** Given five experimental vapor-liquid equilibrium data for the binary system propanol (1)–water (2) at 1.01 bar, predict the $T$-$y$-$x$ diagram for the same system at 1.33 bar.

**Experimental Data at 1.01 bar**
[82]

| $100x_1$ | $100y_1$ | $T$, °C |
|---|---|---|
| 7.5 | 37.5 | 89.05 |
| 17.9 | 38.8 | 87.95 |
| 48.2 | 43.8 | 87.80 |
| 71.2 | 56.0 | 89.20 |
| 85.0 | 68.5 | 91.70 |

**solution** To represent the experimental data, we choose the van Laar equation, as in Example 8-1. Since the temperature range is small, we neglect the effect of temperature on the van Laar constants.
As in Example 8-1, we linearize the van Laar equation as shown in Eq. (8-8.1). To obtain the van Laar constants $A'$ and $B'$, we need, in addition to the data shown above, vapor pressure data for the pure components.
Activity coefficients $\gamma_1$ and $\gamma_2$ are calculated from Eq. (8-4.1) with $\mathcal{F}_i = 1$, and $g^E/RT$ is calculated from Eq. (8-8.2). Antoine constants are from Appendix A. Results are given in Table 8-6. The linearized van Laar plot is shown in Fig. 8-3. From the intercepts in Fig. 8-3 we obtain

$$A' = 2.60 \qquad B' = 1.13 \qquad \frac{A'}{B'} = 2.30 \tag{8-8.8}$$

Activity coefficients $\gamma_1$ and $\gamma_2$ are now given by the van Laar equations

$$\ln \gamma_1 = 2.60\left(1 + 2.30\frac{x_1}{x_2}\right)^{-2} \tag{8-8.9}$$

$$\ln \gamma_2 = 1.13\left(1 + \frac{x_2}{2.30x_1}\right)^{-2} \tag{8-8.10}$$

**TABLE 8-6 Experimental Activity Coefficients for Linearized van Laar Plot, *n*-Propanol (1)–Water (2) at 1.01 Bar**

| $100x_1$ | $T$, °C | $\gamma_1$ | $\gamma_2$ | $\dfrac{x_1x_2}{g^E/RT}$ |
|---|---|---|---|---|
| 7.5 | 89.05 | 6.84 | 1.01 | 0.446 |
| 17.9 | 87.95 | 3.10 | 1.17 | 0.448 |
| 48.2 | 87.80 | 1.31 | 1.71 | 0.615 |
| 71.2 | 89.20 | 1.07 | 2.28 | 0.720 |
| 85.0 | 91.70 | 0.99 | 2.85 | 0.848 |

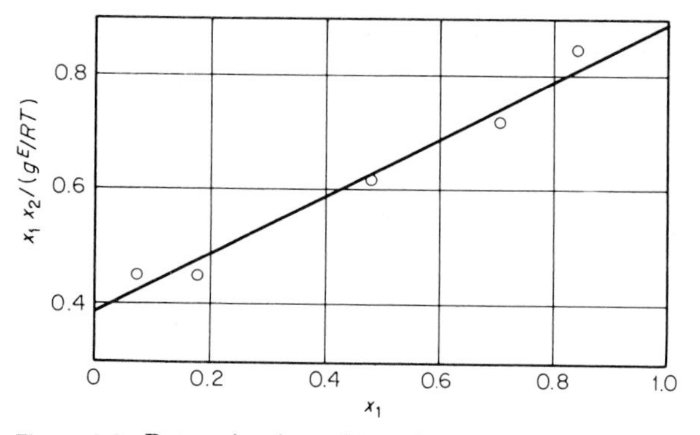


**Figure 8-3** Determination of van Laar constants for the system $n$-propanol (1)–water at 1.01 bar.

To obtain the vapor-liquid equilibrium diagram at 1.33 bar, we must solve simultaneously the two equations of equilibrium

$$y_1 = \frac{\gamma_1 x_1 P_{vp1}(T)}{1.33} \tag{8-8.11}$$

$$1 - y_1 = y_2 = \frac{\gamma_2 x_2 P_{vp2}(T)}{1.33} \tag{8-8.12}$$

In this calculation we assume that $\gamma_1$ and $\gamma_2$ depend only on $x$ (as given by the van Laar equations) and not on temperature. However, $P_{vp1}$ and $P_{vp2}$ are strong functions of temperature.

The two unknowns in the equations of equilibrium are $y_1$ and $T$. To solve for these unknowns, it is also necessary to use the Antoine relations for the two pure components.

The required calculations contain the temperature as an implicit variable; solution of the equations of equilibrium must be by iteration.

While iterative calculations are best performed with a computer, in this example it is possible to obtain results rapidly by hand calculations. Dividing one of the equations of equilibrium by the other, we obtain

$$y_1 = \left(1 + \frac{\gamma_2 x_2}{\gamma_1 x_1} \frac{P_{vp2}}{P_{vp1}}\right)^{-1} \tag{8-8.13}$$

Although $P_{vp2}$ and $P_{vp1}$ are strong functions of temperature, the ratio $P_{vp2}/P_{vp1}$ is a much weaker function of temperature.

For a given $x_1$, find $\gamma_2/\gamma_1$ from the van Laar equations. Choose a reasonable temperature and find $P_{vp2}/P_{vp1}$ from the Antoine relations. Equation (8-8.13) then gives a first estimate for $y_1$. Using this estimate, find $P_{vp1}$ from

$$P_{vp1} = \frac{1.33 y_1}{x_1 \gamma_1} \tag{8-8.14}$$

The Antoine relation for component 1 then gives a first estimate for $T$. By using this $T$, find the ratio $P_{vp2}/P_{vp1}$ and, again using Eq. (8-8.13), find the second estimate for $y_1$. This second estimate for $y_1$ is then used with the Antoine relation to find the second estimate for $T$. Repeat until there is negligible change in the estimate for $T$.

It is clear that Eq. (8-8.14) for component 1 could be replaced with the analogous equation for component 2. Which one should be used? In principle, either one

**TABLE 8-7 Calculated Vapor-Liquid Equilibria for *n*-Propanol (1)–Water (2) at 1.33 Bar**

| $100x_1$ | $\gamma_1$ | $\gamma_2$ | $T$, °C | $100y_1$ |
|---|---|---|---|---|
| 5 | 7.92 | 1.01 | 98.4 | 31.6 |
| 10 | 5.20 | 1.05 | 96.0 | 37.9 |
| 20 | 2.85 | 1.16 | 95.3 | 40.5 |
| 40 | 1.50 | 1.51 | 95.0 | 42.2 |
| 50 | 1.27 | 1.73 | 95.2 | 44.9 |
| 60 | 1.14 | 1.98 | 95.5 | 48.8 |
| 80 | 1.02 | 2.51 | 98.2 | 64.6 |
| 90 | 1.01 | 2.79 | 100.6 | 78.5 |

may be used, but for components of comparable volatility, convergence is likely to be more rapid if Eq. (8-8.14) is used for $x_1 > \frac{1}{2}$ and the analogous equation for component 2 is used when $x_1 < \frac{1}{2}$. However, if one component is much more volatile than the other, the equation for that component is likely to be more useful. Table 8-7 presents calculated results at 1.33 bar. Unfortunately, no experimental results at this pressure are available for comparison.

The two simple examples above illustrate the essential steps for calculating vapor-liquid equilibria from limited experimental data. Because of their illustrative nature, these examples are intentionally simplified, and for more accurate results it is desirable to replace some of the details by more sophisticated techniques. For example, it may be worthwhile to include corrections for vapor phase nonideality and perhaps the Poynting factor, i.e., to relax the simplifying assumption $\mathcal{F}_i = 1$ in Eq. (8-4.1). At the modest pressures encountered here, however, such modifications are likely to have a small effect. A more important change would be to replace the van Laar equation with a better equation for the activity coefficients, e.g., the Wilson equation or the UNIQUAC equation. If this is done, the calculational procedure is the same but the details of computation are more complex. Because of algebraic simplicity, the van Laar equations can easily be linearized, and therefore a convenient graphical procedure can be used to find the van Laar constants.† An equation like UNIQUAC or that of Wilson cannot easily be linearized, and therefore, for practical application, it is necessary to use a computer for data reduction to find the binary constants which appear in the equation.

In Examples 8-1 and 8-2 we have not only made simplifications in the thermodynamic relations but have also neglected to take into quantitative consideration the effect of experimental error.

†The three-suffix Margules equation is also easily linearized, as shown by H. C. Van Ness, "Classical Thermodynamics of Nonelectrolyte Solutions," p. 129, Pergamon, New York, 1964.

It is beyond the scope of this chapter to discuss in detail the highly sophisticated statistical methods now available for optimum reduction of vapor-liquid equilibrium data. Nevertheless, a very short discussion may be useful as an introduction for readers who want to obtain the highest possible accuracy from the available data.

A particularly effective data reduction method is described by Fabries and Renon [33] and by Anderson, Abrams, and Grens [7], who base their analysis on the principle of maximum likelihood while taking into account probable experimental errors in all experimentally determined quantities.

To illustrate the general ideas, we define a calculated pressure (constraining function) by

$$P^c = \exp\left(x_1 \ln \frac{\gamma_1 x_1 f^L_{\text{pure } 1}}{y_1 \phi_1} + x_2 \ln \frac{\gamma_2 x_2 f^L_{\text{pure } 2}}{y_2 \phi_2}\right) \tag{8-8.15}$$

where $f^L_{\text{pure } i}$ is at system temperature and pressure. The most probable values of the parameters (appearing in the function chosen for $g^E$) are those which minimize the function $I$:

$$I = \sum_i \left( \frac{(x_i^0 - x_i^M)^2}{\sigma^2_{xi}} + \frac{(y_i^0 - y_i^M)^2}{\sigma^2_{yi}} + \frac{(P_i^0 - P_i^M)^2}{\sigma^2_{pi}} + \frac{(T_i^0 - T_i^M)^2}{\sigma^2_{Ti}} \right) \tag{8-8.16}$$

Superscript $M$ indicates a measured value, and superscript 0 indicates an estimate of the true value of the variable. The $\sigma^2$'s are estimates of the variances of the measured values, i.e., an indication of the probable experimental uncertainty. These may vary from one point to another but need not.

By using experimental $P$-$T$-$x$-$y$ data and the UNIQUAC equation with estimated parameters $u_{12} - u_{22}$ and $u_{21} - u_{11}$, we obtain estimates of $x_i^0$, $y_i^0$, $T_i^0$, and $P_i^0$. The last of these is found from Eq. (8-8.15). We then evaluate $I$, having previously set average variances $\sigma_x^2$, $\sigma_y^2$, $\sigma_P^2$, and $\sigma_T^2$ from a critical inspection of the data's quality. Upon changing the estimate of UNIQUAC parameters, we calculate a new $I$; with a suitable computer program, we search for the parameters which minimize $I$. Convergence is achieved when, from one iteration to the next, the relative change in $I$ is less than $10^{-5}$. After the last iteration, the variance of fit $\sigma_F^2$ is given by

$$\sigma_F^2 = \frac{I}{D - L} \tag{8-8.17}$$

where $D$ is the number of data points and $L$ is the number of adjustable parameters.

Since all experimental data have some experimental uncertainty, and since any equation for $g^E$ can provide only an approximation to the experimental results, it follows that the parameters obtained from data reduction are not unique; there are many sets of parameters which can equally

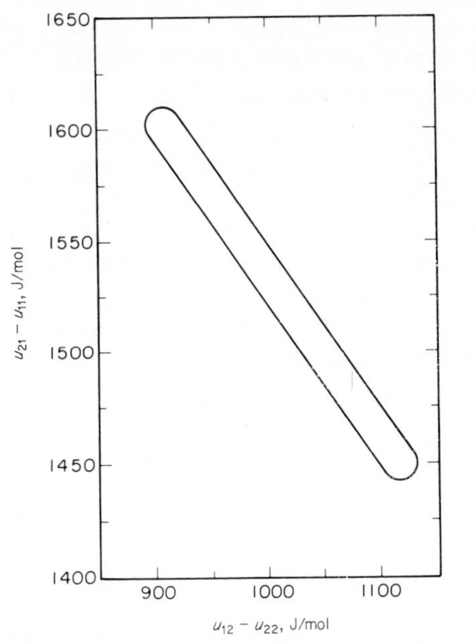


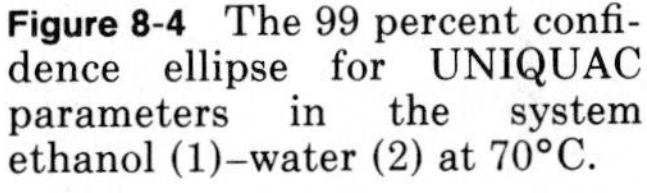
**Figure 8-4** The 99 percent confidence ellipse for UNIQUAC parameters in the system ethanol (1)–water (2) at 70°C.

well represent the experimental data within experimental uncertainty. To illustrate this lack of uniqueness, Fig. 8-4 shows results of data reduction for the binary mixture ethanol (1)–water at 70°C. Experimental data reported by Mertl [76] were reduced by using the UNIQUAC equation with the variances

$$\sigma_x = 10^{-3} \qquad \sigma_y = 10^{-2} \qquad \sigma_p = 6.7 \times 10^{-4} \text{ bar} \qquad \sigma_T = 0.1 \text{ K}$$

For this binary system, the fit is very good; $\sigma_F^2 = 5 \times 10^{-4}$.

The ellipse in Fig. 8-4 clearly shows that, although parameter $u_{21} - u_{11}$ is strongly correlated with parameter $u_{12} - u_{22}$, there are many sets of these parameters that can equally well represent the data. The experimental data used in data reduction are not sufficient to fix a unique set of "best" parameters. Realistic data reduction can determine only a region of parameters.†

While Fig. 8-4 pertains to the UNIQUAC equation, similar results are obtained when other equations for $g^E$ are used; only a region of acceptable parameters can be obtained from $P$-$T$-$y$-$x$ data. For a two-parameter

†Instead of the restraint given by Eq. (8-8.15), it is sometimes preferable to use instead two restraints; first, Eq. (8-8.18), and second,

$$y_1 = \frac{x_1\gamma_1 f_1^0/\phi_1}{x_1\gamma_1 f_1^0/\phi_1 + x_2\gamma_2 f_2^0/\phi_2}$$

or the corresponding equation for $y_2$.

equation this region is represented by an area; for a three-parameter equation it is represented by a volume. If the equation for $g^E$ is suitable for the mixture, the region of acceptable parameters shrinks as the quality and quantity of the experimental data increase. However, considering the limits of both theory and experiment, it is unreasonable to expect this region to shrink to a single point.

As indicated by numerous authors, notably Abbott and Van Ness [3], experimental errors in vapor composition $y$ are usually larger than those in experimental pressure $P$, temperature $T$, and liquid phase composition $x$. Therefore, a relatively simple fitting procedure is provided by reducing only $P$-$x$-$T$ data; $y$ data, even if available, are not used.† The essential point is to minimize the deviation between calculated and observed pressures.

The pressure is calculated according to

$$P_{\text{calc}} = y_1 P + y_2 P = \gamma_1 x_1 P_{\text{vp}1} \mathcal{F}_1 + \gamma_2 x_2 P_{\text{vp}2} \mathcal{F}_2 \tag{8-8.18}$$

where $\mathcal{F}_i$ is given by Eq. (8-4.2).

Thermodynamically consistent equations are now chosen to represent $\gamma_1$ and $\gamma_2$ as functions of $x$ (and perhaps $T$); some are suggested in Table 8-3. These equations contain a number of adjustable binary parameters. With a computer, these parameters can be found by minimizing the deviation between calculated and measured pressures.

At low pressures, we can assume that $\mathcal{F}_1 = \mathcal{F}_2 = 1$. However, at higher pressures, correction factors $\mathcal{F}_1$ and $\mathcal{F}_2$ are functions of pressure, temperature, and vapor compositions $y_1$ and $y_2$; these compositions are calculated from

$$y_1 = \frac{\gamma_1 x_1 P_{\text{vp}1} \mathcal{F}_1(P, T, y)}{P} \quad \text{and} \quad y_2 = \frac{\gamma_2 x_2 P_{\text{vp}2} \mathcal{F}_2(P, T, y)}{P} \tag{8-8.19}$$‡

The data reduction scheme, then, is iterative; to get started, it is necessary first to assume an estimated $y$ for each $x$. After the first iteration, a new set of estimated $y$'s is found from Eq. (8-8.19). Convergence is achieved when, following a given iteration, the $y$'s calculated differ negligibly from those calculated after the previous iteration and when the pressure deviation is minimized.

---

†This technique is commonly referred to as *Barker's method*.

‡If the Lewis fugacity rule is used to calculate vapor phase fugacity coefficients, $\mathcal{F}_1$ and $\mathcal{F}_2$ depend on pressure and temperature but are independent of $y$. The Lewis rule provides mathematical simplification, but, unfortunately, it is a poor rule. If a computer is available, there is no need to use it.

## 8-9 Multicomponent Vapor-Liquid Equilibria

The equations required to calculate vapor-liquid equilibria in multicomponent systems are, in principle, the same as those required for binary systems. In a system containing $N$ components, we must solve $N$ equations simultaneously: Eq. (8-4.1) for each of the $N$ components. We require the saturation (vapor) pressure of each component, as a pure liquid, at the temperature of interest. If all pure component vapor pressures are low, the total pressure also is low. In that event, the factor $\mathcal{F}_i$ [Eq. (8-4.2)] can often be set equal to unity.

Activity coefficients $\gamma_i$ are found from an expression for the excess Gibbs energy, as discussed in Sec. (8-5). For a mixture of $N$ components, the excess Gibbs energy $G^E$ is defined by

$$G^E = RT \sum_{i=1}^{N} n_i \ln \gamma_i \qquad (8\text{-}9.1)$$

where $n_i$ is the number of moles of component $i$. The molar excess Gibbs energy $g^E$ is simply related to $G^E$ by

$$g^E = \frac{G^E}{n_T} \qquad (8\text{-}9.2)$$

where $n_T$, the total number of moles, is equal to $\sum_{i=1}^{N} n_i$.

Individual activity coefficients can be obtained from $G^E$ upon introducing the Gibbs-Duhem equation for a multicomponent system at constant temperature and pressure. That equation is

$$\sum_{i=1}^{N} n_i \, d \ln \gamma_i = 0 \qquad (8\text{-}9.3)$$

The activity coefficient $\gamma_i$ is found by a generalization of Eq. (8-5.3):

$$RT \ln \gamma_i = \left( \frac{\partial G^E}{\partial n_i} \right)_{T,P,n_j} \qquad (8\text{-}9.4)$$

where $n_j$ indicates that all mole numbers (except $n_i$) are held constant in the differentiation.

The key problem in calculating multicomponent vapor-liquid equilibria is to find an expression for $g^E$ which provides a good approximation for the properties of the mixture. Toward that end, the expressions for $g^E$ for binary systems, shown in Table 8-3, can be extended to multicomponent systems. A few of these are shown in Table 8-8.

**TABLE 8-8 Three Expressions for the Molar Excess Gibbs Energy and Activity Coefficients of Multicomponent Systems Using Only Pure Component and Binary Parameters**

Symbols defined in Table 8-3; the number of components is $N$

| Name | Molar excess Gibbs energy | Activity coefficient for component $i$ |
|---|---|---|
| Wilson | $\dfrac{g^E}{RT} = -\sum_i^N x_i \ln\left(\sum_j^N x_j \Lambda_{ij}\right)$ | $\ln \gamma_i = -\ln\left(\sum_j^N x_j \Lambda_{ij}\right) + 1 - \sum_k^N \dfrac{x_k \Lambda_{ki}}{\sum_j^N x_j \Lambda_{kj}}$ |
| NRTL | $\dfrac{g^E}{RT} = \sum_i^N x_i \dfrac{\sum_j^N \tau_{ji} G_{ji} x_j}{\sum_k^N G_{ki} x_k}$ | $\ln \gamma_i = \dfrac{\sum_j^N \tau_{ji} G_{ji} x_j}{\sum_k^N G_{ki} x_k} + \sum_j^N \dfrac{x_j G_{ij}}{\sum_k^N G_{kj} x_k}\left(\tau_{ij} - \dfrac{\sum_k^N x_k \tau_{kj} G_{kj}}{\sum_k^N G_{kj} x_k}\right)$ |
| UNIQUAC† | $\dfrac{g^E}{RT} = \sum_i^N x_i \ln \dfrac{\Phi_i}{x_i} + \dfrac{z}{2} \sum_i^N q_i x_i \ln \dfrac{\theta_i}{\Phi_i} - \sum_i^N q_i x_i \ln\left(\sum_j^N \theta_j \tau_{ji}\right)$ | $\ln \gamma_i = \ln \dfrac{\Phi_i}{x_i} + \dfrac{z}{2} q_i \ln \dfrac{\theta_i}{\Phi_i} + l_i - \dfrac{\Phi_i}{x_i} \sum_j^N x_j l_j - q_i \ln\left(\sum_j^N \theta_j \tau_{ji}\right) + q_i - q_i \sum_j^N \dfrac{\theta_j \tau_{ij}}{\sum_k^N \theta_k \tau_{kj}}$<br>where<br>$\Phi_i = \dfrac{r_i x_i}{\sum_k^N r_k x_k}$ and $\theta_i = \dfrac{q_i x_i}{\sum_k^N q_k x_k}$ |

†Parameters $q$ and $r$ can be calculated from Eq. (8-10.52).

The excess Gibbs energy concept is particularly useful for multicomponent mixtures because in many cases, to a good approximation, extension from binary to multicomponent systems can be made in such a way that only binary parameters appear in the final expression for $g^E$. When that is the case, a large saving in experimental effort is achieved, since experimental data are then required only for the mixture's constituent binaries, not for the multicomponent mixture itself. For example, activity coefficients in a ternary mixture (components 1, 2, and 3) can often be calculated with good accuracy by using only experimental data for the three binary mixtures: components 1 and 2, components 1 and 3, and components 2 and 3.

Many physical models for $g^E$ for a binary system consider only two-body intermolecular interactions, i.e., interactions between two (but not more) molecules. Because of the short range of molecular interaction between nonelectrolytes, it is often permissible to consider only interactions between molecules that are first neighbors and then to sum all the two-body, first-neighbor interactions. A useful consequence of these simplifying assumptions is that extension to ternary (and higher) systems requires only binary, i.e., two-body, information; no ternary (or higher) constants appear. However, not all physical models use this simplifying assumption, and those which do not often require additional simplifying assumptions if the final expression for $g^E$ is to contain only constants derived from binary data.

To illustrate with the simplest case, consider the two-suffix Margules relation for $g^E$ (Table 8-3). For a binary mixture, this relation is given by Eq. (8-5.5), leading to activity coefficients given by Eqs. (8-5.9) and (8-5.10). The generalization to a system containing $N$ components is

$$g^E = \frac{1}{2}\sum_{i=1}^{N}\sum_{j=1}^{N} A_{ij}x_ix_j \tag{8-9.5}$$

where the factor $\frac{1}{2}$ is needed to avoid counting molecular pairs twice. The coefficient $A_{ij}$ is obtained from data for the $ij$ binary. [In the summation indicated in Eq. (8-9.5), $A_{ii} = A_{jj} = 0$ and $A_{ij} = A_{ji}$.] For a ternary system Eq. (8-9.5) becomes

$$g^E = A_{12}x_1x_2 + A_{13}x_1x_3 + A_{23}x_2x_3 \tag{8-9.6}$$

Activity coefficients are obtained by differentiating Eq. (8-9.6) according to Eq. (8-9.4), remembering that $x_i = n_i/n_T$, where $n_T$ is the total number of moles. Upon performing this differentiation, we obtain for component $k$

$$RT \ln \gamma_k = \sum_{i=1}^{N}\sum_{j=1}^{N} (A_{ik} - \tfrac{1}{2}A_{ij})x_ix_j \tag{8-9.7}$$

For a ternary system, Eq. (8-9.7) becomes

$$RT \ln \gamma_1 = A_{12}x_2^2 + A_{13}x_3^2 + (A_{12} + A_{13} - A_{23})x_2x_3 \tag{8-9.8}$$

$$RT \ln \gamma_2 = A_{12}x_1^2 + A_{23}x_3^2 + (A_{12} + A_{23} - A_{13})x_1x_3 \tag{8-9.9}$$

$$RT \ln \gamma_3 = A_{13}x_1^2 + A_{23}x_2^2 + (A_{13} + A_{23} - A_{12})x_1x_2 \tag{8.9.10}$$

All constants appearing in these equations can be obtained from binary data; no ternary data are required.

Equations (8-9.8) to (8-9.10) follow from the simplest model for $g^E$. This model is adequate only for nearly ideal mixtures, where the molecules of the constituent components are similar in size and chemical nature, e.g., benzene-cyclohexane-toluene. For most mixtures encountered in the chemical process industries, more elaborate models for $g^E$ are required.

First it is necessary to choose a model for $g^E$. Depending on the model chosen, some (or possibly all) of the constants in the model may be obtained from binary data. Second, individual activity coefficients are found by differentiation, as indicated in Eq. (8-9.4).

Once we have an expression for the activity coefficients as functions of liquid phase composition and temperature, we can then obtain vapor-liquid equilibria by solving simultaneously *all* the equations of equilibrium. For every component $i$ in the mixture,

$$y_iP = \gamma_i x_i P_{\text{vp}i} \mathcal{F}_i \tag{8-9.11}$$

where $\mathcal{F}_i$ is given by Eq. (8-4.2).

Since the equations of equilibrium are highly nonlinear, simultaneous solution is almost always achieved only by iteration. Such iterations can be efficiently performed with a computer [99].

**Example 8-3** A simple example illustrating how binary data can be used to predict ternary equilibria is provided by Steele, Poling, and Manley [116], who studied the system 1-butene (1)–isobutane (2)–1,3-butadiene (3) in the range 4.4 to 71°C.

**solution** Steele et al. measured isothermal total pressures of the three binary systems as functions of liquid composition. For the three pure components, the pressures are given as functions of temperature by the Antoine equation

$$\ln P_{\text{vp}} = a + b(c + t)^{-1} \tag{8-9.12}$$

where $P_{\text{vp}}$ is in bars and $t$ is in degrees Celsius. Pure component constants $a$, $b$, and $c$ are shown in Table 8-9.

For each binary system the total pressure $P$ is given by

$$P = \sum_{i=1}^{2} y_iP = \sum_{i=1}^{2} x_i\gamma_iP_{\text{vp}i} \exp\frac{(V_i^L - B_{ii})(P - P_{\text{vp}i})}{RT} \tag{8-9.13}$$

where $\gamma_i$ is the activity coefficient of component $i$ in the liquid mixture, $V_i^L$ is the molar volume of pure liquid $i$, and $B_{ii}$ is the second virial coefficient of pure vapor

**TABLE 8-9 Antoine Constants for 1-Butene (1)–Isobutene (2)–1,3-Butadiene (3) at 4.4 to 71°C [Eq. (8-9.12)] [116]**

| Component | $a$ | $-b$ | $c$ |
|---|---|---|---|
| (1) | 9.37579 | 2259.58 | 247.658 |
| (2) | 9.47209 | 2316.92 | 256.961 |
| (3) | 9.43739 | 2292.47 | 247.799 |

$i$, all at system temperature $T$. Equation (8-9.13) assumes that vapor phase imperfections are described by the (volume explicit) virial equation truncated after the second term (see Sec. 3-5). Also, since the components are chemically similar, and since there is little difference in molecular size, Steele et al. used the Lewis fugacity rule $B_{ij} = (\frac{1}{2})(B_{ii} + B_{jj})$. For each pure component, the quantity $(V_i^L - B_{ii})/RT$ is shown in Table 8-10.

For the molar excess Gibbs energy of the binary liquid phase, a one-parameter (two-suffix) Margules equation was assumed:

$$\frac{g_{ij}^E}{RT} = A'_{ij}x_ix_j \tag{8-9.14}$$

From Eq. (8-9.14) we have

$$\ln \gamma_i = A'_{ij}x_j^2 \qquad \text{and} \qquad \ln \gamma_j = A'_{ij}x_i^2 \tag{8-9.15}$$

Equation (8-9.15) is used at each temperature to reduce the binary, total-pressure data yielding the Margules constant $A'_{ij}$. For the three binaries studied, Margules constants are shown in Table 8-11.

To predict ternary phase equilibria, Steele et al. assume that the molar excess Gibbs energy is given by

$$\frac{g^E}{RT} = A'_{12}x_1x_2 + A'_{13}x_1x_3 + A'_{23}x_2x_3 \tag{8-9.16}$$

Activity coefficients $\gamma_1$, $\gamma_2$, and $\gamma_3$ are then found by differentiation. [See Eqs. (8-9.8) to (8-9.10), noting that $A'_{ij} = A_{ij}/RT$.]

Vapor-liquid equilibria are found by writing for each component

$$y_iP = \gamma_ix_iP_{\text{vp}i}\mathcal{F}_i \tag{8-9.17}$$

**TABLE 8-10 Pure Component Parameters for 1-Butene (1)–Isobutane (2)–1,3-Butadiene (3) [116]**

| Temperature, | $10^3(V_i^L - B_{ii})/RT$, bar$^{-1}$ | | |
|---|---|---|---|
| °C | (1) | (2) | (3) |
| 4.4 | 35.13 | 38.62 | 33.92 |
| 21 | 33.04 | 33.04 | 31.85 |
| 38 | 28.30 | 28.82 | 27.60 |
| 54 | 24.35 | 25.02 | 23.32 |
| 71 | 21.25 | 22.12 | 20.33 |

**TABLE 8-11 Margules Constants $A'_{ij}$ for Three Binary Mixtures formed by 1-Butene (1), Isobutane (2), and 1,3-Butadiene (3) [116]**

| Temp. °C | $10^3 A'_{12}$ | $10^3 A'_{13}$ | $10^3 A'_{23}$ |
|---|---|---|---|
| 4.4 | 73.6 | 77.2 | 281 |
| 21 | 60.6 | 64.4 | 237 |
| 38 | 52.1 | 54.8 | 201 |
| 54 | 45.5 | 47.6 | 172 |
| 71 | 40.7 | 42.4 | 147 |

where, consistent with earlier assumptions,

$$\mathcal{F}_i = \exp\frac{(V_i^L - B_{ii})(P - P_{\text{vp}i})}{RT} \tag{8-9.18}$$

Steele and coworkers find that predicted ternary vapor-liquid equilibria are in excellent agreement with their ternary data.

**Example 8-4** A simple procedure for calculating multicomponent vapor-liquid equilibria from binary data is to assume that for the multicomponent mixture

$$g^E = \sum_{\substack{\text{all}\\ \text{binary}\\ \text{pairs}}} g_{ij}^E \tag{8-9.19}$$

**solution** To illustrate Eq. (8-9.19), we consider the ternary mixture acetonitrile–benzene–carbon tetrachloride studied by Clarke and Missen [27] at 45°C.

The three sets of binary data were correlated by the Redlich-Kister expansion, which is equivalent to the Margules equation

$$g_{ij}^E = x_i x_j[A + B(x_i - x_j) + C(x_i - x_j)^2 + D(x_i - x_j)^3] \tag{8-9.20}$$

The constants are given in Table 8-12.

When Eq. (8-9.20) for each binary is substituted into Eq. (8-9.19), the excess Gibbs energy of the ternary is obtained. Clarke and Missen compared excess

**TABLE 8-12 Redlich-Kister Constants for the Three Binaries Formed by Acetonitrile (1), Benzene (2), and Carbon Tetrachloride (3) at 45°C [see Eq. (8-9.20)] [27]**

| Binary system | | J/mol | | | |
|---|---|---|---|---|---|
| $i$ | $j$ | $A$ | $B$ | $C$ | $D$ |
| 1 | 2 | 2691.6 | −33.9 | 293 | 0 |
| 2 | 3 | 317.6 | −3.6 | 0 | 0 |
| 3 | 1 | 4745.9 | 497.5 | 678.6 | 416.3 |

**TABLE 8-13 Calculated and Observed Molar Excess Gibbs Energies for Acetonitrile (1)–Benzene (2)–Carbon Tetrachloride (3) at 45°C [27]**

Calculations from Eq. (8-9.19)

| Composition | | $g^E$, J/mol | |
|---|---|---|---|
| $x_1$ | $x_2$ | Calc. | Obs. |
| 0.156 | 0.767 | 414 | 431 |
| 0.422 | 0.128 | 1067 | 1063 |
| 0.553 | 0.328 | 808 | 774 |
| 0.673 | 0.244 | 711 | 686 |
| 0.169 | 0.179 | 690 | 724 |
| 0.289 | 0.506 | 711 | 707 |

Gibbs energies calculated in this way with those obtained from experimental data for the ternary system according to the definition

$$g^E = RT(x_1 \ln \gamma_1 + x_2 \ln \gamma_2 + x_3 \ln \gamma_3) \tag{8-9.21}$$

Calculated and experimental excess Gibbs energies were in good agreement, as illustrated by a few results shown in Table 8-13. Comparison between calculated and experimental results for more than 60 compositions showed that the average deviation (without regard to sign) was only 16 J/mol. Since the uncertainty due to experimental error is about 13 J/mol, Clarke and Missen conclude that Eq. (8-9.19) provides an excellent approximation for this ternary system.

Since accurate experimental studies on ternary systems are not plentiful, it is difficult to say to what extent the positive conclusion of Clarke and Missen can be applied to other systems. It appears that, for mixtures of typical organic fluids, Eq. (8-9.19) usually gives reliable results, although some deviations have been observed, especially for systems with appreciable hydrogen bonding. In many cases the uncertainties introduced by assuming Eq. (8-9.19) are of the same magnitude as the uncertainties due to experimental error in the binary data.

**Example 8-5** Although the additivity assumption [Eq. (8-9.19)] often provides a good approximation for strongly nonideal mixtures, there may be noticeable deviations between experimental and calculated multicomponent equilibria. Such deviations, however, are significant only if they exceed experimental uncertainty. To detect significant deviations, data of high accuracy are required, and such data are rare, especially for ternary systems; they are nearly nonexistent for quaternary (and higher) systems. To illustrate, we consider the ternary system chloroform-ethanol-heptane at 50°C studied by Abbott et al. [2]. Highly accurate data were first obtained for the three binary systems. The data were reduced by using Barker's method, as explained by Abbott and Van Ness [3] and elsewhere [97]; the essential feature of this method is that it uses only $P$-$x$ data (at constant temperature); it does not use data for vapor composition $y$.

**TABLE 8-14 Binary Parameters in Eq. (8-9.22) or (8-9.23) and rms Deviation in Total Pressure for the Systems Chloroform–Ethanol–*n*-Heptane at 50°C [2]**

| | Chloroform (1), ethanol (2) | Chloroform (1), heptane (2) | Ethanol (1), heptane (2) |
|---|---|---|---|
| $A'_{12}$ | 0.4713 | 0.3507 | 3.4301 |
| $A'_{21}$ | 1.6043 | 0.5262 | 2.4440 |
| $\alpha_{12}$ | | 0.1505 | 11.1950 |
| $\alpha_{21}$ | | 0.1505 | 2.3806 |
| $\eta$ | | 0 | 9.1369 |
| $\lambda_{12}$ | −0.3651 | | |
| $\lambda_{21}$ | 0.5855 | | |
| rms $\Delta P$, bar | 0.00075 | 0.00072 | 0.00045 |

**solution** To represent the binary data, Abbott et al. considered a five-suffix Margules equation and a modified Margules equation

$$\frac{g^E}{RT} = x_1x_2[A'_{21}x_1 + A'_{12}x_2 - (\lambda_{21}x_1 + \lambda_{12}x_2)x_1x_2] \tag{8-9.22†}$$

$$\frac{g^E}{RT} = x_1x_2\left(A'_{21}x_1 + A'_{12}x_2 - \frac{\alpha_{12}\alpha_{21}x_1x_2}{\alpha_{12}x_1 + \alpha_{21}x_2 + \eta x_1x_2}\right) \tag{8-9.23†}$$

If in Eq. (8-9.22), $\lambda_{21} = \lambda_{12} = D$, and if in Eq. (8-9.23) $\alpha_{12} = \alpha_{21} = D$ and $\eta = 0$, both equations reduce to

$$\frac{g^E}{RT} = x_1x_2(A'_{21}x_1 + A'_{12}x_2 - Dx_1x_2) \tag{8-9.24}$$

which is equivalent to the four-suffix Margules equation shown in Table 8-3. If, in addition, $D = 0$, Eqs. (8-9.22) and (8-9.23) reduce to the three-suffix Margules equation.

For the two binaries chloroform-heptane and chloroform-ethanol, experimental data were reduced by using Eq. (8-9.22); however, for the binary ethanol-heptane, Eq. (8-9.23) was used. Parameters reported by Abbott et al. are shown in Table 8-14. With these parameters, calculated total pressures for each binary are in excellent agreement with those measured.

For the ternary, Abbott and coworkers expressed the excess Gibbs energy by

$$\frac{g^E_{123}}{RT} = \frac{g^E_{12}}{RT} + \frac{g^E_{13}}{RT} + \frac{g^E_{23}}{RT} + (C_0 - C_1x_1 - C_2x_2 - C_3x_3)x_1x_2x_3 \tag{8-9.25}$$

where $C_0$, $C_1$, $C_2$, and $C_3$ are ternary constants and $G^E_{ij}$ is given by Eq. (8-9.22) or (8-9.23) for the $ij$ binary. Equation (8-9.25) successfully reproduced the ternary data within experimental error (rms $\Delta P = 0.0012$ bar).

Abbott et al. considered two simplifications:

Simplification *a*: $C_0 = C_1 = C_2 = C_3 = 0$

†The $\alpha$'s and $\lambda$'s are not to be confused with those used in the NRTL and Wilson equations.

Simplification $b$: $C_1 = C_2 = C_3 = 0 \qquad C_0 = \frac{1}{2}\sum\sum_{i \neq j} A'_{ij}$

where the $A'_{ij}$'s are the binary parameters shown in Table 8-14. Simplification $b$ was first proposed by Wohl in 1953 [132] on semitheoretical grounds. When calculated total pressures for the ternary system were compared with experimental results, the deviations exceeded the experimental uncertainty.

| Simplification | rms $\Delta P$, bar |
|---|---|
| $a$ | 0.0517 |
| $b$ | 0.0044 |

These results suggest that Wohl's approximation (simplification $b$) provides significant improvement over the additivity assumption for $g^E$ (simplification $a$). However, one cannot generalize from results for one system. Abbott et al. made similar studies for another ternary (acetone-chloroform-methanol) and found that for this system simplification $a$ gave significantly better results than simplification $b$, although both simplifications produced errors in total pressure beyond the experimental uncertainty.

Although the results of Abbott and coworkers illustrate the limits of predicting ternary (or higher) vapor-liquid equilibria for nonelectrolyte mixtures from binary data only, these limitations are rarely serious for engineering work. As a practical matter, it is common that experimental uncertainties in binary data are as large as the errors which result when multicomponent equilibria are calculated with some model for $g^E$ by using only parameters obtained from binary data.

Although Eq. (8-9.19) provides a particularly simple approximation, the UNIQUAC equation and the Wilson equation can be generalized to multicomponent mixtures without using that approximation but also without requiring ternary (or higher) parameters. Experience has shown that multicomponent vapor-liquid equilibria can usually be calculated with satisfactory engineering accuracy by using the Wilson equation, the NRTL equation, or the UNIQUAC equation provided that care is exercised in obtaining binary parameters.

**Example 8-6** A liquid mixture at 1.013 bar contains 4.7 mole percent ethanol (1), 10.7 mole percent benzene (2), and 84.5 mole percent methylcyclopentane (3). Find the bubble point temperature and the composition of the equilibrium vapor.

**solution** There are three unknowns: the bubble point temperature and two vapor phase mole fractions. To find them, we use three equations of equilibrium:

$$y_i \phi_i P = \gamma_i x_i f_i^{OL} \qquad i = 1, 2, 3 \tag{8-9.26}$$

where $y$ is the vapor phase mole fraction and $x$ is the liquid phase mole fraction. Fugacity coefficient $\phi_i$ is given by the truncated virial equation of state

$$\ln \phi_i = \left(2 \sum_{j=1}^{3} y_j B_{ij} - B_M\right) \frac{P}{RT} \tag{8-9.27}$$

where subscript $M$ stands for mixture and

$$B_M = y_1^2 B_{11} + y_2^2 B_{22} + y_3^2 B_{33} + 2y_1y_2B_{12} + 2y_1y_3B_{13} + 2y_2y_3B_{23} \tag{8-9.28}$$

All second virial coefficients $B_{ij}$ are found from the correlation of Hayden and O'Connell as presented in Ref. 48.

The standard-state fugacity $f_i^{OL}$ is the fugacity of pure liquid $i$ at system temperature and system pressure $P$.

$$f_i^{OL} = P_{vp_i} \phi_i^s \exp \frac{V_i^L (P - P_{vp_i})}{RT} \tag{8-9.29}$$

where $P_{vp_i}$ is the saturation pressure (i.e., the vapor pressure) of pure liquid $i$, $\phi_i^s$ is the fugacity coefficient of pure saturated vapor $i$, and $V_i^L$ is the liquid molar volume of pure $i$, all at system temperature $T$.

Activity coefficients are given by the UNIQUAC equation with the following parameters:

**Pure Component Parameters**

| Component | $r$ | $q$ | $q'$ |
|---|---|---|---|
| 1 | 2.11 | 1.97 | 0.92 |
| 2 | 3.19 | 2.40 | 2.40 |
| 3 | 3.97 | 3.01 | 3.01 |

**Binary Parameters**

$$\tau_{ij} = \exp\left(-\frac{a_{ij}}{T}\right) \quad \text{and} \quad \tau_{ji} = \exp\left(-\frac{a_{ji}}{T}\right)$$

| $i$ | $j$ | $a_{ij}$, K | $a_{ji}$, K |
|---|---|---|---|
| 1 | 2 | −128.9 | 997.4 |
| 1 | 3 | −118.3 | 1384 |
| 2 | 3 | −6.47 | 56.47 |

For a bubble point calculation, a useful objective function $F(1/T)$ is

$$F\left(\frac{1}{T}\right) = \ln\left[\sum_{i=1}^{3} K_i x_i\right] \rightarrow \text{zero}$$

where $K_i = y_i/x_i$. In this calculation, the important unkown is $T$ (rather than $y$) because $P_{vp_i}$ is a strong function of temperature, whereas $\phi_i$ is only a weak function of $y$.

A suitable program for these iterative calculations uses the Newton-Raphson method, as discussed in Ref. [99]. This program requires initial estimates of $T$ and $y$.

The calculated bubble point temperature is 335.99 K. At this temperature, the second virial coefficients ($cm^3$/mole) and liquid molar volumes ($cm^3$/mole) are:

$B_{11} = -1155$

$B_{12} = B_{21} = -587 \qquad V_1^L = 61.1$

$B_{22} = -1086 \qquad V_2^L = 93,7$

$B_{23} = B_{32} = -1134 \qquad V_3^L = 118$

$B_{33} = -1186$

$B_{31} = B_{13} = -618$

$B_M = -957.3$

The detailed results at 335.99 K are:

| Component | $\gamma_i$ | $f_i^{OL}$ (bar) | $\phi_i$ | $100y_i$ | |
|---|---|---|---|---|---|
| | | | | Calculated | Observed |
| 1 | 10.58 | 0.521 | 0.980 | 26.1 | 25.8 |
| 2 | 1.28 | 0.564 | 0.964 | 7.9 | 8.4 |
| 3 | 1.03 | 0.739 | 0.961 | 66.0 | 65.7 |

The experimental bubble point temperature is 336.15K. Experimental results are from J. E. Sinor and J. H. Weber, *J. Chem. Eng. Data,* **5:** 243 (1960).
In this particular case, there is very good agreement between calculated and experimental results. Such agreement is not unusual, but it is, unfortunately, not guaranteed. For many mixtures of nonelectrolyte liquids (including water), agreement between calculated and observed VLE is somewhat less satisfactory than that shown in this example. However, if there is serious disagreement between calculated and observed VLE, do not give up. There may be some error in the calculation, or there may be some error in the data, or both.

## 8-10 Estimation of Activity Coefficients

As discussed in Secs. 8-5 and 8-6, activity coefficients in binary liquid mixtures can often be estimated from a few experimental vapor-liquid equilibrium data for the mixtures by using some empirical (or semiempirical) excess function, as shown in Table 8-3. The excess functions provide a thermodynamically consistent method for interpolating and extrapolating limited binary experimental mixture data and for extending binary data to multicomponent mixtures. Frequently, however, few or no mixture data are at hand, and it is necessary to estimate activity coefficients from some suitable correlation. Unfortunately, few such correlations have been established. Theoretical understanding of liquid mixtures is still in an early stage. There has been progress for simple mixtures containing small, spherical, nonpolar molecules, e.g., argon-xenon, but little useful theory is available for mixtures containing larger molecules, especially if they are polar or form hydrogen bonds. Therefore, the few available correlations are essentially empirical. This means that predictions of activity coefficients can be made only for systems similar to those used to establish the empirical correlation. Even with this restriction, it must be emphasized that, with few exceptions, the accuracy of prediction is not likely to be high whenever predictions for a binary system do not utilize at least some reliable binary data for that system or for another that is closely related. In the following sections we summarize a few of the activity coefficient correlations which are useful for chemical engineering applications.

### Regular solution theory

Following ideas first introduced by van der Waals and van Laar, Hildebrand and Scatchard, working independently [54], showed that for binary mixtures of nonpolar molecules, activity coefficients $\gamma_1$ and $\gamma_2$ can be expressed by

$$RT \ln \gamma_1 = V_1^L \Phi_2^2 (c_{11} + c_{22} - 2c_{12}) \tag{8-10.1}$$

$$RT \ln \gamma_2 = V_2^L \Phi_1^2 (c_{11} + c_{22} - 2c_{12}) \tag{8-10.2}$$

where $V_i^L$ is the liquid molar volume of pure liquid $i$ at temperature $T$, $R$ is the gas constant, and volume fraction $\Phi_1$ and $\Phi_2$ are defined by

$$\Phi_1 = \frac{x_1 V_1^L}{x_1 V_1^L + x_2 V_2^L} \tag{8-10.3}$$

$$\Phi_2 = \frac{x_2 V_2^L}{x_1 V_1^L + x_2 V_2^L} \tag{8-10.4}$$

with $x$ denoting the mole fraction.

For pure liquid $i$, the cohesive energy density $c_{ii}$ is defined by

$$c_{ii} = \frac{\Delta U_i}{V_i^L} \tag{8-10.5}$$

where $\Delta U_i$ is the energy required isothermally to evaporate liquid $i$ from the saturated liquid to the ideal gas. At temperatures well below the critical,

$$\Delta U_i \approx \Delta H_{v_i} - RT \tag{8-10.6}$$

where $\Delta H_{v_i}$ is the molar enthalpy of vaporization of pure liquid $i$ at temperature $T$.

Cohesive energy density $c_{12}$ reflects intermolecular forces between molecules of component 1 and 2; this is the key quantity in Eqs. (8-10.1) and (8-10.2). Formally, $c_{12}$ can be related to $c_{11}$ and $c_{22}$ by

$$c_{12} = (c_{11} c_{22})^{1/2} (1 - \ell_{12}) \tag{8-10.7}$$

where $\ell_{12}$ is a binary parameter, positive or negative, but small compared with unity. Upon substitution, Eqs. (8-10.1) and (8-10.2) can be rewritten

$$RT \ln \gamma_1 = V_1^L \Phi_2^2[(\delta_1 - \delta_2)^2 + 2\ell_{12}\delta_1\delta_2] \tag{8-10.8}$$

$$RT \ln \gamma_2 = V_2^L \Phi_1^2[(\delta_1 - \delta_2)^2 + 2\ell_{12}\delta_1\delta_2] \tag{8-10.9}$$

where solubility parameter $\partial_i$ is defined by

$$\delta_i = (c_{ii})^{1/2} = \left(\frac{\Delta U_i}{V_i^L}\right)^{1/2} \tag{8-10.10}$$

For a first approximation, Hildebrand and Scatchard assume that $\ell_{12} = 0$. In that event, Eqs. (8-10.8) and (8-10.9) contain no binary parameter, and activity coefficients $\gamma_1$ and $\gamma_2$ can be predicted using only pure component data.

Although $\delta_1$ and $\delta_2$ depend on temperature, the theory of regular solutions assumes that the excess entropy is zero. It then follows that, at constant composition,

$$RT \ln \gamma_i = \text{const} \tag{8-10.11}$$

Therefore, the right-hand sides of Eqs. (8-10.8) and (8-10.9) may be evaluated at any convenient temperature provided that all quantities are calculated at the same temperature. For many applications the customary convenient temperature is 25°C. A few typical solubility parameters and molar liquid volumes are shown in Table 8-15, and some calculated vapor-liquid equilibria (assuming $\ell_{12} = 0$) are shown in Figs. 8-5 to 8-7. For typical nonpolar mixtures, calculated results are often in good agreement with experiment.

The regular solution equations are readily generalized to multicomponent mixtures. For component $k$

$$RT \ln \gamma_k = V_k^L \sum_i \sum_j (A_{ik} - \tfrac{1}{2}A_{ij})\Phi_i\Phi_j \tag{8-10.12}$$

where $$A_{ij} = (\delta_i - \delta_j)^2 + 2\ell_{ij}\delta_i\delta_j \tag{8-10.13}$$

If all binary parameters $\ell_{ij}$ are assumed equal to zero, Eq. (8-10.12) simplifies to

$$RT \ln \gamma_k = V_k^L(\delta_k - \bar{\delta})^2 \tag{8-10.14}$$

where $$\bar{\delta} = \sum_i \Phi_i\delta_i \tag{8-10.15}$$

where the summation refers to all components, including component $k$.

The simplicity of Eq. (8-10.14) is striking. It says that, in a multicomponent mixture, activity coefficients for all components can be calculated

**TABLE 8-15 Molar Liquid Volumes and Solubility Parameters of Some Nonpolar Liquids**

| | $V^L$, cm$^3$ mol$^{-1}$ | $\delta$, (J cm$^{-3}$)$^{1/2}$ |
|---|---|---|
| Liquefied gases at 90 K: | | |
| Nitrogen | 38.1 | 10.8 |
| Carbon monoxide | 37.1 | 11.7 |
| Argon | 29.0 | 13.9 |
| Oxygen | 28.0 | 14.7 |
| Methane | 35.3 | 15.1 |
| Carbon tetrafluoride | 46.0 | 17.0 |
| Ethane | 45.7 | 19.4 |
| Liquid solvents at 25°C: | | |
| Perfluoro-*n*-heptane | 226 | 12.3 |
| Neopentane | 122 | 12.7 |
| Isopentane | 117 | 13.9 |
| *n*-Pentane | 116 | 14.5 |
| *n*-Hexane | 132 | 14.9 |
| 1-Hexene | 126 | 14.9 |
| *n*-Octane | 164 | 15.3 |
| *n*-Hexadecane | 294 | 16.3 |
| Cyclohexane | 109 | 16.8 |
| Carbon tetrachloride | 97 | 17.6 |
| Ethyl benzene | 123 | 18.0 |
| Toluene | 107 | 18.2 |
| Benzene | 89 | 18.8 |
| Styrene | 116 | 19.0 |
| Tetrachloroethylene | 103 | 19.0 |
| Carbon disulfide | 61 | 20.5 |
| Bromine | 51 | 23.5 |

at any composition and temperature by using only solubility parameters and molar liquid volumes for the pure components. For mixtures of hydrocarbons, Eq. (8-10.14) often provides a good approximation.

Although binary parameter $\ell_{12}$ is generally small compared with unity in nonpolar mixtures, its importance may be significant, especially if the

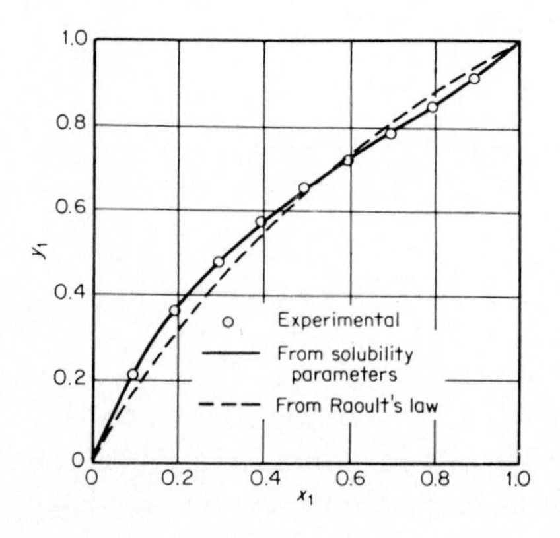


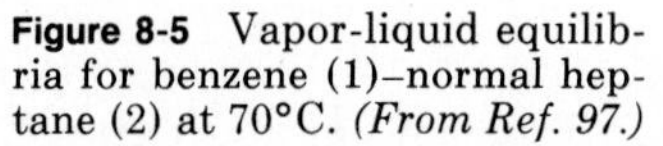
**Figure 8-5** Vapor-liquid equilibria for benzene (1)–normal heptane (2) at 70°C. *(From Ref. 97.)*

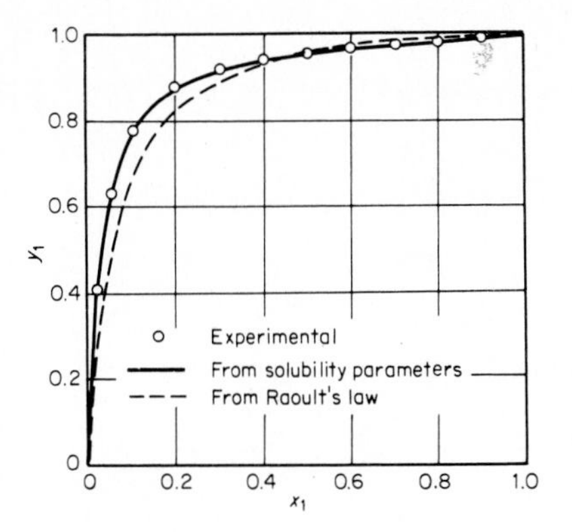


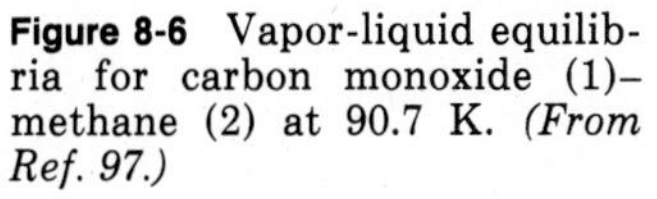

**Figure 8-6** Vapor-liquid equilibria for carbon monoxide (1)–methane (2) at 90.7 K. *(From Ref. 97.)*

difference between $\delta_1$ and $\delta_2$ is small. To illustrate, suppose $T = 300$ K, $V_1^L = 100$ cm$^3$/mol, $\delta_1 = 14.3$, and $\delta_2 = 15.3$ (J/cm$^3$)$^{1/2}$. At infinite dilution ($\Phi_2 = 1$) we find from Eq. (8-10.8) that $\gamma_1^\infty = 1.04$ when $\ell_{12} = 0$. However, if $\ell_{12} = 0.01$, we obtain $\gamma_1^\infty = 1.24$, and if $\ell_{12} = 0.03$, $\gamma_1^\infty = 1.77$. These illustrative results indicate that calculated activity coefficients are often sensitive to small values of $\ell_{12}$ and that much improvement in predicted results can often be achieved when just one binary datum is available for evaluating $\ell_{12}$.

Efforts to correlate $\ell_{12}$ have met with little success. In their study of binary cryogenic mixtures, Bazúa and Prausnitz [13] found no satisfactory variation of $\ell_{12}$ with pure component properties, although some rough trends were found by Cheung and Zander [26] and by Preston and Prausnitz [101]. In many typical cases $\ell_{12}$ is positive and becomes larger as the differences in molecular size and chemical nature of the components increase. For example, for carbon dioxide–paraffin mixtures at low temperatures, Preston found that $\ell_{12} = -0.02$ (methane); $+0.08$ (ethane); $+0.08$ (propane); $+0.09$ (butane).

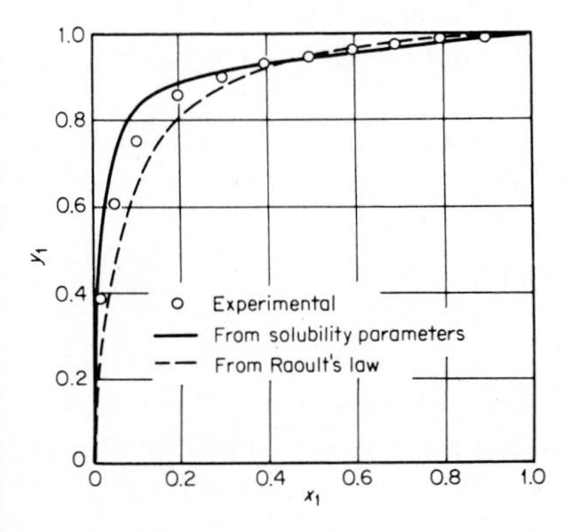


**Figure 8-7** Vapor–liquid equilibria for neopentane (1)–carbon tetrachloride (2) at 0°C. *(From Ref. 97.)*

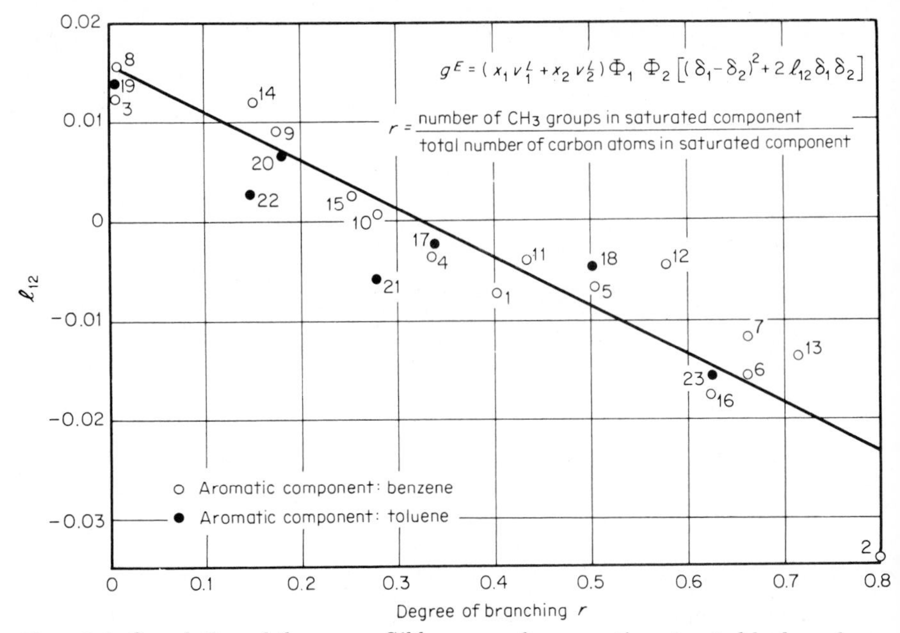


**Figure 8-8** Correlation of the excess Gibbs energy for aromatic–saturated hydrocarbon mixtures at 50°C. Numbers relate to list of binary systems in Ref. 38, table 1. *(From Ref. 38.)*

Since $\ell_{12}$ is an essentially empirical parameter, it depends on temperature. However, for typical nonpolar mixtures over a modest range of temperature, that dependence is usually small.

For mixtures of aromatic and saturated hydrocarbons, Funk and Prausnitz [38] found a systematic variation of $\ell_{12}$ with the structure of the saturated component, as shown in Fig. 8-8. In that case a good correlation could be established because experimental data are relatively plentiful and because the correlation is severely restricted with respect to the chemical nature of the components. Figure 8-9 shows the effect of $\ell_{12}$ on calculating relative volatility in a typical binary system considered by Funk and Prausnitz.

Our inability to correlate $\ell_{12}$ for a wide variety of mixtures follows from our lack of understanding of intermolecular forces, especially between molecules at short separations.

Several authors have tried to extend regular solution theory to mixtures containing polar components; but unless the classes of components considered are restricted, such extension has only semiquantitative significance. In establishing the extensions, the cohesive energy density is divided into separate contributions from nonpolar (dispersion) forces and from polar forces:

$$\left(\frac{\Delta U}{V^L}\right)_{\text{total}} = \left(\frac{\Delta U}{V^L}\right)_{\text{nonpolar}} + \left(\frac{\Delta U}{V^L}\right)_{\text{polar}} \tag{8-10.16}$$

Equations (8-10.1) and (8-10.2) are used with the substitutions

$$c_{11} = \tau_1^2 + \lambda_1^2 \tag{8-10.17}$$

$$c_{22} = \tau_2^2 + \lambda_2^2 \tag{8-10.18}$$

$$c_{12} = \lambda_1\lambda_2 + \tau_1\tau_2 + \psi_{12} \tag{8-10.19}$$

where $\lambda_i$ is the nonpolar solubility parameter $[\lambda_i^2 = (\Delta U_i/V_i^L)_{\text{nonpolar}}]$ and $\tau_i$ is the polar solubility parameter $[\tau_i^2 = (\Delta U_i/V_i^L)_{\text{polar}}]$. The binary parameter $\psi_{12}$ is not negligible, as shown by Weimer and Prausnitz [128] in their correlation of activity coefficients at infinite dilution for hydrocarbons in polar non-hydrogen-bonding solvents.

Further extension of the Scatchard-Hildebrand equation to include hydrogen-bonded components makes little sense theoretically, since the

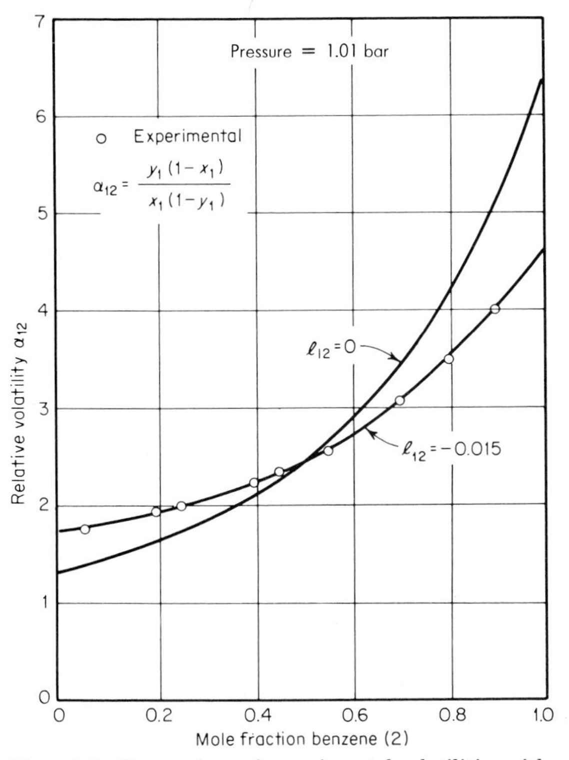


**Figure 8-9** Comparison of experimental volatilities with volatilities calculated by Scatchard-Hildebrand theory for 2,2-dimethylbutane (1)–benzene (2). *(From Ref. 38.)*

assumptions of regular solution theory are seriously in error for mixtures containing such components. Nevertheless, some semiquantitative success has been achieved by Hansen et al. [47] and others [25] interested in establishing criteria for formulating solvents for paints and other surface coatings. Also, Null and Palmer [85] have used extended solubility parameters for establishing an empirical correlation of activity coefficients.

### Activity coefficients at infinite dilution

Experimental activity coefficients at infinite dilution are particularly useful for calculating the parameters needed in an expression for the excess Gibbs energy (Table 8-3). In a binary mixture, suppose experimental data are available for infinite-dilution activity coefficients $\gamma_1^\infty$ and $\gamma_2^\infty$. These can be used to evaluate two adjustable constants in any desired expression for $g^E$. For example, consider the van Laar equation

$$g^E = Ax_1x_2\left(x_1\frac{A}{B} + x_2\right)^{-1} \tag{8-10.20}$$

As indicated in Sec. 8-5, this gives

$$RT \ln \gamma_1 = A\left(1 + \frac{A}{B}\frac{x_1}{x_2}\right)^{-2} \tag{8-10.21}$$

and

$$RT \ln \gamma_2 = B\left(1 + \frac{B}{A}\frac{x_2}{x_1}\right)^{-2} \tag{8-10.22}$$

In the limit, as $x_1 \to 0$ or as $x_2 \to 0$, Eqs. (8-10.21) and (8-10.22) become

$$RT \ln \gamma_1^\infty = A \tag{8-10.23}$$

and

$$RT \ln \gamma_2^\infty = B \tag{8-10.24}$$

Calculation of parameters from $\gamma^\infty$ data is particularly simple for the van Laar equation, but in principle, similar calculations can be made by using any two-parameter equation for the excess Gibbs energy. If a three-parameter equation, e.g., NRTL, is used, an independent method must be chosen to determine the third parameter $\alpha_{12}$.

In recent years, relatively simple experimental methods have been developed for rapid determination of activity coefficients at infinite dilution. These are based on gas-liquid chromatography and on ebulliometry [13, 62, 85, 135, 137, 138].

Schreiber and Eckert [109] have shown that if reliable values of $\gamma_1^\infty$ and $\gamma_2^\infty$ are available, either from direct experiment or from a correla-

**TABLE 8-16 Fit of Binary Data Using Limiting Activity Coefficients in the Wilson Equation [109]**

| | | Average absolute error in calc. $y \times 10^3$ | |
|---|---|---|---|
| System and $\gamma^\infty$ | Temp., °C | All points | $\gamma_1^\infty$ and $\gamma_2^\infty$ only |
| Acetone (1.65)–benzene (1.52) | 45 | 2 | 4 |
| Carbon tetrachloride (5.66)–acetonitrile (9.30) | 45 | 7 | 11 |
| Ethanol (18.1)–*n*-hexane (9.05) | 69–79 | 10 | 12 |
| Chloroform (2.00)–methanol (9.40) | 50 | 10 | 28 |
| Acetone (8.75)–water (3.60) | 100 | 10 | 15 |

tion, it is possible to predict vapor-liquid equilibria over the entire range of composition. For completely miscible mixtures the Wilson equation is particularly useful. Parameters $\Lambda_{12}$ and $\Lambda_{21}$ are found from simultaneous solution of the relations

$$\ln \gamma_1^\infty = -\ln \Lambda_{12} - \Lambda_{21} + 1 \tag{8-10.25}$$

$$\ln \gamma_2^\infty = -\ln \Lambda_{21} - \Lambda_{12} + 1 \tag{8-10.26}$$

Table 8-16 shows some typical results obtained by Schreiber and Eckert. The average error in vapor composition using $\gamma^\infty$ data alone is only slightly larger than that obtained when $\gamma$ data are used over the entire composition range. Schreiber and Eckert also show that reasonable results are often obtained when $\gamma_1^\infty$ or $\gamma_2^\infty$ (but not both) are used. When only one $\gamma^\infty$ is available, it is necessary to use the one-parameter Wilson equation, as discussed earlier. [See Eq. (8-5.12).]

An extensive correlation for $\gamma^\infty$ data in binary systems has been presented by Pierotti, Deal, and Derr [91]. This correlation can be used to predict $\gamma^\infty$ for water, hydrocarbons, and typical organic components, e.g., esters, aldehydes, alcohols, ketones, nitriles, in the temperature region 25 to 100°C. The pertinent equations and tables are summarized by Treybal [123] and, with slight changes, are reproduced in Tables 8-17 and 8-18. The accuracy of the correlation varies considerably from one system to another; provided that $\gamma^\infty$ is not one or more orders of magnitude removed from unity, the average deviation in $\gamma^\infty$ is about 8 percent.

To illustrate use of Table 8-17, an example, closely resembling one given by Treybal, follows.

**Example 8-7** Estimate infinite-dilution activity coefficients for the ethanol-water binary system at 100°C. (Solution appears on p. 299.)

**TABLE 8-17 Correlating Constants for Activity Coefficients at Infinite Dilution; Homologous Series of Solutes and Solvents [91]**

| Solute (1) | Solvent (2) | Temp., °C | $\alpha$ | $\epsilon$ | $\zeta$ | $\eta$ | $\theta$ | Eq. |
|---|---|---|---|---|---|---|---|---|
| $n$-Acids | Water | 25 | −1.00 | 0.622 | 0.490 | . . . | 0 | ($a$) |
| | | 50 | −0.80 | 0.590 | 0.290 | . . . | 0 | ($a$) |
| | | 100 | −0.620 | 0.517 | 0.140 | . . . | 0 | ($a$) |
| $n$-Primary alcohols | Water | 25 | −0.995 | 0.622 | 0.558 | . . . | 0 | ($a$) |
| | | 60 | −0.755 | 0.583 | 0.460 | . . . | 0 | ($a$) |
| | | 100 | −0.420 | 0.517 | 0.230 | . . . | 0 | ($a$) |
| $n$-Secondary alcohols | Water | 25 | −1.220 | 0.622 | 0.170 | 0 | . . . | ($b$) |
| | | 60 | −1.023 | 0.583 | 0.252 | 0 | . . . | ($b$) |
| | | 100 | −0.870 | 0.517 | 0.400 | 0 | . . . | ($b$) |
| $n$-Tertiary alcohols | Water | 25 | −1.740 | 0.622 | 0.170 | . . . | . . . | ($c$) |
| | | 60 | −1.477 | 0.583 | 0.252 | . . . | . . . | ($c$) |
| | | 100 | −1.291 | 0.517 | 0.400 | . . . | . . . | ($c$) |
| Alcohols, general | Water | 25 | −0.525 | 0.622 | 0.475 | 0 | . . . | ($d$) |
| | | 60 | −0.33 | 0.583 | 0.39 | 0 | . . . | ($d$) |
| | | 100 | −0.15 | 0.517 | 0.34 | 0 | . . . | ($d$) |
| $n$-Allyl alcohols | Water | 25 | −1.180 | 0.622 | 0.558 | . . . | 0 | ($a$) |
| | | 60 | −0.929 | 0.583 | 0.460 | . . . | 0 | ($a$) |
| | | 100 | −0.650 | 0.517 | 0.230 | . . . | 0 | ($a$) |
| $n$-Aldehydes | Water | 25 | −0.780 | 0.622 | 0.320 | . . . | 0 | ($a$) |
| | | 60 | −0.400 | 0.583 | 0.210 | . . . | 0 | ($a$) |
| | | 100 | −0.03 | 0.517 | 0 | . . . | 0 | ($a$) |
| $n$-Alkene aldehydes | Water | 25 | −0.720 | 0.622 | 0.320 | . . . | 0 | ($a$) |
| | | 60 | −0.540 | 0.583 | 0.210 | . . . | 0 | ($a$) |
| | | 100 | −0.298 | 0.517 | 0 | . . . | 0 | ($a$) |

|  |  |  |  |  |  |  |  |  |
|---|---|---|---|---|---|---|---|---|
| *n*-Ketones | Water | 25 | −1.475 | 0.622 | 0.500 | 0 | ... | (*b*) |
|  |  | 60 | −1.040 | 0.583 | 0.330 | 0 | ... | (*b*) |
|  |  | 100 | −0.621 | 0.517 | 0.200 | 0 | ... | (*b*) |
| *n*-Acetals | Water | 25 | −2.556 | 0.622 | 0.486 | ... | ... | (*e*) |
|  |  | 60 | −2.184 | 0.583 | 0.451 | ... | ... | (*e*) |
|  |  | 100 | −1.780 | 0.517 | 0.426 | ... | ... | (*e*) |
| *n*-Ethers | Water | 20 | −0.770 | 0.640 | 0.195 | 0 | ... | (*b*) |
| *n*-Nitriles | Water | 25 | −0.587 | 0.622 | 0.760 | ... | 0 | (*a*) |
|  |  | 60 | −0.368 | 0.583 | 0.413 | ... | 0 | (*a*) |
|  |  | 100 | −0.095 | 0.517 | 0 | ... | 0 | (*a*) |
| *n*-Alkene nitriles | Water | 25 | −0.520 | 0.622 | 0.760 | ... | 0 | (*a*) |
|  |  | 60 | −0.323 | 0.583 | 0.413 | ... | 0 | (*a*) |
|  |  | 100 | −0.074 | 0.517 | 0 | ... | 0 | (*a*) |
| *n*-Esters | Water | 20 | −0.930 | 0.640 | 0.260 | 0 | ... | (*b*) |
| *n*-Formates | Water | 20 | −0.585 | 0.640 | 0.260 | ... | 0 | (*a*) |
| *n*-Monoalkyl chlorides | Water | 20 | 1.265 | 0.640 | 0.073 | ... | 0 | (*a*) |
| *n*-Paraffins | Water | 16 | 0.688 | 0.642 | 0 | ... | 0 | (*a*) |
| *n*-Alkyl benzenes | Water | 25 | 3.554 | 0.622 | −0.466 | ... | ... | (*f*) |
| *n*-Alcohols | Paraffins | 25 | 1.960 | 0 | 0.475 | −0.00049 | ... | (*d*) |
|  |  | 60 | 1.460 | 0 | 0.390 | −0.00057 | ... | (*d*) |
|  |  | 100 | 1.070 | 0 | 0.340 | −0.00061 | ... | (*d*) |
| *n*-Ketones | Paraffins | 25 | 0.0877 | 0 | 0.757 | −0.00049 | ... | (*b*) |
|  |  | 60 | 0.016 | 0 | 0.680 | −0.00057 | ... | (*b*) |
|  |  | 100 | −0.067 | 0 | 0.605 | −0.00061 | ... | (*b*) |

**TABLE 8-17 Correlating Constants for Activity Coefficients at Infinite Dilution; Homologous Series of Solutes and Solvents** ***(Continued)***

| Solute (1) | Solvent (2) | Temp., °C | $\alpha$ | $\epsilon$ | $\zeta$ | $\eta$ | $\theta$ | Eq. |
|---|---|---|---|---|---|---|---|---|
| Water | *n*-Alcohols | 25 | 0.760 | 0 | 0 | ... | −0.630 | (*a*) |
| | | 60 | 0.680 | 0 | 0 | ... | −0.440 | (*a*) |
| | | 100 | 0.617 | 0 | 0 | ... | −0.280 | (*a*) |
| Water | *sec*-Alcohols | 80 | 1.208 | 0 | 0 | ... | −0.690 | (*c*) |
| Water | *n*-Ketones | 25 | 1.857 | 0 | 0 | ... | −1.019 | (*c*) |
| | | 60 | 1.493 | 0 | 0 | ... | −0.73 | (*c*) |
| | | 100 | 1.231 | 0 | 0 | ... | −0.557 | (*c*) |
| Ketones | *n*-Alcohols | 25 | −0.088 | 0.176 | 0.50 | −0.00049 | −0.630 | (*g*) |
| | | 60 | −0.035 | 0.138 | 0.33 | −0.00057 | −0.440 | (*g*) |
| | | 100 | −0.035 | 0.112 | 0.20 | −0.00061 | −0.280 | (*g*) |
| Aldehydes | *n*-Alcohols | 25 | −0.701 | 0.176 | 0.320 | −0.00049 | −0.630 | (*h*) |
| | | 60 | −0.239 | 0.138 | 0.210 | −0.00057 | −0.440 | (*h*) |
| Esters | *n*-Alcohols | 25 | 0.212 | 0.176 | 0.260 | −0.00049 | −0.630 | (*g*) |
| | | 60 | 0.055 | 0.138 | 0.240 | −0.00057 | −0.440 | (*g*) |
| | | 100 | 0 | 0.112 | 0.220 | −0.00061 | −0.280 | (*g*) |
| Acetals | *n*-Alcohols | 60 | −1.10 | 0.138 | 0.451 | −0.00057 | −0.440 | (*i*) |
| Paraffins | Ketones | 25 | ... | 0.1821 | ... | −0.00049 | 0.402 | (*j*) |
| | | 60 | ... | 0.1145 | ... | −0.00057 | 0.402 | (*j*) |
| | | 90 | ... | 0.0746 | | −0.00061 | 0.402 | (*j*) |

| Equations |
| --- |
| (a) $\log \gamma_1^\infty = \alpha + \epsilon N_1 + \frac{\zeta}{N_1} + \frac{\theta}{N_2}$ |
| (b) $\log \gamma_1^\infty = \alpha + \epsilon N_1 + \zeta\left(\frac{1}{N_1'} + \frac{1}{N_1''}\right) + \eta(N_1 - N_2)^2$ |
| (c) $\log \gamma_1^\infty = \alpha + \epsilon N_1 + \zeta\left(\frac{1}{N_1'} + \frac{1}{N_1''} + \frac{1}{N_1'''}\right) + \theta\left(\frac{1}{N_2'} + \frac{1}{N_2''}\right)$ |
| (d) $\log \gamma_1^\infty = \alpha + \epsilon N_1 + \zeta\left(\frac{1}{N_1'} + \frac{1}{N_1''} + \frac{1}{N_1'''} - 3\right) + \eta(N_1 - N_2)^2$ |
| (e) $\log \gamma_1^\infty = \alpha + \epsilon N_1 + \zeta\left(\frac{1}{N_1'} + \frac{1}{N_1''} + \frac{2}{N_1'''}\right)$ |
| (f) $\log \gamma_1^\infty = \alpha + \epsilon N_1 + \zeta\left(\frac{1}{N_1} - 4\right)$ |
| (g) $\log \gamma_1^\infty = \alpha + \epsilon \frac{N_1}{N_2} + \zeta\left(\frac{1}{N_1'} + \frac{1}{N_1''}\right) + \eta(N_1 - N_2)^2 + \frac{\theta}{N_2}$ |
| (h) $\log \gamma_1^\infty = \alpha + \epsilon \frac{N_1}{N_2} + \frac{\zeta}{N_1} + \eta(N_1 - N_2)^2 + \frac{\theta}{N_2}$ |
| (i) $\log \gamma_1^\infty = \alpha + \epsilon \frac{N_1}{N_2} + \zeta\left(\frac{1}{N_1'} + \frac{1}{N_1''} + \frac{2}{N_1'''}\right) + \eta(N_1 - N_2)^2 + \frac{\theta}{N_2}$ |
| (j) $\log \gamma_1^\infty = \epsilon \frac{N_1}{N_2} + \eta(N_1 - N_2)^2 + \theta\left(\frac{1}{N_2'} + \frac{1}{N_2''}\right)$ |
| $N_1, N_2$ = total number of carbon atoms in molecules 1 and 2, respectively |
| $N', N'', N'''$ = number of carbon atoms in respective branches of branched compounds, counting the polar grouping; thus, for *t*-butanol, $N' = N'' = N''' = 2$ |

**TABLE 8-18 Correlating Constants for Activity Coefficients at Infinite Dilution; Homologous Series of Hydrocarbons in Specific Solvents [91]**

| Temperature, °C | Solute (1) | Eq. | | Solvent | | | | | | | |
|---|---|---|---|---|---|---|---|---|---|---|---|
| | | | | Heptane | Methyl ethyl ketone | Furfural | Phenol | Ethanol | Triethylene glycol | Diethylene glycol | Ethylene glycol |
| | | | $\eta$ | Value of $\epsilon$ | | | | | | | |
| 25 | | | −0.00049 | 0 | 0.0455 | 0.0937 | 0.0625 | 0.088 | ... | 0.191 | 0.275 |
| 50 | | | −0.00055 | 0 | 0.033 | 0.0878 | 0.0590 | 0.073 | 0.161 | 0.179 | 0.249 |
| 70 | | | −0.00058 | 0 | 0.025 | 0.0810 | 0.0586 | 0.065 | ... | 0.173 | 0.236 |
| 90 | | | −0.00061 | 0 | 0.019 | 0.0686 | 0.0581 | 0.059 | 0.134 | 0.158 | 0.226 |
| | | | | Value of $\theta$ | | | | | | | |
| 25 | | | | 0.2105 | 0.1435 | 0.1152 | 0.1421 | 0.2125 | 0.181 | 0.2022 | 0.275 |
| 70 | | | | 0.1668 | 0.1142 | 0.0836 | 0.1054 | 0.1575 | 0.129 | 0.1472 | 0.2195 |
| 130 | | | | 0.1212 | 0.0875 | 0.0531 | 0.0734 | 0.1035 | 0.0767 | 0.0996 | 0.1492 |
| | | | | Value of $\kappa$ | | | | | | | |
| 25 | | | | 0.1874 | 0.2079 | 0.2178 | 0.2406 | 0.2425 | 0.3124 | 0.3180 | 0.4147 |
| 70 | | | | 0.1478 | 0.1754 | 0.1675 | 0.1810 | 0.1753 | 0.2406 | 0.2545 | 0.3516 |
| 130 | | | | 0.1051 | 0.1427 | 0.1185 | 0.1480 | 0.1169 | 0.1569 | 0.1919 | 0.2772 |
| | Solute (1) | Eq. | $\zeta$ | Value of $\alpha$ | | | | | | | |
| 25 | Paraffins | $(a)$ | 0 | 0 | 0.335 | 0.916 | 0.870 | 0.580 | ... | 0.875 | |
| 50 | | | 0 | 0 | 0.332 | 0.756 | 0.755 | 0.570 | 0.72 | 0.815 | 1.208 |
| 70 | | | 0 | 0 | 0.331 | 0.737 | 0.690 | 0.590 | ... | 0.725 | 1.154 |
| 90 | | | 0 | 0 | 0.330 | 0.771 | 0.620 | 0.610 | 0.68 | [illegible] | [illegible] |

| | | | | | | | | | | | |
|---|---|---|---|---|---|---|---|---|---|---|---|
| 25 | Alkyl cyclohexanes | (a) | −0.200 | 0.18 | 0.70 | [illegible] | [illegible] | [illegible] | . . . | [illegible] | |
| 50 | | | −0.220 | . . . | 0.650 | 1.120 | 1.040 | 1.01 | 1.46 | 1.61 | 2.36 |
| 70 | | | −0.195 | 0.131 | 0.581 | 1.020 | 0.935 | 0.972 | . . . | 1.550 | 2.22 |
| 90 | | | −0.180 | 0.09 | 0.480 | 0.930 | 0.843 | 0.925 | 1.25 | 1.505 | 2.08 |
| 25 | Alkyl benzenes | (a) | −0.466 | 0.328 | 0.277 | 0.67 | 0.694 | 1.011 | . . . | 1.08 | |
| 50 | | | −0.390 | 0.243 | . . . | 0.55 | 0.580 | 0.938 | 0.80 | 1.00 | 1.595 |
| 70 | | | −0.362 | 0.225 | 0.240 | 0.45 | 0.500 | 0.900 | . . . | 0.96 | 1.51 |
| 90 | | | −0.350 | 0.202 | 0.239 | 0.44 | 0.420 | 0.862 | 0.74 | 0.935 | 1.43 |
| 25 | Alkyl naphthalenes | (a) | −0.10 | 0.53 | 0.169 | 0.46 | 0.595 | 1.06 | . . . | 1.00 | |
| 50 | | | −0.14 | 0.53 | 0.141 | 0.40 | 0.54 | 1.03 | 0.75 | 1.00 | 1.92 |
| 70 | | | −0.173 | 0.53 | 0.215 | 0.39 | 0.497 | 1.02 | . . . | 0.991 | 1.82 |
| 90 | | | −0.204 | 0.53 | 0.232 | . . . | 0.445 | . . . | 0.83 | 1.01 | 1.765 |
| 25 | Alkyl tetralins | (a) | 0.28 | 0.244 | 0.179 | 0.652 | 0.378 | . . . | . . . | 1.43 | |
| 50 | | | 0.24 | . . . | . . . | 0.528 | 0.364 | . . . | 1.00 | 1.38 | |
| 70 | | | 0.21 | 0.220 | 0.217 | 0.447 | 0.371 | . . . | . . . | 1.33 | |
| 90 | | | 0.19 | . . . | . . . | 0.373 | 0.348 | . . . | 0.893 | 1.28 | |
| 25 | Alkyl decalins | (a) | −0.43 | . . . | 0.871 | 1.54 | 1.411 | . . . | . . . | 2.46 | |
| 50 | | | −0.368 | . . . | . . . | 1.367 | 1.285 | . . . | 1.906 | 2.25 | |
| 70 | | | −0.355 | 0.356 | 0.80 | 1.253 | 1.161 | . . . | . . . | 2.07 | |
| 90 | | | −0.320 | . . . | . . . | 1.166 | 1.078 | . . . | 1.68 | 2.06 | |
| 25 | Unalkylated aromatics, naphthenes, naphthene aromatics | (b) | 1.176† 1.845‡ | −1.072 | −0.7305 | −0.230 | −0.383 | −0.485 | −0.406 | −0.377 | −0.154 |
| 70 | | | 0.846† 1.362‡ | −0.886 | −0.625 | −0.080 | −0.226 | −0.212 | −0.186 | −0.0775 | −0.0174 |
| 130 | | | 0.544† 0.846‡ | −0.6305 | −0.504 | +0.020 | −0.197 | +0.47 | +0.095 | +0.181 | +0.229 |

**TABLE 8-18 Correlating Constants for Activity Coefficients at Infinite Dilution; Homologous Series of Hydrocarbons in Specific Solvents (*Continued*)**

Equations

(*a*) $\log \gamma_1^{\infty} = \alpha + \epsilon N_p + \dfrac{\zeta}{N_p + 2} + \eta (N_1 + N_2)^2$

(*b*) $\log \gamma_1^{\infty} = \alpha + \theta N_a + \kappa N_n + \xi \left(\dfrac{1}{r} - 1\right)$

where $N_1$, $N_2$ = total number of carbon atoms in molecules 1 and 2, respectively
$N_p$ = number of paraffinic carbon atoms in solute
$N_a$ = number of aromatic carbon atoms, including $=\overset{|}{\text{C}}-$, $=$CH—, ring-juncture naphthenic carbons $-\overset{|}{\underset{|}{\text{C}}}-$H, and naphthenic carbons in the $\alpha$ position to an aromatic nucleus
$N_n$ = number of naphthenic carbon atoms not counted in $N_a$
$r$ = number of rings

*Examples:*

| | | | | | |
|---|---|---|---|---|---|
| Butyl decalin: | $N_p = 4$ | $N_a = 2$ | $N_n = 8$ | $N_1 = 14$ | $r = 2$ |
| Butyl tetralin: | $N_p = 4$ | $N_a = 8$ | $N_n = 2$ | $N_1 = 14$ | $r = 2$ |

**solution** First we find $\gamma^\infty$ for ethanol. Subscript 1 stands for ethanol, and subscript 2 stands for water. From Table 8-17, $\alpha = -0.420$, $\epsilon = 0.517$, $\zeta = 0.230$, $\theta = 0$, and $N_1 = 2$. Using Eq. ($a$) at the end of Table 8-17, we have

$$\log \gamma^\infty = 0.420 + (0.517)(2) + \frac{0.230}{2} = 0.729$$

$$\gamma^\infty \text{ (ethanol)} = 5.875$$

Next, for water, we again use Table 8-17. Now subscript 1 stands for water and subscript 2 stands for ethanol.

$$\alpha = 0.617 \qquad \epsilon = \zeta = 0 \qquad \theta = -0.280 \qquad N_2 = 2$$

$$\log \gamma^\infty = 0.617 - \frac{0.280}{2} = 0.477$$

$$\gamma^\infty \text{ (water)} = 3.0$$

These calculated results are in good agreement with experimental data of Jones et al. [59].

An alternative method for estimating activity coefficients at infinite dilution is provided by a modified regular solution theory. MOSCED (modified separation of cohesive energy density) is a model proposed by Thomas and Eckert [119] for predicting limiting activity coefficients ($\gamma^\infty$'s) from pure component parameters only. It is essentially an extension of regular solution theory to polar and associating systems. The extension is based on the assumption that forces contributing to the cohesive energy density are additive. Forces included are dispersion, orientation, induction, and hydrogen bonding. The five parameters associated with these forces are the dispersion parameter $\lambda$, the induction parameter $q$, the polar parameter $\tau$, and the acidity $\alpha$ and basicity $\beta$ parameters. These modifications affect the magnitudes of the activity coefficients, but they do so in a symmetric way; i.e. both $\gamma_1^\infty$ and $\gamma_2^\infty$ are affected the same way, contrary to experiment. To account for asymmetry, Thomas and Eckert introduce two more parameters, $\psi$ and $\xi$, to account for asymmetry effects resulting from differences in polarity and degree of hydrogen bonding, respectively. These two are functions of the other parameters.

A summary of the equation and a list of the parameters for 144 substances at 20°C are given in Table 8-19. In a binary mixture, the activity coefficient for component 2 at infinite dilution is

$$\ln \gamma_2^\infty = \frac{v_2}{RT}\left[(\lambda_1 - \lambda_2)^2 + \frac{q_1^2 q_2^2(\tau_1 - \tau_2)^2}{\psi_1} + \frac{(\alpha_1 - \alpha_2)(\beta_1 - \beta_2)}{\xi_1}\right] + d_{12} \qquad \text{(8-10.27)}$$

There is a corresponding expression for $\ln \gamma_1^\infty$. Here, $v_2$ is liquid molar volume at 20°C in cm$^3$ mol$^{-1}$ and $d_{12}$ is a Flory-Huggins combinatorial term to account for differences in molecular size:

$$d_{12} = \ln\left(\frac{v_2}{v_1}\right)^{aa} + 1 - \left(\frac{v_2}{v_1}\right)^{aa} \qquad \text{(8-10.28)}$$

**TABLE 8-19 Parameters for MOSCED Model at 20°C**

| RI | $v$ | $\lambda$ | $\tau$ | $q$ | $\alpha$ | $\beta$ | $\psi$ | $\xi$ | $aa$ | ID | Compound |
|---|---|---|---|---|---|---|---|---|---|---|---|
| 1.628 | 72.0 | 9.80 | 0.30 | 1.00 | 0.29 | 0.16 | 1.00 | 1.00 | 0.952 | 1 | Carbon disulfide |
| 1.460 | 96.5 | 8.58 | 0.87 | 1.00 | 0.58 | 0.15 | 1.02 | 1.01 | 0.945 | 1 | Carbon tetrachloride |
| 1.506 | 98.5 | 9.05 | 1.00 | 1.00 | 0.70 | 0.10 | 1.02 | 1.02 | 0.943 | 1 | Bromotrichloromethane |
| 1.446 | 80.7 | 8.43 | 1.95 | 1.00 | 3.05 | 0.06 | 1.16 | 1.11 | 0.914 | 1 | Chloroform |
| 1.496 | 82.7 | 8.95 | 2.00 | 1.00 | 2.90 | 0.10 | 1.17 | 1.12 | 0.911 | 1 | Bromodichloromethane |
| 1.424 | 64.1 | 8.20 | 2.79 | 1.00 | 2.49 | 0.38 | 1.42 | 1.33 | 0.868 | 1 | Dichloromethane |
| 1.542 | 69.6 | 9.41 | 2.90 | 1.00 | 2.00 | 0.10 | 1.45 | 1.31 | 0.870 | 1 | Dibromomethane |
| 1.328 | 40.5 | 7.14 | 2.55 | 1.00 | 7.45 | 7.45 | 1.94 | 3.62 | 0.353 | 1 | Methanol |
| 1.381 | 53.7 | 7.73 | 6.24 | 1.00 | 1.30 | 2.40 | 2.18 | 2.06 | 0.546 | 1 | Nitromethane |
| 1.531 | 79.1 | 9.30 | 1.95 | 1.00 | 0.35 | 0.29 | 1.16 | 1.11 | 0.915 | 1 | Methyl iodide |
| 1.344 | 52.2 | 7.43 | 5.99 | 1.00 | 0.86 | 3.98 | 2.17 | 2.10 | 0.573 | 1 | Acetonitrile |
| 1.438 | 99.6 | 8.35 | 1.40 | 1.00 | 0.50 | 0.20 | 1.06 | 1.04 | 0.933 | 1 | 1,1,1-Trichloroethane |
| 1.471 | 92.7 | 8.70 | 2.40 | 1.00 | 1.50 | 0.20 | 1.28 | 1.20 | 0.984 | 1 | 1,1,2-Trichloroethane |
| 1.445 | 79.1 | 8.42 | 3.13 | 1.00 | 0.79 | 0.51 | 1.53 | 1.37 | 0.854 | 1 | 1,2-Dichloroethane |
| 1.416 | 84.2 | 8.12 | 2.20 | 1.00 | 1.00 | 0.30 | 1.22 | 1.16 | 0.903 | 1 | 1,1-Dichloroethane |
| 1.392 | 72.0 | 7.85 | 4.85 | 1.00 | 0.29 | 2.20 | 2.04 | 1.73 | 0.719 | 1 | Nitroethane |
| 1.513 | 93.6 | 9.12 | 1.66 | 1.00 | 0.31 | 0.27 | 1.10 | 1.07 | 0.926 | 1 | Ethyl iodide |
| 1.424 | 76.5 | 8.20 | 2.04 | 1.00 | 0.32 | 0.23 | 1.18 | 1.12 | 0.912 | 1 | Ethyl bromide |
| 1.361 | 58.4 | 7.51 | 1.36 | 1.00 | 6.19 | 6.19 | 1.48 | 3.43 | 0.564 | 1 | Ethanol |
| 1.366 | 70.4 | 7.57 | 4.82 | 1.00 | 0.38 | 3.31 | 2.04 | 1.78 | 0.716 | 1 | Propionitrile |
| 1.359 | 74.4 | 7.49 | 4.10 | 1.00 | 0.00 | 4.87 | 1.86 | 1.58 | 0.790 | 1 | Acetone |
| 1.360 | 80.4 | 7.50 | 3.32 | 1.00 | 0.00 | 3.83 | 1.60 | 1.41 | 0.846 | 1 | Ethyl formate |
| 1.361 | 79.3 | 7.52 | 3.32 | 1.00 | 0.00 | 3.83 | 1.60 | 1.41 | 0.846 | 1 | Methyl acetate |
| 1.402 | 89.0 | 7.96 | 4.15 | 1.00 | 0.25 | 1.88 | 1.88 | 1.61 | 0.782 | 1 | 1-Nitropropane |
| 1.430 | 77.0 | 8.26 | 4.62 | 1.00 | 0.65 | 10.30 | 2.06 | 2.46 | 0.682 | 1 | Dimethylformamide |
| 1.394 | 90.1 | 7.88 | 4.13 | 1.00 | 0.23 | 1.85 | 1.87 | 1.61 | 0.784 | 1 | 2-Nitropropane |
| 1.503 | 111.6 | 8.83 | 1.40 | 1.00 | 0.14 | 0.15 | 1.06 | 1.04 | 0.934 | 1 | 2-Propyl iodide |
| 1.388 | 88.1 | 7.81 | 1.90 | 1.00 | 0.30 | 0.18 | 1.15 | 1.10 | 0.918 | 1 | *n*-Propyl chloride |
| 1.434 | 90.9 | 8.31 | 1.74 | 1.00 | 0.27 | 0.20 | 1.12 | 1.08 | 0.923 | 1 | *n*-Propyl bromide |
| 1.290 | 88.1 | 6.70 | 0.00 | 1.00 | 0.00 | 0.00 | 1.00 | 1.00 | 0.953 | 1 | Propane |
| 1.386 | 74.8 | 7.79 | 1.16 | 1.00 | 5.28 | 5.28 | 1.34 | 3.34 | 0.670 | 1 | Propanol |
| 1.379 | 89.6 | 7.71 | 3.25 | 1.00 | 0.00 | 4.05 | 1.57 | 1.39 | 0.851 | 1 | Butanone |
| 1.378 | 96.3 | 7.69 | 2.84 | 1.00 | 0.00 | 3.28 | 1.42 | 1.29 | 0.875 | 1 | Methyl propionate |
| 1.407 | 81.1 | 8.02 | 2.30 | 1.00 | 0.00 | 4.58 | 1.25 | 1.17 | 0.902 | 1 | Tetrahydrofuran |
| 1.377 | 97.3 | 7.69 | 2.84 | 1.00 | 0.00 | 3.28 | 1.42 | 1.29 | 0.875 | 1 | Propyl formate |
| 1.372 | 97.8 | 7.64 | 2.84 | 1.00 | 0.00 | 3.28 | 1.42 | 1.29 | 0.875 | 1 | Ethyl acetate |
| 1.422 | 84.2 | 8.08 | 3.32 | 1.00 | 0.00 | 4.14 | 1.60 | 1.41 | 0.846 | 1 | Dioxane |
| 1.397 | 106.0 | 7.91 | 1.62 | 1.00 | 0.25 | 0.15 | 1.09 | 1.06 | 0.927 | 1 | 2-Chlorobutane |
| 1.402 | 104.5 | 7.96 | 1.66 | 1.00 | 0.26 | 0.16 | 1.10 | 1.07 | 0.926 | 1 | *n*-Butyl chloride |
| 1.386 | 109.9 | 7.79 | 1.60 | 1.00 | 0.25 | 0.15 | 1.09 | 1.06 | 0.928 | 1 | *tert*-Butyl chloride |
| 1.440 | 107.4 | 8.37 | 1.52 | 1.00 | 0.24 | 0.17 | 1.08 | 1.05 | 0.930 | 1 | *n*-Butyl bromide |
| 1.333 | 100.4 | 7.20 | 0.00 | 1.00 | 0.00 | 0.00 | 1.00 | 1.00 | 0.953 | 1 | Butane |
| 1.399 | 91.5 | 7.93 | 1.02 | 1.00 | 4.62 | 4.62 | 1.26 | 3.17 | 0.736 | 1 | Butanol |
| 1.510 | 80.5 | 8.57 | 3.16 | 0.90 | 0.68 | 6.70 | 1.40 | 1.72 | 0.812 | 3 | Pyridine |
| 1.430 | 100.8 | 8.27 | 0.58 | 0.80 | 0.00 | 0.37 | 1.00 | 1.00 | 0.950 | 1 | *trans*-1,3-Pentadiene |
| 1.422 | 100.0 | 8.18 | 0.52 | 0.80 | 0.00 | 0.34 | 1.00 | 1.00 | 0.950 | 1 | Isoprene |
| 1.407 | 94.1 | 8.01 | 0.00 | 1.00 | 0.00 | 0.00 | 1.00 | 1.00 | 0.953 | 1 | Cyclopentane |
| 1.387 | 105.9 | 7.80 | 0.26 | 0.90 | 0.00 | 0.21 | 1.00 | 1.00 | 0.952 | 1 | Isopentene |
| 1.383 | 107.0 | 7.76 | 0.25 | 0.90 | 0.00 | 0.21 | 1.00 | 1.00 | 0.952 | 1 | 2-Pentene |
| 1.371 | 109.6 | 7.62 | 0.25 | 0.90 | 0.00 | 0.20 | 1.00 | 1.00 | 0.952 | 1 | 1-Pentene |
| 1.338 | 107.8 | 7.25 | 0.26 | 1.00 | 0.00 | 0.21 | 1.00 | 1.00 | 0.952 | 1 | 3-Methyl-1-butene |
| 1.364 | 111.8 | 7.55 | 0.25 | 0.90 | 0.00 | 0.21 | 1.00 | 1.00 | 0.952 | 1 | 2-Methyl-1-butene |
| 1.388 | 107.3 | 7.81 | 2.50 | 1.00 | 0.00 | 3.20 | 1.31 | 1.21 | 0.893 | 2 | Methyl isopropyl ketone |
| 1.453 | 90.9 | 8.51 | 1.10 | 1.00 | 4.50 | 4.50 | 1.25 | 3.13 | 0.745 | 1 | Cyclopentanol |
| 1.392 | 105.8 | 7.86 | 2.77 | 1.00 | 0.00 | 3.43 | 1.40 | 1.27 | 0.879 | 1 | 3-Pentanone |
| 1.390 | 106.5 | 7.83 | 2.77 | 1.00 | 0.00 | 3.43 | 1.40 | 1.27 | 0.879 | 1 | 2-Pentanone |
| 1.384 | 115.0 | 7.77 | 2.48 | 1.00 | 0.00 | 2.86 | 1.30 | 1.21 | 0.893 | 1 | Propyl acetate |
| 1.388 | 113.7 | 7.81 | 2.48 | 1.00 | 0.00 | 2.86 | 1.30 | 1.21 | 0.893 | 1 | Methyl butylate |
| 1.384 | 114.5 | 7.77 | 2.48 | 1.00 | 0.00 | 2.86 | 1.30 | 1.21 | 0.893 | 1 | Ethyl propionate |
| 1.445 | 124.0 | 8.42 | 1.36 | 1.00 | 0.21 | 0.15 | 1.06 | 1.04 | 0.935 | 1 | *n*-Pentyl bromide |
| 1.358 | 115.3 | 7.48 | 0.00 | 1.00 | 0.00 | 0.00 | 1.00 | 1.00 | 0.953 | 1 | Pentane |
| 1.354 | 116.4 | 7.43 | 0.00 | 0.00 | 0.00 | 0.00 | 1.00 | 1.00 | 0.953 | 1 | Isopentane |
| 1.410 | 108.1 | 8.05 | 0.90 | 1.00 | 4.12 | 4.12 | 1.20 | 2.93 | 0.781 | 1 | 1-Pentanol |
| 1.552 | 102.3 | 8.95 | 3.78 | 0.90 | 0.63 | 1.36 | 1.51 | 1.38 | 0.806 | 3 | Nitrobenzene |
| 1.525 | 101.8 | 8.71 | 1.84 | 0.90 | 0.75 | 0.45 | 1.09 | 1.07 | 0.917 | 3 | Chlorobenzene |
| 1.501 | 89.1 | 8.49 | 1.95 | 0.90 | 0.22 | 0.56 | 1.11 | 1.07 | 0.915 | 3 | Benzene |
| 1.551 | 89.0 | 8.94 | 2.16 | 0.90 | 16.20 | 1.64 | 1.43 | 3.39 | 0.651 | 3 | Phenol |

| RI | $v$ | $\lambda$ | $\tau$ | $q$ | $\alpha$ | $\beta$ | $\psi$ | $\xi$ | $aa$ | ID | Compound |
|---|---|---|---|---|---|---|---|---|---|---|---|
| 1.586 | 91.1 | 9.25 | 4.22 | 0.90 | 3.81 | 2.13 | 1.68 | 2.39 | 0.702 | 3 | Aniline |
| 1.447 | 101.4 | 8.44 | 0.28 | 0.92 | 0.00 | 0.25 | 1.00 | 1.00 | 0.952 | 2 | Cyclohexene |
| 1.451 | 103.6 | 8.48 | 3.05 | 1.00 | 0.00 | 4.82 | 1.50 | 1.34 | 0.863 | 2 | Cyclohexanone |
| 1.392 | 123.8 | 7.85 | 0.23 | 0.92 | 0.00 | 0.18 | 1.00 | 1.00 | 0.952 | 1 | 4-Methyl-1-pentene |
| 1.426 | 108.1 | 8.22 | 0.00 | 1.00 | 0.00 | 0.00 | 1.00 | 1.00 | 0.953 | 2 | Cyclohexane |
| 1.388 | 125.0 | 7.81 | 0.23 | 0.92 | 0.00 | 0.18 | 1.00 | 1.00 | 0.952 | 1 | 1-Hexene |
| 1.383 | 126.7 | 7.75 | 0.23 | 0.92 | 0.00 | 0.18 | 1.00 | 1.00 | 0.952 | 1 | 2-Methyl-1-pentene |
| 1.410 | 112.4 | 8.05 | 0.00 | 1.00 | 0.00 | 0.00 | 1.00 | 1.00 | 0.953 | 1 | Methylcyclopentane |
| 1.400 | 122.6 | 7.95 | 0.23 | 0.92 | 0.00 | 0.18 | 1.00 | 1.00 | 0.952 | 1 | 2-Methyl-2-pentene |
| 1.396 | 125.1 | 7.89 | 2.42 | 1.00 | 0.00 | 3.00 | 1.28 | 1.19 | 0.896 | 1 | 4-Methyl-2-pentanone |
| 1.467 | 105.4 | 8.65 | 0.90 | 1.00 | 4.00 | 4.00 | 1.19 | 2.86 | 0.790 | 2 | Cyclohexanol |
| 1.392 | 132.0 | 7.85 | 2.21 | 1.00 | 0.00 | 2.55 | 1.22 | 1.15 | 0.906 | 1 | Butyl acetate |
| 1.375 | 130.8 | 7.67 | 0.00 | 1.00 | 0.00 | 0.00 | 1.00 | 1.00 | 0.953 | 1 | Hexane |
| 1.375 | 130.3 | 7.67 | 0.00 | 1.00 | 0.00 | 0.00 | 1.00 | 1.00 | 0.953 | 1 | 2,3-Dimethylbutane |
| 1.377 | 129.7 | 7.68 | 0.00 | 1.00 | 0.00 | 0.00 | 1.00 | 1.00 | 0.953 | 1 | 3-Methylpentane |
| 1.369 | 132.9 | 7.60 | 0.00 | 1.00 | 0.00 | 0.00 | 1.00 | 1.00 | 0.953 | 1 | 2,2-Dimethylbutane |
| 1.372 | 131.9 | 7.63 | 0.00 | 1.00 | 0.00 | 0.00 | 1.00 | 1.00 | 0.953 | 1 | 2-Methylpentane |
| 1.418 | 124.7 | 8.14 | 0.82 | 1.00 | 3.73 | 3.73 | 1.17 | 2.68 | 0.812 | 1 | Hexanol |
| 1.401 | 139.0 | 7.52 | 0.53 | 1.00 | 0.00 | 4.98 | 1.00 | 1.00 | 0.950 | 1 | Triethylamine |
| 1.528 | 102.6 | 8.73 | 3.30 | 0.90 | 0.64 | 2.94 | 1.41 | 1.40 | 0.829 | 3 | Benzonitrile |
| 1.539 | 117.2 | 8.84 | 2.71 | 0.90 | 0.74 | 0.93 | 1.26 | 1.20 | 0.875 | 3 | Benzyl chloride |
| 1.564 | 128.4 | 9.06 | 2.20 | 0.90 | 0.80 | 3.06 | 1.17 | 1.30 | 0.882 | 3 | Bromoanisole |
| 1.497 | 106.3 | 8.45 | 1.56 | 0.90 | 0.15 | 1.60 | 1.06 | 1.04 | 0.929 | 3 | Toluene |
| 1.517 | 108.1 | 8.64 | 2.77 | 0.90 | 0.28 | 1.60 | 1.27 | 1.19 | 0.874 | 3 | Anisole |
| 1.420 | 137.2 | 8.16 | 2.90 | 1.00 | 0.22 | 1.99 | 1.45 | 1.32 | 0.867 | 1 | Heptanenitrile |
| 1.423 | 124.4 | 8.19 | 0.00 | 1.00 | 0.00 | 0.00 | 1.00 | 1.00 | 0.953 | 1 | Methylcyclohexane |
| 1.400 | 140.9 | 7.94 | 0.24 | 0.93 | 0.00 | 0.22 | 1.00 | 1.00 | 0.952 | 1 | 1-Heptene |
| 1.444 | 121.3 | 8.40 | 0.00 | 1.00 | 0.00 | 0.00 | 1.00 | 1.00 | 0.953 | 2 | Cycloheptane |
| 1.406 | 139.5 | 8.00 | 2.17 | 1.00 | 0.00 | 2.69 | 1.21 | 1.14 | 0.907 | 1 | Ethyl butyl ketone |
| 1.388 | 146.6 | 7.81 | 0.00 | 1.00 | 0.00 | 0.00 | 1.00 | 1.00 | 0.953 | 1 | Heptane |
| 1.389 | 145.8 | 7.82 | 0.00 | 1.00 | 0.00 | 0.00 | 1.00 | 1.00 | 0.953 | 1 | 3-Methylhexane |
| 1.392 | 148.7 | 7.75 | 0.00 | 1.00 | 0.00 | 0.00 | 1.00 | 1.00 | 0.953 | 1 | 2,2-Dimethylpentane |
| 1.381 | 149.0 | 7.74 | 0.00 | 1.00 | 0.00 | 0.00 | 1.00 | 1.00 | 0.953 | 1 | 2,4-Dimethylpentane |
| 1.534 | 116.9 | 8.79 | 3.02 | 0.90 | 0.86 | 3.13 | 1.35 | 1.45 | 0.839 | 3 | Acetophenone |
| 1.496 | 122.5 | 8.49 | 1.28 | 0.90 | 0.05 | 0.50 | 1.03 | 1.02 | 0.937 | 3 | Ethylbenzene |
| 1.496 | 123.3 | 8.44 | 1.27 | 0.90 | 0.07 | 0.74 | 1.03 | 1.02 | 0.937 | 3 | *p*-Xylene |
| 1.506 | 120.6 | 8.58 | 1.65 | 0.90 | 0.04 | 0.90 | 1.07 | 1.04 | 0.926 | 3 | *o*-Xylene |
| 1.459 | 134.4 | 8.56 | 0.00 | 1.00 | 0.00 | 0.00 | 1.00 | 1.00 | 0.953 | 2 | Cyclooctane |
| 1.409 | 157.0 | 8.04 | 0.23 | 0.94 | 0.00 | 0.21 | 1.00 | 1.00 | 0.952 | 1 | 1-Octene |
| 1.433 | 142.4 | 8.30 | 0.00 | 1.00 | 0.00 | 0.00 | 1.00 | 1.00 | 0.953 | 2 | Ethylcyclohexane |
| 1.391 | 165.1 | 7.84 | 0.00 | 1.00 | 0.00 | 0.00 | 1.00 | 1.00 | 0.953 | 1 | Isooctane |
| 1.397 | 162.6 | 7.91 | 0.00 | 1.00 | 0.00 | 0.00 | 1.00 | 1.00 | 0.953 | 1 | Octane |
| 1.430 | 157.8 | 8.26 | 0.69 | 1.00 | 3.13 | 3.13 | 1.12 | 2.22 | 0.854 | 1 | Octanol |
| 1.627 | 118.1 | 9.60 | 2.80 | 0.90 | 0.34 | 3.10 | 1.28 | 1.24 | 0.867 | 3 | Quinoline |
| 1.527 | 132.9 | 8.73 | 2.74 | 0.90 | 0.69 | 2.91 | 1.28 | 1.33 | 0.861 | 3 | Propiophenone |
| 1.523 | 142.4 | 8.69 | 3.26 | 0.90 | 0.23 | 2.51 | 1.38 | 1.28 | 0.845 | 3 | Benzyl acetate |
| 1.427 | 163.8 | 8.23 | 0.00 | 1.00 | 0.00 | 0.00 | 1.00 | 1.00 | 0.953 | 2 | 1,3,5-Trimethylcyclohexane |
| 1.420 | 173.1 | 8.15 | 1.78 | 1.00 | 0.00 | 2.21 | 1.12 | 1.08 | 0.922 | 1 | Dibutyl ketone |
| 1.412 | 176.6 | 8.07 | 1.60 | 1.00 | 0.00 | 2.40 | 1.09 | 1.06 | 0.928 | 1 | Diisobutyl ketone |
| 1.405 | 178.7 | 8.00 | 0.00 | 1.00 | 0.00 | 0.00 | 1.00 | 1.00 | 0.953 | 1 | Nonane |
| 1.418 | 189.6 | 7.63 | 0.45 | 1.00 | 0.00 | 3.30 | 1.00 | 1.00 | 0.951 | 1 | Tripropylamine |
| 1.658 | 139.6 | 9.85 | 1.95 | 0.90 | 0.57 | 0.38 | 1.11 | 1.08 | 0.914 | 3 | Bromonaphthalene |
| 1.490 | 156.0 | 8.43 | 1.00 | 0.90 | 0.05 | 0.40 | 1.02 | 1.01 | 0.943 | 3 | Butylbenzene |
| 1.481 | 154.0 | 8.80 | 0.00 | 0.90 | 0.00 | 0.00 | 1.00 | 1.00 | 0.953 | 3 | Decalin |
| 1.430 | 186.9 | 8.26 | 2.27 | 1.00 | 0.17 | 1.56 | 1.24 | 1.17 | 0.901 | 1 | Decanenitrile |
| 1.441 | 175.5 | 8.38 | 0.00 | 1.00 | 0.00 | 0.00 | 1.00 | 1.00 | 0.953 | 2 | Butyl cyclohexane |
| 1.421 | 189.3 | 8.17 | 0.21 | 0.95 | 0.00 | 0.19 | 1.00 | 1.00 | 0.953 | 1 | 1-Decene |
| 1.506 | 143.6 | 9.05 | 1.56 | 1.00 | 0.00 | 1.81 | 1.08 | 1.06 | 0.929 | 1 | Ethyloctanoate |
| 1.412 | 194.9 | 8.07 | 0.00 | 1.00 | 0.00 | 0.00 | 1.00 | 1.00 | 0.953 | 1 | Decane |
| 1.437 | 190.8 | 8.34 | 0.60 | 1.00 | 2.72 | 2.72 | 1.09 | 1.89 | 0.878 | 1 | Decanol |
| 1.422 | 227.5 | 8.17 | 0.00 | 1.00 | 0.00 | 0.00 | 1.00 | 1.00 | 0.953 | 2 | Dodecane |
| 1.445 | 286.0 | 8.42 | 1.62 | 1.00 | 0.12 | 1.11 | 1.10 | 1.07 | 0.926 | 1 | Palmitanitrile |
| 1.441 | 287.3 | 8.38 | 0.10 | 0.97 | 0.00 | 0.11 | 1.00 | 1.00 | 0.953 | 1 | Hexadecene |
| 1.451 | 301.5 | 8.49 | 0.72 | 1.00 | 0.11 | 0.07 | 1.01 | 1.01 | 0.948 | 1 | *n*-Hexadecyl chloride |
| 1.462 | 305.6 | 8.60 | 0.66 | 1.00 | 0.10 | 0.07 | 1.01 | 1.00 | 0.949 | 1 | *n*-Hexadecyl bromide |
| 1.435 | 292.8 | 8.32 | 0.00 | 1.00 | 0.00 | 0.00 | 1.00 | 1.00 | 0.953 | 1 | Hexadecane |
| 1.449 | 288.0 | 8.47 | 0.44 | 1.00 | 1.99 | 1.99 | 1.05 | 1.40 | 0.913 | 1 | Hexadecanol |

**TABLE 8-19 Parameters for MOSCED Model at 20°C (*Continued*)**

| RI | $v$ | $\lambda$ | $\tau$ | $q$ | $\alpha$ | $\beta$ | $\psi$ | $\xi$ | $aa$ | ID | Compound |
|---|---|---|---|---|---|---|---|---|---|---|---|
| 1.439 | 327.1 | 8.36 | 0.00 | 1.00 | 0.00 | 0.00 | 1.00 | 1.00 | 0.953 | 1 | Octadecane |
| 1.443 | 358.3 | 8.40 | 0.00 | 1.00 | 0.00 | 0.00 | 1.00 | 1.00 | 0.953 | 1 | Eicosane |
| 1.444 | 374.6 | 8.41 | 0.00 | 0.00 | 0.00 | 0.00 | 1.00 | 1.00 | 0.953 | 1 | Heneicosane |
| 1.448 | 423.8 | 8.45 | 0.00 | 1.00 | 0.00 | 0.00 | 1.00 | 1.00 | 0.953 | 1 | Tetracosane |
| 1.452 | 489.4 | 8.50 | 0.00 | 1.00 | 0.00 | 0.00 | 1.00 | 1.00 | 0.953 | 1 | Octacosane |
| 1.499 | 478.5 | 8.98 | 0.28 | 0.87 | 0.00 | 0.70 | 1.00 | 1.00 | 0.952 | 1 | Squalene |
| 1.454 | 522.2 | 8.51 | 0.00 | 1.00 | 0.00 | 0.00 | 1.00 | 1.00 | 0.953 | 1 | Triacontane |
| 1.453 | 522.0 | 8.50 | 0.00 | 1.00 | 0.00 | 0.00 | 1.00 | 1.00 | 0.953 | 1 | Squalane |
| 1.455 | 555.0 | 8.53 | 0.00 | 1.00 | 0.00 | 0.00 | 1.00 | 1.00 | 0.953 | 1 | Dotriacontane |
| 1.457 | 604.4 | 8.55 | 0.00 | 1.00 | 0.00 | 0.00 | 1.00 | 1.00 | 0.953 | 1 | Pentatriacontane |

Parameters $\tau$, $\alpha$, $\beta$, $\psi$, $\xi$, and $aa$ are temperature-dependent. The temperature dependence is given by

$$\tau_T = \tau_{293}\left(\frac{293}{T}\right)^{0.4} \tag{8-10.29}$$

$$\alpha_T = \alpha_{293}\left(\frac{293}{T}\right)^{0.8} \qquad \beta_T = \beta_{293}\left(\frac{293}{T}\right)^{0.8} \tag{8-10.30}$$

$$\psi = \text{POL} + 0.011\alpha_T\beta_T \tag{8-10.31}$$

$$\xi = 0.68(\text{POL} - 1) + \{3.4 - 2.4\exp[(-0.023)(\alpha_0\beta_0)^{1.5}]\}^{t2} \tag{8-10.32}$$

where

$$\text{POL} = q^4[1.15 - 1.15\exp(-0.020\tau_T^3)] + 1 \tag{8-10.33}$$

and where $t = 293/T$ ($T$ in kelvins). Subscript 0 refers to 20°C (293 K), and subscript $T$ refers to system temperature.

$$aa = (0.953 - 0.00968)(\tau_2^2 + \alpha_2\beta_2) \tag{8-10.34}$$

where $\tau$, $\alpha$, and $\beta$ are at system temperature $T$.

In Eq. (8-10.27), when subscripts 1 and 2 are interchanged, there is no effect in the terms $(\lambda_1 - \lambda_2)^2$, $q_1^2q_2^2(\tau_1 - \tau_2)^2$, and $(\alpha_1 - \alpha_2)(\beta_1 - \beta_2)$. Asymmetry is introduced through parameters $\psi$ and $\xi$ because $\psi_1 \neq \psi_2$ and $\xi_1 \neq \xi_2$.

In Table 8-19, RI refers to refractive index and ID refers to identification number. ID = 1 for alkane, 2 for naphthene, and 3 for aromatic.

The units of the parameters in Table 8-19 are such that, in the equation for $\ln\gamma_2^\infty$, gas constant $R$ has the value 1.987 cal/mol·K.

For substances not listed in the table, parameters for some classes of compounds can be estimated by the following equations:

**Dispersion parameter**

Nonaromatics:†

$$\lambda = 10.3\frac{n_D^2 - 1}{n_D^2 + 2} + 3.02 \tag{8-10.35}$$

†Except for tertiary amines, nitriles, and $CS_2$.

Aromatics:

$$\lambda = 19.5 \frac{n_D^2 - 1}{n_D^2 + 2} + 2.79 \tag{8-10.36}$$

$n_D \equiv$ refractive index for the sodium D line.

**Induction parameter**

$q = 1.0$ for saturated compounds

$q = 0.9$ for aromatics

$q = 1.0 - 0.5$ [(No. C=C bonds)/(No. C atoms)] for unsaturated aliphatics

**Polar and acidity and basicity parameters†**

$$\tau, \alpha, \beta = C_{(\tau,\alpha,\beta)} \left(\frac{4.5}{3.5 + \text{No. C}}\right)\left(1 + \frac{\text{No. C} - 1}{100}\right) \tag{8-10.37}$$

where the constants are as tabulated.

| | $C_\tau$ | $C_\alpha$ | $C_\beta$ | | $C_\tau$ | $C_\alpha$ | $C_\beta$ |
|---|---|---|---|---|---|---|---|
| Chlorides | 2.69 | 0.42 | 0.25 | Nitroalkanes | 5.87 | 0.35 | 2.66 |
| Bromides | 2.47 | 0.39 | 0.38 | Alcohols | 1.65 | 7.49 | 7.49 |
| Iodides | 2.02 | 0.37 | 0.30 | Esters | 4.03 | 0.00 | 4.64† |
| Nitriles | 5.84 | 0.33 | 4.00 | Ketones | 3.93 | 0.00 | 4.87‡ |

†No. C atoms = 1 for methylformate.
‡No. C atoms = 1 for acetone.

Constants $C_\tau$, $C_\alpha$, and $C_\beta$ are restricted to monofunctional, primary alkanes. Not enough information is available to predict the effects of multifunctionalities, secondary or tertiary positioning of the functional group, chain branching, and cyclic or aromatic backbones.

Like any model for mixture properties, MOSCED has both advantages and disadvantages:

**Advantages**

1. Good overall quantitative predictions. For 3357 $\gamma^\infty$'s, an average error of 9.1 percent was achieved with few errors greater than 30 percent. This compares with an average error of 20.5 percent with UNIFAC.‡

†Except for first member of a homologous series and for ethanol.

‡While UNIFAC is often in error at high dilution, it is usually reliable at intermediate mole fractions.

2. Parameters have some physical significance which give a "feel" for the relative magnitudes of the types of forces in a solution.
3. Calculations are tedious but simple.
4. The model applies to a large number of binary mixtures.

**Disadvantages**

1. Since numerous modifications were made to adjust pure component parameters (e.g., molar volumes of aniline, $CS_2$, and iodides and the $\lambda$'s of nitriles, tertiary amines, and $CS_2$), the general applicability of the model is limited.
2. Predictions are poor for systems where steric considerations are significant, e.g., systems containing TEA (triethylamine) with a moderate or strong acid.
3. Cannot be used for aqueous systems.
4. Works only for systems with activity coefficients below about 100.

**Example 8-8** Calculate the $P$-$y$-$x$ diagram at 50.2°C for the system cyclohexane (1)–aniline (2).

**solution** From the MOSCED model, limiting activity coefficients ($\gamma^{\infty}$'s) are predicted for the system. To calculate the diagram over the entire composition range, binary parameters are obtained from a model for the excess Gibbs energy by using $\gamma_1^{\infty}$ and $\gamma_2^{\infty}$ to fix parameters. MOSCED parameters at 20°C are taken from Table 8-19:

| | Cyclohexane (1) | Aniline (2) | | Cyclohexane (1) | Aniline (2) |
|---|---|---|---|---|---|
| $v$ | 108.1 | 91.1 | $\beta$ | 0 | 2.13 |
| $\lambda$ | 8.22 | 9.25 | $\psi$ | 1 | 1.68 |
| $\tau$ | 0 | 4.22 | $\xi$ | 1 | 2.39 |
| $q$ | 1 | 0.90 | $aa$ | 0.953 | 0.702 |
| $\alpha$ | 0 | 3.81 | | | |

$\tau$, $\alpha$, $\beta$, $\psi$, $\xi$, and $aa$ are temperature-dependent. For aniline, these parameters at 50.2°C (323.2 K) are obtained from

$$\tau_{323.2} = (4.22)\left(\frac{293}{323.2}\right)^{0.4} = 4.06$$

$$\alpha_{323.2} = (3.81)\left(\frac{293}{323.2}\right)^{0.8} = 3.52$$

$$\beta_{323.2} = (2.13)\left(\frac{293}{323.2}\right)^{0.8} = 1.97$$

$$\text{POL} = 0.9^4\{1.15 - 1.15 \exp [(-0.020)(4.06)^3]\} + 1 = 1.557$$

$$\psi_{323.2} = 1.557 + (0.011)(3.52)(1.97) = 1.63$$

$$\xi_{323.2} = (0.68)(1.557 - 1) + [3.4 - 2.4 \exp\{(-0.023)[(3.81)(2.13)]^{1.5}\}]^{(293/323.2)^2}$$
$$= 2.14$$
$$aa_{323.2} = 0.953 - (0.00968)[4.06^2 + (3.52)(1.97)] = 0.726$$

The temperature-dependent parameters are obtained similarly for cyclohexane. Summarizing,

| | Cyclohexane (1) | Aniline (2) | | Cyclohexane (1) | Aniline (2) |
|---|---|---|---|---|---|
| $v$ | 108.1 | 91.1 | $\beta$ | 0 | 1.97 |
| $\lambda$ | 8.22 | 9.25 | $\psi$ | 1 | 1.63 |
| $\tau$ | 0 | 4.06 | $\xi$ | 1 | 2.14 |
| $q$ | 1 | 0.9 | $aa$ | 0.953 | 0.726 |
| $\alpha$ | 0 | 3.52 | | | |

Next, we calculate the combinatorial term:

$$d_{12} = \ln\left(\frac{91.1}{108.1}\right)^{0.726} + 1 - \left(\frac{91.1}{108.1}\right)^{0.726} = -0.00741$$
$$d_{12} = \ln\left(\frac{108.1}{91.1}\right)^{0.953} + 1 - \left(\frac{108.1}{91.1}\right)^{0.953} = -0.01405$$

Finally, we calculate the infinite dilution activity coefficients:

$$\gamma_1^{\infty} = \exp\left\{\frac{108.1}{(1.987)(323.2)}\left[(9.25 - 8.22)^2 + \frac{(0.9)^2(4.06)^2}{1.63} + \frac{(3.52)(1.97)}{2.14}\right] - 0.01405\right\} = 8.08$$
$$\gamma_2^{\infty} = \exp\left\{\frac{91.1}{(1.987)(323.2)}\left[(8.22 - 9.25)^2 + \frac{(0.9)^2(4.06)^2}{1.0} + \frac{(3.52)(1.97)}{1.0}\right] - 0.0741\right\}$$
$$= 20.5$$

We choose the Wilson equation for excess Gibbs energy. We find binary parameters $\Lambda_{12}$ and $\Lambda_{21}$ from Eqs. (8-10.25) and (8-10.26). By solving those equations simultaneously, we obtain

$$\Lambda_{12} = 0.3051 \qquad \Lambda_{21} : 0.09773$$

We can now calculate the *P-y-x* diagram. In view of the approximate nature of this calculation, we neglect all vapor phase corrections and the Poynting factor. Recall that

$$\ln \gamma_1 = -\ln(x_1 + \Lambda_{12}x_2) + x_2\left(\frac{\Lambda_{12}}{x_1 + \Lambda_{12}x_2} - \frac{\Lambda_{21}}{\Lambda_{21}x_1 + x_2}\right)$$
$$\ln \gamma_2 = -\ln(x_2 + \Lambda_{21}x_1) - x_1\left(\frac{\Lambda_{12}}{x_1 + \Lambda_{12}x_2} - \frac{\Lambda_{21}}{\Lambda_{21}x_1 + x_2}\right)$$
$$P = \gamma_1 x_1 P_{vp1} + \gamma_2 x_2 P_{vp2}$$
$$y_i = \frac{\gamma_i x_i P_{vpi}}{P}$$

where $P_{vpi}$ is the pure component vapor pressure at temperature $T$. At 50.2°C, $P_{vp1} = 0.3604$ bar and $P_{vp2} = 0.00493$ bar.

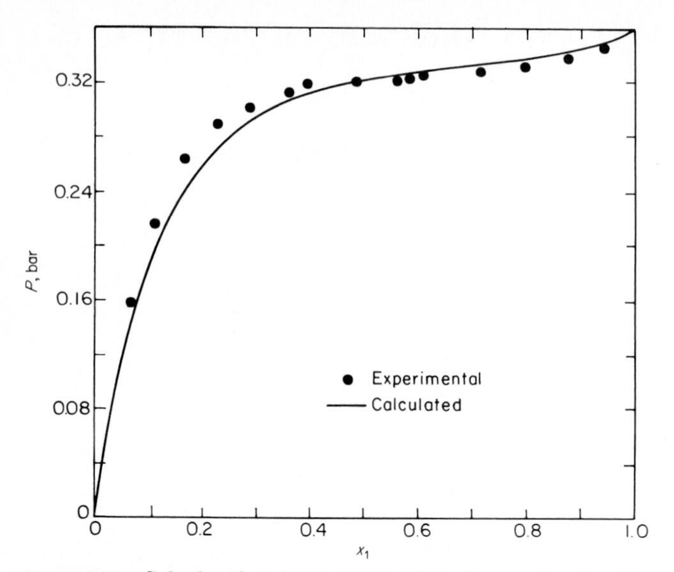


**Figure 8-10** Calculated and experimental total pressures for the system cyclohexane (1)–aniline (2) at 50.2°C. Activity coefficients calculated from Wilson equation with parameters determined from MOSCED correlation for infinite-dilution activity coefficients.

The results are summarized in the accompanying table and Fig. 8-10. The experimental data are taken from Abello et al. [1].

**Vapor-Liquid Equilibrium Composition for Cyclohexane (1)–Aniline (2) at 50.2°C**

| | $100y_1$ | | | $100y_1$ | |
|---|---|---|---|---|---|
| $100x_1$ | Calculated | Observed | $100x_1$ | Calculated | Observed |
| 6.4 | 96.7 | 97.2 | 48.6 | 98.8 | 98.9 |
| 10.9 | 97.7 | 97.9 | 58.2 | 98.8 | 98.9 |
| 16.8 | 98.2 | 98.4 | 71.3 | 98.9 | 98.8 |
| 28.4 | 98.6 | 98.8 | 87.4 | 99.0 | 99.1 |
| 36.0 | 98.7 | 98.8 | 93.9 | 99.2 | 99.3 |

**Example 8-9** Repeat the preceding example for the system methanol (1)–benzene (2) at 35°C.

**solution** At 20°C, the MOSCED parameters are:

| Parameter | Methanol | Benzene | Parameter | Methanol | Benzene |
|---|---|---|---|---|---|
| $v$ | 40.5 | 89.1 | $\beta$ | 7.45 | 0.56 |
| $\lambda$ | 7.14 | 8.49 | $\Psi$ | 1.94 | 1.11 |
| $\tau$ | 2.55 | 1.95 | $\xi$ | 3.62 | 1.07 |
| $q$ | 1 | 0.9 | $aa$ | 0.353 | 0.915 |
| $\alpha$ | 7.45 | 0.22 | | | |

As in the preceding example, the parameters at 35°C are calculated. Summarizing,

| Parameter | Methanol | Benzene | Parameter | Methanol | Benzene |
|---|---|---|---|---|---|
| $v$ | 40.5 | 89.1 | $\beta$ | 7.16 | 0.538 |
| $\lambda$ | 7.14 | 8.49 | $\Psi$ | 1.87 | 1.10 |
| $\tau$ | 2.50 | 1.91 | $\xi$ | 3.24 | 1.07 |
| $q$ | 1 | 0.9 | $aa$ | 0.396 | 0.916 |
| $\alpha$ | 7.16 | 0.211 | | | |

The combinatorial terms are given by

$$d_{12} = \ln\left(\frac{89.1}{40.5}\right)^{0.916} + 1 - \left(\frac{89.1}{40.5}\right)^{0.916} = -0.337$$

$$d_{21} = \ln\left(\frac{40.5}{89.1}\right)^{0.396} + 1 - \left(\frac{40.5}{89.1}\right)^{0.396} = -\ 0.0440$$

Finally, we calculate the infinite-dilution activity coefficients:

$$\gamma_1^\infty = \exp\left\{\frac{40.5}{(1.987)(308)}\left[(8.49 - 7.14)^2 + \frac{(0.9)^2(1.91 - 2.50)^2}{1.10}\right.\right.$$
$$\left.\left. + \frac{(0.211 - 7.16)(0.538 - 7.16)}{1.07}\right] - 0.0440\right\}$$
$$= 18.9$$

$$\gamma_2^\infty = \exp\left\{\frac{89.1}{(1.987)(308)}\left[(7.14 - 8.49)^2 + \frac{(0.9)^2(2.50 - 1.91)^2}{1.87}\right.\right.$$
$$\left.\left. + \frac{(7.16 - 0.211)(7.16\text{-}0.538)}{3.24}\right] - 0.337\right\}$$
$$= 7.52$$

For the Wilson equation, using $\gamma_1^\infty$ and $\gamma_2^\infty$, the binary parameters are

$$\Lambda_{12} = 0.1038 \qquad \Lambda_{21} = 0.3258$$

At 35°C, $P_{vp1} = 0.2759$ bar and $P_{vp2} = 0.1952$ bar

The results are shown in Fig. 8-11. Experimental data are taken from Scatchard et al. [108].

## Azeotropic data

Many binary systems exhibit azeotropy, i.e., a condition in which the composition of a liquid mixture is equal to that of its equilibrium vapor. When the azeotropic conditions (temperature, pressure, composition) are known, activity coefficients $\gamma_1$ and $\gamma_2$ at that condition are readily found. These activity coefficients can then be used to calculate two parameters in some arbitrarily chosen expression for the excess Gibbs energy (Table 8-3). Extensive compilations of azeotropic data are available [56].

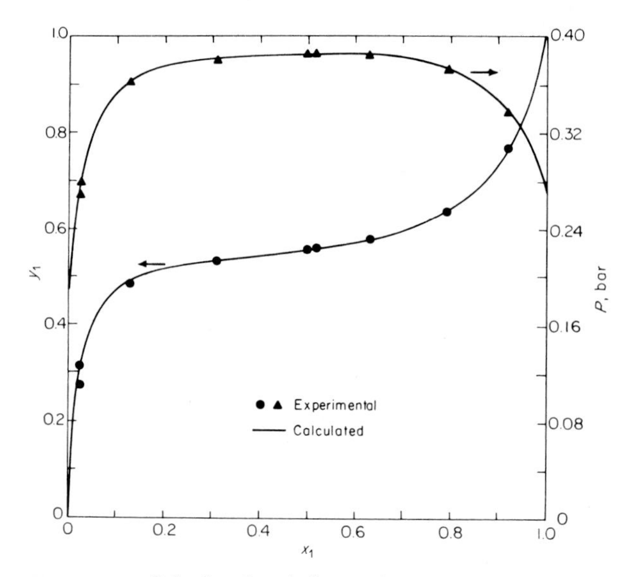


**Figure 8-11** Calculated and observed vapor-liquid equilibria for methanol (1)–benzene (2) at 35°C. Activity coefficients calculated from the Wilson equation with parameters determined from the MOSCED correlation for infinite-dilution activity coefficients.

For a binary azeotrope, $x_1 = y_1$ and $x_2 = y_2$. Therefore, Eq. (8-4.1), with $\mathcal{F}_i = 1$, becomes

$$\gamma_1 = \frac{P}{P_{vp1}} \quad \text{and} \quad \gamma_2 = \frac{P}{P_{vp2}} \tag{8-10.38}$$

Knowing total pressure $P$ and pure component vapor pressures $P_{vp1}$ and $P_{vp2}$, we determine $\gamma_1$ and $\gamma_2$. With these activity coefficients and the azeotropic composition $x_1$ and $x_2$ it is now possible to find two parameters $A$ and $B$ by simultaneous solution of two equations of the form

$$RT \ln \gamma_1 = f_1(x_2, A, B) \tag{8-10.39a}$$

$$RT \ln \gamma_2 = f_2(x_1, A, B) \tag{8-10.39b}$$

where, necessarily, $x_1 = 1 - x_2$ and where functions $f_1$ and $f_2$ represent thermodynamically consistent equations derived from the choice of an expression for the excess Gibbs energy. Simultaneous solution of Eqs. (8-10.39$a$) and (8-10.39$b$) is simple in principle, although the necessary algebra may be tedious if $f_1$ and $f_2$ are complex.

**Example 8-10** To illustrate, consider an example similar to one given by Treybal [123] for the system ethyl acetate (1)–ethanol (2). This system forms an azeotrope at 1.01 bar, 71.8°C, and $x_2 = 0.462$.

**solution** At 1.01 bar and 71.8°C, we use Eq. (8-10.38):

$$\gamma_1 = \frac{1.01}{0.839} = 1.204 \qquad \gamma_2 = \frac{1.01}{0.772} = 1.308$$

where 0.839 and 0.772 bar are the pure component vapor pressures at 71.8°C. For functions $f_1$ and $f_2$ we choose the van Laar equations shown in Table 8-3. Upon algebraic rearrangement, we obtain explicit solutions for $A$ and $B$.

$$\frac{A}{RT} = \ln 1.204 \left(1 + \frac{0.462 \ln 1.308}{0.538 \ln 1.204}\right)^2 = 0.93$$

$$\frac{B}{RT} = \ln 1.308 \left(1 + \frac{0.538 \ln 1.204}{0.462 \ln 1.308}\right)^2 = 0.87$$

and $A/B = 1.07$.
At 71.8°C, the activity coefficients are given by

$$\ln \gamma_1 = \frac{0.93}{(1 + 1.07x_1/x_2)^2}$$

$$\ln \gamma_2 = \frac{0.87}{(1 + x_2/1.07x_1)^2}$$

Figure 8-12 shows a plot of the calculated activity coefficients. Also shown are experimental results at 1.01 bar by Furnas and Leighton [39] and by Griswold, Chu, and Winsauer [43]. Since the experimental results are isobaric, the temperature is not constant. However, in this example, the calculated activity coefficients are assumed to be independent of temperature.

Figure 8-12 shows good overall agreement between experimental and calculated activity coefficients. Generally, fair agreement is found if the azeotropic data are accurate, if the binary system is not highly complex, and, most important, if the azeotropic composition is in the midrange $0.25 < x_1$ (or $x_2$) $< 0.75$. If the azeotropic composition is at either dilute end, azeotropic data are of much less value for estimating activity coefficients over the entire composition range. This negative conclusion follows from the limiting relation $\gamma_1 \to 1$ as $x_1 \to 1$. Thus, if we have an azeotropic mixture where $x_2 \ll 1$, the experimental value of $\gamma_1$ gives us very little information, since $\gamma_1$ is necessarily close to unity. For such a mixture, only $\gamma_2$ supplies significant information, and therefore we cannot expect to calculate two meaningful adjustable parameters when we have only one significant datum. However, if the azeotropic composition is close to unity, we may, nevertheless, use the azeotropic data to find one activity coefficient, namely, $\gamma_2$ (where $x_2 \ll 1$), and then use that $\gamma_2$ to determine the single adjustable parameter in any of the one-parameter equations for the molar excess Gibbs energy, as discussed in Sec. 8-5.

### Mutual solubilities of liquids

When two liquids are only partially miscible, experimental data for the two mutual solubilities can be used to estimate activity coefficients over

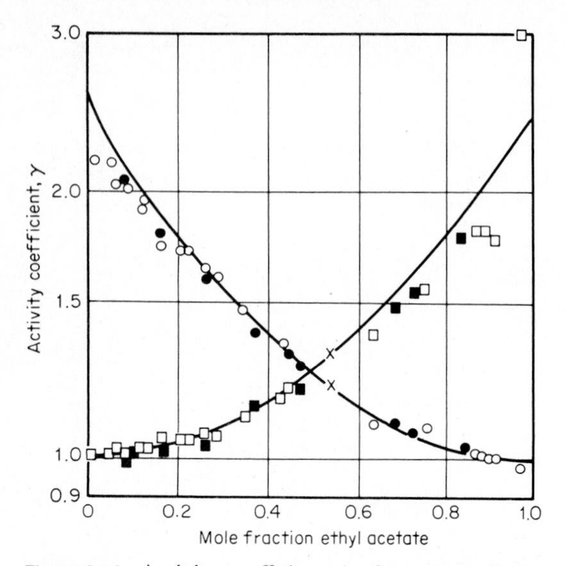


**Figure 8-12** Activity coefficients in the system ethylacetate–ethanol. Calculated lines from azeotropic data (indicated by x) at 1.01 bar. Points are experimental [39, 43]. *(From Ref. 123.)*

the entire range of composition in the homogeneous regions. Suppose the solubility (mole fraction) of component 1 in component 2 is $x_1^{s'}$ and that of component 2 in component 1 is $x_2^{s''}$, where superscript $s$ denotes saturation and the primes designate the two liquid phases. If $x_1^{s'}$ and $x_2^{s''}$ are known at some temperature $T$, it is possible to estimate acitivity coefficients for both components in the homogeneous regions $0 \leqslant x_1' \leqslant x_1^{s'}$ and $0 \leqslant x_2'' \leqslant x_2^{s''}$.

To estimate the activity coefficients, it is necessary to choose some thermodynamically consistent analytical expression which relates activity coefficients $\gamma_1$ and $\gamma_2$ to mole fraction $x$. (See Sec. 8-5.) Such an expression contains one or more constants characteristic of the binary system; these constants are generally temperature-dependent, although the effect of temperature is often not large. From the equations of liquid-liquid equilibrium, it is possible to determine two of these constants. The equations of equilibrium are

$$(\gamma_1 x_1)^{s'} = (\gamma_1 x_1)^{s''} \qquad \text{and} \qquad (\gamma_2 x_2)^{s'} = (\gamma_2 x_2)^{s''} \tag{8-10.40}$$

Suppose we choose a two-constant expression for the molar excess Gibbs energy $g^E$. Then, as discussed in Sec. 8-5,

$$RT \ln \gamma_1 = f_1(x_2, A, B) \qquad \text{and} \qquad RT \ln \gamma_2 = f_2(x_1, A, B) \tag{8-10.41}$$

where $f_1$ and $f_2$ are known functions and the two (unknown) constants are designated by $A$ and $B$. These constants can be found by simultaneous solution of Eqs. (8-10.40) and (8-10.41) coupled with experimental values for $x_1^{s'}$ and $x_2^{s''}$ and the material balances

$$x_2^{s'} = 1 - x_1^{s'} \qquad \text{and} \qquad x_1^{s''} = 1 - x_2^{s''} \tag{8-10.42}$$

In principle, the calculation is simple although the algebra may be tedious, depending on the complexity of the functions $f_1$ and $f_2$.

To illustrate, Table 8-20 presents results obtained by Brian [21] for five binary aqueous systems, where subscript 2 refers to water. Calculations are based on both the van Laar equation and the three-suffix (two-parameter) Margules equation (see Table 8-3). Table 8-20 shows the calculated activity coefficients at infinite dilution, which are easily related to the constants $A$ and $B$. [See Eqs. (8-10.23) and (8-10.24).]

Brian's calculations indicate that results are sensitive to the expression arbitrarily chosen for the molar excess Gibbs energy. Brian found that, compared with experimental vapor-liquid equilibrium data for the homogeneous regions, the Margules equations gave poor results and the van Laar equation gave fair, but not highly accurate, results.

Calculations of this sort can also be made by using a three-parameter equation for $g^E$, but in that event, the third parameter must be estimated independently. A nomogram for such calculations, using the NRTL equation, has been given by Renon and Prausnitz [105].

Generally speaking, mutual solubility data provide only approximate values of activity coefficients, although such estimates may be better than none at all. These estimates are sensitive not only to the choice of an expression for $g^E$ but also to small errors in the experimental mutual solubility data.

## Group contribution methods

For correlating thermodynamic properties, it is often convenient to regard a molecule as an aggregate of functional groups; as a result, some thermodynamic properties of pure fluids, e.g., heat capacity and critical vol-

**TABLE 8-20 Limiting Activity Coefficients as Calculated from Mutual Solubilities in Five Binary Aqueous Systems [21]**

| | | Solubility limits | | $\log \gamma_1^\infty$ | | $\log \gamma_2^\infty$ | |
|---|---|---|---|---|---|---|---|
| Component (1) | Temp., °C | $x_1''$ | $x_2'$ | van Laar | Margules | van Laar | Margules |
| Aniline | 100 | 0.01475 | 0.372 | 1.8337 | 1.5996 | 0.6076 | −0.4514 |
| Isobutyl alcohol | 90 | 0.0213 | 0.5975 | 1.6531 | 0.6193 | 0.4020 | −3.0478 |
| 1-Butanol | 90 | 0.0207 | 0.636 | 1.6477 | 0.2446 | 0.3672 | −4.1104 |
| Phenol | 43.4 | 0.02105 | 0.7325 | 1.6028 | −0.1408 | 0.2872 | −8.2901 |
| Propylene oxide | 36.3 | 0.166 | 0.375 | 1.1103 | 1.0743 | 0.7763 | 0.7046 |

ume, can be calculated by summing group contributions. Extension of this concept to mixtures was suggested long ago by Langmuir, and several attempts have been made to establish group contribution methods for heats of mixing and for activity coefficients. Here we mention only two methods, both for activity coefficients, which appear to be particularly useful for making reasonable estimates for those strongly nonideal mixtures for which data are sparse or totally absent. The two methods, called ASOG and UNIFAC, are similar in principle but differ in detail.

In any group contribution method, the basic idea is that whereas there are thousands of chemical compounds of interest in chemical technology, the number of functional groups which constitute these compounds is much smaller. Therefore, if we assume that a physical property of a fluid is the sum of contributions made by the molecule's functional groups, we obtain a possible technique for correlating the properties of a very large number of fluids in terms of a much smaller number of parameters which characterize the contributions of individual groups.

Any group contribution method is necessarily approximate because the contribution of a given group in one molecule is not necessarily the same as that in another molecule. The fundamental assumption of a group contribution method is additivity: the contribution made by one group within a molecule is assumed to be independent of that made by any other group in that molecule. This assumption is valid only when the influence of any one group in a molecule is not affected by the nature of other groups within that molecule.

For example, we would not expect the contribution of a carbonyl group in a ketone (say, acetone) to be the same as that of a carbonyl group in an organic acid (say, acetic acid). On the other hand, experience suggests that the contribution of a carbonyl group in, for example, acetone, is close to (although not identical with) the contribution of a carbonyl group in another ketone, say 2-butanone.

Accuracy of correlation improves with increasing distinction of groups; in considering, for example, aliphatic alcohols, in a first approximation no distinction is made between the position (primary or secondary) of a hydroxyl group, but in a second approximation such a distinction is desirable. In the limit, as more and more distinctions are made, we recover the ultimate group, namely, the molecule itself. In that event, the advantage of the group contribution method is lost. For practical utility, a compromise must be attained. The number of distinct groups must remain small but not so small as to neglect significant effects of molecular structure on physical properties.

Extension of the group contribution idea to mixtures is attractive because, although the number of pure fluids in chemical technology is already very large, the number of different mixtures is larger by many orders of magnitude. Thousands, perhaps millions, of multicomponent liquid mixtures of interest in the chemical industry can be constituted from perhaps 30, 50, or at most 100 functional groups.

### ASOG method

The analytical solution of groups (ASOG) method was developed by Wilson and Deal [131] and Wilson [130] following earlier work by Redlich, Derr, Pierotti, and Papadopoulos [104]. An introduction to ASOG was presented by Palmer [87].

For component $i$ in a mixture, activity coefficient $\gamma_i$ consists of a configurational (entropic) contribution due to differences in molecular size and a group interaction contribution due primarily to differences in intermolecular forces:

$$\ln \gamma_i = \ln \gamma_i^S + \ln \gamma_i^G \tag{8-10.43}$$

where superscript $S$ designates size and superscript $G$ designates group.

Activity coefficient $\gamma_i^S$ depends only on the number of size groups, e.g., $CH_2$, CO, OH, in the various molecules that constitute the mixture. From the Flory-Huggins theory for athermal mixtures of unequal-size molecules.

$$\ln \gamma_i^S = 1 - \mathcal{R}_i + \ln \mathcal{R}_i \tag{8-10.44}$$

where

$$\mathcal{R}_i = \frac{s_i}{\sum_j s_j x_j} \tag{8-10.45}$$

where $x_j$ = mole fraction of component $j$ in mixture
$s_j$ = number of size groups in molecule $j$

Parameter $s_j$ is independent of temperature. The summation extends over all components, including component $i$.

To calculate $\gamma_i^G$, we need to know the *group* mole fractions $X_k$, where subscript $k$ stands for a particular group in molecule $j$

$$X_k = \frac{\sum_j x_j \nu_{kj}}{\sum_j x_j \sum_k \nu_{kj}} \tag{8-10.46}$$

where $\nu_{kj}$ is the number of interaction groups $k$ in molecule $j$. Activity coefficient $\gamma_i^G$ is given by

$$\ln \gamma_i^G = \sum_k \nu_{ki} \ln \Gamma_k - \sum_k \nu_{ki} \ln \Gamma_k^* \tag{8-10.47}$$

where $\Gamma_k$ = activity coefficient of group $k$ in the mixture
$\Gamma_k^*$ = activity coefficient of group $k$ in the standard state

This standard state depends on molecule $i$.

Activity coefficient $\Gamma_k$ is given by Wilson's equation

$$\ln \Gamma_k = -\ln \sum_\ell X_\ell A_{k\ell} + \left(1 - \sum_\ell \frac{X_\ell A_{\ell k}}{\sum_m x_m A_{\ell m}}\right) \tag{8-10.48}$$

where the summations extend over all groups present in the mixture.

Equation (8-10.48) is also used to find $\Gamma_k^*$ for component $i$, but in that case it is applied to a "mixture" of groups as found in pure component $i$. For example, if $i$ is water, hexane,† or benzene, there is only one kind of group and $\ln \Gamma_k^*$ is zero. However, if $i$ is methanol, $\ln \Gamma_k^*$ has a finite value for both hydroxyl and methyl groups.

Parameters $A_{k\ell}$ and $A_{\ell k}$ ($A_{k\ell} \neq A_{\ell k}$) are group interaction parameters which depend on temperature. These parameters are obtained from reduction of vapor-liquid equilibria, and a substantial number of such parameters have been reported by Derr and Deal [30]. The important point here is that, at a fixed temperature, these parameters depend only on the nature of the groups and, by assumption, are independent of the nature of the molecule. Therefore, group parameters obtained from available experimental data for some mixtures can be used to predict activity coefficients in other mixtures that contain not the same molecules, but the same groups. For example, suppose we wish to predict activity coefficients in the binary system dibutyl ketone–nitrobenzene. To do so, we require group interaction parameters for characterizing interactions between methyl, phenyl, keto, and nitrile groups. These parameters can be obtained from other binary mixtures which contain these groups, e.g., acetone-benzene, nitropropane-toluene, and methyl ethyl ketonenitro–ethane.

### UNIFAC method

The fundamental idea of a solution-of-groups model is to utilize existing phase equilibrium data for predicting phase equilibria of systems for which no experimental data are available. In concept, the UNIFAC method follows the ASOG method, wherein activity coefficients in mixtures are related to interactions between structural groups. The essential features are:

1. Suitable reduction of experimentally obtained activity coefficient data to yield parameters characterizing interactions between pairs of structural groups in nonelectrolyte systems.
2. Use of those parameters to predict activity coefficients for other sys-

†It is assumed here that with respect to group interactions, no distinction is made between groups $CH_2$ and $CH_3$.

tems which have not been studied experimentally but which contain the same functional groups.

The molecular activity coefficient is separated into two parts: one part provides the contribution due to differences in molecular size, and the other provides the contribution due to molecular interactions. In ASOG, the first part is arbitrarily estimated by using the athermal Flory-Huggins equation; the Wilson equation, applied to functional groups, is chosen to estimate the second part. Some of this arbitrariness is removed by combining the solution-of-groups concept with the UNIQUAC equation (see Table 8-3); first, the UNIQUAC model per se contains a combinatorial part, essentially due to differences in size and shape of the molecules in the mixture, and a residual part, essentially due to energy interactions, and second, functional group sizes and interaction surface areas are introduced from independently obtained, pure component molecular structure data.

The UNIQUAC equation often gives good representation of both vapor-liquid and liquid-liquid equilibria for binary and multicomponent mixtures containing a variety of nonelectrolytes such as hydrocarbons, ketones, esters, water, amines, alcohols, nitriles, etc. In a multicomponent mixture, the UNIQUAC equation for the activity coefficient of (molecular) component $i$ is

$$\ln \gamma_i = \underset{\text{combinatorial}}{\ln \gamma_i^C} + \underset{\text{residual}}{\ln \gamma_i^R} \tag{8-10.49}$$

where

$$\ln \gamma_i^C = \ln \frac{\Phi_i}{x_i} + \frac{z}{2} q_i \ln \frac{\theta_i}{\Phi_i} + \ell_i - \frac{\Phi_i}{x_i} \sum_j x_j \ell_j \tag{8-10.50}$$

and

$$\ln \gamma_i^R = q_i \left[ 1 - \ln \left( \sum_j \theta_j \tau_{ji} \right) - \sum_j \frac{\theta_j \tau_{ij}}{\sum_k \theta_k \tau_{kj}} \right] \tag{8-10.51}$$

$$\ell_i = \frac{z}{2}(r_i - q_i) - (r_i - 1) \qquad z = 10$$

$$\theta_i = \frac{q_i x_i}{\sum_j q_j x_j} \qquad \Phi_i = \frac{r_i x_i}{\sum_j r_j x_j} \qquad \tau_{ji} = \exp\left( -\frac{u_{ji} - u_{ii}}{RT} \right)$$

In these equations $x_i$ is the mole fraction of component $i$ and the summations in Eqs. (8-10.50) and (8-10.51) are over all components, including component $i$, $\theta_i$ is the area fraction, and $\Phi_i$ is the segment fraction, which is similar to the volume fraction. Pure component parameters $r_i$ and $q_i$ are, respectively, measures of molecular van der Waals volumes and molecular surface areas.

In UNIQUAC, the two adjustable binary parameters $\tau_{ij}$ and $\tau_{ji}$ appearing in Eq. (8-10.51) must be evaluated from experimental phase equilibrium data. No ternary (or higher) parameters are required for systems containing three or more components.

In the UNIFAC method [36], the combinatorial part of the UNIQUAC activity coefficients, Eq. (8-10.50), is used directly. Only pure component properties enter into this equation. Parameters $r_i$ and $q_i$ are calculated as the sum of the group volume and area parameters $R_k$ and $Q_k$, given in Table 8-21:

$$r_i = \sum_k \nu_k^{(i)} R_k \qquad \text{and} \qquad q_i = \sum_k \nu_k^{(i)} Q_k \tag{8-10.52}$$

where $\nu_k^{(i)}$, always an integer, is the number of groups of type $k$ in molecule $i$. Group parameters $R_k$ and $Q_k$ are obtained from the van der Waals group volume and surface areas $V_{wk}$ and $A_{wk}$, given by Bondi [16]:

$$R_k = \frac{V_{wk}}{15.17} \qquad \text{and} \qquad Q_k = \frac{A_{wk}}{2.5 \times 10^9} \tag{8-10.53}$$

The normalization factors 15.17 and $2.5 \times 10^9$ are determined by the volume and external surface area of a $CH_2$ unit in polyethylene.

The residual part of the activity coefficient, Eq. (8-10.51), is replaced by the solution-of-groups concept. Instead of Eq. (8-10.51), we write

$$\ln \gamma_i^R = \sum_{\substack{k \\ \text{all groups}}} \nu_k^{(i)} (\ln \Gamma_k - \ln \Gamma_k^{(i)}) \tag{8-10.54}$$

where $\Gamma_k$ is the group residual activity coefficient and $\Gamma_k^{(i)}$ is the residual activity coefficient of group $k$ in a reference solution containing only molecules of type $i$. (In UNIFAC, $\Gamma^{(i)}$ is similar to ASOG's $\Gamma^*$.) In Eq. (8-10.54) the term $\ln \Gamma_k^{(i)}$ is necessary to attain the normalization that activity coefficient $\gamma_i$ becomes unity as $x_i \to 1$. The activity coefficient for group $k$ in molecule $i$ depends on the molecule $i$ in which $k$ is situated. For example, $\Gamma_k^{(i)}$ for the COH group† in ethanol refers to a "solution" containing 50 group percent COH and 50 group percent $CH_3$ at the temperature of the mixture, whereas $\Gamma_k^{(i)}$ for the COH group in $n$-butanol refers to a "solution" containing 25 group percent COH, 50 group percent $CH_2$, and 25 group percent $CH_3$.

The group activity coefficient $\Gamma_k$ is found from an expression similar to Eq. (8-10.51):

$$\ln \Gamma_k = Q_k \left[ 1 - \ln \left( \sum_m \theta_m \Psi_{mk} \right) - \sum_m \frac{\theta_m \Psi_{km}}{\sum_n \theta_n \Psi_{nm}} \right] \tag{8-10.55}$$

†COH is shortened notation for $CH_2OH$.

**TABLE 8-21 UNIFAC Group Specifications and Sample Group Assignments**

| Group numbers | | | | | | Sample assignment | |
|---|---|---|---|---|---|---|---|
| Main | Sec. | Name | $R$ | $Q$ | MW | NOGP | IDGP |
| 1 | 1 | $CH_3$ | 0.9011 | 0.848 | 15.03 | 2,2,-4-Trimethylpentane: | |
| 1 | 2 | $CH_2$ | 0.6744 | 0.540 | 14.03 | 5 | 1 |
| 1 | 3 | CH | 0.4469 | 0.228 | 13.02 | 1 | 2 |
| | | | | | | 1 | 3 |
| 1 | 4 | C | 0.2195 | 0 | 12.01 | 1 | 4 |
| 2 | 5 | $CH_2{=}CH$ | 1.3454 | 1.176 | 27.05 | 3-Methyl-1-hexene: | |
| 2 | 6 | CH=CH | 1.1167 | 0.867 | 26.04 | 2 | 1 |
| 2 | 7 | $CH_2{=}C$ | 1.1173 | 0.988 | 26.04 | 2 | 2 |
| 2 | 8 | CH=C | 0.8886 | 0.676 | 25.03 | 1 | 3 |
| 2 | 9 | C=C | 0.6605 | 0.485 | 24.02 | 1 | 5 |
| 3 | 10 | ACH | 0.5313 | 0.400 | 13.02 | Benzene: | |
| 3 | 11 | AC | 0.3652 | 0.120 | 12.01 | 6 | 10 |
| 4 | 12 | $ACCH_3$ | 1.2663 | 0.968 | 27.05 | Xylene: | |
| 4 | 13 | $ACCH_2$ | 1.0396 | 0.660 | 26.04 | 4 | 10 |
| 4 | 14 | ACCH | 0.8121 | 0.348 | 25.03 | 2 | 12 |
| 5 | 15 | OH | 1.0000 | 1.200 | 17.01 | Ethanol: | |
| | | | | | | 1 | 1 |
| | | | | | | 1 | 2 |
| | | | | | | 1 | 15 |
| 6 | 16 | $CH_3OH$ | 1.4311 | 1.432 | 32.04 | Methanol: | |
| | | | | | | 1 | 16 |
| 7 | 17 | $H_2O$ | 0.9200 | 1.400 | 18.02 | Water: | |
| | | | | | | 1 | 17 |
| 8 | 18 | ACOH | 0.8952 | 0.680 | 29.02 | Phenol: | |
| | | | | | | 5 | 10 |
| | | | | | | 1 | 18 |

**TABLE 8-21 UNIFAC Group Specifications and Sample Group Assignments (*Continued*)**

| Group numbers | | | | | | Sample assignment | |
|---|---|---|---|---|---|---|---|
| Main | Sec. | Name | $R$ | $Q$ | MW | NOGP | IDGP |
| 9 | 19 | $CH_3CO$ | 1.6724 | 1.488 | 43.05 | Methylethylketone: | |
| 9 | 20 | $CH_2CO$ | 1.4457 | 1.180 | 42.04 | 1 | 1 |
| | | | | | | 1 | 2 |
| | | | | | | 1 | 19 |
| 10 | 21 | CHO | 0.9980 | 0.948 | 29.02 | Hexanal: | |
| | | | | | | 1 | 1 |
| | | | | | | 4 | 2 |
| | | | | | | 1 | 21 |
| 11 | 22 | $CH_3COO$ | 1.9031 | 1.728 | 59.04 | Butyl acetate: | |
| 11 | 23 | $CH_2COO$ | 1.6764 | 1.420 | 58.04 | 1 | 1 |
| | | | | | | 3 | 2 |
| | | | | | | 1 | 22 |
| 12 | 24 | HCOO | 1.2420 | 1.188 | 45.02 | Ethyl formate: | |
| | | | | | | 1 | 1 |
| | | | | | | 1 | 2 |
| | | | | | | 1 | 24 |
| 13 | 25 | $CH_3O$ | 1.1450 | 1.088 | 31.03 | Ethyl ether: | |
| 13 | 26 | $CH_2O$ | 0.9183 | 0.780 | 30.03 | 2 | 1 |
| 13 | 27 | CH−O | 0.6908 | 0.468 | 29.02 | 1 | 2 |
| 13 | 28 | $FCH_20$ | 0.9183 | 1.100 | 30.03 | 1 | 26 |
| 14 | 29 | $CH_3NH_2$ | 1.5959 | 1.544 | 31.06 | Propyl amine: | |
| 14 | 30 | $CH_2NH_2$ | 1.3692 | 1.236 | 30.05 | 1 | 1 |
| 14 | 31 | $CHNH_2$ | 1.1417 | 0.924 | 29.04 | 1 | 2 |
| | | | | | | 1 | 30 |
| 15 | 32 | $CH_3NH$ | 1.4337 | 1.244 | 30.05 | Diethyl amine: | |
| 15 | 33 | $CH_2NH$ | 1.2070 | 0.936 | 29.04 | 2 | 1 |
| 15 | 34 | CHNH | 0.9795 | 0.624 | 28.03 | 1 | 2 |
| | | | | | | 1 | 33 |

| | | | | | | | |
|---|---|---|---|---|---|---|---|
| 16 | 35 | $CH_3N$ | 1.1865 | 0.940 | 29.04 | Triethyl amine: | |
| 16 | 36 | $CH_2N$ | 0.9597 | 0.632 | 28.03 | 3 | 1 |
| | | | | | | 2 | 2 |
| | | | | | | 1 | 36 |
| 17 | 37 | $ACNH_2$ | 1.0600 | 0.816 | 28.03 | Aniline: | |
| | | | | | | 5 | 10 |
| | | | | | | 1 | 37 |
| 18 | 38 | $C_5H_5N$ | 2.9993 | 2.113 | 79.10 | Methyl pyridine: | |
| 18 | 39 | $C_5H_4N$ | 2.8332 | 1.833 | 78.09 | 1 | 1 |
| 18 | 40 | $C_5H_3N$ | 2.6670 | 1.553 | 77.09 | 1 | 39 |
| 19 | 41 | $CH_3CN$ | 1.8701 | 1.724 | 41.05 | Propionitrile: | |
| 19 | 42 | $CH_2CN$ | 1.6434 | 1.416 | 40.04 | 1 | 1 |
| | | | | | | 1 | 42 |
| 20 | 43 | COOH | 1.3013 | 1.224 | 45.02 | Acetic acid: | |
| 20 | 44 | HCOOH | 1.5280 | 1.532 | 46.03 | 1 | 1 |
| | | | | | | 1 | 43 |
| 21 | 45 | $CH_2Cl$ | 1.4654 | 1.264 | 49.48 | Chloroethane: | |
| 21 | 46 | CHCl | 1.2380 | 0.952 | 48.47 | 1 | 1 |
| 21 | 47 | CCl | 1.0060 | 0.724 | 47.46 | 1 | 45 |
| 22 | 48 | $CH_2Cl_2$ | 2.2564 | 1.988 | 84.93 | 1, 1-Dichloroethane: | |
| 22 | 49 | $CHCl_2$ | 2.0606 | 1.684 | 83.92 | 1 | 1 |
| 22 | 50 | $CCl_2$ | 1.8016 | 1.448 | 82.92 | 1 | 49 |
| 23 | 51 | $CHCl_3$ | 2.8700 | 2.410 | 119.38 | 1, 1, 1-Trichloroethane: | |
| 23 | 52 | $CCl_3$ | 2.6401 | 2.184 | 118.37 | 1 | 1 |
| | | | | | | 1 | 52 |
| 24 | 53 | $CCl_4$ | 3.3900 | 2.910 | 153.82 | Trichloromethane: | |
| | | | | | | 1 | 53 |
| 25 | 54 | ACCl | 1.1562 | 0.844 | 47.46 | Chlorobenzene: | |
| | | | | | | 5 | 10 |
| | | | | | | 1 | 54 |
| 26 | 55 | $CH_3NO_2$ | 2.0086 | 1.868 | 61.04 | Nitroethane: | |
| 26 | 56 | $CH_2NO_2$ | 1.7818 | 1.560 | 60.03 | 1 | 1 |
| 26 | 57 | $CHNO_2$ | 1.5544 | 1.248 | 59.02 | 1 | 56 |

**TABLE 8-21 UNIFAC Group Specifications and Sample Group Assignments (*Continued*)**

| Group numbers | | | | | | Sample assignment | |
|---|---|---|---|---|---|---|---|
| Main | Sec. | Name | $R$ | $Q$ | MW | NOGP | IDGP |
| 27 | 58 | $ACNO_2$ | 1.4199 | 1.104 | 58.02 | Nitrobenzene: | |
| | | | | | | 5 | 10 |
| | | | | | | 1 | 58 |
| 28 | 59 | $CS_2$ | 2.0570 | 1.650 | 76.13 | Carbon disulfide: | |
| | | | | | | 1 | 59 |
| 29 | 60 | $CH_3SH$ | 1.8770 | 1.676 | 48.10 | Ethanethiol: | |
| 29 | 61 | $CH_2SH$ | 1.6510 | 1.368 | 47.09 | 1 | 1 |
| | | | | | | 1 | 61 |
| 30 | 62 | Furfural | 3.1680 | 2.481 | 96.09 | Furfural: | |
| | | | | | | 1 | 62 |
| 31 | 63 | $(CH_2OH)_2$ | 2.4088 | 2.248 | 62.07 | Ethylene glycol: | |
| | | | | | | 1 | 63 |
| 32 | 64 | I | 1.2640 | 0.992 | 126.90 | Iodomethane: | |
| | | | | | | 1 | 1 |
| | | | | | | 1 | 64 |
| 33 | 65 | BR | 0.9492 | 0.832 | 79.90 | Bromomethane: | |
| | | | | | | 1 | 1 |
| | | | | | | 1 | 64 |
| 34 | 66 | CH-trip-C | 1.2920 | 1.088 | 25.03 | Propyne: | |
| 34 | 67 | C-trip-C | 1.0613 | 0.784 | 24.02 | 1 | 1 |
| | | | | | | 1 | 66 |
| 35 | 68 | $Me_2SO$ | 2.8266 | 2.472 | 78.13 | Dimethylsulfoxide: | |
| | | | | | | 1 | 68 |
| 36 | 69 | Acry | 2.3144 | 2.052 | 53.06 | Acrylonitrile: | |
| | | | | | | 1 | 69 |
| 37 | 70 | Cl(C=C) | 0.7910 | 0.724 | 35.45 | Trichloroethylene: | |
| | | | | | | 1 | 8 |
| | | | | | | 3 | 70 |

| | | | | | | | |
|---|---|---|---|---|---|---|---|
| 38 | 71 | ACF | 0.6948 | 0.524 | 31.01 | Fluorobenzene: | |
| | | | | | | 5 | 10 |
| | | | | | | 1 | 71 |
| 39 | 72 | DMF-1 | 3.0856 | 2.736 | 73.09 | Dimethylformamide: | |
| 39 | 73 | DMF-2 | 2.6322 | 2.120 | 43.03 | 1 | 72 |
| | | | | | | Diethylformamide: | |
| | | | | | | 2 | 1 |
| | | | | | | 1 | 73 |
| 40 | 74 | $CF_3$ | 1.4060 | 1.380 | 69.01 | Perfluoroethane: | |
| 40 | 75 | $CF_2$ | 1.0105 | 0.920 | 50.01 | 2 | 74 |
| 40 | 76 | CF | 0.6150 | 0.460 | 31.01 | | |
| 41 | 77 | COO | 1.3800 | 1.200 | 44.01 | Butylacetate: | |
| | | | | | | 2 | 1 |
| | | | | | | 3 | 2 |
| | | | | | | 1 | 77 |
| 42 | 78 | $SiH_3$ | 1.6035 | 1.263 | 31.11 | Methylsilane: | |
| 42 | 79 | $SiH_2$ | 1.4443 | 1.006 | 30.10 | 1 | 1 |
| 42 | 80 | SiH | 1.2851 | 0.749 | 29.09 | 1 | 78 |
| 42 | 81 | Si | 1.0470 | 0.410 | 28.09 | | |
| 43 | 82 | $SiH_2O$ | 1.4338 | 1.062 | 46.10 | Hexamethyldisiloxane: | |
| 43 | 83 | SiHO | 1.3030 | 0.764 | 45.09 | 6 | 1 |
| 43 | 84 | SiO | 1.1044 | 0.466 | 44.09 | 1 | 81 |
| | | | | | | 1 | 84 |
| 44 | 85 | *tert*-N | 0.2854 | 0.092 | 14.01 | Triethylamine: | |
| | | | | | | 3 | 1 |
| | | | | | | 3 | 2 |
| | | | | | | 1 | 85 |
| 45 | 86 | Amide | 1.4660 | 1.336 | 44.03 | Acetamide: | |
| | | | | | | 1 | 1 |
| | | | | | | 1 | 86 |
| 46 | 87 | $CON(Me)_2$ | 2.8590 | 2.428 | 72.09 | *N*,*N*-Methylethylamide: | |
| 46 | 88 | $CONMeCH_2$ | 2.6320 | 2.120 | 71.08 | 2 | 1 |
| 46 | 89 | $CON(CH_2)_2$ | 2.4050 | 1.812 | 70.07 | 1 | 88 |

Equation (8-10.55) also holds for $\ln \Gamma_k^{(i)}$. In Eq. (8-10.55), $\theta_m$ is the area fraction of group $m$, and the sums are over all different groups. $\theta_m$ is calculated in a manner similar to that for $\theta_i$:

$$\theta_m = \frac{Q_m X_m}{\sum_n Q_n X_n} \tag{8-10.56}$$

where $X_m$ is the mole fraction of group $m$ in the mixture. The group interaction parameter $\Psi_{mn}$ is given by

$$\Psi_{mn} = \exp\left(-\frac{U_{mn} - U_{nn}}{RT}\right) = \exp\left(-\frac{a_{mn}}{T}\right) \tag{8-10.57}$$

where $U_{mn}$ is a measure of the energy of interaction between groups $m$ and $n$. The group interaction parameters $a_{mn}$ must be evaluated from experimental phase equilibrium data. Note that $a_{mn}$ has units of kelvins and the $a_{mn} \neq a_{nm}$. Parameters $a_{mn}$ and $a_{nm}$ are obtained from a database using a wide range of experimental results. Some of these are shown in Table 8-22. Efforts toward updating and extending Table 8-22 are in progress in several university laboratories.

The combinatorial contribution to the activity coefficient [Eq. (8-10.50)] depends only on the sizes and shapes of the molecules present. As the coordination number $z$ increases, for large-chain molecules $q_i/r_i \rightarrow 1$ and in that limit, Eq. (8-10.50) reduces to the Flory-Huggins equation used in the ASOG method.

The residual contribution to the activity coefficient [Eqs. (8-10.54) and (8-10.55)] depends on group areas and group interaction. When all group areas are equal, Eqs. (8-10.54) and (8-10.55) are similar to those used in the ASOG method.

The functional groups considered in this work are those given in Table 8-21. Whereas each group listed has its own values of $R$ and $Q$, the subgroups within the same main group, e.g., subgroups 1, 2, and 3 are assumed to have identical energy interaction parameters. We present one example which illustrates (1) the nomenclature and use of Table 8-21 and (2) the UNIFAC method for calculating activity coefficients.

**Example 8-11** Obtain activity coefficients for the acetone (1)–$n$-pentane (2) system at 307 K and $x_1 = 0.047$.

**solution** Acetone has one ($\nu_1 = 1$) $CH_3$ group (main group 1, secondary group 1) and one ($\nu_9 = 1$) $CH_3CO$ group (main group 9, secondary group 19). $n$-Pentane has two ($\nu_1 = 2$) $CH_3$ groups (main group 1, secondary group 1) and three ($\nu_1 = 3$) $CH_2$ groups (main group 1, secondary group 2).

Based on the information in Table 8-21, we can construct the following table:

| Molecule ($i$) | Group identification | | | $\nu_j^{(i)}$ | $R_j$ | $Q_j$ |
|---|---|---|---|---|---|---|
| | Name | Main No. | Sec. No. | | | |
| Acetone (1) | $CH_3$ | 1 | 1 | 1 | 0.9011 | 0.848 |
| | $CH_3CO$ | 9 | 19 | 1 | 1.6724 | 1.488 |
| $n$-Pentane (2) | $CH_3$ | 1 | 1 | 2 | 0.9011 | 0.848 |
| | $CH_2$ | 1 | 2 | 3 | 0.6744 | 0.540 |

We can now write:

$$r_1 = (1)(0.9011) + (1)(1.6724) = 2.5735$$

$$q_1 = (1)(0.848) + (1)(1.488) = 2.336$$

$$\Phi_1 = \frac{(2.5735)(0.047)}{(2.5735)(0.047) + (3.8254)(0.953)} = 0.0321$$

$$\theta_1 = \frac{(2.336)(0.047)}{(2.336)(0.047) + (3.316)(0.953)} = 0.0336$$

$$\ell_1 = (5)(2.5735 - 2.336) - 1.5735 = -0.3860$$

or in tabular form:

| Molecule ($i$) | $r_i$ | $q_i$ | 100 $\Phi_i$ | 100 $\theta_i$ | $\ell_i$ |
|---|---|---|---|---|---|
| Acetone (1) | 2.5735 | 2.336 | 3.21 | 3.36 | −0.3860 |
| $n$-Pentane (2) | 3.8254 | 3.316 | 96.79 | 96.64 | −0.2784 |

We can now calculate the combinatorial contribution to the activity coefficients:

$$\ln \gamma_1^c = \ln \frac{0.0321}{0.047} + (5)(2.336) \ln \frac{0.0336}{0.0321} - 0.3860$$

$$+ \frac{0.0321}{0.047}[(0.047)(0.3860) + (0.953)(0.2784)]$$

$$= -0.0403$$

$$\ln \gamma_2^c = -0.0007$$

Next, we calculate the residual contributions to the activity coefficients. Since only two main groups are represented in this mixture, the calculation is relatively simple. The group interaction parameters $a_{mn}$ are obtained from Table 8-22.

$$a_{1,9} = 476.40$$

$$\Psi_{1,9} = \exp\left(\frac{-476.40}{307}\right)$$

$$= 0.2119$$

$$a_{9,1} = 26.760$$

$$\Psi_{9,1} = \exp\left(\frac{-26.760}{307}\right)$$

$$= 0.9165$$

Note that $\Psi_{1,1} = \Psi_{9,9} = 1.0$, since $a_{1,1} = a_{9,9} = 0$. Let 1 = $CH_3$, 2 = $CH_2$, and 19 = $CH_3CO$.

**TABLE 8-22 UNIFAC Group–Group Interaction Parameters, in Kelvins**

| Main group numbers | 1 | 2 | 3 | 4 | 5 | 6 | 7 |
|---|---|---|---|---|---|---|---|
| 1 | 0 | 86.020 | 61.130 | 76.500 | 986.500 | 697.200 | 1318.000 |
| 2 | −35.360 | 0 | 38.810 | 74.150 | 524.100 | 787.600 | 270.600 |
| 3 | −11.120 | 3.446 | 0 | 167.000 | 636.100 | 637.300 | 903.800 |
| 4 | −69.700 | −113.600 | −146.800 | 0 | 803.200 | 603.200 | 5695.000 |
| 5 | 156.400 | 457.000 | 89.600 | 25.820 | 0 | −137.100 | 353.500 |
| 6 | 16.510 | −12.520 | −50.000 | −44.500 | 249.100 | 0 | −181.000 |
| 7 | 300.000 | 496.100 | 362.300 | 377.600 | −229.100 | 289.600 | 0 |
| 8 | 275.800 | 217.500 | 25.340 | 244.200 | −451.600 | −265.200 | −601.800 |
| 9 | 26.760 | 42.920 | 140.100 | 365.800 | 164.500 | 108.700 | 472.500 |
| 10 | 505.700 | 133.000 | 0 | 0 | −404.800 | −340.200 | 232.700 |
| 11 | 114.800 | 132.100 | 85.840 | −170.000 | 245.400 | 249.600 | 10000.000 |
| 12 | 90.490 | −62.550 | 0 | 0 | 191.200 | 155.700 | 0 |
| 13 | 83.360 | 26.510 | 52.130 | 65.690 | 237.700 | 339.700 | −314.700 |
| 14 | −30.480 | 1.163 | −44.850 | 0 | −164.000 | −481.700 | −330.400 |
| 15 | 65.330 | −28.700 | −22.310 | 223.000 | −150.000 | −500.400 | −448.200 |
| 16 | −83.980 | −25.380 | −223.900 | 109.900 | 28.600 | −406.800 | −598.800 |
| 17 | 5339.000 | 0 | 650.400 | 979.800 | 529.000 | 5.182 | −339.500 |
| 18 | −101.600 | 0 | 31.870 | 49.800 | −132.300 | −378.200 | −332.900 |
| 19 | 24.820 | −40.620 | −22.970 | −138.400 | −185.400 | 157.800 | 242.800 |
| 20 | 315.300 | 1264.000 | 62.320 | 268.200 | −151.000 | 1020.000 | −66.170 |
| 21 | 91.460 | 97.510 | 4.680 | 122.900 | 562.200 | 529.000 | 698.200 |
| 22 | 34.010 | 18.250 | 121.300 | 0 | 747.700 | 669.900 | 708.700 |
| 23 | 36.700 | 51.060 | 288.500 | 33.610 | 742.100 | 649.100 | 826.700 |
| 24 | −78.450 | 160.900 | −4.700 | 134.700 | 856.300 | 860.100 | 1201.000 |
| 25 | −141.300 | −158.800 | −237.700 | 375.500 | 246.900 | 661.600 | 920.400 |
| 26 | −32.690 | −1.996 | 10.380 | −97.050 | 341.700 | 252.600 | 417.900 |
| 27 | 5541.000 | 0 | 1824.000 | −127.800 | 561.600 | 0 | 360.700 |
| 28 | −52.650 | 16.620 | 21.500 | 40.680 | 823.500 | 914.200 | 1081.000 |
| 29 | −7.481 | 0 | 28.410 | 0 | 461.600 | 382.800 | 0 |
| 30 | −25.310 | 0 | 157.300 | 404.300 | 521.600 | 0 | 23.480 |
| 31 | 140.000 | 0 | 221.400 | 150.600 | 267.600 | 0 | 0 |
| 32 | 128.000 | 0 | 58.680 | 0 | 501.300 | 0 | 0 |
| 33 | −31.520 | 0 | 155.600 | 291.100 | 721.900 | 0 | 0 |
| 34 | −72.880 | 41.380 | 0 | 0 | 0 | 0 | 0 |
| 35 | 50.490 | 422.400 | −2.504 | −143.200 | −25.870 | 695.000 | −240.000 |
| 36 | −165.900 | 0 | 0 | 0 | 0 | 0 | 386.600 |
| 37 | 47.410 | 124.200 | 395.800 | 0 | 738.900 | 528.000 | 0 |
| 38 | −5.132 | 0 | −237.200 | −157.300 | 649.700 | 645.900 | 0 |
| 39 | −31.950 | 249.000 | −133.900 | −240.200 | 64.160 | 172.200 | −287.100 |
| 40 | 147.300 | 0 | 0 | 0 | 0 | 0 | 0 |
| 41 | 529.000 | 1397.000 | 317.600 | 615.800 | 88.630 | 171.000 | 284.400 |
| 42 | 97.900 | 0 | −184.300 | 191.600 | 85.190 | 0 | 0 |
| 43 | 109.200 | 0 | 293.800 | 221.800 | 84.850 | 0 | 0 |
| 44 | 272.000 | 0 | −288.000 | −1020.000 | 0 | −668.000 | −1080.000 |
| 45 | 8960.000 | −963.000 | −63.100 | −196.000 | 0 | 0 | 0 |
| 46 | −11.100 | 0 | −11.800 | −36.600 | 0 | 0 | 0 |

| Main group numbers | 8 | 9 | 10 | 11 | 12 | 13 | 14 |
|---|---|---|---|---|---|---|---|
| 1 | 1333.000 | 476.400 | 677.000 | 232.100 | 741.400 | 251.500 | 391.500 |
| 2 | 526.100 | 182.600 | −35.100 | 37.850 | 449.100 | 214.500 | 240.900 |
| 3 | 1329.000 | 25.770 | 0 | 5.994 | 0 | 32.140 | 161.700 |
| 4 | 884.900 | −52.100 | 0 | 5688.000 | 0 | 213.100 | 0 |
| 5 | −259.700 | 84.000 | 441.800 | 101.100 | 193.100 | 28.060 | 83.020 |
| 6 | −101.700 | 23.390 | 306.400 | −10.720 | 193.400 | −180.600 | 359.300 |
| 7 | 324.500 | −195.400 | −257.300 | 14.420 | 0 | 540.500 | 48.890 |
| 8 | 0 | −356.100 | 0 | −449.400 | 0 | 0 | 0 |
| 9 | −133.100 | 0 | −37.360 | −213.700 | 0 | 5.202 | 0 |
| 10 | 0 | 128.000 | 0 | −448.000 | 0 | 304.100 | 0 |
| 11 | −36.720 | 372.200 | 2390.000 | 0 | 372.900 | −235.700 | 0 |
| 12 | 0 | 0 | 0 | −261.100 | 0 | 0 | 0 |
| 13 | 0 | 52.380 | −7.838 | 461.300 | 0 | 0 | 0 |
| 14 | 0 | 0 | 0 | 0 | 0 | 0 | 0 |
| 15 | 0 | 0 | 0 | 136.000 | 0 | −49.300 | 108.800 |
| 16 | 0 | 0 | 0 | 0 | 0 | 0 | 38.890 |
| 17 | 0 | −399.100 | 0 | 0 | 0 | 0 | 0 |
| 18 | −341.600 | −51.540 | 0 | 0 | 0 | 0 | 0 |
| 19 | 0 | −287.500 | 0 | −266.600 | 0 | 0 | 0 |
| 20 | 0 | −297.800 | 0 | −256.300 | 312.500 | −338.500 | 0 |
| 21 | 0 | 286.300 | −47.510 | 0 | 0 | 225.400 | 0 |
| 22 | 0 | 423.200 | 0 | −132.900 | 0 | −197.700 | 0 |
| 23 | 0 | 552.100 | 0 | 176.500 | 488.900 | −20.930 | 0 |
| 24 | 10000.000 | 372.000 | 0 | 129.500 | 0 | 113.900 | 261.100 |
| 25 | 0 | 128.100 | 0 | −246.300 | 0 | 0 | 203.500 |
| 26 | 0 | −142.600 | 0 | 0 | 0 | −94.490 | 0 |
| 27 | 0 | 0 | 0 | 0 | 0 | 0 | 0 |
| 28 | 0 | 303.700 | 0 | 243.800 | 0 | 112.400 | 0 |
| 29 | 0 | 160.600 | 0 | 0 | 239.800 | 63.710 | 106.700 |
| 30 | 0 | 317.500 | 0 | −146.300 | 0 | 0 | 0 |
| 31 | 838.400 | 0 | 0 | 152.000 | 0 | 9.207 | 0 |
| 32 | 0 | 138.000 | 0 | 21.920 | 0 | 476.600 | 0 |
| 33 | 0 | −142.600 | 0 | 0 | 0 | 736.400 | 0 |
| 34 | 0 | 443.600 | 0 | 0 | 0 | 0 | 0 |
| 35 | 0 | 110.400 | 0 | 41.570 | 0 | −122.100 | 0 |
| 36 | 0 | 0 | 0 | 0 | 0 | 0 | 0 |
| 37 | 0 | −40.900 | 0 | 16.990 | 0 | −217.900 | 0 |
| 38 | 0 | 0 | 0 | 0 | 0 | 0 | 0 |
| 39 | 0 | 97.040 | 0 | 0 | 0 | −158.200 | 0 |
| 40 | 0 | 0 | 0 | 0 | 0 | 0 | 0 |
| 41 | −167.300 | 123.400 | 0 | −234.900 | 65.370 | −247.800 | 0 |
| 42 | 0 | 0 | 0 | 0 | 0 | 0 | 0 |
| 43 | 0 | 0 | 0 | 0 | 0 | 0 | 0 |
| 44 | 0 | −435.000 | −686.000 | −463.000 | 0 | 2880.000 | 0 |
| 45 | 0 | −444.000 | −167.000 | 0 | 0 | −74.700 | 0 |
| 46 | 0 | 1530.000 | −60.800 | −466.000 | 0 | 0 | 0 |

**TABLE 8-22 UNIFAC Group–Group Interaction Parameters, in Kelvins (*Continued*)**

| Main group numbers | 15 | 16 | 17 | 18 | 19 | 20 | 21 |
|---|---|---|---|---|---|---|---|
| 1 | 225.700 | 206.600 | 1245.000 | 287.700 | 597.000 | 663.500 | 35.930 |
| 2 | 163.900 | 61.110 | 0 | 0 | 336.900 | 318.900 | 204.600 |
| 3 | 122.800 | 90.490 | 668.200 | −4.449 | 212.500 | 537.400 | −18.810 |
| 4 | −49.290 | 23.500 | 764.700 | 52.800 | 6096.000 | 603.800 | −114.100 |
| 5 | 42.700 | −323.000 | −348.200 | 170.000 | 6.712 | 199.000 | 75.620 |
| 6 | 266.000 | 53.900 | 335.500 | 580.500 | 36.230 | −289.500 | −38.320 |
| 7 | 168.000 | 304.000 | 213.000 | 459.000 | 112.600 | −14.090 | 325.400 |
| 8 | 0 | 0 | 0 | −305.500 | 0 | 0 | 0 |
| 9 | 0 | 0 | 937.900 | 165.100 | 481.700 | 669.400 | −191.700 |
| 10 | 0 | 0 | 0 | 0 | 0 | 0 | 751.900 |
| 11 | −73.500 | 0 | 0 | 0 | 494.600 | 660.200 | 0 |
| 12 | 0 | 0 | 0 | 0 | 0 | −356.300 | 0 |
| 13 | 141.700 | 0 | 0 | 0 | 0 | 664.600 | 301.100 |
| 14 | 63.720 | −41.110 | 0 | 0 | 0 | 0 | 0 |
| 15 | 0 | −189.200 | 0 | 0 | 0 | 0 | 0 |
| 16 | 865.900 | 0 | 0 | 0 | 0 | 0 | 0 |
| 17 | 0 | 0 | 0 | 0 | −216.800 | 0 | 0 |
| 18 | 0 | 0 | 0 | 0 | −169.700 | −153.700 | 0 |
| 19 | 0 | 0 | 617.100 | 134.300 | 0 | 0 | 0 |
| 20 | 0 | 0 | 0 | −313.500 | 0 | 0 | 44.420 |
| 21 | 0 | 0 | 0 | 0 | 0 | 826.400 | 0 |
| 22 | 0 | −141.400 | 0 | 587.300 | 0 | 1821.000 | −84.530 |
| 23 | 0 | −293.700 | 0 | 18.980 | 74.040 | 1346.000 | −157.100 |
| 24 | 91.130 | −126.000 | 1301.000 | 309.200 | 492.000 | 689.000 | 11.800 |
| 25 | −108.400 | 1088.000 | 323.300 | 0 | 356.900 | 0 | −314.900 |
| 26 | 0 | 0 | 0 | 0 | 0 | 0 | 0 |
| 27 | 0 | 0 | 5250.000 | 0 | 0 | 0 | 0 |
| 28 | 0 | 0 | 0 | 0 | 335.700 | 0 | −73.090 |
| 29 | 0 | 0 | 0 | 0 | 125.700 | 0 | −27.940 |
| 30 | 0 | 0 | 0 | 0 | 0 | 0 | 0 |
| 31 | 0 | 0 | 164.400 | 0 | 0 | 0 | 0 |
| 32 | 0 | 0 | 0 | 0 | 0 | 0 | 0 |
| 33 | 0 | 0 | 0 | 0 | 0 | 0 | 1169.000 |
| 34 | 0 | 0 | 0 | 0 | 329.100 | 0 | 0 |
| 35 | 0 | 0 | 0 | 0 | 0 | 0 | 0 |
| 36 | 0 | 0 | 0 | 0 | −42.310 | 0 | 0 |
| 37 | 0 | 0 | 0 | 0 | 304.000 | 898.200 | 428.500 |
| 38 | 0 | 0 | 0 | 0 | 0 | 0 | 0 |
| 39 | 0 | 0 | 335.600 | 0 | 0 | 0 | 0 |
| 40 | 0 | 0 | 0 | 0 | 0 | 0 | 0 |
| 41 | 284.500 | 0 | 0 | 0 | −61.600 | 1179.000 | 0 |
| 42 | 0 | 0 | 0 | 0 | 0 | 2450.000 | 0 |
| 43 | 0 | 0 | 0 | 0 | 0 | 2496.000 | 0 |
| 44 | 0 | 0 | 0 | 0 | 0 | 0 | 0 |
| 45 | 0 | 0 | 0 | 0 | 0 | 0 | 0 |
| 46 | 0 | 0 | 0 | 0 | 0 | 0 | 0 |

| Main group numbers | 22 | 23 | 24 | 25 | 26 | 27 | 28 |
|---|---|---|---|---|---|---|---|
| 1 | 53.760 | 24.900 | 104.300 | 321.500 | 661.500 | 543.000 | 153.600 |
| 2 | 5.892 | −13.990 | −109.700 | 393.100 | 357.500 | 0 | 76.300 |
| 3 | −144.400 | −231.900 | 3.000 | 538.200 | 168.000 | 194.900 | 52.070 |
| 4 | 0 | −12.140 | −141.300 | −126.900 | 3629.000 | 4448.000 | −9.451 |
| 5 | −112.100 | −98.120 | 143.100 | 287.800 | 61.110 | 157.100 | 477.000 |
| 6 | −102.500 | −139.400 | −67.800 | 17.120 | 75.140 | 0 | −31.090 |
| 7 | 370.400 | 353.700 | 497.500 | 678.200 | 220.600 | 399.500 | 887.100 |
| 8 | 0 | 0 | 1827.000 | 0 | 0 | 0 | 0 |
| 9 | −284.000 | −354.600 | −39.200 | 174.500 | 137.500 | 0 | 216.100 |
| 10 | 0 | 0 | 0 | 0 | 0 | 0 | 0 |
| 11 | 108.900 | −209.700 | 54.470 | 629.000 | 0 | 0 | 183.000 |
| 12 | 0 | −287.200 | 0 | 0 | 0 | 0 | 0 |
| 13 | 137.800 | −154.300 | 47.670 | 0 | 95.180 | 0 | 140.900 |
| 14 | 0 | 0 | −99.810 | 68.810 | 0 | 0 | 0 |
| 15 | 0 | 0 | 71.230 | 4350.000 | 0 | 0 | 0 |
| 16 | −73.850 | −352.900 | −8.283 | −86.360 | 0 | 0 | 0 |
| 17 | 0 | 0 | 8455.000 | 699.100 | 0 | −62.730 | 0 |
| 18 | −351.600 | −114.700 | −165.100 | 0 | 0 | 0 | 0 |
| 19 | 0 | −15.620 | −54.860 | 52.310 | 0 | 0 | 230.900 |
| 20 | −183.400 | 76.750 | 212.700 | 0 | 0 | 0 | 0 |
| 21 | 108.300 | 249.200 | 62.420 | 464.400 | 0 | 0 | 450.100 |
| 22 | 0 | 0 | 56.330 | 0 | 0 | 0 | 0 |
| 23 | 0 | 0 | −30.100 | 0 | 0 | 0 | 116.600 |
| 24 | 17.970 | 51.900 | 0 | 475.800 | 490.900 | 534.700 | 132.200 |
| 25 | 0 | 0 | −255.400 | 0 | 154.500 | 0 | 0 |
| 26 | 0 | 0 | −34.680 | 794.400 | 0 | 533.200 | 0 |
| 27 | 0 | 0 | 514.600 | 0 | −85.120 | 0 | 0 |
| 28 | 0 | −26.060 | −60.710 | 0 | 0 | 0 | 0 |
| 29 | 0 | 0 | 0 | 0 | 0 | 0 | 0 |
| 30 | 0 | 48.480 | −133.100 | 0 | 0 | 0 | 0 |
| 31 | 0 | 0 | 0 | 0 | 481.300 | 0 | 0 |
| 32 | −40.820 | 21.760 | 48.490 | 0 | 64.280 | 0 | 0 |
| 33 | 0 | 0 | 225.800 | 224.000 | 125.300 | 0 | 0 |
| 34 | 0 | 0 | 0 | 0 | 174.400 | 0 | 0 |
| 35 | −215.000 | −343.600 | −58.430 | 0 | 0 | 0 | 0 |
| 36 | 0 | 0 | 0 | 0 | 0 | 0 | 0 |
| 37 | 0 | −149.800 | 134.200 | 0 | 379.400 | 0 | 167.900 |
| 38 | 0 | 0 | −124.600 | 0 | 0 | 0 | 0 |
| 39 | 0 | 0 | −186.700 | 0 | 0 | 0 | 0 |
| 40 | 0 | 0 | 0 | 0 | 0 | 0 | 0 |
| 41 | 305.400 | −193.000 | 335.700 | 1107.000 | 0 | 0 | 885.500 |
| 42 | 0 | 0 | 0 | 0 | 0 | 0 | 0 |
| 43 | 0 | 0 | 102.000 | 0 | 0 | 0 | 0 |
| 44 | 0 | 0 | 0 | 0 | 0 | 0 | 0 |
| 45 | 0 | 0 | 0 | 0 | 0 | 0 | 0 |
| 46 | 0 | 0 | 0 | 0 | 0 | 0 | 0 |

**TABLE 8-22 UNIFAC Group–Group Interaction Parameters, in Kelvins** ***(Continued)***

| Main group numbers | 29 | 30 | 31 | 32 | 33 | 34 | 35 |
|---|---|---|---|---|---|---|---|
| 1 | 184.400 | 354.500 | 3025.000 | 335.800 | 479.500 | 298.900 | 526.500 |
| 2 | 0 | 0 | 0 | 0 | 0 | 31.140 | −137.400 |
| 3 | −10.430 | −64.690 | 210.400 | 113.300 | −13.590 | 0 | 169.900 |
| 4 | 0 | −20.360 | 4975.000 | 0 | −171.300 | 0 | 4284.000 |
| 5 | 147.500 | −120.500 | −318.900 | 313.500 | 133.400 | 0 | −202.100 |
| 6 | 37.840 | 0 | 0 | 0 | 0 | 0 | −399.300 |
| 7 | 0 | 188.000 | 0 | 0 | 0 | 0 | −139.000 |
| 8 | 0 | 0 | −687.100 | 0 | 0 | 0 | 0 |
| 9 | −46.280 | −163.700 | 0 | 53.590 | 245.200 | −246.600 | −44.580 |
| 10 | 0 | 0 | 0 | 0 | 0 | 0 | 0 |
| 11 | 0 | 202.300 | −101.700 | 148.300 | 0 | 0 | 52.080 |
| 12 | 4.339 | 0 | 0 | 0 | 0 | 0 | 0 |
| 13 | −8.538 | 0 | −20.110 | −149.500 | −202.300 | 0 | 172.100 |
| 14 | −70.140 | 0 | 0 | 0 | 0 | 0 | 0 |
| 15 | 0 | 0 | 0 | 0 | 0 | 0 | 0 |
| 16 | 0 | 0 | 0 | 0 | 0 | 0 | 0 |
| 17 | 0 | 0 | 125.300 | 0 | 0 | 0 | 0 |
| 18 | 0 | 0 | 0 | 0 | 0 | 0 | 0 |
| 19 | 21.370 | 0 | 0 | 0 | 0 | −203.000 | 0 |
| 20 | 0 | 0 | 0 | 0 | 0 | 0 | 0 |
| 21 | 59.020 | 0 | 0 | 0 | −25.900 | 0 | 0 |
| 22 | 0 | 0 | 0 | 177.600 | 0 | 0 | 215.000 |
| 23 | 0 | −64.380 | 0 | 86.400 | 0 | 0 | 363.700 |
| 24 | 0 | 546.700 | 0 | 247.800 | 41.940 | 0 | 337.700 |
| 25 | 0 | 0 | 0 | 0 | −60.700 | 0 | 0 |
| 26 | 0 | 0 | 139.800 | 304.300 | 10.170 | −27.700 | 0 |
| 27 | 0 | 0 | 0 | 0 | 0 | 0 | 0 |
| 28 | 0 | 0 | 0 | 0 | 0 | 0 | 0 |
| 29 | 0 | 0 | 0 | 0 | 0 | 0 | 31.660 |
| 30 | 0 | 0 | 0 | 0 | 0 | 0 | 0. |
| 31 | 0 | 0 | 0 | 0 | 0 | 0 | −417.200 |
| 32 | 0 | 0 | 0 | 0 | 0 | 0 | 0 |
| 33 | 0 | 0 | 0 | 0 | 0 | 0 | 0 |
| 34 | 0 | 0 | 0 | 0 | 0 | 0 | 0 |
| 35 | 85.700 | 0 | 535.800 | 0 | 0 | 0 | 0 |
| 36 | 0 | 0 | 0 | 0 | 0 | 0 | 0 |
| 37 | 0 | 0 | 0 | 0 | 0 | 0 | 0 |
| 38 | 0 | 0 | 0 | 0 | 0 | 0 | 0 |
| 39 | −71.000 | 0 | −191.700 | 0 | 0 | 6.699 | 136.600 |
| 40 | 0 | 0 | 0 | 0 | 0 | 0 | 0 |
| 41 | 0 | 0 | 0 | 288.100 | 0 | 0 | −29.340 |
| 42 | 0 | 0 | 0 | 0 | 0 | 0 | 0 |
| 43 | 0 | 0 | 0 | 0 | 0 | 0 | 0 |
| 44 | 0 | 0 | 0 | 0 | 0 | 0 | 0 |
| 45 | 0 | 0 | 0 | 0 | 0 | 0 | 0 |
| 46 | 0 | 0 | 0 | 0 | 0 | 0 | 0 |

| Main group numbers | 36 | 37 | 38 | 39 | 40 | 41 | 42 |
|---|---|---|---|---|---|---|---|
| 1 | 689.000 | −4.189 | 125.800 | 485.300 | −2.859 | 387.100 | 407.200 |
| 2 | 0 | −66.460 | 0 | −70.450 | 0 | 48.330 | 0 |
| 3 | 0 | −259.100 | 389.300 | 245.600 | 0 | 103.500 | 551.900 |
| 4 | 0 | 0 | 101.400 | 5629.000 | 0 | 69.260 | 683.300 |
| 5 | 0 | 225.800 | 44.780 | −143.900 | 0 | 190.300 | 269.100 |
| 6 | 0 | 33.470 | −48.250 | −172.400 | 0 | 165.700 | 0 |
| 7 | 160.800 | 0 | 0 | 319.000 | 0 | −197.500 | 0 |
| 8 | 0 | 0 | 0 | 0 | 0 | −494.200 | |
| 9 | 0 | −34.570 | 0 | −61.700 | 0 | −18.800 | 0 |
| 10 | 0 | 0 | 0 | 0 | 0 | 0 | 0 |
| 11 | 0 | −83.300 | 0 | 0 | 0 | 560.200 | 0 |
| 12 | 0 | 0 | 0 | 0 | 0 | −70.240 | 0 |
| 13 | 0 | 240.200 | 0 | 254.800 | 0 | 417.000 | 0 |
| 14 | 0 | 0 | 0 | 0 | 0 | 0 | 0 |
| 15 | 0 | 0 | 0 | 0 | 0 | −38.770 | 0 |
| 16 | 0 | 0 | 0 | 0 | 0 | 0 | 0 |
| 17 | 0 | 0 | 0 | −293.100 | 0 | 0 | 0 |
| 18 | 0 | 0 | 0 | 0 | 0 | 0 | 0 |
| 19 | 81.570 | 3.509 | 0 | 0 | 0 | 120.300 | 0 |
| 20 | 0 | −11.160 | 0 | 0 | 0 | −337.000 | 169.300 |
| 21 | 0 | −245.400 | 0 | 0 | 0 | 0 | 0 |
| 22 | 0 | 0 | 0 | 0 | 0 | −96.870 | 0 |
| 23 | 0 | 111.200 | 0 | 0 | 0 | 255.800 | 0 |
| 24 | 0 | 187.100 | 215.200 | 498.600 | 0 | 256.500 | 639.300 |
| 25 | 0 | 0 | 0 | 0 | 0 | −145.100 | 0 |
| 26 | 0 | 10.760 | 0 | 0 | 0 | 0 | 0 |
| 27 | 0 | 0 | 0 | 0 | 0 | 0 | 0 |
| 28 | 0 | −47.370 | 0 | 0 | 0 | 469.800 | 0 |
| 29 | 0 | 0 | 0 | 78.920 | 0 | 0 | 0 |
| 30 | 0 | 0 | 0 | 0 | 0 | 0 | 0 |
| 31 | 0 | 0 | 0 | 302.200 | 0 | 0 | 0 |
| 32 | 0 | 0 | 0 | 0 | 0 | 68.550 | 0 |
| 33 | 0 | 0 | 0 | 0 | 0 | 0 | 0 |
| 34 | 0 | 0 | 0 | −119.800 | 0 | 0 | 0 |
| 35 | 0 | 0 | 0 | −97.710 | 0 | 153.700 | 0 |
| 36 | 0 | 0 | 0 | 0 | 0 | 423.400 | 0 |
| 37 | 0 | 0 | 0 | 0 | 0 | 730.800 | 0 |
| 38 | 0 | 0 | 0 | 0 | 0 | 0 | 0 |
| 39 | 0 | 0 | 0 | 0 | 0 | 0 | 0 |
| 40 | 0 | 0 | 0 | 0 | 0 | 0 | 0 |
| 41 | −53.910 | −198.000 | 0 | 0 | 0 | 0 | 0 |
| 42 | 0 | 0 | 0 | 0 | 0 | 0 | 0 |
| 43 | 0 | 0 | 0 | 0 | 0 | 0 | 639.300 |
| 44 | 0 | 0 | 0 | 0 | 0 | 0 | 0 |
| 45 | 0 | 0 | 0 | 0 | 0 | 0 | 0 |
| 46 | 0 | 0 | 0 | 0 | 0 | 0 | 0 |

**TABLE 8-22 UNIFAC Group–Group Interaction Parameters, in Kelvins** (*Continued*)

| Main group numbers | 43 | 44 | 45 | 46 |
|---|---|---|---|---|
| 1 | 327.000 | 383.000 | −1380.000 | 729.000 |
| 2 | 0 | 0 | −2340.000 | 0 |
| 3 | 254.300 | 109.000 | 75.900 | 784.000 |
| 4 | 355.500 | 1320.000 | 482.000 | 386.000 |
| 5 | 202.700 | 0 | 0 | 0 |
| 6 | 0 | 214.000 | 0 | 0 |
| 7 | 0 | 365.000 | 0 | 0 |
| 8 | 0 | 0 | 0 | 0 |
| 9 | 0 | 135.000 | −1680.000 | −58.000 |
| 10 | 0 | −7.180 | 333.000 | 6810.000 |
| 11 | 0 | −54.600 | 0 | 6960.000 |
| 12 | 0 | 0 | 0 | 0 |
| 13 | 0 | 5780.000 | 131.000 | 0 |
| 14 | 0 | 0 | 0 | 0 |
| 15 | 0 | 0 | 0 | 0 |
| 16 | 0 | 0 | 0 | 0 |
| 17 | 0 | 0 | 0 | 0 |
| 18 | 0 | 0 | 0 | 0 |
| 19 | 0 | 0 | 0 | 0 |
| 20 | 127.200 | 0 | 0 | 0 |
| 21 | 0 | 0 | 0 | 0 |
| 22 | 0 | 0 | 0 | 0 |
| 23 | 0 | 0 | 0 | 0 |
| 24 | 0 | 0 | 0 | 0 |
| 25 | 0 | 0 | 0 | 0 |
| 26 | 0 | 0 | 0 | 0 |
| 27 | 0 | 0 | 0 | 0 |
| 28 | 0 | 0 | 0 | 0 |
| 29 | 0 | 0 | 0 | 0 |
| 30 | 0 | 0 | 0 | 0 |
| 31 | 0 | 0 | 0 | 0 |
| 32 | 0 | 0 | 0 | 0 |
| 33 | 0 | 0 | 0 | 0 |
| 34 | 0 | 0 | 0 | 0 |
| 35 | 0 | 0 | 0 | 0 |
| 36 | 0 | 0 | 0 | 0 |
| 37 | 0 | 0 | 0 | 0 |
| 38 | 0 | 0 | 0 | 0 |
| 39 | 0 | 0 | 0 | 0 |
| 40 | 0 | 0 | 0 | 0 |
| 41 | 0 | 0 | 0 | 0 |
| 42 | 498.800 | 0 | 0 | 0 |
| 43 | 0 | 0 | 0 | 0 |
| 44 | 0 | 0 | 0 | 0 |
| 45 | 0 | 0 | 0 | 0 |
| 46 | 0 | 0 | 0 | 0 |

Next we compute $\Gamma_k^{(i)}$, the residual activity coefficient of group $k$ in a reference solution containing only molecules of type $i$. For pure acetone (1), the mole fraction of group $m$, $X_m$, is

$$X_1^{(1)} = \frac{\nu_1^{(1)}}{\nu_1^{(1)} + \nu_{19}^{(1)}} = \frac{1}{1+1} = \frac{1}{2} \qquad X_{19}^{(1)} = \frac{1}{2}$$

Hence

$$\theta_1^{(1)} = \frac{\frac{1}{2}(0.848)}{\frac{1}{2}(0.848) + \frac{1}{2}(1.488)}$$

$$= 0.363 \qquad \theta_{19}^{(1)} = 0.637$$

$$\ln \Gamma_1^{(1)} = 0.848 \left\{1 - \ln[0.363 + (0.637)(0.9165)] - \left[\frac{(0.363)}{0.363 + 0.637(0.9165)} + \frac{(0.637)(0.2119)}{(0.363)(0.2119) + 0.637}\right]\right\}$$

$$= 0.409$$

$$\ln \Gamma_{19}^{(1)} = 1.488 \left\{1 - \ln\,[(0.363)(0.2119) + 0.637] - \left[\frac{(0.363)(0.9165)}{0.363 + (0.637)(0.9165)} + \frac{0.637}{(0.363)(0.2119) + 0.637)}\right]\right\}$$

$$= 0.139$$

For pure $n$-pentane (2), the mole fraction of group $m$, $X_m$, is

$$X_1^{(2)} = \frac{\nu_1^{(2)}}{\nu_1^{(2)} + \nu_2^{(2)}} = \frac{2}{2+3} = \frac{2}{5} \qquad X_2^{(2)} = \frac{3}{5}$$

Since only one main group in is $n$-pentane (2),

$$\ln \Gamma_1^{(2)} = \ln \Gamma_2^2 = 0.0$$

The group residual activity coefficients can now be calculated for $x_1 = 0.047$:

$$X_1 = \frac{(0.047)(1) + 0.953(2)}{(0.047)(2) + 0.953(5)} = 0.4019 \qquad X_2 = 0.05884 \qquad X_{19} = 0.097$$

$$\theta_1 = \frac{(0.848)(0.4019)}{(0.848)(0.4019) + (0.540)(0.5884) + (1.488)(0.0097)} = 0.5064$$

$$\theta_2 = 0.4721 \qquad \theta_{19} = 0.0214$$

$$\ln \Gamma_1 = 0.848 \left\{1 - \ln[0.5064 + 0.4721 + (0.0214)(0.9165)] - \left[\frac{0.5064 + 0.4721}{0.5064 + 0.4721 + (0.0214)(0.9165)} + \frac{(0.0214)(0.2119)}{(0.5064 + 0.4721)(0.2119) + 0.0214}\right]\right\}$$

$$= 1.45 \times 10^{-3}$$

$$\ln \Gamma_2 = 0.540 \left\{1 - \ln[0.5064 + 0.4721 + (0.0214)(0.9165)] - \left[\frac{0.5064 + 0.4721}{0.5064 + 0.4721 + (0.0214)(0.9165)} + \frac{(0.0214)(0.2119)}{(0.5064 + 0.4721)(0.2119) + 0.0214}\right]\right\}$$

$$= 9.26 \times 10^{-4}$$

$$\ln \Gamma_{19} = 1.488 \left\{ 1 - \ln[(0.5064 + 0.4721)(0.2119) + 0.214] - \left[ \frac{(0.5064 + 0.4721)(0.9165)}{0.5064 + 0.4721 + (0.0214)(0.9165)} + \frac{0.0214}{(0.5064 + 0.4721)(0.2119) + 0.0214} \right] \right\}$$
$$= 2.21$$

The residual contributions to the activity coefficients follow

$$\ln \gamma_1^R = (1)(1.45 \times 10^{-3} - 0.409) + (1)(2.21 - 0.139) = 1.66$$
$$\ln \gamma_2^R = (2)(1.45 \times 10^{-3} - 0.0) + 3(9.26 \times 10^{-4} - 0.0) = 5.68 \times 10^{-3}$$

Finally, we calculate the activity coefficients:

$$\ln \gamma_1 = \ln \gamma_1^C + \ln \gamma_1^R = -0.403 + 1.66 = 1.62$$
$$\ln \gamma_2 = \ln \gamma_2^C + \ln \gamma_2^R = -0.0007 + 5.68 \times 10^{-3} = 4.98 \times 10^{-3}$$

Hence,

$$\gamma_1 = 5.07 \qquad \gamma_2 = 1.01$$

The corresponding experimental values of Lo et al. (70) are:

$$\gamma_1 = 4.41 \qquad \gamma_2 = 1.11$$

Although agreement with experiment is not as good as we might wish, it is not bad and it is representative of what UNIFAC can do. The main advantage of UNIFAC is its wide range of application.†

## 8-11 Solubilities of Gases in Liquids

At modest pressures, most gases are only sparingly soluble in typical liquids. For example, at 25°C and a partial pressure of 1.01 bar, the (mole fraction) solubility of nitrogen in cyclohexane is $x = 7.6 \times 10^{-4}$ and that in water is $x = 0.18 \times 10^{-4}$. Although there are some exceptions (notably, hydrogen), the solubility of a gas in typical solvents usually falls with rising temperature. However, at higher temperatures, approaching the critical temperature of the solvent, the solubility of a gas usually rises with temperature, as illustrated in Fig. 8-13.

Experimentally determined solubilities have been reported in the chemical literature for over 100 years, but many of the data are of poor quality. Although no truly comprehensive and critical compilation of the available data exists, Table 8-23 gives some useful data sources.

Unfortunately, a variety of units has been employed in reporting gas solubilities. The most common of these are two dimensionless coefficients: *Bunsen coefficient*, defined as the volume (corrected to 0°C and 1 atm) of gas dissolved per unit volume of solvent at system temperature $T$ when the partial pressure of the solute is 1 atm; *Ostwald coefficient*, defined as

†Tables 8-21 and 8-22 are periodically revised and extended. For the latest versions, contact J. M. Prausnitz, Department of Chemical Engineering, University of California, Berkeley.

**TABLE 8-23 Some Sources for Solubilities of Gases in Liquids**

Washburn, E. W., (ed.): *International Critical Tables,* McGraw-Hill, New York, 1926.

Markam, A. E., and K. A. Kobe: *Chem. Rev.,* **28:** 519 (1941).

Seidell, A.: *Solubilities of Inorganic and Metal-Organic Compounds,* Van Nostrand, New York, 1958, and *Solubilities of Inorganic and Organic Compounds, ibid.,* 1952.

Linke, W. L.: *Solubilities of Inorganic and Metal-Organic Compounds,* 4th ed., Van Nostrand, Princeton, N.J., 1958 and 1965, Vols. 1 and 2. (A revision and continuation of the compilation originated by A. Seidell.)

Stephen, H., and T. Stephen: *Solubilities of Inorganic and Organic Compounds,* Vols. 1 and 2, Pergamon Press, Oxford, and Macmillan, New York, 1963 and 1964.

Battino, R., and H. L. Clever: *Chem. Rev.,* **66:** 395 (1966).

Wilhelm, E., and R. Battino: *Chem. Rev.,* **73:** 1 (1973).

Clever, H. L., and R. Battino: "The Solubility of Gases in Liquids," in M. R. J. Dack (ed.), *Solutions and Solubilities,* Vol. 8, Part 1, Wiley, New York, 1975, pp. 379–441.

Kertes, A. S., O. Levy, and G. Y. Markovits: "Solubility," in B. Vodar (ed.), *Experimental Thermodynamics of Nonpolar Fluids,* Vol. II, Butterworth, London, 1975, pp. 725–748.

Gerrard, W.: *Solubility of Gases and Liquids,* Plenum, New York, 1976.

Landolt-Börnstein: 2. Teil, Bandteil b, *Lösungsgleichgewichte I,* Springer, Berlin, 1962; IV. Band, Technik, 4. Teil, Wärmetechnik; Bandteil c, *Gleichgewicht der Absorption von Gasen in Flussigkeiten, ibid.,* 1976.

Wilhelm, E., R. Battino, and R. J. Wilcock: *Chem. Rev.,* **77:** 219 (1977).

Gerrard, W.: *Gas Solubilities, Widespread Applications,* Pergamon, Oxford, 1980.

Battino, R., T. R. Rettich, and T. Tomingaga: *J. Phys. Chem. Ref. Data,* **12:** 163 (1983).

Wilhelm, E.: *Pure Appl. Chem.,* **57**(2):303–322 (1985).

Wilhelm, E.: *CRC Crit. Rev. Anal. Chem.,* **16**(2):129–175 (1985).

IUPAC: *Solubility Data Series,* A. S. Kertes, editor-in-chief, Pergamon, Oxford.

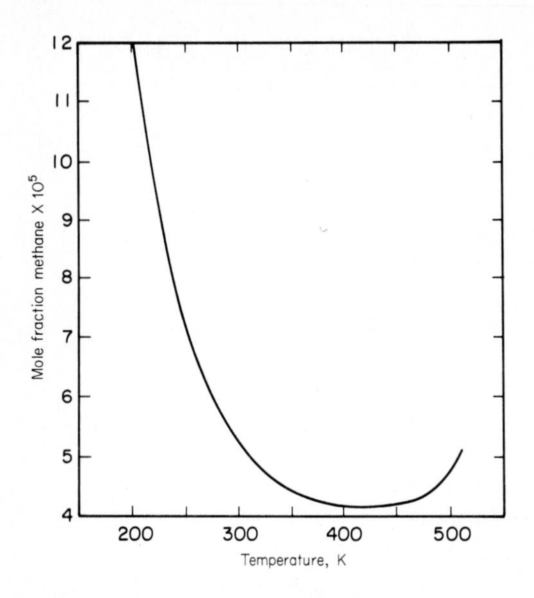


**Figure 8-13** Solubility of methane in *n*-heptane when vapor phase fugacity of methane is 0.0101 bar. *(From Ref. 97.)*

the volume of gas at system temperature $T$ and partial pressure $p$ dissolved per unit volume of solvent. If the solubility is small and the gas phase is ideal, the Ostwald coefficient is independent of $p$ and these two coefficients are simply related by

$$\text{Ostwald coefficient} = \frac{T}{273}\,(\text{Bunsen coefficient})$$

where $T$ is in kelvins. Adler [5], Battino [11], and Friend and Adler [37] have discussed these and other coefficients for expressing solubilities as well as some of their applications for engineering calculations.

These units are often found in older articles. In recent years it has become more common to report solubilities in units of mole fraction or Henry's constants.

When the solubility is small, Henry's law provides a good approximation. In a binary system let subscript 2 refer to the gaseous solute and subscript 1 to the liquid solvent. Henry's law is written

$$f_2 = H_{2,1}^{(P_{\text{vp}1})} x_2 \qquad x_2 \ll 1 \tag{8-11.1}$$

where $x$ is the mole fraction, $f$ is the fugacity, and $H$ is Henry's constant, rigorously defined by

$$H_{2,1}^{(P_{\text{vp}1})} = \lim_{x_2 \to 0} \left(\frac{f}{x}\right)_2 \tag{8-11.2}$$

The subscript 2,1 indicates that Henry's constant $H$ is for solute 2 in solvent 1. Superscript $P_{\text{vp}_1}$ indicates that the pressure of the system (as $x_2 \to$

0) is equal to the saturation (vapor) pressure of solvent 1 at temperature $T$. Henry's constant depends on temperature, and often strongly so.

If the gas pressure is large, the effect of pressure on Henry's constant must be taken into consideration. In that event, Eq. (8-11.1) takes the more general form

$$\ln\left(\frac{f}{x}\right)_2 = \ln\left(\frac{\phi y P}{x}\right)_2 = \ln H_{2,1}^{(P_{vp1})} + \frac{\overline{V}_2^{\infty}(P - P_{vp1})}{RT} \tag{8-11.3}$$

where $\phi$ = vapor phase fugacity coefficient

$P$ = system pressure

$\overline{V}_2^{\infty}$ = partial molar volume of solute 2 at infinite dilution in the liquid phase

$R$ = gas constant

[Equation (8-11.3) assumes that $\overline{V}_2^{\infty}$ is independent of pressure in the interval $P - P_{vp1}$.] An illustration of how it can be used to reduce solubility data is given in Fig. 8-14 which shows the solubility of nitrogen in water. In this case, since the vapor phase is predominantly nitrogen, the fugacity coefficient for nitrogen is calculated by using the Lewis fugacity rule.† The important assumption in Eq. (8-11.3) is that $x_2$ is small com-

†The Lewis fugacity rule assumes that $\phi_i^V(T, P, y_i) = \phi_i^V(T, P, y_i = 1)$, where superscript $V$ refers to the vapor phase.

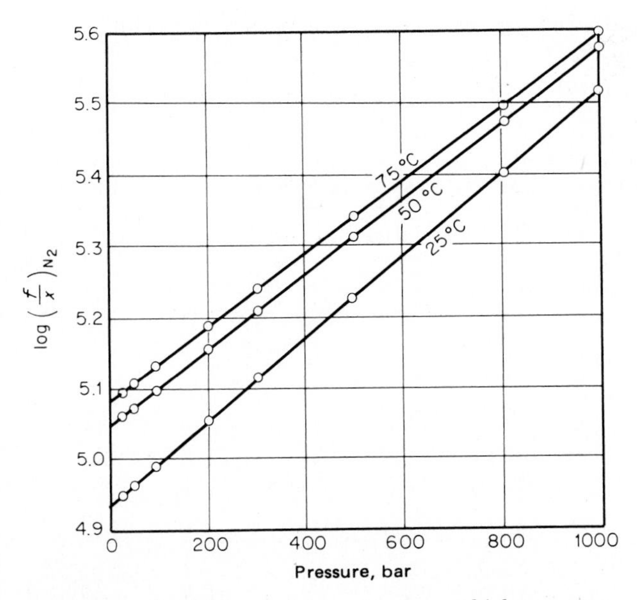


**Figure 8-14** Solubility of nitrogen in water at high pressures. *(From Ref. 97.)*

pared with unity. Just what "small" means depends on the chemical nature of the solute and solvent, as discussed elsewhere [97]. In general, the larger the difference in chemical nature between solute and solvent, the smaller $x_2$ must be for Eq. (8-11.3) to hold.

In Eq. (8-11.3), the first term on the right-hand side is always dominant; the second term is a correction. Correlations for $\overline{V}_2^\infty$ have been presented by Lyckman et al. [71], by Brelvi and O'Connell [20], and by Tiepel and Gubbins [120]. Some typical values for $\overline{V}_2^\infty$ quoted by Hildebrand and Scott are shown in Table 8-24. At temperatures well below the solvent's critical temperature, $\overline{V}^\infty$ is larger than, but in the vicinity of, the solute's molar volume at its normal boiling temperature. However, $\overline{V}^\infty$ may become much larger at temperatures close to the critical temperature of the solvent [97].

Many attempts have been made to correlate gas solubilities, but success has been severely limited because, on the one hand, a satisfactory theory for gas-liquid solutions has not been established and, on the other, reliable experimental data are not plentiful, especially at temperatures remote from 25°C. Among others, Battino and Wilhelm [12] have obtained some success in correlating solubilities in nonpolar systems near 25°C by using concepts from perturbed-hard-sphere theory, but, as yet, these are of limited use for engineering work. A more useful graphical correlation, including polar systems, was prepared by Hayduk et al. [49], and a correlation based on regular solution theory for nonpolar systems was established by Prausnitz and Shair [100] and, in similar form, by Yen and McKetta [136]. The regular solution correlation is limited to nonpolar (or weakly polar) systems, and although its accuracy is not high, it has two advantages: it applies over a wide temperature range, and it requires no mixture data. Correlations for nonpolar systems, near 25°C, are given by Hildebrand and Scott [54].

**TABLE 8-24 Partial Molal Volumes $\overline{V}^\infty$ of Gases in Liquid Solution at 25°C, cm³/mol†**

| | $H_2$ | $N_2$ | CO | $O_2$ | $CH_4$ | $C_2H_2$ | $C_2H_4$ | $C_2H_6$ | $CO_2$ | $SO_2$ |
|---|---|---|---|---|---|---|---|---|---|---|
| Ethyl ether | 50 | 66 | 62 | 56 | 58 | | | | | |
| Acetone | 38 | 55 | 53 | 48 | 55 | 49 | 58 | 64 | ... | 68 |
| Methyl acetate | 38 | 54 | 53 | 48 | 53 | 49 | 62 | 69 | ... | 47 |
| Carbon tetrachloride | 38 | 53 | 53 | 45 | 52 | 54 | 61 | 67 | ... | 54 |
| Benzene | 36 | 53 | 52 | 46 | 52 | 51 | 61 | 67 | ... | 48 |
| Methanol | 35 | 52 | 51 | 45 | 52 | ... | ... | ... | 43 | |
| Chlorobenzene | 34 | 50 | 46 | 43 | 49 | 50 | 58 | 64 | ... | 48 |
| Water | 26 | 40 | 36 | 31 | 37 | ... | ... | ... | 33 | |
| Molar volume of pure solute at its normal boiling point | 28 | 35 | 35 | 28 | 39 | 42 | 50 | 55 | 40 | 45 |

†J. H. Hildebrand and R. L. Scott, "Solubility of Nonelectrolytes," 3d ed., Reinhold, New York, 1950.

A crude estimate of solubility can be obtained rapidly by extrapolating the vapor pressure of the gaseous solute on a linear plot of log $P_{vp}$ vs. $1/T$. The so-called *ideal solubility* is given by

$$x_2 = \frac{y_2 P}{P_{vp2}} \tag{8-11.4}$$

where $P_{vp2}$ is the (extrapolated) vapor pressure of the solute at system temperature $T$. The ideal solubility is a function of temperature, but it is independent of the solvent. Table 8-25 shows that for many typical cases, Eq. (8-11.4) provides an order-of-magnitude estimate.

**TABLE 8-25 Solubilities of Gases in Several Liquid Solvents at 25°C and 1.01 bar Partial Pressure. Mole Fraction × 10⁴**

| | Ideal† | $n$-$C_7F_{16}$ | $n$-$C_7H_{16}$ | $CCl_4$ | $CS_2$ | $(CH_3)_2CO$ |
|---|---|---|---|---|---|---|
| $H_2$ | 8 | 14.01 | 6.88 | 3.19 | 1.49 | 2.31 |
| $N_2$ | 10 | 38.7 | . . . | 6.29 | 2.22 | 5.92 |
| $CH_4$ | 35 | 82.6 | . . . | 28.4 | 13.12 | 22.3 |
| $CO_2$ | 160 | 208.8 | 121 | 107 | 32.8 | |

†See Eq. (8-11.4).

To find the solubility of a gas in a mixed solvent, a first approximation is provided by the expression

$$\ln H_{2,\text{mix}} = \sum_{\substack{i=1,3,\ldots \\ i \neq 2}} x_i \ln H_{2,i} \tag{8-11.5}$$

where $H_{2,\text{mix}}$ is Henry's constant for solute 2 in the solvent mixture and $H_{2,i}$ is Henry's constant for solute 2 in solvent $i$, both at system temperature. The mole fraction $x_i$ is on a solute-free basis.

Equation (8-11.5) is rigorous when the solvent mixture is ideal. However, even for nonideal mixtures, this equation provides a reasonable approximation. For more accurate estimates, it is necessary to add to Eq. (8-11.5) terms which depend on the nonideality of the solvent mixture [86, 120].

## 8-12 Vapor-Liquid Equilibria at High Pressures

Vapor-liquid equilibrium calculations at high pressures are more difficult than those at low or modest pressures for several reasons.

1. The effect of pressure on liquid phase properties is significant only at high pressures. At low or modest pressures this effect can often be

neglected or approximated; the common approximation is to assume in Eq. (8-2.3) that the standard-state fugacity depends on pressure (as given by the Poynting factor) but the activity coefficient is independent of pressure at constant composition and temperature. Since

$$\left(\frac{\partial \ln \gamma_i}{\partial P}\right)_{T,x} = \frac{\overline{V}_i^L - V_{\text{pure } i}^L}{RT} \tag{8-12.1}$$

the assumption that activity coefficient $\gamma_i$ is independent of pressure is equivalent to assuming that in the liquid phase the partial molar volume $\overline{V}_i^L$ is equal to the molar volume of pure liquid $i$. At high pressures, especially in the critical region, this assumption can lead to serious error.

2. The vapor phase fugacity coefficient $\phi_i$ must be found from an equation of state suitable for high pressures, as discussed in Sec. 5-8. Such equations tend to be complex. By contrast, at low pressures we can often set $\phi_i = 1$, and at modest pressures we can often calculate $\phi_i$ with the virial equation truncated after the second term.

3. In high-pressure vapor-liquid equilibria we frequently must deal with supercritical components; we are often concerned with mixtures at a temperature which is larger than the critical temperature of one (or possibly more) of the components. In that event, how do we evaluate the standard-state fugacity of the supercritical component? Normally we use as a standard state the pure liquid at system temperature and pressure. For a supercritical component the pure liquid at system temperature is hypothetical, and therefore there is no unambiguous way to calculate its fugacity. The problem of supercritical hypothetical standard states can be avoided by using the unsymmetric convention for normalizing activity coefficients [97, chap. 6], and some correlations for engineering use have been established on that basis [86, 98]. However, there are computational disadvantages in using unsymmetrically normalized activity coefficients, especially in multicomponent systems, and therefore their use in engineering work is not popular.

4. In high-pressure vapor-liquid equilibria we frequently encounter critical phenomena, including retrograde condensation. Since these phenomena are not well understood, it is difficult to establish simple algebraic equations for representing them.

Vapor-liquid equilibria at high pressures are conveniently calculated by using an equation of state applicable to both phases.

High-pressure vapor-liquid equilibria are often expressed in terms of $K$ factors: For any component $i$.

$$K_i = \frac{y_i}{x_i} \tag{8-12.2}$$

where $y$ is the mole fraction in the vapor phase and $x$ is the mole fraction in the liquid phase. Equation (8-12.2) can be rewritten in terms of fugacity coefficients which can be calculated from an equation of state.

The condition for phase equilibrium is

$$f_i^V = f_i^L \tag{8-12.3}$$

where $f$ is fugacity and superscripts $V$ and $L$ denote vapor phase and liquid phase, respectively. Since, for each phase, fugacity coefficient $\phi_i$ is defined by

$$\phi_i^V = \frac{f_i^V}{y_i P} \tag{8-12.4a}$$

and

$$\phi_i^L = \frac{f_i^L}{x_i P} \tag{8-12.4b}$$

it follows that

$$K_i = \frac{y_i}{x_i} = \frac{\phi_i^L}{\phi_i^V} \tag{8-12.5}$$

Assuming that we have available an equation of state of the form

$$P = F(T, V, z_1, z_2, \ldots) \tag{8-12.6}$$

and assuming further that this equation of state holds for all fluid densities (i.e., gases and liquids) and for all compositions $z_1 z_2, \ldots$, we can then calculate $\phi_i^L$ and $\phi_i^V$ from

$$RT \ln \phi_i^L = \int_{V^L}^{\infty} \left[ \left( \frac{\partial P}{\partial n_i} \right)_{T,V,n_j} - \frac{RT}{V} \right] dV - RT \ln Z^L \tag{8-12.7a}$$

$$RT \ln \phi_i^V = \int_{V^V}^{\infty} \left[ \left( \frac{\partial P}{\partial n_i} \right)_{T,V,n_j} - \frac{RT}{V} \right] dV - RT \ln Z^V \tag{8-12.7b}$$

where compressibility factor $Z$ is given by

$$Z^L = \frac{PV^L}{RT} \tag{8-12.8a}$$

$$Z^V = \frac{PV^V}{RT} \tag{8-12.8b}$$

In the liquid phase, the total volume $V_T^L$ is related to the molar volume $V^L$ by

$$V^L = \frac{V_T^L}{n_T^L} \tag{18-12.9a}$$

where $n_T^L$ is the total number of moles in the liquid phase. Similarly,

$$V^V = \frac{V_T^V}{n_T^V} \tag{8-12.9b}$$

In principle. Eqs. (8-12.5) to (8-12.8) are sufficient to find all $K$ factors in a multicomponent system containing two (or more) fluid phases. However, if a realistic equation of state is used, the required computations are strongly nonlinear and often require extensive iterations.

To fix ideas, consider a two-phase (vapor-liquid) system containing $m$ components at a fixed total pressure $P$. The mole fractions in the liquid phase are $x_1, x_2, \ldots, x_{m-1}$. We want to find the bubble point temperature $T$ and the vapor phase mole fractions $y_1, y_2, \ldots, y_{m-1}$. The total number of unknowns, therefore, is $m$. However, to use Eqs. (8-12.7$a$) and (8-12.7$b$), we also must know the molar volumes $V^L$ and $V^V$. Therefore, the total number of unknowns is $m + 2$.

To find $m + 2$ unknowns, we require $m + 2$ independent equations. These are:

| | |
|---|---|
| Equation (8-12.5) for each component $i$: | $m$ equations |
| Equation (8-12.6), once for the vapor phase and once for the liquid phase: | 2 equations |
| Total number of independent equations: | $m + 2$ |

This case, in which $P$ and $x$ are given and $T$ and $y$ are to be found, is called a bubble point $T$ problem. Other common cases are:

| Given variables | Variables to be found | Name |
|---|---|---|
| $P, y$ | $T, x$ | Dew point $T$ |
| $T, x$ | $P, y$ | Bubble point $P$ |
| $T, y$ | $P, x$ | Dew point $P$ |

However, the most common problem in process design is the flash problem in which we are given $P$, $T$, and the total (feed) composition; we must then find $x$, $y$, and $\alpha$, where $\alpha$ is the fraction of feed which is vaporized in the flash chamber. We cannot go into details here; numerous articles have discussed computational procedures for solving flash problems with an equation of state [50, 78]. A useful discussion is given by Topliss (122).

Knapp et al. [61] have presented a comprehensive (877-page) monograph on calculation of vapor-liquid equilibria by using an equation of state. The monograph contains an exhaustive literature survey (1900–

1980) of experimental data for binary mixtures encountered in natural gas and petroleum technology: hydrocarbons, common gases, freons, and a few oxygenated hydrocarbons. Knapp considered four equations of state:

1. LKP. An equation of state of the Benedict-Webb-Rubin form, proposed by Lee and Kesler [67], based on Pitzer's extended theory of corresponding states where compressibility factor $Z$ of a fluid consists of two parts:

$$Z(P, T) = Z^{(0)}\left(\frac{P}{P_c}, \frac{T}{T_c}\right) + \omega Z^{(1)}\left(\frac{P}{P_c}, \frac{T}{T_c}\right) \tag{8-12.10}$$

Here $P_c$ and $T_c$ are the critical pressure and temperature and $\omega$ is the acentric factor, $Z^{(0)}$ is the generalized dominant contribution corresponding to the properties of a simple fluid ($\omega = 0$), and $Z^{(1)}$ is a generalized correction function. Extension to mixtures follows a procedure described by Plöcker et al. [95] (see also p. 84).

2. BWRS. A modification of the Benedict-Webb-Rubin equation of state proposed by Starling [115] in which all 11 coefficients are generalized as functions of $T_c$, $\rho_c$ (critical density), and an acentric factor. Extension to mixtures follows from mixing rules that give the coefficients as functions of mole fraction.

3. RKS. Soave's [114] modification of the Redlich-Kwong equation of state is

$$Z = \frac{V}{V - b} - \frac{a}{RT}\left(\frac{1}{V + b}\right) \tag{8-12.11}$$

where constant $b$ is found from the critical pressure and temperature and "constant" $a$ is a function of reduced temperature and acentric factor:

$$a = \frac{0.427R^2T_c^2}{P_c}\,[1 + m(1 - \sqrt{T/T_c})]^2 \tag{8-12.12}$$

$$m = 0.48 + 1.574\omega - 0.176\omega^2 \tag{8-12.13}$$

Extension to mixtures is achieved with mixing rules which give $a$ and $b$ as simple functions of mole fraction.

4. PR. Peng and Robinson [90] proposed an equation of state similar to that of Redlich and Kwong:

$$Z = \frac{V}{V - b} - \frac{a}{RT}\left[\frac{1}{(V + b) + (b/V)(V - b)}\right] \tag{8-12.14}$$

Constant $b$ is found from the critical pressure and temperature, and "constant" $a$ depends on reduced temperature and acentric factor:

$$a = \frac{0.457R^2T_c^2}{P_c}[1 + \chi(1 - \sqrt{T/T_c})]^2 \tag{8-12.15}$$

$$\chi = 0.37464 + 1.54266\omega - 0.26992\omega^2 \tag{8-12.16}$$

Extension to mixtures is the same as that in the RKS equation of state.

All four of these equations of state require essentially the same input parameters: for each pure fluid, critical properties and acentric factors, and for each binary mixture, one binary parameter, designated in this chapter by $K_{ij}$ for LKP and by $k_{ij}$ for the others.† For simple mixtures, $K_{ij}$ is close to unity and $k_{ij}$ is close to zero.

To determine binary parameters, Knapp et al. fit calculated vapor-liquid equilibria to experimental ones. The optimum binary parameter is the one which minimizes $DP/P$ defined by

$$\frac{DP}{P} = \frac{100}{N}\sum_{n=1}^{N}\frac{|P_n^e - P_n^c|}{P_n^e} \tag{8-12.17}$$

where $P_n^e$ is the experimental total pressure of point $n$ and $P_n^c$ is the corresponding calculated total pressure, given temperature $T$ and liquid phase mole fraction $x$. The total number of experimental points is $N$.

Similar definitions hold for $Dy_1/y_1$ and for $DK_1/K_1$. Here $y_1$ is the vapor phase mole fraction and $K_1$ is the $K$ factor ($K_1 = y_1/x_1$) for the more volatile component. In addition, Knapp et al. calculated $Df/f$ by

$$\frac{Df}{f} = \frac{100}{N}\sum_{n=1}^{N}\frac{|f_{1n}^V - f_{1n}^L|}{f_{1n}^V} \tag{8-12.18}$$

where $f_1^V$ is the calculated fugacity of the more volatile component in the vapor phase and $f_1^L$ is that in the liquid phase.

When the binary parameter is obtained by minimizing $DP/P$, the other deviation functions are usually close to their minima. However, for a given set of data, it is unavoidable that the optimum binary parameter depends somewhat on the choice of objective function for minimization. Minimizing $DP/P$ is preferred because that objective function gives the sharpest minimum and pressures are usually measured with better accuracy than compositions.

---

†In the original publication by Plöcker, the binary parameter is designated $K_{ij}$. In Knapp's monograph, it is designated by $k_{ij}^*$. In Chap. 4, it is $k'_{ij}$.

Tables 8-26 and 8-27 show some results reported by Knapp et al. Table 8-26, for propylene-propane, concerns a simple system in which the components are similar; in that case, excellent results are obtained by all four equations of state with only very small corrections to the geometric mean assumption for the characteristic potential energy of a dissimilar pair.

However, calculated results are not nearly as good for the system nitrogen-isopentane. Corrrections to the geometric mean assumption are now appreciable, but, even with such corrections, calculated and observed $K$ factors for nitrogen disagree by about 6 percent for LKP, RKS, and PR and by nearly 12 percent for BWRS.

These two examples illustrate the range of results obtained by Knapp et al. for binary mixtures containing nonpolar components. (Disagreement between calculated and observed vapor-liquid equilibria is often larger when polar components are present.) For most nonpolar binary mixtures,

**TABLE 8-26 Comparison of Calculated and Observed Vapor-Liquid Equilibria for the System Propylene (1)–Propane (2)**

**(From Monograph by Knapp et al.†)**

Temperature range: 310 to 344 K
Pressure range: 13 to 31 bar

| Equation of state | Binary constant‡ | Percent | | | |
|---|---|---|---|---|---|
| | | $DP/P$ | $Dy_1/y_1$ | $DK_1/K_1$ | $Df_1/f_1$ |
| LKP | $K_{12} = 0.9919$ | 0.31 | 0.10 | 0.38 | 0.46 |
| BWRS | $k_{12} = 0.0025$ | 0.55 | 0.06 | 0.27 | 0.46 |
| RKS | $k_{12} = 0$ | 0.56 | 0.23 | 0.61 | 0.90 |
| PR | $k_{12} = 0.0063$ | 0.31 | 0.08 | 0.40 | 0.29 |

†Experimental data (77 points) from Laurence and Swift [66].
‡Binary constants obtained by minimizing $DP/P$.

**TABLE 8-27 Comparison of Calculated and Observed Vapor-Liquid Equilibria for the System Nitrogen (1)–Isopentane (2)**

**(From Monograph by Knapp et al.†)**

Temperature range: 277 to 377 K
Pressure range: 1.8 to 207 bar

| Equation of state | Binary constant‡ | Percent | | | |
|---|---|---|---|---|---|
| | | $DP/P$ | $Dy_1/y_1$ | $DK_1/K_1$ | $Df_1/f_1$ |
| LKP | $K_{12} = 1.347$ | 5.14 | 0.87 | 6.12 | 4.99 |
| BWRS | $k_{12} = 0.1367$ | 12.27 | 3.73 | 11.62 | 10.70 |
| RKS | $k_{12} = 0.0867$ | 4.29 | 1.58 | 6.66 | 5.78 |
| PR | $k_{12} = 0.0922$ | 3.93 | 1.61 | 5.98 | 5.26 |

†Experimental data (47 points) from Krishnan et al. [64].
‡Binary constants obtained by minimizing $DP/P$.

the accuracy of calculated results falls between the limits indicated by Tables 8-26 and 8-27.

While Knapp et al. found overall that the BWRS equation did not perform as well as the others, it is not possible to conclude that, of the four equations used, one particular equation is distinctly superior to the others. Further, it is necessary to keep in mind that the quality of experimental data varies appreciably from one set of data to another. Therefore, if calculated results disagree significantly with experimental ones, one must not immediately conclude that the disagreement is due to a poor equation of state.

Knapp's monograph is limited to binary mixtures. If pure component equation-of-state constants are known and if the mixing rules for these constants are simple, requiring only characteristic binary parameters, then it is possible to calculate vapor-liquid equilibria for ternary (and higher) mixtures by using only pure component and binary data. Although few systematic studies have been made, it appears that this "scale-up" procedure usually provides good results for vapor-liquid equilibria, especially in nonpolar systems. (However, this scale-up procedure is usually not successful for ternary liquid-liquid equilibria, unless special precautions are observed.)

Regardless of what equation of state is used, it is usually worthwhile to make an effort to obtain the best possible equation-of-state constants for the fluids that comprise the mixture. Such constants can be estimated from critical data, but it is usually better to obtain them from vapor pressure and density data as discussed, for example, by Panagiotopoulos and Kumar (88*a*).

Although Knapp's monograph is concerned with several popular equations of state, it is sometimes useful for special applications to modify one of the standard equations of state. A particularly useful example is provided by Turek et al. [125], who used a modified Redlich-Kwong equation to correlate phase behavior in carbon dioxide–hydrocarbon systems for miscible enhanced oil recovery.

For a pure component, the equation of state is

$$P = \frac{RT}{V - b} - \frac{a}{T^{1/2}V(V + b)} \tag{8-12.19}$$

where $V$ is the molar volume and constants $a$ and $b$ are found from critical pressure $P_c$ and critical temperature $T_c$:

$$a = \frac{\Omega_a R^2 T_c^{2.5}}{P_c} \tag{8-12.20}$$

$$b = \frac{\Omega_b R T_c}{P_c} \tag{8-12.21}$$

Coefficients $\Omega_a$ and $\Omega_b$ can be found from the critical condition [$\partial P/\partial V = \partial^2 P/\partial V^2 = 0$ at the critical point] yielding $\Omega_a = 0.4278$ and $\Omega_b = 0.0867$. To improve agreement with experimental vapor pressure, liquid density and supercritical density data, Turek uses slightly temperature-dependent values for $\Omega_a$ and $\Omega_b$ for carbon dioxide. For hydrocarbons, he establishes generalized relations for $\Omega_a$ and $\Omega_b$ as functions of reduced temperature and acentric factor.

For a multicomponent mixture M, the composition dependences of constants $a$ and $b$ are given by customary quadratic mixing rules with two adjustable binary parameters, $C_{ij}$ and $D_{ij}$:

$$a_M = \sum_{i=1} \sum_{j=1} z_i z_j (1 - C_{ij})(a_i a_j)^{1/2} \tag{8-12.22}$$

$$b_M = \tfrac{1}{2} \sum_{i=1} \sum_{j=1} z_i z_j \, (1 + D_{ij})(b_i + b_j) \tag{8-12.23}$$

where $z$ is the mole fraction ($z_i = y_i$ for the vapor phase and $z_i = x_i$ for the liquid phase) and parameters $C_{ij}$ and $D_{ij} = 0$ whenever $i = j$.

Phase equilibria are related to the equation of state as discussed previously.

Turek et al. reduced extensive binary VLE data to obtain parameters $C_{ij}$ and $D_{ij}$ ($i \neq j$). For hydrocarbon-hydrocarbon pairs, these parameters are small compared to unity; they depend primarily on the molecular size ratio of components $i$ and $j$. For carbon dioxide–hydrocarbon pairs, parameters $C_{ij}$ and $D_{ij}$ depend on temperature and hydrocarbon acentric factor as shown in Fig. 8-15.

Experimental studies were made for mixtures of carbon dioxide and a synthetic oil whose properties are given in Table 8-28. Experimental $K$

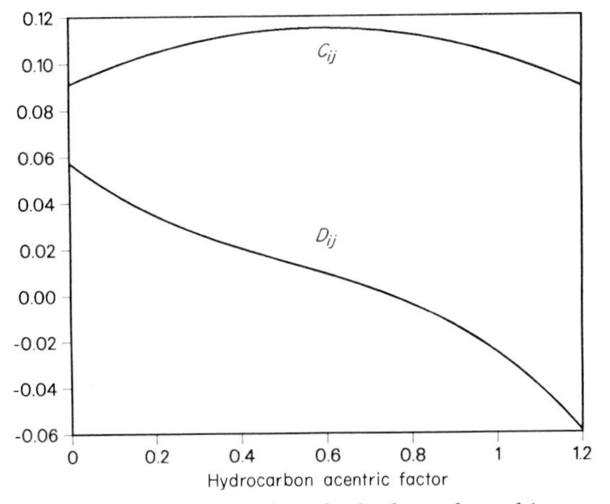


**Figure 8-15** Carbon dioxide–hydrocarbon binary interaction parameters.

**TABLE 8-28 Composition of Synthetic Oil Used by Turek et al. for Experimental Studies of Vapor-Liquid Equilibria with Carbon Dioxide**

| Component | Mole percent | Component | Mole percent |
|---|---|---|---|
| Methane | 34.67 | *n*-Hexane | 3.06 |
| Ethane | 3.13 | *n*-Heptane | 4.95 |
| Propane | 3.96 | *n*-Octane | 4.97 |
| *n*-Butane | 5.95 | *n*-Decane | 30.21 |
| *n*-Pentane | 4.06 | *n*-Tetradecane | 5.04 |

Density at 322.0 K and 15.48 MPa is 637.0 kg/m$^3$.
Density at 338.7 K and 14.13 MPa is 613.5 kg/m$^3$.

factors ($K_i = y_i/x_i$) at 322 K are compared with calculated results in Fig. 8-16. For this case, the modified Redlich-Kwong equation (Amoco Redlich-Kwong), using at most two binary parameters for each binary pair, gives very good agreement with experimental data. (For many of the binary hydrocarbon-hydrocarbon pairs, one or both of the binary parameters can be set equal to zero.) This application of a simple equation of state indicates that, when care is taken to represent thermodynamic properties of pure and binary fluids, it is often possible to scale up to multicomponent vapor-liquid equilibria with good accuracy.

A similar successful scale-up has been established by Mollerup [80], who uses the known thermodynamic properties of a reference fluid (meth-

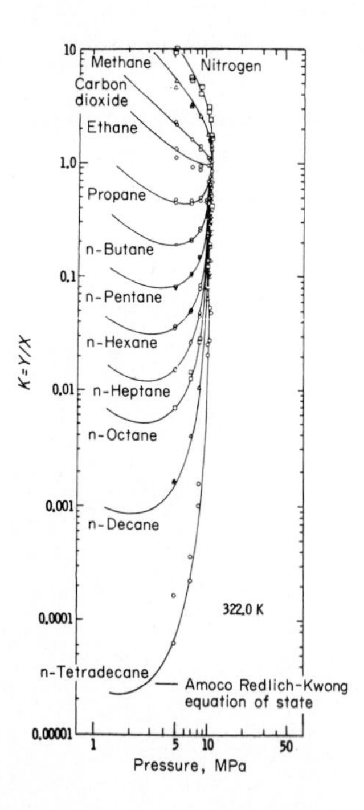


**Figure 8-16** $K$ factors for carbon dioxide–synthetic oil.

ane), coupled with pure component and binary parameters, to calculate vapor-liquid equilibria, densities, and enthalpies for natural gas mixtures at low temperatures.

In a typical equation of state, the pressure is given as a function of temperature, volume, and composition. When pressure, temperature, and composition are specified, it is necessary to find the volume. Finding the correct volume is often not simple.

When a single equation of state is used for both liquid and vapor phases, a common pitfall occurs when one is trying to calculate properties for one phase but generates properties for the other. Figure 8-17 shows the phase envelope for an ethane-heptane mixture calculated by the Soave equation (3-6.1). The largest $Z$ value represents a vapor phase: the smallest, a liquid phase. Note that, at low pressures, there is a broad temperature range (and a correspondingly broad range of composition) for which three roots are generated. At higher pressures, this three-root region becomes smaller, and it disappears at the pseudocritical point. Although the behavior in Fig. 8-17 is for the Soave equation, it would be similar for any equation of state. A number of articles have referred to the trivial root problem [19, 28, 45, 96], but the best way to avoid the problem is to make sufficiently accurate initial guesses in iterative calculations. This is illustrated by the following example.

**Example 8-12** Estimate the $K$ values for a 26.54% ethane (1)–73.46% heptane (2) mixture at 400 K and 15 bar. This might be the first step in a bubble point pressure calculation with an initial guess of 15 bar.

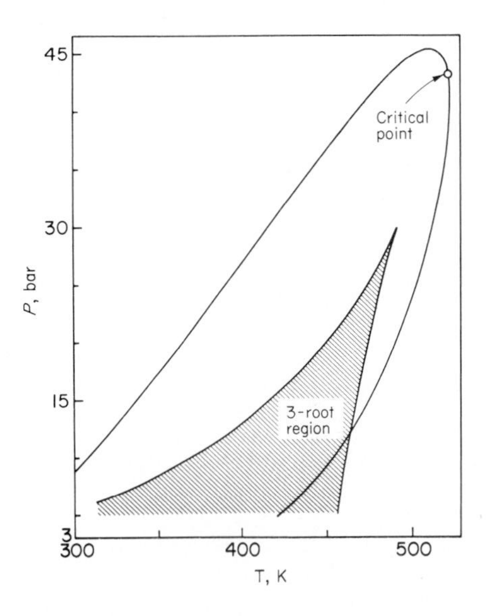


**Figure 8-17** Phase envelope calculated by the Soave equation of state for a 26.54% ethane–73.46% $n$-heptane mixture. In the three-root region, both vapor and liquid properties are calculated.

**solution** We must make an initial guess for the vapor composition. If we guess $y_1 = 0.2654$ and $y_2 = 0.7346$ and use the following properties, Eq. (3-6.2) has only one root, $Z = 0.07952$.

| Compound | $T_c$, K | $P_c$, bar | $\omega$ |
|---|---|---|---|
| Ethane | 305.4 | 48.8 | 0.099 |
| Heptane | 540.3 | 27.4 | 0.349 |

As expected from Fig. 8-17, this corresponds to a liquid value; $K$ values cannot be calculated because no vapor properties are generated. We need to make an initial guess more likely to generate a vapor phase, for example, $y_1 = 0.9$ and $y_2 = 0.1$. Now Eq. (3-6.2) must be solved twice, once with constants from the liquid mole fractions and once with vapor phase constants. Even though both solutions fall in the one root region, the following liquid and vapor properties are generated. Values of $\phi_i$ are from Eq. (5-8.12).

| | Mole fraction of 1 | $Z$ | $\phi_1$ | $\phi_2$ |
|---|---|---|---|---|
| Liquid | 0.2654 | 0.07952 | 5.378 | 0.1484 |
| Vapor | 0.9 | 0.9327 | 0.9607 | 0.7377 |

$K_1 = 5.378/0.9607 = 5.598$
$K_2 = 0.1484/0.7377 = 0.2012$

### Phase envelope construction—dew and bubble point calculations

Calculations at low pressures present little difficulty, but high-pressure calculations can be complicated by both trivial root and convergence difficulties. Trivial root problems can be avoided by starting computations at a low pressure and marching toward the critical point in small increments of temperature or pressure. When the initial guess of each calculation is the result of a previous calculation, trivial roots are avoided. Convergence difficulties are avoided if one does dew or bubble point *pressure* calculations when the phase envelope is flat and dew or bubble point *temperature* calculations when the phase envelope is steep. These conditions have been stated [139] as follows

$$\text{If } \left|\frac{d \ln P}{d \ln T}\right| < 2, \text{ calculate dew or bubble point pressures}$$

$$\text{If } \left|\frac{d \ln P}{d \ln T}\right| > 20, \text{ calculate dew or bubble point temperatures.}$$

This technique allows convergence on a single variable. Several multivariable Newton-Raphson techniques have been presented in the literature

For the above procedure, 10 iterations were required to obtain the condition that

$$\left|\sum_i (y_i - x_i)\right| < 10^{-8}$$

Although well-known equations of state (e.g., Redlich-Kwong-Soave and Peng-Robinson) are suitable for calculating vapor-liquid equilibria for nonpolar mixtures, these equations of state, using conventional mixing rules, are not satisfactory for mixtures containing strongly polar and hydrogen-bonded fluids in addition to the common gases and hydrocarbons. For those mixtures, the assumption of simple (random) mixing is poor because strong polarity and hydrogen bonding can produce significant segregation or ordering of molecules in mixtures. For example, at ordinary temperatures, water and benzene form a strongly nonrandom mixture; the mixture is so far from random that water and benzene are only partially miscible at ordinary temperatures because preferential forces of attraction between water molecules tend to keep these molecules together and prevent their random mixing with benzene molecules.

It is possible to describe deviations from simple mixing by using complex (essentially empirical) mixing rules, as shown, for example, by Vidal [127]. For thermodynamic consistency, however, these mixing rules must be density-dependent because at low densities, the equation of state must give a second virial coefficient which is quadratic in mole fraction. The common mixing rules satisfy that boundary condition, but the mixing rules of Vidal (which are independent of density) do not. Some early work on density-dependent mixing rules has been reported by Whiting and Prausnitz [129], Mollerup [81], Mathias and Copeman [74], and by Won [134], and some promising results have been achieved for binary mixtures. Panagiotopoulos and Reid [88*b*] have reported practical application to ternary systems containing two liquid phases.

A useful technique for describing nonrandom fluid mixtures is provided by the chemical hypothesis which postulates the existence of various chemical species formed by the nominal components. For example, a mixture of fluids A and B is assumed to contain not only monomers A and B but, in addition, dimers, trimers, etc., of A and of B and, further, complexes of A and B with the general formula $A_nB_m$, where $n$ and $m$ are positive integers. Concentrations of the various chemical species are found from chemical equilibrium constants coupled with material balances.

The chemical hypothesis was used many years ago to calculate activity coefficients in liquid mixtures and also to calculate second virial coefficients of pure and mixed gases. However, the early work was restricted to liquids or to gases at moderate densities, and most of that early work assumed that the "true" chemical species form ideal mixtures. It was not until 1976 that Heidemann [51] combined the chemical hypothesis with an equation of state valid for all fluid densities. Unfortunately, Heidemann's work is limited to pure fluids; for extension to mixtures additional assumptions are required as discussed by Hu et al. [58]. However, the chemical hypothesis, coupled with an equation of state, becomes tractable for mixtures provided that association is limited to dimers as shown in 1979 by Gmehling et al. Since then, several other authors have presented similar ideas.

Gmehling et al. [42] used an equation of state of the van der Waals form (in particular, the perturbed-hard-chain equation of state) coupled with a dimerization hypothesis. A binary mixture of nominal components A and B is considered to be a five-species mixture containing two types of monomer ($A_1$ and $B_1$) and three types of dimer ($A_2$, $B_2$, AB).

There are three chemical equilibrium constants:

$$K_{A_2} = \frac{z_{A2}}{z_{A_1}^2}\frac{\phi_{A2}}{\phi_{A_1}^2}\frac{1}{P} \tag{8-12.31a}$$

$$K_{B_2} = \frac{z_{B2}}{z_{B_1}^2}\frac{\phi_{B2}}{\phi_{B_1}^2}\frac{1}{P} \tag{8-12.31b}$$

$$K_{AB} = \frac{z_{AB}}{z_{A_1}z_{B_1}}\frac{\phi_{AB}}{\phi_{A_1}\phi_{B_1}}\frac{1}{P} \tag{8-12.31c}$$

where $z_{A_1}$ is the mole fraction of $A_1$ (etc.) and $\phi_{A_1}$ is the fugacity coefficient of $A_1$ (etc.). The fugacity coefficient is found from the equation of state by using physical interaction parameters to characterize monomer-monomer, monomer-dimer, and dimer-dimer interactions.

Mole fractions $z$ are related to nominal mole fractions $x_A$ and $x_B$ through chemical equilibrium constants and material balances.

To reduce the number of adjustable parameters, Gmehling established physically reasonable relations between parameters for monomers and those for dimers.

The temperature dependence of equilibrium constant $K_{A_2}$ is given by

$$\ln K_{A_2} = \frac{\Delta H^\circ_{A2}}{RT} + \frac{\Delta S^\circ_{A2}}{R} \tag{8-12.32}$$

where $\Delta H^\circ_{A_2}$ is the enthalpy and $\Delta S^\circ_{A_2}$ is the entropy of formation of dimer $A_2$ in the standard state. Similar equations hold for $K_{B_2}$ and $K_{AB}$.

All pure component parameters (including $K_{A_2}$ and $K_{B_2}$) are obtained from experimental density and vapor pressure data.

A reasonable estimate for $\Delta H^\circ_{AB}$ is provided by

$$\Delta H^\circ_{AB} = \tfrac{1}{2}(\Delta H^\circ_{A_2} + \Delta H^\circ_{B_2}) \tag{8-12.33}$$

but a similar relation of $\Delta S^\circ_{AB}$ does not hold. For a binary mixture of A and B, $\Delta S^\circ_{AB}$ must be found from binary data.

The equations for vapor-liquid equilibrium are

$$f_A^V = f_A^L \quad \text{and} \quad f_B^L = f_B^V \tag{8-12.34}$$

where $f$ stands for fugacity and superscripts $V$ and $L$ stand for vapor and liquid, respectively. As shown in Ref. 103, chap. 26, Eq. (8-12.34) can be replaced without loss of generality by

$$f_{A_1}^V = f_{A_1}^L \quad \text{and} \quad f_{B_1}^V = f_{B_1}^L \tag{8-12.35}$$

Figure 8-18 shows calculated and observed vapor-liquid equilibria for methanol-water at modest and advanced pressures. Calculations are based on Gmehling's equation as outlined above. For this mixture, the calculations require only two adjustable binary parameters which are independent of temperature over the indicated temperature range. One of these is $\Delta S^\circ_{AB}$, and the other is $k_{A_1-B_1}$, a physical parameter to characterize $A_1 - B_1$ interactions.

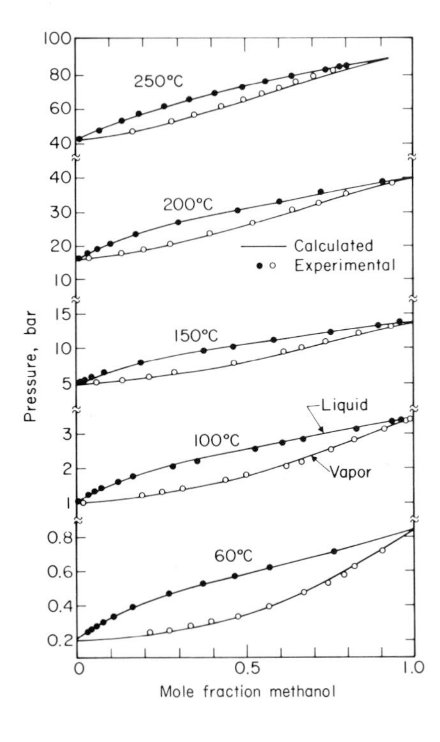


**Figure 8-18** Vapor-liquid equilibria for methanol-water.

Gmehling's equation of state, coupled with a (chemical) dimerization hypothesis, is particularly useful for calculating vapor-liquid equilibria at high pressures for fluid mixtures containing polar and nonpolar components, some subcritical and some supercritical. By using an equation of state valid for both phases, the equations of phase equilibrium avoid the awkward problem of defining a liquid phase standard state for a supercritical component. By superimposing dimerization equilibria onto a "normal" equation of state, Gmehling achieves good representation of thermodynamic properties for both gaseous and liquid mixtures containing polar or hydrogen-bonded fluids in addition to "normal" fluids (such as common gases and hydrocarbons) by using the same characteristic parameters for both phases.

Buck [24] tested Gmehling's method by comparing calculated and observed vapor-liquid equilibria for several ternary systems containing polar and hydrogen-bonded fluids. Encouraged by favorable comparisons, Buck then describes an application of Gmehling's method to an isothermal flash calculation at 200°C and 100 bar. Table 8-29 shows specified feed compositions and calculated compositions for the vapor and for the liquid at equilibrium. All required parameters were obtained from pure component and binary experimental data.

To implement Gmehling's method for multicomponent fluid mixtures, it is necessary to construct a far-from-trivial computer program requiring a variety of iterations. The calculations summarized in Table 8-29 are for seven components, but the number of (assumed) chemical species is much larger. For $H_2$, CO, and $CH_4$ it is reasonable to assume that no dimers are formed; further, it is reasonable to assume that these components do not form cross-dimers with each other or with the other components in the mixture. However, the four polar components form dimers with them-

**TABLE 8-29 Isothermal Flash Calculation Using Gmehling's Equation of State at 200°C and 100 bar [24]**

| Component | Mole percent of | | |
|---|---|---|---|
| | Feed | Vapor | Liquid |
| Hydrogen | 6.0 | 33.86 | 2.03 |
| Carbon monoxide | 5.5 | 24.63 | 2.77 |
| Methane | 0.3 | 1.08 | 0.19 |
| Methyl acetate | 27.2 | 13.34 | 29.18 |
| Ethanol | 39.9 | 19.35 | 42.83 |
| Water | 3.3 | 1.81 | 3.51 |
| 1,4-Dioxane | 17.8 | 5.93 | 19.49 |
| Total moles | 100.00 | 12.49 | 87.51 |

selves and with each other. In Gmehling's method, therefore, this 7-component mixture is considered to be a mixture of 17 chemical species.

**Example 8-15** Use Gmehling's method to calculate the bubble point pressure and vapor phase composition for a mixture of 4.46 mole percent methanol (1) and 95.54 mole percent water (2) at 60°C.

**solution** In the Gmehling model, a water-methanol mixture contains five species: methanol and water monomers, methanol dimers, water dimers, and a methanol-water cross-dimer. The mole fractions of these are, respectively, $z_{M1}$, $z_{M2}$, $z_{D1}$, $z_{D2}$, and $z_{D12}$. There are 13 unknowns and 13 equations. The unknowns include five liquid $z$ values, five vapor $z$ values, and the pressure and molar volumes, of the liquid and vapor phases. The 13 equations are the equation of state for both the liquid and vapor phases, the three reaction equilibrium equations (8-12.31$a$) through (8-12.31$c$), five fugacity equalities (that is, $f_i^L = f_i^V$), $\Sigma z_i = 1$ in both the liquid and vapor, and a material balance accounting for the mixture composition. Pure component parameters from [42] are as follows:

| Component | $T^*$, K | $V^*$, cm$^3$/mol | $\Delta S°/R$ | $\Delta H° R$, K |
|---|---|---|---|---|
| Methanol (1) | 348.09 | 26.224 | −16.47 | −5272 |
| Water (2) | 466.73 | 12.227 | −14.505 | −4313 |

From Eq. (8-12.32),

$$K_1 = \exp\left(-16.47 + \frac{5272}{333.15}\right) = 0.525$$

Similarly, $K_2 = 0.210$

From [42], $\Delta S°_{12}/R = -15.228$ and $k_{12} = 0.0371$ and $K_{12} = \exp\left(-15.228 + \frac{(5272 + 4313)(0.5)}{333.15}\right) = 0.431$

The problem may be solved by the following procedure:

1. Guess $P$.
2. Guess all $\phi_i = 1$.
3. Guess $y_i = x_i$.
4. Solve the reaction equilibria problem for values of $z_i$ in each phase [Eqs. (8-12.31$a$) to (8-12.31$c$)].
5. Calculate mixture parameters with mixing rules from [42].
6. Solve the equation of $V^L$ and $V^V$.
7. Calculate $\phi_i$ for each of the five species in both phases.
8. Go back to step 4 and recalculate $z_i$ values. When $z_i$ values no longer change, reaction equilibria are satisfied, but phase equilibria are not.
9. Calculate $K_i$ by $K_i = \phi_i^L/\phi_i^V$, where $K_i \equiv z_i^V/z_i^L$.
10. See if $\sum_i K_i z_i^L = 1$; if not, adjust P according to $P_{\text{new}} = P_{\text{old}}\left(\sum_i K_i z_i^L\right)$ and go back to step 4.

This procedure converges to the following values:

| | $z_{M1}$ | $z_{M2}$ | $z_{D1}$ | $z_{D2}$ | $z_{D12}$ | $V$ |
|---|---|---|---|---|---|---|
| Liquid | 0.001680 | 0.05252 | 0.003510 | 0.8642 | 0.07808 | 0.03706 |
| Vapor | 0.2641 | 0.6794 | 0.00980 | 0.02603 | 0.0207 | 103.3 |

| | $\phi_{M1}$ | $\phi_{M2}$ | $\phi_{D1}$ | $\phi_{D2}$ | $\phi_{D12}$ |
|---|---|---|---|---|---|
| Liquid | 156.6 | 12.91 | 2.775 | 0.02996 | 0.2636 |
| Vapor | 0.9976 | 0.9983 | 0.9937 | 0.9947 | 0.9942 |

$P = 0.2675$ bar

The above numbers satisfy the material balance and reaction equilibria equations:

$$x_1 = \frac{z^L_{M1} + 2z^L_{D1} + z^L_{D12}}{z^L_{M1} + z^L_{M2} + 2(z^L_{D1} + z^L_{D12} + z^L_{D2})}$$

$$= \frac{0.00168 + (2)(0.00351) + 0.0781}{0.00168 + 0.0525 + (2)(0.00351 + 0.864 + 0.0781)} = 0.0446$$

$$K_1 = \frac{z_{D1}\phi_{D1}}{z^2_{M1}\phi^2_{M1}}\frac{1}{P} = \frac{0.00351}{(0.00168)^2}\frac{2.775}{(156.6)^2}\frac{1}{0.2675} = 0.526$$

$$K_{12} = \frac{z_{D12}}{z_{M1}z_{M2}}\frac{\phi_{D12}}{\phi_{M1}\phi_{M2}}\frac{1}{P} = \frac{0.07808}{(0.00168)(0.05252)}\frac{(0.2336)}{(156.6)(12.91)}\frac{1}{0.2675} = 0.431$$

$$K_2 = \frac{z_{D2}}{z^2_{M2}}\frac{\phi_{D2}}{\phi^2_{M2}}\frac{1}{P} = \frac{0.8642}{(0.05252)^2}\frac{0.02996}{(12.91)^2}\frac{1}{0.2675} = 0.210$$

The reaction expressions are verified above for liquid phase values. They are also satisfied for vapor phase values, since $f_i^L = f_i^V$ is satisfied for each of the five components (as can easily be verified). Mixture parameters to be used can be obtained from mixing rules in [42]. For example, for the liquid phase,

$$\begin{aligned}
<cT^*V^*> &= z_{M1}c_{M1}T^*_{M1}\,[z_{M1}V^*_{M1} + z_{M2}V^*_{M2}(1 - k_{12}) + z_{D1}V^*_{D1} + z_{D2}V^*_{D2}(1 - k_{12}) \\
&\quad + z_{D12}V^*_{D12}(1 - k_{12})^{1/2} + z_{M2}c_{M2}T^*_{M2}(z_{M1}V^*_{M1}(1 - k_{12}) + z_{M2}V^*_{M2} \\
&\quad + z_{D1}V^*_{D1}(1 - k_{12}) \\
&\quad + z_{D2}V^*_{D2} + z_{D12}V^*_{D12}(1 - k_{12})^{1/2}] + z_{D1}c_{D1}T^*_{D1}[z_{M1}V^*_{M1} \\
&\quad + z_{M2}V^*_{M2}(1 - k_{12}) + z_{D1}V^*_{D1} + z_{D2}V^*_{D2}(1 - \mathrm{k}_{12}) \\
&\quad + z_{D12}V^*_{D12}(1 - k_{12})^{1/2}] + z_{D2}c_{D2}T^*_{D2}[z_{M1}V^*_{M1}(1 - k_{12}) \\
&\quad + z_{M2}V^*_{M2} + z_{D1}V^*_{D1}(1 - k_{12}) + z_{D2}V^*_{D2} \\
&\quad + z_{D12}V^*_{D12}(1 - k_{12})^{1/2}] + z_{D12}c_{D12}T^*_{D12}[z_{M1}V^*_{M1}(1 - k_{12})^{1/2} \\
&\quad + z_{M2}V^*_{M2}(1 - k_{12})^{1/2} + z_{D1}V^*_{D1}(1 - k_{12})^{1/2} \\
&\quad + z_{D2}V^*_{D2}(1 - k_{12})^{1/2} + z_{D12}V^*{}_{D12}(1 - k_{12})^{1/2}] \\
&= 12 + 537 + 45 + 15465 + 1200 \\
<cT^*V^*> &= 17{,}259\ (\mathrm{K\cdot cm^3})/\mathrm{mol}
\end{aligned}$$

Values of all parameters as as follows:

| | $\langle c \rangle$ | $\langle V^* \rangle$ | $\langle cT^* V^* \rangle$ | $\langle T^* \rangle^{(2)}$ |
|---|---|---|---|---|
| Vapor | 1.017 | 16.93 | 7503 | 444.92 |
| Liquid | 1.284 | 21.97 | 17259 | 614.72 |

As the last step, values of $y_i$ and $V$ in moles per liter of original monomer may be calculated:

$$y_1 = \frac{z_{M1}^V + 2z_{D1}^V + z_{D12}^V}{z_{M1}^V + z_{M2}^V + 2(z_{D1}^V + z_{D2}^V + z_{D12}^V)} = 0.2881$$

$$y_2 = 0.7119$$

$$V^V = \frac{103.3}{z_{M1}^V + z_{M2}^V + 2(z_{D1}^V + z_{D2}^V + z_{D12}^V)} = 97.8 \text{ mol/L}$$

$$V^L = \frac{0.03706}{z_{M1}^L + z_{M2}^L + 2(z_{D1}^L + z_{D2}^L + z_{D12}^L)} = 0.01905 \text{ mol/L}$$

Experimental values are $P = 0.2625$ bar and $y_1 = 0.2699$.

## 8-13 Liquid-Liquid Equilibria

Many liquids are only partially miscible, and in some cases, e.g., mercury and hexane at normal temperatures, the mutual solubilities are so small that, for practical purposes, the liquids may be considered immiscible. Partial miscibility is observed not only in binary mixtures but also in ternary (and higher) systems, thereby making extraction a possible separation operation. This section introduces some useful thermodynamic relations which, in conjunction with limited experimental data, can be used to obtain quantitative estimates of phase compositions in liquid-liquid systems.

At ordinary temperatures and pressures, it is (relatively) simple to obtain experimentally the compositions of two coexisting liquid phases, and, as a result, the technical literature is rich in experimental results for a variety of binary and ternary systems near 25°C and near atmospheric pressure. However, as temperature and pressure deviate appreciably from those corresponding to normal conditions, the availability of experimental data falls rapidly.

Partial miscibility in liquids is often called *phase splitting*. The thermodynamic criteria which indicate phase splitting are well understood regardless of the number of components [79], but most thermodynamic texts confine discussion to binary systems. Stability analysis shows that, for a binary system, phase splitting occurs when

$$\left(\frac{\partial^2 g^E}{\partial x_1^2}\right)_{T,P} + RT\left(\frac{1}{x_1} + \frac{1}{x_2}\right) < 0 \tag{8-13.1}$$

where $g^E$ is the molar excess Gibbs energy of the binary mixture (see Sec. 8-5). To illustrate Eq. (8-13.1), consider the simplest nontrivial case. Let

$$g^E = Ax_1x_2 \tag{8-13.2}$$

where $A$ is an empirical coefficient characteristic of the binary mixture. Substituting into Eq. (8-13.1), we find that phase splitting occurs if

$$A > 2RT \tag{8-13.3}$$

In other words, if $A < 2RT$, the two components 1 and 2 are completely miscible; there is only one liquid phase. However, if $A > 2RT$, two liquid phases form because components 1 and 2 are only partially miscible.

The condition in which $A = 2RT$ is called *incipient instability,* and the temperature corresponding to that condition is called the *consolute temperature,* designated by $T^c$. Since Eq. (8-13.2) is symmetric in mole fractions $x_1$ and $x_2$, the composition at the consolute point is $x_1^c = x_2^c = 0.5$. In a typical binary mixture, the coefficient $A$ is a function of temperature, and therefore it is possible to have either an upper consolute temperature or a lower consolute temperature, or both, as indicated in Figs. 8-19 and 8-20. Upper consolute temperatures are more common than lower consolute temperatures. Systems with both upper and lower consolute temperatures are rare.†

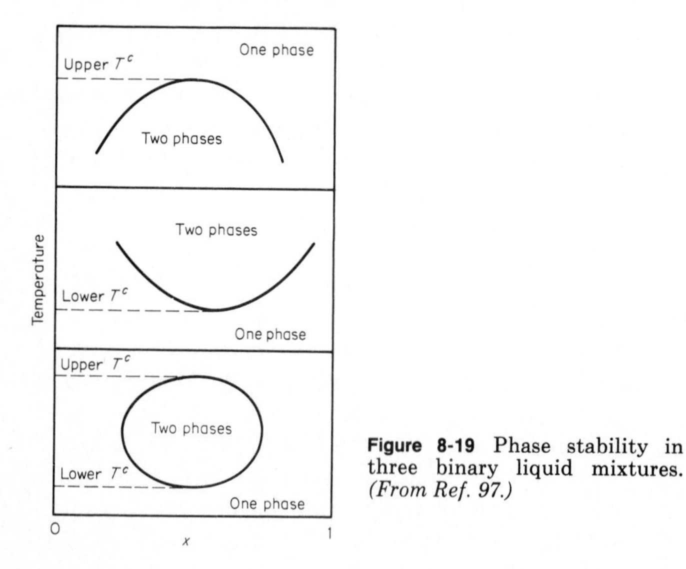


**Figure 8-19** Phase stability in three binary liquid mixtures. *(From Ref. 97.)*

†Although Eq. (8-13.3) is based on the simple two-suffix (one-parameter) Margules equation, similar calculations can be made using other expressions for $g^E$. See, for example, Ref. 111.

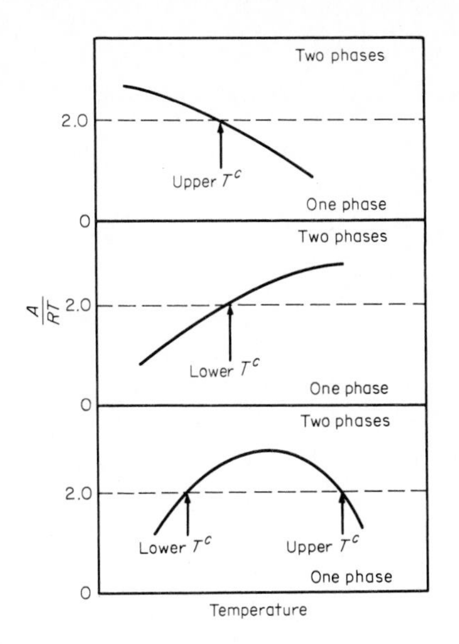


**Figure 8-20** Phase stability in three binary liquid mixtures whose excess Gibbs energy is given by a two-suffix Margules equation. *(From Ref. 97.)*

Stability analysis for ternary (and higher) systems is, in principle, similar to that for binary systems, although the mathematical complexity rises with the number of components. (See, for example, Refs. 14 and 103.) However, it is important to recognize that stability analysis can tell us only whether a system can or cannot *somewhere* exhibit phase splitting at a given temperature. That is, if we have an expression for $g^E$ at a particular temperature, stability analysis can determine whether or not there is *some* range of composition where two liquids exist. It does *not* tell us what that compositon range is. To find the range of compositions in which two liquid phases exist at equilibrium requires a more elaborate calculation. To illustrate, consider again a simple binary mixture whose excess Gibbs energy is given by Eq. (8-13.2). If $A > 2RT$, we can calculate the compositions of the two coexisting equations by solving the two equations of phase equilibrium

$$(\gamma_1 x_1)' = (\gamma_1 x_1)'' \qquad \text{and} \qquad (\gamma_2 x_2)' = (\gamma_2 x_2)'' \tag{8-13.4}$$

where the prime and double prime designate, respectively, the two liquid phases.

From Eq. (8-13.2) we have

$$\ln \gamma_1 = \frac{A}{RT} x_2^2 \tag{8-13.5}$$

and

$$\ln \gamma_2 = \frac{A}{RT} x_1^2 \tag{8-13.6}$$

Substituting into the equation of equilibrium and noting that $x_1' + x_2' = 1$ and $x_1'' + x_2'' = 1$, we obtain

$$x_1' \exp \frac{A(1 - x_1')^2}{RT} = x_1'' \exp \frac{A(1 - x_1'')^2}{RT} \tag{8-13.7}$$

and

$$(1 - x_1') \exp \frac{Ax_1'^2}{RT} = (1 - x_1'') \exp \frac{Ax_1''^2}{RT} \tag{8-13.8}$$

Equations (8-13.7) and (8-13.8) contain two unknowns ($x_1'$ and $x_1''$), which can be found by iteration. Mathematically, several solutions of these two equations can be obtained. However, to be physically meaningful, it is necessary that $1 < x_1' < 1$ and $0 < x_1'' < 1$.

Similar calculations can be performed for ternary (or higher) mixtures. For a ternary system the three equations of equilibrium are

$$(\gamma_1 x_1)' = (\gamma_1 x_1)'' \qquad (\gamma_2 x_2)' = (\gamma_2 x_2)'' \qquad (\gamma_3 x_3)' = (\gamma_3 x_3)'' \tag{8-13.9}$$

If we have an equation relating the excess molar Gibbs energy $g^E$ of the mixture to the overall composition ($x_1$, $x_2$, $x_3$), we can obtain corresponding expressions for the activity coefficients $\gamma_1$, $\gamma_2$, and $\gamma_3$, as discussed elsewhere [see Eq. (8-9.4)]. The equations of equilibrium [Eq. (8-13.9)], coupled with material balance relations, can then be solved to obtain the four unknowns ($x_1'$, $x_2'$ and $x_1''$, $x_2''$).

Systems containing four or more components are handled in a similar manner. An expression for $g^E$ for the multicomponent system is used to relate the activity coefficient of each component in each phase to the composition of that phase. From the equations of equilibrium [$(\gamma_i x_i)' = (\gamma_i x_i)''$ for every component $i$] the phase compositions $x_i'$ and $x_i''$ are found by trial and error.

Considerable skill in numerical analysis is required to construct a computer program that finds the equilibrium compositions of a multicomponent liquid-liquid system from an expression for the excess Gibbs energy for that system. It is difficult to construct a program which always converges to a physically meaningful solution by using only a small number of iterations. This difficulty is especially pronounced in the region near the plait point, where the compositions of the two equilibrium phases become identical.

King [60] has given some useful suggestions for constructing efficient programs toward computation of equilibrium compositions in two-phase systems. See also Ref. 99.

Although the thermodynamics of multicomponent liquid-liquid equilibria is, in principle, straightforward, it is difficult to obtain an expression for $g^E$ which is sufficiently accurate to yield reliable results. Liquid-liquid equilibria are much more sensitive to small changes in activity coefficients than vapor-liquid equilibria. In the latter, activity coefficients play a role which is secondary to the all-important pure component vapor pressures. In liquid-liquid equilibria, however, the activity coefficients are dominant; pure component vapor pressures play no role at all. Therefore, it has often been observed that good estimates of vapor-liquid equilibria can be made for many systems by using only approximate activity coefficients, provided the pure component vapor pressures are accurately known. However, in calculating liquid-liquid equilibria, small inaccuracies in activity coefficients can lead to serious errors.

Renon et al [99] have discussed application of the NRTL equation to liquid-liquid equilibria, Anderson and Prausnitz have discussed application of the UNIQUAC equation [6], and there are many other articles providing similar discussions. Regardless of which equation is used, much care must be exercised in determining parameters from experimental data. Whenever possible, such parameters should come from binary mutual solubility data.

When parameters are obtained from reduction of vapor-liquid equilibrium data, there is always some ambiguity. Unless the experimental data are of very high accuracy, it is usually not possible to obtain a truly unique set of parameters; i.e., in a typical case, there is a range of parameter sets such that any set in that range can equally well reproduce the experimental data within the probable experimental error. (See, for example, Refs. 4 and 33.) When multicomponent vapor-liquid equilibria are calculated, results are often not sensitive to which sets of binary parameters are chosen. However, when multicomponent liquid-liquid equilibria are calculated, results are extremely sensitive to the choice of binary parameters. Therefore, it is difficult to establish reliable ternary (or higher) liquid-liquid equilibria by using only binary parameters obtained from binary liquid-liquid and binary vapor-liquid equilibrium data. For reliable results it is usually necessary to utilize at least some multicomponent liquid-liquid equilibrium data.

To illustrate these ideas, we quote some calculations reported by Bender and Block [15], who considered two ternary systems at 25°C:

System I: Water (1), toluene (2), aniline (3)

System II: Water (1), TCE† (2), acetone (3)

To describe these systems, the NRTL equation was used to relate activity coefficients to composition. The essential problem lies in finding the parameters for the NRTL equation. In system I, components 2 and 3 are completely miscible but components 1 and 2 and components 1 and 3 are only partially miscible. In system II, components 1 and 3 and components 2 and 3 are completely miscible but components 1 and 2 are only partially miscible.

---

†1,1,2-Trichloroethane.

For the completely miscible binaries, Bender and Block set NRTL parameter $\alpha_{ij} = 0.3$. Parameter $\tau_{ij}$ and $\tau_{ji}$ were then obtained from vapor-liquid equilibria. Since it is not possible to obtain unique values of these parameters from vapor-liquid equilibria, Bender and Block used a criterion suggested by Abrams and Prausnitz [4], namely, to choose those sets of parameters for the completely miscible binary pairs which correctly give the limiting liquid-liquid distribution coefficient for the third component at infinite dilution. In other words, NRTL parameters $\tau_{ij}$ and $\tau_{ji}$ chosen were those which not only represent the $ij$ binary vapor-liquid equilibria within experimental accuracy but also give the experimental value of $K_k^\infty$ defined by

$$K_k^\infty = \lim_{\substack{w_k' \to 0 \\ w_k'' \to 0}} \frac{w_k''}{w_k'}$$

where $w$ stands for weight fraction, component $k$ is the third component, i.e., the component *not* in the completely miscible $ij$ binary, and the prime and double prime designate the two equilibrium liquid phases.

For the partially miscible binary pairs, estimates of $\tau_{ij}$ and $\tau_{ji}$ are obtained from mutual-solubility data following an arbitrary choice for $\alpha_{ij}$ in the region $0.20 \leqslant \alpha_{ij} \leqslant 0.40$. When mutual-solubility data are used, the parameter set $\tau_{ij}$ and $\tau_{ji}$ depends only on $\alpha_{ij}$; to find the best $\alpha_{ij}$, Bender and Block used ternary tie-line data. In other words, since the binary parameters are not unique, the binary parameters chosen were those which gave an optimum representation of the ternary liquid-liquid equilibrium data.

Table 8-30 gives mutual solubility data for the three partially miscible binary systems. Table 8-31 gives NRTL parameters following the procedure outlined above. With these parameters, Bender and Block obtained good representation of the ternary phase diagrams, essentially within experimental error. Figures 8-21 and 8-22 compare calculated with observed distribution coefficients for systems I and II.

**TABLE 8-30 Mutual Solubilities in Binary Sysems at 25°C [15]**

| Component | | Weight fraction | |
|---|---|---|---|
| $i$ | $j$ | $i$ in $j$ | $j$ in $i$ |
| Water | TCE | 0.0011 | 0.00435 |
| Water | Toluene | 0.0005 | 0.000515 |
| Water | Aniline | 0.053 | 0.0368 |

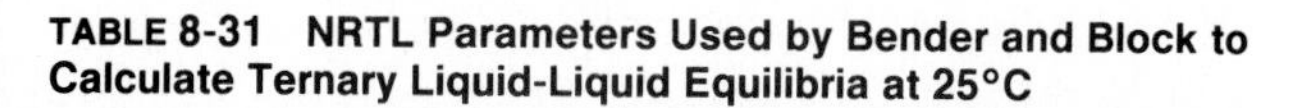

**TABLE 8-31 NRTL Parameters Used by Bender and Block to Calculate Ternary Liquid-Liquid Equilibria at 25°C**

| $i$ | $j$ | $\tau_{ij}$ | $\tau_{ji}$ | $\alpha_{ij}$ |
|---|---|---|---|---|
| System I: water (1), toluene (2), aniline (3) | | | | |
| 1 | 2 | 7.77063 | 4.93035 | 0.2485 |
| 1 | 3 | 4.18462 | 1.27932 | 0.3412 |
| 2 | 3 | 1.59806 | 0.03509 | 0.3 |
| System II: water (1), TCE (2), acetone (3) | | | | |
| 1 | 2 | 5.98775 | 3.60977 | 0.2485 |
| 1 | 3 | 1.38800 | 0.75701 | 0.3 |
| 2 | 3 | −0.19920 | −0.20102 | 0.3 |

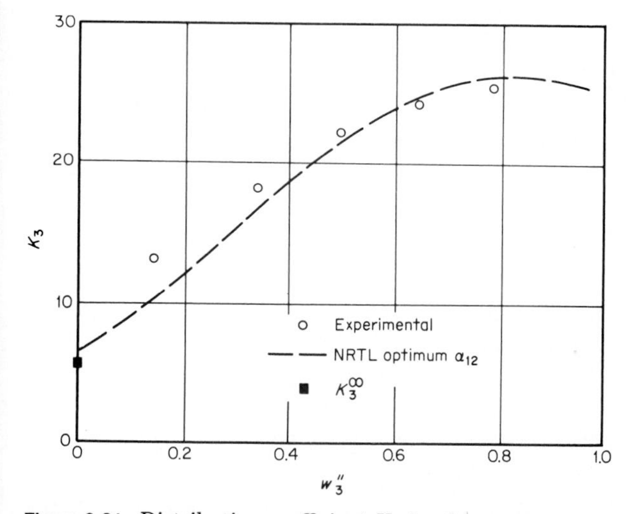


**Figure 8-21** Distribution coefficient $K_3$ for the system water (1)–toluene (2)–aniline (3) at 25°C. Concentrations are weight fractions.

$$K_3 = \frac{w''_3}{w'_3} = \frac{\text{weight fraction aniline in toluene-rich phase}}{\text{weight fraction aniline in water-rich phase}}$$

$$K_3^{\infty} = \frac{\gamma'^{\infty}_3}{\gamma''^{\infty}_3} = \frac{\text{activity coefficient of aniline in water-rich phase at infinite dilution}}{\text{activity coefficient of aniline in toluene-rich phase at infinite dilution}}$$

Activity coefficient $\gamma$ is here defined as the ratio of activity to weight fraction. *(From Ref. 15.)*

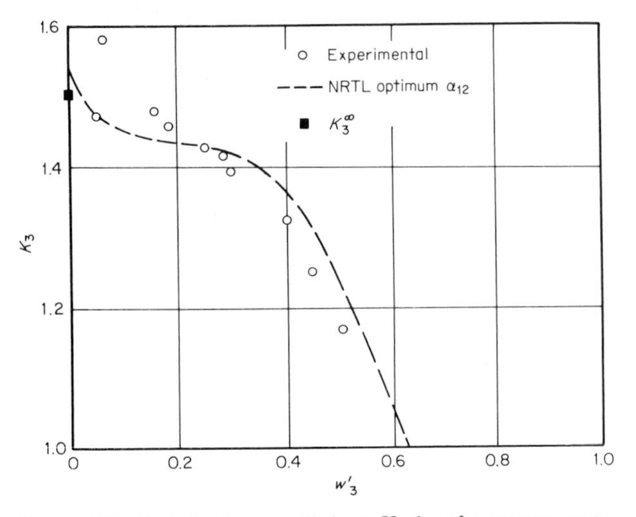


**Figure 8-22** Distribution coefficient $K_3$ for the system water (1)–TCE (2)–acetone (3) at 25°C. Concentrations are in weight fractions.

$$K_3 = \frac{w''_3}{w'_3} = \frac{\text{weight fraction acetone in TCE-rich phase}}{\text{weight fraction acetone in water-rich phase}}$$

$$K_3^\infty = \frac{\gamma'^\infty_3}{\gamma''^\infty_3} = \frac{\text{activity coefficient of acetone in water-rich phase at infinite dilution}}{\text{activity coefficient of acetone in TCE-rich phase at infinite dilution}}$$

Activity coefficient $\gamma$ is here defined as the ratio of activity to weight fraction. *(From Ref. 15.)*

When the NRTL equation is used to represent ternary liquid-liquid equilibria, there are nine adjustable binary parameters; when the UNIQUAC equation is used, there are six. It is tempting to use the ternary liquid-liquid data alone for obtaining the necessary parameters, but this procedure is unlikely to yield a set of *meaningful* parameters; in this context "meaningful" indicates the parameters which also reproduce equilibrium data for the binary pairs. As shown by Heidemann and others [52], unusual and bizarre results can be calculated if the parameter sets are not chosen with care. Experience in this field is not yet plentiful, but all indications are that it is always best to use binary data for calculating binary parameters. Since it often happens that binary parameter sets cannot be determined uniquely, ternary (or higher) data should then be used to fix the best binary sets from the ranges obtained from the binary data. (For a typical range of binary parameter sets, see Fig. 8-4.) It is, of course, always possible to add ternary (or higher) terms to the expression for the excess Gibbs energy and thereby introduce ternary (or higher) constants.

This is sometimes justified, but it is meaningful only if the multicomponent data are plentiful and of high accuracy.

In calculating multicomponent equilibria, the general rule is to use binary data first. Then use multicomponent data for fine tuning.

**Example 8-16** Acetonitrile (1) is used to extract benzene (2) from a mixture of benzene and *n*-heptane (3) at 45°C.

(*a*) 0.5148 mol of acetonitrile is added to a mixture containing 0.0265 mol of benzene and 0.4587 mol of *n*-heptane to form 1 mol of feed.

(*b*) 0.4873 mol of acetonitrile is added to a mixture containing 0.1564 mol of benzene and 0.3563 mol of *n*-heptane to form 1 mol of feed.

For (*a*) and for (*b*), find the composition of the extract phase $E$, the composition of the raffinate phase $R$ and $\alpha$, and the fraction of feed in the extract phase.

**solution** To find the desired quantities, we must solve an isothermal flash problem in which 1 mol of feed separates into $\alpha$ mol of extract and $1 - \alpha$ mol of raffinate.

There are five unknowns: 2 mole fractions in $E$, 2 mole fractions in $R$, and $\alpha$. To find these five unknowns, we require five independent equations. They are three equations of phase equilibrium

$$(\gamma_i x_i)^E = (\gamma_i x_i)^R \qquad i = 1, 2, 3$$

and two material balances

$$z_i = x_i^E \alpha + x_i^R (1 - \alpha) \qquad \text{for any two components}$$

Here $z_i$ is the mole fraction of component $i$ in the feed; $x^E$ and $x^R$ are, respectively, mole fractions in $E$ and in $R$, and $\gamma$ is the activity coefficient.

To solve five equations simultaneously, we use an iterative procedure based on the Newton-Raphson method as described in Ref. 99. The objective function $F$ is

$$F(x^R, x^E, \alpha) = \sum_{i=1}^{3} \frac{(K_i - 1)z_i}{(K_i - 1)\alpha + 1} \rightarrow 0$$

where $$K_i = \frac{x_i^E}{x_i^R} = \frac{\gamma_i^R}{\gamma_i^E}$$

For activity coefficients, we use the UNIQUAC equation with the following parameters:

**Pure Component Parameters**

| Component | $r$ | $q$ |
|---|---|---|
| 1 | 1.87 | 1.72 |
| 2 | 3.19 | 2.40 |
| 3 | 5.17 | 4.40 |

**Binary Parameters**

$$\tau_{ij} = \exp\left(-\frac{a_{ij}}{T}\right) \qquad \tau_{ji} = \exp\left(-\frac{a_{ji}}{T}\right)$$

| Components | | $a_{ij}$, K | $a_{ji}$, K |
|---|---|---|---|
| $i$ | $j$ | | |
| 1 | 2 | 60.28 | 89.57 |
| 1 | 3 | 23.71 | 545.8 |
| 2 | 3 | −135.9 | 245.4 |

In the accompanying table calculated results are compared with experimental data.†

**Liquid-Liquid Equilibria in the System Acetonitrile (1)–Benzene (2)–*n*-Heptane (3) at 45°C**

| | | | $100x_i^R$ | | | $100x_i^E$ | |
|---|---|---|---|---|---|---|---|
| | $i$ | $\gamma_i^R$ | Calc. | Exp. | $\gamma_i^E$ | Calc. | Exp. |
| (*a*) | 1 | 7.15 | 13.11 | 11.67 | 1.03 | 91.18 | 91.29 |
| | 2 | 1.25 | 3.30 | 3.41 | 2.09 | 1.98 | 1.88 |
| | 3 | 1.06 | 83.59 | 84.92 | 12.96 | 6.84 | 6.83 |
| (*b*) | 1 | 3.38 | 25.63 | 27.23 | 1.17 | 73.96 | 70.25 |
| | 2 | 1.01 | 18.08 | 17.71 | 1.41 | 12.97 | 13.56 |
| | 3 | 1.35 | 56.29 | 55.06 | 5.80 | 13.07 | 16.19 |

For (*a*), the calculated $\alpha = 0.4915$; for (*b*), it is 0.4781. When experimental data are substituted into the material balance, $\alpha = 0.5$ for both (*a*) and (*b*).
In this case, there is good agreement between calculated and experimental results because the binary parameters were selected by using binary and ternary data.

While activity coefficient models are commonly used for liquid-liquid equilibria, it is also possible to use an equation of state to correlate liquid-liquid solubility data, as shown, for example, by Tsonopoulos and Wilson [124], who correlated the solubility of water in hydrocarbons by using a variation of the Redlich-Kwong equation of state proposed by Zudkevitch and Joffe. In this variation, for every pure component, equation-of-state constants are given a temperature dependence such that, for temperatures below the critical, the equation of state correctly reproduces experimental vapor pressures and saturated liquid densities. For binary water-hydrocarbon mixtures, the usual mixing rules were used with an adjustable binary constant to correct for deviations from the geometric mean assumption for cross coefficient $a_{12} = (a_1 a_2)^{1/2}(1 - k_{12})$.

†Palmer and Smith [88].

To find $k_{12}$ from solubility data, let subscript 1 stand for water and subscript 2 for hydrocarbon. The equation of equilibrium is

$$f_1' = f_1'' \tag{8-13.10}$$

where $f$ is fugacity, prime stands for the water-rich phase, and double prime stands for the hydrocarbon-rich phase.

For $f_1'$ we write

$$f_1' = f_{\text{pure }1}\,(1 - x_2')\gamma_1' \tag{8-13.11}$$

where $x_2'$ is the experimental solubility of hydrocarbon in water. Since $x_1' <<< 1$, activity coefficient $\gamma_1' = 1$. Since $x_2'$ is very small, the calculations are not at all sensitive to $x_2'$.

For $f_1''$ we write

$$f_1'' = \phi_1'' x_1'' P \tag{8-13.12}$$

where $\phi_1''$ is the fugacity coefficient as calculated from the equation of state applied to the hydrocarbon-rich liquid. $P$ is the experimental system pressure, and $x_1''$ is the experimental solubility of water in the hydrocarbon. Since $x_1''$ and $x_1'$ are very small, $P$ is essentially the sum of the pure component vapor pressures.

For solubilities of water in hydrocarbons in the range 0 to 200°C, Tsonopoulos and Wilson obtained a good correlation with temperature-independent values of $k_{12}$ (0.260 for water in benzene, 0.519 in cyclohexane, and 0.486 in $n$-hexane).

Although Eq. (8-13.10) is the equilibrium condition for water, a similar equation can be written for the hydrocarbon; solubility data for the hydrocarbon in water can then be used to obtain $k_{12}$ or, alternatively, the $k_{12}$ values obtained from solubility data for water in the hydrocarbon can be used to predict solubilities of hydrocarbons in water. That prediction is very poor, or what is equivalent, $k_{12}$ obtained from water-in-hydrocarbon solubility data is very much different from $k_{12}$ obtained from hydrocarbon-in-water solubility data. In other words, for a fixed binary system at constant temperature, $k_{12}$ depends on composition. Such dependence tells us that the quadratic mixing rule for constant $a$ is not correct. However, since the second virial coefficient $B(x)$ of a mixture is related simply to equation-of-state constants† and, since the second virial coefficent must be a quadratic function of $x$, we face a dilemma: at low densities, constant $a$ is quadratic in $x$, but at high densities it is not. In principle, the solution to this dilemma is provided by density-dependent mixing rules as briefly discussed earlier. While equations of state provide an attractive tool for

---

†For an equation of the van der Waals form, $B(x) = b(x) - a(x)/RT$.

describing liquid-liquid equilibria, most published studies are limited to binary systems.

## 8-14 Phase Equilibria in Polymer Solutions

Strong negative deviations from Raoult's law are observed in binary liquid mixtures where one component consists of very large molecules (polymers) and the other consists of molecules of normal size. For mixtures of normal solvents and amorphous polymers, phase equilibrium relations are usually described by the Flory-Huggins theory, discussed fully in a book by Flory [34] and in a monograph by Tompa [121]; a brief introduction is given by Prausnitz et al. [97]. For engineering application, a useful summary is provided by Sheehan and Bisio [112].

There are several versions of the Flory-Huggins theory, and unfortunately, different authors use different notation. The primary composition variable for the liquid phase is the volume fraction, here designated by $\Phi$ and defined by Eqs. (8-10.3) and (8-10.4). In polymer-solvent systems, volume fractions are very different from mole fractions because the molar volume of a polymer is much larger than that of the solvent.

Since the molecular weight of the polymer is often not known accurately, it is difficult to determine the mole fraction. Therefore, an equivalent definition of $\Phi$ is frequently useful:

$$\Phi_1 = \frac{w_1/\rho_1}{w_1/\rho_1 + w_2/\rho_2} \quad \text{and} \quad \Phi_2 = \frac{w_2/\rho_2}{w_1/\rho_1 + w_2/\rho_2} \tag{8-14.1}$$

where $w_i$ is the weight fraction of component $i$ and $\rho_i$ is the mass density (*not* molar density) of pure component $i$.

Let subscript 1 stand for solvent and subscript 2 for polymer. The activity $a_1$ of the solvent, as given by the Flory-Huggins equation, is

$$\ln a_1 = \ln \Phi_1 + \left(1 - \frac{1}{m}\right)\Phi_2 + \chi\Phi_2^2 \tag{8-14.2}$$

where $m$ is defined by $m = V_2^L/V_1^L$ and the adjustable constant $\chi$ is called the *Flory interaction parameter*. In typical polymer solutions $1/m$ is negligibly small compared with unity, and therefore it may be neglected. The parameter $\chi$ depends on temperature, but for polymer-solvent systems in which the molecular weight of the polymer is very large, it is nearly independent of polymer molecular weight. In theory, $\chi$ is also independent of polymer concentration, but in fact it often varies with concentration, especially in mixtures containing polar molecules, for which the Flory-Huggins theory provides only a rough approximation.

In a binary mixture of polymer and solvent at ordinary pressures, only the solvent is volatile; the vapor phase mole fraction of the solvent is unity, and therefore the total pressure is equal to the partial pressure of the solvent.

In a polymer solution, the activity of the solvent is given by

$$a_1 = \frac{P}{P_{vp1}} \frac{1}{\mathcal{F}_1} \tag{8-14.3}$$

where factor $\mathcal{F}_1$ is defined by Eq. (8-4.2). At low or moderate pressures, $\mathcal{F}_1$ is equal to unity.

Equation (8-14.2) holds only for temperatures at which the polymer in the pure state is amorphous. If the pure polymer has appreciable crystallinity, corrections to Eq. (8-14.2) are significant, as discussed elsewhere [34].

Equation (8-14.2) is useful for calculating the volatility of a solvent in a polymer solution, provided that the Flory parameter $\chi$ is known. Sheehan and Bisio [112] report Flory parameters for a large number of binary systems† and present methods for estimating $\chi$ from solubility parameters. Similar data are also given in the *Polymer Handbook* [9]. Table 8-32 shows some $\chi$ values reported by Sheehan and Bisio.

A particularly convenient and rapid experimental method for obtaining $\chi$ is provided by gas-liquid chromatography [44]. Although this experimental technique can be used at finite concentrations of solvent, it is most efficiently used for solutions infinitely dilute with respect to solvent, i.e., at the limit where the volume fraction of polymer approaches unity. Some

†Unfortunately, Sheehan and Bisio use completely different notation; $v$ for $\Phi$, $x$ for $m$, and $\mu$ for $\chi$.

**TABLE 8-32 Flory $\chi$ Parameters for Some Polymer-Solvent Systems near Room Temperature [112]**

| Polymer | Solvent | $\chi$ |
|---|---|---|
| Natural rubber | Heptane | 0.44 |
| | Toluene | 0.39 |
| | Ethyl acetate | 0.75 |
| Polydimethyl siloxane | Cyclohexane | 0.44 |
| | Nitrobenzene | 2.2 |
| Polyisobutylene | Hexadecane | 0.47 |
| | Cyclohexane | 0.39 |
| | Toluene | 0.49 |
| Polystyrene | Benzene | 0.22 |
| | Cyclohexane | 0.52 |
| Polyvinyl acetate | Acetone | 0.37 |
| | Dioxane | 0.41 |
| | Propanol | 1.2 |

solvent volatility data obtained from chromatography [84] are shown in Fig. 8-23. From these data, $\chi$ can be found by rewriting Eq. (8-14.2) in terms of a weight fraction activity coefficient $\Omega$

$$\Omega_1 \equiv \frac{a_1}{w_1} = \frac{P}{P_{vp1} w_1} \frac{1}{\mathcal{F}_1} \tag{8-14.4}$$

Combining with Eq. (8-14.2), in the limit as $\Phi_2 \to 1$, we obtain

$$\chi = \ln \left( \frac{P}{w} \right)_1^\infty - \left( \ln P_{vp1} + \ln \frac{\rho_2}{\rho_1} + 1 \right) \tag{8-14.5}$$

where $\rho$ is the mass density (*not* molar density). Equation (8-14.5) also assumes that $\mathcal{F}_1 = 1$ and that $1/m \ll 1$. Superscript $\infty$ denotes that weight fraction $w_1$ is very small compared with unity. Equation (8-14.5) provides a useful method for finding $\chi$ because $(P/w)_1^\infty$ is easily measured by gas-liquid chromatography.

Equation (8-14.2) was derived for a binary system, i.e., one in which all polymer molecules have the same molecular weight (monodisperse system). For mixtures containing one solvent and one polymer with a variety

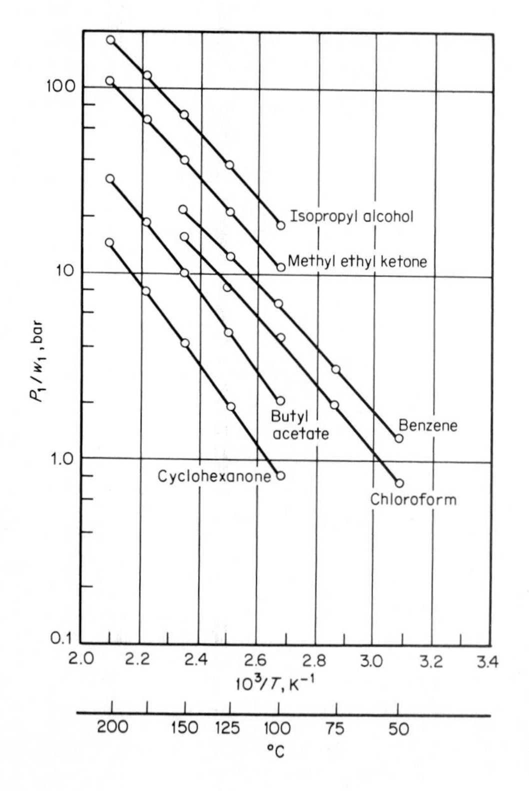


**Figure 8-23** Volatilities of solvents in Lucite 2044 for a small weight fraction of solute. *(From Ref. 84.)*

of molecular weights (polydisperse system), Eq. (8-14.2) can be used provided $m$ and $\Phi$ refer to the polymer whose molecular weight is the number average molecular weight.

The theory of Flory and Huggins can be extended to multicomponent mixtures containing any number of polymers and any number of solvents. No ternary (or higher) constants are required.

Solubility relations (liquid-liquid equilibria) can also be calculated with the Flory-Huggins theory. Limited solubility is often observed in solvent-polymer systems, and it is common in polymer-polymer systems (incompatibility). The Flory-Huggins theory indicates that, for a solvent-polymer system, limited miscibility occurs when

$$\chi > \frac{1}{2}\left(1 + \frac{1}{m^{1/2}}\right)^2 \qquad \text{(8-14.6)}$$

For large $m$, the value of $\chi$ may not exceed $\frac{1}{2}$ for miscibility in all proportions.

Liquid-liquid phase equilibria in polymer-containing systems are described in numerous articles published in journals devoted to polymer science and engineering. The thermodynamics of such equilibria is discussed in Flory's book and in articles by Scott and Tompa [110] and by Hsu and Prausnitz [57]. A comprehensive review of polymer compatibility and incompatibility is given by Krause [63].

For semiquantitative calculations the three-dimensional solubility parameter concept [47] is often useful, especially for formulations of paints, coating, inks, etc.

The Flory-Huggins equation contains only one adjustable binary parameter. For simple nonpolar systems one parameter is often sufficient, but for complex systems, much better representation is obtained by empirical extension of the Flory-Huggins theory using at least two adjustable parameters, as shown by Maron and Nakajima [72] and by Heil and Prausnitz [53]. The latters' extension is a generalization of Wilson's equation. The UNIQUAC equation with two adjustable parameters is also applicable to polymer solutions [4].

The theory of Flory and Huggins is based on a lattice model which ignores free-volume differences; in general, polymer molecules in the pure state pack more densely than molecules of normal liquids. Therefore, when polymer molecules are mixed with molecules of normal size, the polymer molecules gain freedom to exercise their rotational and vibrational motions; at the same time, the smaller solvent molecules partially lose such freedom. To account for these effects, an *equation-of-state theory* of polymer solutions has been developed by Flory [35] and Patterson [89] based on ideas suggested by Prigogine [102]. The newer theory is necessarily more complicated, but, unlike the older one, it can at least semiquantitatively describe some forms of phase behavior commonly observed in polymer solutions. In particular, it can explain the observation that some polymer-solvent systems exhibit lower consolute temperatures as well as upper consolute temperatures (Fig. 8-19).† Engineering applications of the theory are indicated by Bonner [17], Bondi [16], and Tapavicza [117]. Application to phase equilibria in the system polyethylene-ethylene at high pressures is discussed by Liu and Prausnitz [69].

## 8-15 Solubilities of Solids in Liquids

The solubility of a solid in a liquid is determined not only by the intermolecular forces between solute and solvent but also by the melting point and the enthalpy of fusion of the solute. For example, at 25°C, the solid aromatic hydrocarbon phenanthrene is highly soluble in benzene; its solubility is 20.7 mole percent. By contrast, the solid aromatic hydrocarbon anthracene, an isomer of phenanthrene, is only slightly soluble in benzene at 25°C; its solubility is 0.81 mole percent. For both solutes, intermolecular forces between solute and benzene are essentially identical. However, the melting points of the solutes are significantly different: phenanthrene melts at 100°C and anthracene at 217°C. In general, it can be shown that, when other factors are held constant, the solute with the higher melting point has the lower solubility. Also, when other factors are held constant, the solute with the higher enthalpy of fusion has the lower solubility.

These qualitative conclusions follow from a quantitative thermodynamic analysis given in several texts. (See, for example, Refs. 54 and 97.)

In a binary system, let subscript 1 stand for solvent and subscript 2 for solute. Assume that the solid phase is pure. At temperature $T$, the solubility (mole fraction) $x_2$ is given by

$$\ln \gamma_2 x_2 = -\frac{\Delta h_f}{RT}\left(1 - \frac{T}{T_t}\right) + \frac{\Delta C_p}{R}\left(\frac{T_t - T}{T}\right) - \frac{\Delta C_p}{R} \ln \frac{T_t}{T} \tag{8-15.1}$$

where $\Delta h_f$ is the enthalpy of fusion of the solute at the triple-point temperature $T_t$ and $\Delta C_p$ is given by the molar heat capacity of the pure solute:

$$\Delta C_p = C_p \text{ (subcooled liquid solute)} - C_p \text{ (solid solute)} \tag{8-15.2}$$

The standard state for activity coefficient $\gamma_2$ is pure (subcooled) liquid 2 at system temperature $T$.

To a good approximation, we can substitute normal melting temperature $T_m$ for triple-point temperature $T_t$, and we can assume that $\Delta h_f$ is essentially the same at the two temperatures. In Eq. (8-15.1) the first term

---

†However, in polymer-solvent systems, the upper consolute temperature usually is below the lower consolute temperature.

on the right-hand side is much more important than the remaining two terms, and therefore a simplified form of that equation is

$$\ln \gamma_2 x_2 = -\frac{\Delta h_f}{RT}\left(1 - \frac{T}{T_m}\right) \tag{8-15.3}$$

If we substitute

$$\Delta s_f = \frac{\Delta h_f}{T_m} \tag{8-15.4}$$

we obtain an alternative simplified form

$$\ln \gamma_2 x_2 = -\frac{\Delta s_f}{R}\left(\frac{T_m}{T} - 1\right) \tag{8-15.5}$$

where $\Delta s_f$ is the entropy of fusion. A plot of Eq. (8-15.5) is shown in Fig. 8-24.

If we let $\gamma_2 = 1$, we can readily calculate the ideal solubility at temper-

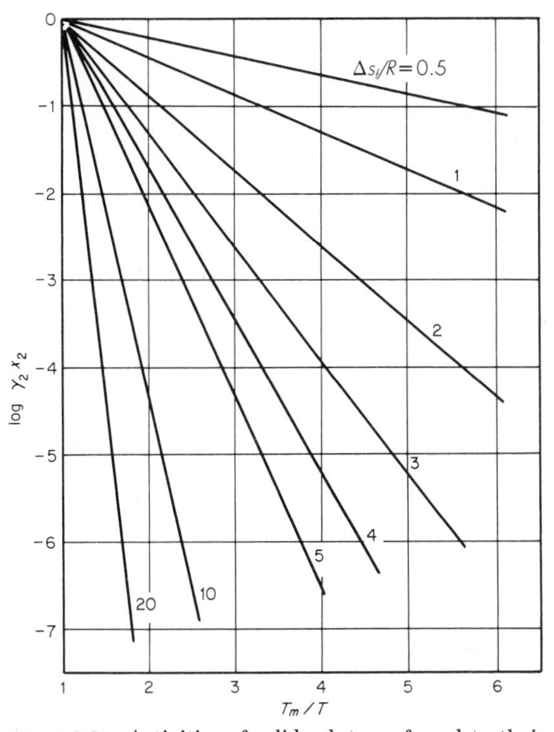


**Figure 8-24** Activities of solid solutes referred to their pure subcooled liquids. *(From Ref. 101.)*

ature $T$, knowing only the solute's melting temperature and its enthalpy (or entropy) of fusion. This ideal solubility depends only on properties of the solute; it is independent of the solvent's properties. The effect of intermolecular forces between molten solute and solvent are reflected in the activity coefficient $\gamma_2$.

To describe $\gamma_2$, we can utilize any of the expressions for the excess Gibbs energy, as discussed in Sec. 8-5. However, since $\gamma_2$ depends on the mole fraction $x_2$, solution of Eq. (8-15.5) requires iteration. For example, suppose that $\gamma_2$ is given by a simple one-parameter Margules equation

$$\ln \gamma_2 = \frac{A}{RT} (1 - x_2)^2 \tag{8-15.6}$$

where $A$ is an empirical constant. Substitution into Eq. (8-15.5) gives

$$\ln x_2 + \frac{A}{RT} (1 - x_2)^2 = -\frac{\Delta s_f}{R} \left( \frac{T_m}{T} - 1 \right) \tag{8-15.7}$$

and $x_2$ must be found by a trial-and-error calculation.

In nonpolar systems, the activity coefficient $y_2$ can often be estimated by using the Scatchard-Hildebrand equation, as discussed in Sec. 8-10. In that event, since $\gamma_2 \geqslant 1$, the ideal solubility ($\gamma_2 = 1$) is larger than that obtained from regular solution theory. As shown by Preston and Prausnitz [101], and as illustrated in Fig. 8-25, regular solution theory is useful for calculating solubilities in nonpolar systems, especially when the geometric mean assumption is relaxed through introduction of an empirical correction $\ell_{12}$ (see Sec. 8-10).

Figure 8-25 shows three lines: the top line is calculated by using the geometric mean assumption ($\ell_{12} = 0$) in the Scatchard-Hildebrand equation. The bottom line is calculated with $\ell_{12} = 0.11$, a value estimated from gas-phase PVT$y$ data. The middle line is calculated with $\ell_{12} = 0.08$, which is the optimum value obtained from solubility data. Figure 8-25 suggests that even an approximate estimate of $\ell_{12}$ usually produces better results than assuming that $\ell_{12}$ is zero. Unfortunately, *some* mixture datum is needed to estimate $\ell_{12}$. In a few fortunate cases one freezing point datum, e.g., the eutectic point, may be available to fix $\ell_{12}$.

In some cases it is possible to use UNIFAC for estimating solubilities of solids, as discussed in Ref. 41.

It is important to remember that the calculations outlined above rest on the assumption that the solid phase is pure, i.e., that there is no solubility of the solvent in the solid phase. This assumption is usually a good one, especially if the two components differ appreciably in molecular size and shape. However, in many cases that are known, the two components are at least partially miscible in the solid phase, and in that event it is necessary to correct for solubility and nonideality in the solid phase as well as in the liquid phase. This complicates the thermodynamic treatment, but, more important, solubility in the solid phase may significantly affect the phase diagram. Figure 8-26 shows results for the solubility of solid argon in liquid nitrogen. The top line presents calculated results assuming that $x^{\mathcal{S}}$ (argon) $= 1$, where superscript $^{\mathcal{S}}$ denotes the solid phase. The bottom line takes into account the experimentally known solubility of nitrogren in solid argon [$x^{\mathcal{S}}$ (argon) $\neq 1$]. In this case is it clear that serious error is introduced by neglecting solubility of the solvent in the solid phase.

## 8-16 Aqueous Solutions of Electrolytes

Physical chemists have given much attention to aqueous mixtures containing solutes that ionize either completely (e.g., strong salts like sodium chloride) or partially (e.g., sulfur dioxide and acetic acid). The thermodynamics of such mixtures is discussed in numerous references, but the

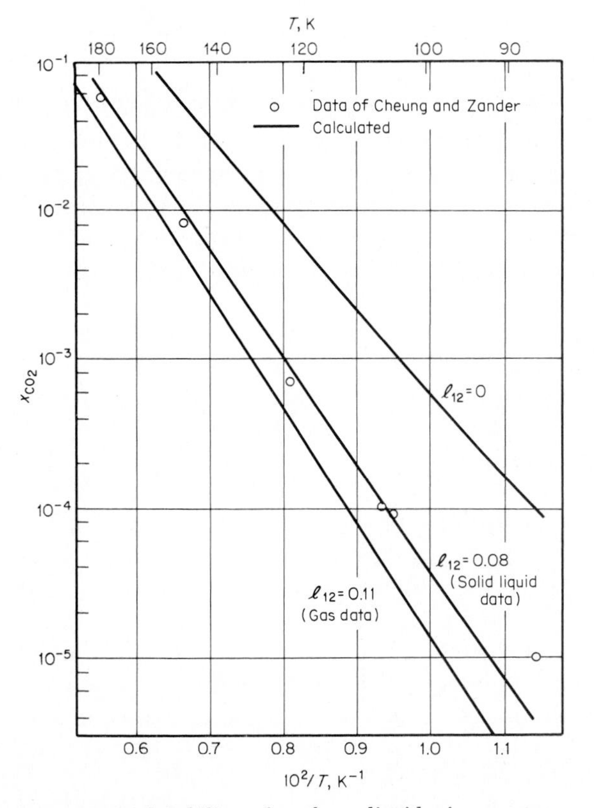


**Figure 8-25** Solubility of carbon dioxide in propane. *(From Ref. 101.)*

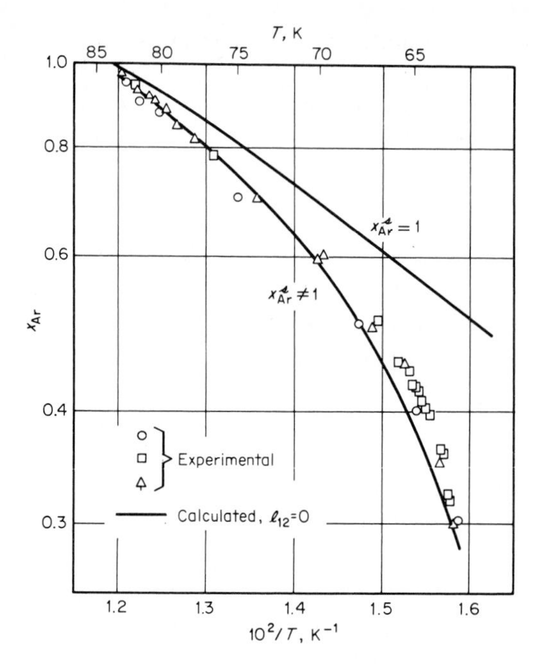


**Figure 8-26** Solubility of argon in nitrogen: effect of solid-phase composition. *(From Ref. 101.)*

most useful general discussions are by Pitzer and Brewer [68] and by Robinson and Stokes [107]. A helpful brief survey is given by Pitzer [93]. Unfortunately, however, these discussions are primarily concerned with single-solute systems and with nonvolatile electrolytes. Further, the discussions are not easily reduced to practice for engineering design, in part because the required parameters are not available, especially at higher temperatures.

Chemical engineers have only recently begun to give careful attention to aqueous mixtures of electrolytes. Much effort is in progress, but as of 1984, little has appeared in the literature. However, see Refs. 18, 22, 29, 46, 75, and 94.

Table 8-33 gives references for four engineering-oriented correlations. The first is designed for systems of strong and/or weak electrolytes in aqueous or mixed solvents; an unpublished version is incorporated in the ASPEN PLUS computer programs. In most cases, the user must regress experimental data to determine the required parameters. The second uses a modified UNIQUAC equation to predict salt effects on vapor-liquid equilibria in simple or mixed solvents. The third and fourth are similar; both focus on multicomponent aqueous solutions of volatile weak electrolytes (e.g., ammonia, carbon dioxide, hydrogen sulfide, sulfur dioxide, and hydrogen cyanide). The fourth also includes nonelectrolyte solutes such as methane, nitrogen, and carbon monoxide.

A monograph edited by Furter [40] discusses salt effects on vapor-liquid equilibria in solvent mixtures.

**TABLE 8-33 Four Engineering-Oriented Correlations for Vapor-Liquid Equilibria for Aqueous Solutions of Electrolytes**

1. C. C. Chen, H. I. Britt, J. F. Boston, and L. B. Evans: *AIChE J.*, **28:** 533 (1982); B. Mock, L. B. Evans, and C. C. Chen: *Proc. 1984 Summer Computer Simulation Conf.*, (1984) p. 558; C. C. Chen, J. F. Boston, and L. B. Evans; also B. Mock, C. C. Chen, and L. B. Evans: papers presented at *AIChE Annual Meeting, San Francisco, November 1984.*

2. B. Sander, A. Fredenslund, and P. Rasmussen: paper presented at *AIChE Annual Meeting, San Francisco, November 1984.*

3. D. Beutier, and H. Renon: *IEC Proc. Des. Dev.*, **17:** 220 (1978): corrections, *ibid.* **19:** 722 (1980). Updated in *Thermodynamics of Aqueous Systems with Industrial Applications, ACS Symposium Series 133,* edited by S. A. Newman, 1980, p. 173.

4. T. J. Edwards, G. Maurer, J. Newman, and J. M. Prausnitz: *AIChE J.*, **24:** 966 (1978).

**Example 8-17** Vapor-liquid equilibria are required for design of a sour water stripper. An aqueous stream at 120°C has the following composition, expressed in molality (moles per kilogram of water):

| | | | | |
|---|---|---|---|---|
| $CO_2$ | 0.4 | $NH_3$ | 2.62 | CO 0.0016 |
| $H_2S$ | 1.22 | $CH_4$ | 0.003 | |

Find the total pressure and the composition of the equilibrium vapor phase.

**solution** This is a bubble point problem with three volatile weak electrolytes and two nonreacting ("inert") gases. The method for solution follows that outlined by Edwards et al. [32].

For the chemical species in the liquid phase, we consider the following equilibria:

$$NH_3 + H_2O \rightleftharpoons NH_4^+ + OH^- \tag{8-16.1}$$

$$CO_2 + H_2O \rightleftharpoons HCO_3^- + H^+ \tag{8-16.2}$$

$$HCO_3^- \rightleftharpoons CO_3^{2-} + H^+ \tag{8-16.3}$$

$$H_2S \rightleftharpoons HS^- + H^+ \tag{8-16.4}$$

$$HS^- \rightleftharpoons S^{2-} + H^+ \tag{8-16.5}$$

$$NH_3 + HCO_3^- \rightleftharpoons NH_2COO^- + H_2O \tag{8.16.6}$$

$$H_2O \rightleftharpoons H^+ + OH^- \tag{8.16.7}$$

We assume that "inert" gases $CH_4$ and CO do not participate in any reactions and that their fugacities are proportional to their molalities (Henry's law).

As indicated by Eqs. (8-16.1) to (8-16.7), we must find the concentrations of thirteen species in the liquid phase (not counting water) and six in the vapor phase (the three volatile weak electrolytes, the two inert gases, and water).

We have the following unknowns:

$m_i$ molality of noninert chemical species $i$ in solution (11 molalities)

$\gamma_i^*$ activity coefficient of noninert chemical species $i$ in solution (molality scale, unsymmetric convention) (11 activity coefficients)

$a_w$ activity of liquid water (one activity)

$p_i$ partial pressure of each volatile component $i$ (six partial pressures)

$\phi_i$ fugacity coefficient of volatile component $i$ in the vapor (six fugacity coefficients)

We have 35 unknowns; we must now find 35 independent relations among these variables.

REACTION EQUILIBRIA. The equilibrium constant is known for each of the seven reactions at 120°C. For any reaction of the form

$$A \rightarrow B^+ + C^-$$

the equilibrium constant $K$ is given by

$$K = \frac{a_{B+}a_{C-}}{a_A} = \frac{\gamma^*_{B+}\gamma^*_{C+}m_{B+}m_{C-}}{a_A} \tag{8-16.8}$$

where $a$ = activity.

ACTIVITY COEFFICIENTS. Activity coefficients for the (noninert) chemical species in solution are calculated from an expression based on the theory of Pitzer [92] as a function of temperature and composition. Binary interaction parameters are given by Edwards et al. [32]

WATER ACTIVITY. By applying the Gibbs-Duhem equation to expressions for the solute activity coefficients, we obtain an expression for the activity of liquid water as a function of temperature and composition. (See Edwards et al. [32]).

VAPOR-LIQUID EQUILIBRIA. For each volatile weak electrolyte and for each "inert" gas, we equate fugacities in the two phases:

$$\phi_i p_i = m_i \gamma_i^* H_i \text{ (PC)} \tag{8-16.9}$$

where $H_i$ is Henry's constant for volatile solute $i$. For the "inert" gases in liquid solution, $m_i$ is given and $\gamma_i^*$ is taken as unity. PC is the Poynting correction.
For water, the equation for phase equilibrium is:

$$\phi_w p_w = a_w P_w^{sat} \phi_w^{sat} \text{ (PC)} \tag{8-16.10}$$

where the superscripts refer to pure, saturated liquid.
Henry's constants are available at 120°C. For the Poynting correction, liquid-phase partial molar volumes are estimated.

VAPOR PHASE FUGACITY COEFFICIENTS. Since some of the components in the gas phase are polar and the total pressure is likely to exceed 1 bar, it is necessary to correct for vapor phase nonideality. The method of Nakamura et al. [83] is used to calculate fugacity coefficients as a function of temperature, pressure, and composition.

MATERIAL BALANCES A material balance is written for each weak electrolyte in the liquid phase. For example, for $NH_3$:

$$m_{NH3} = m_{NH3} + m_{NH_4^+} + m_{NH2COO^-} \tag{8-16.11}$$

where $m$ is the nominal concentration of the solute as specified in the problem statement.

ELECTRONEUTRALITY. Since charge is conserved, the condition for electroneutrality is given by

$$\sum_i z_i m_i = 0 \tag{8-16.12}$$

where $z_i$ is the charge on chemical species $i$.

TOTAL NUMBER OF INDEPENDENT RELATIONS. We have seven reaction equilibria, eleven activity coefficient expressions, one equation for the water activity, six vapor-liquid equilibrium relations, six equations for vapor phase fugacities, three material balances, and one electroneutrality condition, providing a total of 35 independent equations.

**TABLE 8-34 Program Tides Results**

Input specifications: Temperature = 393.15 K

| component | stoichiometric concentration |
|---|---|
| $CO_2$ | 0.40000 |
| $H_2S$ | 1.2200 |
| $NH_3$ | 2.6200 |
| $CH_4$ | 0.003 |
| CO | 0.0016 |

| | Liquid phase | | Vapor phase | | | |
|---|---|---|---|---|---|---|
| Component | Concentration, molality | Activity coef., unitless | Partial pressure, atm. | Fug. coef. | Poynt. corr. | |
| $NH_3$ | 0.97193 | 1.0147 | 0.41741 | 0.95997 | 0.98420 | |
| $NH_4^+$ | 1.4873 | 0.52287 | | | | |
| $CO_2$ | $0.27749 \times 10^{-1}$ | 1.0925 | 3.9433 | 0.97193 | 0.98219 | |
| $HCO_3^-$ | 0.20513 | 0.57394 | | | | |
| $CO_3^{2-}$ | $0.63242 \times 10^{-2}$ | $0.15526 \times 10^{-1}$ | | | | |
| $H_2S$ | 0.11134 | 1.1784 | 4.4096 | 0.95444 | 0.98092 | |
| $HS^-$ | 1.1086 | 0.56771 | | | | |
| $S^{2-}$ | $0.11315 \times 10^{-4}$ | $0.14992 \times 10^{-1}$ | | | | |
| $NH_2COO^-$ | 0.16080 | 0.42682 | | | | |
| $H^+$ | $0.83354 \times 10^{-7}$ | 0.86926 | | | | |
| $OH^-$ | $0.23409 \times 10^{-4}$ | 0.59587 | | | | |
| $CH_4$ | $0.30000 \times 10^{-2}$ | | 3.3627 | 0.99758 | 0.97975 | |
| CO | $0.16000 \times 10^{-2}$ | | 2.0580 | 1.0082 | 0.98053 | |
| | Activity $H_2O$ | 0.93393 | | | $H_2O$ vap. press. | 1.9596 atm. |
| | | Total pressure | 16.154 atm. | | | |

Equilibrium constants at 393.15 K

| Reaction | Equil. constant | units |
|---|---|---|
| $NH_3$ dissociation | $0.11777 \times 10^{-4}$ | molality |
| $CO_2$ dissociation | $0.30127 \times 10^{-6}$ | molality |
| $HCO_3$ dissociation | $0.60438 \times 10^{-10}$ | molality |
| $H_2S$ dissociation | $0.34755 \times 10^{-6}$ | molality |
| HS dissociation | $0.19532 \times 10^{-13}$ | molality |
| $NH_2CO$ formation | 0.55208 | 1/molality |
| $H_2O$ dissociation | $0.10821 \times 10^{-11}$ | molality$^2$ |

| Henry's constants, kg·atm/mol | |
|---|---|
| $CO_2$ | 124.52 |
| $H_2S$ | 31.655 |
| $NH_3$ | 0.40159 |
| $CH_4$ | 1091.3 |
| CO | 1263.9 |

Computer program TIDES† is designed to perform the tedious trial-and-error calculations. For this bubble point problem, the only required inputs are the temperature and the molalities of the nominal (stoichiometric) solutes in the liquid phase. The total pressure is 16.4 bar, and the calculated mole fractions in the vapor phase are:

| | | | |
|---|---|---|---|
| $CO_2$ | 0.243 | $CH_4$ | 0.208 |
| $H_2S$ | 0.272 | CO | 0.127 |
| $NH_3$ | 0.026 | $H_2O$ | 0.122 |

Table 8-34 shows the program output. Note that activities of solutes have units of molality. However, all activity coefficients and the activity of water are dimensionless.

## 8-17 Concluding Remarks

This chapter on phase equilibria has presented no more than a brief introduction to a very broad subject. The variety of mixtures encountered in the chemical industry is extremely large, and, except for general thermodynamic equations, there are no quantitative relations which apply rigorously to all, or even to a large fraction, of these mixtures. Thermodynamics provides only a coarse but reliable framework; the details must be supplied by physics and chemistry, which ultimately rest on experimental data.

For each mixture it is necessary to construct an appropriate mathematical model for representing the properties of that mixture. Whenever possible, such a model should be based on physical concepts, but since our fundamental understanding of fluids is severely limited, any useful model is inevitably influenced by empiricism. While at least some empiricism cannot be avoided, the strategy of the process engineer must be to use enlightened rather than blind empiricism. This means foremost that critical and informed judgment must always be exercised. While such judgment is attained only by experience, we conclude this chapter with a few guidelines.

1. Face the facts: you cannot get something from nothing. Do not expect magic from thermodynamics. If you want reliable results, you will need some reliable experimental data. You may not need many, but you do need some. The required data need not necessarily be for the particular system of interest; sometimes they may come from experimental studies on closely related systems, perhaps represented by a suitable correlation. Only in very simple cases can partial thermodynamic properties in a mixture, e.g., activity coefficients, be found from pure-component data alone.
2. Correlations provide the easy route, but they should be used last, not

†Available from J. M. Prausnitz, Department of Chemical Engineering, University of California, Berkeley.

first. The preferred first step should always be to obtain *reliable* experimental data, either from the literature or from the laboratory. Do not at once reject the possibility of obtaining a few crucial data yourself. Laboratory work is more tedious than pushing a computer button, but ultimately, at least in some cases, you may save time by making a few simple measurements instead of a multitude of furious calculations. A small laboratory with a few analytical instruments (especially a chromatograph or a simple boiling-point apparatus) can often save both time and money. If you cannot do the experiment yourself, consider the possibility of having someone else do it for you.

3. It is always better to obtain a few well-chosen and reliable experimental data than to obtain many data of doubtful quality and relevance. Beware of statistics, which may be the last refuge of a poor experimentalist.
4. Always regard published experimental data with skepticism. Many experimental results are of high quality, but many are not. Just because a number is reported by someone and printed by another, do not automatically assume that it must therefore be correct.
5. When choosing a mathematical model for representing mixture properties, give preference if possible to those which have some physical basis.
6. Seek simplicity; beware of models with many adjustable parameters. When such models are extrapolated even mildly into regions other than those for which the constants were determined, highly erroneous results may be obtained.
7. In reducing experimental data, keep in mind the probable experimental uncertainty of the data. Whenever possible, give more weight to those data which you have reason to believe are more reliable.
8. If you do use a correlation, be sure to note its limitations. Extrapolation outside its domain of validity can lead to large error.
9. Never be impressed by calculated results merely because they come from a computer. The virtue of a computer is speed, not intelligence.
10. Maintain perspective. Always ask yourself: Is this result reasonable? Do other similar systems behave this way? If you are inexperienced, get help from someone who has experience. Phase equilibria in fluid mixtures is not a simple subject. Do not hesitate to ask for advice.

## Notation

| | |
|---|---|
| $a, b, c,$ | empirical coefficients |
| $a_i$ | activity of component $i$ |
| $a_{mn}$ | group interaction parameter, Eq. (8-10.57) |
| $aa$ | empirical parameter in Eq. (8-10.28) |
| $A$ | empirical constant |
| $B$ | empirical constant |

$B_{ij}$ second virial coefficient for the $ij$ interaction

$c, d$ empirical constants in Eq. (8-7.1)

$c_{ij}$ cohesive energy density for the $ij$ interaction in Sec. 8-10.

$c_{12}$ empirical constant in Eq. (8-5.14).

$C$ empirical constant

$C_p$ molar specific heat at constant pressure

$C_{ij}$ binary parameter in Eq. (8-12.22)

$d_{12}$ Flory-Huggins combinatorial term in Eq. (8-10.27)

$D$ empirical constant; number of data points, Eq. (8-8.16)

$D_{ij}$ binary parameter in Eq. (8-12.23)

$f_i$ fugacity of component $i$

$f$ a function

$\mathcal{F}_i$ nonideality factor defined by Eq. (8-4.2)

$g_{ij}$ empirical constant (Table 8-3)

$g^E$ molar excess Gibbs energy

$G^E$ total excess Gibbs energy

$G_{ij}$ empirical constant (Table 8-3)

$h^E$ molar excess enthalpy

$\Delta h_f$ molar enthalpy of fusion

$\overline{h}_i$ partial molar enthalpy of component $i$

$H$ Henry's constant

$\Delta H_v$ enthalpy of vaporization

$I$ defined by Eq. (8-8.17)

ID index (Table 8-19)

$k_{12}$ binary parameter in Sec. 8-13

$K$ $y/x$ in Sec. 8-12; distribution coefficient in Sec. 8-13; chemical equilibrium constant in Sec. 8-16

$\ell_i$ constant defined in Table 8-3

$\ell_{12}$ empirical constant in Sec. 8-10

$m$ defined after Eq. (8-14.2); defined by Eq. (8-12.13)

$m_i$ molality in Sec. 8-16

$n_i$ number of moles of component $i$

$n_T$ total number of moles

$N$ number of components; parameter in Tables 8-17 and 8-18

$p$ partial pressure

$P$ total pressure

$P_{vp}$ vapor pressure

POL defined by Eq. (8-10.33)

$q$ molecular surface parameter, an empirical constant (Table 8-3); induction parameter in Eq. (8-10.27)

$Q_k$ group surface parameter, Eq. (8-10.53)

$r$ molecular-size parameter, an empirical constant (Table 8-3); number of rings (Table 8-18)

$R$ gas constant

$R_k$ group size parameter, Eq. (8-10.53)

$\mathcal{R}$ defined by Eq. (8-10.45)

RI index of refraction (Table 8-19); also known as $n_D$ in Eqs. (8-10.35 and 8-10.36)

$s^E$ molar excess entropy

$\Delta s_f$ molar entropy of fusion

$s_j$ number of size groups in molecule $j$, Eq. (8-10.45)

$t$ temperature

$T$ absolute temperature

$T_m$ melting point temperature

$T_t$ triple-point temperature

$u_{ij}$ empirical constant (Table 8-3)

$\Delta U$ change in internal energy

$V$ molar volume

$V_T$ total volume

$w_k$ weight fraction of component $k$

$x_i$ liquid phase mole fraction of component $i$

$X_k$ group mole fraction for group $k$

$y_i$ vapor phase mole fraction of component $i$

$z$ coordination number (Table 8-3); compressibility factor in Sec. 8-12

$z_i$ overall mole fraction in Sec. 8-12; charge on chemical species $i$ in Eq. (8-16.12)

$Z$ compressibility factor

GREEK

$\alpha$ parameter in Tables 8-17 and 8-18; acidity parameter in Eq. (8-10.27)

$\alpha_{ij}$ empirical constant

$\beta$ proportionality factor in Eq. (8-5.12); basicity parameter in Eq. (8-10.27)

$\gamma_i$ activity coefficient of component $i$

activity coefficient of group $k$ in Eq. (8-10.47)

$\delta$ solubility parameter defined by Eq. (8-10.10); also parameter in Eqs. (8-12.25) and (8-12.30)

$\bar{\delta}$ average solubility parameter defined by Eq. (8-10.15)

$\epsilon$ parameter in Tables 8-17 and 8-18

$\zeta$ parameter in Tables 8-17 and 8-18

$\eta$ empirical constant (Table 8-14) and Eq. (8-9.23); empirical constant in Tables 8-17 and 8-18

$\theta$ parameter in Tables 8-17 and 8-18

$\Theta_i$ surface fraction of component $i$ (Table 8-3)

$\lambda$ nonpolar solubility parameter or dispersion parameter in Sec. 8-10

$\lambda_{ij}$ empirical constant in Eq. (8-5.11) and Table 8-14

$\Lambda_{ij}$ empirical constant (Table 8-3)

$\nu_k^{(i)}$ number of groups of type $k$ in molecule $i$

$\nu_{kj}$ number of interaction groups $k$ in molecule $j$ [Eq. (8-10.46)]

$\xi$ hydrogen-bonding asymmetry parameter in Eq. (8-10.27)

$\rho$ density

$\sigma^2$ variance, Eqs. (8-8.16) and (8-8.17)

$\tau$ polar solubility or polar parameter in Eqs. (8-10.17) and (8.10.18)

$\tau_{ij}$ empirical constant (Table 8-3)

$\phi_i$ fugacity coefficient of component $i$

$\Phi_i$ site fraction (or volume fraction) of component $i$

$\chi$ Flory interaction parameter

$\Psi_{mn}$ group interaction parameter, Eq. (8-10.57)

$\psi$ polarity asymmetry parameter in Eq. (8-10.27)

$\psi_{12}$ binary (induction) parameter in Eq. (8-10.19)

$\omega$ acentric factor

$\Omega$ coefficient in Eq. (8-12.21); weight fraction activity coefficient in Eq. (8-14.4)

SUPERSCRIPTS

| | | | |
|---|---|---|---|
| $c$ | consolute (Sec. 8-13); calculated quantity, Eqs. (8-18.15) and (8.12.17) | $\circ$ | standard state as in $f_i^\circ$ |
| $C$ | configurational | $R$ | residual |
| $e$ | experimental, Eq. (8-12.17) | $\mathcal{S}$ | solid phase |
| $E$ | excess | $s$ | saturation |
| $G$ | group (Sec. 8-10) | $S$ | size (Sec. 8-10) |
| $L$ | liquid phase | $V$ | vapor phase |
| $M$ | measured value, Eq. (8-8.16) | 0 | estimated true value, Eq. (8-8.16) |
| | | $\infty$ | infinite dilution |

## REFERENCES


1. Abello, L., B. Servais, M. Kern, and G. Pannetier: *Bull Soc. Chim. France,* **11:** 4360 (1968).
2. Abbott, M. M., J. K. Floess, G. E. Walsh, and H. C. Van Ness: *AIChE J.,* **21:** 72 (1975).
3. Abbott, M. M., and H. C. Van Ness: *AIChE J.,* **21:** 62 (1975).
4. Abrams, D. S., and J. M. Prausnitz: *AIChE J.,* **21:** 116 (1975); *Joint Mtg. VDI AIChE, Munich, September 1974.*
5. Adler, S. B.: *Hydrocarbon Process. Intern. Ed.,* **62**(5): 109, **62**(6): 93 (1983).
6. Anderson, T. F., and J. M. Prausnitz: *Ind. Eng. Chem. Process Design Develop.,* **17:** 561 (1978).
7. Anderson, T. F., D. S. Abrams, and E. A. Grens: *AIChE J.,* **21:** 116 (1978).
8. Asselineau, L., G. Bogdanic, and J. Vidal: *Fluid Phase Equil.,* **3:** 273 (1980).
9. Bandrup, J., and E. H. Immergut: *Polymer Handbook,* 2d ed., Wiley, New York, 1975.
10. Barton, A. F. M.: *Handbook of Solubility Parameters and Other Cohesion Parameters,* CRC Press, Boca Raton, Fla., 1983.
11. Battino, R.: *Fluid Phase Equil.* **15:** 231 (1984).
12. Battino, R., and E. Wilhelm: *J. Chem. Thermodyn.,* **3:** 379 (1971); L. R. Field, E. Wilhelm, and R. Battino: ibid., **6:** 237 (1974).
13. Bazúa E. R., and J. M. Prausnitz: *Cryogenics,* **11:** 114 (1971).
14. Beegle, B. L., M. Modell, and R. C. Reid: *AIChE J.,* **20:** 1200 (1974).
15. Bender, E., and U. Block: *Verfahrenstechnik,* **9:** 106 (1975).
16. Bondi, A.: *Physical Properties of Molecular Liquids, Crystals and Glasses,* Wiley, New York, 1968.
17. Bonner, D. C. and J. M. Prausnitz: *AIChE J.,* **19:** 943 (1973); errata, **20:** 206 (1974).
18. Boone, J. E., R. W. Rousseau, E. M. Schoenborn: *Advan. Chem. Ser.* **155:** 36–52 (1976).
19. Boston, J. F., and P. M. Mathias: "Phase Equilibria in a Third Generation Process Simulator," *2d Intern. Conf. on Phase Equil. and Fluid Prop. in the Chemical Industry, Berlin (West), March 1980,* pub. by DECHEMA.
20. Brelvi, S. W., and J. P. O'Connell: *AIChE J.,* **18:** 1239 (1972); **21** 157 (1975).
21. Brian, P. L. T.: *Ind. Eng. Chem. Fundam.,* **4:** 101 (1965).
22. Bromley, L. A.: *AIChE* J., **19:** 313 (1973); *J. Chem. Thermodyn.,* **4:** 669 (1972).
23. Bruin, S., and J. M. Prausnitz: *Ind. Eng. Chem. Process Design Develop.,* **10:** 562 (1971); S. Bruin: *Ind. Eng. Chem. Fundam.,* **9:** 305 (1970).
24. Buck, E.: *CHEMTECH.,* September 1984: 570.
25. Burrell, H.: *J. Paint Technol.,* **40:** 197 (1968); J. L. Gardon: ibid., **38:** 43 (1966); R. C. Nelson, R. W. Hemwall, and G. D. Edwards: ibid., **42:** 636 (1970); *Encyclopedia of Chemical Technology* (Kirk-Othmer), 2d ed., vol. 18, pp. 564–588.

26. Cheung, H., and E. H. Zander: *Chem. Eng. Progr. Symp. Ser.*, **64**(88): 34 (1968).
27. Clarke, H. A., and R. W. Missen: *J. Chem. Eng. Data,* **19:** 343 (1974).
28. Coward, I., S. E. Gayle, and D. R. Webb: *Trans. Inst. Chem. Engrs. (London),* **56:** 19 (1978).
29. Cruz, J. L., and H. Renon: *AIChE J.,* **24:** 817 (1978).
30. Derr, E. L., and C. H. Deal: *Inst. Chem. Eng. Symp. Ser. London,* **3**(32): 40 (1969).
31. Eckert, C. A., B. A. Newman, G. L. Nicolaides, and T. C. Long: paper presented at *AIChE Mtg., Los Angeles, November 1975.*
32. Edwards, T. J., G. Maurer, J. Newman, and J. M. Prausnitz: *AIChE J.,* **24:** 966 (1978).
33. Fabries, J. F., and H. Renon: *AIChE J.,* **21:** 735 (1975).
34. Flory, P. J.: *Principles of Polymer Chemistry,* Cornell University Press, Ithaca, N.Y., 1953.
35. Flory, P. J.: *Discussions Faraday Soc.,* **49:** 7 (1970).
36. Fredenslund, A., R. L. Jones, and J. M. Prausnitz: *AIChE J.,* **21:** 1086 (1975); A. Fredenslund, J. Gmehling, and P. Rasmusen: *Vapor-Liquid Equilibria Using UNIFAC,* Elsevier, Amsterdam, 1977.
37. Friend, L., and S. B. Adler: *Chem. Eng. Progr.,* **53:** 452 (1957).
38. Funk, E. W., and J. M. Prausnitz: *Ind. Eng. Chem.,* **62**(9): 8 (1970).
39. Furnas, C. C., and W. B. Leighton: *Ind. Eng. Chem.,* **29:** 709 (1937).
40. Furter, W. F. (ed.): *Advan. Chem. Ser.* **177:** (1979).
41. Gmehling, J., T. F. Anderson, and J. M. Prausnitz: *Ind. Eng. Chem. Fundam.,* **17:** 269 (1978).
42. Gmehling, J., D. D. Liu, and J. M. Prausnitz: *Chem. Eng. Sci.,* **34:** 951 (1979).
43. Griswold, J., P. L. Chu, and W. O. Winsauer: *Ind. Eng. Chem.,* **41:** 2352 (1949).
44. Guillet, J. E.: *Advan. Anal. Chem. Instrum.,* **11:** 187 (1973).
45. Gundersen, T.: *Computers Chem. Eng.,* **6:** 245 (1982).
46. Hala, E.: *Fluid Phase Equil.,* **13:** 311 (1983).
47. Hansen, C. M.: *J. Paint Technol.,* **39:** 104, 505 (1967); C. M. Hansen and K. Skaarup: ibid., **39:** 511 (1967); C. M. Hansen and A. Beerbower: "Solubility Parameters," in H. F. Mark, J. J. McKetta, and D. F. Othmer (eds.), *Encylopedia of Chemical Technology,* 2d ed., suppl. vol., Interscience, New York, 1971.
48. Hayden, J. G., and J. P. O'Connell: *Ind. Eng. Chem. Process Design Develop.,* **14:** 3 (1975).
49. Hayduk, W., and S. C. Cheng: *Can. J. Chem. Eng.,* **48:** 93 (1970); W. Hayduk and W. D. Buckley: ibid., **49:** 667 (1971); W. Hayduk and H. Laudie: *AIChE J.,* **19:** 1233 (1973).
50. Heidemann, R. A.: *Fluid Phase Equil.,* **14:** 55 (1983).
51. Heidemann, R. A., and J. M. Prausnitz: *Proc. Natl. Acad. Sci. USA,* **73:** 1773 (1976).
52. Heidemann, R. A., and J. M. Mandhane: *Chem. Eng. Sci.,* **28:** 1213 (1973); T. Katayama, M. Kato, and M. Yasuda: *J. Chem. Eng. Japan,* **6:** 357 (1973); A. C. Mattelin and L. A. J. Verhoeye: *Chem. Eng. Sci.,* **30:** 193 (1975).
53. Heil, J. F., and J. M. Prausnitz: *AIChE J.,* **12:** 678 (1966).
54. Hildebrand, J. H., and R. L. Scott: *Regular Solutions,* Prentice-Hall, Englewood Cliffs, N.J., 1962.
55. Hiranuma, M., and K. Honma: *Ind. Eng. Chem. Process Design Develop.,* **14:** 221 (1975).
56. Horsley, L. H.: *"Azeotropic Data," Advan. Chem. Ser.,* **6:** (1952), **35:** (1962), **116:** (1973).
57. Hsu, C. C., and J. M. Prausnitz: *Macromolecules,* **7:** 320 (1974).
58. Hu, Y., E. Azevedo, D. Lüdecke, and J. M. Prausnitz: *Fluid Phase Equil.,* **17:** 303 (1984).
59. Jones, C. A., A. P. Colburn, and E. M. Schoenborn: *Ind. Eng. Chem.,* **35:** 666 (1943).
60. King, C. J.,: *Separation Processes,* 2d ed., McGraw-Hill, New York, 1980, chap. 2.
61. Knapp, H., R. Döring, L. Oellrich, U. Plöcker, and J. M. Prausnitz: *Chemistry Data Series,* Vol. VI: *VLE for Mixtures of Low Boiling Substances,* D. Behrens and R. Eckerman (eds.), DECHEMA, Frankfurt a. M., 1982.
62. Kobayashi, R., P. S. Chappelear, and H. A. Deans: *Ind. Eng. Chem.,* **59:** 63 (1967).
63. Krause, S.: *J. Macromol. Sci. Rev. Macromol. Chem.,* **C7:** 251 (1972).

64. Krishnan, T. R., H. Kabra, and D. B. Robinson: *J. Chem. Eng. Data,* **22:** 282 (1977).
65. Ladurelli, A. J., C. H. Eon, and G. Guiochon: *Ind. Eng. Chem. Fundam.,* **14:** 191 (1975).
66. Laurence, D. A., and G. W. Swift: *J. Chem. Eng. Data,* **17:** 333 (1972).
67. Lee, B. I., and Kesler, M. G.: *AIChE J.,* **21:** 510 (1975); errata, 21: 1040 (1975).
68. Lewis, G. N., M. Randall, K. S. Pitzer, and L. Brewer: *Thermodynamics,* 2d ed., McGraw-Hill, New York, 1961.
69. Liu, D. D.,and J. M. Prausnitz: *Ind. Eng. Chem. Process Design Develop.,* **19:** 205 (1980).
70. Lo, T. C., H. H. Bierber, and A. E. Karr: *J. Chem. Eng. Data,* **7:** 327 (1962).
71. Lyckman, E. W., C. A. Eckert, and J. M. Prausnitz: *Chem. Eng. Sci.,* **20:** 685 (1965).
72. Maron, S. H., and N. Nakajima: *J. Polymer Sci.,* **40:** 59 (1959); S. H. Maron: ibid., **38:** 329 (1959).
73. Martin, R. A., and K. L. Hoy: *Tables of Solubility Parameters,* Union Carbide Corp., Chemicals and Plastics, Research and Development Dept., Tarrytown, N.Y., 1975.
74. Mathias, P. M., and T. W. Copeman: *Fluid Phase Equil.,* **13:** 91 (1983).
75. Meissner, H. P., and C. L. Kusik: *AIChE J.,* **18:** 294 (1972); H. P. Meissner and J. W. Tester: *Ind. Eng. Chem. Process Design Develop.,* **11:** 128 (1972); *AIChE J.,* **18:** 661 (1972).
76. Mertl, I.: *Coll. Czech. Chem. Commun.,* **37:** 366 (1972).
77. Michelsen, M. L.: *Fluid Phase Equil.,* **4:** 1 (1980).
78. Michelsen, M. L.: *Fluid Phase Equil.,* **9:** 1 (1982); **9:** 21 (1982).
79. Modell, M., and R. C. Reid: *Thermodynamics and Its Applications,* 2d ed., Prentice-Hall, Englewood Cliffs, N.J., 1983.
80. Mollerup, J.: *Advan. Cryog. Eng.,* **20:** 172 (1975); *Fluid Phase Equil.,* **4:** (1980).
81. Mollerup, J.: *Fluid Phase Equil.,* **7:** 121 (1981); **15:** 189 (1983).
82. Murti, P. S., and M. van Winkle: *Chem. Eng. Data Ser.,* **3:** 72 (1978).
83. Nakamura, R., G. J. F. Breedveld, and J. M. Prausnitz: *Ind. Eng. Chem. Process Design* **15:** 557 (1976).
84. Newman, R. D., and J. M. Prausnitz: *J. Paint Technol.,* **43:** 33 (1973).
85. Null, H. R., and D. A. Palmer: *Chem. Eng. Progr.,* **65:** 47 (1969); H. R. Null: *Phase Equilibrium in Process Design,* Wiley, New York, 1970.
86. O'Connell, J. P.: *AIChE J.,* **12:** 658 (1971).
87. Palmer, D. A.: *Chem. Eng.,* June 9, 1975, p. 80.
88. Palmer, D. A., and B. D. Smith: *J. Chem. Eng. Data,* **17:** 71 (1972).

88*a*. Panagiotopoulos, A. Z., and S. K. Kumar: *Fluid Phase Equil.,* **22:** 77 (1985).

88*b*. Panagiotopoulos, A. Z., and R. C. Reid: *Fluid Phase Equil.,* **29:** 525 (1986).

89. Patterson, D.: *Macromolecules,* **2:** 672 (1969).
90. Peng, D. Y., and D. B. Robinson: *Ind. Eng. Chem. Fundam.,* **15:** 59 (1976).
91. Pierotti, G. J., C. H. Deal, and E. L. Derr: *Ind. Eng. Chem.,* **51:** 95 (1959).
92. Pitzer, K. S.: *J. Phys. Chem.,* **77:** 268 (1973).
93. Pitzer, K. S.: *Accounts Chem. Res.,* **10:** 371 (1977).
94. Pitzer, K. S.: *J. Am. Chem. Soc.,* **102:** 2902 (1980).
95. Plöcker, U., H. Knapp, and J. M. Prausnitz: *Ind. Eng. Chem. Process Design Develop.,* **17:** 324 (1978).
96. Poling, B. E., E. A. Grens II, and J. M. Prausnitz: *Ind. Eng. Chem. Process Design Develop.,* **20:** 127 (1981).
97. Prausnitz, J. M., R. N. Lichtenthaler, and E. G. Azevedo: *Molecular Thermodynamics of Fluid-Phase Equilibria,* 2d ed., Prentice-Hall, Englewood Cliffs, N.J., 1986.
98. Prausnitz, J. M., and P. L. Chueh: *Computer Calculation for High-Pressure Vapor-Liquid Equilibria,* Prentice-Hall, Englewood Cliffs, N.J., 1968.
99. Prausnitz, J. M., T. A. Anderson, E. A. Grens, C. A. Eckert, R. Hsieh, and J. P. O'Connell: *Computer Calculations for Multicomponent Vapor-Liquid and Liquid-Liquid Equilibria,* Prentice-Hall, Englewood Cliffs, N.J., 1980; H. Renon, L. Asselineau, G. Cohen, and C. Raimbault: *Calcul sur ordinateur des équilibres liquide-vapeur et liquide-liquide,* Editions Technip, Paris, 1971.
100. Prausnitz, J. M., and F. H. Shair: *AIChE J.,* **7:** 682 (1961).

101. Preston, G. T., and J. M. Prausnitz: *Ind. Eng. Chem. Process Design Develop.*, **9:** 264 (1970).
102. Prigogine, I.: *The Molecular Theory of Solutions,* North-Holland Publishing, Amsterdam, 1957.
103. Prigogine, I., and R. Defay: *Chemical Thermodynamics,* Longmans, London, 1954.
104. Redlich, O., E. L. Derr, and G. Pierotti: *J. Am. Chem. Soc.*, **81:** 2283 (1959); E. L. Derr and M. Papadopoulous: ibid., **81:** 2285 (1959).
105. Renon, H., and J. M. Prausnitz: *Ind. Eng. Chem. Process Design Develop.*, **8:** 413 (1969).
106. Renon, H., and J. M. Prausnitz: *AIChE J.*, **15:** 785 (1969).
107. Robinson, R. A., and R. H. Stokes: *Electrolyte Solutions,* 2d ed., Academic, New York, 1959.
108. Scatchard, G., S. E. Wood, and J. M. Mochel: *J. Am. Chem. Soc.*, **68:** 1957 (1946).
109. Schreiber, L. B., and C. A. Eckert: *Ind. Eng. Chem. Process Desgin Develop.*, **10:** 572 (1971).
110. Scott, R. L.: *J. Chem. Phys.*, **17:** 279 (1949); H. Tompa: *Trans. Faraday Soc.*, **45:** 1142 (1949).
111. Shain, S. A., and J. M. Prausnitz: *Chem. Eng. Sci.*, **18:** 244 (1963).
112. Sheehan, C. J., and A. L. Bisio: *Rubber Chem. Technol.*, **39:** 149 (1966).
113. Sinor, J. E., and J. H. Weber: *J. Chem. Eng. Data,* **5:** 243 (1960).
114. Soave, G.: *Chem. Eng. Sci.*, **27:** 1197 (1972).
115. Starling, K. E.: *Fluid Thermodynamic Properties of Light Petroleum System,* Gulf Publishing Co., 1973.
116. Steele, K., B. E. Poling, and D. B. Manley: paper presented at *AIChE Mtg., Washington, D.C., December 1974.*
117. Tapavicza, S., and J. M. Prausnitz: *Chem. Ing. Tech.*, **47:** 552 (1975); English translation in *Intern. Chem. Eng.*, **16**(2): 329 (April 1976).
118. Tassios, D.: *AIChe J.*, **17:** 1367 (1971).
Thomas, E. R., and C. A. Eckert: *Ind. Eng. Chem. Process Design Develop.*, **23:** 194 (1984).
120. Tiepel, E. W., and K. E. Gubbins: *Ind. Eng. Chem. Fundam.*, **12:** 18 (1973); *Can. J. Chem. Eng.*, **50:** 361 (1972).
121. Tompa, H.: *Polymer Solutions,* Butterworth, London, 1956.
122. Topliss, R. J.: Ph.D. dissertation, University of California, Berkeley, 1985.
123. Treybal, R. E.: *Liquid Extraction,* 2d ed., McGraw-Hill, New York, 1963.
124. Tsonopoulos, C., and G. M. Wilson: *AIChE J.*, **29:** 990 (1983).
125. Turek, E. A., R. S. Metcalfe, L. Yarborough, and R. L. Robinson: *Soc. Petrol. Engrs. J.*, **24:** 308 (1984).
126. Udovenko, V. V., and T. B. Frid: *Zh. Fiz. Khim.*, **22:** 1263 (1948).
127. Vidal, J.: *Chem. Eng. Sci.*, **31:** 1077 (1978); *Fluid Phase Equil.*, **13:** 15 (1983).
128. Weimer, R. F., and J. M. Prausnitz: *Hydrocarbon Process. Petrol. Refiner,* **44:** 237 (1965).
129. Whiting, W. B., and J. M. Prausnitz: *Fluid Phase Equil.*, **9:** 119 (1982). For a more recent version, see D. Luedecke, and J. M. Prausnitz, *Fluid Phase Equil.*, **22:** 1 (1985).
130. Wilson, G. M.: *J. Am. Chem. Soc.*, **86:** 127, 133 (1964).
131. Wilson, G. M., and C. H. Deal: *Ind. Eng. Chem. Fundam.*, **1:** 20 (1962).
132. Wohl, K.: *Chem. Eng. Progr.*, **49:** 218 (1953).
133. Wohl, K.: *Trans. AIChE,* **42:** 215 (1946).
134. Won, K. W.: *Fluid Phase Equil.*, **10:** 191 (1983).
135. Wong, K. F., and C. A. Eckert: *Ind. Eng. Chem. Fundam.*, **10:** 20 (1971).
136. Yen, L., and J. J. McKetta: *AIChE J.*, **8:** 501 (1962).
137. Yodovich, A., R. L. Robinson, and K. C. Chao: *AIChE J.*, **17:** 1152 (1971).
138. Young, C. L.: *Chromatog. Rev.*, **10:** 129 (1968).
139. Ziervogel, R. G., and B. E. Poling, *Fluid Phase Equil.*, **11:** 127 (1983).

# Chapter 9

# Viscosity

## 9-1 Scope

The first part of this chapter deals with the viscosity of gases and the second with the viscosity of liquids. In each part, methods are recommended for (1) correlating viscosities with temperature, (2) estimating viscosities when no experimental data are available, (3) estimating the effect of pressure on viscosity, and (4) estimating the viscosities of mixtures. The molecular theory of viscosity is considered briefly.

## 9-2 Definition and Units of Viscosity

If a shearing stress is applied to any portion of a confined fluid, the fluid will move and a velocity gradient will be set up within it with a maximum velocity at the point where the stress is applied. If the shear stress per unit area at any point is divided by the velocity gradient, the ratio obtained is defined as the viscosity of the medium. It can be seen, therefore, that viscosity is a measure of the internal fluid friction, which tends to oppose any dynamic change in the fluid motion; i.e., if the friction between layers of fluid is small (low viscosity), an applied shearing force will result in a large velocity gradient. As the viscosity increases, each fluid layer exerts a larger frictional drag on adjacent layers and the velocity gradient decreases.

It is to be noted that viscosity differs in one important respect from the properties discussed previously in this book; namely, viscosity is a nonequilibrium property on a macroscale. Density, for example, is an equilibrium property. On a microscale, both properties reflect the effect of molecular motion and interaction. Even though viscosity is ordinarily referred to as a nonequilibrium property, it is, like temperature, pressure, and volume, a function of the state of the fluid, and it may be used to define the state of the material.† Brulé and Starling [30] have emphasized the desirability of using both viscosity and thermodynamic data to characterize complex fluids and to develop correlations.

The mechanism or theory of gas viscosity has been reasonably well clarified by the application of the kinetic theory of gases, but the theory of liquid viscosity is poorly developed. Brief summaries of both theories will be presented.

Since viscosity is defined as a shearing stress per unit area divided by a velocity gradient, it should have the dimensions of (force)(time)/(length)$^2$ or mass/(length)(time). Both dimensional groups are used, although for most scientific work, viscosities are expressed in poises, centipoises, micropoises, etc. A poise (P) denotes a viscosity of 0.1 N·s/m$^2$ and 1.0 cP = 0.01 P. The following conversion factors apply to the viscosity units:

$$\begin{aligned} 1\ \text{P} &= 1.000 \times 10^2\ \text{cP} = 1.000 \times 10^6\ \mu\text{P} = 0.1\ \text{N s/m}^2 \\ &= 6.72 \times 10^{-2}\ \text{lb-mass/(ft}\cdot\text{s)} = 242\ \text{lb-mass/(ft}\cdot\text{h)} \end{aligned}$$

The *kinematic viscosity* is the ratio of the viscosity to the density. With viscosity in poises and the density of grams per cubic centimeter, the unit of kinematic viscosity is the *stoke*, with the units square centimeters per second. In the SI system of units, viscosities are expressed in N·s/m$^2$ (or Pa·s) and kinematic viscosities in either m$^2$/s or cm$^2$/s.

## 9-3 Theory of Gas Transport Properties

The theory of gas transport properties is simply stated, but it is quite complex to express in equations which can be used directly to calculate viscosities. In simple terms, when a gas undergoes a shearing stress so that there is some bulk motion, the molecules at any one point have the bulk velocity vector added to their own random velocity vector. Molecular collisions cause an interchange of momentum throughout the fluid, and this bulk motion velocity (or momentum) becomes distributed. Near the source of the applied stress, the bulk velocity vector is high, but as the

†This discussion is limited to Newtonian fluids, i.e., fluids in which the viscosity, as defined, is independent of either the magnitude of the shearing stress or velocity gradient (rate of shear).

molecules move away from the source, they are "slowed down" (in the direction of bulk flow), which causes the other sections of the fluid to move in that direction. This random, molecular momentum interchange is the predominant cause of gaseous viscosity.

### Elementary kinetic theory

If the gas is modeled in the simplest manner, it is possible to show the general relations among viscosity, temperature, pressure, and molecular size. More rigorous treatments will yield similar relations which contain important correction factors. The elementary gas model assumes all molecules to be noninteracting rigid spheres of diameter $\sigma$ (with mass $m$) moving randomly at a mean velocity $\upsilon$. The density is $n$ molecules in a unit volume. Molecules move in the gas and collide, and they may transfer momentum or energy if there are velocity or temperature gradients. Such processes also result in a transfer of molecular species if a concentration gradient exists. The net flux of momentum, energy, or component mass between two layers is assumed proportional to the momentum, energy, or mass density gradient, i.e.,

$$\text{Flux} \propto -\frac{d\rho'}{dz} \tag{9-3.1}$$

where the density $\rho'$ decreases in the $+z$ direction and $\rho'$ may be $\rho_i$ (mass density), $nm\upsilon_y$ (momentum density), or $C_\upsilon nT$ (energy density). The coefficient of proportionality for all these fluxes is given by elementary kinetic theory as $\upsilon L/3$, where $\upsilon$ is the average molecular speed and $L$ is the mean free path.

Equation (9-3.1) is also used to define the transport coefficients of diffusivity $D$, viscosity $\eta$, and thermal conductivity $\lambda$; that is,

$$\text{Mass flux} = -Dm\frac{dn_i}{dz} = -\frac{\upsilon L}{3}\frac{d\rho_i}{dz} \tag{9-3.2}$$

$$\text{Momentum flux} = -\eta\frac{d\upsilon_y}{dz} = -\frac{\upsilon L}{3}mn\frac{d\upsilon_y}{dz} \tag{9-3.3}$$

$$\text{Energy flux} = -\lambda\frac{dT}{dz} = -\frac{\upsilon L}{3}C_\upsilon n\frac{dT}{dz} \tag{9-3.4}$$

Equations (9-3.2) to (9-3.4) define the transport coefficients $D$, $\eta$, and $\lambda$. If the average speed is proportional to $(RT/M)^{1/2}$ and the mean free path to $(n\sigma^2)^{-1}$,

$$D = \frac{\upsilon L}{3} = (\text{const})\frac{T^{3/2}}{M^{1/2}P\sigma^2} \tag{9-3.5}$$

$$\eta = \frac{m\rho v L}{3} = (\text{const}) \frac{T^{1/2}M^{1/2}}{\sigma^2} \tag{9-3.6}$$

$$\lambda = \frac{vLC_v n}{3} = (\text{const}) \frac{T^{1/2}}{M^{1/2}\sigma^2} \tag{9-3.7}$$

The constant multipliers in Eqs. (9-3.5) to (9.3-7) are different in each case; the interesting fact to note from these results is the dependency of the various transfer coefficients on $T$, $P$, $M$, and $\sigma$. A similar treatment for rigid, noninteracting spheres having a Maxwellian velocity distribution yields the same final equations but with slightly different numerical constants.

The viscosity relation [Eq. (9-3.6)] for a rigid, noninteracting sphere model is

$$\eta = 26.69 \frac{(MT)^{1/2}}{\sigma^2} \tag{9-3.8}$$

where $\eta$ = viscosity, $\mu$P
$M$ = molecular weight, g/mol
$T$ = temperature, K
$\sigma$ = hard-sphere diameter, Å

Analogous equations for $\lambda$ and $D$ are given in Chaps. 10 and 11.

### Effect of intermolecular forces

If the molecules attract or repel one another by virtue of intermolecular forces, the theory of Chapman and Enskog is normally employed [40, 100]. There are four important assumptions in this development: (1) the gas is sufficiently dilute for only binary collisions to occur, (2) the motion of the molecules during a collision can be described by classical mechanics, (3) only elastic collisions occur, and (4) the intermolecular forces act only between fixed centers of the molecules; i.e., the intermolecular potential function is spherically symmetric. With these restrictions, it would appear that the resulting theory should be applicable only to low-pressure, high-temperature monatomic gases. The pressure and temperature restrictions are valid; but for lack of tractable, alternate models, it is very often applied to polyatomic gases except in the case of thermal conductivity. Then a correction for internal energy transfer and storage must be included (see Chap. 10).

The Chapman-Enskog treatment considers in detail the interactions between colliding molecules with a potential energy $\psi(r)$ included. The equations are well known, but their solution is often very difficult. Each choice of an intermolecular potential $\psi(r)$ must be solved separately. In general terms, the solution for viscosity is written

$$\eta = \frac{(\frac{5}{16})(\pi MRT)^{1/2}}{\pi\sigma^2\Omega_v} = \frac{(26.69)(MT)^{1/2}}{\sigma^2\Omega_v} \tag{9-3.9}$$

which is identical to Eq. (9-3.8) except for the inclusion of the *collision integral* $\Omega_v$. $\Omega_v$ is unity if the molecules do not attract each other. Given a potential energy of interaction $\psi(r)$, $\Omega_v$ can be calculated; results from using the Lennard-Jones potential function are illustrated in Sec. 9-4.

## 9-4 Estimation of Low-Pressure Gas Viscosity

Essentially all gas viscosity estimation techniques are based on either the Chapman-Enskog theory or the law of corresponding states. Both approaches are discussed below, and recommendations are presented at the end of the section.

### Theoretical approach

The Chapman-Enskog viscosity equation was given as (9-3.9).† To use this relation to estimate viscosities, the collision diameter $\sigma$ and the collision integral $\Omega_v$ must be found. In the derivation of Eq. (9-3.9), $\Omega_v$ is obtained as a complex function of a dimensionless temperature $T^*$. The functionality depends upon the intermolecular potential chosen. As shown in Fig. 9-1, let $\psi(r)$ be the potential energy of interaction between two molecules separated by distance $r$. At large separation distances, $\psi(r)$ is negative; the molecules attract each other.‡ At small distances, repulsion occurs. The minimum in the $\psi(r)$ versus $r$ curve is termed the *characteristic energy* $\epsilon$. For any potential curve, the dimensionless temperature $T^*$ is related to $\epsilon$ by

$$T^* = \frac{kT}{\epsilon} \tag{9-4.1}$$

where $k$ is Boltzmann's constant. Referring again to Fig. 9-1, the collision diameter $\sigma$ is defined as the separation distance when $\psi(r) = 0$.

The relation between $\psi(r)$ and $r$ is called an *intermolecular potential function*. Such a function written by using only the parameters $\epsilon$ and $\sigma$ is a two-parameter potential. The Lennard-Jones 12-6 potential given in Eq. (9-4.2) is an example of this type. Many other potential functions with

---

†A correction factor, which is essentially unity, was omitted in Eq. (9-3.9).

‡The negative gradient of $\psi(r)$ is the *force* of interaction.

different or additional parameters also have been proposed. The important element, however, is that one must know how $\psi(r)$ varies with $r$ in order to obtain $\Omega_v$ in Eq. (9-3.9). The working equation for $\eta$ must have as many parameters as were used to define the original $\psi(r)$ relation.

The Lennard-Jones 12-6 potential is

$$\psi(r) = 4\epsilon\left[\left(\frac{\sigma}{r}\right)^{12} - \left(\frac{\sigma}{r}\right)^{6}\right] \tag{9-4.2}$$

This relation is based upon rather tenuous theoretical grounds and has been widely criticized. However, it is one of the more tractable relations for $\psi(r)$; and since $\Omega_v$ is relatively insensitive to the exact form of the $\psi(r)$ relation, Eq. (9-4.2) has been extensively used.

With this potential, the collision integral has been determined by a number of investigators [13, 100, 114, 126, 127, 148]. Neufeld et al. [150] proposed an empirical equation which is convenient for computer application:

$$\Omega_v = [A(T^*)^{-B}] + C[\exp(-DT^*)] + E[\exp(-FT^*)] \tag{9-4.3}$$

where $T^* = kT/\epsilon$, $A = 1.16145$, $B = 0.14874$, $C = 0.52487$, $D = 0.77320$, $E = 2.16178$, and $F = 2.43787$. Equation (9-4.3) is applicable from $0.3 \leq T^* \leq 100$ with an average deviation of only 0.064 percent. A graph of log $\Omega_v$ as a function of log $T^*$ is shown in Fig. 9-2.

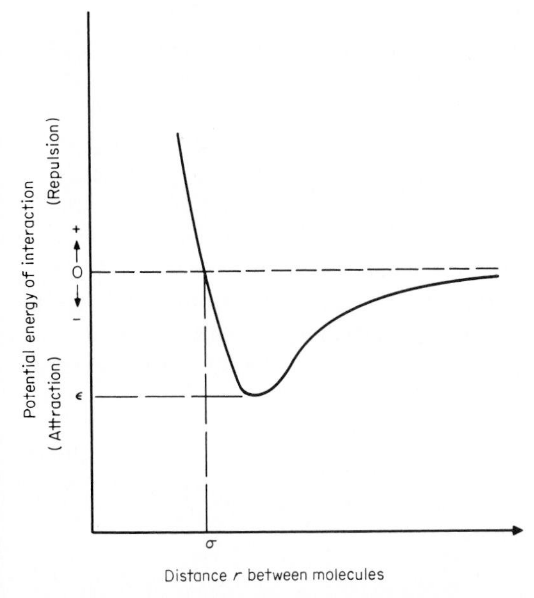


**Figure 9-1** Intermolecular potential function.

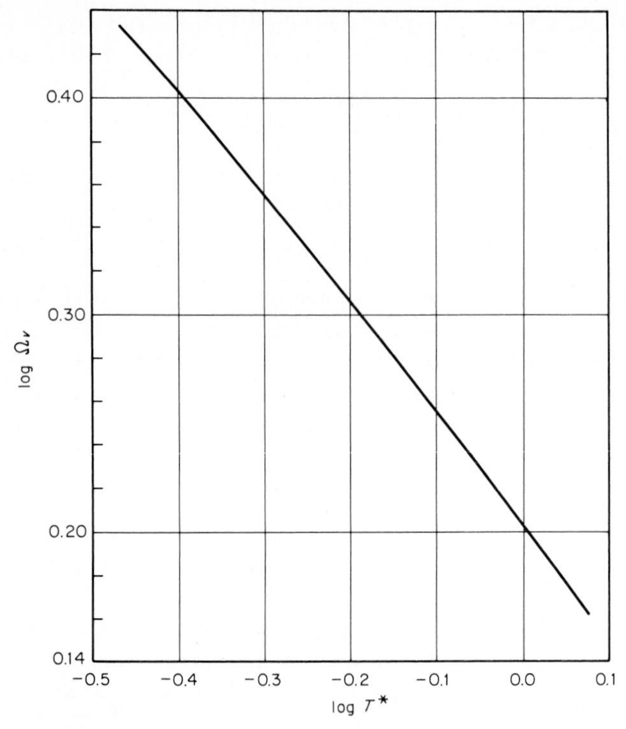


**Figure 9-2** Effect of temperature on the Lennard-Jones viscosity collision integral.

With values of $\Omega_v$ as a function of $T^*$, a number of investigators have used Eq. (9-3.9) and regressed experimental viscosity-temperature data to find the best values of $\epsilon/k$ and $\sigma$ for many substances. Appendix B lists a number of such sets as reported by Svehla [194]. It should be noted, however, that there appears also to be a number of other quite satisfactory *sets* of $\epsilon/k$ and $\sigma$ for any given compound. For example, with $n$-butane, Svehla suggested $\epsilon/k = 513.4$ K, $\sigma = 4.730$ Å, whereas Flynn and Thodos [74] recommend $\epsilon/k = 208$ K and $\sigma = 5.869$ Å.

Both sets, when used to calculate viscosities, yield almost exactly the same values of viscosity as shown in Fig. 9-3. This interesting paradox has been resolved by Reichenberg [165], who suggested that log $\Omega_v$ is essentially a linear function of log $T^*$ (see Fig. 9-2).†

$$\Omega_v = a(T^*)^n \tag{9-4.4}$$

†Kim and Ross [125] do, in fact, propose that:

$$\log \Omega_v = 0.2052 - 0.5 \log T^*$$

where $0.4 < T^* < 1.4$. They note a maximum error of only 0.7 percent.

Equation (9-3.9) may then be written

$$\eta = 26.69 M^{1/2} T^{(0.5-n)} a^{-1} \frac{(\epsilon/k)^n}{\sigma^2} \ \mu\text{P} \tag{9-4.5}$$

Here the parameters $\sigma$ and $\epsilon/k$ are combined as a *single* term $(\epsilon/k)^n/\sigma^2$. There is then no way of delineating individual values of $\epsilon/k$ and $\sigma$ by using experimental viscosity data, at least over the range where Eq. (9-4.4) applies.

The conclusion to be drawn from this discussion is that Eq. (9-3.9) can be used to calculate gas viscosity, although the chosen set of $\epsilon/k$ and $\sigma$ may have little relation to molecular properties. There will be an infinite number of acceptable sets as long as the temperature range is not too broad, e.g., if one limits the estimation to the range of reduced temperatures from about 0.3 to 1.2. In using published values of $\epsilon/k$ and $\sigma$ for a fluid of interest, the two values from the same set must be used—never $\epsilon/k$ from one set and $\sigma$ from another.

The difficulty in obtaining a priori meaningful values of $\epsilon/k$ and $\sigma$ has led most authors to specify rules which relate $\epsilon/k$ and $\sigma$ to better characterized parameters such as the critical constants. One such method is shown below.

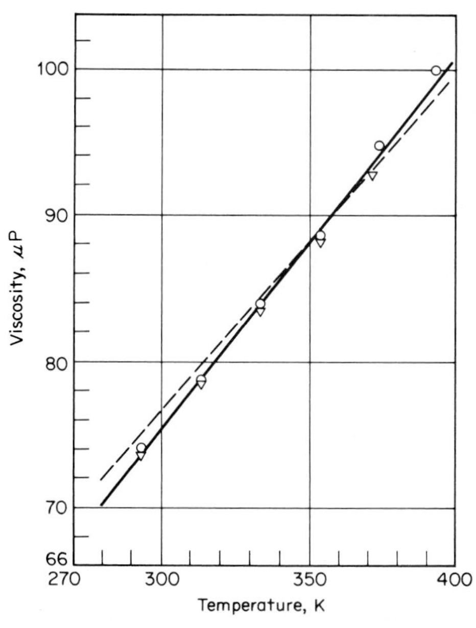


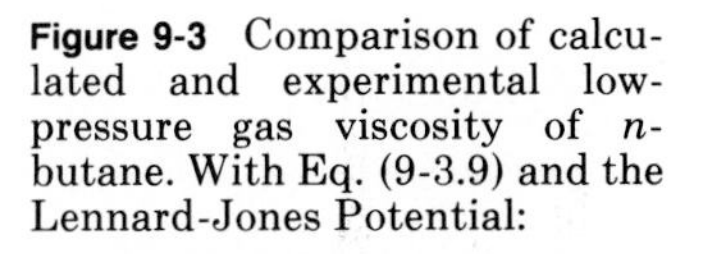

**Figure 9-3** Comparison of calculated and experimental low-pressure gas viscosity of $n$-butane. With Eq. (9-3.9) and the Lennard-Jones Potential:

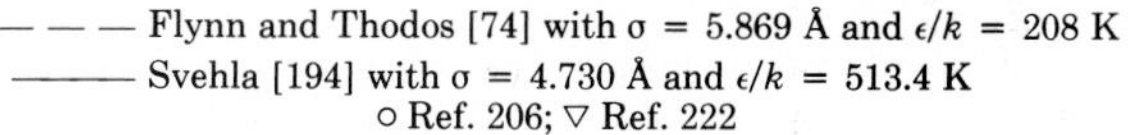

— — — Flynn and Thodos [74] with $\sigma$ = 5.869 Å and $\epsilon/k$ = 208 K
——— Svehla [194] with $\sigma$ = 4.730 Å and $\epsilon/k$ = 513.4 K
○ Ref. 206; ▽ Ref. 222

### Method of Chung et al. [44, 45]

These authors have employed Eq. (9-3.9) with

$$\frac{\epsilon}{k} = \frac{T_c}{1.2593} \tag{9-4.6}$$

$$\sigma = 0.809 V_c^{1/3} \tag{9-4.7}$$

where $\epsilon/k$ and $T_c$ are in kelvins, $\sigma$ is in angstroms, and $V_c$ is in cm$^3$/mol. Then, using Eqs. (9-4.1) and (9-4.6),

$$T^* = 1.2593 T_r \tag{9-4.8}$$

$\Omega_v$ in Eq. (9-3.9) is found from Eq. (9-4.3) with $T^*$ defined by Eq. (9-4.8). Chung et al. also multiply the right-hand side of Eq. (9-3.9) by a factor $F_c$ to account for molecular shapes and polarities of dilute gases. Their final result may be expressed as:

$$\eta = 40.785 \frac{F_c(MT)^{1/2}}{V_c^{2/3}\Omega_v} \tag{9-4.9}$$

where $\eta$ = viscosity, $\mu$P
$M$ = molecular weight, g/mol
$T$ = temperature, K
$V_c$ = critical volume, cm$^3$/mole
$\Omega_v$ = viscosity collision integral from Eq. (9-4.3) and $T^* = 1.2593 T_r$
$F_c = 1 - 0.2756\omega + 0.059035\mu_r^4 + \kappa$ (9-4.10)

In Eq. (9-4.10), $\omega$ is the acentric factor (see Sec. 2-3) and $\kappa$ is a special correction for highly polar substances such as alcohols and acids. Values of $\kappa$ for a few such materials are shown in Table 9-1.† The term $\mu_r$ is a dimensionless dipole moment.‡ When $V_c$ is in cm$^3$/mole, $T_c$ is in kelvins,

**TABLE 9-1 The Association Factor $\kappa$ in Eq. (9-4.10) [44]**

| Compound | $\kappa$ | Compound | $\kappa$ |
|---|---|---|---|
| Methanol | 0.215 | *n*-Pentanol | 0.122 |
| Ethanol | 0.175 | *n*-Hexanol | 0.114 |
| *n*-Propanol | 0.143 | *n*-Heptanol | 0.109 |
| *i*-Propanol | 0.143 | Acetic Acid | 0.0916 |
| *n*-Butanol | 0.132 | Water | 0.076 |
| *i*-Butanol | 0.132 | | |

†Chung et al. [45] suggest that for other alcohols not shown in Table 9-1,

$\kappa = 0.0682 + 0.2767$ [(17)(number of —OH groups)/molecular weight]

‡See the discussion under Eq. (9-4.16) for techniques to nondimensionalize a dipole moment.

and $\mu$ is in debyes,

$$\mu_r = 131.3 \frac{\mu}{(V_c T_c)^{1/2}} \tag{9-4.11}$$

**Example 9-1** Estimate the viscosity of sulfur dioxide gas at atmospheric pressure and 300°C by using the Chung et al. method. The experimental viscosity is 246 $\mu$P [129].

**solution** From Appendix A, $T_c$ = 430.8 K, $V_c$ = 122 cm$^3$/mole, $\omega$ = 0.256, $M$ = 64.063, and the dipole moment is 1.6 debyes. Assume $\kappa$ is negligible. From Eq. (9-4.11),

$$\mu_r = \frac{(131.3)(1.6)}{[(122)(430.8)]^{1/2}} = 0.916$$

and with Eq. (9-4.10),

$$F_c = 1 - (0.2756)(0.256) + (0.059035)(0.916)^4 = 0.971$$

$$T^* = 1.2593 \frac{300 + 273}{430.8} = 1.675$$

Then, with Eq. (9-4.3), $\Omega_v$ = 1.256. The viscosity is determined from Eq. (9-4.9).

$$\eta = (40.785)(0.971) \frac{[(64.063)(300 + 273)]^{1/2}}{(122)^{2/3}(1.256)} = 245.6\ \mu\text{P}$$

$$\text{Error} = \frac{245.6 - 246}{246} \times 100 = -0.2\%$$

Experimental viscosities and those estimated by the Chung et al. method are shown in Table 9-2. The average absolute error was about 1.9 percent. This agrees well with the more extensive comparison by Chung [43], who found an average absolute error of about 1.5 percent.

### Corresponding states methods

From an equation such as (9-3.9), if one associates $\sigma^3$ with $V_c$ [as in Eq. (9-4.7)] and assumes $V_c$ is proportional to $RT_c/P_c$, a dimensionless viscosity can be defined:

$$\eta_r = \xi\eta = f(T_r) \tag{9-4.12}$$

$$\xi = \left[\frac{(RT_c)(N_0)^2}{M^3 P_c^4}\right]^{1/6} \tag{9-4.13}$$

In SI units, if $R$ = 8314 J/(kmol·K) and $N_0$ (Avogadro's number) = 6.023 $\times 10^{26}$ (kmol)$^{-1}$ and with $T_c$ in kelvins, $M$ in kg/kmol, and $P_c$ in N/m$^2$, $\xi$ has the units of m$^2$/(N·s) or inverse viscosity. In more convenient units,

$$\xi = 0.176 \left(\frac{T_c}{M^3 P_c^4}\right)^{1/6} \tag{9-4.14}$$

**TABLE 9-2 Comparison between Calculated and Experimental Low-Pressure Gas Viscosities**

| | | | Percent error† | | |
|---|---|---|---|---|---|
| Compound | $T$, °C | Experimental value, $\mu P$‡ | Chung et al., Eq. (9-4.9) | Lucas, Eq. (9-4.15) | Reichenberg, Eq. (9-4.20) |
| Acetic acid | 150 | 118 | 3.4 | — | 1.7 |
| | 250 | 151 | 0.2 | — | 2.4 |
| Acetylene | 30 | 102 | 0.6 | 0.6 | 2.4 |
| | 101 | 126 | −0.6 | −0.8 | 0.7 |
| | 200 | 155 | −0.5 | −0.6 | 0.7 |
| Ammonia | 37 | 106 | 2.2 | 2.3 | — |
| | 147 | 146 | 0.4 | 0.2 | — |
| | 267 | 189 | −2.1 | −1.9 | — |
| Benzene | 28 | 73.2 | 1.0 | 3.2 | 4.4 |
| | 100 | 92..5 | 0.2 | 1.3 | 2.1 |
| | 200 | 117 | 0.5 | 1.6 | 1.8 |
| Bromotrifluoromethane | 17 | 145 | 7.6 | 8.3 | −0.3 |
| | 97 | 183 | 8.2 | 8.6 | −0.5 |
| Isobutane | 37 | 79 | −1.6 | 1.6 | 1.6 |
| | 155 | 105 | 2.0 | 4.7 | 4.0 |
| | 287 | 132 | 3.9 | 6.7 | 5.6 |
| $n$-Butane | 7 | 74 | −5.8 | −4.1 | −5.6 |
| | 127 | 101 | −1.0 | 0.1 | −2.3 |
| | 267 | 132 | 0.6 | 1.5 | −1.4 |
| 1-Butene | 20 | 76.1 | −1.0 | 1.1 | 1.8 |
| | 60 | 86.3 | −0.5 | 1.4 | 1.7 |
| | 120 | 102 | −0.6 | 0.9 | 0.9 |
| Carbon dioxide | 37 | 154 | −0.6 | 1.6 | — |
| | 127 | 194 | −0.1 | 2.2 | — |
| | 327 | 272 | 0.5 | 3.0 | — |
| Carbon disulfide | 30 | 100 | 0 | 5.9 | — |
| | 98.2 | 125 | −1.3 | 3.5 | — |
| | 200 | 161 | −1.9 | 2.2 | — |
| Carbon tetrachloride | 125 | 133 | 0.8 | 1.9 | −2.8 |
| | 200 | 156 | 2.4 | 3.1 | −2.1 |
| | 300 | 186 | 3.8 | 4.1 | −1.4 |
| Chlorine | 20 | 133 | 2.2 | 3.5 | — |
| | 100 | 168 | 3.3 | 4.1 | — |
| | 200 | 209 | 4.5 | 5.1 | — |
| Chloroform | 20 | 100 | −3.8 | 3.4 | −1.0 |
| | 100 | 125 | −1.3 | 5.4 | 0.6 |
| | 300 | 191 | −0.9 | 5.1 | −0.6 |
| Cyclohexane | 35 | 72.3 | −2.4 | −0.1 | −4.2 |
| | 100 | 87.3 | −1.4 | 0.3 | −4.1 |
| | 300 | 129 | 2.6 | 3.6 | −1.9 |
| Dimethyl ether | 20 | 90.9 | −5.2 | −1.0 | 2.0 |
| | 100 | 117 | −6.1 | −2.4 | 0.2 |

| Compound | $T$, °C | Experimental value, $\mu P$‡ | Percent error† Chung et al., Eq. (9-4.9) | Lucas, Eq. (9-4.15) | Reichenberg, Eq. (9-4.20) |
|---|---|---|---|---|---|
| Ethane | 47 | 100 | 0.1 | 0.7 | −3.2 |
| | 117 | 120 | 0.2 | 0.8 | −3.3 |
| | 247 | 156 | −1.0 | −0.3 | −4.7 |
| Ethyl acetate | 125 | 101 | −2.6 | 9.0 | −1.6 |
| | 200 | 120 | −2.4 | 8.8 | −2.0 |
| | 300 | 146 | −3.3 | 7.4 | −3.5 |
| Ethanol | 110 | 111 | −0.5 | −2.5 | 0.6 |
| | 197 | 137 | −0.8 | 1.6 | −0.4 |
| | 267 | 156 | −0.3 | 2.0 | −0.1 |
| Diethyl ether | 125 | 99.1 | −0.4 | 0.2 | 0.7 |
| | 200 | 118 | −0.8 | −0.5 | −0.3 |
| | 300 | 141 | −0.6 | −0.4 | −0.5 |
| $n$-Hexane | 107 | 81 | −1.1 | 0.9 | 0.3 |
| | 267 | 116 | −2.1 | −0.7 | −2.1 |
| Methane | −13 | 98 | −0.7 | −0.5 | — |
| | 147 | 147 | 0 | −0.9 | — |
| Methyl acetate | 125 | 108 | 0 | 10 | 1.7 |
| | 300 | 157 | −1.6 | 7.6 | −1.0 |
| Methanol | 67 | 112 | −0.4 | −0.9 | 1.1 |
| | 127 | 132 | −0.3 | −1.1 | 1.2 |
| | 277 | 181 | −0.3 | −1.7 | 0.9 |
| Methyl chloride | 50 | 119 | 3.9 | 0.8 | −0.6 |
| | 130 | 147 | 5.0 | 1.4 | 1.9 |
| Nitrogen | 27 | 178 | 0.3 | 0.3 | — |
| | 227 | 258 | 0.3 | 0.4 | — |
| Isopropanol | 157 | 113 | 0.3 | 5.9 | 3.5 |
| | 257 | 139 | 0.1 | 5.4 | 2.7 |
| Propylene | 17 | 83 | 2.4 | 2.4 | 3.0 |
| | 127 | 115 | 1.5 | 0.9 | 0.8 |
| | 307 | 160 | 1.7 | 1.2 | 0.7 |
| Sulfur dioxide | 10 | 120 | 2.8 | 5.5 | — |
| | 100 | 163 | 0.3 | 2.2 | — |
| | 300 | 246 | −0.2 | 1.5 | — |
| | 700 | 376 | 1.5 | 3.4 | — |
| Toluene | 60 | 78.9 | −5.2 | −3.7 | −2.3 |
| | 250 | 123 | −3.5 | −3.1 | −2.5 |
| Average absolute error | | | 1.9 | 3.0 | 1.9 |

†Percent error = [(calc. − exp.)/exp.] × 100.
‡All experimental viscosity values were obtained from Refs. 129, 137, and 188.

where $\xi$ = reduced, inverse viscosity, $(\mu\text{P})^{-1}$, $T_c$ is kelvins, $M$ is in g/mol, and $P_c$ is in bars.

Equation (9-4.12) has been recommended by several authors [73, 83, 141, 143, 210, 225]. The specific form suggested by Lucas [135, 137, 138] is illustrated below.

$$\eta\xi = [0.807T_r^{0.618} - 0.357 \exp(-0.449T_r) + 0.340 \exp(-4.058T_r) + 0.018]\, F_P^\circ F_Q^\circ \tag{9-4.15}$$

$\xi$ is defined by Eq. (9-4.14), $\eta$ is in $\mu$P, $T_r$ is the reduced temperature, and $F_P^\circ$ and $F_Q^\circ$ are correction factors to account for polarity or quantum effects. To obtain $F_P^\circ$, a reduced dipole moment is required. Lucas defines this quantity somewhat differently than did Chung et al. in Eq. (9-4.11), i.e.,

$$\mu_r = 52.46 \frac{\mu^2 P_c}{T_c^2} \tag{9-4.16}$$

where $\mu$ is in debyes, $P_c$ is in bars, and $T_c$ is in kelvins.† Then $F_P^\circ$ values are found as:

$$\begin{aligned} F_P^\circ &= 1 & 0 \le \mu_r &< 0.022 \\ F_P^\circ &= 1 + 30.55(0.292 - Z_c)^{1.72} & 0.022 \le \mu_r &< 0.075 \\ F_P^\circ &= 1 + 30.55(0.292 - Z_c)^{1.72}\, |0.96 + 0.1(T_r - 0.7)| & 0.075 &\le \mu_r \end{aligned} \tag{9-4.17}$$

The factor $F_Q^\circ$ is used only for the quantum gases He, $H_2$, and $D_2$.

$$F_Q^\circ = 1.22Q^{0.15}\{1 + 0.00385[(T_r - 12)^2]^{1/M} \text{ sign}(T_r - 12)\}‡ \tag{9-4.18}$$

where $Q = 1.38$ (He), $Q = 0.76$ ($H_2$), $Q = 0.52$ ($D_2$).

Equation (9-4.15) is similar to an equation proposed by Thodos and coworkers (e.g., [225]). It is interesting to note that, if $T_r \le 1$, the $f(T_r)$ in brackets in Eq. (9-4.15) is closely approximated by $0.606T_r$, that is,

$$\eta\xi \approx (0.606T_r)F_P^\circ F_Q^\circ \qquad T_r \le 1 \tag{9-4.19}$$

---

†Actually the definition used by Lucas was $\mu_r = [\mu^2 P_c/(kT_c)^2]$. If $\mu$ is expressed in debyes, where 1 debye = $10^{-18}$ esu·cm = $3.336 \times 10^{-30}$ C·m, and $(4\pi)(8.854187 \times 10^{-12})$C = N·m², then, if $P_c$ is in N/m², $T_c$ is in kelvins, and $k$ = Boltzmann's constant = $1.3806 \times 10^{-23}$ J/K, $\mu_r$ is dimensionless. Equation (9-4.16) has been presented with the unit conversions already done and $P_c$ is in bars rather than N/m².

‡sign ( ) indicates that one should use +1 or −1 depending on whether the value of the argument ( ) is >0 or <0.

Thus, when $T_r \leq 1$, low-pressure gas viscosities are essentially proportional to the absolute temperature. This might have been anticipated since, in Fig. 9-2, the slope of log $\Omega_v$ versus log $T^*$ is close to $-0.5$. With $n = -0.5$ in Eq. (9-4.5), it would have been predicted that $\eta \propto T$.

The method of Lucas is illustrated in Example 9-2.

**Example 9-2** Estimate the viscosity of methanol vapor at a temperature of 550 K and 1 bar by using Lucas' method. The experimental value is 181 $\mu$P [188].

**solution** From Appendix A, $T_c = 512.6$ K, $P_c = 80.9$ bar, $Z_c = 0.224$, $M = 32.042$, and $\mu = 1.7$ debyes. $T_r = 550/512.6 = 1.07$, and $\mu_r = 52.46[(1.7)^2(80.9)/(512.6)^2] = 4.67 \times 10^{-2}$. From Eq. (9-4.17),

$$F_P^\circ = 1 + (30.55)(0.292 - 0.224)^{1.72} = 1.30$$

With Eq. (9-4.14),

$$\xi = 0.176\left[\frac{512.6}{(32.042)^3(80.9)^4}\right]^{1/6} = 4.71 \times 10^{-3}\ (\mu\text{P})^{-1}$$

Then, with Eq. (9-4.15)

$$\begin{aligned}\eta\xi &= \{(0.807)(1.07)^{0.618} - 0.357 \exp[-(0.449)(1.07)] \\ &\quad + 0.340 \exp[-(4.058)(1.07)] + 0.018\}(1.30) \\ &= 0.836\end{aligned}$$

$$\eta = \frac{0.836}{(4.71 \times 10^{-3)})} = 178\ \mu\text{P}$$

$$\text{Error} = \frac{178 - 181}{181} \times 100 = -1.7\%$$

In Table 9-2, experimental viscosities are compared with those computed by Lucas's method. The average absolute error is 3.0 percent. Even with the correction factor $F_P^\circ$, higher errors are noted for polar compounds compared to nonpolar.

Reichenberg [165, 168] has suggested an alternate corresponding states relation for low-pressure gas viscosity of *organic* compounds.

$$\eta = \frac{M^{1/2}T}{a^*[1 + (4/T_c)][1 + 0.36T_r(T_r - 1)]^{1/6}} \frac{T_r(1 + 270\mu_r^4)}{T_r + 270\mu_r^4} \tag{9-4.20}$$

$\eta$ is in $\mu$P; $M$ is the molecular weight; $T$ is the temperature; $T_c$ is the critical temperature, in kelvins; $T_r$ is the reduced temperature; and $\mu_r$ is the reduced dipole moment defined earlier in Eq. (9-4.16). The parameter $a^*$ is defined as

$$a^* = \Sigma N_i C_i \tag{9-4.21}$$

where $n_i$ represents the number of groups of the $i$th type and $C_i$ is the group contribution shown in Table 9-3.

The term $(1 + 4/T_c)$ in the denominator of Eq. (9-4.20) may be neglected except for treating quantum gases with low values of $T_c$.

**TABLE 9-3 Values of the Group Contributions $C_i$ for the Estimation of $a^*$ in Eq. (9-4.21) [165]**

| Group | Contribution $C_i$ |
|---|---|
| $-CH_3$ | 9.04 |
| $>CH_2$ (nonring) | 6.47 |
| $>CH-$ (nonring) | 2.67 |
| $>C<$ (nonring) | −1.53 |
| $=CH_2$ | 7.68 |
| $=CH-$ (nonring) | 5.53 |
| $>C=$ (nonring) | 1.78 |
| $\equiv CH$ | 7.41 |
| $\equiv C-$ (nonring) | 5.24 |
| $>CH_2$ (ring) | 6.91 |
| $>CH-$ (ring) | 1.16 |
| $>C<$ (ring) | 0.23 |
| $=CH-$ (ring) | 5.90 |
| $>C=$ (ring) | 3.59 |
| $-F$ | 4.46 |
| $-Cl$ | 10.06 |
| $-Br$ | 12.83 |
| $-OH$ (alcohols) | 7.96 |
| $>O$ (nonring) | 3.59 |
| $>C=O$ (nonring) | 12.02 |

**TABLE 9-3 Values of the Group Contributions $C_i$ for the Estimation of $a^*$ in Eq. (9-4.21) [165]** *(Continued)*

| | |
|---|---|
| —CHO (aldehydes) | 14.02 |
| —COOH (acids) | 18.65 |
| —COO— (esters) or HCOO (formates) | 13.41 |
| $—NH_2$ | 9.71 |
| >NH (nonring) | 3.68 |
| =N—(ring) | 4.97 |
| —CN | 18.13 |
| >S (ring) | 8.86 |

A comparison between calculated and experimental low-pressure gas viscosity values is given in Table 9-2, and the method is illustrated in Example 9-3.

**Example 9-3** Estimate the viscosity of ethyl acetate vapor at 125°C and low pressure. The experimental value is reported to be 101 $\mu$P [129].

**solution** From Appendix A, $T_c = 523.2$ K, $M = 88.107$, $P_c = 38.3$ bar, and $\mu = 1.9$ debyes. With Eq. (9-4.16),

$$\mu_r = \frac{(52.46)(1.9)^2(38.3)}{(523.2)^2} = 0.0265$$

$T_r = (125 + 273)/523.2 = 0.761$. With Eq. (9-4.21) and Table 9-3,

$$a^* = 2(-CH_3) + (-CH_2) + (-COO-) = (2)(9.04) + 6.47 + 13.41 = 37.96$$

With Eq. (9-4.20),

$$\eta = \frac{(88.107)^{1/2}(125 + 273)}{37.96[1 + (0.36)(0.761)(0.761 - 1)]^{1/6}} \frac{(0.761)[1 + (270)(0.0265)^4]}{0.761 + (270)(0.0265)^4} = 99.4\ \mu P$$

$$\text{Error} = \frac{99.4 - 101}{101} \times 100 = -1.5\%$$

## Recommendations for estimating low-pressure viscosities of pure gases

Any of the three estimation methods described in this section may be used with the expectation of errors of 0.5 to 1.5 percent for nonpolar compounds and 2 to 4 percent for polar compounds. Lucas's method requires as input data $T_c$, $P_c$, and $M$ as well as $\mu$ and $Z_c$ for polar compounds and is easy to apply. At present, it is not suitable for highly associated gases like acetic acid, but it could probably be extended by multiplication of an appropriate factor as in the Chung et al.'s technique. Chung et al.'s method requires somewhat more input ($T_c$, $V_c$, and $M$ and $\mu$, $\omega$, and $\kappa$ for the polar correction). The critical volume is less readily available than the critical pressure, and the association factor $\kappa$ is an empirical constant that must be determined from viscosity data. The method is not suited for quantum gases. Reichenberg's technique requires $M$, $T_c$, and structural groups as well as $\mu$ for the polar correction. This method is not suitable for inorganic gases and cannot be applied to organic gases for which necessary group contributions have not been determined.

## 9-5 Viscosities of Gas Mixtures at Low Pressures

The rigorous kinetic theory of Chapman-Enskog can be extended to determine the viscosity of low-pressure multicomponent mixtures [23, 24, 25, 27, 40, 100, 119]. The final expressions are quite complicated and are rarely used to estimate mixture viscosities. Three simplifications of the rigorous theoretical expressions are described below. Reichenberg's equations are the most complex, but, as shown later, the most consistently accurate. Wilke's method is simpler, and that of Herning and Zipperer is even more so. All these methods are essentially interpolative; i.e., the viscosity values for the pure components must be available. The methods then lead to estimations showing how the mixture viscosity varies with composition. Later in this section, two corresponding states methods are described; they do not require pure component values as inputs.

### Method of Reichenberg [164, 167, 168]

In this technique, Reichenberg has incorporated elements of the kinetic theory approach of Hirschfelder, Curtiss, and Bird [100] with corresponding states methodology to obtain desired parameters. In addition, a polar correction has been included. The general, multicomponent mixture viscosity equation is:

$$\eta_m = \sum_{i=1}^{n} K_i \left( 1 + 2 \sum_{j=1}^{i-1} H_{ij} K_j \sum_{\substack{j=1 \\ \neq i}}^{n} \sum_{\substack{k=1 \\ \neq i}}^{n} H_{ij} H_{ik} K_j K_k \right) \tag{9-5.1}$$

where $\eta_m$ is the mixture viscosity and $n$ is the number of components.

With $\eta_i$ the viscosity of pure $i$, $M_i$ the molecular weight of $i$, and $y_i$ the mole fraction of $i$ in the mixture,

$$K_i = \frac{y_i \eta_i}{y_i + \eta_i \sum_{\substack{k=1 \\ \neq i}}^{n} y_k H_{ik}[3 + (2M_k/M_i)]} \tag{9-5.2}$$

Two other component properties used are:

$$U_i = \frac{[1 + 0.36T_{r_i}(T_{r_i} - 1)^{1/6}F_{R_i}}{(T_{r_i})^{1/2}} \tag{9-5.3}$$

$$C_i = \frac{M_i^{1/4}}{(\eta_i U_i)^{1/2}} \tag{9-5.4}$$

where $T_{r_i} = T/T_{c_i}$ and $F_{R_i}$ is a polar correction.

$$F_{R_i} = \frac{T_{r_i}^{3.5} + (10\mu_{r_i})^7}{T_{r_i}^{3.5}[1 + (10\mu_{r_i})^7]} \tag{9-5.5}$$

Here $\mu_{r_i}$ is the reduced dipole moment of $i$ and is calculated as shown earlier in Eq. (9-4.16). For the term $H_{ij} = H_{ji}$,

$$H_{ij} = \left[\frac{M_i M_j}{32(M_i + M_j)^3}\right]^{1/2} (C_i + C_j)^2 \times \frac{[1 + 0.36T_{r_{ij}}(T_{r_{ij}} - 1)]^{1/6}F_{R_{ij}}}{(T_{r_{ij}})^{1/2}} \tag{9-5.6}$$

with

$$T_{r_{ij}} = \frac{T}{(T_{c_i}T_{c_j})^{1/2}} \tag{9-5.7}$$

$F_{R_{ij}}$ is found from Eq. (9-5.5) with $T_{r_i}$ replaced by $T_{r_{ij}}$ and $\mu_{r_i}$ by $\mu_{r_{ij}} = (\mu_{r_i}\mu_{r_j})^{1/2}$.

For a binary gas mixture of 1 and 2, these equations may be written as:

$$\eta_m = K_1(1 + H_{12}^2 K_2^2) + K_2(1 + 2H_{12}K_1 + H_{12}^2 K_1^2) \tag{9-5.8}$$

$$K_1 = \frac{y_1\eta_1}{y_1 + \eta_1\{y_2 H_{12}[3 + (2M_2/M_1)]\}} \tag{9-5.9}$$

$$K_2 = \frac{y_2\eta_2}{y_2 + \eta_2\{y_1 H_{12}[3 + (2M_1/M_2)]\}} \tag{9-5.10}$$

$$U_1 = \frac{[1 + 0.36T_{r1}(T_{r1} - 1)]^{1/6}}{T_{r1}^{1/2}} \frac{T_{r1}^{3.5} + 10^7\mu_{r1}^7}{T_{r1}^{3.5}(1 + 10^7\mu_{r1}^7)} \tag{9-5.11}$$

and a comparable expression for $U_2$. The meaning of $C_1$ and $C_2$ is clear from Eq. (9-5.4). Finally, with

$$T_{r12} = \frac{T}{(T_{c1}T_{c2})^{1/2}} \quad \text{and} \quad \mu_{r12} = (\mu_{r1}\mu_{r2})^{1/2}$$

$$H_{12} = \frac{(M_1 M_2/32)^{1/2}}{(M_1 + M_2)^{3/2}} \frac{[1 + 0.36T_{r12}(T_{r12} - 1)]^{1/6}}{(T_{r12})^{1/2}} \times (C_1 + C_2)^2 \frac{T_{r12}^{3.5} + (10\mu_{r12})^7}{T_{r12}^{3.5}[1 + (10\mu_{r12})^7]} \tag{9-5.12}$$

To employ Reichenberg's method, for each component one needs the pure gas viscosity at the system temperature as well as the molecular weight, dipole moment, critical temperature, and critical pressure. The temperature and composition are state variables.

The method is illustrated in Example 9-4. A comparison of experimental and calculated gas-mixture viscosities is shown in Table 9-4.

**Example 9-4** Use Reichenberg's method to estimate the viscosity of a nitrogen-monochlorodifluoromethane (R-22) mixture at 50°C and atmospheric pressure. The mole fraction nitrogen is 0.286. The experimental viscosity is 145 $\mu$P [195].

**solution** The following pure component properties are used:

| | $N_2$ | $CHClF_2$ |
|---|---|---|
| $T_c$, K | 126.2 | 369.3 |
| $P_c$, bar | 33.9 | 49.7 |
| $M$, g/mol | 28.013 | 86.469 |
| $\mu$, debyes | 0 | 1.4 |
| $\eta$, 50°C, $\mu$P | 188 | 134 |

With $T$ = 50°C, $T_r$ ($N_2$) = 2.56, and $T_r$ ($CHClF_2$) = 0.875,

$$T_{r12} = \frac{50 + 273.2}{[(126.2)(369.3)]^{1/2}} = 1.497$$

$\mu_r$ ($N_2$) = 0, and, from Eq. (9-4.16)

$$\mu_r\,(CHClF_2) = \frac{(52.46)(1.4)^2(49.7)}{(369.3)^2} = 0.0375$$

Since $\mu_{r12} = (\mu_{r1}\mu_{r2})^{1/2}$, then for this mixture, $\mu_{r12} = 0$. With Eq. (9-5.11), for $CHClF_2$,

$$U\,(CHClF_2) = \frac{[1 + (0.36)(0.875)(0.875 - 1)]^{1/6}}{(0.875)^{1/2}} \times \frac{(0.875)^{3.5} + (10)^7(0.0375)^7}{(0.875)^{3.5}[1 + (10)^7(0.0375)^7]} = 1.062$$

and $U$ ($N_2$) = 0.725.

Then, $C\ (N_2) = \dfrac{(28.013)^{1/4}}{[(188)(0.725)]^{1/2}} = 0.197$

and $C\ (CHClF_2) = 0.256$

Next,

$$H\ (N_2\text{–}CHClF_2) = (0.197 + 0.256)^2 \frac{[(28.013)(86.469)]^{1/2}}{[32(28.013 + 86.469)^3]^{1/2}} \times \frac{[1 + (0.36)(1.497)(1.497 - 1)]^{1/6}}{(1.497)^{1/2}} \times 1.0 = 1.237 \times 10^{-3}$$

$$K\ (N_2) = \frac{(0.286)(188)}{0.286 + (188)(0.714)(1.237 \times 10^{-3})\{3 + [(2)(86.469)/28.013]\}} = 29.71$$

and $K\ (CHClF_2) = 107.9$. Substituting into Eq. (9-5.8),

$$\eta_m = (29.71)[1 + (1.237 \times 10^{-3})^2(107.9)^2] + (107.9)[1 + (2)(1.237 \times 10^{-3})(29.71) + (1.237 \times 10^{-3})^2(29.71)^2] = 146.2\ \mu P$$

$$\text{Error} = \frac{146.2 - 145}{145} \times 100 = 0.8\%$$

### Method of Wilke

In a further simplification of the kinetic theory approach, Wilke [221] neglected second-order effects and proposed:

$$\eta_m = \sum_{i=1}^{n} \frac{y_i \eta_i}{\sum_{j=1}^{n} y_j \phi_{ij}} \tag{9-5.13}$$

where

$$\phi_{ij} = \frac{[1 + (\eta_i/\eta_j)^{1/2}(M_j/M_i)^{1/4}]^2}{[8(1 + M_i/M_j)]^{1/2}} \tag{9-5.14}$$

$\phi_{ji}$ is found by interchanging subscripts or by

$$\phi_{ji} = \frac{\eta_j}{\eta_i} \frac{M_i}{M_j} \phi_{ij} \tag{9-5.15}$$

For a binary system of 1 and 2, with Eqs. (9-5.13) to (9-5.15),

$$\eta_m = \frac{y_1 \eta_1}{y_1 + y_2 \phi_{12}} + \frac{y_2 \eta_2}{y_2 + y_1 \phi_{21}} \tag{9-5.16}$$

where $\eta_m$ = viscosity of the mixture
$\eta_1, \eta_2$ = pure component viscosities
$y_1, y_2$ = mole fractions

**TABLE 9-4 Comparison of Calculated and Experimental Low-Pressure Gas Mixture Viscosities**

| | | | | | Percent deviation† calculated by method of: | | | | |
|---|---|---|---|---|---|---|---|---|---|
| System | $T$, K | Mole fraction first component | Viscosity (exp.) $\mu P$ | Ref. | Reichenberg, Eq. (9-5.8) | Wilke, Eq. (9-5.16) | Herning and Zipperer, Eq. (9-5.17) | Lucas, Eq. (9-4.15) with Eqs. (9-5.18) through (9-5.23) | Chung et al., Eq. (9-5.24) |
| Nitrogen-hydrogen | 373 | 0.0 | 104.2 | 155, 208 | — | — | — | 0.8 | −11 |
| | | 0.2 | 152.3 | | 4.3 | 12 | 2.0 | 2.1 | −23 |
| | | 0.51 | 190.3 | | 1.8 | 5.6 | −1.0 | −2.0 | −11 |
| | | 0.80 | 205.8 | | 0.1 | 1.4 | −1.2 | 3.6 | −3.3 |
| | | 1.0 | 210.1 | | — | — | — | 0.4 | 0 |
| Methane-propane | 298 | 0.0 | 81.0 | 18 | — | — | — | 3.5 | 1.3 |
| | | 0.2 | 85.0 | | 0.2 | −0.3 | −0.2 | 4.6 | 1.7 |
| | | 0.4 | 89.9 | | 0.1 | −0.8 | −0.6 | 5.0 | 1.7 |
| | | 0.6 | 95.0 | | 0.6 | −0.4 | −0.2 | 5.4 | 1.9 |
| | | 0.8 | 102.0 | | 0.2 | −0.6 | −0.5 | 3.7 | 1.0 |
| | | 1.0 | 110.0 | | — | — | — | 0.3 | 1.0 |
| | 498 | 0.0 | 131.0 | 18 | — | — | — | 4.0 | 2.3 |
| | | 0.2 | 136.0 | | 0.4 | 0.0 | −0.2 | 5.2 | 2.7 |
| | | 0.4 | 142.0 | | 0.6 | 0.0 | −0.5 | 5.6 | 2.6 |
| | | 0.6 | 149.0 | | 0.7 | 0.0 | −0.6 | 5.2 | 2.0 |
| | | 0.8 | 157.0 | | 0.7 | 0.0 | −0.3 | 3.7 | 1.1 |
| | | 1.0 | 167.0 | | — | — | — | −0.2 | 0.2 |
| Carbon tetrafluoride–sulfur hexafluoride | 303 | 0.0 | 159.0 | 162 | — | — | — | 6.6 | 0.7 |
| | | 0.257 | 159.9 | | 2.0 | 2.0 | 1.8 | 9.3 | 3.8 |
| | | 0.491 | 161.5 | | 3.4 | 3.4 | 3.1 | 11.0 | 6.2 |
| | | 0.754 | 164.3 | | 4.6 | 4.6 | 4.3 | 13.0 | 8.4 |
| | | 1.0 | 176.7 | | — | — | — | 7.4 | 5.6 |
| Nitrogen–carbon dioxide | 293 | 0.0 | 146.6 | 120 | — | — | — | 1.6 | −1.2 |
| | | 0.213 | 153.5 | | 0.5 | −1.3 | −1.0 | 0.4 | −0.3 |
| | | 0.495 | 161.8 | | 0.4 | −1.8 | −1.5 | −0.2 | 0.7 |
| | | 0.767 | 172.1 | | −2.0 | −2.8 | −2.5 | −1.7 | −0.7 |

| | | | | | | | | | |
|---|---|---|---|---|---|---|---|---|---|
| Ammonia-hydrogen | 306 | 0.0 | 90.6 | 155 | — | — | — | 2.4 | −9.9 |
| | | 0.195 | 118.4 | | −4.0 | −11 | −18 | −2.7 | 2.1 |
| | | 0.399 | 123.8 | | −4.6 | −12 | −19 | −3.0 | 10.0 |
| | | 0.536 | 122.4 | | −4.5 | −11 | −16 | −2.7 | 10.0 |
| | | 0.677 | 120.0 | | −4.8 | −9.7 | −14 | −3.1 | 7.1 |
| | | 1.0 | 105.9 | | — | — | — | 1.3 | 0.9 |
| Hydrogen sulfide–ethyl ether | 331 | 0.0 | 84.5 | 156 | — | — | — | −1.7 | −4.0 |
| | | 0.204 | 87 | | −2.9 | −3.2 | 0.2 | 2.3 | 0.4 |
| | | 0.500 | 97 | | −2.2 | −2.8 | 3.2 | 3.4 | 1.7 |
| | | 0.802 | 116 | | 0.0 | −0.4 | 4.2 | 0.6 | −0.7 |
| | | 1.0 | 137 | | — | — | — | −3.0 | −4.1 |
| Ammonia–methylamine | 423 | 0.0 | 130.0 | 32 | — | — | — | −2.1 | −8.0 |
| | | 0.25 | 134.5 | | −0.8 | −0.3 | −0.6 | −1.5 | −7.5 |
| | | 0.75 | 142.2 | | −1.0 | −0.3 | −0.7 | 0.1 | −3.4 |
| | | 1.0 | 146.0 | | — | — | — | 1.1 | 1.1 |
| | 673 | 0.0 | 204.8 | 32 | — | — | — | −4.6 | −11 |
| | | 0.25 | 212.8 | | −2.6 | −0.7 | −0.9 | −4.9 | −11 |
| | | 0.75 | 228.3 | | −3.1 | −0.7 | −0.9 | −4.7 | −9.3 |
| | | 1.0 | 236.0 | | — | — | — | −4.3 | −5.4 |
| Nitrogen-monochlorodifluoro-methane | 323 | 0.0 | 134 | 195 | — | — | — | 11.0 | 6.0 |
| | | 0.286 | 145 | | 0.8 | −0.8 | −0.7 | 11.0 | 7.3 |
| | | 0.463 | 153 | | 1.3 | −1.0 | −0.8 | 11.0 | 7.4 |
| | | 0.644 | 164 | | 0.7 | −1.8 | −1.6 | 8.9 | 5.6 |
| | | 0.824 | 177 | | −0.3 | −2.2 | −2.1 | 5.0 | 2.4 |
| | | 1.0 | 188 | | — | — | — | 0.8 | 0.4 |
| Nitrous oxide-sulfur dioxide | 353 | 0.0 | 152.3 | 39 | — | — | — | 3.9 | −1.5 |
| | | 0.325 | 161.7 | | −1.7 | −2.2 | −2.2 | 0.6 | −1.6 |
| | | 0.625 | 167.8 | | −1.6 | −2.2 | −2.1 | −0.6 | −2.3 |
| | | 0.817 | 170.7 | | −0.9 | −1.3 | −1.2 | −0.7 | −1.9 |
| | | 1.0 | 173.0 | | — | — | — | −0.6 | −0.7 |
| Nitrogen–*n*-heptane | 344 | 0.0 | 69.4 | 36,188 | — | — | — | 0.8 | −3.5 |
| | | 0.515 | 104.0 | | 0.7 | −6.2 | 11 | −0.2 | −4.8 |
| | | 0.853 | 154.6 | | 0.9 | −4.7 | 7.4 | −0.5 | −3.2 |
| | | 1.0 | 197.5 | | — | — | — | 0.6 | 0.2 |

†Percent deviation = [(calc. − exp.)/(exp.)] × 100.

and $$\phi_{12} = \frac{[1 + (\eta_1/\eta_2)^{1/2}(M_2/M_1)^{1/4}]^2}{\{8[1 + (M_1/M_2)]\}^{1/2}}$$

$$\phi_{21} = \phi_{12} \frac{\eta_2}{\eta_1} \frac{M_1}{M_2}$$

Equation (9-5.13), with $\phi_{ij}$ from Eq. (9-5.14), has been extensively tested. Wilke [221] compared values with data on 17 binary systems and reported an average deviation of less than 1 percent; several cases in which $\eta_m$ passed through a maximum were included. Many other investigators have tested this method [4, 28, 42, 51, 78, 161, 176, 177, 191, 214, 223]. In most cases, only nonpolar mixtures were compared, and very good results obtained. For some systems containing hydrogen as one component, less satisfactory agreement was noted. In Table 9-4, Wilke's method predicted mixture viscosities that were larger than experimental for the $H_2$-$N_2$ system, but for $H_2$-$NH_3$, it underestimated the viscosities. Gururaja et al. [91] found that this method also overpredicted in the $H_2$-$O_2$ case but was quite accurate for the $H_2$-$CO_2$ system. Wilke's approximation has proved reliable even for polar-polar gas mixtures of aliphatic alcohols [169]. The principal reservation appears to lie in those cases where $M_i \gg M_j$ and $\eta_i \gg \eta_j$.

**Example 9-5** Kestin and Yata [124] report that the viscosity of a mixture of methane and $n$-butane is 93.35 $\mu$P at 293 K when the mole fraction of $n$-butane is 0.303. Compare this result with the value estimated by Wilke's method. For pure methane and $n$-butane, these same authors report viscosities of 109.4 and 72.74 $\mu$P.

**solution** Let 1 refer to methane and 2 to $n$-butane. $M_1 = 16.043$ and $M_2 = 58.124$.

$$\phi_{12} = \frac{[1 + (109.4/72.74)^{1/2}(58.124/16.043)^{1/4}]^2}{\{8[1 + (16.043/58.124)]\}^{1/2}} = 2.268$$

$$\phi_{21} = 2.268 \frac{72.74}{109.4} \frac{16.043}{58.124} = 0.416$$

$$\eta_m = \frac{(0.697)(109.4)}{0.697 + (0.303)(2.268)} + \frac{(0.303)(72.74)}{0.303 + (0.697)(0.416)}$$

$$= 92.26\ \mu\text{P}$$

$$\text{Error} = \frac{92.26 - 93.35}{93.35} \times 100 = -1.2\%$$

### Herning and Zipperer approximation of $\phi_{ij}$

As an approximate expression for $\phi_{ij}$ the following is proposed [98]:

$$\phi_{ij} = \left(\frac{M_j}{M_i}\right)^{1/2} = \phi_{ji}^{-1} \tag{9-5.17}$$

When Eq. (9-5.17) is used with Eq. (9-5.16) to estimate low-pressure binary gas mixture viscosities, quite reasonable predictions are obtained (Table 9-4) except for systems such as $H_2$-$NH_3$. The technique is illustrated in Example 9-6. Note that Examples 9-5 and 9-6 treat the same problem; each provides a viscosity estimate close to the experimental value. But the $\phi_{12}$ and $\phi_{21}$ values employed in the two cases are quite different. Apparently, multiple sets of $\phi_{ij}$ and $\phi_{ji}$ work satisfactorily in Eq. (9-5.13).

**Example 9-6** Repeat Example 9-5 by using the Herning and Zipperer approximation for $\phi_{ij}$.

**solution** As before, with 1 as methane and 2 as $n$-butane,

$$\phi_{12} = \left(\frac{58.124}{16.043}\right)^{1/2} = 1.903$$

$$\phi_{21} = \phi_{12}^{-1} = 0.525$$

$$\eta_m = \frac{(0.697)(109.4)}{0.697 + (0.303)(1.903)} + \frac{(0.303)(72.74)}{0.303 + (0.697)(0.525)}$$

$$= 92.82\ \mu\text{P}$$

$$\text{Error} = \frac{92.82 - 93.35}{93.35} \times 100 = -0.6\%$$

### Corresponding states methods

In this approach, one employs pure component estimation methods with mixing and combining rules to relate critical and other properties of the mixture to pure component properties and to composition.

### Lucas rules

Lucas [135, 137, 138] defined mixture properties as shown below for use in Eqs. (9-4.14) through (9-4.18).

$$T_{cm} = \sum_i y_i T_{ci} \tag{9-5.18}$$

$$P_{cm} = RT_{cm} \frac{\sum_i y_i Z_{ci}}{\sum_i y_i V_{ci}} \tag{9-5.19}$$

$$M_m = \sum_i y_i M_i \tag{9-5.20}$$

$$F^\circ_{Pm} = \sum_i y_i F^\circ_{Pi} \tag{9-5.21}$$

$$F^\circ_{Qm} = \left(\sum_i y_i F^\circ_{Qi}\right) A \tag{9-5.22}$$

and, letting the subscript H denote the mixture component of highest molecular weight and L the component of lowest molecular weight,

$$A = 1 - 0.01\left(\frac{M_H}{M_L}\right)^{0.87} \quad \text{for } \frac{M_H}{M_L} > 9 \text{ and } 0.05 < y_H < 0.7; \tag{9-5.23}$$

otherwise, $A = 1$

The method of Lucas does not necessarily lead to the pure component viscosity $\eta_i$ when all $y_j = 0$ except $y_i = 1$. Thus the method is not interpolative in the same way as are the techniques of Reichenberg, Wilke, and Herning and Zipperer. Nevertheless, as seen in Table 9-4, the method provides reasonable estimates of $\eta_m$ in most test cases.

**Example 9-7** Estimate the viscosity of a binary mixture of ammonia and hydrogen at 33°C and low pressure by using the Lucas corresponding states method.

**solution** Let us illustrate the method for a mixture containing 67.7 mole percent ammonia. The data required (from Appendix A) are:

| | Ammonia | Hydrogen |
|---|---|---|
| $T_c$ K | 405.5 | 33.2 |
| $P_c$, bar | 113.5 | 13.0 |
| $Z_c$ | 0.244 | 0.306 |
| $M$ | 17.031 | 2.016 |
| $\mu$, debyes | 1.47 | 0 |
| $T_r$ | 0.755 | 9.223 |

Using Eqs. (9-5.18) to (9-5.20), $T_{c_m} = 285.2$ K, $P_{c_m} = 89.4$ bar, and $M_m = 12.18$. From these values and Eq. (9-4.14), $\xi_m = 6.4720 \times 10^{-3}\ (\mu\text{P})^{-1}$. With Eq. (9-4.16), $\mu_r\ (NH_3) = 7.825 \times 10^{-2}$ and $\mu_r\ (H_2) = 0$. Then, with Eq. (9-4.17),

$$F_P^\circ\ (NH_3) = 1 + 30.55(0.292 - 0.244)^{1.72}|0.96 + 0.1(0.755 - 0.7)|$$

$$= 1.159$$

$$F_P^\circ\ (H_2) = 1.0$$

$$F_{Pm}^\circ = (1.159)(0.677) + (1)(0.323) = 1.107$$

For the quantum correction, with Eq. (9-5.23), since $\dfrac{M_H}{M_L} = \dfrac{17.031}{2.016} = 8.4 < 9$, then $A = 1$. $F_Q^\circ\ (NH_3) = 1.0$, and with Eq. (9-4.18),

$$F_Q^\circ\ (H_2) = (1.22)(0.76)^{0.15}\{1 + 0.00385[(9.223 - 12)^2]^{1/2.016} \times \text{sign}\ (9.223 - 12)\}$$

$$= (1.171)[1 + (0.01061)(-1)] = 1.158$$

$$F_{Qm}^\circ = (1.158)(0.323) + (1)(0.677) = 1.051$$

Next, from Eq. (9-4.15) with $T_{r_m} = (33 + 273.2)/285.2 = 1.073$

$$\eta_m \xi_m = (0.645)(1.107)(1.051) = 0.750$$

$$\eta_m = \frac{0.750}{6.472 \times 10^{-3}} = 115.9\ \mu\text{P}$$

The experimental value is 120.0 $\mu$P; thus

$$\text{Error} = \frac{115.9 - 120.0}{120.0} \times 100 = -3.4\%$$

The viscosity of the ammonia-hydrogen mixture at 33°C is plotted in Fig. 9-4.

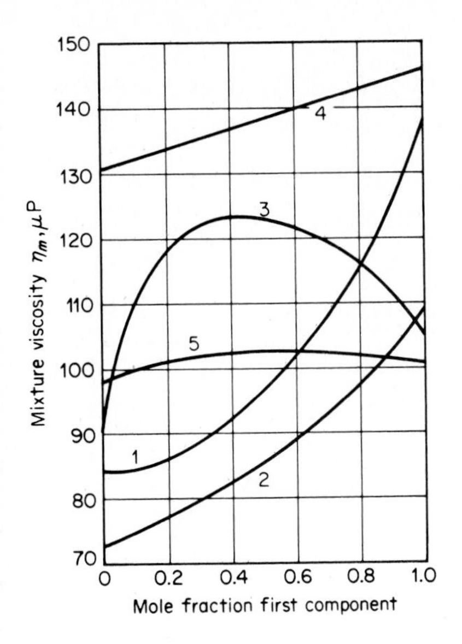


**Figure 9-4** Gas mixture viscosities.

| No. | System | Reference |
|---|---|---|
| 1 | Hydrogen sulfide–ethyl ether | 156 |
| 2 | Methane–$n$-butane | 124 |
| 3 | Ammonia-hydrogen | 155 |
| 4 | Ammonia–methyl amine | 32 |
| 5 | Ethylene-ammonia | 209 |

**Chung et al. rules [44, 45]**

In this case, Eq. (9-3.9) is employed to estimate the mixture viscosity with, however, a factor $F_{cm}$ as used in Eq. (9-4.9) to correct for shape and polarity.

$$\eta_m = \frac{26.69 F_{cm}(M_m T)^{1/2}}{\sigma_m^2 \Omega_v} \tag{9-5.24}$$

where $\Omega_v = f(T_m^*)$. In the Chung et al. approach, the mixing rules are:

$$\sigma_m^3 = \sum_i \sum_j y_i y_j \sigma_{ij}^3 \tag{9-5.25}$$

$$T_m^* = \frac{T}{(\epsilon/k)_m} \tag{9-5.26}$$

$$\left(\frac{\epsilon}{k_m}\right) = \frac{\sum_i \sum_j y_i y_j (\epsilon_{ij}/\mathrm{k}) \sigma_{ij}^3}{\sigma_m^3} \tag{9-5.27}$$

$$M_m = \left[\frac{\sum_i \sum_j y_i y_j (\epsilon_{ij}/k) \sigma_{ij}^2 M_{ij}^{1/2}}{(\epsilon/k)_m \sigma_m^2}\right]^2 \tag{9-5.28}$$

$$\omega_m = \frac{\sum_i \sum_j y_i y_j \omega_{ij} \sigma_{ij}^3}{\sigma_m^3} \tag{9-5.29}$$

$$\mu_m^4 = \sigma_m^3 \sum_i \sum_j \left( \frac{y_i y_j \mu_i^2 \mu_j^2}{\sigma_{ij}^3} \right) \tag{9-5.30}$$

$$\kappa_m = \sum_i \sum_j y_i y_j \kappa_{ij} \tag{9-5.31}$$

and the combining rules are:

$$\sigma_{ii} = \sigma_i = 0.809 V_{ci}^{1/3} \tag{9-5.32}$$

$$\sigma_{ij} = \xi_{ij}(\sigma_i \sigma_j)^{1/2} \tag{9-5.33}$$

$$\frac{\epsilon_{ii}}{k} = \frac{\epsilon_i}{k} = \frac{T_{ci}}{1.2593} \tag{9-5.34}$$

$$\frac{\epsilon_{ij}}{k} = \zeta_{ij} \left( \frac{\epsilon_i}{k} \frac{\epsilon_j}{k} \right)^{1/2} \tag{9-5.35}$$

$$\omega_{ii} = \omega_i \tag{9-5.36}$$

$$\omega_{ij} = \frac{\omega_i + \omega_j}{2} \tag{9-5.37}$$

$$\kappa_{ii} = \kappa_i \tag{9-5.38}$$

$$\kappa_{ij} = (\kappa_i \kappa_j)^{1/2} \tag{9-5.39}$$

$$M_{ij} = \frac{2 M_i M_j}{M_i + M_j} \tag{9-5.40}$$

$\xi_{ij}$ and $\zeta_{ij}$ are binary interaction parameters which are normally set equal to unity. The $F_{c_m}$ term in Eq. (9-5.24) is defined as in Eq. (9-4.10).

$$F_{c_m} = 1 - 0.275\omega_m + 0.059035\mu_{r_m}^4 + \kappa_m \tag{9-5.41}$$

where $\mu_{r_m}$ is given by Eq. (9-4.11)

$$\mu_{r_m} = \frac{131.3\mu_m}{(V_{c_m} T_{c_m})^{1/2}} \tag{9-5.42}$$

$$V_{c_m} = (\sigma_m / 0.809)^3 \tag{9-5.43}$$

$$T_{c_m} = 1.2593 \left( \frac{\epsilon}{k} \right)_m \tag{9-5.44}$$

In these equations, $T_c$ is in kelvins, $V_c$ is in $cm^3$/mol and $\mu$ is in debyes.

The rules suggested by Chung et al. are illustrated for a binary gas mixture in Example 9-8. As with the Lucas approach, the technique is not interpolative between pure component viscosities. Some calculated binary gas mixture viscosities are compared with experimental values in Table 9-4. Errors vary, but they are usually less than about $\pm 5$ percent.

**Example 9-8** Using Chung et al.'s method, estimate the low-pressure gas viscosity of a binary of hydrogen sulfide and ethyl ether containing 20.4 mole percent $H_2S$. The temperature is 331 K.

**solution** For a mixture of hydrogen sulfide and ethyl ether, the following properties are needed (Appendix A):

| | Hydrogen sulfide | Ethyl ether |
|---|---|---|
| $T_c$, K | 373.2 | 466.7 |
| $V_c$, cm³/mol | 98 | 280 |
| $\omega$ | 0.109 | 0.281 |
| $\mu$, debyes | 0.9 | 1.3 |
| $\kappa$ | 0 | 0 |
| $M$, g/mol | 34.080 | 74.123 |
| $y$ | 0.204 | 0.796 |

From Eqs. (9-5.32) and (9-5.33),

$$\sigma\,(\text{H}_2\text{S}) = (0.809)(98)^{1/3} = 3.730\ \text{Å}$$

$$\sigma\,(\text{EE}) = 5.293\ \text{Å}$$

$$\sigma\,(\text{H}_2\text{S-EE}) = 4.443\ \text{Å}$$

then, with Eq. (9-5.25),

$$\sigma_m^3 = (0.204)^2(3.730)^3 + (0.796)^2(5.293)^3 + (2)(0.204)(0.796)(4.443)^3$$
$$= 124.58\ \text{Å}^3$$

From Eqs. (9-5.34) and (9-5.35),

$$\frac{\epsilon}{k}\,(\text{H}_2\text{S}) = \frac{373.2}{1.2593} = 296.4\ \text{K}$$

$$\frac{\epsilon}{k}\,(\text{EE}) = 370.6\ \text{K}$$

$$\frac{\epsilon}{k}\,(\text{H}_2\text{S-EE}) = 331.4\ \text{K}$$

Then, with Eq. (9-5.27),

$$\left(\frac{\epsilon}{k}\right)_m = [(0.204)^2(296.4)(3.730)^3 + (0.796)^2(370.6)(5.293)^3$$
$$+ (2)(0.204)(0.796)(331.4)(4.443)^3]/124.58$$
$$= 360.4\ \text{K}$$

With Eqs. (9-5.28) and (9-5.40),

$$M_m = (\{(0.204)^2(296.4)(3.730)^2(34.080)^{1/2} + (0.796)^2(370.6)(5.293)^2(74.123)^{1/2}$$
$$+ (2)(0.204)(0.796)(331.4)(4.443)^2\,[(2)(34.080)(74.123)/(34.080 + 74.123)]^{1/2}\}$$
$$/(360.4)\,(124.58)^{2/3})^2 = 64.43\ \text{g/mol}$$

With Eq. (9-5.29),

$$\omega_m = \{(0.204)^2(0.109)(3.730)^3 + (0.796)^2(0.281)(5.293)^3$$
$$+ (2)(0.204)(0.796)[(0.109 + 0.281)/2](4.443)^3\}/124.58$$
$$= 0.258$$

and with Eq. (9-5.30),

$$\mu_m^4 = \{[(0.204)^2(0.9)^4/(3.730)^3] + [(0.796)^2(1.3)^4/(5.293)^3] + [(2)(0.204)(0.796)(0.9)^2(1.3)^2/(4.443)^3]\}(124.58) = 2.218$$

$$\mu_m = 1.22 \text{ debyes}$$

so, with Eqs. (9-5.42) to (9-5.44),

$$V_{cm} = \frac{(124.58)}{(0.809)^3} = 235.3 \text{ cm}^3/\text{mol}$$

$$T_{cm} = (1.2593)(360.4) = 453.9 \text{ K}$$

$$\mu_{rm} = \frac{(131.3)(1.22)}{[(235.3)(453.9)]^{1/2}} = 0.490$$

Since $\kappa_m = 0$, with Eq. (9-5.41),

$$F_{cm} = 1 - (0.275)(0.258) + (0.059035)(0.490)^4 = 0.932$$

Using $T_m^*$ from Eq. (9-5.26) [$= 331/360.4 = 0.918$] and Eq. (9-4.3), $\Omega_v = 1.664$. Finally, with Eq. (9-5.24),

$$\eta_m = \frac{(26.69)(0.932)[(64.43)(331)]^{1/2}}{(124.58)^{2/3}(1.664)} = 87.6 \ \mu\text{P}$$

The experimental value is 87 $\mu$P (Table 9-4).

$$\text{Error} = \frac{87.6 - 87}{87} \times 100 = 0.4\%$$

### Discussion and recommendations to estimate the low-pressure viscosity of gas mixtures

As is obvious from the estimation methods discussed in this section, the viscosity of a gas mixture can be a complex function of composition. This is evident from Fig. 9-4. There may be a maximum in mixture viscosity in some cases, e.g., system 3, ammonia-hydrogen. No cases of a viscosity minimum have, however, been reported. Behavior similar to that of the ammonia-hydrogen case occurs most often in polar-nonpolar mixtures in which the pure component viscosities are not greatly different [101, 172]. Maxima are more pronounced as the molecular weight ratio differs from unity.

Of the five estimation methods described in this section, three (Herning and Zipperer, Wilke, and Reichenberg) use the kinetic theory approach and yield interpolative equations between the pure component viscosities. Reichenberg's method is most consistently accurate, but it is the most complex. To use Reichenberg's procedure, one needs, in addition to temperature and composition, the viscosity, critical temperature, critical pressure, molecular weight, and dipole moment of each constituent. Wilke's and Herning and Zipperer's methods require only the pure component viscosities and molecular weights; these latter two yield reasonably accurate predictions of the mixture viscosity.

Arguing that it is rare to have available the pure gas viscosities at the temperature of interest, both Lucas and Chung et al. provide estimation methods to cover the entire range of composition. At the end points where only pure components exist, their methods reduce to those described earlier in Sec. 9-3. Although the errors from these two methods are, on the average, slightly higher than those of the interpolative techniques, they are usually less than $\pm 5$ percent as seen from Table 9-4. Such errors could be reduced even further if pure component viscosity data were available and were employed in a simple linear correction scheme.

Many other estimation methods for determining $\eta_m$ have been proposed [25, 26, 33, 54, 75, 90, 97, 118, 119, 174, 191, 192, 207, 225], but they were judged either less accurate or less general than those discussed in this section.

It is recommended that Reichenberg's method [Eq. (9-5.8)] be used to calculate $\eta_m$ if pure component viscosity values are available. Otherwise, either the Lucas method [Eq. (9-4.15)] or the Chung et al. method [Eq. (9-5.24)] can be employed if critical properties are available for all components.

A compilation of references dealing with gas mixture viscosities (low and high pressure) has been prepared by Sutton [193].

## 9-6 Effect of Pressure on the Viscosity of Pure Gases

The viscosity of a gas is a strong function of pressure near the critical point and at reduced temperatures of about 1 to 2 at high pressures. The complexity of the $T$-$P$-$\eta$ phase diagram is seen in Figs. 9-5 and 9-6 [188]. In Fig. 9-5, the viscosity of carbon dioxide is graphed as a function of temperature with various isobars shown; for $CO_2$, $T_c = 304.1$ K and $P_c = 73.8$ bar. If the viscosity were plotted as a function of pressure with isotherms, one would have a phase diagram as illustrated in Fig. 9-6 for nitrogen ($T_c = 77.4$ K, $P_c = 33.9$ bar). Lucas [136, 137] has generalized the viscosity phase diagrams (for nonpolar gases) as shown in Fig. 9-7. In this case the ordinate is $\eta\xi$ and the temperatures and pressures are reduced values. $\xi$ is the inverse reduced viscosity defined earlier in Eq. (9-4.14).

In Fig. 9-7, the lower limit of the $P_r$ curves would be indicative of the dilute gas state, as described in Sec. 9-4. In such a state, $\eta$ increases with temperature. At high reduced pressures, we see there is a wide range of temperatures where $\eta$ decreases with temperature. In this region the viscosity behavior more closely simulates a liquid state, and, as will be shown in Sec. 9-10, an increase in temperature results in a decrease in viscosity.

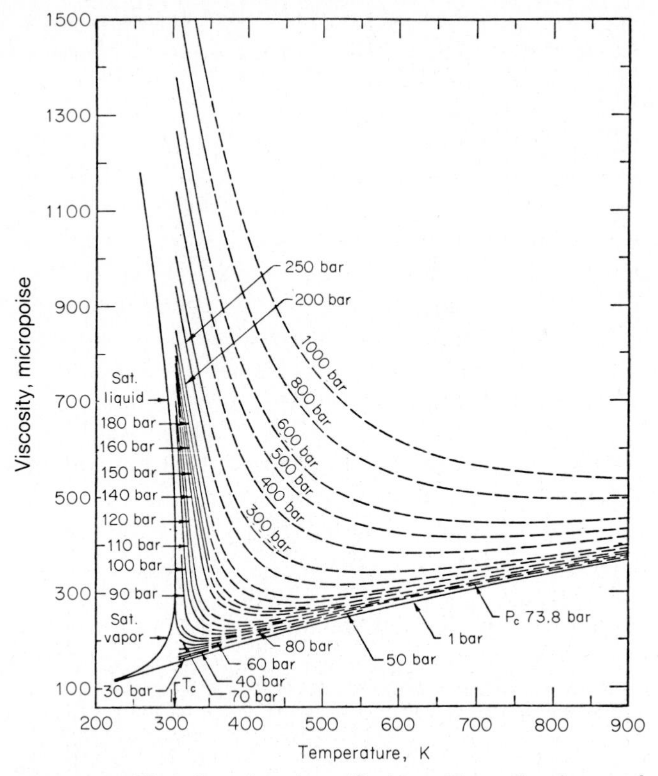


**Figure 9-5** Viscosity of carbon dioxide. *(From Stephan and Lucas, Ref. 188.)*

Finally, at very high reduced temperatures, there again results a condition in which pressure has little effect and viscosities increase with temperature.

## Enskog dense-gas theory

One of the very few theoretical efforts to predict the effect of pressure on the viscosity of gases is due to Enskog and is treated in detail by Chapman and Cowling [40]. The theory has also been applied to dense gas diffusion coefficients, bulk viscosities, and, for monatomic gases, thermal conductivities. The assumption is made that the gas consists of dense, hard spheres and behaves like a low-density hard-sphere system except that all events occur at a faster rate due to the higher rates of collision [1, 2]. The increase in collision rate is proportional to the radial distribution function $\Psi$. The Enskog equation for shear viscosity is

$$\frac{\eta}{\eta^{\circ}} = \Psi^{-1} + 0.8 b_0 \rho + 0.761 \Psi (b_0 \rho)^2 \tag{9-6.1}$$

where $\eta$ = viscosity, $\mu$P
$\eta^\circ$ = low-pressure viscosity, $\mu$P
$b_0$ = excluded volume = $\frac{2}{3}\pi N_0 \sigma^3$, cm$^3$/mol
$N_0$ = Avogadro's number
$\sigma$ = hard-sphere diameter, Å
$\rho$ = molar density, mole/cm$^3$

And $\Psi$ is the radial distribution function at contact and can be related to an equation of state by

$$\Psi = \frac{Z - 1}{\rho b_0} \tag{9-6.2}$$

where $Z$ is the compressibility factor. In the Enskog model, there is no correlation between successive hard-sphere collisions (the molecular chaos approximation). Dymond [61, 62, 63, 64], among others [3, 55, 175], has used molecular dynamics to correct for this deficiency as well as to account for attractive forces. Approaches such as these may lead to reli-

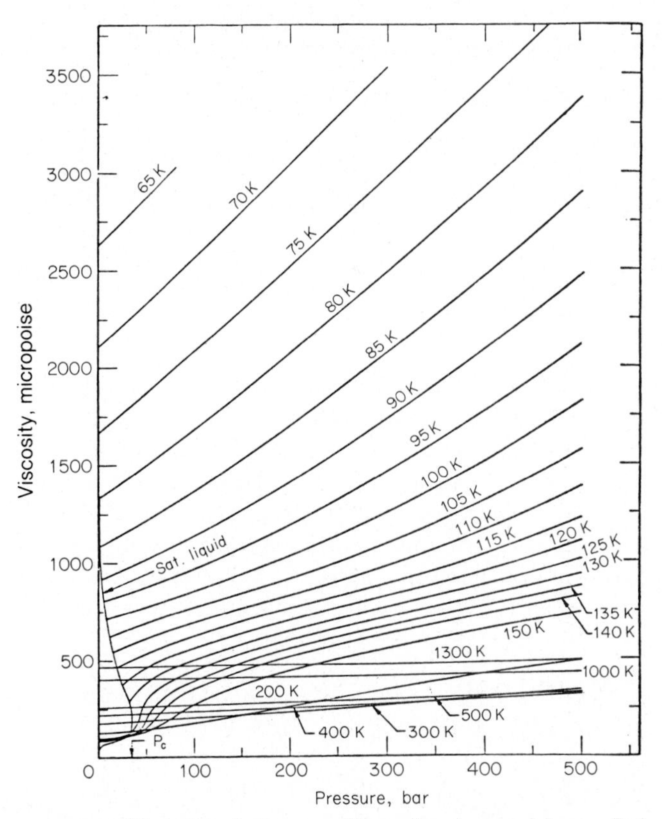


**Figure 9-6** Viscosity of nitrogen. *(From Stephan and Lucas, Ref. 188.)*

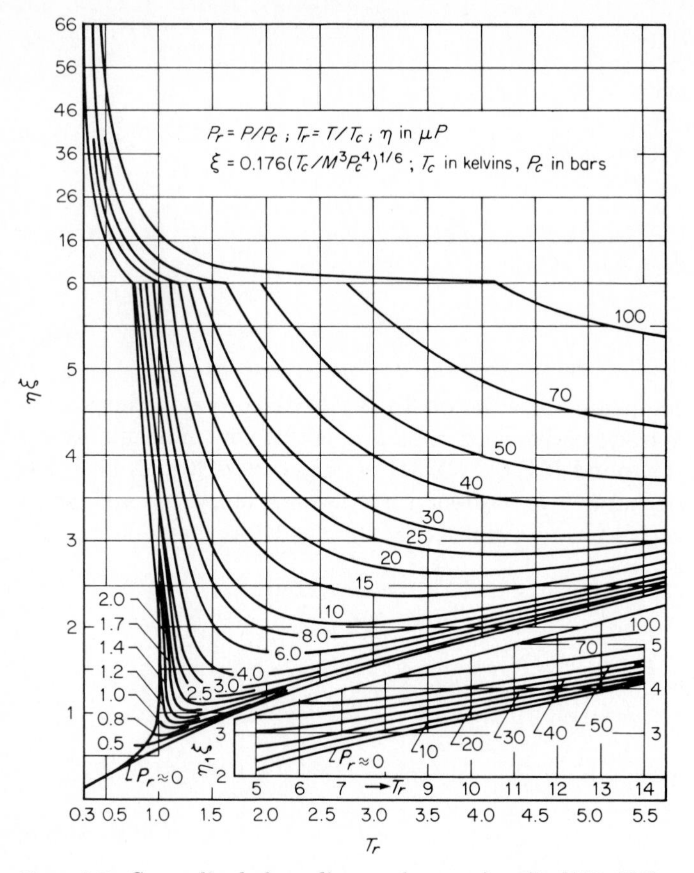


**Figure 9-7** Generalized phase diagram for gas viscosity [136, 137].

able predictive methods, but at the present time the more accurate techniques are based on the concept of corresponding states.

Following the lead of Enskog, many investigators have proposed that the dense-gas viscosity be correlated as $(\eta/\eta^\circ)$, $(\eta/\eta_c^\circ)$, or $(\eta/\eta_c)$ as a function of $T_r$ and $P_r$ or other reduced volumetric properties [10, 14, 20, 22, 28, 37, 38, 46, 47, 48, 49, 76, 83, 88, 133, 134, 171, 178]. In such ratios, $\eta_c$ is the true viscosity at the critical point, $\eta_c^\circ$ is the low-pressure viscosity at $T_c$, and $\eta^\circ$ is the low-pressure viscosity at the system temperature.

### Reichenberg method [165, 166, 168]

In this case, the viscosity ratio $\eta/\eta^\circ$ is given by Eq. (9-6.3)

$$\frac{\eta}{\eta^\circ} = 1 + Q\frac{AP_r^{3/2}}{BP_r + (1 + CP_r^D)^{-1}} \tag{9-6.3}$$

The constants, $A$, $B$, $C$, and $D$ are functions of the reduced temperature $T_r$:

$$A = \frac{\alpha_1}{T_r} \exp \alpha_2 T_r^a \qquad B = A(\beta_1 T_r - \beta_2)$$

$$C = \frac{\gamma_1}{T_r} \exp \gamma_2 T_r^c \qquad D = \frac{\delta_1}{T_r} \exp \delta_2 T_r^d$$

$$\begin{array}{llll} \alpha_1 = 1.9824 \times 10^{-3} & \alpha_2 = 5.2683 & a = -0.5767 & \beta_1 = 1.6552 \\ \beta_2 = 1.2760 & \gamma_1 = 0.1319 & \gamma_2 = 3.7035 & c = -79.8678 \\ \delta_1 = 2.9496 & \delta_2 = 2.9190 & d = -16.6169 & \end{array}$$

and $Q = (1\text{-}5.655\mu_r)$, where $\mu_r$ is defined in Eq. (9-4.16). For nonpolar materials, $Q = 1.0$. Example 9-9 illustrates the application of Eq. (9-6.3), and, in Table 9-6, experimental dense gas viscosities are compared to the viscosities estimated with this method. Errors are generally only a few percent; the poor results for ammonia at 420 K seem to be an anomaly.

**Example 9-9** Use Reichenberg's method to estimate the viscosity of $n$-pentane vapor at 500 K and 101 bar. The experimental value is 546 $\mu$P [188].

**solution** Whereas one could estimate the low-pressure viscosity of $n$-pentane at 500 K by using the methods described in Sec. 9-4, the experimental value is available (114 $\mu$P) [188] and will be used.
The dipole moment of $n$-pentane is zero, so $Q = 1.0$. From Appendix A, $T_c$ = 469.7 K and $P_c$ = 33.7 bar. Thus $T_r = (500/469.7) = 1.065$ and $P_r = (101/33.7) = 3.00$. From the definitions of $A$, $B$, $C$, and $D$ given under Eq. (9-6.3), $A$ = 0.2999, $B$ = 0.1458, $C$ = 1.271, and $D$ = 7.785. With Eq. (9-6.3),

$$\frac{\eta}{\eta^\circ} = 1 + \frac{(0.2999)(3.00)^{3/2}}{(0.1458)(3.00) + [1 + (1.271)(3.00)^{7.785}]^{-1}} = 4.56$$

$$\eta = (4.56)(114) = 520\ \mu\text{P}$$

$$\text{Error} = \frac{520 - 546}{546} \times 100 = -4.7\%$$

If one refers back to Fig. 9-7, at $T_r = 1.065$ and $P_r = 3.00$, the viscosity is changing rapidly with both temperature and pressure. Thus, an error of only 5% is quite remarkable.

### Lucas method

In a technique which, in some aspects, is similar to Reichenberg's, Lucas [135, 136, 137] recommends the following procedure. For the reduced temperature of interest, first calculate a parameter $Z_1$:

$$Z_1 = [0.807\,T_r^{0.618} - 0.357 \exp(-0.449T_r) + 0.340 \exp(-4.058T_r) + 0.018]F_P^\circ F_Q^\circ \qquad (9\text{-}6.4)$$

Note that, from Eq. (9-4.15), $Z_1 = \eta^\circ\xi$, where $\eta^\circ$ refers to the low-pressure

If $T_r \leq 1.0$ and $P_r < (P_{vp}/P_c)$, then

$$Z_2 = 0.600 + 0.760P_r^{\alpha} + (6.990P_r^{\beta} - 0.6)(1 - T_r) \tag{9-6.5}$$

with $\alpha = 3.262 + 14.98P_r^{5.508}$

$\beta = 1.390 + 5.746P_r$

If $(1 < T_r < 40)$ and $(0 < P_r \leq 100)$, then

$$Z_2 = \eta°\xi\left[1 + \frac{aP_r^e}{bP_r^f + (1 + cP_r^d)^{-1}}\right] \tag{9-6.6}$$

where $\eta°\xi$ is found from Eq. (9-6.4). The term multiplying this group is identical to the pressure correction term in Reichenberg's method, Eq. (9-6.3), but the values of the constants are different.

$$a = \frac{a_1}{T_r}\exp a_2T_r^{\gamma}$$

$$b = a(b_1T_r - b_2)$$

$$c = \frac{c_1}{T_r}\exp c_2T_r^{\delta}$$

$$d = \frac{d_1}{T_r}\exp d_2T_r^{\epsilon}$$

$$e = 1.3088$$

$$f = f_1 \exp f_2T_r^{\zeta}$$

and

| | | |
|---|---|---|
| $a_1 = 1.245 \times 10^{-3}$ | $a_2 = 5.1726$ | $\gamma = -0.3286$ |
| $b_1 = 1.6553$ | $b_2 = 1.2723$ | |
| $c_1 = 0.4489$ | $c_2 = 3.0578$ | $\delta = -37.7332$ |
| $d_1 = 1.7368$ | $d_2 = 2.2310$ | $\epsilon = -7.6351$ |
| $f_1 = 0.9425$ | $f_2 = -0.1853$ | $\zeta = 0.4489$ |

After computing $Z_1$ and $Z_2$, we define

$$Y = \frac{Z_2}{Z_1} \tag{9-6.7}$$

and the correction factors $F_P$ and $F_Q$,

$$F_P = \frac{1 + (F_P° - 1)Y^{-3}}{F_P°} \tag{9-6.8}$$

$$F_Q = \frac{1 + (F_Q° - 1)[Y^{-1} - (0.007)(\ln Y)^4]}{F_Q°} \tag{9-6.9}$$

where $F_P^\circ$ and $F_Q^\circ$ are low-pressure polarity and quantum factors determined as shown in Eqs. (9-4.17) and (9-4.18). Finally, the dense gas viscosity is calculated as

$$\eta = \frac{Z_2 F_P F_Q}{\xi} \tag{9-6.10}$$

where $\xi$ is defined in Eq. (9-4.14). At low pressures, $Y$ is essentially unity, and $F_P = 1$, $F_Q = 1$. Also $Z_2$ then equals $\eta^\circ \xi$ so $\eta \rightarrow \eta^\circ$, as expected.

The Lucas method is illustrated in Example 9-10, and calculated dense gas viscosities are compared with experimental data in Table 9-6. In Example 9-10 and Table 9-6, the low-pressure viscosity $\eta^\circ$ was not obtained from experimental data but was estimated by the Lucas method in Sec. 9-4. Except in a few cases, the error was found to be less than 5 percent. The critical temperature, critical pressure, critical compressibility factor, and dipole moment are required, as well as the system temperature and pressure.

**Example 9-10** Estimate the viscosity of ammonia gas at 420 K and 300 bar by using Lucas's method. The experimental values of $\eta$ and $\eta^\circ$ are 571 and 146 $\mu$P [188].

**solution** From Appendix A, for ammonia, $M = 17.03$, $Z_c = 0.244$, $T_c = 405.5$ K, $P_c = 113.5$ bar, and $\mu = 1.47$ debyes. Thus, $T_r = (420/405.5) = 1.036$ and $P_r = (300/113.5) = 2.643$. From Eq. (9-4.14),

$$\xi = (0.176)\left[\frac{405.5}{(17.03)^3(113.5)^4}\right]^{1/6} = 4.96 \times 10^{-3}\ (\mu\text{P})^{-1}$$

and, with Eq. (9-4.16),

$$\mu_r = (52.46)\left[\frac{(1.47)^2(113.5)}{(405.5)^2}\right] = 7.825 \times 10^{-2}$$

$$F_Q^\circ = 1.0$$

and with Eq. (9-4.17),

$$F_P^\circ = 1 + 30.55(0.292 - 0.244)^{1.72}\ |0.96 + (0.1)(1.036 - 0.7)|$$
$$= 1.164$$

From Eq. (9-6.4), $Z_1 = \eta^\circ \xi = 0.7258$

$$\eta^\circ = \frac{0.7258}{4.96 \times 10^{-3}} = 146\ \mu\text{P}$$

The estimation of the low-pressure viscosity of ammonia agrees very well with the experimental value.

Since $T_r > 1.0$, we use Eq. (9-6.6) to determine $Z_2$. The values of the coefficients are $a = 0.1998$, $b = 8.834 \times 10^{-2}$, $c = 0.9764$, $d = 9.235$, $e = 1.3088$, and $f = 0.7808$. Then,

$$Z_2 = \left\{1 + \frac{(0.1998)(2.643)^{1.3088}}{(8.834 \times 10^{-2})(2.643)^{0.7808} + [1 + (0.9764)(2.643)^{9.235}]^{-1}}\right\}(0.7258)$$
$$= (4.776)(0.7258) = 3.466$$

With Eqs. (9-6.7) to (9-6.9),

$$Y = \frac{3.466}{0.7258} = 4.776$$

$$F_P = \frac{1 + (1.164 - 1)(4.776)^{-3}}{1.164} = 0.860$$

$$F_Q = 1.0$$

and, with Eq. (9-6.10),

$$\eta = \frac{(3.466)(0.860)(1.0)}{4.96 \times 10^{-3}} = 601\ \mu\text{P}$$

$$\text{Error} = \frac{601 - 571}{571} \times 100 = 5.2\%$$

The Reichenberg and Lucas methods employ temperature and pressure as the state variables. In most other dense gas viscosity correlations, however, the temperature and density (or specific volume) are used. In those cases, one must have accurate volumetric data or an applicable equation of state to determine the dense gas viscosity. Three different methods are illustrated below.

**Method of Jossi, Stiel, and Thodos [116, 190]**

In this case the *residual* viscosity, $\eta - \eta°$, is correlated with fluid density. All temperature effects are incorporated in the $\eta°$ term. To illustrate the behavior of the $\eta - \eta°$ function, consider Fig. 9-8, which shows $\eta - \eta°$ for $n$-butane graphed as a function of density [59]. Note that there does not appear to be any specific effect of temperature over the range shown. At the highest density, 0.6 g/cm$^3$, the reduced density $\rho/\rho_c$ is 2.63. Similar plots for many other substances are available, for example, He, air, $O_2$, $N_2$, $CH_4$ [121]; ammonia [34, 179]; rare gases [178]; diatomic gases [20]; sulfur dioxide [180]; $CO_2$ [117]; steam [122]; and various hydrocarbons [35, 69, 80, 186, 187]. Other authors have also shown the applicability of a residual viscosity-density correlation [83, 95, 122, 170, 184, 185].

In the Jossi, Stiel, and Thodos method, separate residual viscosity expressions are given for nonpolar and polar gases, but no quantitative criterion is presented to distinguish these classes.

**Nonpolar gases [116]**

$$[(\eta - \eta°)\xi_T + 1]^{1/4} = 1.0230 + 0.23364\rho_r + 0.58533\rho_r^2 - 0.40758\rho_r^3 + 0.093324\rho_r^4 \qquad (9\text{-}6.11)$$

where $\eta$ = dense gas viscosity, $\mu$P
$\eta°$ = low-pressure gas viscosity, $\mu$P
$\rho_r$ = reduced gas density $\rho/\rho_c = V_c/V$
$\xi_T$ = the group $(T_c/M^3P_c^4)^{1/6}$, where $T_c$ is in kelvins and $P_c$ is in atmospheres, $(\mu\text{P})^{-1}$
$M$ = molecular weight, g/mol

This relation is reported by Jossi et al. to be applicable in the range $0.1 \le \rho_r < 3$.

**Polar gases [190]**

$$(\eta - \eta^\circ)\xi_T = 1.656\rho_r^{1.111} \qquad \rho_r \le 0.1 \qquad (9\text{-}6.12)$$

$$(\eta - \eta^\circ)\xi_T = 0.0607(9.045\rho_r + 0.63)^{1.739} \qquad 0.1 \le \rho_r \le 0.9 \qquad (9\text{-}6.13)$$

$$\log\{4 - \log[(\eta - \eta^\circ)\xi_T]\} = 0.6439 - 0.1005\rho_r - \Delta \qquad 0.9 \le \rho_r < 2.6 \qquad (9\text{-}6.14)$$

where $\Delta = 0$ when $0.9 \le \rho_r \le 2.2$

$$\Delta = (4.75 \times 10^{-4})(\rho_r^3 - 10.65)^2 \qquad \text{when } 2.2 < \rho_r < 2.6 \qquad (9\text{-}6.15)$$

and $(\eta - \eta^\circ)\xi_T = 90.0$ and 250 at $\rho_r = 2.8$ and 3.0, respectively. The notation used in Eqs. (9-6.12) to (9-6.15) is defined under Eq. (9-6.11). Note that the parameter $\xi_T$ is *not* the same as $\xi$ defined earlier in Eq. (9-4.14).

An example of the Jossi et al. method is shown below, and calculated dense gas viscosities are compared with experimental values in Table 9-6.

**Example 9-11** Use the Jossi, Stiel, and Thodos method to estimate the viscosity of isobutane at 500 K and 100 bar. The experimental value is 261 $\mu$P [188] and,

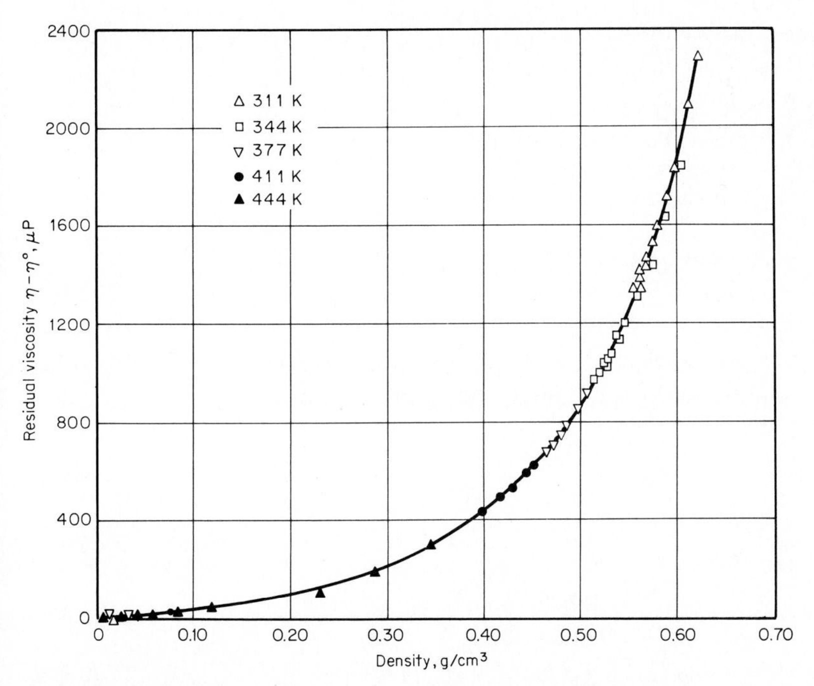


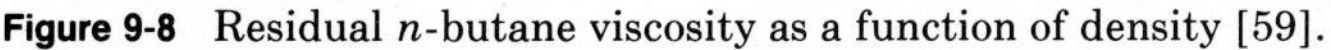
**Figure 9-8** Residual $n$-butane viscosity as a function of density [59].

at 500 K, 100 bar; the specific volume of isobutane is 243.8 $cm^3/mol$ [219]. At low pressures, 500 K, $\eta° = 120\ \mu P$.

**solution** Since isobutane is nonpolar, Eq. (9-6.11) is used. From Appendix A, $T_c = 408.2$ K, $P_c = 36.5$ bar $= 36.0$ atm, $V_c = 263\ cm^3/mol$, and $M = 58.12$. Then

$$\xi_T = \left[\frac{(408.2)}{(58.12)^3(36.0)^4}\right]^{1/6} = 3.277 \times 10^{-2}\ (\mu P)^{-1}$$

The reduced density $= \rho_r = V_c/V = 263/243.8 = 1.079$. With Eq. (9-6.11),

$$[(\eta - 120)(3.277 \times 10^{-2}) + 1]^{1/4} = 1.0230 + (0.23364)(1.079) + (0.58533)(1.079)^2 - (0.40758)(1.079)^3 + (0.093324)(1.079)^4$$

$$= 1.571$$

$$\eta = 275\ \mu P$$

$$\text{Error} = \frac{275 - 261}{261} \times 100 = 5.4\%$$

### Chung et al. method [44]

In an extension of the Chung et al. technique to estimate low-pressure gas viscosities, the authors began with Eq. (9-3.9) and employed empirical correction factors to account for the fact that the fluid has a high density. Their relations are shown below.

$$\eta = \eta^* \frac{36.344(MT_c)^{1/2}}{V_c^{2/3}} \tag{9-6.16}$$

where $\eta$ = viscosity, $\mu P$
$M$ = molecular weight, g/mol
$T_c$ = critical temperature, K
$V_c$ = critical volume, $cm^3/mol$

and

$$\eta^* = \frac{(T^*)^{1/2}}{\Omega_v}\{F_c[(G_2)^{-1} + E_6 y]\} + \eta^{**} \tag{9-6.17}$$

$T^*$ and $F_c$ are defined as in Eqs. (9-4.8) and (9-4.10). $\Omega_v$ is found with Eq. (9-4.3) as a function of $T^*$, and, with $\rho$ in $mol/cm^3$,

$$y = \frac{\rho V_c}{6} \tag{9-6.18}$$

$$G_1 = \frac{1 - 0.5y}{(1 - y)^3} \tag{9-6.19}$$

$$G_2 = \frac{E_1\{[1 - \exp(-E_4 y)]/y\} + E_2 G_1 \exp(E_5 y) + E_3 G_1}{E_1 E_4 + E_2 + E_3} \tag{9-6.20}$$

$$\eta^{**} = E_7 y^2 G_2 \exp[E_8 + E_9 (T^*)^{-1} + E_{10}(T^*)^{-2}] \tag{9-6.21}$$

and the parameters $E_1$ to $E_{10}$ are given in Table 9-5 as linear functions of $\omega$ (the acentric factor), $\mu_r^4$ [as defined in Eq. (9-4.11)], and the association factor $\kappa$ (see Table 9-1). One might note that, at very low densities, $y$ approaches zero, $G_1$ and $G_2$ approach unity, and $\eta^{**}$ is negligible. At these limiting conditions, combining Eqs. (9-6.16) and (9-6.17) and (9-4.8) leads to Eq. (9-4.9), which then applies for estimating $\eta°$.

The application of the Chung et al. method is shown in Example 9-12. Some calculated values of $\eta$ are compared with experimental results in Table 9-6. The agreement is quite good and errors usually are below 5%.

**Example 9-12** With the Chung et al. method, estimate the viscosity of ammonia at 520 K and 600 bar. The experimental value of $\eta$ is 466 $\mu$P [188]. At this temperature, $\eta° = 182\ \mu$P. The specific volume of ammonia at 520 K and 600 bar is 48.2 cm$^3$/mol [93].

**solution** From Appendix A, $T_c = 405.5$ K, $V_c = 72$ cm$^3$/mol, $\omega = 0.250$, $M = 17.03$, and $\mu = 1.47$ debyes. Thus

$$T_r = 520/405.5 = 1.282 \text{ and } \rho = 1/48.2 = 2.07 \times 10^{-2} \text{ mol/cm}^3.$$

With Eq. (9-4.11),

$$\mu_r = \frac{(131.3)(1.47)}{[(72)(405.5)]^{1/2}} = 1.13$$

and with Eq. (9-4.10),

$$F_c = 1 - (0.2756)(0.250) + (0.059035)(1.13)^4 = 1.208$$

$$T^* = (1.2593)(1.282) = 1.615$$

and with Eq. (9-4.3), $\Omega_v = 1.275$. Using Eqs. (9-6.18) and (9-6.19),

$$y = \frac{(2.075 \times 10^{-2})(72)}{6} = 0.249 \qquad \text{and} \qquad G_1 = 2.067$$

From Table 9-5, the following coefficients were computed: $E_1 = -64.82$, $E_2 = -9.218 \times 10^{-3}$, $E_3 = -204.2$, $E_4 = 2.430$, $E_5 = -1.609$, $E_6 = 3.045$, $E_7 = 6.839$,

**TABLE 9-5 Chung et al. Coefficients to Calculate $E_i$**

$E_i = a_i + b_i\omega + c_i\mu_r^4 + d_i\kappa$

| $i$ | $a_i$ | $b_i$ | $c_i$ | $d_i$ |
|---|---|---|---|---|
| 1 | 6.324 | 50.412 | −51.680 | 1189.0 |
| 2 | $1.210 \times 10^{-3}$ | $-1.154 \times 10^{-3}$ | $-6.257 \times 10^{-3}$ | 0.03728 |
| 3 | 5.283 | 254.209 | −168.48 | 3898.0 |
| 4 | 6.623 | 38.096 | −8.464 | 31.42 |
| 5 | 19.745 | 7.630 | −14.354 | 31.53 |
| 6 | −1.900 | −12.537 | 4.985 | −18.15 |
| 7 | 24.275 | 3.450 | −11.291 | 69.35 |
| 8 | 0.7972 | 1.117 | 0.01235 | −4.117 |
| 9 | −0.2382 | 0.06770 | −0.8163 | 4.025 |
| 10 | 0.06863 | 0.3479 | 0.5926 | −0.727 |

**TABLE 9-6 Comparison of Experimental and Calculated Dense Gas Viscosities**

| | | | | | | | | Percent error† by method of | | | | |
|---|---|---|---|---|---|---|---|---|---|---|---|---|
| Compound | $T$, K | $P$, bar | $V$, $cm^3$/mole | Ref. | $\eta$, $\mu P$ | $\eta°$, $\mu$P | Ref. | Reichenberg, Eq. (9-6.3) | Lucas, Eq. (9-6.10) | Jossi et al., Eq. (9-6.11) | Chung et al., Eq. (9-6.16) Table 9-5 | Brulé and Starling, Eq. (9-6.16) Table 9-7 |
| Oxygen | 300 | 30.4 | 806.1 | 189 | 212.8 | 207.2 | 188 | −1.0 | −1.6 | 0.6 | −1.5 | 0.2 |
| | | 81.0 | 295.3 | | 225.7 | | | −1.2 | −1.1 | −0.6 | −1.9 | 0.8 |
| | | 152.0 | 155.3 | | 250.3 | | | −0.3 | −0.2 | −0.8 | −0.2 | 1.6 |
| | | 304.0 | 81.4 | | 319.3 | | | 3.6 | 0.8 | 2.8 | 3.9 | 4.6 |
| Methane | 200 | 40.0 | 282.0 | 84 | 90 | 78.0 | 188 | 7.0 | 0.6 | 5.9 | 3.5 | 1.7 |
| | | 100.0 | 60.2 | | 296 | | | 10.0 | 8.2 | 5.1 | 3.1 | 14 |
| | | 200.0 | 51.1 | | 415 | | | 3.8 | 5.0 | −0.5 | −2.2 | 16 |
| | 500 | 40 | 1039.0 | 84 | 180 | 177 | 188 | −0.4 | −5.6 | 0.9 | −5.3 | −3.8 |
| | | 100 | 417.7 | | 187 | | | −0.6 | −5.1 | 0.3 | −7.2 | −2.3 |
| | | 200 | 213.7 | | 204 | | | −1.0 | −5.0 | −0.5 | −2.9 | −1.0 |
| | | 500 | 98.9 | | 263 | | | 1.5 | −3.3 | 3.3 | 2.5 | 5.3 |
| Isobutane | 500 | 20 | 2396.0 | 219 | 127 | 120 | 188 | 0.9 | 6.3 | 0.2 | 0.8 | 2.2 |
| | | 50 | 620.0 | | 146 | | | 5.7 | 12.0 | 4.5 | 9.3 | 12 |
| | | 100 | 244.0 | | 261 | | | −5.2 | 3.8 | 5.4 | 5.5 | 8.6 |
| | | 200 | 159.0 | | 506 | | | −11 | 2.3 | −9.0 | −7.2 | −5.0 |
| | | 400 | 130.0 | | 794 | | | −19 | −8.2 | −16 | −10 | −9.9 |

| | | | | | | | | | | | | |
|---|---|---|---|---|---|---|---|---|---|---|---|---|
| Ammonia | 420 | 50 | 588 | 93 | 149 | 146 | 188 | 3.0 | −2.4 | 1.7 | 3.1 | −5.2 |
| | | 150 | 61.9 | | 349 | | | −17 | −6.5 | −15 | −13 | 5.1 |
| | | 300 | 39.8 | | 571 | | | −21 | 5.2 | 3.6 | −4.0 | 22 |
| | | 600 | 34.3 | | 752 | | | −24 | 7.8 | 11 | −1.3 | 31 |
| | 520 | 50 | 807.6 | 93 | 185 | 182 | 188 | 0.7 | −5.8 | 0.5 | −0.1 | −9.2 |
| | | 150 | 229.6 | | 196 | | | 4.5 | 0.9 | 2.3 | 4.0 | 1.6 |
| | | 300 | 90.7 | | 296 | | | −1.4 | 5.3 | −2.5 | 0.7 | 13 |
| | | 600 | 48.2 | | 466 | | | −13 | 5.8 | −3.2 | −3.4 | 12 |
| Carbon dioxide | 360 | 50 | 514.6 | 11 | 190 | 177 | 188 | 3.0 | 3.1 | 1.9 | 1.1 | 2.4 |
| | | 100 | 211.2 | | 230 | | | 2.1 | 3.3 | 0.8 | 3.6 | 6.1 |
| | | 400 | 55.0 | | 730 | | | 1.3 | 7.7 | −3.5 | −0.8 | 1.2 |
| | | 800 | 45.8 | | 1104 | | | −7.0 | 1.1 | −9.6 | −2.2 | −1.3 |
| | 500 | 50 | 802.8 | 11 | 243 | 235 | 188 | 0.0 | 3.0 | 1.3 | 0.7 | 1.6 |
| | | 100 | 389.2 | | 254 | | | 1.7 | 5.1 | 1.4 | 3.3 | 5.3 |
| | | 400 | 97.1 | | 411 | | | 9.8 | 7.4 | 2.6 | 3.6 | 9.4 |
| | | 800 | 62.9 | | 636 | | | 10 | 9.6 | 0.9 | −3.2 | 2.8 |
| *n*-Pentane | 600 | 20.3 | 2240 | 52 | 143 | 134 | 188 | 0.0 | 1.4 | 0.0 | 1.2 | 2.1 |
| | | 81.1 | 418.3 | | 242 | | | −7.5 | −5.3 | −11 | −4.6 | −1.0 |
| | | 152 | 237.5 | | 382 | | | 0.9 | 2.3 | −7.9 | −7.0 | −3.7 |

†Percent error = [(calc. − exp.)/exp.] × 100

$E_8 = 1.097$, $E_9 = -1.544$, and $E_{10} = 1.116$. Then, with Eq. (9-6.20), $G_2 = 1.494$ and, from Eq. (9-6.21), $\eta^{**} = 1.118$.
Finally, using Eqs. (9-6.17) and (9-6.16),

$$\eta^* = \frac{(1.615)^{1/2}}{1.275}(1.208)[(1.494)^{-1} + (3.045)(0.249)] + 1.118$$

$$= 2.579$$

$$\eta = \frac{(2.579)(36.344)[(17.03)(405.5)]^{1/2}}{(72)^{2/3}}$$

$$= 450\ \mu\text{P}$$

$$\text{Error} = \frac{450 - 466}{466} \times 100 = -3.4\%$$

### Brulé and Starling Method [30]

In a manner identical in form with that of Chung et al., Brulé and Starling propose a different set of coefficients for $E_1$ to $E_{10}$. These are shown in Table 9-7. Note that no polarity terms are included and the *orientation* parameter $\gamma$ has replaced the acentric factor $\omega$. If values of $\gamma$ are not available, the acentric factor may be substituted.

The Brulé and Starling technique was developed to be more applicable for heavy hydrocarbons rather than for simple molecules as tested in Table 9-6.

### Discussion and recommendations for estimating dense gas viscosities

Five estimation techniques were discussed in this section. Two (Reichenberg's and Lucas's) were developed to use temperature and pressure as

**TABLE 9-7 Brulé and Starling Coefficients to Calculate $E_i$**
**$E_i = a_i + b_i\gamma$**

| $i$ | $a_i$ | $b_i$ |
|---|---|---|
| 1 | 17.450 | 34.063 |
| 2 | $-9.611 \times 10^{-4}$ | $7.235 \times 10^{-3}$ |
| 3 | 51.043 | 169.46 |
| 4 | $-0.6059$ | 71.174 |
| 5 | 21.382 | $-2.110$ |
| 6 | 4.668 | $-39.941$ |
| 7 | 3.762 | 56.623 |
| 8 | 1.004 | 3.140 |
| 9 | $-7.774 \times 10^{-2}$ | $-3.584$ |
| 10 | 0.3175 | 1.1600 |

Note: If $\gamma$ values are not available, use $\omega$, the acentric factor, or, preferably, obtain from multiproperty analysis by using vapor pressure data [30].

the input variables to estimate the viscosity. The other three required temperature and density; thus, an equation of state would normally be required to obtain the necessary volumetric data if not directly available. In systems developed to estimate many types of properties, it would not be difficult to couple the *PVT* and viscosity programs to provide densities when needed. In fact, the Brulé and Starling method [30] is predicated on combining thermodynamic and transport analyses to obtain the characterization parameters most suitable for both types of estimations.

Another difference to be recognized among the methods noted in this section is that Reichenberg's and Jossi et al.'s methods require a low-pressure viscosity at the same temperature. The other techniques bypass this requirement and have imbedded into the methods a low-pressure estimation method; i.e., at low densities they reduce to techniques as described in Sec. 9-4. If the Lucas, Chung et al., or Brulé-Starling method were selected, no special low-pressure estimation method would have to be included in a property estimation package.

With these few remarks, along with the testing in Table 9-6 as well as evaluations by authors of the methods, we recommend that either the Lucas or Chung et al. procedure be used to estimate dense (and dilute) gas viscosities of both polar and nonpolar compounds. The Brulé-Starling method is, however, preferable when complex hydrocarbons are of interest, but even for those materials, the Chung et al. procedure should be used at low reduced temperatures ($T_r < 0.5$).

Except when one is working in temperature and pressure ranges in which viscosities are strong functions of these variables (see Fig. 9-7), errors for the recommended methods are usually only a few percent. Near the critical point and in regions where the fluid density is approaching that of a liquid, higher errors may be encountered.

## 9-7 Viscosity of Gas Mixtures at High Pressures

The most convenient method to estimate the viscosity of dense gas mixtures is to combine, where possible, techniques given previously in Secs. 9-5 and 9-6.

### Lucas Approach [135, 136, 137]

In the (pure) dense gas viscosity approach suggested by Lucas, Eqs. (9-6.4) to (9-6.10) were used. To apply this technique to mixtures, rules must be chosen to obtain $T_c$, $P_c$, $M$, and $\mu$ as functions of composition. For $T_c$, $P_c$, and $M$ of the mixture, Eqs. (9-5.18) to (9-5.20) should be used. The polarity (and quantum) corrections are introduced by using Eqs. (9-6.8) and (9-6.9), where $F_P^\circ$ and $F_Q^\circ$ refer to mixture values from Eqs. (9-5.21) and (9-5.22). The parameter $Y$ in Eqs. (9-6.8) and (9-6.9) must be based on $T_{cm}$ and $P_{cm}$. $F_P^\circ$ and $F_Q^\circ$, for the pure components, were defined in Eqs. (9-4.17) and (9-4.18).

### Chung et al. approach [44]

To use this method for dense gas mixtures, Eqs. (9-6.16) to (9-6.21) are used. The parameters $T_c$, $V_c$ $\omega$, $M$, $\mu$, and $\kappa$ in these equations are given as functions of composition in Sec. 9-5. That is,

| Parameter | Equations to use |
|---|---|
| $T_{cm}$ | (9-5.44), (9-5.27) |
| $V_{cm}$ | (9-5.43), (9-5.25) |
| $\omega_m$ | (9-5.29), (9-5.25) |
| $M_m$ | (9-5.28), (9-5.27), and (9-5.25) |
| $\mu_m$ | (9-5.30), (9-5.25) |
| $\kappa_m$ | (9-5.31) |

### Discussion

Both the Lucas and Chung et al. methods utilize the relations for the estimation of dense gas viscosity and apply a *one-fluid* approximation to relate the component parameters to composition. In the Lucas method, the state variables are $T$, $P$, and composition, whereas in the Chung et al. procedure, $T$, $\rho$, and composition are used. Similar, but less accurate, methods also have been proposed [54, 80].

The accuracy of the Lucas and Chung et al. forms is somewhat less than when applied to pure, dense gases. Also, as noted at the end of Sec. 9-6, the accuracy is often poor when working in the critical region or at densities approaching those of a liquid at the same temperature. The paucity of accurate high-pressure gas mixture viscosity data has limited the testing that could be done, but Chung et al. [44] report absolute average deviations of 8 to 9 percent for both polar and nonpolar dense gas mixtures. A comparable error would be expected from the Lucas form.

As a final comment to the first half of this chapter, it should be noted that, if one were planning a property estimation *system* for use on a high-speed computer, it is recommended that the Lucas, Chung et al., or Brulé and Starling method be used in the dense gas mixture viscosity correlations. Then, at low pressures or for pure components, the relations simplify directly to those described in Secs. 9-4 to 9-6. In other words, it is not necessary—when using these particular methods—to program separate relations for low-pressure pure gases, low-pressure gas mixtures, and high-pressure pure gases. One program is sufficient to cover all those cases as well as high-pressure gas mixtures. Ely and Hanley [70, 71] have also proposed a general viscosity estimation method.

## 9-8 Liquid Viscosity

Most gas and gas mixture estimation techniques for viscosity are modifications of theoretical expressions described briefly in Secs. 9-3 and 9-5. There is no comparable theoretical basis for the estimation of liquid viscosities.

The viscosities of liquids are larger than those of gases at the same temperature. As an example, in Fig. 9-9, the viscosities of liquid and vapor benzene are plotted as functions of temperature. Near the normal boiling point (353.4 K), the liquid viscosity is about 36 times the vapor viscosity, and at lower temperatures, this ratio increases even further. Two vapor

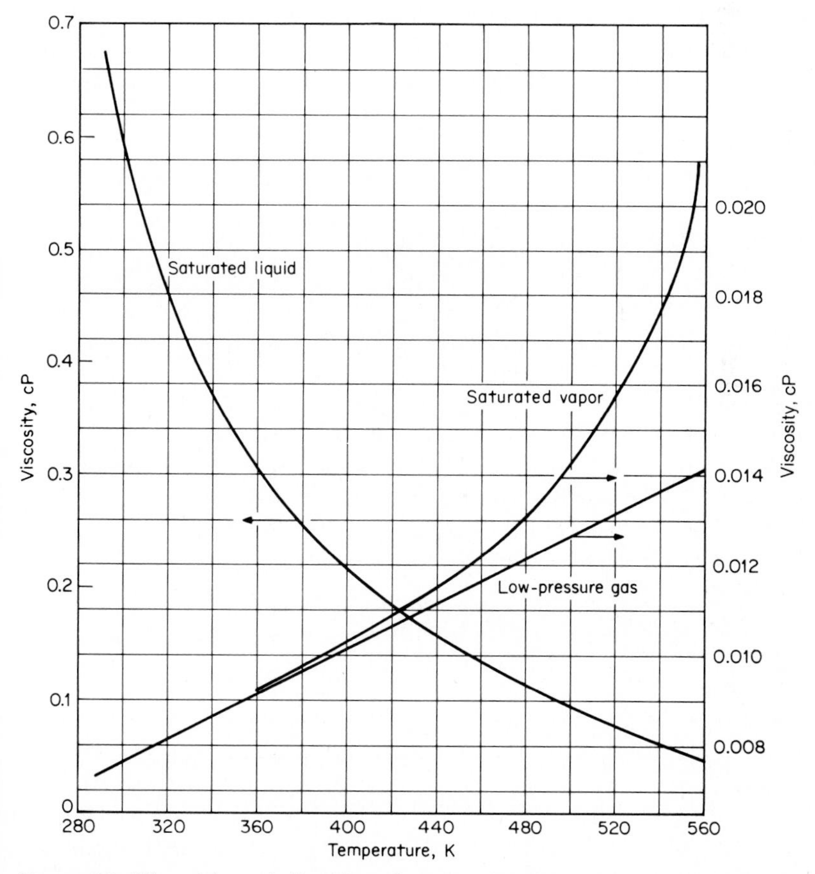


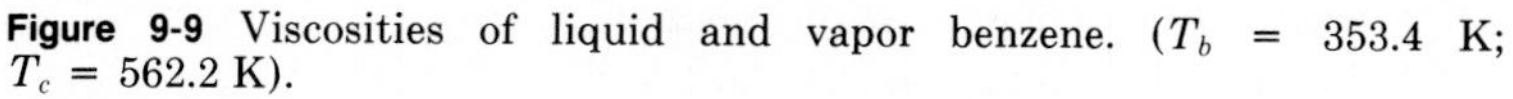
**Figure 9-9** Viscosities of liquid and vapor benzene. ($T_b$ = 353.4 K; $T_c$ = 562.2 K).

viscosities are shown in Fig. 9-9. The low-pressure gas line would correspond to vapor at about 1 bar. As noted earlier in Eq. (9-4.19), below $T_c$, low-pressure gas viscosities vary in a nearly linear manner with temperature. The curve noted as *saturated vapor* reflects the effect of the increase in vapor pressure at higher temperatures. The viscosity of the saturated vapor should equal that of the saturated liquid at the critical temperature (for benzene, $T_c$ = 562.2 K).

Much of the curvature in the liquid viscosity–temperature curve may be eliminated if the logarithm of the viscosity is plotted as a function of reciprocal (absolute) temperature. This change is illustrated in Fig. 9-10 for four saturated liquids: ethanol, benzene, $n$-heptane, and nitrogen. (To allow for variations in the temperature range, the reciprocal of the reduced temperature is employed.) Typically, the normal boiling point would be at a value of $T_r^{-1} \approx 1.5$. For temperatures below the normal boiling point ($T_r^{-1} > 1.5$), the logarithm of the viscosity varies linearly with $T_r^{-1}$. Above the normal boiling point, this no longer holds. In the nonlinear region, several corresponding states estimation methods have been suggested, and they are covered in Sec. 9-12. In the linear region,

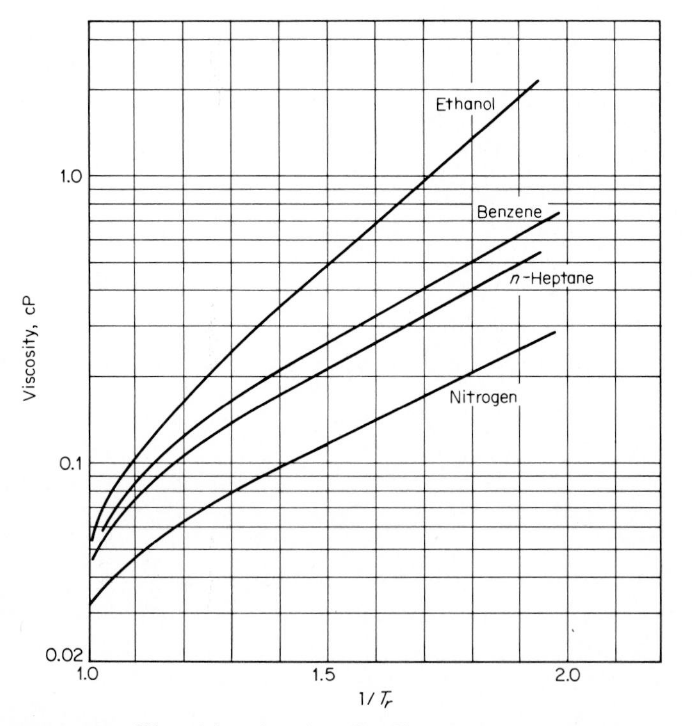


**Figure 9-10** Viscosities of various liquids as functions of temperature [188].

most corresponding states methods have not been found to be accurate, and many estimation techniques employ a group contribution approach to emphasize the effects of the chemical structure on viscosity. The curves in Fig. 9-10 suggest that, at comparable reduced temperatures, viscosities of polar fluids are higher than those of nonpolar liquids such as hydrocarbons, which themselves are larger than those of simple molecules such as nitrogen. If one attempts to replot Fig. 9-10 by using a nondimensional viscosity such as $\eta\xi$ [see, for example, Eqs. (9-4.12) to (9-4.14)] as a function of $T_r$, the separation between curves diminishes, especially at $T_r >$ 0.7. However, at lower values of $T_r$, there are still significant differences between the example compounds.

In the use of viscosity in engineering calculations, one is often interested not in the dynamic viscosity, but, rather, in the ratio of the dynamic viscosity to the density. This quantity, called the *kinematic viscosity,* would normally be expressed in $m^2/s$ or in stokes. One stoke (St) is equivalent to $10^{-4}$ $m^2/s$. When working with the kinematic viscosity $\nu$, this property

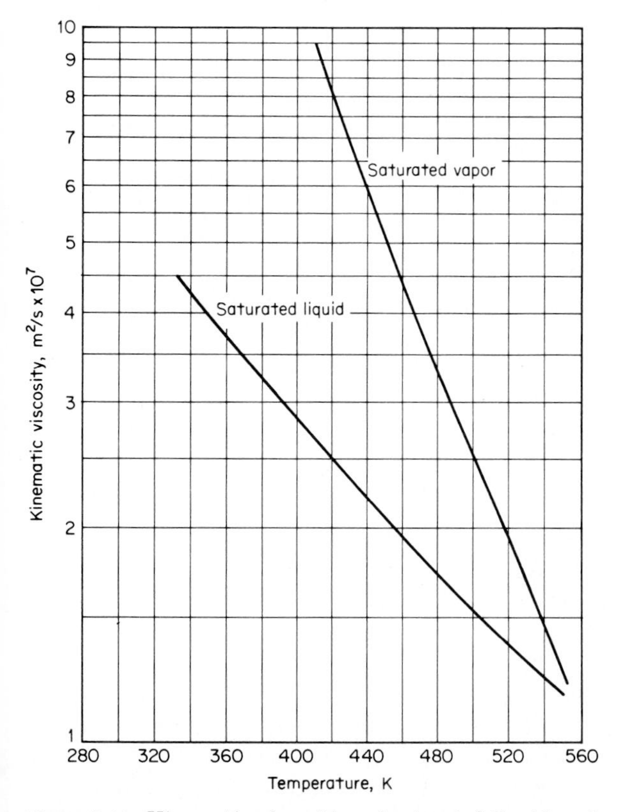


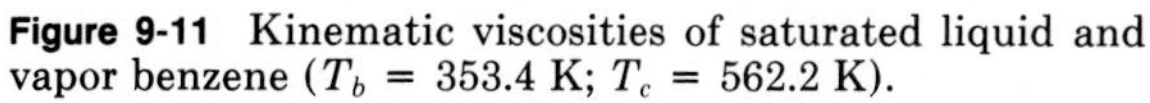
**Figure 9-11** Kinematic viscosities of saturated liquid and vapor benzene ($T_b$ = 353.4 K; $T_c$ = 562.2 K).

decreases in a manner such that ln $\nu$ is nearly linear in temperature for both the saturated liquid and vapor as illustrated in Fig. 9-11 for benzene. As with the dynamic viscosity, the kinematic viscosity of the saturated vapor and liquid become equal at the critical point.

The behavior of the kinematic viscosity with temperature has led to several correlation schemes to estimate $\nu$ rather than $\eta$. However, in most instances, ln $\nu$ is related to $T^{-1}$ rather than $T$. If Fig. 9-11 is replotted by using $T^{-1}$, again there is a nearly linear correlation with some curvature near the critical point (as there is in Fig. 9-10).

In summary, pure liquid viscosities at high reduced temperatures are usually correlated with some variation of the law of corresponding states (Sec. 9-12). At lower temperatures, most methods are empirical and involve a group contribution approach (Sec. 9-11). Current liquid *mixture* correlations are essentially mixing rules relating pure component viscosities to composition (Sec. 9-13). Little theory has been shown to be applicable to estimating liquid viscosities [9, 27, 31, 79, 100].

## 9-9 Effect of High Pressure on Liquid Viscosity

Increasing the pressure over a liquid results in an increase in viscosity. Lucas [136] has suggested that the change may be estimated from Eq. (9-9.1).

$$\frac{\eta}{\eta_{SL}} = \frac{1 + D\,(\Delta P_r/2.118)^A}{1 + C\omega\,\Delta P_r} \tag{9-9.1}$$

where $\eta$ = viscosity of the liquid at pressure

$\eta_{SL}$ = viscosity of the saturated liquid at $P_{vp}$

$\Delta P_r = (P - P_{vp})/P_c$

$\omega$ = acentric factor

$A = 0.9991 - [4.674 \times 10^{-4}/(1.0523T_r^{-0.03877} - 1.0513)]$

$D = [0.3257/(1.0039 - T_r^{2.573})^{0.2906}] - 0.2086$

$C = -0.07921 + 2.1616T_r - 13.4040T_r^2 + 44.1706T_r^3 - 84.8291T_r^4 + 96.1209T_r^5 - 59.8127T_r^6 + 15.6719T_r^7$

In a test with 55 liquids, polar and nonpolar, Lucas found errors less than 10 percent. To illustrate the predicted values of Eq. (9.9.1), Figs. 9-12 and 9-13 were prepared. In both, $\eta/\eta_{SL}$ was plotted as a function of $\Delta P_r$ for various reduced temperatures. In Fig. 9-12, $\omega = 0$, and in Fig. 9-13, $\omega = 0.2$. Except at high values of $T_r$, $\eta/\eta_{SL}$ is approximately proportional to $\Delta P_r$. The effect of pressure is more important at the high reduced temperatures. As the acentric factor increases, there is a somewhat smaller effect of pressure. The method is illustrated in Example 9-13.

**Example 9-13** Estimate the viscosity of liquid methylcyclohexane at 300 K and 500 bar. The viscosity of the saturated liquid at 300 K is 0.68 cP, and the vapor pressure is less than 1 bar.

**solution** From Appendix A, $T_c = 572.2$ K, $P_c = 34.7$ bar, and $\omega = 0.236$. Thus, $T_r = 300/572.2 = 0.524$ and $\Delta P_r = 500/34.7 = 14.4$. ($P_{vp}$ was neglected.) Then

$$A = 0.9991 - \frac{4.674 \times 10^{-4}}{(1.0523)(0.524)^{-0.03877} - 1.0513}$$

$$= 0.9822$$

$$D = \frac{0.3257}{[1.0039 - (0.524)^{2.573}]^{0.2906}} - 0.2086$$

$$= 0.1371$$

$$C = -0.07921 + (2.1616)(0.524) - (13.4040)(0.524)^2$$
$$+ (44.1706)(0.524)^3 - (84.8291)(0.524)^4 + (96.1209)(0.524)^5$$
$$-(59.8127)(0.524)^6 + (15.6719)(0.524)^7$$
$$= 0.0619$$

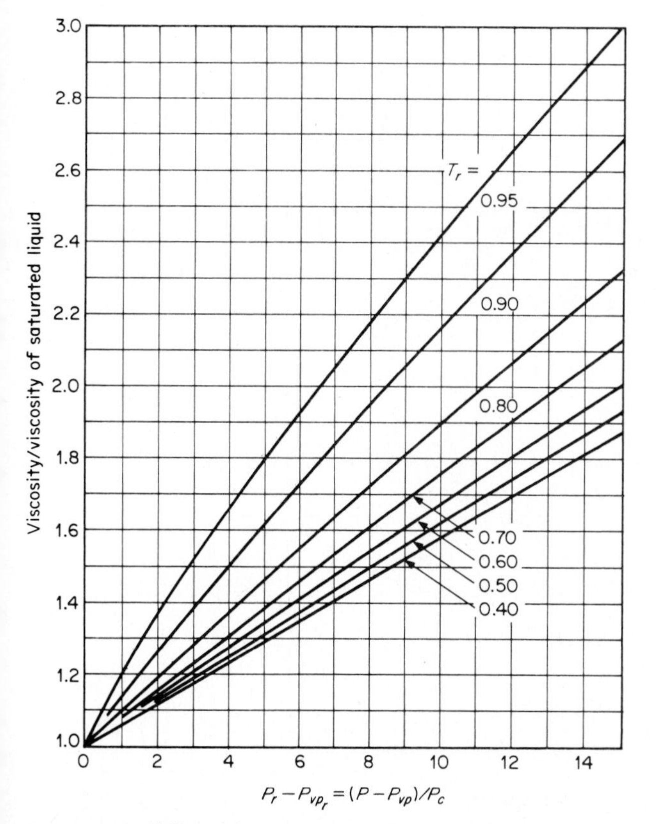


**Figure 9-12** Effect of pressure on the viscosity of liquids; $\omega = 0$.

With Eq. (9-9.1),

$$\frac{\eta}{\eta_{SL}} = \frac{1 + (0.1371)(14.4/2.118)^{0.9822}}{1 + (0.236)(14.4)(0.0619)} = 1.57$$

$$\eta = (1.57)(0.68) = 1.07 \text{ cP}$$

The experimental value of $\eta$ at 300 K and 500 bar is 1.09 cP [188].

$$\text{Error} = \frac{1.07 - 1.09}{1.09} \times 100 = -1.8\%$$

Whereas the correlation by Lucas would encompass most pressure ranges, at pressures over several thousand bar the data of Bridgman suggest that the *logarithm* of the viscosity is proportional to pressure and that the structural complexity of the molecule becomes important. Those who are interested in such high-pressure regions should consult the original publications of Bridgman [21] and others [12, 19, 65, 66, 68, 85, 110, 111, 115, 128, 132, 163].

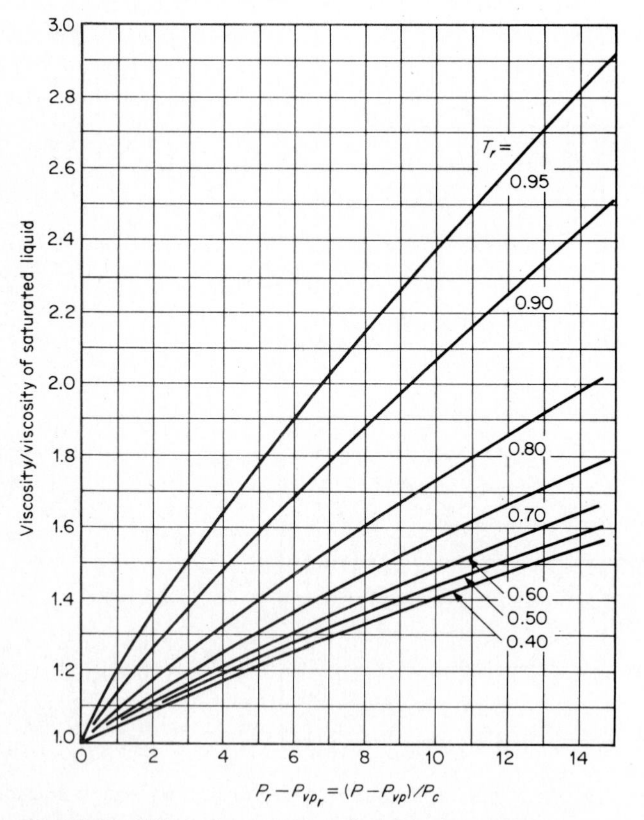


**Figure 9-13** Effect of pressure on the viscosity of liquids; $\omega = 0.2$.

## 9-10 Effect of Temperature on Liquid Viscosity

The viscosities of liquids decrease with increasing temperature either under isobaric conditions or as saturated liquids. This behavior can be seen in Fig. 9-9, where, for example, the viscosity of saturated liquid benzene is graphed as a function of temperature. Also, as noted in Sec. 9-8 and illustrated in Fig. 9-10, for a temperature range from the freezing point to somewhere around the normal boiling temperature, it is often a good approximation to assume $\ln \eta_L$ is linear in reciprocal absolute temperature; i.e.,

$$\ln \eta_L = A + \frac{B}{T} \tag{9-10.1}$$

This simple form was apparently first proposed by de Guzman [53, 153] in 1913, but it is more commonly referred to as the *Andrade* equation [7, 8]. Variations of Eq. (9-10.1) have been proposed to improve upon its correlation accuracy; many include some function of the liquid molar volume in either the $A$ or $B$ parameter [17, 50, 72, 81, 92, 103, 142, 144, 145, 146, 147, 201, 213]. Another variation involves the use of a third constant to obtain the *Vogel* equation [216],

$$\ln \eta_L = A + \frac{B}{T + C} \tag{9-10.2}$$

Goletz and Tassios [82] have used this form (for the kinematic viscosity) and report values of $A$, $B$, and $C$ for many pure liquids.

Equation (9-10.1) requires at least two viscosity-temperature datum points to determine the two constants. If only one datum point is available, one of the few ways to extrapolate this value is to employ the approximate Lewis-Squires chart [131], which is based on the empirical fact that the sensitivity of viscosity to temperature variations appears to depend primarily upon the value of the viscosity. This chart, shown in Fig. 9-14, can be used by locating the known value of viscosity on the ordinate and then extending the abscissa by the required number of degrees to find the new viscosity. Figure 9-14 can be expressed in an equation form as

$$\eta_L^{-0.2661} = \eta_K^{-0.2661} + \frac{T - T_K}{233} \tag{9-10.3}$$

where $\eta_L$ = liquid viscosity at $T$, cP

$\eta_K$ = known value of liquid viscosity at $T_K$, cP

$T$ and $T_K$ may be expressed in either °C or K. Thus, given a value of $\eta_L$ at $T_K$, one can estimate values of $\eta_L$ at other temperatures. Equation (9-10.3) or Fig. 9-14 is only approximate, and errors of 5 to 15 percent (or

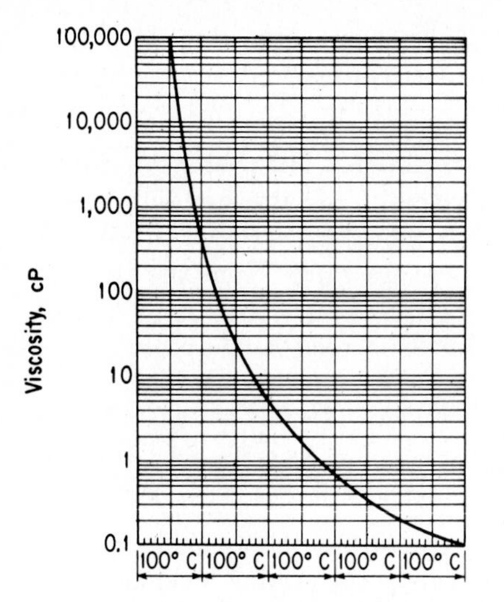


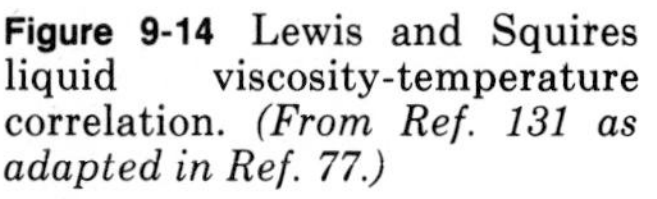
**Figure 9-14** Lewis and Squires liquid viscosity-temperature correlation. *(From Ref. 131 as adapted in Ref. 77.)*

greater) may be expected. This method should not be used if the temperature is much above the normal boiling point.

**Example 9-14** The viscosity of acetone at 30°C is 0.292 cP; estimate the viscosities at −90°C, −60°C, 0°C, and 60°C.

**solution** At −90°C, with Eq. (9-10.3),

$$\eta_L^{-0.2661} = (0.292)^{-0.2661} + \frac{-90 - 30}{233}$$

$$\eta_L = 1.7 \text{ cP}$$

For the other cases,

| $T$, °C | $\eta_L$, cP Eq. (9-10.3) | $\eta_L$, cP Experimental | Percent error |
|---|---|---|---|
| −90 | 1.7 | 2.1 | −19 |
| −60 | 0.99 | 0.98 | 1 |
| 0 | 0.42 | 0.39 | 8 |
| 60 | 0.21 | 0.23 | −9 |

Van Velzen et al. [212], Yaws et al. [224], and Duhne [60] have published constants to allow an estimation of liquid viscosities for most liquids for which experimental data exist. Their results have been modified slightly to yield consistent units and are shown in Table 9-8. Whenever possible, the equation forms shown in the table should be used rather than the less accurate estimation equations given in Sec. 9-11. However, the equations should *not* be used outside the recommended temperature range. A compilation similar to that of Table 9-8 is available in a paper

**TABLE 9-8 Correlation of Experimental Liquid Viscosity Data[1]**

| Formula | Name | Eq. No. | A | B | C | D | Range, °C | η,cP at(T,°C) | Ref. |
|---|---|---|---|---|---|---|---|---|---|
| A | Argon | 3 | -2.851E+01 | 1.057E+03 | 2.429E-01 | -8.096E-04 | -189 to -124 | 0.25 (-185) | 224 |
| $Br_2$ | Bromine | 2 | -3.112E+00 | 9.075E+02 | | | -4 to 29 | 0.99 ( 19.5) | 60 |
| $Cl_2$ | Chlorine | 3 | -1.768E+00 | 3.486E+02 | -1.857E-03 | 7.8 E-07 | -101 to 144 | 0.34 ( 25 ) | 224 |
| $F_2$ | Fluorine | 3 | -3.629E+00 | 1.972E+02 | -9.378E-04 | -6.275E-06 | -219 to -185 | 0.73 (-215) | 224 |
| HBr | Hydrogen bromide | 3 | -2.127E+01 | 1.996E+03 | 7.902E-02 | -1.191E-04 | -88 to 90 | 0.20 ( 25 ) | 224 |
| HCl | Hydrogen chloride | 3 | -3.488E+00 | 4.481E+02 | 7.062E-03 | -3.168E-05 | -110 to 50 | 0.068 ( 25 ) | 224 |
| HF | Hydrogen fluoride | 3 | -1.404E+01 | 1.879E+03 | 2.975E-02 | -3.060E-05 | -80 to 180 | 0.20 ( 25 ) | 224 |
| HI | Hydrogen iodide | 3 | -2.158E+01 | 2.337E+03 | 7.336E-02 | -9.717E-05 | -50 to 150 | 0.60 ( 25 ) | 224 |
| $H_2$ | Hydrogen | 3 | -1.118E+01 | 5.786E+01 | 3.244E-01 | -6.385E-03 | -258 to -240 | 0.016 (-256) | 224 |
| $H_2O$ | Water | 3 | -2.471E+01 | 4.209E+03 | 4.527E-02 | -3.376E-05 | 0 to 370 | 0.90 ( 25 ) | 224 |
| $H_3N$ | Ammonia | 3 | -1.978E+01 | 2.018E+03 | 6.173E-02 | -8.317E-05 | -75 to 130 | 0.13 ( 25 ) | 224 |
| $H_2N_2$ | Hydrazine | 3 | -1.848E+01 | 2.991E+03 | 3.709E-02 | -3.062E-05 | 2 to 370 | 0.90 ( 25 ) | 224 |
| $H_2O_2$ | Hydrogen peroxide | 3 | -3.719E+00 | 1.160E+03 | 8.06 E-04 | -2.689E-06 | 0 to 400 | 1.19 ( 25 ) | 224 |
| $H_2O_4S$ | Sulfuric acid | 2 | -6.178E+00 | 2.736E+03 | | | 0 to 80 | 25.4 ( 20 ) | 60 |
| He | Helium | | | | | | | 0.0034(-270) | 224 |
| $I_2$ | Iodine | 3 | -2.083E+00 | 1.195E+03 | -4.566E-04 | 1.08 E-07 | 114 to 200 | 1.8 (150 ) | 224 |
| NO | Nitric oxide | 3 | -1.150E+01 | 5.487E+02 | 8.448E-02 | -3.092E-04 | -160 to -90 | 0.35 (-160) | 224 |
| $NO_2$ | Nitrogen dioxide | 3 | -1.941E+01 | 2.147E+03 | 6.353E-02 | -8.644E-05 | -11 to 150 | 0.39 ( 25 ) | 224 |
| $N_2$ | Nitrogen | 3 | -2.795E+01 | 8.660E+02 | 2.763E-01 | -1.084E-03 | -205 to -195 | 0.18 (-200) | 224 |
| $N_2O$ | Nitrous oxide | 3 | 1.090E+00 | 5.020E+01 | -1.134E-02 | -9.841E-06 | -100 to 30 | 0.05 ( 25) | 224 |
| Ne | Neon | 3 | -1.929E+01 | 1.990E+02 | 5.453E-01 | -6.675E-03 | -248 to -229 | 0.137 (-247) | 224 |
| $O_2$ | Oxygen | 3 | -4.771E+00 | 2.146E+02 | 1.389E-02 | -6.255E-05 | -218 to -120 | 0.47 (-210) | 224 |
| $O_2S$ | Sulfur dioxide | 3 | -6.148E+00 | 9.365E+02 | 1.414E-02 | -2.887E-05 | -70 to 155 | 0.26 ( 25 ) | 224 |
| $O_3S$ | Sulfur trioxide | 3 | 2.894E+01 | -2.277E+03 | -9.392E-02 | 8.064E-05 | 16 to 210 | 1.60 ( 25 ) | 224 |

**TABLE 9-8 Correlation of Experimental Liquid Viscosity Data** *(Continued)*

| Formula | Name | Eq. No. | A | B | C | D | Range, °C | η,cP at(T,°C) | Ref. |
|---|---|---|---|---|---|---|---|---|---|
| $CCl_4$ | Carbon tetrachloride | 3 | -1.303E+01 | 2.290E+03 | 2.339E-02 | -2.011E-05 | -20 to 283 | 0.86 ( 25 ) | 224 |
| $CO$ | Carbon monoxide | 3 | -5.402E+00 | 2.422E+02 | 1.062E-02 | -4.522E-05 | -200 to -140 | 0.21 (-200) | 224 |
| $CO_2$ | Carbon dioxide | 3 | -3.097E+00 | 4.886E+01 | 2.381E-02 | -7.840E-05 | -56 to 30 | 0.06 ( 25 ) | 224 |
| $CS_2$ | Carbon disulfide | 2 | -3.442E+00 | 7.138E+02 | | | -13 to 40 | 0.36 ( 20 ) | 60 |
| $CHBr_3$ | Bromoform | 2 | -3.405E+00 | 1.195E+03 | | | 5 to 90 | 1.89 ( 25 ) | 212 |
| $CHCl_3$ | Chloroform | 3 | -4.172E+00 | 9.153E+02 | 2.70 E-03 | -4.108E-06 | -63 to 263 | 0.52 ( 25 ) | 224 |
| $CH_2Br_2$ | Methylene bromide | 2 | -3.353E+00 | 9.876E+02 | | | 15 to 40 | 1.09 ( 15 ) | 212 |
| $CH_2Cl_2$ | Methylene chloride | 3 | -8.061E+00 | 1.185E+03 | 1.162E-02 | -1.839E-05 | -97 to 240 | 0.41 ( 25 ) | 224 |
| $CH_2O_2$ | Formic acid | 2 | -5.156E+00 | 1.679E+03 | | | 8 to 110 | 1.80 ( 20 ) | 212 |
| $CH_3Cl$ | Methyl chloride | 2 | -5.073E+00 | 9.819E+02 | | | 0 to 130 | 0.18 ( 20 ) | 212,173 |
| $CH_3I$ | Methyl iodide | 2 | -3.366E+00 | 7.741E+02 | | | 0 to 50 | 0.50 ( 20 ) | 212 |
| $CH_3NO$ | Formamide | 1 | 7.737E+23 | -9.445E+00 | | | 0 to 25 | 3.30 ( 25 ) | 224 |
| $CH_3NO_2$ | Nitromethane | 2 | -3.989E+00 | 1.042E+03 | | | 0 to 90 | 0.63 ( 25 ) | 212 |
| $CH_4$ | Methane | 3 | -2.687E+01 | 1.150E+03 | 1.871E-01 | -5.211E-04 | -180 to -84 | 0.14 (-170) | 224 |
| $CH_4O$ | Methanol | 3 | -3.935E+01 | 4.826E+03 | 1.091E-01 | -1.127E-04 | -40 to 239 | 0.55 ( 25 ) | 224 |
| $CH_5N$ | Methyl amine | | | | | | | 0.24 ( 0 ) | 212 |
| $C_2Cl_3F_3$ | Trichlorotrifluoroethane | 2 | -4.219E+00 | 1.126E+03 | | | 20 to 50 | 0.70 ( 20 ) | 212 |
| $C_2Cl_4$ | Tetrachloroethylene | 2 | -3.334E+00 | 9.464E+02 | | | 0 to 117 | 0.88 ( 22 ) | 60 |
| $C_2HF_3O_2$ | Trifluoroacetic acid | 2 | -4.750E+00 | 1.348E+03 | | | 20 to 70 | 0.87 ( 20 ) | 212 |
| $C_2H_2Cl_4$ | 1,1,2,2-Tetrachloroethane | 2 | -4.505E+00 | 1.490E+03 | | | 0 to 90 | 1.64 ( 25 ) | 212 |
| $C_2H_3N$ | Acetonitrile | 1 | 3.851E+06 | -2.849E+00 | | | 0 to 25 | 0.35 ( 25 ) | 60 |
| $C_2H_4$ | Ethylene | 3 | -1.774E+01 | 1.078E+03 | 8.577E-02 | -1.758E-04 | -169 to 9 | 0.031( 0 ) | 224 |
| $C_2H_4Br_2$ | 1,2-Dibromoethane | 2 | -3.899E+00 | 1.299E+03 | | | 0 to 130 | 1.71 ( 20 ) | 212 |
| $C_2H_4Cl_2$ | 1,1-Dichloroethane | 2 | -3.970E+00 | 9.493E+02 | | | 7 to 60 | 0.49 ( 19 ) | 212 |
| $C_2H_4Cl_2$ | 1,2-Dichloroethane | 2 | -3.926E+00 | 1.091E+03 | | | 0 to 100 | 0.83 ( 20 ) | 212 |
| $C_2H_4F_2$ | 1,2-Difluoroethane | 2 | -3.941E+00 | 7.352E+02 | | | 0 to 70 | 0.25 ( 20 ) | 212 |

| Formula | Name | Eq. No. | A | B | C | D | Range, °C | $\eta$,cP at(T,°C) | Ref. |
|---|---|---|---|---|---|---|---|---|---|
| $C_2H_4O$ | Acetaldehyde | 1 | 5.140E+07 | -3.390E+00 | | | 0 to 20 | 0.22 ( 20 ) | 60 |
| $C_2H_4O$ | Ethylene oxide | 3 | -3.864E+00 | 7.193E+02 | 7.44 E-04 | -1.805E-06 | -112 to 195 | 0.25 ( 25 ) | 224 |
| $C_2H_4O_2$ | Methyl formate | 2 | -3.932E+00 | 8.363E+02 | | | 0 to 40 | 0.35 ( 20 ) | 212 |
| $C_2H_4O_2$ | Acetic acid | 2 | -4.519E+00 | 1.384E+03 | | | 15 to 120 | 1.30 ( 18 ) | 212 |
| $C_2H_5Br$ | Ethyl bromide | 2 | -3.859E+00 | 8.515E+02 | | | -100 to 50 | 0.40 ( 20 ) | 212 |
| $C_2H_5Cl$ | Ethyl Chloride | 2 | -3.873E+00 | 7.390E+02 | | | -20 to 50 | 0.27 ( 20 ) | 212 |
| $C_2H_5I$ | Ethyl iodide | 2 | -3.467E+00 | 8.539E+02 | | | 0 to 80 | 0.59 ( 20 ) | 212 |
| $C_2H_5NO$ | Acetamide | 2 | -5.470E+00 | 2.173E+03 | | | 105 to 120 | 1.32 (105 ) | 60 |
| $C_2H_6$ | Ethane | 3 | -1.023E+01 | 6.680E+02 | 4.386E-02 | -9.588E-05 | -183 to 32 | 0.032( 25 ) | 224 |
| $C_2H_6O$ | Ethanol | 3 | -6.210E+00 | 1.614E+03 | 6.18 E-03 | -1.132E-05 | -105 to 243 | 1.04 ( 25 ) | 224 |
| $C_2H_6O_2$ | Ethylene glycol | 2 | -7.811E+00 | 3.143E+03 | | | 20 to 110 | 19.9 ( 20 ) | 212 |
| $C_2H_7N$ | Ethylamine | | | | | | | 0.44 (-33 ) | 212 |
| $C_3H_3F_3O$ | Trifluoroacetone | 2 | -4.684E+00 | 1.019E+03 | | | 8 to 25 | 0.33 ( 15 ) | 212 |
| $C_3H_4O_2$ | Acrylic acid | 1 | 1.510E+10 | -4.089E+00 | | | 20 to 40 | | 60 |
| $C_3H_5Br$ | Allyl bromide | 2 | -3.782E+00 | 8.971E+02 | | | 0 to 80 | 0.50 ( 18 ) | 212 |
| $C_3H_5Cl$ | Allyl chloride | 2 | -4.015E+00 | 8.457E+02 | | | 0 to 50 | 0.32 ( 22 ) | 212 |
| $C_3H_5F_3O$ | Trifluoro-2-propanol | 2 | -9.540E+00 | 3.115E+03 | | | 15 to 60 | 3.67 ( 15 ) | 212 |
| $C_3H_5I$ | Allyl iodide | 2 | -3.675E+00 | 9.798E+02 | | | 0 to 100 | 0.75 ( 17) | 212 |
| $C_3H_6$ | Propylene | 3 | -1.153E+01 | 9.514E+02 | 4.078E-02 | -7.120E-05 | -160 to 91 | 0.081( 25 ) | 224 |
| $C_3H_6$ | Cyclopropane | 3 | -3.074E+00 | 2.676E+02 | 2.55 E-04 | -8.83 E-08 | -127 to 124 | 0.12 ( 25 ) | 224 |
| $C_3H_6Br_2$ | 1,2-Dibromopropane | 2 | -3.921E+00 | 1.290E+03 | | | 0 to 140 | 1.49 ( 25 ) | 212 |
| $C_3H_6O$ | 2-Propenol | 2 | -5.947E+00 | 1.827E+03 | | | 0 to 100 | 1.36 ( 20 ) | 212 |

**TABLE 9-8 Correlation of Experimental Liquid Viscosity Data** *(Continued)*

| Formula | Name | Eq. No. | A | B | C | D | Range, °C | η,cP at(T,°C) | Ref. |
|---|---|---|---|---|---|---|---|---|---|
| $C_3H_6O$ | Acetone | 2 | -4.033E+00 | 8.456E+02 | | | -80 to 60 | 0.32 ( 25 ) | 212 |
| $C_3H_6O$ | Allyl alcohol | 1 | 3.529E+13 | -5.445E+00 | | | 0 to 70 | 1.36 ( 20 ) | 60 |
| $C_3H_6O$ | Proponal | 2 | -4.817E+00 | 1.153E+03 | | | 10 to 30 | 0.41 ( 20 ) | 60 |
| $C_3H_6O_2$ | Propionic acid | 2 | -4.116E+00 | 1.232E+03 | | | 5 to 150 | 1.10 ( 20 ) | 212 |
| $C_3H_6O_2$ | Methyl acetate | 2 | -4.200E+00 | 9.409E+02 | | | 0 to 70 | 0.38 ( 20 ) | 212 |
| $C_3H_6O_2$ | Ethyl formate | 2 | -4.081E+00 | 9.231E+02 | | | 0 to 70 | 0.37 ( 28 ) | 212 |
| $C_3H_7Br$ | Propyl bromide | 2 | -3.781E+00 | 9.102E+02 | | | 0 to 80 | 0.52 ( 19 ) | 212 |
| $C_3H_7Br$ | Isopropyl bromide | 2 | -3.761E+00 | 8.922E+02 | | | 0 to 50 | 0.49 ( 20 ) | 60 |
| $C_3H_7Cl$ | Propyl chloride | 2 | -4.014E+00 | 8.629E+02 | | | 0 to 50 | 0.35 ( 20 ) | 212 |
| $C_3H_7Cl$ | Isopropyl chloride | 2 | -3.323E+00 | 7.052E+02 | | | 0 to 40 | 0.31 (22.5) | 212 |
| $C_3H_7I$ | Propyl iodide | 2 | -3.715E+00 | 9.995E+02 | | | 0 to 100 | 0.73 ( 21 ) | 212 |
| $C_3H_7I$ | Isopropyl iodide | 2 | -3.743E+00 | 9.841E+02 | | | 0 to 100 | 0.66 ( 23 ) | 212 |
| $C_3H_7N$ | 3-Aminopropene | | | | | | | 0.37 ( 25 ) | 212 |
| $C_3H_7NO_2$ | Ethyl carbamate | 2 | -6.578E+00 | 2.454E+03 | | | 105 to 120 | 0.92 (105 ) | 60 |
| $C_3H_7O$ | Propylene oxide | 3 | -2.717E+00 | 7.000E+02 | -4.384E-03 | 5.363E-06 | -112 to 209 | 0.30 ( 25 ) | 224 |
| $C_3H_8$ | Propane | 3 | -7.764E+00 | 7.219E+02 | 2.381E-02 | -4.665E-05 | -187 to 96 | 0.091( 25 ) | 224 |
| $C_3H_8O$ | n-Propanol | 3 | -1.228E+01 | 2.666E+03 | 2.008E-02 | -2.233E-05 | -72 to 260 | 1.94 ( 25 ) | 224 |
| $C_3H_8O$ | Isopropanol | 2 | -8.114E+00 | 2.624E+03 | | | 0 to 90 | 0.98 ( 52 ) | 212 |
| $C_3H_8O_2$ | Propylene glycol | 2 | -7.577E+00 | 3.233E+03 | | | 40 to 180 | 19.4 ( 40 ) | 212 |
| $C_3H_8O_3$ | Glycerol | 1 | 3.426E+73 | -2.852E+01 | | | 0 to 30 | 954. ( 25 ) | 60 |
| $C_3H_9N$ | Propyl amine | | | | | | | 0.35 ( 25 ) | 212 |
| $C_4H_4N_2$ | Succinonitrile | 2 | -5.183E+00 | 2.060E+03 | | | 59 to 83 | 2.76 ( 59 ) | 60 |
| $C_4H_4S$ | Thiophene | 2 | -4.039E+00 | 1.065E+03 | | | 0 to 83 | 0.64 (22.5) | 60 |
| $C_4H_5Cl$ | Chloroprene | 3 | -3.583E+00 | 7.085E+02 | 1.38 E-03 | -1.841E-06 | -130 to 260 | 0.38 ( 25 ) | 224 |
| $C_4H_6$ | 1,3-Butadiene | 3 | -6.072E+00 | 1.000E+03 | 4.46 E-03 | -6.694E-06 | -108 to 152 | 0.14 ( 25 ) | 224 |
| $C_4H_6O_3$ | Acetic anhydride | 2 | -3.173E+00 | 9.054E+02 | | | 0 to 100 | 0.9 ( 18 ) | 60 |

| Formula | Name | Eq. No. | A | B | C | D | Range, °C | η,cP at(T,°C) | Ref. |
|---|---|---|---|---|---|---|---|---|---|
| $C_4H_7F_3O$ | Trifluoro-2-methylpropanol | 2 | -1.001E+01 | 3.352E+03 | | | 25 to 60 | 3.32 ( 25 ) | 212 |
| $C_4H_7NO_2$ | 2 Nitro-2-butene | 2 | -3.855E+00 | 1.101E+03 | | | 30 to 70 | 0.81 ( 30 ) | 212 |
| $C_4H_8$ | 1-Butene | 3 | -1.063E+01 | 9.816E+02 | 3.525E-02 | -5.593E-05 | -140 to 146 | 0.17 ( 25 ) | 224 |
| $C_4H_8$ | Isobutylene | 3 | -6.447E+00 | 8.135E+02 | 1.320E-02 | -2.438E-05 | -140 to 144 | 0.14 ( 25 ) | 224 |
| $C_4H_8$ | Cyclobutane | 3 | -4.541E+00 | 6.724E+02 | 3.27 E-03 | -3.928E-06 | -90 to 190 | 0.19 ( 25 ) | 224 |
| $C_4H_8Br_2$ | 1,2-Dibromo-2-methylpropane | 2 | -4.335E+00 | 1.497E+03 | | | 0 to 150 | 1.92 ( 27 ) | 212 |
| $C_4H_8O$ | Methylethyl ketone | 2 | -4.213E+00 | 9.759E+02 | | | 0 to 80 | 0.42 ( 21 ) | 212 |
| $C_4H_8O_2$ | Butyric acid | 2 | -4.592E+00 | 1.475E+03 | | | 0 to 160 | 1.54 ( 20 ) | 212 |
| $C_4H_8O_2$ | Isobutyric acid | 2 | -4.355E+00 | 1.355E+03 | | | 4 to 150 | 1.38 ( 17 ) | 212 |
| $C_4H_8O_2$ | Methyl propionate | 2 | -4.173E+00 | 9.872E+02 | | | 0 to 80 | 0.41 ( 30 ) | 212 |
| $C_4H_8O_2$ | Ethyl acetate | 2 | -4.171E+00 | 9.841E+02 | | | 0 to 80 | 0.46 ( 20 ) | 212 |
| $C_4H_8O_2$ | Propyl formate | 2 | -4.238E+00 | 1.043E+03 | | | 0 to 90 | 0.51 ( 23 ) | 212 |
| $C_4H_8O_2$ | Isopropyl formate | | | | | | | 0.57 ( 20 ) | 212 |
| $C_4H_9Br$ | n-Butyl bromide | 2 | -3.805E+00 | 9.830E+02 | | | 0 to 100 | 0.65 ( 20 ) | 212 |
| $C_4H_9Br$ | 1-Bromo-2-methylpropane | 2 | -4.011E+00 | 1.039E+03 | | | 0 to 100 | 0.61 ( 24 ) | 212 |
| $C_4H_9Cl$ | n-Butyl chloride | | | | | | | 0.47 ( 15 ) | 60 |
| $C_4H_9Cl$ | 1-Chloro-2-methylpropane | 2 | -4.240E+00 | 1.007E+03 | | | 0 to 80 | 0.46 ( 19 ) | 212 |
| $C_4H_9Cl$ | 2-Chloro-2-methylpropane | | | | | | | 0.54 ( 15 ) | 60 |
| $C_4H_9I$ | 1-Iodo-2-methylpropane | 2 | -3.859E+00 | 1.089E+03 | | | 0 to 120 | 0.84 ( 22 ) | 212 |
| $C_4H_9I$ | Isobutyl iodide | 2 | -3.783E+00 | 1.072E+03 | | | 0 to 120 | 0.875( 20 ) | 60 |
| $C_4H_{10}$ | n-Butane | 2 | -3.821E+00 | 6.121E+02 | | | -90 to 0 | 0.22 ( -5 ) | 212 |
| $C_4H_{10}$ | 2-Methylpropane | 2 | -4.093E+00 | 6.966E+02 | | | -80 to 0 | 0.27 (-20 ) | 212 |
| $C_4H_{10}O$ | n-Butanol | 3 | -9.722E+00 | 2.602E+03 | 9.53 E-03 | -9.966E-06 | -60 to 289 | 2.61 ( 25 ) | 224 |
| $C_4H_{10}O$ | Isobutanol | 2 | -8.163E+00 | 2.789E+03 | | | 0 to 110 | 3.98 ( 19 ) | 212 |
| $C_4H_{10}O$ | 2-Methylpropanol | 2 | -1.192E+01 | 3.979E+03 | | | 20 to 90 | 5.89 ( 22 ) | 212 |

**TABLE 9-8 Correlation of Experimental Liquid Viscosity Data** (*Continued*)

| Formula | Name | Eq. No. | A | B | C | D | Range, °C | η,cP at(T,°C) | Ref. |
|---|---|---|---|---|---|---|---|---|---|
| $C_4H_{10}O$ | Diethyl ether | 2 | -4.267E+00 | 8.131E+02 | | | -80 to 100 | 0.23 ( 20 ) | 212 |
| $C_4H_{10}O$ | Methyl propyl ether | 2 | -4.189E+00 | 8.149E+02 | | | 0 to 40 | 0.25 ( 20 ) | 212 |
| $C_4H_{10}S$ | Diethyl sulfide | 2 | -3.925E+00 | 9.156E+02 | | | 0 to 88 | 0.43 ( 25 ) | 60 |
| $C_4H_{11}N$ | n-Butyl amine | | | | | | | 0.68 ( 25 ) | 212 |
| $C_4H_{11}N$ | Diethyl amine | 2 | -4.759E+00 | 1.091E+03 | | | -30 to 40 | 0.35 ( 25 ) | 212 |
| $C_4H_{11}N$ | Isobutyl amine | | | | | | | 0.55 ( 25 ) | 212 |
| $C_5H_4O_2$ | Furfural | 1 | 3.628E+14 | -5.815E+00 | | | 0 to 25 | 1.49 ( 25 ) | 60 |
| $C_5H_7F_3O_2$ | Trifluoroacetic acid-2-propyl ester | 2 | -4.780E+00 | 1.301E+03 | | | 20 to 70 | 0.72 ( 20 ) | 212 |
| $C_5H_8$ | Isoprene | 3 | -2.228E+00 | 6.357E+02 | -7.32 E-03 | 7.665E-06 | -146 to 210 | 0.20 ( 25 ) | 224 |
| $C_5H_9NO_2$ | 2-Nitro-2-pentene | 2 | -3.938E+00 | 1.147E+03 | | | 30 to 70 | 0.87 ( 30 ) | 212 |
| $C_5H_9NO_2$ | 3-Nitro-2-pentene | 2 | -4.070E+00 | 1.206E+03 | | | 30 to 70 | 0.93 ( 30 ) | 212 |
| $C_5H_{10}$ | 1-Pentene | 2 | -4.023E+00 | 7.029E+02 | | | -90 to 0 | 0.24 ( 0 ) | 212 |
| $C_5H_{10}$ | 2-Methyl-2-butene | 2 | -4.115E+00 | 7.425E+02 | | | 0 to 40 | 0.21 ( 20 ) | 212 |
| $C_5H_{10}$ | Cyclopentane | 3 | -6.021E+00 | 1.118E+03 | 7.28 E-03 | -8.662E-06 | -90 to 235 | 0.42 ( 25 ) | 224 |
| $C_5H_{10}O$ | Diethyl ketone | 2 | -4.123E+00 | 9.798E+02 | | | 0 to 100 | 0.47 ( 19 ) | 212 |
| $C_5H_{10}O$ | Methyl propyl ketone | 2 | -4.149E+00 | 1.008E+03 | | | 0 to 100 | 0.51 ( 18 ) | 212 |
| $C_5H_{10}O_2$ | Valeric acid | 2 | -4.921E+00 | 1.679E+03 | | | 16 to 100 | 2.30 ( 20 ) | 212 |
| $C_5H_{10}O_2$ | Methyl butyrate | 2 | -4.334E+00 | 1.104E+03 | | | 0 to 110 | 0.58 ( 20 ) | 212 |
| $C_5H_{10}O_2$ | Ethyl propionate | 2 | -4.289E+00 | 1.067E+03 | | | 0 to 100 | 0.535( 20 ) | 212 |
| $C_5H_{10}O_2$ | Propyl acetate | 2 | -4.406E+00 | 1.127E+03 | | | 0 to 110 | 0.58 ( 21 ) | 212 |
| $C_5H_{10}O_2$ | n-Butyl formate | 1 | 4.752E+10 | -4.394E+00 | | | 0 to 20 | 0.69 ( 20 ) | 60 |
| $C_5H_{10}O_2$ | Methyl isobutyrate | 2 | -4.222E+00 | 1.039E+03 | | | 0 to 100 | 0.47 ( 29 ) | 212 |
| $C_5H_{10}O_2$ | Isopropyl acetate | | | | | | | 0.525( 20 ) | 212 |
| $C_5H_{10}O_2$ | Isobutyl formate | | | | | | | 0.65 ( 20 ) | 212 |
| $C_5H_{11}Br$ | n-Amyl bromide | 2 | -3.881E+00 | 1.071E+03 | | | 0 to 100 | 0.81 ( 20 ) | 212 |
| $C_5H_{12}$ | n-Pentane | 2 | -3.958E+00 | 7.222E+02 | | | -130 to 40 | 0.225( 25 ) | 212 |

| Formula | Name | Eq. No. | A | B | C | D | Range, °C | η,cP at(T,°C) | Ref. |
|---|---|---|---|---|---|---|---|---|---|
| $C_5H_{12}$ | 2-Methylbutane | 2 | -4.415E+00 | 8.458E+02 | | | -50 to 30 | 0.21 ( 25 ) | 212 |
| $C_5H_{12}$ | 2,2-Dimethylpropane | 2 | -5.715E+00 | 1.248E+03 | | | -10 to 10 | 0.30 ( 5 ) | 212 |
| $C_5H_{12}O$ | n-Pentanol | 2 | -7.581E+00 | 2.651E+03 | | | 0 to 140 | 4.40 ( 20 ) | 212 |
| $C_5H_{12}O$ | 3-Methylbutanol | 2 | -7.568E+00 | 2.645E+03 | | | 0 to 140 | 3.86 ( 24 ) | 212 |
| $C_5H_{12}O$ | 2-Methylbutanol | 2 | -8.289E+00 | 2.900E+03 | | | 0 to 140 | 5.11 ( 20 ) | 212 |
| $C_5H_{12}O$ | 2-Methylbutanol-2 | 2 | -1.027E+01 | 3.459E+03 | | | 0 to 100 | 5.0 (18.5) | 212 |
| $C_5H_{12}O$ | Ethyl propyl ether | 2 | -4.315E+00 | 9.207E+02 | | | 0 to 70 | 0.32 ( 20 ) | 212 |
| $C_5H_{12}O$ | Methyl isobutyl ether | 2 | -4.270E+00 | 8.972E+02 | | | 0 to 70 | 0.30 ( 21 ) | 212 |
| $C_5H_{13}N$ | Ethyl propyl amine | | | | | | | 0.90 ( 25 ) | 212 |
| $C_6H_5Br$ | Bromobenzene | 2 | -3.869E+00 | 1.170E+03 | | | 0 to 150 | 1.17 ( 18 ) | 212 |
| $C_6H_5Cl$ | Chlorobenzene | 3 | -4.573E+00 | 1.196E+03 | 1.37 E-03 | -1.378E-06 | -45 to 350 | 0.76 ( 25 ) | 224 |
| $C_6H_5ClO$ | o-Chlorophenol | 2 | -6.236E+00 | 2.262E+03 | | | 0 to 160 | 4.21 ( 20 ) | 212 |
| $C_6H_5ClO$ | m-Chlorophenol | 2 | -1.130E+01 | 4.088E+03 | | | 25 to 60 | 11.5 ( 25 ) | 212 |
| $C_6H_5ClO$ | p-Chlorophenol | 2 | -1.169E+01 | 4.287E+03 | | | 45 to 60 | 6.15 ( 45 ) | 212 |
| $C_6H_5F$ | Fluorobenzene | 2 | -4.116E+00 | 1.041E+03 | | | 9 to 100 | 0.58 ( 20 ) | 212 |
| $C_6H_5I$ | Iodobenzene | 2 | -3.933E+00 | 1.303E+03 | | | 4 to 140 | 1.78 ( 17 ) | 212 |
| $C_6H_5NO_2$ | Nitrobenzene | 2 | -4.344E+00 | 1.480E+03 | | | 0 to 200 | 2.02 ( 20 ) | 212 |
| $C_6H_5NO_3$ | o-Nitrophenol | 2 | -5.195E+00 | 1.932E+03 | | | 40 to 90 | 2.75 ( 40 ) | 212 |
| $C_6H_6$ | Benzene | 3 | 4.612E+00 | 1.489E+02 | -2.544E-02 | 2.222E-05 | 6 to 288 | 0.61 ( 25 ) | 224 |
| $C_6H_6BrN$ | m-Bromoaniline | 1 | 7.810E+18 | -7.330E+02 | | | 20 to 80 | 6.81 ( 20 ) | 60 |
| $C_6H_6BrN$ | o-Bromoaniline | | | | | | | 3.19 ( 40 ) | 60 |
| $C_6H_6BrN$ | p-Bromoaniline | | | | | | | 1.81 ( 80 ) | 60 |
| $C_6H_6ClN$ | 2-Chloroaniline | | | | | | | 1.65 ( 55 ) | 212 |
| $C_6H_6ClN$ | 3-Chloroaniline | 2 | -6.283E+00 | 2.242E+03 | | | 25 to 60 | 3.50 ( 25 ) | 212 |

**TABLE 9-8 Correlation of Experimental Liquid Viscosity Data** *(Continued)*

| Formula | Name | Eq. No. | A | B | C | D | Range, °C | η,cP at(T,°C) | Ref. |
|---|---|---|---|---|---|---|---|---|---|
| $C_6H_6ClN$ | 4-Chloroaniline | | | | | | | 1.96 ( 55 ) | 212 |
| $C_6H_6O$ | Phenol | 3 | -1.851E+01 | 4.350E+03 | 2.429E-02 | -1.547E-05 | 41 to 420 | 3.25 ( 50 ) | 224 |
| $C_6H_7N$ | Aniline | 3 | 3.569E+02 | -3.237E+04 | -1.254E+00 | 1.428E-03 | -6 to 50 | 3.93 ( 25 ) | 224 |
| $C_6H_{10}$ | Cyclohexene | 1 | 4.566E+05 | -2.367E+00 | | | 13 to 20 | 0.66 ( 20 ) | 60 |
| $C_6H_{10}$ | Hexadiene-1,5 | 2 | -4.029E+00 | 8.062E+02 | | | 0 to 56 | 0.275( 20 ) | 60 |
| $C_6H_{10}$ | 2,3-Dimethyl-1,3-butadiene | 2 | -4.270E+00 | 8.640E+02 | | | 0 to 70 | 0.22 ( 40 ) | 212 |
| $C_6H_{11}NO_2$ | 2-Nitro-2-hexene | 2 | -4.123E+00 | 1.286E+03 | | | 30 to 70 | 1.14 ( 30 ) | 212 |
| $C_6H_{12}$ | Cyclohexane | 3 | -4.398E+00 | 1.380E+03 | -1.55 E-03 | 1.157E-06 | 7 to 280 | 0.88 ( 25 ) | 224 |
| $C_6H_{12}$ | 1-Hexene | 2 | -4.162E+00 | 8.230E+02 | | | -55 to 70 | 0.25 ( 25 ) | 212 |
| $C_6H_{12}$ | Methylcyclopentane | 2 | -4.170E+00 | 1.014E+03 | | | -25 to 80 | 0.48 ( 25 ) | 212 |
| $C_6H_{12}O$ | Methyl butyl ketone | 2 | -4.642E+00 | 1.218E+03 | | | 20 to 40 | 0.63 ( 20 ) | 212 |
| $C_6H_{12}O$ | Methyl isobutyl ketone | | | | | | | 0.58 ( 20 ) | 212 |
| $C_6H_{12}O_2$ | Caproic acid | 2 | -5.082E+00 | 1.827E+03 | | | 16 to 100 | 3.23 ( 20 ) | 212 |
| $C_6H_{12}O_2$ | Methyl valerate | | | | | | | 0.71 ( 20 ) | 212 |
| $C_6H_{12}O_2$ | Ethyl butyrate | 2 | -4.270E+00 | 1.128E+03 | | | 15 to 80 | 0.69 ( 20 ) | 212 |
| $C_6H_{12}O_2$ | Butyl acetate | 2 | -4.546E+00 | 1.238E+03 | | | 0 to 50 | 0.73 ( 20 ) | 212 |
| $C_6H_{12}O_2$ | Isobutyl acetate | 2 | -4.546E+00 | 1.230E+03 | | | 20 to 100 | 0.72 ( 20 ) | 212 |
| $C_6H_{13}Br$ | n-Hexyl bromide | 2 | -4.001E+00 | 1.172E+03 | | | 0 to 100 | 1.01 ( 20 ) | 212 |
| $C_6H_{14}$ | n-Hexane | 2 | -4.034E+00 | 8.354E+02 | | | -95 to 70 | 0.30 ( 25 ) | 212 |
| $C_6H_{14}$ | 2-Methylpentane | 2 | -4.247E+00 | 8.845E+02 | | | 0 to 70 | 0.285( 25 ) | 212 |
| $C_6H_{14}$ | 2,2-Dimethylbutane | 2 | -4.454E+00 | 1.010E+03 | | | 0 to 40 | 0.35 ( 25 ) | 212 |
| $C_6H_{14}$ | 2,3-Dimethylbutane | 2 | -4.469E+00 | 1.023E+03 | | | 0 to 40 | 0.36 ( 25 ) | 212 |
| $C_6H_{14}O$ | n-Hexanol | 2 | -7.651E+00 | 2.716E+03 | | | 20 to 60 | 4.37 ( 25 ) | 212 |
| $C_6H_{14}O$ | Dipropyl ether | 2 | -4.391E+00 | 1.027E+03 | | | 0 to 100 | 0.40 ( 25 ) | 212 |
| $C_6H_{14}O$ | Ethyl isobutyl ether | 2 | -4.404E+00 | 9.983E+02 | | | 0 to 80 | 0.37 ( 22 ) | 212 |
| $C_6H_{14}O$ | Diisopropyl ether | | | | | | | 0.32 ( 20 ) | 212 |
| $C_6H_{15}N$ | Triethyl amine | 2 | -4.106E+00 | 9.216E+02 | | | -33 to 25 | 0.36 ( 25 ) | 29* |

| Formula | Name | Eq. No. | A | B | C | D | Range, °C | η,cP at(T,°C) | Ref. |
|---|---|---|---|---|---|---|---|---|---|
| $C_7H_5Cl_3$ | Phenyl chloroform | 2 | -6.567E+00 | 2.176E+03 | | | 10 to 20 | 2.55 ( 17 ) | 60 |
| $C_7H_5F_3$ | Trifluorotoluene | 2 | -3.859E+00 | 9.621E+02 | | | 20 to 70 | 0.57 ( 20 ) | 212 |
| $C_7H_5N$ | Benzonitrile | | | | | | | 1.24 ( 25 ) | 60 |
| $C_7H_6O$ | Benzaldehyde | | | | | | | 1.39 ( 25 ) | 60 |
| $C_7H_6O_2$ | Benzoic acid | 2 | -1.478E+01 | 6.027E+03 | | | 122 to 140 | 1.26 (130 ) | 212 |
| $C_7H_6O_2$ | Salicylaldehyde | | | | | | | 1.67 ( 45 ) | 60 |
| $C_7H_6O_3$ | Salicylic acid | 1 | 3.343E+14 | -5.716E+00 | | | 10 to 40 | 2.71 ( 20 ) | 60 |
| $C_7H_7NO_2$ | o-Nitrotoluene | 2 | -5.093E+00 | 1.750E+03 | | | 0 to 70 | 2.37 ( 20 ) | 212 |
| $C_7H_7NO_2$ | m-Nitrotoluene | 1 | 3.085E+13 | -5.320E+00 | | | 20 to 60 | 2.33 ( 20 ) | 60 |
| $C_7H_7NO_2$ | p-Nitrotoluene | 2 | -5.966E+00 | 2.003E+03 | | | 20 to 100 | 2.33 ( 20 ) | 212 |
| $C_7H_8$ | Toluene | 3 | -5.878E+00 | 1.287E+03 | 4.575E-03 | -4.499E-06 | -40 to 315 | 0.55 ( 25 ) | 224 |
| $C_7H_8O$ | Benzyl alcohol | 2 | -6.822E+00 | 2.505E+03 | | | 20 to 60 | 5.58 ( 20 ) | 212 |
| $C_7H_8O$ | o-Cresol | 2 | -9.657E+00 | 3.531E+03 | | | 0 to 120 | 9.56 ( 20 ) | 212 |
| $C_7H_8O$ | m-Cresol | 2 | -1.109E+01 | 4.111E+03 | | | 0 to 120 | 16.4 ( 20 ) | 212 |
| $C_7H_8O$ | p-Cresol | 2 | -1.129E+01 | 4.207E+03 | | | 0 to 120 | 18.9 ( 20 ) | 212 |
| $C_7H_8O$ | Methyl phenyl ether | 2 | -2.748E+00 | 8.953E+02 | | | 0 to 70 | 1.32 ( 20 ) | 212 |
| $C_7H_9N$ | Benzyl amine | 2 | -4.463E+00 | 1.466E+03 | | | 25 to 130 | 1.59 ( 25 ) | 212 |
| $C_7H_9N$ | N-Methylaniline | 2 | -6.333E+00 | 2.107E+03 | | | 0 to 80 | 2.30 ( 20 ) | 212 |
| $C_7H_9N$ | 2-Aminotoluene | 2 | -7.009E+00 | 2.499E+03 | | | 0 to 100 | 4.39 ( 20 ) | 212 |
| $C_7H_9N$ | 3-Aminotoluene | 2 | -6.036E+00 | 2.137E+03 | | | 25 to 130 | 3.31 ( 20 ) | 212 |
| $C_7H_9N$ | 4-Aminotoluene | 2 | -4.779E+00 | 1.701E+03 | | | 40 to 175 | 1.56 ( 55 ) | 212 |
| $C_7H_{11}F_3$ | Trifluoromethylcyclohexane | 2 | -4.239E+00 | 1.224E+03 | | | 20 to 70 | 0.95 ( 20 ) | 212 |
| $C_7H_{14}$ | 1-Heptene | 2 | -3.961E+00 | 8.489E+02 | | | 0 to 100 | 0.34 ( 25 ) | 212 |
| $C_7H_{14}$ | Methylcyclohexane | 2 | -4.480E+00 | 1.217E+03 | | | -25 to 110 | 0.68 ( 25 ) | 212 |
| $C_7H_{14}$ | Ethylcyclopentane | 2 | -4.000E+00 | 9.989E+02 | | | -20 to 110 | 0.53 ( 25 ) | 212 |

**TABLE 9-8 Correlation of Experimental Liquid Viscosity Data** *(Continued)*

| Formula | Name | Eq. No. | A | B | C | D | Range, °C | η,cP at(T,°C) | Ref. |
|---|---|---|---|---|---|---|---|---|---|
| $C_7H_{14}O$ | Methyl amyl ketone | | | | | | | 0.77 ( 25 ) | 212 |
| $C_7H_{14}O$ | Ethyl butyl ketone | | | | | | | 0.84 ( 20 ) | 212 |
| $C_7H_{14}O$ | Dipropyl ketone | | | | | | | 0.69 ( 25 ) | 212 |
| $C_7H_{14}O_2$ | Heptanoic acid | 2 | -5.280E+00 | 1.973E+03 | | | 17 to 100 | 4.34 ( 20 ) | 212 |
| $C_7H_{14}O_2$ | Ethyl valerate | 2 | -4.932E+00 | 1.386E+03 | | | 20 to 60 | 0.84 ( 20 ) | 212 |
| $C_7H_{14}O_2$ | Propyl butyrate | 2 | -4.604E+00 | 1.290E+03 | | | 20 to 60 | 0.83 ( 20 ) | 212 |
| $C_7H_{14}O_2$ | Amyl acetate | 2 | -1.379E+01 | 4.035E+03 | | | 11 to 30 | 0.81 ( 25 ) | 212 |
| $C_7H_{14}O_2$ | Isopropyl propionate | | | | | | | 0.67 ( 25 ) | 212 |
| $C_7H_{14}O_2$ | Isoamyl acetate | 1 | 5.394E+10 | -4.375E+00 | | | 9 to 20 | 0.87 ( 20 ) | 60 |
| $C_7H_{15}Br$ | n-Heptyl bromide | 2 | -4.152E+00 | 1.291E+03 | | | 0 to 100 | 1.29 ( 20 ) | 212 |
| $C_7H_{16}$ | n-Heptane | 2 | -4.325E+00 | 1.006E+03 | | | -90 to 100 | 0.40 ( 25 ) | 212 |
| $C_7H_{16}$ | 2-Methylhexane | 2 | -4.270E+00 | 9.612E+02 | | | 0 to 100 | 0.36 ( 25 ) | 212 |
| $C_7H_{16}O$ | n-Heptanol | 2 | -8.190E+00 | 2.964E+03 | | | 15 to 100 | 5.68 ( 25 ) | 212 |
| $C_8H_8$ | Styrene | 3 | -2.717E+00 | 9.461E+02 | -3.173E-03 | 1.683E-06 | -30 to 360 | 0.71 ( 25 ) | 224 |
| $C_8H_8O$ | Acetophenone | 2 | -4.493E+00 | 1.494E+03 | | | 12 to 100 | 1.62 ( 25 ) | 212 |
| $C_8H_9O$ | Acetanilide | 1 | 2.617E+16 | -6.195E+00 | | | 120 to 140 | 2.22 (120 ) | 60 |
| $C_8H_{10}$ | o-Xylene | 3 | -3.332E+00 | 1.039E+03 | -1.768E-03 | 1.076E-06 | -25 to 350 | 0.76 ( 25 ) | 224 |
| $C_8H_{10}$ | m-Xylene | 3 | -3.820E+00 | 1.027E+03 | -6.38 E-04 | 4.52 E-07 | -47 to 340 | 0.60 ( 25 ) | 224 |
| $C_8H_{10}$ | p-Xylene | 3 | -7.790E+00 | 1.580E+03 | 8.73 E-03 | -6.735E-06 | 13 to 340 | 0.61 ( 25 ) | 224 |
| $C_8H_{10}$ | Ethylbenzene | 3 | -6.106E+00 | 1.353E+03 | 5.112E-03 | -4.552E-06 | -40 to 340 | 0.64 ( 25 ) | 224 |
| $C_8H_{10}O$ | 2-Phenyl ethanol | 2 | -9.191E+00 | 3.339E+03 | | | 25 to 60 | 7.61 ( 25 ) | 212 |
| $C_8H_{10}O$ | Ethyl phenyl ether | 2 | -4.869E+00 | 1.490E+03 | | | 0 to 90 | 1.25 ( 20 ) | 212 |
| $C_8H_{10}O_2$ | Phenylacetic acid | 2 | -5.819E+00 | 2.471E+03 | | | 77 to 140 | 3.54 ( 77 ) | 212 |

| Formula | Name | Eq. No. | A | B | C | D | Range, °C | η,cP at(T,°C) | Ref. |
|---|---|---|---|---|---|---|---|---|---|
| $C_8H_{11}N$ | N-Ethylaniline | 2 | -5.786E+00 | 1.943E+03 | | | 0 to 100 | 2.25 ( 20 ) | 212 |
| $C_8H_{11}N$ | N,N-Dimethylaniline | 2 | -3.979E+00 | 1.273E+03 | | | 0 to 180 | 1.40 ( 20 ) | 212 |
| $C_8H_{11}NO$ | o-Phenetidine | 1 | 2.908E+33 | -1.324E+01 | | | 0 to 30 | 6.08 ( 20 ) | 60 |
| $C_8H_{11}NO$ | p-Phenetidine | 1 | 3.346E+33 | -1.314E+01 | | | 20 to 30 | 12.9 ( 20 ) | 60 |
| $C_8H_{16}$ | Propylcyclopentane | 2 | -3.959E+00 | 1.046E+03 | | | -20 to 120 | 0.64 ( 25 ) | 212 |
| $C_8H_{16}$ | 1-Octene | 2 | -4.058E+00 | 9.644E+02 | | | 0 to 125 | 0.45 ( 25 ) | 212 |
| $C_8H_{16}$ | Ethylcyclohexane | 2 | -4.153E+00 | 1.166E+03 | | | -25 to 125 | 0.785 ( 25 ) | 212 |
| $C_8H_{16}O_2$ | Octanoic acid | 2 | -5.980E+00 | 2.255E+03 | | | 20 to 100 | 5.75 ( 20 ) | 212 |
| $C_8H_{17}Br$ | n-Octyl bromide | 2 | -4.325E+00 | 1.412E+03 | | | 0 to 100 | 1.63 ( 20 ) | 212 |
| $C_8H_{18}$ | n-Octane | 2 | -4.333E+00 | 1.091E+03 | | | -55 to 125 | 0.51 ( 25 ) | 212 |
| $C_8H_{18}O$ | n-Octanol | 2 | -8.166E+00 | 3.021E+03 | | | 15 to 100 | 7.21 ( 25 ) | 212 |
| $C_8H_{18}O$ | Diisobutyl ether | | | | | | | 0.75 ( 20 ) | 212 |
| $C_9H_{10}O_2$ | 3-Phenyl propionic acid | 2 | -6.488E+00 | 2.821E+03 | | | 50 to 140 | 9.8 ( 50 ) | 212 |
| $C_9H_{10}O_2$ | Ethyl benzoate | 2 | -5.078E+00 | 1.719E+03 | | | 20 to 80 | 2.24 ( 20 ) | 212 |
| $C_9H_{12}$ | Propylbenzene | 2 | -4.297E+00 | 1.215E+03 | | | -25 to 160 | 0.80 ( 25 ) | 212 |
| $C_9H_{12}$ | Isopropylbenzene | 3 | -8.292E+00 | 1.700E+03 | 1.003E-02 | -7.829E-06 | -20 to 360 | 0.74 ( 25 ) | 224 |
| $C_9H_{12}$ | 1-Methyl-4-ethylbenzene | 2 | -4.008E+00 | 1.067E+03 | | | 10 to 90 | 0.66 ( 25 ) | 212 |
| $C_9H_{12}$ | 1,2,4-Trimethylbenzene | 2 | -6.749E+00 | 2.010E+03 | | | 25 to 40 | 1.01 ( 25 ) | 212 |
| $C_9H_{12}O$ | Propyl phenyl ether | | | | | | | 1.59 ( 20 ) | 212 |
| $C_9H_{18}$ | 1-Nonene | 2 | -4.189E+00 | 1.085E+03 | | | 0 to 125 | 0.58 ( 25 ) | 212 |
| $C_9H_{18}$ | Propylcyclohexane | 2 | -4.301E+00 | 1.264E+03 | | | -25 to 125 | 0.93 ( 25 ) | 212 |
| $C_9H_{18}$ | Butylcyclopentane | 2 | -4.159E+00 | 1.182E+03 | | | -20 to 120 | 0.83 ( 25 ) | 212 |
| $C_9H_{18}O$ | Dibutyl ketone | | | | | | | 1.28 ( 20 ) | 212 |
| $C_9H_{18}O_2$ | Nonanoic acid | 2 | -6.002E+00 | 2.365E+03 | | | 20 to 100 | 8.08 ( 20 ) | 212 |
| $C_9H_{19}Br$ | n-Nonyl bromide | 2 | -4.531E+00 | 1.535E+03 | | | 0 to 100 | 2.05 ( 20 ) | 212 |
| $C_9H_{20}$ | n-Nonane | 2 | -4.447E+00 | 1.210E+03 | | | -50 to 150 | 0.67 ( 25 ) | 212 |

**TABLE 9-8 Correlation of Experimental Liquid Viscosity Data** *(Continued)*

| Formula | Name | Eq. No. | A | B | C | D | Range, °C | η,cP at(T,°C) | Ref. |
|---|---|---|---|---|---|---|---|---|---|
| $C_{10}H_8$ | Naphthalene | 3 | -1.027E+01 | 2.517E+03 | 1.098E-02 | -5.867E-06 | 81 to 475 | 0.78 (100 ) | 224 |
| $C_{10}H_{12}O_2$ | Eugenol (1,3,4) | 1 | 2.545E+36 | -1.435E+01 | | | 0 to 40 | 9.22 ( 20 ) | 60 |
| $C_{10}H_{14}$ | Butylbenzene | 2 | -4.386E+00 | 1.298E+03 | | | -25 to 160 | 0.96 ( 25 ) | 212 |
| $C_{10}H_{15}N$ | N,N-Diethylaniline | 2 | -5.577E+00 | 1.871E+03 | | | 0 to 100 | 2.18 ( 20 ) | 212 |
| $C_{10}H_{20}$ | 1-Decene | 2 | -4.297E+00 | 1.194E+03 | | | 0 to 125 | 0.75 ( 25 ) | 212 |
| $C_{10}H_{20}$ | n-Butylcyclohexane | 2 | -4.424E+00 | 1.378E+03 | | | -20 to 125 | 1.20 ( 25 ) | 212 |
| $C_{10}H_{20}$ | Pentylcyclopentane | 2 | -4.309E+00 | 1.302E+03 | | | -20 to 120 | 1.06 ( 25 ) | 212 |
| $C_{10}H_{20}O_2$ | Decanoic acid | 2 | -5.591E+00 | 2.275E+03 | | | 50 to 80 | 4.34 ( 50 ) | 212 |
| $C_{10}H_{21}Br$ | n-Decyl bromide | 2 | -4.704E+00 | 1.641E+03 | | | 0 to 100 | 2.55 ( 20 ) | 212 |
| $C_{10}H_{22}$ | n-Decane | 2 | -4.460E+00 | 1.286E+03 | | | -25 to 175 | 0.86 ( 25 ) | 212 |
| $C_{11}H_{16}$ | Pentylbenzene | 2 | -4.658E+00 | 1.464E+03 | | | -20 to 160 | 1.22 ( 25 ) | 212 |
| $C_{11}H_{22}$ | 1-Undecene | 2 | -4.422E+00 | 1.304E+03 | | | 0 to 125 | 0.96 ( 25 ) | 212 |
| $C_{11}H_{22}$ | n-Pentylcyclohexane | 2 | -4.814E+00 | 1.569E+03 | | | -10 to 120 | 1.56 ( 25 ) | 212 |
| $C_{11}H_{22}$ | Hexylcyclopentane | 2 | -4.463E+00 | 1.422E+03 | | | -20 to 120 | 1.36 ( 25 ) | 212 |
| $C_{11}H_{24}$ | n-Undecane | 2 | -4.571E+00 | 1.394E+03 | | | -25 to 200 | 1.09 ( 25 ) | 212 |
| $C_{12}H_{10}$ | Biphenyl | 2 | -4.572E+00 | 1.690E+03 | | | 70 to 450 | 0.95 (100 ) | 212 |
| $C_{12}H_{11}N$ | Diphenyl amine | 2 | -6.629E+00 | 2.676E+03 | | | 55 to 130 | 4.66 ( 55 ) | 212 |
| $C_{12}H_{18}$ | Hexylbenzene | 2 | -4.802E+00 | 1.575E+03 | | | -20 to 160 | 1.52 ( 25 ) | 212 |
| $C_{12}H_{24}$ | 1-Dodecene | 2 | -4.572E+00 | 1.418E+03 | | | 0 to 125 | 1.20 ( 25 ) | 212 |
| $C_{12}H_{24}$ | n-Hexylcyclohexane | 2 | -4.997E+00 | 1.697E+03 | | | -10 to 120 | 1.99 ( 25 ) | 212 |
| $C_{12}H_{24}$ | Heptylcyclopentane | 2 | -4.526E+00 | 1.508E+03 | | | -20 to 120 | 1.72 ( 25 ) | 212 |
| $C_{12}H_{24}O_2$ | Dodecanoic acid | 2 | -6.146E+00 | 2.615E+03 | | | 50 to 100 | 7.30 ( 50 ) | 212 |
| $C_{12}H_{26}$ | n-Dodecane | 2 | -4.562E+00 | 1.454E+03 | | | -5 to 220 | 1.37 ( 25 ) | 212 |

| Formula | Name | Eq. No. | A | B | C | D | Range, °C | η,cP at(T,°C) | Ref. |
|---|---|---|---|---|---|---|---|---|---|
| $C_{13}H_{10}O$ | Benzophenone | 2 | -7.181E+00 | 2.886E+03 | | | 25 to 130 | 13.6 ( 25 ) | 212 |
| $C_{13}H_{13}N$ | Benzyl phenyl amine | 2 | -6.415E+00 | 2.650E+03 | | | 55 to 130 | 5.4 ( 55 ) | 212 |
| $C_{13}H_{13}N$ | N-Methyl diphenyl amine | 2 | -6.328E+00 | 2.428E+03 | | | 10 to 130 | 7.3 ( 20 ) | 212 |
| $C_{13}H_{13}N$ | Benzyl amine | 1 | 5.344E+05 | -2.168E+00 | | | 33 to 130 | 2.18 ( 33 ) | 60 |
| $C_{13}H_{20}$ | Heptylbenzene | 2 | -4.916E+00 | 1.673E+03 | | | -20 to 160 | 1.88 ( 25 ) | 212 |
| $C_{13}H_{26}$ | 1-Tridecene | 2 | -4.682E+00 | 1.516E+03 | | | 0 to 125 | 1.50 ( 25 ) | 212 |
| $C_{13}H_{26}$ | n-Heptylcyclohexane | 2 | -5.164E+00 | 1.814E+03 | | | -10 to 120 | 2.47 ( 25 ) | 212 |
| $C_{13}H_{26}$ | Octylcyclopentane | 2 | -4.628E+00 | 1.602E+03 | | | -20 to 120 | 2.15 ( 25 ) | 212 |
| $C_{13}H_{28}$ | n-Tridecane | 2 | -4.605E+00 | 1.529E+03 | | | -5 to 240 | 1.71 ( 25 ) | 212 |
| $C_{14}H_{12}O_2$ | Diethylphthalate | 2 | -6.589E+00 | 2.606E+03 | | | 25 to 60 | 10.1 ( 25 ) | 212 |
| $C_{14}H_{12}O_2$ | Benzyl benzoate | 2 | -6.589E+00 | 2.606E+03 | | | 5 to 100 | 8.5 ( 25 ) | 212 |
| $C_{14}H_{14}O$ | Benzyl ether | 1 | 1.425E+22 | -8.678E+00 | | | 0 to 40 | 5.33 ( 20 ) | 60 |
| $C_{14}H_{22}$ | Octylbenzene | 2 | -5.190E+00 | 1.816E+03 | | | -20 to 140 | 2.31 ( 25 ) | 212 |
| $C_{14}H_{28}$ | 1-Tetradecene | 2 | -4.778E+00 | 1.606E+03 | | | 0 to 125 | 1.83 ( 25 ) | 212 |
| $C_{14}H_{28}$ | n-Octylcyclohexane | 2 | -5.353E+00 | 1.935E+03 | | | -10 to 120 | 3.10 ( 25 ) | 212 |
| $C_{14}H_{28}$ | Nonylcyclopentane | 2 | -4.732E+00 | 1.693E+03 | | | -20 to 120 | 2.63 ( 25 ) | 212 |
| $C_{14}H_{28}O_2$ | Tetradecanoic acid | 2 | -6.150E+00 | 2.705E+03 | | | 60 ro 100 | 7.43 ( 60 ) | 212 |
| $C_{14}H_{30}$ | n-Tetradecane | 2 | -4.615E+00 | 1.588E+03 | | | 5 to 255 | 2.1 ( 25 ) | 212 |
| $C_{15}H_{24}$ | Nonylbenzene | 2 | -5.352E+00 | 1.924E+03 | | | -20 to 140 | 2.8 ( 25 ) | 212 |
| $C_{15}H_{30}$ | 1-Pentadecene | 2 | -4.898E+00 | 1.702E+03 | | | 0 to 125 | 2.24 ( 25 ) | 212 |
| $C_{15}H_{30}$ | n-Nonylcyclohexane | 2 | -5.511E+00 | 2.042E+03 | | | -10 to 120 | 3.80 ( 25 ) | 212 |
| $C_{15}H_{30}$ | Decylcyclopentane | 2 | -4.825E+00 | 1.777E+03 | | | -20 to 120 | 3.20 ( 25 ) | 212 |
| $C_{15}H_{32}$ | n-Pentadecane | 2 | -4.648E+00 | 1.654E+03 | | | 10 to 280 | 2.56 ( 25 ) | 212 |

**TABLE 9-8 Correlation of Experimental Liquid Viscosity Data** (*Continued*)

| Formula | Name | Eq. No. | A | B | C | D | Range, °C | η,cP at(T,°C) | Ref. |
|---|---|---|---|---|---|---|---|---|---|
| $C_{16}H_{26}$ | Decylbenzene | 2 | -5.251E+00 | 1.944E+03 | | | -15 to 160 | 3.36 ( 25 ) | 212 |
| $C_{16}H_{32}$ | 1-Hexadecene | 2 | -4.938E+00 | 1.767E+03 | | | 5 to 125 | 2.70 ( 25 ) | 212 |
| $C_{16}H_{32}$ | n-Decylcyclohexane | 2 | -5.630E+00 | 2.132E+03 | | | 0 to 120 | 4.60 ( 25 ) | 212 |
| $C_{16}H_{32}$ | Undecylcyclopentane | 2 | -4.992E+00 | 1.882E+03 | | | -10 to 120 | 3.84 ( 25 ) | 212 |
| $C_{16}H_{32}O_2$ | Hexadecanoic acid | 2 | -6.103E+00 | 2.785E+03 | | | 70 to 100 | 7.8 ( 70 ) | 212 |
| $C_{16}H_{34}$ | n-Hexadecane | 2 | -4.643E+00 | 1.700E+03 | | | 20 to 280 | 3.09 ( 25 ) | 212 |
| $C_{16}H_{34}O$ | Cetyl alcohol | | | | | | | 13.4 ( 50 ) | 60 |
| $C_{17}H_{28}$ | Undecylbenzene | 2 | -5.305E+00 | 2.008E+03 | | | -10 to 160 | 4.0 ( 25 ) | 212 |
| $C_{17}H_{34}$ | 1-Heptadecene | 2 | -4.950E+00 | 1.820E+03 | | | 15 to 125 | 3.24 ( 25 ) | 212 |
| $C_{17}H_{34}$ | n-Undecylcyclohexane | 2 | -5.710E+00 | 2.206E+03 | | | 10 to 120 | 5.5 ( 25 ) | 212 |
| $C_{17}H_{34}$ | Dodecylcyclopentane | 2 | -5.100E+00 | 1.966E+03 | | | -5 to 120 | 4.56 ( 25 ) | 212 |
| $C_{17}H_{36}$ | n-Heptadecane | 2 | -4.642E+00 | 1.745E+03 | | | 25 to 200 | 3.7 ( 25 ) | 212 |
| $C_{18}H_{14}$ | o-Terphenyl | 2 | -5.461E+00 | 2.519E+03 | | | 100 to 400 | 4.4 (100 ) | 212 |
| $C_{18}H_{14}$ | m-Terphenyl | 2 | -4.699E+00 | 2.167E+03 | | | 150 to 400 | 1.6 (150 ) | 212 |
| $C_{18}H_{14}$ | p-Terphenyl | 2 | -4.549E+00 | 2.098E+03 | | | 250 to 400 | 0.60 (250 ) | 212 |
| $C_{18}H_{30}$ | Dodecylbenzene | 2 | -5.285E+00 | 2.044E+03 | | | 0 to 160 | 4.78 ( 25 ) | 212 |
| $C_{18}H_{34}O_2$ | Octadecenoic acid | 2 | -5.907E+00 | 2.760E+03 | | | 20 to 200 | 29.4 ( 25 ) | 212 |
| $C_{18}H_{36}$ | 1-Octadecene | 2 | -4.986E+00 | 1.879E+03 | | | 20 to 125 | 3.85 ( 25 ) | 212 |
| $C_{18}H_{36}$ | n-Dodecylcyclohexane | 2 | -5.815E+00 | 2.285E+03 | | | 15 to 120 | 6.5 ( 25 ) | 212 |
| $C_{18}H_{36}$ | Tridecylcyclopentane | 2 | -5.228E+00 | 2.053E+03 | | | 5 to 120 | 5.4 ( 25 ) | 212 |
| $C_{18}H_{36}O_2$ | Octadecanoic acid | 2 | -7.415E+00 | 3.338E+03 | | | 70 to 110 | 7.7 ( 80 ) | 212 |
| $C_{18}H_{38}$ | n-Octadecane | 2 | -4.649E+00 | 1.790E+03 | | | 30 to 300 | 3.88 ( 30 ) | 212 |

| Formula | Name | Eq. No. | A | B | C | D | Range, °C | η,cP at(T,°C) | Ref. |
|---|---|---|---|---|---|---|---|---|---|
| $C_{19}H_{32}$ | Tridecylbenzene | 2 | -5.351E+00 | 2.108E+03 | | | 5 to 160 | 5.63 ( 25 ) | 212 |
| $C_{19}H_{38}$ | 1-Nonadecene | 2 | -5.044E+00 | 1.944E+03 | | | 25 to 125 | 4.56 ( 25 ) | 212 |
| $C_{19}H_{38}$ | n-Tridecylcyclohexane | 2 | -5.904E+00 | 2.357E+03 | | | 20 to 120 | 7.67 ( 25 ) | 212 |
| $C_{19}H_{38}$ | Tetradecylcyclopentane | 2 | -5.328E+00 | 2.129E+03 | | | 10 to 120 | 6.29 ( 25 ) | 212 |
| $C_{19}H_{40}$ | n-Nondecane | 2 | -4.643E+00 | 1.827E+03 | | | 35 to 300 | 4.0 ( 35 ) | 212 |
| $C_{20}H_{34}$ | Tetradecylbenzene | 2 | -5.354E+00 | 2.147E+03 | | | 15 to 160 | 6.6 ( 25 ) | 212 |
| $C_{20}H_{40}$ | 1-Eicosene | 2 | -5.033E+00 | 1.982E+03 | | | 35 to 125 | 4.2 ( 35 ) | 212 |
| $C_{20}H_{40}$ | n-Tetradecylcyclohexane | 2 | -5.998E+00 | 2.428E+03 | | | 25 to 120 | 9.0 ( 25 ) | 212 |
| $C_{20}H_{40}$ | Pentadecylcyclopentane | 2 | -5.387E+00 | 2.189E+03 | | | 20 to 120 | 7.3 ( 25 ) | 212 |
| $C_{20}H_{42}$ | n-Eicosane | 2 | -4.651E+00 | 1.868E+03 | | | 40 to 300 | 4.14 ( 40 ) | 212 |
| $C_{21}H_{36}$ | Pentadecylbenzene | 2 | -5.416E+00 | 2.206E+03 | | | 20 to 160 | 7.72 ( 25 ) | 212 |
| $C_{21}H_{42}$ | n-Pentadecylcyclohexane | 2 | -6.078E+00 | 2.493E+03 | | | 30 to 120 | 8.91 ( 30 ) | 212 |
| $C_{21}H_{42}$ | Hexadecylcyclopentane | 2 | -5.459E+00 | 2.251E+03 | | | 25 to 120 | 8.44 ( 25 ) | 212 |
| $C_{22}H_{38}$ | Hexadecylbenzene | 2 | -5.428E+00 | 2.245E+03 | | | 30 to 160 | 7.74 ( 30 ) | 212 |
| $C_{22}H_{44}$ | n-Hexadecylcyclohexane | 2 | -6.169E+00 | 2.560E+03 | | | 35 to 120 | 8.76 ( 35 ) | 212 |

Footnote:
1. While the literature was often not precise, it should be assumed that the state of the liquid is saturated at the prevailing pressure.

Equation numbers refer to:

Eq. (1): $\eta = AT^B$

Eq. (2): $\ln \eta = A + B/T$

Eq. (3): $\ln \eta = A + B/T + CT + DT^2$

with $\eta$ in cP and $T, T_c$ in kelvins

by Luckas and Lucas [139], who present constants to determine liquid viscosities for a large number of organic compounds. The correlating equations used are Eq. (9-11.2) and a modification of Eq. (9-11.9) with a switching function to change from the former to the latter at about $T_r$ = 0.55. To employ Eq. (9-11.9), accurate liquid volumes are required either from experimental data or from a separate estimation method. Although the Luckas and Lucas table may be slightly more accurate than the correlations in Table 9-8, the latter have been retained because of their simplicity and the fact they require only the state variable of temperature.

Gambill [77] mentions several other approximate one-datum-point extrapolation formulas; the estimation techniques discussed in the next section may also be used by employing a single viscosity point to yield the structural constant.

In summary, from the freezing point to near the normal boiling point, Eq. (9-10.1) is a satisfactory temperature–liquid viscosity function. Two datum points are required. If only one datum point is known, a rough approximation of the viscosity at other temperatures can be obtained from Eq. (9-10.3) or Fig. 9-14. At temperatures above the normal boiling point, Eq. (9-11.9) is preferable.

Liquid viscosities above the normal boiling point are treated in Sec. 9-12.

## 9-11 Estimation of Low-Temperature Liquid Viscosity

Estimation methods for low-temperature liquid viscosity often employ structural-sensitive parameters which are valid only for certain homologous series or are found from group contributions. These methods usually use some variation of Eq. (9-10.1) and are limited to reduced temperatures less than about 0.75. We present two such methods in this section. We also describe a technique which employs corresponding states concepts. None of the three methods considered is particularly reliable, and we recommend that if the compound of interest is given in Table 9-8, the correlating equations presented in the table be used to estimate viscosities rather than the techniques shown below.

### Orrick and Erbar method [154]

This method employs a group contribution technique to estimate $A$ and $B$ in Eq. (9-11.1).

$$\ln \frac{\eta_L}{\rho_L M} = A + \frac{B}{T} \tag{9-11.1}$$

where $\eta_L$ = liquid viscosity, cP
$\rho_L$ = liquid density at 20°C, g/cm$^3$
$M$ = molecular weight
$T$ = temperature, K

The group contributions for obtaining $A$ and $B$ are given in Table 9-9. For liquids that have a normal boiling point below 20°C, use the value of $\rho_L$ at this temperature; for liquids whose freezing point is above 20°C, $\rho_L$ at the melting point should be employed. Compounds containing nitrogen or sulfur cannot be treated. Orrick and Erbar tested this method for 188 organic liquids. The errors varied widely, but they reported an average deviation of 15 percent. This is close to the average value of 16 percent shown in Table 9-12 for a more limited test.

**Example 9-15** Estimate the viscosity of liquid *n*-butyl alcohol at 120°C. The experimental value is 0.394 cP.

**solution** From Table 9-9,

$$A = -6.95 - (0.21)(4) - 3.00 = -10.79$$

$$B = 275 + (99)(4) + 1600 = 2271$$

From Appendix A, at 20°C, $\rho_L$ = 0.809 g/cm$^3$ and $M$ = 74.12. Then, with Eq. (9-11.1),

$$\ln \frac{\eta_L}{(0.809)(74.12)} = -10.79 + \frac{2271}{T}$$

At $T$ = 120°C = 393 K, $\eta_L$ = 0.399 cP

$$\text{Error} = \frac{0.399 - 0.394}{0.394} \times 100 = 1.3\%$$

### Van Velzen, Cardozo, and Langenkamp method

In an unusually detailed study of the effect of structure on liquid viscosities, van Velzen et al. [211, 212] proposed a modification of Eq. (9-10.1),

$$\log \eta_L = B(T^{-1} - T_0^{-1}) \tag{9-11.2}$$

where $\eta_L$ = liquid viscosity, cP
$T$ = temperature, K

and $B$ and $T_0$ are related to structure. To determine these parameters, one must first find the *equivalent chain length* $N^*$, where

$$N^* = N + \Sigma\, \Delta N_i \tag{9-11.3}$$

$N$ is the actual number of carbon atoms in the molecule, and $\Delta N$ represents structural contributions from Table 9-10. If the structural or functional group $\Delta N_i$ appears $n_i$ times in the molecule, $n_i\, \Delta N_i$ corrections must

**TABLE 9-9 Orrick and Erbar Group Contributions for *A* and *B* in Eq. (9-11.1) [154]**

| Group | *A* | *B* |
|---|---|---|
| Carbon atoms† | $-(6.95+0.21n)$ | $275+99n$ |
| $R{-}\overset{\mid}{\underset{R}{\underset{\mid}{C}}}{-}R$ | −0.15 | 35 |
| $R{-}\overset{R}{\overset{\mid}{\underset{R}{\underset{\mid}{C}}}}{-}R$ | −1.20 | 400 |
| Double bond | 0.24 | −90 |
| Five-membered ring | 0.10 | 32 |
| Six-membered ring | −0.45 | 250 |
| Aromatic ring | 0 | 20 |
| Ortho substitution | −0.12 | 100 |
| Meta substitution | 0.05 | −34 |
| Para substitution | −0.01 | −5 |
| Chlorine | −0.61 | 220 |
| Bromine | −1.25 | 365 |
| Iodine | −1.75 | 400 |
| —OH | −3.00 | 1600 |
| —COO— | −1.00 | 420 |
| —O— | −0.38 | 140 |
| $-\overset{\mid}{C}{=}O$ | −0.50 | 350 |
| —COOH | −0.90 | 770 |

†$n$ = number, not including those in groups shown above.

be added. In the tabulation of $\Delta N_i$ contributions, some entries are to be used every time the functional group appears; other entries represent additional *corrections* to be used to modify the basic group contribution. A few examples given below illustrate the technique of calculating $N^*$.

**Example 9-16** Calculate $N^*$ for benzophenone, chloroform, *N*-methyldiphenylamine, *N,N*-diethylaniline, allyl alcohol, and *m*-nitrotoluene.

**solution** For benzophenone, $C_6H_5COC_6H_5$, $N = 13$ (number of carbon atoms). There is a $\Delta N_i$ contribution for ketones, that is, $3.265 - 0.122N = 3.265 - (0.122)(13) = 1.68$. Also, there is a *correction* for aromatic ketones of 2.70 for each aromatic ring; thus,

$$N^* = 13 + 1.68 + (2)(2.70) = 20.08$$

For chloroform, $N = 1$. Each of the three chlorine atoms has a $\Delta N_i$ contribution of 3.21. Also, there is a *correction* term for the $C(Cl)_x$ structure of $1.91 - 1.459x$, where, in this case, $x = 3$. Thus,

$$N^* = 1 + (3)(3.21) + 1.91 - (3)(1.459) = 8.16$$

For *N*-methyldiphenylamine, $(C_6H_5)_2N(CH_3)$, $N = 13$. This is a tertiary amine, so that $\Delta N_i = 3.27$. Also, there is a *correction* for each of the aromatic groups (see note *f* of Table 9-10) of 0.6. Then

$$N^* = 13 + 3.27 + (2)(0.6) = 17.47$$

For *N,N*-diethylaniline, $(C_6H_5)N(C_2H_5)_2$, $N = 10$. Again, a tertiary amine contribution of 3.27 is required, but only one aromatic group is present with a correction of 0.6:

$$N^* = 10 + 3.27 + 0.6 = 13.87$$

For allyl alcohol, $CH_2{=}CHCH_2OH$, $N = 3$. As a primary alcohol, there is a $\Delta N_i$ of $10.606 - 0.276N = 10.606 - (3)(0.276) = 9.778$. In addition, as an alkene, another $\Delta N_i$ of $-0.152 - 0.042N = -0.152 - (3)(0.042) = -0.278$ is necessary. Thus,

$$N^* = 3 + 9.778 - 0.278 = 12.50$$

Finally, for *m*-nitrotoluene, $C_6H_4NO_2(CH_3)$, $N = 7$. The $\Delta N_i$ for the aromatic nitro compound is $7.812 - 0.236N = 7.812 - (7)(0.236) = 6.16$. There is no contribution for an alkyl benzene. For the meta correction (see note *a* of Table 9-10), $\Delta N_i = 0.11$. Thus,

$$N^* = 7 + 6.16 + 0.11 = 13.27$$

The value of $N^*$ is then used to determine the constants $B$ and $T_0$ which appear in Eq. (9-11.2). For $T_0$,

$$T^\circ = \begin{cases} 28.86 + 37.439N^* - 1.3547(N^*)^2 + 0.02076(N^*)^3 & N^* < 20 \quad (9\text{-}11.4) \\ 8.164N^* + 238.59 & N^* > 20 \quad (9\text{-}11.5) \end{cases}$$

**TABLE 9-10 Van Velzen, Cardozo, and Langenkamp Contributions for Liquid Viscosity [211]**

| Structures or functional group | $\Delta N_i$ | $\Delta B_i$ | Example: Compound | $N^*$ | $B$ | $T_0$ | Remarks |
|---|---|---|---|---|---|---|---|
| $n$-Alkanes | 0 | 0 | $n$-Hexane | 6.00 | 377.86 | 209.21 | |
| Isoalkanes | $1.389 - 0.238N$ | 15.51 | 2-Methylbutane | 5.20 | 351.95 | 189.83 | |
| Saturated hydrocarbons with two methyl groups in iso position | $2.319 - 0.238N$ | 15.51 | 2,3-Dimethylbutane | 6.89 | 437.37 | 229.29 | |
| $n$-Alkenes | $-0.152 - 0.042N$ | $-44.94 + 5.410N^*$ | 1-Octene | 7.51 | 446.89 | 242.41 | |
| $n$-Alkadienes | $-0.304 - 0.084N$ | $-44.94 + 5.410N^*$ | 1,3-Butadiene | 3.36 | 211.21 | 140.15 | |
| Isoalkenes | $1.237 - 0.280N$ | $-36.01 + 5.410N^*$ | 2-Methyl-2-butene | 4.84 | 307.40 | 180.68 | |
| Isoalkadienes | $1.085 - 0.322N$ | $-36.01 + 5.410N^*$ | 2-Methyl-1,3-butadiene | 4.48 | 285.89 | 171.26 | |
| Hydrocarbon with one double bond and two methyl groups in iso position | $2.626 - 0.518N$ | $-36.01 + 5.410N^*$ | 2,3-Dimethyl-1-butene | 5.52 | 347.07 | 197.74 | For any additional $CH_3$ groups in iso position, increase $\Delta N$ by $1.389 - 0.238N$ |
| Hydrocarbon with two double bonds and two methyl groups in iso position | $2.474 - 0.560N$ | $-36.01 + 5.410N^*$ | 2,3-Dimethyl-1,3-butadiene | 5.11 | 323.30 | 187.57 | For any additional $CH_3$ groups in iso position, increase $\Delta N$ by $1.389 - 0.238N$ |
| Cyclopentanes | $0.205 + 0.069N$ | $-45.96 + 2.224N^*$ | $n$-Butylcyclopentane | 9.83 | 527.3 | 285.7 | $N \leq 16$; not recommended for $N = 5, 6$ |
| | $3.971 - 0.172N$ | $-339.67 + 23.135N^*$ | Tridecylcyclopentane | 18.87 | 889.40 | 392.45 | $N \geq 16$ |
| Cyclohexanes | 1.48 | $-272.85 + 25.041N^*$ | Ethylcyclohexane | 9.48 | 501.80 | 279.72 | $N < 17$; not recommended for $N = 6, 7$ |
| | $6.517 - 0.311N$ | $-272.85 + 25.041N^*$ | Dodecylcyclohexane | 18.92 | 994.10 | 392.87 | $N \geq 17$ |
| Alkyl benzenes | 0.60 | $-140.04 + 13.869N^*$ | $o$-Xylene | 9.11 | 563.09 | 273.20 | $N < 16$; not recommended for $N = 6, 7$[a,e,f] |
| | $3.055 - 0.161N$ | $-140.04 + 13.869N^*$ | | | | | $N \geq 16$[a,e,f] |
| Polyphenyls | $-5.340 + 0.815N$ | $-188.40 + 9.558N^*$ | $m$-Terphenyl | 27.44 | 1008.7 | 462.58 | [a] |
| Alcohols: | | | | | | | |
| Primary | $10.606 - 0.276N$ | $-589.44 + 70.519N^*$ | 1-Pentanol | 14.23 | 1113.0 | 347.12 | [b] |
| Secondary | $11.200 - 0.605N$ | 497.58 | Isopropyl alcohol | 12.38 | 1141.35 | 324.12 | [b] |
| Tertiary | $11.200 - 0.605N$ | 928.83 | 2-Methyl butanol-2 | 13.42 | 1699.1 | 337.49 | [b] |
| Diols (correction) | See remarks | 557.77 | Propylene glycol | 22.66 | 1399.71 | 423.55 | For $\Delta N$, use alcohol contributions and add $N - 2.50$ |
| Phenols (correction) | $16.17 - N$ | 213.68 | | | | | |
| —OH on side chain to aromatic ring (correction) | $-0.16$ | 213.68 | | | | | [a,c,d] |
| Acids | $6.795 + 0.365N$ | $-249.12 + 22.449N^*$ | $n$-Butyric acid | 12.25 | 665.40 | 322.36 | $N < 11$, not recommended for $N = 1, 2$ |
| | 10.71 | $-249.12 + 22.449N^*$ | | | | | $N \geq 11$ |

| | | | | | | | |
|---|---|---|---|---|---|---|---|
| Iso acids | See remarks | $-249.12 + 22.449N^*$ | Isobutyric acid | 12.01 | 652.02 | 319.06 | Calculate $\Delta B$ as for straight-chain acid; calculate $\Delta N$ for straight-chain acid but reduce $\Delta N$ by 0.24 for each methyl group in iso position |
| Acids with aromatic nucleus in structure (correction) | 4.81 | $-188.40 + 9.558N^*$ | Phenylacetic acid | 22.52 | 1123.29 | 422.41 | |
| Esters | $4.337 - 0.230N$ | $-149.13 + 18.695N^*$ | Ethyl valerate | 9.73 | 580.16 | 284.01 | If hydrocarbon groups have iso configuration, see footnote *e* |
| Esters with aromatic nucleus in structure (correction) | $-1.174 + 0.376N$ | $-140.04 + 13.869N^*$ | Benzyl benzoate | 19.21 | 1133.17 | 395.31 | Add to values of $\Delta N$, $\Delta B$ calculated for ester |
| Ketones | $3.265 - 0.122N$ | $-117.21 + 15.781N^*$ | Methyl *n*-butyl ketone | 8.53 | 514.53 | 262.53 | If hydrocarbon groups have iso configuration, see footnote *e* |
| Ketones with aromatic nucleus in structure (correction) | 2.70 | $-760.65 + 50.478N^*$ | Acetophenone | 12.99 | 645.92 | 322.10 | Add to values of $\Delta N$, $\Delta B$ calculated for ketone |
| Ethers | $0.298 + 0.209N$ | $-9.39 + 2.848N^*$ | Ethyl hexyl ether | 9.97 | 575.96 | 288.04 | If hydrocarbon groups have iso configuration, see footnote *e* |
| Aromatic ethers | $11.5 - N$ | $-140.04 + 13.869N^*$ | Propyl phenyl ether | 11.5 | 656.83 | 311.82 | [c]The $\Delta N$ value is not a correction to regular ether value, but the $\Delta B$ value is a correction to regular ether |
| Amines: | | | | | | | |
| Primary | $3.581 + 0.325N$ | $25.39 + 8.744N^*$ | Propylamine | 7.56 | 545.01 | 243.44 | If hydrocarbon groups have iso configuration, see footnote *e* |
| Primary amine in side chain of aromatic compound (correction) | $-0.16$ | 0 | Benzylamine | 12.70 | 790.47 | 328.36 | Corrections to be added to amine calculation[e] |
| Secondary | $1.390 + 0.461N$ | $25.39 + 8.744N^*$ | Ethylpropylamine | 8.69 | 605.44 | 265.53 | If hydrocarbon groups have iso configuration, see footnote *e* |
| Tertiary | 3.27 | $25.39 + 8.744N^*$ | | ... | ... | ... | If hydrocarbon groups have iso configuration, see footnote *e* |
| Primary amines with $NH_2$ group on aromatic nucleus | $15.04 - N$ | 0 | *m*-Toluidine | 15.04 | 904.08 | 356.13 | [a,c]The $\Delta N$ value is not a correction to regular amine value; to find $\Delta B$ use primary amine value |
| Secondary or tertiary amine with at least one aromatic group attached to amino nitrogen | [f] | [f] | Benzylphenylamine | 21.58 | 1041.18 | 414.74 | |
| Nitro compounds: | | | | | | | |
| 1-nitro | $7.812 - 0.236N$ | $-213.14 + 18.330N^*$ | Nitromethane | 8.57 | 442.82 | 263.28 | |
| 2-nitro | 5.84 | $-213.14 + 18.330N^*$ | 2-Nitro-2-pentene | 10.48 | 567.43 | 296.33 | Note alkene contribution is necessary |
| 3-nitro | 5.56 | $-338.01 + 25.086N^*$ | | | | | |
| 4-nitro; 5-nitro | 5.36 | $-338.01 + 25.086N^*$ | | | | | |
| Aromatic nitro-compounds | $7.812 - 0.236N$ | $-213.14 + 18.330N^*$ | Nitrobenzene | 13.00 | 728.79 | 332.23 | For aromatic correction see footnote *f* |

**TABLE 9-10 Van Velzen, Cardozo, and Langenkamp Contributions for Liquid Viscosity** (*Continued*)

| Structures or functional group | $\Delta N_i$ | $\Delta B_i$ | Example: Compound | Example: $N^*$ | Example: $B$ | Example: $T_0$ | Remarks |
|---|---|---|---|---|---|---|---|
| Halogenated compounds: | | | | | | | |
| Fluoride | 1.43 | 5.75 | | | | | |
| Chloride | 3.21 | −17.03 | Ethyl chloride | 5.21 | 319.94 | 190.08 | [e,f] |
| Bromide | 4.39 | $-101.97 + 5.954N^*$ | 1-Bromo-2-methyl propane | 8.15 | 435.85 | 255.24 | [e,f] |
| Iodide | 5.76 | −85.32 | Iodobenzene | 12.36 | 589.18 | 323.85 | [e,f] |
| Special configurations (corrections): | | | | | | | |
| $C(Cl)_x$ | $1.91 - 1.459x$ | −26.38 | | | | | |
| —CCl—CCl— | 0.96 | 0 | | | | | |
| $—C(Br)_x—$ | 0.50 | $81.34 - 86.850x$ | | | | | |
| —CBr—CBr— | 1.60 | −57.73 | | | | | |
| $CF_3$, in alcohols | −3.93 | 341.68 | | | | | |
| In other compounds | −3.93 | 25.55 | | | | | |
| Aldehydes | 3.38 | $146.45 - 25.11N^*$ | Propionaldehyde | 6.38 | 383.16 | 217.97 | |
| Aldehydes with an aromatic nucleus in structure (correction) | 2.70 | $-760.65 + 50.478N^*$ | Benzaldehyde | 13.08 | 391.19 | 333.25 | |
| Anhydrides | $7.97 - 0.50N$ | −33.50 | Propionic anhydride | 10.79 | 554.87 | 301.19 | |
| Anhydrides with an aromatic nucleus in structure (correction) | 2.70 | $-760.65 + 50.478N^*$ | | | | | |
| Amides | $13.12 + 1.49N$ | $524.63 - 20.72N^*$ | Acetamide | 18.10 | 931.07 | 385.79 | |
| Amides with an aromatic nucleus in structure (correction) | 2.70 | $-760.65 + 50.478N^*$ | | | | | |

[a] For substitutions on an aromatic nucleus in more than one position, additional corrections are required:

| | | |
|---|---|---|
| Ortho: | $\Delta N = 0.51$ | $\Delta B = \begin{cases} -571.94 & \text{with —OH} \\ 54.84 & \text{without —OH} \end{cases}$ |
| Meta: | $\Delta N = 0.11$ | $\Delta B = 27.25$ |
| Para: | $\Delta N = -0.04$ | $\Delta B = -17.57$ |

[b] For alcohols, if there is a methyl group in the iso position, increase $\Delta N$ by 0.24 and $\Delta B$ by 94.23.

[c] If the compound has an aromatic —OH or —$NH_2$, or if it is an aromatic ether, use $\Delta N$ contribution in table but neglect other substituents on the ring such as halogen, $CH_3$, $NO_2$, etc. For the calculation of $\Delta B$, however, such substituents must be taken into account.

[d] For aromatic alcohols and compounds with an —OH on a side chain, the alcohol contribution (primary, etc.) must be included. For example, *o*-chlorophenol:

$$\Delta B = \Delta B \text{ (primary alcohol)} + \Delta B \text{ (chlorine)} + \Delta B \text{ (phenol)} + \Delta B \text{ (ortho correction}^a\text{)}$$

With $N^* = 16.17^c$:

$$\Delta B = (-589.44 + 70.519 \times 16.17) + (-17.03) + (213.68) + (-571.94) = 175.56$$

$$B_a = 745.94 \qquad B = B_a + \Delta B = 921.50$$

2-Phenylethanol:

$$N = 8 \qquad \Delta N = \Delta N \text{ (primary alcohol)} + \Delta N \text{ (correction)} = [10.606 - (0.276)(8)] + (-0.16) = 8.24$$

$$N^* = N + \Delta N = 8 + 8.24 = 16.24$$

$$\Delta B = \Delta B \text{ (primary alcohol)} + \Delta B \text{ (correction)} = [-589.44 + (70.519)(16.24)] + 213.68 = 769.47$$

$$B_a = 747.43 \qquad B = B_a + \Delta B = 1516.9$$

[e] For esters, alkylbenzenes, halogenated hydrocarbons, and ketones, if the hydrocarbon chain has a methyl group in an iso position, decrease $\Delta N$ by 0.24 and increase $\Delta B$ by 8.93 for each such grouping. For ethers and amines, decrease $\Delta N$ by 0.50 and increase $\Delta B$ by 8.93 for each iso group.

[f] For alkylbenzenes, nitrobenzenes, halogenated benzenes, and for secondary or tertiary amines where at least one aromatic group is connected to an amino nitrogen, add the following corrections for each aromatic nucleus. If $N < 16$, increase $\Delta N$ by 0.60; if $N \geq 16$, increase $\Delta N$ by $3.055 - 0.161N$ for each aromatic group. For any $N$, increase $\Delta B$ by $(-140.04 + 13.869N^*)$.

For $B$,

$$B = B_A + \Sigma\, \Delta B_i \tag{9-11.6}$$

where

$$B_A = \begin{cases} 24.79 + 66.885N^* - 1.3173(N^*)^2 - 0.00377(N^*)^3 & N^* < 20 \quad (9\text{-}11.7) \\ 530.59 + 13.740N^* & N^* > 20 \quad (9\text{-}11.8) \end{cases}$$

and $\Sigma\, \Delta B_i$ can be determined by summing contributions as shown in Table 9-10. Even though a functional group may appear more than once in a compound, the $\Delta B_i$ contribution is applied only a single time.

The $B$ and $T_0$ values found for any specific compound are then used in Eq. (9-11.2) and the liquid viscosity is determined. The units of $\eta_L$ and $T$ are centipoises and kelvins.

**Example 9-17** Obtain the constants $B$ and $T_0$ for benzophenone and estimate the liquid viscosity at at 25, 55, 95, and 120°C. Experimental values are 13.61, 4.67, 1.74, and 1.38 cP.

**solution** $N^*$ was calculated in Example 9-16 as 20.08. To determine $T_0$, since $N^* > 20$, Eq. (9-11.5) must be used.

$$T_0 = (8.164)(20.08) + 238.59 = 402.52$$

For $B$, first $B_A$ is determined from Eq. (9-11.8),

$$B_A = 530.59 + (13.740)(20.08) = 806.49$$

The $\Delta B_i$ corrections are

$$\Delta B \text{ (ketone)} = -117.21 + (15.781)(20.08) = 199.67$$

$$\Delta B \text{ (aromatic ketone correction)} = -760.65 + (50.478)(20.08) = 252.95$$

$$B = 806.49 + 199.67 + 252.95 = 1259.11$$

Then, from Eq. (9-11.2),

$$\log \eta_L = 1259.11[T^{-1} - (402.52)^{-1}]$$

| $T$, °C | $\eta_L$, cP Calc. | $\eta_L$, cP Exp. | % Error |
|---|---|---|---|
| 25 | 12.44 | 13.61 | −8.7 |
| 55 | 5.11 | 4.67 | 9.4 |
| 95 | 1.96 | 1.74 | 12 |
| 120 | 1.19 | 1.38 | −14 |

Van Velzen et al. tested their method on 314 different liquids with nearly 4500 datum points, and a careful statistical evaluation was made. Large errors were often noted for the first members of a homologous series. This point is significant because, in the comparison of calculated and experimental liquid viscosities in Table 9-11, many of the test compounds shown are first members of a series. This reflects unfairly on the van Velzen et al. method.

Some care must be used in selecting appropriate $\Delta N_i$ and $\Delta B_i$ contributions. A Euratom report [212] is very helpful in illustrating the rules for complex compounds.

#### Przezdziecki and Sridhar method [160]

In this technique the authors propose using the Hildebrand-modified Batschinski equation [15, 99, 217].

$$\eta_L = \frac{V_0}{E(V - V_0)} \tag{9-11.9}$$

where $\eta_L$ = liquid viscosity, cP
$V$ = liquid molar volume, $cm^3/mol$

and the parameters $E$ and $V_0$ are defined below.

$$E = -1.12 + \frac{V_c}{12.94 + 0.10M - 0.23P_c + 0.0424T_f - 11.58(T_f/T_c)} \tag{9-11.10}$$

$$V_0 = 0.0085\omega T_c - 2.02 + \frac{V_m}{0.342(T_f/T_c) + 0.894} \tag{9-11.11}$$

where $T_c$ = critical temperature, K
$P_c$ = critical pressure, bar
$V_c$ = critical volume, $cm^3/mol$
$M$ = molecular weight, g/mol
$T_f$ = freezing point, K
$\omega$ = acentric factor
$V_m$ = liquid molar volume at $T_f$, $cm^3/mol$

Thus, to use Eq. (9-11.9), one must have values for $T_c$, $P_c$, $V_c$, $T_f$, $\omega$, and $V_m$ in addition to the liquid molar volume $V$ at the temperature of interest. Rarely does one have data for $V_m$ or even $V$, and even if data are available, the authors recommend that these two liquid volumes be estimated from $T_f$ and $T$ by the Gunn-Yamada method [89]. In this procedure, one accurate value of $V$ is required in the temperature range of applicability of Eq. (9-11.9). We define this datum point as $V^R$ at $T^R$; then at any other temperature $T$,

$$V(T) = \frac{f(T)}{f(T^R)} V^R \tag{9-11.12}$$

where

$$f(T) = H_1(1 - \omega H_2) \tag{9-11.13}$$

$$H_1 = 0.33593 - 0.33953T_r + 1.51941T_r^2 - 2.02512T_r^3 + 1.11422T_r^4 \tag{9-11.14}$$

$$H_2 = 0.29607 - 0.09045T_r - 0.04842T_r^2 \tag{9-11.15}$$

**TABLE 9-11 Comparison of Calculated and Experimental Viscosities of Liquids**

| Compound | $T$, K | $\eta_L$, (exp.) cP‡ | Percent error† in liquid viscosity calculated by the method of: Orrick and Erbar | Van Velzen, Cardozo, and Langenkamp | Przezdziecki and Sridhar |
|---|---|---|---|---|---|
| Acetone | 183 | 2.075 | −25 | −20 | −11 |
| | 213 | 0.892 | −6.7 | −0.1 | −4.6 |
| | 273 | 0.389 | −8.3 | 0.1 | −2.3 |
| | 303 | 0.292 | −9.4 | −0.2 | −1.2 |
| | 333 | 0.226 | −8.3 | 1.6 | 0.2 |
| Acetic acid | 283 | 1.45 | −22 | −34 | 8.6 |
| | 313 | 0.901 | −15 | −28 | 0 |
| | 353 | 0.561 | −9.5 | −24 | −1.3 |
| | 383 | 0.416 | −5.3 | −21 | 0.3 |
| Aniline | 263 | 13.4 | — | −52 | — |
| | 293 | 4.38 | — | −23 | — |
| | 333 | 1.52 | — | −2.8 | −49 |
| | 393 | 0.658 | — | −11 | −33 |
| Benzene | 278 | 0.826 | −45 | −42 | 1.1 |
| | 313 | 0.492 | −35 | −30 | 7.3 |
| | 353 | 0.318 | −26 | −20 | 12 |
| | 393 | 0.219 | −46 | −8.5 | 18 |
| | 433 | 0.156 | −7.1 | 5.8 | 23 |
| | 463 | 0.121 | 5.1 | 20 | 28 |
| *n*-Butane | 183 | 0.63 | −14 | −7.2 | −9.0 |
| | 213 | 0.403 | −20 | −9.8 | −8.9 |
| | 273 | 0.210 | −23 | 8.8 | −5.8 |
| 1-Butene | 163 | 0.79 | −22 | −6.1 | −13 |
| | 193 | 0.45 | −20 | 0 | −9.6 |
| | 233 | 0.26 | −18 | 8.7 | −3.3 |
| *n*-Butyl alcohol | 273 | 5.14 | −2.1 | 5.6 | — |
| | 313 | 1.77 | −1.6 | 0 | — |
| | 353 | 0.762 | 0.5 | −2.3 | — |
| | 393 | 0.394 | −1.4 | −5.3 | — |
| Carbon tetrachloride | 273 | 1.369 | 20 | −10 | −24 |
| | 303 | 0.856 | 22 | −6.1 | −15 |
| | 343 | 0.534 | 20 | −4.3 | −6.7 |
| | 373 | 0.404 | 19 | −4.0 | −2.8 |
| Chlorobenzene | 273 | 1.054 | 1.4 | 15 | −8.3 |
| | 313 | 0.639 | −0.6 | 7.0 | −7.0 |
| | 353 | 0.441 | −0.9 | 0.2 | −5.2 |
| | 393 | 0.326 | −5.1 | −5.0 | −3.8 |
| Chloroform | 273 | 0.700 | 40 | 11 | −11 |
| | 303 | 0.502 | 34 | 7.0 | −8.1 |
| | 333 | 0.390 | 27 | 2.1 | −7.9 |
| Cyclohexane | 278 | 1.300 | −51 | −51 | −38 |
| | 333 | 0.528 | −38 | −27 | −22 |
| Cyclopentane | 293 | 0.439 | −32 | −32 | −33 |
| | 323 | 0.323 | −28 | −26 | −29 |
| 2,2-Dimethylpropane | 258 | 0.431 | −3.5 | 7.2 | 20 |
| | 283 | 0.281 | −0.8 | 20 | 30 |

| Compound | $T$, K | $\eta_L$, (exp.) cP‡ | Percent error† in liquid viscosity calculated by the method of: Orrick and Erbar | Van Velzen, Cardozo, and Langenkamp | Przezdziecki and Sridhar |
|---|---|---|---|---|---|
| Ethane | 98 | 0.985 | 30 | 2.6 | −24 |
| | 153 | 0.257 | −12 | 8.2 | −14 |
| | 188 | 0.162 | −22 | 12 | −13 |
| Ethylene chloride | 273 | 1.123 | −43 | −4.0 | — |
| | 313 | 0.644 | −35 | −4.0 | — |
| | 353 | 0.417 | −27 | −3.5 | — |
| Ethyl alcohol | 273 | 1.770 | 27 | 69 | — |
| | 313 | 0.826 | 3.5 | 38 | — |
| | 348 | 0.465 | −5.4 | 27 | — |
| Ethyl acetate | 293 | 0.458 | −4.2 | 3.1 | −16 |
| | 353 | 0.246 | 0.4 | 7.2 | −5.3 |
| | 413 | 0.153 | 7.4 | 14 | −1.8 |
| | 463 | 0.0998 | 27 | 34 | 4.8 |
| Ethylbenzene | 253 | 1.24 | −2.9 | −3.8 | −33 |
| | 313 | 0.535 | −1.2 | −3.3 | −23 |
| | 373 | 0.308 | −1.7 | −4.7 | −16 |
| | 413 | 0.231 | −1.2 | −4.6 | −13 |
| Ethyl bromide | 293 | 0.395 | 27 | 2.9 | −23 |
| | 333 | 0.269 | 32 | 10 | −17 |
| | 373 | 0.199 | 36 | 16 | −16 |
| Ethylene | 103 | 0.70 | −25 | 4.0 | −25 |
| | 133 | 0.31 | −27 | 40 | −17 |
| | 173 | 0.15 | −22 | — | −6.4 |
| Ethyl ether | 273 | 0.289 | 0 | −4.7 | 0 |
| | 293 | 0.236 | 0 | −3.9 | 2.2 |
| | 333 | 0.167 | 2.0 | −1.2 | 4.0 |
| | 373 | 0.118 | 11 | 8.8 | 7.4 |
| Ethyl formate | 273 | 0.507 | −18 | −3.3 | −16 |
| | 303 | 0.362 | −17 | −1.8 | −11 |
| | 328 | 0.288 | −16 | −1.2 | −9.6 |
| *n*-Heptane | 183 | 3.77 | −21 | 18 | −1.7 |
| | 233 | 0.965 | −0.5 | 0.8 | −27 |
| | 293 | 0.418 | −1.9 | −1.8 | −21 |
| | 373 | 0.209 | −3.3 | −4.4 | −17 |
| *n*-Hexane | 213 | 0.888 | 2.9 | 4.2 | −8.3 |
| | 273 | 0.381 | −2.4 | −0.9 | −8.2 |
| | 343 | 0.205 | −4.9 | −3.8 | −7.1 |
| Isobutane | 193 | 0.628 | −23 | −3.5 | −37 |
| | 233 | 0.343 | −25 | −6.4 | −29 |
| | 263 | 0.239 | −24 | −5.4 | −23 |
| Isopropyl alcohol | 283 | 3.319 | −24 | −2.8 | — |
| | 303 | 1.811 | −15 | −3.4 | — |
| | 323 | 1.062 | −10 | −3.6 | — |
| Methane | 88 | 0.226 | 60 | — | −11 |
| | 113 | 0.115 | 23 | — | −4.3 |

(*Continued*)

**TABLE 9-11 Comparison of Calculated and Experimental Viscosities of Liquids (*Continued*)**

| | | | Percent error† in liquid viscosity calculated by the method of | | |
|---|---|---|---|---|---|
| Compound | $T$, K | $\eta_L$, (exp.) cP‡ | Orrick and Erbar | Van Velzen, Cardozo, and Langenkamp | Przezdziecki and Sridhar |
| 2-Methylbutane | 223 | 0.55 | −13 | −4.0 | −30 |
| | 253 | 0.353 | −12 | −2.7 | −21 |
| | 303 | 0.205 | −10 | −1.1 | −12 |
| $n$-Pentane | 153 | 2.35 | −1.0 | −1.7 | 11 |
| | 193 | 0.791 | 3.8 | 5.5 | −7.0 |
| | 233 | 0.428 | −3.3 | 0.6 | −6.0 |
| | 273 | 0.279 | −8.2 | −3.6 | −4.7 |
| | 303 | 0.216 | −11 | −5.2 | −4.9 |
| Phenol | 323 | 3.020 | 0 | 22 | −50 |
| | 373 | 0.783 | 37 | 11 | −5.4 |
| Propane | 133 | 0.984 | −1.5 | −8.4 | −23 |
| | 193 | 0.327 | −22 | −12 | −19 |
| | 233 | 0.205 | −25 | −9.7 | −16 |
| $n$-Propyl alcohol | 283 | 2.897 | −9.1 | 4.1 | — |
| | 313 | 1.400 | −9.8 | 1.4 | — |
| | 373 | 0.443 | −6.5 | 2.0 | — |
| Toluene | 253 | 1.07 | −19 | −19 | −33 |
| | 293 | 0.587 | −13 | −29 | −24 |
| | 333 | 0.380 | −10 | −8.7 | −16 |
| | 383 | 0.249 | −6.8 | −4.6 | −10 |
| $o$-Xylene | 273 | 1.108 | 3.1 | −9.7 | −5.5 |
| | 313 | 0.625 | 5.0 | −13 | −4.8 |
| | 373 | 0.345 | 3.7 | −1.1 | −0.3 |
| | 413 | 0.254 | 3.6 | −21 | 1.9 |
| $m$-Xylene | 273 | 0.808 | 1.1 | 9.9 | 1.9 |
| | 313 | 0.492 | 1.4 | 3.9 | 1.8 |
| | 353 | 0.340 | 0.3 | −1.9 | 2.9 |
| | 413 | 0.218 | 1.4 | −5.8 | 4.6 |

†[(calc. − exp.)/exp.] × 100.
‡Refs. 5 and 129.

Equation (9-11.9) was employed with Eqs. (9-11.10) to (9-11.15) to estimate liquid viscosities for the compounds in Table 9-11. The values of $T_c$, $P_c$, $V_c$, $T_f$, and $\omega$ were obtained from Appendix A. The reference volume for each compound was calculated from the liquid density datum value given in Appendix A. Large errors were noted for alcohols, and those results are not included in the table. For other compounds, the errors varied widely and, except for a few materials, the technique underestimated the liquid viscosity. Larger errors were normally noted at low temperatures, but that might have been expected from the form of Eq. (9-11.9). That is, because $V_0$ is of the order of the volume at the freezing point and $\eta_L \propto (V - V_0)^{-1}$, the estimated value of $\eta_L$ becomes exceedingly sensitive to the choice of $V$. This problem was emphasized by Luckas and Lucas [139], who suggest that Eq. (9-11.9) should not be used below $T_r$ values of about 0.55.

**Example 9-18** Use the Przezdziecki and Sridhar correlation to estimate the liquid viscosity of toluene at 383 K. The experimental value is 0.249 cP [215].

**solution** From Appendix A, for toluene

$$T_c = 591.8 \text{ K}$$
$$P_c = 41.0 \text{ bar}$$
$$V_c = 316 \text{ cm}^3/\text{mol}$$
$$T_f = 178 \text{ K}$$
$$M = 92.14 \text{ g/mol}$$
$$\omega = 0.263$$
$$\rho_L = 0.867 \text{ g/cm}^3 \text{ at } 293 \text{ K}$$

Thus, $V^R = 92.14/0.867 = 106.3 \text{ cm}^3/\text{mol}$ at $T^R = 293$ K. With Eqs. (9-11.12) to (9-11.15),

$$T = T^R;\ T_r^R = \frac{293}{591.8} = 0.495$$

$$H_1(T_r^R) = 0.33593 - (0.33953)(0.495) + (1.51941)(0.495)^2 - (2.02512)(0.495)^3 + (1.11422)(0.495)^4 = 0.361$$

$$H_2(T_r^R) = 0.29607 - (0.09045)(0.495) - (0.04842)(0.495)^2 = 0.239$$

$$f(T^R) = 0.361[1 - (0.263)(0.239)] = 0.338$$

Similarly,

| | $T$, K | $T_r$ | $H_1$ | $H_2$ | $f(T)$ |
|---|---|---|---|---|---|
| $T_f$ | 178 | 0.301 | 0.325 | 0.265 | 0.302 |
| $T$ | 383 | 0.647 | 0.399 | 0.206 | 0.377 |

Then
$$V_m = \frac{0.302}{0.338}(106.3) = 95.0 \text{ cm}^3/\text{mol}$$
$$V = \frac{0.377}{0.338}(106.3) = 118.6 \text{ cm}^3/\text{mol}$$

With Eqs. (9-11.10) and (9-11.11),

$$E = -1.12 + 316/[12.94 + (0.10)(92.14) - (0.23)(41.0) + (0.0424)(178) - (11.58)(178/591.8)] = 17.70$$

$$V_0 = (0.0085)(0.263)(591.8) - 2.02 + \frac{95.0}{[(0.342)(178/591.8) + 0.894]} = 94.6 \text{ cm}^3/\text{mol}$$

Then, with Eq. (9-11.9)

$$\eta_L = \frac{94.6}{17.70(118.6 - 94.6)} = 0.223 \text{ cP}$$

$$\text{Error} = \frac{0.223 - 0.249}{0.249} \times 100 = -10\%$$

In this case Vargaftik [215] gives the value of the liquid molar volume at 383 K as 118.2 $cm^3$/mol. Using this number rather than the estimated value of 118.6 $cm^3$/mol leads to an error in the viscosity of −9%.

### Other correlations

Many viscosity-correlating methods have been proposed, but few are predictive in nature. Of those that are, group contribution approaches have been used [41, 149, 181, 182, 202]. Most other estimation methods relate the liquid viscosity to other physical properties (assumed known) and, in addition, require one or more constants to be determined from experimental data. Luckas and Lucas [139] recently proposed an alternative group contribution method to estimate liquid viscosities. The method is promising, but, at present, only a few group values are available. Other recent correlations are given in Refs. 106, 112, 113, 157, 203, 204, and the earlier literature was reviewed in the third edition of this book.

### Recommendations for estimating low-temperature liquid viscosities

Three estimation methods have been discussed. In Table 9-11 calculated liquid viscosities are compared with experimental values for 35 different liquids (usually of simple structure). Large errors may result, as illustrated for all methods. The method of van Velzen et al. is *not* recommended for first members of a homologous series, and the method of Przezdziecki and Sridhar should not be used for alcohols.

The method of van Velzen et al. assumes that log $\eta_L$ is linear in $T^{-1}$, whereas the Orrick and Erbar method is slightly modified to include the liquid density. Neither is reliable for highly branched structures (van Velzen et al. can treat only iso compounds) or for inorganic liquids or sulfur compounds. Both are limited to a temperature range from somewhat above the freezing point to about $T_r \approx 0.75$. Przezdziecki and Sridhar's method employs the Hildebrand equation, which necessitates knowledge of liquid volumes.

It is recommended that, in general, the method of van Velzen, Cardozo, and Langenkamp be used to estimate low-temperature liquid viscosities except for first members of a homologous series. Errors vary widely, but as judged from extensive testing [211], errors should be less than 10 to 15 percent in most instances.

## 9-12 Estimation of Liquid Viscosity at High Temperatures

Low-temperature viscosity correlations as covered in Sec. 9-10 usually assume that ln $\eta_L$ is a linear function of reciprocal absolute temperature. Above a reduced temperature of about 0.7, this relation is no longer valid, as illustrated in Fig. 9-10. In the region from about $T_r = 0.7$ to near the critical point, many estimation methods are of a corresponding states type that resemble or are identical with those used in the first sections of this chapter to treat gases. For example, Letsou and Stiel [130] proposed, for *saturated liquids,*

$$\eta_{SL}\xi = (\eta_L\xi)^{(0)} + \omega(\eta_L\xi)^{(1)} \tag{9-12.1}$$

where the parameters $(\eta_L\xi)^{(0)}$ and $(\eta_L\xi)^{(1)}$ are functions only of reduced temperature and $\xi$ is defined in Eq. (9-4.14). Letsou and Stiel tabulate these functions; but to a close approximation, from $0.76 < T_r < 0.98$ they can be expressed as

$$(\eta_L\xi)^{(0)} = 10^{-3}(2.648 - 3.725T_r + 1.309T_r^2) \tag{9-12.2}$$

$$(\eta_L\xi)^{(1)} = 10^{-3}(7.425 - 13.39T_r + 5.933T_r^2) \tag{9-12.3}$$

The units have been converted to yield $\eta_{SL}$ in centipoises even though $\xi$ [from Eq. (9-4.14)] has the dimensions of micropoises.

This correlation was developed from data on only 14 liquids, mostly simple hydrocarbons. The authors report average errors of about 3 percent for most materials up to $T_r \approx 0.92$; larger errors were found as the critical point was approached. In our testing of this method, using some compounds other than hydrocarbons, larger errors were found, i.e., up to 15 to 20 percent. In general, however, the technique is a simple one to use and yields estimates that are often surprisingly good.

**Example 9-19** Estimate the saturated liquid viscosity of $n$-propanol at 433.2 K by using the Letsou-Stiel correlation. The experimental value is 0.188 cP.

**solution** From Appendix A, $T_c = 536.8$ K, $P_c = 51.7$ bar, $\omega = 0.623$, and $M = 60.10$. Thus, from Eq. (9-4.14)

$$\xi = (0.176)\left[\frac{(536.8)}{(60.10)^3(51.7)^4}\right]^{1/6} = 4.664 \times 10^{-3}$$

With $T = 433.2$ K and $T_r = 433.2/536.8 = 0.807$ and with Eqs. (9-12.2) and (9-12.3),

$$(\eta_L\xi)^{(0)} = (10)^{-3}[2.648 - (3.725)(0.807) + (1.309)(0.807)^2] = 4.944 \times 10^{-4}$$

$$(\eta_L\xi)^{(1)} = (10)^{-3}[7.425 - (13.39)(0.807) + (5.933)(0.807)^2] = 4.831 \times 10^{-4}$$

$$\eta_{SL}\xi = [4.944 + (0.623)(4.831)](10^{-4}) = 7.954 \times 10^{-4}$$

$$\eta_{SL} = \frac{7.954 \times 10^{-4}}{4.664 \times 10^{-3}} = 0.171 \text{ cP}$$

$$\text{Error} = \frac{0.171 - 0.188}{0.188} \times 100 = -9.0\%$$

A more general estimation method would logically involve the extension of the high-pressure gas viscosity correlations described in Sec. 9-6 into the liquid region. Two techniques have, in fact, been rather widely tested and found reasonably accurate for reduced temperatures above about 0.5. These methods are those of Chung et al. [44] and Brulé and Starling [30]. Both methods use Eq. (9-6.16), but they have slightly different coefficients to compute some of the parameters. The Chung et al. form is preferable for simple molecules and will treat polar as well as nonpolar compounds. The Brulé and Starling relation was developed primarily for complex hydrocarbons, and the authors report their predictions are within 10 percent of experimental values in the majority of cases. The Chung et al. method has a similar accuracy for most nonpolar compounds, but significantly higher errors can occur with polar, halogenated, or high-molecular-weight compounds. In both cases, one needs accurate liquid density data, and the reliability of the methods decreases significantly for $T_r$ less than about 0.5. The liquids need not be saturated; subcooled compressed liquid states simply reflect a higher liquid density.

The Chung et al. technique was illustrated for dense gas ammonia in Example 9-12. The procedure is identical when applied to high-temperature liquids. We present, instead, an example of the use of the Brulé and Starling approach.

**Example 9-20** Estimate the viscosity of liquid acenaphthene (a double-ring aromatic compound found in shale oil and tar sand fluids) at 100°C (373.2 K) by using the Brulé-Starling method. At this temperature, the experimental viscosity is 1.39 cP and the liquid density is about 1.03 g/cm$^3$.

**solution** For acenaphthene, $T_c = 824$ K, $V_c = 460$ cm$^3$/mol, $\omega = 0.36$, and $M = 154.21$. At 373 K, $T_r = (373/824) = 0.453$ and the molar density $\rho_L = 1.03/154.21 = 6.68 \times 10^{-3}$ mol/cm$^3$. We use Eq. (9-6.16) with Table 9-7. $F_c$ is given by Eq. (9-4.10) as

$$F_c = 1 - (0.2756)(0.36) = 0.901$$

$T^* = (1.2593)(0.453) = 0.570$ [Eq. (9-4.8)] and, with Eq. (9-4.3), $\Omega_v = 2.139$. The density is introduced in Eqs. (9-6.18) and (9-6.19),

$$y = \frac{(6.68 \times 10^{-3})(460)}{6} = 0.512$$

$$G_1 = \frac{1 - (0.5)(0.512)}{(1 - 0.512)^3} = 6.40$$

Using Table 9-7, and substituting $\omega$ for $\gamma$, $E_1 = 29.71$, $E_2 = 1.644 \times 10^{-3}$, $E_3 = 112.0$, $E_4 = 25.02$, $E_5 = 20.62$, $E_6 = -9.71$, $E_7 = 24.15$, $E_8 = 2.134$, $E_9 = -1.368$, and $E_{10} = 0.735$. From Eq. (9-6.20), $G_2 = 1.381$; from Eq. (9-6.21), $\eta^{**} = 64.34$; and from Eq. (9-6.17), $\eta^* = 62.99$. Finally, with Eq. (9-6.16),

$$\eta_L = (62.99) \frac{(36.344)[(154.21)(824)]^{1/2}}{(460)^{2/3}}$$

$$= 13{,}700\ \mu\text{P} = 1.37\ \text{cP}$$

$$\text{Error} = \frac{1.37 - 1.39}{1.39} \times 100 = -1.5\%$$

It might be noted that an error of $-15$ percent would have been found if the Chung et al. method had been used.

### Discussion

The quantity of accurate liquid viscosity data at temperatures much above the normal boiling point is not large. In addition, to test estimation methods such as those of Chung et al. or Brulé and Starling, one needs accurate liquid density data under the same conditions as apply to the viscosity data. This matching makes it somewhat difficult to test the methods with many compounds. However, Brulé and Starling developed their technique so that they would be coupled to a separate computation program using a modified BWR equation of state to provide densities. They report relatively low errors (as noted above), and this fact appears to confirm the general approach (see also Ref. 30). Hwang et al. [102] have proposed viscosity (as well as density and surface tension) correlations for coal liquids.

Regardless of what high-temperature estimation method is chosen, there is the problem of joining both high- and low-temperature estimated viscosities should that be necessary.

## 9-13 Liquid Mixture Viscosity

Essentially all correlations for liquid mixture viscosity refer to solutions of liquids below or only slightly above their normal boiling points; i.e., they are restricted to reduced temperatures (of the pure components) to values below about 0.7. The bulk of the discussion below is limited to that temperature range. At the end of the section, however, we suggest approximate methods to treat high-pressure, high-temperature liquid mixture viscosity.

At temperatures below $T_r \approx 0.7$, liquid viscosities are very sensitive to the structure of the constituent molecules (see Sec. 9-11). This generality is also true for liquid mixtures, and even mild association effects between components can often significantly affect the viscosity.

Almost all methods to estimate or correlate liquid mixture viscosities assume that values of the viscosities of the pure components are available. Thus the methods are, in reality, interpolative. Nevertheless, there is no agreement on the best way to carry out the interpolation. Irving [104] surveyed more than 50 equations for binary liquid viscosities and classified them by type. He points out that only very few do not have some adjustable constant that must be determined from experimental data—and the few that do not require such a parameter are applicable only to systems of similar components with comparable viscosities. In a companion report from the National Engineering Laboratory, Irving [105] has also evaluated 25 of the more promising equations with experimental data from the literature. He recommends the one-constant Grunberg-Nissan equation [87] [see Eq. (9-13.1)] as being widely applicable yet reasonably accurate except for aqueous solutions. This NEL report is also an excellent source of viscosity data tabulated from the literature.

### Method of Grunberg and Nissan [87]

In this procedure, the low-temperature liquid viscosity for mixtures is given as

$$\ln \eta_m = \sum_i x_i \ln \eta_i + \sum_{i \neq j}\sum x_i x_j G_{ij} \tag{9-13.1}$$

or, for a binary of 1 and 2,

$$\ln \eta_m = x_1 \ln \eta_1 + x_2 \ln \eta_2 + x_1 x_2 G_{12} \tag{9-13.2}$$

since $G_{ii} = 0$. In Eqs. (9-13.1) and (9-13.2), $x$ is the liquid mole fraction and $G_{ij}$ is an interaction parameter which is a function of the components $i$ and $j$ as well as the temperature (and, in some cases, the composition). This relation has probably been more extensively examined than any other liquid mixture viscosity correlation. Isdale [107] presents the results of a very detailed testing using more than 2000 experimental mixture datum points. When the interaction parameter was regressed from experimental data, nonassociated mixtures and many mixtures containing alcohols, carboxylic acids, and ketones were fitted satisfactorily. The overall root mean square deviation for the mixtures tested was 1.6 percent. More recently, Isdale et al. [109] proposed a group contribution method to estimate the binary interaction parameter $G_{ij}$ at 298 K.

The procedure to be followed is:

1. For a binary of $i$ and $j$, select $i$ by following the priority rules below. ($j$ then becomes the second component.)
   *a.* $i$ = an alcohol, if present
   *b.* $i$ = an acid, if present
   *c.* $i$ = the component with the most carbon atoms
   *d.* $i$ = the component with the most hydrogen atoms
   *e.* $i$ = the component with the most $-CH_3$ groups

$G_{ij}$ = 0 if none of these rules establish a priority.

2. Once the decision has been made which component is $i$ and which is $j$, calculate $\Sigma\,\Delta$ for $i$ and $j$ from the group contributions in Table 9-12.

3. Determine the parameter $W$. (If either $i$ or $j$ contains atoms other than carbon and hydrogen, set $W = 0$ and go to step 4.) Let the number of carbon atoms in $i$ be $N_i$ and that in $j$ be $N_j$.

$$W = \frac{(0.3161)(N_i - N_j)^2}{N_i + N_j} - (0.1188)(N_i - N_j) \tag{9-13.3}$$

4. Calculate $G_{ij}$ from

$$G_{ij} = \Sigma\,\Delta_i - \Sigma\,\Delta_j + W \tag{9-13.4}$$

**TABLE 9-12 Group Contributions for $G_{ij}$ at 298K**

| Group | Notes | Value |
|---|---|---|
| $-CH_3$ | | −0.100 |
| $>CH_2$ | | 0.096 |
| $>CH-$ | | 0.204 |
| $>C<$ | | 0.433 |
| Benzene ring | | 0.766 |
| Substitutions: | | |
| Ortho | | 0.174 |
| Meta | | — |
| Para | | 0.154 |
| Cyclohexane ring | | 0.416 |
| $-OH$ | Methanol | 0.887 |
| | Ethanol | −0.023 |
| | Higher aliphatic alcohols | −0.443 |
| $>C=O$ | Ketones | 1.046 |
| $-Cl$ | | $0.653–0.161N_{Cl}$ |
| $-Br$ | | −0.116 |
| $-COOH$ | Acid with: | |
| | Nonassociated liquids | $-0.411 + 0.06074N_C$ |
| | Ketones | 1.130 |
| | Formic acid with ketones | 0.167 |

$N_{Cl}$ = number of chlorine atoms in the molecule
$N_C$ = *total* number of carbon atoms in the binary set

**Example 9-21** Estimate the viscosity of a liquid mixture of $n$-hexane and $n$-hexadecane as a function of composition. The temperature is 298 K. The viscosities of pure $n$-hexane and $n$-hexadecane at this temperature are 0.298 and 3.078 cP [68].

**solution** We establish that $n$-hexadecane is to be called component $i$ by using the priority rules above (rule $c$ in this instance). With Table 9-12, we can compute $\Sigma\,\Delta_i$ and $\Sigma\,\Delta_j$.

$$\Sigma\,\Delta_i\ (n\text{-hexadecane}) = (2)(-0.100) + (14)(0.096) = 1.144$$

$$\Sigma\,\Delta_j\ (n\text{-hexane}) = (2)(-0.100) + (4)(0.096) = 0.184$$

$N_i$, the number of carbon atoms in $n$-hexadecane, equals 16, and $N_j$ is 6. With Eq. (9-13.3),

$$W = \frac{(0.3161)(16-6)^2}{(16+6)} - (0.1188)(16-6) = 0.249$$

Then, with Eq. (9-13.4),

$$G_{ij} = 1.144 - 0.184 + 0.249 = 1.209$$

and Eq. (9-13.2) becomes

$$\ln \eta_m = x_1 \ln (0.298) + x_2 \ln (3.078) + (x_1 x_2)(1.209)$$

where subscript 1 refers to $n$-hexane. The estimated values of $\eta_m$ are compared with experimental data in the table below. The experimental mixture viscosities were reported by Dymond et al. [68].

| Mole fraction $n$-hexane | Mixture viscosities for $n$-hexane and $n$-hexadecane, cP | |
|---|---|---|
| | Estimated | Experimental |
| 0. | — | 3.078 |
| 0.2 | 2.34 | 2.24 |
| 0.4 | 1.62 | 1.51 |
| 0.6 | 1.01 | 0.991 |
| 0.8 | 0.577 | 0.584 |
| 1.0 | — | 0.298 |

In Fig. 9-15, estimated and experimental mixture viscosities are plotted against composition. An excellent fit is seen.

$G_{ij}$ is sometimes a function of temperature. However, existing data suggest that, for alkane-alkane solutions or for mixtures of an associated component with an unassociated one, $G_{ij}$ is independent of temperature. However, for mixtures of nonassociated compounds (but *not* of only alkanes) or for mixtures of associating compounds, $G_{ij}$ is a mild function of temperature. Isdale et al. [109] suggest for these latter two cases,

$$G_{ij}\,(T) = 1 - [1 - G_{ij}(298)]\,\frac{573 - T}{275} \tag{9-13.5}$$

where $T$ is in kelvins.

**Example 9-22** Estimate the viscosity of a mixture of acetic acid and acetone at 323 K (50°C) that contains 70 mole percent acetic acid. Isdale et al. quote the experimental value to be 0.587 cP, and, at 50°C, the viscosities of pure acetic acid and acetone are 0.798 and 0.241 cP, respectively.

**solution** First we must estimate $G_{ij}$ at 298 K. Component $i$ is acetic acid (priority rule $b$). Since the mixture contains atoms other than carbon and hydrogen (i.e., oxygen), $W = 0$. Then, with Table 9-12,

$$\Sigma\, \Delta_i \text{ (acetic acid)} = -\text{CH}_3 + -\text{COOH} = -0.100 + 1.130 = 1.030$$

$$\Sigma\, \Delta_j \text{ (acetone)} = (2)(-\text{CH}_3) + >\text{C}{=}0 = (2)(-0.100) + 1.046 = 0.846$$

With Eq. (9-13.4),

$$G_{ij} = 1.030 - 0.846 = 0.184 \text{ at } 298 \text{ K}$$

At 50°C = 323 K, we need to adjust $G_{ij}$ with Eq. (9-13.5).

$$G_{ij}(323 \text{ K}) = 1 - \frac{(1 - 0.184)(573 - 323)}{275} = 0.258$$

Then, using Eq. (9-13.2),

$$\ln \eta_m = (0.7) \ln (0.798) + (0.3) \ln (0.241) + (0.7)(0.3)(0.258) = -0.531$$

$$\eta_m = 0.588 \text{ cP}$$

This estimated value is essentially identical with the experimental result of 0.587 cP.

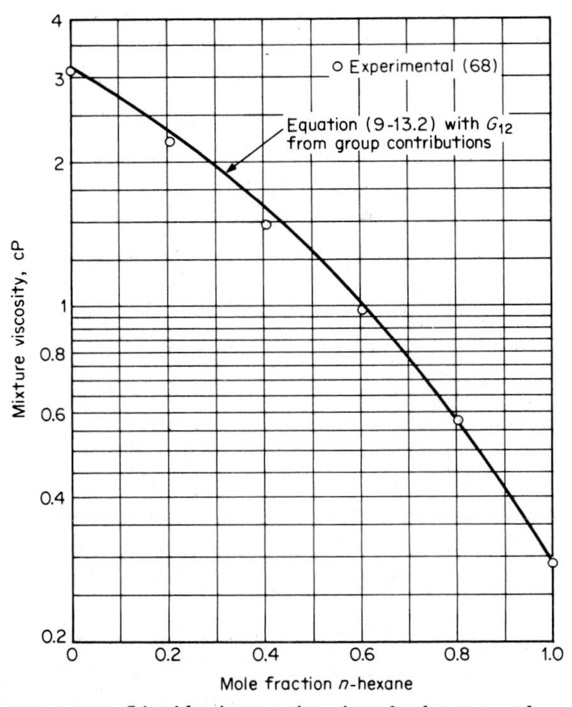


**Figure 9-15** Liquid mixture viscosity of $n$-hexane and $n$-hexadecane at 298 K.

Generally, one assumes the estimated value of $G_{ij}$ is independent of composition. However, for special cases of mixtures of normal and/or branched chain alkanes, the estimated value of $G_{ij}$ should be corrected for composition by Eq. (9-13.6) [109].

$$G_{ij}(x_i) = G_{ij}(1.343 - 0.685x_i) \tag{9-13.6}$$

where $G_{ij}$ is the value of the interaction parameter from Eq. (9-13.4) and $x_i$ is the mole fraction of the component identified as $i$. No temperature correction is necessary because, for alkane mixtures, $G_{ij}$ is independent of temperature. $G_{ij}(x_i)$ is the interaction parameter at a composition $x_i$. Note that $G_{ij}(x_i) = G_{ij}$ when $x_i = 0.5$, so the effect of using Eq. (9-13.6) is more pronounced at either larger or smaller values of $x_i$.

**Example 9-23** Repeat Example 9-21 but allow for variations in $G_{ij}$ with composition.

**solution** In Example 9-21 for the system $i$ = $n$-hexadecane and $j$ = $n$-hexane, $G_{ij}$ was computed to be 1.209. With Eqs. (9-13.6) and (9-13.2), the calculated $G_{ij}(x_i)$ and $\eta_m$ values are as shown below.

| Mole fraction | | $\eta_m$, cP | |
|---|---|---|---|
| $n$-hexane | $G_{ij}(x_i)$ | Estimated | Experimental |
| 0. | — | — | 3.08 |
| 0.2 | 0.961 | 2.19 | 2.24 |
| 0.4 | 1.127 | 1.59 | 1.51 |
| 0.6 | 1.292 | 1.03 | 0.991 |
| 0.8 | 1.458 | 0.60 | 0.584 |
| 1.0 | — | — | 0.298 |

When these results are compared with those of Example 9-21, it is seen that, at low mole fractions of $n$-hexane, the estimated values of $\eta_m$ are somewhat closer to the experimental results than was the case when $G_{ij}$ was assumed independent of composition.

To summarize the Isdale modification of the Grunberg-Nissan equation, for each possible binary pair in the mixture, first decide which component is to be labeled $i$ and which $j$ by the use of the priority rules. Determine $\Sigma\,\Delta_i$ and $\Sigma\,\Delta_j$ by using Table 9-12 and $W$ from Eq. (9-13.3), if necessary. Use Eq. (9-13.4) to calculate $G_{ij}$. Correct for temperatures other than 298 K, if necessary, with Eq. (9-13.5), and, for alkane and/or branched alkane mixtures, for composition by Eq. (9-13.6). With the values of $G_{ij}$ so determined, use either Eq. (9-13.1) or (9-13.2) to determine the viscosity of the liquid mixture. This technique yields quite acceptable estimates of low-temperature liquid mixture viscosities for many systems, but Table 9-12 does not allow one to treat many types of compounds. Also, the method does not cover aqueous mixtures.

**Method of Teja and Rice [197, 198]**

Based on a corresponding states treatment for mixture compressibility factors [196, 199], Teja and Rice propose an analogous form for liquid mixture viscosity.

$$\ln (\eta_m \epsilon_m) = \ln (\eta\epsilon)^{(r1)} + [\ln (\eta\epsilon)^{(r2)} - \ln (\eta\epsilon)^{(r1)}] \frac{\omega_m - \omega^{(r1)}}{\omega^{(r2)} - \omega^{(r1)}} \tag{9-13.7}$$

where the superscripts $(r_1)$ and $(r_2)$ refer to two reference fluids. $\eta$ is the viscosity, $\omega$ the acentric factor, and $\epsilon$ is a parameter similar to $\xi$ in Eq. (9-4.14) but defined here as:

$$\epsilon \frac{V_c^{2/3}}{(T_c M)^{1/2}} \tag{9-13.8}$$

The variable of composition is introduced in four places: the definitions of $\omega_m$, $V_{cm}$, $T_{cm}$, and $M_m$. The rules suggested by the authors to compute these mixture parameters are:

$$V_{cm} = \sum_i \sum_j x_i x_j V_{cij} \tag{9-13.9}$$

$$T_{cm} = \frac{\sum_i \sum_j x_i x_j T_{cij} V_{cij}}{V_{cm}} \tag{9-13.10}$$

$$M_m = \sum_i x_i M_i \tag{9-13.11}$$

$$\omega_m = \sum_i x_i \omega_i \tag{9-13.12}$$

$$V_{cij} = \frac{(V_{ci}^{1/3} + V_{cj}^{1/3})^3}{8} \tag{9-13.13}$$

$$T_{cij} V_{cij} = \psi_{ij} (T_{ci} T_{cj} V_{ci} V_{cj})^{1/2} \tag{9-13.14}$$

$\psi_{ij}$ is an interaction parameter of order unity which must be found from experimental data.

It is important to note that, in the use of Eq. (9-13.7) for a given mixture at a specified temperature, the viscosity values for the two reference fluids $\eta^{(r1)}$ and $\eta^{(r2)}$ are to be obtained *not at T*, but at a temperature equal to $T[(T_c)^{(r1)}/T_{cm}]$ for $(r_1)$ and $T[(T_c)^{(r2)}/T_{cm}]$ for $(r_2)$. $T_{cm}$ is given by Eq. (9-13.10).

Whereas the reference fluids $(r_1)$ and $(r_2)$ may be chosen as different from the actual components in the mixture, it is normally advantageous to select them from the principal components in the mixture. In fact, for a binary of 1 and 2, if $(r_1)$ is selected as component 1 and $(r_2)$ as component 2, then, by virtue of Eq. (9-13.12), Eq. (9-13.7) simplifies to

$$\ln (\eta_m \epsilon_m) = x_1 \ln (\eta\epsilon)_1 + x_2 \ln (\eta\epsilon)_2 \tag{9-13.15}$$

but, as noted above, $\eta_1$ is to be evaluated at $T(T_{c_1}/T_{cm})$ and $\eta_2$ at $T(T_{c2}/T_{cm})$.

Our further discussion of this method will be essentially limited to Eq. (9-13.15), since that is the form most often used for binary liquid mixtures and, by this choice, one is assured that the relation gives correct results when $x_1 = 0$ or 1.0. In addition, the assumption is made that the interaction parameter $\psi_{ij}$ is not a function of temperature or composition.

The authors claim good results for many mixtures ranging from strictly nonpolar to highly polar aqueous-organic systems. For nonpolar mixtures, errors averaged about 1 percent. For nonpolar-polar and polar-polar mixtures, the average rose to about 2.5 percent, whereas for systems containing water, an average error of about 9 percent was reported.

In comparison with the Grunberg-Nissan correlation [Eq. (9-13.1)], with $G_{ij}$ found by regressing data, Teja and Rice show that about the same accuracy is achieved for both methods for nonpolar-nonpolar and nonpolar-polar systems, but their technique was significantly more accurate for polar-polar mixtures, and particularly for aqueous solutions for which Grunberg and Nissan's form should not be used.

**Example 9-24** Estimate the viscosity of a liquid mixture of water and 1,4-dioxane at 60°C when the mole fraction water is 0.83. For this very nonideal solution, Teja and Rice suggest an interaction parameter $\psi_{ij} = 1.37$.

**solution** From Appendix A, for water, $T_c = 647.1$ K, $V_c = 56$ cm$^3$/mol, and $M = 18.02$; for 1,4-dioxane, $T_c = 587$ K, $V_c = 238$ cm$^3$/mol, and $M = 88.11$. Let 1 be water and 2 be 1,4-dioxane. With Eq. (9-13.8), $\epsilon_1 = (56)^{2/3}/[(647.1)(18.02)]^{1/2} = 0.136$; $\epsilon_2 = 0.169$. From Eq. (9-13.9),

$$V_{cm} = (0.830)^2(56) + (0.170)^2(238) + (2)(0.830)(0.170) \times \frac{[(56)^{1/3} + (238)^{1/3}]^3}{8}$$

$$= 80.98 \text{ cm}^3\text{/mol}$$

and, with Eq. (9-13.10),

$$T_{cm} = \{(0.830)^2(647.1)(56) + (0.170)^2(587)(238)$$
$$+ (2)(0.830)(0.170)(1.37)[(647.1)(56)(587)(238)]^{1/2}\}/80.98$$
$$= 697.8 \text{ K}$$

$$M_m = (0.830)(18.02) + (0.170)(88.11) = 29.94$$

so, with Eq. (9-13.8),

$$\epsilon_m = \frac{(80.98)^{2/3}}{[(697.8)(29.94)]^{1/2}} = 0.130$$

Next, we need to know the viscosity of water not at 333.2 K (60°C), but at a temperature of (333.2)(647.1)/697.8 = 309.0 K (35.8°C). This value is 0.712 cP [105]. [Note that, at 60°C, $\eta$ (water) = 0.468 cP.] For 1,4-dioxane, the reference temperature is (333.2)(587)/697.8 = 280.3 K (7.1°C), and at that temperature, $\eta$ = 1.63 cP [105]. Again this value is quite different from the viscosity of 1,4-dioxane at 60°C, which is 0.715 cP.

Finally, with Eq. (9-13.15),

$$\ln[(\eta_m)(0.130)] = (0.830)\ln[(0.712)(0.136)] + (0.170)\ln[(1.63)(0.169)]$$
$$= -2.157$$
$$\eta_m = 0.90 \text{ cP}$$

The experimental viscosity is 0.89 cP.

Although the agreement between the experimental and estimated viscosity in Example 9-24 is excellent, in other composition ranges, higher errors occur. In Fig. 9-16 we have plotted the estimated and experimental values of the mixture viscosity over the entire range of composition. From a mole fraction water of about 0.8 (weight fraction = 0.45) to unity, the method provides an excellent fit to experimental results. At smaller concentrations of water, the technique overpredicts $\eta_m$. Still, for such a nonideal aqueous mixture, the general fit should be considered good.

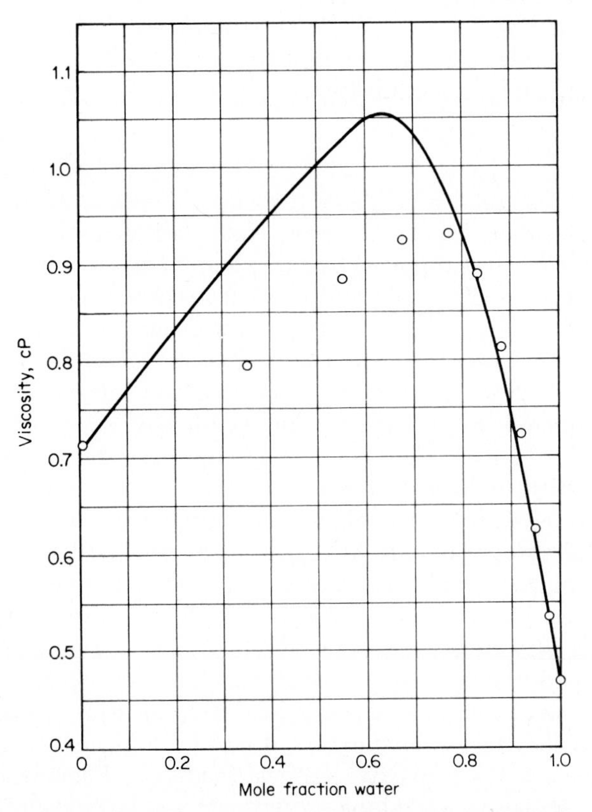


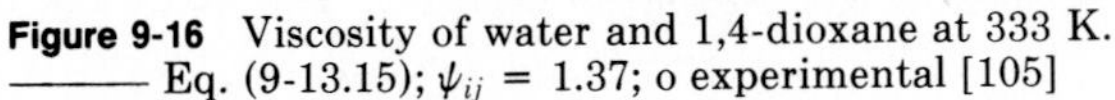

**Figure 9-16** Viscosity of water and 1,4-dioxane at 333 K. ——— Eq. (9-13.15); $\psi_{ij} = 1.37$; o experimental [105]

### Discussion

Two methods have been introduced to estimate the viscosity of liquid mixtures: the Grunberg-Nissan relation [Eq. (9-13.1)] and the Teja-Rice form [Eq. (9-13.15)]. Both contain one adjustable parameter per binary pair in the mixture. An approximate technique is available to estimate the Grunberg-Nissan parameter $G_{ij}$ as a function of temperature [Eq. (9-13.5)] for many types of systems. Teja and Rice suggest that their parameter $\psi_{ij}$ is independent of temperature—at least over reasonable temperature ranges. This latter technique seems to be better for highly polar systems, especially if water is one of the components, and it has also been applied to undefined mixtures [200, 205] with the introduction of reference components [see Eq. (9-13.7)].

Stairs [183] proposed another one-parameter mixture correlation, but it may be shown that his relations reduce to the Grunberg-Nissan form when mole fractions are used as composition variables. Vogel and Weiss [218] correlated their mixture viscosity data with the Grunberg-Nissan equation and, for systems of quite spherical molecules, show $G_{ij}$ is related to the excess entropy of mixing. Pikkarainen [159] studied very nonideal systems of $N$-methylacetamide with various aliphatic alcohols. He determined a parameter $Q$ defined by a modification of Eq. (9-13.2),

$$Q = \ln \nu_m - x_1 \ln \nu_1 - x_2 \ln \nu_2 \tag{9-13.16}$$

where $\nu$ is the kinematic viscosity, $\eta/\rho$. $Q$ was then expressed as a polynomial in $x_1$ for several systems. In a similar way, Noda and Ishida [151] use Eq. (9-13.16) except that $\nu$ was replaced by $\nu M$, where $M$ is the molecular weight. Both Pikkarainen and Noda and Ishida interpret $Q$ as $\Delta G/RT$, where $\Delta G$ is the Gibbs energy of activation for flow. Noda and Ishida then propose that $Q/RT$ be expressed by a local composition model with two adjustable parameters. A viscosity correlation based on the NRTL correlation has been proposed by Wei and Rowley [200]. NRTL parameters are necessary, as is the excess enthalpy of mixing. An adjustable parameter is used for highly nonideal systems.

Several recent studies have correlated binary and ternary liquid mixture viscosities with the two-parameter McAllister relation [140] with good results [6, 57, 58, 152], although the Dizechl form [56] gave a better fit for very polar mixtures. Other investigators [16, 65, 68, 94] have used Eq. (9-11.9) for mixtures. New data were reported [66, 67, 108] for various systems and correlated with the Grunberg-Nissan equation. New mixture correlation techniques were also suggested [86, 96] from liquid theory, and Kestin and Shankland [123] discuss the viscosity of multicomponent electrolyte solutions from data on the respective salt-water binaries. Pedersen et al. [158] present a correlation technique to estimate the viscosity of crude oils.

To finish this section, we again reiterate that the methods proposed should be limited to situations in which the reduced temperatures of the components comprising the mixture are less than about 0.7, although the exact temperature range of the Teja-Rice procedure is as yet undefined.

Should one desire the viscosity of liquid mixtures at high pressures and temperatures, it is possible to employ the Chung et al. method [44] described in Sec. 9-7 to estimate high-pressure gas mixture viscosities. This recommendation is tempered by the fact that such a procedure has been only slightly tested, and usually with rather simple systems where experimental data exist.

### Recommendations to estimate the viscosities of liquid mixtures

To estimate low-temperature liquid mixture viscosities, either the Grunberg-Nissan equation [Eq. (9-13.1) or (9-13.2)] or the Teja-Rice relation [Eq. (9-13.7) or (9-13.15)] may be used. Both require some experimental data to establish the value of an interaction parameter specific for each binary pair in the mixture. In many instances, however, it is possible to estimate the Grunberg-Nissan interaction parameter $G_{ij}$ by a group contribution technique and thus alleviate any necessity to have viscosity data for the mixture. Both methods are essentially interpolative in nature, so viscosities of the pure components comprising the mixture must be known. (Or in the Teja-Rice procedure, one may instead use reference fluids of similar structure rather than the actual mixture components.) The errors to be expected range from a few percent for nonpolar or slightly polar mixtures to 5 to 10 percent for polar mixtures. With aqueous solutions, the Grunberg-Nissan form is not recommended.

## Notation

| | |
|---|---|
| $a^*$ | group contribution sum; Eq. (9-4.21) |
| $b_0$ | excluded volume, $(2/3)\pi N_0\sigma^3$ |
| $B$ | viscosity parameter in Eq. (9-11.2) |
| $C_v$ | heat capacity at constant volume, J/(mol·K); $C_i$, structural contribution in Eq. (9-4.21) and Table 9-3 |
| $D$ | diffusion coefficient, cm$^2$/s or m$^2$/s |
| $F_c$ | shape and polarity factor in Eq. (9-4.10); $F_P^\circ$, low-pressure polar correction factor in Eq. (9-4.17); $F_Q^\circ$, low-pressure quantum correction factor in Eq. (9-4.18); $F_P$, high-pressure polar correction factor in Eq. (9-6.8); $F_Q$, high-pressure quantum correction factor in Eq. (9-6.9) |
| $G_1$, $G_2$ | parameters in Eqs. (9-6.19) and (9-6.20) |
| $\Delta G$ | Gibbs energy of activation for flow, J/mol |

| | |
|---|---|
| $k$ | Boltzmann's constant |
| $L$ | mean free path |
| $m$ | mass of molecule |
| $M$ | molecular weight |
| $n$ | number density of molecules; number of components in a mixture |
| $N$ | number of carbon atoms; $N^*$, equivalent chain length; $\Delta N_i$, structural contribution in Eq. (9-11.3) and Table 9-10; $N_0$; Avogadro's number |
| $P$ | pressure, N/m$^2$ or bar (unless otherwise specified); $P_c$, critical pressure; $P_r$, reduced pressure, $P/P_c$; $P_{vp}$, vapor pressure; $\Delta P_r = (P - P_{vp})/P_c$ |
| $Q$ | polar parameter in Eq. (9-6.3); parameter in Eq. (9-13.16) |
| $r$ | distance of separation |
| $R$ | gas constant, usually 8.314 J/(mol·K) |
| $T^*$ | $kT/\epsilon$ |
| $T$ | temperature, K; $T_c$, critical temperature; $T_r$, reduced temperature, $T/T_c$; $T_0$, parameter in Eq. (9-11.2) |
| $v$ | molecular velocity |
| $V$ | volume, cm$^3$/mol; $V_c$, critical volume; $V_r$, reduced volume, $V/V_c$ |
| $x$ | mole fraction, liquid |
| $y$ | mole fraction, vapor; parameter in Eq. (9-6.18) |
| $Y$ | parameter in Eq. (9-6.7) |
| $Z$ | compressibility factor; $Z_c$, critical compressibility factor; $Z_1$, $Z_2$, parameters in Eqs. (9-6.4) and (9-6.7) |

GREEK

| | |
|---|---|
| $\gamma$ | orientation factor in the Brulé-Starling method, Table 9-7, or obtain from Ref. 30 |
| $\Delta$ | correction term in Eq. (9-6.15) |
| $\epsilon$ | energy-potential parameter; variable defined in Eq. (9-13.8) |
| $\eta$ | viscosity (usually in micropoises for gas and in centipoises for liquids); $\eta^\circ$, denotes value at low-pressure (about 1 bar); $\eta_c$, at the critical point; $\eta_c^\circ$, at the critical temperature but at about 1 bar; $\eta^*$, $\eta^{**}$, parameters in Eqs. (9-6.17) and (9-6.21) |
| $\kappa$ | polar correction factor in Eq. (9-4.10), see Table 9-1 |
| $\lambda$ | thermal conductivity, W/(m·K) |
| $\mu$ | dipole moment, debyes; $\mu_r$, dimensionless dipole moment defined in either Eq. (9-4.11) or Eq. (9-4.16) |
| $\nu$ | kinematic viscosity, $\eta/\rho$, m$^2$/s |
| $\xi$ | inverse viscosity, defined in Eq. (9-4.13) or Eq. (9-4.14); $\xi_T$, inverse viscosity defined in Eq. (9-6.11) |

| | |
|---|---|
| $\rho$ | density (usually mol/cm$^3$); $\rho_c$, critical density; $\rho_r$, reduced density, $\rho/\rho_c$ |
| $\sigma$ | molecular diameter, Å |
| $\Psi$ | radial distribution function |
| $\psi$ | intermolecular potential energy as a function of $r$ |
| $\psi_{ij}$ | interaction parameter in Eq. (9-13.14) |
| $\omega$ | acentric factor, Sec. 2-3 |
| $\Omega_v$ | collision integral for viscosity |

SUBSCRIPTS

| | |
|---|---|
| $i, j, k$ | components $i, j, k$ |
| 1, 2 | components 1, 2 |
| $L$ | liquid |
| $m$ | mixture |
| $SL$ | saturated liquid |

## REFERENCES


1. Alder, B. J.: "Prediction of Transport Properties of Dense Gases and Liquids," *UCRL 14891-T,* University of California, Berkeley, Calif., May 1966.
2. Alder, B. J., and J. H. Dymond: "Van der Waals Theory of Transport in Dense Fluids," *UCRL 14870-T,* University of California, Berkeley, Calif., April 1966.
3. Alder, B. J., D. M. Gass, and T. E. Wainwright: *J. Chem. Phys.,* **75:** 394 (1971).
4. Amdur, I., and E. A. Mason: *Phys. Fluids,* **1:** 370 (1958).
5. American Petroleum Institute, *Selected Values of Physical and Thermodynamic Properties of Hydrocarbons and Related Compounds,* Project 44, Carnegie Press, Pittsburgh, Pa., 1953, and supplements.
6. Aminabhavi, T. M., R. C. Patel, and K. Bridger: *J. Chem. Eng. Data,* **27:** 125 (1982).
7. Andrade, E. N. da C.: *Nature,* **125:** 309 (1930).
8. Andrade, E. N. da C.: *Phil. Mag.,* **17:** 497, 698 (1934).
9. Andrade, E. N. da C.: *Endeavour,* **13:** 117 (1954).
10. Andrussow, L.: *Z. Electrochem.,* **61:** 253 (1957).
11. Angus, S., B. Armstrong, and K. M. deReuck: *International Thermodynamic Tables of the Fluid State—Carbon Dioxide,* Pergamon, New York, 1976.
12. Babb, S. E., and G. J. Scott: *J. Chem. Phys.,* **40:** 3666 (1964).
13. Barker, J. A., W. Fock, and F. Smith: *Phys. Fluids,* **7:** 897 (1964).
14. Baron, J. D., J. G. Root, and F. W. Wells: *J. Chem. Eng. Data,* **4:** 283 (1959).
15. Batschinski, A. J.: *Z. Physik. Chim.,* **84:** 643 (1913).
16. Bertrand, G. L.: *Ind. Eng. Chem. Fundam.,* **16:** 492 (1977).
17. Bingham, E. C., and S. D. Stookey: *J. Am. Chem. Soc.,* **61:** 1625 (1939).
18. Bircher, L. B.: Ph.D. thesis, University of Michigan, Ann Arbor, Mich., 1943.
19. Bradley, R. S. (ed.): *High Pressure Physics and Chemistry,* Academic, New York, 1963, p. 18.
20. Brebach, W. J., and G. Thodos: *Ind. Eng. Chem.,* **50:** 1095 (1958).
21. Bridgman, P. W.: *Proc. Am. Acad. Arts Sci.,* **61:** 57 (1926).
22. Brokaw, R. S.: *NASA Tech. Rept. R-81,* 1961.
23. Brokaw, R. S.: *NASA Tech. Note D-2502,* November 1964.

24. Brokaw, R. S.: *J. Chem. Phys.*, **42:** 1140 (1965).
25. Brokaw, R. S.: *NASA Tech. Note D-4496*, April 1968.
26. Brokaw, R. S.: *Ind. Eng. Chem. Process Design Develop.*, **8:** 240 (1969).
27. Brokaw, R. S., R. A. Svehla, and C. E. Baker: *NASA Tech. Note D-2580,* January 1965.
28. Bromley, L. A., and C. R. Wilke: *Ind. Eng. Chem.*, **43:** 1641 (1951).
29. Brulé, M. R.: private communication, March 1985.
30. Brulé, M. R., and K. E. Starling: *Ind. Eng. Chem. Process Design Develop.*, **23:** 833 (1984).
31. Brush, S. G.: *Chem. Rev.*, **62:** 513 (1962).
32. Burch, L. G., and C. J. G. Raw: *J. Chem. Phys.*, **47:** 2798 (1967).
33. Burnett, D.: *J. Chem. Phys.*, **42:** 2533 (1965).
34. Carmichael, L. T., H. H. Reamer, and B. H. Sage: *J. Chem. Eng. Data,* **8:** 400 (1963).
35. Carmichael, L. T., and B. H. Sage: *J. Chem. Eng. Data,* **8:** 94 (1963).
36. Carmichael, L. T., and B. H. Sage: *AIChE J.*, **12:** 559 (1966).
37. Carr, N. L., R. Kobayashi, and D. B. Burroughs: *J. Petrol. Technol.*, **6**(10): 47 (1954).
38. Carr, N. L., J. D. Parent, and R. E. Peck: *Chem. Eng. Symp. Ser.*, **51**(16): 91 (1955).
39. Chakraborti, P. K., and P. Gray: *Trans. Faraday Soc.*, **61:** 2422 (1965).
40. Chapman, S., and T. G. Cowling: *The Mathematical Theory of Nonuniform Gases,* Cambridge, New York, 1939.
41. Chatterjee, A., and A. K. Vasant: *Chem. Ind.* (London), June 5, 1982, p. 375.
42. Cheung, H.: *UCRL Report 8230,* University of California, Berkeley, Calif., April 1958.
43. Chung, T.-H.: Ph.D. thesis, University of Oklahoma, Norman, Okla., 1980.
44. Chung, T.-H., M. Ajlan, L. L. Lee, and K. E. Starling: *Ind. Eng. Chem. Process Design Develop.*, submitted 1986.
45. Chung, T.-H., L. L. Lee, and K. E. Starling: *Ind. Eng. Chem. Fundam.*, **23:** 8 (1984).
46. Cohen, Y., and S. I. Sandler: *Ind. Eng. Chem. Fundam.*, **19:** 186 (1980).
47. Comings, E. W. and R. S. Egly: *Ind. Eng. Chem.*, **32:** 714 (1940).
48. Comings, E. W., and B. J. Mayland: *Chem. Metall. Eng.*, **42**(3): 115 (1945).
49. Coremans, J. M. J., and J. J. M. Beenakker: *Physica,* **26:** 653 (1960).
50. Cornelissen, J., and H. I. Waterman: *Chem. Eng. Sci.*, **4:** 238 (1955).
51. Dahler, J. S.: *Thermodynamic and Transport Properties of Gases, Liquids, and Solids,* McGraw-Hill, New York, 1959, pp. 14–24.
52. Das, T. R., C. O. Reed, Jr., and P. T. Eubank: *J. Chem. Eng. Data,* **22:** 3 (1977).
53. de Guzman, J., *Anales Soc. espan. fia. y quim.*, **11;** 353 (1913).
54. Dean, D. E. and L. I. Stiel: *AIChe J.*, **11:** 526 (1965).
55. DiPippo, R., J. F. Dorfman, J. Kestin, H. E. Khalifa, and E. A. Mason: *Physica,* **86A:** 205 (1977).
56. Dizechl, M.: Ph.D. thesis, University of California, Santa Barbara, Calif., 1980.
57. Dizechl, M., and E. Marschall: *Ind. Eng. Chem. Process Design Develop.*, **21:** 282 (1982).
58. Dizechl, M., and E. Marschall: *J. Chem. Eng. Data,* **27:** 358 (1982).
59. Dolan, J. P., K. E. Starling, A. L. Lee, B. E. Eakin, and R. T. Ellington: *J. Chem. Eng. Data,* **8:** 396 (1963).
60. Duhne, C. R.: *Chem. Eng.*, **86**(15): 83 (1979).
61. Dymond, J. H.: *Physica,* **75:** 100 (1974).
62. Dymond, J. H.: *Physica,* **79A:** 65 (1975).
63. Dymond, J. H.: *Chem. Physics,* **17:** 101 (1976).
64. Dymond, J. H.: *Physica,* **85A:** 175 (1976).
65. Dymond, J. H., J. Robertson, and J. D. Isdale: *Intern. J. Thermophys.*, **2**(2): 133 (1981).
66. Dymond, J. H., J. Robertson, and J. D. Isdale: *Intern. J. Thermophys.*, **2**(3): 223 (1981).

67. Dymond, J. H., and K. J. Young: *Intern. J. Thermophys.*, **1**(4): 331 (1980).
68. Dymond, J. H., K. J. Young, and J. D. Isdale: *Intern. J. Thermophys.*, **1**(4): 345 (1980).
69. Eakin, B. E., and R. T. Ellington: *J. Petrol. Technol.*, **15:** 210 (1963).
70. Ely, J. F.: *J. Res. Natl. Bur. Stand.*, **86:** 597 (1981).
71. Ely, J. F., and H. J. M. Hanley: *Ind. Eng. Chem. Fundam.*, **20:** 323 (1981).
72. Eversteijn, F. C., J. M. Stevens, and H. I. Waterman: *Chem. Eng. Sci.*, **11:** 267 (1960).
73. Flynn, L. W., and G. Thodos: *J. Chem. Eng. Data*, **6:** 457 (1961).
74. Flynn, L. W., and G. Thodos: *AIChE J.*, **8:** 362 (1962).
75. Francis, W. E.: *Trans. Faraday Soc.*, **54:** 1492 (1958).
76. Gambill, W. R.: *Chem. Eng.*, **65**(23): 157 (1958).
77. Gambill, W. R.: *Chem. Eng.*, **66**(3): 123 (1959).
78. Gandhi, J. M., and S. C. Saxena: *Indian J. Pure Appl. Phys.*, **2:** 83 (1964).
79. Gemant, A. J.: *Appl. Phys.*, **12:** 827 (1941).
80. Giddings, J. D.: Ph.D. thesis, Rice University, Houston, Texas, 1963.
81. Girifalco, L. A.: *J. Chem. Phys.*, **23:** 2446 (1955).
82. Goletz, E., and D. Tassios: *Ind. Eng. Chem. Process Design Develop.*, **16:** 75 (1977).
83. Golubev, I. F.: "Viscosity of Gases and Gas Mixtures: A Handbook," *Natl. Tech. Inf. Serv., TT 70 50022,* 1959.
84. Goodwin, R. D.: "The Thermophysical Properties of Methane from 90 to 500 K at Pressures up to 700 bar," *NBSIR 93-342, Natl. Bur. Stand.*, October 1973.
85. Griest, E. M., W. Webb, and R. W. Schiessler: *J. Chem. Phys.*, **29:** 711 (1958).
86. Grunberg, L.: "An Equation for Predicting the Viscosity of Liquid Mixtures," *National Eng. Lab., Rept. 626,* East Kilbride, Glasgow, Scotland, December 1976.
87. Grunberg, L., and A. H. Nissan: *Nature*, **164:** 799 (1949).
88. Grunberg, L., and A. H. Nissan: *Ind. Eng. Chem.*, **42:** 885 (1950).
89. Gunn, R. D., and T. Yamada: *AIChE J.*, **17:** 1341 (1971).
90. Gupta, G. P., and S. C. Saxena: *AIChE J.*, **14:** 519 (1968).
91. Gururaja, G. J., M. A. Tirunarayanan, and A. Ramachandran: *J. Chem. Eng. Data*, **12:** 562 (1967).
92. Gutman, F., and L. M. Simmons: *J. Appl. Phys.*, **23:** 977 (1952).
93. Haar, L., and J. S. Gallagher: *J. Phys. Chem. Ref. Data*, **7:** 635 (1978).
94. Hafez, M., and S. Hartford: *J. Chem. Eng. Data*, **21:** 179 (1976).
95. Hanley, H. J. M., R. D. McCarty, and J. V. Sengers: *J. Chem. Phys.*, **50:** 857 (1969).
96. Harada, M., M. Tanigaki, and W. Eguchi: *J. Chem. Eng. Japan*, **8:** 1 (1975).
97. Hattikudur, U. R., and G. Thodos: *AIChE J.*, **17:** 1220 (1971).
98. Herning, F., and L. Zipperer: *Gas Wasserfach*, **79:** 49 (1936).
99. Hildebrand, J. H.: *Science*, **174:** 490 (1971).
100. Hirschfelder, J. O., C. F. Curtiss, and R. B. Bird: *Molecular Theory of Gases and Liquids*, Wiley, New York, 1954.
101. Hirschfelder, J. O., M. H. Taylor, and T. Kihara: *Univ. Wisconsin Theoret. Chem. Lab., WIS-OOR-29,* Madison, Wis., July 8, 1960.
102. Hwang, S. C., C. Tsonopoulos, J. R. Cunningham, and G. M. Wilson: *Ind. Eng. Chem. Process Design Develop.*, **21:** 127 (1982).
103. Innes, K. K.: *J. Phys. Chem.*, **60:** 817 (1956).
104. Irving, J. B.: "Viscosities of Binary Liquid Mixtures: A Survey of Mixture Equations," *Natl. Eng. Lab., Rept. 630,* East Kilbride, Glasgow, Scotland, February 1977.
105. Irving, J. B.: "Viscosities of Binary Liquid Mixtures: The Effectiveness of Mixture Equations," *Natl. Eng. Lab., Rept. 631,* East Kilbride, Glasgow, Scotland, February 1977.
106. Isdale, J. D.: Ph.D. thesis, University of Strathclyde, Glasgow, Scotland, 1976.
107. Isdale, J. D.: *Symp. Transp. Prop. Fluids and Fluid Mixtures, Natl. Eng. Lab., East Kilbride, Glasgow, Scotland, 1979.*
108. Isdale, J. D., J. H. Dymond, and T. A. Brawn: *High Temp.—High Press.*, **11:** 571 (1979).
109. Isdale, J. D., J. C. MacGillivray, and G. Cartwright: "Prediction of Viscosity of

Organic Liquid Mixtures by a Group Contribution Method," *Natl. Eng. Lab. Rept.*, East Kilbride, Glasgow, Scotland, 1985.
110. Isdale, J. D., and C. M. Spence: "A Self-centring Falling Body Viscometer for High Pressures," *Natl. Eng. Lab. Rept. 592,* East Kilbride, Glasgow, Scotland, June 1975.
111. Isdale, J. D., and C. M. Spence: "High Pressure Viscosities and Densities of Eight Halogenated Hydrocarbons," *Natl. Eng. Lab. Rept. 604,* East Kilbride, Glasgow, Scotland, December 1975.
112. Islam, N., and M. Ibrahim: *Ind. J. Chem.,* **20A:** 963 (1981).
113. Islam, N., and M. Ibrahim: *Ind. J. Chem.,* **20A:** 969 (1981).
114. Itean, E. C., A. R. Glueck, and R. A. Svehla: *NASA Lewis Research Center, TND-481,* Cleveland, Ohio, 1961.
115. Jobling, A., and A. S. C. Laurence: *Proc. Roy. Soc. London,* **A206:** 257 (1951).
116. Jossi, J. A., L. I. Stiel, and G. Thodos: *AIChE J.,* **8:** 59 (1962).
117. Kennedy, J. T., and G. Thodos: *AIChE J.,* **7:** 625 (1961).
118. Kessel'man, P. M., and A. S. Litvinov. *J. Eng. Phys. USSR,* **10**(3): 385 (1966).
119. Kestin, J., H. E. Khalifa, and W. A. Wakeham: *J. Chem. Phys.,* **65:** 5186 (1976).
120. Kestin, J., and W. Leidenfrost: *Physica,* **25:** 525 (1959).
121. Kestin, J., and W. Leidenfrost: in Y. S. Touloukian (ed.), *Thermodynamic and Transport Properties of Gases, Liquids, and Solids,* ASME and McGraw-Hill, New York, 1959, pp. 321–338.
122. Kestin, J., and J. R. Moszynski: in Y. S. Touloukian (ed.), *Thermodynamic and Transport Properties of Gases, Liquids, and Solids,* ASME and McGraw-Hill, New York, 1959, pp. 70–77.
123. Kestin, J., and I. R. Shankland: in J. V. Sengers (ed.), *Proc. 8th Symp. Thermophys. Prop.,* II, ASME, New York, 1981, p. 352.
124. Kestin, J., and J. Yata: *J. Chem. Phys.,* **49:** 4780 (1968).
125. Kim, S. K., and J. Ross: *J. Chem. Phys.,* **46:** 818 (1967).
126. Klein, M., and F. J. Smith: *J. Res. Natl. Bur. Stand.,* **72A:** 359 (1968).
127. Krieger, F. A.: *Rand Corp. Rept. RM-646,* Santa Monica, Calif.
128. Kuss, E.: *Z. Angew. Phys.,* **7:** 372 (1955).
129. *Landolt-Bornstein Tabellen,* vol. 4, pt. 1, Springer-Verlag, Berlin, 1955.
130. Letsou, A., and L. I. Stiel: *AIChE J.,* **19:** 409 (1973).
131. Lewis, W. K., and L. Squires, *Refiner Nat. Gasoline Manuf.,* **13**(12): 448 (1934).
132. Lowitz, D. A., J. W. Spencer, W. Webb, and R. W. Schiessler: *J. Chem. Phys.,* **30:** 73 (1959).
133. Lucas, K.: *Int. J. Heat Mass Trans.,* **16:** 371 (1973).
134. Lucas, K.: *Chem. Ing. Tech.,* **46:** 157 (1974).
135. Lucas, K.: *Phase Equilibria and Fluid Properties in the Chemical Industry,* Dechema, Frankfurt, 1980, p. 573.
136. Lucas, K.: *Chem. Ing. Tech.,* **53:** 959 (1981).
137. Lucas, K.: personal communications, August 1983, September 1984.
138. Lucas, K.: VDI-Warmeatlas, Abschnitt DA, "Berechnungsmethoden für Stoffeigenschaften," Verin Deutscher Ingenieure, Dusseldorf, 1984.
139. Luckas, M., and K. Lucas: *AIChE J.,* **32:** 139 (1986).
140. McAllister, R. A., *AIChE J.,* **6:** 427 (1960).
141. Malek, K. R., and L. I. Stiel: *Can. J. Chem. Eng.,* **50:** 491 (1972).
142. Marschalko, B., and J. Barna: *Acta Tech. Acad. Sci. Hung.,* **19:** 85 (1957).
143. Mathur, G. P., and G. Thodos: *AIChE J.,* **9:** 596 (1963).
144. Medani, M. S., and M. A. Hasan: *Can. J. Chem. Eng.,* **55:** 203 (1977).
145. Miller, A. A.: *J. Chem. Phys.,* **38:** 1568 (1963).
146. Miller, A. A.: *J. Phys. Chem.,* **67:** 1031 (1963).
147. Miller, A. A.: *J. Phys. Chem.,* **67:** 2809 (1963).
148. Monchick, L., and E. A. Mason: *J. Chem. Phys.,* **35:** 1676 (1961).
149. Morris, P. S.: M.S. thesis, Polytechnic Institute of Brooklyn, Brooklyn, N.Y., 1964.
150. Neufeld, P. D., A. R. Janzen, and R. A. Aziz: *J. Chem. Phys.,* **57:** 1100 (1972).
151. Noda, K., and K. Ishida: *J. Chem. Eng. Japan,* **10:** 478 (1977).
152. Noda, K., M. Ohashi, and K. Ishida: *J. Chem. Eng. Data,* **27:** 326 (1982).

153. O'Loane, J. K.: private communication, June 1979.
154. Orrick, C., and J. H. Erbar: private communication, December 1974.
155. Pal, A. K., and A. K. Barua: *J. Chem. Phys.*, **47:** 216 (1967).
156. Pal, A. K., and P. K. Bhattacharyya: *J. Chem. Phys.*, **51:** 828 (1969).
157. Papadopoulos, C. G.: Ph.D. thesis, Northwestern University, Evanston, Ill., 1977.
158. Pedersen, K. S., A. Fredenslund, P. L. Christensen, and P. Thomassen: *Chem. Eng. Sci.*, **39:** 1011 (1984).
159. Pikkarainen, L.: *J. Chem. Eng. Data*, **28:** 381 (1983).
160. Przezdziecki, J. W., and T. Sridhar: *AIChE J.*, **31:** 333 (1985).
161. Ranz, W. E., and H. A. Brodowsky: *Univ. Minn. OOR Proj. 2340 Tech. Rept. 1*, Minneapolis, Minn., March 15, 1962.
162. Raw, C. J. G., and H. Tang: *J. Chem. Phys.*, **39:** 2616 (1963).
163. Reamer, H. H., G. Cokelet, and B. H. Sage: *Anal. Chem.*, **31:** 1422 (1959).
164. Reichenberg, D.: "The Viscosities of Gas Mixtures at Moderate Pressures," *NPL Rept. Chem. 29*, National Physical Laboratory, Teddington, England, May, 1974.
165. Reichenberg, D.: (*a*) *DCS report 11*, National Physical Laboratory, Teddington, England, August 1971; (*b*) *AIChE J.*, **19:** 854 (1973); (*c*) *ibid.*, **21:** 181 (1975).
166. Reichenberg, D.: "The Viscosities of Pure Gases at High Pressures," *Natl. Eng. Lab., Rept. Chem. 38*, East Kilbride, Glasgow, Scotland, August 1975.
167. Reichenberg, D.: "New Simplified Methods for the Estimation of the Viscosities of Gas Mixtures at Moderate Pressures," *Natl. Eng. Lab. Rept. Chem. 53*, East Kilbride, Glasgow, Scotland, May 1977.
168. Reichenberg, D.: *Symp. Transp. Prop. Fluids and Fluid Mixtures, Natl. Eng. Lab., East Kilbride, Glasgow, Scotland, 1979.*
169. Reid, R. C., and L. I. Belenyessy: *J. Chem. Eng. Data*, **5:** 150 (1960).
170. Rogers, J. D., and F. G. Brickwedde: *AIChE J.*, **11:** 304 (1965).
171. Ross, J. F. and G. M. Brown: *Ind. Eng. Chem.*, **49:** 2026 (1957).
172. Rutherford, R., M. H. Taylor, and J. O. Hirschfelder: *Univ. Wisconsin Theoret. Chem. Lab., WIS-OOR-29a*, Madison, Wis., August 23, 1960.
173. Rutherford, W. M.: *J. Chem. Eng. Data*, **29:** 163 (1984).
174. Saksena, M. P., and S. C. Saxena: *Proc. Natl. Inst. Sci. India*, **31A**(1): 18 (1965).
175. Sandler, S. I., and J. K. Fiszdon: *Physica*, **95A:** 602 (1979).
176. Saxena, S. C., and R. S. Gambhir: *Brit. J. Appl. Phys.*, **14:** 436 (1963).
177. Saxena, S. C., and R. S. Gambhir: *Proc. Phys. Soc. London*, **81:** 788 (1963).
178. Shimotake, H., and G. Thodos: *AIChE J.*, **4:** 257 (1958).
179. Shimotake, H., and G. Thodos: *AIChE J.*, **9:** 68 (1963).
180. Shimotake, H., and G. Thodos: *J. Chem. Eng. Data*, **8:** 88 (1963).
181. Souders, M.: *J. Am. Chem. Soc.*, **59:** 1252 (1937).
182. Souders, M.: *J. Am. Chem. Soc.*, **60:** 154 (1938).
183. Stairs, R. A.: *Can. J. Chem.*, **58:** 296 (1980).
184. Starling, K. E.: M.S. thesis, Illinois Institute of Technology, Chicago, Ill., 1960.
185. Starling, K. E.: Ph.D. thesis, Illinois Institute of Technology, Chicago, Ill., 1962.
186. Starling, K. E., B. E. Eakin, and R. T. Ellington: *AIChE J.*, **6:** 438 (1960).
187. Starling, K. E., and R. T. Ellington: *AIChE J.*, **10:** 11 (1964).
188. Stephan, K., and K. Lucas: *Viscosity of Dense Fluids*, Plenum, New York, 1979.
189. Stewart, R. B.: Ph.D. thesis, University of Iowa, Iowa City, Iowa, June 1966.
190. Stiel, L. I., and G. Thodos: *AIChE J.*, **10:** 275 (1964).
191. Strunk, M. R., W. G. Custead, and G. L. Stevenson: *AIChE J.*, **10:** 483 (1964).
192. Strunk, M. R., and G. D. Fehsenfeld: *AIChE J.*, **11:** 389 (1965).
193. Sutton, J. R.: "References to Experimental Data on Viscosity of Gas Mixtures," *Natl. Eng. Lab. Rept. 613*, East Kilbride, Glasgow, Scotland, May 1976.
194. Svehla, R. A.: "Estimated Viscosities and Thermal Conductivities at High Temperatures," *NASA-TRR-132*, 1962.
195. Tanaka, Y., H. Kubota, T. Makita, and H. Okazaki: *J. Chem. Eng. Japan*, **10:** 83 (1977).
196. Teja, A. S.: *AIChE J.*, **26:** 337 (1980).
197. Teja, A. S., and P. Rice: *Chem. Eng. Sci.*, **36:** 7 (1981).
198. Teja, A. S., and P. Rice: *Ind. Eng. Chem. Fundam.*, **20:** 77 (1981).

199. Teja, A. S., and S. I. Sandler: *AIChE, J.*, **26:** 341 (1980).
200. Teja, A. S., P. A. Thurner, and B. Pasumarti: paper presented at the *Annual AIChE Mtg., Washington, D.C., 1983; Ind. Eng. Chem. Process Design Develop.*, **24:** 344 (1985).
201. Telang, M. S.: *J. Phys. Chem.*, **49:** 579 (1945); **50:** 373 (1946).
202. Thomas, L. H.: *J. Chem. Soc.*, **1946:** 573.
203. Thomas, L. H.: *Chem. Eng. J.*, **11:** 201 (1976).
204. Thomas, L. H., R. Meatyard, H. Smith, and G. H. Davis: *J. Chem. Eng. Data*, **24:** 161 (1979).
205. Thurner, P. A.: S.M. thesis, Georgia Institute of Technology, Atlanta, Ga., 1984.
206. Titani, T.: *Bull. Inst. Phys. Chem. Res. Tokyo*, **8:** 433 (1929).
207. Tondon, P. K., and S. C. Saxena: *Ind. Eng. Chem. Fundam.*, **7:** 314 (1968).
208. Trautz, M., and P. B. Baumann: *Ann. Phys.*, **5:** 733 (1929).
209. Trautz, M., and R. Heberling: *Ann. Phys.*, **10:** 155 (1931).
210. Trautz, M.: *Ann. Phys.*, **11:** 190 (1931).
211. van Velzen, D., R. L. Cardozo, and H. Langenkamp: *Ind. Eng. Chem. Fundam.*, **11:** 20 (1972).
212. van Velzen, D., R. L. Cardozo, and H. Langenkamp: "Liquid Viscosity and Chemical Constitution of Organic Compounds: A New Correlation and a Compilation of Literature Data," *Euratom*, 4735e, Joint Nuclear Research Centre, Ispra Establishment, Italy, 1972.
213. van Wyk, W. R., J. H. van der Veen, H. C. Brinkman, and W. A. Seeder: *Physica*, **7:** 45 (1940).
214. Vanderslice, J. T., S. Weissman, E. A. Mason, and R. J. Fallon: *Phys. Fluids*, **5:** 155 (1962).
215. Vargaftik, N. B.: *Tables on the Thermophysical Properties of Liquids and Gases*, 2d ed., Hemisphere, Washington, D.C., 1975.
216. Vogel, H.: *Physik Z.*, **22:** 645 (1921).
217. Vogel, H., and A. Weiss: *Ber. Bunsenges. Phys. Chem.*, **85:** 539 (1981).
218. Vogel, H., and A. Weiss: *Ber. Bunsenges. Phys. Chem.*, **86:** 193 (1982).
219. Waxman, M., and J. S. Gallagher: *J. Chem. Eng. Data*, **28:** 224 (1983).
220. Wei, I.-Chien, and R. Rowley: *J. Chem. Eng. Data*, **29:** 332, 336 (1984); *Chem. Eng. Sci.*, **40:** 401 (1985).
221. Wilke, C. R.: *J. Chem. Phys.*, **18:** 517 (1950).
222. Wobster, R., and F. Mueller: *Kolloid Beih.*, **52:** 165 (1941).
223. Wright, P. G., and P. Gray: *Trans. Faraday Soc.*, **58:** 1 (1962).
224. Yaws, C. L., J. W. Miller, P. N. Shah, G. R. Schorr, and P. M. Patel: *Chem. Eng.*, **83**(25): 153 (1976).
225. Yoon, P., and G. Thodos: *AIChE J.*, **16:** 300 (1970).

Chapter

# 10

# Thermal Conductivity

## 10-1 Scope

Thermal conductivities of both gases and liquids are considered in this chapter. Some background relevant to the theory of thermal conductivity is given in Secs. 10-2 and 10-3 (for gases) and in Sec. 10-8 (for liquids). Estimation techniques for pure gases at near ambient pressures are covered in Sec. 10-4; the effects of temperature and pressure are discussed in Secs. 10-4 and 10-5. Similar topics for liquids are in Secs. 10-9 to 10-11. Thermal conductivities for gas and for liquid mixtures are covered in Secs. 10-6, 10-7, and 10-12. Thermal conductivities of reacting gas mixtures are not covered. For a recent review, see Ref. 35.

The units used for thermal conductivity are W/(m·K). To convert these to English or cgs units:

$$\mathrm{W/(m\cdot K)} \times 0.5778 = \mathrm{Btu/(hr\cdot ft\cdot °R)}$$
$$\mathrm{W/(m\cdot K)} \times 0.8598 = \mathrm{kcal/(cm\cdot hr\cdot K)}$$
$$\mathrm{W/(m\cdot K)} \times 2.388 \times 10^{-3} = \mathrm{cal/(cm\cdot s\cdot K)}$$

or

$$\mathrm{Btu/(hr\cdot ft\cdot °R)} \times 1.731 = \mathrm{W/(m\cdot K)}$$
$$\mathrm{kcal/(cm\cdot hr\cdot K)} \times 1.163 = \mathrm{W/(m\cdot K)}$$
$$\mathrm{cal/(cm\cdot s\cdot K)} \times 418.7 = \mathrm{W/(m\cdot K)}$$

## 10-2 Theory of Thermal Conductivity

In Sec. 9-3, through rather elementary arguments, the thermal conductivity of an ideal gas was found to be equal to $vLC_v n/3$ [Eq. (9-3.7)], where $v$ is the average molecular velocity, $L$ is the mean free path, $C_v$ is the heat capacity per molecule, and $n$ is the number density of molecules. Similar relations were derived for the viscosity and diffusion coefficients of gases. In the case of the last two properties, this elementary approach yields approximate but reasonable values. For thermal conductivity, it is quite inaccurate. A more detailed treatment is necessary to account for the effect of having a wide spectrum of molecular velocities; also, molecules may store energy in forms other than translational. For *monatomic gases,* which have no rotational or vibrational degrees of freedom, a more rigorous analysis yields

$$\lambda = \frac{25}{32} (\pi mkT)^{1/2} \frac{C_v/m}{\pi \sigma^2 \Omega_v} \tag{10-2.1}$$

or, written for computational ease, with $C_v = \frac{3}{2} k$,

$$\lambda = 2.63 \times 10^{-23} \frac{(T/M')^{1/2}}{\sigma^2 \Omega_v} \tag{10-2.2}$$

where $\lambda$ = thermal conductivity, W/(m·K)
$T$ = temperature, K
$k$ = Boltzmann's constant = $1.3805 \times 10^{-23}$ J/K
$M'$ = molecular weight, kg/mol
$\sigma$ = characteristic dimension of molecule, m
$\Omega_v$ = collision integral, dimensionless

For a hard-sphere molecule, $\Omega_v$ is unity; normally, however, it is a function of temperature, and the exact dependence is related to the intermolecular force law chosen. If the Lennard-Jones 12-6 potential [Eq. (9-4.2)] is selected, $\Omega_v$ is given by Eq. (9-4.3).

If Eq. (10-2.1) is divided by Eq. (9-3.9),

$$\frac{\lambda M'}{\eta C_v} = 2.5 \tag{10-2.3}$$

With $\gamma = C_p/C_v$, the Prandtl number $N_{\mathrm{Pr}}$ is

$$N_{\mathrm{Pr}} = \frac{C_p \eta}{\lambda M'} = \frac{\gamma}{2.5} \tag{10-2.4}$$

Since $\gamma$ for monatomic gases is close to $\frac{5}{3}$ except at very low temperatures, Eq. (10-2.4) would indicate that $N_{\mathrm{Pr}} \approx \frac{2}{3}$, a value close to that found experimentally. To obtain Eq. (10-2.3), the terms $\sigma^2$ and $\Omega_v$ have been eliminated, and the result is essentially independent of the intermolecular potential law chosen.

The dimensionless† group $\lambda M'/\eta C_v$ is known as the *Eucken factor;* it is close to 2.5 for monatomic gases, but it is significantly less for polyatomic gases. Our discussion so far has considered only energy associated with translational motion; since heat capacities of polyatomic molecules exceed those for monatomic gases, a substantial fraction of molecular energy resides in modes other than translational.

## 10-3 Thermal Conductivities of Polyatomic Gases

### Eucken and modified Eucken models

Eucken proposed that Eq. (10-2.3) be modified for polyatomic gases by separating the translational and internal energy contributions into separate terms:

$$\frac{\lambda M'}{\eta C_v} = f_{\text{tr}}\left(\frac{C_{\text{tr}}}{C_v}\right) + f_{\text{int}}\left(\frac{C_{\text{int}}}{C_v}\right) \tag{10-3.1}$$

Thus the translational energy contribution has been decoupled from any internal energy interaction [31, 74, 94, 116, 150, 156, 175, 176], although the validity of this step has been questioned [53, 84, 148, 159]. Invariably, $f_{\text{tr}}$ is set equal to 2.5 to force Eq. (10-3.1) to reduce to Eq. (10-2.3) for a monatomic ideal gas. $C_{\text{tr}}$ is set equal to the classical value of $\frac{3}{2}R$, and $C_{\text{int}}$ is conveniently expressed as $C_v - C_{\text{tr}}$. Then

$$\begin{aligned}\frac{\lambda M'}{\eta C_v} &= \frac{15/4}{C_v/R} + f_{\text{int}}\left(1 - \frac{3/2}{C_v/R}\right) \\ &= \frac{15/4}{(C_p/R) - 1} + f_{\text{int}}\left[1 - \frac{5/2}{(C_p/R) - 1}\right]\end{aligned} \tag{10-3.2}$$

where the ideal-gas relation ($C_p - C_v = R$) has been used.

Eucken chose $f_{\text{int}} = 1.0$, whereby Eq. (10-3.2) reduces to

$$\frac{\lambda M'}{\eta C_v} = 1 + \frac{9/4}{C_v/R} = 1 + \frac{9/4}{(C_p/R) - 1} \tag{10-3.3}$$

the well-known *Eucken correlation* for polyatomic gases.

Many of the assumptions leading to Eq. (10-3.3) are open to question, in particular, the choice of $f_{\text{int}} = 1.0$. Ubbelohde [166], Chapman and

†The group $\lambda M'/\eta C_v$ is dimensionless; with SI units, $\lambda$ is in W/(m·K) = N/(s·K), $\eta$ is in N·s/m$^2$, $C_v$ is in J/(mol·K), and $M'$ is in kg/mol.

Cowling [23], Hirschfelder [53], and Schafer [153] have suggested that molecules with excited internal energy states could be regarded as separate chemical species, and the transfer of internal energy is then analogous to a diffusional process. This concept leads to a result that

$$f_{\text{int}} = \frac{M' \rho D}{\eta} \tag{10-3.4}$$

where $M'$ = molecular weight, kg/mol
$\eta$ = viscosity, $N \cdot s/m^2$†
$\rho$ = molar density, $mol/m^3$
$D$ = diffusion coefficient, $m^2/s$

Most early theories selected $D$ to be equivalent to the molecular self-diffusion coefficient, and $f_{\text{int}}$ is then the reciprocal of the Schmidt number. With Eqs. (9-3.9) and (11-3.2) it can be shown that $f_{\text{int}} \approx 1.32$ and is almost independent of temperature. With this formulation, Eq. (10-3.2) becomes

$$\frac{\lambda M'}{\eta C_v} = 1.32 + \frac{1.77}{C_v/R} = 1.32 + \frac{1.77}{(C_p/R) - 1} \tag{10-3.5}$$

Equation (10-3.5), often referred to as the *modified Eucken* correlation, was used by Svehla [159] in his compilation of high-temperature gas properties.

The modified Eucken relation [Eq. (10-3.5)] predicts larger values of $\lambda$ than the Eucken form [Eq. (10-3.3)], and the difference becomes greater as $C_v$ increases above the monatomic gas value of about 12.6 J/(mol·K). Both yield Eq. (10-2.3) when $C_v = \frac{3}{2}R$. Usually, experimental values of $\lambda$ lie between those calculated by the two Eucken forms except for polar gases, when both predict $\lambda$ values that are too high. For nonpolar gases, Stiel and Thodos [157] suggested a compromise between Eqs. (10-3.3) and (10-3.5) as

$$\frac{\lambda M'}{\eta C_v} = 1.15 + \frac{2.03}{C_v/R} = 1.15 + \frac{2.03}{(C_p/R) - 1} \tag{10-3.6}$$

Equations (10-3.3), (10-3.5), and (10-3.6) indicate that the Eucken factor $(\lambda M'/\eta C_v)$ should decrease with increasing temperature as the heat capacity rises, but experimental data indicate that the Eucken factor is often remarkably constant with temperature, and, if anything, it increases slightly with temperature. In Fig. 10-1 we illustrate the case for ethyl chloride, where the data of Vines and Bennett show the Eucken factor

†*Note:* $1\ N \cdot s/m^2 = 10$ poises $= 10^3$ cP.

increases from only about 1.41 to 1.48 from 40 to 140°C. On this same graph, the predictions of Eqs. (10-3.3), (10-3.5), and (10-3.6) are plotted and, as noted earlier, all predict a small decrease in the Eucken factor as temperature increases. In Fig. 10-2 we have graphed the experimental Eucken factor as a function of *reduced* temperature for 13 quite diverse low-pressure gases. Except for ethane, all show a small rise with an increase in temperature.

### Mason and Monchick analysis

In a pioneer paper published in 1962 [94], Mason and Monchick employed the formal dynamic treatment of Wang Chang and Uhlenbeck [177] and Taxman [160] to derive an approximation to the thermal conductivity of polyatomic gases. In the formalism of Eq. (10-3.1) they found

$$f_{tr} = \frac{5}{2}\left[1 - \frac{10}{3\pi}\left(1 - \frac{2}{5}\frac{M'\rho D}{\eta}\right)\frac{C_{rot}}{RZ_{rot}}\right] \tag{10-3.7}$$

$$f_{int} = \frac{M'\rho D}{\eta}\left[1 + \frac{5}{\pi}\left(1 - \frac{2}{5}\frac{M'\rho D}{\eta}\right)\frac{C_{rot}}{C_{int}Z_{rot}}\right] \tag{10-3.8}$$

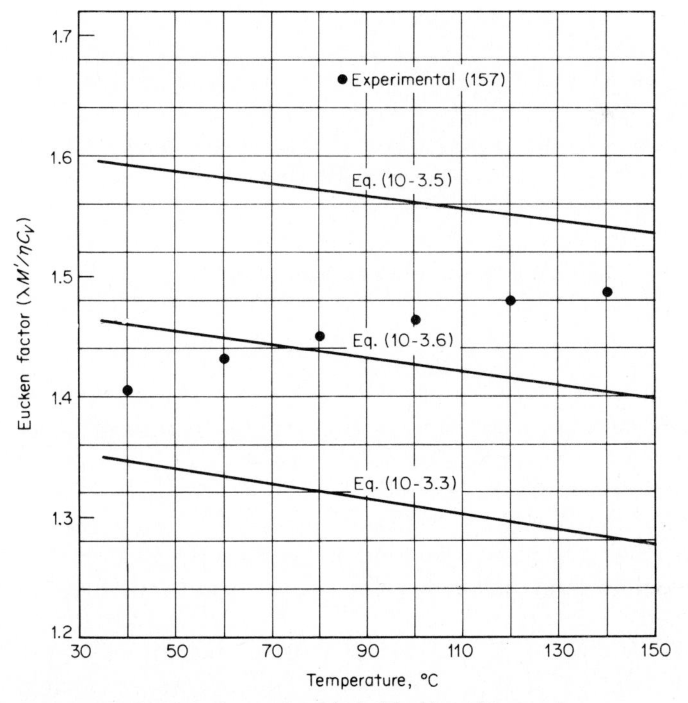


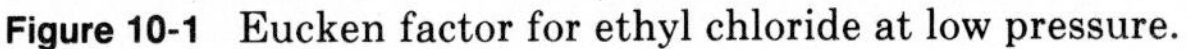
**Figure 10-1** Eucken factor for ethyl chloride at low pressure.

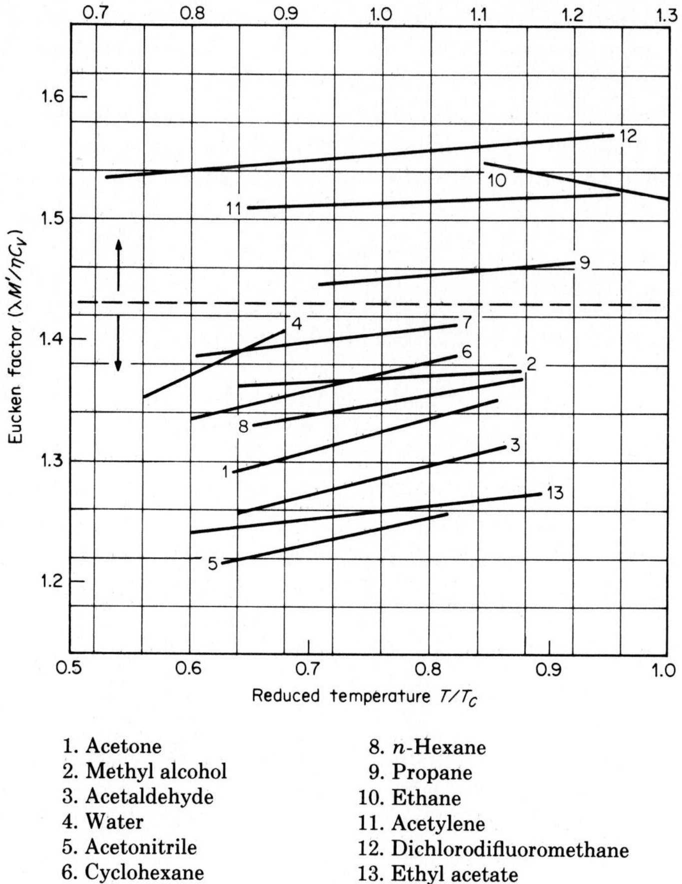


1. Acetone
2. Methyl alcohol
3. Acetaldehyde
4. Water
5. Acetonitrile
6. Cyclohexane
7. Benzene
8. $n$-Hexane
9. Propane
10. Ethane
11. Acetylene
12. Dichlorodifluoromethane
13. Ethyl acetate

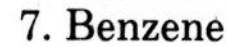

**Figure 10-2** Variation of the Eucken factor with temperature. *(Data primarily from Ref. 157.)*

$C_{\text{rot}}$ is the contribution to the heat capacity from rotational storage modes and is obtained from the classical value $F_rR/2$, where $F_r$ is the number of degrees of freedom for external rotation. $C_{\text{int}} = C_v - C_{\text{tr}}$. $Z_{\text{rot}}$, the collision number, represents the number of collisions required to interchange a quantum of rotational energy with translational energy. [In Eqs. (10-3.7) and (10-3.8) negligible terms involving vibrational energy contributions have been dropped.] For large values of $Z_{\text{rot}}$, Eqs. (10-3.7) and (10-3.8) reduce to $f_{\text{tr}} = 2.5$ and $f_{\text{int}} = M' \rho D/\eta$, that is, those used earlier for the modified Eucken relation. If Eqs. (10-3.7) and (10-3.8) are substituted into Eq. (10-3.1) and $M' \rho D/\eta$ is assumed to be 1.32,

$$\frac{\lambda M'}{\eta C_v} = 1.32 + \frac{1.77}{C_v/R} - \frac{0.886(C_{\text{rot}}/C_v)}{Z_{\text{rot}}} \tag{10-3.9}$$

Equation (10-3.9) should be used only for *nonpolar* polyatomic molecules. The key to its use is the accurate estimation of the rotational collision number. To date, this is not possible, although many authors discuss the problem [8, 14, 16, 33, 94, 95, 116, 117, 125, 146, 148, 149, 150, 156]. In Fig. 10-3, we show the Eucken factor $\lambda M'/\eta C_v$ for hydrogen, nitrogen, and carbon dioxide. As noted earlier, most experimental data indicate values of $\lambda M'/\eta C_v$ between those predicted by the Eucken and the modified

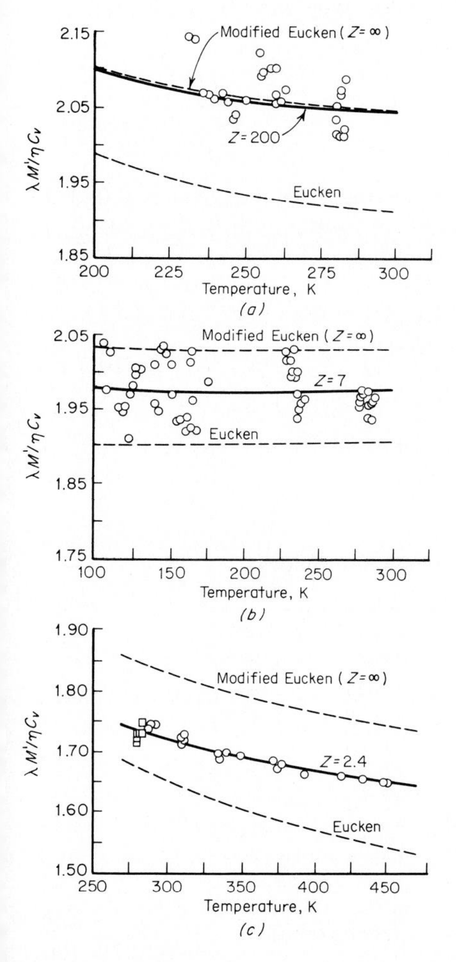


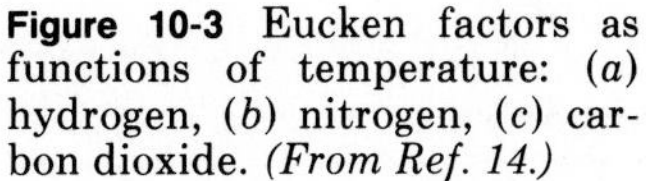

**Figure 10-3** Eucken factors as functions of temperature: (*a*) hydrogen, (*b*) nitrogen, (*c*) carbon dioxide. *(From Ref. 14.)*

Eucken relations. In Fig. 10-3, the $Z_{rot}$ shown is that which gave the best fit to the experimental data. $Z_{rot}$ was assumed temperature-independent.

Although Eq. (10-3.8) is probably the best theoretical equation available for estimating the thermal conductivity of a nonpolar polyatomic gas, without some a priori knowledge of $Z_{rot}$, it is not of much practical value. $Z_{rot}$ values normally lie between 1 and 10 and are probably temperature-dependent [8, 52]; widely differing values are reported in the literature. Attempts to relate $Z_{rot}$ to other, more readily available properties of a molecule have not been successful.

### Roy and Thodos estimation technique

In the same way that the viscosity was nondimensionalized in Eqs. (9-4.12) and (9-4.13), a reduced thermal conductivity may be expressed as

$$\lambda_r = \lambda\Gamma \tag{10-3.10}$$

$$\Gamma = \left[\frac{T_c\,(M')^3\,N_0^2}{R^5 P_c^4}\right]^{1/6} \tag{10-3.11}$$

In SI units, if $R$ = 8314 J/(kmol·K), $N_0$ (Avogadro's number) = 6.023 × $10^{26}$ (kmol)$^{-1}$, and with $T_c$ in kelvins, $M'$ in kg/kmol, and $P_c$ in N/m$^2$, $\Gamma$ has the units of m·K/W or inverse thermal conductivity. In more convenient units,

$$\Gamma = 210\left(\frac{T_c M^3}{P_c^4}\right)^{1/6} \tag{10-3.12}$$

where $\Gamma$ is the reduced, inverse thermal conductivity, [W/(m·K)]$^{-1}$, $T_c$ is in kelvins, $M$ is in g/mol, and $P_c$ is in bars.

**TABLE 10-1 Recommended $f(T_r)$ Equations for the Roy-Thodos Method**

| | |
|---|---|
| Saturated hydrocarbons† | $-0.152T_r + 1.191T_r^2 - 0.039T_r^3$ |
| Olefins | $-0.255T_r + 1.065T_r^2 + 0.190T_r^3$ |
| Acetylenes | $-0.068T_r + 1.251T_r^2 - 0.183T_r^3$ |
| Naphthalenes and aromatics | $-0.354T_r + 1.501T_r^2 - 0.147T_r^3$ |
| Alcohols | $1.000T_r^2$ |
| Aldehydes, ketones, ethers, esters | $-0.082T_r + 1.045T_r^2 + 0.037T_r^3$ |
| Amines and nitriles | $0.633T_r^2 + 0.367T_r^3$ |
| Halides | $-0.107T_r + 1.330T_r^2 - 0.223T_r^3$ |
| Cyclic compounds‡ | $-0.354T_r + 1.501T_r^2 - 0.147T_r^3$ |

†Not recommended for methane.

‡For example, pyridine, thiophene, ethylene oxide, dioxane, piperidine.

The reduced thermal conductivity was employed by Roy and Thodos [142, 143], who, however, separated the $\lambda\Gamma$ product into two parts. The first, attributed only to translational energy, was obtained from a curve fit of the data for the rare gases [141]; this part varies only with the reduced temperature, $T = T/T_c = T_r$. In the second, the contribution from rotational, vibrational, etc., interchange was related to the reduced temperature and a specific constant estimated from group contributions. The final equation may be written

$$\lambda_r = \lambda\Gamma = (\lambda\Gamma)_{tr} + (\lambda\Gamma)_{int} \tag{10-3.13}$$

where $\lambda$ = low-pressure gas thermal conductivity, W/(m·K) and $\Gamma$ is defined in Eq. (10-3.12).

$$(\lambda\Gamma)_{tr} = 8.757[\exp(0.0464T_r) - \exp(-0.2412T_r)] \tag{10-3.14}$$

$$(\lambda\Gamma)_{int} = Cf(T_r) \tag{10-3.15}$$

Relations for $f(T_r)$ are shown in Table 10-1. The constant $C$ is specific for each material, and it is estimated by a group contribution technique as shown below.

**Estimation of Roy-Thodos constant *C*.** T In the discussion to follow, one identifies carbon types as shown:

```
         H          H         H          |
       H—C—       —C—       —C—        —C—
         H          H         |          |
Type:    1          2         3          4
```

| Paraffinic hydrocarbons | $\Delta C$ |
|---|---|
| Base group, methane | 0.73 |
| First methyl substitution | 2.00 |
| Second methyl substitution | 3.18 |
| Third methyl substitution | 3.68 |
| Fourth and successive methyl substitutions | 4.56 |

For example, $C$ for $n$-octane is equal to $[0.73 + 2.00 + 3.18 + 3.68 + 4(4.56)] = 27.8$

*Isoparaffins* are formed by determining the $C$ for the paraffin with the longest possible straight-chain carbon backbone and then making successive substitutions of hydrogen atoms by methyl groups. Values of $\Delta C$

attributable to such substitutions are shown below:

| Type of substitution | $\Delta C$ |
|---|---|
| 1 ← 2 → 1 | 3.64 |
| 1 ← 2 → 2 | 4.71 |
| 1 ← 2 → 3 | 5.79 |
| 2 ← 2 → 2 | 5.79 |
| 1 ← 3 → 1<br>↓<br>1 | 3.39 |
| 1 ← 3 → 1<br>↓<br>2 | 4.50 |
| 1 ← 3 → 1<br>↓<br>3 | 5.61 |

The type of carbon atom from which the arrow points away is the one involved in the methyl substitution. The arrows point toward the types of adjacent atoms. To calculate $C$ for an isoparaffin, beginning with the longest chain, introduce side chains beginning with the left end and proceed in a clockwise direction. To illustrate with 2,2,4-trimethylpentane,

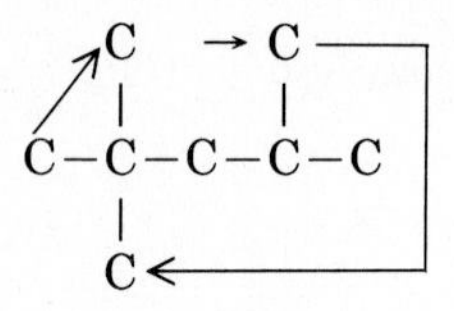

$n$-Pentane = 0.73 + 2.00 + 3.18 + 3.68 + 4.56 = 14.15. For the methyl substitutions

$$1 \leftarrow 2 \rightarrow 2 = 4.71;\ 2 \leftarrow 2 \rightarrow 1 = 4.71;\ \text{and}\ 1 \leftarrow \underset{\substack{\downarrow \\ 1}}{3} \rightarrow 2 = 4.50$$

Thus, $C = 14.15 + 4.71 + 4.71 + 4.50 = 28.07.$

**Olefinic and acetylenic hydrocarbons.** First determine $C$ for the corresponding saturated hydrocarbon, as described above; then insert the unsaturated bond(s) and employ the following $\Delta C$ contributions:

| | | $\Delta C$ |
|---|---|---|
| First double bond | 1 ↔ 1 | −1.19 |
| | 1 ↔ 2 | −0.65 |
| | 2 ↔ 2 | −0.29 |
| Second double bond | 2 ↔ 1 | −0.17 |
| Any acetylenic bond | | −0.83 |

**Naphthenes.** Form the paraffinic hydrocarbon with the same number of carbon atoms as in the naphthene ring. Remove two terminal hydrogens and close the ring. $\Delta C = -1.0$.

**Aromatics.** Benzene has a $C$ value of 13.2. Methyl-substituted benzenes have $C$ values of 13.2 − (5.28) (number of methyl substitutions).

Before discussing the estimation of $C$ for nonhydrocarbons, it is important to note that the simple rules shown above are incomplete and do not cover many types of hydrocarbons. There are, however, no experimental data that can be used to obtain additional $\Delta C$ contributions. In fact, some of the $\Delta C$ values quoted above are based on so few data that they should be considered only approximate. For the 27 *hydrocarbons* studied by Roy and Thodos, one can obtain a rough, but often satisfactory, correlation by using only molecular weight as the correlating parameter, i.e.,

$$C \approx 5.21 \times 10^{-2} M + 1.82 \times 10^{-3} M^2 \qquad M < 120 \qquad (10\text{-}3.16)$$

with $M$ in g/mol and $C$ dimensionless.

For *nonhydrocarbons*, $C$ is again estimated by a group contribution method wherein one mentally synthesizes the final compound by a particular set of rules and employs $\Delta C$ values for each step.

**Alcohols.** Synthesize the corresponding hydrocarbon with the same carbon structure and calculate $C$ as noted above. Replace the appropriate hydrogen atom by a hydroxyl group and correct $C$ as noted:

| Type of −OH substitution | $\Delta C$ |
|---|---|
| On methane | 3.79 |
| $1 \leftarrow 1$ | 4.62 |
| $2 \leftarrow 1$ | 4.11 |
| $3 \leftarrow 1$ | 3.55 |
| $4 \leftarrow 1$ | 3.03 |
| $1 \leftarrow 2 \rightarrow 1$ | 4.12 |

The notation is the same as that used earlier; for example, $3 \leftarrow 1$ indicates that the −OH group is replacing a hydrogen atom on a type 1 carbon which is adjacent to a type 3 carbon:

```
        C                           C
        |                           |
C − C − C − C−H → C − C − C − C−OH
```

These rules apply only to aliphatic alcohols, and they are incomplete even for them.

**Amines.** Estimation of $C$ for amines is similar to that described above for alcohols. First, synthesize the corresponding hydrocarbon segment (with the most complex structure) that is finally to be attached to a nitrogen. For *primary* amines, replace the appropriate terminal hydrogen by a $-NH_2$ group with the following $\Delta C$ contributions:

| Type of substitution | $\Delta C$ |
|---|---|
| On methane | 2.60 |
| $1 \leftarrow 1$ | 3.91 |
| $1 \leftarrow 2 \rightarrow 1$ | 5.08 |
| $2 \leftarrow 2 \rightarrow 1$ | 7.85 |
| $1 \leftarrow 3 \rightarrow 1$ (with $\downarrow 1$ below the 3) | 6.50 |

For *secondary* amines, there are additional $\Delta C$ values:

| | $\Delta C$ |
|---|---|
| $CH_3-NH_2 \rightarrow CH_3-N(H)-CH_3$ | 3.31 |
| $-CH_2-NH_2 \rightarrow -CH_2-N(H)-CH_3$ | 4.40 |

Finally, for *tertiary* amines, Roy and Thodos show three types of corrections applicable to the secondary amines:

| | $\Delta C$ |
|---|---|
| $CH_3-NH-CH_3 \rightarrow (CH_3)_3 \equiv N$ | 2.59 |
| $-CH_2-N(H)-CH_2- \rightarrow -CH_2-N(CH_3)-CH_2-$ | 3.27 |
| $-CH_2-N(H)-CH_3 \rightarrow -CH_2-N-(CH_3)_2$ | 2.94 |

After calculating $C$ for an amine as noted above, any methyl substitutions for a hydrogen on a side chain increase $C$ by 4.56 (the same as shown for fourth and successive methyl substitutions in paraffinic hydrocarbons).

For example,

$$-\overset{|}{N}-CH_3 \rightarrow -\overset{|}{N}-CH_2-CH_3 \qquad \Delta C = 4.56$$

**Nitriles.** Only three $\Delta C$ contributions are shown; they were based on thermal conductivity data for acetonitrile, propionitrile, and acrylic nitrile.

| Type of $-CN$ addition | $\Delta C$ |
|---|---|
| On methane | 5.43 |
| $CH_3-CH_3 \rightarrow CH_3-CH_2-CN$ | 7.12 |
| $-CH=CH_2 \rightarrow -CH=CH-CN$ | 6.29 |

**Halides.** Suggested contributions are shown below; the order of substitution should be F, Cl, Br, I.

| | $\Delta C$ |
|---|---|
| First halogen substitution on methane: | |
| Fluorine | 0.26 |
| Chlorine | 1.38 |
| Bromine | 1.56 |
| Iodine | 2.70 |
| Second and successive substitutions on methane: | |
| Fluorine | 0.38 |
| Chlorine | 2.05 |
| Bromine | 2.81 |
| Substitutions on ethane and higher hydrocarbons: | |
| Fluorine | 0.58 |
| Chlorine | 2.93 |

**Aldehydes and ketones.** Synthesize the hydrocarbon analog with the same number of carbon atoms and calculate $C$ as noted above. Then form the desired aldehyde or ketone by substituting oxygen for two hydrogen atoms:

| | $\Delta C$ |
|---|---|
| $-CH_2-CH_3 \rightarrow -CH_2-CHO$ | 1.93 |
| $-CH_2-CH_2-CH_2- \rightarrow -CH_2-CO-CH_2-$ | 2.80 |

**Ethers.** Synthesize the primary alcohol with the longest carbon chain on one side of the ether oxygen. Convert this alcohol to a methyl ether.

$$-CH_2OH \rightarrow -CH_2-O-CH_3 \qquad \Delta C = 2.46$$

Extend the methyl chain, if desired, to an ethyl.

$$-CH_2-O-CH_3 \rightarrow -CH_2-O-CH_2-CH_3 \qquad \Delta C = 4.18$$

Although Roy and Thodos do not propose extensions beyond the ethyl group, presumably more complex chains could be synthesized by using $\Delta C$ values obtained from paraffinic and isoparaffinic contributions.

**Acids and esters.** Synthesize the appropriate ether so as to allow the following substitutions:

| | $\Delta C$ |
|---|---|
| $-CH_2-O-CH_3 \rightarrow -CH_2-O-\underset{\displaystyle H}{\underset{\mid}{C}}=O$ | 0.75 |
| $-CH_2-O-CH_2- \rightarrow -CH_2-O-\underset{\mid}{C}=O$ | 0.31 |

**Cyclics.** Synthesize the ring, if possible, with the following contributions (not substitutions):

| Group | $\Delta C$ |
|---|---|
| $-CH_2-$ | 4.25 |
| $-CH=$ | 3.50 |
| $-NH-$ | 4.82 |
| $-N=$ | 3.50 |
| $-O-$ | 3.61 |
| $=S=$ | 7.01 |

and determine $C$ as:

$$C = \Sigma\, \Delta C - 7.83$$

The Roy and Thodos group contributions were obtained from limited data and are averaged values. Calculations cannot be made for many compounds by using the rules given above, but an intelligent guess for a missing increment can often be made. The Roy-Thodos method can also be used in a different way. If a single value of $\lambda$ is available at a known temperature, Eq. (10-3.13) can be used with Table 10-1 to yield a value of $C$ that can then be employed to determine $\lambda$ at other temperatures.

### Method of Chung et al. [27, 28]

Chung et al. employed an approach similar to that of Mason and Monchick to obtain a relation for $\lambda$. By using their form and a similar one for low-pressure viscosity [Eq. (9-4.9)], one obtains

$$\frac{\lambda M'}{\eta C_v} = \frac{3.75\Psi}{C_v/R} \tag{10-3.17}$$

where $\lambda$ = thermal conductivity, $W/(m \cdot K)$
$M'$ = molecular weight, kg/mol
$\eta$ = low-pressure gas viscosity, $N \cdot s/m^2$
$C_v$ = heat capacity at constant volume, $J/(mol \cdot K)$
$R$ = gas constant, 8.314 $J/(mol \cdot K)$
$\Psi = 1 + \alpha \{[0.215 + 0.28288\alpha - 1.061\beta + 0.26665Z]/[0.6366 + \beta Z + 1.061\alpha\beta]\}$
$\alpha = (C_v/R) - \frac{3}{2}$
$\beta = 0.7862 - 0.7109\omega + 1.3168\omega^2$
$Z = 2.0 + 10.5T_r^2$

The $\beta$ term is an empirical correlation for $(f_{\text{int}})^{-1}$ [Eq. (10-3.4)] and is said to apply only for nonpolar materials. For polar materials, $\beta$ is specific for each compound; Chung et al. [28] list values for a few materials. If the compound is polar and $\beta$ is not available, use a default value of $(1.32)^{-1} = 0.758$.

$Z$ has the same meaning as $Z_{\text{rot}}$ in the Mason and Monchick analysis. For large values of $Z$, $\Psi$ reduces to

$$\Psi = 1 + 0.2665\frac{\alpha}{\beta} \qquad \text{large } Z \tag{10-3.18}$$

If Eq. (10-3.18) is used in Eq. (10-3.17), the Eucken correlation [Eq. (10-3.3)] is obtained when $\beta$ is set equal to unity. If $\beta = (1.32)^{-1}$, the modified Eucken relation [Eq. (10-3.5)] is recovered. The method is illustrated in Example 10-1.

## Method of Ely and Hanley [40, 51]

An extended corresponding states procedure, the method of Ely and Hanley was developed to estimate the viscosities and thermal conductivities of nonpolar fluids, pure or mixtures, over a wide range of densities and temperatures. As illustrated in this section, the procedure has been simplified to treat the thermal conductivity of low-pressure, pure gases. Later we shall extend the approach to handle fluids at high densities. The estimation technique is based on Eucken's proposal to separate the thermal conductivity into contributions from the interchanges of both translational and internal energy. For the latter, the modified Eucken representation [in Eq. (10-3.5)] was used, but, for the translational component, a corresponding states method using methane as the reference component was selected. Thus, with Eq. (10-3.2), with $f_{int} = 1.32$, we have

$$\frac{\lambda M'}{\eta^* C_v} = \frac{\lambda^* M'}{\eta^* C_v} + 1.32\left(1 - \frac{3/2}{C_v/R}\right) \tag{10-3.19}$$

In the application of Eq. (10-3.2), as noted earlier, the low-pressure, pure gas viscosity $\eta$ was required as an input variable. In the Ely and Hanley method, $\eta^*$ has the same connotation, but it is calculated as a part of the estimation procedure (as shown below). Equation (10-3.19) can be written as

$$\lambda = \lambda^* + \frac{\eta^*}{M'}(1.32)\left(C_v - \frac{3R}{2}\right) \tag{10-3.20}$$

where $\lambda$ = thermal conductivity of the low-pressure gas, W/(m·K)
$M'$ = molecular weight, kg/mol
$C_v$ = heat capacity of the low-pressure gas at constant volume, J/(mol·K)
$R$ = 8.314 J/(mol·K)

$\lambda^*$ and $\eta^*$ are defined below. The calculational procedure is as follows:

1. Determine the reduced temperature $T_r = T/T_c$; then define a parameter $T^+$ such that

$$T^+ = T_r \qquad T_r \leq 2 \qquad \text{or} \qquad T^+ = 2 \qquad T_r > 2 \tag{10-3.21}$$

2. Calculate the compound *shape* factors [78] relative to methane.

$$\Theta = 1 + (\omega - 0.011)\left(0.56553 - 0.86276 \ln T^+ - \frac{0.69852}{T^+}\right) \tag{10-3.22}$$

$$\Phi = [1 + (\omega - 0.011)(0.38560 - 1.1617 \ln T^+)]\frac{0.288}{Z_c} \tag{10-3.23}$$

where the values 0.011 and 0.288 represent the acentric factor and critical compressibility factor of the reference fluid methane.

3. The shape factors are then used to compute scaling parameters for temperature and volume,

$$f = \frac{T_c}{190.4}\Theta \tag{10-3.24}$$

$$h = \frac{V_c}{99.2}\Phi \tag{10-3.25}$$

with 190.4 the critical temperature of methane in kelvins and 99.2 the critical volume of methane in cm$^3$/mol.

4. The temperature scaling parameter $f$ is then used to determine the *equivalent* temperature $T_0$ to estimate the thermal conductivity and viscosity of the methane reference fluid at low pressure.

$$T_0 = \frac{T}{f} \tag{10-3.26}$$

$$\eta_0 = 10^{-7} \sum_{n=1}^{9} C_n T_0^{(n-4)/3} \tag{10-3.27}$$

$$\lambda_0 = 1944\eta_0 \dagger \tag{10-3.28}$$

where $\eta_0$ is the low-pressure methane gas viscosity at $T_0$ (N·s/m$^2$) and $\lambda_0$ is the low-pressure methane gas thermal conductivity [W/(m·K)]. The coefficients $C_n$ for the series in Eq. (10-3.27) are:

| | | |
|---|---|---|
| $C_1 = 2.90774\ E{+}6$ | $C_4 = -4.33190\ E{+}5$ | $C_7 = 4.32517\ E{+}2$ |
| $C_2 = -3.31287\ E{+}6$ | $C_5 = 7.06248\ E{+}4$ | $C_8 = -1.44591\ E{+}1$ |
| $C_3 = 1.60810\ E{+}6$ | $C_6 = -7.11662\ E{+}3$ | $C_9 = 2.03712\ E{-}1$ |

5. To obtain $\lambda^*$ and $\eta^*$ in Eq. (10-3.20),

$$\lambda^* = \lambda_0 H \tag{10-3.29}$$

$$\eta^* = \eta_0 H \frac{M'}{16.04 \times 10^{-3}} \tag{10-3.30}$$

$$H = \left(\frac{16.04 \times 10^{-3}}{M'}\right)^{1/2} f^{1/2}/h^{2/3} \tag{10-3.31}$$

Thus, to use the Ely and Hanley method to estimate the thermal conductivity of low-pressure pure gases as a function of temperature, one needs the critical properties $T_c$, $V_c$, and $Z_c$ as well as the acentric factor, molecular weight, and $C_v$ at low pressure. Some calculated values are compared with experimental data in Table 10-2, and the procedure is illustrated in Example 10-1.

**Example 10-1** Estimate the thermal conductivity of 2-methylbutane (isopentane) vapor at 1 bar and 100°C. The reported value is $2.2 \times 10^{-2}$ W/(m·K) [11].

**solution** From Appendix A, $T_c = 460.4$ K, $P_c = 33.4$ bar, $V_c = 306$ cm$^3$/mol, $Z_c = 0.267$, $\omega = 0.227$, and $M = 72.151$ g/mol $= 72.151 \times 10^{-3}$ kg/mol.

†The coefficient 1944 is obtained by considering the first term of Eq. (10-3.2) after cross-multiplying by $\eta_0 C_v/M'$, that is, the term is then $\frac{15}{4}(R/M')\eta_0$. With $R = 8.314$ J/(mol·K) and $M'$ for methane as $16.04 \times 10^{-3}$, we have $1944\eta_0$, where $\eta_0$ is in N·s/m$^2$.

**TABLE 10-2 Comparison between Calculated and Experimental Values of the Thermal Conductivity of a Pure Gas at 1 bar**

| Compound | $T$, K | λ, exp. W/(m·K) × $10^3$ | $\eta$ N·s/m² × $10^7$ | $C_v$ J/(mol·K) | Ref. | Eucken factor | Percent Error† | | | | | |
|---|---|---|---|---|---|---|---|---|---|---|---|---|
| | | | | | | | Eucken | Mod. Eucken | Stiel and Thodos | Roy and Thodos | Chung et al. | Ely and Hanley |
| Acetaldehyde‡ | 313 | 12.6 | 91.0 | 48.2 | 176 | 1.27 | 8.6 | 27 | 17 | 7.5 | 8.1 | 14 |
| $M' = 44.05 \times 10^{-3}$ | 353 | 15.9 | 103 | 52.3 | | 1.30 | 4.4 | 23 | 13 | 3.9 | 6.4 | 10 |
| | 393 | 19.4 | 115 | 56.5 | | 1.32 | 1.2 | 20 | 10 | 1.6 | 5.3 | 8.8 |
| Acetone‡ | 353 | 15.7 | 90.0 | 77.9 | 176 | 1.30 | −4.7 | 16 | 5.1 | 2.9 | 0 | −0.6 |
| $M' = 58.08 \times 10^{-3}$ | 393 | 19.4 | 100 | 84.2 | | 1.34 | −8.7 | 12 | 0.9 | 1.2 | −2.3 | −3.1 |
| | 457 | 24.7 | 114.5 | 96.1 | 17 | 1.29 | −8.4 | 13 | 1.7 | 4.5 | 0.8 | 1.6 |
| Acetonitrile‡ | 353 | 12.4 | 85 | 49.0 | 176 | 1.22 | 13 | 33 | 22 | 2.7 | 12 | −8.1 |
| $M' = 41.05 \times 10^{-3}$ | 393 | 15.0 | 95 | 52.6 | | 1.23 | 10 | 30 | 19 | 2.9 | 11 | −9.3 |
| Acetylene | 198 | 11.8 | 70.1 | 26.8 | 17 | 1.64 | 3.8 | 14 | 8.8 | 10 | 3.0 | 5.1 |
| $M' = 26.04 \times 10^{-3}$ | 273 | 18.7 | 95.5 | 33.7 | | 1.51 | 2.8 | 16 | 9.1 | 5.7 | 8.3 | 10 |
| | 373 | 29.8 | 126.1 | 40.4 | | 1.52 | −4.0 | 10 | 2.9 | 1.9 | 7.0 | 11 |
| Ammonia‡ | 213 | 16.5 | 73.2 | 25.4 | 17 | 1.51 | 15 | 26 | 20 | | 7.3 | −1.2 |
| $M' = 17.03 \times 10^{-3}$ | 273 | 21.9 | 90.6 | 26.7 | | 1.54 | 10 | 21 | 16 | | 6.6 | 1.4 |
| Benzene | 353 | 14.6 | 90 | 90.4 | 176 | 1.40 | −14 | 5.8 | −4.6 | 1.0 | −1.7 | −2.1 |
| $M' = 78.11 \times 10^{-3}$ | 433 | 22.6 | 109.5 | 114 | | 1.41 | −18 | 2.5 | −8.2 | 0 | −1.6 | −2.7 |
| $n$-Butane | 273 | 13.5 | 68.8 | 84.6 | 39 | 1.35 | −9.4 | 11 | 0.1 | −2.0 | 2.3 | 5.2 |
| $M' = 58.12 \times 10^{-3}$ | 373 | 24.6 | 94.5 | 110 | | 1.38 | −15 | 5.7 | −5.2 | −3.7 | 1.5 | 1.6 |
| Carbon dioxide | 200 | 9.51 | 101.5 | 24.7 | 17 | 1.67 | 5.3 | 15 | 9.8 | | 4.9 | 6.2 |
| $M' = 44.01 \times 10^{-3}$ | 300 | 16.7 | 149.5 | 28.9 | | 1.70 | −3.2 | 7.5 | 1.9 | | 2.8 | 5.4 |
| | 473 | 28.4 | 225.0 | 35.6 | | 1.56 | −2.2 | 11 | 4.1 | | 11 | 20 |
| | 598 | 37.9 | 272.8 | 39.5 | | 1.55 | −4.8 | 9.3 | 1.9 | | 11 | 25 |
| | 1273 | 81.7 | 465.1 | 48.8 | | 1.58 | −13 | 2.4 | −5.6 | | 7.3 | 20 |

| | | | | | | | | | | | | |
|---|---|---|---|---|---|---|---|---|---|---|---|---|
| Carbon tetrachloride | 273 | 5.95 | 91.0 | 72.4 | 17 | 1.39 | −9.4 | 9.7 | −0.4 | −9.6 | −2.4 | 2.9 |
| $M' = 153.82 \times 10^{-3}$ | 373 | 8.58 | 124.3 | 81.7 | | 1.30 | −5.4 | 15 | 4.4 | 5.6 | 6.3 | 11 |
| | 457 | 10.9 | 151.2 | 86.3 | | 1.28 | −5.3 | 16 | 4.7 | 17 | 9.4 | 15 |
| Cyclohexane | 353 | 16.3 | 83.0 | 123 | 176 | 1.34 | −14 | 7.1 | −4.2 | −0.6 | 2.6 | 0 |
| $M' = 84.16 \times 10^{-3}$ | 433 | 25.6 | 100.5 | 155 | | 1.38 | −19 | 2.3 | −9.0 | −1.0 | 0.5 | −2.0 |
| Dichlorodifluoromethane | 273 | 8.29 | 114.2 | 57.8 | 17 | 1.52 | −12 | 4.3 | −4.5 | 0.7 | −5.6 | 0.5 |
| $M' = 120.21 \times 10^{-3}$ | 373 | 13.8 | 155.1 | 69.1 | | 1.56 | −18 | −1.0 | −9.9 | −0.4 | −6.4 | −1.5 |
| | 473 | 19.4 | 193.6 | 77.1 | | 1.57 | −20 | −3.3 | −12 | 3.8 | −5.8 | −0.5 |
| Ethyl acetate‡ | 319 | 12.1 | 81.1 | 105 | 17 | 1.24 | −5.9 | 17 | 4.7 | 6.9 | 1.3 | −0.8 |
| $M' = 88.11 \times 10^{-3}$ | 373 | 16.2 | 95.5 | 121 | | 1.24 | −6.5 | 17 | 4.4 | 7.3 | 3.4 | 0.6 |
| | 457 | 23.8 | 116 | 142 | | 1.27 | −11 | 12 | −0.3 | 8.4 | 1.3 | 0 |
| Ethyl alcohol‡ | 293 | 15.0 | 84.8 | 56.5 | 17 | 1.44 | −7.7 | 9.6 | 0.4 | −2.9 | −8.8 | −25 |
| $M' = 46.07 \times 10^{-3}$ | 401 | 24.9 | 116.6 | 72.4 | | 1.36 | −7.4 | 12 | 1.8 | −1.5 | −2.5 | −18 |
| Ethylene | 273 | 17.4 | 94.2 | 32.8 | 17 | 1.58 | −0.6 | 12 | 5.4 | −0.3 | 3.3 | 8.0 |
| $M' = 28.05 \times 10^{-3}$ | 373 | 27.8 | 124.5 | 43.1 | | 1.46 | −1.3 | 14 | 6.1 | 3.0 | 8.3 | 14 |
| Ethyl ether | 273 | 13.0 | 68.4 | 104 | 176 | 1.35 | −13 | 7.9 | −3.1 | −3.2 | 1.8 | −5.4 |
| $M' = 74.12 \times 10^{-3}$ | 373 | 22.2 | 93.7 | 121 | | 1.45 | −20 | −0.7 | −11 | 2.5 | −3.6 | −9.9 |
| | 486 | 35.1 | 120.8 | 148 | | 1.45 | −23 | −2.5 | −13 | 8.2 | −2.6 | −7.4 |
| *n*-Hexane | 373 | 20.1 | 79.0 | 163 | 176 | 1.34 | −17 | 4.8 | −6.8 | −0.8 | 3.4 | −2.5 |
| $M' = 86.18 \times 10^{-3}$ | 433 | 27.2 | 92.0 | 186 | | 1.37 | −20 | 2.1 | −9.4 | −1.2 | 2.1 | −3.7 |
| Isopropyl alcohol‡ | 304 | 15.2 | 78.5 | 75.4 | 17 | 1.54 | −19 | −1.8 | −11 | −6.6 | −17 | −29 |
| $M' = 60.10 \times 10^{-3}$ | 400 | 25.0 | 105.6 | 101 | | 1.41 | −16 | 4.0 | −6.5 | −8.8 | 8.1 | −22 |
| Sulfur dioxide | 273 | 8.29 | 117 | 30.6 | 17 | 1.48 | 8.6 | 21 | 15 | | 9.0 | 11 |
| $M' = 64.06 \times 10^{-3}$ | | | | | | | | | | | | |

†Percent error = [(calc. − exp.)/exp.] × 100

‡Compounds for which $\beta$ was set equal to $(1.32)^{-1}$ as a default value in the Chung et al. method.

First we need to estimate the viscosity of 2-methylbutane. Using the Chung et al. correlation, Eq. (9-4.9),

$$\eta = \frac{40.785F_c\,(MT)^{1/2}}{V_c^{2/3}\Omega_v}$$

where $F_c = 1 - 0.275\omega$, since $\mu_r$ and $\kappa = 0$, $T^* = 1.2593\ T_r = 1.2593[(100 + 273)/460.4] = 1.020$, and $\Omega = 1.576$ from Eq. (9-4.3). Thus,

$$\eta = \frac{(40.785)\ [1 - (0.275)\ (0.227)]\ [(72.151)\ (373)]^{1/2}}{(306)^{2/3}\ (1.576)}$$

$$= 87.7\ \mu\text{P} = 8.77 \times 10^{-5}\ \text{P} = 8.77 \times 10^{-6}\ \text{N}\cdot\text{s/m}^2$$

The ideal-gas value of $C_v$ is estimated from $(C_p - R)$, where $C_p$ is determined from the polynomial constants in Appendix A,

$$C_p = -9.525 + (0.5066)\ (373) - (2.729 \times 10^{-4})\ (373)^2 + (5.723 \times 10^{-8})\ (373)^3$$

$$= 144.2\ \text{J/(mol}\cdot\text{K)}$$

$$C_v = 144.2 - 8.3 = 135.9\ \text{J/(mol}\cdot\text{K)}$$

$$M' = 72.151 \times 10^{-3}\ \text{kg/mol}$$

EUCKEN METHOD, Eq. (10-3.3)

$$\lambda = \frac{\eta C_v}{M'}\left(1 + \frac{9/4}{C_v/R}\right)$$

$$= \frac{(8.77 \times 10^{-6})\ (135.9)}{72.151 \times 10^{-3}}\left[1 + \frac{9/4}{135.9/8.314}\right]$$

$$= 1.88 \times 10^{-2}\ \text{W/(m}\cdot\text{K)}$$

$$\text{Error} = \frac{1.88 - 2.2}{2.2} \times 100 = -14\%$$

MODIFIED EUCKEN METHOD, Eq. (10-3.5)

$$\lambda = \frac{\eta C_v}{M'}\left(1.32 + \frac{1.77}{C_v/R}\right)$$

$$= \frac{(8.77 \times 10^{-6})\ (135.9)}{72.151 \times 10^{-3}}\left(1.32 + \frac{1.77}{135.9/8.314}\right)$$

$$= 2.36 \times 10^{-2}\ \text{W/(m}\cdot\text{K)}$$

$$\text{Error} = \frac{2.36 - 2.2}{2.2} \times 100 = 7.2\%$$

STIEL AND THODOS METHOD, Eq. (10-3.6)

$$\lambda = \frac{\eta C_v}{M'}\left(1.15 + \frac{2.03}{C_v/R}\right)$$

$$= \frac{(8.77 \times 10^{-6})\ (135.9)}{72.151 \times 10^{-3}}\left(1.15 + \frac{2.03}{135.9/8.314}\right)$$

$$= 2.11 \times 10^{-2}\ \text{W/(m}\cdot\text{K)}$$

$$\text{Error} = \frac{2.11 - 2.2}{2.2} \times 100 = -4.1\%$$

ROY-THODOS METHOD, Eq. (10-3.13)

$$\lambda\Gamma = (\lambda\Gamma)_{tr} + (\lambda\Gamma)_{int}$$

$\Gamma$ is defined in Eq. (10-3.12),

$$\Gamma = 210 \left( \frac{T_c M^3}{P_c^4} \right)^{1/6}$$

$$= 210 \left[ \frac{(460.4)\ (72.151)^3}{(33.4)^4} \right]^{1/6} = 478$$

With a reduced temperature, $T_r = (100 + 273)/460.8 = 0.810$, $(\lambda\Gamma)_{tr}$ is found from Eq. (10-3.14),

$$(\lambda\Gamma)_{tr} = 8.757\{\exp\ [(0.0464)\ (0.810)] - \exp\ [(-0.2412)\ (0.810)]\} = 1.89$$

To find $(\lambda\Gamma)_{int}$, $C$ must first be determined by synthesizing $n$-butane with the recommended $\Delta C$ increments.

$$n\text{-butane} = (0.73 + 2.00 + 3.18 + 3.68) = 9.59$$

Next, to form 2-methylbutane, a $1 \leftarrow 2 \rightarrow 2$ methyl substitution is required. Thus,

$$C\ (\text{2-methylbutane}) = 9.59 + 4.71 = 14.30$$

The appropriate $f(T_r)$ is given in Table 10-1 for saturated hydrocarbons,

$$f(T_r) = -0.152T_r + 1.191T_r^2 - 0.039T_r^3$$

$$= (-0.152)\ (0.810) + (1.191)\ (0.810)^2 - (0.039)\ (0.810)^3 = 0.638$$

Then,

$$(\lambda\Gamma)_{int} = Cf(T_r) = (14.30)\ (0.638) = 9.12$$

$$\lambda = \frac{1.89 + 9.12}{478} = 2.30 \times 10^{-2}\ \text{W/(m·K)}$$

$$\text{Error} = \frac{2.30 - 2.2}{2.2} \times 100 = 4.5\%$$

CHUNG ET AL. METHOD, Eq. (10-3.17)

$$\frac{\lambda M'}{\eta C_v} = \frac{3.75\Psi}{C_v/R}$$

As defined under Eq. (10-3.17),

$$\alpha = \frac{C_v}{R} - 1.5 = \frac{135.9}{8.314} - 1.5 = 14.85$$

$$\beta = 0.7862 - (0.7109)\ (0.227) + (1.3168)\ (0.227)^2 = 0.693$$

$$T_r = \frac{373}{460.4} = 0.810 \text{ and } Z = 2.0 + (10.5)\ (0.810)^2 = 8.89$$

$$\Psi = 1 + 14.85\,\frac{0.215 + (0.28288)\ (14.85) - (1.061)\ (0.693) + (0.26665)\ (8.89)}{0.6366 + (0.693)\ (8.89) + (1.061)\ (14.85)\ (0.693)}$$

$$= 6.073$$

$$\lambda = \frac{(3.75)\ (6.073)}{135.9/8.314}\ \frac{(8.77 \times 10^{-6})\ (135.9)}{72.151 \times 10^{-3}}$$

$$= 2.30 \times 10^{-2}\ \text{W/(m·K)}$$

$$\text{Error} = \frac{2.30 - 2.2}{2.2} \times 100 = 4.5\%$$

Ely and Hanley Method, Eq. (10-3.20). Following the steps outlined in the text:

1. $T_r = 373/460.4 = 0.810$; thus, $T^+ = T_r = 0.810$

2. $\Theta = 1 + (0.227 - 0.011)\left[0.56553 - 0.86276 \ln (0.810) - \dfrac{0.69852}{0.810}\right]$

$= 0.97527$

$\Phi = \{1 + (0.227 - 0.011)\,[0.38560 - 1.1627 \ln (0.810)]\}\dfrac{0.288}{0.267}$

$= 1.2248$

3. $f = \dfrac{460.4}{190.4}\, 0.97527 = 2.356$

$h = \dfrac{306}{99.2}\, 1.2248 = 3.778$

4. $T_0 = \dfrac{373}{2.356} = 158.3 \text{ K}$

$\eta_0 = 10^{-7} \sum C_n\, (158.3)^{(n-4)/3} = 6.22 \times 10^{-6} \text{ N}\cdot\text{s/m}^2$

$\lambda_0 = (1944)\,(6.22 \times 10^{-6}) = 1.21 \times 10^{-2} \text{ W/(m}\cdot\text{K)}$

5. $H = \left(\dfrac{16.04 \times 10^{-3}}{72.151 \times 10^{-3}}\right)^{1/2} \dfrac{(2.356)^{1/2}}{(3.778)^{2/3}} = 0.2984$

$\eta^* = (6.22 \times 10^{-6})\,(0.2984)\,\dfrac{72.151 \times 10^{-3}}{16.04 \times 10^{-3}} = 8.35 \times 10^{-6} \text{ N}\cdot\text{s/m}^2$

$\lambda^* = (1.21 \times 10^{-2})\,(0.2984) = 3.61 \times 10^{-3} \text{ W/(m}\cdot\text{K)}$

Then, with Eq. (10-3.20),

$$\lambda = 3.61 \times 10^{-3} + \frac{8.35 \times 10^{-6}}{72.151 \times 10^{-3}}\,(1.32)\left[135.9 - \frac{(3)\,(8.314)}{2}\right]$$

$$= 3.61 \times 10^{-3} + 1.89 \times 10^{-2} = 2.25 \times 10^{-2} \text{ W/(m}\cdot\text{K)}$$

$$\text{Error} = \frac{2.25 - 2.2}{2.2} \times 100 = 2.2\%$$

**Example 10-2** Use the Roy-Thodos method to estimate the thermal conductivity of ethyl acetate at 184°C and 1 bar. The reported value is $2.38 \times 10^{-2}$ W/(m·K) [17].

**solution** From Appendix A, $T_c = 523.2$ K, $P_c = 38.3$ bar, and $M = 88.107$ g/mol. From Eq. (10-3.12),

$$\Gamma = 210\,(T_c M^3/P_c^4)^{1/6}$$

$$= 210\left[\frac{(523.2)\,(88.107)^3}{(38.3)^4}\right]^{1/6} = 493$$

With $T = 184 + 273 = 457$ K, with $T_r = 457/523.2 = 0.873$, and using Eq. (10-3.14),

$$(\lambda\Gamma)_{tr} = 8.757\,\{\exp\,[(0.0464)\,(0.873)] - \exp\,[(-0.2412)\,(0.873)]\} = 2.02$$

To determine $C$, the synthesis plan is as follows: ethane → ethanol → methyl ethyl ether → diethyl ether → ethyl acetate. For ethane, $C = 0.73 + 2.00 = 2.73$. Converting to ethanol, $\Delta C = 4.62$. Next, to make methyl ethyl ether, $\Delta C = 2.46$ and then on to diethyl ether, $\Delta C = 4.18$. Finally, we form the ester, ethyl acetate, $\Delta C = 0.31$. Summing, $C = 2.73 + 4.62 + 2.46 + 4.18 + 0.31 = 14.30$. With Table 10-1 for esters,

$$f(T_r) = -0.082T_r + 1.045T_r^2 + 0.037T_r^3$$
$$= (-0.082)(0.873) + (1.045)(0.873)^2 + (0.037)(0.873)^3 = 0.750$$

Then, with Eq. (10-3.13),

$$\lambda_r = \lambda\Gamma = (\lambda\Gamma)_{tr} + (\lambda\Gamma)_{int} = 493\lambda = 2.02 + (14.3)(0.750)$$
$$\lambda = 2.58 \times 10^{-2}\ \mathrm{W/(m \cdot K)}$$
$$\text{Error} = \frac{2.58 - 2.38}{2.38} \times 100 = 8.4\%$$

## Discussion

Except for the Roy-Thodos method, all other methods described in this section for estimating the thermal conductivity of a pure gas at ambient pressure correlate the Eucken factor $\lambda M'/\eta C_v$ as a function of other variables such as $C_v$, $T_r$, and $\omega$. To use them, independent values of the gas viscosity and heat capacity are necessary, although the Ely-Hanley procedure estimates its own viscosity. The Roy-Thodos correlation requires only the critical temperature and pressure and employs a group contribution method to account for the effect of internal degrees of freedom. In Table 10-2, we show the percent errors found when applying all of these techniques to estimate $\lambda$ for a variety of compounds. As noted earlier, the Eucken equation (10-3.3) tends to underestimate $\lambda$, whereas the modified Eucken equation overestimates $\lambda$. The Stiel and Thodos equation yields $\lambda$ values between the two Eucken forms. All three of these relations predict that the Eucken factor should decrease with temperature, whereas, in actuality, the factor appears, in most cases, to increase slightly (see Fig. 10-2). The Chung et al. and Ely-Hanley modifications do predict the correct trend of the Eucken factor with temperature and, except for polar compounds, yield $\lambda$ values quite close to those reported experimentally. The Roy-Thodos method generally yields the smallest errors, but it is not applicable to inorganic compounds and, even for many types of organic compounds, group contributions are lacking.

It is recommended that, for *nonpolar* compounds, the Chung et al., the Ely-Hanley, or the Roy-Thodos method be used to estimate $\lambda$ for pure gases at ambient pressure. Errors vary, but, generally, they do not exceed 5 to 7 percent. For *polar* compounds, the Roy-Thodos form is recommended. Other new estimation techniques for $\lambda$ [37, 164] were found to be less accurate than the ones noted above.

## 10-4 Effect of Temperature on the Low-Pressure Thermal Conductivities of Gases

Thermal conductivities of low-pressure gases increase with temperature. The exact dependence of $\lambda$ on $T$ is difficult to judge from the $\lambda$-estimation methods in Sec. 10-3 because other temperature-dependent parameters (e.g., heat capacities and viscosities) are incorporated in the correlations. Generally, $d\lambda/dT$ ranges from $4 \times 10^{-5}$ to $1.2 \times 10^{-4}$ W/(m·K$^2$), with the more complex and polar molecules having the larger values. Several power laws relating $\lambda$ with $T$ have been proposed [30, 107], but they are not particularly accurate. Miller et al. [101] have listed polynomial constants to estimate $\lambda$ as a function of temperature for many gases, and they are shown in Table 10-3. To illustrate the trends, Fig. 10-4 has been drawn to show $\lambda$ as a function of temperature for a few selected gases.

## 10-5 Effect of Pressure on the Thermal Conductivities of Gases

The thermal conductivities of all gases increase with pressure, although the effect is relatively small at low and moderate pressures. Three pressure regions in which the effect of pressure is distinctly different are discussed below.

### Very low pressure

Below pressures of about $10^{-3}$ bar, the mean free path of the molecules is large compared to typical dimensions of a measuring cell, and there $\lambda$ is almost proportional to pressure. This region is called the *Knudsen domain*. In reported thermal conductivity data, the term *zero-pressure value* is often used; however, it refers to values extrapolated from higher pressures (above $10^{-3}$ bar) and not to measured values in the very low pressure domain.

### Low pressure

This region extends from approximately $10^{-3}$ to 10 bar and includes the domain discussed in Secs. 10-3 and 10-4. The thermal conductivity increases about 1 percent or less per bar [68, 174, 175, 176]. Such increases are often ignored in the literature, and either the 1-bar value or the "zero-pressure" extrapolated value may be referred to as the low-pressure conductivity.

**TABLE 10-3 Thermal Conductivities of Some Gases at About 1 Bar**[1]

$\lambda = A + BT + CT^2 + DT^3$; $\lambda$ in W/(m·K) and T in kelvins

| Formula | Name | A | B | C | D | λ(298 K) | Range |
|---|---|---|---|---|---|---|---|
| He | helium | 3.722E-2 | 3.896E-4 | -7.450E-8 | 1.290E-11 | 1.47E-1 | 115 to 1070 |
| Ne | neon | 9.108E-2 | 1.541E-4 | -8.396E-8 | 2.530E-11 | 4.83E-2 | 115 to 1470 |
| A | argon | 2.714E-3 | 5.540E-5 | -2.178E-8 | 5.528E-12 | 1.74E-2 | 115 to 1470 |
| $H_2$ | hydrogen | 8.099E-3 | 6.689E-4 | -4.158E-7 | 1.562E-10 | 1.75E-1 | 115 to 1470 |
| $N_2$ | nitrogen | 3.919E-4 | 9.816E-5 | -5.067E-8 | 1.504E-11 | 2.55E-2 | 115 to 1470 |
| $O_2$ | oxygen | -3.273E-4 | 9.966E-5 | -3.743E-8 | 9.732E-12 | 2.63E-2 | 115 to 1470 |
| $F_2$ | fluorine | 7.812E-4 | 8.287E-5 | 5.193E-8 | -7.441E-11 | 2.81E-2 | 145 to 795 |
| $Cl_2$ | chlorine | 1.361E-3 | 2.429E-5 | 8.794E-9 | -5.235E-12 | 9.24E-3 | 195 to 1470 |
| $Br_2$ | bromine | -6.700E-5 | 1.729E-5 | -1.256E-9 | -3.769E-13 | 4.97E-3 | 195 to 1470 |
| $I_2$ | iodine | 2.638E-4 | 1.143E-5 | -1.256E-9 | 6.281E-13 | 3.58E-3 | 195 to 1470 |
| HF | hydrogen fluoride | 3.857E-3 | 5.276E-5 | 2.261E-8 | -9.841E-13 | 2.13E-2 | 175 to 1670 |
| HCl | hydrogen chloride | -1.089E-4 | 5.306E-5 | -1.047E-8 | 6.700E-13 | 1.48E-2 | 125 to 1670 |
| HBr | hydrogen bromide | -7.915E-4 | 3.836E-5 | -1.089E-8 | 2.219E-12 | 9.73E-3 | 125 to 1670 |
| HI | hydrogen iodide | -2.152E-3 | 3.049E-5 | -9.213E-9 | 1.801E-12 | 6.16E-3 | 125 to 1670 |
| CO | carbon monoxide | 5.067E-4 | 9.125E-5 | -3.524E-8 | 8.199E-12 | 2.48E-2 | 115 to 1670 |
| $CO_2$ | carbon dioxide | -7.215E-3 | 8.015E-5 | 5.477E-9 | -1.053E-11 | 1.69E-2 | 185 to 1670 |
| $SO_2$ | sulfur dioxide | -8.086E-3 | 6.344E-5 | -1.382E-8 | 2.303E-12 | 9.65E-3 | 273 to 1670 |
| $SO_3$ | sulfur trioxide | -6.683E-3 | 7.077E-5 | -1.968E-8 | 1.256E-11 | 1.30E-2 | 175 to 1270 |
| $N_2O$ | nitrous oxide | -7.835E-3 | 8.903E-5 | -8.970E-9 | -2.668E-12 | 1.78E-2 | 175 to 1670 |
| NO | nitric oxide | 5.021E-3 | 7.194E-5 | -0.838E-9 | -3.559E-12 | 2.63E-2 | 85 to 1670 |
| $NO_2$ | nitrogen dioxide | -1.404E-2 | 1.108E-4 | -3.162E-8 | 4.485E-12 | 1.63E-2 | 300 to 1670 |
| $H_2O$ | water | 7.341E-3 | -1.013E-5 | 1.801E-7 | -9.100E-11 | 1.79E-2 | 273 to 1070 |
| $H_2O_2$ | hydrogen peroxide | -8.823E-3 | 7.106E-5 | 7.119E-9 | -6.533E-12 | 1.28E-2 | 273 to 1470 |
| $NH_3$ | ammonia | 3.811E-4 | 5.389E-5 | 1.227E-7 | -3.635E-11 | 2.64E-2 | 273 to 1670 |
| $N_2H_4$ | hydrazine | -2.257E-2 | 1.193E-4 | 8.375E-9 | -7.956E-13 | 1.37E-2 | 273 to 1670 |

**TABLE 10-3 Thermal Conductivities of Some Gases at About 1 Bar *(Continued)***

$\lambda = A + BT + CT^2 + DT^3$; $\lambda$ in W/mK and T in kelvins

| Formula | Name | A | B | C | D | λ(298 K) | Range |
|---|---|---|---|---|---|---|---|
| $CH_4$ | methane | -1.869E-3 | 8.727E-5 | 1.179E-7 | -3.614E-11 | 3.37E-2 | 273 to 1270 |
| $C_2H_6$ | ethane | -3.174E-2 | 2.201E-4 | -1.923E-7 | 1.664E-10 | 2.12E-2 | 273 to 1020 |
| $C_3H_8$ | propane | 1.858E-3 | -4.698E-6 | 2.177E-7 | -8.409E-11 | 1.76E-2 | 273 to 1270 |
| $C_2H_4$ | ethylene | -1.760E-2 | 1.200E-4 | 3.335E-8 | -1.366E-11 | 2.08E-2 | 200 to 1270 |
| $C_3H_6$ | propylene | -7.584E-3 | 6.101E-5 | 9.966E-8 | -3.840E-11 | 1.84E-2 | 175 to 1270 |
| $C_4H_8$ | 1-butene | -1.052E-2 | 5.771E-5 | 1.018E-7 | -4.271E-11 | 1.46E-2 | 175 to 1270 |
| $C_4H_8$ | isobutylene | -2.776E-3 | -2.806E-6 | 2.525E-7 | -1.281E-10 | 1.54E-2 | 273 to 1070 |
| $C_4H_6$ | 1,3-butadiene | -2.844E-2 | 1.255E-4 | 7.286E-8 | -5.109E-11 | 1.41E-2 | 273 to 1270 |
| $C_5H_8$ | isoprene | -2.363E-2 | 1.101E-4 | 5.486E-8 | -3.174E-11 | 1.32E-2 | 273 to 1270 |
| $C_6H_6$ | benzene | -8.455E-3 | 3.618E-5 | 9.799E-8 | -4.058E-11 | 9.96E-3 | 273 to 1270 |
| $C_7H_8$ | toluene | 7.596E-3 | -4.008E-5 | 2.370E-7 | -9.305E-11 | 1.42E-2 | 273 to 1270 |
| $C_8H_{10}$ | ethyl benzene | 6.030E-4 | -5.863E-6 | 2.140E-7 | -8.924E-11 | 1.55E-2 | 273 to 1270 |
| $C_8H_{10}$ | o-xylene | -5.720E-3 | 3.572E-5 | 7.454E-8 | -2.621E-11 | 1.09E-2 | 273 to 1270 |
| $C_8H_{10}$ | m-xylene | 1.320E-2 | -4.196E-5 | 1.662E-7 | -6.106E-11 | 1.39E-2 | 273 to 1270 |
| $C_8H_{10}$ | p-xylene | -8.178E-3 | 3.890E-5 | 7.580E-8 | -2.902E-11 | 9.38E-3 | 273 to 1270 |
| $C_8H_8$ | styrene | 8.752E-4 | -1.926E-6 | 1.244E-7 | -5.071E-11 | 1.00E-2 | 273 to 1270 |
| $C_9H_{12}$ | cumene | -5.590E-3 | 2.253E-5 | 1.813E-7 | -7.504E-11 | 1.52E-2 | 273 to 1270 |
| $C_{10}H_8$ | naphthalene | -9.380E-3 | 4.937E-5 | 3.811E-8 | -1.064E-11 | 8.44E-3 | 273 to 1270 |
| $C_3H_6$ | cyclopropane | -8.568E-3 | 4.079E-5 | 1.579E-7 | -6.817E-11 | 1.58E-2 | 273 to 1070 |
| $C_4H_8$ | cyclobutane | -9.795E-3 | 3.832E-5 | 1.474E-7 | -6.202E-11 | 1.31E-2 | 273 to 1070 |
| $C_5H_{10}$ | cyclopentane | -8.522E-3 | 2.475E-5 | 1.621E-7 | -6.914E-11 | 1.14E-2 | 273 to 1070 |
| $C_6H_{12}$ | cyclohexane | -8.614E-3 | 1.863E-5 | 1.704E-7 | -7.249E-11 | 1.02E-2 | 273 to 1070 |
| $CH_3OH$ | methanol | -7.797E-3 | 4.167E-5 | 1.214E-7 | -5.184E-11 | 1.40E-2 | 273 to 1270 |
| $C_2H_5OH$ | ethanol | -7.797E-3 | 4.167E-5 | 1.214E-7 | -5.184E-11 | 1.40E-2 | 273 to 1270 |
| $C_3H_7OH$ | n-propanol | -7.931E-3 | 3.987E-5 | 1.193E-7 | -5.021E-11 | 1.32E-2 | 273 to 1270 |
| $C_4H_9OH$ | n-butanol | -7.772E-3 | 3.564E-5 | 1.206E-7 | -4.992E-11 | 1.22E-2 | 273 to 1270 |

$$\lambda = A + BT + CT^2 + DT^3; \quad \lambda \text{ in W/mK and T in kelvins}$$

| Formula | Name | A | B | C | D | λ(298 K) | Range |
|---|---|---|---|---|---|---|---|
| $C_2H_4O$ | ethylene oxide | -1.459E-2 | 5.427E-5 | 1.520E-7 | -7.647E-11 | 1.31E-2 | 273 to 1270 |
| $C_3H_7O$ | propylene oxide | -8.204E-3 | 3.664E-5 | 1.072E-7 | -4.569E-11 | 1.10E-2 | 273 to 1270 |
| $C_4H_9O$ | butylene oxide | -9.150E-3 | 3.245E-5 | 9.883E-8 | -4.464E-11 | 8.12E-3 | 273 to 1270 |
| $C_6H_5OH$ | phenol | -1.335E-2 | 6.390E-5 | 7.286E-8 | -1.843E-11 | 1.17E-2 | 273 to 1270 |
| $C_6H_5NH_2$ | aniline | -1.105E-2 | 4.979E-5 | 6.491E-8 | -1.801E-11 | 9.07E-3 | 273 to 1270 |
| $CH_3Cl$ | methyl chloride | -3.191E-3 | 1.579E-5 | 1.181E-7 | -5.406E-11 | 1.06E-2 | 300 to 1270 |
| $CH_2Cl_2$ | methylene chloride | 1.177E-3 | -4.188E-6 | 9.673E-8 | -4.276E-11 | 7.39E-3 | 300 to 1270 |
| $CHCl_3$ | chloroform | -2.400E-3 | 2.634E-5 | 2.472E-8 | -1.404E-11 | 7.27E-3 | 300 to 1270 |
| $CCl_4$ | carbon tetrachloride | -1.742E-4 | 1.703E-5 | 2.561E-8 | -1.493E-11 | 6.78E-3 | 300 to 1270 |
| $C_6H_5Cl$ | chlorobenzene | -6.394E-3 | 2.634E-5 | 7.328E-8 | -2.316E-11 | 7.35E-3 | 273 to 1270 |
| $C_4H_5Cl$ | chloroprene | -1.142E-2 | 4.560E-5 | 1.189E-7 | -5.507E-11 | 1.13E-2 | 273 to 1270 |


1. Miller, J.W. Jr., P.N. Shah, and C.L. Yaws, Chem. Eng. 83 (25), 153 (1976).

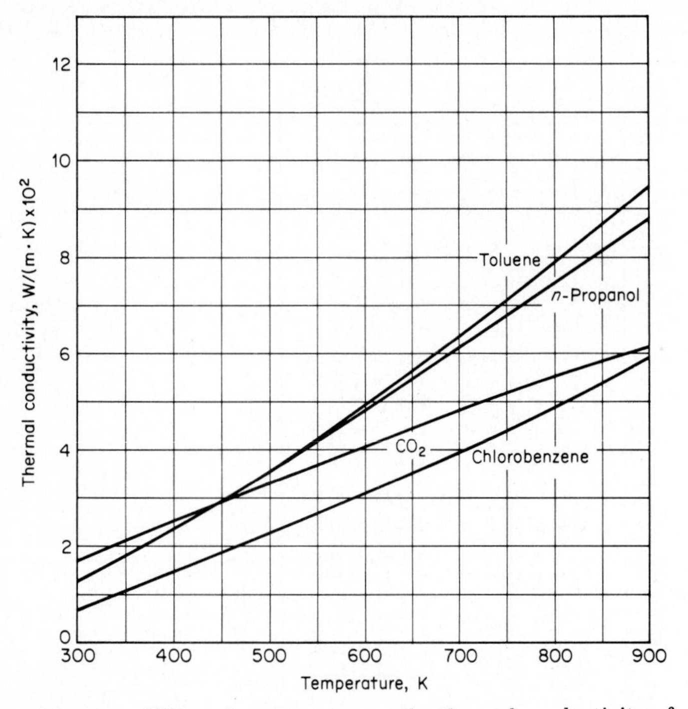


**Figure 10-4** Effect of temperature on the thermal conductivity of some low-pressure gases.

### High pressure

In Fig. 10-5 we show the thermal conductivity of propane over a wide range of pressures and temperatures [56]. The high-pressure gas domain would be represented by the curves on the right-hand side of the graph above the critical temperature (369.8 K). Increasing pressure raises the thermal conductivity, with the region around the critical point being particularly sensitive. Increasing temperature at low pressures results in a larger thermal conductivity, but at high pressure the opposite effect is noted. Similar behavior is shown for the region below $T_c$, where $\lambda$ for liquids decreases with temperature whereas, for gases (see Sec. 10-4), there is an increase of $\lambda$ with $T$. Pressure effects (except at very high pressures) are small below $T_c$. Not shown in Fig. 10-5 is the unusual behavior of $\lambda$ near the critical point. In this region, the thermal conductivity is quite sensitive to both temperature and pressure [9]. Figure 10-6 shows a plot of $\lambda$ for $CO_2$ near the critical point [50]. The explanation for this phenomenon is not clear; it may be due to a transition molecular ordering [73] or to small-scale circulation effects resulting from the migration of clusters of molecules [79]. In any case, when generalized charts of the effect of

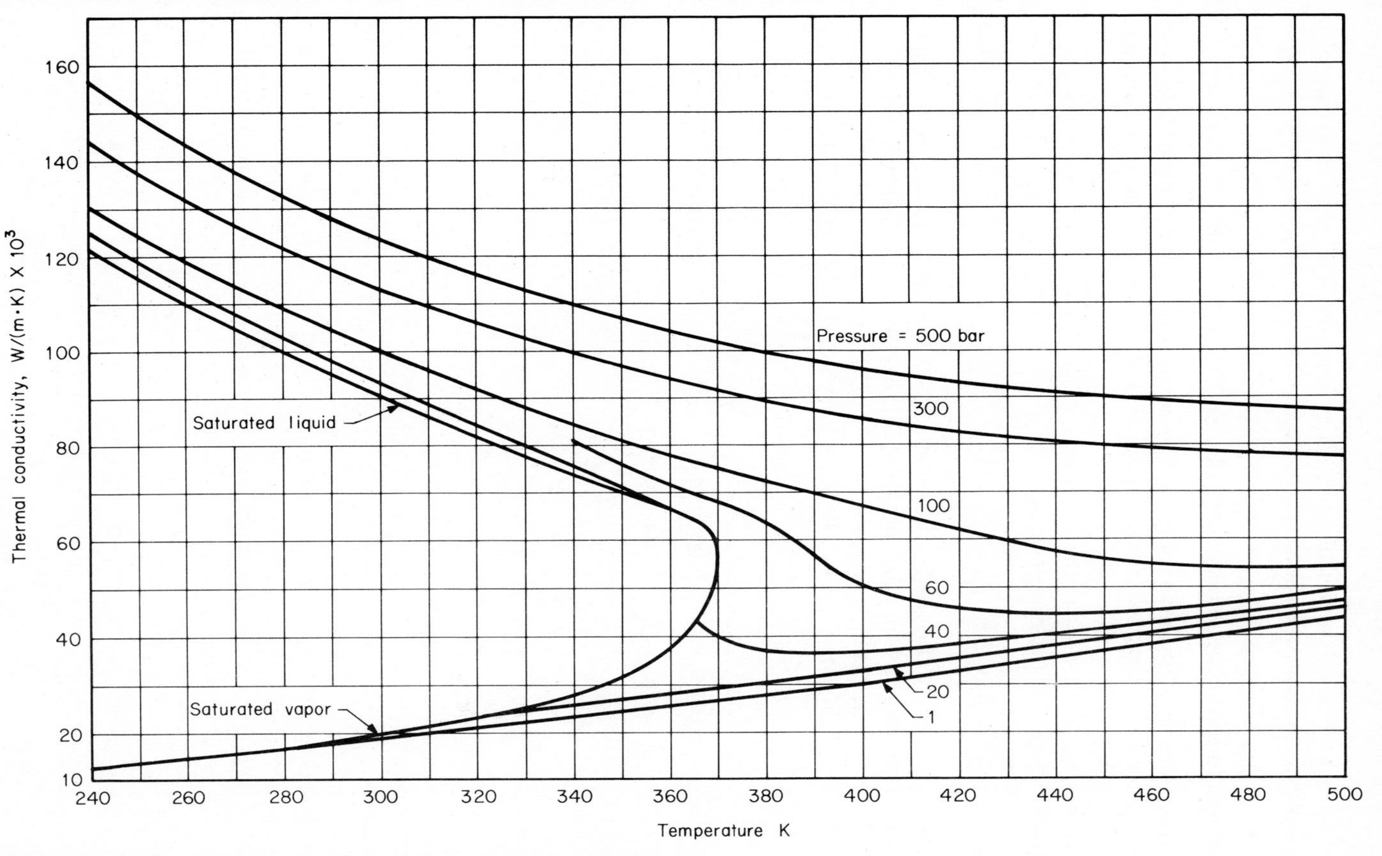


**Figure 10-5** Thermal conductivity of propane. *(Data from Ref. 56.)*

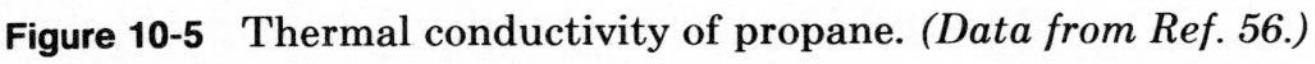

pressure on λ are drawn, these irregularities around $T_c$ and $P_c$ are usually smoothed out and not shown.

### Excess thermal conductivity correlations

Many investigators have adopted the suggestion of Vargaftik [169, 170] that the excess thermal conductivity, $\lambda - \lambda°$, be correlated as a function of the *PVT* properties of the system in a corresponding states manner. (Here $\lambda°$ is the low-pressure thermal conductivity of the gas at the same temperature.) In its simplest form,

$$\lambda - \lambda° = f(\rho) \tag{10-5.1}$$

where $\rho$ is the fluid density. The correlation has been shown to be applicable to ammonia [49, 134], ethane [20], *n*-butane [22, 73], nitrous oxide [133], ethylene [122], methane [21, 91, 121], diatomic gases [103, 152], hydrogen [151], inert gases [120], and carbon dioxide [69]. Temperature

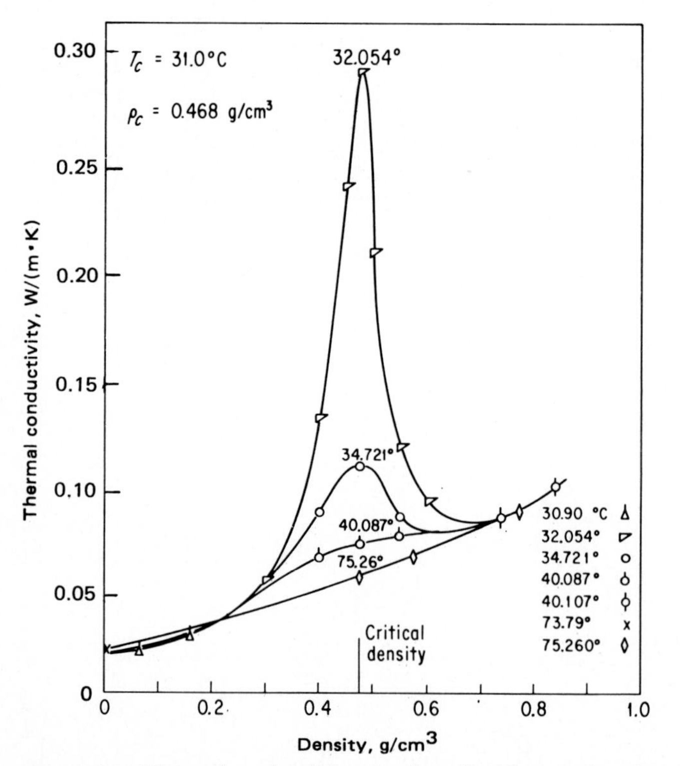


**Figure 10-6** Thermal conductivity of carbon dioxide near the critical point. *(Data from Ref. 50.)*

and pressure do not enter explicitly, but their effects are included in the parameters $\lambda°$ (temperature only) and $\rho$.

Stiel and Thodos [157] have generalized Eq. (10-5.2) by assuming that $f(\rho)$ depends only on $T_c$, $P_c$, $V_c$, $M$, and $\rho$. By dimensional analysis they obtain a correlation between $\lambda - \lambda°$, $Z_c$, $\Gamma$, and $\rho$, where $\Gamma$ was defined in Eq. (10-3.12). From data on 20 nonpolar substances, including inert gases, diatomic gases, $CO_2$, and hydrocarbons, they established the approximate analytical expressions:

$$(\lambda - \lambda°)\Gamma Z_c^5 = 1.22 \times 10^{-2} [\exp(0.535\rho_r) - 1] \qquad \rho_r < 0.5 \tag{10-5.2}$$

$$(\lambda - \lambda°)\Gamma Z_c^5 = 1.14 \times 10^{-2} [\exp(0.67\rho_r) - 1.069] \qquad 0.5 < \rho_r < 2.0 \tag{10-5.3}$$

$$(\lambda - \lambda°)\Gamma Z_c^5 = 2.60 \times 10^{-3} [\exp(1.155\rho_r) + 2.016)] \qquad 2.0 < \rho_r < 2.8 \tag{10-5.4}$$

where $\lambda$ is in W/(m·K), $Z_c$ is the critical compressibility, and $\rho_r$ is the reduced density $\rho/\rho_c = V_c/V$.

Equations (10-5.2) to (10-5.4) should not be used for polar substances or for hydrogen or helium. The general accuracy is in doubt, and errors of $\pm 10$ to 20 percent are possible. The method is illustrated in Example 10-3.

**Example 10-3** Estimate the thermal conductivity of nitrous oxide at 105°C and 138 bar. At this temperature and pressure, the experimental value is reported to be $3.90 \times 10^{-2}$ W/(m·K) [133]. At 1 bar and 105°C, $\lambda° = 2.34 \times 10^{-2}$ W/(m·K) [133]. From Appendix A, $T_c = 309.6$ K, $P_c = 72.4$ bar, $V_c = 97.4$ cm³/mol, $Z_c = 0.274$, and $M = 44.013$ g/mol. At 105°C and 138 bar, $Z$ for $N_2O$ is 0.63 [32].

**solution** With Eq. (10-3.12),

$$\Gamma = 210 \left( \frac{T_c M^3}{P_c^4} \right)^{1/6}$$

$$= 210 \left[ \frac{(309.6)(44.013)^3}{(72.4)^4} \right]^{1/6} = 209$$

$$V = \frac{ZRT}{P}$$

$$= \frac{(0.63)(8.314)(378)}{138 \times 10^5} \times 10^6 = 144 \text{ cm}^3/\text{mol}$$

$$\rho_r = \frac{V_c}{V}$$

$$= \frac{97.4}{144} = 0.676$$

Then, with Eq. (10-5.3),

$$(\lambda - \lambda°)(209)(0.274)^5 = (1.14 \times 10^{-2}) \{\exp[(0.67)(0.676)] - 1.069\}$$

$$\lambda - \lambda° = 1.78 \times 10^{-2} \text{ W/(m·K)}$$

$$\lambda = (2.34 + 1.78) \times 10^{-2} = 4.12 \times 10^{-2} \text{ W/(m·K)}$$

$$\text{Error} = \frac{4.12 - 3.90}{3.90} \times 100 = 5.6\%$$

**Method of Chung et al. [27, 28]**

The low-pressure estimation procedure for pure component thermal conductivities developed by these authors, and given in Eq. (10-3.17), is modified to treat materials at high pressures (or densities).

$$\lambda = \frac{31.2\eta^\circ\Psi}{M'}(G_2^{-1} + B_6 y) + qB_7 y^2 T_r^{1/2} G_2 \tag{10-5.5}$$

where $\lambda$ = thermal conductivity, W/(m·K)
$\eta^\circ$ = *low-pressure* gas viscosity, N·s/m²
$M'$ = molecular weight, kg/mol
$\Psi = f(C_v, \omega, T_r)$ [as defined under Eq. (10-3.17)]
$q = 3.586 \times 10^{-3}\,(Tc/M')^{1/2}/V_c^{2/3}$
$T$ = temperature, K
$T_c$ = critical temperature, K
$T_r$ = reduced temperature, $T/T_c$
$V_c$ = critical volume, cm³/mol

$$y = \frac{V_c}{6V} \tag{10-5.6}$$

$$G_1 = \frac{1 - 0.5y}{(1 - y)^3} \tag{10-5.7}$$

$$G_2 = \frac{(B_1/y)\,[1 - \exp(-B_4 y)] + B_2 G_1 \exp(B_5 y) + B_3 G_1}{B_1 B_4 + B_2 + B_3} \tag{10-5.8}$$

The coefficients $B_1$ to $B_7$ are functions of the acentric factor $\omega$, the reduced dipole moment $\mu_r$ [as defined in Eq. (9-4.11)], and the association factor $\kappa$. Some values of $\kappa$ are shown in Table 9-1.

$$B_i = a_i + b_i\omega + c_i\mu_r^4 + d_i\kappa \tag{10-5.9}$$

with $a_i$, $b_i$, $c_i$, and $d_i$ given in Table 10-4.

The relation for high-pressure thermal conductivities is quite similar to

**TABLE 10-4 Values of $B_i$ in Eq. (10-5.9)**
$B_i = a_i + b_i\omega + c_i\mu_r^4 + d_i\kappa$

| $i$ | $a_i$ | $b_i$ | $c_i$ | $d_i$ |
|---|---|---|---|---|
| 1 | 2.4166 $E$+0 | 7.4824 $E$−1 | −9.1858 $E$−1 | 1.2172 $E$+2 |
| 2 | −5.0924 $E$−1 | −1.5094 $E$+0 | −4.9991 $E$+1 | 6.9983 $E$+1 |
| 3 | 6.6107 $E$+0 | 5.6207 $E$+0 | 6.4760 $E$+1 | 2.7039 $E$+1 |
| 4 | 1.4543 $E$+1 | −8.9139 $E$+0 | −5.6379 $E$+0 | 7.4344 $E$+1 |
| 5 | 7.9274 $E$−1 | 8.2019 $E$−1 | −6.9369 $E$−1 | 6.3173 $E$+0 |
| 6 | −5.8634 $E$+0 | 1.2801 $E$+1 | 9.5893 $E$+0 | 6.5529 $E$+1 |
| 7 | 9.1089 $E$+1 | 1.2811 $E$+2 | −5.4217 $E$+1 | 5.2381 $E$+2 |

the Chung et al. form for high-pressure viscosities [Eqs. (9-7.16) through (9-6.21)].

In Eq. (10-5.5), if $V$ becomes large, $y$ then approaches zero. In such a case, both $G_1$ and $G_2$ are essentially unity and Eq. (10-5.5) reduces to Eq. (10-3.17), the relation for $\lambda$ at low pressures. To use Eq. (10-5.5), it should be noted that the viscosity $\eta°$ is for the *low-pressure,* pure gas. Experimental values may be employed or $\eta°$ can be estimated by the techniques given in Sec. 9-4. The dimensions of $\eta°$ are $N \cdot s/m^2$. The conversion from other viscosity units is $1\ N \cdot s/m^2 = 10\ P = 10^7\ \mu P$.

Chung et al. tested Eq. (10-5.5) with data from a large range of hydrocarbon types and from data for simple gases. Deviations over a wide pressure range were usually less than 5 to 8 percent. For highly polar materials, the correlation for $\beta$ as given under Eq. (10-3.17) is not accurate and, at present no predictive technique to apply such compounds is available. [See the discussion dealing with polar materials under Eq. (10-3.17).]

The high-pressure Chung et al. method is illustrated in Example 10-4.

### Method of Ely and Hanley [40, 51]

This estimation procedure was introduced in Sec. 10-3 for low-pressure, pure gases [Eq. (10-3.20)]. We now extend the treatment to cover pure components at high densities where Eq. (10-3.20) is modified to:

$$\lambda = \lambda^{**} + \frac{\eta^*}{M'}(1.32)\left(C_v - \frac{3R}{2}\right) \tag{10-5.10}$$

$\lambda^{**}$ is defined below, and the second term in Eq. (10-5.10) is identical with that in Eq. (10-3.20) and is determined by the procedure outlined under that equation. It is not a function of the system density.

We adopt a calculational procedure similar to that shown in Sec. 10-3. The data required to estimate $\lambda$ are $T_c$, $V_c$, $Z_c$, $\omega$, $M'$, $C_v$ (at $T$ and low pressure), $T$, and $V$. The dimensions of temperature and volume used here are kelvins and $cm^3/mol$. $Z_c$ and $\omega$ have no dimensions, and $M'$ and $C_v$ are defined under Eq. (10-3.20).

1. Determine the reduced temperature and volume as $T_r = T/T_c$, $V_r = V/V_c$. Then define parameters $T^+$ and $V^+$ such that

$$\begin{aligned} T^+ &= T_r \quad && T_r \le 2 \\ &= 2 && T_r > 2 \end{aligned} \tag{10-5.11}$$

$$\begin{aligned} V^+ &= V_r \quad && 0.5 < V_r < 2 \\ &= 0.5 && V_r \le 0.5 \\ &= 2 && V_r \ge 2 \end{aligned} \tag{10-5.12}$$

2. Calculate *shape* factors for the material relative to methane.

$$\Theta = 1 + (\omega - 0.011)\,[0.09057 - 0.86276 \ln T^{+} + \left(0.31664 - \frac{0.46568}{T^{+}}\right)(V^{+} - 0.5)] \qquad (10\text{-}5.13)$$

$$\Phi = \{1 + (\omega - 0.011)\,[0.39490\,(V^{+} - 1.02355) - (0.93281)\,(V^{+} - 0.75464) \ln T^{+}]\}\,\frac{0.288}{Z_c} \qquad (10\text{-}5.14)$$

Note that, at low pressure, where $V$ is large and, therefore, $V^{+} = 2$, Eqs. (10-5.13) and (10-5.14) reduce to Eqs. (10-3.22) and (10-3.23).

3. Find the compound shape factors *relative* to methane,

$$f = \frac{T_c}{190.4}\,\Theta \qquad (10\text{-}5.15)$$

$$h = \frac{V_c}{99.2}\,\Phi \qquad (10\text{-}5.16)$$

and use them to estimate the equivalent temperature $T_0$ and density $\rho_0$ to determine the thermal conductivity and viscosity of the methane chosen as the *reference fluid.*

$$T_0 = \frac{T}{f} \qquad (10\text{-}5.17)$$

$$\rho_0 = \frac{16.04}{V}\,h \qquad (10\text{-}5.18)$$

4. We now determine $\eta_0$, the low-pressure viscosity of the reference fluid (methane), with $T_0$ from Eq. (10-5.17) and Eq. (10-3.27), where the constants $C_n$ were given below this latter equation.

5. With $\eta_0$, the first component of the thermal conductivity is found from Eq. (10-5.19).

$$\lambda\,(1) = 1944\eta_0 \qquad (10\text{-}5.19)$$

which is identical with the form of Eq. (10-3.28).

6. The second component for $\lambda$ is calculated by

$$\lambda\,(2) = \left\{b_1 + b_2\left[b_3 - \ln\left(\frac{T_0}{b_4}\right)\right]^2\right\}\rho_0 \qquad (10\text{-}5.20)$$

where $T_0$ and $\rho_0$ are given in Eqs. (10-5.17) and (10-5.18) and the coefficients $b_n$ are:

$b_1 = -2.5276\,E{-4}$ $\quad b_2 = 3.3433\,E{-4}$ $\quad b_3 = 1.12$ $\quad b_4 = 1.680\,E{+2}$

7. The third component for $\lambda$ is given as

$$\lambda\,(3) = \exp\left(a_1 + \frac{a_2}{T_0}\right)\left\{\exp\left[\left(a_3 + \frac{a_4}{T_0^{3/2}}\right)\rho_0^{0.1} + \left(\frac{\rho_0}{0.1617} - 1\right)\rho_0^{1/2}\left(a_5 + \frac{a_6}{T_0} + \frac{a_7}{T_0^2}\right)\right] - 1.0\right\} \times 10^{-3} \tag{10-5.21}$$

Again $\rho_0$ and $T_0$ are from Eqs. (10-5.17) and (10-5.28) and the constants $a_n$ are:

$a_1 = -7.19771$ $a_2 = 85.67822$ $a_3 = 12.47183$ $a_4 = -984.6252$

$a_5 = 0.3594685$ $a_6 = 69.79841$ $a_7 = -872.8833$

8. Finally, to determine $\lambda^{**}$ in Eq. (10-5.10),

$$\lambda^{**} = [\lambda\,(1) + \lambda\,(2) + \lambda\,(3)]H \tag{10-5.22}$$

$$H = \frac{(16.04 \times 10^{-3}/M')^{1/2} f^{1/2}}{h^{2/3}} \tag{10-5.23}$$

For a low-pressure gas, $\lambda$ (2) and $\lambda$ (3) approach zero and the method reduces to the form given in Sec. 10-3.

For somewhat higher accuracy, Ely and Hanley suggest an additional correction factor in Eq. (10-5.22) which they call $X$. This parameter multiplies the right-hand side of Eq. (10-5.22).

$$X = \left\{\left[1 - \frac{T}{f}\left(\frac{\partial f}{\partial T}\right)_v\right]\frac{0.288}{Z_c}\right\}^{3/2} \tag{10-5.24}$$

With $f$ from Eq. (10-5.15) and $\Theta$ from Eq. (10-5.13),

$$\left(\frac{\partial f}{\partial T}\right)_v = \frac{T_c}{190.4}\left(\frac{\partial \Theta}{\partial T}\right)_v \tag{10-5.25}$$

$$\left(\frac{\partial \Theta}{\partial T}\right)_v = (\omega - 0.011)\left\{-0.86276\left(\frac{\partial \ln T^+}{\partial T}\right)_v - (V^+ - 0.5)\,(0.46568)\left[\frac{\partial(1/T^+)}{\partial T}\right]_v\right\} \tag{10-5.26}$$

If $T^+$ = constant = 2, by Eq. (10-5.11), then $(\partial\Theta/\partial T)_v = 0$. If $T^+ = T_r = T/T_c$, then

$$\left(\frac{\partial \Theta}{\partial T}\right)_v = (\omega - 0.011)\left[\frac{-0.86276}{T} + (V^+ - 0.5)\frac{0.46568\,T_c}{T^2}\right] \tag{10-5.27}$$

With the value of $(\partial\Theta/\partial T)_v$ either as zero or from Eqs. (10-5.27) and (10-5.25), the correction factor $X$ may be estimated. It is a number of the order of magnitude of unity.

Ely and Hanley tested their method on a large number of hydrocarbons up to densities where the materials were liquids and at temperatures from near the freezing point to above the critical temperature. They report a maximum error of 15 percent with more usual errors in the 3 to 8 percent range. The method does not allow for anomalies in the critical region, and its accuracy for nonhydrocarbons is in doubt. The procedure is illustrated in Example 10-4.

**Example 10-4** Estimate the thermal conductivity of propylene at 473 K and 150 bar by using the methods of (*a*) Chung et al. and (*b*) Ely and Hanley. Under these conditions, Vargaftik [171] reports $\lambda = 6.64 \times 10^{-2}$ W/(m·K) and $V = 172.1$ cm$^3$/mol. Also, the same author lists the low-pressure viscosity and thermal conductivity of propylene at 473 K as $\eta° = 134 \times 10^{-7}$ N·s/m$^2$ and $\lambda° = 3.89 \times 10^{-2}$ W/(m·K).

**solution** For both estimation techniques, we need certain parameters for propylene. From Appendix A:

$$T_c = 364.9 \text{ K}$$
$$P_c = 46.0 \text{ bar}$$
$$V_c = 181 \text{ cm}^3/\text{mol}$$
$$Z_c = 0.274$$
$$\omega = 0.144$$
$$M = 42.081 \text{ g/mol}$$
$$M' = 0.042081 \text{ kg/mol}$$
$$\mu = 0.4 \text{ debye}$$

Also, since propylene is nonpolar, the association factor in Chung et al.'s method $\kappa = 0$.

The low-pressure heat capacity at constant pressure at 473 K is found from the equation and polynomial constants shown in Appendix A as 91.01 J/(mol·K). Thus

$$C_v = C_p - R = 91.01 - 8.31 = 82.70 \text{ J/(mol·K)}.$$

METHOD OF CHUNG ET AL. With the definition of $\Psi$ given under Eq. (10-3.17), where

$$\alpha = \frac{C_v}{R} - \frac{3}{2} = \frac{82.70}{8.314} - \frac{3}{2} = 8.447$$

$$\beta = 0.7862 - 0.7109\omega + 1.3168\omega^2$$
$$= 0.7862 - (0.7109)(0.144) + (1.3168)(0.144)^2 = 0.7111$$

$$T_r = \frac{T}{T_c} = \frac{473}{364.9} = 1.296$$

$$Z = 2.0 + 10.5T_r^2 = 2.0 + (10.5)(1.296)^2 = 19.64$$

then

$$\Psi = 1 + 8.447 \frac{0.215 + (0.28288)(8.447) - (1.061)(0.7111) + (0.26665)(19.64)}{0.6366 + (0.7111)(19.64) + (1.061)(8.447)(0.7111)}$$

$$= 3.854$$

From Eq. (10-5.6)

$$y = \frac{\rho V_c}{6} = \frac{V_c}{6V} = \frac{181}{(6)(172.1)} = 0.1753$$

The values of $B_i$ are found from Table 10-4, where $\omega = 0.144$ and $\kappa = 0$, and with Eq. (9-4.11),

$$\mu_r^4 = \left\{\frac{(131.3)(0.4)}{[(181)(364.9)]^{1/2}}\right\}^4 = 1.74 \times 10^{-3}$$

As an example,

$$B_1 = 2.4166 + (7.4824 \times 10^{-1})(0.144) - (9.1858 \times 10^{-1})(1.74 \times 10^{-3}) = 2.5227$$

$$B_2 = -8.1358 \times 10^{-1} \qquad B_3 = 7.5328 \qquad B_4 = 1.3250 \times 10^1$$

$$B_5 = 9.0964 \times 10^{-1}$$

$$B_6 = -4.0034 \qquad B_7 = 1.0944 \times 10^2$$

With Eqs. (10-5.8) and (10-5.9),

$$G_1 = \frac{1 - (0.5)(0.1753)}{(1 - 0.1753)^3} = 1.627$$

For $G_2$,

$$\frac{B_1}{y}[1 - \exp(-B_4 y)] = \frac{2.5227}{0.1753}\{1 - \exp[-(13.250)(0.1753)]\} = 12.98$$

$$B_2 G_1 \exp(B_5 y) = (-0.81358)(1.627)\exp[(0.90964)(0.1753)] = -1.553$$

$$B_3 G_1 = (7.5328)(1.627) = 12.26$$

$$B_1 B_4 + B_2 + B_3 = (2.5227)(13.250) - 0.81358 + 7.5328 = 40.145$$

Thus

$$G_2 = \frac{(12.98 - 1.553 + 12.26)}{40.145} = 0.5900$$

and

$$q = \frac{3.586 \times 10^{-3}(T_c/M')^{1/2}}{V_c^{2/3}}$$

$$= \frac{3.586 \times 10^{-3}(364.9/0.042081)^{1/2}}{(181)^{2/3}} = 1.044 \times 10^{-2}$$

With Eq. (10-5.5),

$$\lambda = \frac{(31.2)(134 \times 10^{-7})(3.854)}{0.042081}[(0.5900)^{-1} - (4.0034)(0.1753)]$$

$$+ (1.044 \times 10^{-2})(109.44)(0.1753)^2(1.296)^{1/2}(0.5900)$$

$$= (3.829 \times 10^{-2}(0.9931) + 2.36 \times 10^{-2} = 6.16 \times 10^{-2}\ \mathrm{W/(m \cdot K)}$$

$$\text{Error} = \frac{6.16 - 6.64}{6.64} \times 100 = -7\%$$

Note that the first term in the final result ($3.829 \times 10^{-2}$) would represent the estimated value of $\lambda$ at low pressure. This result is in good agreement with the reported value of $3.89 \times 10^{-2}$ W/(m·K) [171].

METHOD OF ELY AND HANLEY. Following the steps outlined in the text, with $T_r = 473/364.9 = 1.296$ and $V_r = 172.1/181 = 0.951$, then, from Eqs. (10-5.11) and (10-5.12), $T^+ = T_r = 1.296$, $V^+ = V_r = 0.951$. With Eqs. (10-5.13) and (10-5.14),

$$\Theta = 1 + (0.144 - 0.011)\{0.09057 - 0.86276 \ln(1.296) + \left[0.31664 - \frac{0.46568}{1.296}\right](0.951 - 0.5)\}$$
$$= 0.9797$$

$$\Phi = \{1 + (0.144 - 0.011)[0.39490(0.951 - 1.02355) - 0.93281(0.951 - 0.75464)\ln(1.296)]\}\frac{0.288}{0.274}$$
$$= 1.040$$

Then, with Eqs. (10-5.15) through (10-5.18)

$$f = \frac{364.9}{190.4}(0.9797) = 1.876$$

$$h = \frac{181}{99.2}(1.040) = 1.898$$

$$T_0 = \frac{473}{1.876} = 252.1 \text{ K}$$

$$\rho_0 = \frac{16.04}{172.1}(1.898) = 0.1769$$

With Eq. (10-3.27) for $T_0 = 252.1$ K, $\eta_0 = 96.35 \times 10^{-7}$ N·s/m². By Eq. (10-5.19),

$$\lambda(1) = (1944)(96.35 \times 10^{-7}) = 1.873 \times 10^{-2}$$

Then, with Eq. (10-5.20)

$$\lambda(2) = \left\{-2.5276 \times 10^{-4} + 3.3433 \times 10^{-4}\left[1.12 - \ln\left(\frac{252.1}{168.0}\right)\right]^2\right\}(0.1769)$$
$$= 9.299 \times 10^{-5}$$

and, from Eq. (10-5.21)

$$\lambda(3) = \left[\exp\left(-7.19771 + \frac{85.67822}{252.1}\right)\right] \times \left\{\exp\left[\left(12.47183 - \frac{984.6252}{252.1^{3/2}}\right)(0.1769)^{0.1} + \left(\frac{0.1769}{0.1617} - 1\right)(0.1769)^{1/2}\left(0.3594685 + \frac{69.79841}{252.1} - \frac{872.8833}{252.1^2}\right)\right] - 1.0\right\} \times 10^{-3}$$
$$= 3.144 \times 10^{-2}$$

Then, with Eqs. (10-5.22) and (10-5.23),

$$H = \frac{(16.04 \times 10^{-3}/0.042081)^{1/2}(1.876)^{1/2}}{(1.898)^{2/3}} = 0.5516$$

$$\lambda^{**} = (1.873 \times 10^{-2} + 9.30 \times 10^{-5} + 3.144 \times 10^{-2})(0.5516) = 2.77 \times 10^{-2} \text{ W/(m·K)}$$

To obtain the second term in Eq. (10-5.10), we follow the same procedure as shown under Eq. (10-3.20). $T_r = 473/364.9 = 1.296$, so $T^+ = T_r = 1.296$. With

Eqs. (10-3.22) and (10-3.23), $\Theta = 0.9738$ and $\Phi = 1.063$. [Note that these are somewhat different from the $\Theta$ and $\Phi$ values calculated from Eqs. (10-5.13) and (10-5.14) because no volume terms are included.] Then, with Eqs. (10-3.24) and (10-3.25), $f = 1.866$ and $h = 1.940$. Continuing, with Eqs. (10-3.26) and (10-3.27), $T_0 = 253.5$ K and $\eta_0 = 9.68 \times 10^{-6}$ N·s/m². With Eq. (10-3.31), $H = 0.542$ and, by Eq. (10-3.30), $\eta^* = 137 \times 10^{-7}$ N·s/m². Note that this term should represent the viscosity of pure propylene at low pressure and at 473 K; it is in agreement with the reported value of $134 \times 10^{-7}$ N·s/m². Returning to Eq. (10-5.10),

$$\lambda = 2.77 \times 10^{-2} + \left(\frac{137 \times 10^{-7}}{0.042081}\right)(1.32)\,[82.70 - (3)(8.314)/2]$$

$$= 2.77 \times 10^{-2} + 3.02 \times 10^{-2} = 5.79 \times 10^{-2}\ \text{W/(m·K)}$$

If the correction term Eq. (10-5.24) is used, by Eqs. (10-5.26) and (10-5.27),

$$\left(\frac{\partial\Theta}{\partial T}\right)_v = (0.144 - 0.011)\left[-\frac{0.86276}{473} + (0.951 - 0.5)\frac{(0.46568)\,(364.9)}{(473)^2}\right] = 10^{-4}$$

$$\left(\frac{\partial f}{\partial T}\right)_v = \frac{364.9}{190.4} \times 10^{-4} = 1.91 \times 10^{-4}$$

$$X = \left\{\left[1 - \left(\frac{473}{1.876}\right)(1.91 \times 10^{-4})\right]\frac{0.288}{0.274}\right\}^{3/2} = 1.001$$

The corrected value of $\lambda$ is then

$$\lambda = (2.77 \times 10^{-2})\,(1.001) + 3.02 \times 10^{-2} = 5.80 \times 10^{-2}\ \text{W/(m·K)}$$

$$\text{Error} = \frac{5.80 - 6.64}{6.64} \times 100 = -12\%$$

## Discussion

Three methods for estimating the thermal conductivity of pure materials in the dense gas region were presented. All use the fluid density rather than pressure as a system variable. The low-density thermal conductivity is required in the Stiel and Thodos method [Eqs. (10.5.2) to (10-5.4)], but it is calculated as a part of the procedure in the Chung et al. [Eq. (10-5.5)]and Ely and Hanley [Eq. (10-5.10)] methods. None of the techniques are applicable for polar gases, and even for nonpolar materials, errors can be large. The Chung et al. and Ely and Hanley procedures are reported to be applicable over a wide density domain even into the liquid phase. No one of the methods appears to have a clear superiority over the others.

A number of alternative estimation methods were considered [26, 29, 34, 87, 112, 113, 124, 147, 165], but their accuracy or generality were significantly less than those of the three described. Riazi and Faghri [131] reduced the thermal conductivity by a term they refer to as the thermal conductivity at the critical point $\lambda_c$. They then relate $\lambda/\lambda_c = f(T_r, P_r, \omega)$. Unfortunately, values of $\lambda_c$ are available for only a few materials, and it is not possible to estimate this parameter from low-pressure thermal conductivity data.

Zheng et al. [181] and Yorizane et al. [179] report high-pressure values of $\lambda$ for several simple gases and light hydrocarbons. Both found the Stiel and Thodos method to correlate their data reasonably well. Other recent papers showing new experimental data include Refs. 44, 127, and 165.

## 10-6 Thermal Conductivities of Low-Pressure Gas Mixtures

The thermal conductivity of a gas mixture is not usually a linear function of mole fraction. Generally, if the constituent molecules differ greatly in polarity, the mixture thermal conductivity is larger than would be predicted from a mole fraction average; for nonpolar molecules, the opposite trend is noted and is more pronounced the greater the difference in molecular weights or sizes of the constituents [48, 102]. Some of these trends are evident in Fig. 10-7, which shows experimental thermal conductivities for four systems. The argon-benzene system typifies a nonpolar case with different molecular sizes, and the methanol–*n*-hexane system is a case representing a significant difference in polarity. The linear systems benzene–*n*-hexane and ether-chloroform represent a balance between the effects of size and polarity.

Papers summarizing various methods for calculating mixture thermal conductivities can be found in Refs. 15, 83, 89, 110, 123, 158, 163, 164, 168. Many theoretical papers discussing the problems, approximations, and limitations of the various methods also have appeared. The theory for calculating the conductivity for *rare-gas* mixtures has been worked out in detail [13, 54, 92, 97, 98, 111]. The more difficult problem, however, is to modify monatomic mixture correlations to apply to polyatomic molecules. Many techniques have been proposed; all are essentially empirical, and most reduce to some form of the Wassiljewa equation. Corresponding states methods for low-pressure thermal conductivities have also been adapted for mixtures, but the results obtained in testing several were not encouraging.

### Wassiljewa equation

In a form analogous to the theoretical relation for mixture viscosity, Eq. (9-5.13),

$$\lambda_m = \sum_{i=1}^{n} \frac{y_i\lambda_i}{\sum_{j=1}^{n} y_j A_{ij}} \tag{10-6.1}$$

where $\lambda_m$ = thermal conductivity of the gas mixture
$\lambda_i$ = thermal conductivity of pure $i$
$y_i, y_j$ = = mole fraction of components $i$ and $j$

$A_{ij}$ = a function, as yet unspecified
$A_{ii}$ = 1.0

This empirical relation was proposed by Wassiljewa in 1904 [178].

### Mason and Saxena modification

Mason and Saxena [96] suggested that $A_{ij}$ in Eq. (10-6.1) could be expressed as

$$A_{ij} = \frac{\epsilon[1 + (\lambda_{tr_i}/\lambda_{tr_j})^{1/2} (M_i/M_j)^{1/4}]^2}{[8(1 + M_i/M_j)]^{1/2}} \tag{10-6.2}$$

where $M$ = molecular weight, g/mol
$\lambda_{tr}$ = monatomic value of the thermal conductivity
$\epsilon$ = numerical constant near unity

Mason and Saxena proposed a value of 1.065 for $\epsilon$, and Tondon and Saxena later suggested 0.85. As used here, $\epsilon$ = 1.0.

From Eq. (10-2.3), noting for monatomic gases that $C_v = C_{tr} = \frac{3}{2}R$,

$$\frac{\lambda_{tr_i}}{\lambda_{tr_j}} = \frac{\eta_i}{\eta_j}\frac{M_j}{M_i} \tag{10-6.3}$$

Substituting Eq. (10-6.3) into Eq. (10-6.2) and comparing with Eq. (9-5.14) gives

$$A_{ij} = \phi_{ij} \tag{10-6.4}$$

where $\phi_{ij}$ is the interaction parameter for gas-mixture viscosity. Thus the relation for estimating mixture viscosities is also applicable to thermal conductivities by simply substituting $\lambda$ for $\eta$. In this approximation, to determine $\lambda_m$, one needs data giving the pure component thermal conductivities and viscosities. An alternative way to proceed is to use Eqs. (10-6.1) and (10-6.2) but obtain the ratio of translational thermal conductivities from Eq. (10-3.14).

$$\frac{\lambda_{tr_i}}{\lambda_{tr_j}} = \frac{\Gamma_j\,[\exp(0.0464T_{r_i}) - \exp(-0.2412T_{r_i})]}{\Gamma_i\,[\exp(0.0464T_{r_j}) - \exp(-0.2412T_{r_j})]} \tag{10-6.5}$$

where $\Gamma$ is defined by Eq. (10-3.12). With Eq. (10-6.5), values of $A_{ij}$ become functions of the reduced temperatures of both $i$ and $j$. However, with this latter approach, pure gas viscosities are not required. Both techniques are illustrated in Example 10-5.

Lindsay and Bromley [85] have also proposed a technique to estimate $A_{ij}$. It is slightly more complex than Eq. (10-6.2), and the results obtained do not differ significantly from the Mason-Saxena approach.

The Wassiljewa equation is capable of representing low-pressure mixture thermal conductivities with either a maximum or minimum as composition is varied. As Gray et al. [48] have shown, if $\lambda_1 < \lambda_2$,

| | |
|---|---|
| $\frac{\lambda_1}{\lambda_2} < A_{12}A_{21} < \frac{\lambda_2}{\lambda_1}$ | $\lambda_m$ varies monotonically with composition |
| $A_{12}A_{21} \geq \frac{\lambda_2}{\lambda_1}$ | $\lambda_m$ has a minimum value below $\lambda_1$ |
| $\frac{\lambda_1}{\lambda_2} \geq A_{12}A_{21}$ | $\lambda_m$ has a maximum value above $\lambda_2$ |

#### Corresponding states methods

Both the Chung et al. [27, 28] and Ely-Hanley [40] methods for estimating low-pressure thermal conductivities [Eqs. (10-3.17) and (10-3.20)] have been adapted to handle mixtures. The emphasis of these authors, however, was, to treat systems at high pressure and, if possible, as liquids. When their methods are used for low-pressure gas mixtures, the accuracy away from the pure components is often not particularly high. However, in their favor is the fact that pure component thermal conductivities are *not* required as input; the methods generate their own values of pure component conductivities.

To illustrate, consider the Chung et al. form. To use this procedure for mixtures, we need to have rules to obtain $M'$, $\eta$, $C_v$, $\omega$, and $T_c$ for the mixture. $\eta_m$ is found from Eq. (9-5.24), and in using this relation, one also obtains $M_m$, $\omega_m$, and $T_{cm}$ [Eqs. (9-5.28), (9-5.29), and (9-5.44)]; $M'_m = M_m/10^3$. For $C_{vm}$, a mole fraction average rule is used, i.e.,

$$C_{vm} = \sum_{i=1}^{n} y_i C_{vi} \tag{10-6.6}$$

With these mixture values, the procedure to compute $\lambda_m$ is identical with that used for the pure component conductivity (see Example 10-1). The method is also illustrated for a mixture in Example 10-5.

#### Discussion

Three techniques were suggested to estimate the thermal conductivity of a gas mixture at low pressure. Two employ the Wassiljewa formulation [Eq. (10-6.1)] and differ only in the manner $\lambda_{tr_i}/\lambda_{tr_j}$ is calculated. The third method (Chung et al.) uses a corresponding states approach. It is the least accurate, but it has the advantage that pure component thermal conductivities do not have to be known. The other two methods require either experimental or estimated values of $\lambda$ for all pure components. All three methods are illustrated in Example 10-5.

For nonpolar gas mixtures, we recommend the Wassiljewa equation with the Mason-Saxena relation for $A_{ij}$, where $\lambda_{tr_i}/\lambda_{tr_j}$ is calculated from Eq. (10-6.5). Errors will generally be less than 3 to 4 percent. For nonpolar-polar and polar-polar gas mixtures, none of the techniques examined were found to be particularly accurate. As an example, in Fig. 10-7, none predicted the maximum in $\lambda_m$ for the methanol–$n$-hexane system. Thus, in such cases, errors greater than 5 to 8 percent may be expected when one employs the procedures recommended for nonpolar gas mixtures. For mixtures in which the sizes and polarities of the constituent molecules are not greatly different, $\lambda_m$ can be estimated satisfactorily by a mole fraction average of the pure component conductivities (e.g., the benzene–$n$-hexane and ether-chloroform cases in Fig. 10-7).

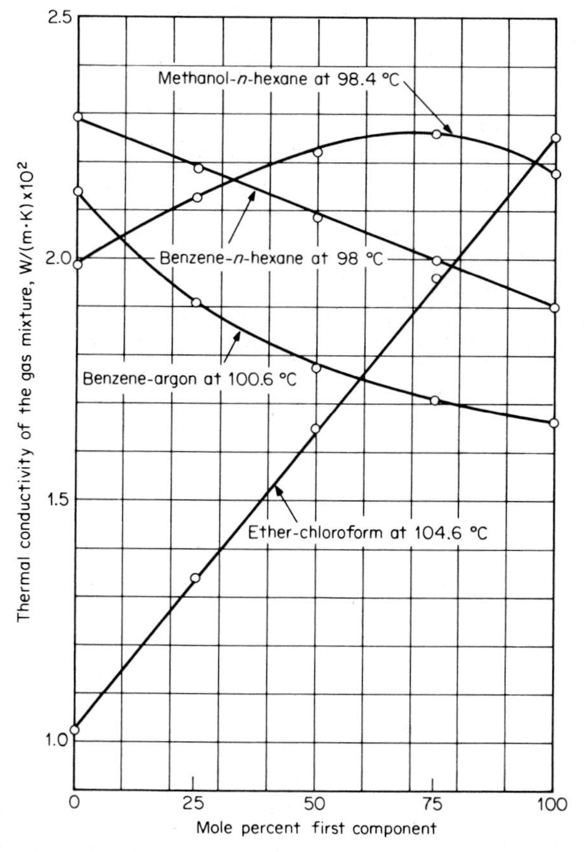


**Figure 10-7** Typical gas-mixture thermal conductivities. *(From Ref. 10.)*

**Example 10-5** Estimate the thermal conductivity of a gas mixture containing 25 mole percent benzene and 75 mole percent argon at 100.6°C and about 1 bar. The experimental value is $1.92 \times 10^{-2}$ W/(m·K)[10].
From Appendix A, the following pure component constants are given:

| | Benzene (1) | Argon (2) |
|---|---|---|
| $T_c$, K | 562.2 | 150.8 |
| $P_c$, bar | 48.9 | 48.7 |
| $V_c$, cm$^3$/mol | 259. | 74.9 |
| $\omega$ | 0.212 | 0.001 |
| $Z_c$ | 0.271 | 0.291 |
| $M$, g/mol | 78.114 | 39.948 |
| $M'$, kg/mol | 0.078114 | 0.039948 |

Using the ideal-gas heat capacity constants for the two pure gases (Appendix A), at 373.8 K,

Benzene $C_p = 104.8$ J/(mol·K):
$C_v = C_p - R = 104.8 - 8.1 = 96.7$ J/(mol·K)

Argon $C_p = 20.8$ J/(mol·K)
$C_v = C_p - R = 20.8 - 8.1 = 12.7$ J/(mol·K)

The pure component viscosities and thermal conductivities at 373.8 K and 1 bar are:

| | Benzene (1) | Argon (2) |
|---|---|---|
| $\eta$, N·s/m$^2$ $\times 10^7$ | 92.5 | 271. |
| $\lambda$, W/(m·K) $\times 10^2$ | 1.66 | 2.14 |

**solution** MASON AND SAXENA. Equation (10-6.1) is used with $A_{12} = \phi_{12}$ and $A_{21} = \phi_{21}$ from Eqs. (9-5.14) and (9-5.15).

$$A_{12} = \frac{[1 + (92.5/271)^{1/2}(39.948/78.114)^{1/4}]^2}{\{[8[1 + (78.114/39.948)]\}^{1/2}} = 0.459$$

$$A_{21} = A_{12}\frac{\eta_2}{\eta_1}\frac{M_1}{M_2}$$

$$= 0.459 \frac{271}{92.5}\frac{78.114}{39.948} = 2.630$$

With Eq. (10-6.1)

$$\lambda_m = \frac{y_1\lambda_1}{y_1 + y_2A_{12}} + \frac{y_2\lambda_2}{y_1A_{21} + y_2}$$

$$= \left[\frac{(0.25)(1.66)}{0.25 + (0.75)(0.459)} + \frac{(0.75)(2.14)}{(0.25)(2.630) + 0.75}\right] \times 10^{-2}$$

$$= 1.84 \times 10^{-2} \text{ W/(m·K)}$$

$$\text{Error} = \frac{1.84 - 1.92}{1.92} \times 100 = -4.2\%$$

MASON AND SAXENA FORM WITH EQ. (10-6.5). In this case, $\lambda_{tr1}/\lambda_{tr2}$ is obtained from Eq. (10-6.5). $\Gamma$ is determined with Eq. (10-3.12).

$$\Gamma_1 = 210\left[\frac{(562.2) \times (78.114)^3}{(48.9)^4}\right]^{1/6} = 398.7$$

$$\Gamma_2 = 210\left[\frac{(150.8) \times (39.948)^3}{(48.7)^4}\right]^{1/6} = 229.6$$

At 373.8 K, $T_{r1} = 373.8/562.2 = 0.665$; $T_{r2} = 373.8/150.8 = 2.479$. Then,

$$\frac{\lambda_{tr1}}{\lambda_{tr2}} = \frac{229.6\{\exp[(0.0464)(0.665)] - \exp(-0.2412)(0.665)]\}}{398.7\{\exp[(0.0464)(2.479)] - \exp[(-0.2412)(2.479)]\}}$$

$$= 0.1808$$

$$\frac{\lambda_{tr2}}{\lambda_{tr1}} = (0.1808)^{-1} = 5.536$$

Inserting these values into Eq. (10-6.2) with $\epsilon = 1.0$ gives

$$A_{12} = \frac{[1 + (0.1808)^{1/2}(78.114/39.948)^{1/4}]^2}{\{8[1+(78.114/39.948)]\}^{1/2}} = 0.4645$$

$$A_{21} = \frac{[1 + (5.536)^{1/2}(39.948/78.114)^{1/4}]^2}{\{8[1 + (39.948/78.114)]\}^{1/2}} = 2.571$$

Then, using Eq. (10-6.1),

$$\lambda_m = \left[\frac{(0.25)(1.66)}{0.25 + (0.75)(0.4645)} + \frac{(0.75)(2.14)}{(0.25)(2.571)+0.75}\right] \times 10^{-2}$$

$$= 1.85 \times 10^{-2} \quad \text{W/(m·K)}$$

$$\text{Error} = \frac{1.85-1.92}{1.92} \times 100 = -3.6\%$$

CHUNG ET AL. With this method, we use the relations described in Chap. 9 to determine the mixture properties $\eta_m$, $M'_m$, $\omega_m$, and $T_{cm}$. (See Example 9-8.) In this case, at 25 mole percent benzene, $\eta_m = 182.2\ \mu P = 182.2 \times 10^{-7}$ N·s/m², $M'_m =$ 0.04631 kg/mol, $\omega_m = 0.0817$, and $T_{cm} = 277.4$ K. From Eq. (10-6.6),

$$C_{vm} = (0.25)(96.7) + (0.75)(12.7) = 33.7 \text{ J/(mol·K)}$$

The mixture thermal conductivity is then found from Eq. (10-3.17).

$$\lambda_m = \frac{(\eta_m C_{vm}/M'_m)(3.75\Psi_m)}{C_{vm}/R} = \frac{\eta_m R}{M'_m}(3.75\Psi_m)$$

$$\Psi_m = 1 + \frac{\alpha_m[0.215 + 0.28288\alpha_m - 1.061\beta_m + 0.26665Z_m]}{0.6366 + \beta_m Z_m + 1.061\alpha_m\beta_m}$$

$$\alpha_m = (C_{vm}/R) - \frac{3}{2} = (33.7/8.31) - \frac{3}{2} = 2.553$$

$$\beta_m = 0.7862 - 0.7109\omega_m + 1.3168\omega_m^2 = 0.7369$$

$$T_{rm} = 373.8/277.4 = 1.348$$

$$Z_m = 2.0 + 10.5\,T_{rm}^2 = 2.0 + (10.5)(1.348)^2 = 21.07$$

$$\Psi_m = 1.812$$

$$\lambda_m = \left[(182.2 \times 10^{-7})\frac{8.314)}{0.04631}\right][(3.75)(1.812)]$$

$$= 2.22 \times 10^{-2} \text{ W/(m·K)}$$

$$\text{Error} = \frac{2.22 - 1.92}{1.92} \times 100 = 16\%$$

## 10-7 Thermal Conductivities of Gas Mixtures at High Pressures

There are few experimental data for gas mixtures at high pressures and, even here, most studies are limited to simple gases and light hydrocarbons. The nitrogen–carbon dioxide system was studied by Keyes [72], and Comings and his colleagues reported on ethylene mixtures with nitrogen and carbon dioxide [66], rare gases [126], and binaries containing carbon dioxide, nitrogen, and ethane [47]. Rosenbaum and Thodos investigated methane-carbon dioxide [139] and methane–carbon tetrafluoride [138] binaries. Binaries containing methane, ethane, nitrogen, and carbon dioxide were also reported by Christensen and Fredenslund [25], and data for systems containing nitrogen, oxygen, argon, methane, ethylene, and carbon dioxide were published by Zheng et al. [181] and Yorizane et al. [180].

Recent theoretical papers [26, 71, 93, 112, 113] have not, as yet, led to accurate predictive techniques.

We present below three estimation methods. All are modifications of procedures developed earlier for low- and high-pressure pure gas thermal conductivities.

### Stiel and Thodos modification

Equations (10-5.2) to (10-5.4) were suggested as a way to estimate the high-pressure thermal conductivity of a pure gas. This procedure may be adapted for mixtures if mixing and combining rules are available to determine $T_{cm}$, $P_{cm}$, $V_{cm}$, $Z_{cm}$ and $M_m$. Yorizane et al. [180] have studied this approach and recommend the following:

$$T_{cm} = \frac{\sum_i \sum_j y_i y_j V_{cij} T_{cij}}{V_{cm}} \tag{10-7.1}$$

$$V_{cm} = \sum_i \sum_j y_i y_j V_{cij} \tag{10-7.2}$$

$$\omega_m = \sum_i y_i \omega_i \tag{10-7.3}$$

$$Z_{cm} = 0.291 - 0.08\omega_m \tag{10-7.4}$$

$$P_{cm} = Z_{cm} R T_{cm} / V_{cm} \tag{10-7.5}$$

$$M_m = \sum_i y_i M_i \tag{10-7.6}$$

$$T_{cii} = T_{ci} \tag{10-7.7}$$

$$T_{cij} = (T_{ci} T_{cj})^{1/2} \tag{10-7.8}$$

$$V_{cii} = V_{ci} \tag{10-7.9}$$

$$V_{cij} = \tfrac{1}{8}[(V_{ci})^{1/3} + (V_{cj})^{1/3}]^3 \tag{10-7.10}$$

Using these simple rules, they found they could correlate their high-pressure thermal conductivity data for $CO_2$-$CH_4$ and $CO_2$-Ar systems quite well. In Fig. 10-8 we show a plot of $\lambda_m$ for the $CO_2$-Ar system at 298 K. This case is interesting because the temperature is slightly below the critical temperature of $CO_2$ and, at high pressure, $\lambda$ for carbon dioxide increases more rapidly than that for argon. The net result is that the $\lambda_m$ composition curves are quite nonlinear. Still, the Stiel-Thodos method, with Eqs. (10-7.1) through (10-7.10), appears to give a quite satisfactory fit to the data. We illustrate the approach in Example 10-6.

**Example 10-6** Estimate the thermal conductivity of a methane (1)–carbon dioxide (2) mixture containing 75.5 mole percent methane at 370.8 K and 174.8 bar. Rosenbaum and Thodos [139] show an experimental value of $5.08 \times 10^{-2}$ W/(m·K); these same investigators report that, for the mixture, $V = 159$ cm$^3$/mol and, at 1 bar, $\lambda_m^\circ = 3.77 \times 10^{-2}$ W/(m·K).

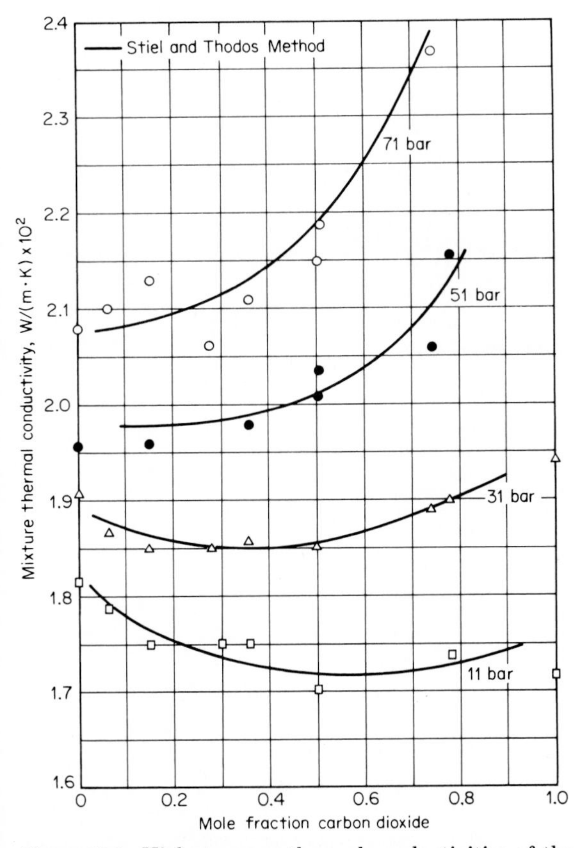


**Figure 10-8** High-pressure thermal conductivities of the argon–carbon dioxide system. *(Experimental Data from Ref. 180.)*

**solution** From Appendix A, we list pure component constants for methane and carbon dioxide that will be used in this example and in Examples 10-7 and 10-8.

| | $CH_4$ (1) | $CO_2$ (2) |
|---|---|---|
| $T_c$, K | 190.4 | 304.1 |
| $P_c$, bar | 46.0 | 73.8 |
| $V_c$, $cm^3/mol$ | 99.2 | 93.9 |
| $Z_c$ | 0.288 | 0.274 |
| $\omega$ | 0.011 | 0.239 |
| $\mu$, debye | 0 | 0 |
| $C_p$ J/(mol·K) | 39.65 | 40.20 |
| $C_v$, J/(mol·K) | 31.33 | 31.89 |
| $M$, g/mol | 16.043 | 44.010 |
| $M'$, kg/mol | 0.01604 | 0.04401 |

The heat capacities were calculated from the equation and polynomial constants in Appendix A at $T = 370.8$ K.
With Eqs. (10-7.1) through (10-7.10),

$$T_{c12} = [(190.4)(304.1)]^{1/2} = 240.6 \text{ K}$$

$$V_{c12} = \tfrac{1}{8}[(99.2)^{1/3} + (93.9)^{1/3}]^3 = 96.5 \text{ cm}^3/\text{mol}$$

$$V_{cm} = (0.755)^2(99.2) + (0.245)^2(93.8) + (2)(0.755)(0.245)(96.5)$$
$$= 97.9 \text{ cm}^3/\text{mol}$$

$$T_{cm} = [(0.755)^2(190.4)\,(99.2) + (0.245)^2(304.1)(93.9)$$
$$+ (2)(0.755)(0.245)(240.6)(96.5)]/97.9$$
$$= 215.2 \text{ K}$$

$$\omega_m = (0.755)(0.011) + (0.245)(0.239) = 0.067$$

$$Z_{cm} = 0.291 - (0.08)(0.067) = 0.286$$

$$P_{cm} = \frac{(0.286)(8.314)(215.2)}{97.9 \times 10^{-6}} = 5.24 \times 10^{-6} \text{ Pa} = 52.4 \text{ bar}$$

$$M_m = (0.755)(16.04) + (0.245)(44.01) = 22.9 \text{ g/mol}$$

With Eq. (10-3.12),

$$\Gamma = (210)\left[\frac{(215.2)(22.9)^3}{(52.4)^4}\right]^{1/6} = 176$$

and

$$\rho_{rm} = \frac{V_{cm}}{V_m} = \frac{97.9}{159} = 0.616$$

Using Eq. (10-5.3),

$$(\lambda_m - \lambda_m^\circ)\,[(176)(0.286)^5] = (1.14\times10^{-2})\{\exp\,[(0.67)(0.616)] - 1.069\}$$

$$\lambda_m - \lambda_m^\circ = 1.50 \times 10^{-2} \text{ W/(m·K)}$$

$$\lambda_m = (1.50 + 3.77)(10^{-2}) = 5.27 \times 10^{-2} \text{ W/(m·K)}$$

$$\text{Error} = \frac{5.27 - 5.08}{5.08} \times 100 = 4\%$$

**Chung et al. method [27, 28]**

To apply this method to estimate the thermal conductivities of high-pressure gas mixtures, one must combine the high-pressure *pure* component relations with the mixing rules given in Secs. 10-6 and in 9-5. To be specific, Eq. (10-5.5) is employed with all variables subscripted with $m$ to denote them as mixture properties. Example 10-7 illustrates the procedure in detail.

**Example 10-7** Repeat Example 10-6 by using the Chung et al. approach.

**solution** For the methane (1)–carbon dioxide (2) system, the required pure component properties were given in Example 10-6.
To use Eq. (10-5.5), let us first estimate $\eta_m^\circ$ with the procedures in Chap. 9. From Eqs. (9-4.6) and (9-4.7),

$$\left(\frac{\epsilon}{k}\right)_1 = \frac{190.4}{1.2593} = 151.2 \text{ K}$$

$$\left(\frac{\epsilon}{k}\right)_2 = \frac{304.1}{1.2593} = 241.5 \text{ K}$$

$$\sigma_1 = (0.809)(99.2)^{1/3} = 3.745 \text{ Å}$$

$$\sigma_2 = (0.809)(93.9)^{1/3} = 3.677 \text{ Å}$$

Interaction values are then found from Eqs. (9-5.32), (9-5.35), (9-5.37), and (9-5.40).

$$\sigma_{12} = [(3.745)(3.677)]^{1/2} = 3.711 \text{ Å}$$

$$(\epsilon/k)_{12} = [(151.2)(241.5)]^{1/2} = 191.1 \text{ K}$$

$$\omega_{12} = \frac{(0.011+0.239)}{2} = 0.125$$

$$M_{12} = \frac{(2)(16.04)(44.01)}{(16.04 + 44.01)} = 23.51$$

With $y_1 = 0.755$ and $y_2 = 0.245$, using Eqs. (9-5.25) to (9-5.29) and Eq. (9-5.41),

$$\sigma_m = 3.728 \text{ Å} \qquad (\epsilon/k)_m = 171.0 \text{ K} \qquad T^* = 2.168$$

$$M_m = 20.89 \text{ g/mol} \qquad \omega_m = 0.066 \qquad F_m = 0.982$$

so, with Eq. (9-4.3), $\Omega_v = 1.144$. Then, with Eq. (9-5.24),

$$\eta_m^\circ = (26.69)(0.982)\frac{[(20.89)(370.8)]^{1/2}}{(3.728)^2(1.144)}$$

$$= 145.1 \ \mu\text{P} = 145.1 \times 10^{-7} \text{ N·s/m}^2$$

With Eqs. (9-5.43) and (9-5.44),

$$V_{cm} = \left(\frac{\sigma_m}{0.809}\right)^3 = 97.85 \text{ cm}^3\text{/mol}$$

$$T_{cm} = (1.2593)\left(\frac{\epsilon}{k}\right)_m = 215.3 \text{ K}$$

$$T_r = \frac{T}{T_{cm}} = \frac{370.8}{215.3} = 1.722$$

$C_v$ for the mixture is found with Eq. (10-6.6) as 31.47 J/(mol·K), and $\Psi$ is determined as indicated under Eq. (10-3.17) with

$$\alpha_m = \frac{31.47}{8.314} - \frac{3}{2} = 2.285$$

$$\beta_m = 0.7862 - (0.7109)(0.066) + (1.3168)(0.066)^2 = 0.745$$

$$Z = 2.0 + (10.5)(1.722)^2 = 33.14$$

$$\Psi = 1.750$$

and $$M' = M/10^3 = 20.89 \times 10^{-3} \text{ kg/mol}$$

With Eqs. (10-5.6) to (10-5.8) and Table 10-4,

$$y_m = 0.1026$$

$$G_1 = 1.313$$

$$G_2 = 0.6522$$

$$B_1 = 2.466,\ B_2 = -0.6089,\ B_3 = 6.982,\ B_4 = 13.95,$$

$$B_5 = 0.8469,\ B_6 = -5.019, \text{ and } B_7 = 99.54$$

$$q = \frac{(3.586 \times 10^{-3})(215.3/20.89 \times 10^{-3})^{1/2}}{(97.85)^{2/3}} = 1.714 \times 10^{-2}$$

Finally, substituting these values into Eq. (10-5.5),

$$\lambda_m = \frac{(31.2)(145.1 \times 10^{-7})(1.750)}{20.89 \times 10^{-3}} [(0.6522)^{-1} - (5.019)(0.1026)]$$
$$+ (1.714 \times 10^{-2})(99.54)(0.1026)^2(1.722)^{1/2}(0.6522)$$
$$= 5.40 \times 10^{-2} \text{ W/(m·K)}$$

$$\text{Error} = \frac{5.40\text{-}5.08}{5.08} \times 100 = 6\%$$

If the pressure were reduced to 1 bar, $y_m$ would become quite small and $G_2$ would be essentially unity. In that case, $\lambda_m^\circ = 3.79 \times 10^{-2}$ W/(m·K), a value very close to the value reported experimentally (3.77 × 10$^{-2}$ W/(m·K).

Whereas the procedure appears tedious, it is readily programmed for computer use. The error found in Example 10-7 is typical for this method when simple gas mixtures are treated. As noted before, the Chung et al. method should not be used for polar gases. Its accuracy for nonpolar gas mixtures containing other than simple gases or light hydrocarbons is in doubt.

### Method of Ely and Hanley [40, 51]

For gas mixtures at high pressure, the thermal conductivity is determined by a combination of the techniques introduced for low-pressure gases (Sec. 10-3) and high-pressure gases (Sec. 10-5) with appropriate mixing rules. The thermal conductivity of the mixture is given by:

$$\lambda_m = \lambda^{**}{}_m + \lambda'_m \qquad (10\text{-}7.11)$$

where
$$\lambda'_m = \sum_i \sum_j y_i y_j \lambda'_{ij} \tag{10-7.12}$$
$$\lambda'_{ii} = \lambda'_i \tag{10-7.13}$$
$$\lambda'_{ij} = \frac{2\lambda'_i \lambda'_j}{\lambda'_i + \lambda'_j} \tag{10-7.14}$$
$$\lambda'_i = (\eta_i^*/M'_i)(1.32)\left(C_{vi} - \frac{3R}{2}\right) \tag{10-7.15}$$

The definition of $\lambda'_i$ is identical with the second term on the right-hand side of Eq. (10-3.20). The procedure described under this equation should be followed to obtain numerical values. $\lambda'_i$ is a function of $T_c$, $C_v$, and $M'$ of each component as well as the system temperature $T$.

The term $\lambda^{**}_m$ is determined in the same manner as $\lambda^{**}$ in Eq. (10-5.10), although appropriate mixing rules must be specified to account for variations in composition. Procedural steps 1, 2, and 3 [under Eq. (10-5.10)] are followed for each component in the mixture to obtain $f_i$ and $h_i$. We then obtain $h_m$, $f_m$, and $M'_m$ from the following mixing and combining rules:

$$h_m = \sum_i \sum_j y_i y_j h_{ij} \tag{10-7.16}$$
$$h_{ii} = h_i \tag{10-7.17}$$
$$h_{ij} = \tfrac{1}{8}\left[(h_i)^{1/3} + (h_j)^{1/3}\right]^3 \tag{10-7.18}$$
$$f_m = \frac{\sum_i \sum_j y_i y_j f_{ij} h_{ij}}{h_m} \tag{10-7.19}$$
$$f_{ii} = f_i \tag{10-7.20}$$
$$f_{ij} = (f_i f_j)^{1/2} \tag{10-7.21}$$
$$M'_m = \left[\sum_i \sum_j y_i y_j M_{ij}^{-1/2} f_{ij}^{1/2}\, h_{ij}^{-4/3}\right]^{-2} (f_m h_m^{-8/3}) \tag{10-7.22}$$
$$M'_{ii} = M'_i \tag{10-7.23}$$
$$M'_{ij} = \frac{2M'_i M'_j}{M'_i + M'_j} \tag{10-7.24}$$

$T_0$ and $\rho_0$ are found from

$$T_0 = \frac{T}{f_m} \tag{10-7.25}$$
$$\rho_0 = \frac{16.04}{V_m} h_m \tag{10-7.26}$$

with $T$ the system temperature, in kelvins, and $V_m$ the molar volume of the mixture, cm$^3$/mol. $\lambda_m$ (1), $\lambda_m$ (2), and $\lambda_m$ (3) are then calculated from

Eqs. (10-5.19) to (10-5.21) with the $T_0$, $\rho_0$ values obtained above. Then,

$$H_m = \left(\frac{16.04 \times 10^{-3}}{M'_m}\right)^{1/2} \frac{f_m^{1/2}}{h_m^{2/3}} \tag{10-7.27}$$

$$\lambda^{**}_m = [\lambda_m(1) + \lambda_m(2) + \lambda_m(3)]H_m \tag{10-7.28}$$

If desired, the correction term $X$, defined in Eq. (10-5.24), may be used to multiply the right-hand side of Eq. (10-7.28). The mixture thermal conductivity is then found from Eq. (10-7.11).

Ely and Hanley tested their method on a number of binary hydrocarbon mixtures over a wide range of densities. They report an average absolute error of about 7 percent, although, in some cases, significantly larger deviations were found. The technique is illustrated in Example 10-8.

**Example 10-8** Repeat Example 10-6 by using the Ely-Hanley procedure.

**solution** The pure component properties for both components of the methane (1), carbon dioxide (2) binary are given at the beginning of the solution of Example 10-6.

We will use Eq. (10-7.11). First $\lambda^{**}_m$ is estimated. $T_c(1) = 190.4$ K, $T_c(2) = 304.1$ K, $V_c(1) = 99.2$ cm$^3$/mol, $V_c(2) = 93.9$ cm$^3$/mol, $T = 370.8$ K, and $V_m = 159.0$ cm$^3$/mol. Then, $T_r(1) = 370.8/190.4 = 1.947$; $T_r(2) = 370.8/304.1 = 1.219$, $V_r(1) = 159.0/99.2 = 1.603$, and $V_r(2) = 159.0/93.9 = 1.693$. With Eqs. (10-5.11) and (10-5.12),

$$T^+(1) = T_r(1) = 1.947 \qquad T^+(2) = T_r(2) = 1.219$$

$$V^+(1) = V_r(1) = 1.603 \qquad V^+(2) = V_r(2) = 1.693$$

Then, with Eqs. (10-5.13) to (10-5.16).†

$$\Theta(1) = 1.0 \qquad \Theta(2) = 0.964$$

$$\phi(1) = 1.0 \qquad \phi(2) = 1.073$$

$$f(1) = 1.0 \qquad f(2) = 1.540$$

$$h(1) = 1.0 \qquad h(2) = 1.016$$

Now, with Eqs. (10-7.16) through (10-7.24) with $y(1) = 0.755$, $y(2) = 0.245$,

$$h(1\text{-}2) = \tfrac{1}{8}[(1.0)^{1/3} + (1.016)^{1/3}]^3 = 1.008$$

$$h_m = (0.755)^2(1.0) + (0.245)^2(1.016) + (2)(0.755)(0.245)(1.008)$$

$$= 1.004$$

$$f(1\text{-}2) = [(1.0)(1.540)]^{1/2} = 1.241$$

$$f_m = (0.755)^2(1.0) + (0.245)^2(1.540) + (2)(0.755)(0.245)(1.241)$$

$$= 1.122$$

†Note that, since methane is the *reference* fluid, all relative terms, $\Theta$, $\phi$, $f$, and $h$, become unity.

$$M'(1\text{-}2) = \frac{(2)(0.01604)(0.04401)}{(0.01604 + 0.04401)} = 0.02351$$

$$M'_m = [(0.755)^2(0.01604)^{-1/2}(1.0)^{1/2}(1.0)^{-4/3} + (0.245)^2(0.04401)^{-1/2}(1.540)^{1/2}(1.016)^{-4/3} + (2)(0.755)(0.245)(0.02351)^{-1/2}(1.241)^{1/2}(1.008)^{-4/3}] \times [(1.122)(1.004)^{-8/3}]$$
$$= 0.01969$$

Thus, with Eqs. (10-7.25) and (10-7.26),

$$T_0 = \frac{370.8}{1.122} = 330.5 \text{ K}$$

$$\rho_0 = \left(\frac{16.04}{159}\right)(1.004) = 0.1013 \text{ g/cm}^3$$

With Eq. (10-3.27), $\eta_0 = 1.225 \times 10^{-5}$ N·s/m² and, by Eq. (10-5.19),

$$\lambda_m(1) = (1944)(1.225 \times 10^{-5}) = 2.38 \times 10^{-2} \text{ W/(m·K)}$$

and, by Eqs. (10-5.20) and (10-5.21),

$$\lambda_m(2) = -1.89 \times 10^{-5} \simeq 0$$

$$\lambda_m(3) = 1.618 \times 10^{-2} \text{ W/(m·K)}$$

Then by Eqs. (10-7.27) and (10-7.28),

$$H_m = \frac{(0.01604/0.01969)^{1/2}(1.122)^{1/2}}{(1.004)^{2/3}} = 0.953$$

$$\lambda_m^{**} = (2.38 + \sim 0 + 1.62)(10^{-2})(0.953) = 3.81 \times 10^{-2} \text{ W/(m·K)}$$

Next, we need to estimate $\lambda'(1)$ and $\lambda'(2)$ as given by Eq. (10-7.15) and calculated as shown under Eq. (10-3.20). We show the steps and results below.

| Parameter | Equation | $CH_4$ (1) | $CO_2$ (2) |
|---|---|---|---|
| $T_r$ | — | 1.947 | 1.219 |
| $T^+$ | (10-3.21) | 1.947 | 1.219 |
| $\Theta$ | (10-3.22) | 1.0 | 0.959 |
| $\phi$ | (10-3.23) | 1.0 | 1.088 |
| $f$ | (10-3.24) | 1.0 | 1.532 |
| $h$ | (10-3.25) | 1.0 | 1.030 |
| $T_0$, K | (10-3.26) | 370.8 | 242.0 |
| $\eta_0$, N·s/m² | (10-3.27) | $1.348 \times 10^{-5}$ | $9.279 \times 10^{-6}$ |
| $H$ | (10-3.31) | 1.0 | 0.7326 |
| $\eta^*$, N·s/m² | (10-3.30) | $1.345 \times 10^{-5}$ | $1.865 \times 10^{-5}$ |
| $C_v$, J/(mol·K) | — | 31.33 | 31.89 |
| $M'$, kg/mol | — | 0.01604 | 0.04401 |
| $\lambda'$, W/(m·K) | (10-7.15) | $2.087 \times 10^{-2}$ | $1.086 \times 10^{-2}$ |

With Eq. (10-7.14),

$$\lambda'(1\text{-}2) = \frac{(2)(2.087 \times 10^{-2})(1.086 \times 10^{-2})}{(2.087 \times 10^{-2} + 1.086 \times 10^{-2})}$$
$$= 1.429 \times 10^{-2} \text{ W/(m·K)}$$

and, by Eq. (10-7.12),

$$\lambda'_m = (0.755)^2(2.087 \times 10^{-2}) + (0.245)^2(1.086 \times 10^2) + (2)(0.755)(0.245)(1.429 \times 10^{-2}) = 1.78 \times 10^{-2}\ \text{W/(m}\cdot\text{K)}$$

So, with Eq. (10-7.11),

$$\lambda_m = 3.81 \times 10^{-2} + 1.78 \times 10^{-2} = 5.59 \times 10^{-2}\ \text{W/(m}\cdot\text{K)}$$

$$\text{Error} = \frac{5.59 - 5.08}{5.08} \times 100 = 10\%$$

No correction [i.e., Eq. (10-5.24)] was used in this example.

### Discussion

Of the three methods presented to estimate the thermal conductivity of high-pressure (or high-density) gas mixtures, all have been tested on available data and shown to be reasonably reliable with errors averaging about 5 to 7 percent. However, the database used for testing is small and comprises, primarily, permanent gases and light hydrocarbons. None are believed applicable to polar fluid mixtures. Chung et al. and Ely and Hanley have also tested their methods on more complex (hydrocarbon) systems at densities which are in the liquid range with quite encouraging results.

It is interesting to note that none of the methods are interpolative in nature, i.e., the pure component conductivities at the system pressure and temperature are not required. Yorizane et al. [180] did investigate the use of the Wassiljewa equation (10-6.1) to account for composition variations while using high-pressure thermal conductivities as "hinge" values. They report that this technique was not particularly accurate.

For simple hand calculation of one or a few values of $\lambda_m$, the Stiel and Thodos method is certainly the simplest. If many values are to be determined, the somewhat more complex, but probably more accurate, methods of Chung et al. or Ely and Hanley should be programmed and used.

## 10-8 Thermal Conductivities of Liquids

For many simple organic liquids, the thermal conductivities are between 10 and 100 times larger than those of the low-pressure gases at the same temperature. There is little effect of pressure, and raising the temperature usually decreases the thermal conductivities. These characteristics are similar to those noted for liquid viscosities, although the temperature dependence of the latter is pronounced and nearly exponential, whereas that for thermal conductivities is weak and nearly linear.

Values of $\lambda_L$ for most common organic liquids range between 0.10 and 0.17 W/(m·K) at temperatures below the normal boiling point, but water, ammonia, and other highly polar molecules have values several times as large. Also, in many cases the dimensionless ratio $M\lambda/R\eta$ is nearly constant (for nonpolar liquids) between values of 2 and 3, so that viscous liquids have a correspondingly larger thermal conductivity. Liquid metals and some organosilicon compounds have large values of $\lambda_L$; the former often are 100 times larger than those for normal organic liquids. The solid thermal conductivity at the melting point is approximately 20 to 40 percent larger than that of the liquid. Liquid thermal conductivity data have been compiled and evaluated by Jamieson et al. [64]. In Table 10-5, we show the thermal conductivities of some liquids as a function of temperature [64].

The difference between transport property values in the gas phase and the values in the liquid phase indicates a distinct change in mechanism of energy (or momentum or mass) transfer, i.e.,

$$\frac{\lambda_L}{\lambda_G} \simeq 10 \text{ to } 100 \qquad \frac{\eta_L}{\eta_G} \simeq 10 \text{ to } 100 \qquad \frac{D_L}{D_G} \simeq 10^{-4}$$

In the gas phase, the molecules are relatively free to move about and transfer momentum and energy by a collisional mechanism. The intermolecular force fields, though not insignificant, do not drastically affect the value of $\lambda$, $\eta$, or $D$. That is, the intermolecular forces are reflected solely in the collision integral terms $\Omega_v$ and $\Omega_D$, which are really ratios of collision integrals for a real force field and an artificial case in which the molecules are rigid, noninteracting spheres. The variation of $\Omega_v$ or $\Omega_D$ from unity then yields a rough quantitative measure of the importance of intermolecular forces in affecting gas phase transport coefficients. Reference to Eq. (9-4.3) (for $\Omega_v$) or Eq. (11-3.6) (for $\Omega_D$) shows that $\Omega$ values are often near unity. One then concludes that a rigid, noninteracting spherical molecular model yields a low-pressure transport coefficient $\lambda$,† $\eta$, or $D$ not greatly different from that computed when intermolecular forces are included.

In the liquid, however, this hypothesis is not even roughly true. The close proximity of molecules to one another emphasizes strongly the intermolecular forces of attraction. There is little wandering of the individual molecules, as evidenced by the low value of liquid diffusion coefficients, and often a liquid is modeled as a lattice with each molecule caged by its nearest neighbors. Energy and momentum are primarily exchanged by oscillations of molecules in the shared force fields surrounding each molecule. McLaughlin [88] discusses in more detail the differences in transport mechanisms between a dense gas or liquid and a low-pressure gas.

---

†$\lambda$ in this case is the monatomic value, *not* including contributions from internal energy transfer mechanisms.

**TABLE 10-5 Thermal Conductivities of Some Liquids**[1]

$\lambda = A + BT + CT^2$; $\lambda$ in W/(m·K) and T in kelvins

| Formula | Name | A | B | C | λ | T | Range |
|---|---|---|---|---|---|---|---|
| He | helium | -3.995E-1 | 6.490E-1 | -2.094E-1 | 8.33E-2 | 1.9 | 1.9 to 2.2 |
| | | 4.118E-2 | -1.833E-2 | 3.789E-3 | 2.11E-2 | 3.2 | 2.2 to 4.9 |
| Ne | neon | 1.374E-2 | 8.392E-3 | -1.726E-4 | 1.01E-1 | 34 | 25 to 43 |
| A | argon | 1.862E-1 | -4.121E-4 | -3.589E-6 | 1.09E-1 | 100 | 84 to 145 |
| $H_2$ | hydrogen | -8.546E-3 | 1.036E-2 | -2.239E-4 | 1.11E-1 | 23 | 14 to 32 |
| $N_2$ | nitrogen | -2.629E-1 | -1.545E-3 | -9.450E-7 | 1.15E-1 | 91 | 64 to 121 |
| $O_2$ | oxygen | 2.444E-1 | -8.813E-4 | -2.023E-6 | 1.49E-1 | 90 | 55 to 138 |
| $F_2$ | fluorine | 2.565E-1 | -6.795E-4 | -4.958E-6 | 1.34E-1 | 103 | 54 to 133 |
| $Cl_2$ | chlorine | 2.508E-1 | -2.022E-4 | -6.381E-7 | 1.37E-1 | 293 | 172 to 405 |
| $Br_2$ | bromine | 1.608E-1 | -1.285E-5 | -3.366E-7 | 1.28E-1 | 293 | 266 to 573 |
| $I_2$ | iodine | 1.340E-1 | 4.296E-5 | -2.031E-7 | 1.15E-1 | 429 | 386 to 785 |
| HF | hydrogen fluoride | 7.100E-1 | -8.622E-4 | -6.440E-7 | 4.02E-1 | 293 | 190 to 438 |
| HCl | hydrogen chloride | 4.487E-1 | -7.721E-5 | -2.756E-6 | 1.89E-1 | 293 | 159 to 304 |
| HBr | hydrogen bromide | 2.428E-1 | 1.605E-4 | -1.721E-6 | 1.42E-1 | 293 | 186 to 343 |
| HI | hydrogen iodide | 2.599E-1 | -4.300E-5 | -9.098E-7 | 1.69E-1 | 293 | 223 to 383 |
| CO | carbon monoxide | 1.991E-1 | 1.386E-5 | -8.971E-6 | 1.52E-1 | 73 | 68 to 128 |
| $CO_2$ | carbon dioxide | 4.070E-1 | -8.438E-4 | -9.626E-7 | 7.69E-2 | 293 | 217 to 299 |
| $SO_2$ | sulfur dioxide | 8.964E-1 | -3.281E-3 | 2.991E-6 | 1.91E-1 | 293 | 223 to 423 |
| $SO_3$ | sulfur trioxide | 9.510E-1 | -3.185E-3 | 2.789E-6 | 2.57E-1 | 293 | 283 to 483 |
| $N_2O$ | nitrous oxide | 3.546E-1 | -8.952E-4 | -1.796E-7 | 7.67E-2 | 293 | 171 to 293 |
| NO | nitric oxide | 1.773E-1 | 1.060E-3 | -8.891E-6 | 1.83E-1 | 113 | 110 to 177 |
| $NO_2$ | nitrogen dioxide | 2.176E-1 | 2.604E-5 | -1.077E-6 | 1.33E-1 | 293 | 262 to 415 |
| $H_2O$ | water | -3.838E-1 | 5.254E-3 | -6.369E-6 | 6.09E-1 | 293 | 273 to 623 |
| $H_2O_2$ | hydrogen peroxide | -1.954E-1 | 3.374E-3 | -3.667E-6 | 4.79E-1 | 293 | 273 to 673 |
| $NH_3$ | ammonia | 1.068E+0 | -1.577E-3 | -1.229E-6 | 5.00E-1 | 293 | 196 to 373 |
| N2H4 | hydrazine | 1.198E+0 | -7.337E-4 | -1.017E-6 | 8.95E-1 | 293 | 275 to 591 |

| Formula | Name | A | B | C | λ | T | Range |
|---|---|---|---|---|---|---|---|
| $CH_4$ | methane | 3.026E-1 | -6.047E-4 | -3.197E-6 | 1.35E-1 | 153 | 90 to 183 |
| $C_2H_6$ | ethane | 2.928E-1 | -6.945E-4 | -2.039E-7 | 7.17E-2 | 293 | 90 to 293 |
| $C_3H_8$ | propane | 2.611E-1 | -5.309E-4 | -8.876E-8 | 9.78E-2 | 293 | 85 to 353 |
| $C_2H_4$ | ethylene | 3.565E-1 | -9.586E-4 | -1.972E-7 | 1.46E-1 | 210 | 104 to 269 |
| $C_3H_6$ | propylene | 2.906E-1 | -6.053E-4 | 1.256E-8 | 1.14E-1 | 293 | 88 to 343 |
| $C_4H_8$ | 1-butene | 2.554E-1 | -3.984E-4 | -1.135E-7 | 1.29E-1 | 293 | 88 to 393 |
| $C_4H_8$ | isobutylene | 2.325E-1 | -5.204E-4 | 2.609E-7 | 1.02E-1 | 293 | 133 to 373 |
| $C_4H_6$ | 1,3-butadiene | 3.007E-1 | -7.837E-4 | 4.916E-7 | 1.13E-1 | 293 | 164 to 393 |
| $C_5H_8$ | isoprene | 2.215E-1 | -3.170E-4 | -5.527E-8 | 1.24E-1 | 293 | 127 to 433 |
| $C_6H_6$ | benzene | 1.776E-1 | 4.773E-6 | -3.781E-7 | 1.47E-1 | 293 | 278 to 533 |
| $C_7H_8$ | toluene | 2.031E-1 | -2.254E-4 | -2.470E-8 | 1.35E-1 | 293 | 178 to 581 |
| $C_8H_{10}$ | ethyl benzene | 2.142E-1 | -3.440E-4 | 1.943E-7 | 1.30E-1 | 293 | 178 to 573 |
| $C_8H_{10}$ | o-xylene | 1.649E-1 | -7.440E-5 | -1.415E-7 | 1.31E-1 | 293 | 248 to 605 |
| $C_8H_{10}$ | m-xylene | 1.643E-1 | -1.466E-5 | -2.387E-7 | 1.39E-1 | 293 | 225 to 603 |
| $C_8H_{10}$ | p-xylene | 1.487E-1 | 2.717E-5 | -2.822E-7 | 1.32E-1 | 293 | 287 to 609 |
| $C_8H_8$ | styrene | 2.696E-1 | -3.384E-4 | 1.675E-8 | 1.72E-1 | 293 | 243 to 623 |
| $C_9H_{12}$ | cumene | 1.973E-1 | -2.421E-4 | 2.052E-8 | 1.28E-1 | 293 | 180 to 578 |
| $C_{10}H_8$ | naphthalene | 1.328E-1 | 5.954E-5 | -1.692E-7 | 1.30E-1 | 393 | 354 to 733 |
| $C_3H_6$ | cyclopropane | 1.661E-1 | -1.763E-4 | -2.814E-7 | 9.03E-2 | 293 | 146 to 390 |
| $C_4H_8$ | cyclobutane | 1.452E-1 | -1.217E-4 | -1.516E-7 | 9.64E-2 | 293 | 183 to 435 |
| $C_5H_{10}$ | cyclopentane | 2.143E-1 | -2.588E-4 | -5.820E-8 | 1.33E-1 | 293 | 180 to 488 |
| $C_6H_{12}$ | cyclohexane | 1.626E-1 | -9.513E-5 | -1.382E-7 | 1.23E-1 | 293 | 267 to 527 |
| $CH_3OH$ | methanol | 3.225E-1 | -4.785E-4 | 1.168E-7 | 1.92E-1 | 293 | 176 to 483 |
| $C_2H_5OH$ | ethanol | 2.629E-1 | -3.847E-4 | 2.211E-7 | 1.69E-1 | 293 | 160 to 463 |
| $C_3H_7OH$ | n-propanol | 1.854E-1 | -3.366E-5 | -2.215E-7 | 1.56E-1 | 293 | 148 to 493 |
| $C_4H_9OH$ | n-butanol | 2.288E-1 | -2.697E-4 | 1.323E-8 | 1.51E-1 | 293 | 184 to 503 |

**TABLE 10-5 Thermal Conductivities of Some Liquids[1] (*Continued*)**

$\lambda = A + BT + CT^2$; $\lambda$ in W/(m·K) and T in kelvins

| Formula | Name | A | B | C | λ | T | Range |
|---|---|---|---|---|---|---|---|
| $C_2H_4O$ | ethylene oxide | 2.624E-1 | -3.329E-4 | -1.193E-7 | 1.55E-1 | 293 | 161 to 453 |
| $C_3H_7O$ | propylene oxide | 2.359E-1 | -2.236E-4 | -2.127E-7 | 1.52E-1 | 293 | 162 to 453 |
| $C_4H_9O$ | butylene oxide | 2.146E-1 | -1.196E-4 | -3.057E-7 | 1.53E-1 | 293 | 123 to 513 |
| $C_6H_5NH_2$ | aniline | 2.251E-1 | -1.274E-4 | -6.239E-8 | 1.82E-1 | 293 | 268 to 680 |
| $CH_3Cl$ | methyl chloride | 3.781E-1 | -6.639E-4 | -1.763E-7 | 1.68E-1 | 293 | 176 to 396 |
| $CH_2Cl_2$ | methylene chloride | 2.252E-1 | -2.532E-4 | -1.126E-7 | 1.41E-1 | 293 | 177 to 459 |
| $CHCl_3$ | chloroform | 1.634E-1 | -8.617E-5 | -2.119E-7 | 1.20E-1 | 293 | 210 to 510 |
| $CCl_4$ | carbon tetrachloride | 1.608E-1 | -1.903E-4 | -1.005E-8 | 1.04E-1 | 293 | 250 to 497 |
| $C_6H_5Cl$ | chlorobenzene | 1.809E-1 | -1.604E-4 | -4.689E-8 | 1.30E-1 | 293 | 228 to 600 |
| $C_4H_5Cl$ | chloroprene | 1.925E-1 | -3.439E-4 | 1.491E-7 | 1.04E-1 | 293 | 143 to 493 |

1. Miller, J.W. Jr., J.J. McGinley, and C.L. Yaws, Chem. Eng. 83 (23), 133 (1976).

To date, theory has not been successful in formulating useful and accurate expressions to calculate liquid thermal conductivities; approximate techniques must be employed for engineering applications.

Only relatively simple organic liquids are considered in the sections to follow. Ho et al. [55] have presented a comprehensive review covering the thermal conductivity of the elements, and Ewing et al. [41] and Gambill [46] consider, respectively, molten metals and molten salt mixtures. Cryogenic liquids are discussed by Preston et al. [128] and Mo and Gubbins [109].

## 10-9 Estimation of the Thermal Conductivities of Pure Liquids

All estimation techniques for the thermal conductivity of pure liquids are empirical; and with only limited examination, they often appear rather accurate. As noted earlier, however, below the normal boiling point, the thermal conductivities of most organic, nonpolar liquids lie between 0.10 and 0.17 W/(m·K). With this fact in mind, it is not too difficult to devise various schemes for estimating $\lambda_L$ within this limited domain.

Many estimation methods were tested; three of the better ones are described below. Others that were considered are noted briefly at the end of the section.

### Latini et al. method

In an examination of the thermal conductivities of many diverse liquids, Latini and his coworkers suggest a correlation of the form [1, 2, 4, 5, 6, 7, 77]:

$$\lambda_L = \frac{A(1 - T_r)^{0.38}}{T_r^{1/6}} \tag{10-9.1}$$

where $\lambda_L$ = thermal conductivity of the liquid, W/(m·K)

$T_b$ = normal boiling temperature (at 1 atm), K

$T_c$ = critical temperature, K

$M$ = molecular weight, g/mol

$T_r$ = $T/T_c$

$$A = \frac{A^* T_b^{\alpha}}{M^{\beta} T_c^{\gamma}} \tag{10-9.2}$$

and the parameters $A^*$, $\alpha$, $\beta$, and $\gamma$ are shown in Table 10-6 for various classes of organic compounds. Specific values of $A$ are given for many compounds in [5]. Equation (10-9.2) is only an approximation of the regressed value of $A$, and this simplification introduces significant error unless $50 < M < 250$.

**TABLE 10-6 Latini et al. Correlation Parameters for Eq. (10-9.2)**

| Family | $A^*$ | $\alpha$ | $\beta$ | $\gamma$ |
|---|---|---|---|---|
| Saturated hydrocarbons | 0.00350 | 1.2 | 0.5 | 0.167 |
| Olefins | 0.0361 | 1.2 | 1.0 | 0.167 |
| Cycloparaffins | 0.0310 | 1.2 | 1.0 | 0.167 |
| Aromatics | 0.0346 | 1.2 | 1.0 | 0.167 |
| Alcohols | 0.00339 | 1.2 | 0.5 | 0.167 |
| Acids (organic) | 0.00319 | 1.2 | 0.5 | 0.167 |
| Ketones | 0.00383 | 1.2 | 0.5 | 0.167 |
| Esters | 0.0415 | 1.2 | 1.0 | 0.167 |
| Ethers | 0.0385 | 1.2 | 1.0 | 0.167 |
| Refrigerants | | | | |
| R20, R21, R22, R23 | 0.562 | 0.0 | 0.5 | −0.167 |
| Others | 0.494 | 0.0 | 0.5 | −0.167 |

Some estimated values of $\lambda_L$ found from Eqs. (10-9.1) and (10-9.2) are compared with experimental results in Table 10-7. Errors vary, but they are usually less than 10 percent. Many types of compounds (e.g., nitrogen- or sulfur-containing materials and aldehydes) cannot be treated, and problems arise if the compound may be fitted into two families. *m*-Cresol (Table 10-7) is an example. It could be considered an aromatic compound or an alcohol. In this case, we chose it to be an aromatic material, but the error would not have been very different if it had been considered an alcohol.

### Boiling-point method

Sato [90] suggested that, at the normal boiling point,

$$\lambda_L\,(T_b) = \frac{1.11}{M^{1/2}} \tag{10-9.3}$$

where $\lambda_L\,(T_b)$ = the thermal conductivity of the liquid at the normal boiling point (at 1 atm), W/(m·K)

$M$ = molecular weight, g/mol

To estimate $\lambda_L$ at other temperatures, the Riedel equation [135] may be used

$$\lambda_L = B[3 + 20(1 - T_r)^{2/3}] \tag{10-9.4}$$

so, combining Eqs. (10-9.3) and (10-9.4),

$$\lambda_L = \frac{(1.11/M^{1/2})[3 + 20(1 - T_r)^{2/3}]}{3 + 20(1 - T_{b_r})^{2/3}} \tag{10-9.5}$$

**TABLE 10-7 Comparison between Calculated and Experimental Values of Liquid Thermal Conductivity**

All values of $\lambda_L$ are in W/(m·K)

| | | | Percent error† calculated by the method of | | |
|---|---|---|---|---|---|
| Compound | $T$, K | $\lambda_L$, exp. | Latini et al. | Sato and Riedel | Missenard and Riedel |
| Propane | 323 | 0.0783 | −19 | 27 | 18 |
| $n$-Pentane | 293 | 0.114 | −5.7 | 20 | 17 |
| | 303 | 0.111 | −5.9 | 20 | 17 |
| $n$-Decane | 314 | 0.127 | −3.2 | −2.0 | 9.5 |
| | 349 | 0.119 | −2.9 | −1.8 | 9.8 |
| Cyclohexane | 293 | 0.124 | −1.2 | 11 | 3.7 |
| Methylcyclopentane | 293 | 0.121 | −3.2 | 13 | 3.8 |
| | 311 | 0.115 | −2.2 | 14 | 4.7 |
| Benzene | 293 | 0.148 | 0 | −3.4 | −5.1 |
| | 323 | 0.137 | 1.9 | −2.1 | −4.0 |
| | 389 | 0.114 | 5.1 | 0 | −1.8 |
| Ethylbenzene | 293 | 0.132 | 2.0 | 2.2 | 4.4 |
| | 353 | 0.118 | 2.9 | 3.2 | 5.3 |
| Ethanol | 293 | 0.165 | −3.3 | 15 | 24 |
| | 313 | 0.152 | 0 | 19 | 28 |
| | 347 | 0.135 | 3.5 | 22 | 32 |
| $n$-Octanol | 293 | 0.166 | −11 | −19 | 5.6 |
| $t$-Butyl alcohol | 311 | 0.116 | 4.5 | 26 | 77 |
| $m$-Cresol | 293 | 0.150 | 10 | −3.6 | 28 |
| | 353 | 0.145 | 3.8 | −8.6 | 21 |
| Aniline | 290 | 0.178 | — | −15 | 10 |
| Propionic acid | 285 | 0.173 | −8.9 | −3.4 | 15 |
| Methylene chloride | 253 | 0.159 | −17 | −13 | −6.3 |
| | 293 | 0.148 | −19 | −15 | −7.9 |
| Carbon tetrachloride | 253 | 0.110 | −6.4 | −0.8 | 15 |
| | 293 | 0.103 | −7.3 | −1.6 | 14 |
| Ethyl bromide | 293 | 0.103 | 2.0 | 7.7 | −6.9 |
| Chlorobenzene | 233 | 0.141 | −0.5 | 0 | 2.6 |
| | 353 | 0.111 | 2.4 | 4.1 | 7.1 |
| Iodobenzene | 253 | 0.106 | −15 | −0.4 | 5.1 |
| | 353 | 0.0938 | −17 | −0.9 | 4.5 |
| Ethyl acetate | 293 | 0.147 | 2.9 | −7.1 | 3.1 |
| | 333 | 0.141 | 2.4 | −12 | −2.7 |
| Butyl acetate | 293 | 0.137 | 2.5 | −4.9 | 9.2 |
| Acetone | 273 | 0.171 | −9.8 | −2.2 | 3.7 |
| | 313 | 0.151 | −6.9 | 0.5 | 6.6 |
| Diethyl ether | 293 | 0.129 | 3.9 | 4.5 | 22 |
| Acetaldehyde | 293 | 0.190 | — | −12 | −11 |

†Percent error = [(calc. − exp.)/exp.] × 100

Experimental values of the thermal conductivity were obtained from Refs. 19, 36, 65, and 173.

Equation (10-9.5) was employed to estimate $\lambda_L$ values in Table 10-7. The data required are $T$, $T_c$, $T_b$, and $M$. Errors varied widely. Poor results were found for low-molecular-weight hydrocarbons and branched hydrocarbons; generally the predicted value was larger than the experimental value. Better results were obtained for nonhydrocarbons.

### Method of Missenard

Missenard has suggested several methods to estimate the thermal conductivity of organic liquids. In one [108], he proposed that

$$\lambda_L = \frac{C(DT_b - T)}{(DT_b - E)} \tag{10-9.6}$$

where $C$, $D$, and $E$ were evaluated for several organic families. $T_b$ is the normal boiling point. In another technique [105], he recommended

$$\lambda_L\,(273\text{ K}) = \frac{(9.0 \times 10^{-3})(T_b\rho'/M)^{1/2}C'_p}{N^{1/4}} \tag{10-9.7}$$

where $\lambda_L$ (273 K) = liquid thermal conductivity at 273 K, W/(m·K)
$T_b$ = normal boiling temperature (at 1 atm), K
$\rho'$ = liquid density, mol/cm$^3$ at 273 K
$C'_p$ = liquid heat capacity at constant pressure at 273 K, J/(mol·K)
$M$ = molecular weight, g/mol
$N$ = number of atoms in the molecule

When this equation is combined with Eq. (10-9.4), one obtains

$$\lambda_L = \frac{\lambda_L\,(273\text{ K})\,[3 + 20(1 - T_r)^{2/3}]}{3 + 20[1 - (273/T_c)]^{2/3}} \tag{10-9.8}$$

Equation (10-9.8) with $\lambda_L$ (273 K) from Eq. (10-9.7) was used to calculate values for Table 10-7. In almost all cases, the estimated $\lambda_L$ exceeded the experimentally reported value. The constant $9.0 \times 10^{-3}$ in Eq. (10-9.7) should, perhaps, be decreased by 10 to 15 percent. In fact, Missenard did suggest a change to $8.4 \times 10^{-3}$ [104].

### Other liquid thermal conductivity estimation techniques

For nonpolar materials, the estimation procedures in Sec. 10-5 may be employed to obtain $\lambda_L$ when temperatures are well above the normal boiling point and accurate fluid densities are available. In particular, the Chung et al. and Ely and Hanley methods were specifically devised to treat liquid systems at high reduced temperatures as well as high-pressure gases.

Teja and Rice [161, 162] have suggested that, in some cases, values of $\lambda_L$ are available for compounds similar to the one of interest, and these data could be employed in an interpolative scheme as follows. Two liquids, similar chemically and with acentric factors bracketing the liquid of interest, are selected. The liquid thermal conductivities of these *reference* liquids should be known over the range of *reduced temperatures* of interest. We denote the properties of one reference fluid by a prime and the other by a double prime. Defining

$$\phi = \frac{V_c^{2/3} M^{1/2}}{T_c^{1/2}} \tag{10-9.9}$$

then $\lambda_L \phi$ is found by an interpolation based on the acentric factor $\omega$ as shown in Fig. 10-9.

$$\lambda_L \phi = (\lambda_L \phi)' + \frac{\omega - \omega'}{\omega'' - \omega'} [(\lambda_L \phi)'' - (\lambda_L \phi)'] \tag{10-9.10}$$

In Eq. (10-9.10) when one selects $\lambda_L'$ and $\lambda_L''$, they should be evaluated at the same reduced temperature as for the compound of interest. The procedure is illustrated in Example 10-9.

Mathur et al. [99] developed a correlation for the thermal conductivity of hydrocarbons as a function of the reduced temperature, critical compressibility factor, and the fluid reduced volume as well as the reduced volume at the freezing point. The procedure is applicable to both light and heavy hydrocarbons, and tests indicated errors of only about 5 to 8 percent. Ogiwara et al. [119] suggested a general estimation relation for

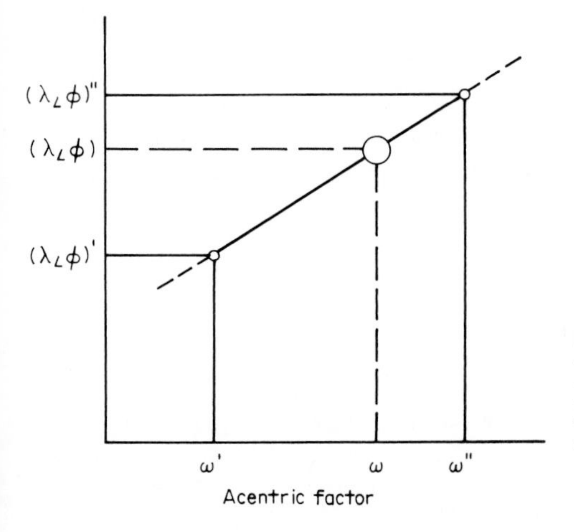


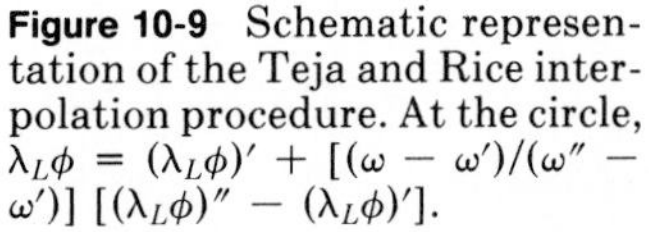
**Figure 10-9** Schematic representation of the Teja and Rice interpolation procedure. At the circle, $\lambda_L\phi = (\lambda_L\phi)' + [(\omega - \omega')/(\omega'' - \omega')] [(\lambda_L\phi)'' - (\lambda_L\phi)']$.

$\lambda_L$ for aliphatic alcohols. Jamieson [58] and Jamieson and Cartwright [60] proposed a general equation to correlate $\lambda_L$ over a wide temperature range (see Sec. 10-10), and they discuss how the constants in their equation vary with structure and molecular size. New liquid thermal conductivity data were reported for alcohols [58, 60, 119], alkyl amines [58, 60], esters [24], hydrocarbons [115, 118], and nitroalkanes [61].

### Discussion and recommendations

The brief comparison shown in Table 10-7 between experimental and estimated values of liquid thermal conductivity would indicate that the Latini et al. and Sato-Riedel methods are somewhat more reliable than the method of Missenard-Riedel, but there are exceptions. A comprehensive testing of the Sato-Riedel form by Baroncini et al. [3] indicated it also was a reasonably accurate estimation method. In many instances, the experimental data are not believed to be particularly reliable and the estimation errors are in the same range as the experimental uncertainity. This is clearly evident from the careful survey of liquid thermal conductivity data provided by the National Engineering Laboratory [58, 60, 65].

For *organic* liquids in the temperature region below the normal boiling point, we recommend either the Latini et al. or Sato-Riedel methods. Errors can vary widely, but they are usually less than 15 percent. There are very few reliable data for liquid thermal conductivities at reduced temperatures exceeding $T_r = 0.65$, and the relations discussed in this section are generally not recommended. If the liquid is nonpolar and is at a reduced temperature greater than about 0.8, one should use the high-pressure fluid correlations given in Sec. 10-5. (The Latini et al. procedure has, however, been applied successfully for refrigerants up to $T_r = 0.9$ [5].) None of the procedures predict the large increase in $\lambda$ near the critical point. The estimation techniques described in this section are illustrated in Examples 10-9 and 10-10.

**Example 10-9** Using the Teja and Rice scheme, estimate the thermal conductivity of liquid *t*-butyl alcohol at 318 K given the thermal conductivities of *n*-propanol and *n*-hexanol reported by Ogiwara et al. [119] as shown below.

*n*-Propanol $\lambda_L = 0.202 - 1.76 \times 10^{-4}\, T$ W/(m·K)

*n*-Hexanol $\lambda_L = 0.190 - 1.36 \times 10^{-4}\, T$ W/(m·K)

**Solution** From Appendix A:

| | $V_c$, cm³/mol | $T_c$, K | $M$, g/mol | $\omega$ |
|---|---|---|---|---|
| *n*-Propanol | 219 | 536.8 | 60.10 | 0.623 |
| *n*-Hexanol | 381 | 611 | 102.18 | 0.560 |
| *t*-Butyl alcohol | 275 | 506.2 | 74.12 | 0.612 |

Thus, with Eq. (10-9.9),

$\phi$ ($n$-propanol) = 12.16

$\phi$ ($n$-hexanol) = 21.49

$\phi$ ($t$-butyl alcohol) = 16.18

At 318 K, for $t$-butyl alcohol, $T_r$ = 318/506.2 = 0.629. At this reduced temperature, the appropriate temperature to use for $n$-propanol is (0.629)(536.8) = 337.6 K, and for $n$-hexanol it is (0.629)(611) = 384 K. With these, using the Ogiwara et al. correlations,

$$\lambda_L\ (n\text{-propanol}) = 0.202 - (1.76 \times 10^{-4})(337.6) = 0.143\ \text{W/(m}\cdot\text{K)}$$
$$\lambda_L\ (n\text{-hexanol}) = 0.190 - (1.36 \times 10^{-4})(384) = 0.138\ \text{W/(m}\cdot\text{K)}$$

Then, using Eq. (10-9.10) with $n$-propanol as the $'$ reference and $n$-hexanol as $''$ reference,

$$\lambda_L\ (16.18) = [(0.143)(12.16)] + \frac{0.612-0.623}{0.560-0.623}$$
$$\times\ [(0.138)(21.49) - (0.143)(12.16)]$$
$$= 1.95$$
$$\lambda_L = \frac{1.95}{16.18} = 0.121\ \text{W/(m}\cdot\text{K)}$$

Ogiwara et al. [119] report the experimental value of $t$-butyl alcohol at 318 K to be 0.128 W/(m·K).

**Example 10-10** Estimate the thermal conductivity of carbon tetrachloride at 293 K. At this temperature, Jamieson and Tudhope [65] list 11 values. Six are given a ranking of A and are considered reliable. They range from 0.102 to 0.107 W/(m·K). Most, however, are close to 0.103 W/(m·K).

**solution** The data (unless otherwise noted from Appendix A) are:

| | 273 K | | 293 K |
|---|---|---|---|
| Heat capacity, J/(mol·K) [136] | 130.7 | | 132.0 |
| Density, mol/cm$^3$ [136] | 0.0106 | | 0.0103 |
| Critical temperature, K | | 556.4 | |
| Normal boiling point (1 atm) | | 349.9 | |
| Molecular weight, g/mol | | 153.823 | |
| Number of atoms, $N$ | | 5 | |

LATINI ET AL. Assuming $CCl_4$ to be a refrigerant, by Eq. (10-9.2) and Table 10-6,

$$A = \frac{0.494(556.4)^{1/6}}{(153.823)^{1/2}} = 0.114$$

Then, with $T_r$ = 293/556.4 = 0.527 and Eq. (10-9.1),

$$\lambda_L = \frac{(0.114)(1 - 0.527)^{0.38}}{(0.527)^{1/6}} = 0.0954\ \text{W/(m}\cdot\text{K)}$$

$$\text{Error} = \frac{0.0954 - 0.103}{0.103} \times 100 = -7.4\%$$

SATO-RIEDEL. With Eq. (10-9.5) with $T_r = 0.527$ and $T_{b_r} = 349.9/556.4 = 0.629$, we have

$$\lambda_L = \frac{1.11}{(153.84)^{1/2}} \frac{3 + 20(1 - 0.527)^{2/3}}{3 + 20(1 - 0.629)^{2/3}} = 0.101 \text{ W/(m·K)}$$

$$\text{Error} = \frac{0.101 - 0.103}{0.103} \times 100 = -1.6\%$$

MISSENARD-RIEDEL. First, from Eq. (10-9.7), $\lambda_L$ (273 K), the thermal conductivity at 273 K is determined:

$$\lambda_L \text{ (273 K)} = (9.0 \times 10^{-3})[(349.9)(0.0106)]^{1/2} \frac{130.7}{(153.84)^{1/2}(5)^{1/4}}$$

$$= 0.122 \text{ W/(m·K)}$$

Then, with Eq. (10-9.8), and 273/556.4 = 0.491

$$\lambda_L = (0.122) \frac{3 + 20(1 - 0.527)^{2/3}}{3 + 20(1 - 0.491)^{2/3}} = 0.118 \text{ W/(m·K)}$$

$$\text{Error} = \frac{0.118 - 0.103}{0.103} \times 100 = 14\%$$

## 10-10 Effect of Temperature on the Thermal Conductivities of Liquids

Except for aqueous solutions, water, and some multihydroxy and multiamine molecules, the thermal conductivities of most liquids decrease with temperature. Below or near the normal boiling point, the decrease is nearly linear and is often represented over small temperature ranges by

$$\lambda_L = A - BT \tag{10-10.1}$$

where $A$ and $B$ are constants and $B$ generally is in the range of 1 to 3 × $10^{-4}$ W/(m·K$^2$). In Fig. 10-10, we show the temperature effect on $\lambda_L$ for a few liquids. Over wider temperature ranges, the correlation suggested by Riedel and given as Eq. (10-9.4) is preferable. Although not suited for water, glycerol, glycols, hydrogen, or helium, Jamieson [57] indicates that the equation represented well the variation of $\lambda_L$ with temperature for a wide range of compounds. Although, as noted earlier, few data for $\lambda_L$ exist over the temperature range from near the melting point to near the critical point, for those that are available, Jamieson [58] has found that neither Eq. (10-10.1) nor Eq. (10-9.4) is suitable, and he recommends

$$\lambda_L = A(1 + B\tau^{1/3} + C\tau^{2/3} + D\tau) \tag{10-10.2}$$

where $A$, $B$, $C$, and $D$ are constants and $\tau = 1 - T_r$. For nonassociating liquids, $C = 1 - 3B$ and $D = 3B$. With these simplifications, Eq. (10-10.2) becomes

$$\lambda_L = A[1 + \tau^{2/3} + B(\tau^{1/3} - 3\tau^{2/3} + 3\tau)] \tag{10-10.3}$$

As an example, in Fig. 10-10, if one fits the data for tributyl amine (a polar, but nonassociating, liquid), to Eq. (10-10.3), approximate values of $A$ and $B$ are $A$ = 0.0590 W/(m·K) and $B$ = 0.875. Using them, one can show by differentiating Eq. (10-10.3) that $d\lambda_L/dT$ decreases with increasing temperature, although, as is obvious from Fig. 10-10, the change in slope is not large in the temperature region shown. For other materials for which data are available over a quite wide temperature range, Eq. (10-10.3) is clearly preferable to Eq. (10-10.1) or (10-9.4) [59].

For associated liquids, $C$ = 1-2.6$B$ and $D \approx 6.5$ for alcohols and 6.0 for alkyd and dialkyd amines. Correlations for $C$ and $D$ for other types of associated molecules are not available. The constants $A$ and $B$ have been correlated, approximately, with carbon number for several homologous series [58, 60, 61].

For saturated liquids at high pressure, variations of $\lambda_L$ with temperature should probably be determined by using the high-pressure correlations in Sec. 10-5.

## 10-11 Effect of Pressure on the Thermal Conductivities of Liquids

At moderate pressures, up to 50 to 60 bar, the effect of pressure on the thermal conductivity of liquids is usually neglected, except near the critical point, where the liquid behaves more like a dense gas than a liquid

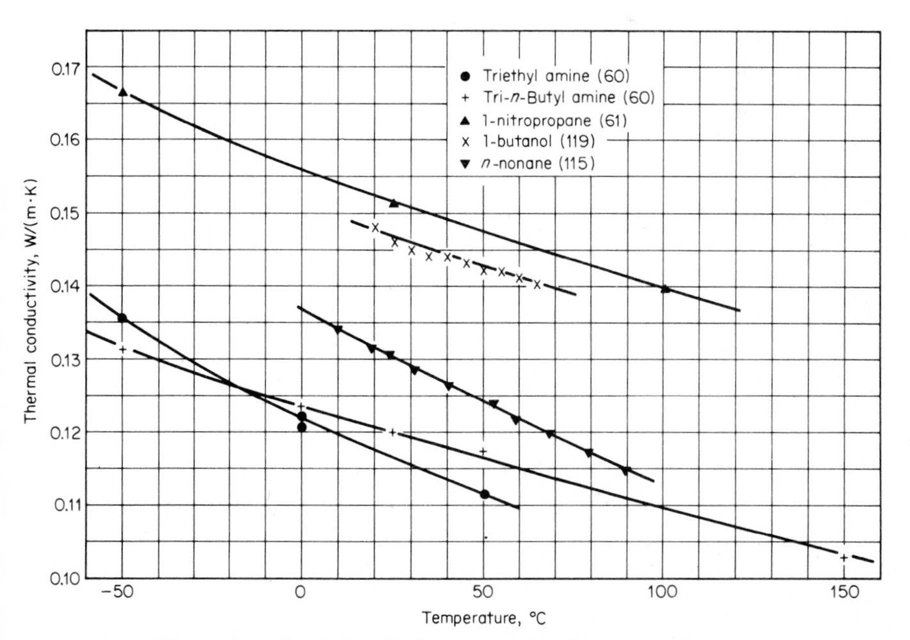


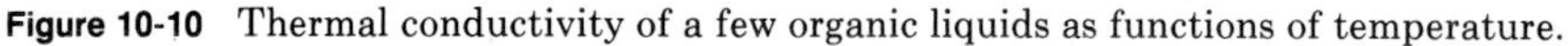

**Figure 10-10** Thermal conductivity of a few organic liquids as functions of temperature.

(see Sec. 10-5). At lower temperatures, $\lambda_L$ increases with pressure. Data showing the effect of pressure on a number of organic liquids are available in Refs. 12 and 64.

A convenient way of estimating the effect of pressure on $\lambda_L$ is by Eq. (10-11.1).

$$\frac{\lambda_2}{\lambda_1} = \frac{L_2}{L_1} \tag{10-11.1}$$

where $\lambda_2$ and $\lambda_1$ refer to liquid thermal conductivities at $T$ and pressures $P_2$ and $P_1$ and $L_2$ and $L_1$ are functions of the reduced temperature and pressure, as shown in Fig. 10-11. This correlation was devised by Lenoir [80]. Testing with data for 12 liquids, both polar and nonpolar, showed errors of only 2 to 4 percent. The use of Eq. (10-11.1) and Fig. 10-11 is illustrated in Example 10-11 with liquid $NO_2$, a material *not* used in developing the correlation.

**Example 10-11** Estimate the thermal conductivity of nitrogen dioxide at 311 K and 276 bar. The experimental value quoted is 0.134 W/(m·K) [132]. The value of $\lambda_L$ for the saturated liquid at 311 K and 2.1 bar is 0.124 W/(m·K) [132].

**solution** From Appendix A, $T_c$ = 431 K, $P_c$ = 101 bar; thus $T_r$ = 311/431 = 0.722, $P_{r1}$ = 2.1/101 = 0.021, and $P_{r2}$ = 276/101 = 2.73. From Fig. 10-11, $L_2$ = 11.75 and $L_1$ = 11.17. With Eq. (10-11.1),

$$\lambda_L \text{ (276 bar)} = (0.124)\,\frac{11.75}{11.17} = 0.130 \text{ W/(m·K)}$$

$$\text{Error} = \frac{0.130 - 0.134}{0.134} \times 100 = -3\%$$

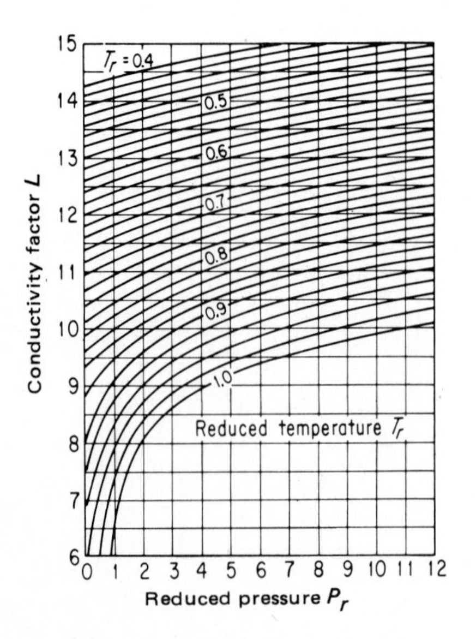


**Figure 10-11** Effect of pressure on liquid thermal conductivities. *(From Ref. 80.)*

Missenard [106] has proposed a simple correlation for $\lambda_L$ that extends to much higher pressures. In analytical form

$$\frac{\lambda_L (P_r)}{\lambda_L \text{ (low pressure)}} = 1 + QP_r^{0.7} \tag{10-11.2}$$

$\lambda_L (P_r)$ and $\lambda_L$ (low pressure) refer to liquid thermal conductivities at high and low, i.e., near saturation, pressure, both at the same temperature. $Q$ is a parameter given in Table 10-8. The correlation is shown in Fig. 10-12.

The correlations of Missenard and Lenoir agree up to a reduced pressure of 12, the maximum value shown for the Lenoir form.

**Example 10-12** Estimate the thermal conductivity of liquid toluene at 6330 bar and 304 K. The experimental value at this high pressure is 0.228 W/(m·K) [67]. At 1 bar and 304 K. $\lambda_L$ = 0.129 W/(m·K) [67].

**TABLE 10-8 Values of $Q$ in Eq. (10-11.2)**

| | Reduced pressure | | | | | |
|---|---|---|---|---|---|---|
| $T_r$ | 1 | 5 | 10 | 50 | 100 | 200 |
| 0.8 | 0.036 | 0.038 | 0.038 | (0.038) | (0.038) | (0.038) |
| 0.7 | 0.018 | 0.025 | 0.027 | 0.031 | 0.032 | 0.032 |
| 0.6 | 0.015 | 0.020 | 0.022 | 0.024 | 0.025 | 0.025 |
| 0.5 | 0.012 | 0.0165 | 0.017 | 0.109 | 0.020 | 0.020 |

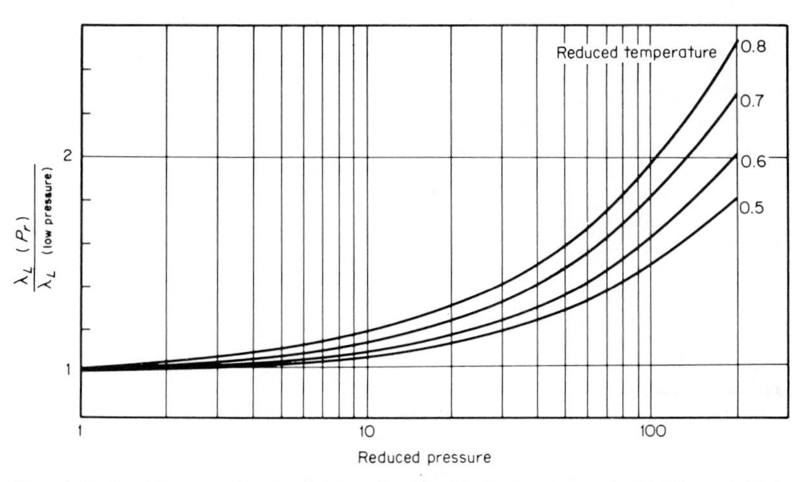


**Figure 10-12** Missenard correlation for liquid thermal conductivities at high pressures. *(From Ref. 106.)*

**solution** From Appendix A, $T_c$ = 591.8 K and $P_c$ = 41.0 bar. Therefore, $T_r$ = 304/591.8 = 0.514 and $P_r$ = 6330/41.0 = 154. From Table 10-8, $Q$ = 0.0205. Then, using Eq. (10-11.2),

$$\lambda_L\,(P_r) = (0.129)[1 + (0.0205)(154)^{0.7}] = 0.219 \text{ W/(m·K)}$$

$$\text{Error} = \frac{0.219 - 0.228}{0.228} \times 100 = -4\%$$

Latini and Baroncini [76] correlated the effect of pressure on liquid thermal conductivity by using Eq. (10-9.1), but they expressed the $A$ parameter as

$$A = A_0 + A_1 P_r \qquad (10\text{-}11.3)$$

Thus, $A_0$ would represent the appropriate $A$ parameter at low pressures, as described in Sec. 10-9 and given by Eq. (10-9.2). Values of $A_1$ were found to range from $6\times10^{-3}$ to $6\times10^{-4}$ W/(m·K); thus the term $A_1P_2$ is negligibly small except at quite high values of $P_r$. The authors have generalized the parameter $A_1$ for hydrocarbons as

$$A_1 = \frac{0.0673}{M^{0.84}} \text{ saturated hydrocarbons} \qquad (10\text{-}11.4)$$

$$A_1 = \frac{102.50}{M^{2.4}} \text{ aromatics} \qquad (10\text{-}11.5)$$

For hydrocarbons the authors found average errors usually less than 6 percent with maximum errors of 10 to 15 percent. The method should not be used for reduced pressures exceeding 50.

**Example 10-13** Rastorguev et al. [130], as quoted in [64], show the liquid thermal conductivity of $n$-heptane at 313 K to be 0.115 W/(m·K) at 1 bar. Estimate the thermal conductivity of the compressed liquid at the same temperature and 490 bar. The experimental value is 0.136 W/(m·K) [130].

**solution** From Appendix A, $T_c$ = 540.3 K, $P_c$ = 27.4 bar, and $M$ = 100.205 g/mol. Since we know the low-pressure value of $\lambda_L$, with Eq. (10-9.1), we can estimate A. With $T_r$ = 313/540.3 = 0.580,

$$0.115 = \frac{A(1 - 0.580)^{0.38}}{(0.580)^{1/6}}$$

$$A = 0.146 \text{ W/(m·K)}$$

This value of $A$ then becomes $A_0$ in Eq. (10-11.3). Using Eq. (10-11.4),

$$A_1 = \frac{0.0673}{(100.205)^{0.84}} = 1.40 \times 10^{-3}$$

Then, using Eqs. (10-9.1) and (10-11.3) with $P_r = 490/27.4 = 17.9$,

$$\frac{\lambda_L(P_r = 17.9)}{\lambda_L \text{ (low pressure)}} = \frac{A_0 + A_1 P_r}{A_0}$$

$$= \frac{[0.146 + (1.40 \times 10^{-3})(17.9)]}{(0.146)} = 1.17$$

$$\lambda_L (P_r = 17.9) = (0.115)(1.17) = 0.135 \text{ W/(m·K)}$$

$$\text{Error} = \frac{0.135 - 0.136}{0.136} \times 100 = -1\%$$

## 10-12 Thermal Conductivities of Liquid Mixtures

The thermal conductivities of most mixtures of organic liquids are usually less than those predicted by either a mole (or weight) fraction average, although the deviations are often small. We show data for several binaries in Fig. 10-13 to illustrate this point.

Many correlation methods for $\lambda_m$ have been proposed [38, 70, 81, 86, 88, 128, 129, 144, 154, 155, 161, 162, 173, 182]. Five were selected for presentation in this section. They are described separately and evaluated later

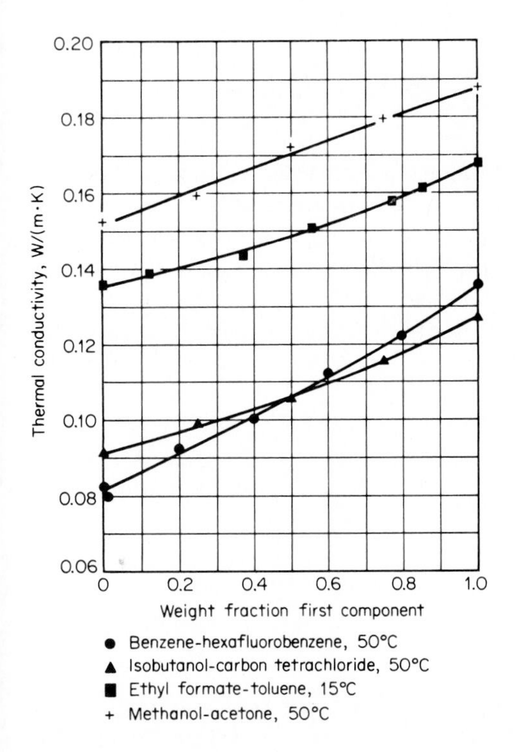


**Figure 10-13** Thermal conductivities of liquid mixtures.

when examples are presented to illustrate the methodology in using each of the methods.

There is a surprisingly large amount of experimental mixture data [7, 45, 62, 63, 64, 118, 119, 129, 154, 167], although most are for temperatures near ambient.

### Filippov equation

This correlation was developed by researchers in the Soviet Union [42, 43], and it has been extensively tested on many types of mixtures.

$$\lambda_m = w_1\lambda_1 + w_2\lambda_2 - 0.72w_1w_2(\lambda_2 - \lambda_1) \tag{10-12.1}$$

where $w_1$, $w_2$ are the weight fractions of components 1 and 2 and $\lambda_1$, $\lambda_2$ are the pure component thermal conductivities. The components were so chosen that $\lambda_2 \geq \lambda_1$. The constant 0.72 may be replaced by an adjustable parameter if binary mixture data are available. The technique is not suitable for multicomponent mixtures.

### Jamieson et al. correlation [64]

Research and data evaluation at the National Engineering Laboratory has suggested, for binary mixtures,

$$\lambda_m = w_1\lambda_1 + w_2\lambda_2 - \alpha(\lambda_2 - \lambda_1)[1 - (w_2)^{1/2}]w_2 \tag{10-12.2}$$

where $w_1$ and $w_2$ are weight fractions and, as in the Filippov method, the components are so selected that $\lambda_2 \geq \lambda_1$. $\alpha$ is an adjustable parameter that is set equal to unity if mixture data are unavailable for regression purposes. The authors indicate that Eq. (10-12.2) enables one to estimate $\lambda_m$ within about 7 percent (with a 95 percent confidence limit) for all types of binary mixtures with or without water. It cannot, however, be extended to multicomponent mixtures.

### Baroncini et al. correlation [1, 6, 7]

The Latini et al. method to estimate pure liquid thermal conductivities [Eq. (10-9.1)] has been adapted to treat binary liquid mixtures as shown in Eq. (10-12.3).

$$\lambda_m = \left[x_1^2A_1 + x_2^2A_2 + 2.2\left(\frac{A_1^3}{A_2}\right)^{1/2} x_1x_2\right]\frac{(1 - T_{rm})^{0.38}}{T_{rm}^{1/6}} \tag{10-12.3}$$

where $x_1$ and $x_2$ are the mole fractions of components 1 and 2. The $A$ parameters, introduced in Eq. (10-9.1), can be estimated from Eq. (10-9.2) and Table 10-6, or they can be calculated from pure component thermal conductivities. (See Example 10-13.) The reduced temperature of the mixture $T_{r_m} = T/T_{c_m}$ where

$$T_{c_m} = x_1 T_{c1} + x_2 T_{c2} \tag{10-12.4}$$

with $T_{c_1}$ and $T_{c_2}$ the pure component critical temperatures. The choice of which component is number 1 is made with criterion $A_1 \leq A_2$.

This correlation was tested [7] with over 600 datum points on 50 binary systems including those with highly polar components. The average error found was about 3 percent. The method is not suitable for multicomponent mixtures.

### Method of Rowley [140]

In this procedure, the liquid phase is modeled by using a two-liquid theory wherein the energetics of the mixture are assumed to favor local variations in composition. The basic relation assumed by Rowley is

$$\lambda_m = \sum_{i=1}^{n} w_i \sum_{j=1}^{n} w_{ji} \lambda_{ji} \tag{10-12.5}$$

where $\lambda_m$ = liquid mixture thermal conductivity, W/(m·K)
$w_i$ = weight fraction of component $i$
$w_{ji}$ = *local* weight fraction of component $j$ relative to a central molecule of component $i$
$\lambda_{ji}$ = characteristic parameter for the thermal conductivity that expresses the interactions between $j$ and $i$, W/(m·K)

Mass fractions were selected instead of mole fractions in Eq. (10-12.5) because it was found that the *excess* mixture thermal conductivity

$$\lambda_m^{\text{ex}} = \lambda_m - \sum_{i=1}^{n} w_i \lambda_i \tag{10-12.6}$$

was more symmetrical when weight fractions were employed.

The two-liquid (or local composition) theory was developed in Chap. 8 to derive several of the liquid activity coefficient-composition models. Rowley develops expressions for $w_{ij}$ and relates this quantity to parameters in the NRTL equation (see Table 8-3). In his treatment, he was able to show that Eq. (10-12.5) could be expressed as

$$\lambda_m = \sum_{i=1}^{n} w_i \frac{\sum_{j=1}^{n} w_j G_{ji} \lambda_{ji}}{\sum_{k=1}^{n} w_k G_{ki}} \tag{10-12.7}$$

where $G_{ji}$ and $G_{ij}$ (or $G_{ki}$ and $G_{ik}$) are the *same* NRTL parameters as used in activity coefficient correlations for the system of interest.

To obtain $\lambda_{ji}$ ($=\lambda_{ij}$), Rowley makes the important assumption that for any binary, say 1 and 2, $\lambda_m = \lambda_{12} = \lambda_{21}$ when the local *mole fractions* are equal, that is, $x_{12} = x_{21}$. Then, after some algebra, the final correlation is obtained.

$$\lambda_m = \sum_{i=1}^{n} w_i\lambda_i + \sum_{i=1}^{n} w_i \frac{\sum_{j=1}^{n} w_j G_{ji}(\lambda_{ji} - \lambda_i)}{\sum_{k=1}^{n} w_k G_{ki}} \qquad (10\text{-}12.8)$$

with

$$\lambda_{ij} = \lambda_{ji} = \frac{w^*_{i\,j}\, w^*_{i\,j}\, \lambda_i + w^*_{j\,i} w^*_{j\,ji}\, \lambda_j}{w^*_{i\,j} w^*_{i\,ij} + w^*_{j\,i} w^*_{j\,ji}} \qquad (10\text{-}12.9)$$

$$w^*_{i\,ij} = \frac{w^*_{i\,j}}{\sum_{k=1}^{n} w^*_{k\,i} G_{ki}} \qquad (10\text{-}12.10)$$

$$w^*_{i\,j} = \frac{M_i (G_{ji})^{1/2}}{M_i (G_{ji})^{1/2} + M_j (G_{ij})^{1/2}} \qquad (10\text{-}12.11)$$

$w^*_{j\,ji}$ and $w^*_{j\,i}$ are defined in a similar manner with the $i$, $j$ subscripts interchanged. Note also that $G_{ii} = G_{jj} = 1$.

For a binary system of 1 and 2, Eqs. (10-12.8) to (10-12.11) become, after some simplification,

$$\lambda_m = w_1\lambda_1 + w_2\lambda_2 + w_1 w_2(\lambda_2 - \lambda_1)R \qquad (10\text{-}12.12)$$

$$R = \frac{G_{21}}{(w_2 G_{21} + w_1)(1 + Y)} - \frac{G_{12}}{(w_2 + w_1 G_{12})(1 + Y^{-1})} \qquad (10\text{-}12.13)$$

$$Y = \left(\frac{w^*_{12}}{w^*_{21}}\right)^2 \frac{w^*_{12} G_{12} + w^*_{21}}{w^*_{12} + w^*_{21} G_{21}} \qquad (10\text{-}12.14)$$

$$w^*_{12} = \frac{M_1 G_{21}^{1/2}}{M_1 G_{21}^{1/2} + M_2 G_{12}^{1/2}} \qquad (10\text{-}12.15)$$

$$w^*_{21} = 1 - w^*_{12} \qquad (10\text{-}12.16)$$

When the Rowley correlation is written in the form of Eq. (10-12.12), it is clear that the entire nonideal effect is included in the $R$ parameter, and the form is quite similar to the Filippov and Jamieson et al. relations described earlier.

To employ this technique, values for the liquid thermal conductivities of all pure components are required. In addition, from data sources or from regressing vapor-liquid equilibrium data, the NRTL parameters, $G_{ij}$ and $G_{ji}$ must be found. The concept of relating transport and thermodynamic properties is an interesting one and bears further study; Brulé and Starling [18] also have advocated such an approach. Rowley's method was devised to treat multicomponent mixtures, although, to date, most of the testing has been on binary systems.

### Li method [82]

In another technique developed for multicomponent systems, Li proposed

$$\lambda_m = \sum_{i=1}^{n} \sum_{j=1}^{n} \phi_i \phi_j \lambda_{ij} \tag{10-12.17}$$

with

$$\lambda_{ij} = 2(\lambda_i^{-1} + \lambda_j^{-1})^{-1} \tag{10-12.18}$$

$$\phi_i = \frac{x_i V_i}{\sum_{j=1}^{n} x_j V_j} \tag{10-12.19}$$

$x_i$ is the mole fraction of component $i$, and $\phi_i$ is the superficial volume fraction of $i$. $V_i$ is the molar volume of the pure liquid. For a binary system of 1 and 2, Eq. (10-12.17) becomes

$$\lambda_m = \phi_1^2 \lambda_1 + 2\phi_1 \phi_2 \lambda_{12} + \phi_2^2 \lambda_2 \tag{10-12.20}$$

The harmonic mean approximation for $\lambda_{ij}$ was chosen over a geometric or arithmetic mean after extensive testing and comparison of calculated and experimental values of $\lambda_m$. Also, it was found that the $V_i$ terms in Eq. (10-12.19) could be replaced by critical volumes for nonaqueous liquid systems without affecting the results significantly.

### Discussion

All five methods for estimating $\lambda_m$ described in this section have been extensively tested by using binary mixture data, and all show approximately the same average error [182]. All require the thermal conductivities of the pure components making up the system (or an estimate of the values), and thus they are interpolative in nature. The Filippov and Jamieson et al. procedures require no addition information for a binary other than the weight fractions and pure component values of $\lambda_L$. The Baroncini et al. method also needs pure component critical properties, and Li's technique utilizes liquid volumes. Rowley's correlation requires the NRTL parameters $G_{ij}$ and $G_{ji}$ from phase equilibrium data. Only Li's and Rowley's methods will treat multicomponent mixtures. In Fig. 10-14 we show some recent measurements of Usmanov and Salikov for very polar systems and illustrate how well Filippov's relation (10-12.1) fits

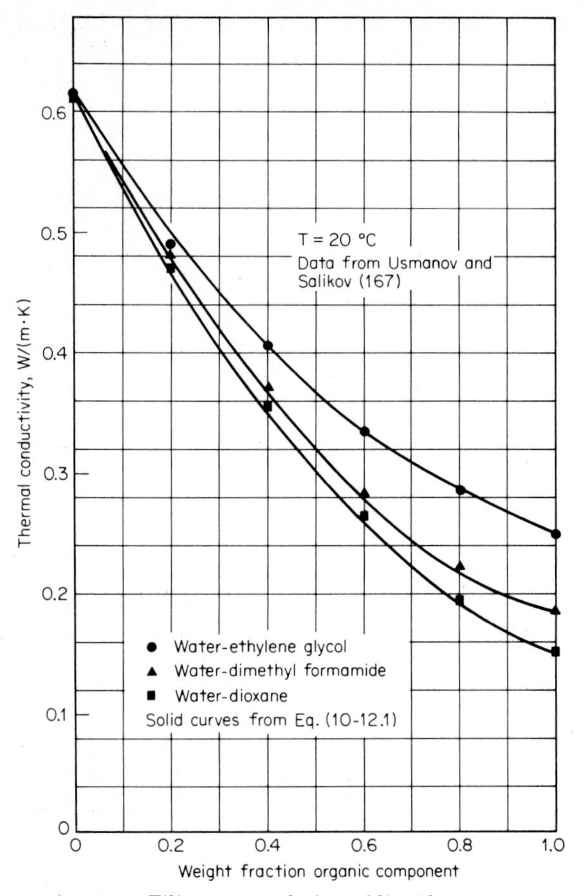


**Figure 10-14** Filippov correlation of liquid mixture thermal conductivity.

these data. The other methods described in this section would have been equally satisfactory. Gaitonde et al. [45] measured $\lambda_m$ for liquid mixtures of alkanes and silicone oils to study systems with large differences in the molecular sizes of the components. They found the Filippov and Jamieson et al. correlations provide a good fit to the data, but they recommended the general form of McLaughlin [88] with the inclusion of an adjustable binary parameter.

In summary, one can use any of the relations described in this section to estimate $\lambda_m$ with the expectation that errors will rarely exceed about 5 percent.

In the case of aqueous *(dilute)* solutions containing electrolytes, the mixture thermal conductivity usually decreases with an increase in the concentration of the dissolved salts. To estimate the thermal conductivity of such mixtures, Jamieson and Tudhope [65] recommend the use of an equation proposed originally by Riedel [135] and tested by Vargaftik and Os'minin [172]. At 293 K:

$$\lambda_m = \lambda\ (H_2O) + \Sigma\ \sigma_i C_i \tag{10-12.21}$$

where $\lambda_m$ = thermal conductivity of the ionic solution at 293 K, W/(m·K)
$\lambda\ (H_2O)$ = thermal conductivity of water at 293 K, W/(m·K)
$C_i$ = concentration of the electrolyte, mol/L
$\sigma_i$ = coefficient that is characteristic for each ion

Values of $\sigma_i$ are shown in Table 10-9. To obtain $\lambda_m$ at other temperatures,

$$\lambda_m\ (T) = \lambda_m\ (293)\ \frac{\lambda\ (H_2O \text{ at } T)}{\lambda\ (H_2O \text{ at 293 K})} \tag{10-12.22}$$

Except for strong acids and bases at high concentrations, Eqs. (10-12.21) and (10-12.22) are usually accurate to within ±5 percent.

**Example 10-14** Using Filippov's, Jamieson et al.'s, and Li's methods, estimate the thermal conductivity of a liquid mixture of methanol and benzene at 273 K. The weight fraction methanol is 0.4. At this temperature, the thermal conductiv-

**TABLE 10-9 Values of $\sigma_i$ for Anions and Cations in Eq. (10-12.21) [65]**

| Anion | $\sigma_i \times 10^5$ | Cation | $\sigma_i \times 10^5$ |
|---|---|---|---|
| $OH^-$ | 20.934 | $H^+$ | −9.071 |
| $F^-$ | 2.0934 | $Li^+$ | −3.489 |
| $Cl^-$ | −5.466 | $Na^+$ | 0.000 |
| $Br^-$ | −17.445 | $K^+$ | −7.560 |
| $I^-$ | −27.447 | $NH_4^+$ | −11.63 |
| $NO_2^-$ | −4.652 | $Mg^{2+}$ | −9.304 |
| $NO_3^-$ | −6.978 | $Ca^{2+}$ | −0.5815 |
| $ClO_3^-$ | −14.189 | $Sr^{2+}$ | −3.954 |
| $ClO_4^-$ | −17.445 | $Ba^{2+}$ | −7.676 |
| $BrO_3^-$ | −14.189 | $Ag^+$ | −10.47 |
| $CO_3^{2-}$ | −7.560 | $Cu^{2+}$ | −16.28 |
| $SiO_3^{2-}$ | −9.300 | $Zn^{2+}$ | −16.28 |
| $SO_3^{2-}$ | −2.326 | $Pb^{2+}$ | −9.304 |
| $SO_4^{2-}$ | 1.163 | $Co^{2+}$ | −11.63 |
| $S_2O_3^{2-}$ | 8.141 | $Al^{3+}$ | −32.56 |
| $CrO_4^{2-}$ | −1.163 | $Th^{4+}$ | −43.61 |
| $Cr_2O_7^{2-}$ | 15.93 | | |
| $PO_4^{3-}$ | −20.93 | | |
| $Fe(CN)_6^{4-}$ | 18.61 | | |
| Acetate$^-$ | −22.91 | | |
| Oxalate$^{2-}$ | −3.489 | | |

ities of pure benzene and methanol are 0.152 and 0.210 W/(m·K) [62], respectively. The experimental mixture value is 0.170 W/(m·K).

**solution** FILIPPOV'S METHOD. We use Eq. (10-12.1). Here methanol is component 2, since $\lambda$ (methanol) > $\lambda$ (benzene). Thus,

$$\lambda_m = (0.6)(0.152) + (0.4)(0.210) - (0.72)(0.6)(0.4)(0.210 - 0.152) = 0.165 \text{ W/(m·K)}$$

$$\text{Error} = \frac{0.165 - 0.170}{0.170} \times 100 = -3\%$$

JAMIESON ET AL. METHOD. Again methanol is chosen as component 2. With Eq. (10-12.2) and $\alpha = 1$,

$$\lambda_m = (0.6)(0.152) + (0.4)(0.210) - (0.210 - 0.152)[1 - (0.4)^{1/2}](0.4)$$
$$= 0.167 \text{ W/(m·K)}$$

$$\text{Error} = \frac{0.167 - 0.170}{0.170} \times 100 = -2\%$$

LI METHOD. With Eq. (10-12.18),

$$\lambda_{12} = 2[(0.152)^{-1} + (0.210)^{-1}]^{-1} = 0.176 \text{ W/(m·K)}$$

At 273 K, $V$ (methanol) = 39.6 cm$^3$/mol and $V$ (benzene) = 88.9 cm$^3$/mol. If the weight fraction methanol in the mixture is 0.4, the mole fraction is 0.619. Then, with Eq. (10-12.19),

$$\phi \text{ (methanol)} = \frac{(0.619)(39.6)}{(0.619)(39.6) + (0.391)(88.9)}$$
$$= 0.414$$

$$\phi \text{ (benzene)} = 1 - 0.414 = 0.586$$

Using Eq. (10-12.20),

$$\lambda_m = (0.414)^2(0.210) + (0.586)^2(0.152) + (2)(0.414)(0.586)(0.176)$$
$$= 0.174 \text{ W/(m·K)}$$

$$\text{Error} = \frac{0.174 - 0.170}{0.170} \times 100 = 2\%$$

**Example 10-15** Estimate the liquid thermal conductivity of a mixture of benzene (1) and methyl formate (2) at 323 K by using the method of Baroncini et al. At this temperature, the values of $\lambda_L$ for the pure components are $\lambda_1 = 0.138$ and $\lambda_2 = 0.179$ W/(m·K) [7].

**solution** We will estimate the values of $\lambda_m$ at 0.25, 0.50, and 0.75 weight fraction benzene. First, however, we need to determine $A_1$ and $A_2$. Although Eq. (10-9.2) and Table 10-6 could be used, it is more convenient to employ the pure component values of $\lambda_L$ with Eq. (10-9.1). From Appendix A, $T_{c_1} = 562.2$ K and $T_{c_2} = 487.2$ K, so $T_{r_1} = 323/562.2 = 0.575$ and $T_{r_2} = 323/487.2 = 0.663$. Then, with Eq. (10-9.1), for benzene,

$$0.138 = \frac{A_1(1 - 0.575)^{0.38}}{(0.575)^{1/6}} \qquad A_1 = 0.174$$

Similarly $A_2 = 0.252$. [Note that, if Eq. (10-9.2) and Table 10-6 had been used, we would have $A_1 = 0.176$ and $A_2 = 0.236$.]
We have selected components 1 and 2 to agree with the criterion $A_1 \leq A_2$.

Consider first a mixture containing 0.25 weight fraction benzene, i.e., $w_1 = 0.25$ and $w_2 = 0.75$. Then, the mole fractions are $x_1 = 0.204$ and $x_2 = 0.796$. Thus,

$$T_{cm} = (0.204)(562.2) + (0.796)(487.2) = 502.5 \text{ K}$$

$$T_{rm} = \frac{323}{502.5} = 0.643$$

With Eq. (10-12.3),

$$\lambda_m = \left\{(0.204)^2(0.174) + (0.796)^2(0.252) + (2.2)\left[\frac{(0.174)^3}{0.252}\right]^{1/2}(0.204)(0.796)\right\}\frac{(1 - 0.643)^{0.38}}{(0.643)^{1/6}}$$

$$= 0.159 \text{ W/(m·K)}$$

Calculated results for this and other compositions are shown below with the experimental values and percent errors.

**Benzene-Methyl Formate Mixtures; $T$ = 323 K**

| Weight fraction benzene | Mole fraction benzene | $T_{cm}$, K | $\lambda_m$, calc., W/(m·K) | $\lambda_m$, exp., W/(m·K) | Percent error |
|---|---|---|---|---|---|
| 0.25 | 0.204 | 502.5 | 0.159 | 0.158 | 0.6 |
| 0.50 | 0.435 | 519.8 | 0.143 | 0.151 | −5.3 |
| 0.75 | 0.698 | 539.6 | 0.135 | 0.140 | −3.6 |

**Example 10-16** Use Rowley's method to estimate the thermal conductivity of a liquid mixture of acetone (1) and chloroform (2) that contains 66.1 weight percent of the former. The temperature is 298 K. As quoted by Jamieson et al. [64], Rodriguez [137] reports $\lambda_1 = 0.161$ W/(m·K), $\lambda_2 = 0.119$ W/(m·K), and for the mixture, $\lambda_m = 0.143$ W/(m·K).

**solution** First, we need the NRTL parameters for this binary at 298 K. Nagata [114] suggests $G_{12} = 1.360$ and $G_{21} = 0.910$. From Appendix A, $M_1 = 58.08$ and $M_2 = 119.38$ g/mol. Using Eqs. (10-12.15) and (10-12.16),

$$w_{12}^* = \frac{(58.08)(0.910)^{1/2}}{(58.08)(0.910)^{1/2} + (119.38)(1.360)^{1/2}}$$

$$= 0.285$$

$$w_{21}^* = 1 - 0.285 = 0.715$$

With Eqs. (10-12.13) and (10-12.14),

$$Y = \left(\frac{0.285}{0.715}\right)^2 \frac{(0.285)(1.360) + 0.715}{0.285 + (0.715)(0.910)}$$

$$= 0.187$$

$$R = \frac{0.910}{[(0.910)(0.339) + 0.661](1 + 0.187)} - \frac{1.360}{[(0.339 + (0.661)(1.360)](1 + 0.187^{-1})}$$

$$= 0.618$$

Then, with Eq. (10-12.12),

$$\lambda_m = (0.661)(0.161) + (0.339)(0.119) + (0.661)(0.339)(0.119 - 0.161)(0.618)$$
$$= 0.141 \text{ W/(m·K)}$$

$$\text{Error} = \frac{0.141 - 0.143}{0.143} \times 100 = -1\%$$

## Notation

| | |
|---|---|
| $A$ | parameter in Eq. (10-9.1), W/(m·K) |
| $A_{ij}$ | Wassiljewa coefficient, Eq. (10-6.1) |
| $B_i$ | parameter in Eq. (10-5.9) |
| $C_i$ | electrolyte concentration, mol/L, Eq. (10-12.21) |
| $C$ | heat capacity, J/(mol·K); $C_v$, at constant volume; $C_p$, at constant pressure; $C_{rot}$, due to rotational degrees of freedom; $C_{int}$, due to internal degrees of freedom; $C_{tr}$, due to translational motion |
| $C$ | group contribution constant in Eq. (10-3.15) |
| $D$ | diffusion coefficient, $m^2$/s |
| $f$ | scaling parameter in Eqs. (10-3.24) and (10-5.15) |
| $f_{tr}$ | translational factor in Eq. (10-3.1) |
| $f_{int}$ | internal energy factor in Eq. (10-3.1) |
| $F_r$ | number of degrees of freedom for external rotation |
| $G_1$ | parameter in Eq. (10-5.7) |
| $G_2$ | parameter in Eq. (10-5.8) |
| $G_{ij}$ | NRTL parameter, Eq. (10-12.7) |
| $h$ | scaling parameter in Eqs. (10-3.25) and (10-5.16) |
| $H$ | parameter in Eqs. (10-3.31) and (10-5.23) |
| $k$ | Boltzmann's constant, J/K |
| $L$ | mean free path, m |
| $L$ | parameter shown in Fig. 10-11 |
| $m$ | molecular mass, g or kg |
| $M$ | molecular weight, g/mol |
| $M'$ | molecular weight, kg/mol unless otherwise noted |
| $M'_m$ | molecular weight defined in Eq. (10-7.22) |
| $n$ | number density of molecules, $m^{-3}$; number of components in a mixture |
| $N$ | number of atoms in a molecule |
| $N_{Pr}$ | Prandtl number, $C_p\eta/\lambda M'$ |
| $N_0$ | Avogadro's number |
| $P$ | pressure, $N/m^2$ or bar; $P_c$, critical pressure; $P_r$, reduced pressure, $P/P_c$ |
| $q$ | parameter defined in Eq. (10-5.5) |

$Q$ parameter in Table 10-8, Fig. 10-12

$R$ gas constant, J/(mol·K)

$R$ parameter in Eq. (10-12.13)

$T$ temperature, K; $T_c$, critical temperature; $T_r$, reduced temperature, $T/T_c$; $T^+$, reduced temperature parameter in Eqs. (10-3.21) and (10-5.11); $T_0$, equivalent temperature in Eqs. (10-3.26) and (10-5.17); $T_b$, normal boiling point (at 1 atm); $T_{b_r} = T_b/T_c$

$v$ molecular velocity, m/s

$V$ molar volume, $cm^3$/mol or $m^3$/mol; $V_c$, critical volume; $V^+$, reduced volume parameter in Eq. (10-5.12)

$w_i$ weight fraction of component $i$

$w^*$ parameter in Eqs. (10-12.10) and (10-12.11)

$x_i$ mole fraction of component $i$ in a liquid mixture

$Y$ parameter in Eq. (10-12.14)

$Z$ compressibility factor $PV/RT$; $Z_c$, critical compressibility factor

$Z$ parameter defined under Eq. (10-3.17)

$Z_{rot}$ collision number

GREEK

$\alpha$ parameter defined under Eq. (10-3.17); parameter in Eq. (10-12.2)

$\beta$ parameter defined under Eq. (10-3.17)

$\gamma$ $C_p/C_v$

$\Gamma$ reduced, inverse thermal conductivity defined in Eq. (10-3.11)

$\epsilon$ interaction energy parameter, J; parameter in Eq. (10-6.2)

$\eta$ viscosity, N·s/$m^2$; $\eta^*$, viscosity parameter in Eq. (10-3.30); $\eta_0$, viscosity parameter in Eq. (10-3.27); $\eta°$, low-pressure gas viscosity

$\Theta$ shape factor in Eqs. (10-3.22) and (10-5.13)

$\kappa$ association constant, see Table 9-1

$\lambda$ thermal conductivity, W/(m·K); $\lambda_r$, reduced thermal conductivity, $\Gamma\lambda$; $\lambda^*$, thermal conductivity parameter in Eq. (10-3.29); $\lambda_0$, thermal conductivity parameter in Eq. (10-3.28); $\lambda°$, low-pressure gas thermal conductivity; $\lambda^{**}$, thermal conductivity parameter in Eqs. (10-5.22) and (10-7.28); $\lambda_{tr}$, monatomic value of thermal conductivity; $\lambda'$, parameter in Eq. (10-7.12)

$\mu$ dipole moment, debye; $\mu_r$, reduced dipole moment defined in Eq. (9-4.11)

$\rho$ molar density, mol/$cm^3$ or mol/$m^3$, $\rho_c$, critical density; $\rho_r$, reduced density, $\rho/\rho_c$; $\rho_0$, density parameter in Eq. (10-5.18); $\rho'$, liquid density, mol $cm^3$, at 273 K

$\sigma$ characteristic dimension of the molecule, m or Å; $\sigma_i$, ion coefficient in Table 10-9

$\tau$ $1 - T_r$

$\phi$ parameter in Eq. (10-9.9); volume fraction in Eq. (10-12.19)

$\Phi$ shape factor in Eqs. (10-3.23) and (10-5.14)

$\Psi$ parameter defined under Eq. (10-3.17)

$\Omega_v$ collision integral for viscosity and thermal conductivity; $\Omega_D$, collision integral for diffusion coefficients

$\omega$ acentric factor

SUBSCRIPTS

$m$ mixture

$L$ liquid

$G$ gas

SUPERSCRIPTS

ex excess property

$'$, $''$ reference properties

## REFERENCES


1. Baroncini, C., P. Di Filippo, and G. Latini: "Comparison between Predicted and Experimental Thermal Conductivity Values for the Liquid Substances and the Liquid Mixtures at Different Temperatures and Pressures," paper presented at the *Workshop on Thermal Conductivity Measurement, IMEKO, Budapest, March 14–16, 1983.*
2. Baroncini, C., P. Di Filippo, and G. Latini: *Intern. J. Refrig.,* **6**(1): 60 (1983).
3. Baroncini, C., P. Di Filippo, G. Latini, and M. Pacetti: *High Temp.—High Press.,* **11:** 581 (1979).
4. Baroncini, C., P. Di Filippo, G. Latini, and M. Pacetti: *Intern. J. Thermophys.,* **1**(2): 159 (1980).
5. Baroncini, C., P. Di Filippo, G. Latini, and M. Pacetti: *Intern. J. Thermophys.,* **2**(1): 21 (1981).
6. Baroncini, C., P. Di Filippo, G. Latini, and M. Pacetti: *Thermal Cond.,* 1981 (pub. 1983), 17th, Plenum Pub. Co, p. 285.
7. Baroncini, C., G. Latini, and P. Pierpaoli: *Intern. J. Thermophys.,* **5**(4): 387 (1984).
8. Barua, A. K., A. Manna, and P. Mukhopadhyay: *J. Chem. Phys.,* **49:** 2422 (1968).
9. Basu, R. S., and J. V. Sengers: "Thermal Conductivity of Fluids in the Critical Region," in *Thermal Conductivity,* D. C. Larson (ed.), Plenum, New York, 1983, p. 591.
10. Bennett, L. A., and R. G. Vines: *J. Chem. Phys.,* **23:** 1587 (1955).
11. Bretsznajder, S., *"Prediction of Transport and Other Physical Properties of Fluids,"* trans. by J. Bandrowski, Pergamon, New York, 1971, p. 251.
12. Bridgman, P. W.: *Proc. Am. Acad. Art Sci.,* **59:** 154 (1923).
13. Brokaw, R. S.: *J. Chem. Phys.,* **29:** 391 (1958).
14. Brokaw, R. S.: *Int. J. Eng. Sci.,* **3:** 251 (1965).
15. Brokaw, R. S.: *Ind. Eng. Chem. Process Design Develop.,* **8:** 240 (1969).
16. Brokaw, R. S., and C. O'Neal, Jr.: "Rotational Relaxation and the Relation between Thermal Conductivity and Viscosity for Some Nonpolar Polyatomic Gases," *9th Intern. Symp. Combust.,* Academic, New York, 1963, p. 725.
17. Bromley, L. A.: "Thermal Conductivity of Gases at Moderate Pressures," *Univ. California Rad. Lab. UCRL-1852,* Berkeley, Calif., June 1952.

18. Brulé, M. R., and K. E. Starling: *Ind. Eng. Chem. Process Design Develop.*, **23:** 833 (1984).
19. Brykov, V. P., G. Kh. Mukhamedzyanov, and A. G. Usmanov: *J. Eng. Phys. USSR*, **18**(1): 62 (1970).
20. Carmichael, L. T., V. Berry, and B. H. Sage: *J. Chem. Eng. Data,* **8:** 281 (1963).
21. Carmichael, L. T., H. H. Reamer, and B. H. Sage: *J. Chem. Eng. Data,* **11:** 52 (1966).
22. Carmichael, L. T., and B. H. Sage: *J. Chem. Eng. Data,* **9:** 511 (1964).
23. Chapman, S., and T. G. Cowling: *The Mathematical Theory of Non-uniform Gases,* Cambridge, New York, 1961.
24. Chase, J. D.: personal communication, Celanese Chemical Co., Corpus Christi, Tex., August 1984.
25. Christensen, P. L., and A. Fredenslund: *J. Chem. Eng. Data,* **24:** 281 (1979).
26. Christensen, P. L., and A. A. Fredenslund: *Chem. Eng. Sci.,* **35:** 871 (1980).
27. Chung, T.-H., M. Ajlan, L. L. Lee, and K. E. Starling: "Generalized Multiparameter Corresponding State Correlation for Polyatomic, Polar Fluid Transport Properties," *Ind. Eng. Chem. Process Design Develop.*, submitted, 1986.
28. Chung, T.-H., L. L. Lee, and K. E. Starling: *Ind. Eng. Chem. Fundam.*, **23:** 8 (1984).
29. Cohen, Y., and S. I. Sandler: *Ind. Eng. Chem. Fundam.,* **19:** 186 (1980).
30. Correla, F. von, B. Schramm, and K. Schaefer: *Ber. Bunsenges. Phys. Chem.,* **72**(3): 393 (1968).
31. Cottrell, T. L., and J. C. McCoubrey: *Molecular Energy Transfer in Gases,* Butterworth, London, 1961.
32. Couch, E. J., and K. A. Dobe: *J. Chem. Eng. Data,* **6:** 229 (1961).
33. Cowling, T. G.: *Brit. J. Appl. Phys.,* **15:** 959 (1964).
34. Crooks, R. G., and T. E. Daubert: *Ind. Eng. Chem. Process Design Develop.,* **18:** 506 (1979).
35. Curtiss, L. A., D. J. Frurip, and M. Blander: *J. Phys. Chem.,* **86:** 1120 (1982).
36. Djalalian, W. H.: *Kaeltechnik,* **18**(11): 410 (1966).
37. Donaldson, A. B.: *Ind. Eng. Chem. Fundam.,* **14:** 325 (1975).
38. Dul'nev, G. N., and Y. P. Zarichnyak: *Thermophysical Properties of Substances and Materials,* 3d issue, Standards Publications, Moscow, 1971, p. 103.
39. Ehya, H., F. M. Faubert, and G. S. Springer: *J. Heat Transfer,* **94:** 262 (1972).
40. Ely, J. F. and H. J. M. Hanley: *Ind. Eng. Chem. Fundam.,* **22:** 90 (1983).
41. Ewing, C. T., B. E. Walker, J. A. Grand, and R. R. Miller: *Chem. Eng. Progr. Symp. Ser.,* **53**(20): 19(1957).
42. Filippov, L. P.: *Vest. Mosk. Univ., Ser. Fiz. Mat. Estestv. Nauk,* (8)**10**(5): 67–69 (1955); *Chem. Abstr.,* **50:** 8276 (1956).
43. Filippov, L. P., and N. S. Novoselova: *Vestn. Mosk. Univ., Ser. Fiz. Mat. Estestv. Nauk,* (3)**10**(2): 37–40(1955); *Chem. Abstr.,* **49:** 11366 (1955).
44. Fleeter, R. J., Kestin, and W. A. Wakeham: *Physica A (Amsterdam),* **103A:** 521 (1980).
45. Gaitonde, U. N., D. D. Deshpande, and S. P. Sukhatme: *Ind. Eng. Chem. Fundam.,* **17:** 321 (1978).
46. Gambill, W. R.: *Chem. Eng.,* **66**(16): 129 (1959).
47. Gilmore, T. F., and E. W. Comings: *AIChE J.,* **12:** 1172 (1966).
48. Gray, P., S. Holland, and A. O. S. Maczek: *Trans. Faraday Soc.,* **66:** 107 (1970).
49. Groenier, W. S., and G. Thodos: *J. Chem. Eng. Data,* **6:** 240 (1961).
50. Guildner, L. A.: *Proc. Natl. Acad. Sci.,* **44:** 1149 (1958).
51. Hanley, H. J. M.: *Cryogenics,* **16**(11): 643 (1976).
52. Healy, R. N., and T. S. Storvick: *J. Chem. Phys.,* **50:** 1419 (1969).
53. Hirschfelder, J. O.: *J. Chem. Phys.,* **26:** 282 (1957).
54. Hirschfelder, J. O., C. F. Curtiss, and R. B. Bird: *Molecular Theory of Gases and Liquids,* Wiley, New York, 1954.
55. Ho, C. Y., R. W. Powell, and P. E. Liley: *J. Phys. Chem. Ref. Data,* **1:** 279 (1972).
56. Holland, P. M., H. J. M. Hanley, K. E. Gubbins, and J. M. Halle: *J. Phys. Chem. Ref. Data,* **8:** 559 (1979).
57. Jamieson, D. T.: private communication, National Engineering Laboratory, East Kilbride, Glasgow, March 1971.

58. Jamieson, D. T.: *J. Chem. Eng. Data,* **24:** 244 (1979).
59. Jamieson, D. T.: private communication, National Engineering Laboratory, East Kilbride, Scotland, October 1984.
60. Jamieson, D. T., and G. Cartwright: *J. Chem. Eng. Data,* **25:** 199 (1980).
61. Jamieson, D. T., and G. Cartwright: *Proc. 8th Symp. Thermophys. Prop.,* Vol. 1, *Thermophysical Properties of Fluids,* J. V. Sengers (ed.), ASME, New York, 1981, p. 260.
62. Jamieson, D. T., and E. H. Hastings: in C. Y. Ho and R. E. Taylor (eds.), *Proc. 8th Conf. Thermal Conductivity,* Plenum, New York, 1969, p. 631.
63. Jamieson, D. T., and J. B. Irving: paper presented in *13th Intern. Thermal Conductivity Conf., Univ. Missouri, 1973.*
64. Jamieson, D. T., J. B. Irving, and J. S. Tudhope: *Liquid Thermal Conductivity: A Data Survey to 1973,* H. M. Stationary Office, Edinburgh, 1975.
65. Jamieson, D. T., and J. S. Tudhope: *Natl. Eng. Lab. Glasgow Rep. 137,* March 1964.
66. Junk, W. A., and E. W. Comings: *Chem. Eng. Progr.,* **49:** 263 (1953).
67. Kandiyoti, R., E. McLaughlin, and J. F. T. Pittman: *Chem. Soc. (London), Faraday Trans.,* **69:** 1953 (1973).
68. Kannuliuk, W. G., and H. B. Donald: *Aust. J. Sci. Res.,* **3A:** 417 (1950).
69. Kennedy, J. T., and G. Thodos: *AIChE J.,* **7:** 625 (1961).
70. Kerr, C. P., and J. Coates: *67th Natl. AIChE Mtg., Atlanta, Ga, February 15–18, 1970,* paper 55c.
71. Kestin, J., and W. A. Wakeham: *Ber. Bunsenges. Phys. Chem.,* **84:** 762 (1980).
72. Keyes, F. G.: *Trans. ASME,* **73:** 597 (1951).
73. Kramer, F. R., and E. W. Comings: *J. Chem. Eng. Data,* **5:** 462 (1960).
74. Lambert, J. D.: in D. R. Bates (ed.), *Atomic and Molecular Processes,* Academic, New York, 1962.
75. Latini, G.: *Chem. Eng. World,* **17**(7): 49 (1982).
76. Latini, G., and C. Baroncini: *High Temp.—High Press.,* **15:** 407 (1983).
77. Latini, G., and M. Pacetti: *Therm. Conduct.,* **15:** 245 (1977); pub. 1978.
78. Leach, J. W., P. S. Chappelear, and T. W. Leland: *AIChE J.,* **14:** 568 (1968).
79. Leng, D. E., and E. W. Comings: *Ind. Eng. Chem.,* **49:** 2042 (1957).
80. Lenoir, J. M.: *Petrol. Refiner,* **36**(8): 162 (1957).
81. Levy, F. L.: *Intern. J. Refrig.,* **4**(4): 223 (1981).
82. Li, C. C.: *AIChE J.,* **22:** 927 (1976).
83. Liley, P. E.: *Symp. Thermal Prop.,* Purdue Univ. Lafayette, Ind., Feb. 23–26, 1959, p. 40.
84. Liley, P. E.: in R. H. Perry and C. H. Chilton (eds.), *Chemical Engineers' Handbook,* 5th ed., McGraw-Hill, New York, 1973, p. 3–216.
85. Lindsay, A. L., and L. A. Bromley: *Ind. Eng. Chem.,* **42:** 1508 (1950).
86. Losenicky, Z.: *J. Phys. Chem.,* **72:** 4308 (1968).
87. McElhannon, W., and E. McLaughlin: *Therm. Conduct.,* **14:** 345 (1975); pub. 1976.
88. McLaughlin, E.: *Chem. Rev.,* **64:** 389 (1964).
89. Maczek, A. O. S., and P. J. Highton: "Calculation of Thermal Conductivity in Dilute Gas Mixtures of Polyatomic Species," *Symp. Transport Properties of Fluids and Fluid Mixtures: Their Measurement, Estimation, Correlation, and Use, Natl. Eng. Lab., East Kilbride, Glasgow, Scotland, April 1979.*
90. Maejima, T., private communication, 1973. Equation (10-9.3) was suggested by Prof. K. Sato, of the Tokyo Institute of Technology.
91. Mani, N., and J. E. S. Venart: *Advan. Cryog. Eng.,* **18:** 280 (1973).
92. Mason, E. A.: *J. Chem. Phys.,* **28:** 1000 (1958).
93. Mason, E. A., H. E. Khalifia, J. Kestin, R. DiPippo, and J. R. Dorfman: *Physica A,* **91A:** 377 (1978).
94. Mason, E. A., and L. Monchick: *J. Chem. Phys.,* **36:** 1622 (1962).
95. Mason, E. A., and L. Monchick: *Theory of Transport Properties of Gases, 9th Intern. Symp. Combust.,* Academic, New York, 1963.
96. Mason, E. A., and S. C. Saxena: *Phys. Fluids,* **1:** 361 (1958).
97. Mason, E. A., and S. C. Saxena: *J. Chem. Phys.,* **31:** 511 (1959).
98. Mason, E. A., and H. von Ubisch: *Phys. Fluids,* **3:** 355 (1960).

99. Mathur, V. K., J. D. Singh, and W. M. Fitzgerald: *J. Chem. Eng. Japan,* **11:** 67 (1978).
100. Miller, J. W., J. J. McGinley, and C. L. Yaws: *Chem. Eng.,* **83**(23): 133 (1976).
101. Miller, J. W., P. N. Shah, and C. L. Yaws: *Chem. Eng.,* **83**(25): 153 (1976).
102. Misic, D., and G. Thodos: *AIChE J.,* **7:** 264 (1961).
103. Misic, D., and G. Thodos: *AIChE J.,* **11:** 650 (1965).
104. Missenard, A.: *C.R.,* **260**(5), 5521 (1965).
105. Missenard, A.: *Conductivite thermique des solides, liquides, gaz et de leurs melanges,* Editions Eyrolles, Paris, 1965.
106. Missenard, A.: *Rev. Gen. Thermodyn.,* **101**(5): 649 (1970).
107. Missenard, A.: *Rev. Gen. Thermodyn.,* **11:** 9 (1972).
108. Missenard, F. A.: *Rev. Gen. Thermodyn.,* **141:** 751 (1973).
109. Mo, K. C., and K. E. Gubbins: *Chem. Eng. Comm.,* **1:** 281 (1974).
110. Monchick, L., A. N. G. Pereira, and E. A. Mason: *J. Chem. Phys.,* **42:** 3241 (1965).
111. Muckenfuss, C., and C. F. Curtiss; *J. Chem. Phys.,* **29:** 1273 (1958).
112. Murad, S., and K. E. Gubbins: *Chem. Eng. Sci.,* **32:** 499 (1977).
113. Murad, S., and K. E. Gubbins: *AIChE J.,* **27:** 864 (1981).
114. Nagata, I.: *J. Chem. Eng. Japan,* **6:** 18 (1973).
115. Nieto de Castro, C. A., J. M. N. A. Fareleira, and J. C. G. Calado: *Proc. 8th Symp. Thermophys. Prop.,* Vol 1, *Thermophysical Prop. of Fluids,* J. V. Sengers (ed.), ASME, New York, 1981, p. 247.
116. O'Neal, C. Jr., and R. S. Brokaw: *Phys, Fluids,* **5:** 567 (1962).
117. O'Neal, C. Jr., and R. S. Brokaw: *Phys. Fluids,* **6:** 1675 (1963).
118. Ogiwara, K., Y. Arai, and S. Saito: *Ind. Eng. Chem. Fundam.,* **19:** 295 (1980).
119. Ogiwara, K., Y. Arai, and S. Saito: *J. Chem. Eng. Japan,* **15:** 335 (1982).
120. Owens, E. J., and G. Thodos: *AIChE J.,* **3:** 454 (1957).
121. Owens, E. J., and G. Thodos: *Proc. Joint Conf. Thermodyn. Transport Prop. Fluids, London, July 1957,* pp. 163–168, Inst. Mech. Engrs., London, 1958.
122. Owens, E. J., and G. Thodos: *AIChE J.,* **6**: 676 (1960).
123. Pandey, J. D., and S. R. Prajapi: *Indian J. Pure Appl. Phys.,* **18:** 815 (1980).
124. Pedersen, K. S., A. Fredenslund, P. L. Christensen, and P. Thomassen: *Chem. Eng. Sci.,* **39:** 1011 (1984).
125. Pereira, A. N. G., and C. J. G. Raw: *Phys. Fluids,* **6:** 1091 (1963).
126. Peterson, J. N., T. F. Hahn, and E. W. Comings: *AIChE J.,* **17:** 289 (1971).
127. Prasad, R. C., and J. E. S. Venart: *Proc. 8th Symp. Thermophys. Prop.,* Vol. 1, *Thermophysical Prop. of Fluids,* J. V. Sengers (ed.), ASME., New York, 1981, p. 263.
128. Preston, G. T., T. W. Chapman, and J. M. Prausnitz: *Cryogenics,* **7**(5): 274 (1967).
129. Rabenovish, B. A.: *Thermophysical Properties of Substances and Materials,* 3d ed., Standards, Moscow, 1971.
130. Rastorguev, Yu. L,, G. F. Bogatov, and B. A. Grigor'ev: *Isv. vyssh. ucheb. Zaved., Neft'i Gaz.,* **11**(12): 59 (1968).
131. Riazi, M. R., and A. Faghri: *AIChE J.,* **31:** 164 (1985).
132. Richter, G. N., and B. H. Sage: *J. Chem. Eng. Data,* **2:** 61 (1957).
133. Richter, G. N., and B. H. Sage: *J. Chem. Eng. Data,* **8:** 221 (1963).
134. Richter, G. N., and B. H. Sage: *J. Chem. Eng. Data,* **9:** 75 (1964).
135. Riedel, L.: *Chem. Ing. Tech.,* **21:** 349 (1949); **23:** 59, 321, 465 (1951).
136. Robbins, L. A., and C. L. Kingrea: *Hydrocarbon Process, Petrol. Refiner,* **41**(5): 133 (1962).
137. Rodriguez, H. V., Ph.D. thesis, Louisiana State Univ., Baton Rouge, La., 1962.
138. Rosenbaum, B. M., and G. Thodos: *Physica,* **37:** 442 (1967).
139. Rosenbaum, B. M., and G. Thodos: *J. Chem. Phys.,* **51:** 1361 (1969).
140. Rowley, R. L.: *Chem. Eng. Sci.,* **37:** 897 (1982).
141. Roy, D.: M.S. thesis, Northwestern University, Evanston, Ill, 1967.
142. Roy, D., and G. Thodos: *Ind. Eng. Chem. Fundam.,* **7:** 529 (1968).
143. Roy, D., and G. Thodos: *Ind. Eng. Chem. Fundam.,* **9:** 71 (1970).
144. Saksena, M. P., and Harminder: *Ind. Eng. Chem. Fundam,* **13:** 245 (1974).
145. Saksena, M. P., and S. C. Saxena: *Appl. Sci. Res.,* **17:** 326 (1966).
146. Sandler, S. I.: *Phys. Fluids,* **11:** 2549 (1968).

147. Sandler, S. I., and J. K. Fizdon: *Physica,* **95A:** 602 (1979).
148. Saxena, S. C., and J. P. Agrawal: *J. Chem. Phys.,* **35:** 2107 (1961).
149. Saxena, S. C., and G. P. Gupta: *Univ. Illinois at Chicago Circle, Dept. Energy Eng., Rept. TR-E-23,* January 1969.
150. Saxena, S. C., M. P. Saksena, and R. S. Gambhir: *Brit. J. Appl. Phys.,* **15:** 843 (1964).
151. Schaefer, C. A., and G. Thodos: *Ind. Eng. Chem,* **50:** 1585 (1958).
152. Schaefer, C. A., and G. Thodos: *AIChE J.,* **5:** 367 (1959).
153. Schafer, K.: *Z. Phys. Chem.,* **B53:** 149 (1943).
154. Shroff, G. H.: Ph.D. thesis, University of New Brunswick, Fredericton, 1968.
155. Shroff, G. H.: in C. Y. Ho and R. E. Taylor (eds.), *Proc, 8th Conf. Thermal Conductivity, Purdue Univ., Oct. 7–10, 1968,* Plenum, New York, 1969, p. 643.
156. Srivastava, B. N., and R. C. Srivastava: *J. Chem. Phys.,* **30:** 1200 (1959).
157. Stiel, L. I., and G. Thodos: *AIChE J.,* **10:** 26 (1964).
158. Sutton, J. R.: "References to Experimental Data on Thermal Conductivity of Gas Mixtures," *Natl. Eng. Lab., Rept. 612,* East Kilbride, Glasgow, Scotland, May 1976.
159. Svehla, R. A.: "Estimated Viscosities and Thermal Conductivities of Gases at High Temperatures," *NASA Tech. Rept. R-132,* Lewis Research Center, Cleveland, Ohio, 1962.
160. Taxman, N.: *Phys. Rev.,* **110:** 1235 (1958).
161. Teja, A. S., and P. Rice: *Chem. Eng. Sci.,* **36:** 417 (1981).
162. Teja, A. S., and P. Rice: *Chem. Eng. Sci.,* **37:** 790 (1982).
163. Tsederberg, N. V.: *Thermal Conductivity of Gases and Liquids,* The M.I.T. Press, Cambridge, Mass., 1965.
164. Tsvetkov, O. B.: *Russ. J. Phys. Chem.,* **56**(5): 680 (1982).
165. Tufeu, R., and B. L. Neindre: *High Temp.—High Press.,* **13:** 31 (1981).
166. Ubbelohde, A. R.: *J. Chem. Phys.,* **3:** 219 (1935).
167. Usmanov, I. U., and A. S. Salikov: *Russ. J. Phys. Chem.,* **51**(10), 1488 (1977).
168. van Dael, W., and H. Cavwenbergh: *Physica,* **40:** 173 (1968).
169. Vargaftik, N. B.: "Thermal Conductivities of Compressed Gases and Steam at High Pressures," *Izv. Vses. Telpotekh. Inst.,* Nov. 7, 1951; private communication, Prof. N. V. Tsederberg, Moscow Energetics Institute.
170. Vargaftik, N. B.: *Proc. Joint Conf. Thermodyn. Transport Prop. Fluids, London, July 1957,* p. 142, Inst. Mech. Engrs., London, 1958.
171. Vargaftik, N. B.: *Tables on the Thermophysical Properties of Liquids and Gases,* 2d. ed., Hemisphere, Washington, D.C., 1975, pp. 315, 316.
172. Vargaftik, N. B., and Y. P. Os'minin: *Teploenergetika,* **3**(7): 11 (1956).
173. Venart, J. E. S., and C. Krishnamurthy: *Proc. 7th Conf. Thermal Conductivity, NBS Spec. Pub. 302,* November 1967, p. 659.
174. Vines, R. G.: *Aust. J. Chem.,* **6:** 1 (1953).
175. Vines, R. G.: *Proc. Joint Conf. Thermodyn. Transport Prop. Fluids, London, July 1957,* Inst. Mech. Engrs., London, 1958, pp. 120–123.
176. Vines, R. G., and L. A. Bennett: *J. Chem. Phys.,* **22:** 360 (1954).
177. Wang Chang, C. S., and G. E. Uhlenbeck: "Transport Phenomena in Polyatomic Gases," *Univ. Michigan Eng. Res. Rept. CM-681,* Ann Arbor, Mich., 1951.
178. Wassiljewa, A.: *Physik. Z.,* **5:** 737 (1904).
179. Yorizane, M., S. Yoshimura, H. Masuoka, and H. Yoshida: *Ind. Eng. Chem. Fundam.,* **22:** 454 (1983).
180. Yorizane, M., S. Yoshimura, H. Masuoka, and H. Yoshida: *Ind. Eng. Chem. Fundam.,* **22:** 458 (1983).
181. Zheng, X.-Y., S. Yamamoto, H. Yoshida, H. Masuoka, and M. Yorizane: *J. Chem. Eng. Japan,* **17:** 237 (1984).
182. Zimmerling, W., and M. Ratzsch: *Chem. Techn.,* **28**(2): 84 (1976).

Chapter

# 11

# Diffusion Coefficients

## 11-1 Scope

In Sec. 11-2 we discuss briefly several frames of reference from which diffusion can be related and define the diffusion coefficient. Low-pressure binary gas diffusion coefficients are treated in Secs. 11-3 and 11-4. The pressure and temperature effects on gas-phase diffusion coefficients are covered in Secs. 11-5 and 11-6, respectively. The theory for liquid diffusion coefficients is introduced in Sec. 11-8, and estimation methods for binary liquid diffusion coefficients at infinite dilution are described in Sec. 11-9. Concentration effects are considered in Sec. 11-10 and temperature and pressure effects in Sec. 11-11. Brief comments on diffusion in multicomponent mixtures are made in Secs. 11-7 (gases) and 11-12 (liquids); ionic solutions are covered in Sec. 11-13.

## 11-2 Basic Concepts and Definitions

The extensive use of the term "diffusion" in the chemical engineering literature is based on an intuitive feel for the concept; i.e., diffusion refers to the net transport of material within a single phase in the absence of mixing (by mechanical means or by convection). Both experiment and theory have shown that diffusion can result from pressure gradients (pressure diffusion), temperature gradients (thermal diffusion), external force fields (forced diffusion), and concentration gradients. Only the last type is considered in this chapter; i.e., the discussion is limited to diffusion in isothermal, isobaric systems with no external force field gradients.

Even with this limitation, confusion can easily arise unless care is taken to define diffusion fluxes and diffusion potentials, e.g., driving forces, clearly. The proportionality constant between the flux and potential is the *diffusion coefficient*, or *diffusivity*.

## Diffusion fluxes

A detailed discussion of diffusion fluxes has been given by Bird et al. [20] and Cussler [47]. Various types originate because different reference frames are employed. The most obvious reference plane is fixed on the equipment in which diffusion is occurring. This plane is designated by $RR'$ in Fig. 11-1. Suppose, in a binary mixture of A and B, that A is diffusing to the left and B to the right. If the diffusion rates of these species are not identical, there will be a net depletion or accumulation of molecules in either side of $RR'$. To maintain the requirements of an isobaric, isothermal system, bulk motion of the mixture occurs. Net movement of A (as measured in the fixed reference frame $RR'$) then results from both diffusion and bulk flow.

Although many reference planes can be delineated, a plane of *no net mole flow* is normally used to define a diffusion coefficient in binary mixtures. If $J_A^M$ represents a mole flux in a mixture of A and B, $J_A^M$ is then the net mole flow of A across the boundaries of a hypothetical (moving) plane such that the total moles of A and B are invariant on both sides of the plane. $J_A^M$ can be related to fluxes across $RR'$ by

$$J_A^M = N_A - x_A (N_A + N_B) \tag{11-2.1}$$

where $N_A$ and $N_B$ are the fluxes of A and B across $RR'$ (relative to the fixed plane) and $x_A$ is the mole fraction of A at $RR'$. Note that $J_A^M$, $N_A$, and $N_B$ are vectorial quantities and a sign convention must be assigned to denote flow directions. Equation (11-2.1) shows that the net flow of A across $RR'$ is due to a diffusion contribution $J_A^M$ and a bulk flow contri-

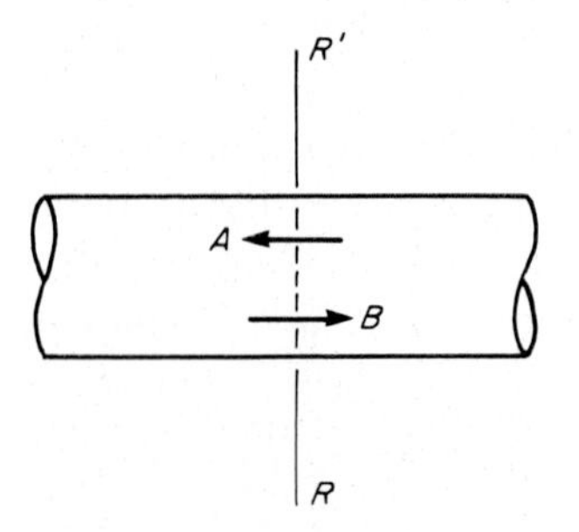


**Figure 11-1** Diffusion across plane $RR'$.

bution $x_A(N_A + N_B)$. For equimolar counterdiffusion, $N_A + N_B = 0$ and $J_A^M = N_A$.

One other flux is extensively used, i.e., one relative to the plane of *no net volume flow*. This plane is less readily visualized. By definition,

$$J_A^M + J_B^M = 0 \tag{11-2.2}$$

and if $J_A^V$ and $J_B^V$ are vectorial molar fluxes of $A$ and $B$ relative to the plane of no net volume flow, then, by definition,

$$J_A^V \overline{V}_A + J_B^V \overline{V}_B = 0 \tag{11-2.3}$$

where $\overline{V}_A$ and $\overline{V}_B$ are the partial molar volumes of A and B in the mixture. It can be shown that

$$J_A^V = \frac{\overline{V}_B}{V} J_A^M \qquad \text{and} \qquad J_B^V = \frac{\overline{V}_A}{V} J_B^M \tag{11-2.4}$$

where $V$ is the volume per mole of mixture. Obviously, if $\overline{V}_A = \overline{V}_B = V$, as in an ideal-gas mixture, then $J_A^V = J_A^M$.

### Diffusion coefficients

Diffusion coefficients for a binary mixture of A and B are defined by

$$J_A^M = -cD_{AB}\frac{dx_A}{dz} \tag{11-2.5}$$

$$J_B^M = -cD_{BA}\frac{dx_B}{dz} \tag{11-2.6}$$

where $c$ is the total molar concentration $(= V^{-1})$ and diffusion is in the $z$ direction. With Eq. (11-2.2), since $(dx_A/dz) + (dx_B/dz) = 0$, we have $D_{AB} = D_{BA}$. The diffusion coefficient then represents the proportionality between the flux of A relative to a plane of no net molar flow and the gradient $c(dx_A/dz)$. From Eqs. (11-2.4) to (11-2.6) and the definition of a partial molar volume it can be shown that, for an isothermal, isobaric binary system,

$$J_A^V = -D_{AB}\frac{dc_A}{dz} \qquad \text{and} \qquad J_B^V = -D_{AB}\frac{dc_B}{dz} \tag{11-2.7}$$

When fluxes are expressed in relation to a plane of no net volume flow, the potential is the concentration gradient. $D_{AB}$ in Eq. (11-2.7) is identical with that defined in Eq. (11-2.5). In many cases $\overline{V}_A \approx \overline{V}_B \approx V$ (ideal gases, ideal solutions), and in such instances $J_A^V \approx J_A^M$, $J_B^V \approx J_B^M$.

### Mutual, self-, and tracer diffusion coefficients

The diffusion coefficient $D_{AB}$ introduced above is termed the *mutual diffusion coefficient*, and it refers to the diffusion of one constituent in a binary system. A similar coefficient $D_{1m}$ would imply the diffusivity of component 1 in a mixture (see Secs. 11-7 and 11-12).

*Tracer diffusion coefficients* (sometimes referred to as *intradiffusion coefficients*) relate to the diffusion of a labeled component within a *homogeneous* mixture. Like mutual diffusion coefficients, tracer diffusion coefficients can be a function of composition. If $D_A^*$ is the tracer diffusivity of A in a mixture of A and B, then as $x_A \rightarrow 1.0$, $D_A^* \rightarrow D_{AA}$, where $D_{AA}$ is the *self-diffusion coefficient* of A in pure A.

In Fig. 11-2, the various diffusion coefficients noted above are shown for a binary liquid mixture of $n$-octane and $n$-dodecane at 60°C [216]. In this case, the mutual diffusion of these two hydrocarbons increases as the mixture becomes richer in $n$-octane. With A as $n$-octane and B as $n$-dodecane, as $x_A \rightarrow 1.0$, $D_{AB} = D_{BA} \rightarrow D_{BA}^\circ$, where this notation signifies that this limiting diffusivity represents the diffusion of B in a medium consisting essentially of A, that is, $n$-dodecane molecules diffusing through almost pure $n$-octane. Similarly, $D_{AB}^\circ$ is the diffusivity of A in essentially pure B. Except in the case of infinite dilution, tracer diffusion coefficients differ from binary-diffusion coefficients, and there is no way to relate the two coefficients [47]. Similarly, there is no relation between quantities such as

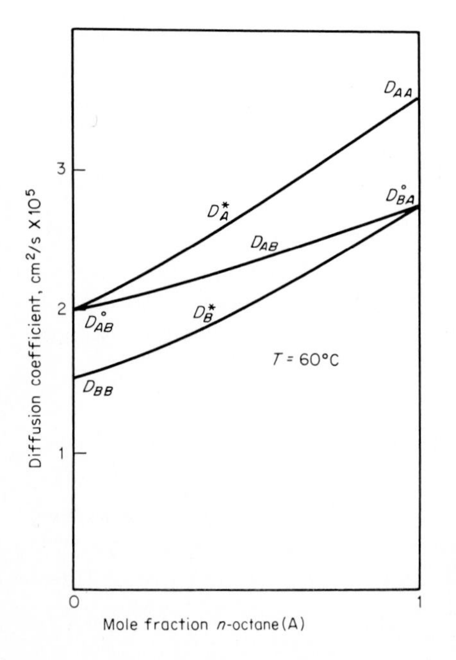


**Figure 11-2** Mutual, self-, and tracer diffusion coefficients in a binary mixture of $n$-octane and $n$-dodecane. *(From Ref. 216.)*

$D_{BB}$ and $D^{\circ}_{AB}$ or $D_{AA}$ and $D^{\circ}_{BA}$. In this chapter, only correlation techniques for $D_{ij}$ (or $D^{\circ}_{ij}$) are considered; corresponding states methods for $D_{ii}$ have, however, been developed [122, 155].

### Chemical potential driving force

The mutual diffusion coefficient $D_{AB}$ in Eq. (11-2.7) indicates that the flux of a diffusing component is proportional to the concentration gradient. Diffusion is, however, affected by more than just the gradient in concentration, e.g., the force fields around molecules [55, 208]. Yet these force fields are some complex function of composition as well as of temperature and pressure. Thus, fluxes should not be expected to be linear in the concentration gradient. Any inadequacy in the defining equation for $D_{AB}$ is reflected by a concentration dependence of experimental diffusion coefficients.

Modern theories of diffusion [75] have adopted the premise that if one perturbs the equilibrium composition of a binary system, the subsequent diffusive flow required to attain a new equilibrium state is proportional to the gradient in chemical potential ($d\mu_A/dz$). Since the diffusion coefficient was defined in Eqs. (11-2.5) and (11-2.6) in terms of a mole fraction gradient instead of a chemical potential gradient, it is argued that one should include a thermodynamic correction in any equation for $D_{AB}$. This correction is

$$\alpha = \left[\frac{(\partial \ln a_A)}{(\partial \ln x_A)}\right]_{T,P} \tag{11-2.8}$$

By virtue of the Gibbs-Duhem equation, $\alpha$ is the same regardless of whether activities and mole fractions of either A or B are used in Eq. (11-2.8). For gases, $\alpha$ is almost always close to unity (except at high pressures), and this correction is seldom used. For liquid mixtures, however, it is widely adopted, as will be illustrated in Sec. 11-10.

## 11-3 Diffusion Coefficients for Binary Gas Systems at Low Pressures: Prediction from Theory

The theory describing diffusion in binary gas mixtures at low to moderate pressures has been well developed. As noted earlier in Chaps. 9 (Viscosity) and 10 (Thermal Conductivity), the theory results from solving the Boltzmann equation, and the results are usually credited to both Chapman and Enskog, who independently derived the working equation

$$D_{AB} = \frac{3}{16}\frac{(4\pi kT/M_{AB})^{1/2}}{n\pi\sigma^2_{AB}\Omega_D} f_D \tag{11-3.1}$$

where $M_A$, $M_B$ = molecular weights of A and B
$M_{AB}$ = $2[(1/M_A) + (1/M_B)]^{-1}$
$n$ = number density of molecules in the mixture
$k$ = Boltzmann's constant
$T$ = absolute temperature

$\Omega_D$, the collision integral for diffusion, is a function of temperature; it depends upon the choice of the intermolecular force law between colliding molecules. $\sigma_{AB}$ is a characteristic length; it also depends upon the intermolecular force law selected. Finally, $f_D$ is a correction term which is of the order of unity. If $M_A$ is of the same order as $M_B$, $f_D$ lies between 1.0 and 1.02 regardless of composition or intermolecular forces. Only if the molecular masses are very unequal and the light component is present in trace amounts is the value of $f_D$ significantly different from unity, and even in such cases, $f_D$ is usually between 1.0 and 1.1 [138].

If $f_D$ is chosen as unity and $n$ is expressed by the ideal-gas law, Eq. (11-3.1) may be written as

$$D_{AB} = \frac{0.00266 T^{3/2}}{P M_{AB}^{1/2} \sigma_{AB}^2 \Omega_D} \tag{11-3.2}$$

where $D_{AB}$ = diffusion coefficient, $cm^2/s$
$T$ = temperature, K
$P$ = pressure, bar
$\sigma_{AB}$ = characteristic length, Å
$\Omega_D$ = diffusion collision integral, dimensionless

and $M_{AB}$ is defined under Eq. (11-3.1). The key to the use of Eq. (11-3.2) is the selection of an intermolecular force law and the evaluation of $\sigma_{AB}$ and $\Omega_D$.

### Lennard-Jones 12-6 potential

As noted earlier [Eq. (9-4.2)], a popular correlation relating the intermolecular energy $\psi$ between two molecules to the distance of separation $r$, is given by

$$\psi = 4\epsilon \left[ \left(\frac{\sigma}{r}\right)^{12} - \left(\frac{\sigma}{r}\right)^{6} \right] \tag{11-3.3}$$

with $\epsilon$ and $\sigma$ as the characteristic Lennard-Jones energy and length, respectively. Application of the Chapman-Enskog theory to the viscosity of pure gases has led to the determination of many values of $\epsilon$ and $\sigma$; some of them are given in Appendix B.

To use Eq. (11-3.2), some rule must be chosen to obtain the interaction value $\sigma_{AB}$ from $\sigma_A$ and $\sigma_B$. Also, it can be shown that $\Omega_D$ is a function only of $kT/\epsilon_{AB}$, where again some rule must be selected to relate $\epsilon_{AB}$ to $\epsilon_A$ and $\epsilon_B$. The simple rules shown below are usually employed:

$$\epsilon_{AB} = (\epsilon_A \epsilon_B)^{1/2} \tag{11-3.4}$$

$$\sigma_{AB} = \frac{\sigma_A + \sigma_B}{2} \tag{11-3.5}$$

$\Omega_D$ is tabulated as a function of $kT/\epsilon$ for the 12-6 Lennard-Jones potential [100], and various analytical approximations also are available [92, 109, 112, 158]. The accurate relation of Neufield et al. [158] is

$$\Omega_D = \frac{A}{(T^*)^B} + \frac{C}{\exp(DT^*)} + \frac{E}{\exp(FT^*)} + \frac{G}{\exp(HT^*)} \tag{11-3.6}$$

where $T^* = kT/\epsilon_{AB}$ $\quad A = 1.06036 \quad B = 0.15610$

$C = 0.19300 \quad D = 0.47635 \quad E = 1.03587$

$F = 1.52996 \quad G = 1.76474 \quad H = 3.89411$

**Example 11-1** Estimate the diffusion coefficient for the system $N_2$-$CO_2$ at 590 K and 1 bar. The experimental value reported by Ellis and Holsen [61] is 0.583 $cm^2/s$.

**solution** To use Eq. (11-3.2), values of $\sigma$ ($CO_2$), $\sigma$ ($N_2$), $\epsilon$ ($CO_2$), and $\epsilon$ ($N_2$) must be obtained. Using the values in Appendix B with Eqs. (11-3.4) and (11-3.5) gives $\sigma$ ($CO_2$) = 3.941 Å, $\sigma$ ($N_2$) = 3.798 Å; $\sigma$ ($CO_2$-$N_2$) = (3.941 + 3.798)/2 = 3.8695 Å; $\epsilon$ ($CO_2$)/$k$ = 195.2 K, $\epsilon$ ($N_2$)/$k$ = 71.4 K; $\epsilon$ ($CO_2$-$N_2$)/$k$ = $[(195.2)(71.4)]^{1/2}$ = 118 K. Then $T^* = kT/\epsilon$ ($CO_2$-$N_2$) = 590/118 = 5.0. With Eq. (11-3.6), $\Omega_D$ = 0.842. Since $M$ ($CO_2$) = 44.0 and $M$ ($N_2$) = 28.0, $M_{AB} = (2)[(1/44.0) + (1/28.0)]^{-1} = 34.22$. With Eq. (11-3.2),

$$D(CO_2\text{-}N_2) = \frac{(0.00266)(590)^{3/2}}{(1)(34.22)^{1/2}(3.8695)^2(0.842)} = 0.52 \text{ cm}^2/\text{s}$$

The error is 11 percent. Ellis and Holsen recommend values of $\epsilon$ ($CO_2$-$N_2$) = 134 K and $\sigma$ ($CO_2$-$N_2$) = 3.660 Å. With these parameters, they predicted $D$ to be 0.56 $cm^2/s$, a value closer to that found experimentally.

Equation (11-3.2) is derived for dilute gases consisting of nonpolar, spherical, monatomic molecules; and the potential function (11-3.3) is essentially empirical, as are the combining rules [Eqs. (11-3.4) and (11-3.5)]. Yet Eq. (11-3.2) gives good results over a wide range of temperatures and provides useful approximate values of $D_{AB}$ [81, 82]. The general nature of the errors to be expected from this estimation procedure is indicated by the comparison of calculated and experimental values shown later in Table 11-2.

The calculated value of $D_{AB}$ is relatively insensitive to the value of $\epsilon_{AB}$ employed and even to the form of the assumed potential function. Values of $\epsilon$ and $\sigma$ are often available from viscosity measurements.

No effect of composition is predicted. A more detailed treatment does indicate that there may be a small effect for cases in which $M_A$ and $M_B$ differ significantly. In a specific study of this effect [234], the low-pressure binary diffusion coefficient for the system He-$CClF_3$ did vary from about 0.416 to 0.430 $cm^2/s$ over the extremes of composition. In another study [154], no effect of concentration was noted for the methyl alcohol–air system, but a small change was observed with chloroform-air.

### Low-pressure diffusion coefficients from viscosity data

Since the equations for low-pressure gas viscosity [Eq. (9-3.9)] and diffusion [Eq. (11-3.2)] have a common basis in the Chapman-Enskog theory, they can be combined to relate the two gas properties. Experimental data on viscosity as a function of composition at constant temperature are required as a basis for calculating the *binary* diffusion coefficient $D_{AB}$ [51, 83, 100, 112, 113]. Weissman and Mason [226, 227] compare the method with a large collection of experimental viscosity and diffusion data and find excellent agreement.

### Polar gases

If one or both components of a gas mixture are polar, a modified Lennard-Jones relation, such as the Stockmayer potential, is often used. A different collision integral relation [rather than Eq. (11-3.6)] is then necessary, and Lennard-Jones $\sigma$ and $\epsilon$ values are not sufficient.

Brokaw [22] has suggested an alternative method for estimating diffusion coefficients for binary mixtures containing polar components. Equation (11-3.1) is still used, but the collision integral $\Omega_D$ is now given as

$$\Omega_D = \Omega_D \text{ [Eq. (11-3.6)]} + \frac{0.19\delta_{AB}^2}{T^*} \tag{11-3.7}$$

where $T^* = \dfrac{kT}{\epsilon_{AB}}$

and
$$\delta = \frac{1.94 \times 10^3 \mu_p^2}{V_b T_b} \tag{11-3.8}$$

$\mu_p$ = dipole moment, debyes

$V_b$ = *liquid* molar volume at the normal boiling point, $cm^3/mol$

$T_b$ = normal boiling point (1 atm), K

$$\frac{\epsilon}{k} = 1.18(1 + 1.3\delta^2)T_b \tag{11-3.9}$$

$$\sigma = \left(\frac{1.585\,V_b}{1 + 1.3\delta^2}\right)^{1/3} \tag{11-3.10}$$

$$\delta_{AB} = (\delta_A\delta_B)^{1/2} \tag{11-3.11}$$

$$\frac{\epsilon_{AB}}{k} = \left(\frac{\epsilon_A}{k}\frac{\epsilon_B}{k}\right)^{1/2} \tag{11-3.12}$$

$$\sigma_{AB} = (\sigma_A\sigma_B)^{1/2} \tag{11-3.13}$$

Note that the polarity effect is related exclusively to the dipole moment; this may not always be a satisfactory assumption [27].

**Example 11-2** Estimate the diffusion coefficient for a mixture of methyl chloride (MC) and sulfur dioxide (SD) at 1 bar and 323 K. The data required to use Brokaw's relation are shown below:

| | Methyl chloride (MC) | Sulfur dioxide (SD) |
|---|---|---|
| Dipole moment, debyes | 1.9 | 1.6 |
| Liquid molar volume at $T_b$, $cm^3/mol$ | 50.6 | 43.8 |
| Normal boiling temperature, K | 249.1 | 263.2 |

**solution** With Eqs. (11-3.8) and (11-3.11),

$$\delta\,(\text{MC}) = \frac{(1.94 \times 10^3)(1.9)^2}{(50.6)(249.1)} = 0.55$$

$$\delta\,(\text{SD}) = \frac{(1.94 \times 10^3)(1.6)^2}{(43.8)(263.2)} = 0.43$$

$$\delta\,(\text{MC-SD}) = [(0.55)(0.43)]^{1/2} = 0.49$$

Also, with Eqs. (11-3.9) and (11-3.12),

$$\frac{\epsilon\,(\text{MC})}{k} = 1.18[1 + 1.3(0.55)^2](249.1) = 412\text{ K}$$

$$\frac{\epsilon\,(\text{SD})}{k} = 1.18[1 + 1.3(0.43)^2](263.2) = 385\text{ K}$$

$$\frac{\epsilon\,(\text{MC-SD})}{k} = [(412)(385)]^{1/2} = 398\text{ K}$$

Then, with Eqs. (11-3.10) and (11-3.13),

$$\sigma\,(\text{MC}) = \left[\frac{(1.585)(50.6)}{1 + (1.3)(0.55)^2}\right]^{1/3} = 3.85\text{ Å}$$

$$\sigma\,(\text{SD}) = \left[\frac{(1.585)(43.8)}{1 + (1.3)(0.43)^2}\right]^{1/3} = 3.82\text{ Å}$$

$$\sigma\,(\text{MC-SD}) = [(3.85)(3.82)]^{1/2} = 3.84\text{ Å}$$

To determine $\Omega_D$, $T^* = kT/\epsilon$ (MC-SD) = 323/398 = 0.811. With Eq. (11-3.6), $\Omega_D = 1.60$. Then with Eq. (11-3.7),

$$\Omega_D = 1.60 + \frac{(0.19)(0.490)^2}{(0.811)} = 1.65$$

With Eq. (11-3.2) and M (MC) = 50.49, M (SD) = 64.60, and $M_{AB} = (2)[(1/50.49) + (1/64.60)]^{-1} = 56.68$

$$D_{\text{MC-SD}} = \frac{(0.00266)(323)^{3/2}}{(1)(56.68)^{1/2}(3.84)^2(1.65)} = 0.084 \text{ cm}^2/\text{s}$$

The experimental value is 0.078 cm$^2$/s and the error is 8 percent.

### Discussion

A comprehensive review of the theory and experimental data for gas diffusion coefficients is available [138]. There have been many studies covering wide temperature ranges, and the applicability of Eq. (11-3.1) is well verified. Most investigators select the Lennard-Jones potential for its convenience and simplicity. The difficult task is to locate appropriate values of $\sigma$ and $\epsilon$. Some values are shown in Appendix B. Brokaw suggests other relations, e.g., Eq. (11-3.9) and (11-3.10). Even after the pure component values of $\sigma$ and $\epsilon$ have been selected, a combination rule is necessary to obtain $\sigma_{AB}$ and $\epsilon_{AB}$. Most studies have employed Eqs. (11-3.4) and (11-3.5) because they are simple and theory suggests no particularly better alternatives. Ravindran et al. [174] have used Eq. (11-3.2) to correlate diffusivities of low-volatile organics in light gases.

It is important to employ values of $\sigma$ and $\epsilon$ obtained from the same source. Published values of these parameters differ considerably, but $\sigma$ and $\epsilon$ from a single source often lead to the same result as the use of a quite different pair from another source.

The estimation equations described in this section were used to calculate diffusion coefficients for a number of different gases, and the results are shown in Table 11-2. The accuracy of the theoretical relations is discussed in Sec. 11-4 after some empirical correlations for the diffusion coefficient have been described.

## 11-4 Diffusion Coefficients for Binary Gas Systems at Low Pressures: Empirical Correlations

Several proposed methods for estimating $D_{AB}$ in low-pressure binary gas systems retain the general form of Eq. (11-3.2), with empirical constants based on experimental data. These include the equations proposed by Arnold [11], Gilliland [77], Wilke and Lee [231], Slattery and Bird [193], Bailey [14], Chen and Othmer [38], Othmer and Chen [163], and Fuller et al. [68, 69, 70]. Values of $D_{AB}$ estimated by these equations generally agree with experimental values to within 5 to 10 percent, although discrepancies of more than 20 percent are possible. We illustrate two methods which have been shown to be quite general and reliable.

### Wilke and Lee [231]

Equation (11-3.2) is rewritten as

$$D_{AB} = \frac{[3.03 - (0.98/M_{AB}^{1/2})](10^{-3})T^{3/2}}{PM_{AB}^{1/2}\sigma_{AB}^{2}\Omega_D} \tag{11-4.1}$$

where $D_{AB}$ = binary diffusion coefficient, $cm^2/s$
$T$ = temperature, K
$M_A$, $M_B$ = molecular weights of A and B, g/mol
$M_{AB} = 2[(1/M_A) + (1/M_B)]^{-1}$
$P$ = pressure, bar

The scale parameter $\sigma_{AB}$ is given by Eq. (11-3.5) where, for each component,

$$\sigma = 1.18V_b^{1/3} \tag{11-4.2}$$

and $V_b$ is the liquid molar volume, $cm^3/mol$, found from experimental data or estimated from the Le Bas contributions in Table 3-8. $\Omega_D$ is determined from Eq. (11-3.6) with $(\epsilon/k)_{AB}$ from Eq. (11-3.4) and, for each component,

$$\frac{\epsilon}{k} = 1.15T_b \tag{11-4.3}$$

with $T_b$ as the normal boiling point (at 1 atm) in kelvins. Note, for systems in which one component is air, $\sigma$ (air) = 3.62 Å and $\epsilon/k$ (air) = 97.0 K. We illustrate this method in Example 11-3.

### Fuller et al. [68, 69, 70]

These authors modified Eq. (11-3.2) to

$$D_{AB} = \frac{0.00143T^{1.75}}{PM_{AB}^{1/2}[(\Sigma_v)_A^{1/3} + (\Sigma_v)_B^{1/3}]^2} \tag{11-4.4}$$

where the terms have been defined under Eq. (11-4.1) and $\Sigma_v$ is found for each component by summing atomic diffusion volumes in Table 11-1 [69]. These atomic parameters were determined by a regression analysis of many experimental data, and the authors report an average absolute error

**TABLE 11-1 Atomic Diffusion Volumes**

| Atomic and Structural Diffusion Volume Increments | | | |
|---|---|---|---|
| C | 15.9 | F | 14.7 |
| H | 2.31 | Cl | 21.0 |
| O | 6.11 | Br | 21.9 |
| N | 4.54 | I | 29.8 |
| Aromatic ring | −18.3 | S | 22.9 |
| Heterocyclic ring | −18.3 | | |
| **Diffusion Volumes of Simple Molecules** | | | |
| He | 2.67 | CO | 18.0 |
| Ne | 5.98 | $CO_2$ | 26.9 |
| Ar | 16.2 | $N_2O$ | 35.9 |
| Kr | 24.5 | $NH_3$ | 20.7 |
| Xe | 32.7 | $H_2O$ | 13.1 |
| $H_2$ | 6.12 | $SF_6$ | 71.3 |
| $D_2$ | 6.84 | $Cl_2$ | 38.4 |
| $N_2$ | 18.5 | $Br_2$ | 69.0 |
| $O_2$ | 16.3 | $SO_2$ | 41.8 |
| Air | 19.7 | | |

of about 4 percent when using Eq. (11-4.4). The technique is illustrated in Example 11-3.

### Discussion

In Table 11-2 we show experimental diffusion coefficients for a number of binary systems and note the errors found when estimating $D_{AB}$ for (*a*) the basic theoretical equation (11-3.2), (*b*) Brokaw's method [Eqs. (11-3.2) and (11-3.7)], (*c*) Wilke and Lee's method [Eq. (11-4.1)], and (*d*) Fuller et al.'s method [Eq. (11-4.4)]. For (*a*), no calculations were made if $\sigma$ and $\epsilon/k$ were not available in Appendix B.

For all methods, there were always a few systems for which large errors were found. These differences may be due to inadequacies of the method or to inaccurate data. In general, however, the Fuller et al. procedure [Eq. (11-4.4) and Table 11-1] yielded the smallest average error, and it is the method recommended for use. Other evaluations [60, 81, 82, 129, 166] have shown both the Fuller et al. and the Wilke-Lee forms to be reliable. The results found when using the Brokaw method were erratic; also, no special advantage was demonstrated for this method when applied to mixtures containing polar components.

A review of experimental data of binary diffusion coefficients is available [80].

**Example 11-3** Estimate the diffusion coefficient of allyl chloride (AC) in air at 298 K and 1 bar. The experimental value reported by Lugg [129] is 0.098 $cm^2/s$.

**solution** WILKE AND LEE METHOD. As suggested in the text, for air $\sigma = 3.62$ Å and $\epsilon/k = 97.0$ K. For allyl chloride, from Appendix A, $T_b = 318.3$ K and, with Table 3-8, $V_b = (3)(14.8) + (5)(3.7) + 24.6 = 87.5$ cm$^3$/mol. Thus, using Eqs. (11-4.2) and (11-4.3),

$$\sigma\,(\text{AC}) = (1.18)(87.5)^{1/3} = 5.24\ \text{Å}$$
$$\epsilon\,(\text{AC})/k = (1.15)(318.3) = 366\ \text{K}$$

Then, with Eqs. (11-3.4) and (11-3.5)

$$\epsilon\,(\text{AC-air})k = [(366)(97.0)]^{1/2} = 188\ \text{K}$$
$$\sigma\,(\text{AC-air}) = (5.24 + 3.62)/2 = 4.43\ \text{Å}$$
$$T^* = \frac{T}{\epsilon\,(\text{AC-air})k} = \frac{298}{188} = 1.59$$

and, with Eq. (11-3.6), $\Omega_D = 1.17$. With $M$ (AC) $= 76.5$ and $M$ (air) $= 29.0$, $M_{AB} = (2)[(1/76.5) + (1/29.0)]^{-1} = 42.0$. Finally, with Eq. (11-4.1) when $P = 1$ bar,

$$D = \frac{\{3.03 - [0.98/(42.0)^{1/2}]\}(10^{-3})(298)^{3/2}}{(1)(42.0)^{1/2}(4.43)^2(1.17)}$$
$$= 0.10\ \text{cm}^2/\text{s}$$
$$\text{Error} = \frac{0.10 - 0.098}{0.098} \times 100 = 2\%$$

FULLER ET AL. METHOD. Equation (11-4.4) is used. $P = 1$ bar; $M_{AB}$ was shown above to be equal to 42.0; and $T = 298$ K. For air $(\Sigma_v) = 19.7$, and for allyl chloride, $C_3H_5Cl$, with Table 11-1, $(\Sigma_v) = (3)(15.9) + (5)(2.31) + 21 = 80.25$. Thus,

$$D = \frac{(0.00143)(298)^{1.75}}{(1)(42.0)^{1/2}[(19.7)^{1/3} + (80.25)^{1/3}]^2}$$
$$= 0.096\ \text{cm}^2/\text{s}$$
$$\text{Error} = \frac{0.096 - 0.098}{0.098} \times 100 = -2\%$$

## 11-5 The Effect of Pressure on the Binary Diffusion Coefficients of Gases

At low to moderate pressures, binary diffusion coefficients vary inversely with pressure or density as suggested by Eqs. (11-3.1) and (11-3.2). At high pressures, the product $DP$ or $D\rho$ is no longer constant but decreases with an increase in either $P$ or $\rho$. Note that it is possible to have a different behavior in the products $DP$ and $D\rho$ as the pressure is raised, since $\rho$ is proportional to pressure only at low pressures, and gas nonidealities—with their concomitant effect on the system density—may become important. Also, as indicated earlier, at low pressures, the binary diffusion coefficient is essentially independent of composition. At high pressures, where the gas phase may deviate significantly from an ideal gas, small, but finite effects of composition have been noted, e.g., ref. 201.

**TABLE 11-2 Comparison of Methods for Estimating Gas Diffusion Coefficients at Low Pressures**

| System | T, K | $D_{AB}P$ (obs.), (cm$^2$/s) bar | Ref. | Errors as percent of observed values | | | |
|---|---|---|---|---|---|---|---|
| | | | | Theoretical | Brokaw | Wilke-Lee | Fuller et al. |
| Air–carbon dioxide | 276 | 0.144 | 103 | −6 | −1 | 6 | −4 |
| | 317 | 0.179 | | −3 | 2 | 10 | −2 |
| Air-ethanol | 313 | 0.147 | 139 | −10 | −13 | −8 | −9 |
| Air-helium | 276 | 0.632 | 103 | 0 | −1 | −2 | −6 |
| | 346 | 0.914 | | 0 | 0 | −1 | −3 |
| Air–$n$-hexane | 294 | 0.081 | 35 | −6 | −5 | −4 | −8 |
| | 328 | 0.094 | | −1 | 0 | 1 | −4 |
| Air–2-methylfuran | 334 | 0.107 | 4 | — | 6 | 10 | 6 |
| Air-naphthalene | 303 | 0.087 | 28 | — | −18 | −17 | −17 |
| Air-water | 313 | 0.292 | 35 | −18 | 2 | −9 | −6 |
| Ammonia–diethyl ether | 288 | 0.101 | 197 | −23 | −13 | −14 | 1 |
| | 337 | 0.139 | | −23 | −13 | −15 | −3 |
| Argon-ammonia | 255 | 0.152 | 196 | 3 | 3 | 3 | 11 |
| | 333 | 0.256 | | 3 | 2 | 1 | 5 |
| Argon-benzene | 323 | 0.085 | 134 | 8 | 9 | 14 | 13 |
| | 373 | 0.112 | | 7 | 8 | 15 | 10 |
| Argon-helium | 276 | 0.655 | 103 | −1 | −6 | −7 | −3 |
| | 418 | 1.417 | 36 | −10 | −13 | −15 | −7 |
| Argon-hexafluorobenzene | 323 | 0.082 | 134 | — | −7 | −5 | −18 |
| | 373 | 0.095 | | — | −5 | −8 | −10 |
| Argon-hydrogen | 295 | 0.84 | 229 | −9 | −2 | −8 | −5 |
| | 628 | 3.25 | | −15 | −10 | −16 | −8 |
| | 1068 | 8.21 | | −19 | −14 | −20 | −7 |
| Argon-krypton | 273 | 0.121 | 195 | −1 | −5 | 2 | −2 |
| Argon-methane | 298 | 0.205 | 36 | 5 | 7 | 13 | 3 |
| Argon–sulfur dioxide | 263 | 0.078 | 139 | 24 | 17 | 23 | 22 |
| Argon-xenon | 195 | 0.052 | 36 | −1 | −3 | 5 | 7 |
| | 378 | 0.180 | | −2 | −4 | 3 | −1 |
| Carbon dioxide–helium | 298 | 0.620 | 188 | −3 | 0 | −2 | −6 |
| | 498 | 1.433 | | −2 | 1 | −1 | 0 |
| Carbon dioxide–nitrogen | 298 | 0.169 | 224 | −9 | −7 | −1 | −4 |
| Carbon dioxide–nitrous oxide | 313 | 0.130 | 6 | −6 | 7 | 16 | [illegible] |

| | | | | | | | |
|---|---|---|---|---|---|---|---|
| | 673 | 0.385 | | −4 | 4 | 12 | −17 |
| Carbon dioxide–water | 307 | 0.201 | 41 | −20 | −5 | −8 | 9 |
| Carbon monoxide–nitrogen | 373 | 0.322 | 5 | −6 | −6 | 0 | −5 |
| Ethylene-water | 328 | 0.236 | 186 | −7 | −6 | −10 | −4 |
| Helium-benzene | 423 | 0.618 | 188 | 8 | 19 | 7 | −6 |
| Helium-bromobenzene | 427 | 0.550 | 69 | — | 30 | 13 | −3 |
| Helium–2-chlorobutane | 429 | 0.568 | 69 | — | 25 | 9 | −3 |
| Helium–*n*-butanol | 423 | 0.595 | 188 | 10 | 3 | 7 | −3 |
| Helium–1-iodobutane | 428 | 0.524 | 69 | — | 26 | 10 | 0 |
| Helium-methanol | 432 | 1.046 | 188 | 7 | 6 | −2 | −3 |
| Helium-nitrogen | 298 | 0.696 | 188 | 1 | 1 | −1 | 1 |
| Helium-water | 352 | 1.136 | 186 | 1 | 20 | −5 | −1 |
| Hydrogen-acetone | 296 | 0.430 | 139 | 0 | 32 | 6 | 1 |
| Hydrogen-ammonia | 263 | 0.58 | 139 | 4 | 23 | 4 | 5 |
| | 358 | 1.11 | | −4 | 11 | −6 | −5 |
| | 473 | 1.89 | | −5 | 6 | −12 | −9 |
| Hydrogen-cyclohexane | 289 | 0.323 | 105 | −3 | 31 | 10 | −4 |
| Hydrogen-naphthalene | 303 | 0.305 | 28 | — | 32 | 8 | 3 |
| Hydrogen-nitrobenzene | 493 | 0.831 | 166 | — | 23 | 3 | 2 |
| Hydrogen-nitrogen | 294 | 0.773 | 187 | −5 | 6 | −2 | −1 |
| | 573 | 2.449 | | −8 | 2 | −5 | 0 |
| Hydrogen-pyridine | 318 | 0.443 | 105 | −5 | 22 | 3 | 4 |
| Hydrogen-water | 307 | 0.927 | 41 | −12 | 24 | −2 | 3 |
| Methane-water | 352 | 0.361 | 186 | −11 | 3 | −3 | −2 |
| Nitrogen-ammonia | 298 | 0.233 | 139 | −5 | −3 | −4 | −3 |
| | 358 | 0.332 | | −6 | −3 | −6 | −6 |
| Nitrogen-aniline | 473 | 0.182 | 166 | — | 5 | 8 | 7 |
| Nitrogen–sulfur dioxide | 263 | 0.105 | 139 | −3 | −2 | 2 | −1 |
| Nitrogen-water | 308 | 0.259 | 186 | −11 | 4 | −2 | 5 |
| | 352 | 0.364 | | −17 | −3 | −10 | −5 |
| Nitrogen–sulfur hexafluoride | 378 | 0.146 | 112 | — | 6 | 11 | −1 |
| Oxygen-benzene | 311 | 0.102 | 105 | −9 | −5 | −2 | −4 |
| Oxygen–carbon tetrachloride | 296 | 0.076 | 139 | −5 | 1 | 3 | −4 |
| Oxygen-cyclohexane | 289 | 0.076 | 105 | −7 | 3 | 4 | −2 |
| Oxygen-water | 352 | 0.357 | 186 | −15 | −1 | −6 | −1 |
| Average absolute error | | | | 7.3 | 9.0 | 7.0 | 5.4 |

There are few experimental studies of binary diffusion coefficients at high pressures, and many of the more recent data involve a trace solute in a supercritical fluid [50, 67, 108, 151, 184, 199, 207, 220] or relate to systems containing helium as one component [13].

With the paucity of reliable data, it is not surprising that few estimation methods have been proposed. Takahashi [200] has suggested a very simple corresponding states method which is satisfactory for the limited data base available. His correlation is

$$\frac{D_{AB}P}{(D_{AB}P)^+} = f(T_r, P_r) \tag{11-5.1}$$

where $D_{AB}$ = diffusion coefficient, $cm^2/s$
$P$ = pressure, bar

The superscript $^+$ indicates that low-pressure values are to be used. The function $f(T_r, P_r)$ is shown in Fig. 11-3, and to obtain pseudocritical properties from which to calculate the reduced temperatures and pressures, Eqs. (11-5.2) to (11-5.5) are used.

$$T_r = \frac{T}{T_c} \tag{11-5.2}$$

$$T_c = y_A T_{cA} + y_B T_{cB} \tag{11-5.3}$$

$$P_r = \frac{P}{P_c} \tag{11-5.4}$$

$$P_c = y_A P_{cA} + y_B P_{cB} \tag{11-5.5}$$

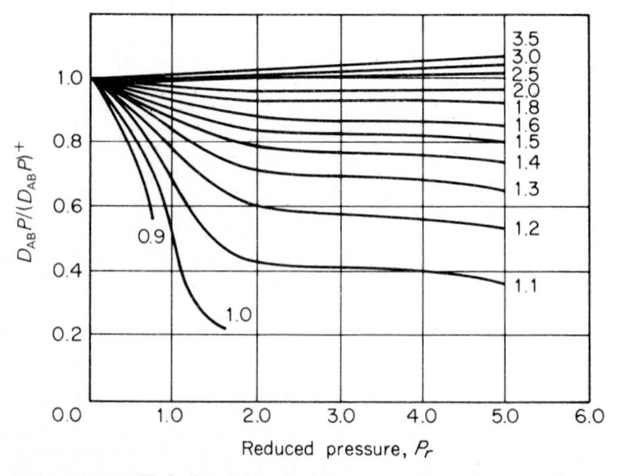


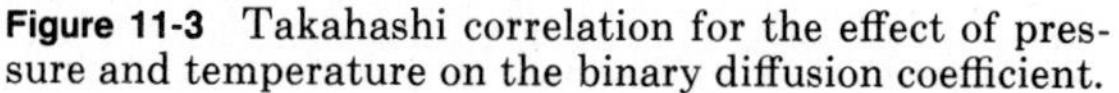

**Figure 11-3** Takahashi correlation for the effect of pressure and temperature on the binary diffusion coefficient.

As an illustration of this technique, in Fig. 11-4, we have plotted the data of Takahashi and Hongo [201] for the system carbon dioxide–ethylene. Two cases are considered, one with a very low concentration of ethylene and the other with a very low concentration of carbon dioxide. Up to about 80 bar, the two limiting diffusion coefficients are essentially identical. Above that pressure, $D_{AB}$ for the trace $CO_2$ system is significantly higher. Plotted as solid curves on this graph are the predicted values of $D_{AB}$ from Fig. 11-3 using the $(D_{AB}P)^+$ product at low pressure to be 0.149 $(cm^2/s)$bar as found by Takahashi and Hongo. Also, the dashed curve has been drawn to indicate the estimated value of $D_{AB}$ if one had assumed that $D_{AB}P$ was a constant. Clearly this assumption is in error above a pressure of about 10 to 15 bar.

Tee et al. [204] suggested a similar corresponding states approach, as have others [156, 189, 190, 201].

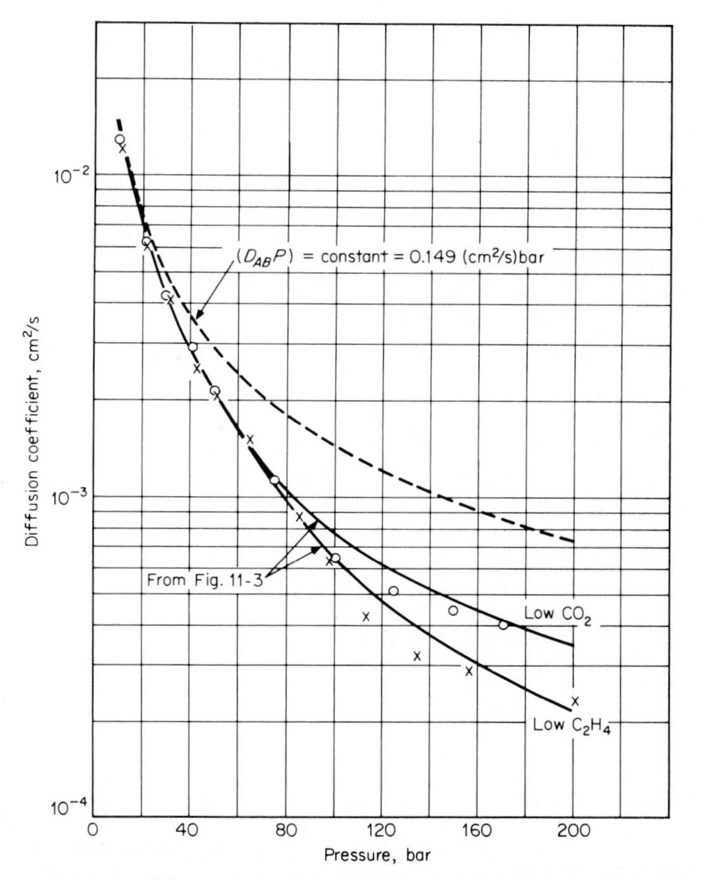


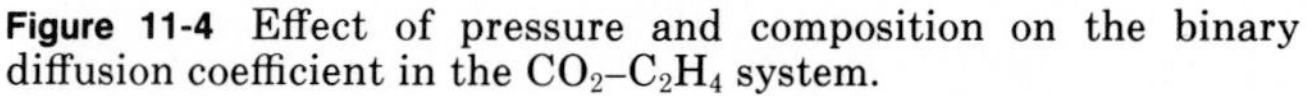
**Figure 11-4** Effect of pressure and composition on the binary diffusion coefficient in the $CO_2$–$C_2H_4$ system.

To illustrate some data for the diffusion coefficient of complex solutes in supercritical fluids, we show Fig. 11-5. There the diffusion coefficient is given as a function of reduced pressure from the ideal-gas range to reduced pressures up to about 6. The solutes are relatively complex molecules, and the solvent gases are $CO_2$, ethylene, and $SF_6$. No temperature dependence is shown, since the temperatures studied (see legend) were such that all the reduced temperatures were similar and were, in most cases, in the range of 1 to 1.05.† Up to about half the critical pressure, $D_{AB}P$ is essentially constant. Above that pressure, the data show the product $D_{AB}P$ decreasing, and at reduced pressures of about 2, it would appear that $D_{AB} \propto P_r^{1/2}$. As supercritical extractions are often carried out in a reduced temperature range of about 1.1 to 1.2 and in a reduced pressure range of 2 to 4, this plot would indicate that $D_{AB} \simeq 10^{-4}$ cm²/s, a value much less than for a low-pressure gas but still significantly higher than

†Since the concentrations of the solutes were quite low, the pressure and temperature were reduced by the pure component values of $P_c$ and $T_c$ of the solvents.

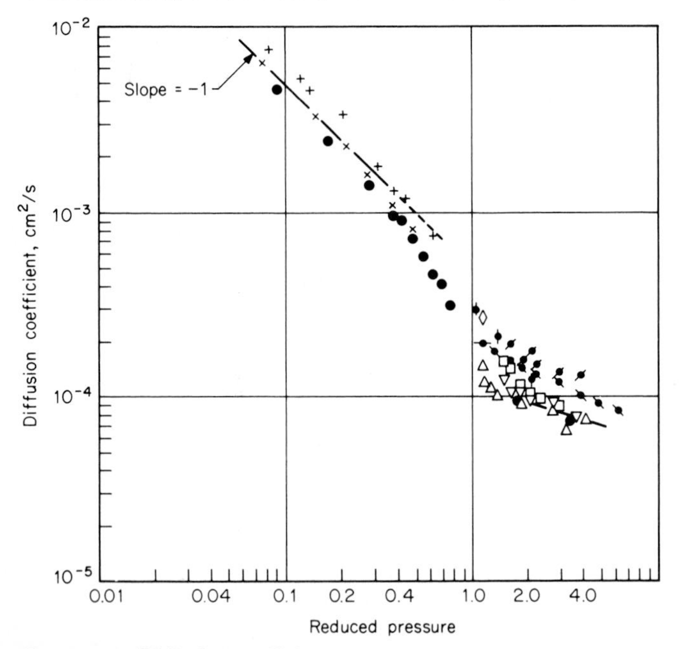


**Figure 11-5** Diffusion coefficients in supercritical fluids.

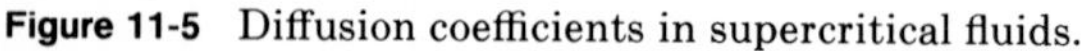

| Key | $T$, °C | System | Key | $T$, °C | System |
|---|---|---|---|---|---|
| ● | 20 | $CO_2$-naphthalene [151, 220] | ⧫ | 40 | $CO_2$-benzene [184] |
| × | 30 | | –●– | 40 | $CO_2$-propylbenzene [199] |
| + | 40 | | ◇ | 40 | $CO_2$-1,2,3-trimethylbenzene [199] |
| △ | 35 | $CO_2$-naphthalene [108, 207] | | | |
| ▽ | 55 | | ◞ | | $SF_6$-naphthalene [50] |
| ◥ | 12 | Ethylene-naphthalene [108, 207] | | 55 | $SF_6$-benzoic acid [50] |
| ◢ | 35 | | | | |

for a typical liquid (see Sec. 11-9).

We illustrate Takahashi's correlation in Example 11-4.

**Example 11-4** Estimate the diffusion coefficient of 2-naphthol in supercritical carbon dioxide at 318 K and 165 bar. The experimental value is reported to be about $7.4 \times 10^{-5}$ cm$^2$/s [50].

**solution** This estimation problem is typical of the problems often encountered. We have no knowledge of the low-pressure diffusion coefficient for this system, and, in fact, experimental values of the critical properties of 2-naphthol are unavailable. However, we suspect that the concentration of 2-naphthol is quite low, so that Takahashi's rules in Eqs. (11-5.2) to (11-5.5) would reduce to $T_c$ and $P_c$ for the pure solvent ($CO_2$). Since $T_c = 304.2$ K and $P_c = 73.8$ bar for $CO_2$, then $T_r = 318/304.2 = 1.05$ and $P_r = 165/73.8 = 2.24$. With Fig. 11-3,

$$\frac{D_{AB}P}{(D_{AB}P)^+} \simeq 0.3$$

To estimate $D^+_{AB}$ (at 1 bar), we will use the Fuller et al. method from Sec. 11-4. For $CO_2$, $M = 44.01$ and $(\Sigma_v) = 26.9$ from Table 11-1. For 2-naphthol, $M = 144$ and $(\Sigma_v) = (10)(C) + (8)(H) + (1)(O) + \text{aromatic ring} = (10)(15.9) + (8)(2.31) + (1)(6.11) - 18.3 = 165.2$. Then, with Eq. (11-4.4) and $P = 1$ bar, $T = 318$ K, $M_{AB} = 2[(1/44.01) + (1/144)]^{-1} = 67.4$,

$$D^+_{AB} = \frac{(0.00143)(318)^{1.75}}{(1)(67.4)^{1/2}[(26.9)^{1/3} + (165.2)^{1/3}]^2} = 5.1 \times 10^{-2}\ \text{cm}^2/\text{s}$$

$$D_{AB} = \frac{(0.30)(5.1 \times 10^{-2})}{165} = 9.2 \times 10^{-5}\ \text{cm}^2/\text{s}$$

This result is about 25 percent higher than that reported experimentally. On the other hand, the estimated value falls well within the band of data in Fig. 11-5 at $P_r \simeq 2$.

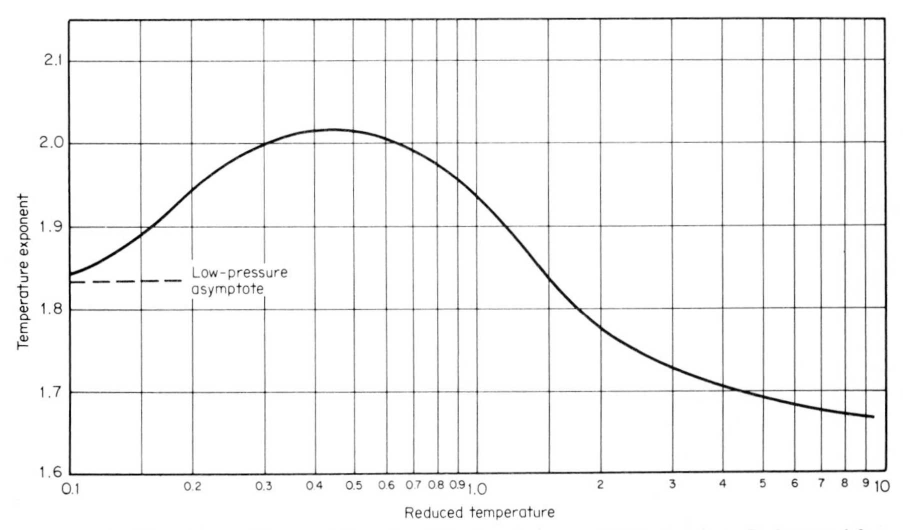


**Figure 11-6** Exponent of temperature for diffusion in gases. *(Adapted from Ref. 138 with the approximation that $\epsilon/k \approx 0.75T_c$.)*

## 11-6 The Effect of Temperature on Diffusion in Gases

At low pressures, where the ideal-gas law approximation is valid, it is seen from Eq. (11-3.2) that

$$D_{AB} \propto \frac{T^{3/2}}{\Omega_D(T)} \tag{11-6.1}$$

or

$$\left(\frac{\partial \ln D_{AB}}{\partial \ln T}\right)_P = \frac{3}{2} - \frac{\mathrm{d} \ln \Omega_D}{\mathrm{d} \ln T} \tag{11-6.2}$$

Marrero and Mason [138] indicate that, in most cases, the term $\mathrm{d} \ln \Omega_D / \mathrm{d} \ln T$ varies from 0 to $-\frac{1}{2}$. Thus $D_{AB}$ varies as $T^{3/2}$ to $T^2$. This result agrees with the empirical estimation methods referred to in Sec. 11-4, e.g., in the Fuller et al. method, $D \propto T^{1.75}$. Over wide temperature ranges, however, the exponent on temperature changes. Figure 11-6 shows the approximate variation of this exponent with reduced temperature. The very fact that the temperature exponent increases and then decreases indicates that empirical estimation techniques with a constant exponent will be limited in their range of applicability. The theoretical and the Wilke-Lee methods are therefore preferable if wide temperature regions are to be covered.

## 11-7 Diffusion in Multicomponent Gas Mixtures

A few general concepts of diffusion in multicomponent liquid mixtures presented later (in Sec. 11-12) are applicable for gas mixtures also. One of the problems with diffusion in liquids is that even the binary diffusion coefficients are often very composition-dependent. For multicomponent liquid mixtures, therefore, it is difficult to obtain numerical values of the diffusion coefficients relating fluxes to concentration gradients.

In gases, $D_{AB}$ is normally assumed independent of composition. With this approximation, multicomponent diffusion in gases can be described by the Stefan-Maxwell equation

$$\frac{dx_i}{dz} = \sum_{j=1}^{n} \frac{c_i c_j}{c^2 D_{ij}} \left(\frac{J_j}{c_j} - \frac{J_i}{c_i}\right) \tag{11-7.1}$$

where $c_i$ = concentration of $i$

$c$ = mixture concentration.

$J_i$, $J_j$ = flux of $i$, $j$

$D_{ij}$ = binary diffusion coefficient of the $ij$ system

$(dx_i/dz)$ = gradient in mole fraction of $i$ in the $z$ direction

This relation is different from the basic binary diffusion relation (11-

2.5), but the employment of common binary diffusion coefficients is particularly desirable. Marrero and Mason [138] discuss many of the assumptions behind Eq. (11-7.1).

Few attempts have been made by engineers to calculate fluxes in multicomponent systems. However, one important and simple limiting case is often cited. If a dilute component $i$ diffuses into a *homogeneous* mixture, then $J_j \approx 0$. With $c_j/c = x_j$, Eq. (11-7.1) reduces to

$$\frac{dx_i}{dz} = -J_i \sum_{\substack{j=1 \\ j \neq i}}^{n} \frac{x_j}{cD_{ij}} \tag{11-7.2}$$

Defining

$$D_{im} = \frac{-J_i}{dx_i/dz} \tag{11-7.3}$$

gives

$$D_{im} = \left( \sum_{\substack{j=1 \\ j \neq i}}^{n} \frac{x_j}{D_{ij}} \right)^{-1} \tag{11-7.4}$$

This simple relation (sometimes called Blanc's law [21, 138]) was shown to apply to several ternary cases in which $i$ was a trace component [140]. Deviations from Blanc's law are discussed by Sandler and Mason [181].

The general theory of diffusion in multicomponent gas systems is covered by Cussler [47] and by Hirschfelder et al. [100]. The problem of diffusion in three-component gas systems has been generalized by Toor [206] and verified by Fairbanks and Wilke [64], Walker et al. [223], and Duncan and Toor [56].

## 11-8 Diffusion in Liquids: Theory

Binary liquid diffusion coefficients are defined by Eq. (11-2.5) or (11-2.7). Since molecules in liquids are densely packed and strongly affected by force fields of neighboring molecules, values of $D_{AB}$ for liquids are much smaller than for low-pressure gases. That does not mean that diffusion rates are necessarily low, since concentration gradients can be large.

Liquid state theories for calculating diffusion coefficients are quite idealized, and none is satisfactory in providing relations for calculating $D_{AB}$. In several cases, however, the *form* of a theoretical equation has provided the framework for several useful prediction methods. A case in point involves the analysis of large spherical molecules diffusing in a dilute solution. Hydrodynamic theory [20, 73] then indicates that

$$D_{AB} = \frac{RT}{6\pi\eta_B r_A} \tag{11-8.1}$$

where $\eta_B$ is the viscosity of the solvent and $r_A$ is the radius of the "spherical" solute. Equation (11-8.1) is the Stokes-Einstein equation. Although this fixed relation was derived for a very special situation, many authors have used the form as a starting point in developing correlations.

Other theories for modeling diffusion in liquids have been based on kinetic theory [10, 18, 31, 33, 48, 52, 91, 114, 148], absolute-rate theory [45, 62, 73, 78, 124, 127, 161, 176], statistical mechanics [17, 18, 111], and other concepts [3, 24, 104, 119, 120, 169]. Several reviews are available for further consideration [53, 75, 76, 99, 128].

Diffusion in liquid metals is not treated, although estimation techniques are available [165].

## 11-9 Estimation of Binary Liquid Diffusion Coefficients at Infinite Dilution

For a binary mixture of solute A in solvent B, the diffusion coefficient $D^\circ_{AB}$ of A diffusing in an infinitely dilute solution of A in B implies that each A molecule is in an environment of essentially pure B. In engineering work, however, $D^\circ_{AB}$ is assumed to be a representative diffusion coefficient even for concentrations of A up to 5 and perhaps 10 mole percent.

In this section, several estimation methods for $D^\circ_{AB}$ are introduced; the effect of concentration for mutual diffusion coefficients is covered in Sec. 11-10.

### Wilke-Chang estimation method [230]

An older but still widely used correlation for $D^\circ_{AB}$, the Wilke-Chang technique is, in essence, an empirical modification of the Stokes-Einstein relation (11-8.1):

$$D^\circ_{AB} = \frac{7.4 \times 10^{-8}(\phi M_B)^{1/2} T}{\eta_B V_A^{0.6}} \tag{11-9.1}$$

where $D^\circ_{AB}$ = mutual diffusion coefficient of solute A at very low concentrations in solvent B, $cm^2/s$

$M_B$ = molecular weight of *solvent* B, g/mol

$T$ = temperature, K

$\eta_B$ = viscosity of *solvent* B, cP

$V_A$ = molar volume of *solute* A at its normal boiling temperature, $cm^3$/mol

$\phi$ = *association factor* of *solvent* B, dimensionless

If experimental data to obtain $V_A$ at $T_b$ do not exist, estimation methods from Chap. 3 may be used, in particular the Le Bas additive volume table (3-8) is convenient.

Wilke and Chang recommend that $\phi$ be chosen as 2.6 if the solvent is water, 1.9 if it is methanol, 1.5 if it is ethanol, and 1.0 if it is unassociated. When 251 solute-solvent systems were tested by these authors, an average error of about 10 percent was noted. Figure 11-7 is a graphical representation of Eq. (11-9.1) with the dashed line representing Eq. (11-8.1); the latter is assumed to represent the maximum value of the ordinate for any value of $V_A$.

A number of authors have suggested modifications of Eq. (11-9.1) particularly to improve its accuracy for systems where water is the solute and the solvent is an organic liquid [7, 29, 93, 94, 97, 123, 130, 133, 160, 183, 191, 232, 233]. However, none of these suggestions have been widely accepted. In Table 11-5, we show a comparison of estimated and experimental values of $D^\circ_{AB}$. The errors vary so greatly that the concept of an *average* error is meaningless. The method should not be used when water is the *solute*.

**Example 11-5** Use the Wilke-Chang correlation to estimate $D^\circ_{AB}$ for ethylbenzene diffusing into water at 293 K. The viscosity of water at this temperature is essentially 1.0 cP. The experimental value of $D^\circ_{AB}$ is $0.81 \times 10^{-5}$ cm²/s [233].

**solution** The normal boiling point of ethylbenzene is 409.3 K (Appendix A). At that temperature, the density is 0.761 g/cm³ [217], so with $M_A = 106.17$, $V_A = 106.17/0.761 = 139.5$ cm³/mol. [If Table 3-8 has been used, $V_A = (14.8)(8) + (3.7)(10) - 15 = 140$ cm³/mol.] Then, using Eq. (11-9.1) with $\phi = 2.6$ and $M_B = 18.0$ for water,

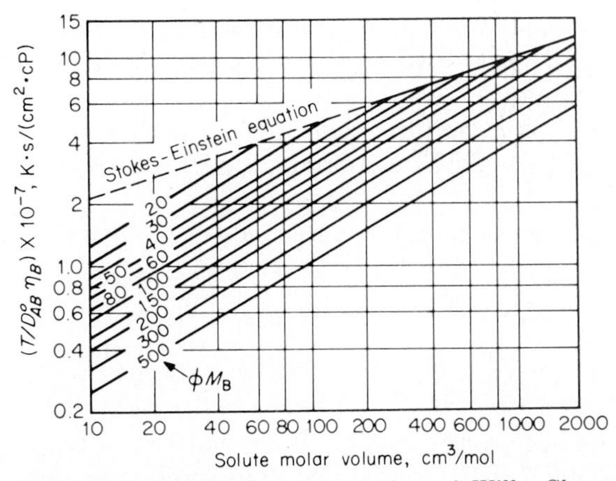


**Figure 11-7** Graphical representation of Wilke-Chang correlation of diffusion coefficients in dilute solutions. *(From Ref. 230.)*

$$D^\circ_{AB} = 7.4 \times 10^{-8} \frac{[(2.6)(18.0)]^{1/2}(293)}{(1.0)(139.5)^{0.6}} = 0.77 \times 10^{-5} \text{ cm}^2/\text{s}$$

$$\text{Error} = \frac{0.77 - 0.81}{0.81} \times 100 = -5\%$$

### Tyn and Calus method [212]

These authors have proposed that $D^\circ_{AB}$ be estimated by the relation

$$D^\circ_{AB} = 8.93 \times 10^{-8} \left(\frac{V_A}{V_B^2}\right)^{1/6} \left(\frac{\mathbf{P}_B}{\mathbf{P}_A}\right)^{0.6} \frac{T}{\eta_B} \tag{11-9.2}$$

where $V_B$ = molar volume of the *solvent* at the normal boiling temperature, cm$^3$/mol, $\mathbf{P}_A$ and $\mathbf{P}_B$ are parachors for the solute and solvent, and the other terms are defined under Eq. (11-9.1).

The parachor is related to the liquid surface tension (see Chap. 12) as

$$\mathbf{P} = V\sigma^{1/4} \tag{11-9.3}$$

where $\sigma$ is the surface tension in dyn/cm = g/s$^2$ = $10^{-3}$ N/m$^2$ and $V$ is the molar volume, cm$^3$/mol, both measured at the same temperature. Thus the units of $\mathbf{P}$ are cm$^3 \cdot$g$^{1/4}$/s$^{1/2} \cdot$mol. Quale [168] has tabulated values of $\mathbf{P}$ for a large number of chemicals; alternatively, $\mathbf{P}$ may be estimated from additive group contributions as shown in Table 11-3. Over moderate temperature ranges, $\mathbf{P}$ is essentially a constant.

When using the correlation shown in Eq. (11-9.2), the authors note several restrictions:

1. The method should *not* be used for diffusion in viscous solvents. Values of $\eta_B$ above about 20 to 30 cP would classify the solvent as viscous.
2. If the solute is water, a *dimer* value of $V_A$ and $\mathbf{P}_A$ should be used. In the calculations for Table 11-5, we used $V_A = V_w = 37.4$ cm$^3$/mol and $\mathbf{P}_A = \mathbf{P}_w = 105.2$ cm$^3 \cdot$g$^{1/4}$/s$^{1/2} \cdot$mol.
3. If the solute is an organic acid and the solvent is other than water, methanol, or butanol, the acid should be considered a *dimer* with twice the expected values of $V_A$ and $\mathbf{P}_A$.
4. For nonpolar solutes diffusing into monohydroxy alcohols, the values of $V_B$ and $\mathbf{P}_B$ should be multiplied by a factor equal to $8\eta_B$, where $\eta_B$ is the solvent viscosity in cP.

By using Eq. (11-9.2) with the restrictions noted above, values of $D^\circ_{AB}$ were estimated for a number of systems. The results are shown in Table 11-5, along with experimentally reported results. In the majority of cases, quite reasonable estimates of $D^\circ_{AB}$ were found and errors normally were less than 10%.

**TABLE 11-3 Structural Contributions for Calculating the Parachor†**

| | | | |
|---|---|---|---|
| Carbon-hydrogen: | | R–[–CO–]–R′ (ketone) | |
| C | 9.0 | R + R′ = 2 | 51.3 |
| H | 15.5 | R + R′ = 3 | 49.0 |
| $CH_3$ | 55.5 | R + R′ = 4 | 47.5 |
| $CH_2$ in $-(CH_2)_n$ | | R + R′ = 5 | 46.3 |
| $n < 12$ | 40.0 | R + R′ = 6 | 45.3 |
| $n > 12$ | 40.3 | R + R′ = 7 | 44.1 |
| | | –CHO | 66 |
| Alkyl groups | | O (not noted above) | 20 |
| 1-Methylethyl | 133.3 | N (not noted above) | 17.5 |
| 1-Methylpropyl | 171.9 | S | 49.1 |
| 1-Methylbutyl | 211.7 | P | 40.5 |
| 2-Methylpropyl | 173.3 | F | 26.1 |
| 1-Ethylpropyl | 209.5 | Cl | 55.2 |
| 1,1-Dimethylethyl | 170.4 | Br | 68.0 |
| 1,1-Dimethylpropyl | 207.5 | I | 90.3 |
| 1,2-Dimethylpropyl | 207.9 | | |
| 1,1,2-Trimethylpropyl | 243.5 | Ethylenic bonds: | |
| $C_6H_5$ | 189.6 | Terminal | 19.1 |
| | | 2,3-position | 17.7 |
| Special groups: | | 3,4-position | 16.3 |
| –COO– | 63.8 | | |
| –COOH | 73.8 | Triple bond | 40.6 |
| –OH | 29.8 | | |
| $-NH_2$ | 42.5 | Ring closure: | |
| –O– | 20.0 | Three-membered | 12 |
| $-NO_2$ | 74 | Four-membered | 6.0 |
| $-NO_3$ (nitrate) | 93 | Five-membered | 3.0 |
| $-CO(NH_2)$ | 91.7 | Six-membered | 0.8 |

†As modified from Ref. 168.

To use the Tyn-Calus form, however, the parachors of both the solute and the solvent must be known. Although the compilation of Quale [168] is of value, it is still incomplete. The structural contributions given in Table 11-3 also are incomplete, and many functional groups are not represented.

A modified form of Eq. (11-9.2) may be developed by combining Eqs. (11-9.2) and (11-9.3) to give

$$D^\circ_{AB} = 8.93 \times 10^{-8} \frac{V_B^{0.267}}{V_A^{0.433}} \frac{T}{\eta_B} \left(\frac{\sigma_B}{\sigma_A}\right)^{0.15} \tag{11-9.4}$$

The definitions of the terms are the same as before except, when substituting Eq. (11-9.3), we must define $V$ and $\sigma$ at $T_b$. Thus $\sigma_B$ and $\sigma_A$ in Eq. (11-9.4) refer to surface tensions at $T_b$. Note also the very low exponent on this ratio of surface tensions. Since most *organic* liquids at $T_b$ have similar surface tensions, one might choose to approximate this ratio as equal to unity. (For example, $0.8^{0.15} = 0.97$ and $1.2^{0.15} = 1.03$.) Then,

$$D^{\circ}_{AB} = 8.93 \times 10^{-8} \frac{V_B^{0.267}}{V_A^{0.433}} \frac{T}{\eta_B} \tag{11-9.5}$$

Alternatively, an approximation to the $\sigma_B/\sigma_A$ ratio may be developed by using one of the correlations shown in Chap. 12. For example, if the Brock and Bird corresponding states method were used, then

$$\sigma = P_c^{2/3} T_c^{1/3} (0.132\alpha_c - 0.278)(1 - T_{b_r})^{11/9} \tag{11-9.6}$$

with $P_c$ in bars and $T_b$ and $T_c$ in kelvins, $T_{b_r} = T_b/T_c$, and

$$\alpha_c = 0.9076 \left[ 1 + \frac{T_{b_r} \ln (P_c/1.013)}{1 - T_{b_r}} \right] \tag{11-9.7}$$

Equation (11-9.6) is only approximate, but it may be satisfactory when used to develop the *ratio* $(\sigma_A/\sigma_B)$. Also, considering the low power (0.15) to which the ratio is raised, estimates of $(\sigma_B/\sigma_A)^{0.15}$ should be quite reasonable.

When Eq. (11-9.5) was employed to estimate $D^{\circ}_{AB}$ for the systems shown in Table 11-5, the results, as expected, were very similar to those found from the original Tyn and Calus form [Eq. (11-9.2)] except when $\sigma_B$ differed appreciably from $\sigma_A$, for example, in the case of water and an organic liquid. In such situations, however, Eq. (11-9.4) with Eqs. (11-9.6) and (11-9.7) still led to results not significantly different from those with Eq. (11-9.2).

The various forms of the Tyn-Calus correlation are illustrated in Example 11-6.

#### Hayduk and Minhas correlation [98]

These authors considered many correlations for the infinite dilution binary diffusion coefficient. By regression analysis, they proposed several depending on the type of solute-solvent system.

For *normal paraffin solutions:*

$$D^{\circ}_{AB} = 13.3 \times 10^{-8} \frac{T^{1.47} \eta_B^{\epsilon}}{V_A^{0.71}} \tag{11-9.8}$$

where $\epsilon = (10.2/V_A) - 0.791$ and the other notation is the same as in Eq. (11-9.1). Equation (11-9.8) was developed from data on solutes ranging from $C_5$ to $C_{32}$ in normal paraffin solvents encompassing $C_5$ to $C_{16}$. An average error of only 3.4 percent was reported.

For *solutes in aqueous solutions:*

$$D^{\circ}_{AB} = 1.25 \times 10^{-8} (V_A^{-0.19} - 0.292) T^{1.52} \eta_w^{\epsilon^*} \tag{11-9.9}$$

with $\epsilon^* = (9.58/V_A) - 1.12$. The rest of the terms are defined in the same manner as under Eq. (11-9.1) except that the subscript $w$ refers to the solvent, water. The authors report that this relation predicted $D^{\circ}_{AB}$ values with an average deviation of slightly less than 10 percent.

For *nonaqueous (nonelectrolyte) solutions:*

$$D^{\circ}_{AB} = 1.55 \times 10^{-8} \frac{T^{1.29}(\mathbf{P}_B^{0.5}/\mathbf{P}_A^{0.42})}{\eta_B^{0.92} V_B^{0.23}} \tag{11-9.10}$$

The notation is the same as in Eq. (11-9.2).

The appropriate equation in the set of (11-9.8) to (11-9.10) was used in computing the errors shown in Table 11-5.

It is important to note that, when using the Hayduk-Minhas correlations, the same restrictions apply as in the Tyn-Calus equations.

If Eq. (11-9.3) is used in Eq. (11-9.10) to eliminate the parachors, one obtains

$$D^{\circ}_{AB} = 1.55 \times 10^{-8} \frac{V_B^{0.27}}{V_A^{0.42}} \frac{T^{1.29}}{\eta_B^{0.92}} \frac{\sigma_B^{0.125}}{\sigma_A^{0.105}} \tag{11-9.11}$$

This relation is remarkably similar to the modified Tyn-Calus equation (11-9.4) except for the larger exponent on temperature. As before, when $\sigma_A$ and $\sigma_B$ are not greatly different, the surface tension ratio may be set equal to unity as was done to obtain Eq. (11-9.5), or if $\sigma_A$ and $\sigma_B$ differ appreciably, Eqs. (11-9.6) and (11-9.7) may be employed.

**Example 11-6** Estimate the infinitely dilute diffusion coefficient of acetic acid into acetone at 313 K. The experimental value is $4.04 \times 10^{-5}$ cm²/s [230].

**solution** The data, from Appendix A and Refs. 168 and 217, are:

| | Acetic acid (solute) A | Acetone (solvent) B |
|---|---|---|
| $T_b$, K | 391.1 | 329.2 |
| $T_c$, K | 592.7 | 508.1 |
| $P_c$, bar | 57.9 | 47.0 |
| $\rho$ (at $T_b$), g/cm³ | 0.939 | 0.749 |
| $M$, g/mol | 60.05 | 58.08 |
| $V$ (at $T_b$), cm³/mol | 64.0 | 77.5 |
| $\mathbf{P}$, cm·g$^{1/4}$/s$^{1/2}$·mol | 129 | 162 |
| $\eta_B$, cP | | 0.270 |

TYN-CALUS, Eq. (11-9.2). By rule 3, acetic acid should be treated as a dimer; thus, $V = (2)(64.0) = 128$ cm³/mol and $\mathbf{P} = (2)(129) = 258$ cm³·g$^{1/4}$/s$^{1/2}$·mol.

$$D^{\circ}_{AB} = 8.93 \times 10^{-8} \left(\frac{128}{(77.5)^2}\right)^{1/6} \left(\frac{162}{258}\right)^{0.6} \frac{313}{0.270}$$

$$= 4.12 \times 10^{-5} \text{ cm}^2/\text{s}$$

$$\text{Error} = \frac{4.12 - 4.04}{4.04} \times 100 = 2\%$$

MODIFIED TYN-CALUS, Eq. (11-9.5)

$$D^\circ_{AB} = 8.93 \times 10^{-8} \frac{(77.5)^{0.267}}{(128)^{0.433}} \frac{313}{0.270}$$

$$= 4.04 \times 10^{-5} \text{ cm}^2\text{/s}$$

Error = 0%

MODIFIED TYN-CALUS, Eqs. (11-9.4), (11-9.6), and (11-9.7). For acetic acid, $T_{b_r} = 391.1/592.7 = 0.660$. With Eq. (11-9.7),

$$\alpha_c = 0.9076 \left\{ 1 + (0.660) \left[ \frac{\ln (57.9/1.013)}{1 - 0.660} \right] \right\} = 8.031$$

Similarly, $\alpha_c$ for acetone = 7.316. Then, with Eq. (11-9.6),

$$\sigma_A = (57.9)^{2/3}(592.7)^{1/3}[(0.132)(8.031) - 0.278](1 - 0.660)^{11/9}$$

$$= 26.0 \text{ erg/cm}^2$$

For acetone, $\sigma_B = 19.9$ erg/cm$^2$ and $(\sigma_B/\sigma_A)^{0.15} = 0.961$; thus,

$$D^\circ_{AB} = (4.04 \times 10^{-5})(0.961)$$

$$= 3.88 \times 10^{-5} \text{ cm}^2\text{/s}$$

$$\text{Error} = \frac{3.88 - 4.04}{4.04} \times 100 = -4\%$$

In this particular case, the use of the $(\sigma_B/\sigma_A)^{0.15}$ factor actually increased the error; in most other cases, however, errors were less when it was employed.

HAYDUK-MINHAS, Eq. (11-9.10)

$$D^\circ_{AB} = 1.55 \times 10^{-8}(313)^{1.29} \frac{(162)^{0.5}/(258)^{0.42}}{(0.270)^{0.92}(77.5)^{0.23}}$$

$$= 3.89 \times 10^{-5} \text{ cm}^2\text{/s}$$

$$\text{Error} = \frac{3.89 - 4.04}{4.04} \times 100 = -4\%$$

### Nakanishi correlation [157]

In this method, empirical parameters were introduced to account for specific interactions between the solvent and the (infinitely dilute) solute. As originally proposed, the scheme was applicable only at 298.2 K. We have scaled the equation assuming $D^\circ_{AB}\eta_B/T$ to be constant.

$$D^\circ_{AB} = \left[ \frac{9.97 \times 10^{-8}}{(I_A V_A)^{1/3}} + \frac{2.40 \times 10^{-8} A_B S_B V_B}{I_A S_A V_A} \right] \frac{T}{\eta_B} \tag{11-9.12}$$

where $D^\circ_{AB}$ is the diffusion coefficient of solute A in solvent B at low concentrations, cm$^2$/s. $V_A$ and $V_B$ are the liquid molar volumes of A and B at the system temperature $T$, cm$^3$/mol, and the factors $I_A$, $S_A$, $S_B$, and $A_B$ are given in Table 11-4. $\eta_B$ is the solvent viscosity, in cP.

Should the solute (pure) not be a liquid at 298 K, it is recommended that the liquid molar volume at the boiling point be obtained either from data or from correlations in Chap. 3. Then,

$$V_A\ (298.2\ \text{K}) = \beta V_A(T_b) \tag{11-9.13}$$

where $\beta = 0.894$ for compounds that are solid at 298 K and $\beta = 1.065$ for compounds that are normally gases at 298 K (and 1 bar). For example, if oxygen is the solute, then, at the normal boiling point of 90.2 K, the molar liquid volume is 27.9 $cm^3$/mol (Appendix A). With Eq. (11-9.13), $V_A = (1.065)(27.9) = 29.7$ $cm^3$/mol.

Values of $D^\circ_{AB}$ were estimated for a number of solute-solvent systems and the results were compared with experimental values in Table 11-5. In this tabulation, $V_A$ for water was set equal to the dimer value of 37.4 $cm^3$/mol to obtain more reasonable results. The poorest estimates were obtained with dissolved gases and with solutes in the more viscous solvents such as $n$-butanol. The use of definite values of $I_A$ to account for solute polarity may cause problems, since it is often difficult to decide whether a compound should be counted as polar ($I_A = 1.5$) or not ($I_A = 1.0$). It might be better to select an average $I_A \simeq 1.25$ if there is doubt about the molecular polarity.

**Example 11-7** Estimate the value of $D^\circ_{AB}$ for $CCl_4$ diffusing into ethanol at 298 K. At this temperature, the viscosity of ethanol is 1.08 cP. The experimental value of $D^\circ_{AB}$ is $1.50 \times 10^{-5}$ $cm^2$/s [133].

**solution** For this system with $CCl_4$ as solute A and ethanol as solvent B, from Table 11-4, $I_A = 1$, $S_A = 1$, $A_B = 2$, and $S_B = 1$. From Appendix A, the liquid densities of $CCl_4$ and ethanol at 298 K are 1.584 and 0.785 g/$cm^3$. With $M_A = 153.82$ and $M_B = 46.07$, $V_A = 153.82/1.584 = 97.1$ $cm^3$/mol and $V_B = 46.07/0.785 = 58.7$ $cm^3$/mol. Then, with Eq. (11-9.12),

**TABLE 11-4 Nakanishi Parameter Values for Liquid Diffusion Coefficients**

| | As solutes (A)† | | As solvents (B) | |
|---|---|---|---|---|
| Compound(s) | $I_A$ | $S_A$ | $A_B$ | $S_B$ |
| Water | 2.8 (1.8)‡ | 1 | 2.8 | 1 |
| Methanol | 2.2 (1.5) | 1 | 2.0 | 1 |
| Ethanol | 2.5 (1.5) | 1 | 2.0 | 1 |
| Other monohydric alcohols | 1.5 | 1 | 1.8 | 1 |
| Glycols, organic acids, and other associated compounds | 2.0 | 1 | 2.0 | 1 |
| Highly polar materials | 1.5 | 1 | 1.0 | 1 |
| Paraffins ($5 \le n \le 12$) | 1.0 | 0.7 | 1.0 | 0.7 |
| Other substances | 1.0 | 1 | 1.0 | 1 |

†If the solute is He, $H_2$, $D_2$, or Ne, the values of $V_A$ should be multiplied by $[1 + (0.85)\Lambda^2]$, where $\Lambda = 3.08$ for $He^3$, 2.67 for $He^4$, 1.73 for $H_2$, 1.22 for $D_2$, and 0.59 for Ne.

‡The values in parentheses are for cases in which these solutes are dissolved in a solvent which is more polar.

$$D^\circ_{AB} = \left[\frac{9.97 \times 10^{-8}}{(97.1)^{1/3}} + \frac{(2.40 \times 10^{-8})(2)(58.7)}{97.1}\right]\frac{298}{1.08}$$

$$= 1.40 \times 10^{-5}\ \text{cm}^2/\text{s}$$

$$\text{Error} = \frac{1.40 - 1.50}{1.50} \times 100 = -7\%$$

Other infinite dilution correlations for diffusion coefficients have been proposed, but after evaluation they were judged either less accurate or less general than the ones noted above [1, 2, 3, 24, 39, 63, 65, 66, 71, 97, 116, 119, 120, 132, 164, 169, 175, 194, 205, 214, 215].

### Effect of solvent viscosity

Most of the estimation techniques introduced in this section have assumed that $D^\circ_{AB}$ varies inversely with the viscosity of the solvent. This inverse dependence originated from the Stokes-Einstein relation for a large (spherical) molecule diffusing through a continuum solvent (small molecules). If, however, the solvent is viscous, one may question whether this simple relation is applicable. Davies et al. [49] found for $CO_2$ that in various solvents, $D^\circ_{AB}\eta_B^{0.45} \simeq$ constant for solvents ranging in viscosity from 1 to 27 cP, and these authors noted that, in 1930, Arnold [12] had proposed an empirical estimation scheme by which $D^\circ_{AB} \propto \eta_B^{-0.5}$. Oosting et al. [162] noted that, for the diffusion of 1-hexanol and 2-butanone in malto-dextrin solutions, the viscosity exponent was close to $-0.5$ over a range of temperatures and concentrations.

Hayduk and Cheng [95] investigated the effect of solvent viscosity more extensively and proposed that, for nonaqueous systems,

$$D^\circ_{AB} = Q\eta_B{}^q \tag{11-9.14}$$

where the constants $Q$ and $q$ are particular for a given solute; some values are listed by these authors. In Fig. 11-8, $CO_2$ diffusion coefficients in various solvents are shown. The solvent viscosity range is reasonably large, and the correlation for organic solvents is satisfactory. In contrast, the data for water as a solvent also are shown [99]. These data fall well below the organic solvent curve and have a slope close to $-1$. Hiss and Cussler [101] measured diffusion coefficients of $n$-hexane and naphthalene in hydrocarbons with viscosities ranging from 0.5 to 5000 cP and report that $D^\circ_{AB} \propto \eta_B^{-2/3}$, whereas Hayduk et al. [94] found that, for methane, ethane, and propane, $D^\circ_{AB}$ was proportional to $\eta_B^{-0.545}$.

These studies and others [73, 131, 225] show clearly that, over wide temperature or solvent viscosity ranges, simple empirical correlations, as presented earlier, are inadequate. The diffusion coefficient does not decrease in proportion to an increase in solvent viscosity, but $D^\circ_{AB} \propto \eta_B{}^q$, where $q$ varies, usually from $-0.5$ to $-1$.

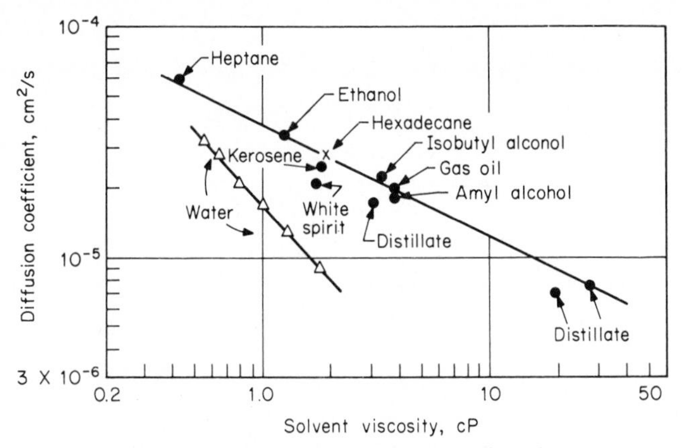


**Figure 11-8** Diffusion coefficients of carbon dioxide in various solvents. ● Ref. 49; × Ref. 95; Δ Ref. 99

## Discussion

Four estimation techniques were described to estimate the infinite dilution diffusion coefficient of a solute A in a solvent B. In Table 11-5, we show comparisons between calculated and experimental values of $D^{\circ}_{AB}$ for a number of binary systems. Several comments are pertinent when analyzing the results. First, the temperature range covered is small; thus, any conclusions based upon this sample may not hold at much higher (or lower) temperatures. Second, while $D^{\circ}_{AB}$ (exp.) is reported to three significant figures, the true accuracy is probably much less because diffusion coefficients are difficult to measure with high precision. Third, all estimation schemes tested showed wide fluctuations in the percent errors. These "failures" may be due to inadequacies in the correlation or to poor data. However, with such wide error ranges, the value of a single *average* percent error is in doubt and was not determined.

With these caveats, it is clearly seen that, in general, the Tyn-Calus and the Hayduk-Minhas correlations usually yield the lowest errors; they are, therefore, recommended for calculating $D^{\circ}_{AB}$. Both require values of the solute and solvent parachors, but using modifications such as Eq. (11-9.5) when $\sigma_A \approx \sigma_B$, or Eq. (11-9.4) [or (11-9.10)] with, say, Eqs. (11-9.6) and (11-9.7) when $\sigma_A$ differs much from $\sigma_B$ obviates the necessity of knowing the parachors.

In special situations such as diffusion in $n$-paraffin solutions, Eq. (11-9.8) is recommended. We did not find a clear advantage for Eq. (11-9.9) over (11-9.2) for solutes diffusing into water, but the former would be more convenient to use.

New experimental data include the systems $H_2S$-$H_2O$ [85], $SO_2$-$H_2O$ [121], $CO_2$ in binary mixtures [202], normal paraffin solutions [96], hydrocarbons in $n$-hexane [57, 58], and rare gases in water [218]. Baldauf and

**TABLE 11-5 Diffusion Coefficients in Liquids at Infinite Dilution**

| Solute A | Solvent B | $T$, K | $D^{\circ}_{AB} \times 10^5$, $cm^2/s$ | Ref. | Percent error† | | | |
|---|---|---|---|---|---|---|---|---|
| | | | | | Tyn and Calus | Hayduk and Minhas | Nakanishi | Wilke and Chang |
| Acetone | Chloroform | 298 | 2.35 | 87 | 2.7 | 1.6 | −7.6 | 39 |
| | | 313 | 2.90 | | 3.0 | 2.0 | −7.3 | 39 |
| Benzene | | 288 | 2.51 | 110 | −22 | −25 | −14 | 0.4 |
| | | 328 | 4.25 | 182 | −24 | −26 | −16 | −1.9 |
| Ethanol | | 288 | 2.20 | 110 | 7.7 | 6.1 | −25 | 53 |
| Ethyl ether | | 298 | 2.13 | 182 | 1.9 | 0.3 | 8.3 | 27 |
| Ethyl acetate | | 298 | 2.02 | 172 | 6.2 | 5.0 | 14 | 34 |
| Methyl ethyl ketone | | 298 | 2.13 | 172 | 3.5 | 2.6 | −11 | 34 |
| Acetic acid | Benzene | 298 | 2.09 | 37 | −11 | −11 | −11 | 26 |
| Aniline | | 298 | 1.96 | 170 | 1.4 | 1.5 | −20 | −3.1 |
| Benzoic acid | | 298 | 1.38 | 37 | −2.9 | −0.8 | −7.2 | 25 |
| Bromobenzene | | 281 | 1.45 | 198 | −8.5 | −7.8 | −4.3 | −10 |
| Cyclohexane | | 298 | 2.09 | 182 | −8.5 | −8.9 | −6.0 | −12 |
| | | 333 | 3.45 | | −3.0 | −3.9 | −0.5 | −7.0 |
| Ethanol | | 288 | 2.25 | 110 | −5.1 | −6.0 | −37 | 3.1 |
| Formic acid | | 298 | 2.28 | 37 | −5.7 | −6.9 | 1.0 | 40 |
| *n*-Heptane | | 298 | 2.10 | 30 | −17 | −18 | −24 | 26 |
| | | 353 | 4.25 | | −7.8 | −9.4 | 6.5 | 26 |
| Methyl ethyl ketone | | 303 | 2.09 | 7 | 7.9 | 7.1 | −11 | 7.1 |
| Naphthalene | | 281 | 1.19 | 198 | 2.7 | 3.7 | 3.2 | −9.2 |
| Toluene | | 298 | 1.85 | 182 | 2.4 | 2.0 | 5.5 | −1.1 |
| 1,2,4-Trichlorobenzene | | 281 | 1.34 | 198 | −7.3 | −7.8 | −13 | −18 |
| Vinyl chloride | | 281 | 1.77 | 198 | −1.8 | −2.1 | 19 | 5.6 |
| Acetic acid | Acetone | 288 | 2.92 | 19 | 6.3 | −0.5 | 2.5 | 39 |
| | | 313 | 4.04 | 230 | 2.1 | −3.7 | −1.6 | 33 |
| Benzoic acid | | 298 | 2.62 | 37 | −4.7 | −7.8 | −15 | 13 |
| Formic acid | | 298 | 3.77 | 37 | 6.2 | −0.6 | 17 | 56 |

| | | | | | | | | |
|---|---|---|---|---|---|---|---|---|
| Nitrobenzene | | 293 | 2.94 | 175 | 8.5 | 3.6 | −11 | 0.7 |
| Water | | 298 | 4.56 | 160 | 6.3 | 4.4 | −16 | — |
| Bromobenzene | *n*-Hexane | 281 | 2.60 | 230 | 18 | 13 | 20 | 16 |
| Carbon tetrachloride | | 298 | 3.70 | 88 | 15 | 8.4 | 12 | 8.6 |
| Dodecane | | 298 | 2.73 | 215 | 9.2 | −2.5 | 1.0 | −13 |
| *n*-Hexane | | 298 | 4.21 | 135 | −7.3 | −2.6 | 2.2 | −16 |
| Methyl ethyl ketone | | 303 | 3.74 | 7 | 24 | 19 | −1.8 | 23 |
| Propane | | 298 | 4.87 | 94 | 2.4 | 22 | 13 | 6.0 |
| Toluene | | 298 | 4.21 | 37 | −4.6 | −9.1 | −4.5 | −6.9 |
| Allyl alcohol | Ethanol | 293 | 0.98 | 107 | 3.6 | 7.6 | 15 | 15 |
| Isoamyl alcohol | | 293 | 0.81 | 107 | 4.3 | 7.5 | 2.3 | 1.2 |
| Benzene | | 298 | 1.81 | 133 | −1.5 | 3.6 | −22 | −40 |
| Iodine | | 298 | 1.32 | 37 | — | — | −1.2 | −1.5 |
| Oxygen | | 303 | 2.64 | 118 | 31 | 34 | 66 | 1.0 |
| Pyridine | | 293 | 1.10 | 107 | −16 | −14 | −20 | −17 |
| Water | | 298 | 1.24 | 123 | 3.4 | 12 | 7.5 | — |
| Carbon tetrachloride | | 298 | 1.50 | 133 | 15 | 21 | −6.7 | −29 |
| Adipic acid | *n*-Butanol | 303 | 0.40 | 7 | 17 | 28 | 9.5 | 5.5 |
| Benzene | | 298 | 1.00 | 133 | 5.2 | 20 | −21 | −48 |
| Butyric acid | | 303 | 0.51 | 7 | 7.0 | 17 | −5.6 | 6.8 |
| *p*-Dichlorobenzene | | 298 | 0.82 | 133 | 12 | 28 | −27 | −49 |
| Methanol | | 303 | 0.59 | 133 | 33 | 46 | 58 | 66 |
| Oleic acid | | 303 | 0.25 | 7 | 25 | 39 | −13 | 1.1 |
| Propane | | 298 | 1.57 | 19 | −23 | −14 | −39 | −61 |
| Water | | 298 | 0.56 | 133 | 4.6 | 23 | 30 | — |
| Benzene | *n*-Heptane | 298 | 3.40 | 30 | 1.0 | −1.6 | 12 | 7.4 |
| | | 372 | 8.40 | | −3.0 | −4.1 | 7.9 | 3.1 |
| Acetic acid | Ethyl acetate | 293 | 2.18 | 192 | 11 | 8.0 | 16 | 68 |
| Acetone | | 293 | 3.18 | 192 | −7.2 | −9.9 | −14 | 2.5 |
| Ethyl benzoate | | 293 | 1.85 | 192 | 16 | 12 | 11 | 7.6 |
| Methyl ethyl ketone | | 303 | 2.93 | 7 | 3.9 | 1.1 | −9.8 | 16 |
| Nitrobenzene | | 293 | 2.25 | 192 | 6.0 | 3.7 | −11 | 9.3 |
| Water | | 298 | 3.20 | 123 | 11 | 13 | −4.0 | — |

**TABLE 11-5 Diffusion Coefficients in Liquids at Infinite Dilution (*Continued*)**

| | | | | | Percent error† | | | |
|---|---|---|---|---|---|---|---|---|
| Solute A | Solvent B | $T$, K | $D^{\circ}_{AB} \times 10^5$, $cm^2/s$ | Ref. | Tyn and Calus | Hayduk and Minhas | Nakanishi | Wilke and Chang |
| Methane | Water | 275 | 0.85 | 233 | −1.6 | 1.0 | 22 | 10 |
| | | 333 | 3.55 | | 2.0 | −2.5 | 27 | 15 |
| Carbon dioxide | | 298 | 2.00 | 222 | −19 | −13 | 13 | 2.0 |
| Propylene | | 298 | 1.44 | 222 | −14 | −11 | 1.1 | −5.6 |
| Methanol | | 288 | 1.26 | 110 | −7.4 | −4.0 | −10 | 7.1 |
| Ethanol | | 288 | 1.00 | 107 | 0 | 1.8 | −7.7 | 11 |
| Allyl alcohol | | 288 | 0.90 | 107 | 2.0 | −0.2 | −10 | 6.7 |
| Acetic acid | | 293 | 1.19 | 126 | −3.3 | −4.6 | −27 | 2.5 |
| Ethyl acetate | | 293 | 1.00 | 126 | −7.8 | −15 | −3.5 | −10 |
| Aniline | | 293 | 0.92 | 126 | −4.0 | −10 | −18 | −4.3 |
| Diethylamine | | 293 | 0.97 | 126 | −5.9 | −14 | −22 | −10 |
| Pyridine | | 288 | 0.58 | 107 | 44 | 22 | 14 | 33 |
| Ethylbenzene | | 293 | 0.81 | 233 | 0.8 | −14 | 1.9 | −6.2 |
| Methylcylopentane | | 275 | 0.48 | 233 | 2.1 | −13 | 6.3 | 0 |
| | | 293 | 0.85 | | 2.5 | −7.5 | 6.8 | 0 |
| | | 333 | 1.92 | | 10 | 9.7 | 15 | 7.3 |
| Vinyl chloride | | 298 | 1.34 | 97 | −5.8 | −3.6 | 10 | 3.0 |
| | | 348 | 3.67 | | −5.5 | 2.5 | 10 | 3.5 |

†Percent error = [(calc. − exp.)/exp.] × 100

Knapp [15] studied a wide variety of polar and nonpolar systems at different temperatures and compositions. Mohan and Srinivasan [150] discuss the effect of solute-solvent complexes on $D^{\circ}_{AB}$ for solutions of nonpolar solvents (benzene, carbon tetrachloride) with small amounts of methanol or ethanol.

## 11-10 Concentration Dependence of Binary Liquid Diffusion Coefficients

In Sec. 11-2 it was suggested that the diffusion coefficient $D_{AB}$ in a binary mixture may be proportional to a thermodynamic correction $\alpha = [(\partial \ln a/\partial \ln x)]_{T,P}$; $a$ is the activity, and $x$ is the mole fraction. From the Gibbs-Duhem equation, the derivative $(\partial \ln a/\partial \ln x)$ is the same whether written for A or B.

Several liquid models purport to relate $D_{AB}$ to composition; e.g., the Darken equation [48, 75] predicts that

$$D_{AB} = (D^{*}_{A} x_A + D^{*}_{B} x_B)\alpha \tag{11-10.1}$$

where $D^{*}_{A}$ and $D^{*}_{B}$ are tracer diffusion coefficients at $x_A$, $x_B$ and $\alpha$ is evaluated at the same composition. Equation (11-10.1) was originally proposed to describe diffusion in metals, but it has been used for organic liquid mixtures by a number of investigators [31, 33, 74, 136, 148, 216, 219] with reasonable success except for mixtures in which the components may associate [89]. The unavailability of tracer diffusion coefficients in most instances has led to a modification of Eq. (11-10.1) as

$$D_{AB} = (D^{\circ}_{BA} x_A + D^{\circ}_{AB} x_B)\alpha = [x_A(D^{\circ}_{BA} - D^{\circ}_{AB}) + D^{\circ}_{AB}]\alpha \tag{11-10.2}$$

That is, $D_{AB}$ is a linear function of composition (see Fig. 11-2) corrected by the thermodynamic factor $\alpha$. Equation (11-10.2) is easier to use because the infinitely dilute diffusion coefficients $D^{\circ}_{BA}$ and $D^{\circ}_{AB}$ may be estimated by techniques shown in Sec. 11-9. The thermodynamic term in Eq. (11-10.2) often overcorrects $D_{AB}$. Rathbun and Babb [173] suggest $\alpha$ be raised to a fractional power; for associated systems, the exponent chosen was 0.6 unless there were *negative* deviations from Raoult's law when an exponent of 0.3 was recommended. It is interesting to note [179] that curves showing $\alpha$ and $D_{AB}$ as a function of $x_A$ tend to have the same curvature, thus providing some credence to the use of $\alpha$ as a correction factor.

Sanchez and Clifton [179] found they could correlate $D_{AB}$ with composition for a wide variety of binary systems by using a modification of Eq. (11-10.2):

$$D_{AB} = (D^{\circ}_{BA} x_A + D^{\circ}_{AB} x_B)(1 - m + m\alpha) \tag{11-10.3}$$

where the parameter $m$ is to be found from one mixture datum point, preferably in the midcompositional range. $m$ varies from system to system and may be either greater or less than unity. When $m = 1$, Eq. (11-10.3) reduces to Eq. (11-10.2). Interestingly, for a number of highly associated systems, $m$ was found to be between 0.8 and 0.9. The temperature dependence of $m$ is not known.

Another theory predicts that the group $D_{AB}\eta/\alpha$ should be a linear function of mole fraction [8, 19, 26]. Vignes [219] shows graphs indicating this is not even approximately true for the systems acetone-water and acetone-chloroform. Rao and Bennett [170] studied several very nonideal mixtures and found that, while the group $D_{AB}\eta/\alpha$ did not vary appreciably with composition, no definite trends could be discerned. One of the systems studied (aniline–carbon tetrachloride) is shown in Fig. 11-9. In this case, $D_{AB}$, $\eta$, $\alpha$, and $D_{AB}\eta$ varied widely; the group $D_{AB}\eta/\alpha$ also showed an unusual variation with composition. Carman and Stein [34] stated that $D_{AB}\eta/\alpha$ is a linear function of $x_A$ for the nearly ideal system benzene–carbon tetrachloride and for the nonideal system acetone-chloroform but not for ethyl alcohol–water. Vignes [219] suggested a convenient way of correlating the composition effect on the liquid diffusion coefficient:

$$D_{AB} = [(D^{\circ}_{AB})^{x_B}(D^{\circ}_{BA})^{x_A}]\alpha \tag{11-10.4}$$

and, therefore, a plot of log $(D_{AB}\eta/\alpha)$ vs. mole fraction should be linear. He illustrated this relation with many systems, and, with the exception of strongly associated mixtures, excellent results were obtained. Figure 11-10 shows the same aniline–carbon tetrachloride system plotted earlier in Fig. 11-9. Although not perfect, there is a good agreement with Eq. (11-10.4).

Dullien [54] carried out a statistical test of the Vignes correlation. It was found to fit experimental data extremely well for ideal or nearly ideal mixtures, but there were several instances when it was not particularly accurate for nonideal, nonassociating solutions. Other authors report that Vignes correlation is satisfactory for benzene and $n$-heptane [30] and toluene and methylcyclohexane [86], not for benzene and cyclohexane [128].

The Vignes relation can be derived from absolute rate theory, and a logical modification of this equation is found to be [124]

$$D_{AB}\eta = [(D^{\circ}_{AB}\eta_B)^{x_B}(D^{\circ}_{BA}\eta_A)^{x_A}]\alpha \tag{11-10.5}$$

A test of 11 systems showed that this latter form was marginally better in fitting experimental data. In Fig. 11-11 we have plotted both log $(D_{AB}\eta/\alpha)$ and log $(D_{AB}\eta/\alpha)$ as a function of composition for the aniline-benzene

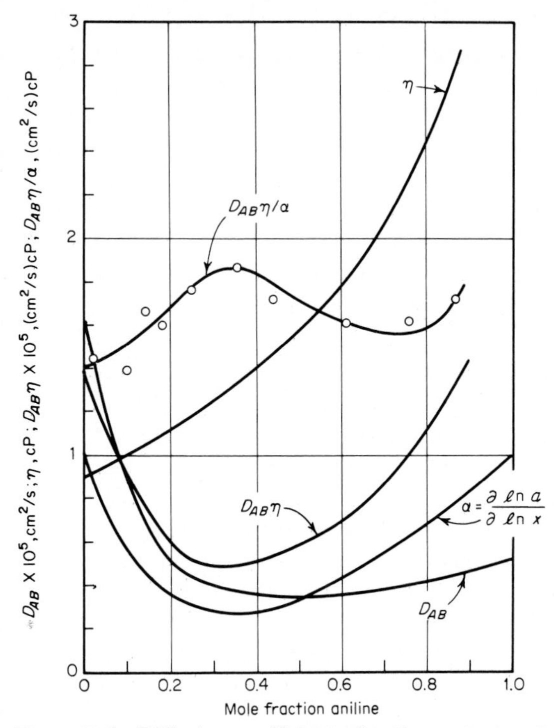


**Figure 11-9** Diffusion coefficients for the system aniline–carbon tetrachloride at 298 K. *(From Ref. 170.)*

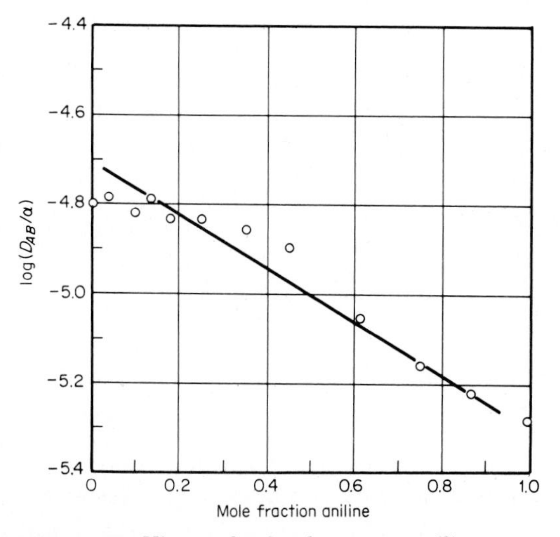


**Figure 11-10** Vignes plot for the system aniline–carbon tetrachloride at 298 K.

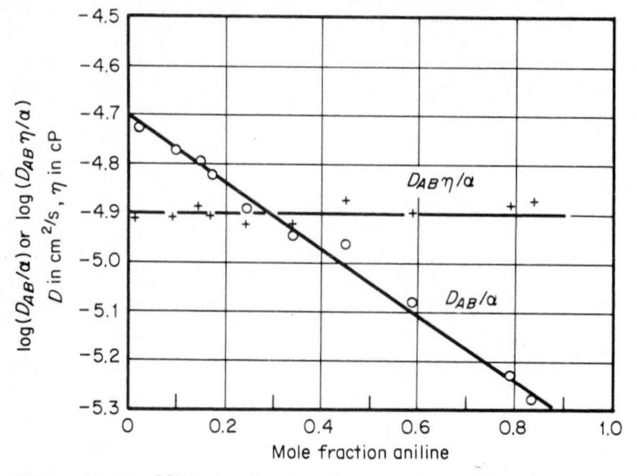


**Figure 11-11** Vignes plot for the system aniline-benzene at 298 K. *(Data from Ref. 170.)*

system. The original Vignes equation fits the data well, but so does Eq. (11-10.5); in fact, for the latter $D_{AB}\eta/\alpha$ is essentially constant.

Tyn and Calus [213] measured the binary diffusion coefficient for several associating systems (ethanol-water, acetone-water, and acetone-chloroform) and found that Eq. (11-10.4) was, generally, preferable to Eq. (11-10.5), although the mean deviation for the Vignes relation was about 14 percent for the three systems studied.

Other correlation methods have been proposed [9, 40, 43, 72, 84, 87, 171, 205], but they are either less accurate or less general than those discussed above.

Baldauf and Knapp [15] present an exceptionally complete data set for eleven binary liquid mixtures giving $D_{AB}$, $\eta_m$, $\rho_m$, and the refractive index as a function of composition.

In summary, no single correlation is always satisfactory for estimating the concentration effect on liquid diffusion coefficients. The Vignes method [Eq. (11-10.4)] is recommended here as a well-tested and generally accurate correlation. It is also the easiest to apply, and no mixture viscosities are necessary. The thermodynamic correction factor $\alpha$ must, however, be known. None of the correlations described work particularly well for liquid mixtures in which the components associate [185].

## 11-11 The Effect of Temperature and Pressure on Diffusion in Liquids

For the Wilke-Chang and Tyn-Calus correlations for $D^\circ_{AB}$ in Sec. 11-9, the effect of temperature was accounted for by assuming

$$\frac{D^\circ_{AB}\eta_B}{T} = \text{constant} \tag{11-11.1}$$

In the Hayduk-Minhas method, the (absolute) temperature was raised to a power > 1, and the viscosity parameter was a function of solute volume. While these approximations may be valid over small temperature ranges, it is usually preferable [180] to assume that

$$D_{AB} \text{ (or } D^{\circ}_{AB}) = A \exp \frac{-B}{T} \tag{11-11.2}$$

Equation (11-11.2) has been employed by a number of investigators [106, 137, 178, 209]. We illustrate its applicability in Fig. 11-12 with the system ethanol-water from about 298 K to 453 K for both infinitely dilute diffusion coefficients and $D_{AB}$ for a 20 mole percent solution [117]. Note that we have not included the thermodynamic correction factor $\alpha$ because it is assumed to be embodied in the $A$ and $B$ parameters. Actually, since the viscosity of liquids is an exponentially decreasing function of temperature, below reduced temperatures of about 0.7, the product $D_{AB}\eta$ might be expected to be temperature-insensitive if the energies of activation for diffusion and viscosity were opposite in sign and of the same magnitude numerically.

Tyn [210, 211] reviewed the various proposed techniques to correlate infinitely dilute binary (and also self-) diffusion coefficients with temperature. He suggested that

$$\frac{D^{\circ}_{AB}(T_2)}{D^{\circ}_{AB}(T_1)} = \left(\frac{T_c - T_1}{T_c - T_2}\right)^n \tag{11-11.3}$$

where $T_c$ is the critical temperature of the *solvent* B. $T_c$, $T_1$, and $T_2$ are in kelvins. The parameter $n$ was related to the heat of vaporization of the solvent at $T_b$ (solvent) as follows:

| $n$ | $\Delta H_v(T_b)$, J/mol |
|---|---|
| 3 | 7,900 to 30,000 |
| 4 | 30,000 to 39,700 |
| 6 | 39,700 to 46,000 |
| 8 | 46,000 to 50,000 |
| 10 | >50,000 |

Typical compounds falling into these categories would be $n = 3$, $n$-pentane, acetone, cyclohexane, chloroform; $n = 4$, benzene, toluene, chlorobenzene, $n$-octane, carbon tetrachloride; $n = 6$, cyclohexane, propanol, butanol, water; $n = 8$, heptanol; and $n = 10$, ethylene and propylene glycols.

Equation (11-11.3), which does not require mixture viscosity data, was tested with a large number of binary systems, and an error of about 9

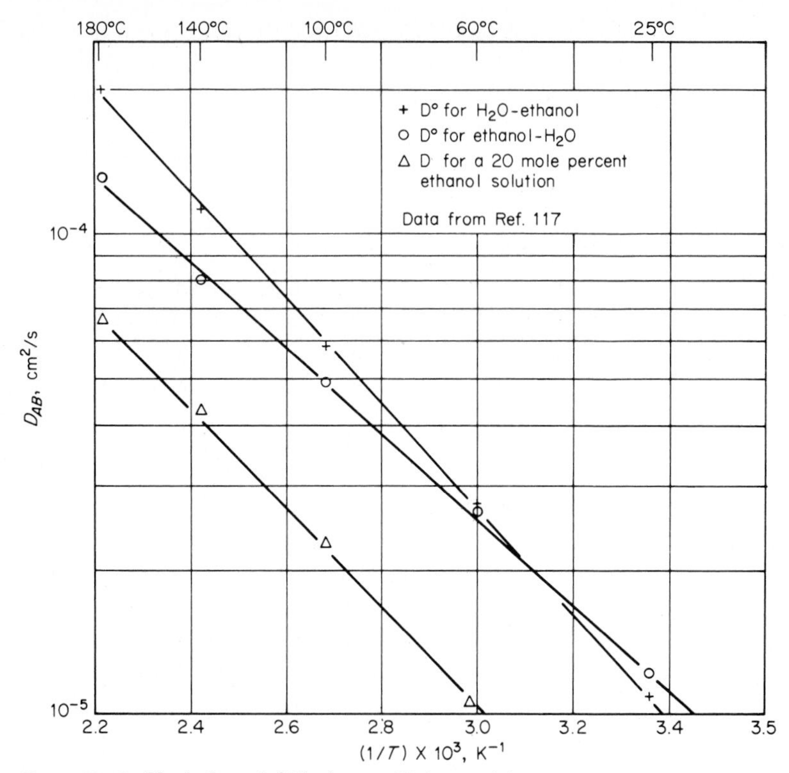


**Figure 11-12** Variation of diffusion coefficients with temperature.

percent was found. When Eq. (11-11.1) also was examined, Tyn reported an error of about 10 percent. The temperature ranges for Eq. (11-11.3) are about 10 K above the freezing point to about 10 K below the normal boiling point. Larger errors were noted if these ranges were exceeded.

The effect of pressure on liquid diffusion coefficients has received little attention. Easteal [59] attempted to correlate tracer or self-diffusion coefficients with pressure and suggested

$$\ln D_j^* = a + bP^{0.75} \tag{11-11.4}$$

where $D_j^*$ is a tracer or self-diffusion coefficient and $a$ and $b$ are constants for a given solute, but they do vary significantly with temperature. $b$ is a negative number, and thus $D_j^*$ decreases with an increase in pressure. As an example, the self-diffusion coefficient for $n$-hexane decreases from about $4.2 \times 10^{-5}$ cm$^2$/s at 1 bar to about $0.7 \times 10^{-5}$ cm$^2$/s at 3500 bar at a temperature of 298 K.

From Eq. (11-11.1), at a given temperature, it could be inferred that

$$D^{\circ}_{AB}\eta_{B} = \text{constant} \tag{11-11.5}$$

If solvent–liquid viscosity data were available at high pressures, it would then be possible to employ Eq. (11-11.5) to estimate $D^{\circ}_{AB}$ at the elevated pressure from low-pressure diffusion coefficient data. Dymond and Woolf [58] show, however, that this proportionality is only approximate for tracer-diffusion coefficients, but they indicate it may be satisfactory for binaries with large solute molecules.

## 11-12 Diffusion in Multicomponent Liquid Mixtures

In a binary liquid mixture, as indicated in Secs. 11-2 and 11-8, a single diffusion coefficient was sufficient to express the proportionality between the flux and concentration gradient. In multicomponent systems, the situation is considerably more complex, and the flux of a given component depends upon the gradient of $n - 1$ components in the mixture. For example, in a ternary system of A, B, and C, the flux of A can be expressed as

$$J_{A} = \mathbf{D}_{AA}\frac{dc_{A}}{dz} + \mathbf{D}_{AB}\frac{dc_{B}}{dz} \tag{11-12.1}$$

Similar relations can be written for $J_B$ and $J_C$. The coefficients $\mathbf{D}_{AA}$ and $\mathbf{D}_{BB}$ are called *main coefficients;* they are *not* self-diffusion coefficients. $\mathbf{D}_{AB}$, $\mathbf{D}_{BA}$, etc., are *cross-coefficients,* because they relate the flux of a component $i$ to a gradient in $j$. $\mathbf{D}_{ij}$ is normally not equal to $\mathbf{D}_{ji}$ for multicomponent systems.

Frames of reference in multicomponent systems must be clearly defined. Yon and Toor [235] discuss some limitations in this context.

One important case of multicomponent diffusion results when a solute diffuses through a homogeneous solution of mixed solvents. When the solute is dilute, there are no concentration gradients for the solvent species and one can speak of a single solute diffusivity with respect to the mixture $D^{\circ}_{Am}$. This problem has been discussed by several authors [46, 102, 167, 203], and empirical relations for $D^{\circ}_{Am}$ have been proposed. Perkins and Geankoplis [167] evaluated several methods and suggested

$$D^{\circ}_{Am}\eta_{m}^{0.8} = \sum_{\substack{j=1 \\ j\neq A}}^{n} x_{j}D^{\circ}_{Aj}\eta_{j}^{0.8} \tag{11-12.2}$$

where $D^\circ_{Am}$ = *effective* diffusion coefficient for a dilute solute A into the mixture, cm²/s

$D^\circ_{Aj}$ = infinite dilution binary diffusion coefficient of solute A into the solvent $j$, cm²/s

$x_j$ = mole fraction of $j$

$\eta_m$ = mixture viscosity, cP

$\eta_j$ = pure component viscosity, cP

When tested with data for eight ternary systems, errors were normally less than 20 percent, except for cases involving $CO_2$. These same authors also suggested that the Wilke-Chang equation (11-9.1) might be modified to include the mixed solvent case, i.e.,

$$D^\circ_{Am} = 7.4 \times 10^{-8} \frac{(\phi M)^{1/2} T}{\eta_m V_A^{0.6}} \tag{11-12.3}$$

$$\phi M = \sum_{\substack{j=1 \\ j \neq A}}^{n} x_j \phi_j M_j \tag{11-12.4}$$

Although not extensively tested, Eq. (11-12.3) provides a rapid, reasonably accurate estimation method.

For $CO_2$ as a solute diffusing into mixed solvents, Takahashi et al. [202] recommend

$$D^\circ(CO_2\text{-}m)\left(\frac{\eta_m}{V_m}\right)^{1/3} = \sum_{\substack{j=1 \\ j \neq CO_2}}^{n} x_j D^\circ(CO_2\text{-}j)\left(\frac{\eta_j}{V_j}\right)^{1/3} \tag{11-12.5}$$

where $V_m$ is the molar volume, cm³/mol, for the mixture at $T$ and $V_j$ applies to the pure component. Tests with a number of ternary systems involving $CO_2$ led to deviations from experimental values usually less than 4 percent.

**Example 11-8** Estimate the diffusion coefficient of acetic acid diffusing into a mixed solvent containing 20.7 mole percent ethyl alcohol in water. The acetic acid concentration is small. Assume the temperature is 298 K.

**data** Let E = ethyl alcohol, W = water, and A = acetic acid. At 298 K, $\eta_E$ = 1.10 cP, $\eta_W$ = 0.894 cP, $D_{AE} = 1.03 \times 10^{-5}$ cm²/s, $D_{AW} = 1.30 \times 10^{-5}$ cm²/s, and for the solvent mixture under consideration, $\eta_m$ = 2.35 cP.

**solution** From Eq. (11-12.2)

$$\begin{aligned} D^\circ_{Am} &= (2.35)^{-0.8}[(0.207)(1.03 \times 10^{-5})(1.10)^{0.8} \\ &\quad + (0.793)(1.30 \times 10^{-5})(0.894)^{0.8}] \\ &= 0.59 \times 10^{-5} \text{ cm}^2/\text{s} \end{aligned}$$

The experimental value reported by Perkins and Geankoplis is $0.571 \times 10^{-5}$ $cm^2/s$. Note that this value is significantly below the two limiting binary values. The decrease in the mixture diffusivity appears to be closely related to the increase in solvent mixture viscosity relative to the pure components. Had the modified Wilke-Chang equation been used, with $V_A = 64.1$ $cm^3/mol$ (Table 3-8), $\phi_E = 1.5$, and $\phi_W = 2.6$, and with $M_E = 46$ and $M_W = 18$, using Eqs. (11-12.3) and (11-12.4),

$$\phi M = (0.207)(1.5)(46) + (0.793)(2.6)(18) = 51.39$$

$$D^{\circ}_{Am} = \frac{(7.4 \times 10^{-8})(51.39)^{1/2}(298)}{(2.35)(64.1)^{0.6}}$$

$$= 0.55 \times 10^{-5} \text{ cm}^2/\text{s}$$

**Example 11-9** Estimate the diffusion coefficient of $CO_2$ (D) into a mixed solvent of *n*-octanol (L) and carbon tetrachloride (C) containing 60 mole percent *n*-octanol. The temperature is 298 K.

**data** From Ref. 202, $D^{\circ}_{DL} = 1.53 \times 10^{-5}$ $cm^2/s$, $D^{\circ}_{DC} = 3.17 \times 10^{-5}$ $cm^2/s$, $\eta_m = 3.55$ cP, $\eta_L = 7.35$ cP, and $\eta_C = 0.88$ cP. With densities and molecular weights given in Appendix A, $V_L = 158$ $cm^3/mol$ and $V_C = 97.1$ $cm^3/mol$ at 298 K. The mixture volume is not known. If we assume that the mole fraction of $CO_2$ in the liquid mixture is small and that *n*-octanol and carbon tetrachloride form ideal solutions,

$$V_m \approx (0.6)(158) + (0.4)(97.1) = 133.6 \text{ cm}^3/\text{mol}$$

**solution** With Eq. (11-12.5)

$$D^{\circ}(CO_2\text{-m}) = (3.55/133.6)^{-1/3}\,[(0.6)(1.53 \times 10^{-5})(7.35/158)^{1/3} + (0.4)(3.17 \times 10^{-5})\,(0.88/97.1)^{1/3}]$$

$$= 1.99 \times 10^{-5} \text{ cm}^2/\text{s}$$

The experimental value [202] is $1.96 \times 10^{-5}$ $cm^2/s$.

When dealing with the general case of multicomponent diffusion coefficients, there are no convenient and simple estimation methods. Cullinan [42, 45] has discussed the possibility of adapting the Vignes correlation [Eq. (11-10.4)] to ternary systems, but the requirement for extensive thermodynamic activity data for the mixture has severely limited the applicability of the method. Kett and Anderson [114, 115] apply hydrodynamic theory to estimate ternary diffusion coefficients. Although some success was achieved, the method again requires extensive data on activities, pure component and mixture volumes, and viscosities, as well as tracer and binary diffusion coefficients. Bandrowski and Kubaczka [16] suggest using the mixture critical volume as a correlating parameter to estimate $D_{Am}$ for multicomponent mixtures.

Other authors [25, 32, 44, 125, 152, 153] also have discussed the problem of obtaining multicomponent liquid diffusion coefficients.

## 11-13 Diffusion in Electrolyte Solutions

When a salt dissociates in solution, ions rather than molecules diffuse. In the absence of an electric potential, however, the diffusion of a single salt may be treated as molecular diffusion.

The theory of diffusion of salts at low concentrations is well developed. At concentrations encountered in most industrial processes, one normally resorts to empirical corrections, with a concomitant loss in generality and accuracy. A comprehensive discussion of this subject is available [159].

For dilute solutions of a *single* salt, the diffusion coefficient is given by the Nernst-Haskell equation

$$D^{\circ}_{AB} = \frac{RT[(1/n_+) + (1/n_-)]}{\mathrm{F}^2[(1/\lambda^{\circ}_+) + (1/\lambda^{\circ}_-)]} \tag{11-13.1}$$

where $D^{\circ}_{AB}$ = diffusion coefficient at infinite dilution, based on molecular concentration, $cm^2/s$

$T$ = temperature, K

$R$ = gas constant, 8.314 J/(mol·K)

$\lambda^{\circ}_+, \lambda^{\circ}_-$ = limiting (zero concentration) ionic conductances, $(A/cm^2)(V/cm)(g\text{-equiv}/cm^3)$

$n_+, n_-$ = valences of cation and anion, respectively

F = faraday = 96,500 C/g-equiv

**TABLE 11-6 Limiting Ionic Conductances in Water at 298 K [90]**
$(A/cm^2)(V/cm)(g\text{-equiv}/cm^3)$

| Anion | $\gamma^{\circ}_-$ | Cation | $\gamma^{\circ}_+$ |
|---|---|---|---|
| $OH^-$ | 197.6 | $H^+$ | 349.8 |
| $Cl^-$ | 76.3 | $Li^+$ | 38.7 |
| $Br^-$ | 78.3 | $Na^+$ | 50.1 |
| $I^-$ | 76.8 | $K^+$ | 73.5 |
| $NO_3^-$ | 71.4 | $NH_4^+$ | 73.4 |
| $CLO_4^-$ | 68.0 | $Ag^+$ | 61.9 |
| $HCO_3^-$ | 44.5 | $Tl^+$ | 74.7 |
| $HCO_2^-$ | 54.6 | $(1/2)Mg^{2+}$ | 53.1 |
| $CH_3CO_2^-$ | 40.9 | $(1/2)Ca^{2+}$ | 59.5 |
| $ClCH_2CO_2^-$ | 39.8 | $(1/2)Sr^{2+}$ | 50.5 |
| $CNCH_2CO_2^-$ | 41.8 | $(1/2)Ba^{2+}$ | 63.6 |
| $CH_3CH_2CO_2^-$ | 35.8 | $(1/2)Cu^{2+}$ | 54 |
| $CH_3(CH_2)_2CO_2^-$ | 32.6 | $(1/2)Zn^{2+}$ | 53 |
| $C_6H_5CO_2^-$ | 32.3 | $(1/3)La^{3+}$ | 69.5 |
| $HC_2O_4^-$ | 40.2 | $(1/3)Co(NH_3)_6^{3+}$ | 102 |
| $(1/2)C_2O_4^{2-}$ | 74.2 | | |
| $(1/2)SO_4^{2-}$ | 80 | | |
| $(1/3)Fe(CN)_6^{3-}$ | 101 | | |
| $(1/4)Fe(CN)_6^{4-}$ | 111 | | |

Values of $\lambda_+^\circ$ and $\lambda_-^\circ$ can be obtained for many ionic species at 298 K from Table 11-6 or from alternative sources [149, 177]. If values of $\lambda_+^\circ$ and $\lambda_-^\circ$ at other temperatures are needed, an approximate correction factor is $T/334\eta_W$, where $\eta_W$ is the viscosity of water at $T$ in centipoises.

As the salt concentration becomes finite and increases, the diffusion coefficient decreases rapidly and then usually rises, often becoming greater than $D_{AB}^\circ$ at high normalities. Figure 11-13 illustrates the typical trend for three simple salts. The initial decrease at low concentrations is proportional to the square root of the concentration, but deviations from this trend are usually significant above 0.1 $N$.

No reliable method has yet been proposed to relate $D_{AB}^\circ$ to concentration. Gordon [79], however, has proposed an empirical equation which has been applied to systems at concentrations up to 2 $N$:

$$D_{AB} = D_{AB}^\circ \frac{\eta_S}{\eta} (\rho_S \overline{V}_S)^{-1} \left(1 + m \frac{\partial \ln \gamma_\pm}{\partial m}\right) \tag{11-13.2}$$

where $D_{AB}^\circ$ = diffusion coefficient at infinite dilution, [Eq. (11-13.1)], cm$^2$/s

$\rho_S$ = molar density of the solvent, mol/cm$^3$

$\overline{V}_S$ = partial molar volume of the solvent, cm$^3$/mol

$\eta_S$ = viscosity of the solvent, cP

$\eta$ = viscosity of the solution, cP

$m$ = molality of the solute, mol/kg solvent

$\gamma_\pm$ = mean ionic activity coefficient of the solute

In many cases, the product $\rho_S \overline{V}_S$ is close to unity, as is the viscosity ratio $\eta_S/\eta$, so that Gordon's relation provides an activity correction to the diffusion coefficient at infinite dilution. Though Harned and Owen [90] tabulate $\gamma_\pm$ as a function of $m$ for many aqueous solutions, there now exist

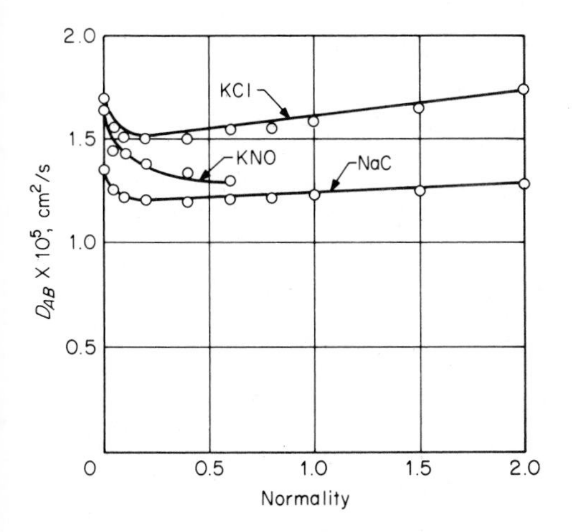


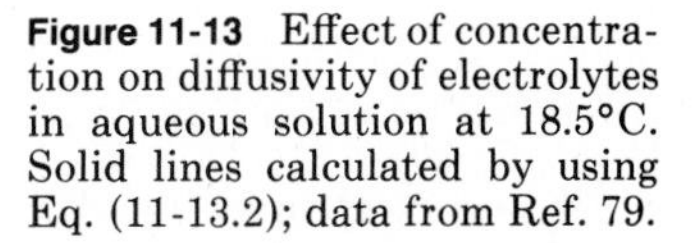

**Figure 11-13** Effect of concentration on diffusivity of electrolytes in aqueous solution at 18.5°C. Solid lines calculated by using Eq. (11-13.2); data from Ref. 79.

several semiempirical correlation techniques to relate $\gamma_{\pm}$ to concentration. Bromley [23] presents an analytical relation, and Meissner et al. [141 to 145] show generalized graphical correlations. The solid lines in Fig. 11-13 were calculated from Eq. (11-13.2).

Data on the diffusion of $CO_2$ into electrolyte solutions have been reported by Ratcliff and Holdcroft [171]. The diffusion coefficient was found to decrease linearly with an increase in salt concentration.

In summary, for very dilute solutions of electrolytes, employ Eq. (11-13.1). When values of the limiting ionic conductances in water are not available at the desired temperature, use those in Table 11-6 for 298 K and multiply $D^{\circ}_{AB}$ at 298 K by $T/334\eta_W$, where $\eta_W$ is the viscosity of water at $T$ in centipoises.

For concentrated solutions, use Eq. (11-13.2). If values of $\gamma_{\pm}$ and $\lambda^{\circ}$ are not available at $T$, calculate $D_{AB}$ at 298 K and multiply it by $(T/298)[(\eta$ at $298)/(\eta$ at $T)]$. If necessary, the ratio of the solution viscosity at 298 K to that at $T$ may be assumed to be the same as the corresponding ratio for water.

**Example 11-10** Estimate the diffusion coefficient of NaOH in a 2 $N$ aqueous solution at 288 K.

**solution** From data on densities of aqueous solutions of NaOH, it is evident that, up to 12 weight percent NaOH (about 3 $N$), the density increases almost exactly in inverse proportion to the weight fraction of water; i.e., the ratio of moles of water per liter is essentially constant at 55.5. Thus both $V/n$ and $\overline{V}_1$ are very nearly 55.5 and cancel in Eq. (11-13.2). In this case, the molality $m$ is essentially identical with the normality.

In plotting the values of $\gamma_{\pm}$ for NaOH at 298 K [90] vs. molality $m$, the slope at $2m$ is approximately 0.047. Hence

$$m \frac{\partial \ln \gamma_{\pm}}{\partial m} = \frac{m}{\gamma_{\pm}} \frac{\partial \gamma_{\pm}}{\partial m} = \frac{2}{0.698}(0.047) = 0.135$$

The value 0.698 is the mean activity coefficient at $m = 2$.

The viscosities of water and 2 $N$ NaOH solution at 298 K are 0.894 and 1.42 cP, respectively. Substituting in Eqs. (11-13.1) and (11-13.2) gives

$$D^{\circ}_{AB} = \frac{(2)(8.314)(298)}{[(1/50) + (1/198)](96{,}500)^2}$$

$$= 2.12 \times 10^{-5} \text{ cm}^2/\text{s}$$

$$D_{AB} = (2.12 \times 10^{-5}) \frac{0.894}{1.42} \frac{55.5}{55.5} [1 + (2)(0.135)]$$

$$= 1.70 \times 10^{-5} \text{ cm}^2/\text{s at 298 K}$$

At 288 K, the viscosity of water is 1.144 cP, and so the estimated value of $D_{AB}$ at 288 K is

$$1.70 \times 10^{-5} \frac{(288)}{(334)(1.144)} = 1.28 \times 10^{-5} \text{ cm}^2/\text{s}$$

which may be compared with the ICT [107] value of $1.36 \times 10^{-5}$ cm$^2$/s.

In a system of *mixed electrolytes,* such as in the simultaneous diffusion of HCl and NaCl in water, the faster-moving $H^+$ ion may move ahead of its $Cl^-$ partner, the electric current being maintained at zero by the lagging behind of the slower-moving $Na^+$ ions. In such systems, the unidirectional diffusion of each ion species results from a combination of electric and concentration gradients:

$$N_+ = \frac{\gamma_+}{\mathrm{F}^2}\left(-RT\frac{\partial c_+}{\partial z} - Fc_+\frac{\partial E}{\partial z}\right) \tag{11-13.3}$$

$$N_- = \frac{\gamma_-}{\mathrm{F}^2}\left(-RT\frac{\partial c_-}{\partial z} - \mathrm{F}c_-\frac{\partial E}{\partial z}\right) \tag{11-13.4}$$

where $N_+, N_-$ = diffusion flux densities of the cation and anion, respectively, g-equiv/cm$^2\cdot$s
$c_+, c_-$ = corresponding ion concentrations, g-equiv/cm$^3$
$\partial E/\partial z$ = gradient in electric potential

The electric field gradient may be imposed externally but is present in the ionic solution even if, owing to the small separation of charges which result from diffusion itself, there is no external electrostatic field. Collision effects, ion complexes, and activity corrections are neglected.

One equation for each cation and one for each anion can be combined with the requirement of zero current at any $z$ to give $\Sigma N_+ = \Sigma N_-$. Solving for the unidirectional flux densities [172],

$$n_+N_+ = \frac{-RT\gamma_+}{\mathrm{F}^2 n_+}(G_+ - n_+c_+Y) \tag{11-13.5}$$

$$n_-N_- = \frac{-RT\gamma_-}{\mathrm{F}^2 n_-}(G_- + n_-c_-Y) \tag{11-13.6}$$

$$Y = \frac{(\Sigma\gamma_+G_+/n_+) - (\Sigma\gamma_-G_-/n_-)}{\Sigma\gamma_+c_+ + \Sigma\gamma_-c_-} \tag{11-13.7}$$

where $G_+$ and $G_-$ are the concentration gradients $\partial c/\partial z$ in the direction of diffusion.

Vinograd and McBain [221] have used these relations to represent their data on diffusion in multi-ion solutions. $D_{AB}$ for the hydrogen ion was found to decrease from 12.2 to 4.9 $\times$ $10^{-5}$ cm$^2$/s in a solution of HCl and $BaCl_2$ when the ratio of $H^+$ to $Ba^{2+}$ was increased from zero to 1.3; $D_{AB}$ at the same temperature is 9.03 $\times$ $10^{-5}$ for the free $H^+$ ion, and 3.3 $\times$ $10^{-5}$ for HCl in water. The presence of the slow-moving $Ba^{2+}$ accelerates the $H^+$ ion, the potential existing with zero current causing it to move in dilute solution even faster than it would as a free ion with no other cation present. That is, electrical neutrality is maintained by the hydrogen ions moving ahead of the chlorine, faster than they would as free ions, while the barium diffuses more slowly than as a free ion.

The interaction of ions in a multi-ion system is important when the several ion conductances differ greatly, as they do when $H^+$ or $OH^-$ is diffusing. When the diffusion of one of these two ions is not involved, no great error is introduced by the use of "molecular" diffusion coefficients for the salt species present.

The relations proposed by Vinograd and McBain are not adequate to represent a ternary system, in which four independent diffusion coefficients must be known to predict fluxes. The subject of ion diffusion in multicomponent systems is covered in detail in the papers by Wendt [228] and Miller [146, 147] in which it is demonstrated how one can obtain multicomponent ion diffusion coefficients, although the data required are usually not available.

## Notation

| | |
|---|---|
| $a_j$ | activity of component $j$ |
| $A_B$ | parameter in Table 11-4 |
| $c$ | concentration, mol/cm$^3$; $c_j$, for component $j$; $c_+$, $c_-$, ion concentrations |
| $D_{AB}$ | binary diffusion coefficient of A diffusing into B, cm$^2$/s; $D^\circ_{AB}$, at infinite dilution of A in B; $\mathbf{D}_{AB}$, cross-coefficient in multicomponent mixtures; $D_{im}$, of $i$ into a homogeneous mixture; $D^+_{AB}$, at a low pressure |
| $D^*_A$ | tracer-diffusion coefficient of A, cm$^2$/s |
| $D_{AA}$ | self-diffusion coefficient of A, cm$^2$/s; $\mathbf{D}_{AA}$, main coefficient for A in multicomponent diffusion |
| $E$ | electric potential |
| $f_D$ | correction term in Eq. (11-3.1) |
| F | faraday = 96,500 C/g-equiv |
| $G_+$, $G_-$ | $\partial c_+/\partial z$ and $\partial c_-/\partial z$ |
| $\Delta H_v$ | heat of vaporization at the normal boiling point, J/mol |
| $I_A$ | parameter in Table 11-4 |
| $J_A$ | flux of A, mol/(cm$^2 \cdot$s); $J^M_A$, flux relative to a plane of no net mole flow; $J^V_A$, flux relative to a plane of no net volume flow |
| $k$ | Boltzmann's constant, $1.3805 \times 10^{-23}$ J/K |
| $m$ | molality of solute, mol/kg solvent; parameter in Eq. (11-10.3) |

| | |
|---|---|
| $M_A$ | molecular weight of A, g/mol; $M_{AB}$, $2[(1/M_A) + (1/M_B)]^{-1}$ |
| $n$ | number density of molecules; parameter in Eq. (11-11.3) |
| $n_+$, $n_-$ | valences of cation and anion, respectively |
| $N_A$ | flux of A relative to a fixed coordinate plane, $mol/(cm^2 \cdot s)$ |
| $N_+$, $N_-$ | diffusion flux of cation and anion, respectively |
| $P$ | pressure, bar; $P_c$, critical pressure; $P_r$, $P/P_c$ |
| $\mathbf{P}_j$ | parachor of component $j$ |
| $q$, $Q$ | parameters in Eq. (11-9.14) |
| $r$ | distance of separation between molecules, Å |
| $r_A$ | molecular radius in the Stokes-Einstein equation |
| $R$ | gas constant, 8.314 J/(mol·K) |
| $S_A$, $S_B$ | parameters in Table 11-4 |
| $T$ | temperature, kelvins; $T_b$, at the normal boiling point (at 1 atm); $T_c$, critical temperature; $T_r$, $T/T_c$ |
| $V$ | volume, $cm^3/mol$; $V_b$, at $T_b$; $\overline{V}_A$, partial molar volume of A |
| $V_j$ | molar volume of component $j$ at either $T_b$ or $T$, $cm^3/mol$ |
| $x_j$ | mole fraction of $j$ |
| $y_j$ | mole fraction of $j$ |
| $z$ | direction coordinate for diffusion |

Greek

| | |
|---|---|
| $\alpha$ | $\partial \ln a / \partial \ln x$, Eq. (11-2.8); $\alpha_c$, parameter in Eq. (11-9.7) |
| $\beta$ | parameter in Eq. (11-9.13) |
| $\gamma$ | activity coefficient; $\gamma_{\pm}$, mean ionic activity coefficient |
| $\gamma^\circ_+$, $\gamma^\circ_-$ | limiting (zero concentration) ionic conductances, $(A/cm^2)(V/cm)$ $(g\text{-}equiv/cm^3)$ |
| $\delta$ | polar parameter defined in Eq. (11-3.8) |
| $\epsilon$ | characteristic energy parameter; $\epsilon_A$, for pure A; $\epsilon_{AB}$, for an A-B interaction |
| $\epsilon$, $\epsilon^*$ | parameters in Eqs. (11-9.8) and (11-9.9) |
| $\eta$ | viscosity, cP; $\eta_A$, for pure A; $\eta_m$, for a mixture |
| $\mu_A$ | chemical potential of A, J/mol |
| $\mu_p$ | dipole moment, debyes |
| $\rho$ | density, $g/cm^3$ |
| $\sigma$ | characteristic length parameter, Å; $\sigma_A$, for pure A; $\sigma_{AB}$, for an A-B interaction; surface tension |

| | |
|---|---|
| $\Sigma_v$ | Fuller et al. volume parameter, Table 11-1 |
| $\phi$ | association parameter for the solvent, Eq. (11-9.1) |
| $\psi$ | intermolecular potential energy of interaction |
| $\Omega_D$ | collision integral for diffusion |

SUPERSCRIPTS

| | |
|---|---|
| $^\circ$ | infinite dilution |
| $*$ | tracer value |
| $^+$ | low pressure |

SUBSCRIPTS

| | |
|---|---|
| A, B | components A and B; usually B is the solvent |
| $m$ | mixture |
| $w$ | water |
| $s$ | solvent |

## REFERENCES


1. Akgerman, A.: *Ing. Eng. Chem. Fundam.*, **15:** 78 (1976).
2. Akgerman, A., and J. L. Gainer: *J. Chem. Eng. Data,* **17:** 372 (1972).
3. Albright, J. G., A. Vernon, J. Edge, and R. Mills: *J. Chem. Soc., Faraday Trans.*, 1, **79:** 1327 (1983).
4. Alvarez, R., I. Medlina, J. L. Bueno, and J. Coca: *J. Chem. Eng. Data,* **28:** 155 (1983).
5. Amdur, I., and L. M. Shuler: *J. Chem. Phys.,* **38:** 188 (1963).
6. Amdur, I., J. Ross, and E. A. Mason: *J. Chem. Phys.,* **20:** 1620 (1952).
7. Amourdam, M. J., and G. S. Laddha: *J. Chem. Eng. Data,* **12:** 389 (1967).
8. Anderson, D. K., and A. L. Babb: *J. Phys. Chem.,* **65:** 1281 (1961).
9. Anderson, D. K., J. R. Hall, and A. L. Babb: *J. Phys. Chem.,* **62:** 404 (1958).
10. Anderson, J. L.: *Ind. Eng. Chem. Fundam.,* **12:** 490 (1973).
11. Arnold, J. H.: *Ind. Eng. Chem.,* **22:** 1091 (1930).
12. Arnold, J. H.: *J. Am. Chem. Soc.,* **52:** 3937 (1930).
13. Arora, P. S., and P. J. Dunlop: *J. Chem. Phys.,* **71:** 2430 (1979).
14. Bailey, R. G.: *Chem. Eng.,* **82**(6), 86 (1975).
15. Baldauf, W., and H. Knapp: *Ber. Bunsenges. Phys. Chem.,* **87:** 304 (1983).
16. Bandrowski, J., and A. Kubaczka: *Chem. Eng. Sci.,* **37:** 1309 (1982).
17. Bearman, R. J.: *J. Chem. Phys.,* **32:** 1308 (1960).
18. Bearman, R. J.: *J. Phys. Chem.,* **65:** 1961 (1961).
19. Bidlack, D. L., and D. K. Anderson: *J. Phys. Chem.,* **68:** 3790 (1964).
20. Bird, R. B., W. E. Stewart, and E. N. Lightfoot: *Transport Phenomena,* Wiley, New York, 1960, chap. 16.
21. Blanc, A.: *J. Phys.,* **7:** 825 (1908).
22. Brokaw, R. S.: *Ind. Eng. Chem. Process Design Develop.,* **8:** 240 (1969).
23. Bromley, L. A.: *AIChE J.,* **19:** 313 (1973).
24. Brunet, J., and M. H. Doan: *Can. J. Chem. Eng.,* **48:** 441 (1970).
25. Burchard, J. K., and H. L. Toor: *J. Phys. Chem.,* **66:** 2015 (1962).
26. Byers, C. H., and C. J. King: *J. Phys. Chem.,* **70:** 2499 (1966).

27. Byrne, J. J., D. Maguire, and J. K. A. Clarke: *J. Phys. Chem.*, **71:** 3051 (1967).
28. Caldwell, L.: *J. Chem. Eng. Data,* **29:** 60 (1984).
29. Caldwell, C. S., and A. L. Babb: *J. Phys. Chem.*, **60:** 14, 56 (1956).
30. Calus, W. F., and M. T. Tyn: *J. Chem. Eng. Data,* **18:** 377 (1973).
31. Carman, P. C.: *J. Phys. Chem.*, **71:** 2565 (1967).
32. Carman, P. C.: *Ind. Eng. Chem. Fundam.*, **12:** 484 (1973).
33. Carman, P. C., and L. Miller: *Trans. Faraday Soc.*, **55:** 1838 (1959).
34. Carman, P. C., and L. H. Stein: *Trans. Faraday Soc.*, **52:** 619 (1956).
35. Carmichael, L. T., B. H. Sage, and W. N. Lacey: *AIChE J.*, **1:** 385 (1955).
36. Carswell, A. J., and J. C. Stryland: *Can. J. Phys.*, **41:** 708 (1963).
37. Chang, P., and C. R. Wilke: *J. Phys. Chem.*, **59:** 592 (1955).
38. Chen, N. H., and D. P. Othmer: *J. Chem. Eng. Data,* **7:** 37 (1962).
39. Chen, S.-H.: *AIChE J.*, **30:** 481 (1984).
40. Cram, R. R., and A. W. Adamson: *J. Phys. Chem.*, **64:** 199 (1960).
41. Crider, W. L.: *J. Am. Chem. Soc.* **78:** 924 (1956).
42. Cullinan, H. T., Jr.: *Can. J. Chem. Eng.*, **45:** 377 (1967).
43. Cullinan, H. T., Jr.: *Can. J. Chem. Eng.*, **49:** 130 (1971).
44. Cullinan, H. T., Jr.: *Can. J. Chem. Eng.*, **49:** 632 (1971).
45. Cullinan, H. T., Jr., and M. R. Cusick: *Ind. Eng. Chem. Fundam.*, **6:** 72 (1967).
46. Cullinan, H. T., Jr., and M. R. Cusick: *AIChE J.*, **13:** 1171 (1967).
47. Cussler, E. L.: *Diffusion: Mass Transfer in Fluid Systems,* Cambridge, Cambridge, 1984, chaps. 3, 7.
48. Darken, L. S.: *Trans. Am. Inst. Mining Metall. Eng.*, **175:** 184 (1948).
49. Davies, G. A., A. B. Ponter, and K. Craine: *Can. J. Chem. Eng.*, **45:** 372 (1967).
50. Debenedetti, P.: Ph.D. thesis, Massachusetts Institute of Technology, Cambridge, Mass., December 1984.
51. Di Pippo, R., J. Kestin, and K. Oguchi: *J. Chem. Phys.*, **46:** 4986 (1967).
52. Dullien, F. A. L.: *Nature,* **190:** 526 (1961).
53. Dullien, F. A. L.: *Trans. Faraday Soc.*, **59:** 856 (1963).
54. Dullien, F. A. L.: *Ind. Eng. Chem. Fundam.*, **10:** 41 (1971).
55. Dullien, F. A. L.: private communication, January 1974.
56. Duncan, J. B., and H. L. Toor: *AIChE J.*, **8:** 38 (1962).
57. Dymond, J. H.: *J. Phys. Chem.*, **85:** 3291 (1981).
58. Dymond, J. H., and L. A. Woolf: *J. Chem. Soc., Faraday Trans.*, 1, **78:** 991 (1982).
59. Easteal, A. J.: *AIChE J.*, **30:** 641 (1984).
60. Elliott, R. W., and H. Watts: *Can. J. Chem.*, **50:** 31 (1972).
61. Ellis, C. S., and J. N. Holsen: *Ind. Eng. Chem. Fundam.*, **8:** 787 (1969).
62. Eyring, H., and T. Ree: *Proc. Natl. Acad. Sci.*, **47:** 526 (1961).
63. Faghri, A., and M.-R. Riazi: *Intern. Comm. Heat Mass Transfer,* **10:** 385 (1983).
64. Fairbanks, D. F., and C. R. Wilke: *Ind. Eng. Chem.*, **42:** 471 (1950).
65. Fedors, R. F.: *AIChE J.*, **25:** 200 (1979).
66. Fedors, R. F.: *AIChE J.*, **25:** 716 (1979).
67. Feist, R., and G. M. Schneider: *Sep. Sci. Tech.*, **17:** 261 (1982).
68. Fuller, E. N., and J. C. Giddings: *J. Gas Chromatogr.*, **3:** 222 (1965).
69. Fuller, E. N., K. Ensley, and J. C. Giddings: *J. Phys. Chem.*, **75:** 3679 (1969).
70. Fuller, E. N., P. D. Schettler, and J. C. Giddings: *Ind. Eng. Chem.*, **58**(5), 18 (1966).
71. Gainer, J. L.: *Ind. Eng. Chem. Fundam.*, **5:** 436 (1966).
72. Gainer, J. L.: *Ind. Eng. Chem. Fundam.*, **9:** 381 (1970).
73. Gainer, J. L., and A. B. Metzner: *AIChE-IChemE Symp. Ser.*, no. 6, 1965, p. 74.
74. Ghai, R. K., and F. A. L. Dullien: private communication, 1976.
75. Ghai, R. K., H. Ertl, and F. A. L. Dullien: *AIChE J.*, **19:** 881 (1973).
76. Ghai, R. K., H. Ertl, and F. A. L. Dullien: *AIChE J.*, **20:** 1 (1974).
77. Gilliland, E. R.: *Ind. Eng. Chem.*, **26:** 681 (1934).
78. Glasstone, S., K. J. Laidler, and H. Eyring: *The Theory of Rate Processes,* McGraw-Hill, New York, 1941, chap. 9.
79. Gordon, A. R.: *J. Chem. Phys.*, **5:** 522 (1937).

80. Gordon, M.: "References to Experimental Data on Diffusion Coefficients of Binary Gas Mixtures," *Natl. Eng. Lab. Rept.*, Glasgow, Scotland, 1977.
81. Gotoh, S., M. Manner, J. P. Sørensen, and W. E. Stewart: *Ind. Eng. Chem.*, **12:** 119 (1973).
82. Gotoh, S., M. Manner, J. P. Sørensen, and W. E. Stewart: *J. Chem. Eng. Data,* **19:** 169, 172 (1974).
83. Gupta, G. P., and S. C. Saxena: *AIChE J.*, **14:** 519 (1968).
84. Haase, R., and H.-J. Jansen: *Z. Naturforsch.*, **35A:** 1116 (1980).
85. Halmour, N., and O. C. Sandall: *J. Chem. Eng. Data,* **29:** 20 (1984).
86. Haluska, J. L., and C. P. Colver: *AIChE J.,* **16:** 691 (1970).
87. Haluska, J. L., and C. P. Colver: *Ind. Eng. Chem. Fundam.,* **10:** 610 (1971).
88. Hammond, B. R., and R. H. Stokes: *Trans. Faraday Soc.,* **51:** 1641 (1955).
89. Hardt, A. P., D. K. Anderson, R. Rathbun, B. W. Mar, and A. L. Babb: *J. Phys. Chem.,* **63:** 2059 (1959).
90. Harned, H. S., and B. B. Owen: "The Physical Chemistry of Electrolytic Solutions," *ACS Monogr. 95,* 1950.
91. Hartley, G. S., and J. Crank: *Trans. Faraday Soc.,* **45:** 801 (1949).
92. Hattikudur, U. R., and G. Thodos: *J. Chem. Phys.,* **52:** 4313 (1970).
93. Hayduk, W., and W. D. Buckley: *Chem. Eng. Sci.,* **27:** 1997 (1972).
94. Hayduk, W., R. Casteñeda, H. Bromfield, and R. R. Perras: *AIChE J.,* **19:** 859 (1973).
95. Hayduk, W., and S. C. Cheng: *Chem. Eng. Sci.,* **26:** 635 (1971).
96. Hayduk, W., and S. Ioakimidis: *J. Chem. Eng. Data,* **21:** 255 (1976).
97. Hayduk, W., and H. Laudie: *AIChE J.,* **20:** 611 (1974).
98. Hayduk, W., and B. S. Minhas: *Can. J. Chem. Eng.,* **60:** 295 (1982).
99. Himmelblau, D. M.: *Chem. Rev.,* **64:** 527 (1964).
100. Hirschfelder, J. O., C. F. Curtiss, and R. B. Bird: *Molecular Theory of Gases and Liquids,* Wiley, New York, 1954.
101. Hiss, T. G., and E. L. Cussler: *AIChE J.,* **19:** 698 (1973).
102. Holmes, J. T., D. R. Olander, and C. R. Wilke: *AIChE J.,* **8:** 646 (1962).
103. Holson, J. N., and M. R. Strunk: *Ind. Eng. Chem. Fundam.,* **3:** 163 (1964).
104. Horrocks, J. K., and E. McLaughlin: *Trans. Faraday Soc.,* **58:** 1367 (1962).
105. Hudson, G. H., J. C. McCoubrey, and A. R. Ubbelohde: *Trans. Faraday Soc.,* **56:** 1144 (1960).
106. Innes, K. K., and L. F. Albright: *Ind. Eng. Chem.,* **49:** 1793 (1957).
107. *International Critical Tables,* McGraw-Hill, New York, 1926–1930.
108. Iomtev, M. B., and Yu. V. Tsekanskaya: *Russ. J. Phys. Chem.,* **38:** 485 (1964).
109. Johnson, D. W., and C. P. Colver: *Hydrocarbon Process. Petrol. Refiner,* **48**(3), 113 (1969).
110. Johnson, P. A., and A. L. Babb: *Chem. Rev.,* **56:** 387 (1956).
111. Kamal, M. R., and L. N. Canjar: *AIChE J.,* **8:** 329 (1962).
112. Kestin, J., H. E. Khalifa, S. T. Ro, and W. A. Wakeham: *Physica,* **88A:** *242 (1977).*
113. Kestin, J., and W. A. Wakeham: *Ber. Bunsenges. Phys. Chem.,* **87:** 309 (1983).
114. Kett, T. K., and D. K. Anderson: *J. Phys. Chem.,* **73:** 1262 (1969).
115. Kett, T. K., and D. K. Anderson: *J. Phys. Chem.,* **73:** 1268 (1969).
116. King, C. J., L. Hsueh, and K.-W. Mao: *J. Chem. Eng. Data,* **10:** 348 (1965).
117. Kircher, K., A. Schaber, and E. Obermeier: *Proc. 8th Symp. Thermophys. Prop.,* vol. 1, ASME, 1981, p. 297.
118. Krieger, I. M., G. W. Mulholland, and C. S. Dickey: *J. Phys. Chem.,* **71:** 1123 (1967).
119. Kuznetsova, E. M., and D. Sh. Rashidova: *Russ. J. Phys. Chem.,* **54:** 1322 (1980).
120. Kuznetsova, E. M., and D. Sh. Rashidova: *Russ. J. Phys. Chem.,* **54:** 1339 (1980).
121. Leaist, D. G.: *J. Chem. Eng. Data,* **29:** 281 (1984).
122. Lee, H., and G. Thodos: *Ind. Eng. Chem. Fundam.,* **22:** 17 (1983).
123. Lees, F. P., and P. Sarram: *J. Chem. Eng. Data,* **16:** 41 (1971).
124. Leffler, J., and H. T. Cullinan, Jr.: *Ind. Eng. Chem. Fundam.,* **9:** 84, 88 (1970).
125. Lenczyk, J. P., and H. T. Cullinan, Jr.: *Ind. Eng. Chem. Fundam.,* **10:** 600 (1971).

126. Lewis, J. B.: *J. Appl. Chem. London,* **5:** 228 (1955).
127. Li, J. C. M., and P. Chang: *J. Chem. Phys.,* **23:** 518 (1955).
128. Loflin, T., and E. McLaughlin: *J. Phys. Chem.,* **73:** 186 (1969).
129. Lugg, G. A.: *Anal. Chem.,* **40:** 1072 (1968).
130. Lusis, M. A.: *Chem. Proc. Eng.,* **5:** 27 (May 1971).
131. Lusis, M. A.: *Chem. Ind. Devel. Bombay,* January 1972, 48; *AIChE J.,* **20:** 207 (1974).
132. Lusis, M. A., and G. A. Ratcliff: *Can. J. Chem. Eng.,* **46:** 385 (1968).
133. Lusis, M. A., and G. A. Ratcliff: *AIChE J.,* **17:** 1492 (1971).
134. Maczek, A. O. S., and C. J. C. Edwards: "The Viscosity and Binary Diffusion Coefficients of Some Gaseous Hydrocarbons, Fluorocarbons and Siloxanes," *Symp. Transport Prop. Fluids and Fluid Mixtures, Natl. Eng. Lab., East Kilbride, Glasgow, Scotland, April 1979.*
135. McCall, D. W., and D. C. Douglas: *Phys. Fluids,* **2:** 87 (1959).
136. McCall, D. W., and D. C. Douglas: *J. Phys. Chem.,* **71:** 987 (1967).
137. McCall, D. W., D. C. Douglas, and E. W. Anderson: *Phys. Fluids,* **2:** 87 (1959); **4:** 162 (1961).
138. Marrero, T. R., and E. A. Mason: *J. Phys. Chem. Ref. Data,* **1:** 3 (1972).
139. Mason, E. A., and L. Monchick: *J. Chem. Phys.,* **36:** 2746 (1962).
140. Mather, G. P., and S. C. Saxena: *Ind. J. Pure Appl. Phys.,* **4:** 266 (1966).
141. Meissner, H. P., and C. L. Kusik: *AIChE J.,* **18:** 294 (1972).
142. Meissner, H. P., and C. L. Kusik: *Ind. Eng. Chem. Process Design Develop.,* **12:** 205 (1973).
143. Meissner, H. P., C. L. Kusik, and J. W. Tester: *AIChE J.,* **18:** 661 (1972).
144. Meissner, H. P., and N. A. Peppas: *AIChE J.,* **19:** 806 (1973).
145. Meissner, H. P., and J. W. Tester: *Ind. Eng. Chem. Process Design Develop.,* **11:** 128 (1972).
146. Miller, D. G.: *J. Phys. Chem.,* **70:** 2639 (1966).
147. Miller, D. G.: *J. Phys. Chem.,* **71:** 616 (1967).
148. Miller, L., and P. C. Carman: *Trans. Faraday Soc.,* **57:** 2143 (1961).
149. Moelwyn-Hughes, E. A.: *Physical Chemistry,* Pergamon, London, 1957.
150. Mohan, V., and D. Srinivasan: *Chem. Eng. Comm.,* **29:** 27 (1984).
151. Morozov, V. S., and E. G. Vinkler: *Russ. J. Phys. Chem.,* **49:** 1404 (1975).
152. Mortimer, R. G.: *Ind. Eng. Chem. Fundam.,* **12:** 492 (1973).
153. Mortimer, R. G., and N. H. Clark: *Ind. Eng. Chem. Fundam.,* **10:** 604 (1971).
154. Mrazek, R. V., C. E. Wicks, and K. N. S. Prabhu: *J. Chem. Eng. Data,* **13:** 508 (1968).
155. Murad, S.: *Chem. Eng. Sci.,* **36:** 1867 (1981).
156. Murad, S., and N. Bhupathiraju: *Int. J. Thermophys.,* **4:** 329 (1983).
157. Nakanishi, K.: *Ind. Eng. Chem. Fundam.,* **17:** 253 (1978).
158. Neufeld, P. D., A. R. Janzen, and R. A. Aziz: *J. Chem. Phys.,* **57:** 1100 (1972).
159. Newman, J. S.: in C. W. Tobias (ed.), *Advances in Electrochemistry and Electrochemical Engineering,* vol. 5, Interscience, New York, 1967.
160. Olander, D. R.: *AIChE J.,* **7:** 175 (1961).
161. Olander, D. R.: *AIChE J.,* **9:** 207 (1963).
162. Oosting, E. M., J. I. Gray, and E. A. Grulke: *AIChE J.,* **31:** 773 (1985).
163. Othmer, D. F., and T. T. Chen: *Ind. Eng. Chem. Process Design Develop.,* **1:** 249 (1962).
164. Othmer, D. F., and M. S. Thakar: *Ind. Eng. Chem.,* **45:** 589 (1953).
165. Pasternak, A. D., and D. R. Olander: *AIChE J.,* **13:** 1052 (1967).
166. Pathak, B. K., V. N. Singh, and P. C. Singh: *Can. J. Chem. Eng.,* **59:** 362 (1981).
167. Perkins, L. R., and C. J. Geankoplis: *Chem. Eng. Sci.,* **24:** 1035 (1969).
168. Quale, O. R.: *Chem. Rev.,* **53:** 439 (1953).
169. Raina, G. K.: *AIChE J.,* **26:** 1046 (1980).
170. Rao, S. S., and C. O. Bennett: *AIChE J.,* **17:** 75 (1971).
171. Ratcliff, G. A. and J. G. Holdcroft: *Trans. Inst. Chem. Eng. London,* **41:** 315 (1963).

172. Ratcliff, G. A., and M. A. Lusis: *Ind. Eng. Chem. Fundam.*, **10:** 474 (1971).
173. Rathbun, R. E., and A. L. Babb: *Ind. Eng. Chem. Process Design Develop.*, **5:** 273 (1966).
174. Ravindran, P., E. J. Davis, and A. K. Ray: *AIChE J.*, **25:** 966 (1979).
175. Reddy, K. A., and L. K. Doraiswamy: *Ind. Eng. Chem. Fundam.*, **6:** 77 (1967).
176. Ree, F. H., T. Ree, and H. Eyring: *Ind. Eng. Chem.*, **50:** 1036 (1958).
177. Robinson, R. A., and R. H. Stokes: *Electrolyte Solutions,* 2d ed., Academic, New York, 1959.
178. Robinson, R. L., Jr., W. C. Edmister, and F. A. L. Dullien: *Ind. Eng. Chem. Fundam.*, **5:** 75 (1966).
179. Sanchez, V., and M. Clifton: *Ind. Eng. Chem. Fundam.*, **16:** 318 (1977).
180. Sanchez, V., H. Oftadeh, C. Durou, and J.-P. Hot: *J. Chem. Eng. Data,* **22:** 123 (1977).
181. Sandler, S., and E. A. Mason: *J. Chem. Phys.*, **48:** 2873 (1968).
182. Sanni, S. A., and P. Hutchinson: *J. Chem. Eng. Data,* **18:** 317 (1973).
183. Scheibel, E. G.: *Ind. Eng. Chem.*, **46:** 2007 (1954).
184. Schneider, G. M.: *Angew. Chem. Intern. Ed. English,* **17:** 716 (1978).
185. Schonert, H.: *Z. Phys. Chem.*, **119:** 63 (1980).
186. Schwartz, F. A., and J. E. Brow: *J. Chem. Phys.*, **19:** 640 (1951).
187. Scott, D. S., and K. E. Cox: *Can. J. Chem. Eng.*, **38:** 201 (1960).
188. Seager, S. L., L. R. Geertson, and J. C. Giddings: *J. Chem. Eng. Data,* **8:** 168 (1963).
189. Semenov, A. V., and A. N. Berezhnoi: *Russ. J. Phys. Chem.*, **52:** 1165 (1978).
190. Shankland, I. R., and P. J. Dunlop: *Physica,* **100A:** 64 (1980).
191. Shrier, A. L.: *Chem. Eng. Sci.*, **22:** 1391 (1967).
192. Sitaraman, R., S. H. Ibrahim, and N. R. Kuloor: *J. Chem. Eng. Data,* **8:** 198 (1963).
193. Slattery, J. C., and R. B. Bird: *AIChE J.*, **4:** 137 (1958).
194. Sridhar, T., and O. E. Potter: *AIChE J.*, **23:** 590, 946 (1977).
195. Srivastava, B. N., and K. P. Srivastava: *J. Chem. Phys.*, **30:** 984 (1959).
196. Srivastava, B. N., and I. B. Srivastava: *J. Chem. Phys.*, **36:** 2616 (1962).
197. Srivastava, B. N., and I. B. Srivastava: *J. Chem. Phys.*, **38:** 1183 (1963).
198. Stearn, A. E., E. M. Irish, and H. Eyring: *J. Phys. Chem.*, **44:** 981 (1940).
199. Swaid, I., and G. M. Schneider: *Ber. Bunsenges. Phys. Chem.*, **83:** 969 (1979).
200. Takahashi, S.: *J. Chem. Eng. Japan,* **7:** 417 (1974).
201. Takahashi, S., and M. Hongo: *J. Chem. Eng. Japan,* **15:** 57 (1982).
202. Takahashi, M., Y. Kobayashi, and H. Takeuchl: *J. Chem. Eng. Data,* **27:** 328 (1982).
203. Tang, Y. P., and D. M. Himmelblau: *AIChE J.*, **11:** 54 (1965).
204. Tee, L. S., G. R. Kuether, R. C. Robinson, and W. E. Stewart: Am. Petrol. Inst. Div. Refining, Houston, Texas, May 1966.
205. Teja, A. S.: private communication, 1982.
206. Toor, H. L.: *AIChE J.*, **3:** 198 (1957).
207. Tsekhanskaya, Yu. V.: *Russ. J. Phys. Chem.*, **45:** 744 (1971).
208. Turner, J. C. R.: *Chem. Eng. Sci.*, **30:** 151 (1975).
209. Tyn, M. T.: *Chem. Eng.*, **82**(12): 106 (1975).
210. Tyn, M. T.: *Chem. Eng. J.*, **12:** 149 (1976).
211. Tyn, M. T.: *Trans. Inst. Chem. Engrs.*, ***59**(2), 112 (1981).*
212. Tyn, M. T., and W. F. Calus: *J. Chem. Eng. Data,* **20:** 106 (1975).
213. Tyn, M. T., and W. F. Calus: *J. Chem. Eng. Data,* **20:** 310 (1975).
214. Umesi, N. O., and R. P. Danner: *Ind. Eng. Chem. Process Design Develop.*, **20:** 662 (1981).
215. Vadovic, C. J., and C. P. Colver: *AIChE J.*, **19:** 546 (1973).
216. Van Geet, A. L., and A. W. Adamson: *J. Phys. Chem.*, **68:** 238 (1964).
217. Vargaftik, N. B.: *Tables on the Thermophysical Properties of Liquids and Gases,* Hemisphere, Washington, D.C., 1975.
218. Verhallen, P. T. H. M., L. J. O. Oomen, A. J. J. M. v. d. Elsen, and A. J. Kruger: *Chem. Eng. Sci.*, 39, 1535 (1984).
219. Vignes, A.: *Ind. Eng. Chem. Fundam.*, **5:** 189 (1966).
220. Vinkler, E. G., and V. S. Morozov: *Russ. J. Phys. Chem.*, **49:** 1405 (1975).

221. Vinograd, J. R., and J. W. McBain: *J. Am. Chem. Soc.*, **63:** 2008 (1941).
222. Vivian, J. E., and C. J. King: *AIChE J.*, **10:** 220 (1964).
223. Walker, R. E., N. de Haas, and A. A. Westenberg: *J. Chem. Phys.*, **32:** 1314 (1960).
224. Walker, R. E., and A. A. Westenberg: *J. Chem. Phys.*, **29:** 1139 (1958).
225. Way, P.: Ph.D. thesis, Massachusetts Institute of Technology, Cambridge, Mass., 1971.
226. Weissman, S.: *J. Chem. Phys.*, **40:** 3397 (1964).
227. Weissman, S., and E. A. Mason: *J. Chem. Phys.*, **37:** 1289 (1962).
228. Wendt, R. P.: *J. Phys. Chem.*, **69:** 1227 (1965).
229. Westenberg, A. A., and G. Frazier: *J. Chem. Phys.*, **36:** 3499 (1962).
230. Wilke, C. R., and P. Chang: *AIChE J.*, **1:** 264 (1955).
231. Wilke, C. R., and C. Y. Lee: *Ind. Eng. Chem.*, **47:** 1253 (1955).
232. Wise, D. L., and G. Houghton: *Chem. Eng. Sci.*, **21:** 999 (1966).
233. Witherspoon, P. A., and L. Bonoli: *Ind. Eng. Chem. Fundam.*, **8:** 589 (1969).
234. Yabsley, M. A., P. J. Carlson, and P. J. Dunlop: *J. Phys. Chem.*, **77:** 703 (1973).
235. Yon, C. M., and H. L. Toor: *Ind. Eng. Chem. Fundam.*, **7:** 319 (1968).

Chapter

# 12

# Surface Tension

## 12-1 Scope

The surface tensions of both pure liquids and liquid mixtures are considered in this chapter. For the former, methods based on the law of corresponding states and upon the parachor are judged most accurate when estimated values are compared with experimental determinations. For mixtures, extensions of the pure component methods are presented, as is a method based upon a thermodynamic analysis of the system. Interfacial tensions for liquid-liquid or liquid-solid systems are not included.

## 12-2 Introduction

The boundary layer between a liquid phase and a gas phase may be considered a third phase with properties intermediate between those of a liquid and a gas. A qualitative picture of the microscopic surface layer shows that there are unequal forces acting upon the molecules; i.e., at low gas densities, the surface molecules are attracted sidewise and toward the bulk liquid but experience less attraction in the direction of the bulk gas. Thus the surface layer is in tension and tends to contract to the smallest area compatible with the mass of material, container restraints, and external forces, e.g., gravity.

A quantitative index of this tension can be presented in various ways; the most common is the surface tension $\sigma$, defined as the force exerted in the plane of the surface per unit length. We can consider a reversible isothermal process whereby surface area $A$ is increased by pulling the surface apart and allowing the molecules from the bulk liquid to enter at constant temperature and pressure. The differential reversible work is $\sigma\, dA$; in this case $\sigma$ is the surface Gibbs energy per unit of area. As equilibrium systems tend to a state of minimum Gibbs energy, the product $\sigma A$ also tends to a minimum. For a fixed $\sigma$, equilibrium is a state of minimum area consistent with the restraints of the system.

Surface tension is usually expressed in dynes per centimeter; surface Gibbs energy per unit area has units of ergs per square centimeter. These units and numerical values of $\sigma$ are identical. In SI units, 1 erg/cm$^2$ = 1 mJ/m$^2$ = 1 mN/m.

The thermodynamics of surface layers furnishes a fascinating subject for study. Guggenheim [23], Gibbs [20], and Modell and Reid [43] have formulated treatments which differ considerably but reduce to similar equations relating macroscopically measurable quantities. In addition to the thermodynamic aspect, treatments of the physics and chemistry of surfaces have been published [1, 3, 4, 8, 57]. These subjects are not covered here; instead, the emphasis is placed upon the few reliable methods available to estimate $\sigma$ from either semitheoretical or empirical equations.

## 12-3 Estimation of the Surface Tension of a Pure Liquid

As the temperature is raised, the surface tension of a liquid in equilibrium with its own vapor decreases and becomes zero at the critical point [53]. In the reduced-temperature range 0.45 to 0.65, $\sigma$ for most organic liquids ranges from 20 to 40 dyn/cm, but for some low-molecular-weight dense liquids such as formamide, $\sigma > 50$ dyn/cm. For water $\sigma = 72.8$ dyn/cm at 293 K, and for liquid metals $\sigma$ is between 300 and 600 dyn/cm; e.g., mercury at 293 K has a value of about 476.

A recent, thorough critical evaluation of experimental surface tensions has been prepared by Jasper [32]. Additional data are given in Ref. 35.

Essentially all useful estimation techniques for the surface tension of a liquid are empirical. Only two are discussed in any detail here, although others are briefly noted at the end of this section.

### Macleod-Sugden correlation

Macleod [38] in 1923 suggested a relation between $\sigma$ and the liquid and vapor densities:

$$\sigma^{1/4} = [P](\rho_L - \rho_v) \tag{12-3.1}$$

**TABLE 12-1 Comparison of Calculated and Experimental Values of Surface Tension of Pure Liquids**

| Compound | $T$, K | $\sigma$ (exp.),† dyn/cm | Percent Error in Method‡: Macleod and Sugden, Eq. (12-3.1) | Percent Error in Method‡: Brock and Bird, Eq. (12-3.6) |
|---|---|---|---|---|
| Acetic acid | 293 | 27.59 | −4.6 | |
| | 333 | 23.62 | −3.2 | |
| Acetone | 298 | 24.02 | −5.4 | 2.4 |
| | 308 | 22.34 | −3.9 | 3.7 |
| | 318 | 21.22 | −4.5 | 2.7 |
| Aniline | 293 | 42.67 | −3.2 | 11 |
| | 313 | 40.50 | −6.0 | 10 |
| | 333 | 38.33 | −7.1 | 9.1 |
| | 353 | 36.15 | −8.8 | 8.0 |
| Benzene | 293 | 28.88 | −5.1 | −2.0 |
| | 313 | 26.25 | −5.6 | −1.9 |
| | 333 | 23.67 | −5.0 | −0.4 |
| | 353 | 21.20 | −3.9 | −1.9 |
| Benzonitrile | 293 | 39.37 | −2.0 | 1.2 |
| | 323 | 35.89 | −3.4 | 1.1 |
| | 363 | 31.26 | −4.2 | 1.3 |
| Bromobenzene | 293 | 35.82 | −0.7 | −0.2 |
| | 323 | 32.34 | −1.4 | 0 |
| | 373 | 26.54 | −0.7 | 0.7 |
| $n$-Butane | 203 | 23.31 | 11 | 1.6 |
| | 233 | 19.69 | 5.2 | 0.9 |
| | 293 | 12.46 | 1.5 | 0.9 |
| Carbon disulfide | 293 | 32.32 | 3.8 | 3.0 |
| | 313 | 29.35 | 3.8 | 3.1 |
| Carbon tetrachloride | 288 | 27.65 | −1.1 | −4.9 |
| | 308 | 25.21 | −1.2 | −5.1 |
| | 328 | 22.76 | −1.0 | −5.0 |
| | 348 | 20.31 | 0.1 | −4.9 |
| | 368 | 17.86 | 2.5 | −4.4 |
| Chlorobenzene | 293 | 33.59 | −0.6 | −1.7 |
| | 323 | 30.01 | 0.7 | −1.7 |
| | 373 | 24.06 | 5.8 | −1.2 |
| $p$-Cresol | 313 | 34.88 | 0.5 | |
| | 373 | 29.32 | −0.3 | |
| Cyclohexane | 293 | 25.24 | −3.9 | −4.8 |
| | 313 | 22.87 | −3.5 | −4.5 |
| | 333 | 20.49 | −2.2 | −4.3 |
| Cyclopentane | 293 | 22.61 | −5.6 | −2.0 |
| | 313 | 19.68 | −2.4 | 0.1 |
| Diethyl ether | 288 | 17.56 | 0 | 0 |
| | 303 | 16.20 | 0.4 | −2.5 |
| 2,3-Dimethylbutane | 293 | 17.38 | −0.9 | −0.2 |
| | 313 | 15.38 | 0.6 | −0.5 |
| Ethyl acetate | 293 | 23.97 | −4.6 | 1.1 |
| | 313 | 21.65 | −4.8 | 0 |
| | 333 | 19.32 | −4.3 | −0.9 |
| | 353 | 17.00 | −2.7 | −1.4 |
| | 373 | 14.68 | 0.5 | −2.0 |

| Compound | $T$, K | $\sigma$ (exp.),† dyn/cm | Percent Error in Method‡ Macleod and Sugden, Eq. (12-3.1) | Brock and Bird, Eq. (12-3.6) |
|---|---|---|---|---|
| Ethyl benzoate | 293 | 35.04 | −1.9 | 2.7 |
| | 313 | 32.92 | −2.7 | 2.9 |
| | 333 | 30.81 | −3.1 | 2.9 |
| Ethyl bromide | 283 | 25.36 | −5.3 | 15 |
| | 303 | 23.04 | −6.1 | 13 |
| Ethyl mercaptan | 288 | 23.87 | −6.7 | 3.5 |
| | 303 | 22.68 | −9.1 | −1.2 |
| Formamide | 298 | 57.02 | −8.8 | § |
| | 338 | 53.66 | −15 | |
| | 373 | 50.71 | −20 | |
| $n$-Heptane | 293 | 20.14 | −0.6 | 0.3 |
| | 313 | 18.18 | 0.7 | 0.1 |
| | 333 | 16.22 | 3.1 | 0.3 |
| | 353 | 14.26 | 6.8 | 0.8 |
| Isobutyric acid | 293 | 25.04 | 1.2 | |
| | 313 | 23.20 | 0.5 | |
| | 333 | 21.36 | −1.2 | |
| | 363 | 18.60 | −3.5 | |
| Methyl formate | 293 | 24.62 | −7.6 | 4.4 |
| | 323 | 20.05 | −7.2 | 4.6 |
| | 373 | 12.90 | −7.4 | 4.3 |
| | 423 | 6.30 | −8.8 | 5.4 |
| | 473 | 0.87 | −14 | 21 |
| Methyl alcohol | 293 | 22.56 | −13 | |
| | 313 | 20.96 | −15 | |
| | 333 | 19.41 | −17 | |
| Phenol | 313 | 39.27 | −6.7 | |
| | 333 | 37.13 | −7.3 | |
| | 373 | 32.86 | −7.8 | |
| $n$-Propyl alcohol | 293 | 23.71 | −0.6 | |
| | 313 | 22.15 | −1.9 | |
| | 333 | 20.60 | −3.3 | |
| | 363 | 18.27 | −4.0 | |
| $n$-Propyl benzene | 293 | 29.98 | 0.2 | −1.1 |
| | 313 | 26.83 | 0.8 | −0.7 |
| | 333 | 24.68 | 2.1 | −0.1 |
| | 353 | 22.53 | 3.9 | 0.7 |
| | 373 | 20.38 | 6.6 | 1.9 |
| Pyridine | 293 | 37.21 | −2.8 | −0.4 |
| | 313 | 34.60 | −3.6 | −0.8 |
| | 333 | 31.98 | −4.1 | −1.2 |

†Experimental values from Ref. 32, except for methyl formate, experimental values for which were taken from Ref. 38. Surface tensions quoted by Jasper are smoothed values obtained after plotting $\sigma$ vs. $T$. Normally a linear relation was assumed over small temperature ranges. See discussion with Eq. (12-4.2).

‡Error = [(calc. − exp.)/exp.] × 100.

§Critical properties not known.

Sugden [67, 68] has called the temperature-independent parameter $[P]$ the *parachor* and indicated how it might be estimated from the structure of the molecule. Quale [50] employed experimental surface tension and density data for many compounds and calculated parachors. From these, he suggested an additive scheme to correlate $[P]$ with structure, and a modified list of his values is shown in Table 11-3. When $[P]$ values determined in this manner are used, the surface tension is given in dynes per centimeter and the densities are expressed in moles per cubic centimeter. The method is illustrated in Example 12-1, and calculated values of $\sigma$ are compared with experimental surface tensions in Table 12-1.

**Example 12-1** Use the Macleod-Sugden correlation to estimate the surface tension of isobutyric acid at 333 K. The experimental value quoted by Jasper [32] is 21.36 dyn/cm.

**solution** At 333 K, the liquid density is 0.912 g/cm³ [31], and, with $M = 88.107$, $\rho_L = 0.912/88.107 = 1.035 \times 10^{-2}$ mol/cm³. At 333 K, isobutyric acid is well below the boiling point, and at this low pressure $\rho_v \ll \rho_L$ and the vapor density term is neglected.

To determine the parachor from Table 11-3,

$$[P] = CH_3{-}CH(CH_3){-} + {-}COOH$$

$$= 133.3 + 73.8 = 207.1$$

Then, with Eq. (12-3.1),

$$\sigma = [(207.1)(1.035 \times 10^{-2})]^4 = 21.10 \text{ dyn/cm}$$

$$\text{Error} = \frac{21.10 - 21.36}{21.36} \times 100 = -1.2\%$$

Since $\sigma$ is proportional to $([P]_{\rho L})^4$, Eq. (12-3.1) is *very* sensitive to the values of the parachor and liquid density chosen. It is remarkable that the estimated values are as accurate as shown in Table 12-1.

Instead of employing experimental densities, correlations relating $\rho$ to $T$, given in Chap. 3, may be used. One technique not covered in that chapter is given by Goldhammer [21] and discussed by Gambill [19]:

$$\rho_L - \rho_v = \rho_{Lb}\left(\frac{1 - T_r}{1 - T_{br}}\right)^n \tag{12-3.2}$$

$\rho_{L_b}$ is the molal liquid density at the normal boiling point in moles per cubic centimeter. The exponent $n$ ranges from 0.25 to 0.31; Fishtine [17] suggests the following values:

| | $n$ |
|---|---|
| Alcohols | 0.25 |
| Hydrocarbons and ethers | 0.29 |
| Other organic compounds | 0.31 |

With Eq. (12-3.2), Eq. (12-3.1) becomes

$$\sigma = ([P]\rho_{Lb})^4 \left(\frac{1 - T_r}{1 - T_{b_r}}\right)^{4n} \tag{12-3.3}$$

where $4n$ varies between 1.0 and 1.24. As shown later, other correlations predict a similar temperature dependence; that is, $\sigma$ decreases with temperature at a rate somewhat exceeding that predicted by a linear relation.

#### Corresponding states correlation

The group $\sigma/P_c^{2/3}T_c^{1/3}$ is dimensionless except for a numerical constant which depends upon the units of $\sigma$, $P_c$, and $T_c$.† Van der Waals suggested in 1894 [73] that this group could be correlated with $1 - T_r$. Brock and Bird [7] developed this idea for nonpolar liquids and proposed that

$$\frac{\sigma}{P_c^{2/3}T_c^{1/3}} = (0.132\alpha_c - 0.279)(1 - T_r)^{11/9} \tag{12-3.4}$$

where $\alpha_c$ is the Riedel parameter [55] at the critical point and $\alpha$ is defined as $d \ln P_{vp_r}/d \ln T_r$. Using a suggestion by Miller [42] to relate $\alpha_c$ to $T_{b_r}$ and $P_c$,

$$\alpha_c = 0.9076 \left[1 + \frac{T_{b_r} \ln (P_c/1.01325)}{1 - T_{b_r}}\right] \tag{12-3.5}$$

it can be shown that

$$\sigma = P_c^{2/3}T_c^{1/3}\, Q(1 - T_r)^{11/9} \tag{12-3.6}$$

$$Q = 0.1196 \left[1 + \frac{T_{b_r} \ln (P_c/1.01325)}{1 - T_{b_r}}\right] - 0.279 \tag{12-3.7}$$

Equations (12-3.6) and (12-3.7) were used to compute $\sigma$ values for nonpolar liquids in Table 12-1. The accuracy is similar to that for the Macleod-Sugden relation discussed earlier. However, the corresponding states method is not applicable to compounds exhibiting strong hydrogen-bonding (alcohols, acids) and quantum liquids ($H_2$, He, Ne).

A similar correlation was proposed by Riedel [56]. More recently, to

†The fundamental dimensionless group is $\sigma V_c^{2/3}/RT_c$ [24, 74], but the proportionality $V_c \propto RT_c/P_c$ has been used and the gas constant has been dropped from the group.

broaden the approach to include polar liquids, Hakim et al. [26] introduced the Stiel polar factor, $X$, and proposed the following equation:

$$\sigma = P_c^{2/3} T_c^{1/3} Q_p \left( \frac{1 - T_r}{0.4} \right)^m \tag{12-3.8}$$

where $\sigma$ = surface tension of polar liquid, dyn/cm
$P_c$ = critical pressure, bar
$T_c$ = critical temperature, K
$Q_p = 0.1560 + 0.365\omega - 1.754X - 13.57X^2 - 0.506\omega^2 + 1.287\omega X$
$m = 1.210 + 0.5385\omega - 14.61X - 32.07X^2 - 1.656\omega^2 + 22.03\omega X$
$X$ = Stiel polar factor
$\omega$ = acentric factor, Sec. 2-3

Values of $X$ can be estimated from the reduced vapor pressure at $T_r = 0.6$, $P_{\mathrm{vp}_r}(0.6)$,

$$X = \log P_{\mathrm{vp}_r}(0.6) + 1.70\omega + 1.552 \tag{12-3.9}$$

The general reliability of Eq. (12-3.8) is not known. The six constants in $Q_p$ and $m$ were obtained from data on only 16 polar compounds, some of which were only slightly polar (diethyl ether, dimethyl ether, ethyl mercaptan, acetone, etc.). $X$ values are available for only a few substances, and estimated values of $\sigma$ are sensitive to the value of $X$ chosen.

**Example 12-2** Using Eqs. (12-3.6) and (12-3.7), estimate the surface tension of liquid ethyl mercaptan at 303 K. The experimental value is 22.68 dyn/cm [32].

**solution** From Appendix A, for ethyl mercaptan, $T_c = 499$ K, $T_b = 308.2$ K, $P_c = 54.9$ bar. Thus $T_{b_r} = 308.2/499 = 0.618$. With Eq. (12-3.7)

$$Q = 0.1196 \left[ 1 + \frac{(0.618) \ln (54.9/1.01325)}{1 - 0.618} \right] - 0.279 = 0.613$$

$$\sigma = (54.9)^{2/3}(499)^{1/3}(0.613) \left( 1 - \frac{303}{499} \right)^{11/9} = 22.4 \text{ dyn/cm}$$

$$\text{Error} = \frac{22.41 - 22.68}{22.68} \times 100 = -1.2\%$$

### Other estimation methods

Statistical-mechanical theories of liquids yield reasonable results for surface tensions of simple liquids [48]. The surface tension has also been correlated with the molar refraction and refractive index [71], liquid compressibility [39, 58], and viscosity [9, 44, 47, 61, 72]. Schonhorn [61] expanded upon an earlier idea of Pelofsky [47] and showed that ln $\sigma$ is linearly related to $(\eta_L - \eta_V)^{-1}$, where $\eta$ is the viscosity. Ramana et al. [51]

proposed a linear relation between log $\sigma_b$ and $T_{b_r}$ for different homologous series ($\sigma_b$ is the surface tension at $T_b$) and Carey et al. [10] have related surface tension to parameters in the Peng-Robinson equation of state.

For rapid estimations of $\sigma$, nomographs have been presented [25, 46] for relating $\sigma$ and $T$ for hydrocarbons. Data for light hydrocarbons and their mixtures have been correlated by Porteous [49].

Surface tensions for several cryogenic liquids have been reported by Sprow and Prausnitz [64]; they were correlated by a modification of Eq. (12-3.4)

$$\sigma = \sigma_0(1 - T_r)^p \tag{12-3.10}$$

where $\sigma_0$ and $p$ were fitted to the data by a least-squares analysis. Values of $p$ were close to $\frac{11}{9}$, but the best value of $\sigma_0$ was often slightly larger than $P_c^{2/3}T_c^{1/3}Q$, as predicted by Eq. (12-3.6).

Gray et al. [22] have correlated surface tensions of coal liquid fractions by

$$\sigma = P_c^{2/3}T_c^{1/3}(0.3993)(1 - T_r)^{0.4} \tag{12-3.11}$$

Rice and Teja [55] have estimated surface tensions with a two-reference-fluid corresponding states method (see Secs. 3-7, 7-5, and 7-9), along with the assumption that the group $\sigma V_c^{2/3}/T_c$ varies linearly with the acentric factor. Murad [45] has used a shape factor approach (Sec. 10-3) to estimate surface tensions; several authors [33, 63] have proposed correlations which relate surface tension to the enthalpy of vaporization. None of these last methods appears to offer any significant advantage over the methods previously described.

### Recommendations

For surface tensions of organic liquids, use the data collection of Jasper [32]. Two estimation methods were presented in this section, and calculated values are compared with experiment in Table 12-1. For non-hydrogen-bonded liquids, use the corresponding states method [Eqs. (12-3.6) and (12-3.7)]. The normal boiling point, critical temperature, and critical pressure are required. Errors are normally less than 5 percent.

For hydrogen-bonded liquids, use the Macleod-Sugden form [Eq. (12-3.1)] with the parachor determined from group contributions in Table 11-3. Either experimental saturated liquid and saturated vapor densities may be used or, with slightly less accuracy, the modified temperature form, Eq. (12-3.3), may be substituted. Errors are normally less than 5 to 10 percent.

If reliable values of $\omega$ and $X$ are available, however, Eq. (12-3.8) is somewhat more accurate for alcohols.

## 12-4 Variation of Surface Tension with Temperature

Equation (12-3.3) indicates that

$$\sigma \propto (1 - T_r)^{4n} \tag{12-4.1}$$

where $n$ varies from 0.25 to 0.31. In Fig. 12-1, log $\sigma$ is plotted against log $(1 - T_r)$ with experimental data for acetic acid, diethyl ether, and ethyl acetate. For the latter two, the slope is close to 1.25 for a value of $n = 0.31$; for acetic acid, the slope is 1.16 or $n = 0.29$. For most organic liquids, this encompasses the range normally found for $n$, although for alcohols, $n$ may be slightly less. The corresponding states correlation predicts a slope of 1.22, giving $n = 0.305$.

For values of $T_r$ between 0.4 and 0.7, Eq. (12-4.1) indicates that $d\sigma/dT$ is almost constant, and often the surface tension–temperature relation is represented by a linear equation

$$\sigma = a + bT \tag{12-4.2}$$

As an example, for nitrobenzene, between 313 and 473 K, data from Jasper [32] are plotted in Fig. 12-2. The linear approximation is satisfactory. Jasper lists values of $a$ and $b$ for many materials.

## 12-5 Surface Tensions of Nonaqueous Mixtures

The surface tension of a liquid mixture is not a simple function of the surface tensions of the pure components because, in a mixture, the composition of the surface is not the same as that of the bulk. In a typical situation, we know the bulk composition but not the surface composition.

The surface tension of a mixture $\sigma_m$ is usually but not always [2, 77] less than that calculated from a mole fraction average of the surface tensions of the pure components. Also, the derivative $d\sigma_m/dx$ usually increases with composition for that component with the largest pure component surface tension.

The techniques suggested for estimating $\sigma_m$ can conveniently be divided into two categories: those based on empirical relations suggested earlier for pure liquids and those derived from thermodynamics.

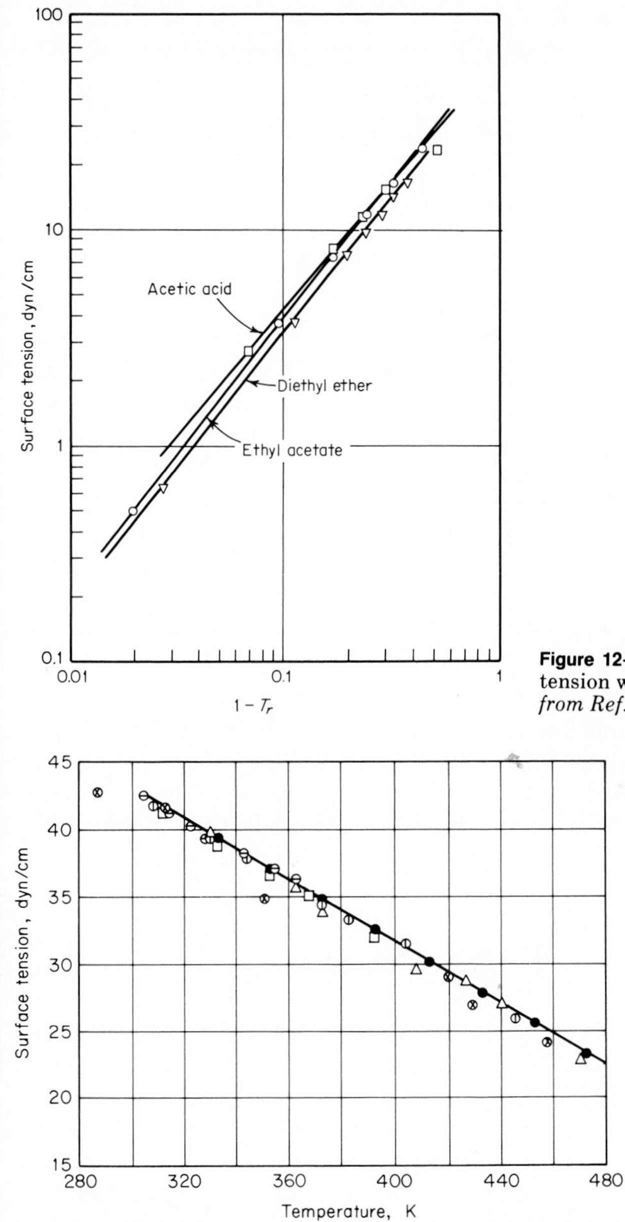


**Figure 12-1** Variation of surface tension with temperature. *(Data from Ref. 38.)*

**Figure 12-2** Surface tension of nitrobenzene: R. B. Badachhope, M. K. Gharpurey, and A. B. Biswas, *J. Chem. Eng. Data,* **10**:143 (1965); △ Ref. 68; ⊖ R. Kremann and R. Meirgast, *Monatsh. für Chemie,* **35**:1332 (1914); ⊖ F. M. Jaeger, *Z. Anorg. Allgem. Chem.,* **101**:1 (1917); □ W. Hückel and W. Jahnentz, *Chem. Ber.,* **75B**:1438 (1942); ⊗ W. Ramsay and J. Shields, *J. Chem. Soc.,* **63**:1089 (1893). *(From Ref. 32.)*

### Macleod-Sugden correlation

Applying Eq. (12-3.1) to mixtures gives

$$\sigma_m^{1/4} = \sum_{i=1}^{n} [P_i](\rho_{Lm}x_i - \rho_{vm}y_i) \tag{12-5.1}$$

where $\sigma_m$ = surface tension of mixture, dyn/cm
$[P_i]$ = parachor of component $i$
$x_i, y_i$ = mole fraction of $i$ in liquid and vapor phases
$\rho_{Lm}$ = liquid mixture density, mol/cm$^3$
$\rho_{vm}$ = vapor mixture density, mol/cm$^3$

At low pressures, the term involving the vapor density and composition may be neglected; when this simplification is possible, Eq. (12-5.1) has been employed to correlate mixture surface tensions for a wide variety of organic liquids [6, 18, 27, 41, 56] with reasonably good results. Most authors, however, do not obtain $[P_i]$ from general group contribution tables such as Table 11-3; instead, they regress data to obtain the best value of $[P_i]$ for each component in the mixture. This same procedure has been used with success for gas-liquid systems under high pressure when the vapor term is significant. Weinaug and Katz [75] showed that Eq. (12-5.1) correlates methane-propane surface tensions from 258 to 363 K and from 2.7 to 103 bar. Deam and Mattox [13] also employed the same equation for the methane-nonane mixture from 239 to 297 K and 1 to 101 bar. Some smoothed data are shown in Fig. 12-3. At any temperature, $\sigma_m$ decreases with increasing pressure as more methane dissolves in the liquid phase. The effect of temperature is more unusual; instead of decreasing

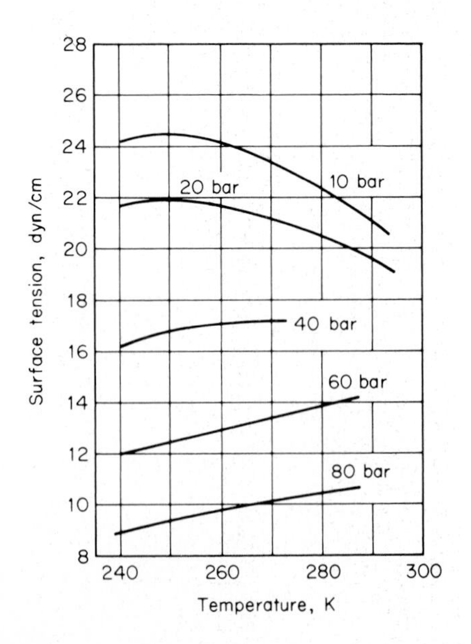


**Figure 12-3** Surface tension for the system methane-nonane.

with rising temperature, $\sigma_m$ increases, except at the lowest pressures. This phenomenon illustrates the fact that at the lower temperatures methane is more soluble in nonane and the effect of liquid composition is more important than the effect of temperature in determining $\sigma_m$.

When correlating these data with Eq. (12-5.1), Deam and Mattox found that the best results were obtained when $[P_{CH_4}] = 81.0$ and $[P_{C_9H_{20}}] = 387.6$. Note that if Table 11-3 is used, $[P_{CH_4}] = 71.0$ and $[P_{C_9H_{20}}] = 391$.

Other authors also have used Eq. (12-5.1) for correlating high-pressure surface tension data; e.g., Stegemeier† [66] studied the methane-pentane and methane-decane systems, Reno and Katz [52] studied the nitrogen-butane (and heptane) systems, and Lefrançois and Bourgeois [36] investigated the effect of rare-gas pressure on many organic liquids as well as the pressure of $N_2$ and $H_2$ on the surface tension of liquid ammonia.

When the Macleod-Sugden correlation is used, mixture liquid and vapor densities and compositions must be known. Errors at low pressures rarely exceed 5 to 10 percent and can be much less if $[P_i]$ values are obtained from experimental data.

### Corresponding states correlation

Equation (12-3.6) has seen little application for mixtures because the composition of the surface is not the same as that of the bulk. Some mixing rule has to be assumed for $P_{c_m}$, $T_{c_m}$, and $Q_m$. A number of possible rules can be obtained from Chap. 4, or simple mole fraction averages can be used. A limited test of the latter technique for nonpolar mixtures indicated that the accuracy was of the same order as that obtained with the Macleod-Sugden mixture relation with $[P_i]$ values from Table 11-3. Murad [45] and Rice and Teja [54] have used their corresponding states methods to estimate mixture surface tensions.

### Other empirical methods

Often, when only approximate estimates of $\sigma_m$ are necessary, one chooses the general form

$$\sigma_m^r = \sum_{j=1}^{n} x_j \sigma_j^r \tag{12-5.2}$$

Hadden [25] recommends $r = 1$ for most hydrocarbon mixtures,‡ but much closer agreement is found if $r = -1$ to $-3$. Equation (12-5.1) may

†In this case, the exponent chosen for $\sigma_m$ was $\frac{3}{11}$ rather than $\frac{1}{4}$.

‡In this reference, an empirical, *very* approximate method is suggested for mixtures when the temperature exceeds the critical temperature of at least one component.

also be written in a form similar to Eq. (12-5.2), using Eq. (12-3.1) to express the parachor $[P_i]$. Neglecting the vapor term (low pressures), we have

$$\sigma_m^{1/4} = \frac{\rho_{Lm} \sum_{i=1}^{n} x_i \sigma_i^{1/4}}{\rho_{Li}} \tag{12-5.3}$$

Equation (12-5.3) has an advantage over Eq. (12-5.1) in that, at the extremes of composition, it yields the correct values ($\sigma_m \rightarrow \sigma_i$ as $x_i \rightarrow 1.0$), whereas, depending upon the $[P_i]$ chosen, Eq. (12-5.1) may or may not reduce to the appropriate limiting value. The $\frac{1}{4}$ power on $\sigma_i$, however, is usually insufficient to provide the necessary curvature.

These empirical rules are illustrated in Example 12-3. The problem is that, with no theoretical basis, it is difficult to generalize, and although one rule may correlate data well for one system, it may fail on one quite similar. Figure 12-4 shows mixture surface tensions for several systems. All illustrate the nonlinearity of the $\sigma_m$-vs.-$x$ relation but to different degrees. The surface tension of the acetophenone-benzene system is almost linear in composition, whereas the nitromethane-benzene and nitrobenzene-carbon tetrachloride systems are decidedly nonlinear and the diethyl ether-benzene case is intermediate.

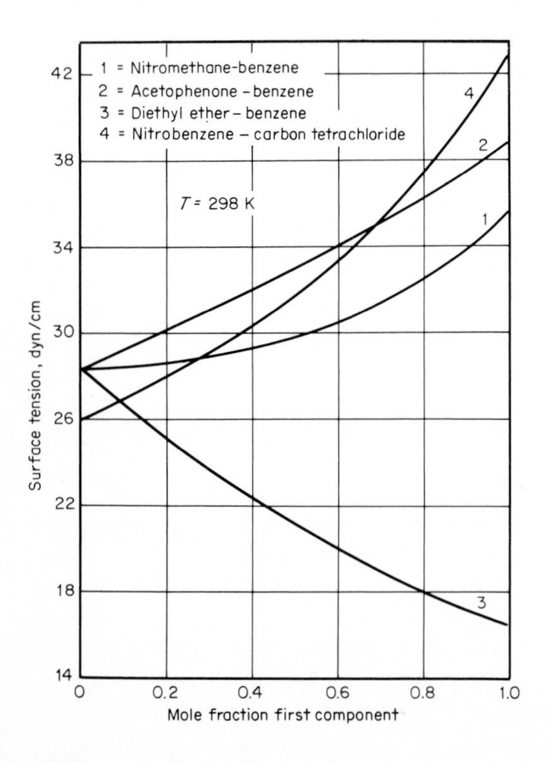


**Figure 12-4** Mixture surface tensions. *(From Ref. 27.)*

**Example 12-3** Estimate the surface tension of a mixture of diethyl ether and benzene containing 42.3 mole percent of the former. The temperature is 298 K. Hammick and Andrew [27] list densities and surface tensions for this system at 298 K; other data are from Appendix A.

| | Benzene | Diethyl ether | Mixture with 42.3 mol % diethyl ether |
|---|---|---|---|
| Density, g/cm$^3$ | 0.8722 | 0.7069 | 0.7996 |
| Surface tension, dyn/cm | 28.23 | 16.47 | 21.81 |
| Molecular weight | 78.114 | 74.123 | 76.426 |
| Critical temperature, K | 562.2 | 466.7 | |
| Critical pressure, bar | 48.9 | 36.4 | |
| Normal boiling point, K | 353.2 | 307.6 | |

**solution** MACLEOD-SUGDEN CORRELATION To employ Eq. (12-5.1), the parachors of benzene and diethyl ether are required. From Table 11-3,

$$[P_{\text{benzene}}] = C_6H_5{-} + H = 189.6 + 15.5 = 205.1$$

$$[P_{\text{ether}}] = (2)(CH_3{-}) + (2)({-}CH_2{-}) + {-}O{-}$$

$$= (2)(55.5) + (2)(40.0) + 20.0 = 211$$

$$M_m = (0.423)(74.123) + (0.577)(78.114) = 76.426$$

When the vapor term is neglected, Eq. (12-5.1) becomes

$$\sigma_m^{1/4} = \frac{0.7996}{76.426}[(0.423)(211) + (0.577)(205.1)]$$

$$\sigma_m = 22.25 \text{ dyn/cm}$$

$$\text{Error} = \frac{22.25 - 21.81}{21.81} \times 100 = 2.0\%$$

With the parachor values estimated above and with pure component densities, Eq. (12-3.1) yields surface tensions of 27.5 dyn/cm for pure benzene and 16.40 dyn/cm for pure diethyl ether, compared with experimental values of 28.23 and 16.47 dyn/cm, respectively.

MODIFIED MACLEOD-SUGDEN CORRELATION. With Eq. (12-5.3)

$$\sigma_m^{1/4} = \frac{0.7996}{76.426}\left[\frac{(0.423)(16.47)^{1/4}}{0.7069/74.123} + \frac{(0.577)(28.23)^{1/4}}{0.8722/78.114}\right]$$

$$\sigma_m = 22.63 \text{ dyn/cm}$$

$$\text{Error} = \frac{22.63 - 21.81}{21.81} \times 100 = 3.8\%$$

Although Eq. (12.5-3) yields the correct pure component values of $\sigma$, it invariably yields larger errors for $\sigma_m$ than Eq. (12-5.1).

CORRESPONDING STATES CORRELATION With Eqs. (12-3.6) and (12-3.7) applied to mixtures, assume, for simplicity, mole fraction averages for $P_{c_m}$, $T_{c_m}$, and $Q_m$:

$$P_{c_m} = (0.423)(36.4) + (0.577)(48.9) = 43.6 \text{ bar}$$

$$T_{c_m} = (0.423)(466.7) + (0.577)(562.2) = 522 \text{ K}$$

With $Q$ (diethyl ether) = 0.674 and $Q$ (benzene) = 0.628, we have

$$Q_m = (0.423)(0.674) + (0.577)(0.628) = 0.647$$

Then, with $T_{r_m} = (298)/522 = 0.571$, we get

$$\sigma_m = (43.6)^{2/3}(522)^{1/3}(0.647)(1 - 0.571)^{11/9}$$
$$= 22.97 \text{ dyn/cm}$$

$$\text{Error} = \frac{22.97 - 21.81}{21.81} \times 100 = 5.3\%$$

POWER RELATION. With Eq. (12-5.2), various values of $r$ can be selected. The results shown below indicate the errors.

| $r$ | Calculated $\sigma_m$, dyn/cm | % error |
|---|---|---|
| 1 | 23.25 | 6.6 |
| 0† | 22.48 | 3.1 |
| −1 | 21.68 | −0.6 |
| −2 | 20.92 | −4.0 |
| −3 | 20.19 | −7.4 |

†When $q = 0$, $\sigma_m = \sigma_1 \exp[x_2 \ln(\sigma_2/\sigma_1)]$.

## Thermodynamic correlations

The estimation procedures introduced earlier in this section are empirical; all employ the bulk liquid (and sometimes vapor) composition to characterize a mixture. However, the "surface phase" differs in composition from that of the bulk liquid and vapor, and it is reasonable to suppose that, in mixture surface tension relations, surface compositions play a more important role than bulk compositions. The fact that $\sigma_m$ is almost always less than the bulk mole fraction average is interpreted as indicating that the component or components with the lower pure component values of $\sigma$ preferentially concentrate in the surface phase. Eberhart [14] assumes that $\sigma_m$ is given by the surface composition–mole fraction average.

Both classical and statistical thermodynamics have been employed to derive expressions for $\sigma_m$ [5, 11, 15, 16, 28, 29, 30, 34, 59, 60, 62, 65, 69, 76]; the results differ in some aspects, but most arrive at a result similar to

$$\sum_{i=1}^{n} \left( \frac{x_i^B \gamma_i^B}{\gamma_i^\sigma} \right) \exp \frac{\mathcal{A}_i(\sigma_m - \sigma_i)}{RT} = 1 \qquad (12\text{-}5.4)$$

where $x_i^B$ = mole fraction of $i$ in bulk liquid
$\gamma_i^B$ = activity coefficient of $i$ in bulk liquid normalized so that $\gamma_i^B \to 1$ as $x_i \to 1$

$\gamma_i^\sigma$ = activity coefficient of $i$ in surface phase normalized so that $\gamma_i^\sigma \to 1$ as surface phase becomes identical with that for pure $i$
$\mathcal{A}_i$ = partial molar surface area of $i$, cm$^2$/mol
$\sigma_m, \sigma_i$ = surface tension of mixture and of component $i$

The activity coefficients $\gamma_i^B$ are usually obtained from vapor-liquid equilibrium data or from some liquid model, as discussed in Chap. 8. For the surface phase, however, no direct measurements appear possible, and a liquid model must be assumed. Hildebrand and Scott [29] treat the case of a "perfect" surface phase, $\gamma_i^\sigma = 1$, whereas Eckert and Prausnitz [15] and Sprow and Prausnitz [65] have used regular solution theory to describe the same phase. In all cases, the partial molal area $\mathcal{A}_i$ has been approximated as $(V_i)^{2/3}(N_0)^{1/3}$, where $V_i$ is the *pure liquid* molal volume of $i$ and $N_0$ is Avogadro's number. Sprow and Prausnitz have successfully applied Eq. (12-5.4) to a variety of nonpolar binary systems and have estimated $\sigma_m$ accurately. Even for polar mixtures, some success was achieved, although in this case a modified Wilson activity coefficient expression was introduced for the surface phase and one empirical constant was retained to give a good fit between experimental and calculated values of $\sigma_m$.

Although the Sprow-Prausnitz approach is more realistic, it is not uncommon to find Eq. (12-5.4) simplified by assuming an ideal-liquid mixture, that is, $\gamma_i^B = \gamma_i^\sigma = 1$. For a binary system, it can then be shown that this equation simplifies to

$$\sigma_m = x_A\sigma_A + x_B\sigma_B - \frac{\mathcal{A}}{2RT}(\sigma_A - \sigma_B)^2 x_A x_B \tag{12-5.5}$$

where the terms are as defined above (with $x_A$, $x_B$ bulk mole fractions) and $\mathcal{A}$ is an average surface area for the molecules constituting the system. This simplified form clearly indicates that $\sigma_m$ is less than a mole fraction average.

**Example 12-4** Repeat Example 12-3 by using the simplified (ideal) form of the thermodynamic correlation.

**solution** Let subscript A stand for diethyl ether and subscript B for benzene. Then $x_A = 0.423$, $\sigma_A = 16.47$, $x_B = 0.577$, and $\sigma_B = 28.23$, giving

$$RT = (8.314 \times 10^7)(298) = 2.478 \times 10^{10} \text{ ergs/mol K}$$

$$\mathcal{A}_A = \left(\frac{74.123}{0.7069}\right)^{2/3}(6.023 \times 10^{23})^{1/3} = 1.775 \times 10^9 \text{ cm}^2/\text{mol}$$

$$\mathcal{A}_B = \left(\frac{78.114}{0.8722}\right)^{2/3}(6.023 \times 10^{23})^{1/3} = 1.690 \times 10^9 \text{ cm}^2/\text{mol}$$

With $\mathcal{A} \approx (\mathcal{A}_A + \mathcal{A}_B)/2 = 1.743 \times 10^9$ cm$^2$/mol, with Eq. (12-5.5) we have

$$\sigma_m = (0.423)(16.47) + (0.577)(28.23)$$
$$- \frac{1.783 \times 10^9}{(2)(2.478 \times 10^{10})} \times (16.47 - 28.23)^2(0.423)(0.577)$$
$$= 22.03 \text{ dyn/cm}$$
$$\text{Error} = \frac{22.03 - 21.81}{21.81} \times 100 = 1\%$$

Although Example 12-4 shows but a small error when Eq. (12-5.5) is applied to the diethyl ether–benzene system at 298 K, this is more fortuitous than representative. The more accurate form, Eq. (12-5.4), is preferable, though considerably more effort must be expended to utilize it.

### Recommendations

For *nonpolar* systems, use Eq. (12-5.4) as prescribed by Sprow and Prausnitz; even polar systems may be treated if at least one mixture value of $\sigma_m$ is available. Errors are typically less than 2 to 3 percent. For less accurate estimates, there is little choice between the Macleod-Sugden correlation (12-5.1), the corresponding states correlation (12-3.6) (see Example 12-3), and the ideal thermodynamic relation (12-5.5). Errors are generally less than 5 to 10 percent.

For mixtures containing one or more polar components, the corresponding states method cannot be used; also, none of the thermodynamic methods is applicable without introducing a mixture constant. For polar (nonaqueous) systems, therefore, the Macleod-Sugden relation (12-5.1) is the only method available. When employed to correlate the mixture surface tension data of Ling and van Winkle [37], only moderate agreement (5 to 15 percent) was achieved for the polar-polar and polar-nonpolar systems studied.

## 12-6 Surface Tensions of Aqueous Solutions

Whereas for nonaqueous solutions the mixture surface tension is often approximated by a linear dependence on mole fraction, aqueous solutions show pronounced nonlinear characteristics. A typical case is shown in Fig. 12-5 for acetone-water at 353 K. The surface tension of the mixture is

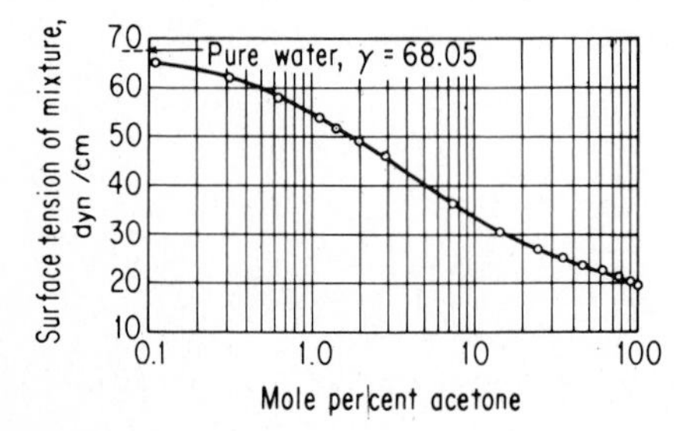


**Figure 12-5** Surface tensions of water-acetone solutions at 353 K. *(From Ref. 40.)*

represented by an approximately straight line on semilogarithmic coordinates. This behavior is typical of organic-aqueous systems, in which small concentrations of the organic material may significantly affect the mixture surface tension. The hydrocarbon portion of the organic molecule behaves like a hydrophobic material and tends to be rejected from the water phase by preferentially concentrating on the surface. In such a case, the bulk concentration is very different from the surface concentration. Unfortunately, the latter is not easily amenable to direct measurement. Meissner and Michaels [41] show graphs similar to Fig. 12-5 for a variety of dilute solutions of organic materials in water and suggest that the general behavior is approximated by the Szyszkowski equation, which they modify to the form

$$\frac{\sigma_m}{\sigma_W} = 1 - 0.411 \log\left(1 + \frac{x}{a}\right) \tag{12-6.1}$$

where $\sigma_W$ = surface tension of pure water
$x$ = mole fraction of organic material
$a$ = constant characteristic of organic material

Values of $a$ are listed in Table 12-2 for a few compounds. This equation should not be used if the mole fraction of the organic solute exceeds 0.01.

The method of Tamura, Kurata, and Odani [70] may be used to estimate surface tensions of aqueous binary mixtures over wide concentration ranges of the dissolved organic material and for both low- and high-molecular weight organic-aqueous systems. Equation (12-5.1) is assumed as a

**TABLE 12-2 Constants for the Szyszkowski Equation (12-6.1) [41]**

| Compound | $a \times 10^4$ | Compound | $a \times 10^4$ |
|---|---|---|---|
| Propionic acid | 26 | Ethyl propionate | 3.1 |
| *n*-Propyl alcohol | 26 | Propyl acetate | 3.1 |
| Isopropyl alcohol | 26 | | |
| Methyl acetate | 26 | *n*-Valeric acid | 1.7 |
| | | Isovaleric acid | 1.7 |
| *n*-Propyl amine | 19 | *n*-Amyl alcohol | 1.7 |
| Methyl ethyl ketone | 19 | Isoamyl alcohol | 1.7 |
| *n*-Butyric acid | 7.0 | Propyl propionate | 1.0 |
| Isobutyric acid | 7.0 | *n*-Caproic acid | 0.75 |
| *n*-Butyl alcohol | 7.0 | *n*-Heptanoic acid | 0.17 |
| Isobutyl alcohol | 7.0 | *n*-Octanoic acid | 0.034 |
| | | *n*-Decanoic acid | 0.0025 |
| Propyl formate | 8.5 | | |
| Ethyl acetate | 8.5 | | |
| Methyl propionate | 8.5 | | |
| Diethyl ketone | 8.5 | | |

starting point, but the significant densities and concentrations are taken to be those characteristic of the surface layer, that is, $(V^\sigma)^{-1}$ replaces $\rho_{Lm}$, where $V^\sigma$ is a hypothetical molal volume of the surface layer. $V^\sigma$ is estimated with

$$V^\sigma = \sum_j x_j^\sigma V_j \tag{12-6.2}$$

where $x_j^\sigma$ is the mole fraction of $j$ in the surface layer. $V_j$, however, is chosen as the pure liquid molal volume of $j$. Then, with Eq. (12-5.1), assuming $\rho_L \gg \rho_v$,

$$V^\sigma \sigma_m^{1/4} = x_W^\sigma[P_W] + x_O^\sigma[P_O] \tag{12-6.3}$$

where the subscripts W and O represent water and the organic component. To eliminate the parachor, however, Tamura et al. introduce Eq. (12-3.1); the result is

$$\sigma_m^{1/4} = \psi_W^\sigma \sigma_W^{1/4} + \psi_O^\sigma \sigma_O^{1/4} \tag{12-6.4}$$

In Eq. (12-6.4), $\psi_W^\sigma$ is the superficial volume fraction water in the surface layer

$$\psi_W^\sigma = \frac{x_W^\sigma V_W}{V^\sigma} \tag{12-6.5}$$

and similarly for $\psi_O^\sigma$.

Equation (12-6.4) is the final correlation. To obtain values of the superficial surface volume fractions $\psi_W^\sigma$ and $\psi_O^\sigma$, equilibrium is assumed between the surface and bulk phases. Tamura's equation is complex, and after rearrangement it can be written in the following set of equations:

$$\mathcal{B} = \log \frac{\psi_W{}^q}{\psi_O} \tag{12-6.6}$$

$$\mathcal{C} = \log \frac{(\psi_W^\sigma)^q}{\psi_O^\sigma} \tag{12-6.7}$$

$$\mathcal{C} = \mathcal{B} + \mathcal{W} \tag{12-6.8}$$

$$\mathcal{W} = 0.441 \frac{q}{T}\left(\frac{\sigma_O V_O^{2/3}}{q} - \sigma_W V_W^{2/3}\right) \tag{12-6.9}$$

where $\psi_W^\sigma$ is defined by Eq. (12-6.5) and $\psi_W$, $\psi_O$ are the superficial bulk volume fractions of water and organic material, i.e.,

| Materials | $q$ | Example |
|---|---|---|
| Fatty acids, alcohols | Number of carbon atoms | Acetic acids, $q = 2$ |
| Ketones | One less than the number of carbon atoms | Acetone, $q = 2$ |
| Halogen derivatives of fatty acids | Number of carbons times ratio of molal volume of halogen derivative to parent fatty acid | Chloroacetic acid, $q = 2\dfrac{V_b \text{ (chloroacetic acid)}}{V_b \text{ (acetic acid)}}$ |

$$\psi_W = \frac{x_W V_W}{x_W V_W + x_O V_O} \qquad \psi_O = \frac{x_O V_O}{x_W V_W + x_O V_O} \tag{12-6.10}$$

where $x_W, x_O$ = bulk mole fraction of pure water and pure organic component
$V_W, V_O$ = molal volume of pure water and pure organic component
$\sigma_W, \sigma_O$ = surface tension of pure water and pure organic component
$T$ = temperature, K
$q$ = constant depending upon type and size of organic constituent

The method is illustrated in Example 12-5. Tamura et al. [70] tested the method with some 14 aqueous systems and 2 alcohol-alcohol systems;† the percentage errors are less than 10 percent when $q$ is less than 5 and within 20 percent for $q$ greater than 5. The method cannot be applied to multicomponent mixtures.

**Example 12-5** Estimate the surface tension of a mixture of methyl alcohol and water at 303K when the mole fraction alcohol is 0.122. The experimental value reported is 46.1 dyn/cm [70].

**solution** At 303 K (O represents methyl alcohol, W water), $\sigma_W$ = 71.18 dyn/cm, $\sigma_O$ = 21.75 dyn/cm, $V_W$ = 18 cm$^3$/mol, $V_O$ = 41 cm$^3$/mol, and $q$ = number of carbon atoms = 1. From Eqs. (12-6.10),

$$\frac{\psi_W}{\psi_O} = \frac{(0.878)(18)}{(0.122)(41)} = 3.16$$

and from Eq. (12-6.6),

$$\mathcal{B} = \log 3.16 = 0.50$$

$$\mathcal{W} \text{ [from Eq. (12-6.9)]} = (0.441)\left(\tfrac{1}{303}\right)[(21.75)(41)^{2/3} - (71.18)(18)^{2/3}]$$

$$= -0.34$$

†For nonaqueous mixtures comprising polar molecules, the method is unchanged except that $q$ = ratio of molal volumes of the solute to solvent.

Hence

$$\mathcal{C} \text{ [from Eq. (12-6.8)]} = \mathcal{B} + \mathcal{W} = 0.50 - 0.34 = 0.16$$

$$= \log \frac{\psi_W^\sigma}{\psi_O^\sigma} \qquad \text{from Eq. (12-6.7)}$$

Since $\psi_W^\sigma + \psi_O^\sigma = 1$, we have

$$\frac{\psi_W^\sigma}{1 - \psi_W^\sigma} = 10^{0.16} = 1.45$$

$$\psi_W^\sigma = 0.59 \qquad \psi_O^\sigma = 0.41$$

Finally, from Eq. (12-6.4)

$$\sigma_m = [(0.59)(71.18)^{1/4} + (0.41)(21.75)^{1/4}]^4 = 46 \text{ dyn/cm}$$

$$\text{Error} = \frac{46 - 46.1}{46.1} \times 100 = -0.2\%$$

## Recommendations

For estimating the surface tensions of binary organic-aqueous mixtures, use the method of Tamura, Kurata, and Odani, as given by Eqs. (12-6.2) to (12-6.10) and illustrated in Example 12-5. The method may also be used for other highly polar solutes such as alcohols. Errors normally do not exceed 10 to 15 percent. If the solubility of the organic compound in water is low, the Szyszkowski equation (12-6.1), as developed by Meissner and Michaels, may be used.

## Notation

| | |
|---|---|
| $a$ | parameter in Eq. (12-6.1) and obtained from Table 12-2 |
| $A$ | area |
| $\mathcal{A}$ | surface area; $\mathcal{A}_i$, partial molal area of component $i$ on the surface |
| $\mathcal{B}$ | parameter in Eq. (12-6.6) |
| $\mathcal{C}$ | parameter in Eq. (12-6.7) |
| $m$ | parameter in Eq. (12-3.8) |
| $n$ | parameter in Eq. (12-3.2) |
| $N_0$ | Avogadro's number |
| $p$ | parameter in Eq. (12-3.10) |
| $[P_i]$ | parachor of component $i$ (see Table 11-3) |
| $P_{vp}$ | vapor pressure, bar |
| $P_c$ | critical pressure, bar; $P_{cm}$, pseudocritical pressure |
| $q$ | parameter in Eq. (12-6.9) |
| $Q$ | parameter in Eq. (12-3.7) |

$Q_p$ parameter in Eq. (12-3.8)

$r$ parameter in Eq. (12-5.2)

$R$ gas constant

$T$ temperature, $K$; $T_c$, critical temperature; $T_b$, normal boiling point, $T_r$, reduced temperature $T/T_c$; $T_{br}$, $T_b/T_c$; $T_{cm}$, pseudocritical mixture temperature

$V$ liquid molal volume, $cm^3/mol$; $V^\sigma$, for the surface phase

$V_c$ critical volume, $cm^3/mol$

$\mathcal{W}$ parameter in Eq. (12-6.9)

$x_i$ liquid mole fraction; $x_i^B$, in the bulk phase; $x_i^\sigma$, in the surface phase

$y_i$ vapor mole fraction

GREEK

$\alpha_c$ Riedel factor

$\gamma_i$ activity coefficient of component $i$; $\gamma_i^B$, in the bulk liquid phase; $\gamma_i^\sigma$, in the surface phase

$\eta$ liquid or vapor viscosity, cP

$\rho$ liquid or vapor density, $mol/cm^3$

$\sigma$ surface tension, dyn/cm; $\sigma_m$, for a mixture; $\sigma_O$, a parameter in Eq. (12-3.10) or representing an organic component

$\psi_i$ volume fraction of $i$ in the bulk liquid; $\psi_i^\sigma$, in the surface phase

SUBSCRIPTS

$b$ normal boiling point

$L$ liquid

$m$ mixture

O organic component in aqueous solution

$r$ reduced value, i.e., the property divided by its value at the critical point

$v$ vapor

W water

## REFERENCES

1. Adamson, A. W.: *Physical Chemistry of Surfaces,* Interscience, New York, 1960.
2. Agarwal, D. K., R. Gopal, and S. Agarwal: *J. Chem. Eng. Data,* **24:** 181 (1979).
3. Aveyard, R., and D. A. Haden: *An Introduction to Principles of Surface Chemistry,* Cambridge, London, 1973.
4. Barbulescu, N.: *Rev. Roum. Chem.,* **19:** 169 (1974).
5. Belton, J. W., and M. G. Evans: *Trans. Faraday Soc.,* **41:** 1 (1945).
6. Bowden, S. T., and E. T. Butler: *J. Chem. Soc.,* **1939:** 79.
7. Brock, J. R., and R. B. Bird: *AIChE J.,* **1:** 174 (1955).

8. Brown, R. C.: *Contemp. Phys.*, **15:** 301 (1974).
9. Buehler, C. A.: *J. Phys. Chem.*, **42:** 1207 (1938).
10. Carey, B. S., L. E. Scriven, and H. T. Davis: *AIChE J.*, **24:** 1076 (1978).
11. Clever, H. L., and W. E. Chase, Jr.: *J. Chem. Eng. Data*, **8:** 291 (1963).
12. Cotton, D. J.: *J. Phys. Chem.*, **73:** 270 (1969).
13. Deam, J. R., and R. N. Mattox: *J. Chem. Eng. Data*, **15:** 216 (1970).
14. Eberhart, J. G.: *J. Phys. Chem.*, **70:** 1183 (1966).
15. Eckert, C. A., and J. M. Prausnitz: *AIChE J.*, **10:** 677 (1964).
16. Evans, H. B., Jr., and H. L. Clever: *J. Phys. Chem.*, **68:** 3433 (1964).
17. Fishtine, S. H.: *Ind. Eng. Chem. Fundam.*, **2:** 149 (1963).
18. Gambill, W. R.: *Chem. Eng.*, **64**(5): 143 (1958).
19. Gambill, W. R.: *Chem. Eng.*, **66**(23): 193 (1959).
20. Gibbs, J. W.: *The Collected Works of J. Willard Gibbs*, vol. I, *Thermodynamics*, Yale University Press, New Haven, Conn., 1957.
21. Goldhammer, D. A.: *Z. Phys. Chem.*, **71:** 577 (1910).
22. Gray, J. A., C. J. Brady, J. R. Cunningham, J. R. Freeman, and G. M. Wilson: *Ind. Eng. Chem. Process Des. Dev.*, **22:** 410 (1983).
23. Guggenheim, E. A.: *Thermodynamics*, 4th ed., North-Holland, Amsterdam, 1959.
24. Guggenheim, E. A.: *Proc. Phys. Soc.*, **85:** 811 (1965).
25. Hadden, S. T.: *Hydrocarbon Process Petrol. Refiner*, **45**(10): 161 (1966).
26. Hakim, D. I., D. Steinberg, and L. I. Stiel: *Ind. Eng. Chem. Fundam.*, **10:** 174 (1971).
27. Hammick, D. L., and L. W. Andrew: *J. Chem. Soc.*, **1929:** 754.
28. Hansen, R. S., and L. Sogor: *J. Colloid Interface Sci.*, **40:** 424 (1972).
29. Hildebrand, J. H., and R. L. Scott: *The Solubility of Nonelectrolytes*, 3d ed., Dover, New York, 1964, chap. 21.
30. Hoar, T. P., and D. A. Melford: *Trans. Faraday Soc.*, **53:** 315 (1957).
31. *International Critical Tables*, vol. III, McGraw-Hill, New York, 1928, p. 28.
32. Jasper, J. J.: *J. Phys. Chem. Ref. Data*, **1:** 841 (1972).
33. Keeney, M., and J. Heicklen: *J. Inorg. Nucl. Chem.*, **41:** 1755 (1979).
34. Kim, S. W., M. S. Jhon, T. Ree, and H. Eyring: *Proc. Natl. Acad. Sci.*, **59:** 336 (1968).
35. Körösi, G., and E. sz. Kováts: *J. Chem. Eng. Data*, **26:** 323 (1981).
36. Lefrançois, B., and Y. Bourgeois: *Chim. Ind. Genie Chim.*, **105**(15): 989 (1972).
37. Ling, T. D., and M. van Winkle: *J. Chem. Eng. Data*, **3:** 88 (1958).
38. Macleod, D. B.: *Trans. Faraday Soc.*, **19:** 38 (1923).
39. Mayer, S. W.: *J. Phys. Chem.*, **67:** 2160 (1963).
40. McAllister, R. A., and K. S. Howard: *AIChE J.*, **3:** 325 (1957).
41. Meissner, H. P., and A. S. Michaels: *Ind. Eng. Chem.*, **41:** 2782 (1949).
42. Miller, D. G.: *Ind. Eng. Chem. Fundam.*, **2:** 78 (1963).
43. Modell, M., and R. C. Reid: *Thermodynamics and Its Applications*, 2d ed., Prentice-Hall, Englewood Cliffs, N.J., 1983.
44. Móritz, P.: *Period. Polytech.*, **3:** 167 (1959).
45. Murad, S: *Chem. Eng. Commun.*, **24:** 353 (1983).
46. Othmer, D. F., S. Josefowitz, and Q. F. Schmutzler, *Ind. Eng. Chem.*, **40:** 886 (1948).
47. Pelofsky, A. H.: *J. Chem. Eng. Data*, **11:** 394 (1966).
48. Plesner, I. W., and O. Platz: *J. Chem. Phys.*, **48:** 5361, 5364 (1968).
49. Porteous, W.: *J. Chem. Eng. Data*, **20:** 339 (1975).
50. Quale, O. R.: *Chem. Rev.*, **53:** 439 (1953).
51. Rao, M. V., K. A. Reddy, and S. R. S. Sastri: *Petrol. Refiner*, **47**(1): 151 (1958).
52. Reno, G. J., and D. L. Katz: *Ind. Eng. Chem.*, **35:** 1091 (1943).
53. Rice, O. K.: *J. Phys. Chem.*, **64:** 976 (1960).
54. Rice, P., and A. S. Teja: *J. Colloid. Interface Sci.*, **86:** 158 (1982).
55. Riedel, L.: *Chem. Ing. Tech.*, **26:** 83 (1954).
56. Riedel, L.: *Chem. Ing. Tech.*, **27:** 209 (1955).
57. Ross, S. (Chairman): *Chemistry and Physics of Interfaces*, American Chemical Society, Washington, D.C., 1965.
58. Sanchez, I. C.: *J. Chem. Phys.*, **79:** 405 (1983).
59. Schmidt, R. L.: *J. Phys. Chem.*, **71:** 1152 (1967).

60. Schmidt, R. L., J. C. Randall, and H. L. Clever: *J. Phys. Chem.*, **70:** 3912 (1966).
61. Schonhorn, H.: *J. Chem. Eng. Data,* **12:** 524 (1967).
62. Shereshefsky, J. L.: *J. Colloid. Interface Sci.*, **24:** 317 (1967).
63. Sivaraman, A., J. Zega, and R. Kobayashi: submitted to *Fluid Phase Equil.* (1985).
64. Sprow, F. B., and J. M. Prausnitz: *Trans. Faraday Soc.*, **62:** 1097 (1966).
65. Sprow, F. B., and J. M. Prausnitz: *Trans. Faraday Soc.*, **62:** 1105 (1966); *Can. J. Chem. Eng.*, **45:** 25 (1967).
66. Stegemeier, G. L.: Ph.D. dissertation, University of Texas, Austin, Tex., 1959.
67. Sugden, S.: *J. Chem. Soc.*, **1924:** 32.
68. Sugden, S.: *J. Chem. Soc.*, **1924:** 1177.
69. Suri, S. K., and V. Ramkrishna: *J. Phys. Chem.*, **72:** 3073 (1968).
70. Tamura, M., M. Kurata, and H. Odani: *Bull. Chem. Soc. Japan,* **28:** 83 (1955).
71. Tripathi, R. C.: *J. Indian Chem. Soc.*, **18:** 411 (1941).
72. Tripathi, R. C.: *J. Indian Chem. Soc.*, **19:** 51 (1942).
73. van der Waals, J. D.: *Z. Phys. Chem.*, **13:** 716 (1894).
74. Watkinson, A. P., and J. Lielmezs: *J. Chem. Phys.*, **47:** 1558 (1967).
75. Weinaug, C. F., and D. L. Katz: *Ind. Eng. Chem.*, **35:** 239 (1943).
76. Winterfeld, P. H., L. E. Scriven, and H. T. Davis: *AIChE J.*, **24:** 1010 (1978).
77. Zihao, W., and F. Jufu: "An Equation for Estimating Surface Tension of Liquid Mixtures," *Proc. Jt. Mtg. of Chem. Ind. and Eng. Soc. of China and AIChE, Beijing, China, September 1982.*

Appendix

# A

# Property Data Bank

The listing of compounds is by the total carbon number. Within each carbon number class, subgroups are indexed by the number of hydrogens and, further, by additional atoms in alphabetical order.

To the best of our knowledge, all tabulated data are experimental; no estimation results are included.

The symbols and equations used are shown below. The enthalpy and Gibbs energy of formation at 298.2 K (DELHF and DELGF) are for the *ideal-gas state.* The reference states chosen for the elements are as follows:

Ideal gases at one atmosphere: Ar, $Cl_2$, $D_2$, $F_2$, He, $H_2$, Kr, Ne, $O_2$, Rn, $T_2$, and Xe. Al (crystal); As (crystal); B (crystal); $Br_2$ (liquid); C (graphite); Hg (liquid); $I_2$ (crystal); P (solid, red); S (crystal, rhombic); Se (crystal); Si (crystal); Ti (crystal, alpha); U (crystal).

MolWt = molecular weight, g/mol
Tfp = normal freezing point, K
Tb = normal boiling point (at 1 atm), K
Tc = critical temperature, K
Pc = critical pressure, bar
Vc = critical volume, $cm^3$/mole
Zc = critical compressibility factor, PcVc/RTc
Omega = Pitzer's acentric factor
Dipm = dipole moment, debyes

CPVAP A, CPVAP B, CPVAP C, CPVAP D = constants to calculate the isobaric heat capacity of the ideal gas, with Cp in J/(mol·K) and T in kelvins:

$$Cp = CPVAP\ A + (CPVAP\ B)T + (CPVAP\ C)T^2 + (CPVAP\ D)T^3$$

DELHF = standard enthalpy of formation for the ideal gas at 298.2 K, J/ mol
(See note above on reference states.)

DELGF = standard Gibbs energy of formation for the ideal gas at 298.2 K and 1 atm, J/ mol
(See note above on reference states.)

Vapor pressure: Pvp = vapor pressure, in bars
Pc = critical pressure, in bars
Tc = critical temperature, in kelvins
T = temperature, in kelvins

There is a choice of equations as noted in the tables:

Equation (1):

$$\ln (Pvp/Pc) = (1 - x)^{-1}[(VP\ A)x + (VP\ B)x^{1.5} + (VP\ C)x^3 + (VP\ D)x^6]$$

$$x = 1 - T/Tc$$

Equation (2):

$$\ln Pvp = VP\ A - (VP\ B)/T + (VP\ C) \ln T + (VP\ D)(Pvp)/T^2$$

Equation (3):

$$\ln Pvp = VP\ A - (VP\ B)/[T + (VP\ C)]$$

LDEN = liquid density, g/cm$^3$
TDEN = temperature for LDEN, K

| No | Formula | Name | MolWt | Tfp K | Tb K | Tc K | Pc bar | Vc $cm^3$/mole | Zc | Omega | Dipm debye |
|---|---|---|---|---|---|---|---|---|---|---|---|
| 1 | AlBr3 | aluminum tribromide | 266.694 | 370.7 | 528. | 763. | 28.9 | 310. | 0.141 | 0.399 | 5.0 |
| 2 | AlCl3 | aluminum trichloride | 133.341 | 467. | 453. | 620. | 26.3 | 259. | 0.132 | 0.660 | 2.0 |
| 3 | AlI3 | aluminum triiodide | 407.697 | 464. | 655. | 983. | | 408. | | | 2.3 |
| 4 | Ar | argon | 39.948 | 83.8 | 87.3 | 150.8 | 48.7 | 74.9 | 0.291 | 0.001 | 0.0 |
| 5 | As | arsenic | 74.922 | | 888. | 1673. | 223. | 34.9 | 0.056 | 0.121 | |
| 6 | AsCl3 | arsenic trichloride | 181.281 | 264.7 | 403. | 654. | | 252. | | | 1.6 |
| 7 | BBr3 | boron tribromide | 250.568 | 227. | 364. | 581. | | 272. | | | 0.0 |
| 8 | BCl3 | boron trichloride | 117.191 | 165.9 | 285.8 | 455. | 38.7 | 239.5 | 0.245 | 0.140 | 0.0 |
| 9 | BF3 | boron trifluoride | 67.805 | 146.5 | 172. | 260.8 | 49.9 | 114.7 | 0.264 | 0.393 | 0.0 |
| 10 | BI3 | boron triiodide | 391.55 | 323.1 | 483. | 773. | | 356. | | | |
| 11 | Br2 | bromine | 159.808 | 266.0 | 331.9 | 588. | 103. | 127.2 | 0.268 | 0.108 | 0.2 |
| 12 | BrI | iodine bromide | 206.813 | 315. | 389. | 719. | | 139. | | | 1.2 |
| 13 | Br3P | phosphorus tribromide | 270.723 | 233. | 446.1 | 711. | | 300. | | | 0.5 |
| 14 | Br4Si | silicon tetrabromide | 347.702 | 278.6 | 427. | 663. | | 382. | | | 0.0 |
| 15 | Br4Ti | titanium tetrabromide | 367.536 | 312. | 503. | 795.7 | | 391. | | | |
| 16 | ClFO3 | perchloryl fluoride | 102.448 | 125.5 | 226.4 | 368.4 | 53.7 | 160.8 | 0.282 | 0.170 | 0.0 |
| 17 | ClF2N | nitrogen chloride difluoride | 87.456 | | 207. | 337.5 | 51.5 | | | 0.154 | |
| 18 | ClF2P | phosphorus chloride difluoride | 104.423 | | 225.9 | 362.4 | 45.2 | | | 0.164 | |
| 19 | ClF2PS | thiophosphoryl chloride difluoride | 136.489 | | 279. | 439.2 | 41.4 | | | 0.202 | |
| 20 | ClF5 | chlorine pentafluoride | 130.443 | | 260.0 | 416. | 52.7 | 233. | 0.355 | 0.216 | |
| 21 | ClNO | nitrosyl chloride | 65.459 | 213.5 | 267.7 | 440. | | | | | 1.8 |
| 22 | Cl2 | chlorine | 70.906 | 172.2 | 239.2 | 416.9 | 79.8 | 123.8 | 0.285 | 0.090 | 0.0 |
| 23 | Cl2FP | phosphorus dichloride fluoride | 120.878 | | 287.0 | 463.0 | 49.6 | | | 0.174 | |
| 24 | Cl3P | phosphorus trichloride | 137.333 | 161. | 349.1 | 563. | | 264. | | | 0.9 |
| 25 | Cl4Si | silicon tetrachloride | 169.898 | 204.3 | 330.8 | 508.1 | 35.9 | 325.7 | 0.277 | 0.232 | 0.0 |

| No | Formula | Name | CPVAP A | CPVAP B | CPVAP C | CPVAP D | DELHF | DELGF |
|---|---|---|---|---|---|---|---|---|
| 1 | AlBr3 | aluminum tribromide | 6.494E+1 | 6.098E-2 | -7.306E-5 | 2.978E-8 | -4.233E+5 | -4.522E+5 |
| 2 | AlCl3 | aluminum trichloride | 5.054E+1 | 1.037E-1 | -1.202E-4 | 4.793E-8 | -5.849E+5 | -5.704E+5 |
| 3 | AlI3 | aluminum triiodide | 6.270E+1 | 6.802E-2 | -8.113E-5 | 3.298E-8 | -2.052E+5 | -2.531E+5 |
| 4 | Ar | argon | 2.080E+1 | | | | 0.0 | 0.0 |
| 5 | As | arsenic | | | | | | |
| 6 | AsCl3 | arsenic trichloride | | | | | | |
| 7 | BBr3 | boron tribromide | 4.331E+1 | 1.160E-1 | -1.267E-4 | 4.849E-8 | -2.043E+5 | -2.312E+5 |
| 8 | BCl3 | boron trichloride | 3.261E+1 | 1.390E-1 | -1.461E-4 | 5.439E-8 | -4.032E+5 | -3.882E+5 |
| 9 | BF3 | boron trifluoride | 1.858E+1 | 1.399E-1 | -1.217E-4 | 3.916E-8 | -1.136E+6 | -1.120E+6 |
| 10 | BI3 | boron triiodide | 4.937E+1 | 1.028E-1 | -1.159E-4 | 4.529E-8 | 7.118E+4 | 2.089E+4 |
| 11 | Br2 | bromine | 3.386E+1 | 1.125E-2 | -1.192E-5 | 4.534E-9 | 3.093E+4 | 3.136E+3 |
| 12 | BrI | iodine bromide | 3.402E+1 | 1.229E-2 | -1.420E-5 | 5.847E-9 | 4.091E+5 | 3.714E+3 |
| 13 | Br3P | phosphorus tribromide | 6.102E+1 | 7.421E-2 | -8.899E-5 | 3.631E-8 | -1.285E+5 | -1.575E+5 |
| 14 | Br4Si | silicon tetrabromide | 7.466E+1 | 1.097E-1 | -1.298E-4 | 5.246E-8 | -4.159E+5 | -4.324E+5 |
| 15 | Br4Ti | titanium tetrabromide | 8.499E+1 | 7.785E-2 | -9.361E-5 | 3.826E-8 | -5.506E+5 | -5.695E+5 |
| 16 | ClFO3 | perchloryl fluoride | 1.245E+1 | 2.390E-1 | -2.346E-4 | 8.321E-8 | -2.144E+4 | 5.062E+4 |
| 17 | ClF2N | nitrogen chloride difluoride | | | | | | |
| 18 | ClF2P | phosphorus chloride difluoride | | | | | | |
| 19 | ClF2PS | thiophosphoryl chloride difluoride | | | | | | |
| 20 | ClF5 | chlorine pentafluoride | 3.098E+1 | 3.203E-1 | -3.685E-4 | 1.462E-7 | -2.386E+5 | -1.469E+5 |
| 21 | ClNO | nitrosyl chloride | 3.410E+1 | 4.472E-2 | -3.340E-5 | 1.015E-8 | 5.263E+4 | 6.699E+4 |
| 22 | Cl2 | chlorine | 2.693E+1 | 3.384E-2 | -3.869E-5 | 1.547E-8 | 0.0 | 0.0 |
| 23 | Cl2FP | phosphorus dichloride fluoride | | | | | | |
| 24 | Cl3P | phosphorus trichloride | 4.849E+1 | 1.131E-1 | -1.334E-4 | 5.380E-8 | -2.713E+5 | -2.577E+5 |
| 25 | Cl4Si | silicon tetrachloride | 5.658E+1 | 1.636E-1 | -1.897E-4 | 7.565E-8 | -6.577E+5 | -6.178E+5 |

| No | Formula | Name | Eq. | VP A | VP B | VP C | VP D | Tmin | Tmax | LDEN | TDEN |
|---|---|---|---|---|---|---|---|---|---|---|---|
| 1 | AlBr3 | aluminum tribromide | | | | | | | | | |
| 2 | AlCl3 | aluminum trichloride | | | | | | | | 1.31 | 473 |
| 3 | AlI3 | aluminum triiodide | | | | | | | | | |
| 4 | Ar | argon | 1 | -5.90501 | 1.12627 | -0.76787 | -1.62721 | 84 | TC | 1.373 | 90 |
| 5 | As | arsenic | | | | | | | | | |
| 6 | AsCl3 | arsenic trichloride | | | | | | | | 2.163 | 293 |
| 7 | BBr3 | boron tribromide | | | | | | | | 2.643 | 291 |
| 8 | BCl3 | boron trichloride | 2 | 46.103 | 4443.16 | -5.404 | 2228. | 230 | TC | 1.349 | 284 |
| 9 | BF3 | boron trifluoride | 2 | 61.138 | 3481.19 | -7.963 | 576. | 160 | TC | 2.811 | 293 |
| 10 | BI3 | boron triiodide | | | | | | | | 3.35 | 323 |
| 11 | Br2 | bromine | 3 | 9.2239 | 2582.32 | -51.56 | | 259 | 354 | 3.119 | 293 |
| 12 | BrI | iodine bromide | | | | | | | | | |
| 13 | Br3P | phosphorous tribromide | | | | | | | | 2.852 | 288 |
| 14 | Br4Si | silicon tetrabromide | | | | | | | | 2.772 | 298 |
| 15 | Br4Ti | titanium tetrabromide | | | | | | | | | |
| 16 | ClFO3 | perchloryl fluoride | | | | | | | | 2.003 | 399 |
| 17 | ClF2N | nitrogen chloride difluoride | | | | | | | | | |
| 18 | ClF2P | phosphorous chloride difluoride | | | | | | | | | |
| 19 | ClF2PS | thiophosphoryl chloride difluoride | | | | | | | | | |
| 20 | ClF5 | chlorine pentafluoride | | | | | | | | | |
| 21 | ClNO | nitrosyl chloride | 2 | 29.760 | 3748.59 | -2.819 | 900. | 230 | TC | 1.42 | 261 |
| 22 | Cl2 | chlorine | 1 | -6.34074 | 1.15037 | -1.40416 | -2.23220 | 206 | TC | 1.563 | 239 |
| 23 | Cl2FP | phosphorous dichloride fluoride | | | | | | | | | |
| 24 | Cl3 P | phosphorus trichloride | | | | | | | | 1.574 | 294 |
| 25 | Cl4Si | silicon tetrachloride | 3 | 9.1817 | 2634.16 | -43.15 | | 238 | 364 | 1.48 | 293 |

| No | Formula | Name | MolWt | Tfp K | Tb K | Tc K | Pc bar | Vc $cm^3/mol$ | Zc | Omega | Dipm debye |
|---|---|---|---|---|---|---|---|---|---|---|---|
| 26 | Cl4Ti | titanium tetrachloride | 189.712 | 243. | 409.6 | 638. | 46.6 | 339.2 | 0.298 | 0.268 | 0.0 |
| 27 | Cl5P | phosphorus pentachloride | 208.260 | 148. | 433. | 646. | | | | | 0.8 |
| 28 | D2 | deuterium (equilibrium) | 4.032 | 18.7 | 23.6 | 38.2 | 16.5 | 60.3 | 0.313 | -0.137 | 0.0 |
| 29 | D2 | deuterium (normal) | 4.032 | 18.6 | 23.5 | 38.4 | 16.6 | | | -0.160 | 0.0 |
| 30 | D2O | deuterium oxide | 20.031 | 277.0 | 374.6 | 644.0 | 216.6 | 56.6 | 0.225 | 0.351 | 1.9 |
| 31 | FNO2 | nitryl fluoride | 65.003 | | 213.2 | 349.5 | | | | | 0.5 |
| 32 | F2 | fluorine | 37.997 | 53.5 | 85.0 | 144.3 | 52.2 | 66.3 | 0.288 | 0.054 | 0.0 |
| 33 | F2N2 | cis-difluorodiazine | 66.010 | | 167.5 | 272. | 70.9 | | | 0.252 | |
| 34 | F2N2 | trans-difluorodiazine | 66.010 | | 161.7 | 260. | 55.7 | | | 0.217 | |
| 35 | F2O | oxygen difluoride | 53.995 | 50. | 128.4 | 215. | 49.6 | | | | 0.2 |
| 36 | F2Xe | xenon difluoride | 169.296 | | 387.5 | 631. | 93.2 | 148.6 | 0.264 | 0.317 | |
| 37 | F3N | nitrogen trifluoride | 71.002 | 66.4 | 144.4 | 234.0 | 45.3 | | | 0.135 | 0.2 |
| 38 | F3NO | trifluoroamine oxide | 87.001 | | 186. | 303. | 64.3 | 146.9 | 0.375 | 0.212 | |
| 39 | F3P | phosphorus trifluoride | 87.968 | | 178. | 271.2 | 43.3 | | | 0.326 | |
| 40 | F3PS | thiophosphoryl trifluoride | 120.034 | | 220.9 | 346.0 | 38.2 | | | 0.187 | 0.6 |
| 41 | F4N2 | tetrafluorohydrazine | 104.016 | 105. | 199. | 309. | 37.5 | | | 0.206 | 0.3 |
| 42 | F4S | sulfur tetrafluoride | 108.058 | 152. | 232.7 | 364. | | | | | 1.0 |
| 43 | F4Si | silicon tetrafluoride | 104.09 | 183.0 | 187. | 259.0 | 37.2 | | | 0.753 | 0.0 |
| 44 | F4Xe | xenon tetrafluoride | 207.292 | 387. | 388.9 | 612. | 70.4 | 188.6 | 0.261 | 0.357 | |
| 45 | F6S | sulfur hexafluoride | 146.054 | 222.5 | 209.6 | 318.7 | 37.6 | 198.8 | 0.282 | 0.286 | 0.0 |
| 46 | F6U | uranium hexafluoride | 352.018 | 337. | 329. | 505.8 | 46.6 | 250.0 | 0.277 | 0.318 | 0.0 |
| 47 | He | helium-3 | 3.017 | | 3.19 | 3.31 | 1.14 | 72.9 | 0.302 | -0.473 | 0.0 |
| 48 | He | helium-4 | 4.003 | | 4.25 | 5.19 | 2.27 | 57.4 | 0.302 | -0.365 | 0.0 |
| 49 | Hg | mercury | 200.61 | 234.3 | 630. | 1765. | 1510. | 42.7 | 0.439 | -0.167 | |
| 50 | I2 | iodine | 253.82 | 386.8 | 457.5 | 819. | | 155. | | | 1.3 |

| No | Formula | Name | CPVAP A | CPVAP B | CPVAP C | CPVAP D | DELHF | DELGF |
|---|---|---|---|---|---|---|---|---|
| 26 | Cl4Ti | titanium tetrachloride | 7.064E+1 | 1.224E-1 | -1.443E-4 | 5.819E-8 | -7.637E+5 | -7.272E+5 |
| 27 | Cl5P | phosphorus pentachloride | 6.946E+1 | 2.079E-1 | -2.455E-4 | 9.914E-8 | -3.429E+5 | -2.785E+5 |
| 28 | D2 | deuterium (equilibrium) | 3.025E+1 | -6.615E-3 | 1.170E-5 | -3.684E-9 | 0.0 | 0.0 |
| 29 | D2 | deuterium (normal) | | | | | 0.0 | 0.0 |
| 30 | D2O | deuterium oxide | 3.182E+1 | 3.045E-3 | 2.033E-5 | -9.737E-9 | -2.494E+5 | -2.348E+5 |
| 31 | FNO2 | nitryl fluoride | 1.778E+1 | 1.416E-1 | -1.254E-4 | 4.140E-8 | -1.089E+5 | -6.649E+4 |
| 32 | F2 | fluorine | 2.322E+1 | 3.657E-2 | -3.613E-5 | 1.204E-8 | 0.0 | 0.0 |
| 33 | F2N2 | cis-difluorodiazine | 1.121E+1 | 1.754E-1 | -1.688E-4 | 5.898E-8 | 6.866E+4 | 1.088E+5 |
| 34 | F2N2 | trans-difluorodiazine | 2.254E+1 | 1.377E-1 | -1.258E-4 | 4.232E-8 | 8.122E+4 | 1.205E+5 |
| 35 | F2O | oxygen difluoride | 2.207E+1 | 9.875E-2 | -1.028E-4 | 3.796E-8 | 2.453E+4 | 4.178E+4 |
| 36 | F2Xe | xenon difluoride | | | | | | |
| 37 | F3N | nitrogen trifluoride | 1.141E+1 | 1.948E-1 | -2.023E-4 | 7.454E-8 | -1.316E+5 | -9.010E+4 |
| 38 | F3NO | trifluoroamine oxide | 1.513E+1 | 2.446E-1 | -2.528E-4 | 9.375E-8 | -1.633E+5 | -9.646E+4 |
| 39 | F3P | phosphorus trifluoride | 2.179E+1 | 1.733E-1 | -1.852E-4 | 6.974E-8 | -9.378E+5 | -9.253E+5 |
| 40 | F3PS | thiophosphoryl trifluoride | 2.492E+1 | 2.326E-1 | -2.472E-4 | 9.275E-8 | -9.923E+5 | -9.743E+5 |
| 41 | F4N2 | tetrafluorohydrazine | 3.553E+0 | 3.509E-1 | -3.637E-4 | 1.338E-7 | -8.374E+3 | 7.988E+4 |
| 42 | F4S | sulfur tetrafluoride | 2.542E+1 | 2.420E-1 | -2.653E-4 | 1.017E-7 | -7.813E+5 | -7.406E+5 |
| 43 | F4Si | silicon tetrafluoride | 2.678E+1 | 2.157E-1 | -2.204E-4 | 8.031E-8 | -1.616E+6 | -1.574E+6 |
| 44 | F4Xe | xenon tetrafluoride | | | | | -1.876E+5 | |
| 45 | F6S | sulfur hexafluoride | -6.599E-1 | 4.639E-1 | -5.089E-4 | 1.953E-7 | -1.222E+6 | -1.118E+6 |
| 46 | F6U | uranium hexafluoride | | | | | -2.139E+6 | -2.060E+6 |
| 47 | He | helium-3 | 2.080E+1 | | | | 0.0 | 0.0 |
| 48 | He | helium-4 | 2.080E+1 | | | | 0.0 | 0.0 |
| 49 | Hg | mercury | 2.080E+1 | | | | 6.134E+4 | 3.186E+4 |
| 50 | I2 | iodine | 3.559E+1 | 6.515E-3 | -6.988E-6 | 2.834E-9 | 6.247E+4 | 1.938E+4 |

| No | Formula | Name | Eq. | VP A | VP B | VP C | VP D | Tmin | Tmax | LDEN | TDEN |
|---|---|---|---|---|---|---|---|---|---|---|---|
| 26 | Cl4Ti | titanium tetrachloride | | | | | | | | 1.70 | 298 |
| 27 | Cl5P | phosphorous pentachloride | | | | | | | | | |
| 28 | D2 | deuterium (equilibrium) | 3 | 6.6752 | 157.89 | 0.0 | | 19 | 25 | 0.165 | 22.7 |
| 29 | D2 | deuterium (normal) | | | | | | | | | |
| 30 | D2O | deuterium oxide | | | | | | | | 1.105 | 298 |
| 31 | FNO2 | nitryl fluoride | | | | | | | | | |
| 32 | F2 | fluorine | 1 | -6.18224 | 1.18062 | -1.16555 | -1.50167 | 64 | TC | 1.51 | 85 |
| 33 | F2N2 | cis-difluorodiazine | | | | | | | | | |
| 34 | F2N2 | trans-difluorodiazine | | | | | | | | | |
| 35 | F2O | oxygen difluoride | | | | | | | | 1.521 | 128 |
| 36 | F2Xe | xenon difluoride | | | | | | | | | |
| 37 | F3N | nitrogen trifluoride | 2 | 32.599 | 1970.37 | -3.81 | 509. | 130 | TC | 1.54 | 144 |
| 38 | F3NO | fluoroamine oxide | | | | | | | | | |
| 39 | F3P | phosphorous trifluoride | | | | | | | | 3.1 | 172 |
| 40 | F3PS | thiophosphoryl trifluoride | | | | | | | | | |
| 41 | F4N2 | tetrafluorohydrazine | | | | | | | | 1.5 | 163 |
| 42 | F4S | sulfur tetrafluoride | 3 | 7.4561 | 1218.59 | -73.24 | | 161 | 224 | 1.936 | 195 |
| 43 | F4Si | silicon tetrafluoride | | | | | | | | | |
| 44 | F4Xe | xenon tetrafluoride | | | | | | | | | |
| 45 | F6S | sulfur hexafluoride | 3 | 12.7583 | 2524.78 | -11.16 | | 159 | 220 | 1.83 | 223 |
| 46 | F6U | uranium hexafluoride | | | | | | | | | |
| 47 | He | helium-3 | | | | | | | | | |
| 48 | He | helium-4 | 1 | -3.97466 | 1.00074 | 1.50056 | -0.43020 | 2 | TC | 0.123 | 4.3 |
| 49 | Hg | mercury | | | | | | | | 13.594 | 293 |
| 50 | I2 | iodine | 3 | 9.5395 | 3709.23 | -68.16 | | 383 | 487 | 3.74 | 453 |

| No | Formula | Name | MolWt | Tfp K | Tb K | Tc K | Pc bar | Vc $cm^3/mol$ | Zc | Omega | Dipm debye |
|---|---|---|---|---|---|---|---|---|---|---|---|
| 51 | I4Si | silicon tetraiodide | 535.706 | 393.7 | 560.5 | 944. | | 558. | | | |
| 52 | I4Ti | titanium tetraiodide | 555.520 | 423. | 650. | 1040. | | 505. | | | |
| 53 | Kr | krypton | 83.800 | 115.8 | 119.9 | 209.4 | 55.0 | 91.2 | 0.288 | 0.005 | 0.0 |
| 54 | NO | nitric oxide | 30.006 | 109.5 | 121.4 | 180. | 64.8 | 57.7 | 0.250 | 0.588 | 0.2 |
| 55 | NO2 | nitrogen dioxide | 46.006 | 261.9 | 294.3 | 431. | 101. | 167.8 | 0.473 | 0.834 | 0.4 |
| 56 | N2 | nitrogen | 28.013 | 63.3 | 77.4 | 126.2 | 33.9 | 89.8 | 0.290 | 0.039 | 0.0 |
| 57 | N2O | nitrous oxide | 44.013 | 182.3 | 184.7 | 309.6 | 72.4 | 97.4 | 0.274 | 0.165 | 0.2 |
| 58 | Ne | neon | 20.183 | 24.5 | 27.1 | 44.4 | 27.6 | 41.6 | 0.311 | -0.029 | 0.0 |
| 59 | O2 | oxygen | 31.999 | 54.4 | 90.2 | 154.6 | 50.4 | 73.4 | 0.288 | 0.025 | 0.0 |
| 60 | O2S | sulfur dioxide | 64.063 | 197.7 | 263.2 | 430.8 | 78.8 | 122.2 | 0.269 | 0.256 | 1.6 |
| 61 | O3 | ozone | 47.998 | 80.5 | 181.2 | 261.1 | 55.7 | 88.9 | 0.228 | 0.691 | 0.6 |
| 62 | O3S | sulfur trioxide | 80.058 | 290. | 318. | 491.0 | 82.1 | 127.3 | 0.256 | 0.481 | 0.0 |
| 63 | P | phosphorous | 30.974 | | 553. | 994. | | | | | |
| 64 | Rn | radon | 222.00 | 202. | 211.4 | 377. | 62.8 | | | -0.008 | |
| 65 | S | sulfur | 32.066 | | 717.8 | 1314. | 207. | | | 0.171 | |
| 66 | Se | selenium | 78.96 | | 1010. | 1766. | 272. | | | 0.346 | |
| 67 | T2 | tritium | 6.32 | | 25.0 | 40.0 | | 58. | | | 0.0 |
| 68 | Xe | xenon | 131.300 | 161.3 | 165.0 | 289.7 | 58.4 | 118.4 | 0.287 | 0.008 | 0.0 |
| 69 | HBr | hydrogen bromide | 80.912 | 187.1 | 206.8 | 363.2 | 85.5 | | | 0.088 | 0.8 |
| 70 | HCl | hydrogen chloride | 36.461 | 159.0 | 188.1 | 324.7 | 83.1 | 80.9 | 0.249 | 0.133 | 1.1 |
| 71 | HD | hydrogen deuteride | 3.023 | 16.6 | 22.1 | 36.0 | 14.8 | 62.7 | 0.310 | -0.179 | 0.0 |
| 72 | HF | hydrogen fluoride | 20.006 | 190. | 293. | 461. | 64.8 | 69.2 | 0.117 | 0.329 | 1.9 |
| 73 | HI | hydrogen iodide | 127.912 | 222.4 | 237.6 | 424.0 | 83.1 | | | 0.049 | 0.5 |
| 74 | HF2N | difluoroamine | 53.011 | | 250. | 403. | | | | | |
| 75 | H2 | hydrogen (equilibrium) | 2.016 | 14.0 | 20.3 | 33.0 | 12.9 | 64.3 | 0.303 | -0.216 | 0.0 |

| No | Formula | Name | CPVAP A | CPVAP B | CPVAP C | CPVAP D | DELHF | DELGF |
|---|---|---|---|---|---|---|---|---|
| 51 | I4Si | silicon tetraiodide | 8.479E+1 | 7.800E-2 | -9.340E-5 | 3.806E-8 | -1.105E+5 | -1.598E+5 |
| 52 | I4Ti | titanium tetraiodide | 9.575E+1 | 4.253E-2 | -5.179E-5 | 2.135E-8 | -2.775E+5 | -3.282E+5 |
| 53 | Kr | krypton | 2.080E+1 | | | | 0.0 | 0.0 |
| 54 | NO | nitric oxide | 2.935E+1 | -9.378E-4 | 9.747E-6 | -4.187E-9 | 9.043E+4 | 8.675E+4 |
| 55 | NO2 | nitrogen dioxide | 2.423E+1 | 4.836E-2 | -2.081E-5 | 0.293E-9 | 3.387E+4 | 5.200E+4 |
| 56 | N2 | nitrogen | 3.115E+1 | -1.357E-2 | 2.680E-5 | -1.168E-8 | 0.0 | 0.0 |
| 57 | N2O | nitrous oxide | 2.162E+1 | 7.281E-2 | -5.778E-5 | 1.830E-8 | 8.160E+4 | 1.037E+5 |
| 58 | Ne | neon | 2.080E+1 | | | | 0.0 | 0.0 |
| 59 | O2 | oxygen | 2.811E+1 | -3.680E-6 | 1.746E-5 | -1.065E-8 | 0.0 | 0.0 |
| 60 | O2S | sulfur dioxide | 2.385E+1 | 6.699E-2 | -4.961E-5 | 1.328E-8 | -2.971E+5 | -3.004E+5 |
| 61 | O3 | ozone | 2.054E+1 | 8.009E-2 | -6.243E-5 | 1.697E-8 | 1.428E+5 | 1.629E+5 |
| 62 | O3S | sulfur trioxide | 1.921E+1 | 1.374E-1 | -1.176E-4 | 3.700E-8 | -3.960E+5 | -3.713E+5 |
| 63 | P | phosphorous | 2.080E+1 | | | | 3.341E+5 | 2.922E+5 |
| 64 | Rn | radon | 2.080E+1 | | | | 0.0 | 0.0 |
| 65 | S | sulfur | | | | | 2.792E+5 | 2.386E+5 |
| 66 | Se | selenium | | | | | | |
| 67 | T2 | tritium | | | | | | |
| 68 | Xe | xenon | 2.080E+1 | | | | 0.0 | 0.0 |
| 69 | HBr | hydrogen bromide | 3.065E+1 | -9.462E-3 | 1.722E-5 | -6.238E-9 | -3.626E+4 | -5.330E+4 |
| 70 | HCl | hydrogen chloride | 3.067E+1 | -7.201E-3 | 1.246E-5 | -3.898E-9 | -9.236E+4 | -9.533E+4 |
| 71 | HD | hydrogen deuteride | 2.947E+1 | -1.329E-3 | 1.311E-6 | 1.279E-9 | 3.220E+2 | -1.465E+3 |
| 72 | HF | hydrogen fluoride | 2.906E+1 | 6.611E-4 | -2.032E-6 | 2.504E-9 | -2.713E+5 | -2.734E+5 |
| 73 | HI | hydrogen iodide | 3.116E+1 | -1.428E-2 | 2.972E-5 | -1.353E-8 | 2.638E+4 | 1.591E+3 |
| 74 | HF2N | difluoroamine | | | | | | |
| 75 | H2 | hydrogen (equilibrium) | 2.714E+1 | 9.274E-3 | -1.381E-5 | 7.645E-9 | 0.0 | 0.0 |

| No | Formula | Name | Eq. | VP A | VP B | VP C | VP D | Tmin | Tmax | LDEN | TDEN |
|---|---|---|---|---|---|---|---|---|---|---|---|
| 51 | I4Si | silicon tetraiodide | | | | | | | | | |
| 52 | I4Ti | titanium tetraiodide | | | | | | | | | |
| 53 | Kr | krypton | 2 | 24.097 | 1408.77 | -2.579 | 336. | 115 | TC | 2.42 | 120 |
| 54 | NO | nitric oxide | 2 | 54.894 | 2465.78 | -7.211 | 209. | 115 | TC | 1.28 | 121 |
| 55 | NO2 | nitrogen dioxide | 2 | 55.242 | 6073.34 | -6.094 | 780. | 270 | TC | 1.45 | 293 |
| 56 | N2 | nitrogen | 1 | -6.09676 | 1.13670 | -1.04072 | -1.93306 | 63 | TC | 0.804 | 78 |
| 57 | N2O | nitrous oxide | 2 | 39.824 | 2867.98 | -4.655 | 557. | 190 | TC | 1.226 | 184 |
| 58 | Ne | neon | 1 | -6.07686 | 1.59402 | -1.06092 | 4.06656 | 25 | TC | 1.204 | 27 |
| 59 | O2 | oxygen | 1 | -6.28275 | 1.73619 | -1.81349 | -2.53645E-2 | 54 | TC | 1.149 | 90 |
| 60 | O2S | sulfur dioxide | 2 | 48.882 | 4552.50 | -5.666 | 990. | 235 | TC | 1.455 | 263 |
| 61 | O3 | ozone | 3 | 9.1225 | 1272.18 | -22.16 | | 109 | 174 | 1.356 | 161 |
| 62 | O3S | sulfur trioxide | 2 | 132.94 | 10420.1 | -17.38 | 1200. | 300 | TC | 1.78 | 318 |
| 63 | P | phosphorous | | | | | | | | | |
| 64 | Rn | radon | | | | | | | | 4.4 | 211 |
| 65 | S | sulfur | | | | | | | | | |
| 66 | Se | selenium | | | | | | | | | |
| 67 | T2 | tritium | | | | | | | | | |
| 68 | Xe | xenon | 2 | 24.809 | 1951.76 | -2.544 | 603. | 170 | TC | 3.06 | 165 |
| 69 | HBr | hydrogen bromide | 2 | 21.482 | 2394.35 | -1.843 | 653. | 200 | TC | 2.16 | 216 |
| 70 | HCl | hydrogen chloride | 2 | 31.994 | 2626.67 | -3.443 | 538. | 180 | TC | 1.193 | 188 |
| 71 | HD | hydrogen deuteride | | | | | | | | | |
| 72 | HF | hydrogen fluoride | 1 | -9.74369 | 4.68946 | -2.98358 | 9.65825 | 273 | TC | 0.967 | 293 |
| 73 | HI | hydrogen iodide | 2 | 27.264 | 3013.08 | -2.673 | 923. | 235 | TC | 2.80 | 237 |
| 74 | HF2N | difluoroamine | | | | | | | | | |
| 75 | H2 | hydrogen (equilibrium) | 1 | -5.57929 | 2.60012 | -0.85506 | 1.70503 | 14 | TC | 0.071 | 20 |

| No | Formula | Name | MolWt | Tfp K | Tb K | Tc K | Pc bar | Vc $cm^3/mol$ | Zc | Omega | Dipm debye |
|---|---|---|---|---|---|---|---|---|---|---|---|
| 76 | H2 | hydrogen (normal) | 2.016 | 14.0 | 20.4 | 33.2 | 13.0 | 65.1 | 0.306 | -0.218 | 0.0 |
| 77 | H2O | water | 18.015 | 273.2 | 373.2 | 647.3 | 221.2 | 57.1 | 0.235 | 0.344 | 1.8 |
| 78 | H2S | hydrogen sulfide | 34.080 | 189.6 | 213.5 | 373.2 | 89.4 | 98.6 | 0.284 | 0.081 | 0.9 |
| 79 | H3As | arsine | 77.946 | 159.7 | 218. | 373.1 | | | | | 0.2 |
| 80 | H3N | ammonia | 17.031 | 195.4 | 239.8 | 405.5 | 113.5 | 72.5 | 0.244 | 0.250 | 1.5 |
| 81 | H3P | phosphine | 33.998 | 140. | 185.4 | 324.5 | 65.4 | | | 0.038 | 0.6 |
| 82 | H4ClN | ammonium chloride | 53.492 | | 793. | 882. | 16.4 | | | 3.92 | |
| 83 | H4ClP | phosphonium chloride | 70.459 | | 246. | 322.3 | 73.7 | | | 1.64 | |
| 84 | H4N2 | hydrazine | 32.045 | 274.7 | 386.7 | 653. | 147. | | | 0.316 | 3.0 |
| 85 | H4Si | silane | 32.122 | 88.2 | 161. | 269.7 | 48.4 | | | 0.068 | 0.0 |
| 86 | H6B2 | diborane | 27.668 | 108. | 185.6 | 289.8 | 40.5 | | | 0.217 | 0.0 |
| 87 | CBrClF2 | bromochlorodifluoromethane | 165.364 | | 269. | 426.9 | 42.5 | 245.5 | 0.294 | 0.184 | |
| 88 | CBr2F2 | dibromodifluoromethane | 209.815 | 132. | 298. | 471.3 | 41.3 | | | | 0.7 |
| 89 | CBrF3 | trifluorobromomethane | 148.910 | | 215.3 | 340.2 | 39.7 | 195.9 | 0.275 | 0.171 | 0.7 |
| 90 | CClF3 | chlorotrifluoromethane | 104.459 | 92.0 | 193.2 | 302.0 | 38.7 | 180.4 | 0.278 | 0.198 | 0.5 |
| 91 | CCl2F2 | dichlorodifluoromethane | 120.914 | 115.4 | 245.2 | 385.0 | 41.4 | 216.7 | 0.280 | 0.204 | 0.5 |
| 92 | CCl2O | phosgene | 98.916 | 145.0 | 281. | 455. | 56.7 | 190.1 | 0.285 | 0.205 | 1.1 |
| 93 | CCl3F | trichlorofluoromethane | 137.368 | 162.0 | 296.9 | 471.2 | 44.1 | 247.8 | 0.279 | 0.189 | 0.5 |
| 94 | CCl4 | carbon tetrachloride | 153.823 | 250. | 349.9 | 556.4 | 45.6 | 275.9 | 0.272 | 0.193 | 0.0 |
| 95 | CD4 | deuteromethane | 20.071 | | 111.7 | 189.2 | 46.6 | 98.2 | 0.291 | 0.032 | 0.0 |
| 96 | CF4 | carbon tetrafluoride | 88.005 | 86.4 | 145.1 | 227.6 | 37.4 | 139.6 | 0.276 | 0.177 | 0.0 |
| 97 | CO | carbon monoxide | 28.010 | 68.1 | 81.7 | 132.9 | 35.0 | 93.2 | 0.295 | 0.066 | 0.1 |
| 98 | COS | carbonyl sulfide | 60.070 | 134.3 | 223. | 378.8 | 63.5 | 136.3 | 0.275 | 0.105 | 0.7 |
| 99 | CO2 | carbon dioxide | 44.010 | 216.6 | | 304.1 | 73.8 | 93.9 | 0.274 | 0.239 | 0.0 |
| 100 | CS2 | carbon disulfide | 76.131 | 161.3 | 319. | 552. | 79.0 | 160. | 0.276 | 0.109 | 0.0 |

| No | Formula | Name | CPVAP A | CPVAP B | CPVAP C | CPVAP D | DELHF | DELGF |
|---|---|---|---|---|---|---|---|---|
| 76 | H2 | hydrogen (normal) | | | | | | |
| 77 | H2O | water | 3.224E+1 | 1.924E-3 | 1.055E-5 | -3.596E-9 | -2.420E+5 | -2.288E+5 |
| 78 | H2S | hydrogen sulfide | 3.194E+1 | 1.436E-3 | 2.432E-5 | -1.176E-8 | -2.018E+4 | -3.308E+4 |
| 79 | H3As | arsine | | | | | 1.825E+5 | 1.578E+5 |
| 80 | H3N | ammonia | 2.731E+1 | 2.383E-2 | 1.707E-5 | -1.185E-8 | -4.572E+4 | -1.616E+4 |
| 81 | H3P | phosphine | 2.323E+1 | 4.401E-2 | 1.303E-5 | -1.593E-8 | 2.290E+4 | 2.541E+4 |
| 82 | H4ClN | ammonium chloride | | | | | | |
| 83 | H4ClP | phosphonium chloride | | | | | | |
| 84 | H4N2 | hydrazine | 9.768E+0 | 1.895E-1 | -1.657E-4 | 6.025E-8 | 9.525E+4 | 1.586E+5 |
| 85 | H4Si | silane | 1.118E+1 | 1.220E-1 | -5.548E-5 | 6.840E-9 | 3.266E+4 | 5.518E+4 |
| 86 | H6B2 | diborane | | | | | 3.140E+4 | 8.332E+4 |
| 87 | CBrClF2 | bromochlorodifluoromethane | | | | | | |
| 88 | CBr2F2 | dibromodifluoromethane | | | | | | |
| 89 | CBrF3 | trifluorobromomethane | 2.188E+1 | 2.159E-1 | -2.114E-4 | 7.464E-8 | -6.494E+5 | -6.975E+5 |
| 90 | CClF3 | chlorotrifluoromethane | 2.281E+1 | 1.911E-1 | -1.576E-4 | 4.459E-8 | -6.950E+5 | -6.544E+5 |
| 91 | CCl2F2 | dichlorodifluoromethane | 3.160E+1 | 1.782E-1 | -1.509E-4 | 4.342E-8 | -4.815E+5 | -4.425E+5 |
| 92 | CCl2O | phosgene | 2.809E+1 | 1.361E-1 | -1.374E-4 | 5.070E-8 | -2.211E+5 | -2.069E+5 |
| 93 | CCl3F | trichlorofluoromethane | 4.098E+1 | 1.668E-1 | -1.416E-4 | 4.146E-8 | -2.847E+5 | -2.455E+5 |
| 94 | CCl4 | carbon tetrachloride | 4.072E+1 | 2.049E-1 | -2.270E-4 | 8.843E-8 | -1.005E+5 | -5.828E+4 |
| 95 | CD4 | deuteromethane | 1.249E+1 | 1.010E-1 | -2.199E-5 | -8.458E-9 | -8.830E+4 | -5.954E+4 |
| 96 | CF4 | carbon tetrafluoride | 1.398E+1 | 2.026E-1 | -1.625E-4 | 4.513E-8 | -9.337E+5 | -8.890E+5 |
| 97 | CO | carbon monoxide | 3.087E+1 | -1.285E-2 | 2.789E-5 | -1.272E-8 | -1.106E+5 | -1.374E+5 |
| 98 | COS | carbonyl sulfide | 2.357E+1 | 7.984E-2 | -7.017E-5 | 2.453E-8 | -1.385E+5 | -1.658E+5 |
| 99 | CO2 | carbon dioxide | 1.980E+1 | 7.344E-2 | -5.602E-5 | 1.715E-8 | -3.938E+5 | -3.946E+5 |
| 100 | CS2 | carbon disulfide | 2.744E+1 | 8.127E-2 | -7.666E-5 | 2.673E-8 | 1.171E+5 | 6.695E+4 |

| No | Formula | Name | Eq. | VP A | VP B | VP C | VP D | Tmin | Tmax | LDEN | TDEN |
|---|---|---|---|---|---|---|---|---|---|---|---|
| 76 | H2 | hydrogen (normal) | | | | | | | | | |
| 77 | H2O | water | 1 | -7.76451 | 1.45838 | -2.77580 | -1.23303 | 275 | TC | 0.998 | 293 |
| 78 | H2S | hydrogen sulfide | 2 | 36.067 | 3132.31 | -3.985 | 653. | 205 | TC | 0.993 | 214 |
| 79 | H3As | arsine | | | | | | | | 1.604 | 209 |
| 80 | H3N | ammonia | 2 | 45.327 | 4104.67 | -5.146 | 615. | 220 | TC | 0.639 | 273 |
| 81 | H3P | phosphine | | | | | | | | 1.529 | 298 |
| 82 | H4ClN | ammonium chloride | | | | | | | | | |
| 83 | H4ClP | phosphonium chloride | | | | | | | | | |
| 84 | H4N2 | hydrazine | 2 | 49.476 | 6951.84 | -5.286 | 1222. | 350 | TC | 1.008 | 293 |
| 85 | H4Si | silane | | | | | | | | 0.68 | 88 |
| 86 | H6B2 | diborane | 3 | 8.0390 | 1200.78 | -31.22 | | 118 | 181 | 0.470 | 153 |
| 87 | CBrClF2 | bromochlorodifluoromethane | 3 | 9.1295 | 2154.39 | -32.87 | | 178 | 283 | | |
| 88 | CBr2F2 | dibromodifluoromethane | 3 | 9.8485 | 2720.78 | -19.35 | | 247 | 296 | 2.462 | 288 |
| 89 | CBrF3 | trifluorobromomethane | | | | | | | | 1.538 | 298 |
| 90 | CClF3 | chlorotrifluoromethane | 1 | -6.78845 | 1.24435 | -2.32601 | 1.45543 | 233 | TC | 1.298 | 243 |
| 91 | CCl2F2 | dichlorodifluoromethane | 1 | -7.01657 | 1.73224 | -2.97909 | -0.37723 | 155 | TC | 1.750 | 158 |
| 92 | CCl2O | phosgene | 1 | -7.08177 | 1.60461 | -2.57153 | -1.88377 | 216 | TC | 1.381 | 293 |
| 93 | CCl3F | trichlorofluoromethane | 2 | 42.089 | 4464.14 | -4.753 | 2138. | 260 | TC | 1.494 | 290 |
| 94 | CCl4 | carbon tetrachloride | 1 | -7.07139 | 1.71497 | -2.89930 | -2.49466 | 250 | TC | 1.584 | 298 |
| 95 | CD4 | deuteromethane | | | | | | | | | |
| 96 | CF4 | carbon tetrafluoride | 3 | 9.4341 | 1244.55 | -13.06 | | 93 | 148 | 1.33 | 193 |
| 97 | CO | carbon monoxide | 1 | -6.20798 | 1.27885 | -1.34533 | -2.56842 | 71 | TC | 0.803 | 81 |
| 98 | COS | carbonyl sulfide | 1 | -6.40952 | 1.21015 | -1.54976 | -2.10074 | 162 | TC | 1.274 | 174 |
| 99 | CO2 | carbon dioxide | 1 | -6.95626 | 1.19695 | -3.12614 | 2.99448 | 217 | TC | | |
| 100 | CS2 | carbon disulfide | 1 | -6.63896 | 1.20395 | -0.37653 | -4.32820 | 277 | TC | 1.293 | 273 |

| No | Formula | Name | MolWt | Tfp K | Tb K | Tc K | Pc bar | Vc $cm^3/mol$ | Zc | Omega | Dipm debye |
|---|---|---|---|---|---|---|---|---|---|---|---|
| 101 | CHClF2 | chlorodifluoromethane | 86.469 | 113. | 232.4 | 369.3 | 49.7 | 165.6 | 0.268 | 0.221 | 1.4 |
| 102 | CHCl2F | dichloromonofluoromethane | 102.923 | 138. | 282.1 | 451.6 | 51.8 | 196.4 | 0.271 | 0.210 | 1.3 |
| 103 | CHCl3 | chloroform | 119.378 | 209.6 | 334.3 | 536.4 | 53.7 | 238.9 | 0.293 | 0.218 | 1.1 |
| 104 | CHF3 | fluoroform | 70.013 | 110. | 191.0 | 299.3 | 48.6 | 132.7 | 0.259 | 0.260 | 1.6 |
| 105 | CHN | hydrogen cyanide | 27.026 | 259.9 | 298.9 | 456.7 | 53.9 | 138.8 | 0.197 | 0.388 | 3.0 |
| 106 | CH2Br2 | dibromomethane | 173.835 | 220.6 | 370. | 583. | 71. | | | | 1.9 |
| 107 | CH2Cl2 | dichloromethane | 84.933 | 178.1 | 313.0 | 510. | 63. | | | 0.199 | 1.8 |
| 108 | CH2F2 | difluoromethane | 52.023 | | 221.5 | 351.6 | 58.3 | 120.8 | 0.241 | 0.271 | 2.0 |
| 109 | CH2O | formaldehyde | 30.026 | 156. | 254. | 408. | 65.9 | | | 0.253 | 2.3 |
| 110 | CH2O2 | formic acid | 46.025 | 281.5 | 373.8 | 580. | | | | | 1.5 |
| 111 | CH3Br | methyl bromide | 94.939 | 179.5 | 276.6 | 464. | 66.1 | | | | 1.8 |
| 112 | CH3Cl | methyl chloride | 50.488 | 175.4 | 249.1 | 416.3 | 67.0 | 138.9 | 0.269 | 0.153 | 1.9 |
| 113 | CH3F | methyl fluoride | 34.033 | 131.4 | 194.7 | 315.0 | 56. | 113.2 | 0.240 | 0.187 | 1.8 |
| 114 | CH3I | methyl iodide | 141.939 | 206.7 | 315.7 | 528. | 65.9 | | | | 1.6 |
| 115 | CH3NO2 | nitromethane | 61.041 | 244.6 | 374.3 | 588. | 63.1 | 173.2 | 0.208 | 0.310 | 3.1 |
| 116 | CH4 | methane | 16.043 | 90.7 | 111.6 | 190.4 | 46.0 | 99.2 | 0.288 | 0.011 | 0.0 |
| 117 | CH4O | methanol | 32.042 | 175.5 | 337.7 | 512.6 | 80.9 | 118.0 | 0.224 | 0.556 | 1.7 |
| 118 | CH4S | methyl mercaptan | 48.107 | 150. | 279.1 | 470.0 | 72.3 | 144.8 | 0.268 | 0.153 | 1.3 |
| 119 | CH5N | methyl amine | 31.058 | 179.7 | 266.8 | 430.0 | 74.3 | | | 0.292 | 1.3 |
| 120 | CH6N2 | methyl hydrazine | 46.072 | | 362. | 567. | 82.4 | 271.2 | 0.474 | 0.425 | 1.7 |
| 121 | CH6Si | methyl silane | 46.145 | 116.7 | 215.6 | 352.5 | | | | | 0.7 |
| 122 | C2Br2ClF3 | 1,2-dibromo-1-chlorotrifluoroethane | 276.277 | 182.7 | 366. | 560.7 | 36.1 | 368. | 0.285 | 0.248 | |
| 123 | C2Br2F4 | 1,2-dibromotetrafluoroethane | 259.822 | 163. | 320.4 | 487.8 | 33.9 | 341. | 0.285 | 0.245 | |
| 124 | C2ClF3 | chlorotrifluoroethene | 116.469 | 116. | 245.3 | 379. | 40.5 | 212. | 0.272 | 0.252 | 0.4 |
| 125 | C2ClF5 | chloropentafluoroethane | 154.467 | 167. | 235.2 | 353.2 | 32.3 | 251.8 | 0.277 | 0.279 | 0.3 |

| No | Formula | Name | CPVAP A | CPVAP B | CPVAP C | CPVAP D | DELHF | DELGF |
|---|---|---|---|---|---|---|---|---|
| 101 | CHClF2 | chlorodifluoromethane | 1.730E+1 | 1.618E-1 | -1.170E-4 | 3.058E-8 | -5.020E+5 | -4.709E+5 |
| 102 | CHCl2F | dichloromonofluoromethane | 2.366E+1 | 1.581E-1 | -1.200E-4 | 3.264E-8 | -2.989E+5 | -2.684E+5 |
| 103 | CHCl3 | chloroform | 2.400E+1 | 1.893E-1 | -1.841E-4 | 6.657E-8 | -1.013E+5 | -6.858E+4 |
| 104 | CHF3 | fluoroform | 8.156E+0 | 1.813E-1 | -1.379E-4 | 3.938E-8 | -6.975E+5 | -6.628E+5 |
| 105 | CHN | hydrogen cyanide | 2.186E+1 | 6.062E-2 | -4.961E-5 | 1.815E-8 | 1.306E+5 | 1.202E+5 |
| 106 | CH2Br2 | dibromomethane | 2.500E+1 | 2.517E-1 | -1.833E-4 | 5.646E-8 | -3.890E+4 | -1.059E+4 |
| 107 | CH2Cl2 | dichloromethane | 1.295E+1 | 1.623E-1 | -1.302E-4 | 4.208E-8 | -9.546E+4 | -6.891E+4 |
| 108 | CH2F2 | difluoromethane | 1.179E+1 | 1.181E-1 | -4.843E-5 | 2.125E-9 | -4.509E+5 | -4.229E+5 |
| 109 | CH2O | formaldehyde | 2.348E+1 | 3.157E-2 | 2.985E-5 | -2.300E-8 | -1.160E+5 | -1.100E+5 |
| 110 | CH2O2 | formic acid | 1.171E+1 | 1.358E-1 | -8.411E-5 | 2.017E-8 | -3.789E+5 | -3.512E+5 |
| 111 | CH3Br | methyl bromide | 1.443E+1 | 1.091E-1 | -5.401E-5 | 1.000E-8 | -3.768E+4 | -2.818E+4 |
| 112 | CH3Cl | methyl chloride | 1.388E+1 | 1.014E-1 | -3.889E-5 | 2.567E-9 | -8.637E+4 | -6.293E+4 |
| 113 | CH3F | methyl fluoride | 1.382E+1 | 8.616E-2 | -2.071E-5 | -1.985E-9 | -2.340E+5 | -2.101E+5 |
| 114 | CH3I | methyl iodide | 1.081E+1 | 1.389E-1 | -1.041E-4 | 3.486E-8 | 1.398E+4 | 1.566E+4 |
| 115 | CH3NO2 | nitromethane | 7.423E+0 | 1.978E-1 | -1.081E-4 | 2.085E-8 | -7.478E+4 | -6.950E+3 |
| 116 | CH4 | methane | 1.925E+1 | 5.213E-2 | 1.197E-5 | -1.132E-8 | -7.490E+4 | -5.087E+4 |
| 117 | CH4O | methanol | 2.115E+1 | 7.092E-2 | 2.587E-5 | -2.852E-8 | -2.013E+5 | -1.626E+5 |
| 118 | CH4S | methyl mercaptan | 1.327E+1 | 1.457E-1 | -8.545E-5 | 2.075E-8 | -2.299E+4 | -9.923E+3 |
| 119 | CH5N | methyl amine | 1.148E+1 | 1.427E-1 | -5.334E-5 | 4.752E-9 | -2.303E+4 | 3.228E+4 |
| 120 | CH6N2 | methyl hydrazine | | | | | 8.541E+4 | 1.780E+5 |
| 121 | CH6Si | methyl silane | | | | | | |
| 122 | C2Br2ClF3 | 1,2-dibromo-1-chlorotrifluoroethane | | | | | | |
| 123 | C2Br2F4 | 1,2-dibromotetrafluoroethane | | | | | -7.662E+5 | |
| 124 | C2ClF3 | chlorotrifluoroethene | | | | | -5.317E+5 | |
| 125 | C2ClF5 | chloropentafluoroethane | 2.783E+1 | 3.492E-1 | -2.891E-4 | 8.139E-8 | | |

| No | Formula | Name | Eq. | VP A | VP B | VP C | VP D | Tmin | Tmax | LDEN | TDEN |
|---|---|---|---|---|---|---|---|---|---|---|---|
| 101 | CHClF2 | chlorodifluoromethane | 1 | -6.99913 | 1.23014 | -2.49377 | -2.21052 | 170 | TC | 1.23 | 289 |
| 102 | CHCl2F | dichloromonofluoromethane | 2 | 47.943 | 4629.02 | -5.590 | 1665. | 250 | TC | 1.38 | 282 |
| 103 | CHCl3 | chloroform | 1 | -6.95546 | 1.16625 | -2.13970 | -3.44421 | 215 | TC | 1.489 | 293 |
| 104 | CHF3 | fluoroform | 1 | -7.41994 | 1.65884 | -3.14962 | -0.84938 | 125 | TC | 1.246 | 239 |
| 105 | CHN | hydrogen cyanide | 2 | 31.122 | 4183.37 | -3.004 | 1635. | 280 | TC | 0.688 | 293 |
| 106 | CH2Br2 | dibromomethane | | | | | | | | 2.50 | 293 |
| 107 | CH2Cl2 | dichloromethane | 1 | -7.35739 | 2.17546 | -4.07038 | 3.50701 | 233 | TC | 1.317 | 298 |
| 108 | CH2F2 | difluoromethane | 1 | -7.44206 | 1.51914 | -2.75319 | -0.97949 | 155 | TC | | |
| 109 | CH2O | formaldehyde | 1 | -7.29343 | 1.08395 | -1.63882 | -2.30677 | 184 | TC | 0.815 | 253 |
| 110 | CH2O2 | formic acid | 3 | 10.3680 | 3599.58 | -26.09 | | 271 | 409 | 1.226 | 288 |
| 111 | CH3Br | methyl bromide | 1 | -7.43951 | 3.15408 | -4.67922 | 2.33796 | 184 | TC | 1.737 | 268 |
| 112 | CH3Cl | methyl chloride | 1 | -6.86672 | 1.52273 | -1.92919 | -2.61459 | 175 | TC | 0.915 | 293 |
| 113 | CH3F | methyl fluoride | 1 | -6.78099 | 0.828379 | -1.41137 | -2.41700 | 135 | TC | 0.843 | 213 |
| 114 | CH3I | methyl iodide | 1 | -6.51125 | 0.888786 | -1.36624 | -3.03652 | 259 | TC | 2.279 | 293 |
| 115 | CH3NO2 | nitromethane | 1 | -8.41688 | 2.76466 | -3.65341 | -1.01376 | 328 | TC | 1.138 | 293 |
| 116 | CH4 | methane | 1 | -6.00435 | 1.18850 | -0.83408 | -1.22833 | 91 | TC | 0.425 | 112 |
| 117 | CH4O | methanol | 1 | -8.54796 | 0.76982 | -3.10850 | 1.54481 | 288 | TC | 0.791 | 293 |
| 118 | CH4S | methyl mercaptan | 1 | -6.79300 | 1.52687 | -2.45989 | -1.34839 | 222 | TC | 0.866 | 293 |
| 119 | CH5N | methyl amine | 1 | -7.52772 | 1.81615 | -4.20677 | -1.22275 | 200 | TC | 0.703 | 260 |
| 120 | CH6N2 | methyl hydrazine | 3 | 8.5222 | 2319.84 | -91.70 | | 270 | 400 | | |
| 121 | CH6Si | methyl silane | | | | | | | | | |
| 122 | C2Br2ClF3 | 1,2-dibromo-1-chlorotrifluoroethane | 1 | -7.75667 | 2.65450 | -4.26722 | -0.10090 | 184 | TC | | |
| 123 | C2Br2F4 | 1,2-dibromotetrafluoroethane | 1 | -7.30588 | 1.65554 | -3.20770 | -1.65654 | 225 | TC | 2.175 | 294 |
| 124 | C2ClF3 | chlorotrifluoroethene | 1 | -7.73622 | 2.58699 | -4.21453 | -0.15430 | 116 | TC | 1.305 | 293 |
| 125 | C2ClF5 | chloropentafluoroethane | 1 | -7.69084 | 2.41233 | -4.48383 | 1.92058 | 174 | TC | 1.26 | 303 |

| No | Formula | Name | MolWt | Tfp K | Tb K | Tc K | Pc bar | Vc $cm^3/mol$ | Zc | Omega | Dipm debye |
|---|---|---|---|---|---|---|---|---|---|---|---|
| 126 | C2Cl2F4 | 1,1-dichlorotetrafluoroethane | 170.922 | 179. | 277.0 | 418.6 | 33.0 | 294.2 | 0.279 | 0.263 | |
| 127 | C2Cl2F4 | 1,2-dichlorotetrafluoroethane | 170.922 | 179. | 276.2 | 418.9 | 32.6 | 293.8 | 0.275 | 0.246 | 0.5 |
| 128 | C2Cl3F3 | 1,2,2-trichlorotrifluoroethane | 187.380 | 238.2 | 320.8 | 487.3 | 34.1 | 325.5 | 0.274 | 0.256 | |
| 129 | C2Cl4 | tetrachloroethene | 165.834 | 251. | 394.4 | 620.2 | 47.6 | 289.6 | 0.250 | | 0.0 |
| 130 | C2Cl4F2 | 1,1,2,2-tetrachlorodifluoroethane | 203.831 | 298.0 | 366.0 | 551. | 38.7 | | | | |
| 131 | C2F3N | trifluoroacetonitrile | 95.023 | | 205.5 | 311.1 | 36.2 | 202. | 0.283 | 0.267 | |
| 132 | C2F4 | perfluoroethene | 100.016 | 130.7 | 197.2 | 306.5 | 39.4 | 172. | 0.267 | 0.223 | |
| 133 | C2F6 | perfluoroethane | 138.012 | 172.4 | 194.9 | 293.0 | 30.6 | 222. | 0.279 | | 0.0 |
| 134 | C2N2 | cyanogen | 52.035 | 245.3 | 252.0 | 400. | 59.8 | | | 0.278 | 0.2 |
| 135 | C2HClF2 | 1-chloro-2,2-difluoroethene | 98.479 | | 254.6 | 400.6 | 44.6 | 197. | 0.264 | 0.220 | |
| 136 | C2HClF4 | chloro-1,1,2,2,-tetrafluoroethane | 136.475 | | 263. | 399.9 | 37.2 | 244. | 0.273 | 0.281 | |
| 137 | C2HCl3 | trichloroethene | 131.389 | 186.8 | 360.4 | 572. | 50.5 | 256. | 0.265 | 0.213 | 0.9 |
| 138 | C2HCl5 | pentachloroethane | 202.295 | 244. | 435. | 646. | 34.8 | | | | 1.0 |
| 139 | C2HF3O2 | trifluoroacetic acid | 114.024 | 257.9 | 346. | 491.3 | 32.6 | 204. | 0.163 | 0.540 | 2.3 |
| 140 | C2H2 | acetylene | 26.038 | | 188.4 | 308.3 | 61.4 | 112.7 | 0.270 | 0.190 | 0.0 |
| 141 | C2H2Cl2 | cis-1,2-dichloroethene | 96.944 | 192.7 | 333.3 | 537. | 56. | | | | 1.8 |
| 142 | C2H2Cl2 | trans-1,2-dichloroethene | 96.944 | 223. | 321.9 | 513. | 48.1 | | | 0.232 | 0.0 |
| 143 | C2H2Cl4 | 1,1,2,2-tetrachloroethane | 167.850 | 237. | 419.4 | 661.2 | 58.4 | | | | 1.5 |
| 144 | C2H2F2 | 1,1-difluoroethene | 64.035 | 129. | 187.5 | 302.9 | 44.6 | 154.1 | 0.273 | 0.140 | 1.4 |
| 145 | C2H2O | ketene | 42.038 | 138. | 232. | 380. | 65. | 145. | 0.30 | 0.21 | 1.4 |
| 146 | C2H3Cl | vinyl chloride | 62.499 | 119.4 | 259.8 | 425. | 51.5 | 169. | 0.265 | 0.122 | 1.5 |
| 147 | C2H3ClF2 | 1-chloro-1,1-difluoroethane | 100.496 | 142. | 263.4 | 409.6 | 43.3 | 231. | 0.294 | 0.251 | 2.1 |
| 148 | C2H3ClO | acetyl chloride | 78.498 | 160.2 | 323.9 | 508.0 | 58.7 | 204. | 0.280 | 0.344 | 2.4 |
| 149 | C2H3Cl3 | 1,1,1-trichloroethane | 133.405 | 240. | 347.2 | 545. | 43.0 | | | 0.217 | 1.7 |
| 150 | C2H3Cl3 | 1,1,2-trichloroethane | 133.405 | 235.8 | 386.7 | 606. | 51.4 | | | | 1.4 |

| No | Formula | Name | CPVAP A | CPVAP B | CPVAP C | CPVAP D | DELHF | DELGF |
|---|---|---|---|---|---|---|---|---|
| 126 | C2Cl2F4 | 1,1-dichlorotetrafluoroethane | 4.045E+1 | 3.278E-1 | -2.752E-4 | 7.821E-8 | | |
| 127 | C2Cl2F4 | 1,2-dichlorotetrafluoroethane | 3.878E+1 | 3.440E-1 | -2.950E-4 | 8.508E-8 | -8.985E+5 | |
| 128 | C2Cl3F3 | 1,2,2-trichlorotrifluoroethane | 6.114E+1 | 2.874E-1 | -2.420E-4 | 6.904E-8 | -7.457E+5 | |
| 129 | C2Cl4 | tetrachloroethene | 4.597E+1 | 2.255E-1 | -2.294E-4 | 8.382E-8 | -1.214E+4 | 2.261E+4 |
| 130 | C2Cl4F2 | 1,1,2,2-tetrachlorodifluoroethane | | | | | | |
| 131 | C2F3N | trifluoroacetonitrile | 2.213E+1 | 2.519E-1 | -2.361E-4 | 8.207E-8 | -4.957E+5 | -4.622E+5 |
| 132 | C2F4 | perfluoroethene | 2.901E+1 | 2.277E-1 | -2.036E-4 | 6.778E-8 | -6.590E+5 | -6.241E+5 |
| 133 | C2F6 | perfluoroethane | 2.682E+1 | 3.458E-1 | -2.869E-4 | 8.135E-8 | -1.344E+6 | -1.258E+6 |
| 134 | C2N2 | cyanogen | 3.594E+1 | 9.253E-2 | -8.148E-5 | 2.950E-8 | 3.092E+5 | 2.974E+5 |
| 135 | C2HClF2 | 1-chloro-2,2-difluoroethene | | | | | | |
| 136 | C2HClF4 | chloro-1,1,2,2,-tetrafluoroethane | | | | | | |
| 137 | C2HCl3 | trichloroethene | 3.017E+1 | 2.287E-1 | -2.229E-4 | 8.244E-8 | -5.862E+3 | 1.989E+4 |
| 138 | C2HCl5 | pentachloroethane | 4.394E+1 | 3.374E-1 | -3.356E-4 | 1.213E-7 | -1.424E+5 | -6.670E+4 |
| 139 | C2HF3O2 | trifluoroacetic acid | | | | | | |
| 140 | C2H2 | acetylene | 2.682E+1 | 7.578E-2 | -5.007E-5 | 1.412E-8 | 2.269E+5 | 2.093E+5 |
| 141 | C2H2Cl2 | cis-1,2-dichloroethene | 1.161E+1 | 2.358E-1 | -2.100E-4 | 7.242E-8 | 1.880E+3 | 2.437E+4 |
| 142 | C2H2Cl2 | trans-1,2-dichloroethene | 1.828E+1 | 2.100E-1 | -1.764E-4 | 5.804E-8 | 4.190E+3 | 2.660E+4 |
| 143 | C2H2Cl4 | 1,1,2,2-tetrachloroethane | 2.767E+1 | 3.251E-1 | -2.974E-4 | 1.028E-7 | -1.528E+5 | -8.583E+4 |
| 144 | C2H2F2 | 1,1-difluoroethene | 3.073E+0 | 2.445E-1 | -2.099E-4 | 7.021E-8 | -3.454E+5 | -3.217E+5 |
| 145 | C2H2O | ketene | 6.385E+0 | 1.638E-1 | -1.084E-4 | 2.698E-8 | -6.113E+4 | -6.033E+4 |
| 146 | C2H3Cl | vinyl chloride | 5.949E+0 | 2.019E-1 | -1.536E-4 | 4.773E-8 | 3.517E+4 | 5.154E+4 |
| 147 | C2H3ClF2 | 1-chloro-1,1-difluoroethane | 1.682E+1 | 2.757E-1 | -1.992E-4 | 5.305E-8 | | |
| 148 | C2H3ClO | acetyl chloride | 2.502E+1 | 1.711E-1 | -9.856E-5 | 2.219E-8 | -2.441E+5 | -2.064E+5 |
| 149 | C2H3Cl3 | 1,1,1-trichloroethane | | | | | | |
| 150 | C2H3Cl3 | 1,1,2-trichloroethane | 6.322E+0 | 3.431E-1 | -2.958E-4 | 9.793E-8 | -1.386E+5 | -7.754E+4 |

| No | Formula | Name | Eq. | VP A | VP B | VP C | VP D | Tmin | Tmax | LDEN | TDEN |
|---|---|---|---|---|---|---|---|---|---|---|---|
| 126 | C2Cl2F4 | 1,1-dichlorotetrafluoroethane | 1 | -7.33582 | 1.62482 | -3.06234 | -2.42281 | 217 | TC | 1.455 | 298 |
| 127 | C2Cl2F4 | 1,2-dichlorotetrafluoroethane | 1 | -7.15825 | 1.10752 | -2.12022 | -4.54857 | 180 | TC | 1.48 | 277 |
| 128 | C2Cl3F3 | 1,2,2-trichlorotrifluoroethane | 1 | -7.26519 | 1.39273 | -2.50843 | -5.26657 | 238 | TC | 1.580 | 289 |
| 129 | C2Cl4 | tetrachloroethene | 1 | -7.36067 | 1.82732 | -3.47735 | -1.00033 | 252 | TC | 1.62 | 293 |
| 130 | C2Cl4F2 | 1,1,2,2-tetrachlorodifluoroethane | 1 | -7.80715 | 1.69009 | -3.12042 | -3.29269 | 299 | TC | 1.65 | 298 |
| 131 | C2F3N | trifluoroacetonitrile | 3 | 9.7917 | 1781.77 | -23.28 | | 142 | 206 | | |
| 132 | C2F4 | perfluoroethene | 1 | -6.74371 | 0.62458 | -1.94752 | -3.78881 | 145 | TC | 1.519 | 197 |
| 133 | C2F6 | perfluoroethane | 1 | -7.32301 | 1.50248 | -2.64678 | -4.93429 | 173 | TC | 1.590 | 195 |
| 134 | C2N2 | cyanogen | 2 | 51.703 | 4390.80 | -6.185 | 1130. | 250 | TC | 0.954 | 252 |
| 135 | C2HClF2 | 1-chloro-2,2-difluoroethene | 1 | -7.19815 | 1.77543 | -3.50534 | -0.68772 | 136 | TC | | |
| 136 | C2HClF4 | chloro-1,1,2,2,-tetrafluoroethane | 1 | -7.56490 | 1.81516 | -3.54300 | -1.04102 | 157 | TC | | |
| 137 | C2HCl3 | trichloroethene | 1 | -7.38190 | 1.94817 | -3.03294 | -5.34536 | 291 | TC | 1.462 | 293 |
| 138 | C2HCl5 | pentachloroethane | 1 | -7.50052 | 1.16078 | -3.48149 | -1.04212 | 312 | TC | 1.671 | 298 |
| 139 | C2HF3O2 | trifluoroacetic acid | 3 | 7.5356 | 2828.94 | -57.11 | | 285 | 345 | 1.535 | 273 |
| 140 | C2H2 | acetylene | 1 | -6.90128 | 1.26873 | -2.09113 | -2.75601 | 192 | TC | 0.615 | 189 |
| 141 | C2H2Cl2 | cis-1,2-dichloroethene | 1 | -6.97612 | 1.11972 | -1.88483 | -3.32030 | 274 | TC | 1.282 | 298 |
| 142 | C2H2Cl2 | trans-1,2-dichloroethene | 1 | -6.69776 | 1.08543 | -2.90387 | -0.25533 | 258 | TC | 1.255 | 293 |
| 143 | C2H2Cl4 | 1,1,2,2-tetrachloroethane | 1 | -7.98542 | 2.49931 | -4.07076 | -0.69180 | 303 | TC | 1.600 | 293 |
| 144 | C2H2F2 | 1,1-difluoroethene | 1 | -6.58895 | 0.90734 | -0.82882 | 0.11779 | 130 | TC | 0.617 | 297 |
| 145 | C2H2O | ketene | 3 | 9.3995 | 1849.21 | -35.15 | | 170 | 255 | | |
| 146 | C2H3Cl | vinyl chloride | 1 | -6.50008 | 1.21422 | -2.57867 | -2.00937 | 208 | TC | 0.969 | 259 |
| 147 | C2H3ClF2 | 1-chloro-1,1-difluoroethane | 1 | -7.83556 | 2.79382 | -4.42364 | 0.06334 | 143 | TC | 1.10 | 303 |
| 148 | C2H3ClO | acetyl chloride | 1 | -7.94455 | 1.81437 | -2.09194 | -1.98959 | 267 | TC | 1.104 | 293 |
| 149 | C2H3Cl3 | 1,1,1-trichloroethane | 1 | -7.31317 | 2.04642 | -3.77747 | -0.45475 | 247 | TC | 1.339 | 298 |
| 150 | C2H3Cl3 | 1,1,2-trichloroethane | 1 | -7.71341 | 2.15518 | -3.96435 | -0.54604 | 323 | TC | 1.441 | 293 |

| No | Formula | Name | MolWt | Tfp K | Tb K | Tc K | Pc bar | Vc $cm^3/mol$ | Zc | Omega | Dipm debye |
|---|---|---|---|---|---|---|---|---|---|---|---|
| 151 | C2H3F | vinyl fluoride | 46.044 | 130.0 | 201.0 | 327.9 | 52.4 | 144. | 0.277 | 0.157 | 1.4 |
| 152 | C2H3F3 | 1,1,1-trifluoroethane | 84.041 | 161.9 | 225.6 | 346.3 | 37.6 | 194. | 0.253 | 0.251 | 2.3 |
| 153 | C2H3N | acetonitrile | 41.053 | 229.3 | 354.8 | 545.5 | 48.3 | 173. | 0.184 | 0.327 | 3.5 |
| 154 | C2H3NO | methyl isocyanate | 57.052 | | 312. | 491. | 55.7 | | | 0.278 | |
| 155 | C2H4 | ethylene | 28.054 | 104.0 | 169.3 | 282.4 | 50.4 | 130.4 | 0.280 | 0.089 | 0.0 |
| 156 | C2H4Br2 | 1,2-dibromoethane | 187.862 | 283.3 | 404.7 | 646. | 53.5 | | | 0.795 | 1.0 |
| 157 | C2H4Cl2 | 1,1-dichloroethane | 98.960 | 176.2 | 330.5 | 523. | 50.7 | 236. | 0.275 | 0.240 | 2.0 |
| 158 | C2H4Cl2 | 1,2-dichloroethane | 98.960 | 237.5 | 356.7 | 566. | 53.7 | 225. | 0.259 | 0.278 | 1.8 |
| 159 | C2H4F2 | 1,1-difluoroethane | 66.051 | 156.2 | 248.2 | 386.7 | 45.0 | 181. | 0.253 | 0.256 | 2.3 |
| 160 | C2H4O | acetaldehyde | 44.054 | 150.2 | 294. | 461. | 55.7 | 154. | 0.220 | 0.303 | 2.5 |
| 161 | C2H4O | ethylene oxide | 44.054 | 161. | 283.7 | 469. | 71.9 | 140. | 0.259 | 0.202 | 1.9 |
| 162 | C2H4O2 | acetic acid | 60.052 | 289.8 | 391.1 | 592.7 | 57.9 | 171. | 0.201 | 0.447 | 1.3 |
| 163 | C2H4O2 | methyl formate | 60.052 | 174.2 | 304.9 | 487.2 | 60.0 | 172. | 0.255 | 0.257 | 1.8 |
| 164 | C2H5Br | ethyl bromide | 108.966 | 154.6 | 311.5 | 503.9 | 62.3 | 215. | 0.320 | 0.229 | 2.0 |
| 165 | C2H5Cl | ethyl chloride | 64.515 | 136.8 | 285.5 | 460.4 | 52.7 | 199. | 0.274 | 0.191 | 2.0 |
| 166 | C2H5F | ethyl fluoride | 48.060 | 129.9 | 235.5 | 375.3 | 50.2 | 169. | 0.272 | 0.215 | 2.0 |
| 167 | C2H5I | ethyl iodide | 155.967 | 165. | 345.6 | 554. | 47.0 | | | | 1.7 |
| 168 | C2H6 | ethane | 30.070 | 89.9 | 184.6 | 305.4 | 48.8 | 148.3 | 0.285 | 0.099 | 0.0 |
| 169 | C2H6O | dimethyl ether | 46.069 | 131.7 | 248.3 | 400.0 | 52.4 | 178. | 0.287 | 0.200 | 1.3 |
| 170 | C2H6O | ethanol | 46.069 | 159.1 | 351.4 | 513.9 | 61.4 | 167.1 | 0.240 | 0.644 | 1.7 |
| 171 | C2H6O2 | ethylene glycol | 62.069 | 260.2 | 470.5 | (645.) | (77.) | | | | 2.2 |
| 172 | C2H6S | ethyl mercaptan | 62.134 | 125.3 | 308.2 | 499. | 54.9 | 207. | 0.274 | 0.191 | 1.5 |
| 173 | C2H6S | dimethyl sulfide | 62.130 | 174.9 | 310.5 | 503.0 | 55.3 | 201. | 0.266 | 0.191 | 1.5 |
| 174 | C2H7N | ethyl amine | 45.085 | 192. | 289.7 | 456.4 | 56.4 | 182. | 0.270 | 0.289 | 1.3 |
| 175 | C2H7N | dimethylamine | 45.085 | 181.0 | 280.0 | 437.7 | 53.1 | | | 0.302 | 1.0 |

| No | Formula | Name | CPVAP A | CPVAP B | CPVAP C | CPVAP D | DELHF | DELGF |
|---|---|---|---|---|---|---|---|---|
| 151 | C2H3F | vinyl fluoride | | | | | -1.172E+5 | |
| 152 | C2H3F3 | 1,1,1-trifluoroethane | 5.744E+0 | 3.141E-1 | -2.597E-4 | 8.415E-8 | -7.461E+5 | -6.792E+5 |
| 153 | C2H3N | acetonitrile | 2.048E+1 | 1.196E-1 | -4.492E-5 | 3.203E-9 | 8.792E+4 | 1.057E+5 |
| 154 | C2H3NO | methyl isocyanate | 3.576E+1 | 1.040E-1 | -5.820E-6 | -1.687E-8 | -9.000E+4 | |
| 155 | C2H4 | ethylene | 3.806E+0 | 1.566E-1 | -8.348E-5 | 1.755E-8 | 5.234E+4 | 6.816E+4 |
| 156 | C2H4Br2 | 1,2-dibromoethane | 2.500E+1 | 2.517E-1 | -1.833E-4 | 5.646E-8 | -3.894E+4 | -1.060E+4 |
| 157 | C2H4Cl2 | 1,1-dichloroethane | 1.247E+1 | 2.696E-1 | -2.050E-4 | 6.301E-8 | -1.300E+5 | -7.314E+4 |
| 158 | C2H4Cl2 | 1,2-dichloroethane | 2.049E+1 | 2.310E-1 | -1.438E-4 | 3.389E-8 | -1.298E+5 | -7.390E+4 |
| 159 | C2H4F2 | 1,1-difluoroethane | 8.675E+1 | 2.396E-1 | -1.457E-4 | 3.394E-8 | -4.940E+5 | -4.365E+5 |
| 160 | C2H4O | acetaldehyde | 7.716E+0 | 1.823E-1 | -1.007E-4 | 2.380E-8 | -1.644E+5 | -1.334E+5 |
| 161 | C2H4O | ethylene oxide | -7.519E+0 | 2.222E-1 | -1.256E-4 | 2.592E-8 | -5.267E+4 | -1.310E+4 |
| 162 | C2H4O2 | acetic acid | 4.840E+0 | 2.549E-1 | -1.753E-4 | 4.949E-8 | -4.351E+5 | -3.769E+5 |
| 163 | C2H4O2 | methyl formate | 1.432E+0 | 2.700E-1 | -1.949E-4 | 5.702E-8 | -3.500E+5 | -2.974E+5 |
| 164 | C2H5Br | ethyl bromide | 6.657E+0 | 2.348E-1 | -1.472E-4 | 3.804E-8 | -6.406E+4 | -2.633E+4 |
| 165 | C2H5Cl | ethyl chloride | -5.527E-1 | 2.606E-1 | -1.840E-4 | 5.548E-8 | -1.118E+5 | -6.004E+4 |
| 166 | C2H5F | ethyl fluoride | 4.346E+0 | 2.180E-1 | -1.166E-4 | 2.410E-8 | -2.617E+5 | -2.097E+5 |
| 167 | C2H5I | ethyl iodide | 1.011E+1 | 2.253E-1 | -1.382E-4 | 3.531E-8 | -8.370E+3 | 2.135E+4 |
| 168 | C2H6 | ethane | 5.409E+0 | 1.781E-1 | -6.938E-5 | 8.713E-9 | -8.474E+4 | -3.295E+4 |
| 169 | C2H6O | dimethyl ether | 1.702E+1 | 1.791E-1 | -5.234E-5 | -1.918E-9 | -1.842E+5 | -1.130E+5 |
| 170 | C2H6O | ethanol | 9.014E+0 | 2.141E-1 | -8.390E-5 | 1.373E-9 | -2.350E+5 | -1.684E+5 |
| 171 | C2H6O2 | ethylene glycol | 3.570E+1 | 2.483E-1 | -1.497E-4 | 3.010E-8 | -3.896E+5 | -3.047E+5 |
| 172 | C2H6S | ethyl mercaptan | 1.492E+1 | 2.351E-1 | -1.356E-4 | 3.162E-8 | -4.614E+4 | -4.670E+3 |
| 173 | C2H6S | dimethyl sulfide | 2.430E+1 | 1.875E-1 | -6.875E-5 | 4.099E-9 | -3.756E+4 | 6.950E+3 |
| 174 | C2H7N | ethyl amine | 3.693E+0 | 2.752E-1 | -1.583E-4 | 3.808E-8 | -4.605E+4 | 3.730E+4 |
| 175 | C2H7N | dimethylamine | -1.717E-1 | 2.695E-1 | -1.329E-4 | 2.339E-8 | -1.880E+4 | 6.800E+4 |

| No | Formula | Name | Eq. | VP A | VP B | VP C | VP D | Tmin | Tmax | LDEN | TDEN |
|---|---|---|---|---|---|---|---|---|---|---|---|
| 151 | C2H3F | vinyl fluoride | 1 | -6.80471 | 1.67182 | -3.29094 | -0.69493 | 114 | TC | 0.681 | 263 |
| 152 | C2H3F3 | 1,1,1-trifluoroethane | 1 | -7.87141 | 2.78418 | -4.55799 | 0.56876 | 163 | TC | | |
| 153 | C2H3N | acetonitrile | 2 | 40.774 | 5392.43 | -4.357 | 2615. | 300 | TC | 0.782 | 293 |
| 154 | C2H3NO | methyl isocyanate | 3 | 9.7056 | 2480.37 | -56.31 | | 230 | 340 | 0.958 | 293 |
| 155 | C2H4 | ethylene | 1 | -6.32055 | 1.16819 | -1.55935 | -1.83552 | 105 | TC | 0.577 | 163 |
| 156 | C2H4Br2 | 1,2-dibromoethane | 1 | -7.45007 | 2.22849 | -3.97795 | -0.24734 | 290 | TC | 2.180 | 293 |
| 157 | C2H4Cl2 | 1,1-dichloroethane | 2 | 49.613 | 5422.68 | -5.726 | 2380. | 280 | TC | 1.168 | 298 |
| 158 | C2H4Cl2 | 1,2-dichloroethane | 1 | -7.36864 | 1.76727 | -3.34295 | -1.43530 | 260 | TC | 1.250 | 289 |
| 159 | C2H4F2 | 1,1-difluoroethane | 1 | -7.40625 | 1.76980 | -3.44560 | -1.09392 | 157 | TC | 1.012 | 247 |
| 160 | C2H4O | acetaldehyde | 1 | -7.04687 | 0.12142 | -2.66037E-2 | -5.90300 | 273 | TC | 0.778 | 293 |
| 161 | C2H4O | ethylene oxide | 1 | -6.56234 | 0.42696 | -1.25638 | -3.18133 | 238 | TC | 0.899 | 273 |
| 162 | C2H4O2 | acetic acid | 1 | -7.83183 | 5.51929E-4 | 0.24709 | -8.50462 | 304 | TC | 1.049 | 293 |
| 163 | C2H4O2 | methyl formate | 1 | -6.99601 | 0.89328 | -2.52294 | -3.16636 | 220 | TC | 0.974 | 293 |
| 164 | C2H5Br | ethyl bromide | 1 | -9.14807 | 5.49831 | -6.68657 | 6.27287 | 301 | TC | 1.451 | 298 |
| 165 | C2H5Cl | ethyl chloride | 1 | -7.23667 | 2.11017 | -3.53882 | 0.34775 | 217 | TC | 0.896 | 293 |
| 166 | C2H5F | ethyl fluoride | 1 | -6.82738 | 0.59267 | -0.73934 | -3.69185 | 266 | TC | | |
| 167 | C2H5I | ethyl iodide | 1 | -6.50172 | 1.05321 | -3.16148 | -0.64188 | 290 | TC | 1.950 | 293 |
| 168 | C2H6 | ethane | 1 | -6.34307 | 1.01630 | -1.19116 | -2.03539 | 133 | TC | 0.548 | 183 |
| 169 | C2H6O | dimethyl ether | 1 | -7.12597 | 1.81710 | -3.10058 | -0.91638 | 194 | TC | 0.667 | 293 |
| 170 | C2H6O | ethanol | 1 | -8.51838 | 0.34163 | -5.73683 | 8.32581 | 293 | TC | 0.789 | 293 |
| 171 | C2H6O2 | ethylene glycol | 3 | 13.6299 | 6022.18 | -28.25 | | 364 | 494 | 1.114 | 293 |
| 172 | C2H6S | ethyl mercaptan | 1 | -6.96578 | 1.50970 | -2.73740 | -1.73828 | 273 | TC | 0.839 | 293 |
| 173 | C2H6S | dimethyl sulfide | 1 | -6.94973 | 1.43646 | -2.51444 | -2.47611 | 222 | TC | 0.848 | 293 |
| 174 | C2H2N | ethyl amine | 1 | -7.20059 | 1.20679 | -3.71972 | -4.33511 | 215 | TC | 0.683 | 293 |
| 175 | C2H7N | dimethylamine | 1 | -7.90295 | 2.81577 | -6.31338 | -0.22407 | 240 | TC | 0.656 | 293 |

| No | Formula | Name | MolWt | Tfp K | Tb K | Tc K | Pc bar | Vc $cm^3/mol$ | Zc | Omega | Dipm debye |
|---|---|---|---|---|---|---|---|---|---|---|---|
| 176 | C2H7NO | monoethanolamine | 61.084 | 283.5 | 443.5 | 614. | 44.5 | 196. | 0.17 | | 2.6 |
| 177 | C2H8N2 | ethylenediamine | 60.099 | 284. | 390.4 | 593. | 62.8 | 206. | 0.26 | 0.51 | 1.9 |
| 178 | C3ClF5O | chloropentafluoroacetone | 182.475 | | 281.0 | 410.6 | 28.8 | | | 0.347 | |
| 179 | C3F6O | perfluoroacetone | 166.020 | | 245.7 | 357.1 | 28.4 | 329. | 0.314 | 0.365 | |
| 180 | C3F8 | perfluoropropane | 188.017 | 90. | 236.5 | 345.1 | 26.8 | 299.8 | 0.280 | 0.325 | |
| 181 | C3H3F3 | trifluoropropene | 96.051 | | 244. | 376.2 | 38.0 | 211. | 0.256 | 0.238 | |
| 182 | C3H3F5 | 1,1,1,2,2-pentafluoropropane | 134.047 | | 255.7 | 380.1 | 31.4 | 273. | 0.271 | 0.308 | |
| 183 | C3H3N | acrylonitrile | 53.064 | 189.5 | 350.5 | 536. | 45.6 | 210. | 0.21 | 0.35 | 3.5 |
| 184 | C3H3NO | isoxazole | 69.063 | | 368. | 552. | | | | | 2.8 |
| 185 | C3H4 | propadiene | 40.065 | 136.9 | 238.7 | 393. | 54.7 | 162. | 0.271 | 0.313 | 0.2 |
| 186 | C3H4 | methyl acetylene | 40.065 | 170.5 | 249.9 | 402.4 | 56.3 | 164. | 0.275 | 0.215 | 0.7 |
| 187 | C3H4O | acrolein | 56.064 | 186. | 326. | 506. | 51.6 | | | 0.33 | 2.9 |
| 188 | C3H4O2 | acrylic acid | 72.064 | 285. | 414. | 615. | 56.7 | 210. | 0.23 | 0.56 | |
| 189 | C3H4O2 | vinyl formate | 72.064 | 215.5 | 319.6 | 475. | 57.7 | 210. | 0.31 | 0.55 | |
| 190 | C3H5Cl | allyl chloride | 76.526 | 138.7 | 318.3 | 514. | 47.6 | 234. | 0.26 | 0.13 | 2.0 |
| 191 | C3H5Cl3 | 1,2,3-trichloropropane | 147.432 | 258.5 | 429. | 651. | 39.5 | 348. | 0.25 | 0.31 | 1.6 |
| 192 | C3H5N | propionitrile | 55.080 | 180.3 | 370.3 | 564.4 | 41.8 | 229. | 0.205 | 0.313 | 3.7 |
| 193 | C3H6 | cyclopropane | 42.081 | 145.7 | 240.3 | 397.8 | 54.9 | 163. | 0.274 | 0.130 | 0.0 |
| 194 | C3H6 | propylene | 42.081 | 87.9 | 225.5 | 364.9 | 46.0 | 181. | 0.274 | 0.144 | 0.4 |
| 195 | C3H6Cl2 | 1,2-dichloropropane | 112.987 | 172.7 | 369.5 | 577. | 44.5 | 226. | 0.21 | 0.24 | 1.9 |
| 196 | C3H6O | acetone | 58.080 | 178.2 | 329.2 | 508.1 | 47.0 | 209. | 0.232 | 0.304 | 2.9 |
| 197 | C3H6O | allyl alcohol | 58.080 | 144. | 370.2 | 543.0 | | | | | |
| 198 | C3H6O | propionaldehyde | 58.080 | 193. | 321. | 515.3 | 63.3 | | | 0.313 | 2.7 |
| 199 | C3H6O | 1,2-propylene oxide | 58.080 | 161. | 308. | 482.2 | 49.2 | 186. | 0.229 | 0.269 | 2.0 |
| 200 | C3H6O | vinyl methyl ether | 58.080 | 151.5 | 278. | 436. | 47.6 | 205. | 0.27 | 0.34 | |

| No | Formula | Name | CPVAP A | CPVAP B | CPVAP C | CPVAP D | DELHF | DELGF |
|---|---|---|---|---|---|---|---|---|
| 176 | C2H7NO | monoethanolamine | 9.311E+0 | 3.009E-1 | -1.818E-4 | 4.656E-8 | -2.017E+5 | |
| 177 | C2H8N2 | ethylenediamine | 3.830E+1 | 2.407E-1 | -4.338E-5 | -3.948E-8 | | |
| 178 | C3ClF5O | chloropentafluoroacetone | | | | | | |
| 179 | C3F6O | perfluoroacetone | | | | | | |
| 180 | C3F8 | perfluoropropane | | | | | | |
| 181 | C3H3F3 | trifluoropropene | | | | | | |
| 182 | C3H3F5 | 1,1,1,2,2-pentafluoropropane | | | | | | |
| 183 | C3H3N | acrylonitrile | 1.069E+1 | 2.208E-1 | -1.565E-4 | 4.601E-8 | 1.851E+5 | 1.954E+5 |
| 184 | C3H3NO | isoxazole | | | | | | |
| 185 | C3H4 | propadiene | 9.906E+0 | 1.977E-1 | -1.182E-4 | 2.782E-8 | 1.923E+5 | 2.025E+5 |
| 186 | C3H4 | methyl acetylene | 1.471E+1 | 1.864E-1 | -1.174E-4 | 3.224E-8 | 1.856E+5 | 1.946E+5 |
| 187 | C3H4O | acrolein | 1.197E+1 | 2.106E-1 | -1.071E-4 | 1.906E-8 | -7.092E+4 | -6.519E+4 |
| 188 | C3H4O2 | acrylic acid | 1.742E+0 | 3.191E-1 | -2.352E-4 | 6.975E-8 | -3.365E+5 | -2.863E+5 |
| 189 | C3H4O2 | vinyl formate | 2.781E+1 | 1.839E-1 | -3.560E-5 | -2.335E-7 | | |
| 190 | C3H5Cl | allyl chloride | 2.529E+0 | 3.047E-1 | -2.278E-4 | 7.293E-8 | -6.280E+2 | 4.363E+4 |
| 191 | C3H5Cl3 | 1,2,3-trichloropropane | 2.688E+1 | 3.622E-1 | -2.787E-4 | 8.788E-8 | -1.859E+5 | -9.785E+4 |
| 192 | C3H5N | propionitrile | 1.540E+1 | 2.245E-1 | -1.100E-4 | 1.954E-8 | 5.066E+4 | 9.621E+4 |
| 193 | C3H6 | cyclopropane | -3.524E+1 | 3.813E-1 | -2.881E-4 | 9.035E-8 | 5.334E+4 | 1.045E+5 |
| 194 | C3H6 | propylene | 3.710E+0 | 2.345E-1 | -1.160E-4 | 2.205E-8 | 2.043E+4 | 6.276E+4 |
| 195 | C3H6Cl2 | 1,2-dichloropropane | 1.045E+1 | 3.655E-1 | -2.604E-4 | 7.741E-8 | -1.660E+5 | -8.315E+4 |
| 196 | C3H6O | acetone | 6.301E+0 | 2.606E-1 | -1.253E-4 | 2.038E-8 | -2.177E+5 | -1.532E+5 |
| 197 | C3H6O | allyl alcohol | -1.105E+0 | 3.146E-1 | -2.032E-4 | 5.321E-8 | -1.321E+5 | -7.130E+4 |
| 198 | C3H6O | propionaldehyde | 1.172E+1 | 2.614E-1 | -1.300E-4 | 2.126E-8 | -1.922E+5 | -1.305E+5 |
| 199 | C3H6O | 1,2-propylene oxide | -8.457E+0 | 3.257E-1 | -1.989E-4 | 4.823E-8 | -9.282E+4 | -2.580E+4 |
| 200 | C3H6O | vinyl methyl ether | 1.563E+1 | 2.341E-1 | -9.697E-5 | 1.062E-8 | | |

| No | Formula | Name | Eq. | VP A | VP B | VP C | VP D | Tmin | Tmax | LDEN | TDEN |
|---|---|---|---|---|---|---|---|---|---|---|---|
| 176 | C2H7NO | monoethanolamine | 1 | -10.8842 | 3.03743 | -7.21939 | -2.99322 | 379 | TC | 1.016 | 293 |
| 177 | C2H8N2 | ethylenediamine | 1 | -8.82254 | 2.27867 | -3.52636 | -6.97579 | 285 | TC | 0.896 | 293 |
| 178 | C3ClF5O | chloropentafluoroacetone | | | | | | | | | |
| 179 | C3F6O | perfluoroacetone | | | | | | | | | |
| 180 | C3F8 | perfluoropropane | 3 | 9.3122 | 1901.54 | -31.97 | | 194 | 237 | 1.350 | 293 |
| 181 | C3H3F3 | trifluoropropene | | | | | | | | | |
| 182 | C3H3F5 | 1,1,1,2,2-pentafluoropropane | | | | | | | | | |
| 183 | C3H3N | acrylonitrile | 3 | 9.3051 | 2782.21 | -51.15 | | 255 | 385 | 0.806 | 293 |
| 184 | C3H3NO | isoxazole | | | | | | | | 1.078 | 293 |
| 185 | C3H4 | propadiene | 3 | 6.5361 | 1054.72 | -77.08 | | 174 | 257 | 0.658 | 238 |
| 186 | C3H4 | methyl acetylene | 1 | -7.43860 | 2.62026 | -5.76535 | 7.55261 | 178 | TC | 0.706 | 223 |
| 187 | C3H4O | acrolein | 3 | 9.2855 | 2606.53 | -45.15 | | 235 | 360 | 0.839 | 293 |
| 188 | C3H4O2 | acrylic acid | 3 | 9.9415 | 3319.18 | -80.15 | | 315 | 450 | 1.051 | 293 |
| 189 | C3H4O2 | vinyl formate | 3 | 10.0329 | 2569.68 | -63.15 | | 240 | 350 | 0.963 | 293 |
| 190 | C3H5Cl | allyl chloride | 1 | -6.76334 | 2.50730 | -7.64033 | 11.6666 | 286 | TC | 0.937 | 293 |
| 191 | C3H5Cl3 | 1,2,3-trichloropropane | 3 | 9.5044 | 3417.27 | -69.15 | | 315 | 470 | 1.389 | 293 |
| 192 | C3H5N | propionitrile | 1 | -7.27719 | 0.46035 | -0.45714 | -10.1636 | 309 | TC | 0.782 | 293 |
| 193 | C3H6 | cyclopropane | 1 | -7.98411 | 4.38160 | -5.72309 | 3.40444 | 183 | TC | 0.563 | 288 |
| 194 | C3H6 | propylene | 1 | -6.64231 | 1.21857 | -1.81005 | -2.48212 | 140 | TC | 0.612 | 223 |
| 195 | C3H6Cl2 | 1,2-dichloropropane | 1 | -6.82259 | 0.54655 | -1.59982 | -5.05429 | 318 | TC | 1.15 | 293 |
| 196 | C3H6O | acetone | 1 | -7.45514 | 1.20200 | -2.43926 | -3.35590 | 259 | TC | 0.790 | 293 |
| 197 | C3H6O | allyl alcohol | 3 | 10.2864 | 2928.20 | -85.15 | | 286 | 400 | 0.855 | 288 |
| 198 | C3H6O | propionaldehyde | 1 | -7.18479 | 1.00298 | -1.49247 | -5.13288 | 235 | TC | 0.797 | 293 |
| 199 | C3H6O | 1,2-propylene oxide | 1 | -6.97569 | 0.63650 | -1.49187 | -6.37743 | 249 | TC | 0.829 | 293 |
| 200 | C3H6O | vinyl methyl ether | 3 | 7.8400 | 1980.22 | -25.15 | | 190 | 315 | 0.750 | 293 |

| No | Formula | Name | MolWt | Tfp K | Tb K | Tc K | Pc bar | Vc $cm^3$/mol | Zc | Omega | Dipm debye |
|---|---|---|---|---|---|---|---|---|---|---|---|
| 201 | C3H6O2 | propionic acid | 74.080 | 252.5 | 414.5 | 612. | 54. | 222. | 0.183 | 0.520 | 1.5 |
| 202 | C3H6O2 | ethyl formate | 74.080 | 193.8 | 327.5 | 508.5 | 47.4 | 229. | 0.257 | 0.285 | 2.0 |
| 203 | C3H6O2 | methyl acetate | 74.080 | 175. | 330.4 | 506.8 | 46.9 | 228. | 0.254 | 0.326 | 1.7 |
| 204 | C3H7Cl | propyl chloride | 78.542 | 150.4 | 320.4 | 503. | 45.8 | 254. | 0.278 | 0.235 | 2.0 |
| 205 | C3H7Cl | isopropyl chloride | 78.542 | 156.0 | 308.9 | 485.0 | 47.2 | 230. | 0.269 | 0.232 | 2.1 |
| 206 | C3H8 | propane | 44.094 | 85.5 | 231.1 | 369.8 | 42.5 | 203. | 0.281 | 0.153 | 0.0 |
| 207 | C3H8O | 1-propanol | 60.096 | 146.9 | 370.3 | 536.8 | 51.7 | 219. | 0.253 | 0.623 | 1.7 |
| 208 | C3H8O | isopropyl alcohol | 60.096 | 184.7 | 355.4 | 508.3 | 47.6 | 220. | 0.248 | 0.665 | 1.7 |
| 209 | C3H8O | methyl ethyl ether | 60.096 | 134. | 280.6 | 437.8 | 44.0 | 221. | 0.267 | 0.244 | 1.2 |
| 210 | C3H8O2 | methylal | 76.096 | 168. | 315. | 480.6 | 39.5 | 213. | 0.211 | 0.286 | 1.0 |
| 211 | C3H8O2 | 1,2-propanediol | 76.096 | 213. | 460.5 | 625. | 60.7 | 237. | 0.28 | | 3.6 |
| 212 | C3H8O2 | 1,3-propanediol | 76.096 | 246.4 | 487.6 | 724. | 89.5 | | | | 3.7 |
| 213 | C3H8O3 | glycerol | 92.095 | 291. | 563. | 726. | 66.8 | 255. | 0.28 | | 3.0 |
| 214 | C3H8S | methyl ethyl sulfide | 76.157 | 167.2 | 339.8 | 533. | 42.6 | | | 0.216 | |
| 215 | C3H9BO3 | trimethyl borate | 103.912 | | 342. | 501.7 | 35.9 | | | 0.415 | 0.8 |
| 216 | C3H9N | n-propyl amine | 59.112 | 190. | 321.7 | 497.0 | 48.1 | 233. | 0.271 | 0.303 | 1.3 |
| 217 | C3H9N | isopropyl amine | 59.112 | 177.9 | 305.6 | 471.8 | 45.4 | 221. | 0.255 | 0.291 | |
| 218 | C3H9N | trimethyl amine | 59.112 | 156. | 276.0 | 433.3 | 40.9 | 254. | 0.288 | 0.205 | 0.6 |
| 219 | C4F8 | perfluorocyclobutane | 200.028 | | 267.2 | 388.5 | 27.8 | 324. | 0.279 | 0.356 | |
| 220 | C4F10 | perfluorobutane | 238.024 | 145. | 271.2 | 386.4 | 23.2 | 378. | 0.274 | 0.374 | |
| 221 | C4H4 | vinylacetylene | 52.076 | 227.6 | 278.1 | 455. | 49.6 | 202. | 0.26 | 0.092 | |
| 222 | C4H4O | furan | 68.075 | 187.5 | 304.5 | 490.2 | 55.0 | 218. | 0.295 | 0.209 | 0.7 |
| 223 | C4H4S | thiophene | 84.136 | 234.9 | 357.2 | 579.4 | 56.9 | 219. | 0.258 | 0.196 | 0.5 |
| 224 | C4H5N | allyl cyanide | 67.091 | 186.7 | 392. | 585. | 39.5 | 265. | 0.22 | 0.39 | 3.4 |
| 225 | C4H5N | pyrrole | 67.091 | | 403.0 | 639.8 | | | | | 1.8 |

| No | Formula | Name | CPVAP A | CPVAP B | CPVAP C | CPVAP D | DELHF | DELGF |
|---|---|---|---|---|---|---|---|---|
| 201 | C3H6O2 | propionic acid | 5.669E+0 | 3.689E-1 | -2.865E-4 | 9.877E-8 | -4.554E+5 | -3.696E+5 |
| 202 | C3H6O2 | ethyl formate | 2.467E+1 | 2.316E-1 | -2.120E-5 | -5.359E-8 | -3.715E+5 | |
| 203 | C3H6O2 | methyl acetate | 1.655E+1 | 2.245E-1 | -4.342E-5 | 2.914E-8 | -4.097E+5 | |
| 204 | C3H7Cl | propyl chloride | -3.345E+0 | 3.626E-1 | -2.508E-4 | 7.448E-8 | -1.302E+5 | -5.070E+4 |
| 205 | C3H7Cl | isopropyl chloride | 1.842E+0 | 3.488E-1 | -2.244E-4 | 5.862E-8 | -1.465E+5 | -6.255E+4 |
| 206 | C3H8 | propane | -4.224E+0 | 3.063E-1 | -1.586E-4 | 3.215E-8 | -1.039E+5 | -2.349E+4 |
| 207 | C3H8O | 1-propanol | 2.470E+0 | 3.325E-1 | -1.855E-4 | 4.296E-8 | -2.566E+5 | -1.619E+5 |
| 208 | C3H8O | isopropyl alcohol | 3.243E+1 | 1.885E-1 | 6.406E-5 | -9.261E-8 | -2.726E+5 | -1.735E+5 |
| 209 | C3H8O | methyl ethyl ether | 1.867E+1 | 2.685E-1 | -1.025E-4 | 8.951E-9 | -2.166E+5 | -1.177E+5 |
| 210 | C3H8O2 | methylal | | | | | | |
| 211 | C3H8O2 | 1,2-propanediol | 6.322E-1 | 4.212E-1 | -2.981E-4 | 8.951E-8 | -4.242E+5 | |
| 212 | C3H8O2 | 1,3-propanediol | 8.269E+0 | 3.676E-1 | -2.162E-4 | 5.053E-8 | -4.091E+5 | |
| 213 | C3H8O3 | glycerol | 8.424E+0 | 4.442E-1 | -3.159E-4 | 9.378E-8 | -5.853E+5 | |
| 214 | C3H8S | methyl ethyl sulfide | 1.953E+1 | 2.891E-1 | -1.209E-4 | 1.287E-8 | -5.966E+4 | 1.140E+4 |
| 215 | C3H9BO3 | trimethyl borate | | | | | | |
| 216 | C3H9N | n-propyl amine | 6.691E+0 | 3.498E-1 | -1.822E-4 | 3.586E-8 | -7.243E+4 | 3.982E+4 |
| 217 | C3H9N | isopropyl amine | -7.486E+0 | 4.175E-1 | -2.826E-4 | 8.348E-8 | -8.382E+4 | |
| 218 | C3H9N | trimethyl amine | -8.206E+0 | 3.972E-1 | -2.219E-4 | 4.622E-8 | -2.386E+4 | 9.898E+4 |
| 219 | C4F8 | perfluorocyclobutane | | | | | | |
| 220 | C4F10 | perfluorobutane | | | | | | |
| 221 | C4H4 | vinylacetylene | 6.757E+0 | 2.841E-1 | -2.265E-4 | 7.461E-8 | 3.048E+5 | 3.062E+5 |
| 222 | C4H4O | furan | -3.553E+1 | 4.321E-1 | -3.455E-4 | 1.074E-7 | -3.470E+4 | 8.790E+2 |
| 223 | C4H4S | thiophene | -3.061E+1 | 4.480E-1 | -3.772E-4 | 1.253E-7 | 1.158E+5 | 1.269E+5 |
| 224 | C4H5N | allyl cyanide | 2.170E+1 | 2.572E-1 | -1.192E-4 | 1.229E-8 | | |
| 225 | C4H5N | pyrrole | | | | | 1.084E+5 | |

| No | Formula | Name | Eq. | VP A | VP B | VP C | VP D | Tmin | Tmax | LDEN | TDEN |
|---|---|---|---|---|---|---|---|---|---|---|---|
| 201 | C3H6O2 | propionic acid | 1 | -8.69958 | 1.49460 | -4.50355 | 1.06898 | 345 | TC | 0.993 | 293 |
| 202 | C3H6O2 | ethyl formate | 1 | -7.16968 | 1.13188 | -3.37309 | -3.53058 | 277 | TC | 0.927 | 289 |
| 203 | C3H6O2 | methyl acetate | 1 | -8.05406 | 2.56375 | -5.12994 | 0.16125 | 275 | TC | 0.934 | 293 |
| 204 | C3H7Cl | propyl chloride | 1 | -7.55764 | 2.60153 | -5.06041 | 3.31163 | 248 | TC | 0.891 | 293 |
| 205 | C3H7Cl | isopropyl chloride | 3 | 9.4182 | 2490.48 | -43.15 | | 225 | 340 | 0.862 | 293 |
| 206 | C3H8 | propane | 1 | -6.72219 | 1.33236 | -2.13868 | -1.38551 | 145 | TC | 0.582 | 231 |
| 207 | C3H8O | 1-propanol | 1 | -8.05594 | 4.25183E-2 | -7.51296 | 6.89004 | 260 | TC | 0.804 | 293 |
| 208 | C3H8O | isopropyl alcohol | 1 | -8.16927 | -9.43213E-2 | -8.10040 | 7.85000 | 250 | TC | 0.786 | 293 |
| 209 | C3H8O | methyl ethyl ether | 1 | -7.64466 | 2.88475 | -6.32922 | 0.33736 | 224 | TC | 0.700 | 293 |
| 210 | C3H8O2 | methylal | 3 | 9.2035 | 2415.92 | -52.58 | | 270 | 315 | 0.888 | 291 |
| 211 | C3H8O2 | 1,2-propanediol | 3 | 13.9122 | 6091.95 | -22.46 | | 357 | 483 | 1.036 | 293 |
| 212 | C3H8O2 | 1,3-propanediol | 1 | -10.20156 | 2.93938 | -6.69889 | 5.49989 | 332 | TC | 1.053 | 293 |
| 213 | C3H8O3 | glycerol | 3 | 10.6190 | 4487.04 | -140.2 | | 440 | 600 | 1.261 | 293 |
| 214 | C3H8S | methyl ethyl sulfide | 3 | 9.3563 | 2722.95 | -48.37 | | 250 | 360 | 0.837 | 293 |
| 215 | C3H9BO3 | trimethyl borate | | | | | | | | 0.915 | 293 |
| 216 | C3H9N | n-propyl amine | 1 | -7.23587 | 1.22853 | -3.75004 | -4.33990 | 235 | TC | 0.717 | 293 |
| 217 | C3H9N | isopropyl amine | 1 | -7.40866 | 1.79229 | -4.75675 | -1.70138 | 235 | TC | 0.688 | 293 |
| 218 | C3H9N | trimethyl amine | 1 | -6.88066 | 1.15962 | -2.18332 | -2.94707 | 200 | TC | 0.633 | 293 |
| 219 | C4F8 | perfluorocyclobutane | 3 | 9.0726 | 1985.95 | -48.01 | | 241 | 274 | 1.654 | 253 |
| 220 | C4F10 | perfluorobutane | 3 | 9.5788 | 2280.18 | -32.82 | | 233 | 272 | 1.517 | 293 |
| 221 | C4H4 | vinylacetylene | 3 | 9.3898 | 2203.57 | -43.15 | | 200 | 305 | 0.710 | 273 |
| 222 | C4H4O | furan | 3 | 9.4410 | 2442.70 | -45.41 | | 238 | 363 | 0.938 | 293 |
| 223 | C4H4S | thiophene | 1 | -7.05208 | 1.69640 | -3.17778 | -1.57742 | 312 | TC | 1.071 | 289 |
| 224 | C4H5N | allyl cyanide | 3 | 9.3817 | 3128.75 | -58.15 | | 400 | 430 | 0.835 | 293 |
| 225 | C4H5N | pyrrole | 3 | 10.1764 | 3457.47 | -62.73 | | 330 | 440 | 0.967 | 294 |

| No | Formula | Name | MolWt | Tfp K | Tb K | Tc K | Pc bar | Vc $cm^3/mol$ | Zc | Omega | Dipm debye |
|---|---|---|---|---|---|---|---|---|---|---|---|
| 226 | C4H6 | 1-butyne | 54.092 | 147.4 | 281.2 | 463.7 | 47.1 | 220. | 0.27 | 0.050 | 0.8 |
| 227 | C4H6 | 2-butyne | 54.092 | 240.9 | 300.1 | 488.7 | 50.8 | 221. | 0.277 | 0.124 | 0.8 |
| 228 | C4H6 | 1,2-butadiene | 54.092 | 137.0 | 284.0 | 443.7 | 44.9 | 219. | 0.267 | 0.255 | 0.4 |
| 229 | C4H6 | 1,3-butadiene | 54.092 | 164.2 | 268.7 | 425. | 43.3 | 221. | 0.270 | 0.195 | 0.0 |
| 230 | C4H6O2 | vinyl acetate | 86.091 | 173. | 346. | 525. | 43.5 | 265. | 0.26 | 0.34 | 1.7 |
| 231 | C4H6O3 | acetic anhydride | 102.089 | 199. | 413.2 | 569. | 46.8 | | | 0.908 | 3.0 |
| 232 | C4H6O4 | dimethyl oxalate | 118.090 | 327. | 436.5 | 628. | 39.8 | | | 0.556 | |
| 233 | C4H6O4 | succinic acid | 118.090 | 456. | 508. | | | | | | 2.2 |
| 234 | C4H7N | butyronitrile | 69.107 | 161.0 | 391.1 | 582.2 | 37.9 | | | 0.373 | 3.8 |
| 235 | C4H7O2 | methyl acrylate | 86.091 | 196.7 | 353.5 | 536. | 43. | 265. | 0.25 | 0.35 | |
| 236 | C4H8 | 1-butene | 56.108 | 87.8 | 266.9 | 419.6 | 40.2 | 240. | 0.277 | 0.191 | 0.3 |
| 237 | C4H8 | 2-butene,cis | 56.108 | 134.3 | 276.9 | 435.6 | 42.0 | 234. | 0.271 | 0.202 | 0.3 |
| 238 | C4H8 | 2-butene,trans | 56.108 | 167.6 | 274.0 | 428.6 | 39.9 | 238. | 0.266 | 0.205 | 0.0 |
| 239 | C4H8 | cyclobutane | 56.108 | 182.4 | 285.7 | 460.0 | 49.9 | 210. | 0.274 | 0.181 | |
| 240 | C4H8 | isobutylene | 56.108 | 132.8 | 266.2 | 417.9 | 40.0 | 239. | 0.275 | 0.194 | 0.5 |
| 241 | C4H8O | n-butyraldehyde | 72.107 | 176.8 | 348.0 | 545.4 | 53.8 | | | 0.352 | 2.6 |
| 242 | C4H8O | isobutyraldehyde | 72.107 | 208.2 | 337. | 513. | 41.5 | 274. | 0.27 | 0.35 | |
| 243 | C4H8O | methyl ethyl ketone | 72.107 | 186.5 | 352.7 | 536.8 | 42.1 | 267. | 0.252 | 0.320 | 3.3 |
| 244 | C4H8O | tetrahydrofuran | 72.107 | 164.7 | 338. | 540.1 | 51.9 | 224. | 0.259 | 0.217 | 1.7 |
| 245 | C4H8O | vinyl ethyl ether | 72.107 | 157.9 | 308.7 | 475. | 40.7 | | | 0.268 | 1.3 |
| 246 | C4H8O2 | n-butyric acid | 88.107 | 267.9 | 437.2 | 628. | 52.7 | 290. | 0.292 | 0.683 | 1.5 |
| 247 | C4H8O2 | isobutyric acid | 88.107 | 227.2 | 427.9 | 609. | 40.5 | 292. | 0.234 | 0.623 | 1.3 |
| 248 | C4H8O2 | 1,4-dioxane | 88.107 | 285. | 374.6 | 587. | 52.1 | 238. | 0.254 | 0.281 | 0.4 |
| 249 | C4H8O2 | ethyl acetate | 88.107 | 189.6 | 350.3 | 523.2 | 38.3 | 286. | 0.252 | 0.362 | 1.9 |
| 250 | C4H8O2 | methyl propionate | 88.107 | 185.7 | 352.8 | 530.6 | 40.0 | 282. | 0.256 | 0.350 | 1.7 |

| No | Formula | Name | CPVAP A | CPVAP B | CPVAP C | CPVAP D | DELHF | DELGF |
|---|---|---|---|---|---|---|---|---|
| 226 | C4H6 | 1-butyne | 1.255E+1 | 2.744E-1 | -1.545E-4 | 3.450E-8 | 1.653E+5 | 2.022E+5 |
| 227 | C4H6 | 2-butyne | 1.593E+1 | 2.381E-1 | -1.070E-4 | 1.753E-8 | 1.464E+5 | 1.856E+5 |
| 228 | C4H6 | 1,2-butadiene | 1.120E+1 | 2.724E-1 | -1.468E-4 | 3.089E-8 | 1.623E+5 | 1.986E+5 |
| 229 | C4H6 | 1,3-butadiene | -1.687E+0 | 3.419E-1 | -2.340E-4 | 6.335E-8 | 1.102E+5 | 1.508E+5 |
| 230 | C4H602 | vinyl acetate | 1.516E+1 | 2.795E-1 | -8.805E-5 | -1.660E-8 | -3.160E+5 | |
| 231 | C4H603 | acetic anhydride | -2.313E+1 | 5.087E-1 | -3.580E-4 | 9.835E-8 | -5.761E+5 | -4.770E+5 |
| 232 | C4H604 | dimethyl oxalate | | | | | | |
| 233 | C4H604 | succinic acid | 1.507E+1 | 4.689E-1 | -3.143E-4 | 7.938E-8 | | |
| 234 | C4H7N | butyronitrile | 1.521E+1 | 3.206E-1 | -1.638E-4 | 2.982E-8 | 3.410E+4 | 1.087E+5 |
| 235 | C4H702 | methyl acrylate | 1.516E+1 | 2.796E-1 | -8.805E-5 | -1.660E-8 | | |
| 236 | C4H8 | 1-butene | -2.994E+0 | 3.532E-1 | -1.990E-4 | 4.463E-8 | -1.260E+2 | 7.134E+4 |
| 237 | C4H8 | 2-butene,cis | 4.396E-1 | 2.953E-1 | -1.018E-4 | -0.616E-9 | -6.990E+3 | 6.590E+4 |
| 238 | C4H8 | 2-butene,trans | 1.832E+1 | 2.564E-1 | -7.013E-5 | -8.989E-9 | -1.118E+4 | 6.301E+4 |
| 239 | C4H8 | cyclobutane | -5.025E+1 | 5.024E-1 | -3.558E-4 | 1.047E-7 | 2.667E+4 | 1.101E+5 |
| 240 | C4H8 | isobutylene | 1.605E+1 | 2.804E-1 | -1.091E-4 | 9.098E-9 | -1.691E+4 | 5.811E+4 |
| 241 | C4H80 | n-butyraldehyde | 1.408E+1 | 3.457E-1 | -1.723E-4 | 2.887E-8 | -2.052E+5 | -1.148E+5 |
| 242 | C4H80 | isobutyraldehyde | 2.446E+1 | 3.356E-1 | -2.057E-4 | 6.368E-8 | -2.159E+5 | -1.214E+5 |
| 243 | C4H80 | methyl ethyl ketone | 1.094E+1 | 3.559E-1 | -1.900E-4 | 3.920E-8 | -2.385E+5 | -1.462E+5 |
| 244 | C4H80 | tetrahydrofuran | 1.910E+1 | 5.162E-1 | -4.132E-4 | 1.454E-7 | -1.843E+5 | |
| 245 | C4H80 | vinyl ethyl ether | 1.728E+1 | 3.236E-1 | -1.471E-4 | 2.150E-8 | -1.403E+5 | |
| 246 | C4H802 | n-butyric acid | 1.174E+1 | 4.137E-1 | -2.430E-4 | 5.531E-8 | -4.762E+5 | |
| 247 | C4H802 | isobutyric acid | 9.814E+0 | 4.668E-1 | -3.720E-4 | 1.350E-7 | -4.842E+5 | |
| 248 | C4H802 | 1,4-dioxane | -5.357E+1 | 5.987E-1 | -4.085E-4 | 1.062E-7 | -3.153E+5 | -1.809E+5 |
| 249 | C4H802 | ethyl acetate | 7.235E+0 | 4.072E-1 | -2.092E-4 | 2.855E-8 | -4.432E+5 | -3.276E+5 |
| 250 | C4H802 | methyl propionate | 1.820E+1 | 3.140E-1 | -9.353E-5 | -1.828E-8 | | |

| No | Formula | Name | Eq. | VP A | VP B | VP C | VP D | Tmin | Tmax | LDEN | TDEN |
|---|---|---|---|---|---|---|---|---|---|---|---|
| 226 | C4H6 | 1-butyne | 1 | -6.29693 | 2.12358 | -6.42124 | 4.11543 | 194 | TC | 0.650 | 289 |
| 227 | C4H6 | 2-butyne | 3 | 9.6669 | 2536.78 | -37.34 | | 240 | 320 | 0.691 | 293 |
| 228 | C4H6 | 1,2-butadiene | 3 | 9.4837 | 2397.26 | -30.88 | | 245 | 305 | 0.652 | 293 |
| 229 | C4H6 | 1,3-butadiene | 1 | -7.12563 | 1.73913 | -2.70805 | -1.68376 | 197 | TC | 0.621 | 293 |
| 230 | C4H602 | vinyl acetate | 1 | -7.80478 | 1.80668 | -4.48160 | 1.70357 | 295 | TC | 0.932 | 293 |
| 231 | C4H603 | acetic anhydride | 1 | -18.1529 | 18.3036 | -20.0953 | 16.6970 | 336 | TC | 1.087 | 293 |
| 232 | C4H604 | dimethyl oxalate | | | | | | | | 1.15 | 288 |
| 233 | C4H604 | succinic acid | | | | | | | | | |
| 234 | C4H7N | butyronitrile | 2 | 49.985 | 6476.68 | -5.599 | 3770. | 320 | TC | 0.792 | 293 |
| 235 | C4H702 | methyl acrylate | 3 | 9.4886 | 2788.43 | -59.15 | | 260 | 390 | 0.956 | 293 |
| 236 | C4H8 | 1-butene | 1 | -6.88204 | 1.27051 | -2.26284 | -2.61632 | 170 | TC | 0.595 | 293 |
| 237 | C4H8 | 2-butene,cis | 1 | -6.88706 | 1.15941 | -2.19304 | -3.12758 | 203 | TC | 0.621 | 293 |
| 238 | C4H8 | 2-butene,trans | 2 | 43.517 | 4174.56 | -5.041 | 1995. | 240 | 400 | 0.604 | 293 |
| 239 | C4H8 | cyclobutane | 1 | -7.40011 | 2.37997 | -3.12269 | -0.34310 | 213 | TC | 0.694 | 293 |
| 240 | C4H8 | isobutylene | 1 | -6.95542 | 1.35673 | -2.45222 | -1.46110 | 170 | TC | 0.594 | 293 |
| 241 | C4H80 | n-butyraldehyde | 1 | -7.01403 | 0.12265 | -0.00073 | -8.50911 | 304 | TC | 0.802 | 293 |
| 242 | C4H80 | isobutyraldehyde | 1 | -7.53679 | 1.08548 | -1.52929 | -8.48589 | 286 | TC | 0.789 | 293 |
| 243 | C4H80 | methyl ethyl ketone | 1 | -7.71476 | 1.71061 | -3.68770 | -0.75169 | 255 | TC | 0.805 | 293 |
| 244 | C4H80 | tetrahydrofuran | 3 | 9.4867 | 2768.38 | -46.90 | | 270 | 370 | 0.889 | 293 |
| 245 | C4H80 | vinyl ethyl ether | 1 | -7.33727 | 1.50878 | -3.30376 | -1.10728 | 256 | TC | 0.793 | 293 |
| 246 | C4H802 | n-butyric acid | 1 | -10.0392 | 3.15679 | -7.72604 | 5.27630 | 364 | TC | 0.958 | 293 |
| 247 | C4H802 | isobutyric acid | 2 | 76.037 | 9222.72 | -8.986 | 3863. | 320 | TC | 0.968 | 293 |
| 248 | C4H802 | 1,4-dioxane | 3 | 9.5125 | 2966.88 | -62.15 | | 275 | 410 | 1.033 | 293 |
| 249 | C4H802 | ethyl acetate | 1 | -7.68521 | 1.36511 | -4.08980 | -1.75342 | 289 | TC | 0.901 | 293 |
| 250 | C4H802 | methyl propionate | 1 | -8.23756 | 2.71406 | -5.35097 | -2.34114 | 294 | TC | 0.915 | 293 |

| No | Formula | Name | MolWt | Tfp K | Tb K | Tc K | Pc bar | Vc $cm^3/mol$ | Zc | Omega | Dipm debye |
|---|---|---|---|---|---|---|---|---|---|---|---|
| 251 | C4H8O2 | n-propyl formate | 88.107 | 180.3 | 354.1 | 538.0 | 40.6 | 285. | 0.259 | 0.314 | 1.9 |
| 252 | C4H8S | tetrahydrothiophene | 88.172 | 177.0 | 394.2 | 632. | | | | | 1.9 |
| 253 | C4H9Cl | 1-chlorobutane | 92.569 | 150.1 | 351.6 | 542. | 36.8 | 312. | 0.255 | 0.218 | 2.0 |
| 254 | C4H9Cl | 2-chlorobutane | 92.569 | 141.8 | 341.4 | 520.6 | 39.5 | 305. | 0.28 | 0.30 | 2.1 |
| 255 | C4H9Cl | tert-butyl chloride | 92.569 | 247.8 | 324. | 507. | 39.5 | 295. | 0.28 | 0.19 | 2.1 |
| 256 | C4H9N | pyrrolidine | 71.123 | | 359.6 | 568.6 | 56.1 | 249. | 0.295 | 0.274 | 1.6 |
| 257 | C4H9NO | morpholine | 87.122 | 268.4 | 401.4 | 618. | 54.7 | 253. | 0.27 | 0.37 | 1.5 |
| 258 | C4H10 | n-butane | 58.124 | 134.8 | 272.7 | 425.2 | 38.0 | 255. | 0.274 | 0.199 | 0.0 |
| 259 | C4H10 | isobutane | 58.124 | 113.6 | 261.4 | 408.2 | 36.5 | 263. | 0.283 | 0.183 | 0.1 |
| 260 | C4H10O | n-butanol | 74.123 | 183.9 | 390.9 | 563.1 | 44.2 | 275. | 0.259 | 0.593 | 1.8 |
| 261 | C4H10O | 2-butanol | 74.123 | 158.5 | 372.7 | 536.1 | 41.8 | 269. | 0.252 | 0.577 | 1.7 |
| 262 | C4H10O | isobutanol | 74.123 | 165.2 | 381.0 | 547.8 | 43.0 | 273. | 0.257 | 0.592 | 1.7 |
| 263 | C4H10O | tert-butanol | 74.123 | 298.8 | 355.5 | 506.2 | 39.7 | 275. | 0.259 | 0.612 | 1.7 |
| 264 | C4H10O | diethyl ether | 74.123 | 156.9 | 307.6 | 466.7 | 36.4 | 280. | 0.262 | 0.281 | 1.3 |
| 265 | C4H10O | methyl propyl ether | 74.123 | | 311.7 | 476.3 | 38.0 | | | 0.271 | 1.2 |
| 266 | C4H10O | methyl isopropyl ether | 74.123 | | 303.9 | 464.5 | 37.6 | | | 0.266 | |
| 267 | C4H10O2 | 1,2-dimethoxyethane | 90.123 | 202. | 358. | 536. | 38.7 | 271. | 0.235 | 0.358 | 0.0 |
| 268 | C4H10O3 | diethylene glycol | 106.122 | 265. | 519. | 681. | 47. | | | | |
| 269 | C4H10S | diethyl sulfide | 90.184 | 169.2 | 365.3 | 557. | 39.6 | 318. | 0.272 | 0.292 | 1.6 |
| 270 | C4H10S2 | diethyl disulfide | 122.244 | 171.7 | 427.1 | 642. | | | | | 2.0 |
| 271 | C4H11N | n-butyl amine | 73.139 | 224.1 | 349.5 | 531.9 | 42.0 | | | 0.329 | 1.3 |
| 272 | C4H11N | isobutyl amine | 73.139 | 188.0 | 336.2 | 514.3 | 41.0 | | | 0.368 | 1.2 |
| 273 | C4H11N | diethyl amine | 73.139 | 223.4 | 328.6 | 496.5 | 37.1 | 301. | 0.271 | 0.291 | 1.1 |
| 274 | C5F12 | perfluoropentane | 288.031 | | 302.4 | 420.6 | 20.5 | 473. | 0.276 | 0.432 | 0.0 |
| 275 | C5H2F6O2 | hexafluoroacetylacetone | 208.059 | | 327.3 | 485.1 | 27.7 | | | 0.278 | |

| No | Formula | Name | CPVAP A | CPVAP B | CPVAP C | CPVAP D | DELHF | DELGF |
|---|---|---|---|---|---|---|---|---|
| 251 | C4H8O2 | n-propyl formate | | | | | | |
| 252 | C4H8S | tetrahydrothiophene | | | | | | |
| 253 | C4H9Cl | 1-chlorobutane | -2.613E+0 | 4.497E-1 | -2.937E-4 | 8.081E-8 | -1.474E+5 | -3.881E+4 |
| 254 | C4H9Cl | 2-chlorobutane | -3.433E+0 | 4.559E-1 | -2.981E-4 | 8.256E-8 | -1.616E+5 | -5.351E+4 |
| 255 | C4H9Cl | tert-butyl chloride | -3.931E+0 | 4.652E-1 | -2.886E-4 | 7.871E-8 | -1.834E+5 | -6.414E+4 |
| 256 | C4H9N | pyrrolidine | -5.153E+1 | 5.338E-1 | -3.240E-4 | 7.528E-8 | -3.600E+3 | 1.148E+5 |
| 257 | C4H9NO | morpholine | -4.280E+1 | 5.388E-1 | -2.666E-4 | 4.199E-8 | | |
| 258 | C4H10 | n-butane | 9.487E+0 | 3.313E-1 | -1.108E-4 | -2.822E-9 | -1.262E+5 | -1.610E+4 |
| 259 | C4H10 | isobutane | -1.390E+0 | 3.847E-1 | -1.846E-4 | 2.895E-8 | -1.346E+5 | -2.090E+4 |
| 260 | C4H10O | n-butanol | 3.266E+0 | 4.180E-1 | -2.242E-4 | 4.685E-8 | -2.749E+5 | -1.509E+5 |
| 261 | C4H10O | 2-butanol | 5.753E+0 | 4.245E-1 | -2.328E-4 | 4.773E-8 | -2.928E+5 | -1.677E+5 |
| 262 | C4H10O | isobutanol | -7.708E+0 | 4.689E-1 | -2.884E-4 | 7.231E-8 | -2.834E+5 | -1.674E+5 |
| 263 | C4H10O | tert-butanol | -4.861E+1 | 7.172E-1 | -7.084E-4 | 2.920E-7 | -3.128E+5 | -1.778E+5 |
| 264 | C4H10O | diethyl ether | 2.142E+1 | 3.359E-1 | -1.035E-4 | -9.357E-9 | -2.524E+5 | -1.224E+5 |
| 265 | C4H10O | methyl propyl ether | 2.131E+1 | 3.390E-1 | -1.127E-4 | -2.855E-9 | -2.379E+5 | -1.100E+5 |
| 266 | C4H10O | methyl isopropyl ether | 1.353E+1 | 3.697E-1 | -1.481E-4 | 1.205E-8 | -2.522E+5 | -1.210E+5 |
| 267 | C4H10O2 | 1,2-dimethoxyethane | 3.223E+1 | 3.567E-1 | -1.336E-4 | 8.399E-9 | | |
| 268 | C4H10O3 | diethylene glycol | 7.306E+1 | 3.461E-1 | -1.468E-4 | 1.846E-8 | -5.715E+5 | |
| 269 | C4H10S | diethyl sulfide | 1.359E+1 | 3.959E-1 | -1.780E-4 | 2.649E-8 | -8.353E+4 | 1.780E+4 |
| 270 | C4H10S2 | diethyl disulfide | 2.690E+1 | 4.601E-1 | -2.710E-4 | 5.970E-8 | -7.469E+4 | 2.227E+4 |
| 271 | C4H11N | n-butyl amine | 5.079E+0 | 4.476E-1 | -2.407E-4 | 7.599E-8 | -9.211E+4 | 4.924E+4 |
| 272 | C4H11N | isobutyl amine | 9.491E+0 | 4.430E-1 | -2.110E-4 | 2.333E-8 | | |
| 273 | C4H11N | diethyl amine | 2.039E+0 | 4.430E-1 | -2.183E-4 | 3.653E-8 | -7.243E+4 | 7.214E+4 |
| 274 | C5F12 | perfluoropentane | | | | | | |
| 275 | C5H2F6O2 | hexafluoroacetylacetone | | | | | | |

| No | Formula | Name | Eq. | VP A | VP B | VP C | VP D | Tmin | Tmax | LDEN | TDEN |
|---|---|---|---|---|---|---|---|---|---|---|---|
| 251 | C4H8O2 | n-propyl formate | 1 | -7.48563 | 1.71260 | -5.16404 | 1.64290 | 299 | TC | 0.911 | 289 |
| 252 | C4H8S | tetrahydrothiophene | 3 | 9.3870 | 3160.1 | -57.2 | | 308 | 473 | 1.000 | 293 |
| 253 | C4H9Cl | 1-chlorobutane | 1 | -6.79852 | 0.78511 | -2.31047 | -5.83223 | 256 | TC | 0.886 | 293 |
| 254 | C4H9Cl | 2-chlorobutane | 3 | 9.3705 | 2753.43 | -47.15 | | 250 | 375 | 0.873 | 293 |
| 255 | C4H9Cl | tert-butyl chloride | 3 | 9.1919 | 2567.15 | -44.15 | | 235 | 360 | 0.842 | 293 |
| 256 | C4H9N | pyrrolidine | 1 | -7.73658 | 2.33495 | -4.20213 | -3.71251 | 316 | TC | 0.852 | 295 |
| 257 | C4H9NO | morpholine | 3 | 9.6162 | 3171.35 | -71.15 | | 300 | 440 | 1.000 | 293 |
| 258 | C4H10 | n-butane | 1 | -6.88709 | 1.15157 | -1.99873 | -3.13003 | 170 | TC | 0.579 | 293 |
| 259 | C4H10 | isobutane | 1 | -6.95579 | 1.50090 | -2.52717 | -1.49776 | 165 | TC | 0.557 | 293 |
| 260 | C4H10O | n-butanol | 1 | -8.00756 | 0.53783 | -9.34240 | 6.68692 | 275 | TC | 0.810 | 293 |
| 261 | C4H10O | 2-butanol | 1 | -7.80578 | 0.32456 | -9.41265 | 2.64643 | 265 | TC | 0.807 | 293 |
| 262 | C4H10O | isobutanol | 3 | 10.2510 | 2874.73 | -100.3 | | 293 | 388 | 0.802 | 293 |
| 263 | C4H10O | tert-butanol | 3 | 10.2346 | 2658.29 | -95.50 | | 293 | 376 | 0.787 | 293 |
| 264 | C4H10O | diethyl ether | 1 | -7.29916 | 1.24828 | -2.91931 | -3.36740 | 250 | TC | 0.713 | 293 |
| 265 | C4H10O | methyl propyl ether | 1 | -7.59830 | 2.01601 | -3.70390 | -1.64710 | 258 | TC | 0.738 | 293 |
| 266 | C4H10O | methyl isopropyl ether | 1 | -7.06696 | 0.86497 | -2.16269 | -4.72211 | 252 | TC | 0.724 | 288 |
| 267 | C4H10O2 | 1,2-dimethoxyethane | 3 | 9.4039 | 2869.79 | -53.15 | | 262 | 393 | 0.867 | 293 |
| 268 | C4H10O3 | diethylene glycol | 3 | 10.4124 | 4122.52 | -122.5 | | 402 | 560 | 1.116 | 293 |
| 269 | C4H10S | diethyl sulfide | 3 | 9.3329 | 2896.27 | -54.49 | | 260 | 390 | 0.837 | 293 |
| 270 | C4H10S2 | diethyl disulfide | 3 | 9.4405 | 3421.57 | -64.19 | | 312 | 455 | 0.998 | 293 |
| 271 | C4H11N | n-butyl amine | 1 | -7.91668 | 2.36401 | -5.01170 | -2.54215 | 255 | TC | 0.739 | 293 |
| 272 | C4H11N | isobutyl amine | 1 | -8.41366 | 3.12108 | -5.70064 | -1.83920 | 248 | TC | 0.722 | 295 |
| 273 | C4H11N | diethyl amine | 1 | -7.26796 | 1.15810 | -3.91125 | -1.17981 | 240 | TC | 0.707 | 293 |
| 274 | C5F12 | perfluoropentane | 3 | 9.5390 | 2470.33 | -43.20 | | 282 | 338 | | |
| 275 | C5H2F6O2 | hexafluoroacetylacetone | | | | | | | | | |

| No | Formula | Name | MolWt | Tfp<br>K | Tb<br>K | Tc<br>K | Pc<br>bar | Vc<br>$cm^3$/mol | Zc | Omega | Dipm<br>debye |
|---|---|---|---|---|---|---|---|---|---|---|---|
| 276 | C5H4O2 | furfural | 96.085 | 234.5 | 434.9 | 670. | 58.9 | | | 0.383 | 3.6 |
| 277 | C5H5N | pyridine | 79.102 | 231.5 | 388.4 | 620.0 | 56.3 | 254. | 0.277 | 0.243 | 2.3 |
| 278 | C5H6N2 | 2-methyl pyrazine | 94.117 | | 410. | 634.3 | 50.1 | 283. | 0.268 | 0.315 | |
| 279 | C5H6O | 2-methyl furan | 82.102 | | 338. | 527. | 47.2 | 247. | 0.266 | 0.270 | 0.7 |
| 280 | C5H8 | cyclopentene | 68.119 | 138.1 | 317.4 | 506.0 | | | | | 0.9 |
| 281 | C5H8 | 1,2-pentadiene | 68.119 | 135.9 | 318.0 | 503. | 40.7 | 276. | 0.269 | 0.173 | |
| 282 | C5H8 | 1,3-pentadiene,trans | 68.119 | 185.7 | 315.1 | 496. | 39.9 | 275. | 0.266 | 0.175 | 0.7 |
| 283 | C5H8 | 1,4-pentadiene | 68.119 | 124.9 | 299.1 | 478. | 37.9 | 276. | 0.263 | 0.104 | 0.4 |
| 284 | C5H8 | 1-pentyne | 68.119 | 167.5 | 313.3 | 493.5 | 40.5 | 278. | 0.275 | 0.164 | 0.9 |
| 285 | C5H8 | 2-methyl-1,3-butadiene | 68.119 | 127.2 | 307.2 | 484. | 38.5 | 276. | 0.264 | 0.164 | 0.3 |
| 286 | C5H8 | 3-methyl-1,2-butadiene | 68.119 | 159.5 | 314.0 | 496. | 41.1 | 267. | 0.266 | 0.160 | |
| 287 | C5H8O | cyclopentanone | 84.118 | 222.5 | 403.9 | 634.6 | 51.1 | 268. | 0.260 | 0.35 | 3.0 |
| 288 | C5H8O | dihydropyran | 84.118 | | 359. | 561.7 | 45.6 | 268. | 0.262 | 0.247 | 1.4 |
| 289 | C5H8O2 | ethyl acrylate | 100.118 | 201. | 373. | 552. | 37.4 | 320. | 0.261 | 0.400 | |
| 290 | C5H10 | cyclopentane | 70.135 | 179.3 | 322.4 | 511.7 | 45.1 | 260. | 0.275 | 0.196 | 0.0 |
| 291 | C5H10 | 1-pentene | 70.135 | 107.9 | 303.1 | 464.8 | 35.3 | 300. | 0.31 | 0.233 | 0.4 |
| 292 | C5H10 | 2-pentene,cis | 70.135 | 121.7 | 310.1 | 476. | 36.5 | | | 0.251 | |
| 293 | C4H10 | 2-pentene,trans | 70.135 | 132.9 | 309.5 | 475. | 36.6 | | | 0.259 | |
| 294 | C5H10 | 2-methyl-1-butene | 70.135 | 135.6 | 304.3 | 465. | 34.5 | | | 0.236 | 0.5 |
| 295 | C5H10 | 2-methyl-2-butene | 70.135 | 139.3 | 311.7 | 470. | 34.5 | | | 0.244 | |
| 296 | C5H10 | 3-methyl-1-butene | 70.135 | 104.7 | 293.3 | 450. | 35.1 | | | 0.209 | |
| 297 | C5H10O | valeraldehyde | 86.134 | 182. | 376. | 554. | 35.4 | 333. | 0.26 | 0.40 | 2.6 |
| 298 | C5H10O | methyl n-propyl ketone | 86.134 | 196. | 375.4 | 561.1 | 36.9 | 301. | 0.238 | 0.346 | 2.5 |
| 299 | C5H10O | methyl isopropyl ketone | 86.134 | 181. | 367.5 | 553.4 | 38.5 | 310. | 0.259 | 0.331 | 2.8 |
| 300 | C5H10O | diethyl ketone | 86.134 | 234.2 | 375.1 | 561.0 | 37.3 | 336. | 0.269 | 0.344 | 2.7 |

| No | Formula | Name | CPVAP A | CPVAP B | CPVAP C | CPVAP D | DELHF | DELGF |
|---|---|---|---|---|---|---|---|---|
| 276 | C5H402 | furfural | | | | | | |
| 277 | C5H5N | pyridine | 3.979E+1 | 4.928E-1 | -3.558E-4 | 1.004E-7 | 1.403E+5 | 1.903E+5 |
| 278 | C5H6N2 | 2-methyl pyrazine | | | | | | |
| 279 | C5H60 | 2-methyl furan | | | | | | |
| 280 | C5H8 | cyclopentene | -4.151E+1 | 4.631E-1 | -2.579E-4 | 5.434E-8 | 3.295E+4 | 1.109E+5 |
| 281 | C5H8 | 1,2-pentadiene | 8.826E+0 | 3.880E-1 | -2.280E-4 | 5.246E-8 | 1.457E+5 | 2.106E+5 |
| 282 | C5H8 | 1,3-pentadiene,trans | 3.069E+1 | 2.811E-1 | -6.711E-5 | -2.352E-8 | 7.787E+4 | 1.468E+5 |
| 283 | C5H8 | 1,4-pentadiene | 6.996E+0 | 3.952E-1 | -2.374E-4 | 5.598E-8 | 1.055E+5 | 1.704E+5 |
| 284 | C5H8 | 1-pentyne | 1.807E+1 | 3.511E-1 | -1.913E-4 | 4.098E-8 | 1.444E+5 | 2.104E+5 |
| 285 | C5H8 | 2-methyl-1,3-butadiene | -3.412E+0 | 4.585E-1 | -3.337E-4 | 1.000E-7 | 7.578E+4 | 1.460E+5 |
| 286 | C5H8 | 3-methyl-1,2-butadiene | 1.469E+1 | 3.598E-1 | -1.976E-4 | 4.262E-8 | 1.298E+5 | 1.987E+5 |
| 287 | C5H80 | cyclopentanone | -4.064E+1 | 5.255E-1 | -3.124E-4 | 7.130E-8 | -1.928E+5 | |
| 288 | C5H80 | dihydropyran | | | | | | |
| 289 | C5H802 | ethyl acrylate | 1.681E+1 | 3.690E+0 | -1.382E-4 | -5.732E-9 | | |
| 290 | C5H10 | cyclopentane | -5.362E+1 | 5.426E-1 | -3.031E-4 | 6.485E-8 | -7.729E+4 | 3.860E+4 |
| 291 | C5H10 | 1-pentene | -1.340E-1 | 4.329E-1 | -2.317E-4 | 4.681E-8 | -2.093E+4 | 7.917E+4 |
| 292 | C5H10 | 2-pentene,cis | -1.429E+1 | 4.601E-1 | -2.541E-4 | 5.455E-8 | -2.809E+4 | 7.189E+4 |
| 293 | C4H10 | 2-pentene,trans | 1.947E+0 | 4.182E-1 | -2.178E-4 | 4.405E-8 | -3.178E+4 | 6.996E+4 |
| 294 | C5H10 | 2-methyl-1-butene | 1.057E+1 | 3.997E-1 | -1.946E-4 | 3.314E-8 | -3.634E+4 | 6.565E+4 |
| 295 | C5H10 | 2-methyl-2-butene | 1.180E+1 | 3.509E-1 | -1.117E-4 | -5.807E-9 | -4.258E+4 | 5.970E+4 |
| 296 | C5H10 | 3-methyl-1-butene | 2.174E+1 | 3.890E-1 | -2.007E-4 | 4.011E-8 | -2.897E+4 | 7.482E+4 |
| 297 | C5H100 | valeraldehyde | 1.424E+1 | 4.329E-1 | -2.107E-4 | 3.162E-8 | -2.280E+5 | -1.084E+5 |
| 298 | C5H100 | methyl n-propyl ketone | 1.147E+0 | 4.802E-1 | -2.818E-4 | 6.661E-8 | -2.588E+5 | -1.372E+5 |
| 299 | C5H100 | methyl isopropyl ketone | -2.914E+0 | 4.991E-1 | -2.935E-4 | 6.665E-8 | | |
| 300 | C5H100 | diethyl ketone | 3.001E+1 | 3.939E-1 | -1.907E-4 | 3.398E-8 | -2.588E+5 | -1.354E+5 |

| No | Formula | Name | Eq. | VP A | VP B | VP C | VP D | Tmin | Tmax | LDEN | TDEN |
|---|---|---|---|---|---|---|---|---|---|---|---|
| 276 | C5H4O2 | furfural | 3 | 8.5214 | 2760.09 | -110.4 | | 328 | 434 | 1.159 | 293 |
| 277 | C5H5N | pyridine | 1 | -7.07689 | 1.21511 | -2.76681 | -2.87472 | 340 | TC | 0.983 | 293 |
| 278 | C5H6N2 | 2-methyl pyrazine | | | | | | | | 1.044 | 273 |
| 279 | C5H6O | 2-methyl furan | | | | | | | | 0.913 | 293 |
| 280 | C5H8 | cyclopentene | 3 | 9.3154 | 2583.07 | -39.70 | | 244 | 378 | 0.772 | 293 |
| 281 | C5H8 | 1,2-pentadiene | 3 | 9.3095 | 2544.34 | -44.30 | | 250 | 340 | 0.693 | 293 |
| 282 | C5H8 | 1,3-pentadiene,trans | 3 | 9.2980 | 2541.69 | -41.43 | | 250 | 340 | 0.676 | 293 |
| 283 | C5H8 | 1,4-pentadiene | 3 | 9.1190 | 2344.02 | -41.69 | | 240 | 320 | 0.661 | 293 |
| 284 | C5H8 | 1-pentyne | 3 | 9.4227 | 2515.62 | -45.97 | | 230 | 335 | 0.690 | 293 |
| 285 | C5H8 | 2-methyl-1,3-butadiene | 1 | -6.59262 | 1.28930 | -3.89168 | 1.70215 | 257 | TC | 0.681 | 293 |
| 286 | C5H8 | 3-methyl-1,2-butadiene | 1 | -6.71441 | 1.53531 | -4.64262 | 2.99854 | 274 | TC | 0.686 | 293 |
| 287 | C5H8O | cyclopentanone | 1 | -7.19551 | 1.16379 | -2.52546 | -3.28861 | 273 | TC | 0.950 | 293 |
| 288 | C5H8O | dihydropyran | | | | | | | | | |
| 289 | C5H8O2 | ethyl acrylate | 3 | 9.4688 | 2974.94 | -58.15 | | 274 | 409 | 0.921 | 293 |
| 290 | C5H10 | cyclopentane | 1 | -6.51809 | 0.38442 | -1.11706 | -4.50275 | 289 | TC | 0.745 | 293 |
| 291 | C5H10 | 1-pentene | 1 | -7.04875 | 1.17813 | -2.45105 | -2.21727 | 190 | TC | 0.640 | 293 |
| 292 | C5H10 | 2-pentene,cis | 1 | -6.80160 | 0.54458 | -1.55279 | -5.68029 | 275 | TC | 0.656 | 293 |
| 293 | C4H10 | 2-pentene,trans | 1 | -6.99461 | 1.00724 | -2.42146 | -2.51692 | 274 | TC | 0.649 | 293 |
| 294 | C5H10 | 2-methyl-1-butene | 1 | -6.82990 | 0.72660 | -2.15363 | -3.62225 | 274 | TC | 0.650 | 293 |
| 295 | C5H10 | 2-methyl-2-butene | 1 | -7.71438 | 1.95946 | -3.15710 | -2.22515 | 276 | TC | 0.662 | 293 |
| 296 | C5H10 | 3-methyl-1-butene | 1 | -7.18870 | 1.42502 | -2.27292 | -2.04323 | 273 | TC | 0.627 | 293 |
| 297 | C5H10O | valeraldehyde | 3 | 9.5421 | 3030.20 | -58.15 | | 277 | 412 | 0.810 | 293 |
| 298 | C5H10O | methyl n-propyl ketone | 3 | 9.3829 | 2934.87 | -62.25 | | 275 | 410 | 0.806 | 293 |
| 299 | C5H10O | methyl isopropyl ketone | 3 | 7.5577 | 1993.12 | -103.2 | | 271 | 406 | 0.803 | 293 |
| 300 | C5H10O | diethyl ketone | 1 | -7.70542 | 1.44422 | -3.60173 | -2.88141 | 330 | TC | 0.814 | 293 |

| No | Formula | Name | MolWt | Tfp K | Tb K | Tc K | Pc bar | Vc $cm^3/mol$ | Zc | Omega | Dipm debye |
|---|---|---|---|---|---|---|---|---|---|---|---|
| 301 | C5H10O | 2-methyl tetrahydrofuran | 86.134 | | 351. | 537. | 37.6 | 267. | 0.225 | 0.264 | |
| 302 | C5H10O | tetrahydropyran | 86.134 | | 361. | 572.2 | 47.7 | 263. | 0.263 | 0.218 | 1.6 |
| 303 | C5H10O2 | n-valeric acid | 102.134 | 239.0 | 459.5 | 651. | | | | | |
| 304 | C5H10O2 | isovaleric acid | 102.134 | | 449.7 | 634. | | | | | 1.0 |
| 305 | C5H10O2 | isobutyl formate | 102.134 | 178. | 371.4 | 554. | 37.3 | 352. | 0.285 | 0.396 | 1.9 |
| 306 | C5H10O2 | n-propyl acetate | 102.134 | 178. | 374.7 | 549.4 | 33.3 | 345. | 0.252 | 0.391 | 1.8 |
| 307 | C5H10O2 | ethyl propionate | 102.134 | 199.3 | 372.2 | 546.0 | 33.6 | 345. | 0.256 | 0.391 | 1.8 |
| 308 | C5H10O2 | methyl butyrate | 102.134 | 188.4 | 375.9 | 554.4 | 34.8 | 340. | 0.257 | 0.380 | 1.7 |
| 309 | C5H10O2 | methyl isobutyrate | 102.134 | 185.4 | 365.5 | 540.8 | 34.3 | 339. | 0.259 | 0.362 | 2.0 |
| 310 | C5H11N | piperidine | 85.150 | 262.7 | 379.6 | 594.0 | 47.6 | 289. | 0.280 | 0.251 | 1.2 |
| 311 | C5H12 | n-pentane | 72.151 | 143.4 | 309.2 | 469.7 | 33.7 | 304. | 0.263 | 0.251 | 0.0 |
| 312 | C5H12 | 2-methyl butane | 72.151 | 113.3 | 301.0 | 460.4 | 33.9 | 306. | 0.271 | 0.227 | 0.1 |
| 313 | C5H12 | 2,2-dimethlypropane | 72.151 | 256.6 | 282.6 | 433.8 | 32.0 | 303. | 0.269 | 0.197 | 0.0 |
| 314 | C5H12O | 1-pentanol | 88.150 | 195.0 | 411.1 | 588.2 | 39.1 | 326. | 0.26 | 0.579 | 1.7 |
| 315 | C5H12O | 2-methyl-1-butanol | 88.150 | 203. | 401.9 | 571.0 | 33.4 | | | | |
| 316 | C5H12O | 3-methyl-1-butanol | 88.150 | 156. | 405.2 | 579.4 | | | | | 1.8 |
| 317 | C5H12O | 2-methyl-2-butanol | 88.150 | 264.4 | 375.5 | 545.0 | 39.5 | | | | 1.9 |
| 318 | C5H12O | 2,2-dimethyl-1-propanol | 88.150 | 327. | 386.3 | 549.0 | | | | | |
| 319 | C5H12O | ethyl propyl ether | 88.150 | 146.4 | 336.4 | 500.2 | 33.7 | 339. | 0.275 | 0.333 | 1.2 |
| 320 | C5H12O | butyl methyl ether | 88.150 | 157.7 | 343.3 | 512.8 | 33.7 | 329. | 0.260 | 0.316 | 1.3 |
| 321 | C5H12O | tert-butyl methyl ether | 88.150 | | 328.3 | 496.4 | 33.7 | | | 0.269 | 1.2 |
| 322 | C6BrF5 | bromopentafluorobenzene | 246.960 | | 410.0 | 601. | 30.4 | | | 0.355 | |
| 323 | C6ClF5 | chloropentafluorobenzene | 202.509 | | 391.1 | 570.8 | 32.4 | 376. | 0.256 | 0.400 | |
| 324 | C6Cl2F4 | dichlorotetrafluorobenzene | 218.964 | | 430.9 | 626. | 53.2 | | | 0.622 | |
| 325 | C6Cl3F3 | 1,3,5-trichlorotrifluorobenzene | 235.419 | | 471.5 | 684.9 | 32.7 | 448. | 0.257 | 0.426 | |

| No | Formula | Name | CPVAP A | CPVAP B | CPVAP C | CPVAP D | DELHF | DELGF |
|---|---|---|---|---|---|---|---|---|
| 301 | C5H10O | 2-methyl tetrahydrofuran | | | | | | |
| 302 | C5H10O | tetrahydropyran | | | | | | |
| 303 | C5H10O2 | n-valeric acid | 1.339E+1 | 5.033E-1 | -2.931E-4 | 6.619E-8 | -4.907E+5 | -3.574E+5 |
| 304 | C5H10O2 | isovaleric acid | | | | | | |
| 305 | C5H10O2 | isobutyl formate | 1.985E+1 | 4.034E-1 | -1.436E-4 | -7.402E-9 | | |
| 306 | C5H10O2 | n-propyl acetate | 1.542E+1 | 4.501E-1 | -1.686E-4 | -1.439E-8 | -4.660E+5 | |
| 307 | C5H10O2 | ethyl propionate | 1.985E+1 | 4.034E-1 | -1.437E-4 | -7.394E-9 | -4.702E+5 | -3.237E+5 |
| 308 | C5H10O2 | methyl butyrate | | | | | | |
| 309 | C5H10O2 | methyl isobutyrate | | | | | | |
| 310 | C5H11N | piperidine | -5.307E+1 | 6.289E-1 | -3.358E-4 | 6.427E-8 | -4.903E+4 | |
| 311 | C5H12 | n-pentane | -3.626E+0 | 4.873E-1 | -2.580E-4 | 5.305E-8 | -1.465E+5 | -8.370E+3 |
| 312 | C5H12 | 2-methyl butane | -9.525E+0 | 5.066E-1 | -2.729E-4 | 5.723E-8 | -1.546E+5 | -1.482E+4 |
| 313 | C5H12 | 2,2-dimethlypropane | -1.659E+1 | 5.552E-1 | -3.306E-4 | 7.633E-8 | -1.661E+5 | -1.524E+4 |
| 314 | C5H12O | 1-pentanol | 3.869E+0 | 5.045E-1 | -2.639E-4 | 5.120E-8 | -2.989E+5 | -1.461E+5 |
| 315 | C5H12O | 2-methyl-1-butanol | -9.483E+0 | 5.677E-1 | -3.481E-4 | 8.637E-8 | -3.027E+5 | -1.657E+5 |
| 316 | C5H12O | 3-methyl-1-butanol | -9.542E+0 | 5.681E-1 | -3.485E-4 | 8.650E-8 | -3.023E+5 | |
| 317 | C5H12O | 2-methyl-2-butanol | -1.209E+1 | 6.096E-1 | -4.204E-4 | 1.228E-7 | -3.299E+5 | -1.654E+5 |
| 318 | C5H12O | 2,2-dimethyl-1-propanol | 1.215E+1 | 5.397E-1 | -3.160E-4 | 7.122E-8 | -2.931E+5 | -1.255E+5 |
| 319 | C5H12O | ethyl propyl ether | | | | | | |
| 320 | C5H12O | butyl methyl ether | | | | | | |
| 321 | C5H12O | tert-butyl methyl ether | 2.534E+0 | 5.136E-1 | -2.596E-4 | 4.303E-8 | -2.931E+5 | -1.255E+5 |
| 322 | C6BrF5 | bromopentafluorobenzene | | | | | | |
| 323 | C6ClF5 | chloropentafluorobenzene | | | | | | |
| 324 | C6Cl2F4 | dichlorotetrafluorobenzene | | | | | | |
| 325 | C6Cl3F3 | 1,3,5-trichlorotrifluorobenzene | | | | | | |

| No | Formula | Name | Eq. | VP A | VP B | VP C | VP D | Tmin | Tmax | LDEN | TDEN |
|---|---|---|---|---|---|---|---|---|---|---|---|
| 301 | C5H10O | 2-methyl tetrahydrofuran | | | | | | | | 0.855 | 293 |
| 302 | C5H10O | tetrahydropyran | | | | | | | | 0.886 | 288 |
| 303 | C5H10O2 | n-valeric acid | 3 | 11.0104 | 4092.15 | -86.55 | | 350 | 495 | 0.939 | 293 |
| 304 | C5H10O2 | isovaleric acid | 3 | 2.4671 | 588.09 | -261.9 | | 359 | 378 | 0.925 | 293 |
| 305 | C5H10O2 | isobutyl formate | 1 | -8.01454 | 2.05091 | -4.38201 | -2.85582 | 270 | TC | 0.885 | 293 |
| 306 | C5H10O2 | n-propyl acetate | 1 | -7.85524 | 1.43936 | -4.30187 | -3.04832 | 312 | TC | 0.887 | 293 |
| 307 | C5H10O2 | ethyl propionate | 1 | -8.55094 | 3.10067 | -6.99241 | 3.45112 | 307 | TC | 0.895 | 289 |
| 308 | C5H10O2 | methyl butyrate | 1 | -7.77600 | 1.32028 | -3.93963 | -3.53112 | 275 | TC | 0.898 | 293 |
| 309 | C5H10O2 | methyl isobutyrate | 1 | -7.65814 | 1.29248 | -3.85632 | -3.49858 | 270 | TC | 0.891 | 293 |
| 310 | C5H11N | piperidine | 1 | -7.56707 | 2.15002 | -3.89030 | -3.70363 | 316 | TC | 0.862 | 293 |
| 311 | C5H12 | n-pentane | 1 | -7.28936 | 1.53679 | -3.08367 | -1.02456 | 195 | TC | 0.626 | 293 |
| 312 | C5H12 | 2-methyl butane | 1 | -7.12727 | 1.38996 | -2.54302 | -2.45657 | 220 | TC | 0.620 | 293 |
| 313 | C5H12 | 2,2-dimethlypropane | 1 | -6.89153 | 1.25019 | -2.28233 | -4.74891 | 260 | TC | 0.591 | 293 |
| 314 | C5H12O | 1-pentanol | 1 | -8.97725 | 2.99791 | -12.9596 | 8.84205 | 290 | TC | 0.815 | 293 |
| 315 | C5H12O | 2-methyl-1-butanol | 1 | -9.26305 | 3.86947 | -15.3562 | 12.1464 | 308 | TC | 0.819 | 293 |
| 316 | C5H12O | 3-methyl-1-butanol | 3 | 10.0925 | 3026.43 | -104.1 | | 298 | 426 | 0.810 | 293 |
| 317 | C5H12O | 2-methyl-2-butanol | 1 | -8.66602 | 3.46689 | -14.1750 | 10.9679 | 298 | TC | 0.809 | 293 |
| 318 | C5H12O | 2,2-dimethyl-1-propanol | 3 | 11.5134 | 3694.96 | -65.00 | | 328 | 406 | 0.783 | 327 |
| 319 | C5H12O | ethyl propyl ether | 1 | -8.05820 | 2.35916 | -4.51822 | 0.92352 | 275 | TC | 0.733 | 293 |
| 320 | C5H12O | butyl methyl ether | 1 | -7.75110 | 1.87213 | -3.80629 | -1.81410 | 285 | TC | 0.744 | 293 |
| 321 | C5H10O | tert-butyl methyl ether | 1 | -7.82516 | 2.95493 | -6.94079 | 12.17416 | 287 | TC | | |
| 322 | C6BrF5 | bromopentafluorobenzene | | | | | | | | | |
| 323 | C6ClF5 | chloropentafluorobenzene | 1 | -8.02172 | 1.54665 | -3.78361 | -2.99849 | 309 | TC | | |
| 324 | C6Cl2F4 | dichlorotetrafluorobenzene | | | | | | | | | |
| 325 | C6Cl3F3 | 1,3,5-trichlorotrifluorobenzene | 1 | -8.20940 | 1.68886 | -4.17824 | -1.54115 | 364 | TC | | |

| No | Formula | Name | MolWt | Tfp K | Tb K | Tc K | Pc bar | Vc $cm^3/mol$ | Zc | Omega | Dipm debye |
|---|---|---|---|---|---|---|---|---|---|---|---|
| 326 | C6F6 | perfluorobenzene | 186.056 | | 353.4 | 516.7 | 33.0 | 335. | 0.255 | 0.396 | |
| 327 | C6F12 | perfluorocyclohexane | 300.047 | | 326.0 | 457.2 | 24.3 | 459. | 0.270 | 0.432 | 0.0 |
| 328 | C6F14 | perfluoro-n-hexane | 338.044 | 186.0 | 329.8 | 448.8 | 18.7 | 606. | 0.303 | 0.514 | |
| 329 | C6F14 | perfluoro-2-methylpentane | 338.044 | | 330.8 | 453. | 18.2 | 550. | 0.266 | 0.464 | |
| 330 | C6F14 | perfluoro-3-methylpentane | 338.044 | | 331.5 | 450. | 16.9 | | | 0.476 | |
| 331 | C6F14 | perfluoro-2,3-dimethylbutane | 338.044 | | 332.9 | 463. | 18.7 | 525. | 0.256 | 0.394 | |
| 332 | C6HF5 | pentafluorobenzene | 168.064 | | 358.9 | 531.0 | 35.3 | 324. | 0.260 | 0.373 | |
| 333 | C6HF5O | pentafluorophenol | 184.063 | | 418.8 | 609. | 40.0 | 348. | 0.275 | 0.502 | |
| 334 | C6H2F4 | 1,2,3,4-tetrafluorobenzene | 150.074 | | 367.5 | 550.8 | 37.9 | 313. | 0.259 | 0.344 | |
| 335 | C6H2F4 | 1,2,3,5-tetrafluorobenzene | 150.074 | | 357.6 | 535.3 | 37.5 | | | 0.346 | |
| 336 | C6H2F4 | 1,2,4,5-tetrafluorobenzene | 150.074 | | 363.4 | 543.4 | 38.0 | | | 0.355 | |
| 337 | C6H4Cl2 | o-dichlorobenzene | 147.004 | 256.1 | 452.0 | 729. | 41.0 | 360. | 0.244 | 0.272 | 2.3 |
| 338 | C6H4F2 | 1,4-difluorobenzene | 114.094 | | 362.0 | 556. | 44.0 | | | 0.299 | |
| 339 | C6H5Br | bromobenzene | 157.010 | 242.3 | 429.2 | 670.0 | 45.2 | 324. | 0.263 | 0.251 | 1.5 |
| 340 | C6H5Cl | chlorobenzene | 112.559 | 227.6 | 404.9 | 632.4 | 45.2 | 308. | 0.265 | 0.249 | 1.6 |
| 341 | C6H5F | fluorobenzene | 96.104 | 234.0 | 357.9 | 560.1 | 45.5 | 269. | 0.263 | 0.244 | 1.4 |
| 342 | C6H5I | iodobenzene | 204.011 | 241.8 | 461.6 | 721.0 | 45.2 | 351. | 0.265 | 0.249 | 1.4 |
| 343 | C6H6 | benzene | 78.114 | 278.7 | 353.2 | 562.2 | 48.9 | 259. | 0.271 | 0.212 | 0.0 |
| 344 | C6H6O | phenol | 94.113 | 314.0 | 455.0 | 694.2 | 61.3 | 229. | 0.240 | 0.438 | 1.6 |
| 345 | C6H7N | aniline | 93.129 | 267.0 | 457.6 | 699. | 53.1 | 274. | 0.250 | 0.384 | 1.6 |
| 346 | C6H7N | 2-methylpyridine | 93.129 | 207. | 402.6 | 621. | 46.0 | | | 0.299 | 1.9 |
| 347 | C6H7N | 3-methylpyridine | 93.129 | | 417.3 | 645. | | | | | 2.4 |
| 348 | C6H7N | 4-methylpyridine | 93.129 | 276.9 | 418.5 | 646.0 | 44.6 | 311. | 0.260 | 0.301 | |
| 349 | C6H10 | 1,5-hexadiene | 82.146 | 132.0 | 332.6 | 507.0 | 34.4 | | | 0.160 | |
| 350 | C6H10 | cyclohexene | 82.146 | 169.7 | 356.1 | 560.5 | 43.4 | | | 0.210 | 0.6 |

| No | Formula | Name | CPVAP A | CPVAP B | CPVAP C | CPVAP D | DELHF | DELGF |
|---|---|---|---|---|---|---|---|---|
| 326 | C6F6 | perfluorobenzene | 3.628E+1 | 5.267E-1 | -4.547E-4 | 1.456E-7 | -9.573E+5 | -8.800E+5 |
| 327 | C6F12 | perfluorocyclohexane | | | | | | |
| 328 | C6F14 | perfluoro-n-hexane | | | | | | |
| 329 | C6F14 | perfluoro-2-methylpentane | | | | | | |
| 330 | C6F14 | perfluoro-3-methylpentane | | | | | | |
| 331 | C6F14 | perfluoro-2,3-dimethylbutane | | | | | | |
| 332 | C6HF5 | pentafluorobenzene | | | | | | |
| 333 | C6HF5O | pentafluorophenol | | | | | | |
| 334 | C6H2F4 | 1,2,3,4-tetrafluorobenzene | | | | | | |
| 335 | C6H2F4 | 1,2,3,5-tetrafluorobenzene | | | | | | |
| 336 | C6H2F4 | 1,2,4,5-tetrafluorobenzene | | | | | | |
| 337 | C6H4Cl2 | o-dichlorobenzene | -1.430E+1 | 5.506E-1 | -4.513E-4 | 1.429E-7 | 3.000E+4 | 8.273E+4 |
| 338 | C6H4F2 | 1,4-difluorobenzene | -2.596E+1 | 5.722E-1 | -4.677E-4 | 1.475E-7 | -3.074E+5 | -2.530E+5 |
| 339 | C6H5Br | bromobenzene | -2.881E+1 | 5.351E-1 | -4.080E-4 | 1.212E-7 | 1.051E+5 | 1.386E+5 |
| 340 | C6H5Cl | chlorobenzene | -3.389E+1 | 5.631E-1 | -4.522E-4 | 1.426E-7 | 5.187E+4 | 9.923E+4 |
| 341 | C6H5F | fluorobenzene | -3.873E+1 | 5.669E-1 | -4.434E-4 | 1.355E-7 | -1.166E+5 | -6.908E+4 |
| 342 | C6H5I | iodobenzene | -2.927E+1 | 5.564E-1 | -4.509E-4 | 1.443E-7 | 1.627E+5 | 1.879E+5 |
| 343 | C6H6 | benzene | -3.392E+1 | 4.739E-1 | -3.017E-4 | 7.130E-8 | 8.298E+4 | 1.297E+5 |
| 344 | C6H6O | phenol | -3.584E+1 | 5.983E-1 | -4.827E-4 | 1.527E-7 | -9.642E+4 | -3.290E+4 |
| 345 | C6H7N | aniline | -4.052E+1 | 6.385E-1 | -5.133E-4 | 1.633E-7 | 8.692E+4 | 1.668E+5 |
| 346 | C6H7N | 2-methylpyridine | -3.626E+1 | 5.584E-1 | -3.704E-4 | 9.663E-8 | 9.902E+4 | 1.772E+5 |
| 347 | C6H7N | 3-methylpyridine | -3.709E+1 | 5.600E-1 | -3.719E-4 | 9.685E-8 | 1.062E+5 | 1.844E+5 |
| 348 | C6H7N | 4-methylpyridine | -1.743E+1 | 4.882E-1 | -2.798E-4 | 5.451E-8 | 1.023E+5 | |
| 349 | C6H10 | 1,5-hexadiene | | | | | 8.374E+4 | |
| 350 | C6H10 | cyclohexene | -6.865E+1 | 7.252E-1 | -5.414E-4 | 1.644E-7 | -5.360E+3 | 1.069E+5 |

| No | Formula | Name | Eq. | VP A | VP B | VP C | VP D | Tmin | Tmax | LDEN | TDEN |
|---|---|---|---|---|---|---|---|---|---|---|---|
| 326 | C6F6 | perfluorobenzene | 1 | -7.97271 | 1.43798 | -3.62195 | -4.79241 | 278 | TC | | |
| 327 | C6F12 | perfluorocyclohexane | 3 | 7.2885 | 1374.07 | -136.8 | | 280 | 400 | | |
| 328 | C6F14 | perfluoro-n-hexane | 1 | -9.16184 | 2.97539 | -7.17322 | 5.50684 | 270 | TC | | |
| 329 | C6F14 | perfluoro-2-methylpentane | 3 | 9.6896 | 2760.0 | -45.70 | | 259 | 346 | 1.733 | 293 |
| 330 | C6F14 | perfluoro-3-methylpentane | 3 | 9.2670 | 2565.44 | -54.23 | | 255 | 333 | | |
| 331 | C6F14 | perfluoro-2,3-dimethylbutane | 3 | 9.9846 | 2933.85 | -38.70 | | 262 | 333 | | |
| 332 | C6HF5 | pentafluorobenzene | 1 | -7.79730 | 1.35271 | -3.50409 | -3.76856 | 322 | TC | | |
| 333 | C6HF5O | pentafluorophenol | 1 | -8.69734 | 2.03071 | -5.32619 | -3.28915 | 379 | TC | | |
| 334 | C6H2F4 | 1,2,3,4-tetrafluorobenzene | 1 | -7.71223 | 1.48262 | -3.55699 | -2.83189 | 301 | TC | | |
| 335 | C6H2F4 | 1,2,3,5-tetrafluorobenzene | 1 | -7.71193 | 1.46356 | -3.49452 | -3.04916 | 288 | TC | | |
| 336 | C6H2F4 | 1,2,4,5-tetrafluorobenzene | 1 | -7.79740 | 1.57406 | -3.82060 | -2.45398 | 294 | TC | | |
| 337 | C6H4Cl2 | o-dichlorobenzene | 1 | -8.23991 | 6.34949 | -13.24326 | 17.25417 | 403 | TC | 1.306 | 293 |
| 338 | C6H4F2 | 1,4-difluorobenzene | | | | | | | | | |
| 339 | C6H5Br | bromobenzene | 1 | -7.54985 | 2.09359 | -3.57864 | -1.82558 | 329 | TC | 1.495 | 293 |
| 340 | C6H5Cl | chlorobenzene | 1 | -7.58700 | 2.26551 | -4.09418 | 0.17038 | 335 | TC | 1.106 | 293 |
| 341 | C6H5F | fluorobenzene | 2 | 48.521 | 5819.21 | -5.489 | 2910. | 300 | TC | 1.024 | 293 |
| 342 | C6H5I | iodobenzene | 2 | 51.071 | 7589.50 | -5.646 | 4845. | 380 | TC | 1.855 | 277 |
| 343 | C6H6 | benzene | 1 | -6.98273 | 1.33213 | -2.62863 | -3.33399 | 288 | TC | 0.885 | 289 |
| 344 | C6H6O | phenol | 1 | -8.75550 | 2.92651 | -6.31601 | -1.36889 | 380 | TC | 1.059 | 313 |
| 345 | C6H7N | aniline | 1 | -7.65517 | 0.85386 | -2.51602 | -5.96795 | 376 | TC | 1.022 | 293 |
| 346 | C6H7N | 2-methylpyridine | 3 | 9.5725 | 3259.83 | -61.58 | | 352 | 442 | 0.950 | 288 |
| 347 | C6H7N | 3-methylpyridine | 3 | 9.6136 | 3411.91 | -61.95 | | 347 | 458 | 0.961 | 288 |
| 348 | C6H7N | 4-methyl pyridine | 1 | -7.13732 | 0.93444 | -2.93708 | -2.65045 | 348 | TC | 0.955 | 293 |
| 349 | C6H10 | 1,5-hexadiene | 1 | -7.72848 | 2.21648 | -2.23190 | -8.51382 | 273 | TC | 0.692 | 293 |
| 350 | C6H10 | cyclohexene | 3 | 9.2041 | 2813.53 | -49.98 | | 300 | 360 | 0.816 | 289 |

| No | Formula | Name | MolWt | Tfp K | Tb K | Tc K | Pc bar | Vc $cm^3$/mol | Zc | Omega | Dipm debye |
|---|---|---|---|---|---|---|---|---|---|---|---|
| 351 | C6H10O | cyclohexanone | 98.145 | 242.0 | 428.8 | 629. | 39. | | | | 3.1 |
| 352 | C6H11N | capronitrile | 97.161 | 194. | 436.8 | 622. | 32.5 | | | 0.524 | 3.5 |
| 353 | C6H12 | cyclohexane | 84.162 | 279.6 | 353.8 | 553.5 | 40.7 | 308. | 0.273 | 0.212 | 0.3 |
| 354 | C6H12 | methylcyclopentane | 84.162 | 130.7 | 345.0 | 532.7 | 37.8 | 319. | 0.272 | 0.231 | 0.0 |
| 355 | C6H12 | 1-hexene | 84.163 | 133.3 | 336.6 | 504.0 | 31.7 | 350. | 0.26 | 0.285 | 0.4 |
| 356 | C6H12 | 2-hexene,cis | 84.162 | 132.0 | 342.0 | 518. | 32.8 | 351. | 0.27 | 0.256 | |
| 357 | C6H12 | 2-hexene,trans | 84.162 | 140. | 341.0 | 516. | 32.7 | 351. | 0.27 | 0.242 | |
| 358 | C6H12 | 3-hexene,cis | 84.162 | 135.3 | 339.6 | 517. | 32.8 | 350. | 0.27 | 0.225 | 0.3 |
| 359 | C6H12 | 3-hexene,trans | 84.162 | 159.7 | 340.3 | 519.9 | 32.5 | 350. | 0.26 | 0.227 | 0.0 |
| 360 | C6H12 | 2-methyl-2-pentene | 84.162 | 138.1 | 340.5 | 518. | 32.8 | 351. | 0.27 | 0.229 | |
| 361 | C6H12 | 3-methyl-2-pentene,cis | 84.162 | 138.3 | 340.9 | 518. | 32.8 | 351. | 0.27 | 0.269 | |
| 362 | C6H12 | 3-methyl-2-pentene,trans | 84.162 | 134.7 | 343.6 | 521. | 32.9 | 350. | 0.27 | 0.207 | |
| 363 | C6H12 | 4-methyl-2-pentene,cis | 84.162 | 139. | 329.6 | 490. | 30.4 | 360. | 0.27 | 0.29 | |
| 364 | C6H12 | 4-methyl-2-pentene,trans | 84.162 | 132. | 331.7 | 493. | 30.4 | 360. | 0.27 | 0.29 | |
| 365 | C6H12 | 2,3-dimethyl-1-butene | 84.162 | 115.9 | 328.8 | 501. | 32.4 | 343. | 0.27 | 0.221 | |
| 366 | C6H12 | 2,3-dimethyl-2-butene | 84.162 | 198.9 | 346.4 | 524. | 33.6 | 351. | 0.27 | 0.239 | |
| 367 | C6H12 | 3,3-dimethyl-1-butene | 84.162 | 158. | 314.4 | 490. | 32.5 | 340. | 0.27 | 0.121 | |
| 368 | C6H12O | cyclohexanol | 100.160 | 298. | 434.3 | 625. | 37.5 | | | 0.528 | 1.7 |
| 369 | C6H12O | ethyl propyl ketone | 100.160 | | 396.6 | 582.8 | 33.2 | | | 0.378 | |
| 370 | C6H12O | methyl butyl ketone | 100.160 | 216. | 400.7 | 587.0 | 33.2 | | | 0.392 | |
| 371 | C6H12O | methyl isobutyl ketone | 100.160 | 189. | 389.6 | 571. | 32.7 | | | 0.385 | 2.8 |
| 372 | C6H12O2 | n-butyl acetate | 116.160 | 199.7 | 399.3 | 579. | 31.4 | 400. | 0.26 | 0.417 | 1.8 |
| 373 | C6H12O2 | isobutyl acetate | 116.160 | 174.3 | 389.7 | 564. | 30.2 | 414. | 0.267 | 0.455 | 1.9 |
| 374 | C6H12O2 | ethyl butyrate | 116.160 | 180. | 394.7 | 569. | 29.6 | 421. | 0.263 | 0.461 | 1.8 |
| 375 | C6H12O2 | ethyl isobutyrate | 116.160 | 185. | 383.2 | 555. | 29.7 | 421. | 0.271 | 0.431 | 2.1 |

| No | Formula | Name | CPVAP A | CPVAP B | CPVAP C | CPVAP D | DELHF | DELGF |
|---|---|---|---|---|---|---|---|---|
| 351 | C6H10O | cyclohexanone | -3.781E+1 | 5.539E-1 | -1.953E-4 | -1.534E-8 | -2.303E+5 | -9.081E+4 |
| 352 | C6H11N | capronitrile | | | | | | |
| 353 | C6H12 | cyclohexane | -5.454E+1 | 6.113E-1 | -2.523E-4 | 1.321E-8 | -1.232E+5 | 3.178E+4 |
| 354 | C6H12 | methylcyclopentane | -5.011E+1 | 6.381E-1 | -3.642E-4 | 8.014E-8 | -1.068E+5 | 3.580E+4 |
| 355 | C6H12 | 1-hexene | -1.746E+0 | 5.309E-1 | -2.903E-4 | 6.054E-8 | -4.170E+4 | 8.750E+4 |
| 356 | C6H12 | 2-hexene,cis | -9.810E+0 | 5.309E-1 | -2.717E-4 | 4.827E-8 | -5.238E+4 | 7.628E+4 |
| 357 | C6H12 | 2-hexene,trans | -3.292E+1 | 6.929E-1 | -5.619E-4 | 2.005E-7 | -5.393E+4 | 7.649E+4 |
| 358 | C6H12 | 3-hexene,cis | -2.173E+1 | 5.811E-1 | -3.362E-4 | 7.457E-8 | -4.765E+4 | 8.307E+4 |
| 359 | C6H12 | 3-hexene,trans | -4.338E+0 | 5.510E-1 | -3.282E-4 | 8.047E-8 | -5.447E+4 | 7.767E+4 |
| 360 | C6H12 | 2-methyl-2-pentene | -1.475E+1 | 5.669E-1 | -3.341E-4 | 7.963E-8 | -6.653E+4 | 7.126E+4 |
| 361 | C6H12 | 3-methyl-2-pentene,cis | -1.475E+1 | 5.669E-1 | -3.341E-4 | 7.963E-8 | -6.222E+4 | 7.327E+4 |
| 362 | C6H12 | 3-methyl-2-pentene,trans | -1.475E+1 | 5.669E-1 | -3.341E-4 | 7.963E-8 | -6.314E+4 | 7.134E+4 |
| 363 | C6H12 | 4-methyl-2-pentene,cis | -1.675E+0 | 5.376E-1 | -3.044E-4 | 6.753E-8 | -5.748E+4 | 8.219E+4 |
| 364 | C6H12 | 4-methyl-2-pentene,trans | 1.263E+1 | 5.154E-1 | -3.007E-4 | 7.327E-8 | -6.150E+4 | 7.967E+4 |
| 365 | C6H12 | 2,3-dimethyl-1-butene | 7.025E+0 | 5.585E-1 | -3.696E-4 | 1.063E-7 | -6.636E+4 | 7.909E+4 |
| 366 | C6H12 | 2,3-dimethyl-2-butene | 2.294E+0 | 4.827E-1 | -2.199E-4 | 3.042E-8 | -6.984E+4 | 7.591E+4 |
| 367 | C6H12 | 3,3-dimethyl-1-butene | -1.256E+1 | 5.485E-1 | -2.915E-4 | 5.208E-8 | -6.155E+4 | 9.822E+4 |
| 368 | C6H12O | cyclohexanol | -5.553E+1 | 7.214E-1 | -4.086E-4 | 8.235E-8 | -2.948E+5 | -1.180E+5 |
| 369 | C6H12O | ethyl propyl ketone | | | | | | |
| 370 | C6H12O | methyl butyl ketone | | | | | | |
| 371 | C6H12O | methyl isobutyl ketone | 3.894E+0 | 5.656E-1 | -3.318E-4 | 8.231E-8 | -2.840E+5 | |
| 372 | C6H12O2 | n-butyl acetate | 1.362E+1 | 5.489E-1 | -2.278E-4 | 0.791E-9 | -4.868E+5 | |
| 373 | C6H12O2 | isobutyl acetate | 7.310E+0 | 5.740E-1 | -2.576E-4 | 1.101E-8 | -4.955E+5 | |
| 374 | C6H12O2 | ethyl butyrate | 2.151E+1 | 4.928E-1 | -1.938E-4 | 3.559E-9 | | |
| 375 | C6H12O2 | ethyl isobutyrate | | | | | | |

| No | Formula | Name | Eq. | VP A | VP B | VP C | VP D | Tmin | Tmax | LDEN | TDEN |
|---|---|---|---|---|---|---|---|---|---|---|---|
| 351 | C6H10O | cyclohexanone | | | | | | | | 0.951 | 288 |
| 352 | C6H11N | capronitrile | 3 | 9.7814 | 3677.63 | -60.40 | | 363 | 438 | 0.809 | 288 |
| 353 | C6H12 | cyclohexane | 1 | -6.96009 | 1.31328 | -2.75683 | -2.45491 | 293 | TC | 0.779 | 293 |
| 354 | C6H12 | methylcyclopentane | 1 | -7.15937 | 1.48017 | -2.92482 | -1.98377 | 288 | TC | 0.754 | 289 |
| 355 | C6H12 | 1-hexene | 1 | -7.76467 | 2.29843 | -4.44302 | 0.89947 | 289 | TC | 0.673 | 293 |
| 356 | C6H12 | 2-hexene,cis | 3 | 9.5855 | 2897.97 | -39.30 | | 245 | 370 | 0.687 | 293 |
| 357 | C6H12 | 2-hexene,trans | 2 | 53.818 | 5734.51 | -6.348 | 3548. | 280 | TC | 0.678 | 293 |
| 358 | C6H12 | 3-hexene,cis | 3 | 9.2182 | 2680.52 | -48.40 | | 245 | 365 | 0.680 | 293 |
| 359 | C6H12 | 3-hexene,trans | 3 | 9.3086 | 2718.68 | -47.77 | | 245 | 365 | 0.677 | 293 |
| 360 | C6H12 | 2-methyl-2-pentene | 3 | 9.3221 | 2725.89 | -47.64 | | 245 | 370 | 0.691 | 289 |
| 361 | C6H12 | 3-methyl-2-pentene,cis | 3 | 9.2922 | 2731.79 | -46.47 | | 248 | 364 | 0.694 | 293 |
| 362 | C6H12 | 3-methyl-2-pentene,trans | 3 | 9.3282 | 2750.50 | -48.33 | | 250 | 366 | 0.698 | 293 |
| 363 | C6H12 | 4-methyl-2-pentene,cis | 3 | 9.1325 | 2580.52 | -46.56 | | 238 | 352 | 0.669 | 293 |
| 364 | C6H12 | 4-methyl-2-pentene,trans | 3 | 9.2223 | 2631.57 | -46.00 | | 240 | 354 | 0.669 | 293 |
| 365 | C6H12 | 2,3-dimethyl-1-butene | 3 | 9.1810 | 2612.69 | -43.78 | | 235 | 360 | 0.678 | 293 |
| 366 | C6H12 | 2,3-dimethyl-2-butene | 1 | -7.15852 | 1.36868 | -4.12890 | 1.53046 | 302 | TC | 0.708 | 293 |
| 367 | C6H12 | 3,3-dimethyl-1-butene | 1 | -6.54633 | 1.50412 | -4.54855 | 2.96466 | 264 | TC | 0.653 | 293 |
| 368 | C6H12O | cyclohexanol | 1 | -8.77758 | 3.11622 | -12.3555 | 7.50610 | 367 | TC | 0.942 | 303 |
| 369 | C6H12O | ethyl propyl ketone | 3 | 9.5000 | 3144.85 | -65.19 | | 347 | 408 | 0.813 | 295 |
| 370 | C6H12O | methyl butyl ketone | | | | | | | | 0.816 | 288 |
| 371 | C6H12O | methyl isobutyl ketone | 1 | -8.54349 | 2.92801 | -5.27311 | -2.54507 | 295 | TC | 0.801 | 293 |
| 372 | C6H12O2 | n-butyl acetate | 1 | -8.36658 | 2.40985 | -6.42511 | 4.85939 | 333 | TC | 0.898 | 273 |
| 373 | C6H12O2 | isobutyl acetate | 1 | -8.12456 | 1.66934 | -4.20511 | -3.72813 | 290 | TC | 0.875 | 293 |
| 374 | C6H12O2 | ethyl butyrate | 1 | -8.00073 | 1.34045 | -3.99843 | -3.74347 | 290 | TC | 0.879 | 293 |
| 375 | C6H12O2 | ethyl isobutyrate | 1 | -8.08582 | 1.61436 | -4.14816 | -3.80720 | 280 | TC | 0.869 | 293 |

| No | Formula | Name | MolWt | Tfp K | Tb K | Tc K | Pc bar | Vc $cm^3/mol$ | Zc | Omega | Dipm debye |
|---|---|---|---|---|---|---|---|---|---|---|---|
| 376 | C6H12O2 | n-propyl propionate | 116.160 | 197.3 | 395.8 | 571. | 30.2 | | | | 1.8 |
| 377 | C6H12O2 | n-amyl formate | 116.160 | 199.7 | 403.6 | 576. | 34.6 | | | 0.538 | |
| 378 | C6H12O2 | isoamyl formate | 116.160 | | 396.7 | 578. | | | | | |
| 379 | C6H14 | n-hexane | 86.178 | 177.8 | 341.9 | 507.5 | 30.1 | 370. | 0.264 | 0.299 | 0.0 |
| 380 | C6H14 | 2-methyl pentane | 86.178 | 119.5 | 333.4 | 497.5 | 30.1 | 367. | 0.267 | 0.278 | |
| 381 | C6H14 | 3-methyl pentane | 86.178 | 155. | 336.4 | 504.5 | 31.2 | 367. | 0.273 | 0.272 | |
| 382 | C6H14 | 2,2-dimethyl butane | 86.178 | 173.3 | 322.8 | 488.8 | 30.8 | 359. | 0.272 | 0.232 | |
| 383 | C6H14 | 2,3-dimethyl butane | 86.178 | 144.6 | 331.1 | 500.0 | 31.3 | 358. | 0.269 | 0.247 | |
| 384 | C6H14O | 1-hexanol | 102.177 | 229.2 | 430.2 | 611. | 40.5 | 381. | 0.300 | 0.560 | 1.8 |
| 385 | C6H14O | 2-hexanol | 102.177 | | 411. | 586.2 | | | | | |
| 386 | C6H14O | ethyl butyl ether | 102.177 | 170. | 365.4 | 531. | 30.4 | 390. | 0.27 | 0.40 | 1.2 |
| 387 | C6H14O | methyl amyl ether | 102.177 | | 372. | 546.5 | 30.4 | 392. | 0.262 | 0.347 | |
| 388 | C6H14O | dipropyl ether | 102.177 | 151. | 363.2 | 530.6 | 30.3 | | | 0.369 | 1.2 |
| 389 | C6H14O | diisopropyl ether | 102.177 | 187.7 | 341.7 | 500.3 | 28.8 | 386. | 0.262 | 0.331 | 1.2 |
| 390 | C6H15N | dipropylamine | 101.193 | 233.6 | 382.5 | 555.8 | 29.9 | | | 0.471 | 1.0 |
| 391 | C6H15N | diisopropylamine | 101.193 | 212.2 | 357.1 | 523.1 | 30.2 | | | 0.360 | 1.0 |
| 392 | C6H15N | triethylamine | 101.193 | 158.4 | 362.5 | 535. | 30.3 | 389. | 0.265 | 0.320 | 0.9 |
| 393 | C7F8 | perfluorotoluene | 236.061 | | 377.7 | 534.5 | 27.1 | 428. | 0.260 | 0.475 | |
| 394 | C7F14 | perfluoromethylcyclohexane | 350.055 | | 349.5 | 486.8 | 23.3 | | | 0.491 | |
| 395 | C7F16 | perfluoro-n-heptane | 388.051 | 195. | 355.6 | 474.8 | 16.2 | 664. | 0.273 | 0.556 | |
| 396 | C7H3F5 | 2,3,4,5,6-pentafluorotoluene | 182.091 | | 390.7 | 566.5 | 31.3 | 384. | 0.255 | 0.415 | |
| 397 | C7H5N | benzonitrile | 103.124 | 260. | 464.3 | 699.4 | 42.2 | | | 0.362 | 3.5 |
| 398 | C7H6O | benzaldehyde | 106.124 | 216. | 452.2 | 694.8 | 45.4 | | | 0.316 | 2.8 |
| 399 | C7H6O2 | benzoic acid | 122.124 | 395.6 | 523. | 752. | 45.6 | 341. | 0.25 | 0.62 | 1.7 |
| 400 | C7H8 | toluene | 92.141 | 178. | 383.8 | 591.8 | 41.0 | 316. | 0.263 | 0.263 | 0.4 |

| No | Formula | Name | CPVAP A | CPVAP B | CPVAP C | CPVAP D | DELHF | DELGF |
|---|---|---|---|---|---|---|---|---|
| 376 | C6H12O2 | n-propyl propionate | | | | | | |
| 377 | C6H12O2 | n-amyl formate | | | | | | |
| 378 | C6H12O2 | isoamyl formate | | | | | | |
| 379 | C6H14 | n-hexane | -4.413E+0 | 5.820E-1 | -3.119E-4 | 6.494E-8 | -1.673E+5 | -1.670E+2 |
| 380 | C6H14 | 2-methyl pentane | -1.057E+1 | 6.184E-1 | -3.573E-4 | 8.085E-8 | -1.744E+5 | -5.020E+3 |
| 381 | C6H14 | 3-methyl pentane | -2.386E+0 | 5.690E-1 | -2.870E-4 | 5.033E-8 | -1.717E+5 | -2.140E+3 |
| 382 | C6H14 | 2,2-dimethyl butane | -1.663E+1 | 6.293E-1 | -3.481E-4 | 6.850E-8 | -1.857E+5 | -9.630E+3 |
| 383 | C6H14 | 2,3-dimethyl butane | -1.461E+1 | 6.150E-1 | -3.376E-4 | 6.820E-8 | -1.779E+5 | -4.100E+3 |
| 384 | C6H14O | 1-hexanol | 4.811E+0 | 5.891E-1 | -3.010E-4 | 5.426E-8 | -3.178E+5 | -1.357E+5 |
| 385 | C6H14O | 2-hexanol | | | | | | |
| 386 | C6H14O | ethyl butyl ether | 2.363E+1 | 5.367E-1 | -2.528E-4 | 4.157E-8 | | |
| 387 | C6H14O | methyl amyl ether | | | | | | |
| 388 | C6H14O | dipropyl ether | 1.862E+1 | 5.335E-1 | -2.285E-4 | 2.442E-8 | -2.931E+5 | -1.056E+5 |
| 389 | C6H14O | diisopropyl ether | 7.505E+0 | 5.849E-1 | -3.027E-4 | 5.845E-8 | -3.190E+5 | -1.220E+5 |
| 390 | C6H15N | dipropylamine | 6.460E+0 | 6.293E-1 | -3.390E-4 | 7.072E-8 | | |
| 391 | C6H15N | diisopropylamine | | | | | | |
| 392 | C6H15N | triethylamine | -1.843E+1 | 7.155E-1 | -4.392E-4 | 1.092E-7 | -9.965E+4 | 1.104E+5 |
| 393 | C7F8 | perfluorotoluene | | | | | | |
| 394 | C7F14 | perfluoromethylcyclohexane | | | | | -2.898E+6 | |
| 395 | C7F16 | perfluoro-n-heptane | | | | | -3.387E+6 | -3.089E+6 |
| 396 | C7H3F5 | 2,3,4,5,6-pentafluorotoluene | | | | | | |
| 397 | C7H5N | benzonitrile | -2.605E+1 | 5.732E-1 | -4.430E-4 | 1.349E-7 | 2.190E+5 | 2.610E+5 |
| 398 | C7H6O | benzaldehyde | -1.214E+1 | 4.961E-1 | -2.845E-4 | 5.167E-8 | -3.680E+4 | 2.240E+4 |
| 399 | C7H6O2 | benzoic acid | -5.129E+1 | 6.293E-1 | -4.237E-4 | 1.062E-7 | -2.904E+5 | -2.106E+5 |
| 400 | C7H8 | toluene | -2.435E+1 | 5.125E-1 | -2.765E-4 | 4.911E-8 | 5.003E+4 | 1.221E+5 |

| No | Formula | Name | Eq. | VP A | VP B | VP C | VP D | Tmin | Tmax | LDEN | TDEN |
|---|---|---|---|---|---|---|---|---|---|---|---|
| 376 | C6H12O2 | n-propyl propionate | 1 | -8.00913 | 1.33297 | -3.97513 | -3.83674 | 290 | TC | 0.881 | 293 |
| 377 | C6H12O2 | n-amyl formate | | | | | | | | 0.902 | 273 |
| 378 | C6H12O2 | isoamyl formate | | | | | | | | 0.882 | 293 |
| 379 | C6H14 | n-hexane | 1 | -7.46765 | 1.44211 | -3.28222 | -2.50941 | 220 | TC | 0.659 | 293 |
| 380 | C6H14 | isohexane | 1 | -7.31728 | 1.33940 | -3.06807 | -1.99255 | 240 | TC | 0.653 | 293 |
| 381 | C6H14 | 3-methyl pentane | 1 | -7.27084 | 1.26113 | -2.81741 | -2.17642 | 235 | TC | 0.664 | 293 |
| 382 | C6H14 | 2,2-dimethyl butane | 1 | -7.24296 | 1.66876 | -3.23718 | -0.53171 | 225 | TC | 0.649 | 293 |
| 383 | C6H14 | 2,3-dimethyl butane | 1 | -7.27870 | 1.56349 | -3.05387 | -1.57752 | 235 | TC | 0.662 | 293 |
| 384 | C6H14O | 1-hexanol | 3 | 11.4792 | 4055.45 | -76.49 | | 308 | 430 | 0.819 | 293 |
| 385 | C6H14O | 2-hexanol | 3 | 10.0989 | 3158.53 | -99.98 | | 295 | 418 | 0.816 | 293 |
| 386 | C6H14O | ethyl butyl ether | 1 | -8.30292 | 2.02889 | -3.26245 | -6.32274 | 311 | TC | 0.749 | 293 |
| 387 | C6H14O | methyl amyl ether | | | | | | | | 0.75 | 298 |
| 388 | C6H14O | dipropyl ether | 1 | -8.22229 | 2.22110 | -3.90291 | -3.77431 | 288 | TC | 0.736 | 293 |
| 389 | C6H14O | diisopropyl ether | 1 | -7.62613 | 1.29308 | -2.90101 | -6.14467 | 297 | TC | 0.724 | 293 |
| 390 | C6H15N | dipropylamine | 1 | -8.56471 | 2.93461 | -5.56089 | 0.56571 | 275 | TC | 0.738 | 293 |
| 391 | C6H15N | diisopropylamine | 1 | -7.84319 | 1.80097 | -4.66547 | -0.29364 | 257 | TC | 0.722 | 295 |
| 392 | C6H15N | triethylamine | 1 | -11.3617 | 10.0092 | -13.4750 | -9.36035 | 323 | TC | 0.728 | 293 |
| 393 | C7F8 | perfluorotoluene | | | | | | | | | |
| 394 | C7F14 | perfluoromethylcyclohexane | 1 | -10.5469 | 6.38028 | -10.6940 | 11.6006 | 306 | TC | 1.789 | 298 |
| 395 | C7F16 | perfluoro-n-heptane | 1 | -9.13392 | 2.75328 | -8.33813 | 6.82085 | 271 | TC | 1.733 | 293 |
| 396 | C7H3F5 | 2,3,4,5,6-pentafluorotoluene | 1 | -8.05688 | 1.46673 | -3.82439 | -2.78727 | 313 | TC | | |
| 397 | C7H5N | benzonitrile | 2 | 53.154 | 7912.31 | -5.881 | 4898. | 340 | TC | 1.010 | 288 |
| 398 | C7H6O | benzaldehyde | 1 | -7.16527 | 0.52710 | -1.51484 | -7.92908 | 300 | TC | 1.045 | 293 |
| 399 | C7H6O2 | benzoic acid | 3 | 10.5432 | 4190.70 | -125.2 | | 405 | 560 | 1.075 | 403 |
| 400 | C7H8 | toluene | 1 | -7.28607 | 1.38091 | -2.83433 | -2.79168 | 309 | TC | 0.867 | 293 |

| No | Formula | Name | MolWt | Tfp K | Tb K | Tc K | Pc bar | Vc $cm^3/mol$ | Zc | Omega | Dipm debye |
|---|---|---|---|---|---|---|---|---|---|---|---|
| 401 | C7H8O | methyl phenyl ether | 108.140 | 235.7 | 426.8 | 645.6 | 42.5 | | | 0.347 | 1.2 |
| 402 | C7H8O | benzyl alcohol | 108.140 | 257.8 | 478.6 | 720.2 | 44.0 | | | | 1.7 |
| 403 | C7H8O | o-cresol | 108.140 | 304.1 | 464.2 | 697.6 | 50.1 | | | 0.433 | 1.6 |
| 404 | C7H8O | m-cresol | 108.140 | 285.4 | 475.4 | 705.8 | 45.6 | 309. | 0.240 | 0.454 | 1.8 |
| 405 | C7H8O | p-cresol | 108.140 | 307.9 | 475.1 | 704.6 | 51.5 | | | 0.505 | 1.6 |
| 406 | C7H9N | 2,3-dimethylpyridine | 107.156 | | 434.4 | 655.4 | | | | | 2.2 |
| 407 | C7H9N | 2,4-dimethylpyridine | 107.156 | | 431.6 | 647. | | | | | 2.3 |
| 408 | C7H9N | 2,5-dimethylpyridine | 107.156 | | 430.2 | 644.2 | | | | | 2.2 |
| 409 | C7H9N | 2,6-dimethylpyridine | 107.156 | 267. | 417.2 | 623.8 | | | | | 1.7 |
| 410 | C7H9N | 3,4-dimethylpyridine | 107.156 | | 452.3 | 683.8 | | | | | 1.9 |
| 411 | C7H9N | 3,5-dimethylpyridine | 107.156 | | 445.1 | 667.2 | | | | | 2.6 |
| 412 | C7H9N | N-methylaniline | 107.156 | 216. | 469.4 | 701. | 52.0 | | | 0.475 | 1.7 |
| 413 | C7H9N | o-toluidine | 107.156 | 258.4 | 473.5 | 694. | 37.5 | | | 0.438 | 1.6 |
| 414 | C7H9N | m-toluidine | 107.156 | 242.8 | 476.6 | 709. | 41.5 | | | 0.410 | 1.5 |
| 415 | C7H9N | p-toluidine | 107.156 | 316.9 | 473.7 | 667. | 23.8 | | | 0.443 | 1.6 |
| 416 | C7H14 | cycloheptane | 98.189 | 265. | 391.6 | 604.2 | 38.1 | 353. | 0.268 | 0.237 | |
| 417 | C7H14 | 1,1-dimethylcyclopentane | 98.189 | 203.4 | 361.0 | 547. | 34.4 | 360. | 0.27 | 0.273 | |
| 418 | C4H14 | 1,2-dimethylcyclopentane-cis | 98.189 | 219.3 | 372.7 | 564.8 | 34.4 | 368. | 0.27 | 0.269 | |
| 419 | C7H14 | 1,2-dimethylcyclopentane-trans | 98.189 | 155.6 | 365.0 | 553.2 | 34.4 | 362. | 0.27 | 0.269 | |
| 420 | C7H14 | ethylcyclopentane | 98.189 | 134.7 | 376.6 | 569.5 | 34.0 | 375. | 0.269 | 0.271 | |
| 421 | C7H14 | methylcyclohexane | 98.189 | 146.6 | 374.1 | 572.2 | 34.7 | 368. | 0.268 | 0.236 | 0.0 |
| 422 | C7H14 | 1-heptene | 98.189 | 154.3 | 366.8 | 537.3 | 28.3 | 440. | 0.28 | 0.358 | 0.3 |
| 423 | C7H14 | 2,3,3-trimethyl-1-butene | 98.189 | 163.3 | 351.0 | 533. | 28.9 | 400. | 0.26 | 0.192 | |
| 424 | C7H14O | methyl amyl ketone | 114.188 | | 424.2 | 611.5 | 34.4 | | | 0.483 | |
| 425 | C7H14O2 | n-propyl butyrate | 130.187 | 176.0 | 416.2 | 590. | 27.1 | | | | 1.8 |

| No | Formula | Name | CPVAP A | CPVAP B | CPVAP C | CPVAP D | DELHF | DELGF |
|---|---|---|---|---|---|---|---|---|
| 401 | C7H80 | methyl phenyl ether | | | | | | |
| 402 | C7H80 | benzyl alcohol | -7.398E+0 | 5.481E-1 | -3.357E-4 | 7.771E-8 | -9.408E+4 | |
| 403 | C7H80 | o-cresol | -3.228E+1 | 7.005E-1 | -5.924E-4 | 2.124E-7 | -1.287E+5 | -3.300E+4 |
| 404 | C7H80 | m-cresol | -4.501E+1 | 7.264E-1 | -6.029E-4 | 2.077E-7 | -1.324E+5 | -4.057E+4 |
| 405 | C7H80 | p-cresol | -4.063E+1 | 7.055E-1 | -5.757E-4 | 1.967E-7 | -1.255E+5 | -3.090E+4 |
| 406 | C7H9N | 2,3-dimethylpyridine | | | | | 6.829E+4 | |
| 407 | C7H9N | 2,4-dimethylpyridine | | | | | | |
| 408 | C7H9N | 2,5-dimethylpyridine | | | | | 6.644E+4 | |
| 409 | C7H9N | 2,6-dimethylpyridine | | | | | | |
| 410 | C7H9N | 3,4-dimethylpyridine | | | | | 7.005E+4 | |
| 411 | C7H9N | 3,5-dimethylpyridine | | | | | 7.281E+4 | |
| 412 | C7H9N | N-methylaniline | | | | | 8.541E+4 | 1.993E+5 |
| 413 | C7H9N | o-toluidine | | | | | | |
| 414 | C7H9N | m-toluidine | -1.599E+1 | 5.681E-1 | -3.033E-4 | 4.643E-8 | | |
| 415 | C7H9N | p-toluidine | | | | | | |
| 416 | C7H14 | cycloheptane | -7.619E+1 | 7.867E-1 | -4.204E-4 | 7.561E-8 | -1.194E+5 | 6.305E+4 |
| 417 | C7H14 | 1,1-dimethylcyclopentane | -5.789E+1 | 7.670E-1 | -4.501E-4 | 1.010E-7 | -1.384E+5 | 3.906E+4 |
| 418 | C4H14 | 1,2-dimethylcyclopentane-cis | -5.564E+1 | 7.616E-1 | -4.484E-4 | 1.014E-7 | -1.296E+5 | 4.576E+4 |
| 419 | C7H14 | 1,2-dimethylcyclopentane-trans | -5.452E+1 | 7.591E-1 | -4.480E-4 | 1.017E-7 | -1.368E+5 | 3.839E+4 |
| 420 | C7H14 | ethylcyclopentane | -5.531E+1 | 7.511E-1 | -4.396E-4 | 1.004E-7 | -1.272E+5 | 4.459E+4 |
| 421 | C7H14 | methylcyclohexane | -6.192E+1 | 7.842E-1 | -4.438E-4 | 9.366E-8 | -1.549E+5 | 2.730E+4 |
| 422 | C7H14 | 1-heptene | -3.303E+0 | 6.297E-1 | -3.512E-4 | 7.607E-8 | -6.234E+4 | 9.588E+4 |
| 423 | C7H14 | 2,3,3-trimethyl-1-butene | | | | | -8.654E+4 | |
| 424 | C7H140 | methyl amyl ketone | | | | | | |
| 425 | C7H1402 | n-propyl butyrate | | | | | | |

| No | Formula | Name | Eq. | VP A | VP B | VP C | VP D | Tmin | Tmax | LDEN | TDEN |
|---|---|---|---|---|---|---|---|---|---|---|---|
| 401 | C7H8O | methyl phenyl ether | 1 | -7.87545 | 1.83291 | -4.06977 | -2.18906 | 357 | TC | 0.996 | 293 |
| 402 | C7H8O | benzyl alcohol | 1 | -7.09506 | 1.18389 | -9.14255 | 5.56311 | 303 | TC | 1.041 | 298 |
| 403 | C7H8O | o-cresol | 1 | -8.82061 | 3.14917 | -6.63041 | -0.84857 | 393 | TC | 1.028 | 313 |
| 404 | C7H8O | m-cresol | 1 | -8.58506 | 2.82624 | -8.57418 | 8.74822 | 423 | TC | 1.034 | 293 |
| 405 | C7H8O | p-cresol | 1 | -9.23951 | 3.29880 | -7.17725 | -0.48000 | 401 | TC | 1.019 | 313 |
| 406 | C7H9N | 2,3-dimethylpyridine | 3 | 10.5290 | 4219.74 | -33.04 | | 420 | 440 | 0.942 | 298 |
| 407 | C7H9N | 2,4-dimethylpyridine | 3 | 10.2785 | 3991.27 | -42.79 | | 418 | 438 | 0.949 | 273 |
| 408 | C7H9N | 2,5-dimethylpyridine | 3 | 9.6844 | 3545.14 | -63.59 | | 350 | 435 | 0.938 | 273 |
| 409 | C7H9N | 2,6-dimethylpyridine | 3 | 9.6286 | 3385.20 | -65.19 | | 350 | 420 | 0.923 | 298 |
| 410 | C7H9N | 3,4-dimethylpyridine | 3 | 10.3315 | 4237.04 | -41.65 | | 400 | 460 | 0.954 | 298 |
| 411 | C7H9N | 3,5-dimethylpyridine | 3 | 10.2648 | 4106.95 | -44.45 | | 400 | 460 | 0.939 | 298 |
| 412 | C7H9N | N-methylaniline | 3 | 9.6864 | 3756.28 | -80.71 | | 320 | 480 | 0.989 | 293 |
| 413 | C7H9N | o-toluidine | 1 | -8.68458 | 2.72553 | -5.94620 | -1.09185 | 392 | TC | 0.998 | 293 |
| 414 | C7H9N | m-toluidine | 1 | -8.43741 | 2.58101 | -6.00776 | -1.52856 | 395 | TC | 0.989 | 293 |
| 415 | C7H9N | p-toluidine | 3 | 10.0766 | 4041.04 | -72.15 | | 350 | 500 | 0.964 | 323 |
| 416 | C7H14 | cycloheptane | 3 | 9.1616 | 3066.05 | -56.80 | | 330 | 435 | 0.810 | 293 |
| 417 | C7H14 | 1,1-dimethylcyclopentane | 1 | -7.56029 | 1.82906 | -2.90303 | -3.11433 | 289 | TC | 0.759 | 289 |
| 418 | C4H14 | 1,2-dimethylcyclopentane-cis | 1 | -7.67242 | 2.20160 | -3.86394 | -1.16796 | 299 | TC | 0.777 | 289 |
| 419 | C7H14 | 1,2-dimethylcyclopentane-trans | 1 | -7.19675 | 1.03696 | -1.93618 | -5.30531 | 299 | TC | 0.756 | 289 |
| 420 | C7H14 | ethylcyclopentane | 1 | -7.68089 | 2.28014 | -4.40365 | 0.54338 | 302 | TC | 0.771 | 289 |
| 421 | C7H14 | methylcyclohexane | 1 | -7.01915 | 1.09615 | -2.37009 | -3.37562 | 299 | TC | 0.774 | 289 |
| 422 | C7H14 | 1-heptene | 1 | -8.26875 | 3.02688 | -6.18709 | 4.33049 | 295 | TC | 0.697 | 293 |
| 423 | C7H14 | 2,3,3-trimethyl-1-butene | 3 | 9.0334 | 2719.47 | -49.56 | | 253 | 375 | 0.705 | 293 |
| 424 | C7H14O | methyl amyl ketone | | | | | | | | 0.820 | 288 |
| 425 | C7H14O2 | n-propyl butyrate | 1 | -8.28062 | 1.40511 | -4.19323 | -3.70158 | 300 | TC | 0.879 | 288 |

| No | Formula | Name | MolWt | Tfp K | Tb K | Tc K | Pc bar | Vc $cm^3/mol$ | Zc | Omega | Dipm debye |
|---|---|---|---|---|---|---|---|---|---|---|---|
| 426 | C7H14O2 | n-propyl isobutyrate | 130.187 | | 408.6 | 581. | 28.3 | | | | |
| 427 | C7H14O2 | isoamyl acetate | 130.187 | 194.7 | 415.7 | 599. | | | | | 1.8 |
| 428 | C7H14O2 | isobutyl propionate | 130.187 | 201.8 | 410.0 | 583. | 27.7 | | | | |
| 429 | C7H16 | n-heptane | 100.205 | 182.6 | 371.6 | 540.3 | 27.4 | 432. | 0.263 | 0.349 | 0.0 |
| 430 | C7H16 | 2-methylhexane | 100.205 | 154.9 | 363.2 | 530.4 | 27.3 | 421. | 0.261 | 0.329 | 0.0 |
| 431 | C7H16 | 3-methylhexane | 100.205 | 100.0 | 365.0 | 535.3 | 28.1 | 404. | 0.255 | 0.323 | 0.0 |
| 432 | C7H16 | 2,2-dimethylpentane | 100.205 | 149.4 | 352.4 | 520.5 | 27.7 | 416. | 0.266 | 0.287 | 0.0 |
| 433 | C7H16 | 2,3-dimethylpentane | 100.205 | | 362.9 | 537.4 | 29.1 | 393. | 0.256 | 0.296 | 0.0 |
| 434 | C7H16 | 2,4-dimethylpentane | 100.205 | 154. | 353.6 | 519.8 | 27.4 | 418. | 0.264 | 0.302 | 0.0 |
| 435 | C7H16 | 3,3-dimethylpentane | 100.205 | 138.7 | 359.2 | 536.4 | 29.5 | 414. | 0.273 | 0.267 | 0.0 |
| 436 | C7H16 | 3-ethylpentane | 100.205 | 154.6 | 366.6 | 540.6 | 28.9 | 416. | 0.267 | 0.310 | 0.0 |
| 437 | C7H16 | 2,2,3-trimethylbutane | 100.205 | 248.3 | 354.0 | 531.2 | 29.5 | 398. | 0.266 | 0.250 | 0.0 |
| 438 | C7H16O | 1-heptanol | 116.204 | 239.2 | 449.8 | 633. | 30.4 | 435. | 0.251 | 0.560 | 1.7 |
| 439 | C8H4O3 | phthalic anhydride | 148.118 | 404. | 560. | 810. | 47.6 | 368. | 0.26 | | 5.3 |
| 440 | C8H8 | styrene | 104.152 | 242.5 | 418.3 | 647. | 39.9 | | | 0.257 | 0.1 |
| 441 | C8H8O | methyl phenyl ketone | 120.151 | 292.8 | 474.9 | 714.0 | 40.6 | 376. | 0.257 | 0.42 | 3.0 |
| 442 | C8H8O2 | methyl benzoate | 136.151 | 260.8 | 472.2 | 692. | 36.4 | 396. | 0.25 | 0.43 | 1.9 |
| 443 | C8H8O3 | methyl salicylate | 152.149 | 264.6 | 496.1 | 709. | | | | | 2.4 |
| 444 | C8H10 | o-xylene | 106.168 | 248.0 | 417.6 | 630.3 | 37.3 | 369. | 0.262 | 0.310 | 0.5 |
| 445 | C8H10 | m-xylene | 106.168 | 225.3 | 412.3 | 617.1 | 35.4 | 376. | 0.259 | 0.325 | 0.3 |
| 446 | C8H10 | p-xylene | 106.168 | 286.4 | 411.5 | 616.2 | 35.1 | 379. | 0.260 | 0.320 | 0.1 |
| 447 | C8H10 | ethylbenzene | 106.168 | 178.2 | 409.3 | 617.2 | 36.0 | 374. | 0.262 | 0.302 | 0.4 |
| 448 | C8H10O | o-ethylphenol | 122.167 | 269.8 | 477.7 | 703.0 | | | | | |
| 449 | C8H10O | m-ethylphenol | 122.167 | 269. | 491.6 | 718.8 | | | | | |
| 450 | C8H10O | p-ethylphenol | 122.167 | 318. | 491.1 | 716.4 | | | | | |

| No | Formula | Name | CPVAP A | CPVAP B | CPVAP C | CPVAP D | DELHF | DELGF |
|---|---|---|---|---|---|---|---|---|
| 426 | C7H14O2 | n-propyl isobutyrate | | | | | | |
| 427 | C7H14O2 | isoamyl acetate | | | | | | |
| 428 | C7H14O2 | isobutyl propionate | | | | | | |
| 429 | C7H16 | n-heptane | -5.146E+0 | 6.762E-1 | -3.651E-4 | 7.658E-8 | -1.879E+5 | 8.000E+3 |
| 430 | C7H16 | 2-methylhexane | -3.939E+1 | 8.642E-1 | -6.289E-4 | 1.836E-7 | -1.951E+5 | 3.220E+3 |
| 431 | C7H16 | 3-methylhexane | -7.046E+0 | 6.837E-1 | -3.734E-4 | 7.834E-8 | -1.924E+5 | 4.600E+3 |
| 432 | C7H16 | 2,2-dimethylpentane | -5.010E+1 | 8.956E-1 | -6.360E-4 | 1.736E-7 | -2.063E+5 | 8.400E+2 |
| 433 | C7H16 | 2,3-dimethylpentane | -7.046E+0 | 6.837E-1 | -3.734E-4 | 7.834E-8 | -1.994E+5 | 6.700E+2 |
| 434 | C7H16 | 2,4-dimethylpentane | -7.046E+0 | 6.837E-1 | -3.734E-4 | 7.834E-8 | -2.021E+5 | 3.100E+3 |
| 435 | C7H16 | 3,3-dimethylpentane | -7.046E+0 | 6.837E-1 | -3.734E-4 | 7.834E-8 | -2.017E+5 | 2.640E+3 |
| 436 | C7H16 | 3-ethylpentane | -7.046E+0 | 6.837E-1 | -3.734E-4 | 7.834E-8 | -1.898E+5 | 1.100E+4 |
| 437 | C7H16 | 2,2,3-trimethylbutane | -2.294E+1 | 7.519E-1 | -4.421E-4 | 1.005E-7 | -2.049E+5 | 4.270E+3 |
| 438 | C7H16O | 1-heptanol | 4.907E+1 | 6.778E-1 | -3.447E-4 | 6.046E-8 | -3.320E+5 | -1.210E+5 |
| 439 | C8H4O3 | phthalic anhydride | -4.455E+0 | 6.540E-1 | -4.283E-4 | 1.009E-7 | -3.718E+5 | |
| 440 | C8H8 | styrene | -2.825E+1 | 6.159E-1 | -4.023E-4 | 9.935E-8 | 1.475E+5 | 2.139E+5 |
| 441 | C8H8O | methyl phenyl ketone | -2.958E+1 | 6.410E-1 | -4.071E-4 | 9.722E-8 | -8.692E+4 | 1.840E+3 |
| 442 | C8H8O2 | methyl benzoate | -2.121E+1 | 5.501E-1 | -1.799E-4 | 4.425E-8 | -2.541E+5 | |
| 443 | C8H8O3 | methyl salicylate | | | | | | |
| 444 | C8H10 | o-xylene | -1.585E+1 | 5.962E-1 | -3.443E-4 | 7.528E-8 | 1.900E+4 | 1.222E+5 |
| 445 | C8H10 | m-xylene | -2.917E+1 | 6.297E-1 | -3.747E-4 | 8.478E-8 | 1.725E+4 | 1.189E+5 |
| 446 | C8H10 | p-xylene | -2.509E+1 | 6.042E-1 | -3.374E-4 | 6.820E-8 | 1.796E+4 | 1.212E+5 |
| 447 | C8H10 | ethylbenzene | -4.310E+1 | 7.072E-1 | -4.811E-4 | 1.301E-7 | 2.981E+4 | 1.307E+5 |
| 448 | C8H10O | o-ethylphenol | | | | | -1.458E+5 | |
| 449 | C8H10O | m-ethylphenol | | | | | -1.466E+5 | |
| 450 | C8H10O | p-ethylphenol | | | | | -1.447E+5 | |

| No | Formula | Name | Eq. | VP A | VP B | VP C | VP D | Tmin | Tmax | LDEN | TDEN |
|---|---|---|---|---|---|---|---|---|---|---|---|
| 426 | C7H14O2 | n-propyl isobutyrate | 1 | -8.52052 | 2.10660 | -4.44053 | -3.90420 | 300 | TC | 0.884 | 273 |
| 427 | C7H14O2 | isoamyl acetate | 3 | 10.5011 | 3699.29 | -57.54 | | 311 | 369 | 0.876 | 288 |
| 428 | C7H14O2 | isobutyl propionate | 1 | -8.32761 | 1.56574 | -3.97739 | -4.71845 | 300 | TC | 0.888 | 273 |
| 429 | C7H16 | n-heptane | 1 | -7.67468 | 1.37068 | -3.53620 | -3.20243 | 240 | TC | 0.684 | 293 |
| 430 | C7H16 | 2-methylhexane | 1 | -7.62477 | 1.47806 | -3.53616 | -2.70794 | 230 | TC | 0.679 | 293 |
| 431 | C7H16 | 3-methylhexane | 1 | -7.58592 | 1.47394 | -3.52511 | -2.35419 | 235 | TC | 0.687 | 293 |
| 432 | C7H16 | 2,2-dimethylpentane | 1 | -7.45564 | 1.56232 | -3.44620 | -1.80802 | 225 | TC | 0.674 | 293 |
| 433 | C7H16 | 2,3-dimethylpentane | 1 | -7.46078 | 1.47778 | -3.37079 | -1.88997 | 230 | TC | 0.695 | 293 |
| 434 | C7H16 | 2,4-dimethylpentane | 1 | -7.46358 | 1.43203 | -3.42422 | -2.20238 | 225 | TC | 0.673 | 293 |
| 435 | C7H16 | 3,3-dimethylpentane | 1 | -7.49199 | 1.83146 | -3.57292 | -0.89448 | 225 | TC | 0.693 | 293 |
| 436 | C7H16 | 3-ethylpentane | 1 | -7.58305 | 1.58587 | -3.56732 | -2.42625 | 265 | TC | 0.698 | 293 |
| 437 | C7H16 | 2,2,3-trimethylbutane | 1 | -7.22017 | 1.44914 | -3.11808 | -1.10598 | 250 | TC | 0.690 | 293 |
| 438 | C7H16O | 1-heptanol | 3 | 8.6866 | 2626.42 | -146.6 | | 333 | 449 | 0.822 | 293 |
| 439 | C8H4O3 | phthalic anhydride | 3 | 9.3782 | 4467.01 | -83.15 | | 409 | 615 | | |
| 440 | C8H8 | styrene | 1 | -7.15981 | 1.78861 | -5.10359 | 1.63749 | 303 | TC | 0.906 | 293 |
| 441 | C8H8O | methyl phenyl ketone | 1 | -7.63896 | 1.20432 | -3.60753 | -1.55754 | 298 | TC | 1.032 | 288 |
| 442 | C8H8O2 | methyl benzoate | 3 | 9.6070 | 3751.83 | -81.15 | | 350 | 516 | 1.086 | 293 |
| 443 | C8H8O3 | methyl salicylate | 3 | 9.6897 | 3943.86 | -86.19 | | 350 | 495 | 1.182 | 298 |
| 444 | C8H10 | o-xylene | 1 | -7.53357 | 1.40968 | -3.10985 | -2.85992 | 337 | TC | 0.880 | 293 |
| 445 | C8H10 | m-xylene | 1 | -7.59222 | 1.39441 | -3.22746 | -2.40376 | 332 | TC | 0.864 | 293 |
| 446 | C8H10 | p-xylene | 1 | -7.63495 | 1.50724 | -3.19678 | -2.78710 | 331 | TC | 0.861 | 293 |
| 447 | C8H10 | ethylbenzene | 1 | -7.48645 | 1.45488 | -3.37538 | -2.23048 | 330 | TC | 0.867 | 293 |
| 448 | C8H10O | o-ethylphenol | 3 | 11.3408 | 4928.36 | -45.75 | | 350 | 500 | 1.037 | 273 |
| 449 | C8H10O | m-ethylphenol | 3 | 10.5753 | 4272.77 | -86.08 | | 370 | 500 | 1.025 | 273 |
| 450 | C8H10O | p-ethylphenol | 3 | 12.4703 | 5579.62 | -44.15 | | 370 | 500 | | |

| No | Formula | Name | MolWt | Tfp K | Tb K | Tc K | Pc bar | Vc $cm^3$/mol | Zc | Omega | Dipm debye |
|---|---|---|---|---|---|---|---|---|---|---|---|
| 451 | C8H10O | ethyl phenyl ether | 122.167 | 243. | 443.0 | 647. | 34.2 | | | 0.418 | 1.2 |
| 452 | C8H10O | 2,3-xylenol | 122.167 | 348. | 490.1 | 722.8 | | | | | |
| 453 | C8H10O | 2,4-xylenol | 122.167 | 298. | 484.1 | 707.6 | | | | | 2.0 |
| 454 | C8H10O | 2,5-xylenol | 122.167 | 348. | 484.3 | 706.9 | | | | | 1.5 |
| 455 | C8H10O | 2,6-xylenol | 122.167 | 322. | 474.2 | 701.0 | | | | | |
| 456 | C8H10O | 3,4-xylenol | 122.167 | 338. | 500.2 | 729.8 | | | | | 1.7 |
| 457 | C8H10O | 3,5-xylenol | 122.167 | 337. | 494.9 | 715.6 | | | | | 1.8 |
| 458 | C8H11N | N,N-dimethylaniline | 121.183 | 275.6 | 467.3 | 687. | 36.3 | | | 0.411 | 1.6 |
| 459 | C8H11N | N-ethylaniline | 121.183 | 207.4 | 476.2 | 698. | | | | | 1.7 |
| 460 | C8H14O4 | diethylsuccinate | 174.196 | 251.9 | 490.9 | 663. | | | | | 2.3 |
| 461 | C8H16 | 1,1-dimethylcyclohexane | 112.216 | 239.7 | 392.7 | 591. | 29.6 | 416. | 0.25 | 0.238 | |
| 462 | C8H16 | 1,2-dimethylcyclohexane-cis | 112.216 | 223.1 | 402.9 | 606. | 29.6 | | | 0.236 | |
| 463 | C8H16 | 1,2-dimethylcyclohexane-trans | 112.216 | 185.0 | 396.6 | 596. | | | | 0.242 | |
| 464 | C8H16 | 1,3-dimethylcyclohexane-cis | 112.216 | 197.6 | 393.3 | 591. | 29.6 | | | 0.224 | |
| 465 | C8H16 | 1,3-dimethylcyclohexane-trans | 112.216 | 183.0 | 397.6 | 598. | 29.7 | | | 0.189 | |
| 466 | C8H16 | 1,4-dimethylcyclohexane-cis | 112.216 | 185.7 | 397.5 | 598. | 29.7 | | | 0.234 | |
| 467 | C8H16 | 1,4-dimethylcyclohexane-trans | 112.216 | | 392.5 | 587.7 | 29.7 | | | 0.242 | |
| 468 | C8H16 | ethylcyclohexane | 112.216 | 161.8 | 404.9 | 609. | 30. | 450. | 0.27 | 0.243 | 0.0 |
| 469 | C8H16 | 1,1,2-trimethylcyclopentane | 112.216 | | 386.9 | 579.5 | 29.4 | | | 0.252 | |
| 470 | C8H16 | 1,1,3-trimethylcyclopentane | 112.216 | | 378.0 | 569.5 | 28.3 | | | 0.211 | |
| 471 | C8H16 | 1,2,4-trimethylcyclopentane-c,c,t | 112.216 | | 391. | 579. | 29. | | | 0.277 | |
| 472 | C8H16 | 1,2,4-trimethylcyclopentane-c,t,c | 112.216 | | 382.4 | 571. | 28. | | | 0.246 | |
| 473 | C8H16 | 1-methyl-1-ethylcyclopentane | 112.216 | | 394.7 | 592. | 30. | | | 0.250 | |
| 474 | C8H16 | n-propylcyclopentane | 112.216 | 155.8 | 404.1 | 603. | 30. | 425. | 0.25 | 0.335 | |
| 475 | C8H16 | isopropylcyclopentane | 112.216 | 160.5 | 399.6 | 601. | 30. | | | 0.240 | |

| No | Formula | Name | CPVAP A | CPVAP B | CPVAP C | CPVAP D | DELHF | DELGF |
|---|---|---|---|---|---|---|---|---|
| 451 | C8H10O | ethyl phenyl ether | | | | | | |
| 452 | C8H10O | 2,3-xylenol | | | | | -1.573E+5 | |
| 453 | C8H10O | 2,4-xylenol | | | | | -1.628E+5 | |
| 454 | C8H10O | 2,5-xylenol | | | | | -1.615E+5 | |
| 455 | C8H10O | 2,6-xylenol | | | | | -1.619E+5 | |
| 456 | C8H10O | 3,4-xylenol | | | | | -1.565E+5 | |
| 457 | C8H10O | 3,5-xylenol | | | | | -1.615E+5 | |
| 458 | C8H11N | N,N-dimethylaniline | | | | | 8.415E+4 | 2.314E+5 |
| 459 | C8H11N | N-ethylaniline | | | | | | |
| 460 | C8H14O4 | diethylsuccinate | | | | | | |
| 461 | C8H16 | 1,1-dimethylcyclohexane | -7.211E+1 | 8.997E-1 | -5.020E-4 | 1.030E-7 | -1.811E+5 | 3.525E+4 |
| 462 | C8H16 | 1,2-dimethylcyclohexane-cis | -6.837E+1 | 8.972E-1 | -5.137E-4 | 1.099E-7 | -1.723E+5 | 4.124E+4 |
| 463 | C8H16 | 1,2-dimethylcyclohexane-trans | -6.848E+1 | 9.123E-1 | -5.355E-4 | 1.181E-7 | -1.801E+5 | 3.450E+4 |
| 464 | C8H16 | 1,3-dimethylcyclohexane-cis | -6.516E+1 | 8.838E-1 | -4.932E-4 | 1.020E-7 | -1.849E+5 | 2.985E+4 |
| 465 | C8H16 | 1,3-dimethylcyclohexane-trans | -6.415E+1 | 8.826E-1 | -5.016E-4 | 1.068E-7 | -1.767E+5 | 3.634E+4 |
| 466 | C8H16 | 1,4-dimethylcyclohexane-cis | -6.415E+1 | 8.826E-1 | -5.016E-4 | 1.068E-7 | -1.768E+5 | 3.797E+4 |
| 467 | C8H16 | 1,4-dimethylcyclohexane-trans | -7.036E+1 | 9.131E-1 | -5.309E-4 | 1.155E-7 | -1.847E+5 | 3.174E+4 |
| 468 | C8H16 | ethylcyclohexane | -6.389E+1 | 8.893E-1 | -5.108E-4 | 1.103E-7 | -1.719E+5 | 3.927E+4 |
| 469 | C8H16 | 1,1,2-trimethylcyclopentane | | | | | | |
| 470 | C8H16 | 1,1,3-trimethylcyclopentane | | | | | | |
| 471 | C8H16 | 1,2,4-trimethylcyclopentane-c,c,t | | | | | | |
| 472 | C8H16 | 1,2,4-trimethylcyclopentane-c,t,c | | | | | | |
| 473 | C8H16 | 1-methyl-1-ethylcyclopentane | | | | | | |
| 474 | C8H16 | n-propylcyclopentane | -5.597E+1 | 8.449E-1 | -4.924E-4 | 1.117E-7 | -1.482E+5 | 5.263E+4 |
| 475 | C8H16 | isopropylcyclopentane | | | | | | |

| No | Formula | Name | Eq. | VP A | VP B | VP C | VP D | Tmin | Tmax | LDEN | TDEN |
|---|---|---|---|---|---|---|---|---|---|---|---|
| 451 | C8H10O | ethyl phenyl ether | 1 | -8.50867 | 2.56997 | -5.78999 | 0.10899 | 371 | TC | 0.979 | 277 |
| 452 | C8H10O | 2,3-xylenol | 3 | 9.6222 | 3724.58 | -102.4 | | 420 | 500 | | |
| 453 | C8H10O | 2,4-xylenol | 3 | 9.6254 | 3655.26 | -103.8 | | 410 | 500 | | |
| 454 | C8H10O | 2,5-xylenol | 3 | 9.6166 | 3667.32 | -102.4 | | 410 | 490 | | |
| 455 | C8H10O | 2,6-xylenol | 3 | 9.6607 | 3749.35 | -85.55 | | 400 | 480 | | |
| 456 | C8H10O | 3,4-xylenol | 3 | 9.6802 | 3733.53 | -113.9 | | 430 | 520 | | |
| 457 | C8H10O | 3,5-xylenol | 3 | 9.7990 | 3775.91 | -109.0 | | 410 | 500 | | |
| 458 | C8H11N | N,N-dimethylaniline | 3 | 10.3445 | 4276.08 | -52.80 | | 345 | 480 | 0.956 | 293 |
| 459 | C8H11N | N-ethylaniline | 3 | 10.4715 | 4382.63 | -58.88 | | 321 | 481 | 0.963 | 293 |
| 460 | C8H14O4 | diethylsuccinate | | | | | | | | 1.041 | 293 |
| 461 | C8H16 | 1,1-dimethylcyclohexane | 1 | -6.92810 | 1.01872 | -3.04857 | -1.70684 | 314 | TC | 0.785 | 289 |
| 462 | C8H16 | 1,2-dimethylcyclohexane-cis | 1 | -7.01944 | 1.31860 | -3.96577 | 0.08142 | 322 | TC | 0.796 | 293 |
| 463 | C8H16 | 1,2-dimethylcyclohexane-trans | 2 | 46.903 | 6162.66 | -5.245 | 4785. | 320 | TC | 0.776 | 293 |
| 464 | C8H16 | 1,3-dimethylcyclohexane-cis | 3 | 9.1268 | 3081.95 | -55.08 | | 284 | 420 | 0.766 | 293 |
| 465 | C8H16 | 1,3-dimethylcyclohexane-trans | 2 | 49.477 | 6271.67 | -5.615 | 4718. | 320 | TC | 0.785 | 293 |
| 466 | C8H16 | 1,4-dimethylcyclohexane-cis | 2 | 46.951 | 6219.26 | -5.233 | 4718. | 320 | TC | 0.783 | 293 |
| 467 | C8H16 | 1,4-dimethylcyclohexane-trans | 2 | 46.289 | 6071.72 | -5.163 | 4650. | 320 | TC | 0.763 | 293 |
| 468 | C8H16 | ethylcyclohexane | 3 | 9.1923 | 3183.25 | -58.15 | | 293 | 433 | 0.788 | 293 |
| 469 | C8H16 | 1,1,2-trimethylcyclopentane | 1 | -7.01985 | 1.06194 | -3.15886 | -1.64858 | 309 | TC | | |
| 470 | C8H16 | 1,1,3-trimethylcyclopentane | 1 | -6.97215 | 1.62353 | -4.90587 | 2.76293 | 302 | TC | | |
| 471 | C8H16 | 1,2,4-trimethylcyclopentane-c,c,t | 3 | 9.1341 | 3073.95 | -54.20 | | 283 | 418 | | |
| 472 | C8H16 | 1,2,4-trimethylcyclopentane-c,t,c | 3 | 9.1554 | 3009.70 | -53.23 | | 282 | 417 | | |
| 473 | C8H16 | 1-methyl-1-ethylcyclopentane | 1 | -7.09092 | 1.31715 | -3.96332 | 0.30332 | 316 | TC | | |
| 474 | C8H16 | n-propylcyclopentane | 1 | -7.82031 | 2.88785 | -6.85367 | 6.03561 | 325 | TC | 0.781 | 289 |
| 475 | C8H16 | isopropylcyclopentane | 1 | -7.10096 | 1.54495 | -4.66594 | 2.34067 | 320 | TC | 0.776 | 293 |

| No | Formula | Name | MolWt | Tfp K | Tb K | Tc K | Pc bar | Vc $cm^3/mol$ | Zc | Omega | Dipm debye |
|---|---|---|---|---|---|---|---|---|---|---|---|
| 476 | C8H16 | cyclooctane | 112.216 | 287.6 | 422. | 647.2 | 35.6 | 410. | 0.271 | 0.236 | |
| 477 | C8H16 | 1-octene | 112.216 | 171.4 | 394.4 | 566.7 | 26.2 | 464. | 0.26 | 0.386 | 0.3 |
| 478 | C8H16 | 2-octene-trans | 112.216 | 185.4 | 398.1 | 580. | 27.7 | | | 0.350 | |
| 479 | C8H16O2 | isoamyl propionate | 144.214 | | 433.4 | 611. | | | | | |
| 480 | C8H16O2 | isobutyl butyrate | 144.214 | | 430.1 | 603. | 24.5 | | | | |
| 481 | C8H16O2 | isobutyl isobutyrate | 144.214 | | 421.8 | 594. | 24.6 | | | | |
| 482 | C8H16O2 | n-propyl isovalerate | 144.214 | | 429.1 | 609. | | | | | |
| 483 | C8H18 | n-octane | 114.232 | 216.4 | 398.8 | 568.8 | 24.9 | 492. | 0.259 | 0.398 | 0.0 |
| 484 | C8H18 | 2-methylheptane | 114.232 | 164. | 390.8 | 559.6 | 24.8 | 488. | 0.261 | 0.378 | |
| 485 | C8H18 | 3-methylheptane | 114.232 | 152.7 | 392.1 | 563.7 | 25.5 | 464. | 0.252 | 0.370 | |
| 486 | C8H18 | 4-methylheptane | 114.232 | 152.2 | 390.9 | 561.7 | 25.4 | 476. | 0.259 | 0.371 | |
| 487 | C8H18 | 2,2-dimethylhexane | 114.232 | 152. | 380.0 | 549.9 | 25.3 | 478. | 0.264 | 0.338 | |
| 488 | C8H18 | 2,3-dimethylhexane | 114.232 | | 388.8 | 563.5 | 26.3 | 468. | 0.263 | 0.346 | |
| 489 | C8H18 | 2,4-dimethylhexane | 114.232 | | 382.6 | 553.5 | 25.6 | 472. | 0.262 | 0.343 | |
| 490 | C8H18 | 2,5-dimethylhexane | 114.232 | 181.9 | 382.3 | 550.1 | 24.9 | 482. | 0.262 | 0.356 | |
| 491 | C8H18 | 3,3-dimethylhexane | 114.232 | 147. | 385.1 | 562.0 | 26.5 | 443. | 0.251 | 0.320 | |
| 492 | C8H18 | 3,4-dimethylhexane | 114.232 | | 390.9 | 568.9 | 26.9 | 466. | 0.265 | 0.338 | |
| 493 | C8H18 | 3-ethylhexane | 114.232 | | 391.7 | 565.5 | 26.1 | 455. | 0.252 | 0.361 | |
| 494 | C8H18 | 2,2,3-trimethylpentane | 114.232 | 160.9 | 383.0 | 563.5 | 27.3 | 436. | 0.254 | 0.297 | |
| 495 | C8H18 | 2,2,4-trimethylpentane | 114.232 | 165.8 | 372.4 | 544.0 | 25.7 | 468. | 0.266 | 0.303 | |
| 496 | C8H18 | 2,3,3-trimethylpentane | 114.232 | 172.5 | 387.9 | 573.6 | 28.2 | 455. | 0.269 | 0.290 | |
| 497 | C8H18 | 2,3,4-trimethylpentane | 114.232 | 163.9 | 386.6 | 566.4 | 27.3 | 461. | 0.267 | 0.315 | |
| 498 | C8H18 | 2-methyl-3-ethylpentane | 114.232 | 158.2 | 388.8 | 567.1 | 27.0 | 443. | 0.254 | 0.330 | |
| 499 | C8H18 | 3-methyl-3-ethylpentane | 114.232 | 182.3 | 391.4 | 576.6 | 28.1 | 455. | 0.267 | 0.303 | |
| 500 | C8H18 | 2,2,3,3-tetramethylbutane | 114.232 | 374. | 379.6 | 567.8 | 28.7 | 461. | 0.280 | 0.251 | |

| No | Formula | Name | CPVAP A | CPVAP B | CPVAP C | CPVAP D | DELHF | DELGF |
|---|---|---|---|---|---|---|---|---|
| 476 | C8H16 | cyclooctane | | | | | | |
| 477 | C8H16 | 1-octene | -4.099E+0 | 7.239E-1 | -4.036E-4 | 8.675E-8 | -8.298E+4 | 1.043E+5 |
| 478 | C8H16 | 2-octene-trans | -1.282E+1 | 7.532E-1 | -4.442E-4 | 1.050E-7 | -9.458E+4 | 9.274E+4 |
| 479 | C8H1602 | isoamyl propionate | | | | | | |
| 480 | C8H1602 | isobutyl butyrate | | | | | | |
| 481 | C8H1602 | isobutyl isobutyrate | | | | | | |
| 482 | C8H1602 | n-propyl isovalerate | | | | | | |
| 483 | C8H18 | n-octane | -6.096E+0 | 7.712E-1 | -4.195E-4 | 8.855E-8 | -2.086E+5 | 1.640E+4 |
| 484 | C8H18 | 2-methylheptane | -8.970E+1 | 1.242E+0 | -1.176E-3 | 4.618E-7 | -2.156E+5 | 1.277E+4 |
| 485 | C8H18 | 3-methylheptane | -9.215E+0 | 7.859E-1 | -4.400E-4 | 9.697E-8 | -2.128E+5 | 1.373E+4 |
| 486 | C8H18 | 4-methylheptane | -9.215E+0 | 7.859E-1 | -4.400E-4 | 9.697E-8 | -2.122E+5 | 1.675E+4 |
| 487 | C8H18 | 2,2-dimethylhexane | -9.215E+0 | 7.859E-1 | -4.400E-4 | 9.697E-8 | -2.249E+5 | 1.072E+4 |
| 488 | C8H18 | 2,3-dimethylhexane | -9.215E+0 | 7.859E-1 | -4.400E-4 | 9.697E-8 | -2.141E+5 | 1.771E+4 |
| 489 | C8H18 | 2,4-dimethylhexane | -9.215E+0 | 7.859E-1 | -4.400E-4 | 9.697E-8 | -2.196E+5 | 1.172E+4 |
| 490 | C8H18 | 2,5-dimethylhexane | -9.215E+0 | 7.859E-1 | -4.400E-4 | 9.697E-8 | -2.228E+5 | 1.047E+4 |
| 491 | C8H18 | 3,3-dimethylhexane | -9.215E+0 | 7.859E-1 | -4.400E-4 | 9.697E-8 | -2.203E+5 | 1.327E+4 |
| 492 | C8H18 | 3,4-dimethylhexane | -9.215E+0 | 7.859E-1 | -4.400E-4 | 9.697E-8 | -2.131E+5 | 1.733E+4 |
| 493 | C8H18 | 3-ethylhexane | -9.215E+0 | 7.859E-1 | -4.400E-4 | 9.697E-8 | -2.110E+5 | 1.694E+4 |
| 494 | C8H18 | 2,2,3-trimethylpentane | -9.215E+0 | 7.859E-1 | -4.400E-4 | 9.697E-8 | -2.203E+5 | 1.712E+4 |
| 495 | C8H18 | 2,2,4-trimethylpentane | -7.461E+0 | -7.779E-1 | -4.287E-4 | 9.173E-8 | -2.243E+5 | 1.369E+4 |
| 496 | C8H18 | 2,3,3-trimethylpentane | -9.215E+0 | 7.859E-1 | -4.400E-4 | 9.697E-8 | -2.166E+5 | 1.892E+4 |
| 497 | C8H18 | 2,3,4-trimethylpentane | -9.215E+0 | 7.859E-1 | -4.400E-4 | 9.697E-8 | -2.176E+5 | 1.892E+4 |
| 498 | C8H18 | 2-methyl-3-ethylpentane | -9.215E+0 | 7.859E-1 | -4.400E-4 | 9.697E-8 | -2.113E+5 | 2.127E+4 |
| 499 | C8H18 | 3-methyl-3-ethylpentane | -9.215E+0 | 7.859E-1 | -4.400E-4 | 9.697E-8 | -2.151E+5 | 1.993E+4 |
| 500 | C8H18 | 2,2,3,3-tetramethylbutane | | | | | | |

| No | Formula | Name | Eq. | VP A | VP B | VP C | VP D | Tmin | Tmax | LDEN | TDEN |
|---|---|---|---|---|---|---|---|---|---|---|---|
| 476 | C8H16 | cyclooctane | 3 | 9.1799 | 3310.62 | -63.18 | | 367 | 470 | 0.834 | 293 |
| 477 | C8H16 | 1-octene | 2 | 57.867 | 6883.34 | -6.765 | 5235. | 320 | TC | 0.715 | 293 |
| 478 | C8H16 | 2-octene-trans | 3 | 9.2352 | 3134.97 | -58.00 | | 289 | 425 | 0.720 | 293 |
| 479 | C8H16O2 | isoamyl propionate | | | | | | | | 0.870 | 293 |
| 480 | C8H16O2 | isobutyl butyrate | 1 | -8.32597 | 1.42350 | -4.25376 | -3.09772 | 310 | TC | 0.863 | 291 |
| 481 | C8H16O2 | isobutyl isobutyrate | 1 | -8.18677 | 1.32200 | -3.94343 | -3.68833 | 310 | TC | 0.875 | 273 |
| 482 | C8H16O2 | n-propyl isovalerate | | | | | | | | 0.863 | 293 |
| 483 | C8H18 | n-octane | 1 | -7.91211 | 1.38007 | -3.80435 | -4.50132 | 260 | TC | 0.703 | 293 |
| 484 | C8H18 | 2-methylheptane | 1 | -7.80701 | 1.38191 | -3.78286 | -3.50395 | 250 | TC | 0.702 | 289 |
| 485 | C8H18 | 3-methylheptane | 1 | -7.82876 | 1.50656 | -3.86146 | -3.52377 | 255 | TC | 0.706 | 293 |
| 486 | C8H18 | 4-methylheptane | 1 | -7.78757 | 1.40709 | -3.76234 | -3.50643 | 250 | TC | 0.705 | 293 |
| 487 | C8H18 | 2,2-dimethylhexane | 1 | -7.69898 | 1.56083 | -3.75189 | -3.01869 | 245 | TC | 0.695 | 293 |
| 488 | C8H18 | 2,3-dimethylhexane | 1 | -7.75180 | 1.58578 | -3.80794 | -2.58547 | 250 | TC | 0.712 | 293 |
| 489 | C8H18 | 2,4-dimethylhexane | 1 | -7.65152 | 1.41393 | -3.62789 | -3.06548 | 245 | TC | 0.700 | 293 |
| 490 | C8H18 | 2,5-dimethylhexane | 1 | -7.76508 | 1.51236 | -3.78809 | -3.07843 | 245 | TC | 0.693 | 293 |
| 491 | C8H18 | 3,3-dimethylhexane | 1 | -7.59847 | 1.50336 | -3.49912 | -2.38236 | 245 | TC | 0.710 | 293 |
| 492 | C8H18 | 3,4-dimethylhexane | 1 | -7.72976 | 1.61174 | -3.75756 | -2.62874 | 250 | TC | 0.719 | 293 |
| 493 | C8H18 | 3-ethylhexane | 1 | -7.75246 | 1.42908 | -3.68445 | -3.46671 | 250 | TC | 0.718 | 289 |
| 494 | C8H18 | 2,2,3-trimethylpentane | 1 | -7.48839 | 1.52208 | -3.44481 | -2.12538 | 245 | TC | 0.716 | 293 |
| 495 | C8H18 | 2,2,4-trimethylpentane | 1 | -7.38890 | 1.25294 | -3.16606 | -2.22001 | 265 | TC | 0.692 | 293 |
| 496 | C8H18 | 2,3,3-trimethylpentane | 1 | -7.41747 | 1.42778 | -3.19166 | -1.81367 | 245 | TC | 0.726 | 293 |
| 497 | C8H18 | 2,3,4-trimethylpentane | 1 | -7.62000 | 1.60334 | -3.57834 | -2.04401 | 245 | TC | 0.719 | 293 |
| 498 | C8H18 | 2-methyl-3-ethylpentane | 1 | -7.65393 | 1.54032 | -3.64686 | -2.52380 | 250 | TC | 0.719 | 293 |
| 499 | C8H18 | 3-methyl-3-ethylpentane | 1 | -7.56484 | 1.58810 | -3.40610 | -1.71546 | 250 | TC | 0.727 | 293 |
| 500 | C8H18 | 2,2,3,3-tetramethylbutane | 3 | 11.4937 | 3856.39 | -42.42 | | 270 | 343 | | |

| No | Formula | Name | MolWt | Tfp K | Tb K | Tc K | Pc bar | Vc $cm^3/mol$ | Zc | Omega | Dipm debye |
|---|---|---|---|---|---|---|---|---|---|---|---|
| 501 | C8H18O | 1-octanol | 130.231 | 257.7 | 468.3 | 652.5 | 28.6 | 490. | 0.258 | 0.587 | 2.0 |
| 502 | C8H18O | 2-octanol | 130.231 | 241.2 | 452. | 637. | | | | | 1.6 |
| 503 | C8H18O | 4-methyl-3-heptanol | 130.231 | | 443. | 623.5 | | | | | |
| 504 | C8H18O | 5-methyl-3-heptanol | 130.231 | | 445. | 621.2 | | | | | |
| 505 | C8H18O | 2-ethyl-1-hexanol | 130.231 | 203.2 | 457.8 | 640.2 | | | | | 1.8 |
| 506 | C8H18O | dibutyl ether | 130.231 | 175. | 413.4 | 580. | 25.3 | | | 0.502 | 1.2 |
| 507 | C8H18O | di-tert-butyl ether | 130.231 | | 382.2 | 550. | 24.2 | | | | |
| 508 | C8H19N | dibutyl amine | 129.247 | 211. | 432.8 | 607.5 | 26.4 | | | 0.580 | 1.1 |
| 509 | C8H19N | diisobutyl amine | 129.247 | 203. | 412.8 | 584.4 | 27.2 | | | 0.548 | |
| 510 | C9H7N | quinoline | 129.162 | 258. | 510.8 | 782. | | | | | |
| 511 | C9H7N | isoquinoline | 129.162 | 300. | 516.4 | 803. | | | | | |
| 512 | C9H10 | indane | 118.179 | | 451.1 | 684.9 | 39.5 | | | 0.308 | |
| 513 | C9H10 | alpha-methylstyrene | 118.179 | | 438.5 | 654. | 34. | | | | |
| 514 | C9H10O2 | ethyl benzoate | 150.178 | 238.3 | 485.9 | 668.7 | 23.2 | | | 0.48 | |
| 515 | C9H12 | n-propylbenzene | 120.195 | 173.7 | 432.4 | 638.2 | 32.0 | 440. | 0.265 | 0.344 | |
| 516 | C9H12 | isopropylbenzene | 120.195 | 177.1 | 425.6 | 631.1 | 32.1 | | | 0.326 | |
| 517 | C9H12 | 1-methyl-2-ethylbenzene | 120.195 | 192.3 | 438.3 | 651. | 30.4 | 460. | 0.26 | 0.294 | |
| 518 | C9H12 | 1-methyl-3-ethylbenzene | 120.195 | 177.6 | 434.5 | 637. | 28.4 | 490. | 0.26 | 0.360 | |
| 519 | C9H12 | 1-methyl-4-ethylbenzene | 120.195 | 210.8 | 435.2 | 640. | 29.4 | 470. | 0.26 | 0.322 | |
| 520 | C9H12 | 1,2,3-trimethylbenzene | 120.195 | 247.7 | 449.3 | 664.5 | 34.5 | | | 0.366 | |
| 521 | C9H12 | 1,2,4-trimethylbenzene | 120.195 | 227. | 442.5 | 649.2 | 32.3 | | | 0.376 | |
| 522 | C9H12 | 1,3,5-trimethylbenzene | 120.195 | 228.4 | 437.9 | 637.3 | 31.3 | | | 0.399 | 0.1 |
| 523 | C9H13N | N,N-dimethyl-o-toluidine | 135.210 | 212. | 467.3 | 668. | 31.2 | | | 0.484 | 0.9 |
| 524 | C9H18 | n-propylcyclohexane | 126.243 | 178.7 | 429.9 | 639.0 | 28.0 | | | 0.258 | |
| 525 | C9H18 | isopropylcyclohexane | 126.243 | 183.4 | 427.7 | 640.0 | 28.3 | | | 0.237 | 0.0 |

| No | Formula | Name | CPVAP A | CPVAP B | CPVAP C | CPVAP D | DELHF | DELGF |
|---|---|---|---|---|---|---|---|---|
| 501 | C8H18O | 1-octanol | 6.171E+0 | 7.607E-1 | -3.797E-4 | 6.263E-8 | -3.601E+5 | -1.202E+5 |
| 502 | C8H18O | 2-octanol | 2.588E+1 | 7.641E-1 | -4.224E-4 | 9.064E-8 | | |
| 503 | C8H18O | 4-methyl-3-heptanol | | | | | | |
| 504 | C8H18O | 5-methyl-3-heptanol | | | | | | |
| 505 | C8H18O | 2-ethyl-1-hexanol | -1.499E+1 | 8.654E-1 | -5.280E-4 | 1.285E-7 | -3.655E+5 | |
| 506 | C8H18O | dibutyl ether | 6.054E+0 | 7.729E-1 | -4.085E-4 | 8.085E-8 | -3.341E+5 | -8.859E+4 |
| 507 | C8H18O | di-tert-butyl ether | | | | | | |
| 508 | C8H19N | dibutyl amine | 9.764E+0 | 8.081E-1 | -4.392E-4 | 9.249E-8 | | |
| 509 | C8H19N | diisobutyl amine | | | | | | |
| 510 | C9H7N | quinoline | | | | | | |
| 511 | C9H7N | isoquinoline | | | | | | |
| 512 | C9H10 | indane | | | | | | |
| 513 | C9H10 | alpha-methylstyrene | -2.433E+1 | 6.933E-1 | -4.530E-4 | 1.181E-7 | | |
| 514 | C9H10O2 | ethyl benzoate | 2.067E+1 | 6.887E-1 | -3.608E-4 | 5.062E-8 | | |
| 515 | C9H12 | n-propylbenzene | -3.129E+1 | 7.486E-1 | -4.601E-4 | 1.081E-7 | 7.830E+3 | 1.373E+5 |
| 516 | C9H12 | isopropylbenzene | -3.936E+1 | 7.842E-1 | -5.087E-4 | 1.291E-7 | 3.940E+3 | 1.371E+5 |
| 517 | C9H12 | 1-methyl-2-ethylbenzene | -1.645E+1 | 6.996E-1 | -4.120E-4 | 9.328E-8 | 1.210E+3 | 1.312E+5 |
| 518 | C9H12 | 1-methyl-3-ethylbenzene | -2.900E+1 | 7.293E-1 | -4.363E-4 | 9.998E-8 | -1.930E+3 | 1.265E+5 |
| 519 | C9H12 | 1-methyl-4-ethylbenzene | -2.731E+1 | 7.176E-1 | -4.224E-4 | 9.542E-8 | -2.050E+3 | 1.268E+5 |
| 520 | C9H12 | 1,2,3-trimethylbenzene | -6.942E+0 | 6.335E-1 | -3.326E-4 | 6.611E-8 | -9.590E+3 | 1.246E+5 |
| 521 | C9H12 | 1,2,4-trimethylbenzene | -4.668E+0 | 6.238E-1 | -3.263E-4 | 6.376E-8 | -1.394E+4 | 1.170E+5 |
| 522 | C9H12 | 1,3,5-trimethylbenzene | -1.959E+1 | 6.724E-1 | -3.692E-4 | 7.700E-8 | -1.608E+4 | 1.180E+5 |
| 523 | C9H13N | N,N-dimethyl-o-toluidine | | | | | | |
| 524 | C9H18 | n-propylcyclohexane | -6.252E+1 | 9.889E-1 | -5.795E-4 | 1.291E-7 | -1.934E+5 | 4.735E+4 |
| 525 | C9H18 | isopropylcyclohexane | | | | | | |

| No | Formula | Name | Eq. | VP A | VP B | VP C | VP D | Tmin | Tmax | LDEN | TDEN |
|---|---|---|---|---|---|---|---|---|---|---|---|
| 501 | C8H18O | 1-octanol | 1 | -9.71763 | 4.22514 | -12.9222 | -3.59254 | 325 | TC | 0.826 | 293 |
| 502 | C8H18O | 2-octanol | 3 | 8.0906 | 2441.66 | -150.7 | | 345 | 453 | 0.821 | 293 |
| 503 | C8H18O | 4-methyl-3-heptanol | | | | | | | | | |
| 504 | C8H18O | 5-methyl-3-heptanol | | | | | | | | | |
| 505 | C8H18O | 2-ethyl-1-hexanol | 3 | 8.7412 | 2773.46 | -140.0 | | 348 | 458 | 0.833 | 293 |
| 506 | C8H18O | dibutyl ether | 1 | -9.04970 | 2.78734 | -5.11686 | -3.97104 | 362 | TC | 0.768 | 293 |
| 507 | C8H18O | di-tert-butyl ether | 1 | -7.47062 | 1.33672 | -4.00322 | -1.89122 | 300 | TC | | |
| 508 | C8H19N | dibutyl amine | 1 | -9.14853 | 2.93179 | -6.02092 | 0.93342 | 315 | TC | 0.767 | 293 |
| 509 | C8H19N | diisobutyl amine | 1 | -8.95962 | 2.85335 | -5.81427 | 0.65701 | 300 | TC | 0.741 | 298 |
| 510 | C9H7N | quinoline | 3 | 9.0779 | 3842.40 | -86.94 | | 437 | 515 | 1.095 | 293 |
| 511 | C9H7N | isoquinoline | 3 | 9.2957 | 3968.37 | -88.94 | | 437 | 517 | 1.091 | 303 |
| 512 | C9H10 | indane | | | | | | | | | |
| 513 | C9H10 | alpha-methylstyrene | 3 | 9.7106 | 3644.30 | -67.15 | | 348 | 493 | 0.911 | 293 |
| 514 | C9H10O2 | ethyl benzoate | 1 | -9.32936 | 2.89807 | -6.54758 | 5.56703 | 317 | TC | 1.046 | 293 |
| 515 | C9H12 | n-propylbenzene | 1 | -7.92198 | 1.97403 | -4.27504 | -1.28568 | 346 | TC | 0.862 | 293 |
| 516 | C9H12 | isopropylbenzene | 1 | -7.46042 | 1.14486 | -3.19082 | -3.62628 | 343 | TC | 0.862 | 293 |
| 517 | C9H12 | 1-methyl-2-ethylbenzene | 1 | -7.58007 | 2.20412 | -6.68027 | 6.06587 | 354 | TC | 0.881 | 293 |
| 518 | C9H12 | 1-methyl-3-ethylbenzene | 1 | -7.86301 | 2.47961 | -6.98644 | 6.35609 | 351 | TC | 0.865 | 293 |
| 519 | C9H12 | 1-methyl-4-ethylbenzene | 1 | -7.68892 | 1.92605 | -5.51788 | 2.76399 | 351 | TC | 0.861 | 293 |
| 520 | C9H12 | 1,2,3-trimethylbenzene | 1 | -8.44191 | 2.92198 | -5.66712 | 2.28086 | 363 | TC | 0.894 | 293 |
| 521 | C9H12 | 1,2,4-trimethylbenzene | 1 | -8.50002 | 2.98227 | -6.02665 | 3.51307 | 358 | TC | 0.880 | 289 |
| 522 | C9H12 | 1,3,5-trimethylbenzene | 1 | -8.37150 | 2.41166 | -5.30321 | 2.67635 | 355 | TC | 0.865 | 293 |
| 523 | C9H13N | N,N-dimethyl-o-toluidine | | | | | | | | 0.929 | 293 |
| 524 | C9H18 | n-propylcyclohexane | 1 | -7.37782 | 2.13149 | -6.45979 | 5.82529 | 349 | TC | 0.793 | 293 |
| 525 | C9H18 | isopropylcyclohexane | 1 | -7.24565 | 2.09643 | -6.35158 | 5.5038 | 344 | TC | 0.802 | 293 |

| No | Formula | Name | MolWt | Tfp K | Tb K | Tc K | Pc bar | Vc $cm^3/mol$ | Zc | Omega | Dipm debye |
|---|---|---|---|---|---|---|---|---|---|---|---|
| 526 | C9H18 | 1,trans-3,5-trimethylcyclohexane | 126.243 | | 413.7 | 602.2 | | | | | |
| 527 | C9H18 | 1-nonene | 126.243 | 191.8 | 420.0 | 592. | 23.4 | 580. | 0.28 | 0.430 | |
| 528 | C9H18O | dibutyl ketone | 142.242 | 267.3 | 461.6 | 640. | | | | | 2.7 |
| 529 | C9H20 | n-nonane | 128.259 | 219.7 | 424.0 | 594.6 | 22.9 | 548. | 0.26 | 0.445 | |
| 530 | C9H20 | 2-methyloctane | 128.242 | 192.8 | 416.4 | 587.0 | 23.1 | | | 0.423 | |
| 531 | C9H20 | 2,2-dimethylheptane | 128.242 | 160. | 405.9 | 576.8 | 23.5 | | | 0.390 | |
| 532 | C9H20 | 2,2,3-trimethylhexane | 128.259 | | 406.8 | 588. | 24.9 | | | 0.332 | |
| 533 | C9H20 | 2,2,4-trimethylhexane | 128.259 | 153. | 399.7 | 573.7 | 23.7 | | | 0.321 | |
| 534 | C9H20 | 2,2,5-trimethylhexane | 128.259 | 167.4 | 397.2 | 568. | 23.3 | 519. | 0.260 | 0.357 | |
| 535 | C9H20 | 3,3-diethylpentane | 128.259 | 240.1 | 419.3 | 610. | 26.7 | | | 0.338 | 0.0 |
| 536 | C9H20 | 2,2,3,3-tetramethylpentane | 128.259 | 263. | 413.4 | 607.7 | 27.4 | | | 0.303 | |
| 537 | C9H20 | 2,2,3,4-tetramethylpentane | 128.259 | 152. | 406.1 | 592.7 | 26.0 | | | 0.313 | |
| 538 | C9H20 | 2,2,4,4-tetramethylpentane | 128.259 | 206.0 | 395.4 | 574.7 | 24.9 | | | 0.312 | |
| 539 | C9H20 | 2,3,3,4-tetramethylpentane | 128.259 | 171.1 | 414.7 | 607.7 | 27.2 | | | 0.313 | |
| 540 | C9H20O | 1-nonanol | 144.258 | 268. | 486.7 | 671. | | 546. | | | 1.7 |
| 541 | C10F8 | perfluoronaphthalene | 272.094 | | 482. | 673.1 | | | | | |
| 542 | C10F18 | perfluorodecalin | 462.074 | | 415. | 566. | 15.2 | | | 0.392 | |
| 543 | C10H8 | naphthalene | 128.174 | 353.5 | 491.1 | 748.4 | 40.5 | 413. | 0.269 | 0.302 | 0.0 |
| 544 | C10H12 | 1,2,3,4-tetrahydronaphthalene | 132.206 | 242. | 480.7 | 719. | 35.1 | | | 0.303 | |
| 545 | C10H14 | n-butylbenzene | 134.222 | 185.2 | 456.5 | 660.5 | 28.9 | 497. | 0.261 | 0.393 | 0.4 |
| 546 | C10H14 | isobutylbenzene | 134.222 | 221.7 | 445.9 | 650. | 31.4 | 480. | 0.28 | 0.380 | 0.3 |
| 547 | C10H14 | sec-butylbenzene | 134.222 | 197.7 | 446.5 | 664. | 29.4 | | | 0.274 | 0.4 |
| 548 | C10H14 | tert-butylbenzene | 134.222 | 215.3 | 442.3 | 660. | 29.6 | | | 0.265 | 0.5 |
| 549 | C10H14 | 1-methyl-2-isopropylbenzene | 134.222 | | 451.5 | 670. | 28.9 | | | 0.277 | |
| 550 | C10H14 | 1-methyl-3-isopropylbenzene | 134.222 | | 448.3 | 666. | 29.3 | | | 0.279 | |

| No | Formula | Name | CPVAP A | CPVAP B | CPVAP C | CPVAP D | DELHF | DELGF |
|---|---|---|---|---|---|---|---|---|
| 526 | C9H18 | 1,trans-3,5-trimethylcyclohexane | | | | | | |
| 527 | C9H18 | 1-nonene | -3.718E+0 | 8.122E-1 | -4.509E-4 | 9.705E-8 | -1.036E+5 | 1.128E+5 |
| 528 | C9H18O | dibutyl ketone | | | | | | |
| 529 | C9H20 | n-nonane | -8.374E+0 | 8.729E-1 | -4.823E-4 | 1.031E-7 | -2.292E+5 | 2.483E+4 |
| 530 | C9H20 | 2-methyloctane | -1.011E+1 | 8.805E-1 | -4.936E-4 | 1.083E-7 | -2.292E+5 | 2.483E+4 |
| 531 | C9H20 | 2,2-dimethylheptane | -2.089E+1 | 9.668E-1 | -6.120E-4 | 1.570E-7 | -2.470E+5 | 1.675E+4 |
| 532 | C9H20 | 2,2,3-trimethylhexane | -4.563E+1 | 1.055E+0 | -7.172E-4 | 1.987E-7 | -2.414E+5 | 2.453E+4 |
| 533 | C9H20 | 2,2,4-trimethylhexane | -6.031E+1 | 1.104E+0 | -7.712E-4 | 2.188E-7 | -2.434E+5 | 2.252E+4 |
| 534 | C9H20 | 2,2,5-trimethylhexane | -5.411E+1 | 1.095E+0 | -7.746E-4 | 2.255E-7 | -2.542E+5 | 1.344E+4 |
| 535 | C9H20 | 3,3-diethylpentane | -6.727E+1 | 1.126E+0 | -7.988E-4 | 2.306E-7 | -2.321E+5 | 3.509E+4 |
| 536 | C9H20 | 2,2,3,3-tetramethylpentane | -5.458E+1 | 1.089E+0 | -7.570E-4 | 2.142E-7 | -2.374E+5 | 3.433E+4 |
| 537 | C9H20 | 2,2,3,4-tetramethylpentane | -5.458E+1 | 1.089E+0 | -7.570E-4 | 2.142E-7 | -2.371E+5 | 3.266E+4 |
| 538 | C9H20 | 2,2,4,4-tetramethylpentane | -6.740E+1 | 1.168E+0 | -8.612E-4 | 2.574E-7 | -2.421E+5 | 3.404E+4 |
| 539 | C9H20 | 2,3,3,4-tetramethylpentane | -5.492E+1 | 1.091E+0 | -7.603E-4 | 2.158E-7 | -2.364E+5 | 3.412E+4 |
| 540 | C9H20O | 1-nonanol | 1.280E+0 | 8.817E-1 | -4.791E-4 | 9.801E-8 | -3.872E+5 | -1.183E+5 |
| 541 | C10F8 | perfluoronaphthalene | | | | | | |
| 542 | C10F18 | perfluorodecalin | | | | | | |
| 543 | C10H8 | naphthalene | -6.880E+1 | 8.499E-1 | -6.506E-4 | 1.981E-7 | 1.511E+5 | 2.237E+5 |
| 544 | C10H12 | 1,2,3,4-tetrahydronaphthalene | | | | | 2.760E+3 | 1.671E+5 |
| 545 | C10H14 | n-butylbenzene | -2.299E+1 | 7.934E-1 | -4.396E-4 | 8.570E-8 | -1.382E+4 | 1.448E+5 |
| 546 | C10H14 | isobutylbenzene | | | | | -2.156E+4 | |
| 547 | C10H14 | sec-butylbenzene | -6.515E+1 | 9.893E-1 | -7.214E-4 | 2.152E-7 | -1.746E+4 | |
| 548 | C10H14 | tert-butylbenzene | -8.600E+1 | 1.102E+0 | -8.746E-4 | 2.827E-7 | -2.269E+4 | |
| 549 | C10H14 | 1-methyl-2-isopropylbenzene | | | | | | |
| 550 | C10H14 | 1-methyl-3-isopropylbenzene | -4.876E+1 | 9.064E-1 | -6.054E-4 | 1.627E-7 | -2.931E+4 | |

| No | Formula | Name | Eq. | VP A | VP B | VP C | VP D | Tmin | Tmax | LDEN | TDEN |
|---|---|---|---|---|---|---|---|---|---|---|---|
| 526 | C9H18 | 1,trans-3,5-trimethylcyclohexane | | | | | | | | 0.722 | 293 |
| 527 | C9H18 | 1-nonene | 1 | -8.30824 | 2.03357 | -5.42753 | 0.95331 | 340 | TC | 0.745 | 273 |
| 528 | C9H18O | dibutyl ketone | | | | | | | | 0.827 | 286 |
| 529 | C9H20 | n-nonane | 1 | -8.24480 | 1.57885 | -4.38155 | -4.04412 | 343 | TC | 0.718 | 293 |
| 530 | C9H20 | 2-methyloctane | 3 | 9.3089 | 3246.64 | -67.20 | | 323 | 448 | 0.713 | 293 |
| 531 | C9H20 | 2,2-dimethylheptane | 3 | 9.1710 | 3120.00 | -65.20 | | 313 | 438 | 0.711 | 293 |
| 532 | C9H20 | 2,2,3-trimethylhexane | 3 | 9.1815 | 3164.17 | -61.66 | | 297 | 436 | 0.729 | 293 |
| 533 | C9H20 | 2,2,4-trimethylhexane | 3 | 9.1437 | 3084.08 | -61.94 | | 291 | 428 | 0.720 | 289 |
| 534 | C9H20 | 2,2,5-trimethylhexane | 1 | -7.80573 | 1.68023 | -4.50859 | -0.78808 | 319 | TC | 0.717 | 289 |
| 535 | C9H20 | 3,3-diethylpentane | 1 | -7.98732 | 2.15446 | -4.25035 | -0.09787 | 336 | TC | 0.752 | 293 |
| 536 | C9H20 | 2,2,3,3-tetramethylpentane | 1 | -7.40615 | 1.23976 | -2.94462 | -3.35833 | 331 | TC | 0.757 | 293 |
| 537 | C9H20 | 2,2,3,4-tetramethylpentane | 1 | -7.60624 | 1.62208 | -3.71777 | -1.50403 | 325 | TC | 0.739 | 293 |
| 538 | C9H20 | 2,2,4,4-tetramethylpentane | 1 | -7.71570 | 1.89775 | -4.08940 | -0.75421 | 316 | TC | 0.719 | 293 |
| 539 | C9H20 | 2,3,3,4-tetramethylpentane | 1 | -7.65000 | 1.71897 | -3.82026 | -0.95911 | 332 | TC | 0.755 | 293 |
| 540 | C9H20O | 1-nonanol | 3 | 8.7513 | 2939.54 | -150.1 | | 363 | 487 | 0.828 | 293 |
| 541 | C10F8 | perfluoronaphthalene | | | | | | | | | |
| 542 | C10F18 | perfluorodecalin | | | | | | | | | |
| 543 | C10H8 | naphthalene | 1 | -7.85178 | 2.17172 | -3.70504 | -4.81238 | 399 | TC | 0.971 | 363 |
| 544 | C10H12 | 1,2,3,4-tetrahydronaphthalene | 3 | 9.5883 | 4009.49 | -64.89 | | 365 | 500 | 0.973 | 293 |
| 545 | C10H14 | n-butylbenzene | 1 | -8.39978 | 2.61916 | -5.80532 | 2.11591 | 369 | TC | 0.860 | 293 |
| 546 | C10H14 | isobutylbenzene | 1 | -8.13153 | 1.58186 | -2.37146 | -7.46781 | 360 | TC | 0.853 | 293 |
| 547 | C10H14 | sec-butylbenzene | 1 | -7.49482 | 2.23440 | -6.77346 | 6.31118 | 360 | TC | 0.862 | 293 |
| 548 | C10H14 | tert-butylbenzene | 1 | -7.45802 | 2.33227 | -7.07129 | 6.72178 | 357 | TC | 0.867 | 293 |
| 549 | C10H14 | 1-methyl-2-isopropylbenzene | 3 | 9.3607 | 3564.52 | -70.00 | | 330 | 481 | 0.876 | 293 |
| 550 | C10H14 | 1-methyl-3-isopropylbenzene | 2 | 61.106 | 8033.58 | -7.076 | 6293. | 330 | TC | 0.861 | 293 |

| No | Formula | Name | MolWt | Tfp K | Tb K | Tc K | Pc bar | Vc $cm^3/mol$ | Zc | Omega | Dipm debye |
|---|---|---|---|---|---|---|---|---|---|---|---|
| 551 | C10H14 | 1-methyl-4-isopropylbenzene | 134.222 | 200. | 450.3 | 651. | 27.3 | | | 0.373 | 0.0 |
| 552 | C10H14 | 1,4-diethylbenzene | 134.222 | 231. | 456.9 | 657.9 | 28.0 | | | 0.404 | 0.1 |
| 553 | C10H14 | 1,2,3,5-tetramethylbenzene | 134.222 | 249. | 471.2 | 679. | | | | | |
| 554 | C10H14 | 1,2,4,5-tetramethylbenzene | 134.212 | 352. | 470.0 | 675. | 29.4 | | | 0.435 | |
| 555 | C10H14O | thymol | 150.221 | 323. | 505.7 | 698. | | | | | |
| 556 | C10H15N | n-butylaniline | 149.236 | 259. | 513.9 | 721. | 28.3 | | | | |
| 557 | C10H18 | cis-decalin | 138.254 | 230. | 468.9 | 702.3 | 32.0 | | | 0.286 | 0.0 |
| 558 | C10H18 | trans-decalin | 138.254 | 242.8 | 460.5 | 687.1 | 31.4 | | | 0.270 | 0.0 |
| 559 | C10H18 | 1,3-decadiene | 138.254 | | 442. | 615. | | | | | |
| 560 | C10H19N | caprylonitrile | 153.269 | 255.3 | 516. | 622.0 | 32.5 | | | | |
| 561 | C10H20 | butylcyclohexane | 140.260 | 198.4 | 454.1 | 667. | 31.5 | | | 0.362 | |
| 562 | C10H20 | isobutylcyclohexane | 140.270 | | 444.5 | 659. | 31.2 | | | 0.319 | |
| 563 | C10H20 | sec-butylcyclohexane | 140.270 | | 452.5 | 669. | 26.7 | | | 0.264 | |
| 564 | C10H20 | tert-butylcyclohexane | 140.270 | 232.0 | 444.7 | 659. | 26.6 | | | 0.252 | 0.0 |
| 565 | C10H20 | 1-decene | 140.270 | 206.9 | 443.7 | 615. | 22.0 | 650. | 0.28 | 0.491 | |
| 566 | C10H20O | menthol | 156.269 | 316. | 489.5 | 694. | | | | | |
| 567 | C10H22 | n-decane | 142.286 | 243.5 | 447.3 | 617.7 | 21.2 | 603. | 0.249 | 0.489 | 0.0 |
| 568 | C10H22 | 3,3,5-trimethylheptane | 142.286 | | 428.9 | 609.7 | 23.2 | | | 0.382 | |
| 569 | C10H22 | 2,2,3,3-tetramethylhexane | 142.286 | | 433.5 | 623.2 | 25.1 | | | 0.364 | |
| 570 | C10H22 | 2,2,5,5-tetramethylhexane | 142.286 | | 410.6 | 581.6 | 21.9 | | | 0.375 | |
| 571 | C10H22O | 1-decanol | 158.285 | 280.1 | 506.1 | 687. | 22.2 | 600. | 0.230 | | 1.8 |
| 572 | C11H10 | 1-methylnaphthalene | 142.201 | 242.7 | 517.9 | 772. | 36. | 462. | 0.234 | 0.310 | 0.5 |
| 573 | C11H10 | 2-methylnaphthalene | 142.201 | 307.7 | 514.3 | 761. | 35. | 462. | 0.26 | 0.382 | 0.4 |
| 574 | C11H14O2 | butyl benzoate | 178.232 | 251. | 523. | 723. | 26. | 561. | 0.25 | 0.58 | |
| 575 | C11H16 | pentamethylbenzene | 148.249 | 327.5 | 504.6 | 719. | | | | | |

| No | Formula | Name | CPVAP A | CPVAP B | CPVAP C | CPVAP D | DELHF | DELGF |
|---|---|---|---|---|---|---|---|---|
| 551 | C10H14 | 1-methyl-4-isopropylbenzene | | | | | | |
| 552 | C10H14 | 1,4-diethylbenzene | -3.742E+1 | 8.671E-1 | -5.560E-4 | 1.411E-7 | -2.227E+4 | 1.380E+5 |
| 553 | C10H14 | 1,2,3,5-tetramethylbenzene | 3.923E+0 | 7.131E-1 | -3.711E-4 | 6.840E-8 | -4.484E+4 | 1.188E+5 |
| 554 | C10H14 | 1,2,4,5-tetramethylbenzene | 1.652E+1 | 6.519E-1 | -2.879E-4 | 3.257E-8 | -4.530E+4 | 1.195E+5 |
| 555 | C10H14O | thymol | | | | | | |
| 556 | C10H15N | n-butylaniline | -3.407E+1 | 9.144E-1 | -5.560E-4 | 1.287E-7 | | |
| 557 | C10H18 | cis-decalin | -1.125E+2 | 1.118E+0 | -6.607E-4 | 1.437E-7 | -1.691E+5 | 8.587E+4 |
| 558 | C10H18 | trans-decalin | -9.767E+1 | 1.045E+0 | -5.476E-4 | 8.981E-8 | -1.824E+5 | 7.348E+4 |
| 559 | C10H18 | 1,3-decadiene | | | | | | |
| 560 | C10H19N | caprylonitrile | | | | | | |
| 561 | C10H20 | butylcyclohexane | -6.296E+1 | 1.081E+0 | -6.305E-4 | 1.400E-7 | -2.133E+5 | 5.648E+4 |
| 562 | C10H20 | isobutylcyclohexane | | | | | | |
| 563 | C10H20 | sec-butylcyclohexane | | | | | | |
| 564 | C10H20 | tert-butylcyclohexane | | | | | | |
| 565 | C10H20 | 1-decene | -4.664E+0 | 9.077E-1 | -5.058E-4 | 1.095E-7 | -1.242E+5 | 1.211E+5 |
| 566 | C10H20O | menthol | | | | | | |
| 567 | C10H22 | n-decane | -7.913E+0 | 9.609E-1 | -5.288E-4 | 1.131E-7 | -2.498E+5 | 3.324E+4 |
| 568 | C10H22 | 3,3,5-trimethylheptane | -7.037E+1 | 1.232E+0 | -8.646E-4 | 2.455E-7 | -2.587E+5 | 3.358E+4 |
| 569 | C10H22 | 2,2,3,3-tetramethylhexane | -5.883E+1 | 1.231E+0 | -8.834E-4 | 2.585E-7 | | |
| 570 | C10H22 | 2,2,5,5-tetramethylhexane | -6.234E+1 | 1.245E+0 | -8.956E-4 | 2.618E-7 | | |
| 571 | C10H22O | 1-decanol | 1.457E+1 | 8.947E-1 | -3.921E-4 | 3.451E-8 | -4.019E+5 | -1.043E+5 |
| 572 | C11H10 | 1-methylnaphthalene | -6.482E+1 | 9.387E-1 | -6.942E-4 | 2.016E-7 | 1.169E+5 | 2.178E+5 |
| 573 | C11H10 | 2-methylnaphthalene | -5.652E+1 | 8.997E-1 | -6.469E-4 | 1.840E-7 | 1.162E+5 | 2.163E+5 |
| 574 | C11H14O2 | butyl benzoate | -1.737E+1 | 8.675E-1 | -4.610E-4 | 7.235E-8 | | |
| 575 | C11H16 | pentamethylbenzene | | | | | | |

| No | Formula | Name | Eq. | VP A | VP B | VP C | VP D | Tmin | Tmax | LDEN | TDEN |
|---|---|---|---|---|---|---|---|---|---|---|---|
| 551 | C10H14 | 1-methyl-4-isopropylbenzene | 2 | 56.605 | 7800.97 | -6.432 | 6308. | 360 | TC | 0.857 | 293 |
| 552 | C10H14 | 1,4-diethylbenzene | 1 | -8.11413 | 1.77697 | -4.43960 | -1.47477 | 370 | TC | 0.862 | 293 |
| 553 | C10H14 | 1,2,3,5-tetramethylbenzene | 3 | 9.6750 | 3854.53 | -72.26 | | 368 | 513 | 0.890 | 293 |
| 554 | C10H14 | 1,2,4,5-tetramethylbenzene | 2 | 57.519 | 8300.92 | -6.478 | 6600. | 360 | TC | 0.838 | 354 |
| 555 | C10H14O | thymol | | | | | | | | | |
| 556 | C10H15N | n-butylaniline | 3 | 9.7792 | 4079.72 | -96.15 | | 385 | 560 | 0.932 | 293 |
| 557 | C10H18 | cis-decalin | 3 | 9.2110 | 3671.61 | -69.74 | | 368 | 495 | 0.897 | 293 |
| 558 | C10H18 | trans-decalin | 3 | 9.1787 | 3610.66 | -66.49 | | 363 | 470 | 0.870 | 293 |
| 559 | C10H18 | 1,3-decadiene | | | | | | | | 0.750 | 293 |
| 560 | C10H19N | caprylonitrile | | | | | | | | 0.820 | 293 |
| 561 | C10H20 | butylcyclohexane | 3 | 9.2914 | 3542.57 | -72.32 | | 332 | 485 | 0.799 | 293 |
| 562 | C10H20 | isobutylcyclohexane | 1 | -8.05035 | 2.67134 | -5.49473 | 2.06044 | 358 | TC | 0.795 | 293 |
| 563 | C10H20 | sec-butylcyclohexane | 1 | -7.49250 | 2.47712 | -7.51526 | 7.69513 | 369 | TC | 0.813 | 293 |
| 564 | C10H20 | tert-butylcyclohexane | 1 | -7.34348 | 2.13810 | -6.48025 | 5.89241 | 357 | TC | 0.813 | 293 |
| 565 | C10H20 | 1-decene | 1 | -9.05778 | 3.06154 | -7.07236 | 4.20695 | 360 | TC | 0.741 | 293 |
| 566 | C10H20O | menthol | | | | | | | | | |
| 567 | C10H22 | n-decane | 1 | -8.56523 | 1.97756 | -5.81971 | -0.29982 | 368 | TC | 0.730 | 293 |
| 568 | C10H22 | 3,3,5-trimethylheptane | 3 | 9.1646 | 3305.20 | -67.66 | | 313 | 458 | | |
| 569 | C10H22 | 2,2,3,3-tetramethylhexane | 3 | 9.1396 | 3371.05 | -64.09 | | 314 | 463 | | |
| 570 | C10H22 | 2,2,5,5-tetramethylhexane | 3 | 9.2244 | 3172.92 | -66.15 | | 300 | 438 | | |
| 571 | C10H22O | 1-decanol | 1 | -8.62283 | 1.39315 | -8.24774 | -19.21149 | 400 | TC | 0.830 | 293 |
| 572 | C11H10 | 1-methylnaphthalene | 1 | -7.56390 | 1.19577 | -3.38134 | -2.86388 | 415 | TC | 1.020 | 293 |
| 573 | C11H10 | 2-methylnaphthalene | 1 | -8.43595 | 2.88433 | -5.70017 | 2.50897 | 412 | TC | 0.990 | 313 |
| 574 | C11H14O2 | butyl benzoate | 3 | 9.7161 | 4158.47 | -94.15 | | 390 | 570 | 1.006 | 293 |
| 575 | C11H16 | pentamethylbenzene | 3 | 9.8147 | 4222.48 | -74.20 | | 398 | 543 | | |

| No | Formula | Name | MolWt | Tfp K | Tb K | Tc K | Pc bar | Vc $cm^3/mol$ | Zc | Omega | Dipm debye |
|---|---|---|---|---|---|---|---|---|---|---|---|
| 576 | C11H22 | n-hexylcyclopentane | 154.297 | 200.2 | 476.3 | 660.1 | 21.3 | | | 0.476 | |
| 577 | C11H22 | 1-undecene | 154.297 | 224.0 | 465.8 | 637. | 19.9 | | | 0.518 | |
| 578 | C11H24 | n-undecane | 156.313 | 247.6 | 469.1 | 638.8 | 19.7 | 660. | 0.24 | 0.535 | 0.0 |
| 579 | C12H10 | diphenyl | 154.212 | 342.4 | 529.3 | 789. | 38.5 | 502. | 0.295 | 0.372 | |
| 580 | C12H10O | diphenyl ether | 170.211 | 300. | 531.2 | 766. | 31.4 | | | 0.44 | 1.1 |
| 581 | C12H18 | hexamethylbenzene | 162.276 | | 536.6 | 758. | | | | | |
| 582 | C12H24 | n-heptylcyclopentane | 168.324 | 220. | 497.3 | 679. | 19.4 | | | 0.515 | |
| 583 | C12H24 | 1-dodecene | 168.324 | 238.0 | 486.5 | 657. | 18.5 | | | 0.558 | |
| 584 | C12H26 | dodecane | 170.340 | 263.6 | 489.5 | 658.2 | 18.2 | 713. | 0.24 | 0.575 | 0.0 |
| 585 | C12H26O | dihexylether | 186.339 | 230.0 | 499.6 | 657. | 18.2 | 720. | 0.24 | 0.70 | |
| 586 | C12H26O | dodecanol | 186.339 | 297.1 | 533.1 | 679. | 19.2 | 718. | 0.24 | | 1.6 |
| 587 | C12H27N | tributylamine | 185.355 | | 486.6 | 643. | 18.2 | | | | 0.8 |
| 588 | C13H12 | diphenylmethane | 168.239 | 300. | 538.2 | 770. | 28.6 | | | 0.442 | 0.4 |
| 589 | C13H26 | n-octylcyclopentane | 182.351 | 229. | 516.9 | 694. | 17.9 | | | 0.564 | |
| 590 | C13H26 | 1-tridecene | 182.351 | 250.1 | 505.9 | 674. | 17.0 | | | 0.598 | |
| 591 | C13H28 | n-tridecane | 184.367 | 267.8 | 508.6 | 676. | 17.2 | 780. | 0.240 | 0.619 | |
| 592 | C14H10 | anthracene | 178.234 | 489.7 | 613.1 | 869.3 | | 554. | | | 0.0 |
| 593 | C14H10 | phenanthrene | 178.234 | 373.7 | 613. | 873. | | 554. | | | 0.0 |
| 594 | C14H28 | n-nonylcyclopentane | 196.378 | 244. | 535.3 | 710.5 | 16.5 | | | 0.610 | |
| 595 | C14H28 | 1-tetradecene | 196.378 | 260.3 | 524.3 | 689. | 15.6 | | | 0.644 | |
| 596 | C14H30 | n-tetradecane | 198.394 | 279.0 | 526.7 | 693.0 | 14.4 | 830. | 0.23 | 0.581 | |
| 597 | C15H30 | n-decylcyclopentane | 210.405 | 251.1 | 552.5 | 723.8 | 15.2 | | | 0.654 | |
| 598 | C15H30 | 1-pentadecene | 210.405 | 269.4 | 541.5 | 704. | 14.5 | | | 0.682 | |
| 599 | C15H32 | n-pentadecane | 212.421 | 283. | 543.8 | 707. | 15.2 | 880. | 0.23 | 0.706 | |
| 600 | C16H22O4 | dibutyl-o-phthalate | 278.350 | 238. | 608. | | | | | | |

| No | Formula | Name | CPVAP A | CPVAP B | CPVAP C | CPVAP D | DELHF | DELGF |
|---|---|---|---|---|---|---|---|---|
| 576 | C11H22 | n-hexylcyclopentane | -5.832E+1 | 1.128E+0 | -6.536E-4 | 1.473E-7 | -2.096E+5 | 7.825E+4 |
| 577 | C11H22 | 1-undecene | -5.585E+0 | 1.003E+0 | -5.602E-4 | 1.216E-7 | -1.449E+5 | 1.295E+5 |
| 578 | C11H24 | n-undecane | -8.395E+0 | 1.054E+0 | -5.799E-4 | 1.237E-7 | -2.705E+5 | 4.162E+4 |
| 579 | C12H10 | diphenyl | -9.707E+1 | 1.106E+0 | -8.855E-4 | 2.790E-7 | 1.822E+5 | 2.803E+5 |
| 580 | C12H10O | diphenyl ether | -6.073E+1 | 9.282E-1 | -5.870E-4 | 1.359E-7 | 4.999E+4 | |
| 581 | C12H18 | hexamethylbenzene | | | | | | |
| 582 | C12H24 | n-heptylcyclopentane | -5.926E+1 | 1.223E+0 | -7.084E-4 | 1.596E-7 | -2.303E+5 | 8.667E+4 |
| 583 | C12H24 | 1-dodecene | -6.544E+0 | 1.098E+0 | -6.155E-4 | 1.341E-7 | -1.655E+5 | 1.380E+5 |
| 584 | C12H26 | dodecane | -9.328E+0 | 1.149E+0 | -6.347E-4 | 1.359E-7 | -2.911E+5 | 5.007E+4 |
| 585 | C12H26O | dihexylether | 3.354E+1 | 1.073E+0 | -5.535E-4 | 1.678E-7 | | |
| 586 | C12H26O | dodecanol | 9.224E+0 | 1.103E+0 | -5.338E-4 | 7.779E-8 | -4.431E+5 | -8.713E+4 |
| 587 | C12H27N | tributylamine | 7.993E+0 | 1.198E+0 | -6.703E-4 | 1.449E-7 | | |
| 588 | C13H12 | diphenylmethane | | | | | | |
| 589 | C13H26 | n-octylcyclopentane | -5.995E+1 | 1.317E+0 | -7.612E-4 | 1.708E-7 | -2.509E+5 | 9.512E+4 |
| 590 | C13H26 | 1-tridecene | -7.118E+0 | 1.191E+0 | -6.674E-4 | 1.451E-7 | -1.861E+5 | 1.464E+5 |
| 591 | C13H28 | n-tridecane | -1.046E+1 | 1.245E+0 | -6.912E-4 | 1.490E-7 | -3.117E+5 | 5.849E+4 |
| 592 | C14H10 | anthracene | -5.898E+1 | 1.006E+0 | -6.594E-4 | 1.606E-7 | 2.248E+5 | |
| 593 | C14H10 | phenanthrene | -5.898E+1 | 1.006E+0 | -6.594E-4 | 1.606E-7 | | |
| 594 | C14H28 | n-nonylcyclopentane | -6.081E+1 | 1.412E+0 | -8.156E-4 | 1.830E-7 | -2.715E+5 | 1.035E+5 |
| 595 | C14H28 | 1-tetradecene | -7.967E+0 | 1.286E+0 | -7.210E-4 | 1.569E-7 | -2.067E+5 | 1.549E+5 |
| 596 | C14H30 | n-tetradecane | -1.098E+1 | 1.338E+0 | -7.423E-4 | 1.598E-7 | -3.323E+5 | 6.686E+4 |
| 597 | C15H30 | n-decylcyclopentane | -6.192E+1 | 1.508E+0 | -8.717E-4 | 1.959E-7 | -2.922E+5 | 1.119E+5 |
| 598 | C15H30 | 1-pentadecene | -9.203E+0 | 1.382E+0 | -7.783E-4 | 1.703E-7 | -2.274E+5 | 1.632E+5 |
| 599 | C15H32 | n-pentadecane | -1.192E+1 | 1.433E+0 | -7.972E-4 | 1.720E-7 | -3.530E+5 | 7.528E+4 |
| 600 | C16H22O4 | dibutyl-o-phthalate | 1.880E+0 | 1.254E+0 | -6.121E-4 | 6.971E-8 | | |

| No | Formula | Name | Eq. | VP A | VP B | VP C | VP D | Tmin | Tmax | LDEN | TDEN |
|---|---|---|---|---|---|---|---|---|---|---|---|
| 576 | C11H22 | n-hexylcyclopentane | 3 | 9.3938 | 3702.56 | -81.55 | | 351 | 507 | 0.797 | 293 |
| 577 | C11H22 | 1-undecene | 2 | 71.675 | 9105.75 | -8.489 | 8596. | 350 | TC | 0.751 | 293 |
| 578 | C11H24 | n-undecane | 2 | 73.501 | 9305.80 | -8.729 | 8813. | 350 | TC | 0.740 | 293 |
| 579 | C12H10 | diphenyl | 1 | -7.67400 | 1.23008 | -3.67908 | -2.29172 | 342 | TC | 0.990 | 347 |
| 580 | C12H10O | diphenyl ether | 1 | -8.59849 | 2.46297 | -5.62029 | -1.23996 | 477 | TC | 1.066 | 303 |
| 581 | C12H18 | hexamethylbenzene | | | | | | | | | |
| 582 | C12H24 | n-heptylcyclopentane | 3 | 9.4387 | 3850.38 | -88.75 | | 368 | 529 | 0.810 | 293 |
| 583 | C12H24 | 1-dodecene | 2 | 76.348 | 9846.99 | -9.073 | 9826. | 360 | TC | 0.758 | 293 |
| 584 | C12H26 | dodecane | 2 | 77.628 | 10012.5 | -9.236 | 10030. | 360 | TC | 0.748 | 293 |
| 585 | C12H26O | dihexylether | 3 | 9.7170 | 3982.78 | -89.15 | | 373 | 545 | 0.794 | 293 |
| 586 | C12H26O | dodecanol | 3 | 8.6436 | 3242.04 | -157.1 | | 407 | 580 | 0.835 | 293 |
| 587 | C12H27N | tributylamine | 3 | 9.6676 | 3865.58 | -86.15 | | 362 | 531 | 0.779 | 293 |
| 588 | C13H12 | diphenylmethane | 3 | 7.8654 | 2902.44 | -167.9 | | 473 | 563 | 1.006 | 293 |
| 589 | C13H26 | n-octylcyclopentane | 3 | 9.4739 | 3983.01 | -95.85 | | 385 | 549 | 0.805 | 293 |
| 590 | C13H26 | 1-tridecene | 2 | 81.389 | 10609.4 | -9.709 | 11250. | 370 | TC | 0.766 | 293 |
| 591 | C13H28 | n-tridecane | 3 | 9.5153 | 3892.91 | -98.93 | | 380 | 540 | 0.756 | 293 |
| 592 | C14H10 | anthracene | 3 | 11.0499 | 6492.44 | -26.13 | | 490 | 655 | | |
| 593 | C14H10 | phenanthrene | 3 | 10.0985 | 5477.94 | -69.39 | | 450 | 655 | | |
| 594 | C14H28 | n-nonylcyclopentane | 3 | 9.4887 | 4096.30 | -103.0 | | 400 | 569 | 0.808 | 293 |
| 595 | C14H28 | 1-tetradecene | 2 | 85.854 | 11329.2 | -10.27 | 12800. | 380 | TC | 0.786 | 273 |
| 596 | C14H30 | n-tetradecane | 2 | 84.552 | 11322.9 | -10.07 | 12500. | 380 | TC | 0.763 | 293 |
| 597 | C15H30 | n-decylcyclopentane | 3 | 9.5059 | 4203.94 | -109.7 | | 413 | 586 | 0.811 | 293 |
| 598 | C15H30 | 1-pentadecene | 2 | 92.300 | 12205.3 | -11.09 | 14370. | 400 | TC | 0.791 | 273 |
| 599 | C15H32 | n-pentadecane | 2 | 88.380 | 11995.6 | -10.54 | 13840. | 400 | TC | 0.769 | 293 |
| 600 | C16H22O4 | dibutyl-o-phthalate | 3 | 10.3337 | 4852.47 | -138.1 | | 469 | 657 | 1.047 | 293 |

| No | Formula | Name | MolWt | Tfp K | Tb K | Tc K | Pc bar | Vc $cm^3/mol$ | Zc | Omega | Dipm debye |
|---|---|---|---|---|---|---|---|---|---|---|---|
| 601 | $C_{16}H_{32}$ | n-decylcyclohexane | 224.432 | 271. | 570.8 | 750. | 13.5 | | | 0.583 | |
| 602 | $C_{16}H_{32}$ | 1-hexadecene | 224.432 | 277.3 | 558. | 717. | 13.3 | | | 0.721 | |
| 603 | $C_{16}H_{34}$ | hexadecane | 226.448 | 291. | 560. | 722. | 14.1 | | | 0.742 | |
| 604 | $C_{17}H_{34}$ | n-dodecylcyclopentane | 238.459 | 268. | 584.1 | 750. | 12.9 | | | 0.719 | |
| 605 | $C_{17}H_{36}O$ | heptadecanol | 256.474 | 327. | 597. | 736. | 14.1 | | | | |
| 606 | $C_{17}H_{36}$ | n-heptadecane | 240.475 | 295. | 575.2 | 733. | 13. | 1000. | 0.22 | 0.77 | |
| 607 | $C_{18}H_{14}$ | o-terphenyl | 230.310 | 330. | 605. | 891.0 | 39.0 | 753. | 0.396 | 0.431 | |
| 608 | $C_{18}H_{14}$ | m-terphenyl | 230.310 | 360. | 638. | 924.9 | 35.1 | 768. | 0.358 | 0.449 | |
| 609 | $C_{18}H_{14}$ | p-terphenyl | 230.310 | 485. | 649. | 926.0 | 33.2 | 763. | 0.329 | 0.523 | 0.7 |
| 610 | $C_{18}H_{36}$ | 1-octadecene | 252.486 | 290.8 | 588.0 | 739. | 11.3 | | | 0.807 | |
| 611 | $C_{18}H_{36}$ | n-tridecylcyclopentane | 252.486 | 278. | 598.6 | 761. | 12.0 | | | 0.755 | |
| 612 | $C_{18}H_{38}$ | octadecane | 254.504 | 301.3 | 589.5 | 748. | 12.0 | | | 0.790 | |
| 613 | $C_{18}H_{38}O$ | 1-octadecanol | 270.501 | 331. | 608. | 747. | 14.1 | | | | 1.7 |
| 614 | $C_{19}H_{38}$ | 1-cyclopentyltetradecane | 266.513 | 282. | 599. | 772. | 11.2 | | | 0.789 | |
| 615 | $C_{19}H_{40}$ | n-nonadecane | 268.529 | 305. | 603.1 | 756. | 11.1 | | | 0.827 | |
| 616 | $C_{20}H_{40}$ | 1-cyclopentylpentadecane | 280.540 | 290. | 625. | 780. | 10.2 | | | 0.833 | |
| 617 | $C_{20}H_{42}$ | n-eicosane | 282.556 | 310. | 617. | 767. | 11.1 | | | 0.907 | |
| 618 | $C_{20}H_{42}O$ | 1-eicosanol | 298.555 | 339. | 629. | 770. | 12. | | | | |

| No | Formula | Name | CPVAP A | CPVAP B | CPVAP C | CPVAP D | DELHF | DELGF |
|---|---|---|---|---|---|---|---|---|
| 601 | C16H32 | n-decylcyclohexane | -6.902E+1 | 1.654E+0 | -9.613E-4 | 2.143E-7 | | |
| 602 | C16H32 | 1-hexadecene | -9.705E+0 | 1.475E+0 | -8.298E-4 | 1.810E-7 | -2.480E+5 | 1.716E+5 |
| 603 | C16H34 | hexadecane | -1.302E+1 | 1.529E+0 | -8.537E-4 | 1.850E-7 | -3.736E+5 | 8.374E+4 |
| 604 | C17H34 | n-dodecylcyclopentane | -6.326E+1 | 1.695E+0 | -9.768E-4 | 2.186E-7 | -3.361E+5 | 1.260E+5 |
| 605 | C17H36O | heptadecanol | -7.792E+0 | 1.653E+0 | -9.345E-4 | 2.044E-7 | -5.463E+5 | -4.467E+4 |
| 606 | C17H36 | n-heptadecane | -1.397E+1 | 1.624E+0 | -9.081E-4 | 1.972E-7 | -3.942E+5 | 9.215E+4 |
| 607 | C18H14 | o-terphenyl | | | | | | |
| 608 | C18H14 | m-terphenyl | | | | | | |
| 609 | C18H14 | p-terphenyl | | | | | | |
| 610 | C18H36 | 1-octadecene | -1.133E+1 | 1.664E+0 | -9.374E-4 | 2.049E-7 | -2.892E+5 | 1.884E+5 |
| 611 | C18H36 | n-tridecylcyclopentane | -6.421E+1 | 1.790E+0 | -1.032E-3 | 2.309E-7 | -3.540E+5 | 1.371E+5 |
| 612 | C18H38 | octadecane | -1.447E+1 | 1.717E+0 | -9.592E-4 | 2.078E-7 | -4.148E+5 | 1.006E+5 |
| 613 | C18H38O | 1-octadecanol | -8.704E+0 | 1.748E+0 | -9.881E-4 | 2.157E-7 | -5.669E+5 | -3.622E+4 |
| 614 | C19H38 | 1-cyclopentyltetradecane | -6.493E+1 | 1.884E+0 | -1.085E-3 | 2.426E-7 | -3.746E+5 | 1.456E+5 |
| 615 | C19H40 | n-nonadecane | -1.549E+1 | 1.812E+0 | -1.015E-3 | 2.205E-7 | -4.354E+5 | 1.090E+5 |
| 616 | C20H40 | 1-cyclopentylpentadecane | -6.609E+1 | 1.980E+0 | -1.140E-3 | 2.550E-7 | -3.953E+5 | 1.540E+5 |
| 617 | C20H42 | n-eicosane | -2.238E+1 | 1.939E+0 | -1.117E-3 | 2.528E-7 | -4.561E+5 | 1.174E+5 |
| 618 | C20H42O | 1-eicosanol | -1.258E+1 | 1.950E+0 | -1.118E-3 | 2.516E-7 | -6.081E+5 | -1.943E+4 |

| No | Formula | Name | Eq. | VP A | VP B | VP C | VP D | Tmin | Tmax | LDEN | TDEN |
|---|---|---|---|---|---|---|---|---|---|---|---|
| 601 | C16H32 | n-decylcyclohexane | 3 | 9.5425 | 4373.37 | -111.8 | | 463 | 573 | 0.819 | 293 |
| 602 | C16H32 | 1-hexadecene | 2 | 99.280 | 13117.0 | -11.99 | 16260. | 400 | TC | 0.788 | 283 |
| 603 | C16H34 | hexadecane | 2 | 89.060 | 12411.3 | -10.58 | 15200. | 400 | TC | 0.773 | 293 |
| 604 | C17H34 | n-dodecylcyclopentane | 3 | 9.5713 | 4395.87 | -124.2 | | 441 | 619 | 0.816 | 293 |
| 605 | C17H360 | heptadecanol | 3 | 8.9959 | 3672.62 | -188.1 | | 464 | 656 | 0.848 | 327 |
| 606 | C17H36 | n-heptadecane | 3 | 9.5308 | 4294.55 | -124.0 | | 434 | 610 | 0.778 | 298 |
| 607 | C18H14 | o-terphenyl | | | | | | | | | |
| 608 | C18H14 | m-terphenyl | | | | | | | | | |
| 609 | C18H14 | p-terphenyl | | | | | | | | | |
| 610 | C18H36 | 1-octadecene | 3 | 9.6019 | 4416.13 | -127.3 | | 444 | 623 | 0.789 | 293 |
| 611 | C18H36 | n-tridecylcyclopentane | 3 | 9.6068 | 4483.13 | -131.3 | | 453 | 634 | 0.818 | 293 |
| 612 | C18H38 | octadecane | 3 | 9.5030 | 4361.79 | -129.9 | | 445 | 625 | 0.777 | 301 |
| 613 | C18H380 | 1-octadecanol | 3 | 9.0696 | 3757.82 | -193.1 | | 474 | 658 | 0.812 | 332 |
| 614 | C19H38 | 1-cyclopentyltetradecane | 3 | 9.6430 | 4439.38 | -138.1 | | 465 | 648 | 0.820 | 293 |
| 615 | C19H40 | n-nonadecane | 3 | 9.5331 | 4450.44 | -135.6 | | 456 | 639 | 0.789 | 305 |
| 616 | C20H40 | 1-cyclopentylpentadecane | 3 | 9.6890 | 4692.01 | -145.1 | | 476 | 661 | 0.821 | 293 |
| 617 | C20H42 | n-eicosane | 3 | 9.8483 | 4680.46 | -141.1 | | 471 | 652 | 0.775 | 313 |
| 618 | C20H420 | 1-eicosanol | 3 | 9.2031 | 3912.10 | -203.1 | | 492 | 679 | | |

Appendix

# B

# Lennard-Jones Potentials as Determined from Viscosity Data†

| Substance | | $b_0$,‡ cm$^3$/g-mol | $\sigma$, Å | $\epsilon/k$, K |
|---|---|---|---|---|
| Ar | Argon | 56.08 | 3.542 | 93.3 |
| He | Helium | 20.95 | 2.551§ | 10.22 |
| Kr | Krypton | 61.62 | 3.655 | 178.9 |
| Ne | Neon | 28.30 | 2.820 | 32.8 |
| Xe | Xenon | 83.66 | 4.047 | 231.0 |
| Air | Air | 64.50 | 3.711 | 78.6 |
| $AsH_3$ | Arsine | 89.88 | 4.145 | 259.8 |
| $BCl_3$ | Boron chloride | 170.1 | 5.127 | 337.7 |
| $BF_3$ | Boron fluoride | 93.35 | 4.198 | 186.3 |
| $B(OCH_3)_3$ | Methyl borate | 210.3 | 5.503 | 396.7 |
| $Br_2$ | Bromine | 100.1 | 4.296 | 507.9 |
| $CCl_4$ | Carbon tetrachloride | 265.5 | 5.947 | 322.7 |
| $CF_4$ | Carbon tetrafluoride | 127.9 | 4.662 | 134.0 |
| $CHCl_3$ | Chloroform | 197.5 | 5.389 | 340.2 |
| $CH_2Cl_2$ | Methylene chloride | 148.3 | 4.898 | 356.3 |
| $CH_3Br$ | Methyl bromide | 88.14 | 4.118 | 449.2 |
| $CH_3Cl$ | Methyl chloride | 92.31 | 4.182 | 350 |
| $CH_3OH$ | Methanol | 60.17 | 3.626 | 481.8 |
| $CH_4$ | Methane | 66.98 | 3.758 | 148.6 |
| CO | Carbon monoxide | 63.41 | 3.690 | 91.7 |
| COS | Carbonyl sulfide | 88.91 | 4.130 | 336.0 |
| $CO_2$ | Carbon dioxide | 77.25 | 3.941 | 195.2 |
| $CS_2$ | Carbon disulfide | 113.7 | 4.483 | 467 |
| $C_2H_2$ | Acetylene | 82.79 | 4.033 | 231.8 |
| $C_2H_4$ | Ethylene | 91.06 | 4.163 | 224.7 |
| $C_2H_6$ | Ethane | 110.7 | 4.443 | 215.7 |
| $C_2H_5Cl$ | Ethyl chloride | 148.3 | 4.898 | 300 |
| $C_2H_5OH$ | Ethanol | 117.3 | 4.530 | 362.6 |
| $C_2N_2$ | Cyanogen | 104.7 | 4.361 | 348.6 |
| $CH_3OCH_3$ | Methyl ether | 100.9 | 4.307 | 395.0 |

| Substance | | $b_0$,‡ $cm^3$/g-mol | $\sigma$, Å | $\epsilon/k$, K |
|---|---|---|---|---|
| $CH_2CHCH_3$ | Propylene | 129.2 | 4.678 | 298.9 |
| $CH_3CCH$ | Methylacetylene | 136.2 | 4.761 | 251.8 |
| $C_3H_6$ | Cyclopropane | 140.2 | 4.807 | 248.9 |
| $C_3H_8$ | Propane | 169.2 | 5.118 | 237.1 |
| $n$-$C_3H_7OH$ | $n$-Propyl alcohol | 118.8 | 4.549 | 576.7 |
| $CH_3COCH_3$ | Acetone | 122.8 | 4.600 | 560.2 |
| $CH_3COOCH_3$ | Methyl acetate | 151.8 | 4.936 | 469.8 |
| $n$-$C_4H_{10}$ | $n$-Butane | 130.0 | 4.687 | 531.4 |
| *iso*-$C_4H_{10}$ | Isobutane | 185.6 | 5.278 | 330.1 |
| $C_2H_5OC_2H_5$ | Ethyl ether | 231.0 | 5.678 | 313.8 |
| $CH_3COOC_2H_5$ | Ethyl acetate | 178.0 | 5.205 | 521.3 |
| $n$-$C_5H_{12}$ | $n$-Pentane | 244.2 | 5.784 | 341.1 |
| $C(CH_3)_4$ | 2,2-Dimethylpropane | 340.9 | 6.464 | 193.4 |
| $C_6H_6$ | Benzene | 193.2 | 5.349 | 412.3 |
| $C_6H_{12}$ | Cyclohexane | 298.2 | 6.182 | 297.1 |
| $n$-$C_6H_{14}$ | $n$-Hexane | 265.7 | 5.949 | 399.3 |
| $Cl_2$ | Chlorine | 94.65 | 4.217 | 316.0 |
| $F_2$ | Fluorine | 47.75 | 3.357 | 112.6 |
| HBr | Hydrogen bromide | 47.58 | 3.353 | 449 |
| HCN | Hydrogen cyanide | 60.37 | 3.630 | 569.1 |
| HCl | Hydrogen chloride | 46.98 | 3.339 | 344.7 |
| HF | Hydrogen fluoride | 39.37 | 3.148 | 330 |
| HI | Hydrogen iodide | 94.24 | 4.211 | 288.7 |
| $H_2$ | Hydrogen | 28.51 | 2.827 | 59.7 |
| $H_2O$ | Water | 23.25 | 2.641 | 809.1 |
| $H_2O_2$ | Hydrogen peroxide | 93.24 | 4.196 | 289.3 |
| $H_2S$ | Hydrogen sulfide | 60.02 | 3.623 | 301.1 |
| Hg | Mercury | 33.03 | 2.969 | 750 |
| $HgBr_2$ | Mercuric bromide | 165.5 | 5.080 | 686.2 |
| $HgCl_2$ | Mercuric chloride | 118.9 | 4.550 | 750 |
| $HgI_2$ | Mercuric iodide | 224.6 | 5.625 | 695.6 |
| $I_2$ | Iodine | 173.4 | 5.160 | 474.2 |
| $NH_3$ | Ammonia | 30.78 | 2.900 | 558.3 |
| NO | Nitric oxide | 53.74 | 3.492 | 116.7 |
| NOCl | Nitrosyl chloride | 87.75 | 4.112 | 395.3 |
| $N_2$ | Nitrogen | 69.14 | 3.798 | 71.4 |
| $N_2O$ | Nitrous oxide | 70.80 | 3.828 | 232.4 |
| $O_2$ | Oxygen | 52.60 | 3.467 | 106.7 |
| $PH_3$ | Phosphine | 79.63 | 3.981 | 251.5 |
| $SF_6$ | Sulfur hexafluoride | 170.2 | 5.128 | 222.1 |
| $SO_2$ | Sulfur dioxide | 87.75 | 4.112 | 335.4 |
| $SiF_4$ | Silicon tetrafluoride | 146.7 | 4.880 | 171.9 |
| $SiH_4$ | Silicon hydride | 85.97 | 4.084 | 207.6 |
| $SnBr_4$ | Stannic bromide | 329.0 | 6.388 | 563.7 |
| $UF_6$ | Uranium hexafluoride | 268.1 | 5.967 | 236.8 |

†R. A. Svehla, *NASA Tech. Rep.* R-132, Lewis Research Center, Cleveland, Ohio, 1962.
‡$b_0 = \frac{2}{3}\pi N_0 \sigma^3$, where $N_0$ is Avogadro's number.
§The parameter $\sigma$ was determined by quantum-mechanical formulas.

# Index

## ABOUT THE AUTHORS

ROBERT C. REID is an Emeritus Professor of Chemical Engineering at the Massachusetts Institute of Technology, where he received his Sc.D. degree. A consultant for Nestlé, Cabot, and Arthur D. Little, Inc., he is the author or coauthor of over 130 technical papers and textbooks in thermodynamics and crystal growth from solution. He is a member of the National Academy of Engineering and received the Founders Award from the American Institute of Chemical Engineers.

JOHN M. PRAUSNITZ holds a Ph.D. from Princeton University and an honorary D.Eng. from the University of L'Aquila (Italy). Currently he is Professor of Chemical Engineering at the University of California at Berkeley. He is a consultant to several chemical companies. Dr. Prausnitz is the author or coauthor of more than 350 technical articles and 4 monographs in applied phase-equilibrium thermodynamics. He has won a number of awards and fellowships. He is a member of the National Academy of Sciences and of the National Academy of Engineering.

BRUCE E. POLING is a Chemical Engineering Professor at the University of Missouri-Rolla, where he has taught courses and conducted research in the areas of physical properties, thermodynamics, and equations of state. He received his Ph.D. from the University of Illinois.